The Environment

The second edition of this fully integrated introductory text for courses in environmental studies and physical geography builds on the resounding success of the first edition, providing a comprehensive account of modern environmental issues and the physical and socio-economic framework in which they are set. It explains the principles and applications of the different parts of the Earth's system: the lithosphere, atmosphere, hydrosphere and the biosphere, and explains the interrelationships within and between these systems. It explores the present environmental crisis, examines how the planet Earth fits into the wider universe and explores human–environment interactions.

New features of the second edition include:

- Two entirely new chapters on 'population' and 'retrospect and prospect'
- Updated case studies and examples bringing completely fresh perspectives to recent environmental events and issues
- New boxes introducing emerging approaches, techniques or perspectives
- Additional student-friendly textbook features, including learning outcomes and end-of-chapter summaries and a glossary of key terms
- New colour and black and white plates
- Updated annotated reading
- Web links to sites illustrating themes and material covered in the book
- An associated website structured by chapter with a complete bibliography, links to imagery and an updated lecturer's manual

Chris Park is Senior Lecturer in Geography and Principal of the Graduate College at Lancaster University

What readers said about the second edition

'It is so important that the global perspective is painted as a holistic picture for the citizen of tomorrow . . . Chris Park shows how the planet ticks, and hence how local autonomy is possible even on this dynamic and unstable globe.'

Tim O'Riordan, University of East Anglia

'Ideally suited to the student of the 21st century. It provides a cornucopia of environmental information that will both challenge and stimulate the reader. The principles underlying key issues are presented in a readable manner and are reinforced by the effective presentation of relevant case studies . . . Should appeal to students of environmental science as well as undergraduates on physical geography and ecology programmes.'

Jennifer Jones, Liverpool John Moores University

'I find the content superior to virtually all I have seen on the market.'

Beverley Wemple, University of Vermont

'An excellent text . . . ample thought-provoking material to stimulate student groups, accompanied by clear and concise figures, diagrams and charts . . . A most useful addition to the introductory course material for environmental science students at a price accessible to most cohort groups.'

Clive Roberts, University of Wolverhampton

What readers said about the first edition

'Written by an experienced lecturer, this book has an impressive breadth of topics and presents concepts and processes in an accurate and balanced way. The structure of the book is logical and easily understandable, and allows ample opportunity for discussion of processes and the way systems work, which is important for the understanding of any present-day environmental problem. The examples are very useful as introductions to the themes, as are the references – for example those to recent international conferences (Rio).'

Anders Lundberg, University of Bergen, Norway

'A succinct yet remarkably comprehensive account of modern environmental issues and the physical and socio-economic framework in which they are set. Interdisciplinary in nature, this is a book that can be browsed to obtain course information or used to build a global perspective on the environment.'

David Kemp, Lakehead University, Canada

'Exciting first year text books in the Environmental Sciences are a rare event but Chris Park has produced one which should stand the test of time. The author's approach is fresh and the work can be described in four words: comprehensive, integrative, informative and relevant. The liberal use of text boxes and figures breaks up the page, the work is extensively referenced through notes and lists of additional reading in each chapter. Examples and case studies have been well selected for their practical human significance and are drawn from all over the world. *The Environment: Principles and Applications* deserves to find a ready market in both the upper levels of secondary schools and first and second year university.'

Peter Mitchell, Macquarie University, Australia

'This is an excellent textbook, particularly well suited to introductory courses in geography and environmental studies. I am impressed by the blending of scientific theory and contemporary environmental issues. The text has a broad scope which is focused by the liberal use of well chosen case studies. The writing is clear and concise, and the text is well supported by figures and tables. This is a most accessible text, and should be appreciated by students on both sides of the Atlantic. I recommend it thoroughly.

Michael Day, University of Milwaukee

The Environment *2nd edition*

principles *and* applications **Chris Park**

London *and* New York

First published 1997
by Routledge, an imprint of the Taylor & Francis Group
11 New Fetter Lane, London EC4P 4EE

Simultaneously published in the USA and Canada
by Routledge
29 West 35th Street, New York, NY 10001

Second edition published 2001 by Routledge

Routledge is an imprint of the Taylor & Francis Group

Typeset in Perpetua and Bell Gothic by
Florence Production Ltd, Stoodleigh, Devon
Printed and bound in Great Britain by
St Edmundsbury Press, Bury St Edmunds, Suffolk

British Library Cataloguing in Publication Data
A catalogue record for this book is available
from the British Library

Library of Congress Cataloguing in Publication Data
Park, Chris C.
The environment : principles and applications / Chris Park.
— 2nd ed.
p. cm.
Includes bibliographical references and index.
GE105. P37 2001
363.7—dc21 00–045731

ISBN 0–415–21770–9 (hbk)
ISBN 0–415–21771–7 (pbk)

For Emma-Jane, Andrew, Sam and Elizabeth

May they enjoy the world they inherit,
and pass on a world worth living in.

Brief Contents

Contents

Colour Plates

These appear in a plate section between pages 348 and 349.

Black and White Plates

Figures

Tables

Boxes

Preface to the Second Edition

I am probably as surprised to be writing this second edition of the book, only four years after the first edition, as you are to be reading a new edition so soon! It is a measure of how fast the environmental field is changing, and how necessary it is to keep the 'stories' up to date, that this new edition is necessary. The book has also sold well, and both I and Routledge are heartened to know there is a market 'out there'!

In preparing this new edition, I have tried to preserve all that was good about the first edition, and to add some new material and new features that I hope will make the book even more useful. Two new chapters have been added – one on population and environment, and a more substantive final chapter ('Retrospect and prospect'), which replaces the original Chapter 19 – to make the coverage more rounded, tie major themes together more fully and point to some of the emerging themes and issues in this richly rewarding field of study. All other chapters have been refreshed with new or updated material, including case studies and examples, sometimes in the form of new boxes but often just where appropriate in the main body of the text.

Many of the updates relate to global warming, its causes and consequences, because this is an area that is fast unfolding and whose relevance is obvious. I have also further improved the cross-referencing between chapters in an attempt to underline even more clearly how these major environmental systems and sub-systems interact with one another.

The further reading sections in each chapter have been updated, and all of these items are now listed in a new bibliography at the end of the book. The material in the further reading sections and the bibliography is almost without exception post-1990. The extensive referencing used throughout the first edition has been replaced by the bibliography, and citations have been removed from the text. New learning tools have been provided for each chapter, which now opens with a series of learning objectives and closes with a chapter summary. A glossary of important terms has been included at the back of the book, and items listed in it are marked in bold when first encountered in the text.

A website has been created especially for this book, with its own memorable domain name
(www.park-environment.com)
and each chapter includes a pointer to it. The website is designed to help readers to get more out of the book, and it will be updated regularly. It contains chapter outlines, links to relevant websites, comprehensive chapter-based bibliographies and learning tools.

This new edition also has a brand new cover and a new layout inside. I find the cover – based on Jeff Wall's *A sudden gust of wind* – intriguing, enigmatic and in some ways quite disturbing, but that makes it particularly suitable for this book!

Acknowledgements

I am very grateful to the numerous people (almost all of them anonymous) who have given me valuable feedback on the first edition, and I hope that I have been able to take on board their views and suggestions as I prepared this new edition. It is a humbling experience to receive such positive feedback, and it also makes me a little nervous about changing anything at all in a book which so many people obviously find very useful.

My 'family' at Routledge have, as ever, been solidly supportive, and I owe them a great thanks. Andrew Mould, my new editor, has had the sense to enquire little but hope much about the progress of this new edition; his quiet support and unerring commitment have been important to me. This time round I have also had the great pleasure of working with Moira Taylor, Routledge's energetic textbook supremo, whose attention to detail is superb, and whose prompting has been a great spur to complete on time. I am particularly indebted to Dennis Hodgson for his excellent attention to detail during copy editing.

Revisiting this material some four years after completing the writing of the first edition has been an interesting experience. I realise now just how much work went into the first edition, and even now I wonder how on earth that book ever got finished. This edition remains as a testimony to the invaluable contributions of all those listed in the first Acknowledgements, though with great sadness I must record the untimely death of Sandra Irish after a gallant struggle against cancer. Chris Beacock drew the three new figures for this edition, in his customary calm and quiet manner, and to great effect.

Nietzsche said that 'life is lived forwards, but only understood looking backwards', and this is true of my life as much as everyone else's. What I described in the Acknowledgements to the first edition as a 'turning point in my life' is now past, and my partner Penny has truly brought the sunshine back into my life. Her support, warmth, love and genuine goodness have already brought me many blessings; without her I doubt if I would have found it possible to produce this second edition. My close group of friends – Robin, Jill, Steve, Audrey, Cyril and Muriel – have each given me so much, and asked so little in return, and I am truly thankful to them and for them. My parents have, as always, been supportive and encouraging, and I could not have asked for more. The greatest gift I have ever been given is my four wonderful children – Emma-Jane, Andrew, Sam and Elizabeth – and I dedicate this book to them as a token of my love for them.

Chris Park
Lancaster
July 2000

Part One
Introduction

Chapter 1

Environment in Crisis

LEARNING OBJECTIVES

When you have finished studying this chapter, you should be able to

- *Understand the importance of adopting a systems framework based on perspectives that are interdisciplinary and global*
- *Appreciate the relevance and context of the so-called 'environmental crisis', and explain why environmental problems are symptoms of what is wrong, not root causes*
- *Evaluate environmental problems in historical context and understand the new and emerging dimensions of the environmental crisis*
- *Describe why the 1992 Rio Earth Summit was so important, and why its products and outcomes are so challenging*
- *Appreciate the importance of environmental monitoring and analysis, and the relevance of new technologies such as remote sensing and satellite systems*
- *Explain why coping with environmental risks and hazards is such a major challenge*
- *Chart the emergence of sustainable development and outline its main objectives*
- *Account for the development of environmentalism and understand its diversity*

This book is about the environment, and it focuses on the ways in which people and environments interact, and particularly on the environmental impacts of human activities. The subtitle – 'Principles and applications' – is very deliberate, because it emphasises the two main themes around which the material is structured. These are the ways in which the environment works and the ways in which an understanding of those basic principles helps us to use the environment and its resources.

EMPHASES

Many books have been written about the environment over the last 20 years. This one differs from most others in the three particular emphases that are adopted:

- *A systems framework*: this is useful because it stresses interaction and interrelationships between different parts of the environment.
- *An interdisciplinary perspective*: this requires us to step back from the particular focus of individual subject areas and take a broader view. It is essential if we are to understand how the environment works and appreciate its complexity and diversity.
- *A global perspective*: many of the most serious environmental problems today are global, and they require global solutions. Throughout this book examples are drawn from many parts of the world, and they are included to illustrate broad global patterns and how they vary across the planet.

A recurring theme throughout the book is the way in which we are pushing the Earth, our only home, to the very limits of its capacity. Indeed, many scientists now argue that the very survival of the planet is at risk because of human misuse of its natural resources and disturbance of its natural environmental systems.

We abuse the Earth at our peril, because damaging the environment causes serious if not irreversible damage to the planet's life-support systems, the ecological processes that shape climate, cleanse air and water, regulate water flow, recycle essential elements, create and generate soil, and keep the planet fit for life.

OVERVIEW

The book is divided into six parts, comprising an introduction, four parts that focus on particular parts of the Earth's system, and a final chapter that reflects on the implications of human misuse of environmental systems. While the parts are dealt with separately, this is simply a matter of convenience and in reality the parts overlap and interact (Figure 1.1) to create the world as we know it.

In the first part of the book, we explore the context for today's rising concern about the state of the environment (Chapter 1), examine the interactions between population and environment (Chapter 2), consider how useful a systems perspective is in understanding how the environment functions (Chapter 3) and examine how the planet Earth fits into the wider universe (Chapter 4). In Part II, we focus on the lithosphere – the solid parts of the Earth's environmental systems – by looking at the internal structure and composition of the Earth (Chapter 5), the surface dynamics of the planet (Chapter 6) and the materials that are created and transported by geological processes (Chapter 7). Part III looks at the atmosphere in terms of its structure and composition (Chapter 8), atmospheric processes (Chapter 9), weather systems (Chapter 10) and climate (Chapter 11). Part IV deals with the hydrosphere – the movement of water through the environment – particularly the water cycle (Chapter 12), water resources (Chapter 13), drylands and deserts (Chapter 14), ice and glaciers (Chapter 15) and oceans and coasts (Chapter 16). In Part V, we explore the biosphere – the living parts of the environment – by looking at biodiversity and ecosystems (Chapter 17), at biomes and ecological succession (Chapter 18), and soils (Chapter 19). Finally, in Part VI we explore the implications of the people–environment relationship, examine where we stand at the opening of the new millennium and reflect on what the future might have in store (Chapter 20).

We begin by considering the so-called environmental crisis.

THE ENVIRONMENTAL CRISIS

Interest in the environment has grown a great deal since the early 1970s (Box 1.1), initially mainly among scientists but more recently increasingly among the general public and politicians. Over the last two decades, our reliance on the environment has been graphically displayed in a series of major environmental disasters – like the North African droughts of the 1970s and 1980s (p. 422), the nuclear accident at Chernobyl in 1986 (see Box 8.23), and the mounting problems of air pollution by greenhouse gases (Box 1.2). Since the 1970s, scientists have been writing about what they term the 'environmental crisis', and more recently this debate has evolved and broadened, incorporating some other leading social issues of the day, into the wider *green* debate. Moreover, the general public has woken up to the possible environmental impacts of its own actions, and we now understand that it is *our* problem, not someone else's.

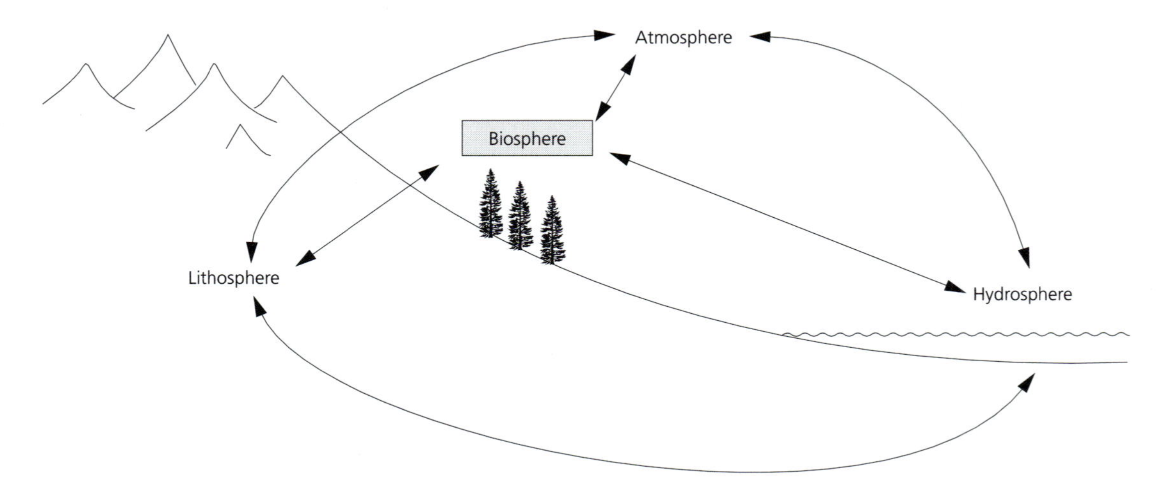

Figure 1.1 *The Earth as an interacting system. The central theme of this book is the interaction between four main environmental systems – the lithosphere, atmosphere, hydrosphere and biosphere.*

> **BOX 1.1 TRIGGERS FOR RISING INTEREST IN THE ENVIRONMENT**
>
> The main triggers for rising interest in the environment have been growing awareness that:
>
> - Most human activities affect the environment in one way or another, usually for the worst.
> - The environment is our basic life-support system. It provides the air we breathe, the water we drink, the food we eat, the land we live on.
> - We rely on the environment to provide us with natural resources (such as wood, energy and minerals).
> - Environmental hazards (natural and caused by human activities) cause much disruption, damage, death, injury and hardship.
> - Many parts of the environment have been badly damaged by over-use or unwise use.
> - If we continue to treat the environment as we have done up to now, the damage will grow worse, the costs will be higher, and the consequences will be more serious.

The UNEP *Global environment outlook 2000* report highlights two critical themes at the opening of the third millennium:

1. The global human ecosystem is threatened by grave imbalances in productivity and in the distribution of goods and services. A significant proportion of humanity still lives in dire poverty, and projected trends are for increasing divergence between those who benefit from economic and technological development and those who do not.
2. The world is undergoing accelerating change, with internationally co-ordinated environmental stewardship lagging behind economic and social development. Environmental gains from new technology are being overtaken by the pace and scale of population growth and economic development.

Key contemporary environmental problems include the greenhouse effect and global warming (p. 261), the hole in the ozone layer (p. 249), acid rain (p. 251) and the destruction of tropical forests (p. 586). But, while the problems appear to be largely physical (environmental), the causes and solutions lie much more in people's attitudes, values and expectations. We expect the environment to be all things to all people – it is to be life-supporting, it is to be useful, and it is to be beautiful. Unfortunately, the environment cannot meet all of these needs at the same time.

SYMPTOMS

Few people doubt that the world is in a mess, that the mess is getting worse, and that we are responsible for both. Humankind seems to be the only species that knowingly continues to foul its own nest, when it already has a fair idea of what damage it is causing. Many of the problems stem from the interactions between people, resources and pollution (Figure 1.2). A complete checklist of what we are doing wrong, and where we are going wrong, would be enormous and would touch upon just about every aspect of modern life. Some of the more obvious symptoms of the crisis are listed in Table 1.1.

There is no shortage of evidence that things are going badly wrong; but remember, these are really *symptoms* of what is wrong, not root causes. Some examples reinforce the point.

AIR POLLUTION

UNEP's *Global environment outlook 2000* highlights climate change associated with air pollution as the key environmental problem in the early 2000s. The major causes of air

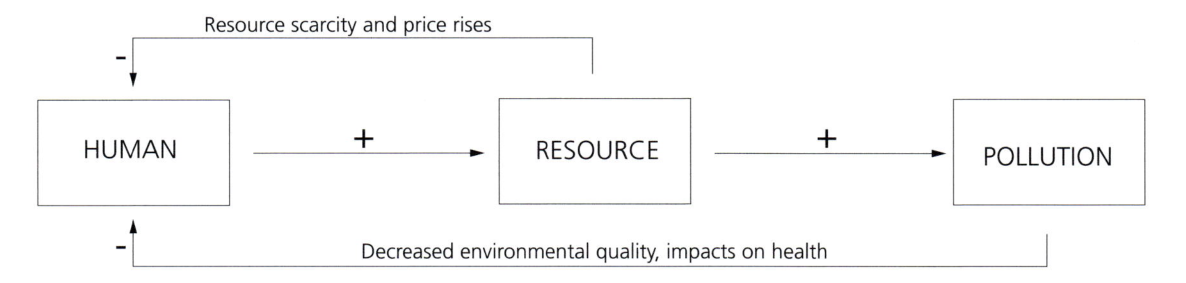

Figure 1.2 *Interactions between people, resources and pollution.*

BOX 1.2 GREENHOUSE GASES AND GLOBAL WARMING

The UNEP *Global environment outlook 2000* report describes global warming as the most serious environmental problem today. The theme of global warming is examined in detail in other chapters (particularly pp. 261–268), but it is useful to highlight here some of the issues and uncertainties that make this such a difficult problem to solve.

Evidence: Until the late 1990s, it was difficult to state categorically that the Earth's atmosphere is getting warmer, but several years of record-breaking temperatures now add strong support to the contention. Average temperatures across the world in 1998 were the highest ever recorded. Thermometer records (which stretch back 130 years) and evidence from tree rings and ice cores (see p. 340) confirm that the Earth is the hottest it has been since at least the Middle Ages. The hottest years on record are (in decreasing average temperature) 1997, 1995, 1990, 1991, 1994, 1998, 1983, 1987, 1996 and 1989. This means that the ten hottest years during the twentieth century were all during the 1980s and 1990s.

Causes: New evidence continues to emerge about what factors and processes might be causing the observed warming. For example, a previously unnoticed greenhouse gas – fluoroform (HFC-23) – with a global warming potential (see p. 261) 10,000 times greater than carbon dioxide and a life-expectancy of 260 years is increasing in the atmosphere at a rate of around 5 per cent a year and may contribute significantly to climate change. Another recently identified source of greenhouse gases is uncontrolled natural fires burning up to 200 million tonnes of coal a year in China's coalfields (which could contribute up to 3 per cent of worldwide emissions of CO_2). Recent studies have also shown that cement production, which is rising by 5 per cent a year worldwide, creates much more CO_2 than previously thought. Those who contest that human activities are causing global warming are losing ground, thanks partly to a recent study which shows that natural forces (high solar activity and low volcanic activity) caused most climate change until about 1950, since when human inputs have become much more important. Also now discredited is satellite-based evidence of slight cooling of the lower atmosphere, which has been accounted for by the previously undetected decrease in altitude of the instruments in question. The third assessment report of the IPCC (see p. 256), published in 2001, argues that the evidence now unequivocally points to humans as the main cause of observed warming.

Impacts: There is mounting evidence that global warming is causing a worldwide resurgence in mosquito-borne diseases such as malaria and dengue-fever, resulting in epidemics among people who have no immunity to them. Studies in Japan have also shown that the summer death toll in that country rises by nearly 600 if temperature rises by 1 °C. New studies have also shown that temperatures have risen sharply in Alaska and north-western Canada, causing a melting of the patchy permafrost (see p. 444) there and in turn triggering more than 2,000 landslides in recent years. Global warming in Alaska has increased tree growth but also increased populations of insect pests, which are damaging forests. A recent study suggests that the cost of protecting the UK coastline from sea-level rise associated with global warming over the next 50 years could be as high as £1.2 billion.

Solutions: The search for effective solutions is gathering pace as scientists and politicians realise that waiting for more hard evidence to emerge is not a viable option. Harnessing market forces could be one way forward, for example by setting up a global trading system to buy and sell permits to pollute the atmosphere with the greenhouse gas CO_2. One possible approach might be to issue permits according to the amount by which each nation has agreed to cut its emissions, so that countries that do better than meet their target could sell spare permits to other countries.

pollution are road traffic, the burning of coal and high-sulphur fuels, and forest fires. Air pollutants – particularly invisible gases – can be blown by the wind over vast areas and can damage human health as well as wildlife. An estimated 625 million people worldwide live in areas (mainly industrial cities) where the air is unhealthy. Increased levels of ozone in the lower atmosphere (see p. 247) decrease crop yields in the USA by up to 10 per cent and damage lung and respiratory tissues in humans. At the same time, ozone depletion in the upper atmosphere (see p. 249),

Plate 1.1 *In many areas, human activities have altered landscapes and the environment throughout history. This pastoral scene of goat herding in Yazir Golü, south-west Turkey, masks the long-term impacts of such traditional practices on vegetation change, particularly through deforestation.*
Photo: Philip Barker.

Table 1.1 Symptoms of the environmental crisis

According to the UNEP *Global environment outlook 2000* report:

- there will be a billion cars by 2025, up from 40 million in 1945
- a quarter of the world's 4,630 types of mammals and 11 per cent of the 9,675 bird species are at serious risk of extinction
- more than half of the world's coral is at risk from dredging, diving and global warming
- 80 per cent of forests have been cleared
- a billion city dwellers are exposed to health-threatening levels of air pollution
- the global population will reach 8.9 billion in 2050, up from 6 billion now
- global warming will raise temperatures by up to 3.6 °C, triggering a 'devastating' rise in sea levels and more severe natural disasters
- global pesticide use is causing up to 5 million acute poisoning incidents each year

caused mainly by CFCs, is allowing more incoming harmful ultraviolet radiation from the Sun to reach the Earth's surface, threatening to increase the incidence of skin cancers and eye damage in humans.

Air pollution by greenhouse gases (including carbon dioxide, nitrous oxide and methane) is already starting to cause global warming (see Box 1.2). The Intergovernmental Panel on Climate Change has concluded that 'the balance of evidence suggests that there is a discernible human influence on global climate'. Scientists predict that average global temperature will rise by 1.5–4.5 °C by 2030, causing patterns of temperature and precipitation to change significantly, sea level to rise, and droughts and storms to become increasingly common and serious. Coping with the expected impacts of global warming will not be easy and will require carefully thought-out responses (Figure 1.3).

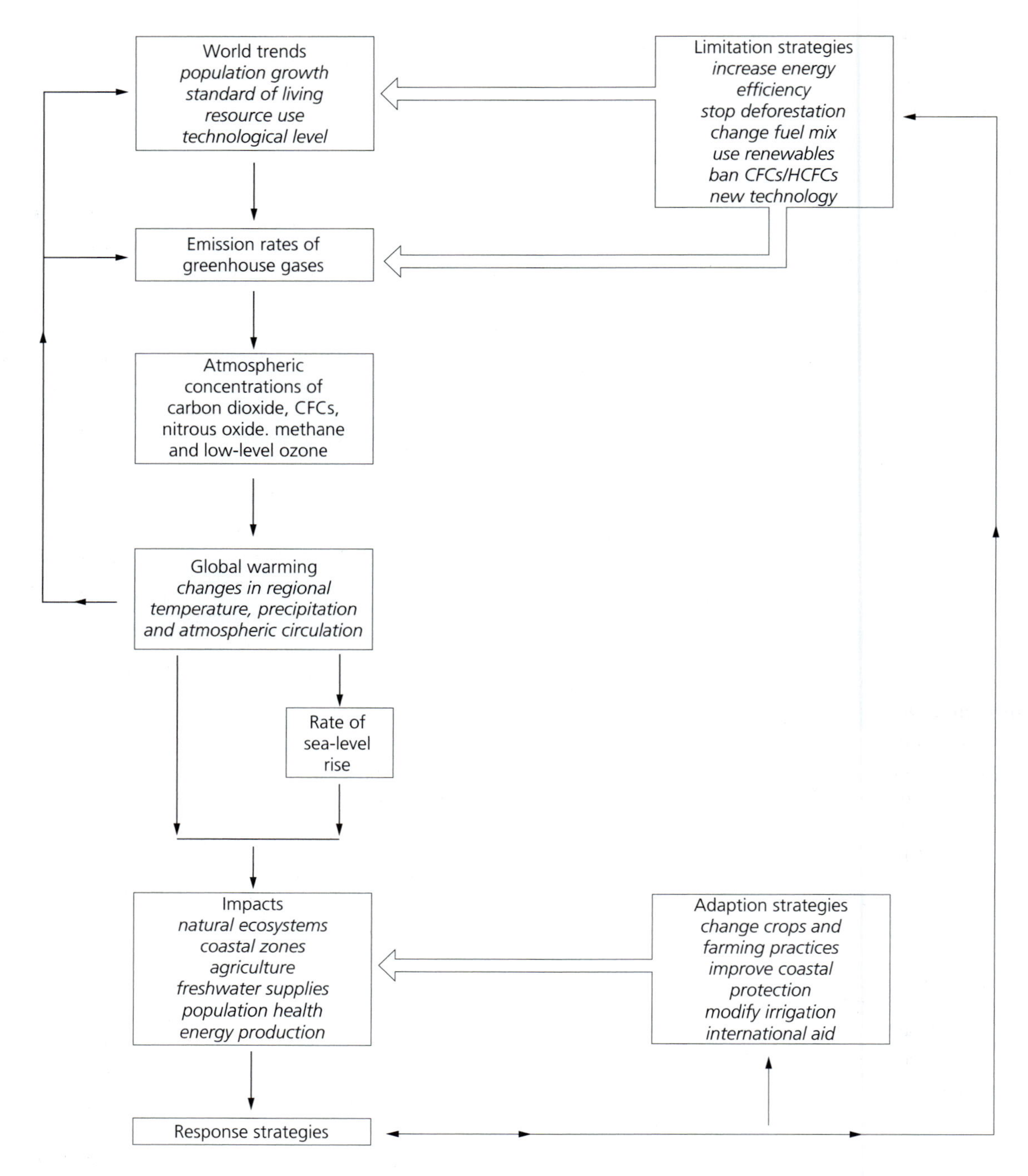

Figure 1.3 *Impacts and responses to the greenhouse effect. The problem of coping with greenhouses gases illustrates many of the complexities of all environmental problems, where scientific understanding of processes is incomplete, assessment of likely impacts is difficult, and choice of coping strategies is never straightforward.*

Source: Figure 12.2 in Elsom, D.M. (1992) Atmospheric pollution. Blackwell, Oxford.

TROPICAL DEFORESTATION

One of the most serious environmental problems today is the continued destruction of tropical rainforests. Four-fifths of the forest area is cleared for farming, and most of the rest is selectively logged. Although the tropical forests cover only about 6 per cent of the world's land surface, they are an essential part of our life-support system (see p. 586). They help to regulate climate, protect soils from erosion and provide habitats for millions of species of plants and animals. Up to nine-tenths of all the species of wildlife on Earth live in the tropical forests. UNEP's *Global environment outlook 2000* notes that Africa lost 39 million hectares of tropical forest during the 1980s and a further 10 million hectares by 1995.

BIODIVERSITY

Beyond the tropics, many species of plants and animals are under threat because their natural habitats are being destroyed (see p. 536). Other wildlife is threatened by excessive hunting and trapping for trade (especially of species that are rare and endangered, like rhinos and African elephants). UNEP's *Global environment outlook 2000* indicates that, in 1996, 25 per cent of the world's approximately 4,630 mammal species and 11 per cent of the 9,675 bird species were at significant risk of total extinction. The oceans are being over-fished (catches rose from about 30 million tonnes in 1958 to 50 million tonnes in 1975, and to more than 97 million tonnes in 1995), putting at risk sustainable yields of some species. Pollution of the oceans is also reducing fish harvests in many areas (see p. 486).

POPULATION

Another critical part of the equation is population pressure. Chapter 2 explores the relationships between population and environment.

THE CRISIS IN CONTEXT

HISTORICAL CONTEXT

While a great deal has been written in recent years about the environmental crisis, there is nothing new about the idea that we are damaging the very environment on which we depend. For example, it is known that forest clearance is a traditional form of land management that has been carried out throughout the settled world for at least the last 4,000 years. Forests have been felled in the present-day Sahara and Arabia since 5000 BC, in China since 2000 BC and in the USA since about AD 1800. Figure 1.4 shows how much natural woodland was cleared from an area in Wisconsin, USA, during the first 150 years of European settlement.

A number of ancient civilisations declined or disappeared completely as a result of their unsustainable use of natural resources. In Mesopotamia, for example, intensive irrigation agriculture and the associated salinisation (see p. 401) and waterlogging of the fields destroyed the basis of Sumerian society. The Mediterranean regions suffered from long-term environmental decline associated with deforestation and soil erosion caused by continuous human occupation. In the Indus valley (India) and Mayan lowland jungle (Central America), large-scale deforestation leading to soil erosion caused the collapse of highly advanced civilisations.

Human activities have caused the extinction of wildlife since early prehistory. For example, many large American mammals (including mammoths and many species of horse) became extinct towards the close of the last ice age, possibly because early Americans used fire drives to encourage whole herds of big game over cliffs for hunting.

Air pollution is not new either. In 1306, a London manufacturer was tried and executed for disobeying a law

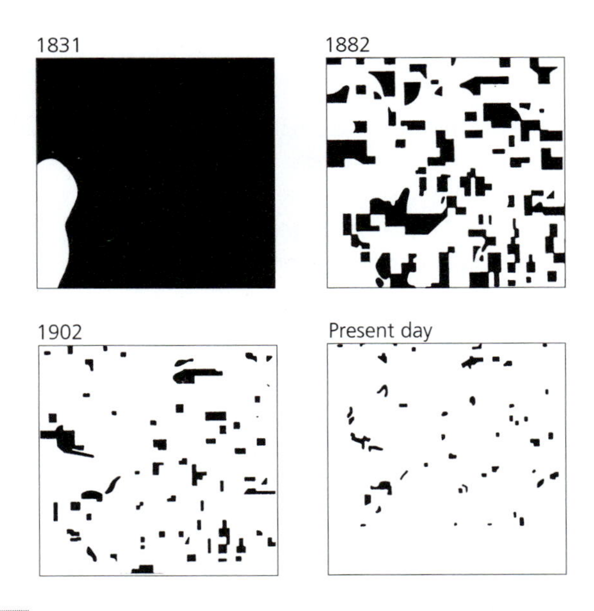

Figure 1.4 *Woodland clearance in Cadiz Township, Wisconsin, USA, during the period of European settlement. When the first European settlers arrived, in the early nineteenth century, most of the area was covered by natural woodland. The woods were cleared to make way for agriculture, so that today relatively little remains. Each small rectangle is about 90 km².*

Source: Figure 5.10 in Simmons, I.G. (1996) Changing the face of the Earth. Blackwell, Oxford.

forbidding the burning of coal in the city. Queen Elizabeth I refused to enter the city in 1578 because of the smoke problem. By 1700, air pollution was causing serious damage by killing vegetation, corroding buildings, and ruining clothes and soft furnishings.

ROOT CAUSES

Inevitably, there has been heated debate about what has caused the environmental crisis, and many different (and at times conflicting) suggestions have been made about what the most important root causes might be. There are certainly a number of important factors (Box 1.3) that, added together, help to account for observed changes in the way in which the environment has been used.

While in recent years there have been great advances in scientific and public understanding of the relevance and seriousness of the environmental crisis, the debate about root causes remains largely unresolved. This is perhaps inevitable, given the many different factors involved and the wide variety of disciplines (including history, social science, economics, psychology and religious studies) that have taken part in the debate.

BOX 1.3 ROOT CAUSES OF THE ENVIRONMENTAL CRISIS

A number of factors are involved, including:

- developments in technology throughout history, which have given people a better ability to use the environment and its resources for their own ends (particularly since the Industrial Revolution);
- the rapid increase in human population in recent centuries, which has significantly increased population densities in many countries;
- a significant rise in human use of natural resources, particularly over the last century;
- the emergence of free-market economies, in which economic factors play a central role in decision making about production, consumption, use of resources and treatment of wastes;
- attitudes towards the environment, particularly among Western cultures, which regard it as freely available for people to do whatever they like with;
- the short-term time horizon over which many people, companies and countries make decisions, which means that short-term profit maximisation has generally been taken more seriously than long-term sustainable use of the environment.

COMMON RESOURCES

An important perspective on the environmental crisis, which sets it into context even if it does not fully explain it, is the idea of the environment as a commons (common property resource) which is freely available – at least in theory – to anybody who wants to use it (Box 1.4).

The way in which herdsmen treated their medieval commons has many similarities with the way we treat the environment today:

- we think of it as freely available without restriction;
- we regard it as our right to be able to use it;
- each of us wants to get as much as we can out of it;
- it can tolerate only a certain level of use before it is damaged;
- we compete for a larger share of it than other people.

The 'tragedy of the commons' analogy applies just as well to common property resources, including atmosphere outside 'national air space', the oceans outside 'territorial limits' and wilderness (unspoiled land that is not privately owned). Growth, competition and private ownership are hallmarks of modern Western society, yet these are the very forces that threaten and damage the commons. UNEP's *Global environment outlook 2000* stresses the need for a much better understanding of the interrelationships between economic, social and environmental change as a basis for a more sustainable future.

Little wonder that many scientists and politicians have recently argued that these forces must be replaced by co-operation and sustainable use, and that the commons debate was rekindled at the 1992 Rio Earth Summit (see p. 13). Critics of modern development practices often argue that many environmental and socio-economic problems (including poverty) can be traced to enclosure of the commons and the domination and dispossession of local communities by enclosure. The principle of permanent sovereignty over natural resources (PSNR), which emerged in the 1950s as a legal concept to formally recognise the sovereignty of developing countries over their resources, is also part of this wider debate.

BOX 1.4 THE TRAGEDY OF THE COMMONS

In 1968, an American biologist called Garrett Hardin published a paper called 'The tragedy of the commons', in which he compared present-day attitudes towards the environment with the attitudes of medieval villagers to common grazing land in their village. In this typical medieval English village, all of the herdsmen had a right of access to the common pasture, which (in these pre-enclosure times) had no field boundaries and was open to all. Each herdsman would be inclined to graze as many cows as possible on the common land, because that way he would get the maximum return. Such an arrangement – of use without restriction – would work fine so long as the number of people and cattle remained below the carrying capacity of the common land (perhaps because of disease and poaching). Once the carrying capacity had been exceeded, each cow got relatively less grass to eat, so its yield declined, and over-grazing decreased environmental quality and led to soil erosion.

Because each person was trying to maximise their income, they were all trapped in a system that could tolerate only certain levels of use, competing against each other for a larger share of the available resources (space and grass in this case). As Hardin concludes:

> each man is locked into a system that compels him to increase his herd without limit – in a world that is limited. Ruin is the destination toward which all men rush, each pursuing his own best interest in a society that believes in the freedom of the commons. Freedom in a commons brings ruin to all.

NEW DIMENSIONS TO THE CRISIS

Many people ask 'Is the situation really as critical as some scientists are saying it is?' There are problems, but is it really a crisis? Optimists say 'There have been environmental scare stories before, and we are still here – so who is kidding whom?' But such an argument overlooks several important aspects of today's crisis that make it different from past situations.

New threats: the UNEP *Global environment outlook 2000* highlights global warming as the biggest threat to the planet, followed by the scarcity of fresh water, deforestation and desertification. But, to make matters worse, new threats are also emerging. One, which is threatening to be a 'largely uncontrolled experiment' on a global scale, is the increase in nitrogen loading (see p. 89) on the world's environment. Nitrogen is used as a fertiliser in intensive agriculture and is released when fossil fuels are burned by industry, power stations and cars (see p. 255). Excessive nitrogen levels in the environment are triggering the growth of unwanted plants, which damage the ecology of areas like the Black Sea, the Baltic Sea and Chesapeake Bay (see Box 16.38).

Global problems: we now have the power to change the environment on a global scale for the first time ever. Many of today's pressing problems are affecting the whole world. For example, if the hole in the atmosphere's ozone layer continues to grow (see p. 249), then people around the world will suffer from excessive amounts of damaging ultraviolet radiation. If felling of tropical rainforests continues (see p. 586), we will all be affected by the resultant climatic changes. Many environmental problems – especially air and water pollution – are international, because they cross national frontiers. The nuclear fallout from Chernobyl (see p. 246) was spread across Europe within a week, and the acid rain (see p. 251) exported from some countries (like Britain) is quickly imported elsewhere (like Scandinavia). Serious political as well as scientific problems are posed by such uncontrollable movements.

Rapid build-up: a second new cause for concern is the speed with which serious problems are building up. Rates of change are accelerating. In a typical day, 300 km^2 of tropical forest is destroyed or badly damaged; deserts advance over a similar area; 200 million tonnes of valuable topsoil is washed or blown away, one more species becomes extinct, and 100,000 people (nearly half of them children) die from starvation.

Persistence: the long-lasting effects of many environmental problems aggravate the situation even further and mean that we are passing on to future generations problems that we have created but for which we have no solutions. Some waste products, such as toxic chemicals and radioactive wastes from nuclear power stations (Figure

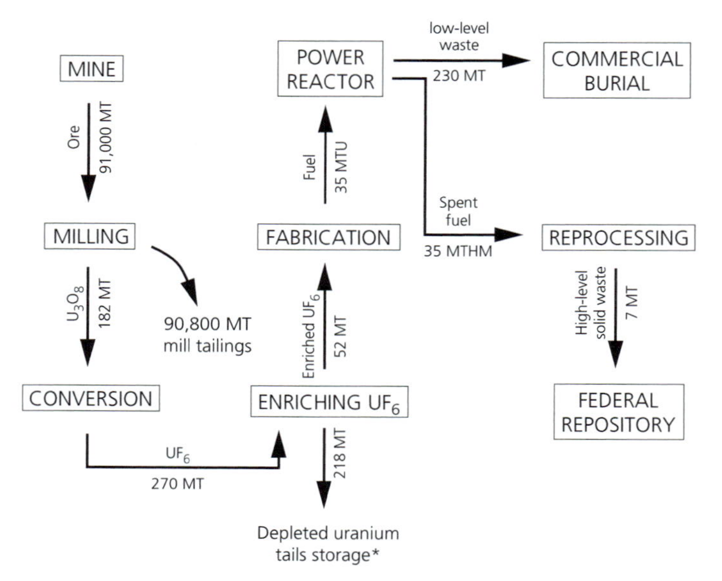

Figure 1.5 *The nuclear fuel cycle. There are many different processes involved in the nuclear fuel cycle, from initial mining and milling of the ore, through conversion and enrichment of the nuclear fuel, its use in nuclear power stations, and the ultimate management of nuclear wastes. Quantities are shown in metric tons, and they represent the average annual fuel requirements for a typical 1,000 MW light water reactor.*

Source: Figure 19.6 in Cunningham, W.P. and B.W. Saigo (1992) Environmental science: a global concern. Wm. C. Brown Publishers, Dubuque, Iowa.

1.5), will still be around in thousands of years' time. Today's land-use changes, such as forest felling and the building of cities, might trigger climatic changes that affect the next century or even further ahead. Activities that cause the extinction of wildlife species have a cost that extends to eternity, because extinction is for ever.

Thresholds: many scientists are worried that we are now stretching ecological systems to their breaking point (critical thresholds beyond which irreversible changes can occur). This makes the present situation more serious and more critical than ever before, and it means that the options available to generations that follow us (including our own children) will depend largely on what actions we take. The planet has a finite ability to absorb our wastes and renew its resources, and to knowingly approach those natural limits is to deliberately play Russian roulette with our environmental life-support system.

Uncertainty: another cause for concern is growing awareness that we simply do not know how many of our actions are affecting the environment and people's health. There are many uncertainties in linking observed effects with possible or suspected causes. Examples include the widespread controversy and anxiety surrounding the suspected links between leukaemia and radiation pollution, and uncertainties over the possible impacts on human health and on ecosystems of growing genetically modified crops (see p. 532). New risks are being created every day through the development of new technologies (such as genetic engineering) or the careless use of existing ones (such as the 1984 explosion at the Bhopal pesticide factory, which killed 3,600 people by methyl isocyanate poisoning).

The climate change debate, where there are clearly many uncertainties (see p. 265), shows just how difficult it is to make reliable predictions of how environmental systems are likely to change over the next 100 years or so. There are also major uncertainties in predicting how these changes are likely to affect national economies. Inevitably, a major change in climate will have wide-ranging impacts, particularly on the agriculture, transport, energy and insurance sectors of the economy (Table 1.2). While some sectors (such as agriculture and transport) can adapt quite quickly to environmental change, others (such as the energy and construction industries) cannot, and they require significant advance warning and long-term investment.

Table 1.2 Some links between environmental change and national economies

Sector	*Potential significance of environmental change*
Agriculture, forestry and fisheries	*Very significant*: changes in climate, ocean circulation and effects of pollution in regional seas, acid precipitation, etc. would have a major influence on all aspects of these activities. Agriculture and forestry are key components of biogeochemical pathways.
Energy and water supply	*Very significant*: changes in climate will have a direct effect on demand for energy and water. Nuclear power and groundwater supply raise issues of biogeochemical pathways in terms of waste disposal and groundwater effects.
Manufacturing	*Some significance*: depends on the supply of primary products. Climate change could have substantial effects on transport systems and vehicle designs.
Construction	*Significant*: sea-level rise will require more coastal defences, and climate affects building design.
Tourism	*Some significance*: tourism will be susceptible to changing climate.
Transport and communication	*Some significance*: sea and air transport will be affected by changing climate.
Banking, finance, insurance, etc.	*Significant:* insurance is particularly affected by changing environmental conditions.
Defence	*Significant*: changing the oceans and atmosphere affects defence decisions.
Education and research	*Very significant*: research on environmental change, and education about it, are both crucial.
Medical and health services	*Very significant*: environmental change will have direct impacts on health and health care.

THE RIO EARTH SUMMIT

Without doubt the most significant environmental event of the last two decades was the United Nations Conference on Environment and Development (UNCED) – widely referred to as the Earth Summit – which took place in Rio de Janeiro, Brazil, in June 1992. The conference (Box 1.5) was the culmination of more than a decade of preparatory work, and it established the tone, pace and direction of the international environmental agenda for the foreseeable future.

PREPARATION

The seeds for the Earth Summit were sown in Stockholm in 1972, where the United Nations held the first international Conference on the Human Environment. One outcome of the Stockholm meeting was the establishment of the United Nations Environment Programme (UNEP) to spearhead international initiatives designed to protect the environment. A major step forward was taken in 1984, when UNEP (jointly with the World Wide Fund for Nature and the International Union for the Conservation of Nature) published the *World Conservation Strategy*. This document was effectively a prospectus for environmental conservation, with a sharp focus on the need to protect nature and natural resources, but it largely ignored the question of development and indeed the whole question of human needs for food and resources.

Several years later, the United Nations appointed a World Commission on Environment and Development (WCED), chaired by Gro Harland Brundtland (then prime minister of Norway). The commission produced the now famous report *Our Common Future* – commonly referred to as the Brundtland Report – which set out the idea of sustainable development (see p. 26). The axis of the debate switched firmly from nature conservation *per se* to sustainable development of all of the Earth's resources.

The United Nations was keen to follow up the Brundtland Report and to implement its recommendations. In 1989, it announced that plans were being made to hold a major international conference on environment and development in 1992, which Brazil offered to host. A series of preparatory meetings was held between 1990 and 1992 in various countries, at which governments, non-governmental organisations (NGOs) and expert scientists discussed and largely reached agreement on a series of basic documents which would be formally debated at Rio.

AGREEMENTS

Agreement was reached at the United Nations Conference on Environment and Development on a wide variety of

BOX 1.5 THE RIO EARTH SUMMIT

The United Nations Conference on Environment and Development, based at the Rio Centro, was accompanied by a series of events held in the city between 3 and 14 June 1992, all designed to raise awareness of environmental issues. There were many unofficial events, displays, concerts and fringe meetings, which added to the colourful carnival atmosphere.

The conference, described by some as 'the biggest show on Earth', was a record breaker in many ways:

- 172 countries were represented and took part;
- the summit at the end was attended by 120 heads of state, making it the largest ever gathering of world leaders;
- there were more than 350 scheduled meetings;
- 1,600 non-government organisations (NGOs) were represented;
- the conference and associated events attracted an estimated 450,000 visitors;
- it was by far the largest conference ever held about the environment;
- there were almost 100 press conferences, and nearly 9,000 journalists covered the events.

Although most of the world's media attention was focused on the conference itself, the debates that took place were effectively open forums where the views of different countries were reported and the draft documents were finally approved. Most of the hard work of drafting documents and position papers, preparing agreements for signing, ironing out many political and bureaucratic differences between the countries represented, and agreeing on the main agenda items had been completed during the preparatory meetings.

environmental objectives. The key outcomes of the Earth Summit are incorporated in five formal documents – two conventions, two statements of principle and an action programme:

- a convention on climate change
- a convention on biodiversity
- a statement on forest principles
- the Rio Declaration
- Agenda 21 (the action programme).

The Convention on Climate Change: one of the most important products of UNCED, this convention calls on countries to commit themselves to stabilise emissions of greenhouse gases by the year 2000 (see p. 15). Oil-producing countries were concerned that limitations on the burning of fossil fuels (a principal source of carbon dioxide) would be a direct threat to their economies, so there was heated debate, and a great deal of behind-the-scenes negotiating was required to get the convention accepted. Five years after Rio, in 1997, the debate was still raging (Box 1.6).

The Convention on Biological Diversity: this convention is designed to preserve plant and animal life (see p. 541). It was also hotly debated, partly because it defined wild (as well as improved) species as having economic value, and it proposed that genetic resources be recognised as national assets that can only be exploited on the basis of agreements between signatory states. US President George Bush did not sign the convention, allegedly out of consideration for the country's biotechnology industry.

The forest principles document: the proper title of this document is the 'Non-legally binding authoritative statement of principles for a global consensus on the management, conservation and sustainable development of all types of forest'. It includes important principles on replanting, establishing new forests and protecting the rights of indigenous people.

The Rio Declaration: this declaration is a general statement of intent that centres on the well-being of people, on the right of states to control their own natural resources and on their obligation not to damage the environment of other countries. Among other things, it recommends the precautionary principle, internalisation of environmental costs, use of environmental impact statements and the polluter-pays principle.

Agenda 21: this is the action plan, which 179 states agreed to, that is designed to turn the theory into practice. Agenda 21 is a national and international 'blueprint for action' on environment and development, an agenda for the twenty-first century. It is over 500 pages long and contains forty chapters, each of which focuses on a specific theme and is divided into programme areas setting out goals, activities and provisions on funding and implementation. The total estimated cost of Agenda 21 is nearly

BOX 1.6 THE 1997 KYOTO PROTOCOL

Under the Climate Change Convention agreed at the Rio Earth Summit in 1992, industrialised countries agreed to stabilise their emissions of CO_2 at 1990 levels by the year 2000. But by 1996, most of these countries had accepted the scientific case for significant reductions in their emissions and promised to set reduction targets by the end of 1997. It then became very obvious that most countries would not be able to meet the 2000 target, and many countries abandoned the 1992 agreements. President Clinton postponed the stabilisation of US emissions to 2012 and ruled out real cuts before 2017. The European Union called for a 15 per cent cut by all industrialised countries by 2010.

In December 1997, scientists and diplomats from 160 countries gathered in Kyoto, Japan, for the UN Climate Convention. The aim was to find acceptable ways forward. Many countries, including the United States, demanded 'flexibility measures', which would allow them to bank, borrow or trade spare emissions 'permissions'. It was also argued that flexibility would allow countries to emit more greenhouse gases if they planted more trees (which act as CO_2 sinks) or could demonstrate that they had reduced pollution in other countries (a procedure known as 'joint implementation'). The European Union negotiators were against flexibility measures and wanted to implement its targets collectively. The Kyoto conference eventually settled for stability targets rather than cuts, and it adopted flexibility measures.

Under the Kyoto Protocol, rich industrial countries agreed to reduce their emissions of greenhouse gases by an average of 5.2 per cent by 2010. Developing countries were not set formal emission limits, partly because to do so would unfairly inhibit their pursuit of economic growth. But the political debate between countries continues, and the United States has insisted that some developing countries accept emission targets before it is willing to reduce its own greenhouse gas emissions. Most other industrialised countries (including the European Union and Japan) have promised to bring the Kyoto Protocol into law by 2002.

A new proposal, unveiled in June 1999, is designed to break this deadlock. The new plan requires developing countries to improve the carbon efficiency of their economies (by reducing the amount of carbon produced per dollar of their gross national product) rather than requiring them to accept absolute limits on emissions. Some developing countries are already making great progress in this direction. China, for example, has increased the carbon efficiency of its economy by 47 per cent over the past 20 years, mainly as a result of measures designed to reduce urban air pollution. Supporters of the new plan point out that it reduces the scope for trading in rights to emit greenhouse gases.

Trading of rights to emit greenhouse gases is a controversial theme, but one taken seriously under the Kyoto Protocol. Under current plans, developing countries would receive fixed targets above their current emission levels to allow for economic growth. Rich countries could buy this excess capacity, and thus avoid the need to reduce their own output. Critics of the trading approach argue that it could well serve to increase overall emissions of greenhouse gases.

An alternative approach is the so-called clean development mechanism (CDM), an incentive scheme proposed at the 1998 Buenos Aires climate summit. The CDM approach seeks to give 'carbon credits' (credit to industrialised countries towards the emission targets agreed in the Kyoto Protocol) in developing countries by investing in renewable energy projects or by enlarging carbon sinks such as forests. But countries are divided over the question of which energy technologies should count (nuclear energy does not, for example) and the tendency to promote coal-fired power stations at the expense of renewable energy technologies.

A number of key issues surround the process and outcome of the Kyoto meeting:

- Global negotiations on climate control will never be easy, but they must be undertaken and within a unified framework committed to seeking effective solutions.
- It must be understood that the main objective of the Kyoto Protocol – stabilising global climate at 'non-dangerous levels' – is a long-term objective that will inevitably take some time to achieve.
- Much more detailed assessment is required about what constitutes 'dangerous' climate change; the Kyoto Protocol proposals are designed without reference either to the level of climate change to which we can adapt or to the level of climate change that will cause significant damage.
- Given the history of global greenhouse gas emissions and the inertia of climate systems, we are already committed to further global warming.

US$600,000 million (Table 1.3), and a number of sources of funding were identified (Box 1.7).

Table 1.3 The cost of Agenda 21

Item	Cost ($ billion)
Protecting the atmosphere	21.00
Mountains and deserts	21.70
Forests	31.20
Rivers and seas	67.90
Biodiversity	23.50
Population growth	7.10
Land use and agriculture	31.85
Human settlements	218.00
Poverty and consumption	30.00
Health	51.00
Waste	28.10
Government and United Nations	20.70
Major groups	0.37
Trade and money	8.90
Total	561.32

GLOBAL ENVIRONMENT FACILITY

A mechanism and a framework for handling international financial transfers in four principal environmental areas, including climate change and biodiversity, were first proposed in 1990 by the governments of some leading developed countries. The proposals were taken seriously, and in 1991 the World Bank and the United Nations set up the Global Environment Facility (GEF) as a funding mechanism within the umbrella of UNCED.

Discussions leading up to and at the Earth Summit, at which the climate change and biodiversity conventions were agreed, explored the likely role of the GEF in helping to finance their adoption in developing countries. It was agreed that the GEF would be entrusted with financial transfers under both conventions, but only on an interim basis to allow reform of the structure and governance of the GEF that will move it away from the World Bank and closer to the UN system and institutions. The climate

BOX 1.7 FUNDING AGENDA 21

In the past, the environment has been viewed as a free resource and largely taken for granted. But it is now recognised that natural resources must be paid for, that restoring, maintaining and enhancing environmental quality is costly, and that money to pay for all this will have to be set aside. Sacrifices will inevitably have to be made as we decide what other activities to cut back in order to generate the money required to protect the environment.

The estimated total cost of implementing all of the recommendations in Agenda 21 is more than US$500,000 million (at 1992 prices) (Table 1.3). The Earth Summit proposed that US$460,000 million (77 per cent) would come from local sources and the remaining US$140,000 million (23 per cent) from international sources, including foreign aid and the World Bank. It was agreed at Rio that each country should meet the costs of implementing Agenda 21 from its own resources, but that additional funding would be needed for developing nations. Four funding mechanisms were discussed to help the developing countries to pay for Agenda 21:

1. Official development assistance: a target was set of each developing country giving 0.7 per cent of its gross national product (GNP) to official development assistance, which compares with the 0.35 per cent on average that developed countries currently give.
2. The Global Environment Facility (GEF) set up by the World Bank, the United Nations Development Program (UNDP) and the United Nations Environment Program (UNEP).
3. The International Development Association, the branch of the World Bank group that provides interest-free loans to the lowest-income countries.
4. The peace dividend – defence allocation freed as a result of *détente* (particularly since the end of the Cold War). Global military spending totals around US$900,000 million a year, and it is hoped that at least some of this money can be redirected into environmental protection. The total cost of UNEP between 1982 and 1992 was about US$500 million, which is the equivalent of about 5 hours of military spending!

change and biodiversity conventions have been significantly strengthened by giving them control (which would otherwise rest with the GEF) of policy, programme priorities and eligibility criteria for financial transfers.

Given the large amounts of money involved, and the significance of the conventions, the pilot phase of the GEF (1990–93) was monitored very closely. A total of fifty biodiversity grants were awarded, totalling over US$300 million, but the projects to which they were awarded suffer from a lack of participation by local communities and NGOs. Critics conclude that neither the outlook for the next phase nor its likely impact on biodiversity conservation is promising. Critics also argue that policy development within the GEF is not fully objective because it is heavily influenced by government agencies and NGOs, and that particular social groups can have a disproportionate input into the policy-making process.

Many scientists are also disappointed that the GEF has not had a more positive impact on agencies like the World Bank, which gives financial assistance to major schemes including dam construction, road building and chemical-intensive agricultural projects. Between 1993 and 1997, for example, the World Bank invested US$9.4 billion in fossil fuel projects that will accelerate climate change, and less than US$300 million on schemes to prevent it.

GEF may still be a young and developing organisation, but it is rapidly emerging as a key player in international environmental politics because it manages many billions of dollars' worth of 'global environmental finance' each year, and through this it can steer the global environmental agenda.

CRITIQUE

Despite the huge amount of media coverage it attracted, and the promising tone of the conventions and other items that were agreed, the Rio Earth Summit has had a mixed reaction. Environmentalists from developed countries hailed the conference as the 'last chance to save the planet', while delegates from developing countries saw it as an opportunity to redress longstanding economic grievances.

Supporters argue that although the conference attracted a great deal of bad press coverage and was widely criticised by environmental pressure groups, it did mark a welcome and substantial change in international political attitudes towards the environment. They also applaud the recognition formally given at Rio of the need to tackle over-consumption in industrialised states and poverty and resource scarcity in the developing world. Agenda 21 recognised that growth, poverty alleviation, population policy and environmental protection were mutually reinforcing and supporting.

Critics argue that while UNCED delivered an impressive number of international agreements on a variety of topics, it hardly began to address the fundamental driving forces of global environmental change, such as trade, population growth and institutional change. They also point out that the legal agreements signed at Rio are relatively weak and lack binding commitments and timetables.

On balance, even though some aims were not achieved at Rio, it was nonetheless an important step on the long-term path towards environmentally sustainable development. While Agenda 21 is not legally binding, it is based on political and moral commitments from those who signed it. An urgent priority is to implement the UNCED decisions at national levels within a spirit of international co-operation. Building consensus is only a first step, and it is widely recognised that much more work needs to be done as to how, at the national level, environmental considerations can be integrated into policies, programmes and development projects.

Many challenges lie ahead, including the question of how to retain long-term environmental perspectives while responding to the short-term political and economic challenges that are so urgent in many countries today. UNCED exposed and ultimately suffered from a number of important tensions – including the balance between present and future, developed and developing countries, people and environment, development and conservation. There are no easy answers.

MONITORING AND ANALYSIS

The rise in concern about the state of the environment, both nationally and globally, has partly prompted and partly been a response to more detailed and comprehensive monitoring of the environment. One hallmark of environmental research since the start of the 1990s has been the development of new technologies, procedures and protocols for collecting, analysing, interpreting and reporting environmental information. Great improvements have been made in harmonising practices and standards between countries, so that it is now becoming both possible and meaningful to collate national statistics into global summaries.

A useful product of this new interest in producing and interpreting environmental information is the state-of-the-environment reports (Box 1.8) that many countries now

> **BOX 1.8 STATE-OF-THE-ENVIRONMENT REPORTS**
>
> State-of-the-environment reports are important because they provide up-to-date information relating to environmental quality and natural resources, with a particular emphasis on:
>
> - an inventory of what is there;
> - an assessment of their state and quality;
> - a baseline against which to compare changes; and
> - the prospect of monitoring changes through time.

produce regularly. A great deal of effort is now being invested in the design, implementation and effective operation of environmental information systems. The development of new approaches and technologies – such as geographical information systems (GIS) (Figure 1.6), remote sensing, simulation modelling and knowledge-based expert systems – is opening up new analytical possibilities and making it much easier to simulate how the environment works and how it might respond to different management scenarios. Such analyses rely heavily on powerful computers, which are used to store and manipulate databases and to run complex predictive models. As computers become more and more powerful – between 1950 and 1990, for example, computers increased in speed by a factor of ten roughly every five years – it becomes possible to analyse bigger and bigger data sets and run more complex models in more sophisticated ways.

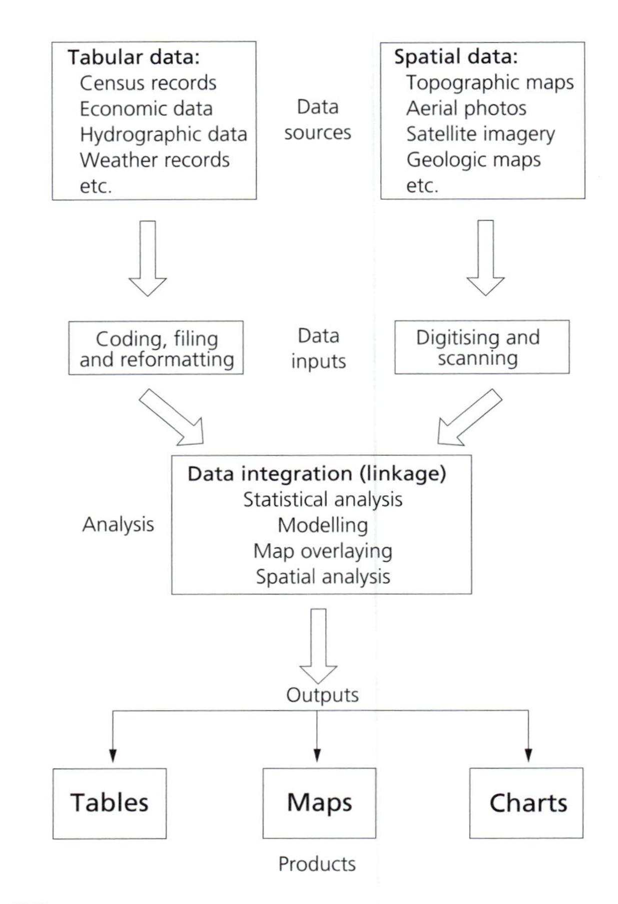

Figure 1.6 *The basis of a geographical information system. All GIS systems contain four essential components – data sources, data inputs, analysis or data integration, and products such as maps, tables and charts.*

Source: Figure 16.13 in Marsh, W.M. and J.M. Grossa (1996) Environmental geography: science, land use and Earth systems. John Wiley & Sons, New York.

ENVIRONMENTAL INDICATORS

The rise in interest in the state of the environment has prompted the search for ways of measuring and expressing environmental quality that are meaningful and measurable. Much attention is being devoted to the selection of appropriate environmental indicators – such as which threshold concentrations of different air pollutants are relevant to human health and to environmental stability, how best to measure the diversity of species of wildlife in natural habitats, and how to express land-use change in the most useful ways.

To date, there have been relatively few attempts to produce composite environmental indices for individual nations, partly because the choice of which indicators are most appropriate depends heavily on the purpose of the analysis. At the international scale the problems are even bigger. One interesting initiative is the CORINE (Co-ordinated Information on the Environment in the European Community) project, which aims to build a common environmental database for the EU using geographical information systems. Early work in compiling the database is promising, but different countries use different data standards and have their own cartographic styles and conventions, which sometimes makes comparisons between countries difficult.

ENVIRONMENTAL MONITORING

Recent years have also seen a significant increase in environmental monitoring (Box 1.9), which is designed to provide detailed, up-to-date and reliable information about

BOX 1.9 DEVELOPMENTS IN ENVIRONMENTAL MONITORING

Significant developments in environmental monitoring during the 1990s include:

- harmonisation of data collection between countries;
- global coverage (often using remote sensing and satellite technologies);
- continuous monitoring (often linked to telemetric systems in which remote instruments can be interrogated live by computer from a central control office); and
- computerised data-handling systems (which can automatically read data from electronic monitoring equipment, store it and analyse it).

the state of the environment. Monitoring provides the data that are needed to facilitate the development, testing and operation of complex global models. It also provides baselines against which to evaluate rates and patterns of environmental change, and it can be used to give early indications of natural environmental adjustments and possible risks.

Many countries have been operating local, regional and national environmental monitoring schemes for some time, but until quite recently there was little harmonisation of procedures or standards. Different analytical techniques were used to monitor air and water chemistry, for example, and samples were collected much more regularly in some countries than in others. This made it very difficult to compare environmental information between countries in any meaningful way, which therefore inhibited scientific research into the scale and pattern of many particular environmental problems.

Many countries are now introducing effective national monitoring programmes that collect coherent natural resource information at regional and national scales. In the USA, for example, the Environmental Protection Agency has launched an Environmental Monitoring and Assessment Program (EMAP) based on an integrated family of monitoring designs that could be applied at a variety of scales, and a Long-Term Monitoring (LTM) project to establish general environmental baselines.

The United Nations Environment Program (UNEP) is responsible for co-ordinating the monitoring networks that provide invaluable information about the global environment. The data are collected mainly from surface stations around the world as part of the Global Environmental Monitoring System (GEMS), which was established by UNEP, covers 142 countries and collects data relating to atmosphere, climate, pollution and renewable resources.

REMOTE SENSING AND SATELLITES

Ground monitoring provides much of the environmental information collected under the GEMS and similar systems, but it is inevitably constrained by spatial coverage and access to remote locations. Developments in remote sensing and satellite technologies (Box 1.10) are opening up exciting new horizons, particularly for monitoring environmental quality and environmental change across the entire globe.

The technology already exists and is widely used to observe important aspects of global change from satellites (Figure 1.7), including desertification, weather systems and environmental management. Satellite monitoring of the Earth's radiation balance, atmospheric dynamics, the oceans and ocean/atmosphere interactions, land resources, glaciers and ice caps is proving to be particularly useful in building up a detailed picture of how the environment works and how it is being affected by human activities.

Satellites are also being increasingly used to monitor natural hazards and disasters. They already provide operational capability for storm warnings and search-and-rescue efforts, and other capabilities such as improved flood prediction and global mobile communications during relief are being developed rapidly. Other applications, such as earthquake prediction, show great promise but still require considerable research. One promising development is the proposed World Environment and Disaster Observation System (WEDOS). This is based on launching twenty-six Earth observation satellites at low altitude, which would allow any area of the world to be observed at least once a day with a 20 m resolution. Particular locations could be monitored several times a day with a 2 m resolution. This would make it possible to detect any irregularities or environmental changes almost immediately, even in remote places.

IGBP

Much of the large-scale environmental research carried out since the start of the 1990s has been done under the umbrella of the International Geosphere–Biosphere

BOX 1.10 MONITORING THE EARTH FROM SATELLITES

By the mid-1990s, a number of satellite surveillance systems were operational, including SPOT (Système Probatoire d'Observation de la Terre), which was launched in 1986 and provides information for farmers, geologists and land-use planners, and ERS-1 (Earth Resources Satellite), which monitors ice patterns and surface temperatures. New satellites with capabilities to monitor the Earth's environment include Meteor 3M-1 (launched in 1998), which monitors atmospheric aerosols and chemical compounds, and ADEOS II (1999), which monitors surface wind speeds and directions over the oceans. Between 1998 and 2003, the USA plans to launch eight new satellites, and in 2001 a new space station will be launched with powerful equipment to monitor the Earth's environment.

In December 1999, NASA launched a new Earth observation satellite (called TERRA) that will circle the Earth sixteen times a day for the next six years in a polar orbit and send back information on how the oceans, continents and atmosphere interact. Its sensors will scan the entire planet every one to two days. TERRA is the first of ten satellites designed to monitor the effects of human activity on the global environment over the next 15 years.

A new commercial market is also opening up for detailed satellite images, the availability of which until recently has been tightly controlled by governments. New commercial satellites – useful for the transmission of television news programmes but also for observing natural hazards on Earth, as well as military movements and vulnerable targets – will have lower resolution than the best spy satellites, but they will deliver images more quickly. A pioneer of this new use for existing technology was the Ikonos satellite, launched by a company from Colorado, USA, in December 1997.

Programme (IGBP). The IGBP was established by the International Council of Scientific Unions (ICSU) in 1986 to describe and understand the interactions between physical, chemical and biological processes that regulate the total Earth system. Information generated by the IGBP is intended to assist the world's decision makers in managing the global environment.

Much of the work of the IGBP centres on a co-ordinated programme of scientific research that is designed to examine and evaluate present and future causes and effects of global environmental change. The programme is built around a series of core projects (Table 1.4) which focus on important aspects of environmental systems and their links to global change.

Table 1.4 Core projects within the IGBP

- The International Global Atmospheric Chemistry Project
- Stratosphere–Troposphere Interactions and the Biosphere
- Joint Global Ocean Flux Study
- Global Ocean Euphotic Zone Study
- Land–Ocean Interactions in the Coastal Zone
- Biospheric Aspects of the Hydrological Cycle
- Global Change and Terrestrial Ecosystems
- Past Global Changes
- Global Analysis, Interpretation and Modelling

Mission to Planet Earth is an important component of the IGBP. This is a co-ordinated international plan to provide the necessary satellite platforms and instruments, data and information systems, and related scientific research for the programme. It examines critical interactions between the Earth's physical, chemical, biological and social systems.

CENTRAL THEMES

The following chapters in this book focus on particular aspects of the environment, and four important threads run through them:

1. the ways in which the environment creates hazards and from time to time gives rise to disasters that endanger people, cause damage and bring economic losses and hardship;
2. growing awareness that many environmental problems are now global in scale and significance and are best studied at the global scale;
3. mounting concern about the need to find more sustainable ways of using the Earth and its natural resources;
4. rising interest in the state of the environment, and in environmental change and environmental futures, among the public at large as well as among specialists (scientists and politicians).

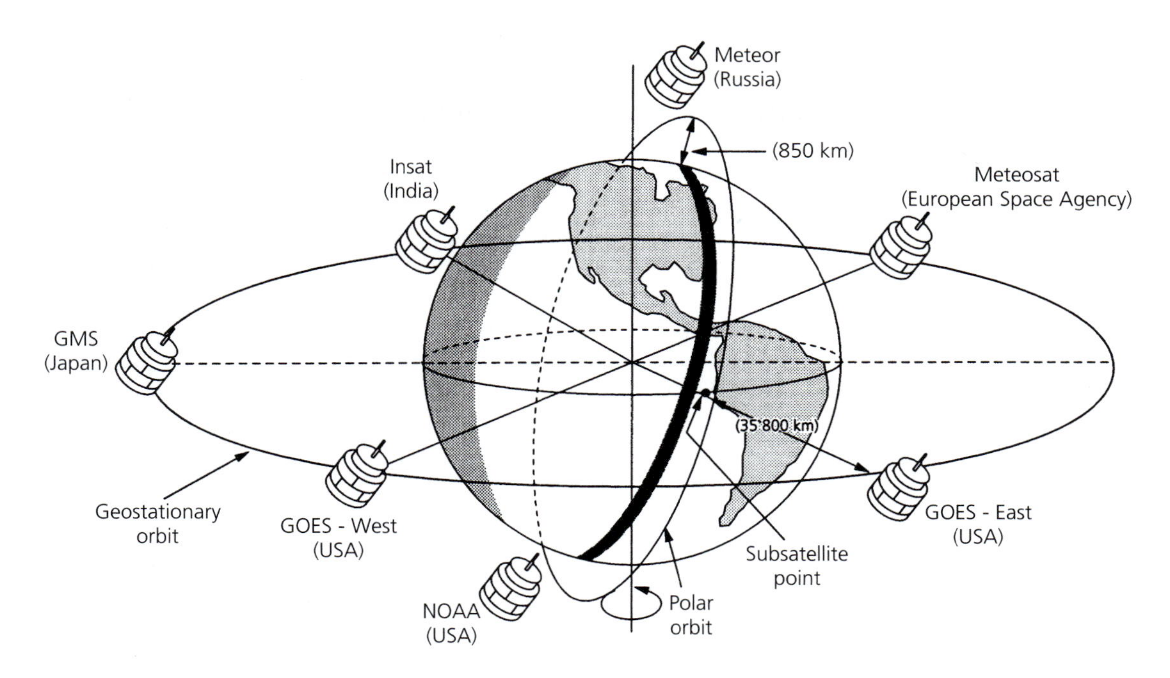

Figure 1.7 *Meteorological satellites operating over the Earth during the mid-1990s.*
Source: Figure 16.7 in Marsh, W.M. and J.M. Grossa (1996) Environmental geography: science, land use and Earth systems. John Wiley & Sons, New York.

ENVIRONMENTAL HAZARDS AND DISASTERS

NATURE OF RISK

Since the earliest times, people have had to cope with environmental hazards, many of which are simply the result of natural environmental processes at work. Hazards are a basic part of the two-way relationship between people and environment (Figure 1.8). There are many different types of environmental hazard, including volcanoes (p. 173), earthquakes (p. 160), violent storms (p. 312), hurricanes (p. 317) and tornadoes (p. 315), river flooding (p. 411), storm surges (p. 508), droughts (p. 419), avalanches and landslides (p. 213), glacier advances (p. 546), and sea-level rise (p. 511), which all affect people in different ways.

Three trends have increased the problems of coping with environmental hazards:

- *Population increase*: this has significantly increased the density of people in areas that have a long history of hazard events (exposing more people to risk and increasing the prospect of major disasters). It has also encouraged the movement of people into areas that in the past were largely unsettled, sometimes because the hazard risk was too great. Many of the critical factors that create vulnerability and result in disasters are related to fundamental human issues such as access to power and resources, lack of training and escalating debts (Figure 1.9).
- *Human impacts on environmental systems*: human activities influence natural environmental systems in many ways, both directly and indirectly, and this often changes the magnitude and frequency with which natural environmental processes operate. Thus, for example, forest clearance can increase downstream river flood risk (see p. 411), and intensive grazing of grasslands in arid climates can promote desertification (see p. 438).
- *Technological hazards*: modern technology creates a new set of hazards that did not previously exist, including the release of air pollutants such as CFCs, which do not exist naturally in the environment (see Box 8.27); the risk of serious industrial accidents such as oil spills at sea (see p. 494), explosions at nuclear power stations (see Box 8.23) and toxic chemical plants (see p. 247); and the creation of waste materials such as nuclear wastes (see p. 137), which are toxic and persistent and which natural environmental systems are incapable of breaking down.

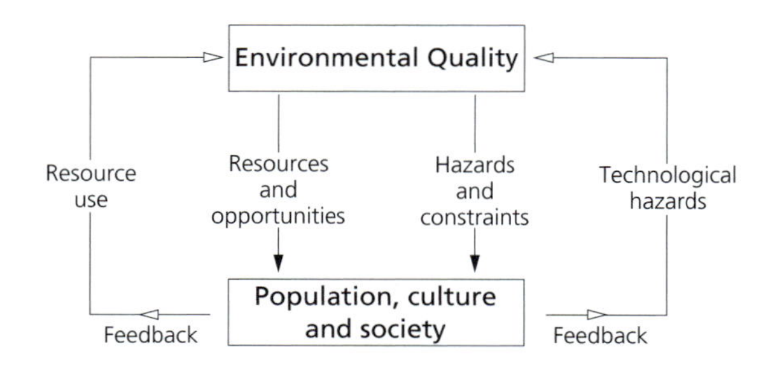

Figure 1.8 *Two-way relationship between people and environment. The relationship between people and their environment is symbiotic, involving both resources (opportunities) and hazards (constraints).*

Source: Figure 2 in Park, C.C. (1991) Environmental hazards. Macmillan Education, London.

CHANGING RISK

The net effect of these changes, particularly during the second half of the twentieth century, is an increase in exposure to many hazards and increased potential for catastrophic losses. More than 1.4 million people have died as a result of natural disasters over the past 50 years, with earthquakes (see p. 160) by far the biggest killers (Table 1.5). Hazards can cause major financial problems, as well as killing many people and damaging much property. In 1998, for example, financial losses from natural disasters worldwide were greater than US$93 billion. Insurance companies have estimated that there were around eighty separate catastrophes related directly to the 1998 El Niño event (see p. 483), which caused economic losses of around US$14 billion.

Analysis of recent trends shows significant regional disparities in losses, particularly between developed and developing regions. Financial losses associated with natural hazards are highest among the developed countries, such as the USA, where natural hazard losses exceed those of many other national social problems, including fire and crime. In the developing world, in contrast, the costs are largely measured in terms of human suffering and hardship. Many low-income populations are forced to occupy illegal settlements on low-lying lands, steep hillsides, floodplains or other hazard-prone areas. They are very vulnerable to significant health risks from flooding, landslides, mud slides and other natural hazards, and their dwellings and infrastructure are subject to accidents, massive damage and collapse. Table 1.6 summarises the major natural hazards worldwide that occurred between 1996 and 1998.

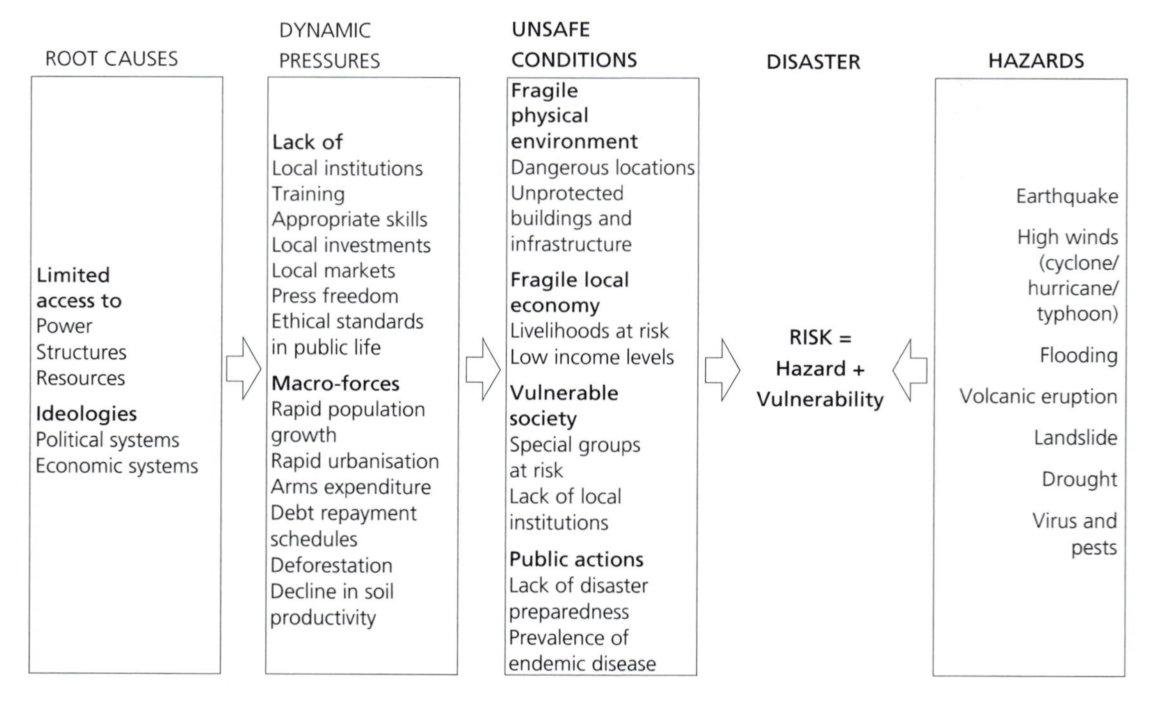

Figure 1.9 *Progression of human vulnerability to disasters. Disasters are caused by hazard events, but the impacts on people are strongly influenced by degree of vulnerability. This in turn is a product of unsafe conditions, which result from some important root causes and are compounded by dynamic pressures.*

Source: A figure in Blaikie, P., T. Cannon, L. Davis and B. Wisner (1994) At risk: natural hazards, people's vulnerability and disasters. Routledge, London.

Table 1.5 Major natural disasters during the twentieth century

Year	Location	Disaster	Deaths
1906	San Francisco, USA	earthquake	3,000
1915	Avezzano, Italy	earthquake	32,610
1920	Gansu, China	earthquake	235,000
1923	Tokyo, Japan	earthquake	142,800
1931	Yangtze kiang, China	flood	140,000
1935	Quetta, Pakistan	earthquake	35,000
1938	USA	hurricane	600
1939	Erzican, Turkey	earthquake	32,740
1942	India and Bangladesh	cyclone	61,000
1954	Dongting area, China	flood	40,000
1960	Agadir, Morocco	earthquake	12,000
1965	Florida and Louisiana, USA	Hurricane Betsy	75
1970	Bangladesh	cyclone	300,000
1970	Chimbote, Peru	earthquake and landslide	67,000
1976	Guatemala City, Guatemala	earthquake	22,084
1976	Tangshan, China	earthquake	290,000
1985	Mexico City, Mexico	earthquake	10,000
1985	Mt Nevado del Ruiz, Colombia	volcano	24,740
1988	Caribbean and Central America	Hurricane Gilbert	355
1988	Spitak, Armenia	earthquake	25,000
1989	San Francisco, USA	earthquake	68
1990	Western Europe	winter storms	230
1991	Bangladesh	Cyclone Gorky and storm surge	139,000
1991	Kyushu and Hokkaido, Japan	Typhoon Mirielle	62
1992	Florida and Louisiana, USA	Hurricane Andrew	62
1993	Maharashtra, India	earthquake	6,348
1995	Kobe, Japan	earthquake	100,000
1998	Nicaragua, Honduras	Hurricane Mitch	9,200
1999	Izmit, Turkey	earthquake	17,000
1999	Taichung, Taiwan	earthquake	2,400

Worldwide losses from natural disasters are increasing rapidly (Figure 1.10). According to the UNEP *Global environment outlook 2000*, losses from natural disasters over the decade 1986–95 were eight times higher than in the 1960s. One response to this was the designation by the United Nations of the 1990s as the International Decade for Natural Disaster Reduction (IDNDR) (Box 1.11).

GLOBAL ENVIRONMENTAL CHANGE

GLOBAL PERSPECTIVE

One hallmark of recent interest in the environment has been the adoption of a global perspective (Box 1.12). This is important because it reflects a growing awareness that the future of the planet is at stake. The catalogue of regional and local environmental problems is immense, and it grows bigger day by day. Graphic examples include damage to ecosystems, the atmosphere, oceans, forests, agricultural systems and water supplies. But, while these are vitally important to the people who live there and the environmental systems that are affected, they must be viewed in a wider context as part of a global problem. This is not simply a matter of scientific curiosity, because it threatens both the stability of industrialised nations and the growth prospects of the developing world.

The growth of global environmental research reflects growing interest in two sets of processes. First, there are the natural environmental processes that are global in

BOX 1.11 THE INTERNATIONAL DECADE FOR NATURAL DISASTER REDUCTION

More than 150 member states of the UN signed the IDNDR resolution calling on all nations to develop programmes to reduce loss of life, economic impact and human suffering caused by natural disasters. This is a significant challenge, and achieving it will require the application of new knowledge and technology to minimise losses in regions of high risk, coupled with the adoption of multidisciplinary efforts at the global scale. Some scientists point out that many of the IDNDR targets complement the objectives of Agenda 21 (p. 14), although few planners and decision makers yet build upon the links between successful disaster management and sustainable development.

scale. These include plate tectonics (p. 145), atmospheric circulation (p. 284), ocean currents (p. 479), the water cycle (p. 352) and the biogeochemical cycles (p. 86). Understanding how these large-scale systems and cycles operate would be impossible without studying them at a global scale, even though many important details and processes within them are perhaps best studied at smaller scales.

Second, there are local and regional phenomena that are repeated around the world and are becoming issues of global concern. Obvious examples include air pollution

Plate 1.2 *The Leaning Tower of Pisa, Pisa, Italy. This is a good example of human adjustment to natural hazards, because construction of the tower in several phases between 1173 and 1355 was altered to try to compensate for subsidence caused by underlying clay. The tower was closed in 1989 to allow remedial work (to improve safety and stability of the structure) to be carried out.*

Photo: Chris Park.

Figure 1.10 *Increasing scale of losses from natural disasters, 1969–89. Economic and insured losses from natural hazards have varied a great deal from year to year, but losses averaged out over successive decades are clearly rising.*

Source: Figure 2.5 in Jones, D.K.C. (1991) Environmental hazards. In Bennett, R. and R. Estall (eds) Global change and challenge. Routledge, London, pp. 27–55.

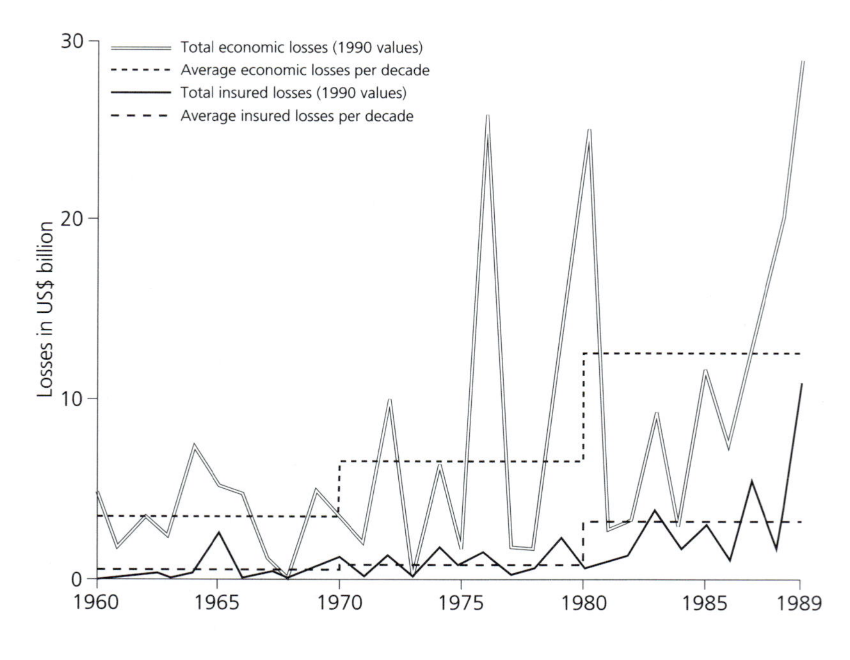

Table 1.6 Major natural disasters between 1996 and 1998

Date	*Disaster*	*Impact*
Late June to mid-August 1996	Flooding of the Yangtze River in China	Affected 20 million people, caused economic losses of more than US$20,000 million
1997	Flooding in Central Europe	Caused economic damage estimated at US$900 million in Poland and US$800 million in the Czech Republic
1997	Severe floods in Kenya, Burma, Somalia, the United States, and along the Pacific coast of Latin America	
1997	Earthquakes in Iran	Killed more than 2,300 people
1997	Earth tremors in central Italy	Major damage and destruction in many towns and villages
1998	Mudslides in central Italy	Widespread damage and destruction
June 1988	Cyclone in the Indian state of Gujarat	Killed more than 10,000 people
September 1998	Hurricane George in the Caribbean	Damage estimated at US$10 million
October 1998	Hurricane Mitch, Nicaragua and Honduras	Killed more than 9,000 people; major setback to development plans

(p. 241), desertification (p. 434), tropical deforestation (p. 586), ocean pollution (p. 493), soil erosion (p. 609) and groundwater contamination (p. 406).

Global environmental research is increasingly being focused on culturally induced environmental change as awareness grows about how widespread this is and how it has evolved generally over the last three million years but more particularly over the last century. Many of the themes that are attracting widespread attention and interest, including desertification, deforestation, acid deposition, stratospheric ozone depletion and climate change, are of vital importance to the future of the planet and its people.

MULTIDISCIPLINARY RESEARCH

As the scale of interest has broadened (from local and regional problems towards global problems), approaches have also progressed from subject-specific disciplinary emphases towards increasingly multidisciplinary and interdisciplinary research programmes. The two biggest international global research programmes are the International Geosphere–Biosphere Programme (IGBP) and the Human Dimension of Global Change Programme (HDGCP). The IGBP (see p. 20) is a natural science programme focusing on the processes and consequences of environmental change.

The HDGCP is a large-scale international social science research programme designed to run through the 1990s. It seeks to obtain a much better understanding of the human causes of global environmental change and to formulate appropriate responses for reconciling economic development and the maintenance of environmental quality.

BOX 1.12 BENEFITS OF A GLOBAL PERSPECTIVE ON THE ENVIRONMENT

Adopting a global perspective emphasises:

- how widespread many environmental problems are;
- how what happens in one place can readily affect the environment elsewhere;
- the fact that the survival and stability of global environmental systems are crucial to the future of humanity; and
- the reality that many local and regional environmental problems are in essence symptoms of a much broader underlying problem, which is the way we view, value and treat the natural environment and its resources.

SUSTAINABLE DEVELOPMENT

Another leading focus within the environmental debate since the start of the 1990s has been the search for more sustainable ways of using the Earth and its resources (Box 1.13). Many of the symptoms of the environmental crisis (see p. 4) illustrate what can go wrong if environmental resources are over-used or used in ways that create environmental damage or instability. This is particularly true in developing countries, where developed-world notions of development have often been transplanted with dire consequences. A central objective of the 1992 Rio Earth Summit was to establish the need to replace existing exploitative and environmentally damaging forms of economic development with more sustainable and environmentally friendly forms of development.

Sustainability is an in-built feature of all natural environmental systems, provided that human interference is absent or minimised. It relates to the capacity of a system to maintain a continuous flow of whatever each part of that system needs for a healthy existence (see p. 78). Human use of environmental resources and interference with environmental systems disturbs this in-built capacity, which can make it unsustainable. Economists argue that resource use and depletion can stimulate research and development, substitution of new materials and the effective creation of new resources (Figure 1.11), but there are limits to what is possible.

Traditional ways of harvesting natural renewable resources, such as fish from the oceans, wood from the forests, and plants and products from natural ecosystems, have usually been sustainable so long as the quantities extracted were not too large. Sustainable yield means taking no more from an ecosystem than it can create and at the same time remain healthy, diverse and self-perpetuating. Many types of natural resource use are now intensive and unsustainable without deliberate management intervention to limit the quantities removed to below threshold sustainable yields.

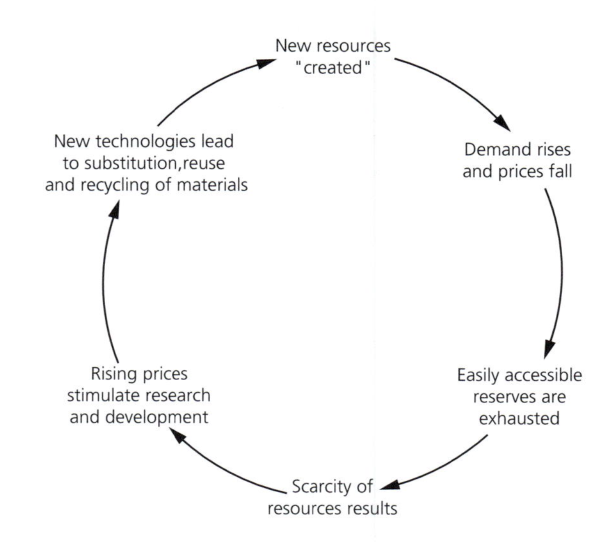

Figure 1.11 *The scarcity–development cycle. Many economists argue that resource use and depletion stimulate research and development, the substitution of new materials and the effective creation of new resources.*

Source: Figure 8.12 in Cunningham, W.P. and B.W. Saigo (1992) Environmental science: a global concern. Wm. C. Brown Publishers, Dubuque, Iowa.

BOX 1.13 THE CONTEXT OF SUSTAINABLE DEVELOPMENT

It has long been recognised that the future of the planet and its people is intimately tied up with our ability to maintain and preserve the life-support systems that nature provides. This makes it our duty to ensure that:

- all uses of renewable resources are sustainable;
- the diversity of life on Earth is conserved; and
- damage to natural environmental systems is minimised.

DEFINITION

Conservationists and ecologists have long been aware of the significance of sustainability within natural environmental systems. However, it was not until the late 1980s that the broader concept of sustainable development was introduced by the World Commission on Environment and Development. In its 1987 report *Our Common Future*, the WCED defined sustainable development as 'meeting the needs of the present without compromising the ability of future generations to meet their needs'. This definition has since been widely adopted.

The search for a single definition of 'sustainability' seems elusive partly because it embodies a number of ideas imported from different disciplines, including economics (no growth or slow growth), ecology (integrity of the biosphere, carrying capacity), sociology (critique of technology) and environmental studies (eco-development, resource–environment links). While the general definition has been widely accepted, there has nonetheless been widespread debate about what sustainability might actually mean in practice and about how it might best be applied to different cultures and economies. It is generally accepted that development can be sustainable only if it is based on sound ecological principles and practices, but beyond that there is little real consensus. Some environmentalists argue that too much attention has been focused on defining the meaning of sustainability and not enough on exploring the implications of sustainability as it is likely to affect the status quo within and between countries.

IMPLICATIONS

One significant effect of the sustainable development debate has been to raise awareness of the need to adopt wider perspectives (Box 1.14) when making decisions that affect the environment. Moving towards a more sustainable way of living will inevitably require some radical changes in attitudes, values and practices. There is no clear answer to the question of how we can create a vibrant world economy that does not destroy the ecosystem on which it is based. The environmentally sustainable 'brave new world' will have to be less polluting, probably heavily reliant on solar energy, make extensive use of new ways of using and reusing materials, adopt less resource-intensive means of growing food, and develop effective strategies for preserving forests. There would also need to be radical changes in energy systems, tax systems, international economic structures and provision of international aid.

A number of different scenarios of what a sustainable society might look like have been formulated, including a particularly wide-ranging one by the Swedish Environmental Protection Agency (Box 1.15). A strategy for building a sustainable society was proposed in *Caring for the Earth,* published in 1991 by the International Union for the Conservation of Nature, jointly with UNEP and the World Wide Fund for Nature. The strategy is based on nine key principles (Table 1.7), the first of which provides an ethical base, the next four define criteria that should be met, and the last four define directions that should be taken.

ENVIRONMENTALISM

'Environmentalism' is a collective term to describe ways in which people express their concern about the state and future of the environment (Box 1.16). Environmentalism is really a social movement, which individuals are free to join if they wish. It is founded on a number of concerns (Box 1.17).

BOX 1.14 ISSUES WITHIN SUSTAINABLE DEVELOPMENT

Three key issues that are now taken much more seriously are:

1. *Inter-generational implications of patterns of resource use*: how effectively do decisions about the use of natural resources preserve an environmental heritage or estate for the benefit of future generations?
2. *Equity concerns*: who has access to resources? How fairly are available resources allocated between competing claimants?
3. *Time horizons*: how much are resource allocation decisions oriented towards short-term economic gain or long-term environmental stability?

EMERGENCE

Western environmentalism can be traced back to the emergence of concern about nature and natural landscapes in the USA towards the close of the nineteenth century. The modern environmental movement emerged during the 1960s, first in the USA and Britain. A number of books were particularly influential in orienting people's views and attitudes during this formative period, including:

- *Silent Spring* by Rachel Carson (published in 1962), which described the loss of birds poisoned by agricultural insecticides.
- *Blueprint for Survival*, published by *The Ecologist* in 1971 as a manifesto for radical changes in lifestyle and patterns of economic development.

Table 1.7 Principles of sustainable development

- Respect and care for the community of life
- Improve the quality of human life
- Conserve the Earth's vitality and diversity
- Minimise the depletion of non-renewable resources
- Keep within the Earth's carrying capacity
- Change personal attitudes and practices
- Enable communities to care for their own environments
- Provide a national framework for integrating development and conservation
- Create a global alliance

BOX 1.15 A SWEDISH PERSPECTIVE ON SUSTAINABLE SOCIETY

In the view of the Swedish Environmental Protection Agency, the sustainable society of the future should look like this:

- The stratospheric ozone layer will be preserved intact and the changes in climate caused by people will be small enough to allow for natural adjustment.
- Trans-boundary water and air pollution will be on such a small scale that every country will independently be able to determine the state of its own environment.
- Levels of air pollution and noise will not impair people's health or well-being, and the threats to our cultural heritage will have been eliminated.
- Lakes and seas will support viable, balanced populations of naturally occurring species, and their value for fishing, recreation or water supplies will not be impaired by pollution.
- The productivity of agricultural and forest soils will be sustainable on a long-term basis. Pollution will not be permitted to disturb natural biological soil processes or restrict the use of groundwater.
- Land and water will be used in ways that husband natural resources. Renewable resources will be used within the limits of ecosystem productivity, non-renewable resources sparingly and responsibly.
- Natural species and populations will be able to survive in viable numbers. Particular care will be taken where native populations represent an important share of the world population of a species.
- The country's most representative and valuable natural habitats and cultural landscapes will enjoy protection and be managed in accordance with that protection.
- The potential of biotechnology will be harnessed for the benefit of environmental protection, and its many applications in other areas will be scrutinised and controlled so that harm to the environment is avoided.
- The flow of goods will be characterised by producer liability 'from cradle to grave', and economic growth will be used for consumption that spares natural resources and the environment.

BOX 1.16 EXPRESSION OF ENVIRONMENTAL CONCERN

Environmental concern is reflected in a number of different ways, including:

- membership of environmental pressure groups and campaigning organisations (such as Friends of the Earth and Greenpeace);
- sympathy with and activism in environmental politics;
- green consumerism (in which people deliberately buy goods that are environment-friendly, such as cars that run on lead-free petrol, aerosols that do not contain CFC propellants, wooden products that do not contain tropical hardwoods, and paper that is recycled);
- local environmental activism (such as membership of nature conservation groups);
- adoption of environment-friendly lifestyles (such as restricting family size to a maximum of two children, using public rather than private transport, using renewable materials wherever possible).

- *Limits to Growth*, published on behalf of the Club of Rome in 1971 (see p. 622).

The strength of public interest in the environment is not static but can vary a great deal through time. It has risen and fallen in a number of cycles since the early 1960s – rising during times of relative prosperity (for example during the 1960s and early 1970s, and again in the early 1980s) and falling during periods of economic recession and rising unemployment (for example during the late 1970s and late 1980s). These cycles of rising and falling interest are reflected in the amount of media coverage given to environmental issues such as air pollution (Figure 1.12).

Public interest in the environment also changes in

BOX 1.17 ROOT CONCERNS WITH ENVIRONMENTALISM

Environmentalism is founded on a number of concerns, including:

- A reaction against technocracy: some environmentalists describe modern Western society as a technocracy, in which scientists and technologists are in control, techno-fix solutions (which rely on engineering and technology) are applied to most environmental problems, and many important decisions are ruled by the technological imperative (if we *can* do something, we *should* do it – like generating energy from nuclear power – even if it has many risks and unknowns associated with it).
- A concern for the welfare of deprived groups of people (particularly in developing countries), and a concern about both practical and ethical implications of the so-called North–South divide (between the developed and developing countries of the world).
- A concern for wider issues of equity and justice: since the early 1970s, two important streams of social activism (environmental concern and civil rights) have grown alongside each other, and in the 1990s they appear to have converged in what is now called the environmental justice movement.
- A sense of personal responsibility to leave a worthwhile environmental heritage for future generations. This trend has promoted public environmentalism through, for example, green consumerism, passive membership of environmental groups and domestic recycling.

emphasis through time as different issues rise to prominence. A good example is the emergence of scientific evidence of the hole in the stratospheric ozone layer over Antarctica in the mid-1980s (see p. 249), which triggered a significant rise in public concern about the issue. Since the early 1970s, the US environmental movement has become much more diverse as the range of serious environmental problems has grown.

DIVERSITY

Because environmentalism is founded on a number of concerns, it is not a monolithic social movement. It is better described as a loose coalition of like-minded people who share many central concerns but probably choose to express them in a variety of different ways, and who have many other beliefs and concerns which are not in themselves necessarily environmental.

Some public debates on environmental issues attract support from most environmentalists. This is certainly true

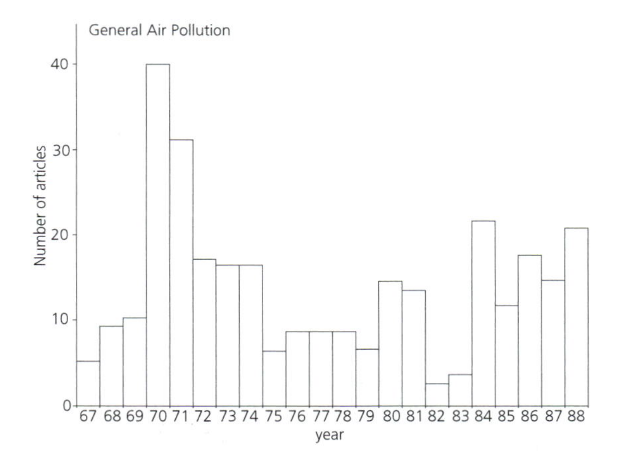

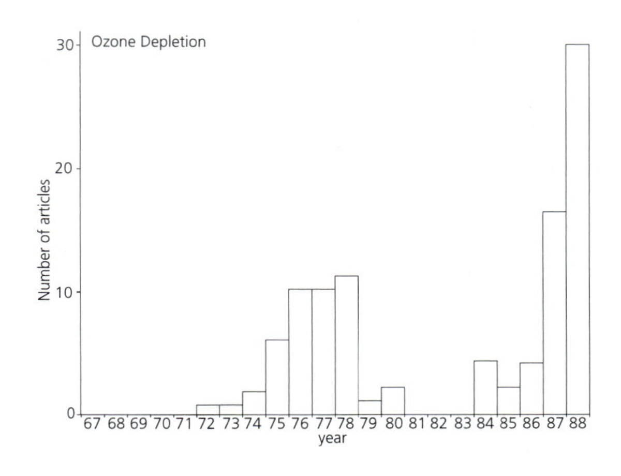

Figure 1.12 *Changing levels of public interest in air pollution. Simple counts of the number of articles in The Times (London) that were devoted to particular environmental problems varied a great deal from year to year between 1967 and 1988, with some marked phases of higher than normal interest.*

Source: Figure 2 in Park, C.C. (1991) Trans-frontier air pollution: some geographical issues. Geography 76(1): 21–35.

SUMMARY

This chapter is designed to set into context the following chapters by outlining the main ingredients of the debate about the so-called 'environmental crisis'. After introducing the general idea of the crisis, we looked at some of the key symptoms of what has gone wrong (including air pollution, tropical deforestation and loss of biodiversity). It was pointed out that most of these problems are not new but have become much worse in recent decades, and it was suggested that attention should really be focused on root causes, not just environmental symptoms. One root cause is the way we view common resources. New ingredients to the 'crisis' were listed, to show how scientific understanding and general awareness have evolved through time. We also explored the 1992 Rio Earth Summit as a major landmark in international co-operation and outlined its main products (including Agenda 21 and the Global Environment Facility). Recent years have seen important developments in monitoring and analysis, including the adoption of new technologies (particularly remote sensing and satellites), and these were explained and set into context. The final section in the chapter outlined four key themes which underpin the remaining chapters of the book – environmental hazards and disasters, global environmental change, sustainable development, and environmentalism. Each theme is evolving and challenging, and each makes an important contribution to understanding how and why interest in the environment is such an important part of life at the start of the new millennium.

of debates in recent years concerning issues such as nuclear energy, fossil fuels, renewable energy, the 'greenhouse' effect and global warming, the hole in the ozone layer, and the conservation of whales and seals. Other environmental debates attract interest and concern only from particular sub-groups of the environmental movement, either because they have only local interest or relevance, their environmental focus is not clearly defined, or they represent an area of controversy even among environmentalists.

WEBSITE

Links to relevant websites, a comprehensive bibliography, tools for teaching and learning, and downloadable images relevant to this chapter can be found at the website specially designed to accompany this book at:
http://www.park-environment.com

FURTHER READING

Adams, W.M. (1992) *Green development: environment and sustainability in the Third World.* Routledge, London. A clear introduction to the environment–development debate, with examples drawn from around the world.

Baarschers, W.H. (1996) *Eco-facts and eco-fiction: understanding the environmental debate.* Routledge, London. Seeks to distinguish between the rhetoric and the reality within the environmental debate.

Barrow, C.J. (1995) *Developing the environment: problems and management.* Longman, Harlow. A comprehensive overview of the nature and evolution of global environmental problems and the need for appropriate environmental management.

Brack, D., M. Grubb and C. Vrolijk (1999) *The Kyoto Protocol: a guide and assessment.* Earthscan, London. A detailed analysis of the agreement to limit greenhouse gas emissions and control climate change.

Brown, L.R. (ed.) (2000) *State of the world 2000.* Earthscan, London. Up-to-date review of the state of the global environment, the factors leading to environmental change and responses to change.

Brown, L.R., M. Renner and B. Halweil (eds) (2000) *Vital signs 2000–2001: the environmental trends that are shaping our future.* Worldwatch Institute, Washington. Excellent source of up-to-date information on the state of the environment and sustainable development.

Burton, I., R.W. Kates and G.F. White (1994) *The environment as hazard.* Longman, Harlow. The second edition of a classic text on environmental hazards, which portrays hazards as extreme events in nature that are made even more dangerous by human neglect.

Carley, M. and I. Christie (2000) *Managing sustainable development.* Earthscan, London. Explores what kind of organisations and management and policy-making approaches are best suited to the complex challenges of sustainable development.

Dodds, F. (ed.) (2000) *Earth Summit 2002: a new deal.* Earthscan, London. A retrospective on the Rio Earth Summit and reflections on its successes and failures by many of the leading figures who took part.

Farmer, A. (1997) *Managing environmental pollution.* Routledge, London. A comprehensive introduction to the nature of pollution, its impacts on the environment, and the practical options and regulatory frameworks for pollution control.

Goudie, A. (ed.) (1997) *The human impact reader: readings and case studies.* Blackwell, Oxford. A carefully chosen and well-structured set of readings on the impacts of human activities on the environment at a variety of spatial scales.

Goudie, A. (1999) *The human impact on the natural environment.* Blackwell, Oxford. A clear review of the main ways in which human activities affect the natural environment, with examples drawn from around the world.

Goudie, A. and H. Viles (1997) *The Earth transformed: an introduction to human impacts on the environment.* Blackwell, Oxford. A wide-ranging non-technical introduction to the ways in which the natural environment has been and is being affected by human activities.

Harvey, D. (2000) *Climate and global environmental change.* Longman, Harlow. Reviews the importance of climate as a major driver of global environmental change and explores the contributions of both human-induced and natural changes.

Hewitt, K. (1997) *Regions of risk: a geographical introduction to disasters.* Longman, Harlow. Wide-ranging exploration of natural hazards, human vulnerability and response, drawing on examples from different countries and contexts.

Heywood, I., S. Cornelius and S. Carver (1998) *Introduction to Geographical Information Systems.* Longman, Harlow. Up-to-date introduction to the theory and practice of GIS.

Hidore, J.J. (1996) *Global environmental change: its nature and impact.* Prentice Hall, Hemel Hempstead. Outlines the nature of the basic processes that produce environmental change, the effects of these changes on people and the impacts of human activities on the environment.

Jensen, J.R. (2000) *Remote sensing of the environment.* Prentice Hall, London. Introduces and illustrates the basic fundamentals of remote sensing from an Earth resources perspective.

Keulartz, J. (1998) *The struggle for nature: a critique of environmental philosophy.* Routledge, London. Outlines and examines the main components of contemporary environmental philosophy, including deep ecology, social and political ecology, eco-feminism and eco-anarchy.

Keys, D. (1999) *Catastrophe: a quest for the origins of the modern world.* Ballantine, New York. Explores the impact of large-scale natural disasters on the rise and fall of civilisations and societies, drawing particularly on examples from the Americas.

Kondratyev, K.Y., A. Buznikov and O. Pokrovsky (1996) *Global change and remote sensing.* John Wiley & Sons, London. Detailed survey of the international space year programme and its results.

Lamb, R. (1996) *Promising the Earth.* Routledge, London. An informative review of the shape and emergence of the environmental movement, written to celebrate the twenty-fifth anniversary of Friends of the Earth.

Lillesand, T.M. and R.W. Kiefer (1999) *Remote sensing and image interpretation.* John Wiley & Sons, London. Clear introduction to the principles and practice of remote sensing and image interpretation.

McCormick, J. (1995) *The global environmental movement.* John Wiley & Sons, London. Traces the roots of environmental activism and shows how concern for the environment has emerged from relative obscurity to centre stage.

Mannion, A.M. (1999) *Natural environmental change.* Routledge, London. Clear introduction to natural environmental change at a range of temporal and spatial scales, drawing on examples from around the world.

Mannion, A. and S.R. Bowlby (eds) (1992) *Environmental issues in the 1990s.* John Wiley & Sons, London. A wide-ranging collection of essays that review the nature and evolution of critical contemporary environmental problems and trace the emergence of modern environmentalism.

Markham, A. (1994) *A brief history of pollution.* Earthscan, London. Charts the history of pollution and examines current dilemmas and trends.

Mather, A.S. and K. Chapman (1995) *Environmental resources.* Longman, Harlow. Explores the meaning of natural and environmental resources and reviews a range of resources, including forests, agriculture, energy, minerals and environmental resources.

Middleton, N. (1999) *The global casino: an introduction to environmental issues.* Edward Arnold, London. Clear introduction to major environmental issues, with an emphasis on underlying causes, human factors that have contributed to the problems, and possible solutions.

Mitchell, B. (1997) *Resource and environmental management.* Longman, Harlow. Wide-ranging overview of concepts, strategies and methods of resource management, with an emphasis on change, complexity, uncertainty and conflict.

Moss, N. (2000) *Managing the planet: the politics of the new millennium.* Earthscan, London. Explores the implications for politics and management of resource scarcity and declining environmental quality, and maps out a possible future shape of politics.

O'Riordan, T. (ed.) (1999) *Environmental science for environmental management.* Longman, Harlow. Explores issues and challenges within environmental management, set against a backdrop of changing environmentalism and emerging views about science.

O'Riordan, T. (ed.) (2000) *Globalism, localism and identity: new perspectives on the transition to sustainability.* Earthscan, London. An overview of the major global economic and social forces that affect people and the environment.

Owen, O.S., D.D. Chiras and J.P. Reganold (1998) *Natural resource conservation: management for a sustainable future*. Prentice Hall, Hemel Hempstead. Outlines the value to humans of each major natural resource, examines the ways in which it is exploited or degraded, and considers how it can be restored and managed sustainably.

Park, C.C. (1991) *Environmental hazards*. Macmillan, London. A brief overview of how people cope with environmental hazards.

Pentecost, A. (1999) *Analysing environmental data*. Longman, Harlow. Useful review of common ways of analysing environmental data, with a particular emphasis on statistical techniques and their strengths and weaknesses.

Pepper, D. (1996) *Modern environmentalism*. Routledge, London. A detailed review of the origins and evolution of ideas within environmentalism, and how these relate to modern environmental ideologies.

Pickering, K.T. and L.A. Owen (1997) *An introduction to global environmental issues*. Routledge, London. The second edition of this non-technical, up-to-date introduction to the most pressing global environmental issues in the 1990s.

Roberts, N. (ed.) (1993) *The changing global environment*. Blackwell, Oxford. A wide-ranging collection of essays on contemporary environmental change, which includes useful case studies from most parts of the world.

Saiko, T. (2000) *Environmental crises*. Prentice Hall, London. Outlines the causes, magnitudes and implications of different types of environmental crisis in the countries of the former Soviet Union and Eastern Europe.

Seitz, J.L. (1995) *Global issues: an introduction*. Blackwell, Oxford. Sets environmental issues into context alongside wealth/poverty, population, food, energy, technology and government.

Simmons, I.G. (1993) *Interpreting nature: cultural constructions of the environment*. Routledge, London. Explores different ways of seeing, thinking about and assigning values to the environment.

Simmons, I.G. (1996) *Changing the face of the Earth: culture, environment, history*. Blackwell, Oxford. A readable overview of how human impact on the natural environment has evolved through time, from early hunter-gatherers to today's nuclear societies.

Slack, P. (ed.) (1999) *Environments and historical change*. Oxford University Press, Oxford. A series of essays that explore the impacts of environmental change on human society, from a variety of disciplinary perspectives.

Smith, J. (ed.) (2000) *The daily globe: environmental change, the public and the media*. Earthscan, London. A series of essays that explore how the media treat and how the public understand key environmental issues.

Smith, K. (1996) *Environmental hazards: assessing risk and reducing disaster*. Routledge, London. Clearly written, comprehensive assessment of environmental risk and the policy responses required to achieve a safer world.

United Nations Environment Program (1999) *Global environment outlook 2000*. UNEP/Earthscan, London. Comprehensive and up-to-date review and analysis of the state of the environment at global and regional scales.

Vogler, J. (2000) *The global commons: environmental and technological governance*. John Wiley & Sons, London. Reviews the pressures on global commons – areas that are beyond national control, such as outer space, the atmosphere, the oceans and Antarctica – and suggests approaches to their sustainable use and development.

Wadsworth, R. and J. Treweek (1999) *GIS for ecology*. Longman, Harlow. Clear introductory guide to GIS and how it can be applied within ecology and environmental sciences, with an emphasis on data acquisition, handling and analysis.

Wall, D. (ed.) (1993) *Green history: a reader in environmental literature, philosophy and politics*. Routledge, London. Annotated extracts from writings throughout history that chart the long-term emergence and evolution of environmental concern.

Chapter 2

Population and Environment

LEARNING OBJECTIVES

When you have finished studying this chapter, you should be able to

- *Define the term demography and explain the three key demographic forces (fertility, mortality and migration)*
- *Understand the principal features of population composition, both globally and for particular countries*
- *Describe the 'population balancing equation' and say why it is important*
- *Outline historical trends in population and explain the 'demographic transition'*
- *Describe and account for the principal features of population distribution, at the global and regional scales*
- *Appreciate the relevance of 'carrying capacity' and 'limits to growth'*
- *Summarise the ongoing debate between optimists and pessimists in the context of global futures and sustainable development*
- *Evaluate likely future changes in population, given recent trends and projections*
- *Defend the need for population policies and appreciate the challenges of devising appropriate sustainable policies*

A central component of the environmental debate is the interaction between people and environment. We have seen (see Figure 1.8) that the relationship is two-way and noted how the emergence of sustainable development (see p. 26) during the 1990s acted as a catalyst for new thinking and new initiatives designed to protect the environment and its resources and systems.

In this chapter, we explore the core theme of population, with a particular emphasis on how and why human populations change (in size and composition) through time. An appreciation of demographic processes allows us to understand critical themes such as population size, growth and distribution. Concerns about population growth and change generally stress the need to recognise and stay within the carrying capacity of the environment, and in recent decades there has been lively debate about how many people the Earth could possibly house, and about whether or not these limits are being reached. Forecasting likely future changes in population is never easy, but it has to be done to inform national and international population policies and initiatives, and to allow the assessment of potential environmental stresses.

DEMOGRAPHIC FACTORS AND PROCESSES

Many environmental problems arise from the simple fact that, through time, people put more and more pressure on environmental resources and environmental systems. This is partly because of rising numbers, but it also reflects rising expectations and changing lifestyles. In this section, we explore the key processes that give rise to population growth and consider the evidence for and implications of continued growth worldwide.

POPULATION COMPOSITION

To assume that the human population is uniform and stable, both through time and from place to place, would clearly be wrong, because the population contains huge variety. While our main concern in this chapter is the number and distribution of people, and the impacts they have on the environment, we need to appreciate this diversity because it is related to population levels, distributions and impacts. The population – of a place, a country or the whole world – contains individuals with many different characteristics, including gender, age, race and ethnicity, occupation, education, religion, marital status and lifestyle preferences. Age, gender and ethnicity (Box 2.1) are determined by the main demographic factors (fertility, mortality and migration), which are considered in the next section.

AGE

The age composition of a population can be illustrated by a population pyramid (Figure 2.1), which shows the proportion of the population in each age group. Slow-growth populations (such as the United States) typically have a relatively large elderly population, so they have a population pyramid of reasonably uniform width from bottom (young) to top (old). Rapidly growing populations (such as Kenya) have a proportionately large number of young people, thus more of a true pyramid shape, becoming narrower towards the top. The pyramid for a population that is not growing, or is decreasing, often has a bulge in the middle-age range, with relatively few young and elderly people.

GENDER

Population pyramids also show the relative proportion of males and females within each age group. One useful way of expressing this is the sex ratio – the ratio of the number of males to females, usually expressed as the number of males for every 100 females. During the 1990s, the global sex ratio was 102 (102 males for every 100 females); the ratio was higher (104) in developing countries and lower (95) in developed countries.

DEMOGRAPHIC FORCES

The level, composition and distribution of population reflect the interplay of three key demographic forces – fertility, mortality and migration.

FERTILITY

Fertility refers to the number of births that occur to an individual or in a population. A number of different measures of fertility are widely used (Box 2.2), one of the most widely used being the average number of births per woman within a defined geographical area (usually a country). For example, in 1998 the average fertility in the world was 2.9 children per woman; the average in the United States was 2.0, the averages in Italy and Spain were 1.2, and in the West African state of Niger it was 7.4. During 1997 roughly 137 million children were born world-wide, which is nearly 261 every minute. Nearly 4 million babies were born in the United States, and 21 million in China, in that single year.

A population stops growing when it reaches the replacement-level fertility (Box 2.2), but this requires a total fertility rate (TFR) higher than 2, mainly because some children die before they grow up to have their own children. The critical TFR is around 2.1 in a country with low mortality (such as the USA), but it can be higher than 3 in a high-mortality country (such as Sierra Leone).

Fertility (actual birth rate) is different from fecundity (the physiological ability of individuals or couples to have children). It is thought that the maximum natural fecundity of a population, assuming that all women engage in regular sexual intercourse between the ages of about 12

> **BOX 2.1 RACE AND ETHNICITY**
>
> Some characteristics of individuals (such as age, marital status and educational attainment) change through time, whereas others (particularly gender, race and ethnicity) are fixed at birth. Race is culturally defined, although it may have genetic roots and sometimes has physical expression. Thus, for example, physical characteristics such as facial features, fair texture and skin colour are often used to identify racial groups. The cultural definition of race can change through time. The US Census Bureau classifies people into six racial groups – white, black or African American, Asian, American Indian and Alaska native, native Hawaiian and other Pacific islander, and other race (the rest). Ethnicity is also culturally defined, by such things as cultural practices, language, food preferences and tradition. The US Census Bureau defines two ethnic groups – Hispanics (Latinos) and non-Hispanics. Hispanics are defined as individuals who trace their ancestry to Spain, the Spanish-speaking countries of Latin America and the Caribbean, or any other Spanish culture. Hispanics can be of any race, but most report themselves as white.

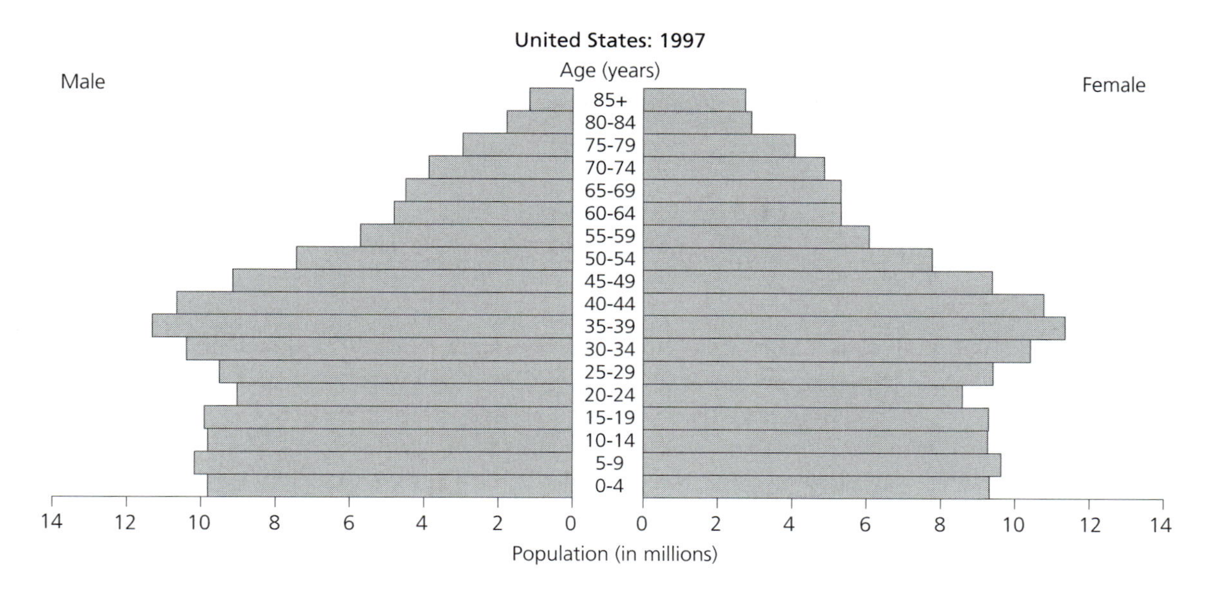

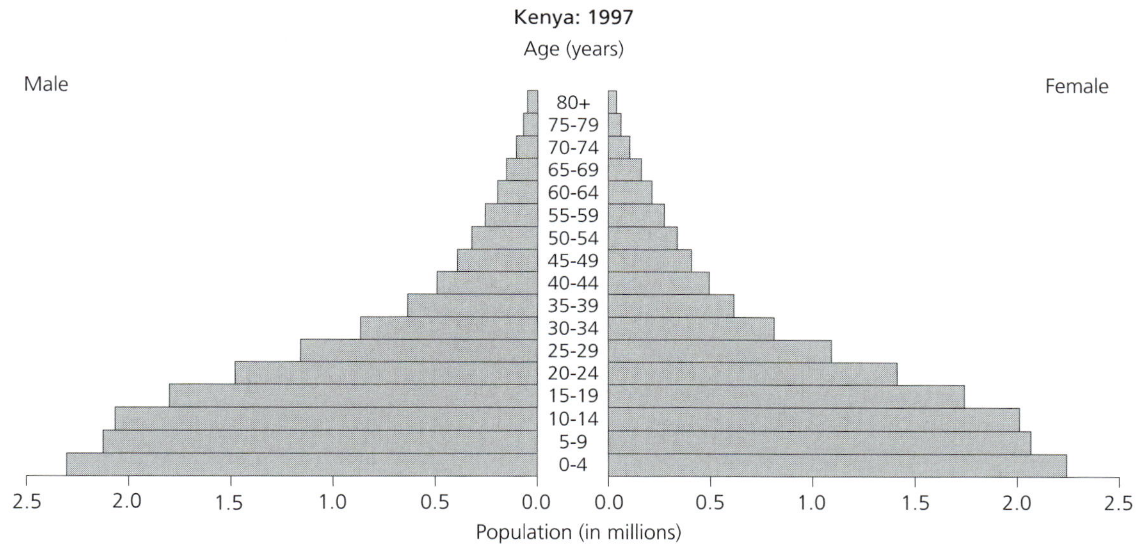

Figure 2.1 *Typical population pyramids.*

and 50, without using any form of birth control, is about 15 children (in single births) per women. But for a variety of reasons (Box 2.3) this maximum is rarely reached. Even in high-fertility countries the average is usually less than 8 children per woman.

A number of cultural and economic factors – so-called intermediate variables (Box 2.4) – affect fertility indirectly, via the factors that determine the rate and level of child-bearing. These factors operate in every society, although they vary a great deal in relative importance. In Africa, for example, women often marry young and rarely use contraceptives, but cultural factors keep fertility lower than it would otherwise be (6–7 births per woman). In many African societies women breast-feed their babies until they are 2–3 years old (which naturally prolongs the infecund period), and they are expected to refrain from sex for up to 2 years after childbirth.

Studies have shown that most of the variation in levels of fertility between different populations can be accounted for by four variables:

- the proportion of women who are married or in a sexual union
- the percentage of women who are using contraceptives

BOX 2.2 MEASURES OF FERTILITY

Fertility can be measured and expressed in a variety of ways, the most common of which are total fertility rate (TFR) and crude birth rate.

Total fertility rate This measures the average number of children in a family. Replacement-level fertility refers to the level of child-bearing at which couples have an average of two children, in which they just replace themselves in the population.

Crude birth rate The most widely reported measure of fertility, calculated from the number of babies born in a given year divided by the mid-year population, and expressed as the number of births per 1,000 population. For example, in 1998 the estimated world crude birth rate was 23 births per 1,000 people; in the United States, the crude birth rate was 15 births per 1,000 people. One problem with the crude birth rate is that it depends strongly on the age structure of the population, so it is not always comparing like with like. Within the total population, the critical group are the women of reproductive age (because these are the ones who produce the babies!), and younger women usually have higher birth rates than older women. So, crude birth rates will tend to be higher in a population with a large proportion of younger women.

Other measures of fertility, designed to remove the distorting effect of age distribution, include:

- *general fertility rate*: measures the number of births per woman of child-bearing age (15–49 years).
- *net reproduction rate*: measures the number of daughters born to a woman given current birth rates and her chances of living to the end of her child-bearing years.

- the proportion of women who are infecund (for example, because they are currently breast-feeding)
- the level of induced abortion.

In North America and most developed countries, key determinants of fertility are contraceptive use and abortion. In 1997, for example, the fertility level for Spain was 1.15 births per woman – one of the lowest recorded fertility rates for a country – mainly because of relatively high rates of contraceptive use.

BOX 2.3 ACTUAL VERSUS POTENTIAL FERTILITY

Reasons why fertility is always lower than fecundity include:

- *Cultural values*: does the society value large or small families?
- *Social roles*: is the wife primarily a child-bearer and child-rearer?
- *Economic realities*: do parents rely on children to look after them in old age?
- *Disease*: some diseases, such as the sexually transmitted disease gonorrhoea, inhibit fecundity.

Fertility levels reported for particular countries mask a great deal of variability between regions, and between individuals. For individuals the most important influences appear to be:

- age: this includes the postponement of child-bearing to older age, perhaps driven by life-style choices but promoted by medical advances;
- income and socio-economic status: in most countries, the poor have more children than the rich;
- race and religion: in many countries, racial and ethnic minorities have higher fertility than the majority, often reflecting religious beliefs and cultural traditions. In many countries, people who practise religion tend to have more children than those who don't.

But a range of other social, economic and cultural factors can also have an effect.

Birth rates in most developed countries have varied a great deal over the last century. In the United States, for example, fertility declined after the early 1950s to reach an all-time low of 1.74 children per woman in 1976. It has remained slightly above 2.0 since (in 1996 it was 2.03).

BOX 2.4 FACTORS THAT DETERMINE THE RATE AND LEVEL OF CHILD-BEARING

Three sets of factors determine the rate and level of child-bearing:

Fecundity

- ability to have intercourse
- ability to conceive
- ability to carry a pregnancy to full term.

Sexual unions (including marriage)

- the formation and dissolution of unions
- age at first intercourse
- proportion of women who are married or in a union
- time spent outside a union (e.g. separated, divorced or widowed)
- frequency of intercourse
- sexual abstinence (e.g. for religious reasons or as cultural custom)
- temporary separations (e.g. during military service).

Birth control

- use of contraceptives
- contraceptive sterilisation
- induced abortion.

Fertility levels were low through the 1970s, mainly because of delayed marriage and the widespread use of contraception and abortion.

Long-term fertility trends and current social trends (including postponement of marriage and child-bearing to older ages, high divorce rates, and the large and growing proportion of women in the labour force) suggest that birth rates in developed countries are likely to remain low for the foreseeable future.

MORTALITY

Mortality – the second cause of population change – relates to death rates. The death rate for a population is usually expressed as the number of deaths per 1,000 persons in a given year. For example, in 1997 world population was estimated at 5.84 billion, and an estimated 53 million people died that year; this produced a global death rate of 9 per 1,000 people. In the United States, 2.3 million people died from the population of 267.7 million, giving a crude death rate of 9 per 1,000 people. Death rates in the late 1990s varied between a low of about 2 per 1,000 in Kuwait, Qatar and the United Arab Emirates, and a high of around 30 per 1,000 in Sierra Leone.

Death rate measures the proportion of a population that dies in a given year, but the index is strongly influenced by the age structure of the population. Comparing death rates between different populations does not show whether one population is healthier than another, or lives longer than another. Life expectancy (Box 2.5) and life span (Box 2.6) are useful ways of describing how long people live.

BOX 2.5 LIFE EXPECTANCY

Life expectancy is a measure of how long individuals live, usually measured as life expectancy at birth. This offers a useful means of comparing mortality conditions between countries. Life expectancy is determined by both biological and social factors. For example, in 1996 the average life expectancy at birth in the United States was 76 years (up from 47 years in 1900); in Japan it was 80 years; for Sierra Leone it was 34 years and for Malawi 36 years. Most individuals live longer than the life expectancy for their society (in 1996, there were more than 12 million Americans older than 76, for example), because it is an average – many people die young, for various reasons, and many live longer than the average.

BOX 2.6 LIFE SPAN

Life span is the theoretical maximum number of years that the most healthy humans can possibly live. Life expectancy is lower than this maximum because it reflects the real-life conditions within a population, including stresses such as disease, malnutrition, environmental risk and social tension. Although some individuals have confirmed ages of over 110, few people have a genetic make-up that allows them to live past 100 years. Continued decline in mortality among the elderly, largely because of better health care, has doubtless increased the life span for many people. It remains to be seen whether advances in medical technology and bioengineering will increase it even further in the future.

Mortality rates are much higher in developing countries than in developed countries. Much of the difference reflects the widespread death toll among children from preventable diseases such as diarrhoea, respiratory infections, measles and neonatal tetanus. Many experts argue that mortality rates in developing countries would drop sharply if adequate and accessible health services were more widely available throughout the developing world – including provision of clean drinking water, immunisation and antibiotics.

In most countries people live longer now than in the past, even though mortality levels are still high in many countries because of curable diseases. Average life expectancy (Box 2.5) in the world in 1900 was less than 30 years; by the late 1990s, it had risen to 66 years. This striking improvement in life expectancy reflects a number of factors, including:

- better understanding of how diseases spread;
- improvements in personal hygiene; and
- better and more widely available public health practices.

Despite the good news about rising life expectancy, the battle against communicable diseases continues. The most prominent is the spread of HIV (Box 2.8), but others that kill large numbers each year include measles and diarrhoea. Modern life-styles promote the spread of communicable diseases in many different ways, including via migration of people, international air travel, and the import and export of foodstuffs (particularly fruit and vegetables). Climate change associated with greenhouse gases and global warming (see p. 261) could add to the problem by allowing the spread of disease-bearing agents into new areas.

The environment also poses its own threats. Natural disasters tend to have minor impacts on long-term mortality rates for individual countries, because disaster events are uncommon and are usually well spaced out through time and dispersed geographically. But major disasters – including earthquakes (see p. 160), volcanic eruptions (see p. 173), hurricanes (see p. 317), tsunamis (see p. 485) and storm surges (see p. 509) – can kill many

BOX 2.7 MORTALITY IN THE UNITED STATES

In 1995, the major causes of death in the United States were heart disease (32 per cent of all deaths) and cancer (23 per cent), and these mostly affected people over 50 years old. Death rates vary by age, gender, socio-economic status, race, ethnicity and religion. But genetic factors are also important, because individuals can inherit a predisposition to develop a potentially fatal disease such as some types of cancer.

The most important factors are:

- *Age*: death rates are much higher among older people (more than about 50 years old) than younger people.
- *Gender*: death rates are lower among women than men in each age group.
- *Socio-economic status*: death rates are lower among higher-status individuals than lower-status individuals, measured by occupation, income and education.
- *Ethnicity*: death rates are usually much higher among racial and ethnic minorities, mostly because they are also economically disadvantaged.

Age is important, particularly for those aged less than about 15 and more than about 50. Infant mortality (death in the first year of life) in the United States in 1996 was 7 per 1,000 live births, down from about 120 per 1,000 in 1900. Within the age group 15–24 years old the most common causes of death are accidents (particularly motor vehicle crashes), homicide (murder) and suicide. Heart disease and cancer cause most deaths in the over 50s age group.

Gender is also important, and the differences are most pronounced among young adults; the death rate for 15–24-year-old males in the United States is nearly three times higher than that for females. Reasons include:

- *HIV/AIDS*: until recently, this was dominant among young men.
- *suicides*: more than four times as many young men commit suicide as women.
- *homicide (murder)*: more men than women are killed and more black men than white men.
- *accidents*: more common among young men, who tend to engage in more risk-taking behaviour and activities.

BOX 2.8 HIV AND AIDS

First identified in 1981, when there were an estimated 200,000 new infections worldwide, HIV has been infecting more people each year since. In 1998, there were 5.8 million new infections. Death rates are rising and life expectancy is falling in the countries most affected by HIV/AIDS. Many experts stress that published figures are estimates which may well underestimate the full impact of the problem, because many countries are unwilling to acknowledge how seriously they are affected. Most of the worst-affected countries are among the poorest in the world.

The disease affects the most sexually active groups within the population, and many young adults have died from it. Worldwide, an estimated 8.2 million children have lost their mothers to AIDS, and many have lost both parents. In addition to increased mortality, HIV/AIDS drives demographic change by reducing the fertility of women who are infected, and by influencing age at marriage, sexual behaviour and contraceptive use.

All industrial countries have held HIV infection rates of their adult populations below 1 per cent, but in some countries in sub-Saharan Africa rates have risen above 20 per cent. For example, the adult infection rate in Botswana is 26 per cent; in Zimbabwe it is 25 per cent, and in South Africa it is 22 per cent. Around 85 per cent of the currently infected people, and 91 per cent of all AIDS deaths to date, are in thirty-four countries (twenty-nine of them in Africa). In the twenty-nine African countries, average life expectancy at birth has fallen by 7 years since the AIDS epidemic started.

Although Africa is the main epicentre of the epidemic, the virus is also widespread in southern and western India, particularly in cities such as Bombay (Mumbai) and Madras. Other countries with high infection rates include Haiti (5 per cent of the population) and Cambodia (2.5 per cent of the population).

Death rates are falling among people infected with HIV in the developed world, largely because of medical advances (particularly the availability of life-extending drugs), but these are too expensive for most people in developing countries. Even without costly medication, transmission of HIV can be reduced or prevented by measures such as surveillance, education, increased reproductive health services and adoption of safer health-care practices.

people in particular places at particular times. The suffering and hardship are not evenly distributed, because most lives are lost through disasters in developing countries, where many people live in low-quality (and often inappropriate) housing, and where the public health and safety systems are much less developed than elsewhere.

MIGRATION

Migration refers to the movement of people into (immigrants) or out of (emigrants) a specific area (Box 2.9), so it increases or decreases that area's population depending on whether more people move in or out. The term 'migration' is not normally used to describe residential relocation (moving house) within an area, or temporary moves for work or leisure. It refers specifically to the movement of people across a territorial boundary – usually a national border – for the purpose of changing their place of usual residence (international migration, see Box 2.9).

Most large-scale migration movements are caused by a combination of push and pull factors – people are pushed from their homelands by difficult conditions (including ethnic and cultural persecution) and pulled to a new country by the prospect of better conditions and a better quality of life.

Migration sometimes occurs in large waves of people responding to particular events. These include political events such as the reunification of East and West Germany in 1990, which triggered a mass exodus of people from the former East into the former West Germany. Environmental

BOX 2.9 MIGRATION TERMS

- *Immigration*: the movement of people into a country or area;
- *Emigration*: the movement of people out of a country or area;
- *Net migration*: the difference between the number of people moving into and out of a country or area;
- *Internal migration*: the movement of people within a country;
- *International migration*: the movement of people across a national border, from one country to another.

disasters and stress, such as the wholesale migration of people in North Africa associated with prolonged drought there since the early 1970s (see p. 422), can also trigger large-scale migration flows.

A great deal of migration also occurs by small-scale migration streams, such as individuals and families migrating into small towns from surrounding farms and villages. Migration within countries is often associated with the growth of large cities (Table 2.6). Urban growth is driven by a combination of natural population growth in cities and rural-to-urban migration.

Some people and groups of people are more likely to migrate than others, particularly where migration is voluntary rather than forced. As a result, migration is usually

- selective: better educated and more adventurous individuals tend to move more readily and more often than others;
- life-stage dependent: people are most likely to move at particular stages in their lives, such as when they marry, have children, divorce or retire.

One of the great unknowns about possible future climate change (see p. 265) is how it will affect populations in different countries. Some countries are likely to benefit from global warming, which will increase their farming and forestry yields and promote economic development. Other countries, particularly in tropical environments, will be net losers. It is highly likely that the number of environmental refugees (people displaced from their homelands by environmental stress such as drought, rising sea level or increased flooding) will increase dramatically. This could trigger a major redistribution of global population, further increasing pressure on available resources in the countries where the refugees settle, as well as causing great hardship and suffering for those forced to move. If sea level rises dramatically (see p. 511), some entire nation states – such as the low-lying Pacific Islands of Tuvalu – could disappear altogether, giving the local population no option but to relocate elsewhere. Under international law, people who flee their own country to escape from persecution, armed conflict or violence can be granted refugee status, but those forced to leave for social or environmental reasons do not qualify.

POPULATION SIZE AND GROWTH

Populations rarely remain fixed in size over time, because the three main factors which determine population size and growth (fertility, mortality and migration) each changes through time in response to cultural and socio-economic drivers.

POPULATION BALANCING EQUATION

The so-called population *balancing equation* (Box 2.10) shows how these three factors interact to produce net change over time, and worked examples illustrate how the equation can be used to estimate world population (Table 2.1) and the population of the United States (Table 2.2) in any given year. Fertility effectively serves to increase population, and mortality to decrease it. The net number of people added to that population through natural increase is thus the annual number of births minus the annual number of deaths.

The rate of natural increase is the birth rate minus the death rate. For the world overall in 1997 the birth rate was 23.1 per 1,000 and the death rate was 9.0 per 1,000. Thus the global rate of natural increase in 1997 was 14 per 1,000 (1.4 per cent). For individual countries, the rate of natural increase plus the rate of net migration (immigration rate minus emigration rate, both expressed as number of individuals per 1,000) gives the overall population growth rate.

Table 2.1 Estimating world population in January 1998 using the demographic balancing equation

Starting population (1 January 1997)	5,800,202,000
+ births during 1997	+ 136,967,000
– deaths during 1997	– 53,282,000
Ending population (1 January 1998)	5,883,887,000

Table 2.2 Estimating the population of the United States in January 1998 using the demographic balancing equation

Starting population (1 January 1997)		266,487,000
+ births during 1997	+ 3,882,000	
– deaths during 1997	– 2,294,000	
Natural increase	+ 1,588,000	+ 1,588,000
+ immigrants during 1997	+ 1,124,000	
– emigrants during 1997	– 277,000	
Net migration	+ 847,000	+ 847,000
Ending population (1 January 1998)		268,922,000

BOX 2.10 THE DEMOGRAPHIC BALANCING EQUATION

The *demographic balancing equation* allows the calculation of population changes from one year to the next, based on number of births, deaths and migrations. The general form of the equation is a mass balance equation:

Population at start of 2005 = population at start of year 2004 + births during 2004 − deaths during 2004 + immigrants during 2004 − emigrants during 2004

or

End population = starting population ± natural increase ± net migration

where

Natural increase = births − deaths

Net migration = immigrants − emigrants

At the global scale there is no net migration (unless the Earth is invaded by aliens from another planet, or people leave the Earth permanently to set up home in space), so the balancing equation is then simply:

End population = starting population ± natural increase

Most populations increase through time, because of natural increase and – where it occurs – in-migration. Some populations decline at particular points in time through a combination of net out-migration (more people leave than move in) and demographic change (death rates rise in an increasingly elderly population, for example).

Population growth rates can be used to estimate how long it would take for a given population to double in size (Table 2.3), assuming that the present rate of growth continues. This is known as the doubling time (Box 2.11).

Table 2.3 Population doubling times associated with different annual growth rates

Annual growth rate (per cent)	*Doubling time (years)*
0.5	140
1	70
2	35
3	23.3
4	17.5
5	14
10	7
15	4.7
20	3.5

BOX 2.11 POPULATION DOUBLING TIME

Doubling time is the time it takes (usually expressed in years) for a population to double in size. It can be estimated by dividing the number 70 by the annual growth rate expressed as a percentage. Thus, for example, a population growing at 5 per cent a year would double in 14 years, while a population growing at 1 per cent a year would double in 70 years. Table 2.3 shows doubling times for a range of annual growth rates.

The importance of doubling time as an indicator of rates of change is that it shows very clearly how quickly population grows as the growth rate changes even a relatively small amount. This is because the change is geometric rather than arithmetic (see p. 623), in the same way that money accumulates in a bank savings account with compound interest.

Population cannot double if growth rates are zero (there is no growth) or negative (population is declining, not expanding).

POPULATION GROWTH

HISTORICAL TRENDS

Throughout most of human history, world population never reached 10 million people (Figure 2.2) but, for comparison, by the late 1990s there were more than ten cities around the world with populations greater than 12 million (see Table 2.6). Population remained quite low, and rates of increase were very small because death rates matched birth rates. The first significant growth in population began about 8000 BC, when people began to grow crops and raise animals.

By 1650, world population had reached around 500 million. It then doubled over the next 150 years, to a billion (1,000 million) by 1800. The second billion was reached within the next 130 years, by 1930. Thirty more years saw the total rise a further billion, to 3 billion by 1960. The fourth billion took only 15 years – by 1975, world population stood at 4 billion. The fifth billion was reached in 1987 (12 years later), and the sixth on 12 October 1999 (another 12 years). Demographers estimate that nearly half of the 6 billion in late 1999 were under 25 years old, and over a billion of them were aged between 15 and 24.

World population has therefore doubled since 1960. Most – up to 95 per cent – of the current growth is in developing countries. In many countries, including the United States, Japan and most of Europe, population growth has slowed or stopped. The United States is the only industrial country where large population increases are projected, and this is mainly because of immigration.

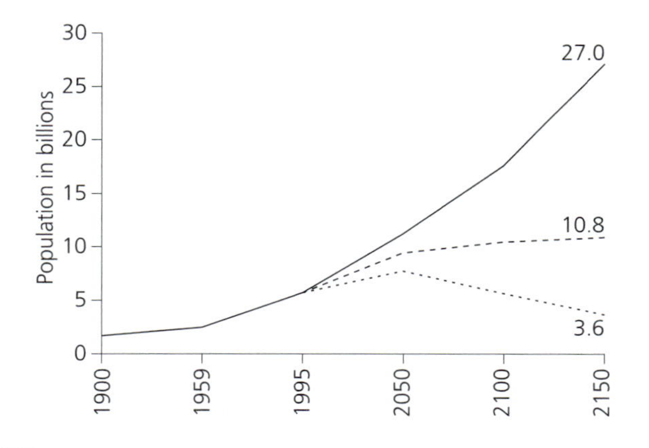

Figure 2.2 *World population growth throughout history. World population has expanded progressively since the development of agriculture about 12,000 years ago, and the pace of increase has speeded up over time.*

Source: Figure 5.1 in Marsh, W.M. and J.M. Grossa (1996) Environmental geography: science, land use and Earth systems. John Wiley & Sons, New York.

The current rate of global population increase can be put into context – it amounts to an extra three people every second, an extra quarter of a million every day and an extra 87 million (roughly the population of Mexico) every year. The world will have an extra population the size of the United States within 2.5 years and an extra population the size of China over the next decade. A number of factors help to explain the rapid population increase since the 1960s (Table 2.4).

The pace of change has been staggering, and it cannot continue unchecked without creating major stresses on environmental life-support systems (see p. 10). The first billion population took most of human history to achieve; each new billion currently takes little more than a decade! Most demographers expect world population to continue rising by a billion people every 11 to 13 years, at least until the middle of the twenty-first century, after which the rate of rise is expected to slow down.

PROBLEMS AND SOLUTIONS

This rapid rate of population growth – sometimes referred to as a *population explosion* – generally brings increased environmental damage, through increased resource use (see p. 26), waste production (see p. 96), pollution emission (see p. 241) and land-use change. But it is also widely associated with social impacts such as:

- low living standards
- low education standards

Table 2.4 Factors responsible for the rapid population increase since 1960

Improved healthcare (which has greatly reduced infant and child mortality, in particular), such as:

- the use of DDT in the battle to eliminate malaria
- childhood immunisation programmes against cholera, diphtheria and other often fatal diseases
- antibiotics

Increased food production (the so-called Green Revolution), involving:

- cultivation of new disease-resistant rice and other food crops
- more extensive use of chemical fertilisers
- introduction and widespread adoption of more effective farming methods

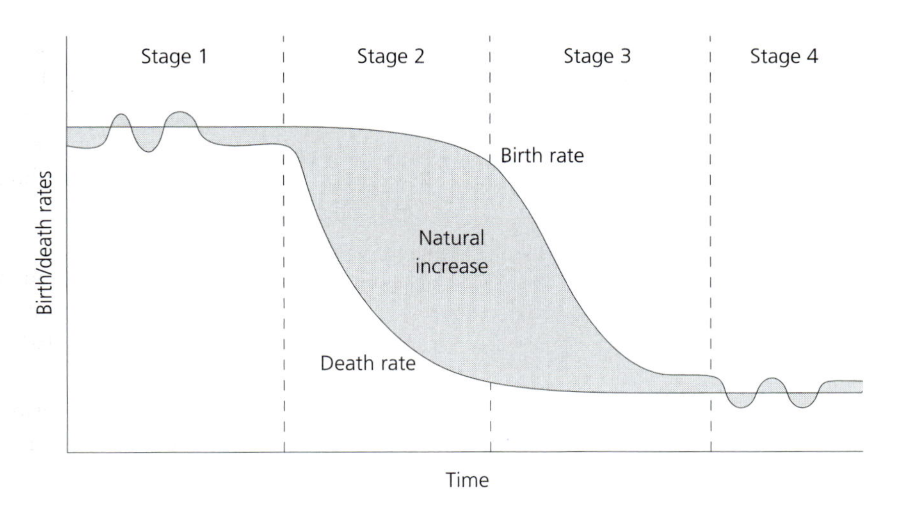

Figure 2.3 *The demographic transition model.*

- unemployment and poverty
- malnutrition and starvation
- civil unrest
- large-scale migration, both internal and international
- the growth of huge mega-cities and associated shanty-towns
- unsustainable pressures on government institutions and national economies.

Many of the problems that rapid population growth gives rise to are interrelated. For example, environmental damage reflects not just the overall number of people but also how much they consume and the ways in which their consumption damages the environment. Poverty is caused not just by large numbers of people but also by the uneven distribution of income within and between countries.

Solutions to these pressing problems, which cause extensive and persistent human suffering and hardship and damage environmental life-support systems, will require direct action by policy makers in many different areas. Population control by itself will not solve the problems; it must be part of a larger package of integrated measures, which should include:

- sustainable development;
- better education;
- empowerment of marginal and landless people; and
- a more co-operative spirit among developed and developing countries.

THE DEMOGRAPHIC TRANSITION

Many populations, including those in Europe and the United States, pass through a number of stages through time. One useful way of defining these is in terms of the so-called demographic transition model, which is illustrated in Figure 2.3 and described in Box 2.12. By the close of the twentieth century, most developed countries had completed all four stages, while most developing countries lagged behind at stage 2 or early in stage 3.

A leading question is whether all developing countries will eventually complete the demographic transition to low fertility and mortality. The evidence suggests that some developing countries have already followed a different path. After about 1950, for example, death rates in many developing countries fell sharply because of the availability of medical supplies and technology. This fall was much quicker than stage 2 occurred in Europe. But it was not matched by a fall in birth rates, so many developing countries experienced unprecedented rates of natural population increase beginning in the 1960s and accelerating during the 1970s. Growth slowed down in many countries during the 1980s and 1990s. Birth rates fell sharply in Brazil, Mexico, South Korea and Thailand during this period.

Developing countries are following different paths and have different timetables for development. The ultimate size of the world population will depend heavily on when and by how much fertility declines in developing countries. Hence the interest in developing countries as well as developed countries in finding effective strategies to promote declining birth rates, including improving the educational levels of women and making contraceptives more readily available.

BOX 2.12 THE DEMOGRAPHIC TRANSITION MODEL

This model (Figure 2.3) shows how birth and death rates change through time in a particular country or area, giving rise to phases of population stability and change. There are four stages in the model:

Stage 1: During this early stage, the death rate is typically very high because of poor health and difficult living conditions. Life expectancy at birth is less than 30 years, but the population does not decrease and die out during this stage because birth rates are also high (mainly for cultural reasons, including religious teaching and social pressure; large families are practically useful and bring power and prestige). Population growth rate is zero or close to zero during this initial stage, in which birth rates and death rates are both high and similar.

Stage 2: This stage – the transitional phase – begins when the death rate begins to fall, probably because of improved living conditions, better food supplies and better health practices (such as immunisation). The birth rate remains high and may even increase, because women are healthier. A high population growth rate is triggered by the excess of births over deaths. It takes time, generations sometimes, for social attitudes (such as the high value attached to having children) to change, so during this phase birth rates remain high.

Stage 3: In this stage the birth rate declines (because of better education, better family planning, more career options for women and reduced infant mortality) and will eventually catch up with the death rate. During the early part of this stage, population growth remains relatively high. It falls close to zero towards the end of the stage.

Stage 4: By now the birth rate and death rate have converged, and they oscillate around a relatively low level. Population growth rate is once again zero or close to zero during this phase, in which birth rates and death rates are once again similar but now both low.

ZERO POPULATION GROWTH

If 1998 rates of increase continue, world population will keep rising to 12 billion by 2050, 24 billion by 2100 and 96 billion by 2200. Such numbers are unthinkable, as well as unsustainable, and many demographers argue that zero population growth (ZPG) must be the target.

ZPG was typical of human populations throughout most of history (see Figure 2.2), and it must be achieved again – albeit at a much higher level of overall population – to be sustainable. But even if that were achievable, it would bring its own problems, because a zero-growth population quickly becomes an ageing population, reducing the labour force and causing heavy increases in demand for medical and other public services.

Assuming that mortality remains low, or declines even further as many experts predict, global total fertility rate (TFR) would need to fall from its present average level of 2.9 children per woman to a two-child average. This is replacement-level fertility, but it is unlikely to happen quickly for two main reasons: the social beliefs and cultural practices that promote high fertility will change only slowly; and the world's current age structure (almost one in three people alive today is less than 15 years old) will generate great growth in the near future even if family sizes decline.

Many experts predict that, once the coming population explosion has run its course (probably by the close of the twenty-first century), population growth rates both globally and in most countries will be close to zero. When growth stops, world population could be between 11 billion and 13 billion.

POPULATION DISTRIBUTION

The spatial pattern of population is strongly influenced by fertility, mortality and migration (see p. 34), is markedly uneven, changes through time and can significantly affect rates and patterns of environmental change.

WORLD POPULATION

GROWTH AND CHANGE

By 2000 there were more than 6 billion people living on the Earth, but they were by no means evenly distributed.

Population distribution is very uneven, and it is becoming even more so because some regions are growing much faster than others. In 1950, for example, developing countries accounted for 68 per cent of the world population. By 1960, the proportion had risen to 70 per cent, and by 1998 it was 80 per cent. The imbalance is expected to grow further, because the United Nations predicts that developing countries will house 88 per cent of the world's population by 2050 and 89 per cent by 2100.

Growth is uneven mainly because of variations in fertility from country to country. But migration has also played a significant role, as large numbers of people migrate from developing countries to developed ones (for example, from Mexico to the United States), and from less affluent countries to more affluent countries within the developed regions (for example, from Portugal to France).

A shifting distribution of population between regions is not necessarily a major problem in itself, provided that development progresses everywhere and that population growth is balanced by the development of social and economic capacity. It would create problems if the main areas of population growth were close to or at the carrying capacity (see p. 48), and if the groups which were growing the fastest were placing disproportionately large pressures on environmental resource systems.

REGIONAL DIFFERENCES

The distribution of population across the globe is very uneven, as also is the distribution of rates of change. Figure 2.4 shows the distribution of global population, and Table 2.5 summarises the variations in population growth rates by region. Population growth rates are highest in Africa but are still high in South America, Asia and Oceania. Growth rates are much more modest in Europe and North America.

Rapid population growth does put great pressure on the environment, through such factors as the need for more food, building of more settlements, creation of more waste materials, generation of more pollution, and greater loss of habitat and changes in land use. But the population–environment equation is not simple, because some of the most serious pollution problems and highest levels of energy use are concentrated in areas with high population densities (Asia and Europe), not necessarily in areas with high population growth rates (Africa and South America).

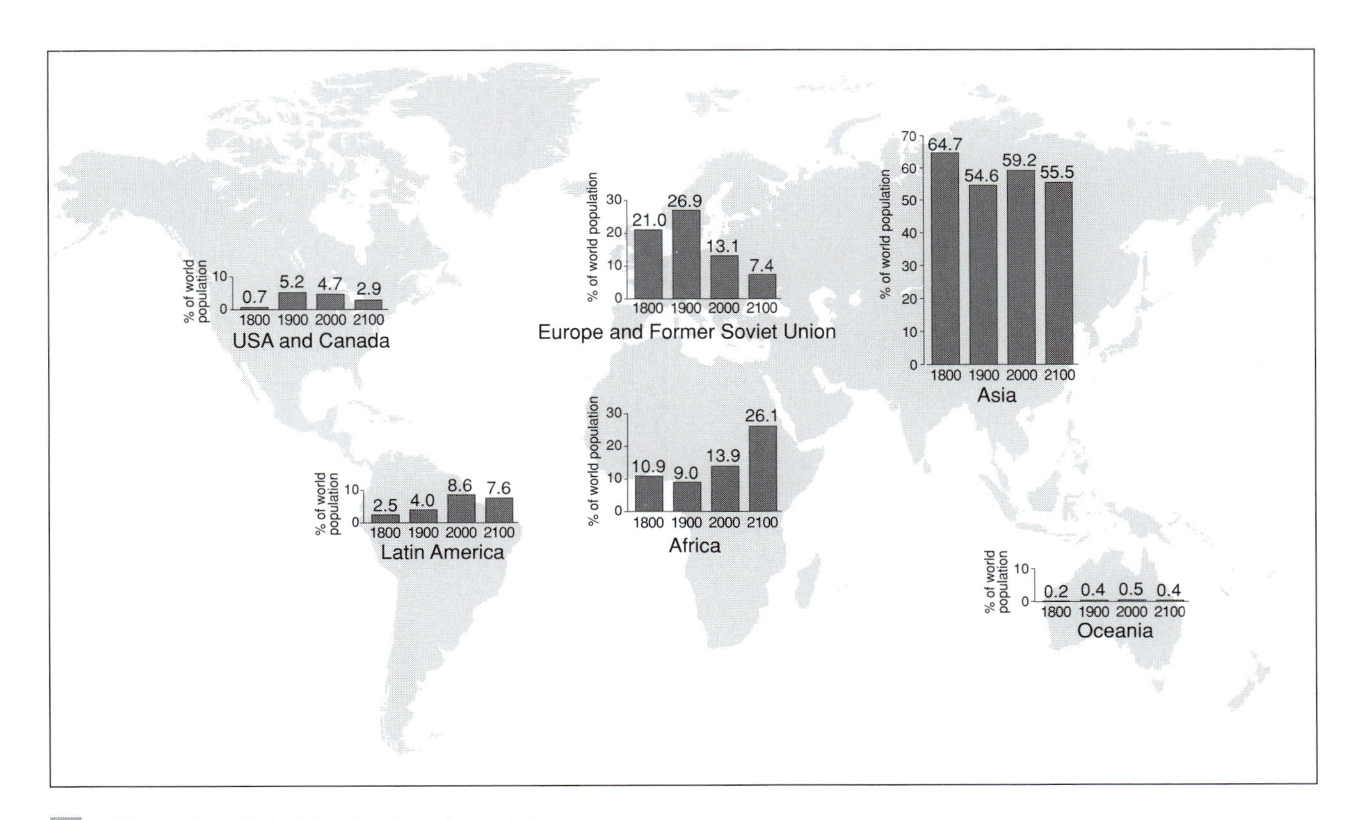

Figure 2.4 *Global distribution of population.*

Table 2.5 Population growth rate by region

Region	Annual growth rate (per cent)	Number of people per km² of agricultural land
Asia	1.8	423
Europe	0.2	213
Africa	3.0	80
Former Soviet Union	0.7	69
South America	1.9	58
North America	0.7	55
Oceania	1.4	15

The distribution of environmental problems will inevitably change in the future, because around 95 per cent of the predicted extra 3.6 billion people between the late 1990s and 2030 will be located in developing countries. The population in developing countries is growing at a rate of 2.1 per cent a year: a doubling time of 33 years (see Table 2.3). If current trends continue, the population of Africa is likely to double within about 23 years, Latin America within 30 years and Asia within 36 years. At current rates of increase, the population of sub-Saharan Africa (Box 2.13) is likely to double by 2016. These are regional averages, which mask a great deal of local variation. Many areas – particularly the major cities – in Africa, Latin America and Asia are growing much faster than the average.

BOX 2.13 RAPID POPULATION CHANGE IN SUB-SAHARAN AFRICA

Sub-Saharan Africa has:

- the highest birth rate of any region in the world;
- the highest rate of population increase of any region in the world (average annual population increase is 3 per cent);
- the lowest use of contraceptives of any major region in the world;
- very large family sizes (in twelve countries in the region, women have on average more than seven children); and
- only limited increases in food supply (increasing at around 1 per cent a year).

INEQUALITIES IN POVERTY AND WEALTH

Many environmental problems and a great deal of human suffering arise because of the strongly uneven distribution of income between countries, and between groups of people within countries.

In 1960, the richest 20 per cent of people in the world accounted for 70 per cent of total global income. By 1989, their share had grown to 83 per cent. Over the same period, the income of the poorest 20 per cent fell from 2.3 per cent to 1.4 per cent. The rich get richer, and the poor get poorer; the poverty gap widens and becomes more and more unbridgeable for growing numbers of people.

This inequality of income distribution (Box 2.14) is serious and must be addressed if there is to be any hope of achieving sustainable development within a reasonable period of time. It has humanitarian consequences because poverty is perpetuated, with all the human hardship and suffering that this brings. But it also has environmental consequences, because it promotes further excess consumption, waste and pollution by the rich (Box 2.15) and further over-exploitation of diminished resources (usually in marginal and vulnerable environments) by the poor. The poor often have no alternative but to over-graze land and grow crops on impoverished land, cut down forest for fuelwood and increase grazing lands and farmland (Box 2.16).

The imbalance between rich and poor is difficult enough as it is, but the signs are that it will get much worse in the years ahead. The most rapid increases in population are expected to occur in the poorest countries, such as Ivory Coast, Togo, Comoros and Kenya, whose per capita gross national product lies between US$340 and US$690 (the equivalent figure for the USA is US$22,560).

BOX 2.14 INEQUALITIES IN INCOME AND WEALTH

According to a 1998 report from the United Nations Development Program:

- The richest 20 per cent of people control 86 per cent of world income, while the poorest 20 per cent control only 1.1 per cent.
- The richest 225 people control as much wealth as the poorest 2.5 billion people.
- The richest three people control more wealth than all the least developed countries (which contain 600 million people) put together.

BOX 2.15 RESOURCE USE BY THE RICH AND THE POOR

Many different comparisons in resource use between the rich and the poor have been made. For example, it is said that – compared with an average person in India – a typical person in the United States consumes 50 times more steel, 56 times more energy, 170 times more synthetic rubber and newsprint, 250 times more motor fuel and 300 times more plastic. The United States houses 4.7 per cent of the world's population, but it uses 25 per cent of the world's resources and produces up to 30 per cent of the world's waste materials.

URBAN GROWTH

A prominent feature of population redistribution, particularly within developing countries, is the growth of major cities. In 1960, about a third of the total world population lived in cities; by 1999, this had grown to 47 per cent, accounting for 2.8 billion people.

The world's urban population is currently growing by 60 million people a year, which is about three times the increase in the rural population. Most (about 60 per cent) of this urban growth reflects natural increase among current city dwellers (those born in the city and those who migrated there); the rest is accounted for by rural–urban migration. Urban growth has been particularly fast in developing countries. In 1950, there were 304 million people living in cities in developing countries, and by 1996 this had risen to 1.8 billion. The proportion of people living in cities in developing countries nearly doubled, from 22 per cent in 1960 to more than 40 per cent in 1999. Urbanisation is projected to continue well into the next century. By 2030, for example, it is expected that nearly 5 billion (61 per cent) of the world's total 8.1 billion people will live in cities.

Rapid urban growth, particularly if sustained, creates huge cities, which are sometimes referred to as *mega-cities*. In 1950, there were only two cities in the developing world with populations of more than 5 million each – Shanghai (China) and Buenos Aires (Argentina). The United Nations estimates that by 2000 there were around thirty-five such cities, including sixteen with more than 10 million people (Table 2.6).

The number of mega-cities continues to rise. In 1960, only New York and Tokyo housed more than 10 million people each, yet by 1997 there were seventeen mega-cities (thirteen of them in developing regions). If recent trends continue, there are likely to be twenty-six mega-cities by 2015, twenty-two of them in developing regions (and eighteen of these twenty-two will be in Asia). Between them,

BOX 2.16 POVERTY, POPULATION AND ENVIRONMENT

More than a fifth of the global population today lives in poverty, subsisting on less than US$1 a day. There are clear links between poverty, population growth and environmental problems. These are most evident at the level of individual households or communities, but they can also be detected at the national scale.

Some key ingredients of these interrelationships:

- Poverty is often accompanied by illiteracy, poor nutrition and health, low status of women, and exposure to environmental hazards.
- Poverty and lack of economic opportunities can force people to exploit marginal resources (e.g. by over-grazing land or over-harvesting forests); this sets in motion a repeating cycle of environmental deterioration, marginalisation, hardship and poverty.
- Poverty is associated with a wide variety of health risks and problems, including inadequate sanitation, unsafe drinking water, air pollution and crowding.
- Poverty is linked to fertility, and in many places women from low-income families have more children than women from richer families in the same society.
- Problems of rapid population growth are compounded by poverty and inequality, and efforts are being made in many countries to reduce poverty as a means of improving the health of children and mothers, reducing the problems of rapid urban development and ensuring adequate nutrition for everyone.

Table 2.6 The ten largest cities in the world. Populations of metropolitan areas, in millions, mid-1998

Rank	City	Population
1	Tokyo, Japan	27.3
2	Mexico City, Mexico	24
3	São Paulo, Brazil	22
4	New York, United States	19.8
5	Bombay, India	16.6
6	Shanghai, China	16.2
7	Los Angeles, United States	15.3
8	Beijing, China	13.3
9	Jakarta, Indonesia	12.8
10	Calcutta, India	12
	Total	179.3

these twenty-six mega-cities are expected to house more than a tenth of the world's population.

Explosive urban growth creates many social, economic and environmental problems. Rapid population growth can outstrip the supply of housing, jobs and social services. Many people – particularly the poor, dispossessed and landless – survive by building makeshift shelters for themselves and their families on open land, trying to make a living on the streets (by such means as begging, rag-picking and waste recycling, and low-paid casual employment). Huge shantytowns have grown up in this way around many major cities – such as Rio de Janeiro in Brazil and Manila in the Philippines – usually on very marginal land prone to environmental hazards (such as landslides, flash floods and storm surges), thus placing the residents and their property at risk. People forced to live under such conditions share a disproportionately large amount of the hardship, suffering and losses associated with natural hazards (see p. 21).

CARRYING CAPACITY AND LIMITS TO GROWTH

A central question in the population–environment debate is this: how many people can the Earth be expected to support? What is the maximum number of people the planet can accommodate and meet the needs of? This is not an easy question to answer, because it all depends on what is regarded as an acceptable quality of life, and on how sustainable that population might be in the long term, particularly in terms of resource use, production of wastes and pollution.

CARRYING CAPACITY

One way of approaching the question of upper limits is via the ecological concept of carrying capacity, which refers to the largest number of individuals in any given species that a habitat can support indefinitely. There are numerous ways of trying to define the Earth's carrying capacity for humans, including:

- *Space*: how much land area is there on the planet, and how many people could that meet the needs of? See Box 5.1 for a brief discussion of this theme.
- *Food*: how much food could be grown on the planet, and how many people could it feed? See Box 2.17.

BOX 2.17 FOOD PRODUCTION ON EARTH

One way of gauging the Earth's carrying capacity is to estimate the total food resource. A good measure of this is total net primary productivity (NPP) – the total amount of solar energy that is converted into biochemical energy via plant photosynthesis, minus the energy needed by those plants for their own life processes. One estimate is that before humans changed vegetation, NPP was about 150 billion tons of organic matter per year. Human activities (such as deforestation, other land-use changes) have destroyed about 12 per cent of the terrestrial (land-based) NPP, and we now use a further 27 per cent directly (for food and fibre) and indirectly (by converting productive land to other uses). This means that humans already use about 40 per cent of the terrestrial food supply, leaving 60 per cent as yet unused by humans. But all of that 60 per cent cannot be used simply for food production; much of it is needed for conserving biodiversity (see p. 527), protecting natural habitats and landscapes (see p. 542) and protecting vital environmental services (such as water supplies – see p. 388; limiting soil erosion – see p. 609; and preventing the build-up of atmospheric CO_2 – see p. 255). Some of it is likely to be damaged in the future by pollution and the build-up of wastes. As a result, many scientists conclude that the likely total carrying capacity of the Earth is nowhere near 15 billion (if today's 6 billion represents 40 per cent of what is possible). Many insist that the sustainable carrying capacity is probably lower than the current population total of 6 billion.

Table 2.7 Population and availability of renewable resources

	1990	2010	Total change (per cent)	Per capita change (per cent)
Population (millions)	5,290	7,030	33	
Fish catch (million tons)	85	102	20	−10
Irrigated land (million hectares)	237	277	17	−12
Cropland (million hectares)	1,444	1,516	5	−21
Rangeland and pasture (million hectares)	3,402	3,540	4	−22
Forests (million hectares)	3,413	3,165	−7	−30

Source: Postel, S. (1994) *Carrying capacity: Earth's bottom line.* State of the World, Washington DC.

- *Damage*: how much damage has been caused to the Earth's environmental support systems, and how much more change could those systems tolerate without wholesale disruption or decline? See Box 20.8.

Each approach leads to different estimates of the world's carrying capacity (published estimates vary between 3 billion and 14 billion), and this is one reason for the lack of consensus among experts about how serious the problem really is. Yet, while this uncertainty persists, it makes it very difficult for politicians and decision makers to agree on the most appropriate and effective ways of tackling the problems of population growth and excessive resource use.

GLOBAL AND REGIONAL CARRYING CAPACITY

It is important to distinguish between global carrying capacity and regional/national carrying capacity. Global carrying capacity refers to the total number of people that the Earth and its environmental systems can support on a sustainable basis. Regional and national carrying capacity refers to the total number of people that a particular region or country and its environmental systems can support on a sustainable basis. Some regions have clearly already exceeded their national carrying capacity – the Sahel region in Africa (Box 2.18) is a vivid example. The same is true of some other countries, such as India (Box 2.19). Japan illustrates a very different situation, because it has a very successful economy and a very high population density (331 people per km^2 on average), but its land resources are small and steep. Thus its ability to grow food is severely limited; the amount of cropland per person is only one-seventh of the global average. As a result, Japan must import many of the resources it needs – nearly three-quarters of its grain, and over two-thirds of its wood. International trade in agricultural and forestry products allows Japan to survive, and it gives opportunities for the exporting countries to generate revenue.

The notion of a global carrying capacity makes it clear that it would be impossible for all countries to exceed their carrying capacity; otherwise, critical environmental systems would be over-stretched, resources would be depleted, and wastes would be produced in unmanageable amounts. The survival of the whole Earth system (see Chapter 3) depends on living within the global carrying capacity and sharing out the planet's environmental resources and opportunities in ways that are fair and sustainable. But who decides what is fair and sustainable? This is the magnitude of the challenge that decision makers in meetings like the Rio Earth Summit

BOX 2.18 CARRYING CAPACITY OF THE SAHEL

Sub-Saharan Africa is one area where population pressure is close to if not above the carrying capacity of that fragile environment. Mortality rates continue to increase, birth rates are falling but not as fast as elsewhere, aquifers (which supply groundwater) are being depleted, and cropland per person is declining. Within sub-Saharan Africa, the problems are most acute in the Sahel, which has a rapidly increasing population (see Box 2.13) and is suffering from very rapid desertification (see p. 422). Prolonged droughts have been common in the Sahel since the 1970s, causing massive human suffering and hardship and requiring international aid and assistance. Rapid population growth promotes desertification in this dry environment, by clearing forests and trees for fuelwood and agriculture, and by over-grazing of the impoverished soils.

BOX 2.19 CARRYING CAPACITY OF INDIA

In August 1999, the population of India reached 1 billion. This is an almost threefold increase since the country achieved independence in 1948. The population of India is currently growing at a rate of about 1.0 per cent a year, and within a few decades India is expected to overtake China to become the world's most populous country. By 2016, India is expected to house more people than all of the industrialised countries in the world. Population density in India is more than double that of China and more than ten times that of the United States.

This rapid population explosion has had significant impacts on the environment, and continues to do so – through deforestation, soil erosion, water shortages, pollution of air and water, and declining biodiversity.

India has struggled for many years to reduce population growth, but it has fallen short of the extreme approach adopted by China. Many experts insist that the future stability and prosperity of India will depend largely on the country's ability to stabilise its population.

BOX 2.20 THOMAS MALTHUS ON HUMAN POPULATION PROBLEMS

In 1798, Thomas Malthus published *An essay on the principle of population*. In that book, he argued that populations tend to grow geometrically, while the resources available to support them tend to grow arithmetically. As a result, population grows much faster than food supply, so a population will almost inevitably outgrow the supply of food available to feed it. Indeed, Malthus stressed that population growth was already outpacing the production of food supplies in eighteenth-century England. He predicted that this imbalance of population and resources would lead to degradation of the land, and this in turn would lead to massive famine, disease and war. The disaster that Malthus believed imminent did not occur, mainly because of improvements in agriculture and the technological innovations introduced by the Industrial Revolution. Nonetheless, many experts argue that his ideas still have great value today.

(see p. 13) and the Cairo Population Conference (see p. 54) must confront.

OPTIMISTS AND PESSIMISTS

There are no clear-cut answers to the question of what the Earth's carrying capacity really is – published values range between about 3 billion and 14 billion people. Experts expect there to be natural limits, but there is no consensus over precisely what those limits are likely to be. The debate, which began in the 1970s (see p. 48), continues.

MALTHUS AND MARX

There is a long history of interest in population problems. For example, Aristotle warned that populations could outstrip the resource base on which they depended, causing poverty and social unrest. In the late eighteenth century, Thomas Malthus (Box 2.20) arrived at much the same conclusion, arguing that the inevitable results of population growth are poverty and misery because the population will eventually outstrip the food supply. This Malthusian view was roundly rejected in the nineteenth century by Karl Marx and Friedrich Engels, who blamed poverty on the evils of social organisation in capitalist societies rather than on the poor or on over-population. They saw over-population as a natural consequence of capitalism that would not exist in socialist societies, which would provide enough resources for each person. Marx and Engels argued that resource scarcity would encourage people to reduce family size and thus keep population and resources in check with one another.

Thankfully, Malthus's predictions and more recent warnings that human population growth would eventually outstrip food supply have not come true. Or rather, not yet. Food supplies have continued to grow faster than population increase, largely due to human ingenuity and continued improvements in agricultural technology.

At the opening of the twenty-first century, many experts are still asking whether there are environmental limits to the number of people the planet can support and to the quality of life those people might rightly expect to enjoy. The debate (to which we return in Chapter 20) has tended to be heated, with views polarised between so-called optimists and pessimists.

OPTIMISTS

Optimists – mainly economists – stress that the development of new technologies and improvement in resource management techniques will further expand the planet's carrying capacity, making the ultimate maximum difficult to define and making it possible that Earth could support an almost infinite number of people. To support their contention, they point to such indicators as the increased food yields associated with the 'Green Revolution', increased efficiency of energy use and increasing life expectancy in many countries. Many optimists believe that an upper limit of about 10 billion may be realistic if people are to have a reasonable quality of life.

PESSIMISTS

Pessimists – many of them biologists and environmentalists – insist that the world's biological systems (forests, grasslands, croplands and fisheries) and energy resources could not cope with a population as high as 10 billion, and they argue that human numbers have already exceeded the Earth's carrying capacity. Pessimists point to the growing rate of extinction of species, global climate change and more people living in poverty than ever before as evidence that this is the case. One thing that is clear is that the Earth would not be able to support its current population if every person enjoyed a standard of living and quality of life comparable with those in the most developed countries, such as the United States (see Box 2.15).

POPULATION PROSPECTS TO 2050

MILLENNIUM MILESTONES

In 1998, total world population was 5.9 billion, and it was growing at a rate of about 1.4 per cent a year (a doubling time of 50 years). Population growth continues because the global birth rate exceeds the global death rate – by more than 80 million people in 1998 alone.

The most heavily populated country in the world is China, which in 1998 had a population of 1.2 billion and a net increase of around 1.0 per cent a year (a 70-year doubling time), assuming minimal net migration. India has a smaller population (989 million) but much faster growth (annual growth rate of around 1.9 percent – a doubling time of 37 years), so it is likely to overtake China as the world's most populated country by around 2050.

The country with the third largest population is the United States, which housed 269 million people in 1998. The US population grew by 2.5 million during 1997 (see Table 2.2), with the number of births greatly exceeding the number of deaths and immigration significantly outpacing emigration. The US Census Bureau predicts that the US population could reach 394 million by 2050.

The most rapid rates of population growth in the 1990s were in the Middle East and Africa. In 1998, Kuwait had a population of 1.9 million, which was growing at a rate of around 3.7 per cent a year (a 19-year doubling time) including net migration. The population of the African continent overall is growing at a rate of 2.5 per cent a year, producing a doubling time of 27 years. Population is growing fastest in the poorest countries, which are least able to provide for basic needs and create opportunities. Fertility is highest in countries in Africa and South Asia, and the fastest-growing regions of the world are sub-Saharan Africa (see Box 2.13), parts of South Asia and West Asia.

Population growth is very slow in some countries, and in others death rates are higher than birth rates, so there is a net decrease in population. Death rates were higher than birth rates during the late 1990s in thirteen European countries (including Russia, Germany and the Czech Republic), and the only population growth was related to net migration. There are more than sixty countries in which fertility is now at or below replacement level, which means that their populations could decline over the long term. Death rates will increase though time in these countries as the populations grow progressively older (Box 2.21).

TRENDS AND CHANGES

The twentieth century saw unparalleled changes in population – world population more than tripled in size, average life expectancy increased by two-thirds, and there were major declines in child-bearing and major shifts in population distribution. However, at the same time, there have been unprecedented improvements in technology, communications, education and agriculture. There have also been major improvements in family planning, health care and education, and a major expansion in social and economic opportunities, especially for women.

The coming century will see further change, as fertility continues to decline and population redistribution between developed and developing countries intensifies. Nearly all future world population growth will be concentrated in developing countries. Some likely trends for the future are already clear:

- Population growth will be concentrated in particular regions, and elsewhere populations will stabilise or even decline.
- Within countries, populations will continue to move from rural to urban areas and will become increasingly older and better educated.

BOX 2.21 IMPACTS OF AN AGEING POPULATION

One consequence of the progress made in many countries during the twentieth century – including reduced infant and child mortality, and better nutrition, education, health care and access to family planning – is increased life expectancy. This has promoted an unprecedented growth in the elderly, both in absolute numbers and as a proportion of overall population. All the signs suggest that this growth will continue during the next century. It is estimated that by 2025 there will be more than 1 billion people aged 60 or more, and by 2050 this will have risen to nearly 2 billion.

The world's population has never been so healthy or lived so long. This is a blessing for the people involved, but it does come at a cost. The stresses that this brings will not be evenly distributed, because three-quarters of this huge elderly population will live in developing countries. Not surprisingly, the largest percentage increases in the elderly population will occur in the poorest regions of the world – South Asia and sub-Saharan Africa.

An ageing population brings a range of problems. Economic problems include the high costs of health care, housing provision and support services for the old and frail, and the outflow of savings from pension funds and a corresponding decrease in the amount of capital available for investment. Social problems include the need to have effective care and support systems in place for people who are generally less mobile and more dependent than when they were younger.

- Migration between countries will continue and will be an increasingly important aspect of international relations and the composition of national populations.

UNITED NATIONS POPULATION PROJECTIONS

Every two years, the UN Population Division produces a set of population projections for every country, based on updated population data and revised projection methodology. The most recent projections (Table 2.8), published in 1998, incorporate these updated data and revisions in methodology. The projections are based on three growth scenarios – low, medium and high – and by 2050 the UN expects total world population to grow to between 7.3 billion and 10.7 billion. Under the high-growth scenario, world population will continue to increase after 2050; under the low-growth scenario, it will have slowly begun to decline. The medium-growth projection – 8.9 billion people by 2050 – is considered the most likely.

Key assumptions underlying the projections are global fertility rates, which are projected to fall to between 2.5 and 1.6 births per woman by 2050. The medium-growth scenario assumes that the fertility rate for all developing countries will fall to 2.1 by 2050, and national fertility rates in developed countries will lie between 1.7 and 1.9

Table 2.8 United Nations population projections (1998) for world regions; three scenarios for 2050. Population in millions

Region or country	*2000*	*2050*		
		High growth	*Medium growth*	*Low growth*
World	6,055	10,674	8,909	7,343
■ more developed	1,188	1,361	1,155	990
■ less developed	4,867	9,313	7,754	6,353
Africa	784	2,102	1,766	1,467
Sub-Saharan Africa	641	1,804	1,522	1,272
Asia	3,583	5,316	5,268	4,312
■ China	1,278	1,686	1,478	1,250
■ Japan	127	117	105	92
Latin America and Caribbean	519	994	809	654
North America	310	464	392	324
Europe	729	746	628	550
Oceania	30	52	46	36

births per woman. Assumptions about declining fertility and rising mortality (particularly from AIDS) allowed the UN demographers to reduce the projected global population for 2050 by 500 million people, from 9.4 billion to 8.9 billion. One-third of the fall was related to increased mortality, two-thirds to reduced fertility.

Under even the low-growth forecast, the world will have to meet the needs of an extra 1.3 billion people, because:

- Fertility in developing countries is twice as high as that in developed countries, overall.
- Developing countries have a younger age structure, which creates a momentum for continued population growth for several decades ahead.
- Continued improvements in mortality will add extra growth, particularly in countries with relatively low life expectancies.

Continued population growth will inevitably affect other environmental trends (Table 2.9). And global warming (see p. 261) threatens to make matters much worse within the next century.

POPULATION POLICIES

NATIONAL INITIATIVES

Concern about the impacts of rapid population growth are not new – the ancient Greeks and Egyptians voiced concern about over-population in lean times and promoted population growth in times of plenty. During the latter half of the twentieth century, the need for effective population planning became more pressing, promoted initially by widespread food shortages and famines in some developing countries during the 1960s, which were widely associated with rapid population growth. Progress was slow, even though some developed countries (particularly the United States) supported and promoted efforts to strengthen family planning programmes in developing countries.

Table 2.9 Population pressure and resource use

Continued population growth will inevitably affect other environmental trends, including

- declining fisheries
- increasing pressure on croplands and grazing lands
- forest clearance and destruction
- loss of plant and animal species
- increased exposure to environmental hazards
- increased production of wastes
- increased pollution of air, land and water resources

Concern about the impacts of rapid population growth on economic development deepened during the 1970s, and many countries started to believe that governments could and should take action and introduce policies that would slow down population growth. By the early 1980s, there was mounting evidence of:

- high rates of population growth;
- high rates of infant and maternal mortality;
- stagnant economic and social development; and
- a widespread desire by women to limit child-bearing.

During the 1980s, many sub-Saharan African countries adopted regional declarations on population and development, and these were followed in the 1990s by the adoption of national population policies. By 1994, more than half of the developing countries had adopted national population policies designed to reduce growth, and most of the rest planned to do so as a matter of priority.

CHINA

Continued rapid population growth in China in the second half of the twentieth century was creating major social and economic problems, and the country adopted a strict regime in an effort to reduce the rate of expansion. A 'one child per couple' policy was adopted in 1979 to try to limit the population to 1.2 billion by 2000, and it has been in operation since then. The policy (Box 2.22) allows couples to have a second child only under special circumstances, and it does not allow more than two children per couple. Many people outside China think that the policy has been pursued too aggressively, in violation of the human rights of Chinese citizens.

The goal of stabilizing the Chinese population at 1.2 billion in the year 2000 was not reached – by 2000 it had grown to 1.3 billion, and a further increase to 1.5 billion by 2025 seems largely inevitable. The policy has been successful in reducing family size (from 5.01 children per woman in 1970 to 1.84 in 1995), but continued population growth was driven by the large number of children born in the earlier phase of high fertility now having one or two children of their own.

BOX 2.22 THE CHINESE 'ONE CHILD PER COUPLE' STRATEGY

The Chinese population control strategy is very much a 'carrot-and-stick' affair, involving both incentives (carrots) and enforcement (sticks). Carrots include granting rewards (including grants of money) for having only one child, allowing additional maternity leave to mothers, and increased allocation of land to farmers. Children from single-child families are also given preferential treatment in education, housing and employment. The sticks include compulsory birth control or sterilisation after a first child is born, and punishment (including heavy fines, loss of land grants and jobs, and expulsion from the Communist Party) for refusal to terminate unapproved pregnancies, for giving birth when under the legal age for marriage, and for having an approved second child too soon. Enforcement of the policy (particularly by abortion) was particularly strong during the early 1980s, but birth rates began to rise again after enforcement was moderated (with education often replacing abortion) in the late 1980s.

INTERNATIONAL INITIATIVES

The most prominent and effective international initiative relating to population was the Fifth United Nations International Conference on Population and Development (ICPD) held in Cairo in September 1994.

INTERNATIONAL CONFERENCE ON POPULATION AND DEVELOPMENT, CAIRO, 1994

This conference followed in the wake of the 1992 Rio Earth Summit (see p. 13), and like UNCED, it was a major international forum. The Cairo conference also gave great prominence to non-governmental organisations (NGOs) – more than 1,200 NGOs participated as delegates or observers – as well as government officials.

A 'Program of Action' (Box 2.23) was drafted by government officials and NGOs and endorsed by leaders of 179 nations. The Program calls on governments to provide universal access to reproductive health care by 2015 as a basic human right.

The conference set goals for 2015 (Box 2.24) that are designed to improve individual and family well-being and enhance women's status. These include:

- universal access to family planning and other reproductive health services;
- universal access to primary school education;
- increased access by girls and women to secondary and higher education; and
- reductions in infant, child and maternal mortality.

Like UNCED, the Cairo conference provided further momentum for a radical re-evaluation of the people–environment relationship, and it reached an important consensus on the relationship between population and development.

BOX 2.23 THE 1994 CAIRO 'PROGRAM OF ACTION'

The Program of Action is built on five key principles:

1. Provide universal access to family-planning and reproductive health programmes, and to information and education regarding these programmes. It is estimated that 125 million women want family-planning services but do not have access to them.
2. Recognise that environmental protection and economic development are not necessarily incompatible, but that economic development is essential for environmental protection. Promote free trade, private investment and development assistance.
3. Make women equal participants in all parts of society by increasing their health, education and employment.
4. Increase access to education. There are close links between inadequate education (which prevents individuals reaching their full potential) and high birth rates. The goal is universal primary education by 2015. Provide information and services for adolescents to prevent unwanted pregnancies, unsafe abortions and the spread of AIDS and sexually transmitted diseases.
5. Ensure that men fulfil their responsibility to ensure healthy pregnancies, proper child care, promotion of women's worth and dignity, prevention of unwanted pregnancies, and prevention of the spread of AIDS and sexually transmitted diseases.

BOX 2.24 ICPD QUANTITATIVE GOALS

The ICPD adopted specific quantitative goals in three areas:

1. *Universal education*: elimination of the gender gap in primary and secondary education by 2005, and complete access to primary school or the equivalent for both girls and boys as quickly as possible and no later than 2015.
2. *Mortality reduction*: reduction in infant and under-5 mortality rates by at least one-third, to no more than 50 and 70 per 1,000 live births, respectively, by 2000, and to below 35 and 45, respectively, by 2015; reduction in maternal mortality to half the 1990s levels by 2000 and by a further one-half by 2015 (specifically, in countries with the highest levels of mortality, to below 60 per 100,000 live births).
3. *Reproductive health*: provision of universal access to a full range of safe and reliable family-planning methods and to related reproductive and sexual health services by 2015.

NEW APPROACH

The action plan agreed at the 1994 Cairo conference redefined the world's view of population growth and the best way to address this growth. It rejected the use of demographic targets by family planning programmes (the traditional approach) and adopted a new approach based on integrating family planning into a broader women's health agenda. This outcome reflects the important role played by women's groups who took part in the conference, as well as mounting research evidence of the links between fertility declines and reductions in infant mortality, increased use of family planning, and improvement in women's education and other aspects of women's status. The action plan goes beyond the single issue of population growth and seeks to identify and resolve many of the interrelated social problems that contribute to poverty and population growth. It recognised the need to provide a range of services, including family planning, and at the same time take action to guarantee rights, inform and empower women in all aspects of their lives, and involve men as supportive partners.

The Program recognises that the key to future population levels lies in providing for individual reproductive health needs. This would enable couples to make informed choices about the number and spacing of their children, which would promote smaller families (by voluntary means) and lead in turn to reduced population growth. The real hope is that this strategy will stabilise population at both global and national scales, and prevent further growth. The target set at Cairo is to stabilise world population at 7.8 billion by 2050.

ICPD+5 REVIEW, THE HAGUE, 1999

A five-year review of progress, based on an international forum held in The Hague in February 1999, showed that the goals of ICPD remain both practical and realistic, but also necessary for individual advancement and balanced development. The 'ICPD+5' review included analysis of surveys carried out by the UN Population Division and

BOX 2.25 POPULATION BENCHMARKS AGREED AT ICPD+5

- The 1990 illiteracy rate for women and girls should be halved by 2005. By 2010, the net primary school enrolment ratio for children of both sexes should be at least 90 per cent.
- By 2005, 60 per cent of primary health care and family planning facilities should offer the widest achievable range of safe and effective family-planning methods, essential obstetric (child-related) care, prevention and management of reproductive tract infections (including sexually transmitted diseases) and barrier methods to prevent infection; 80 per cent of facilities should offer such services by 2010, and all should do so by 2015.
- At least 40 per cent of all births should be assisted by skilled attendants where the maternal mortality rate is very high, and 80 per cent globally, by 2005; these figures should be 50 and 85 per cent, respectively, by 2010 and 60 and 90 per cent by 2015.
- Any gap between the proportion of individuals using contraceptives and the proportion expressing a desire to space or limit their families should be reduced by half by 2005, 75 per cent by 2010 and 100 per cent by 2015.

Table 2.10 Environmental factors that affect human health, according to the World Health Organisation

	Polluted air	*Poor sanitation and waste*	*Polluted water or poor water*	*Polluted food*	*Unhealthy housing*	*Global environmental change*
Acute respiratory infections	✓				✓	
Diarrhoeal diseases		✓	✓	✓		✓
Other infections		✓	✓	✓	✓	
Malaria and other vector-borne diseases		✓	✓		✓	✓
Injuries and poisonings	✓		✓	✓	✓	✓
Mental health conditions					✓	
Cardiovascular diseases	✓					✓
Cancer	✓			✓		✓
Chronic respiratory diseases	✓					

various NGOs, and it allowed governments, parliamentarians, NGOs and private donors to share their experiences and understanding and consider action for the future. The five-year review agreed on new benchmarks (Box 2.25).

The main focus of this chapter has been on demographic issues, particularly rates, patterns and causes of population change and the ways in which these change through time and vary from place to place. But we must not overlook the human health issue, because it has such a major impact on quality of life for many people, as well as on life expectancies and reproduction. Demographic factors affect and are affected by environmental quality, but there are more direct causal links between human health and environment (Table 2.10). Put simply, low environmental quality impairs human health.

SUMMARY

This chapter has focused on the interactions between people and the environment, not just as a root cause of the 'environmental crisis' but also because of the ways in which this affects the quality of life for many people around the world. We began by examining the key demographic forces that determine population composition – fertility, mortality and migration – and exploring what controls them, how they operate and what impact they have. This led in naturally to a consideration of population size and growth, both current and historical, in which we looked at the population balancing equation and the demographic transition model and considered the prospects of zero population growth. Our review of population growth at global and regional scales included the challenges of rapid urbanisation. Carrying capacity and limits to growth were identified as critical factors in the population–environment debate, because they determine quality of life and maximum populations at different scales. Looking ahead, we also examined population prospects to 2020 based on UN estimates of rates and patterns of population growth, and considered the likelihood of global population stabilising at around 9 billion by 2050. Finally, we considered the need for population policies and looked at the major change in attitudes and directions signalled at the 1994 Cairo conference. People are a critical part of the 'environmental crisis', both as victims and perpetrators, but without more determined efforts to manage population growth in equitable and sustainable ways, the 'crisis' will inevitably become much worse, with long-term consequences for everyone.

WEBSITE

Links to relevant websites, a comprehensive bibliography, tools for teaching and learning, and downloadable images relevant to this chapter can be found at the website specially designed to accompany this book at:
http://www.park-environment.com

FURTHER READING

Beatley, T. (2000) *Green urbanism: learning from European cities.* Island Press, Washington. Outlines best practice for confronting unplanned urban growth, based on the experience of twenty-five cities in eleven countries in Europe.

Black, R. (1998) *Refugees, environment and development.* Longman, Harlow. Explores the complex interrelationships between forced migration, natural resource management and sustainable development and challenges the view that refugees 'cause' environmental degradation.

Brown, L.R. (ed.) (2000) *State of the world 2000.* Earthscan, London. Up-to-date review of the state of the global environment, the factors leading to environmental change and responses to change.

Brown, L., G. Gardner and B. Halweil (1999) *Beyond Malthus: nineteen dimensions of the population challenge.* Worldwatch Institute, Washington. Examines the impact of population growth on nineteen global resources and services.

Brown, L.R., M. Renner and B. Halweil (eds) (2000) *Vital signs 2000–2001: the environmental trends that are shaping our future.* Worldwatch Institute, Washington. Excellent source of up-to-date information on the state of the environment and sustainable development.

Buckingham-Hatfield, S. (2000) *Environment and gender.* Routledge, London. Wide-ranging exploration of how gender relations affect the natural environment, and of how environmental issues have a differential impact on men and women.

Demeny, P. and G. McNicholl (eds) (1996) *The Earthscan reader in population and development.* Earthscan, London. A collection of key readings from the 1960s to the 1990s that explore the links between population and development.

Dyson, T. (1999) *Population and food.* Routledge, London. Describes global trends and future prospects in food production and availability, and the implications of this for human populations.

Graham, D.T. and N.K. Poku (1999) *Migration, globalisation and human security.* Routledge, London. Theoretical and empirical treatment of how the displacement of people affects human security.

Honari, M. and T. Boleyn (1999) *Health ecology; nature, culture and human–environment interaction.* Routledge, London. Innovative introduction to human ecology, with an emphasis on how the environment affects health at various scales.

Jackson, S. (1998) *Britain's population: demographic issues in a contemporary society.* Routledge, London. Explores social issues such as single parents, pensioners and the welfare state within the context of demographic trends.

Jones, A. (1997) *Environmental biology.* Routledge, London. Sound introduction to key biological terms and concepts relating to populations and communities.

Livi-Bacci, M. (1997) *A concise history of world population.* Blackwell, Oxford. Interesting review of how and why the human population has changed over time.

Livi-Bacci, M. (2000) *The population of Europe.* Blackwell, Oxford. Traces the cultural history of populations within Europe and stresses the interrelationships between population, land, resources and disease.

Preston, S., P. Heiveline and M. Guillot (2000) *Demography: measuring and modelling population processes.* Blackwell, Oxford. Outlines and illustrates the basic methods and models used by demographers to study the behaviour of human populations.

Pugh, C. (2000) *Sustainable cities in developing countries.* Earthscan, London. Examines the challenges of tackling sustainability alongside the economic, social and ecological challenges of many of the world's most under-developed areas.

Satterthwaite, D. (ed.) (1999) *The Earthscan reader in sustainable cities.* Earthscan, London. A collection of readings on the key issues surrounding sustainable cities and sustainable urban development.

Skeldon, R. (1997) *Migration and development: a global perspective.* Longman, Harlow. Explores the links between internal and international migration, and development, at a global scale.

United Nations Environment Program (1999) *Global environment outlook 2000.* UNEP/Earthscan, London. Comprehensive and up-to-date review and analysis of the state of the environment at global and regional scales.

Wang, G.T. (1999) *China's population: problems, thoughts and policies.* Ashgate, London. Comprehensive review of the history and future of China's population problems and policies.

Chapter 3

Environmental Systems

LEARNING OBJECTIVES

When you have finished studying this chapter, you should be able to

- *Understand the nature and relevance of interrelationships within environmental systems, on the basis of the River Nile case study*
- *Define what is meant by a 'system' and explain the importance of boundaries and scale*
- *Describe the four main environmental systems and show how they are interrelated*
- *Outline the importance of flows, cycles and structures in environmental systems*
- *Appreciate how and why systems respond to change, by adjustment and feedback*
- *Evaluate the importance of equilibrium, resilience, thresholds and lags*
- *Explain why reducing system uncertainty is such a major challenge*
- *Outline the principal components and dynamics of biogeochemical cycles*
- *Describe the nitrogen cycle and explain its relevance*
- *Describe the carbon cycle and explain its relevance*
- *Describe the sulphur cycle and explain its relevance*

A useful approach to looking at and analysing the environment is by way of systems theory, which emphasises structure and interrelationships within and between different parts of the environment. This chapter provides an overview of systems and the systems approach by first examining in some detail a case study of water resource management on the River Nile in Egypt. This is designed to introduce many of the concepts and ideas of environmental systems in a real-world setting, where their relevance and application can be fully appreciated. This leads on to an exploration of the structure and operation of environmental systems, where the emphasis is on how the different parts of a system adjust to each other and to external factors. The final section deals with biogeochemical cycles, which describe the flow of chemical elements through the environment and illustrate many of the more important themes within the systems approach.

RIVER NILE CASE STUDY

THE RIVER NILE

The Nile's statistics are impressive – at nearly 6,700 km it is the world's longest river (Table 3.1), it drains an area of nearly 4.5 million km^2 (a tenth of Africa), and before it was dammed it carried an estimated 85 km^3 of water to the sea every year. The Nile basin stretches across 35 degrees of latitude, embraces several climatic zones and houses 50 million people.

Table 3.1 The world's longest rivers

	River	Continent	Length (km)
1	Nile	Africa	6,648
2	Amazon	South America	6,275
3	Mississippi–Missouri–Red Rock	North America	6,210
4	Ob–Irtysh	Asia	5,569
5	Yangtze	Asia	5,519

But the significance of this mighty river rests not so much on its sheer size as on its relevance to people in this semi-arid region. Throughout history it has been a lifeline, providing an important corridor for travel. Its water resources have been the lifeblood for human settlement and agriculture, and increasingly sophisticated schemes have been used to store, manage and distribute this precious resource. Egypt and Sudan, in particular, rely heavily on the waters of the Nile. Perhaps inevitably in a river that flows through ten dry countries – Egypt, Sudan, Ethiopia, Uganda, Zaire, Tanzania, Kenya, Rwanda, Burundi and Eritrea – there has been great competition for available water supplies. Hydropolitics (the politics of water) has been a powerful force in shaping the history of this area.

THE RIVER

The Nile rises in the East African Plateau and Ethiopian Highlands, a tropical area that receives abundant rainfall. Once it leaves the highlands, flowing northwards, it crosses the much drier savannah country of the Sudan and then the eastern Sahara Desert on its way to the Mediterranean Sea (Figure 3.1).

Most of the water in the Nile comes from the two main headwater rivers, the White Nile and the Blue Nile. The White Nile starts in the mountains and flows through Lake Victoria and Lake Albert, but up to half of its water is lost by evaporation and transpiration in the swampy Sudd area in southern Sudan. The Blue Nile starts in the mountains above Lake Tana in Ethiopia and provides most of the water in the main Nile below its confluence with the White Nile at Khartoum. The historical annual cycle of flow in the River Nile was determined largely by flows in the Blue Nile (Table 3.2).

For most of its lowest 2,000 km, downstream from Khartoum, the River Nile flows through relatively flat, dry country and receives no water from tributary rivers. From the border with Sudan northwards to Cairo (a downstream distance of 1,200 km) the Nile flows through a narrow valley hemmed in by cliffs. Its water resources become more and more important downstream, particularly in the Sahara Desert (see p. 437), where temperatures are high (50 °C in the daytime is common), rainfall is limited (around 120 mm a year), natural vegetation is sparse, and hot, bare ground is the norm. Over 96 per cent of the land area of Egypt is almost uninhabitable desert, and 99 per cent of the country's population lives in the 4 per cent of land accounted for by the fertile narrow corridor of the Nile valley. In 1993, Egypt's population was around 60

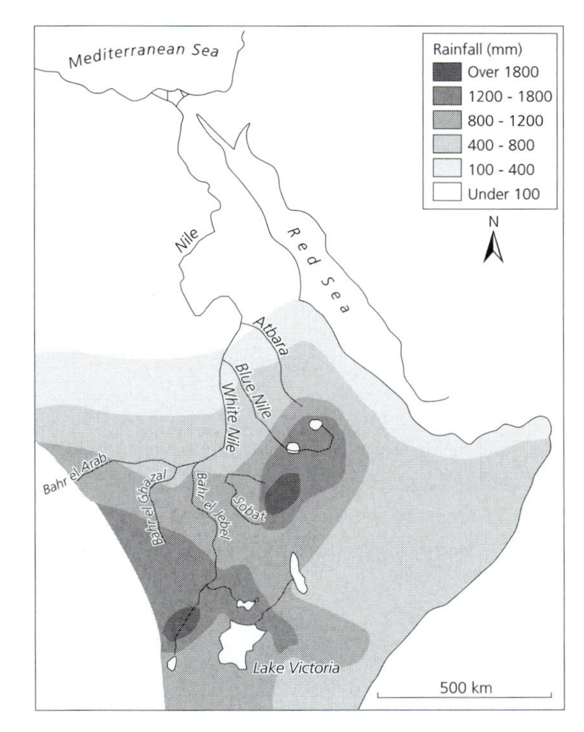

Figure 3.1 *Mean annual rainfall in the Nile basin. The Nile headwaters receive abundant rainfall, but below the confluence of the Blue and While Niles the region is extremely dry.*

Source: Figure 5.4 in Newson, M.D. (1992) Land, water and development. Routledge, London and New York.

Table 3.2 Monthly average discharge (m^3 s^{-1}) in the River Nile

Month	White Nile	Blue Nile	Nile at Aswan
October	1,200	3,040	5,200
November	1,200	1,030	2,270
December	1,100	499	1,400
January	829	282	1,100
February	634	188	1,020
March	553	156	834
April	525	138	819
May	574	182	698
June	742	461	1,340
July	897	2,080	1,910
August	1,030	5,950	6,570
September	1,130	5,650	8,180
Annual average	868	1,638	2,610

million, and during the early 1990s population was growing at the relatively fast rate of 2.3 per cent a year (see p. 46).

The Nile flows into the Mediterranean through its delta, where it splits into several large distributary channels, including the Rosetta and the Damietta. The Nile delta is a fan-shaped flat plain, around 25,000 km^2 in size, stretching 250 km along the Mediterranean coast. This fertile area has long been a major centre for agriculture and settlement. About half of it is now irrigated through the year, and lakes, swamps and desert comprise the rest.

CULTURAL AND ENVIRONMENTAL HISTORY

Regular natural flooding by the Nile endowed its floodplain with valuable water and nutrient-rich silts. The Nile valley has been a focus of settlement since earliest times, and the centre of a major civilisation since at least 3000 BC. The Pharaohs (ancient Egyptian kings) built great cities and temples along the Nile, including the huge pyramid tombs. They also constructed elaborate water management schemes, including dams and canals, to irrigate the sandy soils. The Nile delta is one of the world's oldest agricultural areas, dependent on careful management of available water resources.

Water management became more important within the last 5,000 years when climate change brought drier conditions to northern tropical Africa. Desertification of the Sahara region began, Nile flows decreased and flood levels fell, droughts became more common and more serious in the area, and forests along the Nile were cleared or died. Population levels rose, settlements were more concentrated along the Nile, and the precious Nile water resources were increasingly used for irrigation agriculture. Reliance on the Nile has grown steadily stronger ever since, despite some substantial variations in river flow (Box 3.1).

Plate 3.1 *Nile hydrometer. This is one of a number of sites along the lower Nile where variations in discharge are recorded continuously. The hydrometer is connected to the adjacent river by a horizontal pipe, and changes in river water level are reflected in changes in the level of the water with the gauge.*

Photo: Philip Barker.

BOX 3.1 FLOW VARIATIONS ON THE RIVER NILE

Records have been kept of high and low flows in the Nile at Cairo since the ninth century. Although the record is incomplete, it shows several phases of between 40 and 150 years when river flows have been much higher than those during the twentieth century. This seems to have been the case during Roman times, for example. Higher rainfalls in North Africa would explain why a much larger area of the Sahara Desert was cultivated then than now. The long-term flow record preserves evidence of climatic changes (including the Little Ice Age) and shows how year-to-year variations, particularly in summer flows, have been affected by El Niño events (see p. 482). The flow record at Aswan over the last century (Figure 3.2) also shows a great deal of variability, with flows generally lower during the twentieth century.

TRADITIONAL WATER MANAGEMENT

Agriculture in Egypt depends heavily on irrigation water from the Nile. The potential is vast, because the river's 85 km^3 of water in a typical year could flood 42,500 km^2 to a depth of 2 m. Natural flooding irrigation was used initially. This involved growing crops on low-lying soils after the flood waters went down every autumn. The annual flood on the unregulated Nile started in late summer and reached a peak in early September.

Through time more refined irrigation systems were developed, using specially built embankments and gravity-flow canals to control where and when flooding occurred. The floodplain was divided into flood basins ranging in size from 2 to 200 km^2, which were flooded to a depth of up to 2 m during August and September (the natural flood period on the Nile). Silt, rich in nutrients, was deposited by the flood water as a natural fertiliser. The area that could be farmed was progressively expanded through time by extending the artificial irrigation system and draining marshes. Such gravity-fed basin irrigation systems were the main form of water resource management along the Nile for more than 6,000 years.

Shallow wells were dug to make use of groundwater in the floodplain soils, and this augmented river flow and extended the crop-growing season to include spring and early summer. The most important crops are cotton (Egypt is among the world's leading cotton producers), rice, sugar cane, grains, beans and corn.

By the middle of the nineteenth century, as population levels continued to grow, new approaches were needed to further expand the area under cultivation. In the 1840s, small dams were built on the delta to store water, thus allowing irrigation throughout the year and removing the traditional reliance on the river's natural seasonal flood cycle. Barrages were later built in other places along the Nile in order to increase seasonal storage.

INTEGRATED WATER MANAGEMENT

Seasonal storage played an important role in exploiting the Nile and making its waters available for irrigation agriculture throughout the year. But it offered no solution to the recurrent problem of year-to-year variations in river flow, which brought occasional prolonged drought or severe flooding. Moreover, the growing number of local water management initiatives were creating problems in ensuring that all schemes received enough water and reducing the risk of upstream schemes taking out too much. The need to develop a more rational water management strategy was compounded by the multinational nature of the Nile, because flows in Egypt and Sudan were significantly affected by the activities of upstream countries.

Co-ordinated development of the entire Nile river system, as a single unit, offered the prospect of optimising the availability of water resources, reducing the risk and managing the timing of river flooding, and making use of the river as a source of hydroelectric power. It would not be an easy task to achieve, because of the competing and sometimes conflicting claims of different countries. It would require vast investment in huge engineering schemes, along with great political and diplomatic skill and goodwill, carefully controlled project management and continued co-operation in running an integrated scheme on such a vast scale. But given the importance of reliable water supplies to all of the countries involved, it had to be taken seriously.

In the early 1900s, plans were drawn up to make more efficient use of the River Nile (Box 3.2) by reducing water losses from evaporation and seepage and building a series of storage dams and hydroelectric power stations. The plan could largely ignore the multinational nature of the river, because most of the Nile basin was still under British colonial rule at the time. For example, the 1929 Nile Waters Agreement between Egypt and Sudan shaped the use that both countries made of the river for a period of 25 years

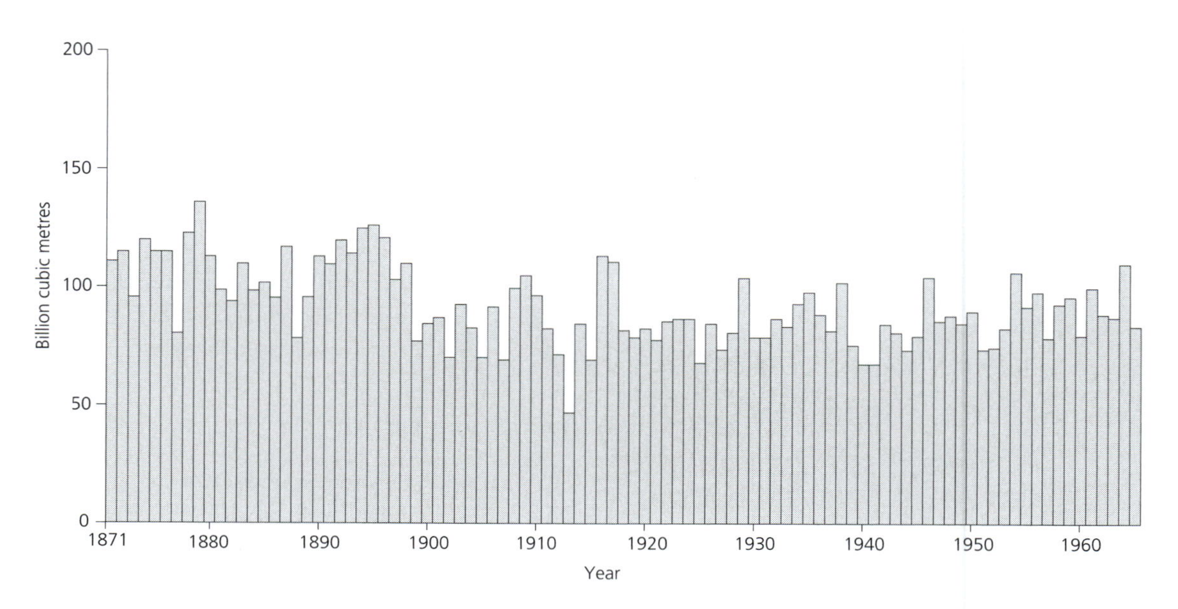

Figure 3.2 *Annual variations in discharge in the River Nile at Aswan, 1871–1965. Over the long term, discharge has varied from year to year around an average of about 100 billion m*^{3} *per year, with flows generally lower since about 1900.*

Source: Figure 6.2 in Beaumont, P. (1993) Drylands: environmental management and development. Routledge, London.

(a new agreement came into force in 1959). Water management became much more complex after the 1950s, when all of the countries concerned became independent.

Work began in late 1999 on one of Egypt's most ambitious water management schemes yet. The aim of the New Valley project is to irrigate thousands of square kilometres around the fringes of the Nile delta, and in the barren desert west of the Nile. Some of the new farms will draw water from the Nile, but others will make use of groundwater and agricultural runoff. A major component of the scheme is the construction of the world's largest pumping station, at a cost of around £300 million, to be completed in 2002, designed to lift 25 million cubic metres of water from Lake Nasser each day and send it down a 70 km trunk canal to feed four large farming areas downstream.

THE ASWAN HIGH DAM

The original Aswan Dam proved very successful in engineering, economic and water supply terms. In 1949, proposals were made for a large new storage dam 6 km further upstream.

BOX 3.2 EARLY ENGINEERING SCHEMES ALONG THE RIVER NILE

A number of large engineering schemes to control the River Nile were proposed or implemented between 1900 and 1960 (Figure 3.3). In 1925, for instance, the Egyptian government proposed the cutting of a new channel for the Nile through the wetlands of the Sudd in southern Sudan (to reduce water loss by evaporation), but political barriers made the scheme impossible. The Jonglei Canal Diversion Scheme, involving the digging of a 260 km diversion canal, produced the required effect. New dam constructions included the Sennar Dam on the Blue Nile, and the Jebel Aulia Dam and the Owen Falls Dam on the White Nile.

One such scheme was the Aswan Dam, built in 1902 on the lowest cataract (waterfall) on the Nile. The dam, 105 m high and nearly 2 km wide, was built from locally quarried granite and had a storage capacity of 1 km^3. It was raised twice (to 140 m in 1912 and 190 m 1934) to increase its storage capacity to 5.7 km^3. The scheme was designed to store water for perennial (year-long) irrigation in Egypt, and allowed controlled release of water downstream to the Assiut Barrage (an irrigation dam more than 500 km downstream). Equipment to generate hydroelectricity was installed in 1960.

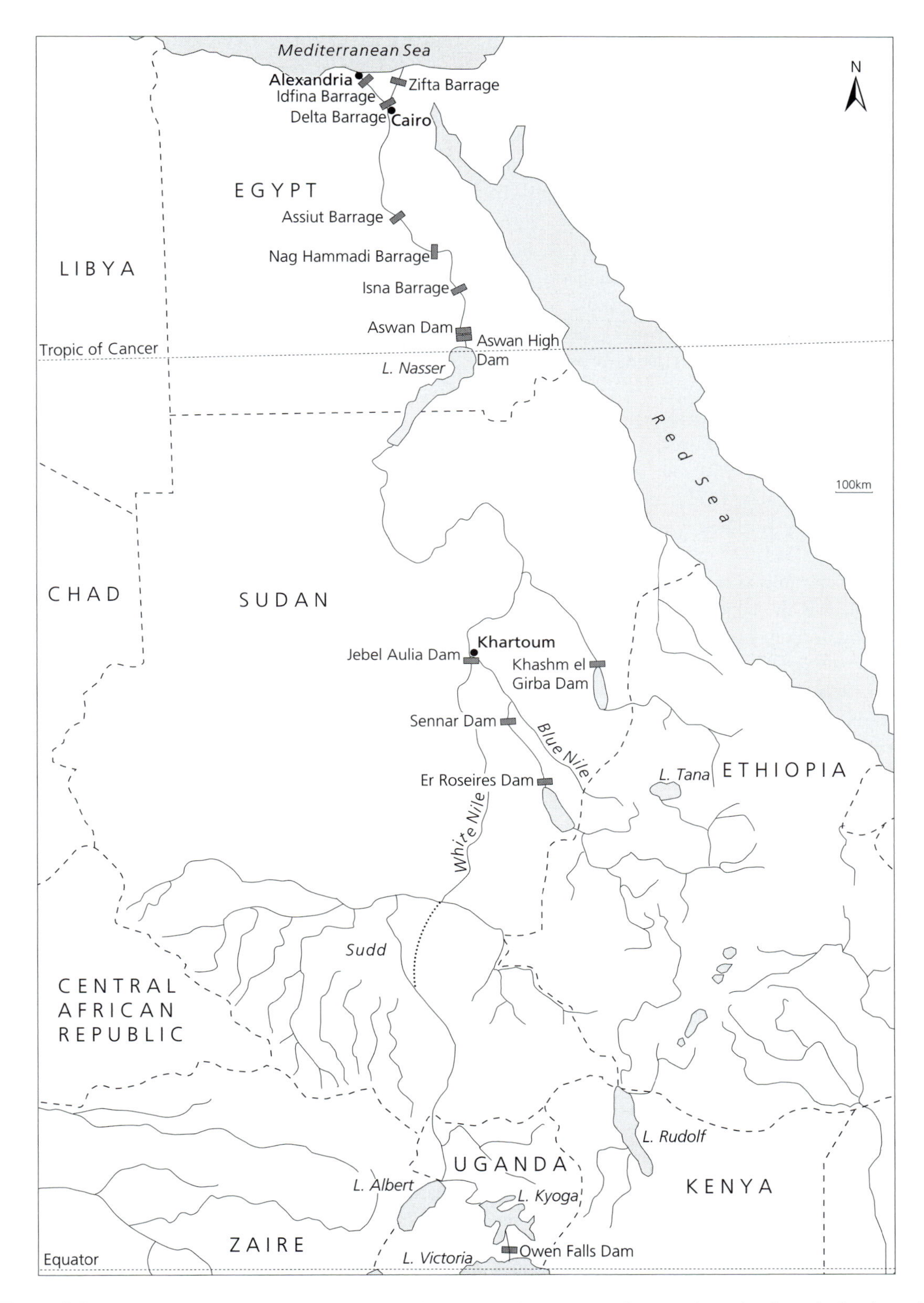

Figure 3.3 *Major water engineering works along the River Nile. A series of major dams has been built along the Nile and its main tributaries, and drainage has been diverted through the Sudd marshlands.*

Source: Figure 6.1 in Beaumont, P. (1993) Drylands: environmental management and development. Routledge, London.

OBJECTIVES AND DESIGN

The new Aswan High Dam was designed to reduce the problems of drought and low river flows, control river flooding downstream, provide irrigation water to increase agricultural productivity, and provide much-needed hydroelectric power for Egypt's industries in Cairo and elsewhere. Located within – and thus controlled by – Egypt, the new dam would be vital to the country's economic security and prosperity. Among the largest dams in the world, it would be a jewel in the crown and a source of immense national pride. President Nasser, Egypt's leader at the time, saw it as 'a source of everlasting prosperity' for his country.

Building the dam was a massive undertaking. It is 111 m high, 1 km wide at the base and nearly 4 km wide at the top. Work began in 1960 and finished in 1969. It became fully operational in 1971, when the last of the twelve hydroelectricity turbines was installed. The total cost was more than US$1,000 million (at 1960s prices; it would have cost ten times as much at mid-1990s prices).

Behind the dam is the huge artificial Lake Nasser. The reservoir is 480 km long, extending upstream beyond the Egyptian border into Sudan (two-thirds of its area lies within Egypt). It is up to 16 km wide and has a storage capacity of about 160 km^3 (160,000 million m^3) which is the equivalent of about two years' flow of the Nile. The reservoir began filling in 1964, and the water level continued rising until the late 1970s. Water levels fell steadily between 1981 and 1987 as the severe North African drought (see Box 13.4) reduced rainfall across the region.

BENEFITS

The Aswan High Dam scheme has brought important practical and economic benefits to modern Egypt:

Flood and drought control: particularly important has been the effective taming of the river, because throughout history Egypt has been at the mercy of the highly variable river flow. The dam now controls variations in the Nile flood and largely removes the risk of years of unusually high or low flow (Figure 3.4). The removal of sustained drought and potentially catastrophic floods has greatly helped Egypt's economy. Valuable rice and cotton crops were not ruined by the droughts of 1972 and 1973, for example.

Irrigation: flood waters trapped in Lake Nasser can now be released as and when engineers decide, and this makes rational use of water resources much easier and more cost-effective. Great benefits have been created by extending irrigation agriculture and increasing agricultural production (Figure 3.5). Nearly two-thirds of the water released from the High Dam is used for irrigation (Table 3.3). More than 800 km of irrigated Nile floodplain have been converted from a one-crop system to a three-crop rotation under year-round irrigation, and an extra 4,000 km^2 of desert has been brought under cultivation by new irrigation schemes.

Hydroelectricity: the huge energy potential of the Nile is also being tapped, and the dam now provides more than half of Egypt's electrical power. Although initial forecasts of 10,000 million kWh (kilowatt hours) of electricity a year have proved over-optimistic, the scheme regularly produces at least 7,000 million kWh a year. This energy source is renewable, reliable, sustainable, pollution-free and environmentally sound.

So in terms of flood and drought control, increased irrigation and crop production, and electricity production the Aswan High Dam must be judged a success. It is bringing the practical benefits it was designed to, and is proving vital to the country's economy. The dam scheme is estimated to have increased Egypt's national income by more than US$500 million a year, and increased agricultural production and hydropower generation paid for the building costs within the first two years. Agriculture and electricity continue to provide significant economic benefits (Table 3.4).

PROBLEMS

Despite the benefits that the Aswan High Dam has brought, there has been great concern over some of the environmental and social impacts of the scheme. Since the early 1960s, a series of impacts have come to light, not all of which were expected and some of which have been unwelcome and costly:

Water losses: although greater and more reliable supplies of irrigation water are available, thanks to Lake Nasser, by the mid-1980s the amount of water available in the fields was only about half of what had been expected and planned for. Water losses turned out to be much higher than had been expected, because of evaporation from the reservoir, seepage into the sandstone rock beneath the reservoir and evaporation and seepage in the extensive network of unlined irrigation canals that deliver water to the fields.

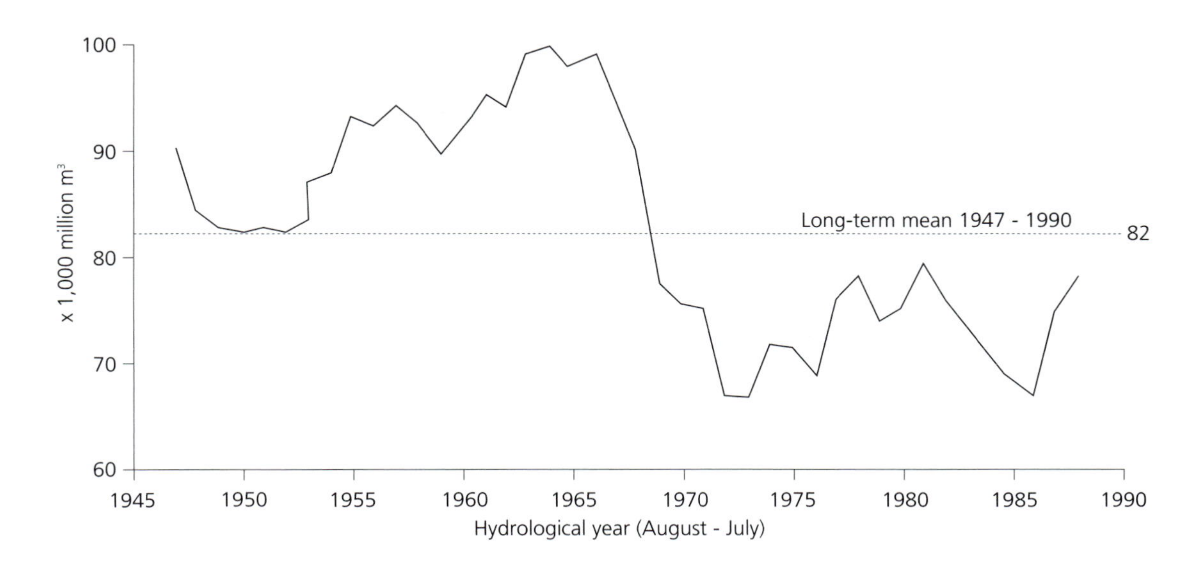

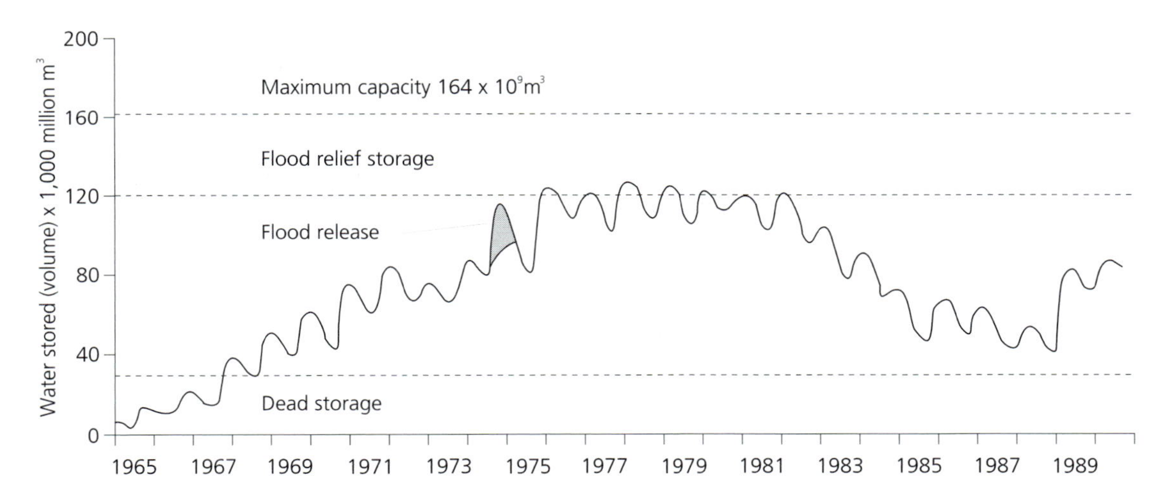

Figure 3.4 *Changes in annual flow of the River Nile at Aswan, 1945–1988. The upper figure shows variations in natural flow (the volume before withdrawal of irrigation water) between 1945 and 1990, based on five-year running averages. The lower figure shows variations in the amount of water stored in Lake Nasser between 1965 and 1990.*
Source: Figures 3 and 5 in Smith, G. (1991) The Aswan Dam. Geography Review 5 (2): 35–41.

Salinisation: soils in arid areas often suffer from salt saturation (salinisation; see p. 401) as high rates of evaporation (in the hot, dry air) remove moisture, and the dissolved salts that are left behind become more and more concentrated (Box 3.3).

Groundwater changes: groundwater conditions have been radically altered since construction of the Aswan High Dam, in two ways. Seepage of surface water from Lake Nasser into the underlying sandstone has raised the water table in the area around the reservoir, making shallow groundwater more accessible. In the year-round irrigated area, seepage of water into the soils from fields and unlined canals has raised the water table, thus contributing to the problems of salinisation.

Displacement of people: an estimated 100,000 indigenous people (Nubians) were displaced in Egypt and Sudan when Lake Nasser filled and their villages on the Nile floodplain upstream were drowned. This meant leaving their homelands and familiar environments, involuntary relocation elsewhere, and great human suffering and hardship. Many chose to stay close to Lake Nasser rather than move to the government-built villages far from the river.

Drowning of archaeological sites: because of its long history of settlement, the Nile valley contains many sites of great archaeological importance. Some of these were flooded when Lake Nasser was created, but measures were taken to protect some of the most important sites. The huge sandstone monuments of Ramesses II and Nefertari, carved into the cliff at Abu Simbel in Egypt 3,200 years ago, were

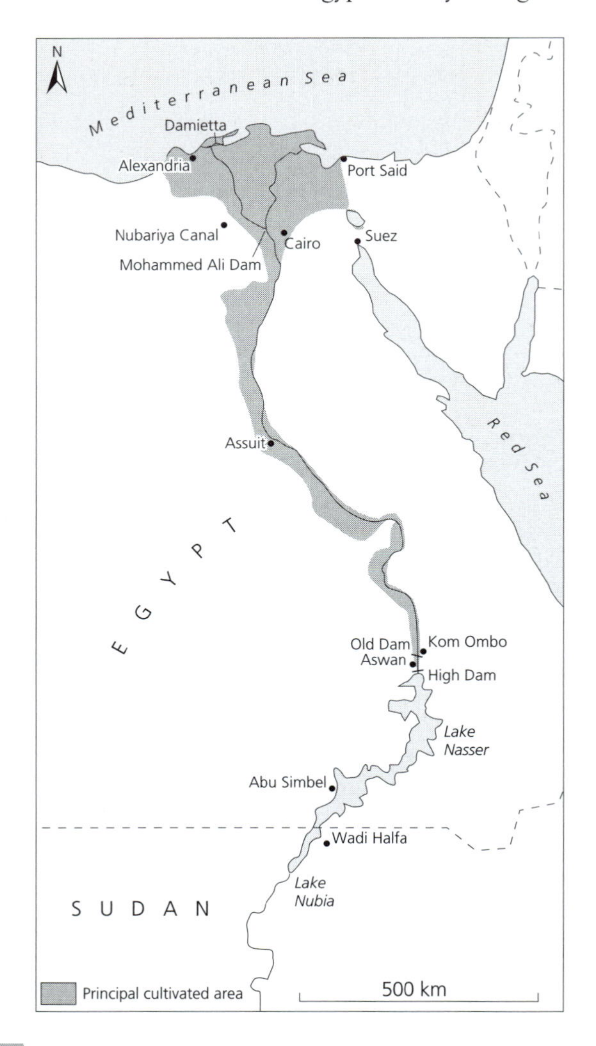

Figure 3.5 *Main areas of irrigation cultivation along the Nile valley. Much of the low-lying land near the river downstream from the Aswan High Dam (including the extensive fertile Nile Delta) is under irrigation agriculture.*

Source: Figure 1 in White, G.F. (1988) The environmental effects of the high dam at Aswan. Environment 30 (7): 4–11.

Table 3.3 Estimated use of water from the Aswan High Dam, 1990

Use	*Volume (km³)*	*Per cent*
Inflow from Aswan Dam	55.5	100
Outflow to sea and inland drainage	17.5	32
Municipal and industrial	2.4	4
Irrigation	33.6	60
Evaporation loss	2.0	4

Table 3.4 Estimated annual economic benefits of the Aswan High Dam scheme

Benefit	*Value (million Egyptian £, early 1980s prices)*	*Per cent*
Agricultural development	140	55
Hydroelectric power	100	39
Flood control	10	4
Improved navigation	5	2
Total annual benefit	255	100

BOX 3.3 SALINISATION OF THE NILE FLOODPLAIN

The annual flood cycle on the pre-dam Nile served the valuable function of flushing away mineral salts from the floodplain soils, thus reducing the risk of salinisation (see p. 401). Eliminating the flood has helped the build-up of salts. But the problem is made worse by the use of irrigation water, which has much higher salinity (a greater salt content) than natural water in the area. The salinity of the water leaving Lake Nasser is about a tenth higher than that of the water that flows into the reservoir (because of evaporation from the lake surface), and salinity increases even more through the irrigation canal network (as evaporation removes some of the water and thus concentrates the salts that remain). Problems are compounded by groundwater, which is drawn up to the soil surface by capillary action by the hot air above the ground. When this water is evaporated, the salts are left behind as a hard crust on the soil surface. Salt damages the root hairs on plants (through which they take in water and nutrients from soil moisture) and reduces yields. Crop yields have declined in up to a third of the area irrigated with water from Lake Nasser.

cut into pieces and reassembled 65 m higher up, above the water level.

Seismic stress: the weight of water stored in a large reservoir can increase pressure on the rocks below enough to trigger small earthquakes in susceptible areas (see p. 172). This appears to have been the case in Lake Nasser in 1981, after the rising water level reached a geologically unstable area. A magnitude 5.3 earthquake occurred on 14 November 1981 and was followed by a series of smaller movements. As the water level fell between 1981 and 1987, in response to prolonged drought, seismic activity died down.

Mediterranean flow circulation: before the dam was constructed, the outflow of water from the Nile strongly influenced water circulation and flow conditions in the south-east Mediterranean. The influence was strongest during the flood period between August and November and extended outwards up to 80 km from the coast and to a depth of about 150 m. These natural circulation patterns have been disturbed and disrupted by the flood-control effects of the Aswan High Dam, thus extending the impacts across a much bigger area.

Lake infilling: as well as storing water, reservoirs trap much of the sediment that would previously have been washed downstream by the river. This decreases the availability of nutrients to land below the dam, but the silting also decreases the storage capacity of the reservoir and so reduces water yield. An estimated 100 million tonnes of silt is trapped each year. Through time, as the silt accumulates, the life span of the reservoir decreases. Lake Nasser was designed with nearly 25 per cent spare capacity for storage of silt, enough to last an estimated 500 years. Rates of deposition on the lake bed are high because the Nile flows over rock that is easily eroded, and the semi-arid climate and lack of vegetation encourage rapid weathering of hill slopes. Early fears were that the lake would fill completely within about 360 years, but the estimated life span of the lake has been revised to 535 years to take into account rates of silt compaction and recent changes in climate and river flow.

Scouring below the dam: because most of the sediment load in the Nile is now deposited behind Lake Nasser, the water released downstream from the dam is quite clear. This has led to erosion of the river bed and some channel instability. Scouring has undermined or threatened some bridges and small dams downstream from Aswan. Compared with many large dams, however, rates of lowering of the river bed have been low (an average of 25 mm over the first 18 years), and they appear to be decreasing through time.

Erosion of the Nile delta: another casualty of silt deposition behind the Aswan High Dam has been the Nile delta, now starved of the sediment that traditionally replenished it. Without a continued supply of silt from upstream, the delta is under attack from coastal erosion and is retreating. This has great economic significance, because it is putting at risk the productivity and survival of the rich agricultural land on the delta and threatening the tourism industry based on Alexandria. The problems posed by this induced change are compounded by the natural subsidence and tilting of the Nile delta, at rates of up to 2.5 cm a year. This is increasing the risk of coastal erosion and salt water intrusion.

Loss of nutrients: in the past, the yearly Nile flood cycle brought nutrient-rich silt as well as water to the soils in the floodplain. Nutrients deposited as flood waters went down provided a free and annually renewed source of fertiliser, which helped the development and survival of Egypt's civilisation. Flood peaks have been reduced by three-quarters since the High Dam was built, so a much smaller area receives nutrients. The silt load below the dam is also much lower than before the dam was built, so nutrient supply to areas that are flooded is much reduced. Around US$100 million of commercial fertilisers must now be bought each year to make up the nutrient shortfall.

Reduced fish catches: the loss of nutrients has had a knock-on effect on fish populations in the eastern Mediterranean around the Nile delta, which have shrunk dramatically in response to the declining availability of food. Sardine catches fell by 95 per cent after the dam was built, and Egypt's sardine, mackerel, shrimp and lobster industries have suffered badly. Around 3,000 jobs in the fishing industry have disappeared, foreign exchange earnings from fish exports have declined sharply, and the country has lost a valuable food source that was high in protein. A fishing industry is being developed in Lake Nasser, although it will not replace what has been lost at sea.

Water-borne diseases: the changed river flow patterns caused by the Aswan High Dam scheme have also seriously affected the spread of some infectious diseases (Box 3.4).

BOX 3.4 WATER-BORNE DISEASES ALONG THE RIVER NILE

The most notable water-borne disease along the Nile, as along many low-latitude rivers, is schistosomiasis (also known as bilharzia) – a painful, debilitating, sometimes incurable and often fatal disease that attacks the liver, kidneys and other vital organs. The disease is spread by snails that live in shallow water and carry the parasites that infect humans. The annual flood cycle previously swept away most of the snails along the Nile, but this natural cleansing process no longer exists. Snail populations have grown along the shores of Lake Nasser and in the extended network of permanent irrigation canals, causing a dramatic rise in the incidence of the disease. Problems have been most acute among people living close to the shores of the reservoir, and among farmers and those who work in the irrigation ditches.

LESSONS FROM ASWAN

The Aswan High Dam has clearly brought both benefits and problems, which are not shared equally. Some people and places are winners, while others are losers. Most of the effects expected from the dam were benefits (Table 3.5), but some of the costs were unexpected.

Many scientists argue that the planning of all major development projects such as the Aswan High Dam must taken into account possible long-term environmental impacts, costs and benefits. Alternative ways of meeting the objectives of such schemes should also be evaluated alongside the proposed techno-fix engineering solutions. One view is that the time, effort and money invested in designing and building the dam scheme could have been better invested in a broader and less technology-led programme of population control, rural development and agricultural production. Critics argue that if all of the real costs of the Aswan High Dam scheme were fully recognised and taken into account, building it in the first place would have been regarded as an economic as well as an environmental mistake.

In the 1990s, it also became clear that despite many initiatives over the past century, Egypt still faces a genuine crisis in terms of water supply. This has arisen partly because of greatly increased demand for water resources by all countries along the Nile, but the problem is made worse by natural variations in climate, particularly serious droughts in Ethiopia in 1972–73, 1982–83 and 1987–88 (see p. 422). Analysis of Nile flow records shows that recent droughts are part of a long-term dry phase that started in the mid-1960s. Global warming associated with the

Table 3.5 Expected effects of the Aswan High Dam

Physical effects	*Resulting social effects*
Submerge reservoir lands	Displace local populations
	Change nomadic grazing patterns
Create a new water body (Lake Nasser)	Provide fishing grounds
	Change nomadic grazing patterns
	Allow navigation
Change the pattern of river flow	Expand agricultural production
	Increase irrigated area
	Change crop patterns
	Change water scheduling
	Improve navigation
	Decrease sediment flow
	Increase channel cutting
	Reduce Mediterranean fisheries
Generate hydroelectric power	Increase electricity supply
	Support fertiliser industry
	Reduce electricity prices

greenhouse effect (see p. 261) poses an even greater threat, with some climate change models forecasting a decline in rainfall over the Nile headwaters and increasing evaporation losses (particularly in the Sudd marshland and the lakes at the head of the White Nile).

Conflicts within and between countries over the allocation of available water resources look likely to continue. The need to locate new sources of water, reallocate some existing water supplies, price water more realistically and manage water demand becomes increasingly urgent. Hydropolitics, so central to the history of this arid region, will remain a critical issue in the twenty-first century.

SYSTEMS

The Nile case study illustrates some important features of the natural environment, and the ways in which human activities can alter the environment and how the environment can affect people both directly and indirectly. The complex history of water management in the Nile shows how:

- people depend a great deal on the environment and its resources;
- allocation of scarce resources fairly between different users is extremely complex; and
- sustainable practices that rely on natural resources are difficult to develop.

Some of the Nile water problems arise because human activities rarely coincide with physical regions, and political boundaries can exert stronger control over human activities than environmental boundaries. Put another way, human-use systems make use of environmental systems, but not always in the original locations. Hence, for example, water that has flowed into Lake Nasser is eventually used for irrigation some distance downstream.

The Nile case study provides a good example of an environmental system. It is helpful to reflect on what systems are and how they work, because many of the concepts and approaches of systems theory have been used in environmental management. The systems approach offers a useful framework for the rest of this book.

WHAT IS A SYSTEM?

A system is in effect any collection of components that work together to perform a function – a computer (Figure 3.6) is a familiar example. This definition highlights three important qualities of systems:

1. The system is made up of component parts.
2. The parts work together.
3. The whole thing serves some purpose.

It follows from quality 1 that systems can be studied at different scales, such as looking at the overall system without bothering about the detail of its component parts (e.g. the overall Nile river system), or looking at what components are present and how they seem to fit together (such as the individual tributary rivers, the delta, the dams and irrigation networks). Quality 1 also creates the possibility of movement or dynamics within systems if one or more of the components is altered (e.g. damming the Nile at Aswan) or removed (e.g. decline of nutrients washed down the Nile into the Mediterranean).

The fact that the parts work together (quality 2) shows that systems have structures to link the various components together (e.g. the flow of water and sediment down the natural Nile, which flooded low-lying areas, deposited silt, built up the delta, provided nutrients, flushed out snails). This in turn implies interrelationships, so that changes in one component might trigger or be associated with changes

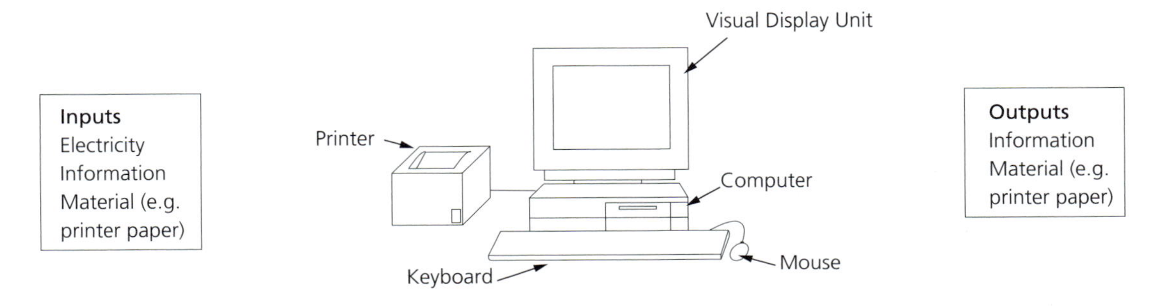

Figure 3.6 *A computer as a system. Like most systems, a computer system has inputs and outputs, and the internal workings of the system convert the inputs into outputs.*

BOX 3.5 DIFFERENT TYPES OF SYSTEM

The term 'system' is widely used, and it can be defined in many different ways. People sometimes speak of 'the system', by which they mean the bureaucratic and political structure that governs the way society operates. Computer specialists speak of 'computer systems' as the hardware (computer) and software (programs) (Figure 3.6). There are social systems, which include groups of various sizes including the family and the community. Managers use accounting systems (ways of recording and analysing financial data), transport planners use road systems (networks of roads), and so on. Quite how a system is defined depends largely on what type of system we are dealing with.

in other components (e.g. damming the Nile has caused erosion downstream, silting in Lake Nasser and the spread of schistosomiasis (see Box 3.4)).

Quality 3 means that a system is not a meaningless jumble of components but rather is goal-directed (that is, it works towards some goal or purpose). For example, the natural River Nile was an important artery along which water, sediment and nutrients were redistributed in time and space across parts of North Africa. There are clear purposes behind the various water resource schemes on the river, so this quality applies equally well to managed systems.

Environmental systems include living (organic) and non-living (inanimate) components, which interact to create the world around us.

SYSTEM BOUNDARIES

A particularly important quality of systems is that they operate within boundaries that can be identified and defined. The boundary defines the limits within which the components interact, and thus it defines the scale of the system and the way in which systems are related together.

Environmental systems are physical systems with physical boundaries (Figure 3.7). Some boundaries are sharp, such as the coastline or a lake shore, which mark boundaries between aquatic (water-based) and terrestrial (land-based) systems, or a frontal weather system, which marks the boundary between warm and cold air masses. Other boundaries within environmental systems are less sharp and more transitional, such as the gradual change in vegetation towards a desert margin. Few environmental boundaries are totally uncrossable, although they can create sharp changes in environmental conditions.

There are clear boundaries to the Nile river system overall – its watershed, separating the area that contributes water to the Nile from surrounding areas that feed other river systems, and the edge of the Nile delta, which marks the boundary with the Mediterranean Sea. Within the Nile, there are clear boundaries to the sub-systems, such as Lake Nasser (its shoreline), the Nile delta (its shoreline at the seaward edge and the former estuary limits at the landward edge) and the Sudd wetlands (the marsh edge).

EXCHANGES ACROSS BOUNDARIES

System boundaries are not just important for defining the limits to a particular system; they also determine what type of system it is. Some systems have inputs and outputs (that is, they exchange energy and/or material with the environment outside), so they are affected by and can in turn affect other systems. Three types of system can be identified on this basis (Box 3.6) – isolated, closed and open systems (Figure 3.8).

BOX 3.6 ISOLATED, CLOSED AND OPEN SYSTEMS

Isolated systems: an isolated system has no exchange of energy or material across the system boundaries. Such systems are rare in the natural world and occur mostly in laboratory experiments under tightly controlled conditions.

Closed systems: these can exchange energy but not material across the system boundaries. They are more common than isolated systems. The global water cycle is a closed system (Figure 3.9) because there is a fixed amount of material (water) within the system (so no water is exported or imported across its boundaries), and the circulation of water is maintained by energy from the Sun (which is imported across the system boundary).

Open systems: open systems exchange energy and materials quite freely across system boundaries. Most environmental systems – such as drainage basins, lakes and forests – are open in this way.

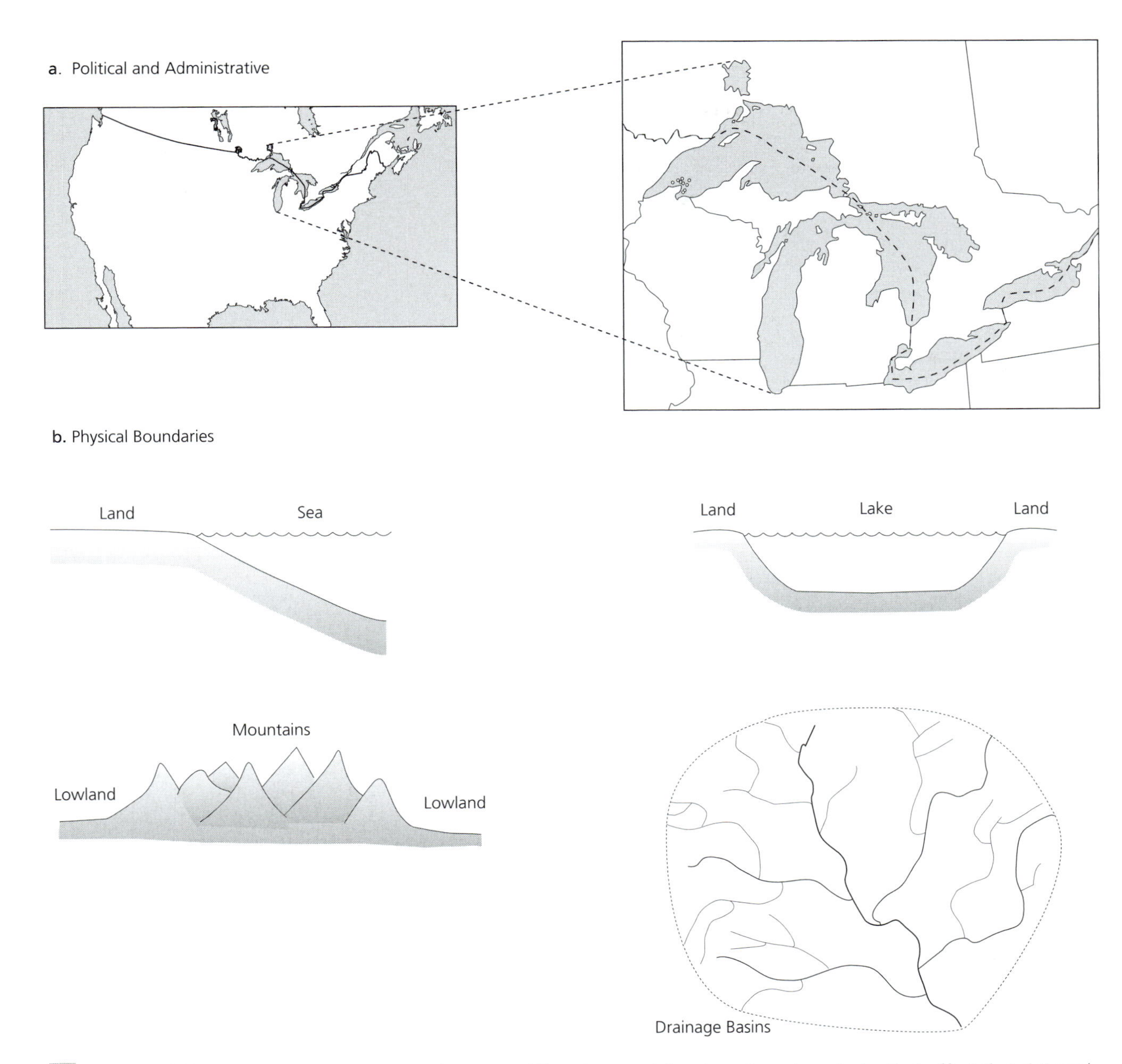

Figure 3.7 *Boundaries in environmental systems. All environmental systems have boundaries that affect their internal operations and the relationships between systems. Some boundaries, such as the frontier between the USA and Canada (a), are political and administrative, but the most important boundaries in environmental systems are physical (b).*

Outputs from one open system can become inputs to another. Examples from the Nile case study include the flow of water and sediment from the Blue and White Niles into the main Nile, the output of water from Lake Nasser, which becomes the input to the irrigation network, and the output of nutrients from the pre-dam Nile to the Mediterranean Sea.

All environmental systems on Earth are interrelated. As a consequence, any action that affects one component in one system might ultimately influence every other system in one way or another. Thus, for example, current changes in the Pacific Ocean can be associated with changing rainfall and runoff in parts of North and South America, drought conditions in North Africa, and air quality and

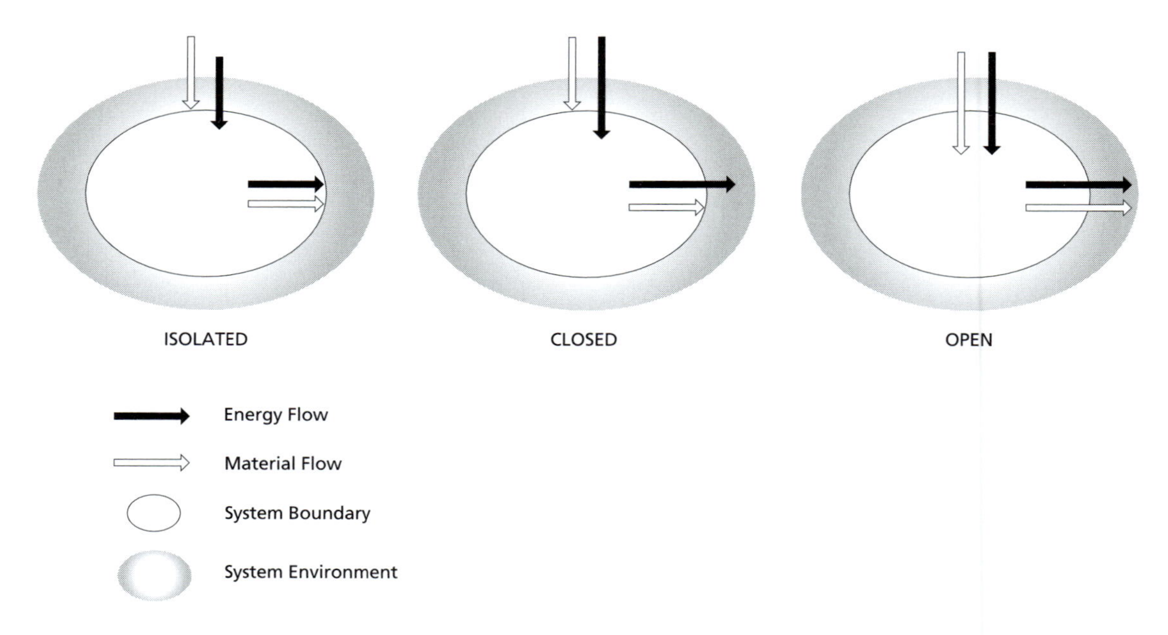

Figure 3.8 *Isolated, closed and open systems. The nature of a system is heavily influenced by the flow of energy and material across the system boundaries. See Box 3.6 for explanation.*

food supplies in Europe (via the El Niño teleconnection; see p. 482).

SYSTEM SCALE

Sometimes the distinction between closed and open systems is a matter of scale. Thus, for example, the global water cycle (Figure 3.9) is usually defined as a closed system, but individual drainage basins within it are clearly open systems. Similarly, the global rock cycle is a closed system, but sediment movements on individual hill slopes or along specific rivers are part of smaller-scale open systems.

Every organism, from the smallest bacterium through the full range of plants and animals (including humans), is an integrated whole and hence a living system in its own right. Collectively, all living material on Earth comprises the biosphere (see p. 519), which is a global system. Ecological systems can be defined at all scales, from the biosphere down to the individual organism. A forest is an open system that has flows of materials and energy into and out of the system through its boundary (Figure 3.10).

Because they can be defined at a variety of scales, systems can overlap. Indeed, some systems exist within other systems, like a set of stacking Russian dolls. This type of structure (a *nested hierarchy*) is typical of river systems (see p. 355): small headwater streams, which are systems in their own right, are part of (nested within) larger streams, and they in turn are part of whole river systems. Similarly, an individual tree (again, a system in its own right) is part of a forest, which is part of the biosphere.

To simplify matters, given that systems can be defined at a variety of scales, and the choice of scale adopted depends on the purpose of any given study, we could use a threefold structure to describe a system and its sub-divisions:

- *System*: this refers to the total environmental system in question (such as a drainage basin).
- *Sub-system*: this refers to major sub-divisions within that system (such as the hill slopes, floodplain and river channel within the drainage basin, each of which displays signs of the system at work, including the flow of energy and material).
- *System component* (or *element*): this refers to a portion of the system or sub-system that has specific properties (such as the slope of a hill, the water flow along the channel or the amount of sediment carried by the river).

ENVIRONMENTAL SYSTEMS

Planet Earth is itself an integrated system, but it is useful to identify its four main sub-systems, which are explored in the following four sections of this book (Box 3.7). These

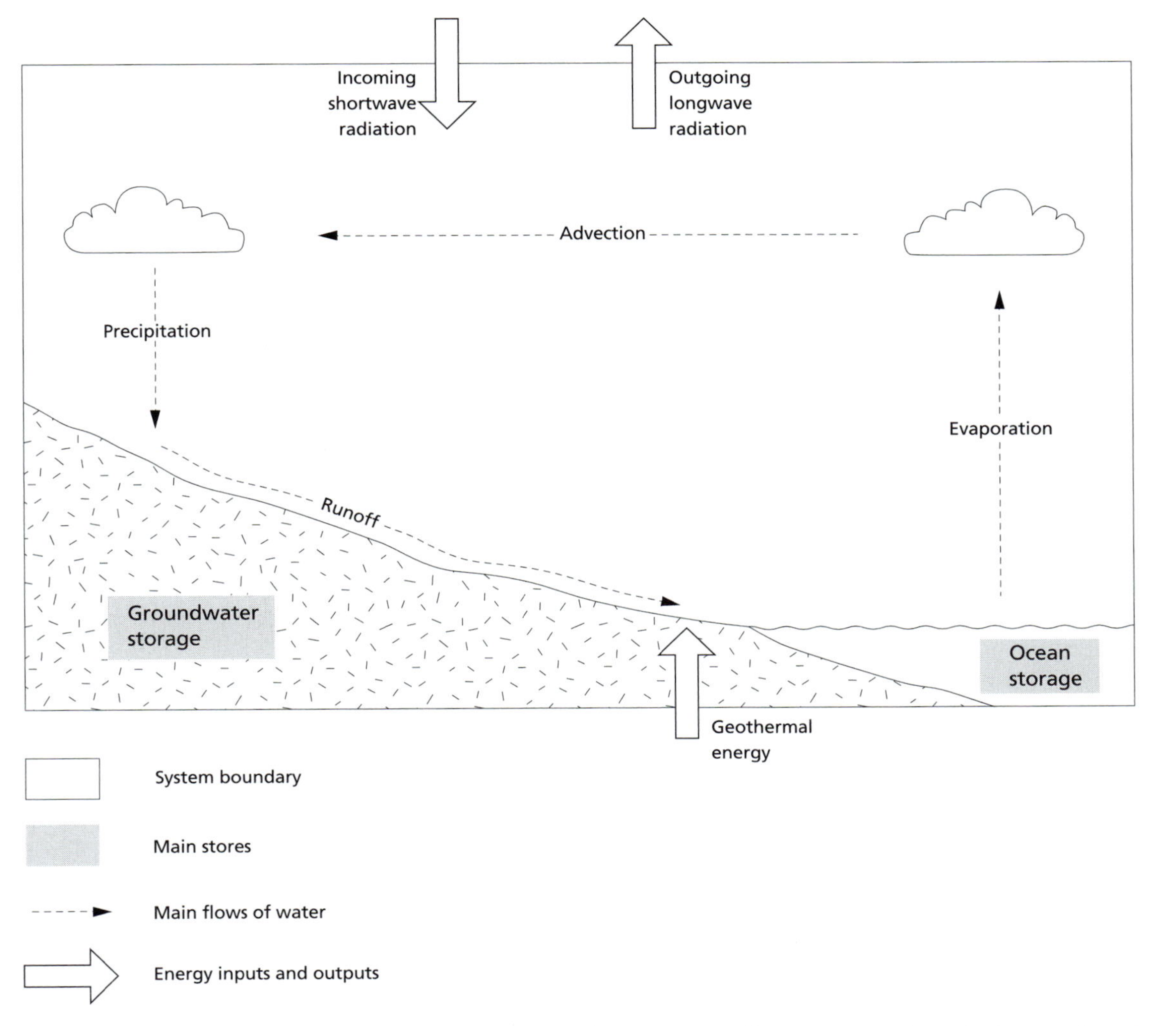

Figure 3.9 *General structure of the global water cycle. The water cycle illustrates some important properties of environmental systems, including the transfer of material (water) and energy across the system boundaries, storage of material and energy (water) within the system, flows and movement between stores, and changes in the character of the material (e.g. water is evaporated into water vapour, then falls again as rain) as it flows through the system.*

BOX 3.7 THE FOUR MAIN ENVIRONMENTAL SYSTEMS

1. *lithosphere*: the rocks and minerals that form the solid body of the Earth (Part II);
2. *atmosphere*: the layer of air surrounding the Earth's surface (Part III);
3. *hydrosphere*: water on and near the Earth's surface (Part IV);
4. *biosphere*: the layer of living organisms of which humans are part (Part V).

sub-systems are also systems in their own right. All four are tightly interconnected and respond to each other and to the flow of materials and energy through the Earth system. The resulting distributions, patterns and dynamics are what give rise to and sustain life on Earth.

If the planet were uninhabited, we could concentrate on the natural processes and structures that make environmental systems work. But, given the long history of human use and abuse of the environment, the human impact cannot be ignored. This creates added complexity, because it is often difficult to determine which changes in the environmental are 'natural' and which are 'induced'. Environmental change is increasingly becoming a mixture of the two.

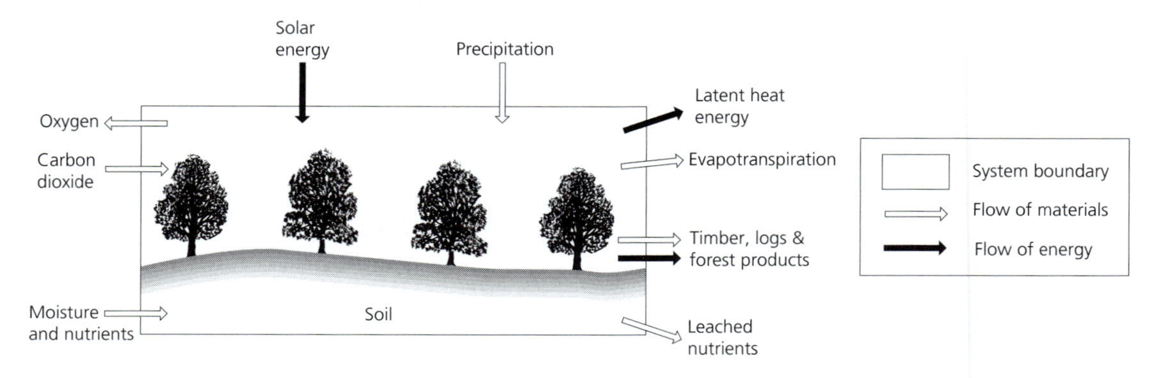

Figure 3.10 *A forest as an open system. A forest, like other ecosystems, displays system properties similar to those of the water cycle (see Figure 3.9).*

A proper understanding of environmental systems must take into account the complex interactions between the Earth system and people. Some of these interactions are physical, such as the different types of structure we erect (large dams, roads, towns, sea defence walls, and so on) and the ways we use the Earth's physical resources. The human influence is also evident in environmental management, and in the institutional and organisational structures that underpin modern society.

FLOWS, CYCLES AND STRUCTURES

Most environmental systems are open systems, and they are usually defined in terms of the movement of energy and materials. This movement (the system dynamics) occurs because the system components work together, in much the same way as the hands on a clock move round because the clock's component parts are working properly. It determines the character of the system, the ability of the system to adapt or adjust, and the ways in which the system interacts with other systems. System dynamics, in turn, are generally strongly influenced by factors outside the system itself. For example, sunlight from the Sun drives the Earth's energy system, rain over East Africa drives the water system in the natural Nile, and waves from across the ocean drive coastal currents and sediment processes.

STRUCTURE

The basic structure of open environmental systems has four key parts – inputs, outputs, flows and stores (Box 3.8). The challenge in understanding environmental systems is to identify the most important system components and to measure or monitor them wherever possible. In a drainage basin, for example, the watershed is the system boundary, and this can be identified on a map or on the ground. The important system components in a natural drainage basin can all be measured with appropriate instrumentation and sampling:

- *inputs*: water and dissolved chemicals from the atmosphere; sediment from hill slopes;
- *outputs*: water and sediment that flow out of the basin into another river, a lake or the sea; water evaporated back to the atmosphere; water transpired by plants;
- *flows*: water and sediment that move over and through hill slopes, then downstream along channel networks to the outlet of the basin;
- *stores*: water is stored short-term on vegetation, on the soil surface, within the soil, in the river; water is stored

BOX 3.8 KEY COMPONENTS IN OPEN ENVIRONMENTAL SYSTEMS

- *inputs*: they import material and energy across the system boundary;
- *outputs*: they export material and energy across the system boundary;
- *flows*: they have flows and pathways (fluxes) within the system along which the energy and materials pass;
- *stores*: they have storage areas within the system where energy and material can be stored for various lengths of time before being released back into the flows.

long-term in groundwater and large lakes; sediment is stored on hill slopes, in channels, on floodplains, on lake floors, in deltas.

System structure determines how these components fit together and operate as a single united system. Each depends on the others. Systems cannot be dissected or they cease to function properly; they are integrated functional units. Because of this, changes in one part of a system can promote changes in other parts of the system. For example, changes in flow can trigger changes in stores, which in turn trigger changes in exports and thus affect downstream systems.

Because of the import and export of energy and material across system boundaries, each environmental system interacts with its surrounding environment (which might be a larger system within which it is located). Systems are affected by that environment (by the imports) and can affect it (by the exports).

The Earth itself is a vast and complex open system powered by two energy sources, one internal and one external. The external source is solar energy (from the Sun; see p. 231); the internal source is heat generated by the decay of radioactive elements inside the Earth itself (see p. 128). These sources provide a relatively constant supply of energy to the Earth, which effectively drives all of the planet's environmental systems.

Flows of matter and energy drive all the processes in environmental systems. A city can be viewed as an open system, driven by and reflected in internal adjustments between a series of inputs and a series of outputs (Figure 3.11).

FLOWS OF ENERGY

Energy drives environmental systems. There are a number of different types of energy (Table 3.6), and any one type can be converted into another in the presence of matter.

Plate 3.2 *The urban system of Hong Kong. This is a largely artificial system in which flows of energy and materials are heavily controlled for human purposes. The urban system interacts with the natural environmental systems of the area – the atmosphere, hydrosphere, lithosphere and biosphere – to produce opportunities (resources) and constraints (hazards).*

Photo: Chris Park.

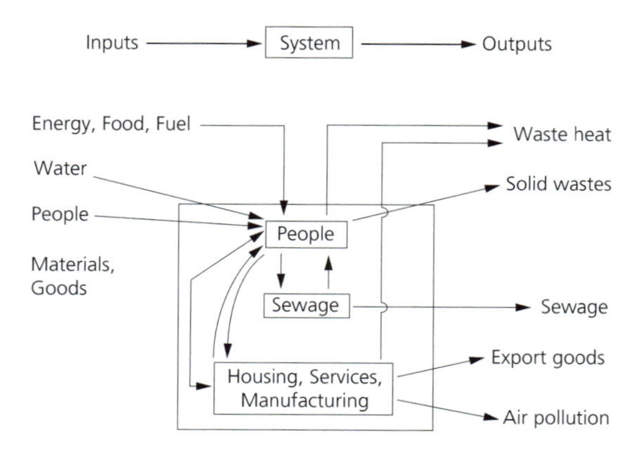

Figure 3.11 *The city as an open system. Like the water cycle (Figure 3.9) and a forest (Figure 3.10), there are inputs into, throughputs within and outputs from a city. The character of the system reflects the nature and balance of the inputs and outputs and the ways in which internal processes convert inputs into outputs.*

Thus, for example, a particle of rock perched on the top of a cliff has *potential* energy (because of its height), but this becomes *kinetic* energy when the rock falls from the cliff. The food we eat provides us with *chemical* energy, some of which we then release as *heat* energy as we move around. The *kinetic* energy of water flow in the Nile is converted to *electrical* energy in the hydroelectric plant at Aswan (see p. 64).

Studying the way in which different environmental systems obtain, store, use and release energy can reveal a great deal about how they operate and how their components are related. It is possible to establish energy budgets for many systems, particularly ecosystems. Comparisons of how a particular energy budget changes through time provide important information on system stability and adjustment. Comparing energy budgets between different systems throws light on issues such as the efficiency of energy use, the complexity of energy relations and the amount of interaction with surrounding systems.

Table 3.6 Different types of energy in the environment

chemical energy: possessed by a substance in its atoms or molecules, and released in chemical reaction

electrical energy: associated with electric charges

heat energy: possessed by a body because of the motion of its atoms or molecules (it is really a form of kinetic energy)

kinetic energy: possessed by a body because of its motion

potential energy: possessed by a body because of its position

Energy flows through environmental systems. It cannot be recycled, unlike materials, but it is used to do the work involved in recycling matter. The use of energy in any system is governed by the first two laws of thermodynamics (Box 3.9).

A useful indicator of energy use in systems is the amount of original energy available that has been dispersed as heat (in accordance with the second law of thermodynamics). This is referred to as *entropy*, and it really describes the amount of energy in the system that is not available to do useful work. It is a measure of wastage, or disorder. Entropy naturally increases through time in any system that does not receive its required input of energy in a usable form, just as a mechanical clock slowly runs down if it is not wound up regularly. Ecosystems (see p. 551) and indeed the Earth system itself are maintained – and thus have low levels of entropy – by the continuous input of energy from the Sun (see p. 231). Human use of energy and environmental resources serves to increase global entropy.

CYCLES OF MATERIALS

As we have seen, energy can be used but not reused, and it flows through environmental systems but cannot be recycled. Material flows are different, because materials can be used and reused, and while they flow through systems they can also be recycled.

While environmental systems have an infinite supply of energy (solar energy), they have a finite supply of materials (such as water or carbon), which must be used over and over again. Hence the material cycles that transform

BOX 3.9 THE LAWS OF THERMODYNAMICS

Law 1 states that 'in a system of constant mass, energy cannot be created or destroyed, it can only be transformed'. This means that energy is converted from one type to another (Table 3.6) as it performs work and explains why, for example, people need to eat regularly to have a proper intake of energy to enable them to grow, move and function.

Law 2 states that 'energy is dispersed in heat energy (thermal waste) as work is done'. This means that a lot of energy is lost from environmental systems as waste heat and explains why, for example, people get hot and perspire when they do physical exercise such as running.

and redistribute materials so that they again become available for use in their original form. These transformations use energy as it flows through the system.

The three main types of material that flow through environmental systems are water, nutrients and sediment. Water is the integrating link that ties together most of the Earth's environmental systems, and it plays important roles in most physical, chemical and biological processes in the environment. Hence the significance of the water cycle (see Figure 3.9) and hydrosphere.

Nutrients provide the chemical energy (food) needed by living organisms. The most needed chemicals are oxygen, hydrogen, carbon, nitrogen, phosphorus, potassium, calcium, magnesium and sulphur. These are moved around the biogeochemical cycles, which interact with the major systems in the lithosphere, atmosphere, hydrosphere and biosphere.

Cycling and recycling are natural features of environmental systems. The air we breathe has been breathed many times before, and the water we drink has been drunk and used many times before. Nature's economy is highly efficient in this respect!

Differences in the rates at which materials move through environmental cycles (Box 3.10 and Table 3.7) are important to humans in a number of ways. They determine the speed with which pollutants are dispersed through air and water, for example, and this in turn affects pollutant concentrations, health risk to humans and prospects for interacting with other systems. The pace of natural environmental recycling should also be taken into account when making decisions about the storage and disposal of waste materials. This is particularly true in the context of toxic and nuclear wastes (see p. 137), which pose great health risks to humans, and where safe long-term storage is a necessity.

BOX 3.10 RESIDENCE TIMES IN ENVIRONMENTAL SYSTEMS

The different environmental cycles operate at different speeds. For example, the typical residence time for smoke particles in the lower atmosphere is a few weeks before rain washes them back to the ground. Water remains in the air for 9–10 days on average. Dust thrown into the upper atmosphere from volcanic eruptions may remain there for several years. Groundwater may remain in place for centuries. Deep water in the Pacific and Atlantic Oceans may return to the surface after perhaps a thousand years. Table 3.7 shows typical residence times of some materials that circulate within the Earth's environmental systems.

SYSTEM DYNAMICS

Much of the complexity of environmental systems can be understood in terms of four principles, which relate mainly to energy flows.

Table 3.7 Typical residence times of materials in some environmental systems

Material	*Typical residence time*
Water vapour in the lower atmosphere	10 days
Carbon dioxide in the atmosphere	5–10 days (with sea)
Aerosol particles, lower atmosphere	1 week to several weeks
Aerosol particles, upper atmosphere	several months to several years
Water in the biosphere	2 million years
Oxygen in the biosphere	2,000 years
Carbon dioxide in the biosphere	300 years
Groundwater	150 years (above 760 m depth)
Atlantic Ocean, surface water	10 years
Atlantic Ocean, deep water	600 years
Pacific Ocean, surface water	25 years
Pacific Ocean, deep water	1,300 years

CONVERSION OF FORM

Imported energy and materials are not always exported in the same form. While open systems are defined in terms of the input and output of material across the system boundaries, it does not necessarily mean that inputs and outputs are of the same form (i.e. what comes out is not necessarily identical to what went in). All of the Earth's environmental systems are ultimately driven by energy from the Sun, but short-wave solar energy imported into natural systems is transformed into outgoing long-wave energy, which is radiated back into space (see p. 235). Trees removed from forests and crops removed from fields are exports of material and energy in the form of biomass, grown from inputs of solar energy, nutrients and water. Conversion of energy in the Nile river flow into hydroelectricity for industrial use (see p. 64) is another example.

TIMESCALES OF BALANCE

Rates of input and output of energy and materials in a system do not have to balance in the short term (hours, days or even months), because both can be stored within the system. But in the long term (over years, decades, centuries or longer) inputs and outputs are often balanced, particularly if the environment outside the system is stable and unchanging. All environmental systems have stores where energy and material can be locked up for varying periods. The storage of water in Lake Nasser behind the Aswan High Dam (see p. 64) is an obvious example. Groundwater flowing into a lake or river originally fell over the area as rainfall weeks, months or even years earlier and has since been stored in permeable rocks.

Storage is a key aspect of system stability, and the alteration or removal of stores within a system can trigger instability and change. Stopping the annual deposition of nutrient-rich silt along the Nile floodplain after closure of the Aswan High Dam (see p. 67) represents removal of a valuable store, and alternative means had to be found to restore what had been lost in order to maintain stability (continued irrigation agriculture).

AVAILABILITY OF ENERGY

All environmental systems (closed and open) require constant supplies of energy to keep going. Throughout the system, as energy is converted from one form to another in performing work, heat energy is released. This is exported from the system, usually into the atmosphere, from where it is ultimately exported into space. New energy must be imported to replace the energy that is lost. A simple example is any type of battery-driven machine (such as a portable radio, a watch or a toy car); once the energy in the battery has been used up, the machine will stop working unless a new battery (energy source) is installed. Environmental systems require energy supplies too. Solar radiation powers ecosystems; relief, water and sunlight provide the energy that drives river systems; and coastal systems get their energy from winds, waves, tides and currents.

ENERGY EFFICIENCY

Although they usually have access to constant inputs of potential energy from outside, most environmental systems are very inefficient in their use of available energy. This is largely because a great deal of useless heat energy is created when energy is transformed as it works within the system. The ratio of useful work done to potential energy supplied is usually very small.

RESPONSE TO CHANGE – ADJUSTMENT AND FEEDBACK

A system changes when it is forced to, otherwise it tends to remain stable and unchanging. Many different factors force such change, and the net effect is for the system to adjust in such a way that it absorbs the change and produces a new equilibrium (Figure 3.12). Because the various elements of a system are related, often by means of the flows of energy and materials that effectively drive the system, it follows that changes to one element might trigger changes to at least some of the other elements in the same system. Clearing a patch of forest, for example, not only removes the trees but also lets in more sunlight, which allows shrubs and bushes to grow in the clearing. But internal system changes like this are rarely linear (where a change in element A causes a change in element B and then a change

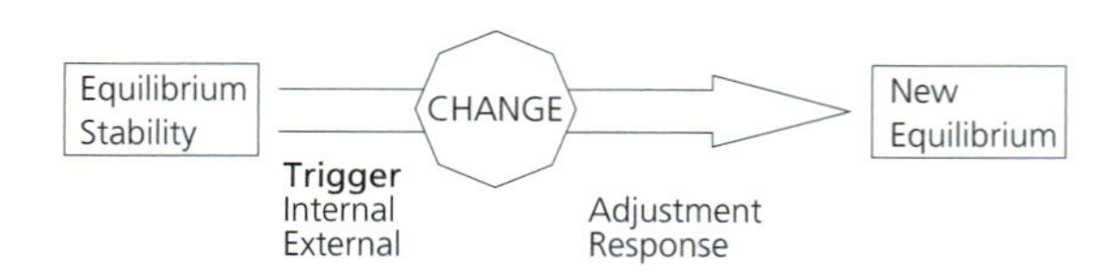

Figure 3.12 *Triggers to change in environmental systems. In most systems, internal and external triggers promote change, which is reflected in a system response and adjustment towards a new equilibrium.*

in element C, then D, and so on, with the changes getting further and further away through time from the initial change in A).

FEEDBACK

As a result of the many ways in which the elements are usually interrelated, a change in one element can often promote complex adjustment, which includes a feedback link. Feedback occurs when a change in one system element produces a sequence of changes in other elements, which ultimately leads back to the element whose initial change set off the sequence in the first place.

Feedback in environmental systems is complex, because many more variables are involved and the feedback loops interlock, overlap and interact. Unravelling all the links in this tangled web of interrelationships is far from easy. Yet this tightly knit structure is what gives environmental systems structure and makes them function. It also helps to explain why even a small change in the system can have many consequences as the adjustment gets passed along the interacting feedback loops.

BOX 3.11 POSITIVE AND NEGATIVE FEEDBACK

Positive feedback: this exists when the feedback serves to further change the element that originally changed, in the same direction. It amplifies or reinforces the initial change within the system, which was triggered by external factors, so the changes get bigger and bigger. It tends to destroy any stability that previously existed in the system. Thankfully positive feedback, which enhances change, is relatively rare in environmental systems.

Negative feedback: this promotes the opposite change to positive feedback, namely it damps down or suppresses system changes promoted by external factors. Such feedback serves to maintain and regulate system stability by self-regulation (homeostasis). Most environmental systems are dominated by negative feedback, which opposes change. This is why such systems appear stable in the absence of external change, and why they can usually adjust to external change.

TYPES OF FEEDBACK

In reality, there are two different types of feedback loop, positive and negative (Box 3.11), and these are illustrated in Figure 3.13. They have radically different impacts on systems.

A large-scale example of positive feedback is the way in which the growth of large ice sheets is a self-reinforcing mechanism that tends to promote further expansion of ice cover by reducing mean global temperature (as happened during the Pleistocene ice age) (see p. 451). However the process starts, ice caps grow and this changes the *albedo* (reflectivity) of the land surface. As a result, more solar energy is reflected back to space, and mean global temperature falls. More ice forms, and the feedback loop continues. Ice covers an increasingly large area, the albedo declines further, temperatures fall even more, and the whole process feeds itself.

Positive feedback loops like this are the ultimate vicious circles and are broken only by a change in external factors. In the case of ice caps, this might be a change in the Earth's position relative to the Sun, or perhaps a change in the frequency of sunspot activity on the surface of the Sun, which heats the Earth and raises temperatures enough to more than compensate for the progressive cooling feedback loop.

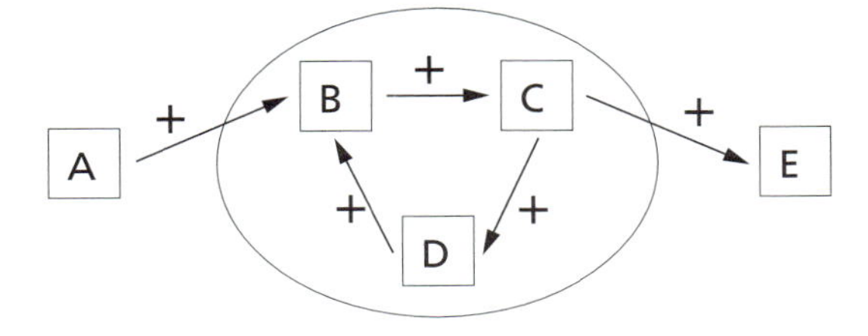

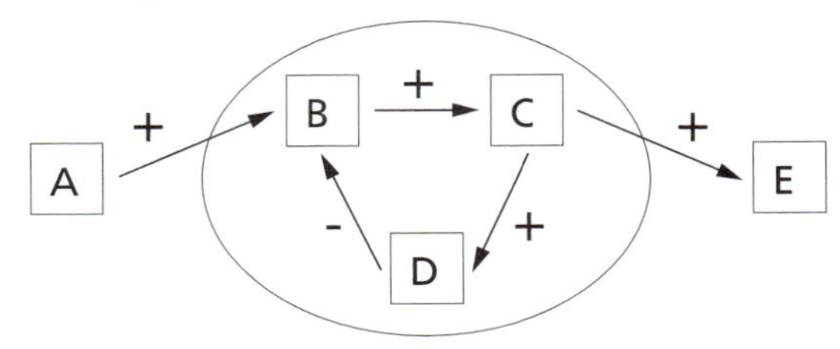

Figure 3.13 *Feedback in environmental systems. Internal adjustment within a system is usually promoted by feedback, involving a sequence of changes between system components. Positive feedback (a) promotes further change, whereas negative feedback (b) damps down change. Negative feedback tends to dominate most environmental systems, which is why stability can exist in the absence of triggers to change.*

EQUILIBRIUM

Most environmental systems show signs of equilibrium (Box 3.12), meaning a state of balance between the system and its surrounding environment. If equilibrium did not exist, or if a system loses equilibrium, natural stability is lost and chaos can follow. In some ways, this was the case with erosion of the river bed and delta downstream from the Aswan High Dam (see p. 67). There are many different types of equilibrium, but the most important ones for environmental systems are steady-state equilibrium and dynamic equilibrium (Figure 3.14).

STEADY-STATE EQUILIBRIUM

In a system in steady state the balance between inflow and output of energy is constant over time, and there is no net change in storage within the system through time. Energy and materials enter, flow through and exit the system, effectively making the system work as they pass through.

BOX 3.12 TIMESCALES AND EQUILIBRIUM

Timescale is important when describing system equilibrium. Equilibrium is usually stable but not necessarily immovable (fixed), and short-term variations within the system can be superimposed on a more stable longer-term pattern. Weather can vary a great deal from day to day, for example, while climate changes much more slowly. Map and air photo analysis might show that a particular river channel has been stable for a long time (with unchanging width, meander size and location), but detailed analysis might show many small-scale and short-term channel adjustments to individual floods. Similarly, winter storms can cause large short-term variations in the character and appearance of a beach, yet over a longer timescale that beach could appear to be remarkably stable.

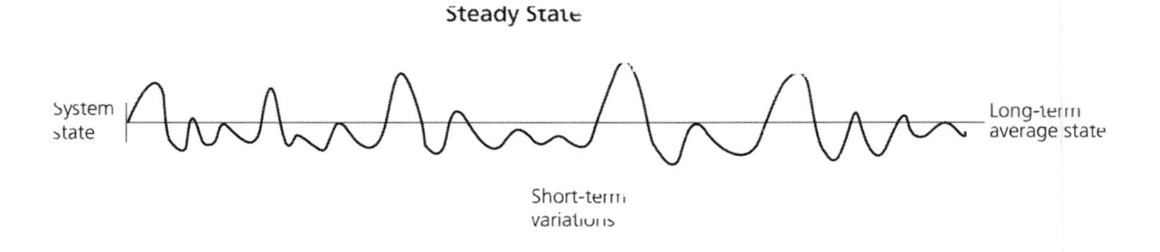

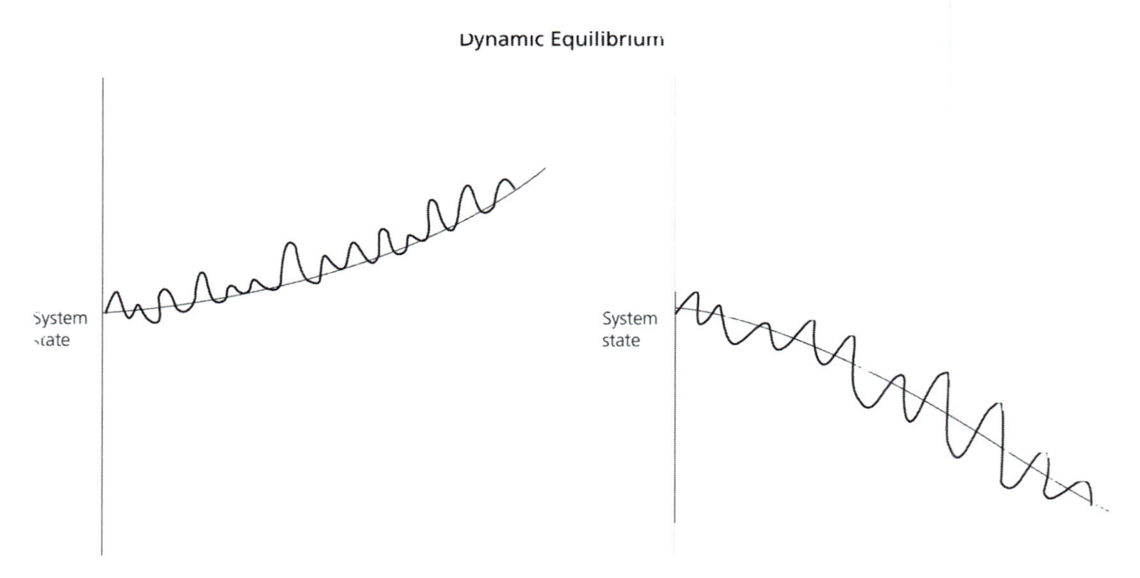

Figure 3.14 *Steady-state and dynamic equilibrium. Steady-state equilibrium is defined by the persistence of short-term variations around an unchanging average state, whereas in dynamic equilibrium the short-term changes are superimposed on a background state that is itself changing.*

In the pre-dam Nile below the confluence of the Blue Nile and White Nile, the main material flows were water and sediment, which flowed down the river into the sea. Water and sediment would be stored temporarily on the floodplain during the yearly flood cycle. But – under stable climatic conditions and without human interference – there would be no net change in storage within the river system. Inputs would match outputs. This does not mean that each grain of sediment simply passes through the system, because some will be deposited and others will be eroded and removed. It means that a similar amount (but not necessarily the same material) was exported as imported over a given period.

While steady-state equilibrium applies to a stable open system, it is possible to maintain steady state under changing conditions provided that the system can adjust quickly enough to the change (which is often promoted by external factors, like climate change or sea-level change). When the Nile was originally dammed at Aswan in 1902, for example, the flows of water and sediment along the lower river were both reduced (see p. 67). But the lower Nile system was not necessarily made unstable, because the river could adjust to the new situation. After a period of adjustment (involving some scouring and deposition of the river bed) – spontaneous adjustment involving negative feedback mechanisms that automatically created new internal distributions of energy and material within the system – steady state was re-established and outputs more or less matched inputs.

DYNAMIC EQUILIBRIUM

Open systems in dynamic equilibrium remain stable over long periods, and they have the ability to adapt to external change (unless the change is extremely fast) but still maintain stability (see Figure 3.14). Hill slopes in temperate climates appear to evolve over long periods of time by retreating backwards parallel to themselves. Thus, although the location of the hill slope within a landscape changes progressively through time, the slope profile (cross-section) remains relatively unchanged. This implies that the slope maintains equilibrium – dynamic equilibrium – with the forces that operate on it and promote change.

Many environmental systems show signs of dynamic equilibrium and progressive adjustment over long timescales. Much of the concern over human impacts on the environment centres on the rapid rate of change induced by many human activities, which can disturb system stability and bring unwanted and unforeseen side-effects. The environmental impacts of damming the River Nile (see Table 3.5) illustrate the point well.

RESILIENCE

Closely related to the idea of equilibrium is the ability of a system to recover after a shock or sudden change (Figure 3.15). This is the system's resilience. Environmental systems are complex, with many interconnecting parts, and they can usually recover fairly quickly from natural interruptions. Thus, for example, vegetation usually grows again after prolonged drought, rivers can usually re-establish equilibrium after major floods, and beaches can stabilise fairly quickly after storms and cyclones.

Natural systems seem, on the whole, to have greater resilience than simple or simplified systems (such as monoculture, where one crop is grown over a wide area). Simple systems can sometimes be seriously affected or even totally destroyed by single failures. This is one reason why ecologists argue for the conservation and protection of entire ecosystems and the preservation of biodiversity (p. 539).

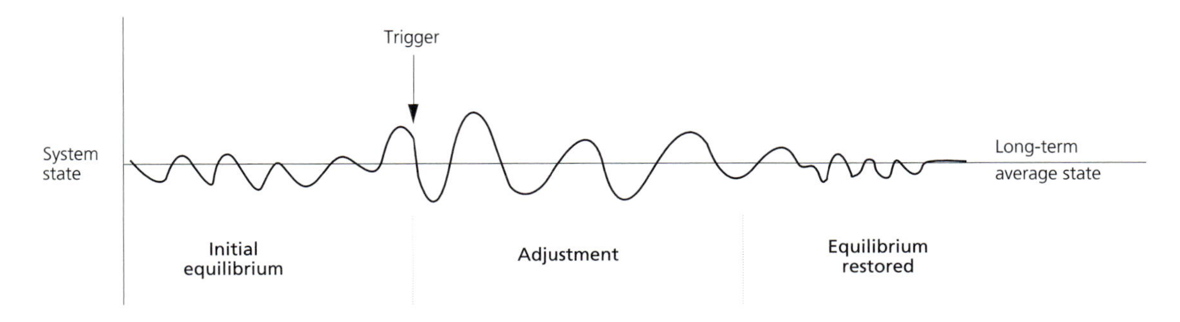

Figure 3.15 *Resilience. Resilience means resistance to change, and in a resilient system (or part of a system) a trigger promotes adjustment; but over time the underlying equilibrium is restored (generally by negative feedback; see Figure 3.13).*

THRESHOLDS AND LAGS

Predicting how an environmental system is likely to respond to imposed change, such as global warming, is generally very difficult. It requires a good understanding of many factors, such as how that system is structured, how its component elements interact, how energy flows and materials are cycled through it, and what determines stability and equilibrium within it. Two complicating factors are thresholds (Box 3.13) and lags (Box 3.14).

IMPLICATIONS

A number of implications follow from what has been said about the structure and functioning of environmental systems:

Interdependence: the key to understanding environmental systems is the interrelationships that give them structure and that affect and are affected by flows of energy and materials. Systems are functioning wholes, and while they often show signs of stability and equilibrium these can be disturbed if any part of the system is damaged, altered or removed. An environmental example would be a small woodland drainage basin, in which climate affects vegetation and soils, vegetation affects water movement and soils, soils affect vegetation and water movement, and all (directly or indirectly) affect river flow and sediment load. If any of these major components changes, the others would change in response. Maintaining system stability is vitally important in environmental management, sustainable development and coping with hazards.

BOX 3.13 THRESHOLDS IN ENVIRONMENTAL SYSTEMS

Thresholds are critical points or turning points at which system response and behaviour change abruptly. Some adjustment in environmental systems is linear and incremental (continuing progressively at a similar pace), but some is sporadic or discontinuous. Discontinuity is often associated with the existence of thresholds (Figure 3.16). An obvious and easily defined threshold is the bankfull state in a river. Water level rises within the channel until it reaches bankfull, then water spills out on to the adjacent floodplain (see p. 410). Conditions change after this point (threshold). Other thresholds that are relatively easy to identify include the slope conditions under which sediment begins to become unstable and move (as in landslides; see p. 213), moisture levels in the atmosphere at which clouds form and rainfall is generated (see p. 296), and levels of exposure to some important air pollutants at which humans start to suffer ill-health or die (p. 245).

The problem is that many thresholds in environmental systems are not easily identified. When they are reached, system response can change without warning and in ways that are difficult to forecast. This is why some catastrophes occur. It poses real problems in environmental management, particularly when the slow build-up of environmental change (such as global warming) produces incremental adjustments, then – possibly without warning – there is a sudden displacement of equilibrium and dramatic system responses when the threshold is reached.

BOX 3.14 LAGS IN ENVIRONMENTAL SYSTEMS

Lags are time delays in a system's response to change (Figure 3.17). Some systems, and some parts of other systems, can adjust rapidly to external change and so there is no time delay in system response. Lags can be caused by system complexity. It might take some time for a chain reaction of adjustments to progress through a system with many elements and interrelationships.

Lags are sometimes created by increasing scale of the system. A small stream, for example, responds rapidly to a thunderstorm overhead as the rainwater is quickly drained into the river and flows downstream. A major river, in contrast, has a much slower response to a similar event because of the time it takes water to drain into all of the tributary streams and then flow downstream. Distance itself can cause a lag. Some lags are associated with thresholds, because incremental change can take time to build up to threshold levels at which system behaviour alters. Whatever the cause, lags can make it difficult to establish cause-and-effect associations within and between systems, and this further complicates the problem of environmental management.

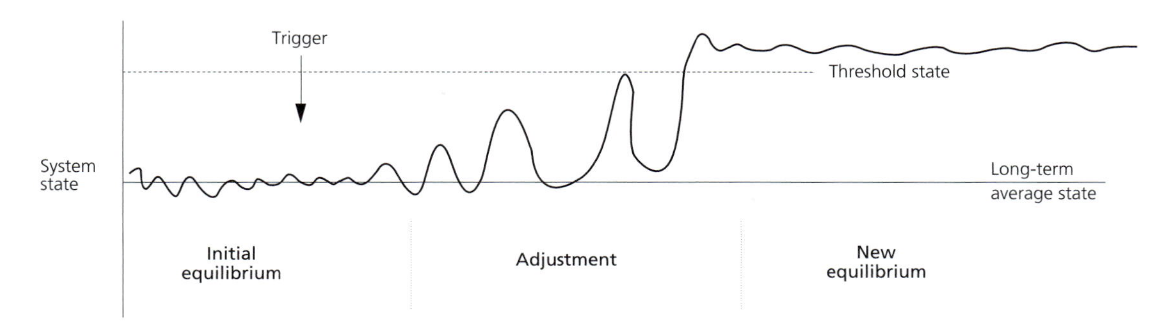

Figure 3.16 *Thresholds. If a system (or part of a system) is forced to adjust, the adjustment may be so great that it crosses a critical point (threshold), and the system settles down at a new equilibrium that is different from the original equilibrium.*

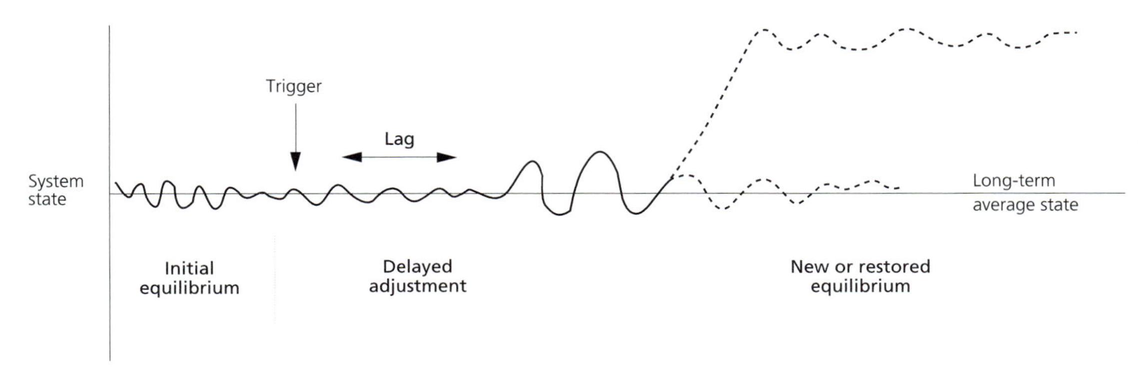

Figure 3.17 *Time lags. A time lag is a delayed adjustment in a system (or part of a system) and is reflected in a time delay in system response.*

BOX 3.15 THE SYSTEMS APPROACH

The systems approach emphasises interrelationships, synergy and holism and it differs from traditional science, which is more concerned with single cause-and-effect relationships. As such, it offers a broader perspective, is interdisciplinary and focuses on the need to understand and preserve the natural environment.

The systems approach is really a way of looking at the environment, a synthesis based on trying to see the whole picture. It is a perspective rather than hard fact. But this is precisely its attraction, because the systems perspective offers a valuable framework for studying, describing, interpreting and modelling the environment. Since the early 1970s, scientists have used the systems approach to study many aspects of the environment. As will become clear in the chapters that follow, systems ideas have been widely used to model and explain environmental stability, adjustment and relationships.

Synergy: different parts of many environmental systems are often studied separately, but this is largely for practical convenience. The essence of an environmental system is its entire structure and operation, rather than its component parts or individual flow pathways or stores. This is why many environmentalists stress the need for holistic approaches, which emphasise unity and completeness. Synergy results from the interaction of components within a system. The *gestalt* principle, which states that 'the whole system is greater than the sum of its component parts', is important here. To take a simple example, the human body is made up of many different organs (such as the heart, lungs and kidneys) within a physical frame (the skeleton, muscles and skin), linked together by flows (of blood in veins and arteries). But a living human being amounts to much more than the right number of pieces stuck together in the right way! Similarly, an entire drainage basin amounts to much more than simply a collection of hillslopes, a covering of vegetation, a column of soil and a course for water to flow along. Ecosystems offer among the most important examples of environmental systems where unity is of paramount importance, and the maintenance of unity and stability are the principal goals of ecosystem management.

Complexity: a simple system, with few component parts and clearly defined pathways for the flow of energy and materials, is relatively easy to describe and understand. But such systems are rare. Most environmental systems are extremely complex, involving many different flows and cycles operating over different timescales, a series of stores of different sizes that operate over different timescales, and numerous links with surrounding systems. It is not uncommon for parts of one system to be in equilibrium at the same time as other parts are adjusting or unstable, or for parts of the system to be adjusting much faster and more obviously than others. Complex response, in which an initial adjustment to an external change in turn triggers other internal adjustments, seems to be inherent in many environmental systems. This all makes it generally very difficult to predict how an environmental system will respond to change, and it explains why the management and sustainable use of environmental systems is such a complex task.

Equifinality: this means that any particular system state or condition can often be brought about in more than one way. As a result, it is often difficult to work out cause-and-effect links between many parts of a system, because the same end-result can be arrived at by a number of different permutations of changes. A given effect can have several possible causes, and any particular environmental change can have several different effects. Global warming can occur because of changes in the distance between the Earth and the Sun, changes in sunspot activity on the Sun, changes in the Earth's rotation, changes in Earth surface conditions and changes in the distribution of oceans and continents, as well as being a result of air pollution by greenhouse gases. In many environmental systems, it is not easy to distinguish changes promoted by external factors from changes promoted within the system itself, and this has great significance for developing coping strategies and remedial measures. A second consequence of equifinality in environmental systems is that it is often difficult to anticipate how a system will respond to any given change.

Timescale: the timescale over which a particular system is considered is important, because the system might appear to be in equilibrium at a geological timescale (thousands or millions of years) but also appear to be highly responsive to short-term changes (perhaps less than tens of years).

Two key themes must be addressed before we leave this overview of environmental systems – increasing awareness of the problems and relevance of global systems, and increasing awareness of environmental uncertainty.

GLOBAL SYSTEMS

A systems perspective is particularly useful when applied at the global scale, because it helps our understanding of planetary balance and change. According to the Gaia hypothesis (see p. 119), the Earth can be viewed as a single system or super-organism. It follows that disturbing parts of that planetary system could trigger unexpected changes elsewhere in the system. Such a perspective forces us to think much more carefully about protecting the entire global environment rather than concentrating on particular hot-spots of damage or change.

A global perspective is also important, because it emphasises large-scale interactions between the Earth's main environmental systems (the biosphere, lithosphere, atmosphere and hydrosphere). One useful approach in this context is to focus on large-scale flows of materials through the planetary environment, such as the major biogeochemical cycles (see p. 86).

Looking at the big picture rather than focusing on small details also helps us to appreciate that:

- The biosphere is an indivisible whole.
- People and their activities are now an integral part of the biosphere.
- Nothing is static, and change is both inevitable and often beneficial.

Whether this realisation forces changes in human behaviour (individual and collective) towards more environmentally sustainable lifestyles remains to be seen, but the prospect lies at the very heart of the sustainable development debate (see p. 26).

ENVIRONMENTAL UNCERTAINTY

Despite great advances over recent decades in understanding how environmental systems work, there are still large areas of uncertainty. As we learn more about the world around us and how it operates, we also become more aware of what we don't know. Some particular causes of uncertainty are summarised in Table 3.8. Sometimes the gaps in understanding are so large that scientists can do no more than offer informed judgements rather than hard facts

Table 3.8 Major sources of environmental uncertainty

Source of uncertainty	*Complications*
Complexity of natural systems	Environmental change is the result of complex interactions between several closely coupled systems. Some changes are less easy to predict than others, partly because they are the result of complex responses.
Thresholds	The presence of thresholds in many environmental systems means that effects may increase suddenly, dramatically and without warning.
Non-linear response to change	Many environmental systems are non-linear, which means that the size of their response is not necessarily directly proportional to the size of the external change that promoted the response.
Imperfection of existing models	Environmental change is the result of complex interactions between a large number of variables (elements), and most models can cope with only a limited number of variables. Many existing models are highly simplified.
Lack of good knowledge about background conditions	Identification of future trends requires good knowledge of background trends, and long-term data are often not available. Such data would help in determining whether observed trends have happened before, and whether they are linear or cyclical in nature.
Potential role of catastrophes	Extreme events are part of natural environmental systems, they will inevitably recur in the future, and they can either increase or decrease the impacts of human activities.
Potential role of unsuspected mechanisms	There are bound to be factors that have not yet been taken into account, either because they are difficult to measure or because they have not even been identified.
Problems of predicting speed of response	Although there is fairly reliable information that many environmental changes are taking place (e.g. in response to global warming), in many cases it is much more difficult to predict the speed of response.
Problems of definition	For many types of environmental change there may be more than one way of defining what is going on. This makes it difficult to identify some trends, or to compare different situations.

or confident predictions. This distinction is important, and it explains why many environmental debates create more heat than light.

Where there is uncertainty there is ample scope for differing opinions and judgements, even among experienced and respected scientists. Hence, for example, there are no clear answers in the ongoing debates about nuclear energy, biodiversity (p. 524), population (p. 51), climate change (p. 261), ozone depletion (p. 249), acid rain (p. 251), desertification (p. 434), over-fishing (p. 486) and rainforest clearance (p. 586).

The climate change debate, where there are clearly many uncertainties, shows just how difficult it is to make reliable forecasts or predictions of how environmental systems are likely to change over the next 100 years or so. Environmental change is being promoted directly by atmospheric warming and related climatic changes (in precipitation, for example), and indirectly via the impacts of warming on other factors (such as vegetation). Likely consequences of these changes can often be worked out in general terms, but many uncertainties remain (Table 3.9).

Managing global environmental change, moving towards more sustainable forms of development and coping with an increasingly hazardous environment are effectively challenges in managing an Earth system that is beset with uncertainty. Ever-present uncertainty, coupled with ignorance (there are many areas about how the environment functions that we simply know next to nothing about), set against a backdrop of incessant environmental change, adds up to a major challenge, the sheer scale and significance of which we have never seen before!

Table 3.9 Major areas of uncertainty about the impacts of climate change

- the hydrological response in rivers and lakes
- the frequency of tropical cyclones
- the pace and pattern of permafrost melting
- the response of glaciers and ice caps
- the extent of sea-level rise
- the reaction of beaches to rising sea levels
- the state of wetlands, deltas and coral reefs

BIOGEOCHEMICAL CYCLES

The availability and transfer of a wide variety of chemical elements in the environment are key factors that affect life on Earth. Indeed, according to the Gaia hypothesis (p. 119), the global chemical balance both shapes and is shaped by the interaction between life and planet, and it helps to maintain global equilibrium.

All of the nutrition elements on which life depends are involved in large-scale cycles involving constituents of soil, water, air, rock and living organisms. The sub-system of the Earth system that is responsible for these stores and transfers is a series of interlocking and delicately balanced cycles. The biogeochemical cycles function at the global scale, and nutrient cycles are those biogeochemical cycles that involve elements necessary for life.

The most important elements that cycle through the major biogeochemical cycles are carbon, oxygen, hydrogen, nitrogen, phosphorus and sulphur. Despite recent advances in understanding how these great cycles work (partly a result of the IGBP studies; see p. 20), many uncertainties and gaps remain. Many of the more pressing global and regional environmental problems we face today – such as greenhouse gases (p. 255), eutrophication (p. 409), acid rain (p. 251) and ozone depletion (p. 249) – are closely related to these cycles. So the need for better understanding becomes greater and more important.

Although we are considering these cycles here as though they were largely self-contained systems, we must not overlook the fact that they are components of the global system and that they interact with the other major environmental systems – the atmosphere, the lithosphere, the hydrosphere and the biosphere (see Figure 1.1).

BOX 3.16 BIOGEOCHEMISTRY – DEFINITIONS

Bio refers to living organisms and *geo* refers to the rocks, air, soil and water of the Earth. *Geochemistry* deals with the chemical composition of the Earth and the exchange of elements between different parts of the Earth's crust and its oceans, rivers and other physical stores. *Biogeochemistry* is the study of the exchange of materials between living (biotic) and non-living (abiotic) components of the biosphere.

BOX 3.17 GEOCHEMISTRY AND HUMAN HEALTH

Geochemistry affects human health in many different ways. People lacking a balanced intake of trace elements – including such exotic materials as gold, platinum, copper, lead, zinc, aluminium, silicon, mercury, cadmium, selenium, arsenic and iodine – are more susceptible to a range of ailments, including multiple sclerosis, various cancers, arthritis, Down's Syndrome and mental retardation.

The reverse side of the coin is also true, because there are significant health risks (including heart attacks and cancer) associated with exposure to high levels of many chemicals and materials in soils, air and food. Lead is a well-documented example, and attempts have been made to reduce human exposure to lead in the environment in many countries, including those in the European Union.

Pollution of freshwater resources by trace metals and other chemicals poses risks for human health but also has a variety of other ecological and environmental impacts.

CHEMICAL ELEMENTS

Organisms – including plants, animals and humans – require a large number of nutrients for survival, but not all in equal quantities. The total list includes carbon (C), hydrogen (H), oxygen (O), nitrogen (N), calcium (Ca), chlorine (Cl), copper (Cu), iron (Fe), magnesium (Mg), potassium (K), sodium (Na), sulphur (S) and small traces of aluminium (Al), boron (B), bromine (Br), chromium (Cr), cobalt (Co), fluorine (F), gallium (Ga), iodine (I), manganese (Mn), molybdenum (Mo), selenium (Se), silicon (Si), strontium (Sr), tin (Sn), titanium (Ti), vanadium (V) and zinc (Zn).

Living cells are composed mainly of carbon, hydrogen and oxygen, and these are required in larger amounts than the other elements (Table 3.10). But although the quantities of trace elements required are much smaller, they are no less vital.

GENERAL STRUCTURE OF THE CYCLES

The biogeochemical cycles are based on the cyclic flow of elements passing between organisms and their environment. Flow through the environment is cyclic because of the finite supply and the relatively constant form of the individual elements.

Individual cycles can be identified for each of the elements, but they all have in common a basic two-part structure involving:

BOX 3.18 GREENHOUSE GASES AND BIOGEOCHEMICAL COMPLEXITY

The likely biogeochemical impact of continued reliance on burning fossil fuels is a major concern, and this important area of controversy clearly illustrates the complexity of many environmental systems. Records show that levels of carbon dioxide (CO_2) in the atmosphere are higher than at any time over the past 150,000 years and are rising faster than at any time over the last 500 million years (see p. 259). The main cause is the burning of fossil fuels, which releases carbon stored in sedimentary deposits into the atmosphere.

Rates and patterns of accumulation and release differ a great deal, and this leads to problems of overload in natural environmental systems. Carbon that has accumulated slowly in ecosystems of the geological past was stored in the energy deposits and is now being released rapidly (on combustion) into today's environment. The carbon accumulated in depositional basins of regional extent, but it is now being released from a large number of specific places (point sources), particularly factories and power stations.

As well as affecting present and future global climate (p. 261), this biogeochemical overload is expected to have significant impacts on agricultural productivity. This will put available food supplies under great pressure, greatly increase the risk of massive and widespread famine and malnutrition, and could trigger huge migration flows of environmental refugees.

These human costs will be high in themselves, but they may be small set alongside the impacts of greenhouse gases on natural ecosystems and biodiversity (habitats and species). Climate change and ecosystems are intimately interlinked, and plants are the only organisms capable of removing significant amounts of CO_2 from the atmosphere. This natural biogeochemical mechanism offers the prospect of environmental self-repair, using plants to mop up (sequester) the additional CO_2 produced by air pollution.

But the good news is overshadowed by bad news, because there is clear evidence that natural responses for sequestering carbon have been outstripped by the continued rise in atmospheric CO_2 and continued clearance of natural vegetation (more than half the world's prehistoric forest cover has now been removed, for example). Quite simply, the plants cannot keep pace with the air pollution. Coping with the greenhouse effect will be much simpler if steps are taken to increase the natural ability of the biosphere to sequester the additional carbon. This would require initiatives such as massive reforestation programmes and large-scale ecosystem engineering schemes.

Table 3.10 Major and minor chemical elements in the environment

Major elements, which plants require large amounts of	Oxygen (O) Carbon (C) Hydrogen (H)
Minor elements, which plants require relatively large amounts of	Nitrogen (N) Phosphorus (P) Potassium (K) Calcium (Ca) Magnesium (Mg) Sulphur (S)
Trace elements: plants also require small amounts of over 100 materials, including	Iron (Fe) Manganese (Mn) Cobalt (Co)

1. an inorganic component (comprising the abiotic or non-living parts of the environment, with sedimentary and atmospheric phases); and
2. an organic component (comprising plants and animals including humans, living and dead, and their physical and chemical interactions).

STORES AND FLOWS

Biogeochemical cycles illustrate many of the properties of environmental systems. They are based on the flow of chemicals between stores (reservoirs) within the environment. The main stores are rocks, the oceans, the atmosphere and biota (living plants and animals). The water cycle (see p. 73) plays a vital role in creating many of the flows between stores.

Elements within the cycles move between stores, and transfers between stores operate at different speeds even within one nutrient cycle. For example, decomposition of dead organic matter quickly releases nutrients into soil (see p. 602), whereas the same nutrients cycle much more slowly in the atmosphere and exceedingly slowly in groundwater and the oceans. Some stores operate over geological timescales. This is true, for example, of carbon, which can be locked in coal deposits for perhaps 300 million years before being released as gas into the atmosphere on burning (see p. 255).

Stores also vary in size. The atmosphere stores a very large amount of nitrogen and carbon, whereas the oceans store vast amounts of hydrogen. Rocks are the main stores for many nutrients.

The general structure of a terrestrial biogeochemical cycle is shown in Figure 3.18. Because it is a cycle, we could begin to describe it at any point in the looped structure. For simplicity, we start with plants within the organic phase.

ORGANIC PHASE

Plants take in chemical elements in solution (via root osmosis) from the soil and convert them into forms that they can use in plant growth by biochemical processes (including photosynthesis). The elements can be released from the plants in a number of ways. When individual parts of a plant are separated from the main plant (such as leaf fall from trees during autumn), they are decomposed by bacterial activity in the soil. This decomposition releases the organic compounds, generally in a soluble inorganic form that can then be stored in the soil or another reservoir (store) within that particular biogeochemical cycle. When whole plants die, the same form of bacterial activity releases elements to the reservoir. If vegetation is burned (naturally as a result of lightning, or accidentally or deliberately by humans), then some of the elements are released directly to the atmosphere in the smoke, gas and ash, and the remaining elements in the ash residue can be decomposed by bacteria in the soil.

Animals also play significant roles in the cycling of chemical elements. Elements are released to grazing animals (herbivores) when they eat and digest plants (see p. 552). Herbivores in turn pass on to carnivores (meat-eating animals) a proportion of the elements they have taken in, when they are eaten by them! Throughout their life-cycle, animals release elements by producing waste material and faeces, which can then be decomposed by bacterial activity. Grazing animals also release elements to the soil and atmosphere when they die (by decomposition or combustion).

INORGANIC PHASE

The elements released from the organic phase are in one of two forms:

1. combustion material released to the atmosphere;
2. decomposition material released by bacteria into soils.

The combustion material can be washed out of the atmosphere by precipitation. It then becomes part of the reservoir of soluble inorganic forms of that element which is stored in the sedimentary phase of the cycle. If it is not washed out, it becomes part of the reservoir stored in the atmospheric phase. From here it can be released by bacterial activity and stored longer in the reservoir, or (by interacting with the water and rock cycles) it can enter the sedimentary phase of the cycle and be stored in sediments, soils and sedimentary rocks for a variable but generally long period of time. This sedimentary storage can later be released to the general reservoir by normal processes of weathering and erosion (see p. 204), which can also release chemical elements from igneous rocks (see p. 194).

The inorganic phase is very important in many biogeochemical cycles because it provides the main storage reservoirs for most nutrients. Flows through the inorganic phase are usually much slower than those through the organic phase.

The sedimentary phase is important to all of the major chemical cycles, but the atmospheric phase is important to only some. However, between them these two sets of system linkages form strong ties with the energy system and the rock cycle. The water cycle (see p. 73) binds all three – the energy, rock and biogeochemical cycles – together. The strength of this binding depends partly on the solubility of the chemical element concerned. With relatively soluble substances, such as nitrogen, the amounts of the elements present in different phases of the cycle are related to the water budget. With insoluble substances, such as phosphorus, there is relatively little removal of the element from the soil (except by soil erosion), so the amount present is determined largely by the rock cycle (see p. 203).

The relative importance of the atmospheric and sedimentary phases differs between chemical elements. For example, sulphur follows a cycle in which both phases are present and important. In the nitrogen cycle, on the other hand, the atmospheric phase is much more important than the sedimentary phase. This is because the atmosphere offers a large reservoir of nitrogen in gas form; decomposers contribute directly to this by denitrification processes, and plants extract nitrogen directly from it by biological fixation through leaf surfaces.

INDUCED CHANGES

Life on Earth depends on the global cycles of carbon, nitrogen, phosphorus and sulphur. But the links between life and biogeochemical cycles are two-way, because living organisms (particularly humans) have significantly altered the natural distributions and circulation of these elements. Human activities can influence the stability and balance of the biogeochemical cycles in many different ways, particularly by altering the initial inputs into the cycle, extracting materials from the cycle, and changing the speed and direction of operation of processes within the cycle.

Inputs: inputs into many biogeochemical cycles can be altered either deliberately or accidentally. Alterations to the inputs into the atmospheric phase might involve changing the quantities of elements that are already in the natural cycle (such as greenhouse gases released from air pollution). They might also involve releasing into the atmosphere 'alien' or exotic elements that are not normally found there (such as through the production of gases by combustion of some materials, particularly synthetic chemicals and compounds). Inputs to the organic phase can be increased by the application of organic fertilisers and other natural elements, as well as by introducing toxic materials (such as DDT) for which stable biogeochemical cycles do not naturally exist (and which are often damaging to plants and animals). Inputs into the sedimentary phase can also be affected by deliberate action, such as the dumping of industrial chemicals and wastes.

The human impact is greatest in terms of increasing inputs into biogeochemical cycles. Human activities are now generating flows of carbon, nitrogen, phosphorus and sulphur that are comparable in magnitude and importance with natural flows, thereby upsetting the delicate balance within natural cycles.

Extraction: many elements are removed from biogeochemical cycles, usually on purpose. Examples include extraction of material from the sedimentary phase (such as the extraction of nitrates and phosphorus, for use as agricultural fertilisers), and the cropping of biomass (living material) in farming and forestry.

Such induced changes in inputs and outputs often have ecological and environmental impacts at a relatively local scale. This is because the overall supplies of the elements are fixed, and human activities effectively redistribute them in time and space, rather than producing net change. It is also because the form in which the chemical elements are used cannot readily be changed, so much so that the material cannot be reused in the organic phases of the cycles. This is the very basis of recycling in natural environmental systems.

Process changes: the third main area of human modification lies in affecting the speed at which the cyclic processes operate. Burning, for example – which might be intentional (such as the periodic and controlled firing of heather moorlands to encourage fresh new growth) or accidental (such as forest fires started by campers) – can destroy all of the vegetation and most of the relatively immobile grazing animals and decomposers from the organic phase of a local cycle, and at the same time increase the atmospheric store of some chemical elements at the local scale. The atmospheric phase can also be influenced by planned weather modification schemes, because cloud seeding can influence precipitation chemistry and thus input into the soluble reservoir of some elements.

Rates of cycling within the sedimentary phase can be significantly affected by land management practices, and accelerated soil erosion can disturb local biogeochemical balances. Total supplies and the speed and pathways of certain elements can be disturbed by many different human activities, including agriculture, water resource management schemes, industrial activities and pollution.

Natural biogeochemical cycles are being disrupted by a range of human activities, including land-use changes and the burning of fossil fuels. The consequences could be very far-reaching, including global warming. The nitrogen, carbon and sulphur cycles are particularly important to the functioning of the biosphere, and they are also closely linked to the climate system.

THE NITROGEN CYCLE

THE NATURAL CYCLE

The nitrogen cycle (Figure 3.18; Box 3.19) involves the circulation of the element nitrogen through the biosphere. Compared with the other major biogeochemical cycles, it is relatively fast but also very complex. The atmosphere is composed mostly of nitrogen gas (N_2). This is chemically unreactive, which means that it cannot be used directly by plants. In order to make it available to plants, some of the

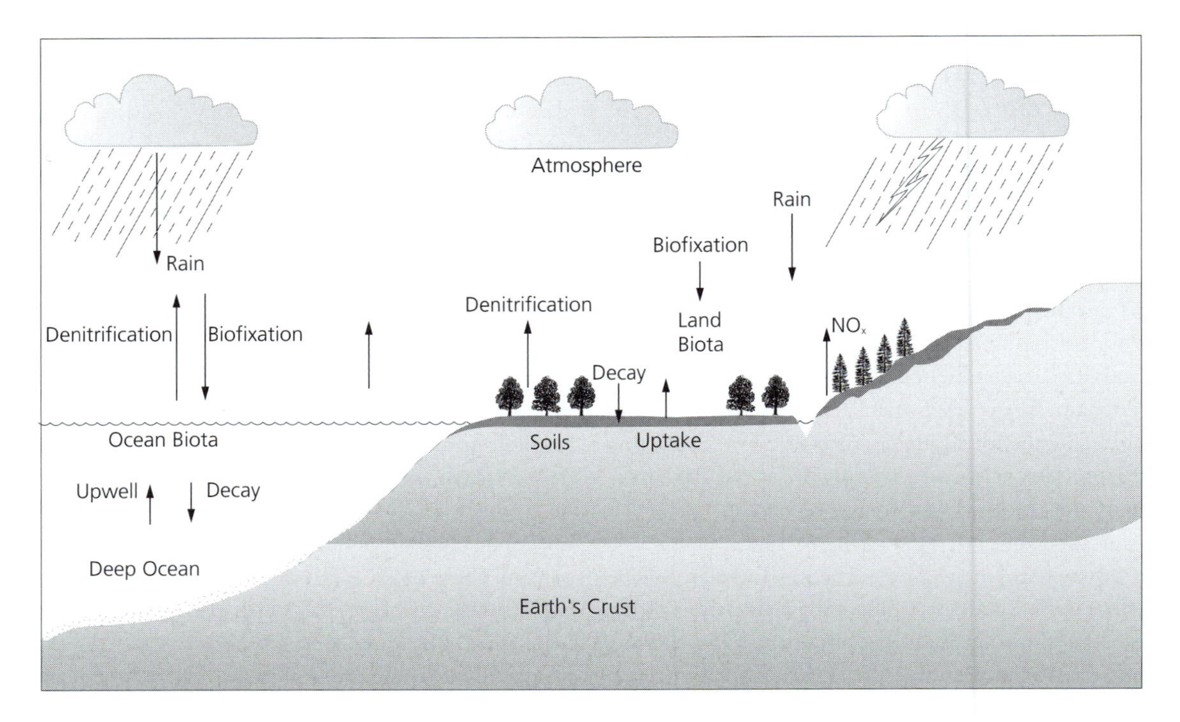

Figure 3.18 *The nitrogen cycle. The global nitrogen cycle involves the storage of nitrogen in a series of reservoirs (including soils, plants, the atmosphere and the oceans) and the transfer of nitrogen between reservoirs by biological and physical processes.*

Source: Natural Environment Research Council (1989) Our future world. NERC, Swindon, p. 14.

nitrogen is oxidised by energy released in lightning, and it then dissolves in raindrops (to form nitric acid – HNO_3).

Most (more than 90 per cent) of nitrogen in soils is fixed – and thus made available to plants – biologically. The nitrogen gas (N_2) is used by soil bacteria (particularly the bacteria living in nodules attached to the roots of legumes such as peas and beans), which convert it into ammonia (NH_3). Inorganic nitrate (NO_3^-) and ammonia in the soil are absorbed by plants (via their roots) and turned into organic compounds (such as proteins) in plant tissue. A proportion of this nitrogen is eaten by herbivores, which use nitrogen in the form of amino acids from which they synthesise proteins. Some of the nitrogen is passed on in the form of proteins to the carnivores that feed on the grazing animals. The nitrogen is ultimately returned to the soil as waste products (in urine and faeces) and when organisms die and decompose. Bacteria convert the organic nitrogen into ammonia or ammonium (NH_4^+) compounds. Other bacteria convert these into nitrites (NO_2^-) and then into nitrates (NO_3^-), which can be taken up again by plants.

Some of the nitrate and ammonia that is not absorbed by plants is leached (washed) from the soil into groundwater and surface water (lake and rivers), where it provides nutrient supplies for aquatic plants. The oceans are also part of the nitrogen cycle. Some nitrogen accumulates as organic sediment, which might through time be compacted and converted into sedimentary rock. The nitrogen can be released back into water by weathering of the deposits and rocks. Ammonia evaporates quickly when it is released, thus recycling some nitrogen back into the atmosphere. From there it quickly dissolves in raindrops and is washed back to

BOX 3.19 NITROGEN

Nitrogen (N) is extremely important for life on Earth. It is a constituent of all plant and animal tissues (where it occurs in proteins and nucleic acids), and it forms nearly 80 per cent of the Earth's atmosphere by volume. Nitrogen availability controls plant growth and thus rates of photosynthesis in many terrestrial ecosystems. It is a non-metallic element with no colour, smell or taste but many different uses. Its compounds are used in the manufacture of foods, drugs, fertilisers, dyes and explosives.

BOX 3.20 UNCERTAINTIES ABOUT THE NITROGEN CYCLE

The nitrogen cycle appears relatively simple, but there are still many uncertainties about details of the cycle. We do not know for certain, for example, how much nitrogen is fixed (made usable) in many biological systems. Neither is it clear what the ratio is between fixation from natural processes and emission of nitrogen from human activities. Many details of denitrification processes remain only partially understood.

the ground. Special denitrifying bacteria complete the cycle, converting nitrates to nitrite, nitrites to ammonia, and nitrates to gaseous nitrogen or nitrogen oxides. This effectively recycles all of the nitrogen compounds.

INDUCED CHANGES

The two most important ways in which human activities alter the natural nitrogen cycle are through the use in agriculture of nitrogen fertilisers and the release of nitrogen oxides in air pollution (Box 3.21).

Nitrogen fertilisers: shortage of available nitrogen can often be a limiting factor in ecosystems, because of its mobility across ecosystem boundaries. This is true particularly where decomposition and thus release of organic nitrogen is slow. When it happens it can limit biological productivity. Sustainable traditional farming practices have replenished the nitrogen in agricultural soils by crop rotations (growing legumes within the cycle) and by applying natural organic fertilisers such as nitrogen-rich animal wastes. Modern intensive farming requires the regular addition of artificial nitrogen fertiliser to soils to improve or maintain soil fertility.

The quantities involved are relatively very high. Natural biological processes fix around 54 million tonnes of nitrogen a year, and annual production of nitrogen fertiliser worldwide in the early 1990s was around 50 million tonnes. This additional nitrogen loading is not evenly spread around the world, and where it does occur it puts natural nitrogen cycles under stress and threatens to destabilise them. It often causes chronic pollution of groundwater, lakes and rivers (see p. 409). This in turn raises nitrate levels in estuaries and coastal waters, which affects primary productivity in the sea. Ocean systems and fisheries can in turn be affected.

BOX 3.21 NITROGEN OXIDES AND AIR POLLUTION

Nitrogen is oxidised in high-compression internal combustion engines and high-temperature industrial furnaces, and released in exhaust gases. Nitrogen gas is itself a pollutant, but its oxides are very reactive in the atmosphere, where they can interact with other gases and aerosols and create complex pollution problems. For example, nitrous oxide acts as a greenhouse gas and is thus part of the wider problem of global warming (see p. 259).

Estimates have been made of likely atmospheric nitrogen loads in 2020, assuming a global population of 8.5 billion people and a doubling in per capita energy consumption (much from the burning of fossil fuels) between 1980 and 2020. They predict that rates of deposition of nitrogen from the atmosphere will increase by a quarter, deposition over developing regions will at least double, and deposition over oceans in the northern hemisphere will rise by at least 50 per cent. This increased deposition apparently has the potential to fertilise both terrestrial and marine ecosystems, resulting in the sequestering (locking up into biogeochemical storage) of carbon. This would help to reduce the contribution of carbon to the greenhouse gas problem, although the nitrogen is itself part of the problem.

THE CARBON CYCLE

THE NATURAL CYCLE

The carbon cycle (Figure 3.19; Box 3.22) is a series of reactions in which carbon is continuously circulated and recycled in the environment, involving interactions between living organisms and their surrounding environment. The organic phase of the cycle has many similarities with that in the nitrogen cycle and reflects the dynamics of food webs within ecosystems (see p. 552).

Carbon is stored in the atmosphere as carbon dioxide (CO_2). Plants take in carbon dioxide gas from the air and can use it directly. They convert it by the process of photosynthesis into carbohydrates (carbon compounds), which can be consumed by animals and thus passed along food chains. Oxygen is released back to the atmosphere in photosynthesis, making plants natural air purifiers. Grazing animals eat the plants, and carnivores eat the grazers. The carbon compounds in these food sources are broken

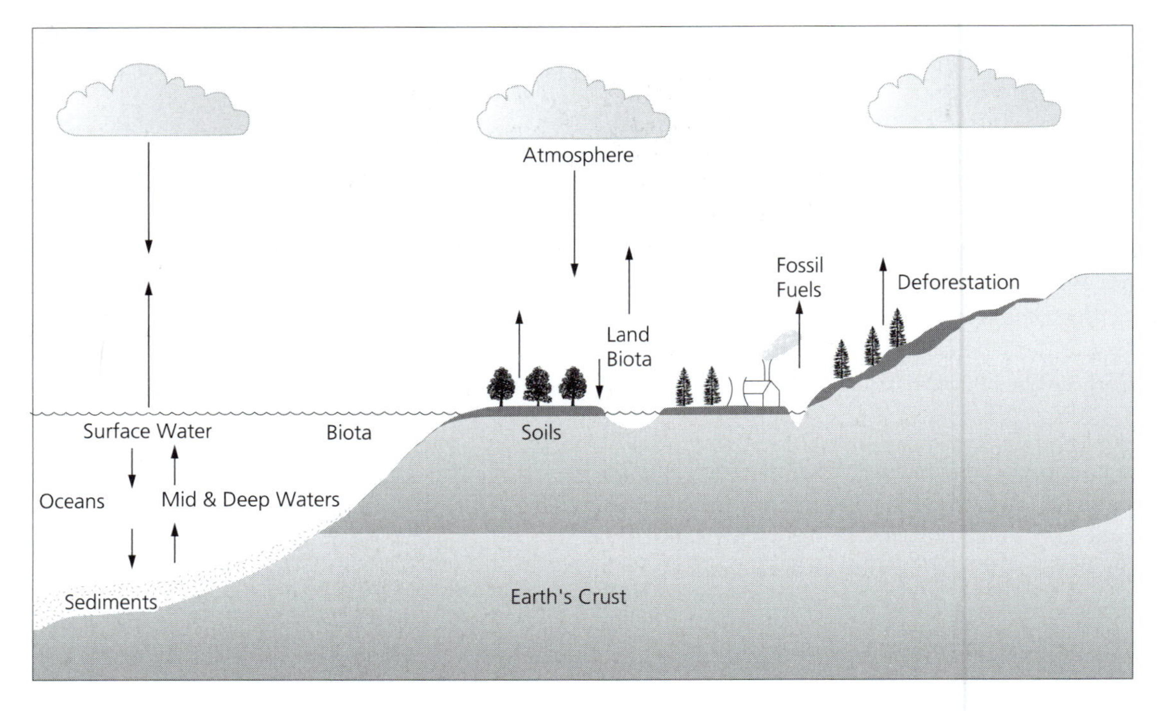

Figure 3.19 *The carbon cycle. Like the nitrogen cycle (Figure 3.18), the global carbon cycle involves the storage of carbon in a series of reservoirs (including soils, plants, the atmosphere and the oceans) and the transfer of carbon between reservoirs by biological and physical processes.*

Source: Natural Environment Research Council (1989) Our future world. NERC, London, p. 13.

BOX 3.22 CARBON

Carbon, a non-metallic element, is one of the most widely distributed elements on Earth. It occurs in combination with other elements in all plants and animals, and is basic to life. In the natural environment, carbon occurs in four basic forms:

- in the pure state in the crystalline forms of graphite and diamond;
- as calcium carbonate ($CaCO_3$) in carbonaceous rocks such as chalk and limestone;
- as carbon dioxide (CO_2) in the atmosphere;
- as hydrocarbons in fossil fuels (petroleum, coal and natural gas).

This is perhaps the most important biogeochemical cycle in terms of human impacts and environmental change, because carbon dioxide is a greenhouse gas that has major impacts on climate change (p. 255).

down in the animals' bodies during respiration. Energy is released, and carbon dioxide is produced as a result of respiration. Carbon dioxide is returned to the air when the animals breathe out and when organic waste and dead organisms decay. This decomposition involves the oxidation of the carbon, releasing it again as carbon dioxide through respiration.

Soils provide important reservoirs for organic carbon, storing twice as much carbon as is stored in the atmosphere and nearly three times as much as is stored in above-ground biomass (plants). Nearly a third of the total soil store of carbon is found in the tropics. Carbon dioxide is also released into the atmosphere when vegetation is burned (naturally after lightning strikes, deliberately in moorland firing or accidentally in forest and grassland fires).

Atmospheric CO_2 is soluble and can enter water directly. It dissolves to form carbonic acid (H_2CO_3). This can break down into hydrogen (H^+) and bicarbonate (HCO_3^-) ions, which then break down further into more hydrogen and carbonate (CO_3^-) ions. The carbonate ions combine with positively charged ions (such as calcium, Ca^{++}) to form salts. Some of these salts (for example, calcium carbonate, $CaCO_3$) are insoluble. The formation of

BOX 3.23 UNCERTAINTIES ABOUT THE CARBON CYCLE

The carbon cycle involves a number of complex reservoirs and flows (fluxes). As with nitrogen, major gaps remain in scientific understanding of some parts of the global cycle. There are still a number of uncertainties surrounding quantities of carbon stored and flowing through some parts of the cycle. For example, estimates of the amount of organic carbon in the world's soils range from 700,000 billion tonnes to 3,000,000 billion tonnes. While data on the flow of carbon along major river systems are becoming available, they refer mainly to dissolved organic loads, and few reliable measurements of particulate loads are yet available. Uncertainty also surrounds the potential effects of rising atmospheric concentrations of CO_2 on natural ecosystems.

The role of the oceans as a sink for CO_2 is not yet fully clear, although it is believed to be significant. Carbon sequestration by marine organisms over geological time has significantly affected the equilibrium of global biogeochemical cycles, but we do not yet have an adequate understanding of how this biological carbon pump is affected by the climatic impacts of rising atmospheric CO_2. Oceans provide a significant sink for carbon dioxide, taking up about a third of the emissions arising from burning fossil fuels and clearing tropical forests. Increases in atmospheric carbon dioxide account for most of the remaining emissions, but there still appears to be a 'missing sink', which may be located in the terrestrial biosphere.

calcium carbonate in the oceans in this way is exploited by animals and some single-cell plants. Calcium carbonate tends to accumulate in sediments in shallow water. At depths greater than about 4,000 m calcium carbonate dissolves because temperatures are lower and seawater tends to be saturated with carbon dioxide. If shallow-water carbonate sediments are converted into sedimentary rocks (see p. 196), the carbon is stored until released by chemical weathering (see p. 208).

Like all biogeochemical cycles, the carbon cycle adapts to changing climate. Models have been made of changes in terrestrial carbon storage over the past 18,000 years (linking together a biome model and a general circulation model). These show the importance of vegetation changes, ecosystem adjustment and the flooding of lowland areas caused by a sea-level rise of 100 m.

Parts of the carbon cycle also respond to short-term climate variations. Studies of carbon cycling in the equatorial Pacific, for example, show that biological productivity varies from year to year in response to fluctuations in the recycling of ocean nutrients. El Niño events (p. 482) appear to be important triggers.

INDUCED CHANGE

The natural carbon cycle can be disturbed in a variety of ways. Most human activities release carbon from storage (particularly long-term storage in the sedimentary phase) and return it to the rapidly flowing parts of the cycle. This effectively represents a short-circuit and can produce overloading by releasing more carbon than the natural cycle is able to handle by recycling. Burning of fossil fuels (Box 3.24) and clearance of vegetation are the two most serious forms of human impact.

Vegetation removal: the carbon cycle is also being disrupted by land-use changes, particularly the removal of natural vegetation. Natural sinks of atmospheric CO_2 are decreasing as a direct result of vegetation clearance and land-use change. Levels of carbon dioxide are building up in the atmosphere as a result, thus contributing to the greenhouse effect. Replacing natural grassland and forest

BOX 3.24 FOSSIL FUELS AND THE CARBON CYCLE

The most obvious and most important human impact on the carbon cycle is the burning of fossil fuels (coal, oil and gas). This quickly releases carbon into the atmosphere from storage in organic sediments. Carbon dioxide is a greenhouse gas, and we will explore the links between increasing atmospheric concentration of carbon dioxide and climatic change in Chapter 8. There is no doubt that the carbon cycle is in danger of being disrupted by the increased burning of fossil fuels, but the consequences of this are still not fully clear. Because of the complexity of ecosystems, and the large number of possible interactions within and between environmental systems, there is still much uncertainty in predicting how the biosphere is likely to respond to rising CO_2 concentrations.

with agriculture brings a sharp decline in soil carbon storage, and inevitably the local carbon cycle is altered when farm products are harvested and exported. Clearance of tropical rainforests (see p. 586) significantly disturbs flows and storage of carbon. Temperate forests are also important carbon stores, and when forests are harvested or cleared the carbon cycle is effectively short-circuited. Forest regrowth has the opposite effect, expanding the terrestrial carbon sink (see p. 585).

While greenhouse gases (including carbon dioxide) drive climate change, the system is complex because of feedback loops within the carbon cycle. There is concern, for example, that climate warming might seriously disturb the carbon cycle in the permafrost zone. Increased release of carbon from storage, vegetation and litter decay seems likely. This might promote further warming, but it might be largely offset by increased carbon uptake by plants as the warmer climate triggers increased productivity and forest migration further north.

Stabilising atmospheric CO_2 is technically feasible, and it is cheaper than adapting to climate change. Making drastic cuts in fossil fuel use is unavoidable but will not be enough. Energy changes must be complemented by appropriate land-use changes. There is a strong case for establishing and managing forests in many areas (boreal, temperate and tropical forests), and for developing agro-forestry systems (combining forestry and agriculture in an integrated system) to enhance the sequestration of carbon on land. This would be a long-term investment, over perhaps a 50-year period, but could be cost-effective. Forest planting on a large scale would bring other environmental benefits too, such as reducing soil erosion and creating new habitats for wildlife.

THE SULPHUR CYCLE

Sulphur is another important biogeochemical cycle (Figure 3.20), which plays a key role in creating acid rain.

NATURAL CYCLE

The sedimentary and atmospheric phases are both important in the sulphur cycle, unlike the nitrogen and carbon cycles, and they are of comparable size. One source of sulphur is volcanoes. Although major volcanic eruptions are relatively infrequent, and therefore release relatively

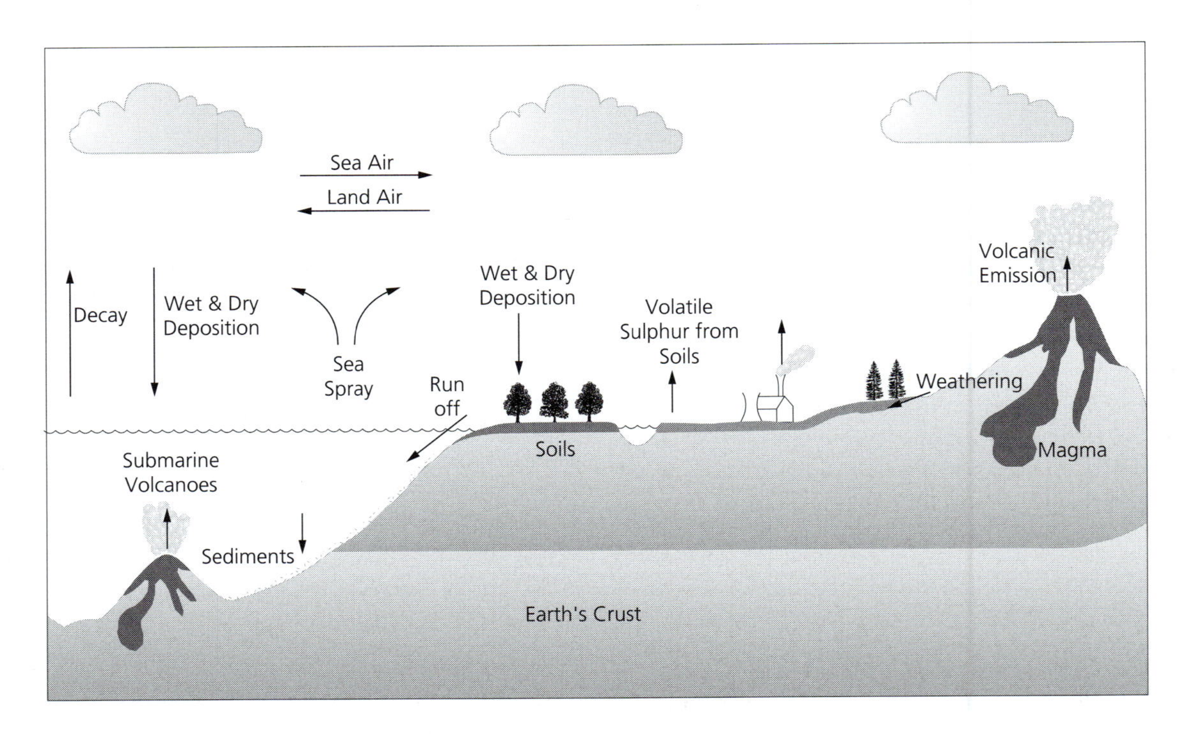

Figure 3.20 *The sulphur cycle. Like the nitrogen cycle (Figure 3.18), the global sulphur cycle involves the storage of sulphur in a series of reservoirs (including soils, plants, the atmosphere and the oceans) and the transfer of sulphur between reservoirs mainly by physical processes.*

Source: Natural Environment Research Council (1989) Our future world. NERC, London, p. 15.

small amounts of sulphur overall, when they do occur they produce sudden increases in atmospheric sulphur. Major eruptions – like the explosive eruption of Mount Pinatubo in the Philippines in 1991 (see Box 6.28) – can throw vast amounts of sulphur (as gas, dust and particles) high into the atmosphere, and these are then distributed widely by wind systems and atmospheric circulation. Sulphur compounds are also released by weathering of rocks. Roughly similar amounts of sulphur reach lakes and rivers from the atmosphere and from rock weathering.

The oceans also play an important role in the sulphur cycle. Several species of marine phytoplankton produce dimethyl sulphide (DMS), much of which breaks down in water. But some of the DMS enters the air, where it is oxidised to sulphur dioxide (SO_2) and sulphate aerosols. During the summer, 20–40 per cent of natural sulphur emissions along the west coast of the USA come from marine sources. The oxidisation of DMS to sulphate has implications for the control of acid rain in many coastal areas, because it can be as important a source of sulphur as air pollution. Sulphate aerosols can act as cloud condensation nuclei, further complicating the situation.

In terrestrial ecosystems, bacteria break down sulphate and release gases, mainly hydrogen sulphide (H_2S).

INDUCED CHANGES

Human activities have doubled the amount of sulphur in the atmosphere since the Industrial Revolution in the eighteenth and nineteenth centuries. By far the most significant human impact is the burning of fossil fuels (Box 3.26), particularly coal (which contains on average 1–5 per cent sulphur) and oil (containing 2–3 per cent on average).

BOX 3.25 UNCERTAINTIES IN THE SULPHUR CYCLE

As in the nitrogen (Box 3.20) and carbon cycles (Box 3.23), significant uncertainties surround some parts of the sulphur cycle. Indeed, the overall global sulphur budget is known in general terms only. Gaps in knowledge include the role of sulphur from natural sources (such as volcanic gases), and the production of hydrogen sulphide and dimethyl sulphide (DMS) by marine organisms. Much also remains to be discovered about sulphate in soils, and about sulphides and sulphates in mineral deposits.

BOX 3.26 FOSSIL FUELS AND THE SULPHUR CYCLE

When sulphur-bearing fossil fuels are burned, sulphur is released into the atmosphere as sulphur dioxide (SO_2) in exhaust gases. It can be blown great distances, across oceans and national boundaries. Hence it is a transfrontier pollution problem, exported by producer countries and unwillingly imported by other countries downwind (see p. 251). Where the sulphur is deposited (as dry fallout or dissolved in rain) it damages lakes, forests and buildings. Not all acid-related damage can be blamed directly on air pollution, because analysis of lake deposits has shown that in some places natural acidification has taken place for up to 10,000 years (related to vegetation changes, particularly the growth of coniferous forest and peat). But it is clear that steps must be taken to remove the SO_2 from exhaust gases if further environmental damage is to be avoided.

The amount of sulphur that enters the atmosphere from fossil fuel burning is similar to that produced from natural processes, but it is not so widely distributed. Combustion emissions are spatially concentrated in major industrial regions, so they often overload regional biogeochemical cycles.

IMPLICATIONS OF BIOGEOCHEMICAL CYCLES

Maintaining the integrity of the major biogeochemical cycles is clearly important, because they affect and are affected by so many aspects of modern life. All forms of pollution in all environmental media (air, water and soil) illustrate the overloading of natural cycles, and most if not all use of natural resources (renewable and non-renewable) has impacts on the stores and flows that are integral parts of the cycles.

CHAIN REACTIONS

Because the biogeochemical cycles are interrelated with the other global environmental systems (particularly the rock cycle, the water cycle and the energy system), any disturbances here will almost inevitably trigger changes in the other systems. Feedback loops within and between systems can amplify initial changes and quickly produce major problems with significant costs and risks. Hence, for example, the chain reaction of fossil fuel burning, emission of greenhouse gases, disturbance of the atmospheric energy system, rising temperatures, response of ecosystems and environmental systems to global warming, and further induced changes in the biogeochemical cycles. Breaking this cycle of self-reinforcing change is not easy, particularly in ways that are sustainable in character and global in scale, and that reduce hazard risks.

PERSISTENCE AND POLLUTION

Some materials break down quickly in the environment (Box 3.27), but others – the persistent ones – take much longer. The more stable a chemical is, the longer it will remain in an ecosystem before it breaks down and becomes relatively harmless.

Many pesticides and industrial wastes include very persistent ingredients. While they are not all particularly toxic, none of them breaks down in the biosphere. They can still be redistributed through natural environmental systems (particularly through the water cycle and through atmospheric circulation), and because they persist for long periods they can be found great distances away from where they started out. This explains why, for example, traces of DDT (dichlorodiphenyltrichloroethane), a highly persistent insecticide which has been banned in some countries, have been found in the fatty tissues of Antarctic penguins many thousands of kilometres away – washed there by ocean currents and dropped into the food chain by migrating species.

One of the best-documented examples of toxic persistent chemicals is dioxin (a family of more than 200 organic chemicals). Exposure to dioxin affects humans in various ways, including disfiguring skin complaints (chloracne), birth defects, miscarriages and cancer. Dioxins are regularly released into the environment in fairly small amounts when chlorinated materials (such as treated wood, plastics and some specially treated fuels) are burned. Recent European Union directives have significantly decreased dioxin emissions from large incinerators. More dramatic and concentrated releases occur from time to time, particularly in chemical accidents. The 1976 accident at a chemical plant in Seveso, Italy, spread dioxin over a wide area, and contamination only came to light slowly as animals died and people showed signs of dioxin exposure.

BOX 3.27 BIODEGRADABILITY AND WASTE MANAGEMENT

Nothing disappears when we throw it away, but at least we get rid of it from our own patch. Much waste management is driven by an 'out of sight, out of mind' mentality; once our waste materials have disappeared (from view) we don't concern ourselves too much about them.

Nature works differently. The biogeochemical cycles show how natural biological and environmental processes constantly move, store and recycle material. Recall (from p. 75) that materials are recycled within but energy flows through environmental systems.

In the natural world, many substances can be broken down by living organisms (mainly bacteria and fungi). Breakdown of these biodegradable materials (such as food and sewage) keeps them part of the biogeochemical system and thus makes them harmless. The process releases nutrients, which are then recycled by ecosystems. Up to a third of the household waste in many developed countries is organic matter, which is easily degraded into valuable compost.

Many substances (particularly plastics) that are described by manufacturers as 'biodegradable' are not, or they cause serious pollution problems as they break down. Non-biodegradable substances (such as glass, heavy metals and most types of plastic) cannot be broken down by living organisms. As a result, they cannot be recycled by natural processes within the biogeochemical system and have to be dealt with in other ways. They present substantial problems in waste management.

Traditional approaches to waste management tended to rely heavily on disposal of wastes, often by burial as landfill. The non-biodegradable wastes would simply be stored long-term. While this approach has often produced solid fill for land reclamation, concern is growing about likely shortages of suitable sites for the future, as land shortages and waste mountains continue to grow.

SUMMARY

This chapter provides a foundation for Chapters 4 to 19, because it introduces the basic ideas about environmental systems and how they operate. The River Nile water management case study sets the scene, allowing us to explore issues such as interconnectedness and interrelationships, cause and effect, direct and indirect impacts, human-induced changes in natural systems, and the impacts of natural processes on humans, in a real-life setting. Next, we explored some basic issues in systems analysis, including defining a system and its boundaries, the nature and importance of flows and cycles in environmental systems, and key aspects of system structure. A great deal of attention was focused on system dynamics – the ways in which systems reach a state of equilibrium and then respond to change by adjustment and feedback. Complications – such as thresholds, resilience and lags – help us to understand why system response is often difficult to predict. The third major theme in this chapter is biogeochemical cycles, through which chemical elements are transferred through the environment, between organic and inorganic phases. After describing the basic components of a typical biogeochemical cycle, we looked specifically at the cycles for nitrogen, carbon and sulphur, considering for each the importance of the cycle, the basis of the natural cycle and the ways in which human activities induce changes. While the world might seem a very complicated place, it does follow predictable patterns, and it is the challenge of science to unravel these mysteries and the responsibility of all people to disturb them as little as possible.

WEBSITE

Links to relevant websites, a comprehensive bibliography, tools for teaching and learning, and downloadable images relevant to this chapter can be found at the website specially designed to accompany this book at:
http://www.park-environment.com

FURTHER READING

Adams, W.M. (1990) *Green development: environment and sustainability in the Third World.* Routledge, London. An excellent introduction to the environment–development debate.

Beaumont, P. (1993) *Drylands: environmental management and development.* Routledge, London. A comprehensive and well-illustrated review of the problems of managing the world's dryland environments.

Bradbury, M., J. Boyle and A. Morse (2000) *Science for physical geography and environment.* Longman, Harlow. Outlines and explains the basic scientific principles necessary to an understanding of key processes occurring in the natural environment.

Briggs, D.J., P. Smithson, K. Addison and K. Atkinson (1997) *Fundamentals of the physical environment.* Routledge, London. Comprehensive introduction to the complex interactions of the natural environment.

Butcher, S.S., G.H. Orians, R.J. Charlson and G.V. Wolfe (1992) *Global biogeochemical cycles.* Academic Press, London. A useful introduction to the transformation and movement of chemical substances within global environmental systems.

Christopherson, R.W. (2000) *Geosystems.* Prentice Hall, Hemel Hempstead. Wide-ranging and clearly written review of the Earth's physical systems.

Collins, R.O. (1990) *The waters of the Nile: hydropolitics and the Jonglei Canal, 1900–1988.* Clarendon Press, Oxford. A detailed exploration of the political debate surrounding water management along the River Nile.

Degens, E.T., S. Kempe and J.E. Richey (1990) *Biogeochemistry of major world rivers.* John Wiley & Sons, London, for Scientific Committee on Problems of the Environment of ICSU/UNEP; SCOPE 42. Detailed review of recent scientific findings.

Faure, G. (1998) *Principles and applications of geochemistry.* Prentice Hall, Hemel Hempstead. Shows how chemical principles can be used in solving geological problems.

Goudie, A. and H. Viles (1997) *The Earth transformed: an introduction to human impacts on the environment.* Blackwell, Oxford. A wide-ranging, non-technical introduction to the ways in which the natural environment has been and is being affected by human activities.

Graedel, T.E. and P.J. Crutzen (1993) *Atmospheric change: an Earth system perspective.* W.H. Freeman, Oxford. A comprehensive introduction to the atmospheric system and its interactions with other environmental systems.

Holland, C. (1999) *The idea of time.* John Wiley & Sons, London. Explores, from a geological perspective, what we mean by 'time' and what we understand when we think about time.

Kemp, D. (1998) *The environment dictionary.* Routledge, London. Useful interdisciplinary, non-technical definitions of scientific terms and concepts.

Mannion, A.M. (1999) *Natural environmental change*. Routledge, London. Clear introduction to natural environmental change at a range of temporal and spatial scales, drawing on examples from around the world.

Marsh, W.M. and J.M. Grossa (1996) *Environmental geography*. John Wiley & Sons, London. A broad introduction to the Earth's physical processes and systems, and the interactions between them and human activities.

Murck, B.W., B.J. Skinner and S.C. Porter (1996) *Dangerous Earth*. John Wiley & Sons, London. A clear introduction to geological hazards, how they occur, what controls them and how they affect people and property.

Newson, M. (1997) *Land, water and development: sustainable management of river basin systems*. Routledge, London. Clearly written and abundantly illustrated overview of the theory and practice of catchment management planning in both developed and developing countries.

Schlesinger, W.H. (1991) *Biogeochemistry: an analysis of global change*. Academic Press, London. Examines the changes taking place in water, in the air and on land and relates them to the global cycles of water, carbon, nitrogen, phosphorus and sulphur.

Skinner, B.J. and S.C. Porter (1995) *The blue planet: an introduction to Earth systems science*. John Wiley & Sons, London. An introduction to the global environmental system and its main component parts – the atmosphere, hydrosphere, biosphere and lithosphere.

Slaymaker, O. and T. Spencer (1998) *Physical geography and global environmental change*. Longman, Harlow. Examines principles and concepts of environmental systems, and the implications of these to society.

Strahler, A.H. and A.N. Strahler (1994) *Introducing physical geography*. John Wiley & Sons, London. A clear introduction to the Earth and its major environmental systems.

Voet, D., J.G. Voet and C. Pratt (1999) *Fundamentals of biochemistry*. John Wiley & Sons, London. A wide-ranging introduction to the principles and practice of biochemistry.

Watts, S. (ed.) (1996) *Essential environmental science: methods and techniques*. Routledge, London. A manual of techniques, methods and basic tools used in the study of the environment.

Chapter 4

Spaceship Earth

LEARNING OBJECTIVES

When you have finished studying this chapter, you should be able to

- *Appreciate how the way we view and describe the universe changes through time*
- *Define the Earth and the universe in systems terms*
- *Describe and account for the structure and composition of the universe*
- *Explain the relevance of the Sun and the Moon to conditions on Earth*
- *Understand some major theories about the origin and evolution of the universe*
- *Outline the four main motions of the Earth (rotation, revolution, precession and galactic rotation), and say why they are important*
- *Appreciate the principles and implications of Lovelock's Gaia theory of the Earth as a living organism*

The Earth is the only planet known to support life, at least in forms that we can recognise. It is our home, and from it we get all the materials and resources we use, everything we eat and drink. The Earth receives energy from the Sun, and this drives its environmental systems and makes life possible.

In this chapter, we examine the origin, age and character of the Earth in the context of the wider universe of which it is part. This is not a needless detour into astronomy and cosmology, because without some basic understanding of how our planet came into being and how it relates to other things in the universe, making sense of important themes like energy flows (p. 75) is difficult. Neither will we have a realistic-scale framework to fit the Earth into, nor an appreciation of how vitally important it is to protect our home planet from further environmental damage and destruction. With existing technology, it remains the best place for us to live, despite what some science fiction films would have us believe!

IMAGES OF THE UNIVERSE

The way we view the world and how we interpret its role and place in the great order of things are important. But this importance goes beyond philosophical mind games, because the way we view the Earth inevitably shapes the way we treat it. Belief and behaviour are closely linked. If we see the Earth – and, indeed, the entire universe – as a random collection of bits and pieces, we are encouraged to treat it as a machine. If, on the other hand, we see in it some notion of design, possibly for some purpose, then we are more likely to adopt a sense of responsibility for it and treat it more carefully.

SPACESHIP EARTH

One way in which many people have viewed the Earth, in recent decades, is as a spaceship. The idea of a *Spaceship Earth* was suggested by US Ambassador Adlai Stevenson in a speech before the United Nations in Geneva on 9 July 1965. He referred to the Earth as 'a little spaceship in which we all travel together, dependent on its vulnerable supplies of air and soil'. This metaphor suggests some important things about the Earth (Box 4.1).

Since 1965, the usefulness of this Spaceship Earth metaphor has become even more obvious. Our image of the universe and the place of the Earth in it was radically

BOX 4.1 IMPLICATIONS OF THE SPACESHIP EARTH METAPHOR

- Spaceships are self-contained and as such must carry all the resources they require.
- Spaceships are small relative to the vastness of space through which they travel.
- Spaceships are complex systems (involving many computer control systems, air conditioning, wiring, and so on) with many components fitted together into a stable, working whole.
- No spaceship would be able to operate for long without a constant supply of energy to power all of its systems.
- In many spaceships, some of the energy comes from within (from batteries and other non-renewable fuel sources) and some from outside (they use solar energy by means of solar panels, for example).
- Conditions in spaceships are carefully controlled to suit human life.
- If conditions in a spaceship change too much or too fast they can become unsuitable for human survival.
- Spaceships don't last for ever!

altered when the first good-quality photographs of the planet came back from space. The Earth, which feels so large when we stand on it, looked like a shiny coloured marble floating through space. It is difficult not to be moved, at least a little, by such a perspective (Box 4.2).

EARTH AS A CLOSED SYSTEM

The Earth receives a non-stop supply of energy from the Sun. This drives the Earth's biological and environmental systems and warms its lower atmosphere enough to sustain life. If the Earth also received input of materials from space, it would make it an open system (see p. 74). But in material terms the Earth is self-contained, so it is really a closed system (Figure 4.1). There are some exchanges of material between Earth and space, but these are minor exceptions which do not call into question the basic definition.

Inputs: these include the dust particles and occasional meteorites (Box 4.3) that reach Earth from space. In the order of 10,000 tonnes a year of this material (mostly dust) might reach the outer edge of the Earth's atmosphere, but most of it is burned up by friction at a height of around 100 km as it falls towards the ground at supersonic speeds. This causes flashes of light in the sky, which we see as 'shooting stars' or meteors.

Outputs: these are confined largely to the artificial satellites sent into space for scientific purposes, communications, weather forecasting and military applications since Sputnik 1 was launched into orbit around the Earth by the former Soviet Union in 1957. More recent examples include the Landsat series of satellites used for monitoring the Earth's resources, launched since 1972. There is now much redundant space hardware floating about in space – an estimated 150,000 small objects in orbit (mostly less than 10 cm across; bits and pieces left behind by the Cold War and the growth of communications and navigation by satellite), and up to 10,000 larger objects (including spent satellites and rocket stages) – which strictly speaking represents a material output from Earth.

VIEWS OF THE UNIVERSE

People have had a curiosity about whatever lies beyond our planet since the dawning of human consciousness. Gazing

BOX 4.2 EARTH IN SPACE

On Christmas Eve 1968, the *Apollo 8* spacecraft was three days out into space, orbiting at a speed of around 9,000 km an hour 55,000 km above the surface of the Moon. This was the first time that humans had travelled around the Moon, and the three US astronauts aboard – Colonel Frank Bormann, Major Bill Anders and Captain James Lovell – were taking photographs of the lunar surface and looking for suitable landing sites.

Millions of television viewers around the world joined them live as they watched the Earth rising over the lunar surface. It was an amazing spectacle. The Earth, seen from a point over South America over 400,000 km away, looked like a small blue marble floating freely in space. Oceans appeared royal blue, and land showed up as brown. The South Pole and its ice cap could clearly be seen, and Major Anders claimed that he could see the mouth of the Mississippi River. Clouds and weather systems were clearly visible too.

The Earth looked like a tiny spaceship adrift in the vastness of the universe, dependent on its own life-support system.

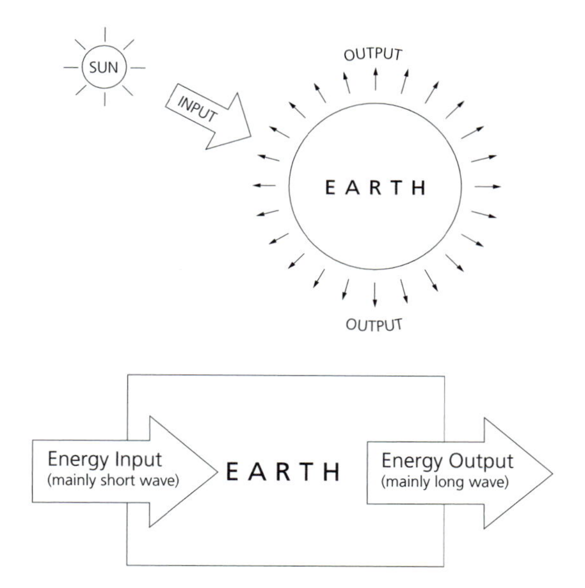

Figure 4.1 *Earth as a closed system. Like other closed systems, the Earth has inputs and outputs of energy but not materials. The energy is received from the Sun mainly as short-wave radiation and is output back to space mainly as long-wave radiation.*

Source: Figure 3.1 in White, I.D., D.N. Mottershead and S.J. Harrison (1984) Environmental systems. George Allen & Unwin, London.

at the Sun, the Moon, the planets, and the stars in the sky takes us beyond ourselves, makes us reflect on the big questions of life and causes us to wonder whether there is any underlying purpose to this apparent cosmic design. This fascination has encouraged the development of astronomy (the scientific study of the universe). It has caused us to invest phenomenal amounts of money since the 1960s in launching artificial satellites, landing men on the Moon, putting observatories into orbit around the Earth, sending probes off to distant planets, searching the night skies for new stars and planets, and looking for signs of extraterrestrial life.

Our view of the universe and our place in it inevitably changes through time, as reliable evidence grows and understanding expands.

PLANETS AND THE PASSAGE OF TIME

Since prehistory, people have been interested in the skies. Early farmers worked out the best times for sowing and harvesting crops by watching the rising and setting of the constellations (groups of stars). Early seafarers used the stars to plan voyages and to navigate around the then known world. Large, elaborate stone structures were erected as the original observatories, designed to predict eclipses

BOX 4.3 METEORITES

Meteorites are small (by planetary standards) masses of rock or metal, possibly fragments from asteroids (minor planets), that enter the Earth's atmosphere. If they are not burned up as they fall through the atmosphere, they crash into the ground at great speed, showering fragments over a wide area and creating large craters on impact. In February 1947, a meteorite 10 m across when it reached the Earth's atmosphere, weighing an estimated 1,000 tonnes, fell on the Siberian village of Novoprovska. Fragments were scattered over several kilometres, leaving more than a hundred craters up to 12 m deep and 25 m wide. Much larger meteorites have fallen in the geological past, and they might explain some of the mass extinctions of species that are recorded in the fossil record. According to one theory, for example, the dinosaurs were killed off 65 million years ago when an asteroid between 8 and 12 km in diameter smashed into the Earth. Inputs from space must have been arriving since the Earth was formed, because the surface of the planet is dotted with the remains of craters formed by the impact of asteroids, comets and meteorites. It is estimated that there may be as many as 1,000 asteroids large enough to pose a risk to Earth.

It is unlikely that the survival of humanity is threatened by the prospect of a major catastrophe associated with such extraterrestrial phenomena, but the risk is always there. The Torino scale (Table 4.1), first proposed in 1999, offers a useful way of assessing the risk of an asteroid impact. Recent research suggests that if a large comet or asteroid smashes into the Earth it would trigger an 'ultraviolet spring' (as the asteroid ionises nitrogen and oxygen in the atmosphere, which then destroys the stratospheric ozone that shields the Earth from damaging ultraviolet light; see p. 238). High ultraviolet levels would suppress plant photosynthesis (see p. 554) and damage DNA (see p. 532), and it would be followed by a devastatingly cold winter. It is estimated that an asteroid 10 km wide would destroy more than 85 per cent of the ozone in the atmosphere, causing peak ultraviolet levels to more than double.

Table 4.1 The Torino scale for assessing the risk of an asteroid impact on Earth

Torino value	*Risk assessment*
0	Likelihood of collision is zero
1	Chance of collision is extremely unlikely
2	Somewhat close, but not unusual, encounter. Collision very unlikely
3	Close encounter, with a 1 per cent or greater chance of collision causing localised destruction
4	Close encounter, with 1 per cent or greater chance of collision causing regional devastation
5	Close encounter, with a significant threat of a collision capable of regional destruction
6	Close encounter, with a significant threat of collision causing global catastrophe
7	Close encounter, with an extremely significant threat of collision causing global catastrophe
8	Collision capable of causing localised destruction. Such events occur somewhere between once per 1,000 years and once per 100,000 years
9	Collision capable of causing regional devastation. Such events occur between once per 1,000 years and once per 100,000 years
10	Collision capable of causing global climatic catastrophe. Such events occur once per 100,000 years, or less often

(particularly of the Sun) and record the passage of time. Stonehenge in southern England was probably built for this purpose. The Great Pyramid of Cheops (Khufu), built by the ancient Egyptians around 2550 BC, was both a tomb and an astronomical clock.

The basis of astronomy was developed when ancient cultures mapped the heavens and charted the courses of the planets across the sky. The Egyptians and Babylonians, in particular, were very skilful in monitoring both cyclical and more complex changes. They built up detailed records of periodic changes in sunrise and sunset during the year, the phases of the Moon, the seasonal appearance and disappearance of constellations, and the more complex motions of the planets. These planetary cycles allowed them to divide time into months and years and to devise reliable calendars. Time was captured by the movement of the Sun, Moon and stars.

BOX 4.4 THE UNIVERSE ACCORDING TO PYTHAGORAS

- The Earth is not flat but round (a sphere).
- The planets move along circular paths.
- The Earth is stationary and does not move.
- The Earth lies at the centre of the universe.
- The planets are unchanging and eternal.
- The Earth is subject to change and decay.

EXPLAINING MOVEMENT OF THE PLANETS

The ancient Egyptians and Babylonians recorded the movements of the planets, but the Greeks tried to explain them. The roots of modern astronomy were laid down by the Greek scientist Pythagoras (*c.* 582–500 BC) (Box 4.4). Seven centuries later, the Greek astronomer Ptolemy (*c.* AD 90–168) echoed the interpretation of Pythagoras with his model of the universe with the stationary Earth at the centre. As he saw it, the Sun, Moon and five known planets (Mercury, Venus, Mars, Jupiter and Saturn) all move around the Earth in concentric rings.

The so-called Ptolemaic system remained unchallenged for 1,400 years, until the Polish priest Nicolaus Copernicus proposed a revolutionary new model in 1543 (Box 4.5) (Figure 4.2). Johannes Kepler revised the Copernican model in 1609, pointing out that the planets move round the Sun in ellipses (flattened circles) not circles.

BOX 4.5 THE UNIVERSE ACCORDING TO COPERNICUS

- The Sun, not the Earth, lies at the centre of the universe.
- The Moon travels around the Earth.
- The Earth and its planets move around the Sun in concentric circular orbits.
- The Earth moves in a number of directions at the same time: it rotates around its axis once a day; it revolves around the Sun once a year; and because it rotates it has a wobbly motion (precession) like a spinning top.

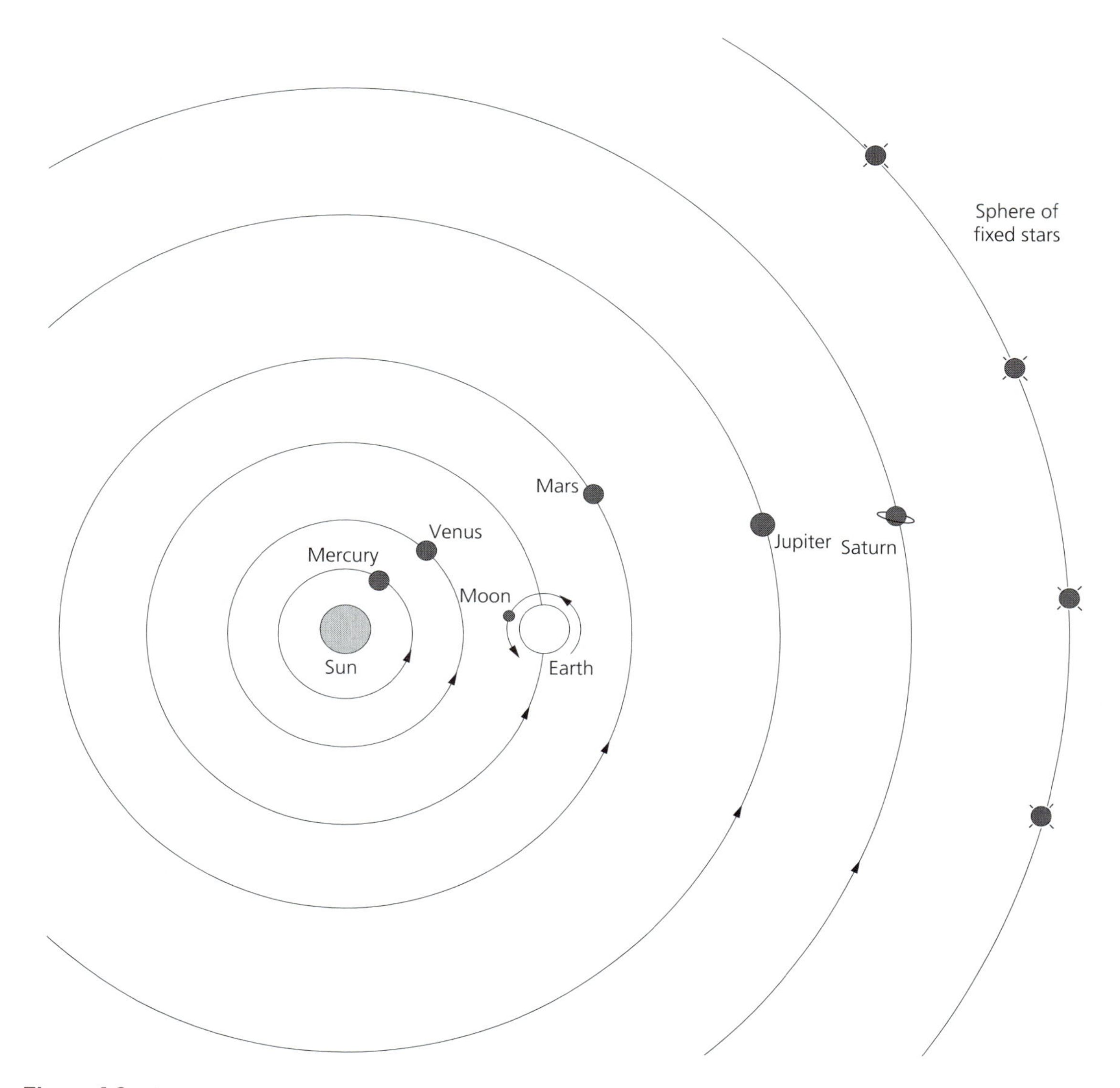

Figure 4.2 *The Sun-centred universe of Copernicus. Copernicus established that the Sun lies at the centre of the universe, the Moon travels around the Earth, and the Earth and the other planets move about the Sun.*
Source: Figure 3.7 in Merken, M. (1989) Physical science: with modern applications. Saunders College Publishing, Philadelphia.

Until about 1610, all that was known about the universe was based on watching the movements of planets across the sky with the naked eye. From the early seventeenth century, with the development of telescopes, much closer observation and clearer images were possible. Galileo studied the Sun, Moon, planets and stars through a telescope, and what he found confirmed Copernican theory. He took things further. The rough surface of the Moon was seen to resemble the Earth, with valleys and mountains. Planets were seen to be round in shape. Many more stars could be seen through the telescope than with the naked eye.

The modern development of radio astronomy has expanded our horizons even further. It is now possible to observe the universe beyond the visible region of the electromagnetic spectrum (see p. 233), effectively 'seeing' new parts of the cosmos. The more detail we observe the universe in, the more it begins to reveal its secrets. Modern interpretations of the age, origins and evolution of the universe (see p. 109) reflect this fact.

STRUCTURE OF THE UNIVERSE

THE UNIVERSE: STRUCTURE AND COMPOSITION

SPACE

Most of the universe (cosmos) consists of emptiness, so space is well named! Or so it seemed until quite recently. Modern astronomers suggest that more than 99 per cent of the universe is made of dark or unseen matter, which might include as yet unknown atomic particles, vast numbers of extremely faint stars and enormous black holes.

STARS

The black emptiness of space is punctuated by tiny lights, the 3,000 or so stars we can see in the sky on a clear, moonless night. These stars are not evenly distributed, and the brightest ones are grouped into eighty-eight constellations named after mythical characters (such as Taurus the bull, Aquarius the water carrier and Aries the ram). A group of twelve constellations (called the zodiac) lies in the same plane as the Earth's orbit around the Sun (Table 4.2). Astrologers believe that these constellations influence human affairs, but the belief has no basis in science.

GALAXIES

On a clear, moonless night, a faint, milky band of light can be seen crossing the sky through certain constellations. Ancient astronomers gave it the name 'Milky Way' (or galaxy, after *gala*, the Greek word for milk). All of the stars visible to the naked eye belong to the Milky Way system (Box 4.6), but it is much bigger than that. It contains more than 100,000 million stars, most of them too far away to be seen from Earth.

Table 4.2 The signs of the zodiac, with meanings of the named constellations

1. Aries (the ram)
2. Taurus (the bull)
3. Gemini (the twins)
4. Cancer (the crab)
5. Leo (the lion)
6. Virgo (the virgin)
7. Libra (the balance)
8. Scorpio (the scorpion)
9. Sagittarius (the archer)
10. Capricorn (the sea goat)
11. Aquarius (the water carrier)
12. Pisces (the fish)

BOX 4.6 THE MILKY WAY GALAXY

The Milky Way is spiral-shaped and large by Earth standards – an estimated 100,000 light years in diameter at its widest, about 10,000 light years at its narrowest. Viewed from outside, it would probably appear as a flat disc with spiral arms rotating around a small, brilliant central core. The core is believed to be a black hole, about a million times the mass of our Sun. Gas and dust are pulled into it by its own gravity. The material accelerates nearly to the speed of light, is strongly heated by friction, and matter shines brightly before it disappears into the black hole (where the gravitational pull is so strong that not even light can escape from it). The spiral arms consist of concentrations of stars, dust and gas. The Sun, about 25,000 light years from the centre, is on one such arm.

Distances in space are measured in light years: a light year is the distance travelled by a beam of light in a vacuum in one year, roughly 9,460,000 million km.

Until the 1920s, it was thought that the Milky Way was the entire universe. We now know that it is only one of many galaxies (perhaps hundreds of millions), all of which formed at the very beginning of the universe, an estimated 15,000 million years ago. Indeed, galaxies appear to be the basic building blocks of the universe.

Ours is one of the larger of a few dozen galaxies known as the Local Group. The Andromeda galaxy (2.5 million light years away) is another large galaxy in our group. Even further away, at about 50 million light years, is the larger group of about 2,500 galaxies called the Virgo cluster. More distant still, at the edge of the universe, are the quasars. According to some scientists, these are the burning remnants of galaxies, which can shine thousands of times more brightly than normal galaxies. Born in the early years of the universe, their light is only now reaching us.

There is evidence of large-scale structures in the universe, because most galaxies appear to be clustered into sheets, with massive voids (expanses of apparent nothingness) between them. No light appears to shine from and no matter appears to exist within these voids. To add to the complexity, our part of the universe is seemingly being

Table 4.3 The main components of the solar system

Component	Description
Sun	A star that radiates energy (particularly heat and light) derived from thermonuclear reactions in its interior
Planet	A large body made of rock, metal or gas, which orbits a star
Moon	A natural satellite of a planet, that is, a large body that orbits a planet
Asteroid	A small planet-like body that orbits the Sun. There are probably more than 100,000 asteroids in the solar system, which might be remnants of a former planet or planets that disintegrated, or may be part of the original matter of the solar system that never became a planet
Comet	A small, icy body that orbits the Sun. It consists of a central nucleus a few kilometres across. As the comet approaches the Sun the nucleus heats up, releasing gas and dust that stream out in a series of tails, which can be millions of kilometres long
Interplanetary matter	Gas and dust spread thinly through the solar system. Low-energy particles flow outwards from the Sun as the solar wind. Fine dust scatters sunlight to cause the zodiacal light. Dust shed by comets enters the Earth's atmosphere to cause meteor showers

pulled towards a mysterious gravitational source whose location remains unknown. The Milky Way, our entire Local Group and thousands of other galaxies nearby are being pulled towards a huge concentration of galactic matter called the Great Attractor. The Great Attractor is itself moving, in an unknown direction and pulled by an unknown force.

So although the universe appears to be empty, it contains many planets, stars and galaxies and appears to be structured. This neatly illustrates the idea of nested hierarchy systems (p. 70) and indicates that the Earth is part of something much bigger. The Spaceship Earth metaphor looks highly appropriate! Earth systems are driven mainly by energy from the Sun, but it is clear that this tightly knit relationship is but a speck in the vastness of the universe. Scales of time and distance look very different from a cosmic perspective. The fact that all we see might be a mere 1 per cent of the real universe simply serves to perpetuate the mystery of it all.

THE SOLAR SYSTEM

The solar system is part of the Milky Way galaxy. If the Earth is our home, the solar system is our neighbourhood, the Milky Way galaxy our city and the universe our world. The edge of the solar system is not clearly defined; it is marked only by the limit of the Sun's gravitational influence. This extends about 1.5 light years, almost halfway to the nearest star, Alpha Centauri, which is 4.3 light years away.

Table 4.4 The size of the Sun and its planets

Planet	Equatorial diameter (km)
Sun	1,392,900
Jupiter	142,800
Saturn	120,000
Uranus	51,800
Neptune	49,500
Earth	12,756
Venus	12,104
Mars	6,787
Pluto	6,000
Mercury	4,878

COMPONENTS OF THE SOLAR SYSTEM

The solar system is made up of the Sun and all the bodies that orbit it. These are the nine planets (Mercury, Venus, Earth, Mars, Jupiter, Saturn, Uranus, Neptune and Pluto), their moons, asteroids and comets, and interplanetary matter (Table 4.3). The Sun is by far the largest body in the solar system (Table 4.4), nearly ten times bigger than Jupiter, the second largest. Earth is relatively small, even within the solar system.

Geological studies of a number of the planets have shown great similarities between them. The Moon and the Earth-like planets (Venus and Mars) appear to be of similar composition (made of silicates and metals that are relatively rare throughout the rest of the universe) and structure (they appear to have undergone separation of material, on cooling, into a core, a mantle and a crust). These types of similarity suggest similar origins and have helped to create a better understanding of the early evolution of the Earth.

Most space travel between the 1960s and the late 1990s was confined to the Moon and areas of space close to the Earth. However, space scientists were excited when NASA's Pathfinder mission landed safely on Mars – the so-called Red Planet – in 1997 and sent back detailed still photographic images of the Martian landscape. Pathfinder carried a six-wheeled rover (called Sojourner) that could move across the surface of Mars and probe rocks and soils with special equipment to detect the presence of different geochemical elements (other than oxygen) in the material.

PLANETS IN MOTION

The nine planets orbit the Sun, and the planetary orbits are elliptical, not circular. This means that they have a minimum (perihelion) and a maximum (aphelion) distance from the Sun (Table 4.5). Each planet receives most solar energy at perihelion (when it is closest to the energy source, the Sun) and least at aphelion (when it is furthest away).

Absolute distance from the Sun is also an important control on solar energy receipts. Mercury orbits closest to the Sun and is thus hottest (surface temperatures range from 350 to −160 °C). Pluto orbits furthest from the Sun and is much colder as a result (the surface temperature is around −240 °C, and the planet is permanently frozen). Earth remains relatively close to the Sun throughout its orbit and thus remains warm enough for life as we know it.

The cyclical movement of planets through the solar system is part of the underlying rhythm of time, and the regularity of the movements attracted the attention of the early astronomers and inspired the construction of the original observatories. As the planets move through their orbits around the Sun, their position relative to each other changes. These changes are cyclical too, but the cycles are not of similar lengths or synchronised with each other. This complex sequence of planetary changes brings predictable changes through time in the alignment of planets relative to one another.

Distances between the Earth and other planets vary through time. Venus is usually closest (Table 4.6), closely followed by Mars and Mercury. Distant planets, such as Pluto and Neptune, are significantly further away. The main relevance of these interplanetary distances to people on Earth is the prospect of space travel and exploration. Perhaps inevitably, the closest planets have been explored first and in more detail. Space probes aimed at Venus during the 1960s and 1970s, for example, sent back much information about the planet's surface and atmosphere. Close-up pictures of Mars were first taken by *Mariner 4* in 1965; Mercury was first photographed at close quarters by *Mariner 10* in 1974.

THE SUN

The Sun is the star at the centre of the solar system, which the planets (including the Earth) orbit. Earth is on average about 150 million km away from the Sun, but this varies between 147 million and 152 million (see Table 4.4) during the elliptical orbit (which goes from west to east relative to the Earth's surface). In galactic terms our Sun is quite insignificant, one of perhaps 100,000 million stars. But it is the most important single member of the solar system and dominates the system in two senses – size and energy output. With a diameter of nearly 1.4 million km, the Sun is more than 100 times bigger than the Earth and has a

Table 4.5 Distances between the planets and the Sun

Planet	Distance in millions of km		
	mean	minimum	maximum
Mercury	58	46	70
Venus	109	108	109
Earth	150	147	152
Mars	229	206	249
Jupiter	779	740	816
Saturn	1,427	1,347	1,506
Uranus	2,871	2,739	3,003
Neptune	4,496	4,451	4,541
Pluto	5,913	4,445	7,382

Table 4.6 Mean distance of the other planets from Earth

Planet	Distance (millions of km)
Venus	40.2
Mars	56.3
Mercury	80.5
Jupiter	591.0
Saturn	1,197.0
Uranus	2,585.0
Pluto	4,297.0
Neptune	4,308.0

volume equivalent to a million Earths. Almost 99 per cent of the mass of the solar system is contained within this one giant star.

The Sun is composed mostly of hydrogen (about 70 per cent) and helium (30 per cent); other elements make up less than 1 per cent. It generates energy by nuclear fusion reactions, which convert hydrogen into helium in its interior. These reactions produce immense heat – temperatures are about 5,530 °C at the surface of the Sun and about 15,000,000 °C at the centre. The reactions also produce vast amounts of energy, and cause sunspots and solar flares that are huge relative to the size of the Earth. Some scientists even propose that the global climate change we are experiencing today is caused more by variations in solar activity than by greenhouse gases. Output from one computer model suggests that the solar influence on global warming between 1890 and 1997 was between 20 and 40 per cent.

Energy: export of energy from the Sun affects conditions throughout the entire solar system, particularly the surface temperatures of the planets (including Earth). This large cosmic battery is a seemingly endless source of energy. Its nuclear reactions have been releasing energy into the solar system over the past 4,700 million years, and the Sun has enough hydrogen to burn continuously for another 5,000 million years.

Sunspots: like the Earth, the Sun has an atmosphere, but its composition is very different. The surface of the lowest part of the solar atmosphere – the photosphere – is marked by sunspots. These are irregular dark spots marking regions where the gases (though hotter than the surface of many stars) are cooler than the surrounding areas. Sunspots vary in size from about 1,500 km to groups up to 200,000 km in diameter, and large ones are sometimes visible to the naked eye. They constantly change size and appearance (Box 4.7). Most spots appear and disappear within a day or so, but some last longer than a month. Quite why sunspots vary so much is not known, but they might be caused by strong magnetic fields that block the outward flow of heat to the Sun's surface.

Solar flares: these are violent eruptions of white-hot gases in the outer part of the Sun's atmosphere (the chromosphere), caused by the disintegration of massive arcs of gas. The explosive process shoots huge streams of burning gas far out into space (further than the distance between the Earth and the Moon).

THE MOON

The Moon is a natural satellite of the Earth. A satellite is a body that orbits a planet or star. So the Moon orbits the Earth, which orbits the Sun. The rhythm of this celestial (heavenly) machine strongly influences conditions on Earth and makes life on Earth possible. Unlike the Sun and the Earth, the Moon has no atmosphere or liquid water.

The Moon is much smaller than the Earth, barely a quarter of the size, with a diameter of 3,476 km. Its surface gravity is only about one-sixth that of the Earth. The Moon

BOX 4.7 SUNSPOT CYCLES

While sunspots vary a great deal in the short term, there is evidence of longer-term solar rhythms linked to environmental changes on Earth. Sunspots follow a semi-regular 11-year cycle, for example, during which they increase in number towards a maximum (more than 100) then fall off to a minimum (none). While 11 is the long-term average, cycles can last between 9 and 13 years. Sunspot peaks were recorded in 1958, 1969, 1980 and 1990. Some long-term weather cycles follow 11-year cycles synchronised with the sunspot cycle. For example, although rainfall varies a great deal within and between years, long-term annual rainfall records often display an underlying 11-year cycle, with years of maximum rainfall corresponding to times of greatest sunspot activity.

There is also evidence of even longer-term patterns of sunspot activity, which appear to be associated with long-term climate change on Earth (see p. 340). Short cycles (less than 11 years) appear to be associated with warm climate and long cycles with colder climate. During the twentieth century, the cycle seems to have shortened from 11.7 to 9.7 years on average, while global land temperatures have risen by 0.6 °C.

Some scientists argue that recent climate change may be associated more with sunspot cycles than with the greenhouse effect. Temperatures fell between 1940 and 1970 while carbon dioxide levels were rising rapidly, but the sunspot cycle slowed from 10.2 to 10.7 years. Since 1970, sunspot cycles have been shorter, and temperatures have been increasing.

orbits in a west–east direction, about 385,000 km from the Earth, and each orbit takes 27.32 days (a sidereal month) to complete. It also spins on its axis, with one side permanently turned towards the Earth. On Earth, the tides (cyclic rise and fall of sea level) are caused by the gravitational pull of the Moon and Sun, so the movements of the Moon affect the oceans that cover two-thirds of the Earth's surface (see p. 125).

The Moon might also be implicated in global climate change, because recent research has shown that the Moon affects temperatures on Earth through the influence of the alignment of Earth, Moon and Sun on the tides (see p. 483). According to some scientists, not only did lunar changes trigger a Little Ice Age 500 years ago (see p. 345), but it threatens to warm the Earth for several thousand years ahead.

We know much more about the Moon than about other planets – except Earth – in the solar system. Photographs and measurements taken by US and Soviet Moon probes (crewless spacecraft) since the 1960s reveal a great deal about the nature and composition of the lunar surface. But the big breakthrough came when the US Apollo space project to land a person on the Moon came to fruition on 20 July 1969. Neil Armstrong was the first human to set foot on another world (Box 4.8). His historic 'one small step for man, one giant leap for mankind' was the first of what was to become a series of Moon landings between 1969 and 1972, which set up experiments, collected many measurements and photographs, and brought back geological samples.

In universal terms the Moon is on our doorstep, and some of the ingenuity that put man on the Moon has been directed to the prospects of mining the Moon. Exploring lunar minerals and mining methods by sophisticated machines driven by solar power is now a practical if rather expensive possibility. The brightest prospects apparently centre on mining oxygen, which is abundant in lunar rocks.

Ice and water were discovered on the Moon during the late 1990s, particularly on the basis of information collected by the Lunar Prospector in 1998. The discovery has excited aspiring lunar colonists but puzzled planetary scientists, who are curious to know its origin. One theory is that the ice is derived from comets and asteroids that have bombarded the lunar surface over millions of years.

For the time being, however, the Earth remains both our home and our sole resource base, despite what science fiction writers and film makers would have us believe.

THE EARTH

The Earth, as we have seen, is the fifth largest of the nine planets in the solar system (see Table 4.4). It rotates around its axis once every 24 hours and orbits the Sun 150 million km away once a year, and the Moon orbits it 385,000 km away once every 27 days 7 hours and 43 minutes. These rhythmic motions create night and day and the seasons of the year.

By planetary standards the Earth is relatively small, with a diameter around the equator of 12,756 km (Table 4.4). Like other planets, the Earth is not a perfect sphere. Because it rotates around its axis it is flattened slightly at the top and bottom (its diameter measured through the poles is only 12,713 km). This shape, a slightly flattened sphere, is described as an oblate spheroid.

BOX 4.8 MAN ON THE MOON

On 20 July 1969, people around the world watched their television sets with enthusiasm. They were watching history being made, live, from a distant world. They heard Edwin (Buzz) Aldrin announce 'Tranquillity base; the Eagle has landed' and watched in awe and amazement as Neil Armstrong started to climb the few steps down from the lunar lander *Eagle* on to the surface of the Moon, in the Sea of Tranquillity. As Armstrong took that momentous first step on the Moon, he proudly proclaimed that it was 'one small step for man, one giant leap for mankind'.

His footprint could be seen clearly on the lunar surface, which he described as 'a very fine-grained powder', like charcoal. 'It is very pretty out here,' he noted as he walked about. 'There are quite a few rocks and boulders in the near area which are going to have some interesting colours in them.'

After twenty minutes, Aldrin joined Armstrong on the Moon, while Collins stayed in orbit around the Moon in the mother ship *Columbus*. Together they unveiled a plaque that has remained on the surface of the Moon. It was signed by US President Richard Nixon and has the message 'Here men from the planet Earth first set foot upon the Moon, July 1969 AD. We came in peace for all mankind.'

Although we cannot feel it, the Earth is moving through space at great speed. It orbits the Sun at an average speed of about 107,000 km h^{-1} and rotates on its axis at a speed of 1,673 km h^{-1}. Compared with fast-moving objects with which we are familiar (such as Concorde, which travels at 2,320 km h^{-1} at full speed), the Earth is a fast machine!

While space exploration has greatly increased our understanding of distant parts of the universe, it has also thrown light on things much closer to home, such as the geological properties of the solid Earth. In Chapter 5 we explore the internal composition and structure of the Earth, and in Chapter 6 the ways in which these shape large-scale features on the Earth's surface. Space research has also revealed a great deal about possible sources of energy and material resources in parts of the solar system that are most accessible from the Earth. Our local patch of the solar system appears to offer exciting possibilities, based on using space materials to produce propellants, structural materials, refractories, life-support fluids and other materials on site in space.

The Earth is surrounded by an atmosphere (Chapter 8), which interacts with conditions on the planet's surface to create the environmental systems (see p. 4) we are concerned with in this book. But the atmosphere is even more important than that, because it supports life. Ours is the only planet on which life is known to exist, thanks to the mediating influence of the atmosphere (Box 4.9), which filters out the most harmful rays from the Sun, creates suitable temperatures and plays important roles in the biogeochemical cycles (see p. 86) and the water cycle (see p. 352).

The Earth is unique, not only in having life on it but also in terms of its relationship with the Sun. Its size, its distance from the Sun, its orbit around the Sun, the speed at which it rotates (around its axis) and revolves (around the Sun) all have a strong influence on conditions on Earth, which in turn make life possible. The Gaia hypothesis (see p. 119) goes even further than this, arguing that the entire planet is a giant organic system in which living organisms and inanimate (non-living) parts of the environment have evolved and remain in harmony with each other. Change one, and the other must inevitably change. This poses immense challenges to us as caretakers of our planetary home.

The Earth is unique in other ways too. Compared with other bodies floating about in the solar system the Earth has very few impact craters (from things like meteorites). Earth's crustal features and dynamics, involving large-scale plate tectonic movements, are also unusual. Climate on Earth is different from that on other planets with atmospheres, particularly the abundance of water and the character of the Earth's ice.

BOX 4.9 ATMOSPHERES ON OTHER PLANETS

Conditions on other planets are not suitable for life as we know it. Mercury and Venus orbit closest to the Sun, have no atmospheres (because of the intense heat) and receive large amounts of lethal ultraviolet (short-wave) radiation from the Sun. Earth comes next, followed by Mars, which has a poorly developed atmosphere and large temperature variations between day and night. Moving even further away from the Sun, the planets (Jupiter, Saturn, Uranus, Neptune and Pluto) are simply too cold for life.

DYNAMICS OF THE UNIVERSE

ORIGIN AND EVOLUTION OF THE UNIVERSE

COSMOLOGY

Cosmology is the study of the origin and nature of the universe. Big questions like this have puzzled people across the ages, as we saw earlier (p. 101). Modern science has greatly expanded our knowledge of the structure and composition of the universe, although the basic framework put forward by Copernicus more than four centuries ago has stood the test of time.

There are now several competing theories about how the universe may have been formed, the most popular of which is the now notorious Big Bang theory. Even the most recent evidence (see Box 4.12) is compatible with the notion of a gigantic explosion and subsequent expansion.

BIG BANG THEORY

The birth of modern cosmology is generally associated with the discovery by Edwin Hubble (at Mount Wilson Observatory in California in 1929) that the universe appears to be expanding, rather like a balloon. Hubble interpreted this to mean that the universe probably began in an explosion (the Big Bang) and has been expanding outwards ever since. The evidence suggested that the universe must once have

BOX 4.10 COSMIC BACKGROUND RADIATION

Edwin Hubble's proposed Big Bang theory required an initial explosion to trigger the process, but at that stage it was simply a logical proposition. Direct evidence in support of the theory first appeared in 1965, when a faint signal was detected from the universe at short radio wavelengths. This so-called cosmic background radiation was interpreted as the heat from the Big Bang explosion, which has been cooled by the subsequent expansion of the universe.

The evidence was extremely important, but it could not provide an adequate explanation of the fact that the universe appears to be extremely lumpy. Galaxies and other objects in the universe appear to be grouped or clustered, with vast areas of nothingness between the clusters. Data relayed back to Earth from the high-precision *COBE* (*Cosmic Background Explorer*) satellite in 1992, showing minute temperature variations across the expanse of the universe, have helped to provide a better picture of this universal lumpiness.

been compressed into a super-dense mass that exploded for some reason (Box 4.10).

Like all theories, the Big Bang theory has its critics, but it seems to be the best explanation of the facts available to date. An alternative way of interpreting what Hubble originally discovered – the steady-state theory – proposes that the universe has no origin but is expanding because new matter is continuously being created.

BEFORE SPACE AND TIME

Understanding of the Big Bang is based on theory, mathematics and experiments using high-energy particle accelerators. Little is known of the earliest instant of the universe, and inevitably there is much speculation (or, rather, inference) surrounding key parts of the model because the event itself cannot be revisited.

At the time of the explosion, the entire universe must have been squeezed into an unbelievably hot, super-dense state. Neither space nor time existed. Seemingly nothing existed other than this highly concentrated speck of material (smaller than a speck of dust). Somehow the entire universe – all matter, energy, space and time – exploded from this nothingness.

There is no trace of what conditions were like before the Big Bang, at least according to current understanding in physics. Neither do we have any clues about where that original material came from, or what caused the huge explosion. The origins of the universe are literally shrouded in mystery. Little wonder, therefore, that big questions of ultimate meaning and purpose have engaged the curiosity and engaged the minds from earliest times.

THE EXPLOSION

The cause of the giant Big Bang explosion may remain unknown, but its consequences have been studied in some detail. The initial speck of material – Stephen Hawking estimates it to have been about the size of a pea – would have had an inconceivably high temperature (estimated in the region of a million trillion trillion degrees). The explosion, which must have been of immense power, blasted all of the material in the universe outwards in every direction. As the speck expanded, it cooled.

This whole process took place within a tiny fraction of a second. By the time the universe was 1 second old the basic building blocks of atoms (protons, neutrons and electrons) had come into being. Within the same instant photons were created. These units of electromagnetic radiation with no mass are the basis of light. But the photons were unable to move in this dense miniature universe, in which matter and energy were still incredibly heavily concentrated.

So, well within its first second and seemingly from nothing, the universe came into being, bringing with it space and time, matter and energy. But there was no light in this dark cosmos. Only when the universe was 300,000 years old was it possible for light to break away from matter and begin to travel freely through the expanding universe.

It is impossible for the human mind to imagine the enormity of this Big Bang explosion, or to appreciate that it happened in less than an instant. Time, as we know and experience it, has little relevance to the origin of the universe. Well within the first second of time the universe expanded more than it has done in the 15,000 million years since then. Within the first 3 minutes, when the explosive material had already cooled a great deal, the nuclei of what were to become atoms (the building blocks of matter) were formed as protons and neutrons (elementary particles with positive and neutral charges, respectively) bound together. Slowly, on this exceedingly fast cosmic clock, the visible universe began to take shape. Gravity in the universe was generated by radiant energy, and matter began to join together and form structures within the first 3,000 years.

THE UNIVERSE TAKES SHAPE

It took a relatively long time for large-scale structures to appear in the evolving universe. The explosion threw out a vast cloud of gas and dust. This matter continued to join together in areas of concentration, and over vast periods of time it was condensed by gravity (Box 4.11). The most recent evidence suggests that the universe is, in fact, more flat than truly three-dimensional (Box 4.12).

The process of evolution continues, and some galaxies, stars and planets are still forming from the original gas and dust thrown out in the Big Bang explosion (Figure 4.3). The universe is in a perpetual state of change. New stars are constantly being formed and existing ones are constantly disappearing (by exploding, or being sucked by gravity into mysterious black holes).

BOX 4.12 THE FLAT UNIVERSE

In 2000, new evidence emerged from a ten-day balloon flight around the Antarctic, in late 1998, in which instruments measured tiny variations in the background radiation. The telescope captured the most detailed picture (a map of radiation released in the Big Bang) ever of the early universe. The results are among the most important in cosmology for many years, because they confirm that the geometry of space is effectively flat rather than three-dimensional – the universe appears to be disc-shaped, like a huge CD! The results also support the idea that in the first second after the Big Bang there was a period of rapid expansion (known as *inflation*). They also rule out the theory that eventually the universe will stop expanding and go into reverse (ending in a 'Big Crunch'; see p. 112). But the research has created a major mystery – 95 per cent of the matter in the universe exists in forms we cannot see!

A BRIEF HISTORY OF TIME

Modern scientific methods of dating have pushed the age of the universe back much further than the date proposed by James Ussher, Archbishop of Armagh, in 1648. He declared that the universe was created on Sunday 23 October 4004 BC!

Observations of the current rate of expansion of the universe suggest that it took place about 15,000 million years ago (Figure 4.4). Galaxies began to form perhaps 14,000 million years ago, and stars developed in those galaxies around 11,000 million years ago. The oldest rocks found on the Moon are estimated to be about 4,600 million years old, and this is now widely accepted as the approximate age of the Earth and solar system. The oldest dated rocks on Earth (Box 4.13), from western Greenland, are believed to be around 3,800 million years old.

The geological timescale shown in Table 4.7, based on millions of years (Ma), reflects present understanding; some dates are revised from time to time.

BOX 4.11 GASES, DUST, STARS AND PLANETS

After about the first 200 million years of the universe (14,800 million years ago), gravity was strong enough to start forming galaxies, stars, planets and asteroids by condensation within the spinning cloud of gases released by the Big Bang. As time passed the universe continued to expand, and galaxies within it began to group together in clusters. Most of the universe comprises the huge voids between these galactic clusters.

Fusion processes in stars convert hydrogen into helium, and in larger stars these processes form all the heavier metals up to iron. As our part of the cloud of cosmic dust and gas released in the Big Bang condensed, its mass was greatest near the centre. This formed the Sun, and the planets were formed from the remaining material in a disc surrounding the Sun. The whole system rotated, and the inner planets formed by accretion (small particles were drawn together by mutual gravitational attraction).

HOW WILL IT ALL END?

The ultimate fate of the universe is no more clear than its origin. Despite much research by physicists, astronomers and cosmologists there is still no consensus about how and when it is likely to end. Some scientists even question whether it is logical to expect some ultimate end-point.

One theory – the so-called oscillating universe theory – predicts that, because of gravitational attraction, the ongoing expansion of the universe will eventually slow

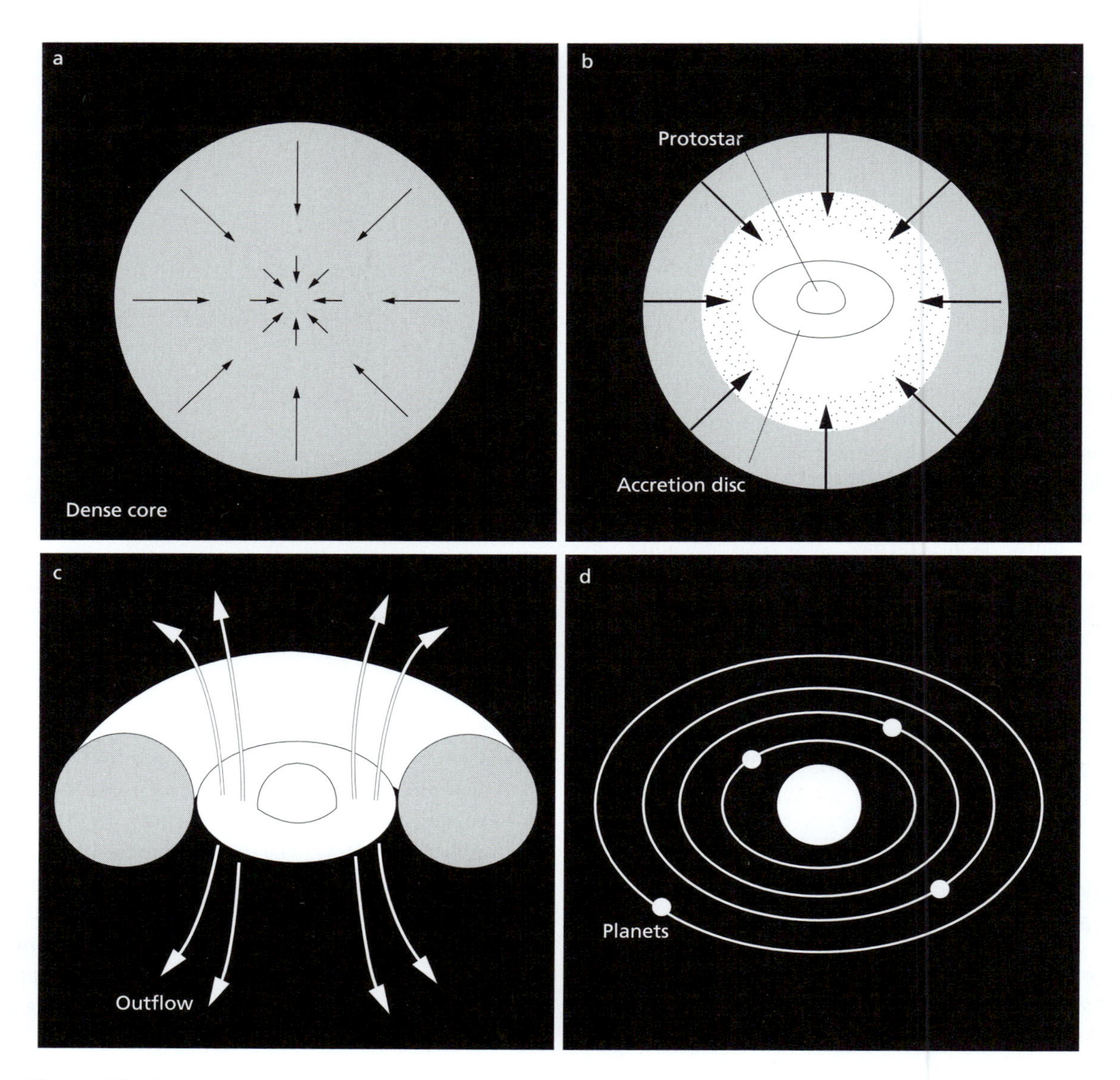

Figure 4.3 *The birth of a star. According to modern science, a collapsing cloud of gas and dust (a) forms a dense core (called a protostar) surrounded by a disc of gas and dust (b). Outflows of hot gas then drive away the remains of the original cloud (c). As nuclear reactions begin in the protostar, it becomes a star, and the matter in the disc condenses into planets (d).*

Source: Henbest, N. (1992) Life of a star. In R. Fifield (ed.) Inside science. Penguin, Harmondsworth, pp. 66–78 (at p. 68).

down, stop and then go into reverse. This raises the prospect of a Big Crunch (reverse Big Bang), in which all the matter in the universe would be contracted into a very small volume of very high density. A subsequent Big Bang could follow some time in the future (whatever time actually means in this context). Repeated cycles of universal expansion and contraction could follow, hence the idea of an 'oscillating universe'.

Other theories stretch the imagination to a similar degree! Various causes of the end of the universe have been predicted, including the progressive destruction of matter by antimatter (in ways as yet unknown). Closer to home, various theories have suggested the fate of planet Earth. One contender is death by freezing, if the solar fires are extinguished (again by unknown processes). Much more likely in the short term, at least in terms of probability, is the risk of a large (at least 2 km diameter) asteroid or comet colliding with the Earth, seriously disrupting biological and environmental systems, and killing a large fraction of the world's population.

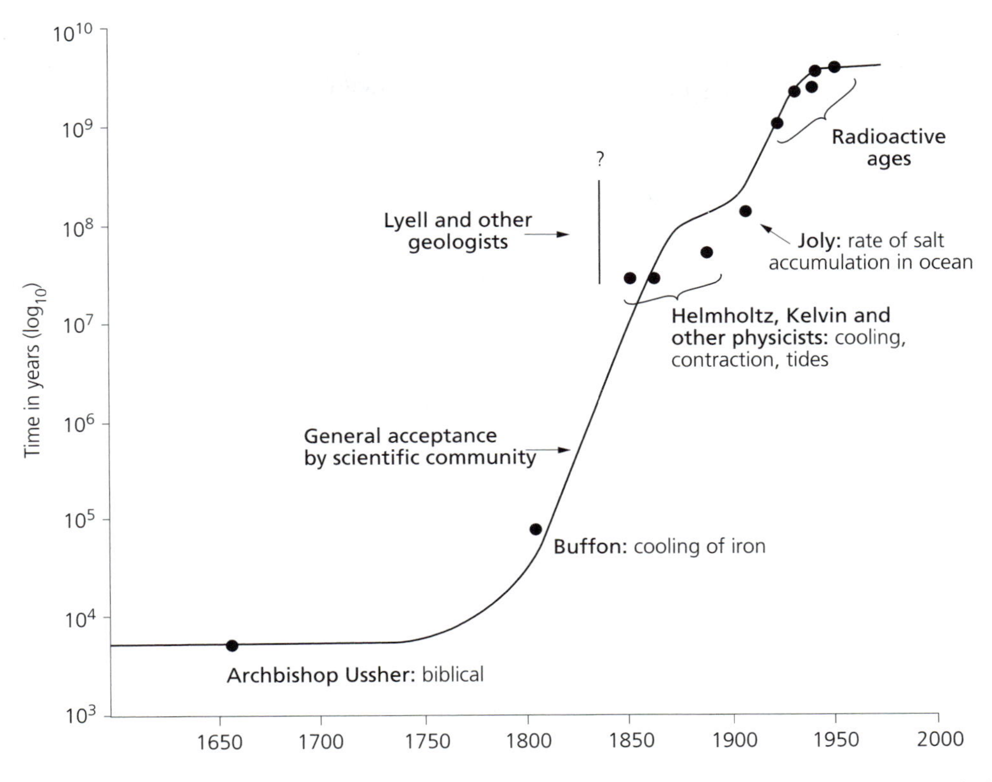

Figure 4.4 *Increasing estimated age of the Earth between 1650 and 1990. Modern scientific methods for dating rocks have greatly extended the estimated age of the Earth, particularly during the twentieth century. The 6,000-year age accepted in the seventeenth century has been pushed back to more than 1,000 million years.*

Source: Figure 2.25 in Press, F. and R. Siever (1978) Earth. W.H. Freeman, San Francisco.

BOX 4.13 DATING THE EARTH

Long-term changes in the Earth's environment are usually defined on the basis of the geological timescale. The relative ages of rocks on the Earth can be determined from fossils within them (palaeontology) and from the relative positions of the different rock strata or layers (stratigraphy). Since the development of absolute dating techniques based on the rate of decay of radioactive substances, it has been possible to give reasonably accurate dates for the geological time divisions. Radiocarbon dating is reliable back to about 30,000 years, but much longer timescales can be measured using potassium–argon dating (applicable from about 100,000 to 30 million years).

A new theory was proposed by cosmological mathematician Stephen Hawking in 1998, called the Open Inflation theory. This explains the formation of the universe from a tiny pea-sized object suspended in a timeless void, which went through a period of rapid expansion (*inflation*) immediately before the Big Bang. The theory predicts that the universe will continue to expand for ever.

PLANET IN MOTION

The Earth appears, from our direct experiences, to be stationary. We feel no movement of the planet as we go about our daily routines, and while clouds appear to float across the sky high above us, this relative movement appears to be caused more by wind than planetary motion. But this sense of a stationary Earth is an illusion, because gravity keeps us (and everything else) in place on its surface as the planet moves continuously through space, and we are so small relative to the size of the Earth that we cannot detect planetary motion ourselves.

BOX 4.14 THE EARTH BOOK

Life is a relatively recent arrival on Earth. Imagine that the entire history of the Earth was written in a book containing 1,000 pages, with each page representing roughly 4.5 million years:

- The first quarter of the book (about 220 pages) would describe the emergence of conditions suitable for the appearance of life.
- Gases condense to form the planet, and eventually primitive simple forms of life emerge in tidal pools of the warm oceans that covered the early Earth.
- Three-quarters of the way through the book, around 500 million years ago, life in the seas begins to appear in forms that we might be able to recognise; this is the Age of Fishes.
- About thirty pages further on, around 350 million years ago, the first land creatures appear as direct descendants of the early fishes.
- Another thirty pages later (225 million years ago) comes the Age of Dinosaurs.
- The Age of Dinosaurs ends suddenly about 65 million years ago, followed by the Age of Mammals (when the first primates appear) about 14 pages on.
- The earliest identifiable ancestors of modern humans (the first hominids) appear about three pages from the end of the book, around 12 million years ago.
- The first modern humans (Homo) appear at the bottom of the last page but one.
- The first stone tools are described halfway down the last page.
- The rise of modern humans appears on the very last line!

Planetary movement is far from random and chaotic, and the order and pattern within it are vital for life on Earth. The Earth moves within the solar system, and the solar system itself is moving through the vastness of space. The Earth is affected by four main motions – rotation, revolution, precession and galactic rotation.

ROTATION

The Earth spins continuously around its axis (an imaginary line joining the North Pole, the centre of the Earth and the South Pole). Because the planet is roughly spherical, the speed of rotation varies across the Earth's surface – at the equator it is about 532 km h^{-1} (km per hour), but near the poles it is much lower.

Rotation has the effect of making objects in space (such as the Sun, Moon, stars and other planets) appear to travel in the opposite direction to that in which they really travel. The Earth rotates eastwards, so the Sun appears to move westwards across the sky (rising in the east and setting in the west). Some artificial satellites circle the Earth in geostationary orbits, circular paths 35,900 km above the equator, taking 24 hours to complete each orbit. The geostationary orbit moves from west to east, so the satellites appears to remain stationary over one place on the Earth's surface. Such orbits are used particularly for weather and communications satellites.

Each rotation of the Earth around its axis takes about 24 hours (23 hours 56 minutes 4.1 seconds to be precise). This motion is important in a number of ways. It defines the length of a day, so it is important in keeping time. Rotation creates day and night on Earth, as the planet is repeatedly lit by the Sun then darkened in its shadow (Figure 4.5), so it shapes patterns of activity for all life (including humans) and for many environmental processes that depend on sunlight as an energy source. It also helps to define location and relative time on the Earth's continuously curved surface (Figure 4.6). The Earth rotates 360° (a complete circle) every 24 hours, which is 15° every four minutes. This is the basis on which time zones (Box 4.15) are defined (Figure 4.7).

The axis around which the Earth rotates is not vertical. It is inclined at an angle of about 23.5° from the vertical. This puts it at an angle of about 66.5° relative to the plane (level) of the Earth's orbital path around the Sun. This tilt is an important factor in determining the seasons through the year.

The speed at which the Earth rotates appears to be extremely conservative and has apparently been quite uniform throughout geological time. There is evidence that

Table 4.7 The geological timescale

Eon	*Era*	*Sub-era*	*Period*	*Epoch*	*Began (Ma)*
Priscoan					4,600
Archaean					4,000
Proterozoic					2,500
Phanerozoic					
	Palaeozoic				
		Lower Palaeozic			
			Cambrian		590
			Ordovician		505
			Silurian		438
		Upper Palaeozoic			
			Devonian		408
			Carboniferous		
			■ Mississippian	360	
			■ Pennsylvanian	320	
			Permian		286
	Mesozoic				
			Triassic		248
			Jurassic		213
			Cretaceous		144
	Cenozoic				
		Tertiary			
			Palaeogene		
				Palaeocene	65
				Eocene	55
				Oligocene	38
			Neogene		
				Miocene	24.6
				Pliocene	5.1
		Quaternary			
			Pleistogene		
				Pleistocene	2.0
				Holocene	0.01

the Earth might be slowing down a little through time, and that it is spinning more slowly than it was 90 years ago. At least in theory, global environmental change could trigger changes in the Earth's rotation by causing wholesale changes in the distribution of material within the atmosphere, hydrosphere and lithosphere. One recent study suggests that global warming has caused a slowdown of the Earth's rotation of about 0.56 milliseconds a century (by altering the angular momentum of the atmosphere); this implies that global warming is responsible for almost a third of the slowdown that scientists have measured.

There is also evidence of small variations in rotation speed over very long timescales, which might cause long-term climatic change. Possible processes include shifts of

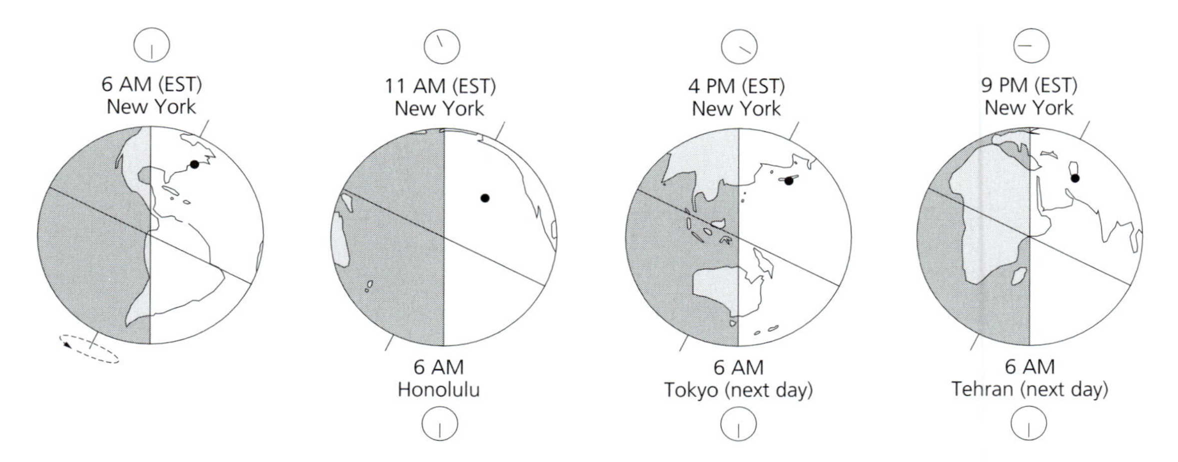

Figure 4.5 *The effects of the Earth's rotation on day and night. The figure shows the effects of the Earth's rotation on the distribution of day and night on a typical summer day in the northern hemisphere.*

Source: Figure 2.1 in Doerr, A.H. (1990) Fundamentals of physical geography. Wm. C. Brown Publishers, Dubuque, Iowa.

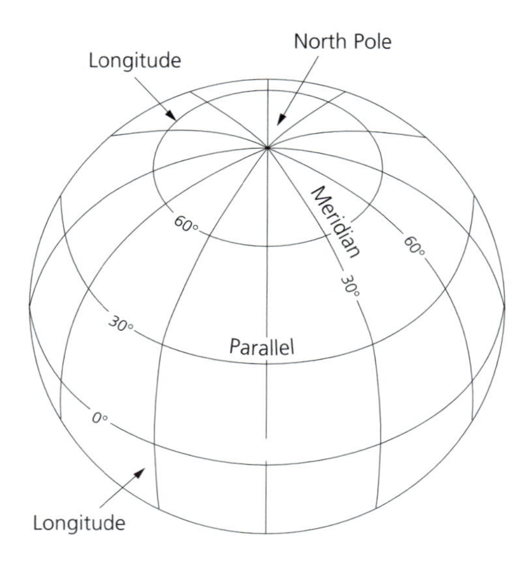

Figure 4.6 *Definition of latitude and longitude on the Earth. Longitude defines location around the Earth's circumference (east to west), like segments of an orange (based on meridians), while latitude defines locations relative to the equator (north to south), like parallel slices through the orange.*

mass within the Earth's core and changes in relationships between the Earth and the Moon, which affect the tides. Such variations cause difficulties in keeping time precisely. For example, at midnight on 31 December 1998, the clocks that keep time for the world had to be stopped for 1 second – a leap second – to bring Universal Time (based on the Earth's rotation) into alignment with Co-ordinated Universal Time (based on atomic clocks). But these rotation changes are believed to be too small to have a significant impact on life on Earth.

REVOLUTION

The Earth revolves around the Sun (Figure 4.8). It travels a total distance of roughly 943 million km per revolution, at an average speed of around 30 km a second (107,000 km h^{-1}). Speed varies through the elliptical orbit, depending on the distance between the Earth and the Sun. The Earth is closest to the Sun – about 147,250,000 km – in early January (perihelion) and furthest away – about 152,079,000 km – in July (aphelion).

Each complete orbit takes a year (more correctly, 365 days 5 hours 48 minutes 46 seconds). This is called the Earth's sidereal period. The fact that it is 365.25 days explains why leap years (every fourth year) have 366 days, to keep calendars on time. The Earth's revolution around the Sun thus defines the calendar year.

As we saw above, the axis around which the Earth rotates is 23.5° from the vertical. It remains at this constant angle throughout its revolution around the Sun. This causes the northern and southern hemispheres to point directly towards the Sun in successive seasons and explains why summer in the northern hemisphere is matched by winter in the southern hemisphere (Table 4.8). The rhythm of the seasons (Box 4.16) is defined by systematic variations at a given place in:

- the height of the Sun in the sky (which determines the angle at which the Sun's rays intersect the ground);

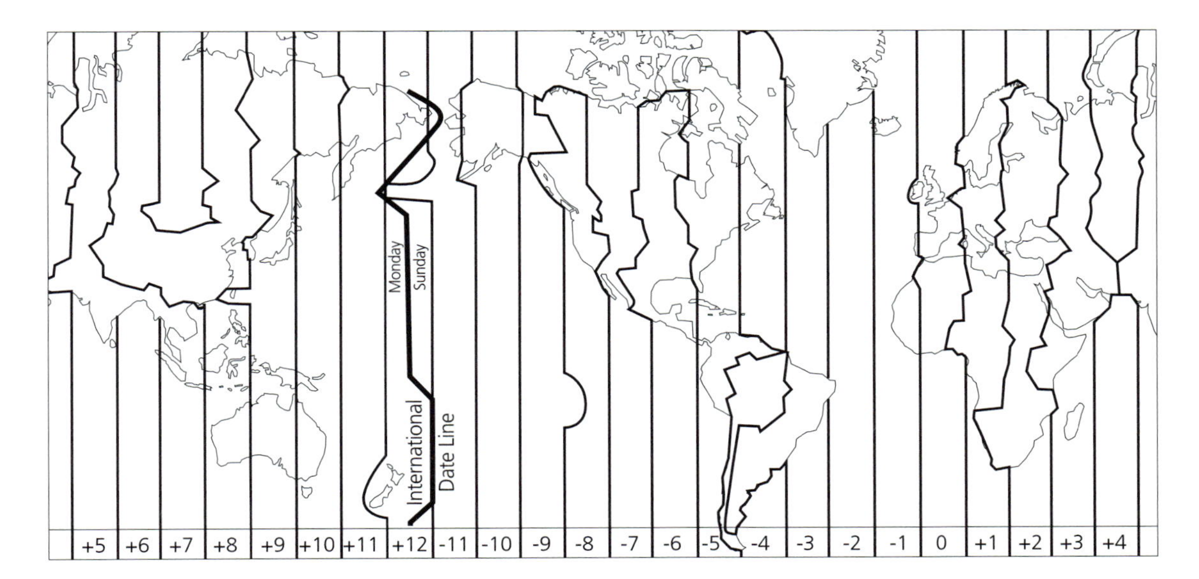

Figure 4.7 *World time zones. The 360° circumference of the Earth is divided into twenty-four time zones, each one roughly 15° wide and defined as 1 hour different from the adjacent time zones. Clocks everywhere within a given time zone are set at the same hour. The boundaries of some zones are adjusted to keep relatively small political units within a given zone. The International Date Line is the crossover from one day to the next.*

BOX 4.15 TIME ZONES

The globe is divided into time zones, each 15° of longitude wide and spanning 1 hour. Each country chooses the standard time or times most convenient to it, and large countries such as the USA spread over a number of time zones. The time zones are relative to Greenwich Mean Time (GMT: local time on the zero line of longitude, which passes through the old Royal Observatory at Greenwich, London). Successive zones to the east of the Greenwich meridian are 1 hour in advance, and successive zones to the west are 1 hour behind. Thus, for example, while people in London are having lunch (1 pm), people in New York are having breakfast (8 am local time), people in Moscow are having tea (4 pm), it is dinner time (8 pm) in Beijing and midnight in Wellington (New Zealand).

In the past these time differences were not really important, but in our increasingly global village relative time matters much more. Trade in the world's leading money markets – London, New York, Zurich, Los Angeles, Tokyo, Hong Kong – is taking place at different times of the day, all day long. Supersonic air travel, on Concorde, allows people to arrive in New York an hour before they left London! Sports events televised live in Los Angeles are shown three hours earlier in the day in New York and eight hours earlier in London.

- the length of day (thus number of hours of sunlight in the day); and
- the amount of solar radiation received (which depends partly on how close the Earth is to the Sun).

During the summer, the Sun is more directly overhead and daylight lasts longer than in the winter months, so solar heating brings hotter conditions. Winters are colder because the Sun is lower in the sky (thus rays are more spread out) and daylight is shorter.

Because the Earth's axis is tilted from the vertical, the relative lengths of day and night vary through the year as the Earth revolves around the Sun. The critical points along this continuous cycle of change are the solstices and equinoxes, the timing of which is offset by six months between the two hemispheres (Table 4.9).

PRECESSION

Precession is a secondary motion that occurs because the Earth rotates around its axis and the axis is not vertical. If a child's top is set spinning upright, it will continue to

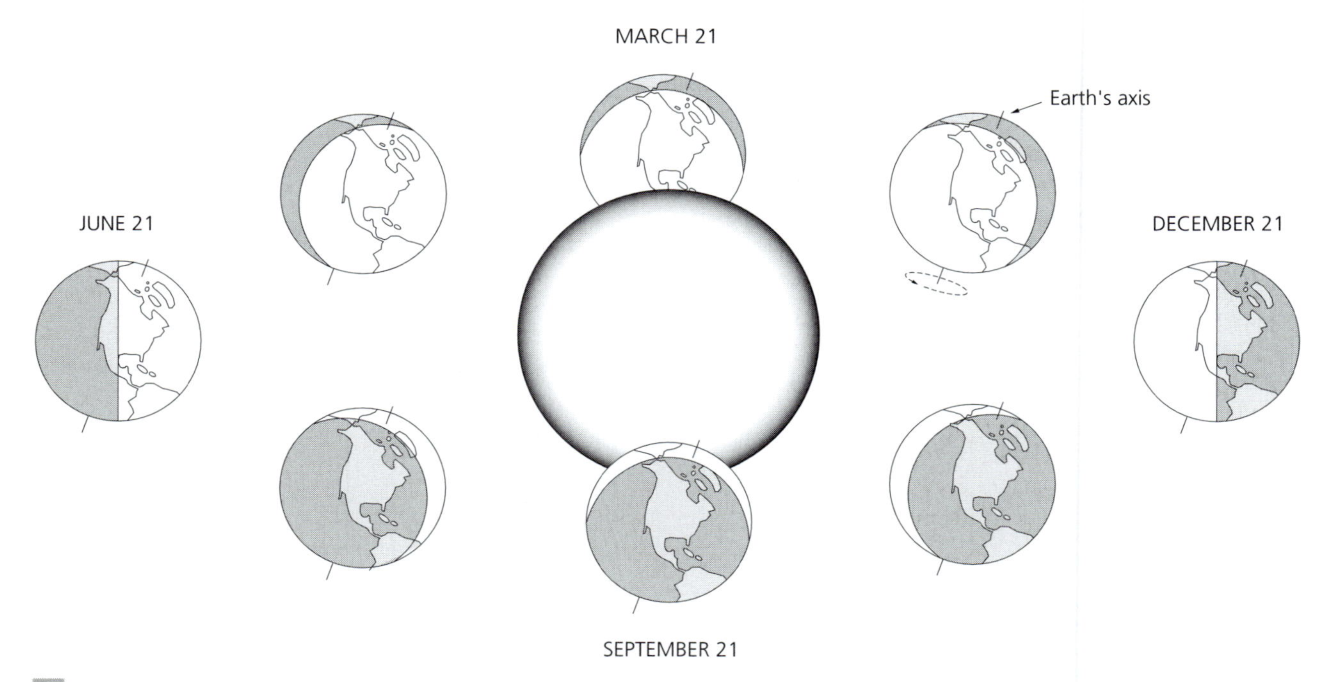

Figure 4.8 *Revolution of the Earth around the Sun. The Earth revolves around the Sun over a yearly cycle, and this defines the passage of the seasons in both hemispheres. See text for explanation.*

Source: Figure 2.2 in Doerr, A.H. (1990) Fundamentals of physical geography. Wm. C. Brown Publishers, Dubuque, Iowa.

Table 4.8 The march of the seasons

Northern hemisphere	*Southern hemisphere*
summer	winter
autumn (fall)	spring
winter	summer
spring	autumn (fall)

Table 4.9 Timing of the equinoxes and solstices in each hemisphere

Date	*Northern hemisphere*	*Southern hemisphere*
21 June	summer solstice	winter solstice
23 September	autumn equinox	vernal (spring) equinox
22 December	winter solstice	summer solstice
21 March	vernal (spring) equinox	autumn equinox

rotate around its axis. If the top is spun at a slight angle, it will still continue to rotate around its axis, but the axis itself will start to turn in a slow circular motion opposite to the main direction of the spinning. So, if the top is spinning clockwise, the axis turns slowly anticlockwise, and vice versa.

Exactly the same process affects the Earth's rotation around its inclined axis. This slow counter-rotation motion is called *precession*, and a complete rotation cycle takes 21,000 years. It is the result of the gravitational pull of the Sun and the Moon operating on the equatorial bulge of the Earth.

One consequence of precession is a continuous slow shift in the position of the equinoxes and a gradual westward drift in the ecliptic (the path that the Sun appears to follow). As a result, the dates of the astrological signs of the zodiac no longer correspond to the times of year when the Sun actually passes through those constellations.

Precession also causes a slight progressive change in the start of the seasons on Earth, which apparently start earlier by up to 20 minutes a year. There is a belief that, in the longer term (perhaps over periods of tens of thousands to millions of years), precession might also be associated with climate change and the onset of glacial conditions.

GALACTIC ROTATION

The Earth rotates around its axis, revolves around the Sun and precesses around its axis (Table 4.10). There is a fourth, and even bigger, motion, and this is the movement of the Milky Way galaxy as part of the ongoing evolution of the

BOX 4.16 THE FOUR SEASONS

Solstices: these occur when the Sun appears to be overhead at midday at the maximum distance north and south of the equator. Days are longest and nights are shortest at the summer solstice, and days are shortest and nights longest at the winter solstice.

Equinoxes: the Sun appears directly overhead at midday on the equator at the equinoxes in spring (vernal) and autumn. Day and night are then roughly the same length (12 hours each).

Summer comes to the northern hemisphere from June to September and to the southern hemisphere from December to March. This seasonal shift has an impact on the movement of people around the world. It encourages long-haul holiday trips by sun-starved Europeans seeking comfort and winter relief in sunnier southern climates, for example, and allows English cricketers to escape the northern hemisphere winter and tour in Australia and New Zealand during their summer. It is also vital to the year-round provision of supplies of fresh food (particularly fruit and vegetables), because southern hemisphere produce is available in northern hemisphere markets during winter in the north, and vice versa.

Table 4.10 The four principal cyclical motions of the Earth

Motion	*Time taken*	*Effects on Earth*
Rotation	24 hours	Day and night
Revolution	365.25 days	Seasons
Precession	21,000 years	Climate change
Galactic rotation	225 million years	Climate change?

universe after the Big Bang. The entire galaxy and everything in it are travelling through space at a speed of about 322 km per second. In addition, the galaxy itself rotates, with a period of 225 million years. Its possible effects on the Earth are as yet unknown, though scientists suspect a link with long-term climate change.

EARTH AS A LIVING ORGANISM

For centuries, scientists have viewed the Earth and its environmental systems as a sort of mechanical machine, driven by physical forces like volcanoes, rock weathering and the water cycle. While it is clear that organic activity plays an important role in some environmental systems, such as the biogeochemical cycles (see p. 86), until quite recently biological factors were seen as secondary to physical and chemical ones.

THE GAIA HYPOTHESIS

In the early 1970s, a revolutionary new theory was put forward by James Lovelock. He called it the *Gaia hypothesis*, after the Greek Earth goddess (or, as some prefer, Mother Earth). The theory is revolutionary because it treats the Earth as a single living organism, in which biological, chemical and physical factors all play important roles. Lovelock argued that the Earth's living and non-living systems form an inseparable whole that is regulated and kept adapted for life by living organisms themselves. He sees Gaia as 'a complex entity involving the Earth's biosphere, atmosphere, oceans and soil; the totality constituting a feedback or cybernetic system which seeks an optimal physical and chemical environment for life on this planet'.

SELF-REGULATION AND SELF-REPAIR

According to the Gaia hypothesis, the planet is alive and operates as a single organism (some scientists refer to it as a super-organism) (Box 4.17). It has an in-built ability to maintain its own equilibrium and an ability to recover from damage, similar in principle to the ways in which any other organism (including humans) can often self-repair, but obviously involving totally different mechanisms. The Gaia super-organism system is designed to maintain conditions suitable for life on Earth, and the planetary self-repair

BOX 4.17 GAIA – ACCIDENT OR DESIGN?

Lovelock's original idea was that Gaia followed some purposeful design that organised living things to stabilise the atmosphere and climate. The notion that the Earth acted with some sense of purpose in seeking to achieve a predefined goal (survival of life) was heavily criticised because it implied some ability of the planet to think and act (rather like humans), or it required an external agent (such as God or gods) to think and act. Lovelock later refined the idea to allow for regulation via internal feedback (p. 79).

potential means that life is likely to continue after a global disaster. But it would not necessarily be life in its present forms, and there is no guarantee that the human species would survive.

Gaia is a self-regulating system. It is highly organised and hierarchical, and all of its components play a part in creating and maintaining stability, which is the ultimate goal. If one part of the system changes, or is changed, the rest of the system will adapt so as to re-establish stability and preserve life on Earth.

CLIMATE CONTROL

A key part of the Gaia model is the way in which climate is moderated or even controlled by biological factors. According to the model, the Earth's atmosphere would be unsuitable for life if it were not regulated by the biosphere, that is by the planet's ecological and environmental systems. Lovelock developed a model called Daisyworld (Box 4.18) to show how atmosphere and plant growth might be closely interdependent.

Geological evidence shows that the Earth's temperature has varied through time, but within limits. If the Gaia hypothesis is correct, the Earth's climate has been improved by living organisms, which operate as a planetary thermostat (heating regulator), serving as either heaters or air conditioners as appropriate. If this is so, then the ecological tolerance of living organisms (that is, their ability to withstand changes in their external environment) (see Box 18.18) is the main sensor in the global thermostat.

GREENHOUSE EFFECT

Gaia is based on natural feedback loops involving organisms that always operate in such a way as to stabilise conditions enough for life to continue. The greenhouse effect (see p. 238), of major concern now because it threatens to bring wholesale climate change and knock-on adjustment in many environmental systems, is an integral part of the Gaia model.

It seems that when the first photosynthetic bacteria appeared on Earth, about 3,500 million years ago, the carbon required to build their shells came from atmospheric carbon dioxide. As CO_2 was appropriated to build the shells, atmospheric concentrations fell and so did average temperature (from about 28 to 15°C). Biological feedback then stabilised temperature, because the bacteria produced methane (another greenhouse gas) and rising atmospheric concentrations of methane pushed up temperatures again. The growth rate of bacteria depends partly on temperature, so this biological feedback turned out to be a remarkably effective thermostat that kept the Earth's temperatures within tolerable limits at the same time as organic life was emerging and becoming more important within the Gaian system.

The idea that life somehow manipulates the Earth's atmosphere for its own purposes is attractive to scientists who do not accept that climate change triggered by greenhouse gases and ozone depletion (pp. 255 and 249) is a major threat to the Earth, its environments and people. There is evidence linking variations in temperature on Earth over the last 1,000 million years with changing atmospheric levels of CO_2 (an important greenhouse gas), but it is not clear whether organic or inorganic processes, or both, were responsible.

Some models of how the biosphere might respond to global warming associated with the continued release of greenhouse gases recognise the possible feedback links (see p. 79) between temperature, CO_2 concentrations and plant growth, and these produce quite optimistic forecasts for the future. One such model reckons that the Earth's biosphere

BOX 4.18 DAISYWORLD

Daisyworld is a computer model developed by James Lovelock to illustrate his Gaia hypothesis on any planet that supports life. It is a good example of the ways in which feedback mechanisms create and maintain equilibrium (in this case the physical and chemical conditions necessary for life) in open systems.

The only living things in this computer world are daisies. Some of them are white and the rest black. Variations in solar radiation favour one group or the other. When the temperature rises, white daisies reflect more heat and thus stay comfortably cool. Conversely, when the temperature falls, black daisies absorb more heat and remain comfortably warm. A change in the temperature of the planet promotes a change in the relative abundance of black and white daisies on it.

Feedback occurs via changes in the albedo (reflectivity; see p. 234) of the planet's surface. White daisies reflect more radiation when radiation increases, and they cover a larger area. Black daisies absorb more radiation when radiation decreases, and they cover a larger area. Albedo changes then raise or lower surface temperature, and thus air temperature. This prevents climate becoming so warm or so cool that neither type of daisy can survive.

could survive for up to 1,500 million years, depending on whether temperature or CO_2 was the limiting factor.

EMERGENCE OF LIFE

Lovelock has used his Gaia model to explain how life on Earth has evolved through time. The process began after the Earth cooled and the first signs of life started to appear, possibly associated with particular clay minerals in the as yet lifeless planet. Over time, the two-way (symbiotic) process of adaptation began, between organisms and environment. Organisms had to adapt to changes in the Earth's system (such as the emergence of oxygen) and to external changes (such as meteorite impacts, which might have caused massive extinction of species).

Lovelock has coined the term 'geophysiology' to describe how the evolution of organisms and the evolution of their material environment is effectively a single process. Geophysiology, as a new evolutionary theory, contrasts with the more widely accepted co-evolutionary or biogeochemical view, in which organisms and their environment evolve more or less separately (see p. 529).

IMPLICATIONS OF GAIA

An interesting corollary of the Gaia hypothesis is that the existence of the conditions that are necessary for life on Earth was not and is still not a matter of coincidence. This has profound implications for how we view the Earth, the life on it, and indeed our own position in the grand design of things. It might radically affect the way we view, value and treat the Earth and its environment.

One of the many implications of the Gaian view is that, since life and environment are so closely linked, we must understand and maintain the physical environment and living things around us if we are to survive on this planet. Environmental protection and conservation of species are thus central to the survival of the human race, not luxuries that we can choose to ignore. Other implications of the Gaia hypothesis (Box 4.19) echo some of our earlier conclusions about environmental systems (p. 82).

Adopting a broad global perspective is vital, because this is the scale at which the Gaia processes function and at which stability is developed and maintained. Local changes and impacts are important, but the planetary scale is what really matters.

Adopting an interdisciplinary perspective is also important. Individual scientific disciplines have boundaries to the questions they ask and the type of information they regard as valid and relevant. This tendency to focus on the specific and ignore the general can make us extremely short-sighted and thus unable to understand how different environmental systems or parts of systems link together.

BOX 4.19 IMPLICATIONS OF THE GAIA HYPOTHESIS

- The whole system matters much more than its component parts.
- Complexity in the environment is inevitable and fundamental, and much of it we simply do not understand.
- Feedback links in environmental systems are critical but are not always fully understood; they should form the basis of any human intervention (deliberate or accidental) in environmental systems.
- Biotic and abiotic aspects of the environment affect each other; the symbiosis is important but easy to disturb or destroy.
- Global environmental systems exist in stability that has evolved over long periods of time; short-term changes must be seen against a background of long-term geological change and progression.
- It is unwise radically to modify global environmental systems and equilibrium, because we cannot always properly predict the outcomes and consequences.
- The Earth has a great natural capacity to look after itself (self-repair), so long as change remains within critical limits (thresholds), many of which we simply do not know enough about.
- The Earth will probably survive, no matter what humans do to it, but its survival might not include humans. Quite simply, it is bigger than we are and can probably survive well without us!

SUMMARY

This chapter takes a much broader perspective than the others in this book, by examining the origin, age and character of the Earth in the context of the wider universe. The chapter title ('Spaceship Earth') is drawn from a metaphor suggested in the 1960s, which likens the Earth to a small spaceship floating about in space but dependent on external sources of energy. We considered how explanations of the origin and structure of the universe have changed through time, and looked at the most recent evidence on the structure and composition of the universe. The Earth sits in the solar system, which sits in the Milky Way galaxy, which is part of the universe, and the Sun and Moon play critical roles within our solar system. The thorny question of how and when the universe began, and how it has evolved, has exercised the minds of scientists for centuries, and we considered the nature and relevance of the Big Bang theory. Of critical importance to life on Earth, and to environmental systems on the planet, are the four main motions of the Earth (rotation, revolution, precession and galactic rotation). Finally, we considered James Lovelock's controversial Gaia ideas about the Earth as a super-organism and their application to global warming and implications for the emergence of life on our planet. Looked at from this grand universal scale, people and indeed the whole planet are very small, but this should not blind us to the responsibilities we have to look after the Earth, our only home.

WEBSITE

Links to relevant websites, a comprehensive bibliography, tools for teaching and learning, and downloadable images relevant to this chapter can be found at the website specially designed to accompany this book at
http://www.park-environment.com

FURTHER READING

Condie, K.C. and R.E. Sloan (1998) *Origin and evolution of the Earth: principles of historical geology*. Prentice Hall, Hemel Hempstead. Explores the complexities of Earth's history and life, with a special emphasis on the role and significance of plate tectonics.

Emiliani, C. (1992) *Planet Earth: cosmology, geology, and the evolution of life and environment.* Cambridge University Press, Cambridge. A comprehensive review of how the planet was formed and how it functions.

Kaufmann, W.J. (1993) *Discovering the universe.* W.H. Freeman, Oxford. A splendidly illustrated review of the development of modern astronomy.

Lewis, J.S., M.S. Matthews and M.L. Guerriri (1993) *Resources of near-Earth space.* University of Arizona Press, Tucson. Suggests that space – the final frontier – has much to offer, if we can find ways of exploiting what is there!

Lovelock, J.E. (1979) *Gaia: a new look at life on Earth.* Oxford University Press, Oxford. Lovelock outlines his key ideas about Gaia in this classic book.

Lovelock, J.E. (1988) *The ages of Gaia: a biography of our living Earth.* W.W. Norton, New York. Lovelock applies his Gaia theory to the evolution of the Earth and the life on it and in it.

Mursky, G. (1996) *Introduction to planetary volcanism.* Prentice Hall, Hemel Hempstead. Outlines the nature, impacts and dynamics of volcanic phenomena throughout the solar system.

Pasachoff, J.M. (1995) *Astronomy.* Saunders College Publishing, Philadelphia. A clear, up-to-date and well-illustrated introduction to the structure and evolution of the universe.

Press, F. and R. Seiver (1993) *Understanding Earth.* W.H. Freeman, Oxford. A clear, comprehensive and well-illustrated introduction to the principles and applications of modern geology.

Stanley, S.M. (1993) *Exploring Earth and life through time.* W.H. Freeman, Oxford. Reviews the history of physical environments on Earth and the evolution and extinction of life up to the present time.

Teisseyre, R., J. Leliwa Kopystynski and B. Lang (1992) *Evolution of the Earth and other planetary bodies.* Elsevier, Amsterdam. How did it all begin, how does it work, and where is it all heading?

Wood, J.A. (1997) *The solar system.* Prentice Hall, London. Up-to-date review of theories and evidence about how and when the solar system was formed, how it functions, and what questions remain unanswered.

Part Two
The lithosphere

Chapter 5

Structure of the Earth

LEARNING OBJECTIVES

When you have finished studying this chapter, you should be able to

- *Outline the distribution of land and sea, and relief, across the Earth's surface*
- *Describe and account for the structure and composition of the interior of the Earth*
- *Appreciate the meaning and relevance of seismic waves*
- *Distinguish between oceanic and continental crust*
- *Understand the environmental and health problems associated with radon gas*
- *Discuss the environmental problems associated with the safe underground storage of nuclear waste*
- *Explain why geothermal energy is such a useful renewable energy source*

We saw in Chapter 4 how the Earth was formed, probably more than 4,000 million years ago, out of a swirling mass of cosmic gas and dust in the solar system. Through time the dust and gas coalesced (clumped together) into solid forms – the Earth and other planets – which have been undergoing change ever since. Nothing in the universe is static, and the universe itself is still evolving from the initial Big Bang more than 15,000 million years ago. Change is the norm in this dynamic cosmos of ours!

In this chapter, we explore the nature of the Earth's surface and the structure of the Earth's interior. The surface is very dynamic and subject to great change through time, as we shall see in Chapter 6, largely because of what goes on in the interior.

SURFACE OF THE EARTH

Geodesy is the branch of science that deals with measuring the size, shape and curvature of the Earth, and modern geodetic survey is greatly assisted by developments in remote sensing.

SIZE, SHAPE AND CHARACTER

SIZE AND SHAPE

Compared with the other planets in our solar system, the Earth is not particularly large (see Table 4.4). It has a surface area of 510 million km^2, a mean radius of 6,371 km and a total mass estimated at $5{,}976 \times 10^{24}$ g (million million million grams). The Earth is almost but not entirely spherical; it is an oblate spheroid. The circumference around the equator is 40,077 km, compared with 40,009 km around the poles.

LAND AND WATER

Under a third of the Earth's surface (29 per cent, 149 million km^2) is occupied by land. Most (71 per cent, 361 million km^2) is covered by water. There is more than twice as much sea as land, and one single ocean (the Pacific, which is over 160 million km^2) covers an area larger than all the land put together. Little wonder that some scientists suggest that the planet be renamed Water instead of Earth!

The distribution of land and water is not constant through time. It can be altered by various long-term changes, including:

BOX 5.1 LAND AND PEOPLE

If the total land area were to be shared equally among the world's population in the year 2000 (estimated at 6,400 million people), each person would have a plot roughly 150 m by 150 m (0.02 km^2) in size. There would be about 43 people per km^2 in this world with a uniform population distribution. But, of course, many parts of the Earth's land surface are constrained by climate and/or relief, and up to 80 per cent of the plots would be relatively useless. This would significantly increase population density (to nearly 215 people per km^2) and decrease plot sizes (to roughly 70 m by 70 m each).

- *climate*: sea level is lower hence water area smaller during ice ages, for example, when much of the Earth's water is stored as ice caps and ice sheets;
- *tectonic activity*: including earthquakes and volcanic eruptions, which can build up or destroy land surfaces;
- *plate tectonics*: the gradual relocation of the continents around the surface of the Earth.

Land and water are not evenly distributed around the Earth (Figure 5.1). There is much more land in the northern hemisphere than in the southern hemisphere; the southern hemisphere is largely water. This distribution has an influence on global patterns of climate, because oceans are giant heat stores, whereas the continents heat up and cool down relatively rapidly. Oceans are also sources of moisture for precipitation. The distribution of land and sea also influences global patterns of pollution (because almost all of the air and water pollution comes from land-based sources), and oceans provide some of the important stores in global biogeochemical cycles (see p. 86).

The continental land masses vary a great deal in size (Table 5.1), and they are the result of plate tectonic processes operating over geological timescales. Perhaps inevitably, the continental areas that are closest to the original land mass (Asia and Africa) are largest, and those that

Table 5.1 Area of the continents

Continent	*Area (million km^2)*	*Per cent*
Asia	44.250	29.82
Africa	30.264	20.40
North America	24.398	16.45
South America	17.793	12.00
Antarctica	13.209	8.90
Europe	9.907	6.68
Oceania	8.534	5.75
Total	148.355	100.00

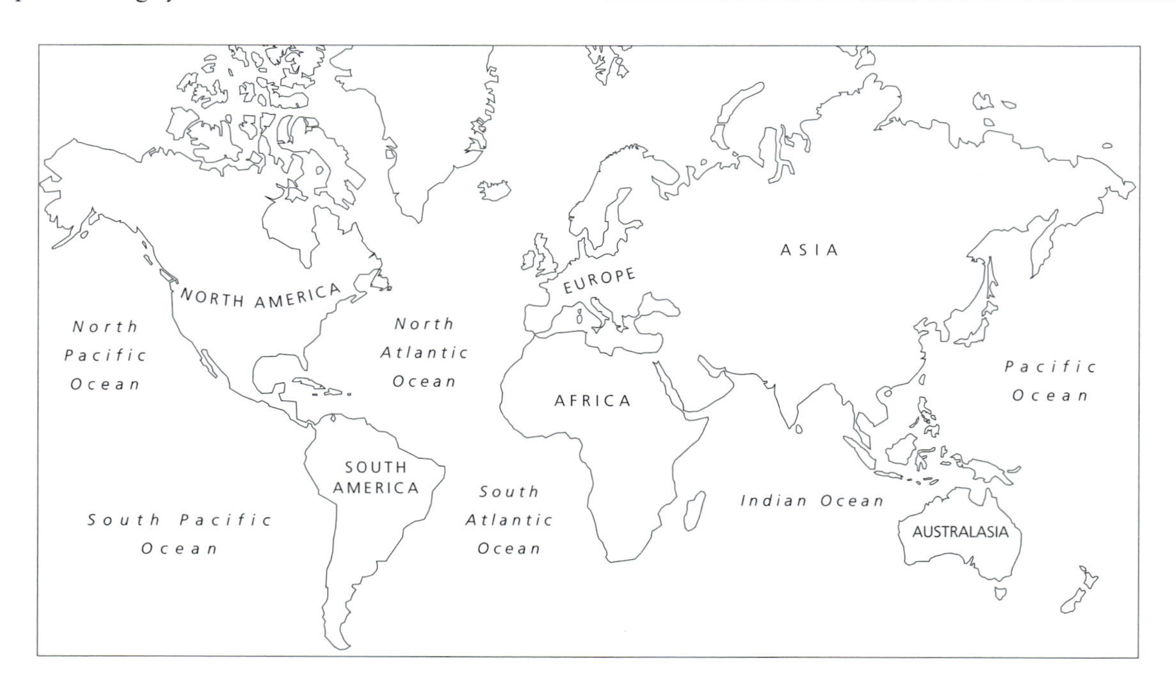

Figure 5.1 *The Earth's major oceans and continents.*

Table 5.2 Area of the oceans and major seas

Ocean or sea	*Area (million km²)*	*Per cent*
Pacific Ocean	165.242	46.91
Atlantic Ocean	82.362	23.38
Indian Ocean	73.556	20.87
Arctic Ocean	13.986	3.97
Malay Sea	8.143	2.31
Caribbean Sea	2.753	0.80
Mediterranean Sea	2.505	0.71
Bering Sea	2.269	0.64
Gulf of Mexico	1.544	0.44
Total (listed)	352.36	100.00

have been dragged further away by plate tectonic forces (such as Europe and the Americas) are much smaller.

The oceans also vary greatly in size (Table 5.2), although the Pacific – which accounts for nearly half of the total area of water – is clearly in a league of its own. This again partly reflects the interplay of plate tectonic forces and the history of plate tectonic movements through geological time. The Atlantic Ocean is much smaller, but it is stretching wider much more quickly. In the long-term future, the relative sizes of the main oceans might look rather different.

The boundaries of the continents and oceans, determined largely by present-day coastlines, provide convenient natural ways of sub-dividing the world. This is effectively the face of the Earth when viewed from space: it ignores political and administrative boundaries constructed by humans; and it offers a meaningful basis for some important aspects of environmental management.

RELIEF

From our human perspective the Earth has a very variable surface, punctuated by high mountains and deep valleys. These topographic variations are indeed large compared with the size of people, and throughout history they have provided major barriers to the movement of people and materials. But compared with the size of the Earth, such features (Box 5.2) are extremely small.

Heights are normally measured relative to mean sea level, which provides a consistent baseline around the world (except in places undergoing uplift or subsidence as a result of tectonic activity; see p. 154). Although global sea-level change accompanies and is a result of global warming, long-term (50-year) average sea level is well recorded in many places and is the baseline used (see p. 511).

BOX 5.2 EXTREMES OF RELIEF ON THE EARTH

The highest point on the Earth's surface is the summit of Mount Everest (Chomolungma), in the Himalaya mountain range, which is 8,840 m above sea level. The lowest point is the bottom of the Marianas Trench, in the Pacific Ocean, which plunges to a depth of 11,034 m below sea level. The total relief of the Earth is thus 19,874 m. This 20 km relief, superimposed on a planet with a radius of 6,371 km, is microscopic – it amounts to less than a third of 1 per cent of the radius.

Deep depressions in the land surface are less of a barrier or inconvenience to humans than high mountains are, but they are still important details of the Earth's topography. The deepest land depression is below the ice in Marie Byrd Land, Antarctica, which dips to 2,469 m below sea level. The deepest land depression at surface level is the Dead Sea in Israel and Jordan, the surface of which lies at 395 m below sea level. The world's deepest lake is Lake Baikal, which is 1,940 m deep, more than 1,480 m of which is below sea level.

Absolute relief, which is the maximum elevation above sea level, varies much less between the continents than area does (Table 5.3). The explanation for this also lies within plate tectonics and mountain building (see p. 154), particularly at the leading edges of the major crustal plates, where massive forces lift up mountain ranges. The forces probably do not vary directly with the size of the continents or plates involved.

One way of examining the distribution of relief, across the globe or within a particular continent or country, is to plot the hypsometric curve (Figure 5.2). This shows the

Table 5.3 Highest peaks in each continent

Continent	*Peak*	*Height (m)*
Asia	Everest	8,840
South America	Aconcagua	6,959.8
North America	McKinley	6,193.5
Africa	Kilimanjaro	5,894.8
Europe	Elbrus	5,632.7
Antarctica	Vinson Massif	5,139.8
Oceania	Wilhelm	4,693.9

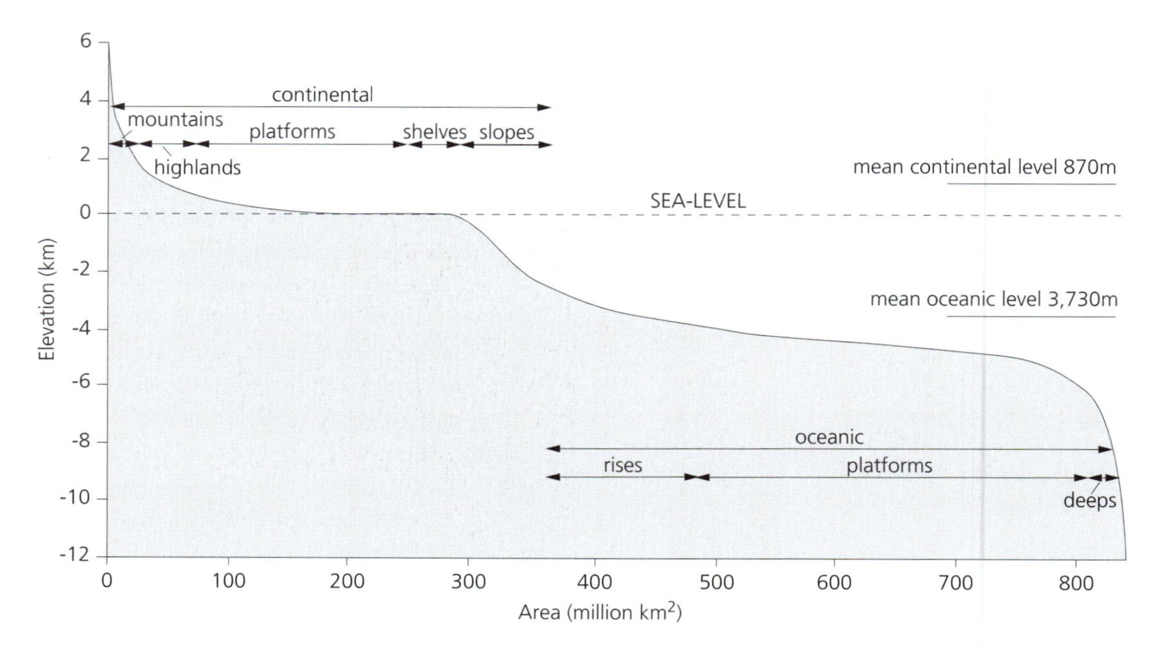

Figure 5.2 *Hypsometric curve of relief on the Earth. The hypsometric curve, which displays the relative distribution of relief, shows that most continental areas are below about 2,000 m above sea level, but most of the ocean floor is deeper than about 4,000 m below sea level.*

Source: Figure 5.3 in White, I.D., D.N. Mottershead and S.J. Harrison (1984) Environmental systems. George Allen & Unwin, London.

proportion of the land surface that is higher (or lower) than a given level. It provides a means of exploring how uniform or variable relief is in an area and of comparing the distribution of relief between areas.

INTERIOR OF THE EARTH

STRUCTURE AND BEHAVIOUR

The internal structure and composition of the Earth is a product of how the planet formed out of the swirling mass of cosmic dust and gas. Early in its life as something approaching a solid object, the different materials within the dust and gas sorted themselves out in terms of density (Figure 5.3). The sorting process was assisted by the heat of compression and by the radioactive decay of many of the elements present. As a result, the Earth's interior is not uniform but – rather like the concentric rings inside an onion – is composed of a series of layers. Unlike the onion rings, however, the layers inside the Earth are made of different materials, have different properties and behave in different ways.

To understand what happens on the face of the Earth, we must appreciate what goes on inside it. The outer shell

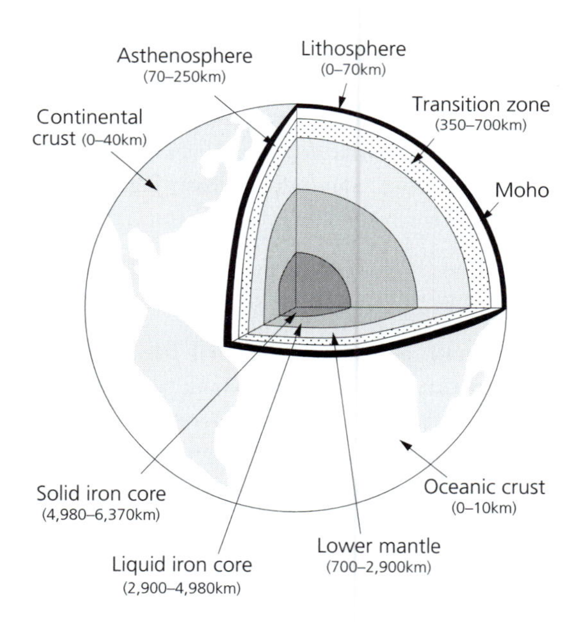

Figure 5.3 *Internal structure of the Earth. The vertical zones of the Earth's interior include the inner core, the outer core, the mantle and the crust.*

Source: Figure 2.1 in Norris, R.E. (1990) World regional geography. West Publishing Co., St Paul, Minnesota.

of the planet is solid, as we know from direct experience, but many of the materials inside it are viscous (thick, sticky) fluids, or at least they behave as if they were – they can move and deform, and they behave like soft melted plastic. Movement of this plastic material and interactions between it and the solid crust have significant impacts on many of the details of the Earth's surface via plate tectonics, mountain building and earth movements (Chapter 6).

EVIDENCE

Most of what we know about the interior of the Earth comes from indirect evidence. Direct evidence is restricted to the extremely thin outer part of the crust with which we have direct contact – we live on it, extract resources from it, tunnel into it, measure it and monitor it with scientific instruments. Given that the deepest wells extend down only about 11 km, and the centre of the Earth is 6,371 km beneath the surface (sea level), these direct sources yield a lot of useful information but only about a tiny part of the whole system. We barely scratch the surface, literally.

Much of our understanding of the internal composition of the Earth comes from the way in which seismic waves (caused by earthquakes or major explosions) are created and move. Other valuable information comes from gravity studies, particularly the pattern of variations of gravitational strength from place to place, and from studies of how magnetic fields vary. New technologies, particularly in remote sensing, are opening up new areas of discovery and exploration at great speed, and this enhanced capacity to study the Earth's interior promises to greatly increase understanding of its composition, properties and behaviour.

SEISMIC WAVES

Seismic waves are shock waves generated by earthquakes. They travel downwards from the surface of the Earth towards the centre and then return to the surface at various points. Geologists monitor these waves continuously on seismographs at a number of sites around the world.

The shock waves travel at different speeds, depending on the density of the material they pass through. There are two main types of seismic wave, S-waves (secondary waves) and P-waves (primary waves), which work in different ways and reveal different things (Box 5.3; Figure 5.4). The behaviour of these waves, particularly the time it takes for them to arrive at different monitoring sites, and the pattern of their arrival reveal a great deal about the properties of the material they have travelled through (such as its thickness, and whether is behaves like a plastic or a solid).

BOX 5.3 TWO MAIN TYPES OF SEISMIC WAVES

S-waves: these never reach seismograph stations on the side of the Earth opposite the point of origin of the earthquake. S-waves can only travel through rigid, non-liquid materials, so this suggests that the Earth's outer core is liquid, or at least partially liquid.

P-waves: these are refracted (bent) as they pass through the liquid outer core. As a result, seismograph stations located at certain distances from the earthquake receive neither S- nor P-waves; they are in the so-called *shadow zone*.

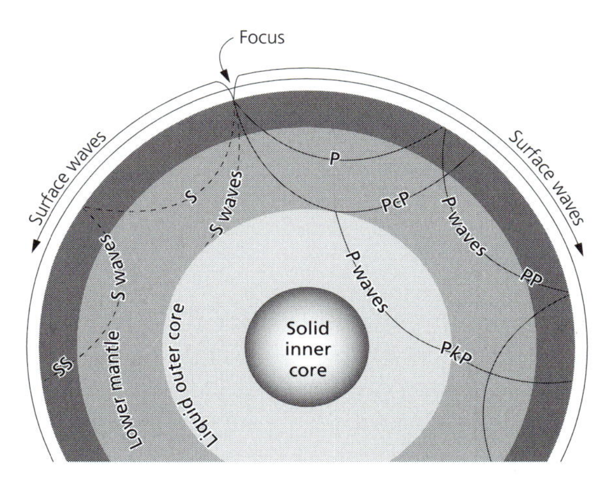

Figure 5.4 *Movement of seismic waves through the interior of the Earth. P-waves and S-waves radiate outwards in different directions from an earthquake focus. Waves reflected from the Earth's surface are known as PP or SS waves. PcP is a wave that bounces off the core, and PKP is a P-wave transmitted through the liquid core. S-waves cannot travel in a liquid.*

Source: Figure 17.33 in Press, F. and R. Siever (1978) Earth. W.H. Freeman, San Francisco.

ZONES AND MATERIALS

Detailed studies of seismic waves suggest that the Earth is made up of a series of concentric zones, each with different properties – the core, mantle and crust (see Figure 5.3). Each of the three zones has a different density (or specific gravity, meaning the ratio of that material's weight to that

Table 5.4 Internal composition of the Earth

Zone	Distance from surface (km)	Thickness (km)	Density (g cm^{-3})
Crust	0–60	5–32	2.8–3.0
Mantle	2,900	2,900	3.3–5.6
Outer core	5,000	2,100	9.5–12
Inner core	6,371	1,400	13+

of an equal volume of water), with density increasing towards the core (Table 5.4). The zones are composed of different materials, hence their different densities and properties. The mantle, for example, appears to be mainly peridotite, which consists almost entirely of the mineral olivine (an iron magnesium silicate), whereas the core appears to be mainly nickel and iron.

CORE

The Earth's core is divided into a relatively small inner core and a much thicker outer core. Each has different properties.

INNER CORE

The inner core has a radius of about 1,255 km, which is about 20 per cent of the total radius of the Earth. It is solid and is made up mainly of iron molecules with some nickel (a silvery-white magnetic metal) and possibly some silicon and sulphur. This material is extremely dense (more than thirteen times the density of surface water), extremely hot (temperatures are estimated at between 4,500 and 5,500 °C, almost as hot as the surface of the Sun), and under extremely high pressure (estimated at 3.5 million times atmospheric pressure). These extreme conditions mean that the iron material in the inner core behaves rather differently from how it would at the Earth's surface.

OUTER CORE

Surrounding the inner core is an outer core about 2,220 km thick. This is also made up mostly of iron and nickel, but this material is molten (in a liquid not a solid state). Conditions at the boundary of the outer core and the inner core are similar to those in the inner core, but heat and pressure decrease outwards. Thus temperatures drop from around 4,500 °C to less than 2,000 °C, and pressure drops by half. Density of material also decreases with distance from the inner core.

The Earth's magnetic field is believed to be generated by this molten iron–nickel outer core rotating around the solid inner core, like a dynamo.

MANTLE

Outside the outer core lies the mantle, which is about 2,900 km thick. It is made from rock that is dense by surface standards, but only as dense as material in the core. The mantle is often described as solid, but in reality it behaves like a plastic and can flow. It is composed mainly of minerals that are silicate compounds (combining silica and oxygen). Iron is the main metal, but there is also a great deal of magnesium.

The composition and behaviour of material in the mantle varies with distance from the core (below) and crust (above), and two main zones can be distinguished within it.

ASTHENOSPHERE

The inner part of the mantle is called the asthenosphere (see Figure 5.3). Its materials behave more like flowing plastic than solid rock. These soft, partially melted, semi-fluid rocks are constantly redistributed by slow-moving but large-scale convection currents generated by the immense heat and pressure.

The boundary between the core and the mantle is sharply defined but variable in character, and the way in which core and mantle materials interact has a strong influence on the Earth's wobble around its axis (precession; see p. 117). It also gives rise to the huge convection currents within the mantle, which ultimately drive plate tectonics and redistribute land and sea around the planet.

LITHOSPHERE

The outer part of the mantle is the lithosphere (see Figure 5.3), which is about 75 km thick. Unlike the asthenosphere, this material is rigid and behaves as a solid. It mostly merges with the crust, and many geologists take the lithosphere to consist of the crust and part of the upper mantle.

CRUST

A very thin crust surrounds the mantle, varying in thickness between about 8 and 65 km. This is effectively the Earth's outer skin, occupying much less than 1 per cent of the radius of the planet. The surface of the crust is the

BOX 5.4 THE MOHO

The sharp and well-defined junction between the mantle and the crust is called the *Mohorovicic discontinuity* (Moho for short), after the Yugoslav geophysicist who first suspected its presence in 1909. It follows variations in the thickness of the crust and is found roughly 32 km below the continents and about 10 km below the oceans. Seismic waves speed up noticeably when they reach the Moho.

surface of the Earth (Figure 5.5). All of the Earth's landforms (mountains, plains and plateaux) are contained within it, along with the oceans and seas. While the crust appears to be solid, it is subjected to repeated movement (including bending, folding and breaking) associated with the movement of material in the mantle below.

This is the part of the planet that we know most about, from direct experience. It is also the part that is most readily damaged by human impacts and the part that interacts with the other major environmental systems (the atmosphere, hydrosphere and biosphere). It sustains life on Earth.

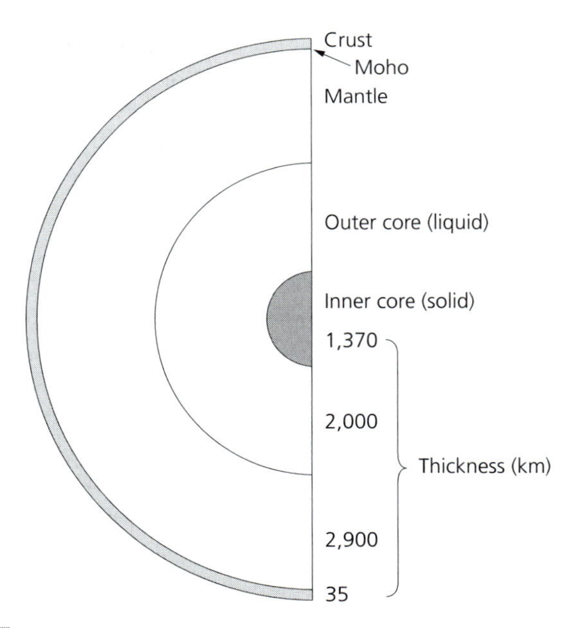

Figure 5.5 *The Earth's mantle and crust. Beyond the thin crust is the relatively thick mantle. The Earth's core is solid, with a liquid outer part.*

Source: Figure 1.1a in Goudie, A. (1993) The nature of the environment. Blackwell, Oxford.

MATERIAL

The crust is composed of a range of materials, many of them in very small amounts (Table 5.5). It contains the least dense rock material in the Earth (see Table 5.4) and appears to float on the denser material beneath it in the mantle. Crustal thickness varies a great deal from place to place, reflecting two very different types of crustal material – oceanic and continental crust (Box 5.5; Figure 5.6).

The material in the oceanic crust is more ductile and deforms without fracturing when subjected to geological stress. It flows and stretches, rather than cracks or breaks. Material in the continental crust, on the other hand, is more brittle. Under stress it fractures along individual faults, with blocks in between that slide against each other, or it is deformed (for example by folding) like a crumpled sheet of paper.

Because of the movements of plate tectonics, the oceanic crust is nowhere older than about 200 million years. Seafloor spreading is continuously creating new oceanic crust, and subduction (sinking) is continuously destroying existing crust, like some huge non-stop geological conveyor belt. Overall, the continental crust is more stable and thus older; parts of it – the shield areas (Figure 5.7) – are more than 3,000 million years old. The continental crust is forever

Table 5.5 Composition of the Earth's crust

Element	*Symbol*	*Percentage*	*Cumulative percentage*
Oxygen	O	46.60	46.60
Silicon	Si	27.72	74.32
Aluminium	Al	8.13	82.45
Iron	Fe	5.00	87.45
Calcium	Ca	3.63	91.08
Sodium	Na	2.83	93.91
Potassium	K	2.59	96.50
Magnesium	Mg	2.09	98.59
Titanium	Ti	0.44	99.03
Hydrogen	H	0.14	99.17
Phosphorus	P	0.12	99.29
Manganese	Mn	0.10	99.39
Fluorine	F	0.08	99.47
Sulphur	S	0.05	99.52
Chlorine	Cl	0.04	99.56
Carbon	C	0.03	99.59
Others		0.41	100.00

BOX 5.5 OCEANIC AND CONTINENTAL CRUST

Oceanic crust: the crust is thinner (thus denser and heavier) beneath the oceans, where it is up to about 8 km thick. It consists mostly of basaltic types of rock. Geologists refer to this oceanic crustal material as *sima*, after its two main chemical constituents, which are silica (si) and magnesium (ma).

Continental crust: the crust is generally much thicker (up to 65 km thick) beneath the continents, but it is also less dense and lighter than the oceanic crust. The lighter but discontinuous continental crust appears to float on top of the denser but more or less continuous oceanic crust, like a series of rafts floating on the sea. It is mainly granitic in composition and is more complex in structure than the oceanic crust. Geologists refer to the continental crustal material as *sial* because its main components are silica (si) and alumina (al).

subjected to destruction and rifting (splitting) at plate margins, and to new growth by the addition of igneous material from the mantle.

Both sial and sima rocks are often found at or close to the Earth's surface. They are broken down by weathering processes (p. 204), producing weathered products, which are then deposited elsewhere in the rock cycle as sedimentary rocks (p. 202).

APPLICATIONS

There are a number of applications that relate to the material and composition of the Earth's interior, particularly the crust. We look at a selection of four of them here – radon gas, long-term underground storage of nuclear waste, hot springs and geysers, and the exploitation of geothermal energy resources.

RADON GAS

There is mounting concern in many countries about the health risks associated with exposure to radioactivity in the environment. It is known to cause cancers and premature death, and the health risks appear to be on the increase. Pollution from the nuclear power industry and similar sources is a major source of radioactivity, and one that can be controlled much more tightly than at present if not totally eliminated. But most people in most places are exposed to more radioactivity from natural than from artificial sources. About 60 per cent of background radiation comes from radioactivity that is naturally present in Earth materials, and much of the rest comes from space.

One of the more important natural sources of radioactivity in the environment is the generation of radon from the natural decay of radium, a rare radioactive element that occurs in uranium ores (such as pitchblende). Radon (Rn) is a colourless, odourless, radioactive gas. It is one of the inert or rare gases, like helium and neon, which are unreactive (meaning that they do not readily take part in chemical reactions). The rare gases are thus very stable.

BOX 5.6 RAW MATERIALS IN THE EARTH'S CRUST

The Earth's crust provides the basic raw materials – including building materials, fossil fuel energy resources, metals and minerals – on which society depends. Nearly three-quarters of the crust is oxygen and silicon, and the other important elements are present in much smaller quantities. About ninety elements make up Earth's crust, but over 98 per cent of it is accounted for by just eight of them (Table 5.5). Some of these play important roles in the global biogeochemical cycles (see p. 86).

BOX 5.7 RADIOACTIVE ISOTOPES

All radioactive elements have more than one isotope (atoms, or types, with identical chemical behaviour but differing physical behaviour), and radon is no exception. The most stable radon isotope is radon-222. Radioisotopes give off radiation and revert through time to their natural form. This is known as radioactive decay, and each isotope has its own rate of decay, expressed by its *half-life* (the time taken for the radioactivity to fall by half). Isotopes with long half-lives remain in the environment (and often pose a threat to people) for a long time, those with short half-lives disappear fast. Radon-222 has a very short half-life of 3.82 days. Compared with radium-226 (half-life of 1,622 years) or thorium-232 (half-life of 14 million years), radon gas disappears very quickly indeed.

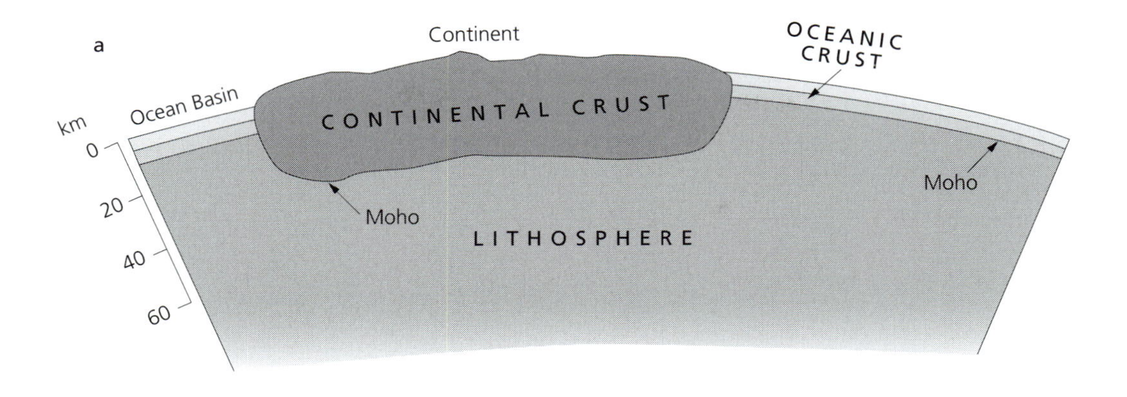

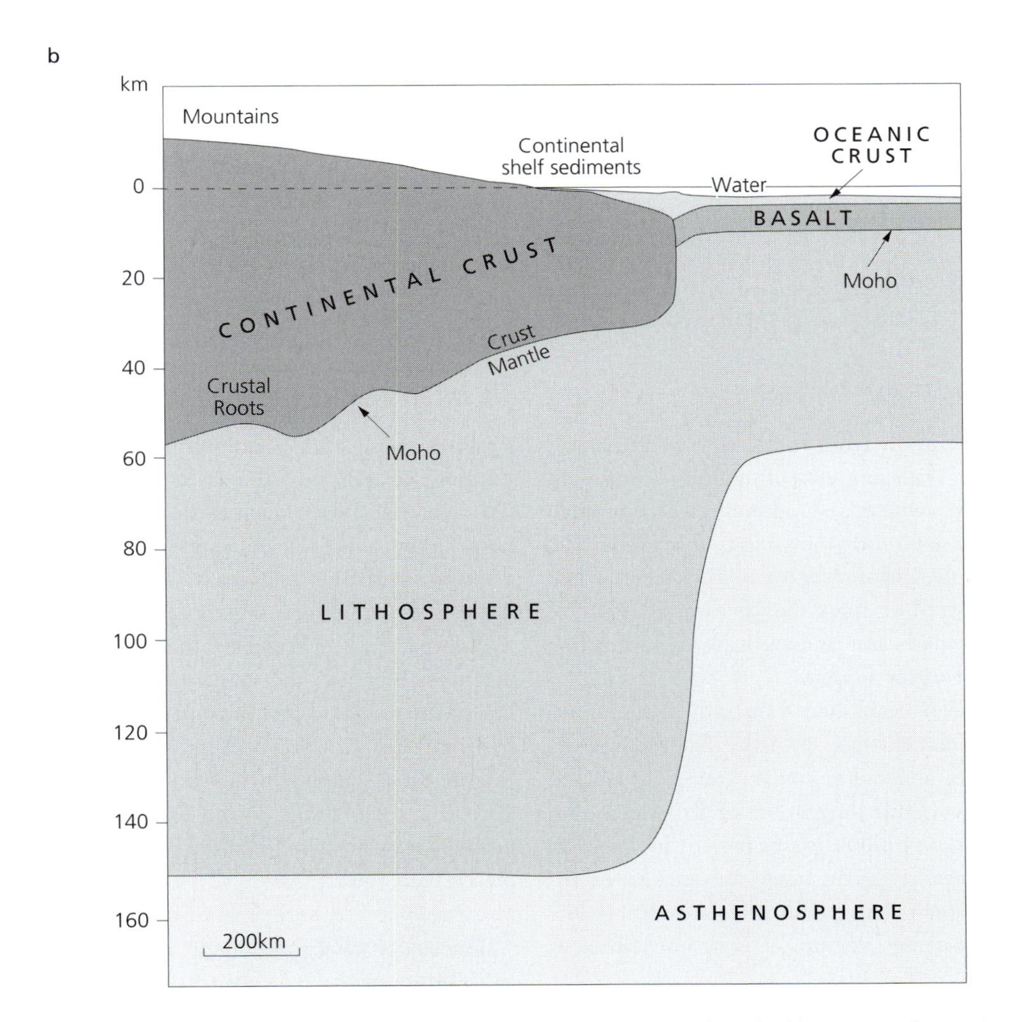

Figure 5.6 *Materials in the Earth's crust. The thick continental crust and much thinner oceanic crust are separated from the mantle (lithosphere) below by the Moho (a), and the boundary between the lithosphere and the asthenosphere is depressed below the blocks of continental crust (b).*

Source: Figure 11.2 in Strahler, A.H. and A.N. Strahler (1994) Introducing physical geography. John Wiley & Sons, New York.

Figure 5.7 *Major shield areas on the Earth's crust. The shield areas are remnants of the oldest rocks on the Earth's surface, having cooled and solidified from molten magma.*

Source: Figure 1.4 in Goudie, A. (1993) The nature of the environment. Blackwell, Oxford.

PRODUCTION AND RELEASE

Radon is released naturally from rocks, soils and minerals that contain radium. Trace amounts of uranium are present in rocks and soils everywhere, and radon is released in small concentrations from most soil grains and rock surfaces. The gas is normally trapped in underground rocks, but when the rocks are weathered or crack the gas escapes. Some of it is dissolved in groundwater and can cause a health risk by contamination of water supplies.

Links have also been established – along the San Andreas fault system in California (see p. 161), for example – between earthquakes and radon concentrations in air and groundwater. It appears that the amount of dissolved radon in water from wells, and radon gas in the air, increases as strain increases in nearby rocks, suggesting that an earthquake may be imminent. This might offer a quite reliable earthquake early warning system, if instrumentation is in place.

Most of the radon escapes directly into the lower atmosphere, where it is mixed, dispersed and diluted. It normally disappears quickly, because of its short half-life and rapid dispersion by winds, preventing the build-up of dangerously high concentrations. Traces of radon appear in the air in many places, but in concentrations that are barely detectable.

HEALTH RISKS

Problems arise when some of the radon is prevented from escaping directly into the air. It readily collects in places like non-coal mines (such as tin and uranium mines) and homes (Figure 5.8). If not moved outside, usually by some artificial ventilation system, it can accumulate and pose a health risk for humans who breathe it in.

The gas and its daughter (decay) products emit alpha particles (Box 5.8) and may cause lung cancer. More than half of the radiation that most people living in typical parts of North America are exposed to comes from radon gas (Figure 5.9). Radon is believed to be responsible for up to 20,000 cases of lung cancer in the USA each year. The problem is by no means confined to the USA: it affects most places with suitable geological conditions, including New Mexico (Box 5.9) and parts of Britain.

Unusually (and potentially dangerously) high radon concentrations have been recorded in houses in parts of Britain such as Devon, Cornwall and south-west Scotland, usually adjacent to granite rocks or old tin-mining regions. Radon levels up to 300 becquerels m^{-3} have been found, which are one and a half times the levels at which action is recommended to reduce concentrations. The risk of at least some residents contracting lung cancer seems quite real.

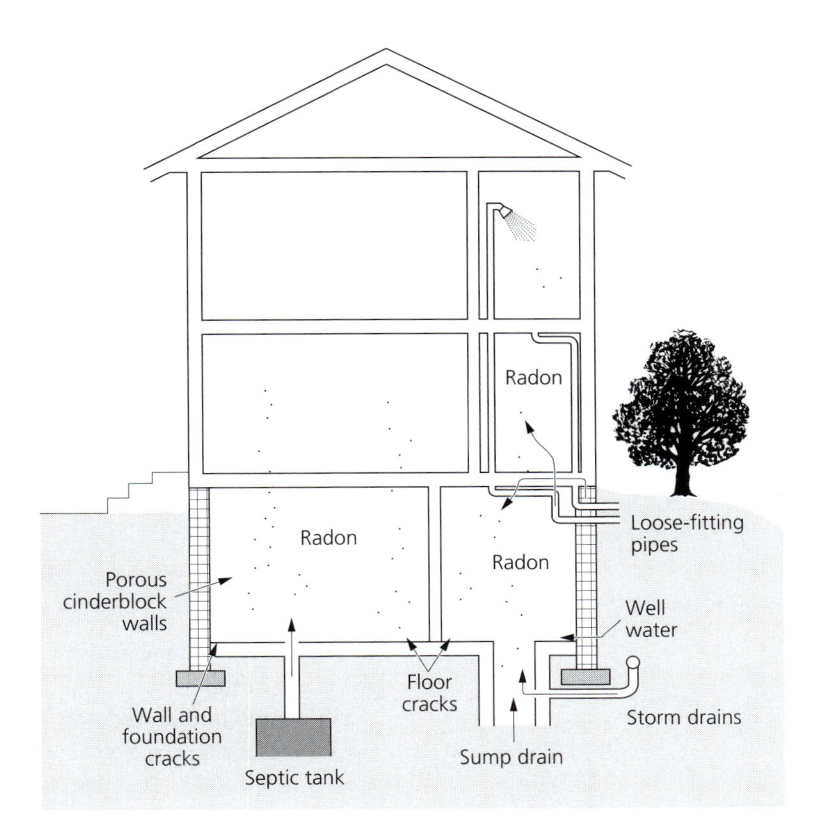

Figure 5.8 *Main entry points for radon gas into a house. Radon gas, which is naturally present in some rocks, can enter buildings in a variety of ways. Most involve seepage of the gas through cracks, porous material, pipes and drains.*

Source: Figure 9.12 in Marsh, W.M. and J.M. Grossa (1996) Environmental geography: science, land use and Earth systems. John Wiley & Sons, New York.

BOX 5.8 ALPHA PARTICLES

Alpha particles:

- have a positive charge;
- are emitted from the nucleus of an atom;
- cause high ionisation;
- are very large compared with other radiating particles;
- lose energy very quickly;
- cannot penetrate very far into materials, even soft tissue (where they penetrate to a depth of 0.001–0.007 cm); but once they are inside a body (by inhalation or through a wound) alpha particles are biologically very damaging.

Radon levels in buildings can usually be reduced by increasing the circulation of air, either by opening windows and doors or by installing relatively simply ventilation systems. In most cases, the capital and operating costs of preventive or remedial measures are low.

UNDERGROUND STORAGE OF NUCLEAR WASTE

Another environmental problem related to the Earth's interior is the safe management of toxic or hazardous wastes. Waste management is based on the management of wastes at all stages – from production, handling, storage, transport, processing and ultimate disposal – in such a way as to minimise the risks to human health, wildlife and environmental systems.

Traditional approaches to the management of non-toxic waste rely heavily on sanitary landfill, in which material is simply dumped in holes in the ground and then covered with soil. Disposing of toxic wastes in this cavalier manner

BOX 5.9 INDOOR RADON IN NEW MEXICO

The build-up of radon in homes is widely recognised around the world as a potentially serious health hazard. Up to 30,000 people in the USA die in a typical year because of indoor radon, according to the US Environmental Protection Agency (EPA).

In order to understand better how the problem arises, a detailed study was carried out in Albuquerque, New Mexico. This area has relatively young soils, and granite bedrock with a higher than normal uranium content. Indoor radon levels were measured in 180 houses during the winter of 1986–87. Radon levels were found to be higher than the EPA's suggested maximum permissible level in 28 per cent of houses. High radon concentrations are associated with:

- closeness to the Sandia mountains (source of the granite);
- places with high radon concentrations in the soil;
- houses with excessive insulation (where there is no natural ventilation of air);
- houses with solar capacities (such as solar panels and photovoltaic cells).

Some building materials may provide very minor sources of some indoor radon.

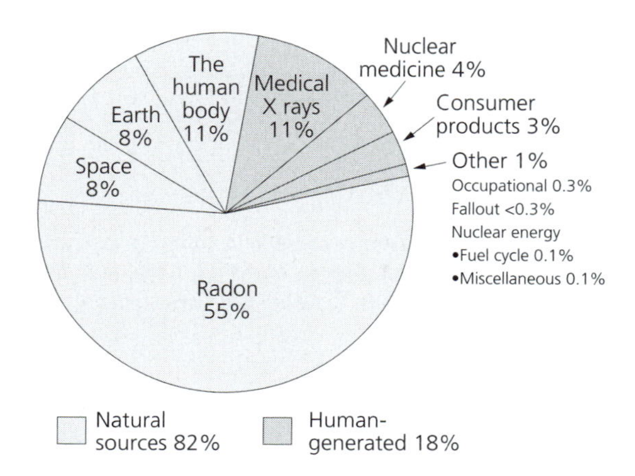

Figure 5.9 *Sources of radiation to which humans are exposed in the USA. Most (82 per cent) of the radiation to which people are exposed comes from natural sources, including radioactive elements in the Earth and in our bodies, cosmic radiation from space, and radon gas. Of the 18 per cent that comes from human activities, most is from medical X-rays and nuclear medicine.*

Source: Figure 10.5 in Cunningham, W.P. and B.W. Saigo (1992) Environmental science: a global concern. Wm. C. Brown Publishers, Dubuque, Iowa.

is unsuitable, largely because of the risk of contaminating groundwater.

The nuclear industry produces sizeable amounts of waste (see Figure 1.5). Much of it is perfectly safe and can be disposed of in traditional ways, but some nuclear waste is highly radioactive, poses serious health risks to people and wildlife, and remains dangerously radioactive for long periods of time (hundreds of thousands of years or longer for many isotopes).

When the nuclear industry was launched in the 1950s, no serious attempt was made to deal with the radioactive wastes it produces. Only now are the long-term problems being taken seriously. There is mounting concern in the West about the ultimate fate of much of the radioactive waste from commercial reactors in the former Soviet Union, where storage practices might not have been as safe as those in the West. Large areas appear to be threatened by radioactive pollution of soils, groundwater and air. This will require a vast clean-up strategy and heavy investment to put right.

This type of waste requires special management. Reprocessing of most of the radioactive waste is a long-term objective and occurs to a limited extent already. Spent nuclear fuel from most commercial reactors in Western Europe is now reprocessed at Sellafield in the UK and La Hague and Marcoule in France, and all three plants import material from other European countries. But the quantities produced are so large, reprocessing costs are so high, and not all radioactive waste can be reprocessed, that the only realistic strategy currently available (and economically viable) is long-term safe storage. Deep underground storage is a preferred option, and vast amounts of resources have been invested in the search for suitable repository sites in the USA and the UK (Box 5.10).

Nuclear waste management in the USA requires long-term underground storage of radioactive waste, but efforts to set up such repositories have been hampered by public opposition. The controversy surrounding the proposed repository at Yucca Mountain in Nevada (Box 5.11; Figure 5.10) illustrates some of the complexity of the issues involved. By the close of the 1990s there were still many problems surrounding the Yucca Mountain scheme, including fierce public opposition and continued uncertainty about long-term safety issues. By early 2000, proposals were being evaluated for international nuclear waste repositories at Krasnoyarsk and Mayak in Russia, in

BOX 5.10 KEY SITE REQUIREMENTS FOR NUCLEAR WASTE REPOSITORIES

- The site must be away from major centres of population.
- The site must be accessible by road and rail.
- Geological conditions must be suitable (it must be possible to tunnel, for example).
- Prospects of minimising groundwater contamination must be good.
- It must be possible to store radioactive waste there safely for at least 10,000 years.
- The site must be geologically stable (away from fault lines and earthquake zones).

BOX 5.11 THE YUCCA MOUNTAIN NUCLEAR REPOSITORY PROJECT

The problems of building an underground storage facility for the long-term storage of radioactive nuclear waste had never been confronted in the USA (or anywhere else) before the late 1980s. Finding a site for the safe storage of high-level nuclear waste for up to 10,000 years is no simple task! It was clear from the outset that the first attempts would be controversial and would bring into focus many of the key issues and difficulties.

Because the need to use such a repository was becoming more urgent, as the stocks of nuclear waste continued to pile up, some short-cuts in normal planning procedures were regarded as acceptable. The Nuclear Waste Policy Act of 1982, for example, exempted repository siting from some of the important requirements for environmental review that had been established by the 1969 National Environmental Policy Act (NEPA) and had become normal practice.

In December 1987, Yucca Mountain in Nevada was named as the first site for the development of a high-level nuclear waste repository. It was selected from a short-list of suitable sites that included Hanford Reservation in Washington and seven sites in Louisiana, Mississippi, Texas and Utah. Yucca Mountain borders the Nevada nuclear test site.

The Yucca Mountain repository proposal sparked great controversy among specialists and raised widespread concern among the general public (not just local residents, but across the USA). The debate centres on a number of issues, including:

- The repository technology is totally new and unproven.
- Some geological questions remain about the suitability of the site itself.
- No proven methods of risk assessment are available for this type of development, so it is difficult to establish just how hazardous it is.

There are still too many unknowns for the Yucca Mountain project to be welcomed by everyone concerned. A better understanding of the geology and environment of the site would probably help to reduce anxiety about issues such as possible bedrock instability and groundwater contamination. Inevitably, a scheme like this involves a great deal of scientific and technical uncertainty, but one of the unknowns is what level of uncertainty is acceptable. There is also the question: 'acceptable to whom?'

Critics argue that the approach used to select and develop Yucca Mountain represents a significant departure from the more traditional, comprehensive and interdisciplinary environmental review for siting nuclear projects in the USA. They recognise that something has to be done about the growing nuclear waste stockpile and accept that we are dealing with complex technologies and experimental approaches, but they insist that corners are being cut, risks are being ignored or downplayed, and the approach is treating the symptoms (disposal of nuclear waste) not the causes (generation of nuclear waste) of the problem.

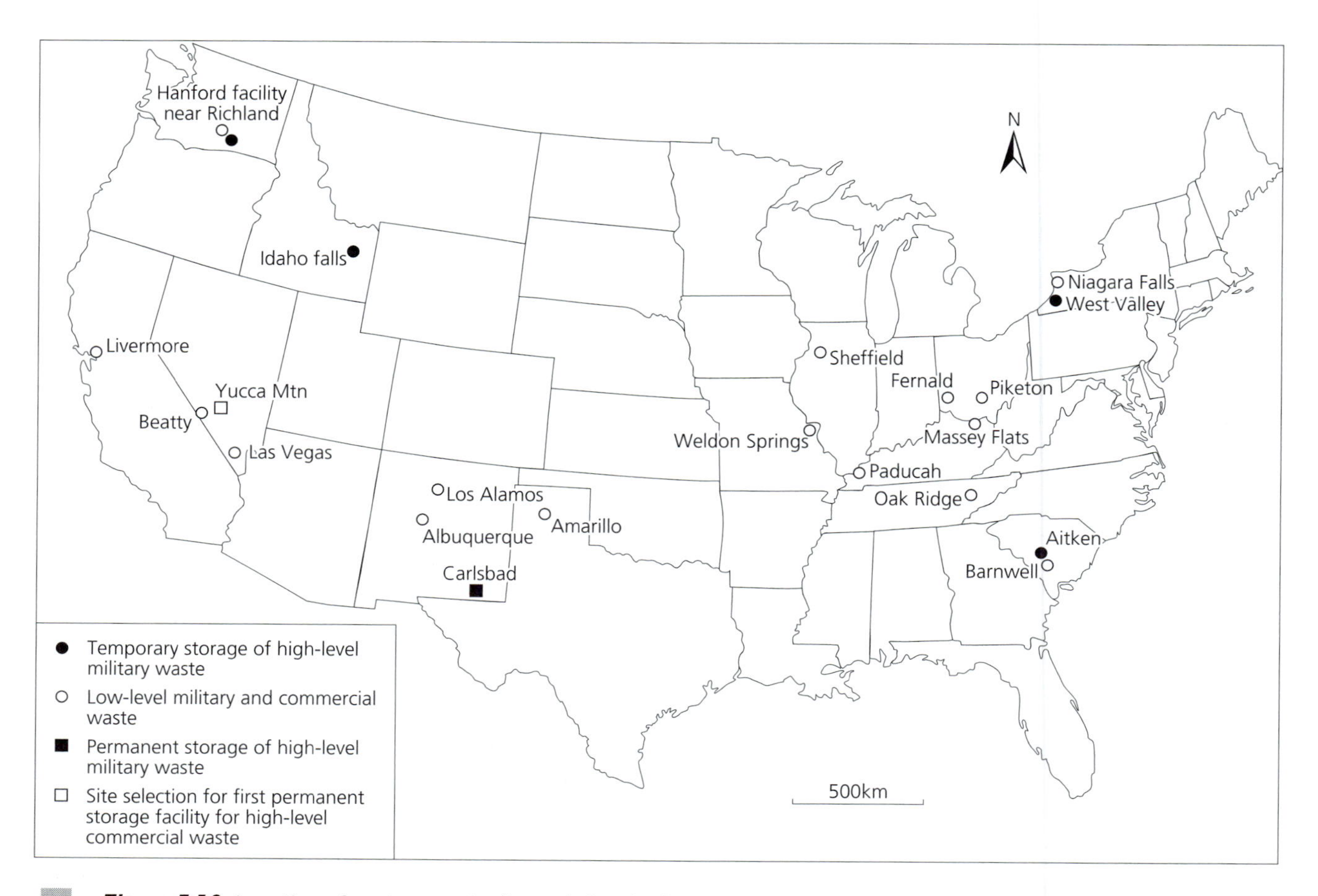

Figure 5.10 *Location of nuclear waste disposal sites in the USA. Most of the sites store low-level radioactive wastes, but Carlsbad (New Mexico) and Yucca Mountain (Nevada) are designed for the safe long-term underground storage of high-level waste.*

Source: Figure 19.14 in Cunningham, W.P. and B.W. Saigo (1992) Environmental science: a global concern. Wm. C. Brown Publishers, Dubuque, Iowa.

the Australian outback, and in South Africa. Public opposition to each scheme has been very strong and very vocal.

Similar public concern has been voiced in Britain over plans for an underground repository at Sellafield in Cumbria. United Kingdom Nirex Limited (Nirex) is responsible for developing and managing a facility for the safe deep disposal of intermediate (ILW) and some low-level (LLW) radioactive waste. In October 1992, the company announced plans to investigate the proposed site seriously to a depth of 800 m below sea level in the underlying volcanic rocks, using seismic surveys, geophysical sensing and geological mapping. Public opposition centred on the health risks and pollution potential of the repository, and on its impacts on the landscape at the edge of the Lake District National Park, and the planning application was refused.

The Sellafield complex faces other nuclear safety challenges too, because it stores about 1,000 m^3 of high-level waste, which comes from the reprocessing of spent fuel from nuclear power stations and military reactors. High-level waste is the longest-lasting, hottest and most dangerous of all the radioactive waste produced by the industry, and by early 2000 the plant's safety record was falling under extremely close scrutiny.

Since the 1960s, more than 200,000 tonnes of spent fuel has been produced by 400 reactors in thirty countries. Each year, an extra 10,000 tonnes of nuclear waste is added to the stockpile. Britain's Royal Commission on Environmental Pollution concluded in 1976 that 'radioactive waste management is a profoundly serious issue . . . there has been a diffuse pattern of responsibility with the result that too little has been done'. Sadly, this is still true today, worldwide.

HOT SPRINGS AND GEYSERS

A geyser is a natural spring that discharges steam and hot water, intermittently releasing an explosive column into the air. The conditions that give rise to this are created in the outer part of the Earth's crust. Geysers tend to be given exotic names and, because they are unusual and dramatic when they erupt, they often attract large numbers of visitors and sightseers. Old Faithful, in Yellowstone National Park in Wyoming, is a good example on both counts. Geyser systems are effectively the steam vents on natural pressure cookers, in which water in a parcel of permeable rock is superheated and turns to steam, with intermittent explosive hydrothermal eruptions. They occur most commonly in heavily fractured permeable rocks surrounded by less permeable rock.

Most geysers erupt at irregular intervals, displaying chaotic (disorderly) behaviour. Regular eruption cycles are much less common, and even the behaviour of Old Faithful appears to be chaotic despite folklore to the contrary. Eruption frequency varies through time in most geysers, apparently controlled largely by lateral recharge rates. This is the speed at which water that has been released is replaced by seepage from surrounding rocks. Recharge rates are partly dependent on local seismic stress, which explains why eruption frequency appears to vary with strains triggered by seismic events (such as earthquakes), atmospheric loading (which changes with air pressure and thus weather systems), and even the Earth's tides.

Geysers provide valuable indicators of the state and likely activity patterns of other environmental systems. Between 1973 and 1991, for example, a periodic geyser in northern California regularly erupted between one and three days before the three largest earthquakes between 130 km and 250 km away (including the major Loma Prieta earthquake in 1989; see p. 170). Such advance warning might be extremely useful in earthquake prediction.

BOX 5.12 IMPORTANCE OF GEYSERS

Geysers have practical application, as well as scientific curiosity and tourist potential, because they occur in areas with particularly active hydrothermal geological conditions. Extraction of geothermal heat and energy often occurs close to geyser sites, and in parts of Iceland the hot water is used directly in district heating schemes. Extraction of geothermal heat for energy use can seriously disturb geyser systems by reducing heat flow and aquifer (groundwater) pressure. Over-extraction from the Rotorua geothermal field in New Zealand during the 1980s caused hot springs to stop flowing and geysers to behave abnormally, for example.

In some places, such as Yellowstone National Park, the distribution in rocks of valuable elements like gold seems to be closely related to known networks of fossil hot spring systems. The belief is that the hot springs transported these elements along their underground networks and deposited them close to the ground surface when they were active in the past.

GEOTHERMAL ENERGY

GEOTHERMAL GRADIENT

Direct observation in places such as deep mines and wells has shown that temperature rises with increased depth below the Earth's surface. This geothermal gradient varies from place to place but is on average 20–40 °C for every km of depth.

EXPLOITATION OF GEOTHERMAL ENERGY

In some places, the geothermal gradient is much steeper than the norm, and this can be exploited as a source of geothermal energy. In a few places, igneous rocks, usually within 10 km of the Earth's surface, are in a molten or partly molten state. These so-called 'hot rocks' conduct heat from the Earth's interior up to the surface, where it can be extracted and used as a source of energy (Box 5.13), which can be used for a variety of purposes (Box 5.14). Areas where the geothermal gradient is steep are referred to as geothermal belts (Figure 5.11).

By the early 1990s, geothermal energy fields were being exploited in more than twenty countries, including New Zealand, Iceland, the USA and the former Soviet Union. Only four European countries were then using geothermal resources for power generation – Italy, Turkey, Portugal and Greece. Italy had by far the largest capacity, at around 500 Mw of electricity, compared with less than 20 Mw in the other three countries. Geothermal energy is an important source of energy in some areas, particularly in Iceland and parts of New Zealand. Direct heating use of geothermal energy in the USA (Table 5.6) occurs mainly in the western states, particularly southern California.

Geothermal power plants are usually relatively small and serve local rather than regional or national needs. The fraction of total world energy needs that is provided by geothermal sources is currently very small but is rising rapidly (Table 5.7). Installed capacity in 1994 was nearly 6,300 Mw of electricity, but demand was expected to rise by about 4 per cent a year for electricity and 10 per cent a year for direct use. Installed capacity worldwide rose by 13.5 per cent a year between 1980 and 1984, by 9.6 per cent a year between 1985 and 1989, and by 0.5 per cent a year between 1990 and 1992. Further increases, in the order of 8 per cent a year, were expected up to the year 2000.

BOX 5.13 GENERATION OF GEOTHERMAL HEAT

Heat within the Earth is generated by a number of processes, including:

- radioactive decay of rocks, particularly at depths of between 40 and 48 km;
- heat generated by compression of material at great depths;
- some residual heat, surviving from the initial formation of the Earth and now trapped inside the planet.

Temperatures in the crust are much lower than those in the mantle and core (see p. 130), and crustal rocks are normally solid and behave as solids. Rock has a very low thermal conductivity, so little of the heat reaches the surface. It has no known direct effect on present climate; nor can humans detect the subtle differences in rock temperature from place to place.

BOX 5.14 USE OF GEOTHERMAL HEAT

There are many different direct uses of geothermal energy, including direct heating (Iceland), industrial processing (New Zealand), greenhouse heating (Hungary) and heating bathing water (Japan). Space and district heating accounted for almost half of the US direct heat use in the late 1980s (Table 5.6). In some parts of China, geothermal energy is used to heat greenhouses, in fish farming, poultry incubation, irrigation, leather processing and animal bathing. In Russia, geothermal energy is used mainly for space and district heating, and for a range of industrial and agricultural purposes. Six Russian towns, housing about 100,000 people, were being heated by geothermal district heating systems in the early 1990s. Geothermal energy provides heating and hot water for over 85 per cent of Iceland's housing.

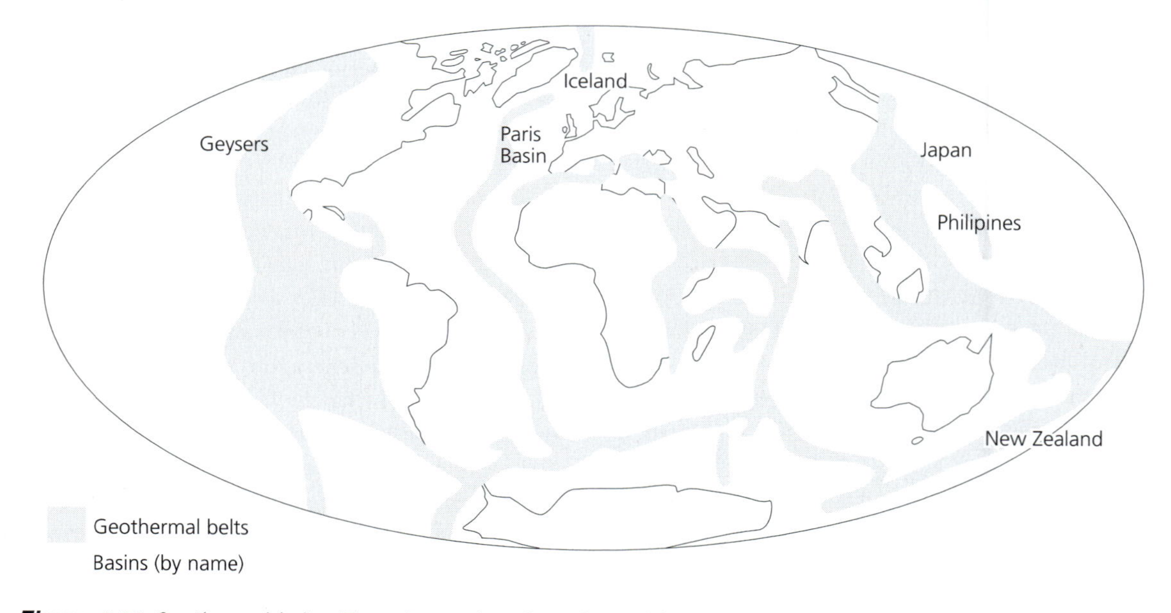

***Figure** 5.11 Geothermal belts. There is a series of geothermal belts around the Earth's crust, where the natural geothermal gradient is unusually steep. The basins in these belts are particularly suitable for the extraction of geothermal energy.*

Source: Figure 1 in Harrison, R. (1992) Applications of geothermal energy. Endeavour 16 (1): 29–36.

Table 5.6 Direct heat use of geothermal energy in the USA, 1988

Use	*Percentage*
Space and district heating	45.8
Greenhouses	20.2
Fish farming	20.7
Industrial processing	6.7
Pools and spas	6.6
Total	100

Table 5.7 Geothermal electricity production, 1980–2000

Year	*Installed capacity (megawatts)*
1980	1,960
1985	3,968
1990	5,835
1993	5,915
2000	10,200 (projected)

Great investments are being made to exploit commercially the geothermal energy potential of these fields, and to discover new fields. Central America, parts of South-east Asia and the western USA have great geothermal energy potential, and promising sites also exist in southern Europe and East Africa.

As a viable energy source geothermal power has a bright future, but large-scale exploitation is unlikely in the foreseeable future beyond areas with high thermal gradients (which tend to be the seismically active zones; see p. 160). Most areas that have normal thermal gradients will be incapable of yielding enough heat with existing technologies, and at acceptable costs, to generate electricity. Despite this, many such areas are suitable for developing geothermal heating, where heat is used directly to warm buildings.

Geothermal power also offers a sustainable energy source, provided that it is managed appropriately. While in theory it is a renewable source of energy, in practice heat is removed from the rock faster than it is replaced by natural processes (particularly the radioactive decay of rocks).

BOX 5.15 CONSERVATION AND GEOTHERMAL ENERGY DEVELOPMENT IN HAWAII

The US state of Hawaii consists of over twenty volcanic islands and atolls in the central Pacific. It has no fossil fuel reserves of its own, although its location endows it with abundant potential for developing renewable energy sources, including solar, tidal, wave and wind power. Its geological basement, an active volcanic zone (see p. 173), also offers great potential for geothermal energy development. Kilauea – a crater on the east side of the volcano of Mauna Loa, on the south-east of the island of Hawaii – is one of only two basaltic volcanoes in the world where geothermal power has been produced commercially.

Since the early 1970s, the Hawaiian authorities have explored ways of making use of the area's apparently massive geothermal energy resources to produce electricity. They hope to promote private investment in developing the energy resource by offering commercial incentives, streamlining the permit process and researching how much energy could realistically be exploited.

One such scheme would generate about 500 Mw of electricity and cost an estimated US$4 billion. Energy would be generated from boreholes located on the east rift zone of Kilauea, and power would be distributed from there to the island of Oahu about 230 km away by a deep undersea cable. Cable laying itself would be no easy task, because it would have to negotiate some of the most unstable trenches in the Pacific at depths of over 2,000 m.

The proposed scheme has attracted a great deal of public opposition, for a number of reasons:

- It would involve destroying America's only lowland tropical forest (Nao Kele O Puna), which houses a rich and unique range of plants and animals.
- It would release hydrogen sulphide, a toxic gas, into the atmosphere.
- It would seriously disturb if not totally destroy traditional local cultures.

Critics point out that the energy problem in Hawaii is probably better tackled by managing energy demand (by better energy conservation and increasing efficiency of energy use) than by increasing energy supply.

APPROACHES

Geothermal heat is normally trapped at depth and does not naturally reach the surface. It can be tapped if a borehole is drilled into the thermal reservoir (hot rock). To be useful, the geothermal energy must be available in the form of superheated water or steam. Hot water is pumped to the surface and converted to steam or run through a heat exchanger (Figure 5.12). Dry steam is pumped to the surface and can then be directed through turbines to generate electricity.

In volcanic regions – such as New Zealand, Japan and Iceland – water heated beneath the ground may erupt at the surface as geysers, hot springs or boiling mud. This heated material can sometimes be used directly, saving the cost of pumping to the surface. There are many different uses of geothermal heat (Figure 5.13).

An alternative approach is the 'hot dry rock' system, which at least in theory allows geothermal energy to be exploited. Cold water is pumped through fractured hot dry rock deep underground. It is heated by contact with the rock and is then extracted and used for heating and electricity generation. Experiments in Cornwall in south-west England have shown that the approach has real potential, but there are technical and economic problems to be resolved before it can be regarded as commercially viable.

ENVIRONMENTAL IMPACTS

Geothermal energy plants can cause heat pollution (which damages fish in local rivers, for example) if they are not carefully controlled. Underground heat transfer systems can be used to reduce the risk of upsetting local natural thermal regimes. In the 'hot dry rock' approach the heated water often contains toxic compounds that have been dissolved from the rocks it has passed through. This waste material would damage rivers, lakes and the sea if it were allowed to seep from the site, so it has to be stored safely.

Other possible environmental impacts of geothermal energy development include:

- surface disturbances
- physical effects of fluid withdrawal (including subsidence)
- noise
- thermal effects
- emission of chemicals.

Some impacts can be minimised by re-injecting the used fluids underground.

On balance, this form of energy use is much less damaging to the environment than continued burning of fossil fuels, which emit greenhouse gases, and continued

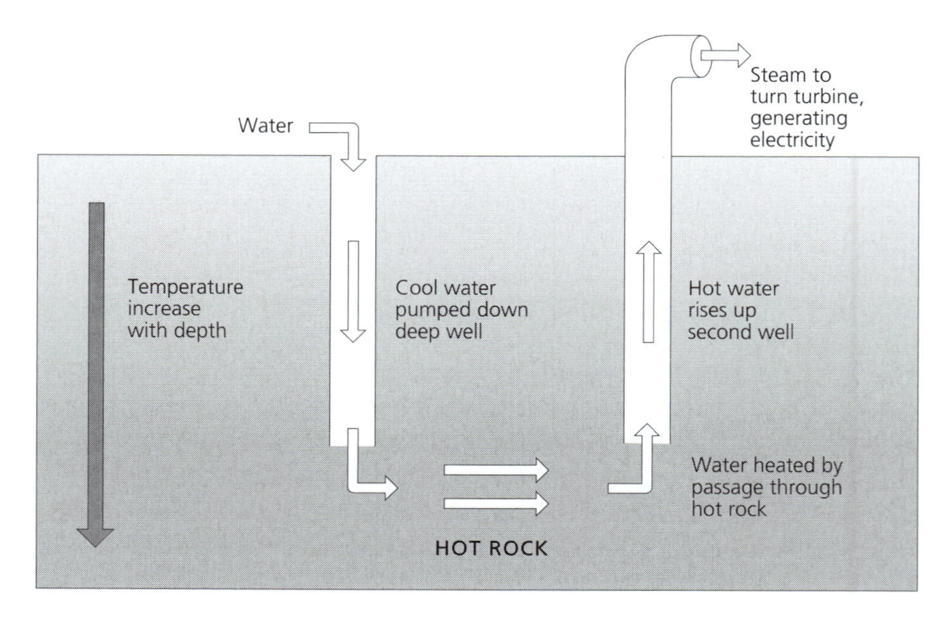

Figure 5.12 *Geothermal heat extraction for electricity production. Cool water is pumped down deep wells to the relatively hot rocks below, where it is heated before being pumped to the surface again as steam to drive turbines to produce electricity.*

Source: Figure 14.13 in Doerr, A.H. (1990) Fundamentals of physical geography. Wm. C. Brown Publishers, Dubuque, Iowa.

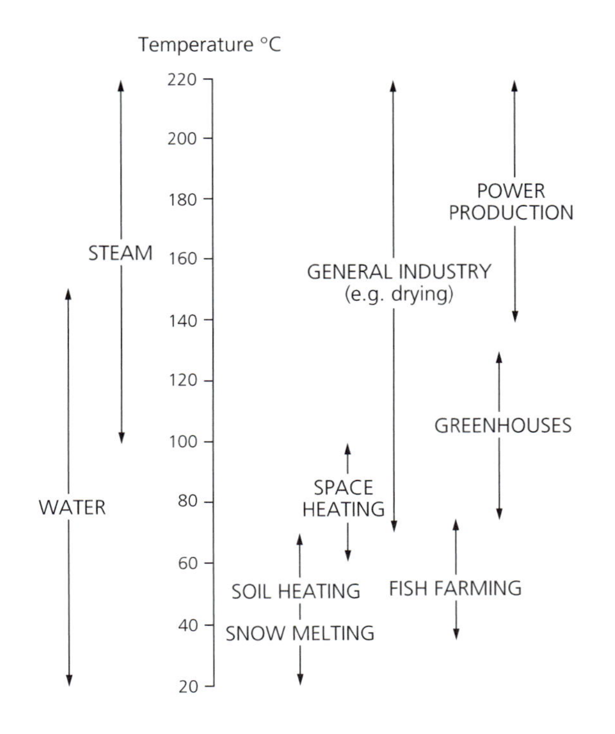

development of nuclear power. Its impacts are smaller, less hazardous to people, less damaging to the environment and much more localised. With careful planning and management, most if not all of them can be avoided.

WEBSITE

Links to relevant websites, a comprehensive bibliography, tools for teaching and learning, and downloadable images relevant to this chapter can be found at the website specially designed to accompany this book at
http://www.park-environment.com

Figure 5.13 *Uses of geothermal heat. Geothermal heat can be used in a variety of ways, depending on its temperature. At low temperatures, uses include heating snow and soil; at high temperatures, uses include industrial drying and power production.*

Source: Figure 5.9 in Armansson, H. and H. Kristmannsdottir (1992) Geothermal environmental impact. Geothermics 21 (5–6): 869–80.

SUMMARY

This chapter examined the nature of the Earth's surface and the structure of its interior, both of which reveal important clues to the origin and evolution of the planet. Both also exert controls on the major environmental systems (lithosphere, atmosphere, hydrosphere and biosphere) described in this book. We saw how most of the Earth's surface is covered by water and its relief is relatively subdued. Looking beneath the surface, we noted how the different layers in the Earth's interior have different properties and play different roles in transferring seismic waves. The crust is particularly important to conditions at the Earth's surface. To illustrate some of the ways in which sub-surface materials and processes help or hinder human activity, we looked at four application case studies – radon gas, underground storage of nuclear waste, hot springs and geysers, and geothermal energy. Although we have very little direct contact with the interior of the Earth, it nonetheless offers valuable opportunities and imposes serious constraints on what humans do on the surface.

FURTHER READING

Bell, F. (1999) *Geological hazards: their assessment, avoidance and mitigation.* E & FN Spon, London. Wide-ranging review of how natural hazards are caused, experienced and coped with.

Berkhaut, F. (1991) *Radioactive waste: politics and technology.* Routledge, London. Introduces and explores critical issues surrounding the search for safe methods of managing radioactive waste materials, with examples drawn from many different countries.

Egorov, N.N., V.M. Novikov, F.L. Parker and V.K. Popov (eds) (2000) *The radiation legacy of the Soviet nuclear complex.* IIASA/Earthscan, London. Outlines the problems associated with nuclear contamination from civil and military sources in the former Soviet Union.

Hasan, S.E. (1996) *Geology and hazardous waste management.* Prentice Hall, Hemel Hempstead. Useful overview of the relationship between geology and hazardous waste management, with a focus on site selection, regulatory frameworks, waste treatment and disposal.

Holland, C. (1999) *The idea of time.* John Wiley & Sons, London. Explores, from a geological perspective, what we mean by 'time' and what we understand when we think about time.

Keller, E.A. (2000) *Environmental geology*. Prentice Hall, London. Comprehensive review of themes, including geological hazards and processes, impacts of human activities on the environment, and environmental decision making.

Levin, H.L. (1995) *The Earth through time*. Saunders College Publishing, Philadelphia. A comprehensive survey of Earth's history, its major landmarks and implications for the evolution of life on the planet.

Park, C.C. (1989) *Chernobyl: the long shadow*. Routledge, London. Explains the causes and consequences of the 1986 nuclear accident at Chernobyl and explores the wider implications for nuclear safety around the world.

Porteous, A. (1992) *Dictionary of environmental science and technology*. John Wiley & Sons, Chichester. A useful source of information, particularly about energy and materials.

Savage, D. (ed.) (1995) *The scientific and regulatory basis for the geological disposal of nuclear waste*. John Wiley & Sons, London. An up-to-date summary of the debate surrounding the safe disposal of radioactive wastes.

Summerfield, M.A. (1991) *Global geomorphology: an introduction to the study of landforms*. Longman, Harlow. Focuses on the significance and development of large-scale landforms and explores the relevance to landform development of global plate tectonics.

Chapter 6

Dynamic Earth

LEARNING OBJECTIVES

When you have finished studying this chapter, you should be able to

- *Appreciate the relevance of the dynamics of the Earth's crust in shaping surface features and creating a range of geological hazards*
- *Describe the principal elements of the theory of plate tectonics*
- *Outline the nature of the evidence for continental drift, sea-floor spreading and palaeomagnetism*
- *Distinguish between processes of warping, folding and faulting, and describe typical landforms associated with each*
- *Explain the causes and mechanisms of earthquakes, and outline how earthquakes are experienced and coped with*
- *Give examples of ways in which human activities can influence earthquake magnitude, frequency, timing and location*
- *Explain the causes and mechanisms of volcanoes, and outline how volcanoes are experienced and coped with*
- *Distinguish between intrusive and extrusive vulcanism, and describe how each affects the Earth's surface*
- *Give examples of the geological hazards and products associated with volcanic events*

The Earth's crust is constantly moving, although most of the time we are unaware of it. Just as we cannot sense the Earth spinning around its axis, revolving around the Sun and precessing around itself, we have no direct experience of this incessant motion. It is simply too slow for us to detect with our five senses. From time to time there is sudden or violent movement, as happens in an earthquake or a volcanic eruption, which cannot escape our attention because of the damage and suffering it can cause. Such periodic disruptions can cause massive and widespread change to the surface of the Earth. Long periods of quiescence, when little seems to happen, are punctuated by short phases of intense change. This pattern seems to be equally true in the long-term evolution of species (see p. 529).

It is easy to concentrate on the sudden, dramatic changes and overlook the progressive, long-term, incremental changes. Yet both are vitally important in understanding how the Earth's surface operates and why it is the way that it is. In this chapter, we examine both timescales of change – the long-term changes associated with plate tectonics and mountain building, and the short-term changes associated with earthquakes and volcanic eruptions.

All of these are affected by conditions in the interior of the Earth (see p. 128), involve the redistribution of crustal materials, give rise to large-scale landforms and alter the face of the Earth. The long-term changes explain the formation and distribution of the oceans and continents, and of major features like mountain ranges and ocean trenches. The short-term changes have more dramatic and direct impacts on people and property, as well as creating and destroying smaller-scale landforms. Just as we are at the mercy of the large-scale cosmic dynamics that affect the movement of the planets and galaxies (Chapter 4), we also cannot escape the Earth's internal dynamics (Chapter 5) and the impacts that these have on the surface of the Earth.

PLATE TECTONICS

We know from geological evidence that the continents have moved around through time, in what appears to be a series

BOX 6.1 RELEVANCE OF THE DRIFTING CONTINENTS

The kinds of question to which new evidence about drifting continents provides answers include:

- *Palaeontology* (fossils): why, for example, are fossils of similar plant and animal species found on both sides of major oceans? Why are fossils of tropical species found in areas that now have cold climates?
- *Biogeography* (plant and animal distributions): why is the distribution of many species of plants and animals similar between areas that have very different climates today? Why do some species that are rare worldwide actually appear scattered between several continents, long distances apart?
- *Geomorphology* (landforms): why do coastlines on opposite sides of major oceans appear to fit together so neatly? Why are glacial landforms found in areas that now have tropical climates?
- *Geology* (rocks and deposits): why are rock types similar in many areas now separated by wide oceans? Why are there coal seams in Antarctica? Why are glacial deposits so widespread in Australia, South America, South Africa, India and Antarctica? Why are the distributions of earthquakes and volcanoes so similar? Why do we rarely find active volcanoes and earthquakes in the middle of large continents?

of slow but continuous movements over many millions of years (Box 6.1). These movements are still going on, and there is every indication that they will continue into the future. Since the 1960s in particular, geologists have been able to piece together various strands of evidence, which all support the notion of drifting continents. The evidence is of many different types, much of which is difficult to explain without allowing for substantial changes in the location of land masses.

DRIFTING CONTINENTS

CONTINENTAL JIGSAW

Photographs of the Earth taken from space show clearly the layout of the continents on the spherical surface of the planet. Even a quick look at the coastlines of South America and Africa suggests that they look like pieces of a giant jigsaw that would fit together remarkably well – if they were not thousands of kilometres apart!

Before the 1960s, we could get the same impression by looking at a large-scale map of the world, accepting that the map projection might introduce distortion into the pattern. As far back as the seventeenth century, well before modern surveying and map-making techniques gave us reliable and accurate maps of the world, Francis Bacon was struck by how well the western coastline of Africa seemed to fit the eastern coastline of South America. He spotted the pattern, but he could not explain it.

CONTINENTAL CRUST

Since the early twentieth century, some geologists have explained the pattern of the continents in terms of continental drift, which assumes that the continents float (literally) on the oceanic crust. The idea was first put forward in 1915 by Alfred Wegener, a German meteorologist.

The continents are the large land masses – Asia, Africa, North America, South America, Europe, Australia and Antarctica (see Table 5.1, Figure 5.1). Satellite photographs and maps show the boundary of the continents at sea level, but the physical reality of the continents does not end at the coastline. Their limits lie at the edge of the shallow continental shelf, which might extend several hundred kilometres out to sea. Present-day sea level is not a useful reference point, because sea level changes through time as climate changes. More importantly, the shallow continental shelves are part of the continental crust (*not* the oceanic crust).

Recall (p. 130) that the oceanic crust is thinner, denser and younger than the continental crust and is composed of different material – sima (silicon and magnesium) rather than sial (silicon and aluminium). The continental crust is not uniform but can be divided into *cratons* and *orogens* (Box 6.2).

CONTINENTAL DRIFT

Wegener suggested that all of the continents were originally joined together as one huge land mass. He called this super-continent Pangaea, and it probably existed between 200 and 250 million years ago. The rest of the Earth was covered by ocean (Panthalassa).

The theory of continental drift is based on the idea that Pangaea broke up and moved apart to form the continents we have today. The northern part of Pangaea (which Wegener called Laurasia) would break up to create North America, Greenland, Europe and Asia. The southern part

> ## BOX 6.2 CRATONS AND OROGENS
>
> **Cratons:** these are the very old, stable core areas in the continental crust and are also known as shields. They are large areas of highly deformed metamorphic rocks, effectively the roots around which continents have been built. Intense mountain-building phases shaped these areas in pre-Cambrian times, before stable conditions were established. About 30 per cent of the Earth is covered by continents, but there are only about ten shield areas (the Archaean cratons), which were formed more than 2.5 billion years ago and have survived. One of the oldest and largest is the Kaapvaal craton in South Africa, which formed and stabilised between 3.7 billion and 2.7 billion years ago.
>
> **Orogens:** these comprise the areas of more mobile crust around the cratons, which have been much more affected by earth movements and intense heating. Many of the orogens have been deformed into fold mountains.

of Pangaea (Gondwanaland) would provide the continents of South America, Africa, Australia and Antarctica. Between Laurasia and Gondwanaland, in this original continent, lay the Tethys Sea (of which the Mediterranean is a surviving remnant) (Figure 6.1). This pattern of evolution of the continents helps to explain the remarkably close fit, not only of the continental margins but also of old mountain belts and shield areas of the Earth's crust (Figure 6.2).

The geological evidence suggests that Gondwanaland began to break away from the rest of Pangaea around 180 Ma, when the Atlantic Ocean opened up. By 150 Ma, it had separated into two distinct land masses – Africa–South America, and Antarctica–India–Madagascar–Australia–New Zealand. Africa broke away from South America and Antarctica–Australia separated from India–Madagascar at about 135–120 Ma. Australia drifted away from Antarctica, New Zealand broke away from both of them, and India finally separated from Madagascar at about 96 Ma. India collided with Asia about 50 million years ago, raising the Himalayan mountain range as a result. While all this was happening, smaller continental blocks drifted away from the northern margins of Gondwanaland from time to

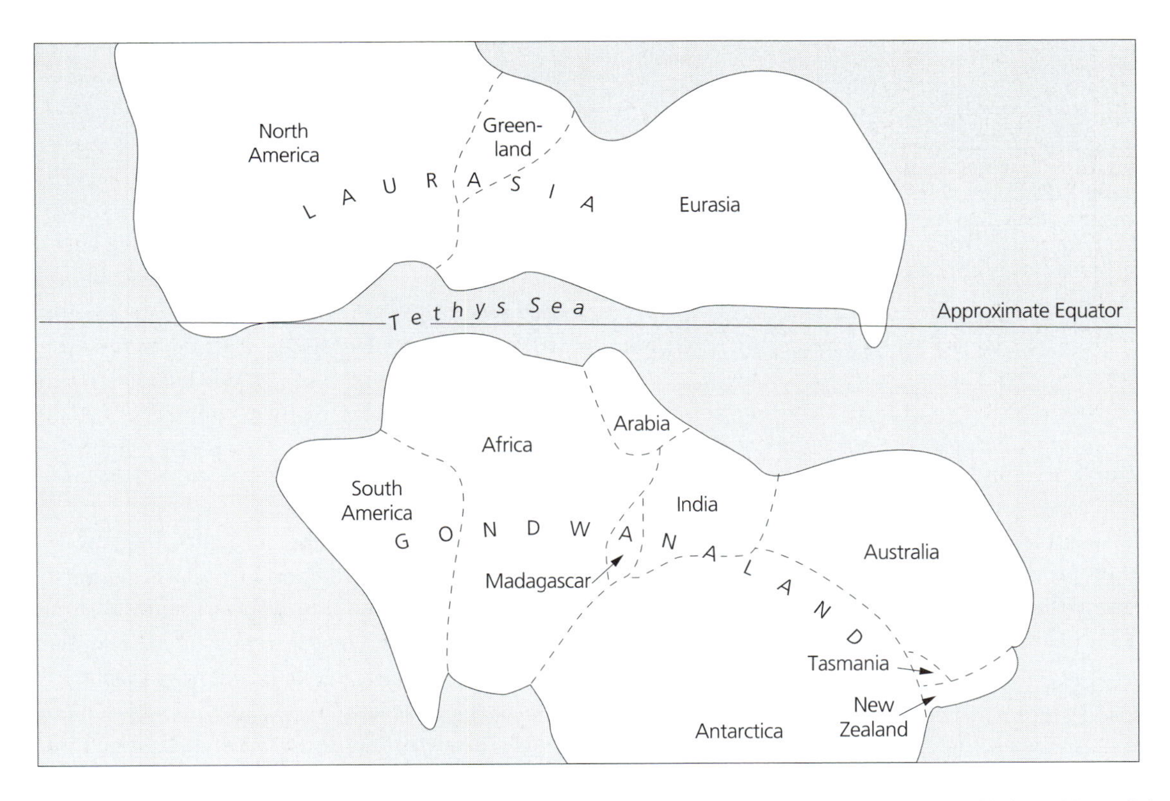

Figure 6.1 Reconstruction of the original land masses of Laurasia and Gondwanaland. The map shows the likely positions of the two main land masses, and the modern continents that have evolved from their breaking up and drifting apart.
Source: Figure 3.10 in Dury, G.H. (1981) Environmental systems. Heinemann, London.

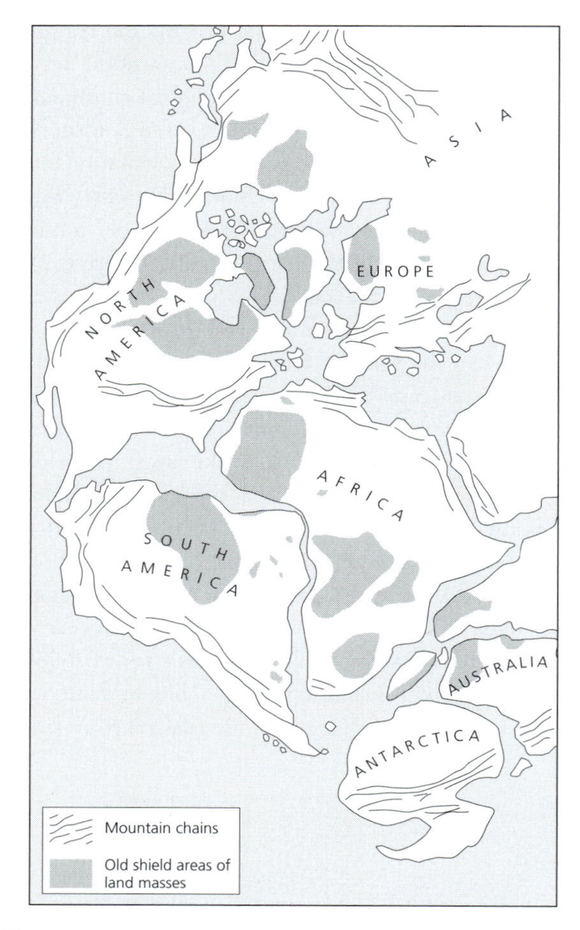

Figure 6.2 *Mountain belts and crustal shield areas in Pangaea about 200 million years ago. The distribution of ancient shield rocks and the orientation of old mountain chains provide important clues about how the modern continents fitted together in the ancient past.*

Source: Figure 9.16 in Doerr, A.H. (1990) Fundamentals of physical geography. Wm. C. Brown Publishers, Dubuque, Iowa.

time and collided with Laurasia. These now form places like southern Europe, Turkey, Afghanistan, Tibet, central China and Malaysia.

Wegener proposed that the lighter continental (sial) crust floats on the more plastic oceanic (sima) crustal rocks (see p. 132) below, rather like a log floating on water. In this way, he argued, the continents more or less drifted from their original locations.

SEA-FLOOR SPREADING

What Wegener was unable to explain was what caused the super-continent to split, what mechanisms drove the continental fragments to drift apart, and why they drifted the way they did. Answers to these fundamental questions started to emerge only in the 1960s, when the plate tectonic model was starting to take shape.

PALAEOMAGNETISM AND POLAR WANDERING

New perspectives were greatly assisted by the development of new technologies in geophysics, particularly those that allowed accurate and reliable measurement of the magnetic properties of rocks. Palaeomagnetism is the science of reconstructing the Earth's ancient magnetic field and the former positions of the continents from the evidence of remanent magnetisation in ancient rocks, based on traces left by the Earth's magnetic field in igneous rocks before they cool.

The palaeomagnetism evidence shows that the Earth's magnetic field has reversed itself (the magnetic north pole becomes the magnetic south pole and vice versa) at roughly half-million-year intervals, with shorter reversal periods in between the major ones. The record of polar wandering in European and North American rocks is different, which suggests that the continents have moved relative to each other. This confirms the principle at least of continental drift.

MID-OCEAN RIDGES

A significant breakthrough in understanding how the continents move came when geologists started to measure magnetic properties of rocks beneath the oceans, particularly around the mid-ocean ridges (Figure 6.3), which run through the Atlantic and Pacific (Box 6.3). These ridges have long been known to be geologically active; submarine volcanic activity regularly takes place along them, and rocks in and around the ridges are among the youngest on the planet.

This curious submarine magnetic mosaic was explained as the product of sea-floor spreading (Figure 6.4). New magma is regularly pushed upwards to the surface of the ocean floor in the central trough of the mid-ocean ridge, where it cools. It mirrors and preserves the magnetic polarity of the Earth at the time of emplacement (formation). Through time, more magma is pushed up, and the existing material is pushed sideways by lateral pressure. The mid-ocean ridges are constantly being generated, as hot mantle material wells up there and spreads laterally from the ridge.

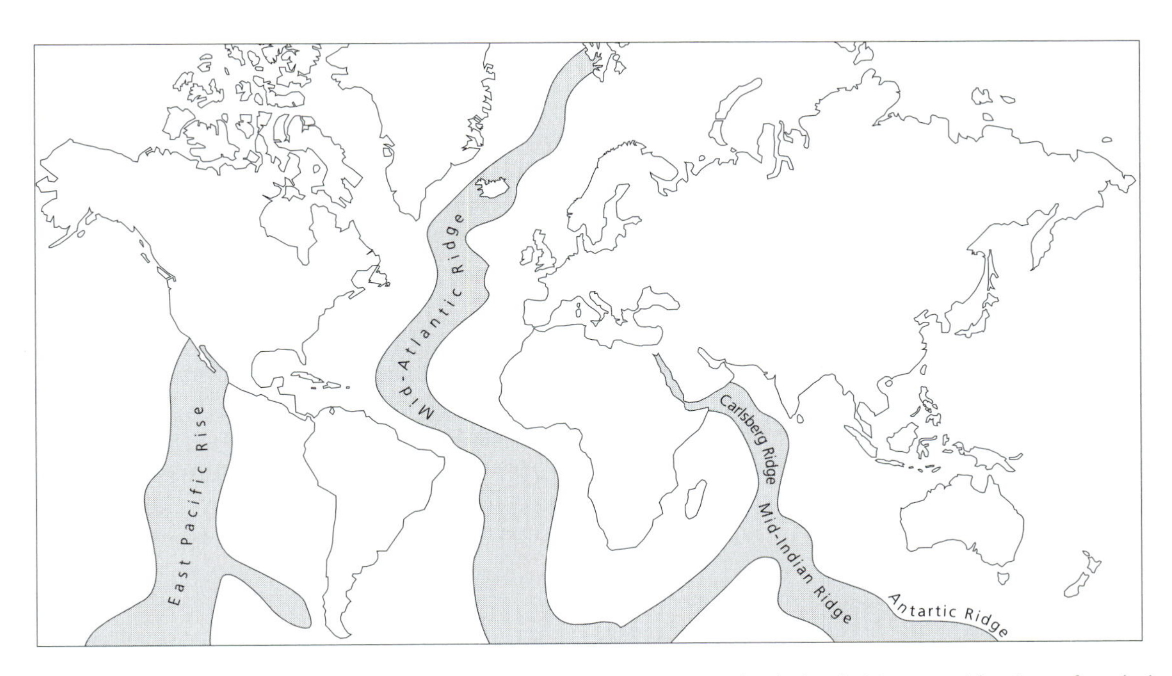

Figure 6.3 *Location of the mid-ocean ridges. Major submarine mountain chains (mid-ocean ridges) are found along the central part of the main oceans.*

Source: Figure 1.3a in Goudie, A. (1993) The nature of the environment. Blackwell, Oxford.

BOX 6.3 CHARACTERISTICS OF THE MID-OCEAN RIDGES

Detailed analysis of the rocks adjacent to the mid-ocean ridges, initially in the Atlantic and Pacific, revealed interesting facts, including:

- *Zonation*: linear bands of rock hundreds of kilometres long and generally between 20 and 30 km wide appeared to lie parallel to one another, each band displaying different magnetic properties from its neighbours.
- *Similarity*: even more striking, the zonation of bands on both sides of the ridges appeared to be remarkably similar.
- *Age*: dating of the material on both sides of the ridge established that the youngest material was at the centre and material was progressively older with increasing distance from the ridge.
- *Temperature*: rocks in the mid-ocean ridges were found to be hotter than those further away.

This process continues incessantly, like a giant submarine conveyor belt. The polar wanderings and magnetic reversals are preserved in the ocean-floor rocks, equally on both sides of the ridge (hence the parallel bands). The variable width of magnetic zones on each side of the ridge reflects different lengths of time of each successive epoch when magnetic polarity remained the same between reversals.

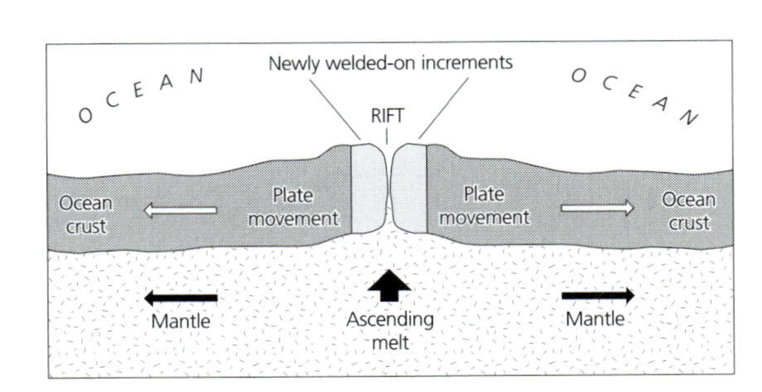

Figure 6.4 *Sea-floor spreading caused by the upward movement of magma. Magma from the mantle is pushed upwards and solidifies at the edges of adjacent ocean crustal plates, pushing them apart and causing sea-floor spreading.*

Source: Figure 3.3 in Dury, G.H. (1981) Environmental systems. Heinemann, London.

IMPLICATIONS OF SEA-FLOOR SPREADING

As well as explaining the pattern of magnetic variations in ocean-floor rocks, this sea-floor mechanism provides the key to explaining continental drift, because it drives the movement of the rafts of continental crust over the ocean crust. This was Wegener's missing link.

Sea-floor spreading might even offer a valuable renewable energy resource. Recent research suggests that hydrothermal vents along the mid-ocean ridges may prove to be the most accessible and concentrated source of geothermal energy yet found. Harnessing this energy will not be easy because of the remoteness of the ridges from the continental land masses, but the challenge is there.

CRUSTAL PLATES

Continental drift offers a useful description of the movement of continents across the surface of the Earth, but without being able to explain how or why. Sea-floor spreading provides a mechanism to drive the movement. The final piece in the puzzle came with the formulation of plate tectonic theory in the late 1960s, which offered an explanation of the processes and dynamics involved.

PLATES

Continental drift and sea-floor spreading are both associated with the continuous formation and destruction of the Earth's outer layer, the crust. The evidence suggests that large sections or blocks of the crust (the so-called plates) are pushed over the plastic-like rocks of the more mobile asthenosphere (p. 133). The plates are curved to the spherical shape of the Earth, and they fit together like a jigsaw with no large gaps between pieces (Figure 6.5). They effectively float on the asthenosphere, so they move relative to one another.

At least seven major plates are recognised – the North American, South American, African, Eurasian, Indo-Australian, Pacific and Antarctic plates. In addition, there are at least twelve minor plates and many so-called microplates.

The plates are moving towards, away from or alongside each other at an average speed of about 70 km per million years (7 cm a year). Speed of movement varies a great deal from one plate to another. These large-scale crustal movements shape the Earth's surface and create all of its major natural features, including mountains, ocean basins, deep ocean trenches, volcanoes and earthquakes. Most of the geological activity (particularly the earthquakes and

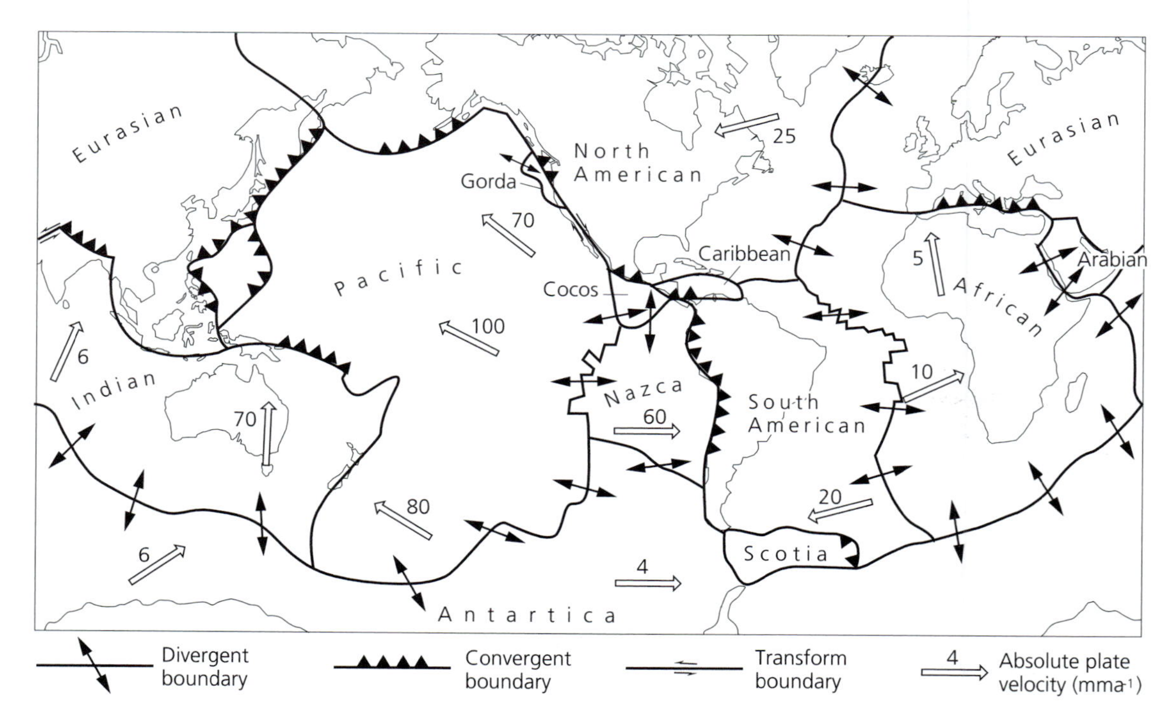

Figure 6.5 *Distribution of the main crustal plates. The Earth's crust is composed of a series of major and minor crustal plates. The map shows the location of the main types of plate boundary and estimated current rates (mm a^{-1}) and directions (arrows) of plate movements.*

Source: Figure 2.14 in Goudie, A. (1993) The nature of the environment. Blackwell, Oxford.

volcanic eruptions, but also mountain building, folding and faulting) takes place along the margins of the plates, where the crust appears to be much more unstable.

CURRENTS

Material in the mantle behaves like a plastic (see p. 130), and it is believed that huge, slow-moving convection currents within it can move sections of the overlying crustal rocks with them (Figure 6.6). This drives the movements of the plates. The convection currents produce upwellings of new material at the mid-ocean ridges. This new material extends the plates and moves them away from the ridges. This effectively sets in motion the crustal conveyor belt, which pushes and drags the plates (and continents) around (Box 6.4).

It is still not entirely clear why these convection cells exist in the mantle, or what drives them or determines their speed or location. Cell and thus plate speeds are believed to have varied through time, because the complete break-up of an assumed original continent could probably not have happened at current plate speeds.

> **BOX 6.4 RATES OF SEA-FLOOR SPREADING**
>
> Sea-floor spreading also explains present-day rates and patterns of continental movements, which are known from geodetic surveys from satellites. Rates of spreading over the past few million years, estimated from the ocean-floor rocks, agree remarkably well with rates actually measured between the mid-1980s and mid-1990s from space surveys. The similarity in rates, based on different dating techniques, strongly supports the idea of sea-floor spreading as the driving force for continental drift.
>
> Much remains to be discovered about the sea-floor spreading mechanism, particularly the processes responsible for creating new oceanic crust. By the mid-1990s, less than 10 per cent of the mid-ocean ridge system around the world had been studied in detail, and a 10-year research programme – Ridge Interdisciplinary Global Experiments (RIDGE) – was designed to integrate observational, experimental and theoretical studies and improve understanding.

PLATE BOUNDARIES

The movement of the plates redistributes crustal material around the face of the Earth, and this in turn determines

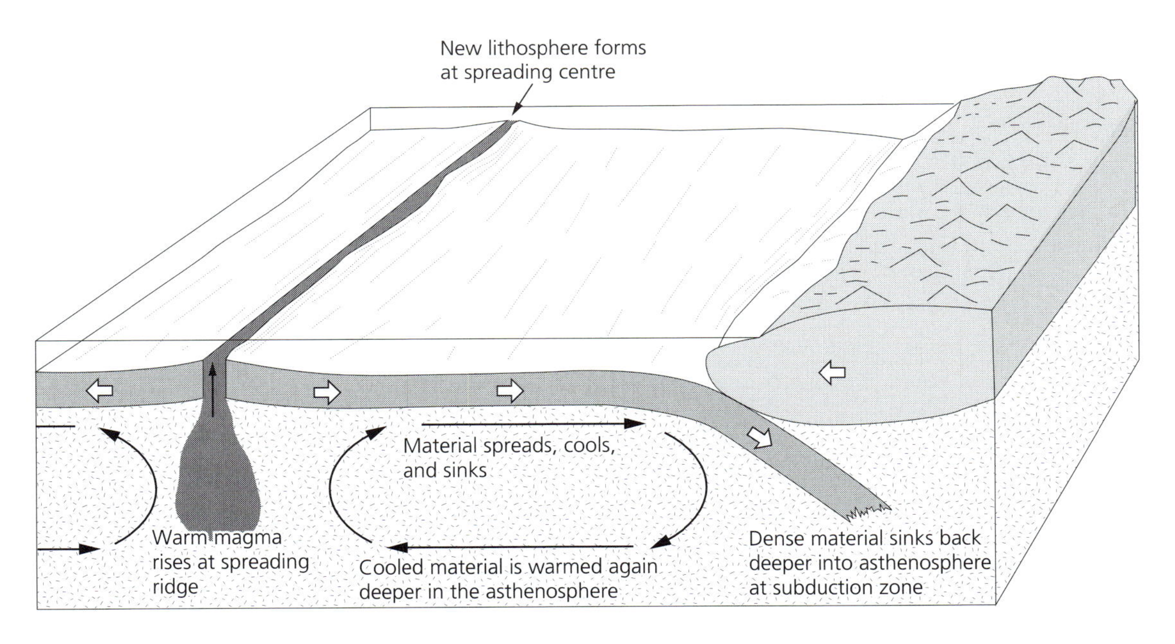
New lithosphere forms at spreading centre
Material spreads, cools, and sinks
Warm magma rises at spreading ridge
Cooled material is warmed again deeper in the asthenosphere
Dense material sinks back deeper into asthenosphere at subduction zone

Figure 6.6 *Sea-floor spreading and plate movement. The lateral movement of crustal plates is driven largely by sea-floor spreading, which is caused by convection currents within the molten magma in the upper mantle.*
Source: Figure 9.12 in Doerr, A.H. (1990) Fundamentals of physical geography. Wm. C. Brown Publishers, Dubuque, Iowa.

the position, size, shape and orientation of the continents and oceans (see Figure 5.1). Most of these plate movements simply relocate land masses without necessarily altering their appearance. The analogy of a conveyor belt (see Figure 6.6) appears to be quite appropriate.

Most of the changes in the appearance of land masses occur at the margins or boundaries of the plates. This where most of the action takes place, and where geological instability is concentrated. Little wonder, therefore, that there is a close match between the distribution of earthquakes and volcanoes around the world and the location of the major plate boundaries.

Boundaries between plates can be constructive, destructive or conservative (Box 6.5). Subduction at destructive plate margins (Figure 6.7) creates a series of large-scale landforms, including mountain belts, volcanoes and ocean trenches (Figure 6.8). Conditions at plate boundaries can be very different, depending on what type of boundary is involved. The outcome of interaction between adjacent plates also depends on whether we are dealing with two ocean plates, two continental plates or one of each. Examples of the more common permutations are given in Table 6.1.

REFLECTIONS

The theory of plate tectonics has revolutionised the way geologists explain the movement of continents, the evolution of oceans, the development of mountain ranges, and the distribution of earthquakes and volcanoes. Plate tectonics is very much a unifying theory, which ties together many geological patterns and processes in a coherent and logically consistent framework.

Questions do remain, however, that the theory – or the real-world evidence supporting it – cannot fully answer. The most fundamental outstanding question is why and how the continents originated in the first place. Plate tectonics can explain how continents are moved around the face of the Earth, but not how the continental and oceanic crustal materials were first formed. Less important, but still puzzling, is the question of why some earthquake zones are found away from the margins of known plates.

Table 6.1 Examples of plate boundary conditions

	Ocean–Ocean	*Ocean–Continent*	*Continent–Continent*
Divergent	Mid-Atlantic		Arabia (Red Sea)
Convergent	Island arcs	Andes	Himalaya
Transform		California (San Andreas)	

BOX 6.5 TYPES OF BOUNDARY BETWEEN CRUSTAL PLATES

Constructive boundary: this is associated with the divergence of plates. Two plates move apart, and new material emerges from the mantle, cools as crustal rock and fills the gap. Ridges, like those found near the centres of all the oceans, usually mark such constructive plate boundaries.

Destructive boundary: this is associated with the convergence of plates. Two plates move towards one another, pushed together under great pressure. One plate normally overrides the other, forcing it to dip downwards underneath it. This process is called *subduction*. Deep ocean trenches like the Marianas Trench (see Figure 6.8) are created where one plate sinks (is subducted) beneath the other and is absorbed back into the mantle. Rocks that were previously at or near the Earth's surface are subducted, subjected to great heating in the mantle, melt and become plastic, and can provide new raw materials (for example, for sea-floor spreading or extrusive or intrusive volcanic activity; pp. 151 and 173). Ocean trenches often have associated island arcs (series of volcanoes) located on the side closer to the nearby continent, for this very reason.

Conservative boundary: at conservative margins two plates move past each other in opposite directions. It is described as a *transform* movement.

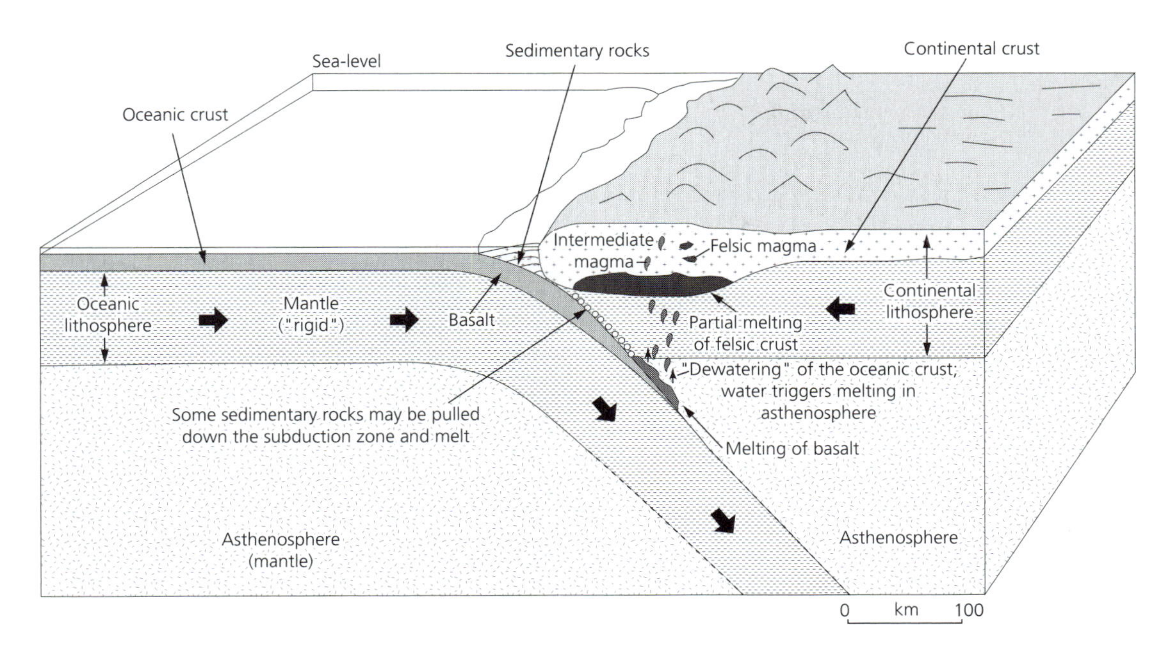

Figure 6.7 *Forms and processes at a typical subduction zone. One crustal plate is pushed downwards into the mantle beneath an adjacent one at a subduction zone, where some of the rock material is melted.*

Source: Figure 10.16 in Doerr, A.H. (1990) Fundamentals of physical geography. Wm. C. Brown Publishers, Dubuque, Iowa.

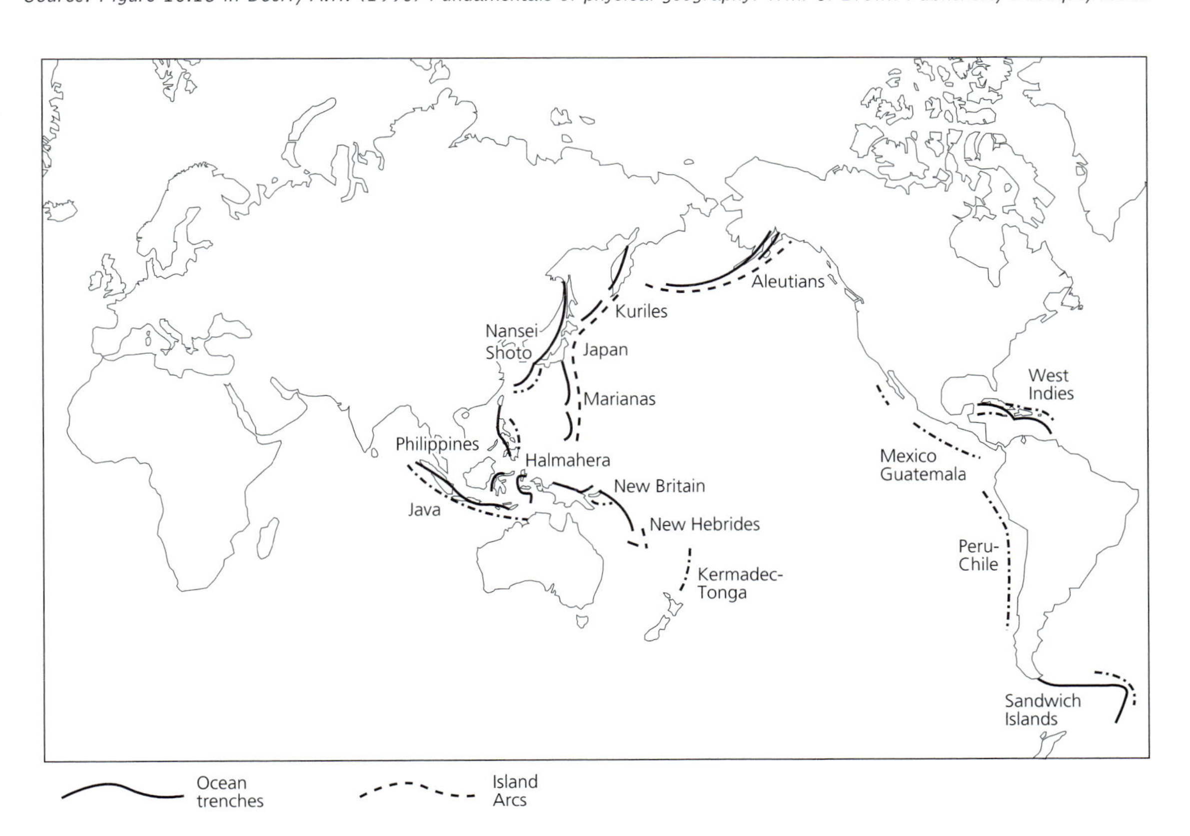

Figure 6.8 *Location of the main ocean trenches. Many of the most active subduction zones at the margins of the continents are marked by deep ocean trenches, which sometimes run parallel to the coast.*

Source: Figure 1.3b in Goudie, A. (1993) The nature of the environment. Blackwell, Oxford.

EARTH MOVEMENTS AND MOUNTAIN BUILDING

THE FACE OF THE EARTH

The face of the Earth is quite varied and can be described at different scales. At the largest scale, the Earth is divided into the major continental land masses and the major oceans between them (see Figure 5.1), but superimposed on this basic framework is the detailed expression of topography, such as the mountains, valleys, flat plains and gently sloping surfaces. At this scale, geological and topographic features affect human activities, both directly (such as creating barriers to communication) and indirectly (such as via the influence of mountains on local weather and climate).

MOUNTAINS

Mountains are a mixed blessing to humans! They provide some of the world's most striking and dramatic scenery, offer great tourism potential, provide some of the most important remaining wilderness areas, offer important habitats to wildlife, have inspired artists and writers through the ages, sometimes provide valuable minerals and materials, and remind us just how relatively small and insignificant we are. On the other hand, mountains threaten people with a range of hazards (including earthquakes, landslides and avalanches, major river flooding, and severe storms).

Throughout history, the inaccessibility and inhospitability of many high mountains has also been a mixed blessing – offering solitude and space for reflection (by religious communities, for example) but also offering refuges where hard drugs (such as coca in South America and opium poppies in South Asia) can be grown for the international trade.

Plate 6.1 *Road bridge over the Haast River, western side of the Southern Alps, New Zealand. Uplift rates of up to 6 mm a year mean that the rivers draining the Alps are extremely dynamic, and permanent river crossings are not possible.*
Photo: Alistair Kirkbride.

Mountain ranges are often created when two crustal plates collide (see p. 150). Collision and upthrust (forcing upwards) can raise a mountain chain. The Himalayan chain is being raised in this way, by collision between the Indian and Eurasian plates. The Himalaya started to form about 50 million years ago, when plate tectonics closed the Tethys Sea (between Laurasia and Gondwanaland). Geology and relief vary a great deal in the Himalaya, depending partly on plate tectonic setting but also on weathering processes (which are themselves heavily influenced by interactions between climate and topography).

BOX 6.6 THE EUROPEAN ALPS

The Alps in southern Europe, linked to the Himalaya by plate tectonics, were formed by intense folding during the Tertiary sub-era (see Table 4.7). The Alpine massif (mountain range) has great variety of rock type, relief and topography, and most of what we see today is the result of Quaternary glaciation. Glacial erosion, which continues today, has carved deep, sheltered valleys and deposited soils and sediments. The massif's relief has endowed it with great hydroelectric power potential (see p. 393) but severely restricted access and transport in many areas.

> **BOX 6.7 GEOLOGICAL TERMS RELATING TO MOUNTAIN BUILDING**
>
> **Tectonics:** tectonics is the study of the movements of rocks on the Earth's surface. At the large scale, plate tectonics is concerned with movements at the global scale. On a smaller scale, tectonics deals with processes that bend, fold or fracture rocks. Orogenesis and diastrophism are studied within tectonics.
>
> **Orogenesis:** orogenesis, or orogeny, describes the formation of mountain ranges by geological processes (particularly large-scale compression and intense upward displacement).
>
> **Diastrophism:** diastrophism refers to the deformation of the Earth's crust by internal forces. Crustal response gives rise to continents, mountains and major geological features (such as folds and faults). There are three main types of diastrophic force – warping, folding and faulting.

NEW ZEALAND

Many of the features associated with plate tectonics and mountain building are clearly displayed in New Zealand, which is effectively a micro-continent straddling the boundary of the Pacific and Indo-Australian plates. This is a particularly dynamic part of the global plate system, and as a result the face of New Zealand is being continuously transformed and altered in response to seismic pressures and processes.

The Pacific and Indo-Australian plates are converging at rates of up to 58 mm a year (over half a metre each century), and this results in compression and extremely high shear forces. New Zealand is seismically active as a result, with a great deal of fault movement, folding and uplift at the present time. Parts of the Southern Alps are being lifted up by between 7 and 12 mm a year, although other parts of the country appear to be subsiding.

A number of faults (including the Matada, Edgecumbe, Obepu and Rotoitopakau faults) have moved at least once during the last 800 years, and some parts of the Rangitaiki Plains face a major seismic hazard of moderate-magnitude earthquakes (magnitude 6 to 6.5). Parts of the Southern Alps are particularly susceptible to seismic damage, because a number of major fault lines cross areas that regularly experience intense rainstorms and snow/ice avalanches. On Saturday 14 December 1991, the top 10 m of New Zealand's highest mountain – Mount Cook – disappeared in such a rock avalanche.

Plate tectonics has created the striking mountain scenery for which New Zealand is renowned, and which is now the basis of a growing tourism industry. It has also endowed the country with valuable geothermal energy resources (see p. 139), which are being used directly (in various types of heating system) and indirectly (to generate electricity).

WARPING

Warping means large-scale bending, and the Earth's crust can be bent upwards (upwarping) or downwards (downwarping) as a response to great pressure. The process operates continuously over long periods of time at a regional scale, affecting very large areas (thousands of km^2 in size) to create major landforms. Smaller-scale features (such as mountains, lakes and seas, folds and faults) are generally superimposed on top of the regional warps.

The Earth's crust is warped in a number of ways. Sometimes it is a response to the convergence and collision of adjacent tectonic plates. Crustal material at or near the edge of one or both of the plates can be warped by the immense pressures of the convergence. This is, in essence, a form of folding (see p. 157).

Warping also occurs as a response to increased weight on the ocean floor caused by the deposition and accumulation of sediment, much of which has been washed from the continents by major rivers that drain into the sea. Such downwarping of the ocean floor creates a geosyncline (a depositional basin on the sea floor). Because it lowers the sea bed it encourages further deposition, so the cycle is self-reinforcing (this is an example of positive feedback; p. 79).

COASTAL CHANGES

By changing relative relief (the height of one area relative to another, usually adjacent, area), warping can significantly alter landforms, weathering and erosion processes, and the general appearance of landscapes. These changes are most obvious along coastlines (see p. 497). The Pacific coast of the USA is showing signs of upwarping, for example, so it is an emerging coastline. Although rates of upwarp are relatively low, it means that through time the threat of coastal flooding and storm surges is reducing, as are rates of beach erosion and cliff attack. The Atlantic coast and Gulf of Mexico, in contrast, appear to be downwarping. Beach erosion rates along this submerging coastline are much higher as a result, and the progressive drowning of river mouths is extending estuary conditions further inland.

ISOSTATIC ADJUSTMENT

Large-scale glaciation and deglaciation can trigger warping (called isostatic adjustment) by redistributing weight on that portion of the Earth's crust. Glaciation (see p. 451), which causes the build-up of ice over wide areas and promotes the extension of continental glaciers, increases crustal weight and this triggers downwarping. The thicker the ice cap, ice sheet or glacier, the heavier the crustal loading and thus the greater the downwarping (in terms of both the depth of depression and the area affected by it).

The Earth's crustal material behaves rather like plastic in the sense that if additional weight is removed from it, it can rebound to its original position. This isostatic rebound happens when ice caps, ice sheets and glaciers melt during deglaciation, and upwarping is the normal crustal response to decreased overlying weight. As melting continues, loading is reduced and uplift continues (Figure 6.9).

The rate of crustal rebound is generally greatest as soon as the weight is removed, and the rate declines through time. Rate of rebound depends partly on the amount of weight removed, and partly on the nature of the rocks concerned (some materials respond faster than others).

Parts of North America and northern Europe are slowly rebounding upwards in response to the melting of large glaciers at the end of the Pleistocene ice age (see p. 45). Ice left these northern latitude areas between 10,000 and 20,000 years ago, and the crustal material is still relaxing under the reduced weight of continental glaciers. Rates of rebound are by no means uniform across the zone, and they vary across even quite small areas. In Sweden, for example, uplift has been faster under the Gulf of Bothnia than in land around the Gulf, and local differentials like this can distort topography and might even promote earthquakes. Rebound is likely to continue for a long time yet, and this gives a slowly but continuously moving baseline (ground level) to all environmental systems in the area.

Uplift associated with warping affects people in a variety of ways. In Finland, for instance, uplift has meant the emergence of new land which is a valuable economic asset in an agricultural society. The downside, literally, has been reduced water depths in harbours and along shipping routes, and sea fishing has also suffered.

Coastal changes (land relative to sea; see p. 511) are likely to be particularly complex in these areas of isostatic

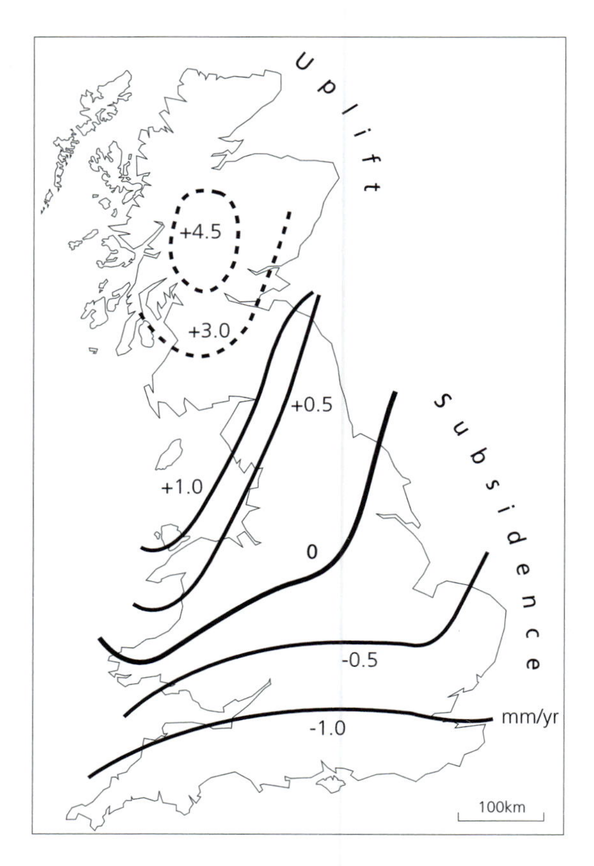

Figure 6.9 *Continued isostatic uplift of Britain after the end of the Pleistocene ice age. The Earth's crust beneath Britain has been adjusting to the reduced pressure on it since thick ice caps melted at the close of the Pleistocene ice age. Central Scotland (where the ice was thickest and survived longest) has been rising at a rate of more than 4.5 mm a^{-1}, and this upward tilt has caused a compensating sinking of southern England at a rate of around 1 mm a^{-1}.*

Source: Brown, E.H. (1979) The shape of Britain. Transactions, Institute of British Geographers New Series 4: 449–62.

rebound, because sea-level change (associated with global warming; see p. 512) will be superimposed on the underlying crustal warping. Responses (such as whether coastal erosion increases or decreases) will depend on the balance between, and relative speeds of, sea-level change and crustal change.

Extensive flooding will occur where sea-level rise coincides with downwarping; equilibrium could be maintained where sea-level rise balances upwarping; emergence will be the case where upwarping is faster than sea-level rise.

FOLDING

Warping is a vertical movement of crustal material, and folding is a horizontal movement. It is usually associated with the convergence of two adjacent tectonic plates that are forced together by lateral (sideways) movement of crustal material driven by convection currents in the mantle.

Lateral pressure in the collision forces a vertical response, in much the same way that the metal shell of a moving vehicle buckles and crumples on impact with a stationary or moving object. Fold mountains, like the Alps and the Himalaya, are formed by strong and sustained compression over long periods of time. The high relief of southern Tibet is explained partly by rapid uplift about 20 million years ago, followed by erosion and valley cutting, which was fastest up to about 8 million years ago (high relief promotes rapid erosion, and as erosion reduces relief, rates of erosion decline).

ROCK RESPONSES TO PRESSURE

Because of plate tectonics, most crustal materials are subjected to pressure, as the crustal plates move together, apart and side by side against each other. Compression pressure has the result of squeezing rocks together, and tensional pressure stretches them apart. Rocks can change size (expand or contract) and shape in response to pressure.

The way in which a particular type of rock responds to pressure depends largely on the properties of the rock and its resistance to change. There are three types of response – elastic strain, plastic strain and fracturing (Box 6.8).

TYPES OF FOLD

Any bend in stratified (layered) rocks is a fold. The line along which a bed of rock folds is called its axis, and the limbs of the fold are the tilted beds that extend outwards from the axis. The fold can be upwards or downwards. Folds vary a great deal in size and type (Figure 6.10), depending mainly on:

- the amount of pressure exerted;
- the symmetry (balance in different directions) of that pressure; and
- the nature of the rocks being compressed.

The simplest type of fold is a *monocline*, a one-sided fold caused by relatively weak pressure. Monoclines are usually found on the outer edges of tightly folded areas, like fold mountains. Two-sided folds are more common, and they can be either symmetrical or asymmetrical. Common multi-sided folds include anticlines, synclines, overturned folds and domes (Box 6.9).

FAULTING

Faulting occurs when rocks that are subjected to pressure (compressional or tensional) fracture or break. This happens

BOX 6.8 ROCK RESPONSES TO PRESSURE

Elastic strain: this occurs when rock returns to its original form after the release of applied pressure, without any overall change. Isostatic rebound caused by the melting of a continental ice sheet is an example of elastic strain. The property of a material that makes it return to its original shape after a force deforming it is removed is its elasticity. Hooke's law states that, within elastic limits, extension (stress) is proportional to the force (strain) producing it. Hence, larger folds are caused by greater pressure in a given type of rock.

Plastic strain: this occurs where rock does not return to its original shape after the release of applied pressure, as occurs with folding. In this case, the material has been stressed beyond its elastic limit. Most rocks are subject to plastic deformation when exposed to slow movements in plate tectonics.

Fracturing: when applied forces exceed the plastic limits of rocks (that is, the threshold at which rocks can absorb pressure by folding) they fracture or break. Some rocks are more brittle and liable to fracturing than others. Fracturing tends to occur most commonly when there are rapid movements along fault lines during earthquakes, or during some volcanic eruptions (when the heat and pressures involved can alter the rocks by metamorphosis; see p. 201).

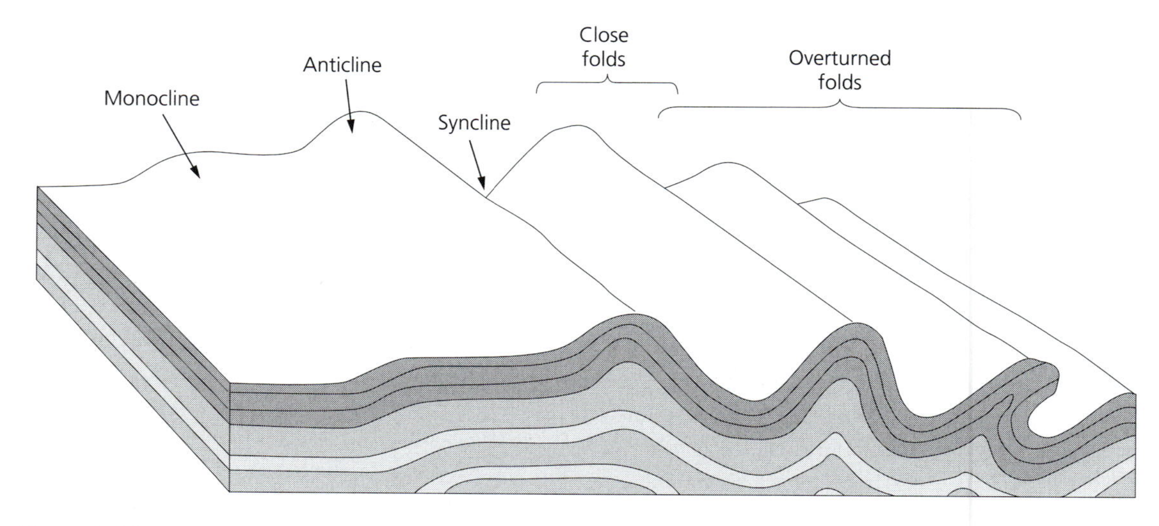

Figure 6.10 *Simple types of fold. The most common types of simple fold are shown; see text for explanation.*
Source: Figure 10.3 in Doerr, A.H. (1990) Fundamentals of physical geography. Wm. C. Brown Publishers, Dubuque, Iowa.

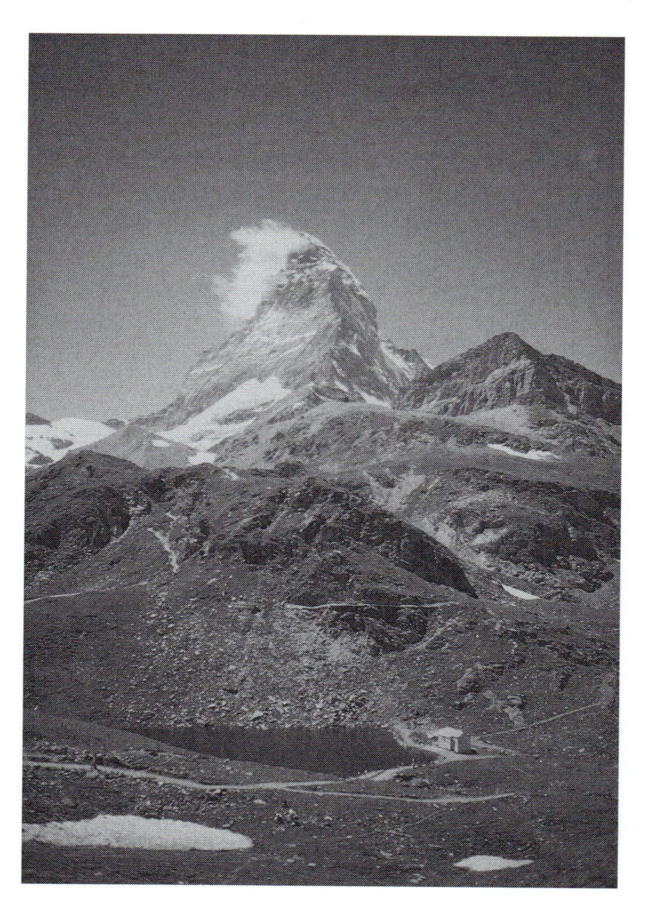

Plate 6.2 *The Matterhorn in the Alps, Switzerland. This dramatic mountain peak is the product of the interaction of forces of uplift (particularly intense folding) and forces of downwearing (weathering and erosion).*
Photo: Chris Park.

when the plastic limit of the rocks is exceeded. When this threshold condition (see p. 82) is reached, the system responds rapidly and abruptly (in contrast to the continuous adjustment reflected in most warping and folding).

Small-scale fractures or breaks in rocks subjected to pressures beyond their plastic limits are referred to as joints. These create natural lines of weakness in the rocks, which can be exploited by weathering processes that concentrate on them. This can produce some interesting and dramatic landforms, and physical and chemical weathering etches details into the surface of exposed rocks.

DISPLACEMENT

Faults are large-scale fractures in rocks at the Earth's crust, along which the rocks on each side have moved relative to each other (they have been displaced). Displacement takes place as a result of differential pressure (strain) in the adjacent bodies of rock; it can be in a vertical direction, a horizontal direction or both at the same time (Table 6.2).

Displacement can be so small as to be almost undetectable, causing no visible or short-term change and posing no threat to people, buildings and structures. Sometimes there is massive displacement, and this causes major earthquakes (see p. 160). Signs of movement can often be seen along recently active fault lines, such as the San Andreas fault, where the fault line is indicated on the ground by such things as displaced lines of trees, non-matching fence lines, sharp changes in the direction of rivers and streams, and repaired highways.

BOX 6.9 SOME COMMON TYPES OF FOLD

Anticline: this is a symmetrical fold that is arched upwards in the middle. Anticlines are rarely preserved intact, because the rocks at the top of the arch have been stretched and weakened and are liable to weather rapidly on exposure at the surface. An *anticlinorium* is a complex anticline in which a large arch contains many small folds superimposed on it.

Syncline: this is a symmetrical fold in which the beds dip inwards, forming a trough-like structure with a dip in the middle. A *synclinorium* is a complicated downfold, with many small folds superimposed on the major syncline. Synclines can form depositional basins into which geological deposits are transported and deposited. Over long periods of time the deposits accumulate, are compacted and turn into sedimentary rocks. Valuable fossil fuels (coal, oil and gas) accumulate in such basins. *Geosynclines* can form in such symmetrical down-folds.

Overturned folds: this is created where compressional forces are asymmetrical, so that one limb of the fold is pushed over the other limb and overrides it. If the force is too great the rock breaks and overthrusting occurs, in which one limb is pushed a great distance over the other one, from which it has been detached. The broken fold formed in this way is called a *nappe*.

Domes: this is a symmetrical anticline that is circular rather than linear when viewed from above. The ground slopes downwards in all directions from the crest. While most domes are formed by the intrusion (forcing) of material like magma or salt into horizontal sedimentary beds from below, they can also be created by compressional folding operating from several directions at the same time.

Table 6.2 Directions of movement along fault lines

Name	*Movement*
Dip-slope fault	Vertical movement up and down the dip of the fault line
Strike-slip fault	Horizontal movement along the strike of the fault line
Oblique-slip fault	Vertical (dip-slope) and horizontal (strike-slip) movements

Faulting takes place along fault zones or fault planes; a fault line occurs where such a zone intersects the Earth's surface. One of the world's most active faults is the Xianshuihe fault of western Sichuan province, China. During the twentieth century, it has produced four major earthquakes (magnitude 7 or greater) along a 350 km segment, and similar events are likely to happen again in the future.

INDIVIDUAL FAULTS

There are four common types of fault, produced in different ways (Figure 6.11). Each produces different landscapes.

Normal fault: this is the simplest type of vertical displacement fault, in which one side rises or falls relative to the other. It is often difficult to work out which side actually moved, because what is seen on the ground is the relative movement. The higher side survives as a cliff-like scarp, or escarpment. Normal faults can develop from either compressional (convergent) or tensional (divergent) forces.

Reverse fault: this is an overhanging type of vertical displacement fault, in which one side is pushed upwards over the other at an angle (the dip of the fault line). Such faults are the result of strong compressional forces. The overhang rarely survives for long, because weathering and erosion wear it back, and gravity tends to pull it down. The end result is often a scarp that looks similar to a normal fault scarp (but the surface expression hides the different internal structure).

Transcurrent (strike-slip) fault: this is a horizontal displacement fault in which movement occurs along the fault line as a result of shearing. The San Andreas fault in California (Box 6.10) is of this type.

Thrust fault: this is a vertical displacement fault associated with low-angle fault planes. Strong compressional forces push blocks of material up along the fault plane, and they ride over the material beneath them (rather like overturned folds).

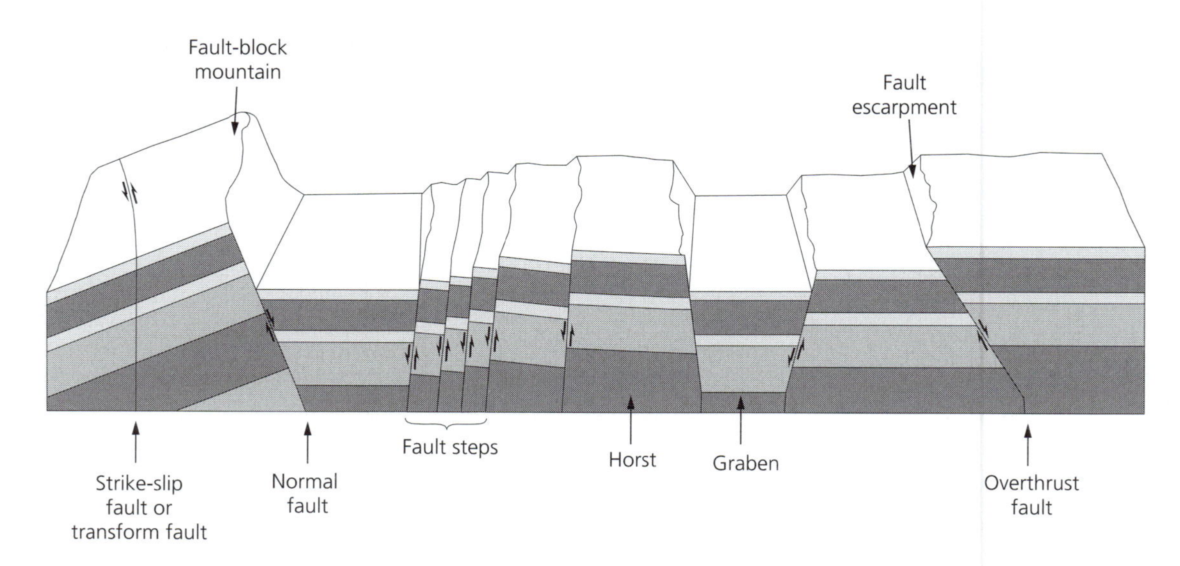

Figure 6.11 *The main types of common fault. The most common types of fault are shown; see text for explanation.*
Source: Figure 10.7 in Doerr, A.H. (1990) Fundamentals of physical geography. Wm. C. Brown Publishers, Dubuque, Iowa.

MULTIPLE FAULT SYSTEMS

Some distinctive large-scale landforms are created when two or more fault systems intersect one another.

Grabens: a graben is a large depression between two faults that are more or less parallel to each other. It can be created by the sinking (downthrow) of the central block, or the lifting (upthrow) of the two outer blocks, but the end-result is the same – the area between the faults sinks relative to the surrounding land. Large grabens, which extend over long distances, are called rift valleys. These tend to have steep side walls, they form where the Earth's crust is being pulled apart, and many rift margins are the focus of earthquakes and volcanic activity. Mid-ocean ridges, which play a central role in plate tectonics (see p. 150), are submarine rift valleys. The Great Rift Valley of East Africa (Figure 6.13) was formed in the same way, as were the rift valleys now occupied by the Red Sea, the Dead Sea and the Jordan valley in the Middle East.

Horst: a horst, an upthrown block between two parallel faults, is the opposite of a graben. Sometimes called a fault-block mountain, it is formed by vertical movements between the two sets of faults. Upthrow creates long, ridge-like block mountains, which often have steep slopes along the fault lines and more gentle slopes running away from them. Many of the western mountains of the Great Basin and the Sierra Nevada range in North America are fault-block mountains.

EARTHQUAKES

Earthquakes are vibrations of the Earth's crust that occur when strain in the crust is suddenly released by displacement along a fault. They vary a great deal in the magnitude of the energy released.

The point at which an earthquake originates inside the Earth is the focus, which may lie many kilometres beneath the surface (Figure 6.14). The strongest shock is normally felt at the earthquake epicentre – the point on the Earth's surface directly above the focus – which is also where most damage usually occurs.

SHOCKS, WAVES AND ZONES

Geologists record Earth tremors and earthquakes on seismographs (Box 6.11). These record the passage of seismic waves through the ground, and analysis of the pattern of these waves reveals a great deal about the nature of underlying rock structures. Seismograph records show that the Earth's crust and mantle are always on the move. The global seismic monitoring unit in Edinburgh, Scotland, picks up about 30,000 tremors a year on its seismograph equipment. Few of them trigger large earthquakes, cause great destruction or are reported in the media.

CAUSES AND MECHANISMS

Earthquakes can be explained in terms of plate tectonics and the associated movement of fault lines. The San Andreas fault (see Box 6.10) is a plate-boundary fault located between the Pacific plate and the North American plate. The Pacific plate is sliding north-west at a speed of a few centimetres a year. Where the movement is restricted, as it is in northern and southern California, the fault is effectively locked and stores up strain energy. This energy is ultimately released as an earthquake.

The immediate cause of most earthquakes is a sudden breaking and displacement of large masses of rock. As seismic energy accumulates in rocks subjected to strain,

BOX 6.10 THE SAN ANDREAS FAULT

One of the best-known and most closely observed fault systems in the world is the San Andreas fault, which runs for 1,040 km up the west coast of California (Figure 6.12). The fault and its tributaries pass close to Los Angeles and through the densely populated communities of Berkeley and Oakland near San Francisco.

It is a long, linear transcurrent fault, marking the point of contact between the Pacific plate (to the west) and the North American plate (to the east). Plate tectonic pressures are moving the Pacific plate north-westwards at a rate of a few centimetres a year, pushing with it past the mainland a narrow strip of the California coast. Recent Global Positioning System (GPS) measurements put the rate of movement along the San Andreas fault between 1990 and 1993 at about 46 mm a year. This is probably a reliable measure of background rates of movement, when no major earthquakes or sudden slippage occurs. Recent evidence suggests that high-pressure carbon dioxide pumped up from deep within the Earth acts as a lubricant that helps the plates to slide past each other.

The San Andreas fault is the source of a number of major destructive earthquakes along western California, including the 1906 San Francisco earthquake, a similarly large earthquake at Fort Tejon in 1857 and the smaller (magnitude 7.1) Loma Prieta earthquake of 1989. Apart from the damage that such large earthquakes cause to buildings and structures, they are also associated with major movements along fault lines. The 1857 Fort Tejon earthquake (and associated seismicity and afterslip) brought lateral displacement of at least 9.5 m along the main fault line.

Sediment sequences often preserve evidence of the timing and size of past earthquake events, and this allows the reconstruction of earthquake history and dynamics. Large earthquakes apparently occurred along the San Andreas fault zone 70 km north-east of Los Angeles in 1470, 1610, about 1700, 1812 and 1857, indicating that they are more frequent and regular than previously thought.

Geologists, seismologists and geophysicists have studied the San Andreas fault system intensively since the 1960s. Their object is to learn more about damaging earthquakes and about the behaviour of major strike-slip faults. Understanding these aspects of fault zone dynamics should help in developing methods of reducing earthquake hazards in populated areas.

The San Andreas fault is not all bad news, because, although it does cause damage and take lives, it has left an indelible imprint on the whole of western California. It has created the spectacular mountain landscape and shoreline, the major valley systems with rich young soils (which help the wine growing) and the deposits of oil, for example. One geologist reckons that the fault earns the state between US$10 billion and US$20 billion a year, whereas major earthquakes occur once every 50 to 100 years at a cost of around US$100 billion each. If these figures are of the right order of magnitude, the fault earns the state five–ten times what it costs!

BOX 6.11 SEISMOGRAPH RECORDS OF EARTHQUAKE ACTIVITY

Earthquakes usually produce a pulse-like pattern on seismograph charts. If these charts are interpreted correctly, the distance to the origin (focus) of the earthquake can be calculated.

The first earthquake waves to be recorded on a seismograph are the primary (P) waves (see Box 5.3), which travel fastest through the upper mantle, at around 7.5 km a second. Secondary (S) waves are bigger, and they move much more slowly through the upper mantle (around 4.3 km a second). The time difference between the P-waves and the S-waves reflects the distance between the seismograph and the epicentre of the earthquake (see Figure 5.4). If this distance can be calculated for at least three widely spaced seismograph sites, the precise location of the epicentre can be determined.

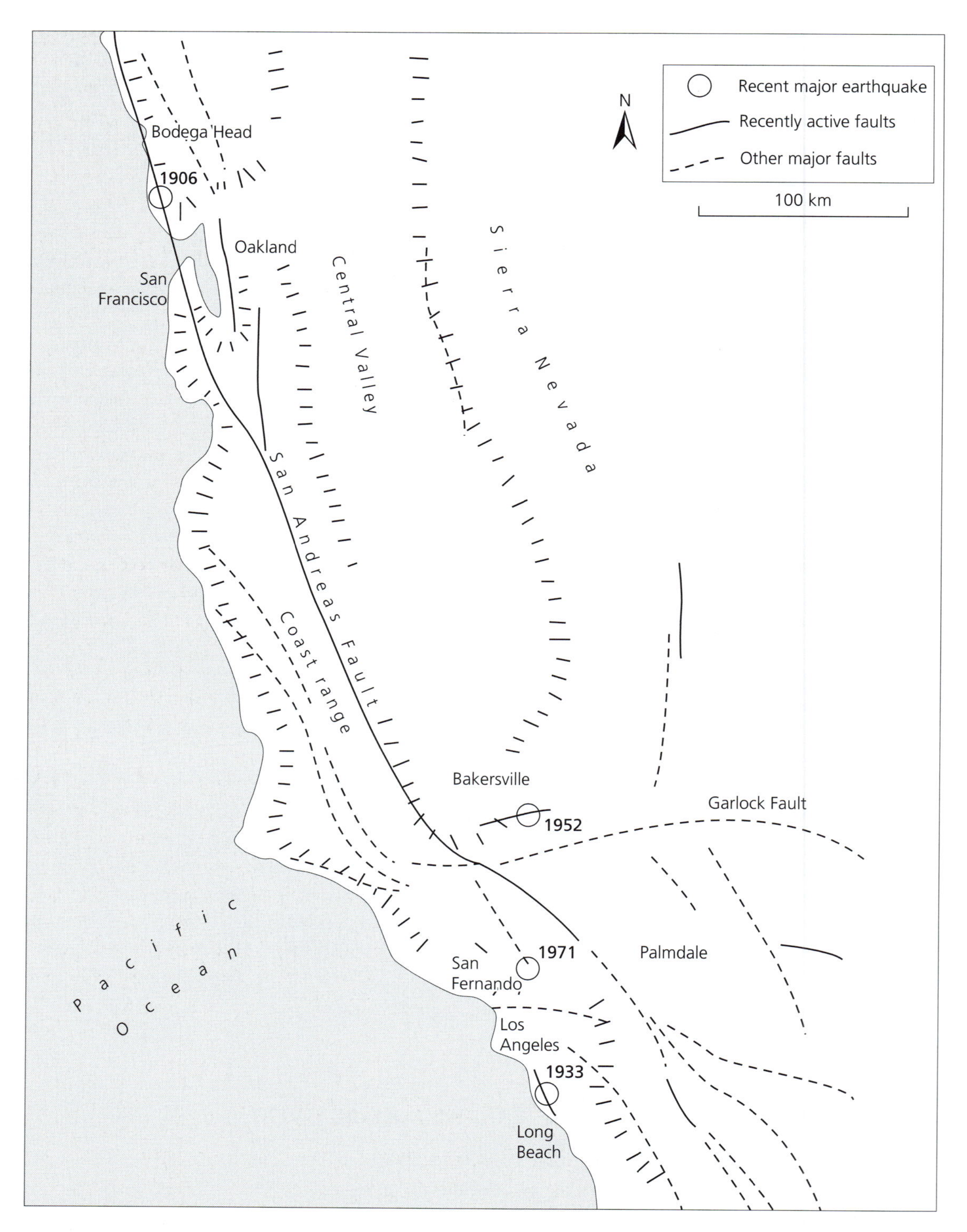

Figure 6.12 *The San Andreas fault system in California. The San Andreas fault runs along western California, through or close to some of the largest centres of population. It is part of a regional network of active faults that cross the area and that have caused some very damaging earthquakes during the twentieth century.*

Source: Waltham, A. (1978) Catastrophe – the violent Earth. George Allen & Unwin, London.

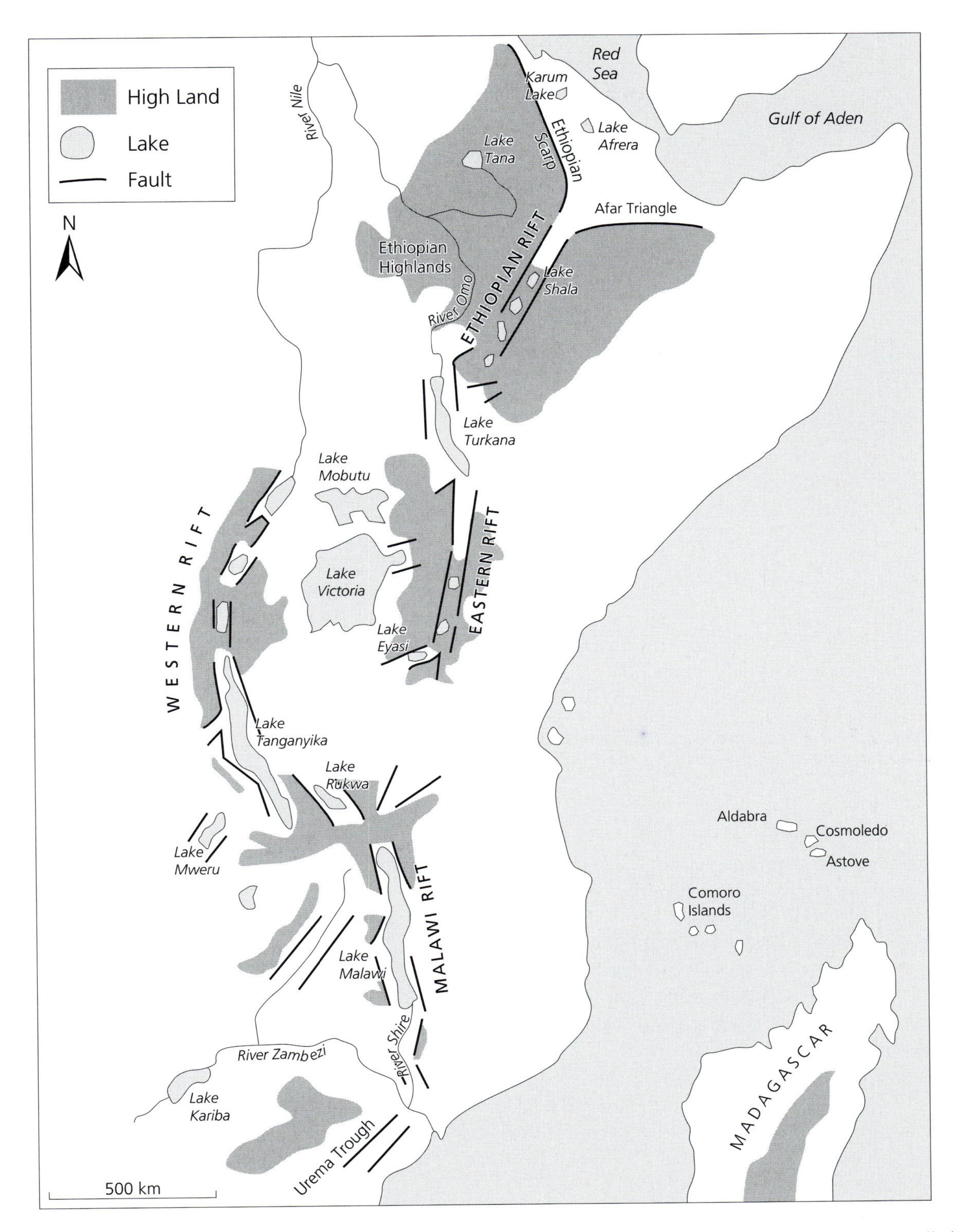

Figure 6.13 *The East African Rift System. The East African Rift System is a series of deep rift valleys controlled by fault lines, which dissect some of the highest parts of East Africa. Large lakes are found in some of the rifts.*

Source: Figure 2.4 in Buckle, C. (1978) Landforms in Africa. Longman, Harlow.

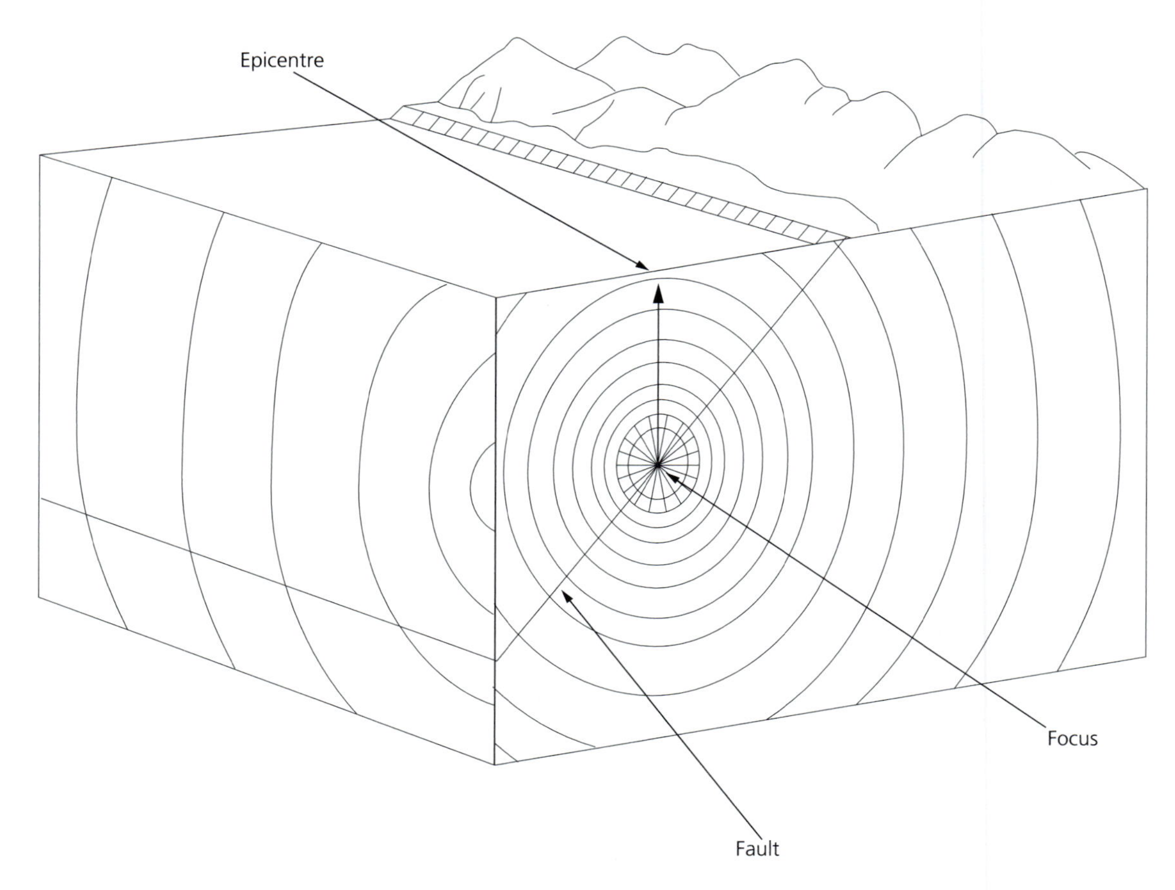

Figure 6.14 *The focus and epicentre of an earthquake. Earthquakes are associated with the movement of faults; the focus is the centre of the earthquake underground, and the epicentre is found on the ground surface directly above the focus.*

they bend slowly until they can no longer withstand the strain. They then split apart along a fault. The sudden failure and displacement produces an earthquake, which might vary in intensity from barely detectable trembling to violent and damaging shaking of the ground. As the stressed area ruptures, this relieves the strain and causes a sudden ground movement on one or both sides of the fault. Aftershocks occur as the fault settles back, and as slower-moving seismic waves affect the area.

Stress builds up through time, and the more accumulated stress is released, the larger the earthquake. It follows that if stress is relieved before it builds up to critical levels (thresholds), a series of minor earthquakes occur rather than one major event. Stress release is one strategy for managing earthquake risk (see p. 170).

Seismic stress can build up to critical levels, and if this is not released (normally by massive fault movement and seismic adjustment) it continues to build up. It is rather like a pressure cooker from which steam is not allowed to escape. This was believed to have been the situation along part of the San Andreas fault near San Bernardino in California in the early 1990s. The area had not received a major seismic shock since 1812, but a major (magnitude 7.4) earthquake at Landers on 28 June 1992 brought the fault very close to failure. The Landers earthquake also increased stress on the San Jacinto fault near San Bernardino and on the San Andreas fault south-east of Palm Springs. Geologists believe that unless creep or moderate earthquakes relieve these stress changes, the next great earthquake on the southern San Andreas fault is likely to arrive early by one to two decades.

The seismograph evidence suggests that major earthquakes are not random in timing, and the pattern of major earthquakes around the world appears to have some periodicity. Between 1910 and 1990, the energy released in great earthquakes has occurred in alternating cycles of 20 to 30 years. The increase in seismic activity in California since the 1980s has been interpreted as a late stage in one

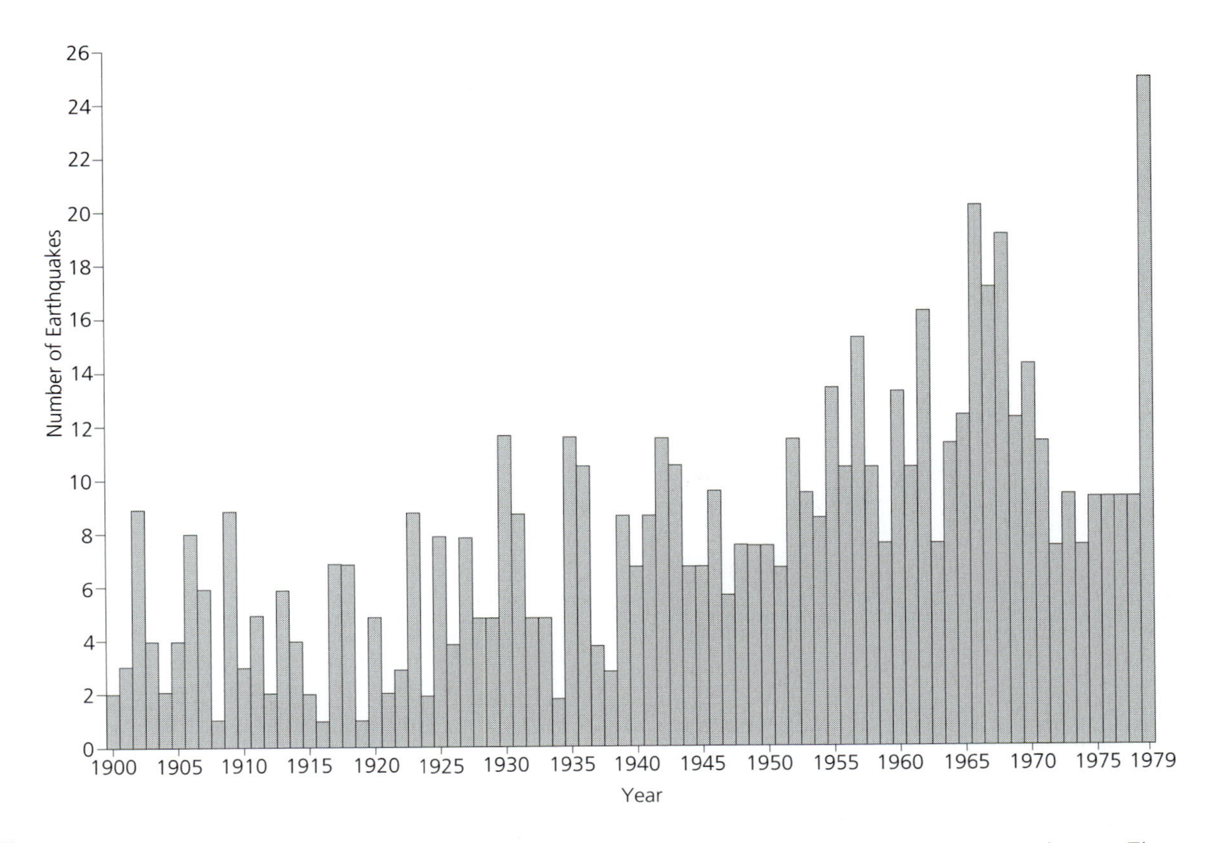

Figure 6.15 *The increasing number of destructive earthquakes around the world between 1900 and 1979. The graph shows great variability in the frequency of destructive earthquakes from year to year, but superimposed on a progressive increase in frequency through the twentieth century.*

Source: Figure 2.6 in Jones, D.K.C. (1991) Environmental hazards. In Bennett, R. and R. Estall (eds) Global change and challenge. Routledge, London, pp. 27–55.

such cycle that began in the 1960s. The number of destructive earthquakes around the world seems to have increased a great deal during the twentieth century (Figure 6.15), possibly in part because of the significant increase in the development of settlements and concentrations of people in seismically active areas.

EARTHQUAKE ZONES

A map of the major earthquakes and volcanic eruptions around the world shows an interesting continuous pattern around the Pacific – extending along the Andes in South America, up California, through Alaska, across the Aleutian Islands, down through Japan and across to New Zealand (Figure 6.16). They mark the boundaries of tectonic plates (see Box 6.5).

Earthquake activity commonly occurs where two plates slide laterally alongside each other. One is normally stationary and the other glides past it, but sometimes both move. The San Andreas fault (see Box 6.10) in California is one of the best-known examples. Another is the Great African Rift system, at the margins of the African and Arabian plates.

But not all earthquakes occur along plate boundaries or in areas with a history of earthquake activity. Even seismically quiet areas, like Western Europe, have had their surprises. Much of northern France, Britain probably as far north as Edinburgh, and the Low Countries and Germany beyond Cologne and Duisburg felt the aftershocks of an earthquake on 6 April 1580 whose epicentre lay offshore in the Straits of Dover, between England and France. Much more recently, on 19 July 1984, much of North Wales was shaken by a moderate-sized earthquake (magnitude 4.5) with an epicentre on the Lleyn Peninsula.

MAGNITUDE AND INTENSITY

The magnitude of earthquakes is usually expressed using the scale devised by Charles Richter in 1935 and named

Plate 6.3 *The Latin American Tower in Mexico City, Mexico. This high-rise tower block is built on special earthquake-proof foundations, including long concrete piles driven deep into the lake clays on which the city has been built. It survived the 1990 Mexico City earthquake when many surrounding buildings were badly damaged.*
Photo: Chris Park.

after him (Box 6.12). Intensity measures the effect of an earthquake on people and structures. The Mercalli scale is widely used to describe earthquake intensity. The zone of maximum intensity is the epicentre, and seismic waves gradually decrease in size outwards from there.

Each individual earthquake has one magnitude (hence one Richter number), but the intensity with which that earthquake is experienced varies from place to place (hence the Mercalli number changes depending on the location relative to the epicentre). The figures shown in Table 6.3 show how visible effects increase with increasing intensity, and indicative Richter magnitude numbers are shown.

BOX 6.12 THE RICHTER SCALE

The Richter scale describes the amplitude of the earthquake waves radiating out in all directions from the focus of the earthquake, which is closely related to the amount of energy released. It is a measure of ground motion as recorded on seismographs.

Magnitude on the Richter scale is proportional to the logarithm of the amplitude of a recorded seismic wave. Because it is a logarithmic scale, simple comparison of numbers can be deceptive. On a scale from 1 to 10, magnitude 6 is ten times the amount of energy of magnitude 5, 100 times magnitude 4, 1,000 times magnitude 3, and so on. Thus an earthquake of magnitude 6 involves a million times as much energy as an earthquake of magnitude 1. An earthquake of magnitude 2 is barely noticeable, whereas a magnitude 7 earthquake (100,000 times as strong) is severe. Total destruction can be expected at the upper end of the scale.

EXPERIENCE AND DAMAGE

Most earthquakes occur with little if any advance warning, despite recent advances in earthquake prediction and forecasting (see p. 171). Some places with a history of coping with earthquakes – such as California – have developed quite effective contingency plans and introduced information programmes to raise residents' awareness of what to do in the event of an earthquake.

TYPES OF DAMAGE

Most problems for people are associated with the damage that earthquakes cause to buildings, structures and transport systems. The collapse of structures is the direct cause of most injuries and deaths, and it also hinders relief operations.

Quite often more damage is caused by the series of aftershocks that shake the area after the event than is caused by the earthquake itself. This is because the aftershocks tend to be more subdued but longer-lasting and more frequent than the initial energy release during the earthquake. Buildings and structures that were weakened or partly damaged during the earthquake are often completely flattened or destroyed by the aftershocks.

Some but not all earthquakes involve surface displacement, usually along fault lines. This creates problems for

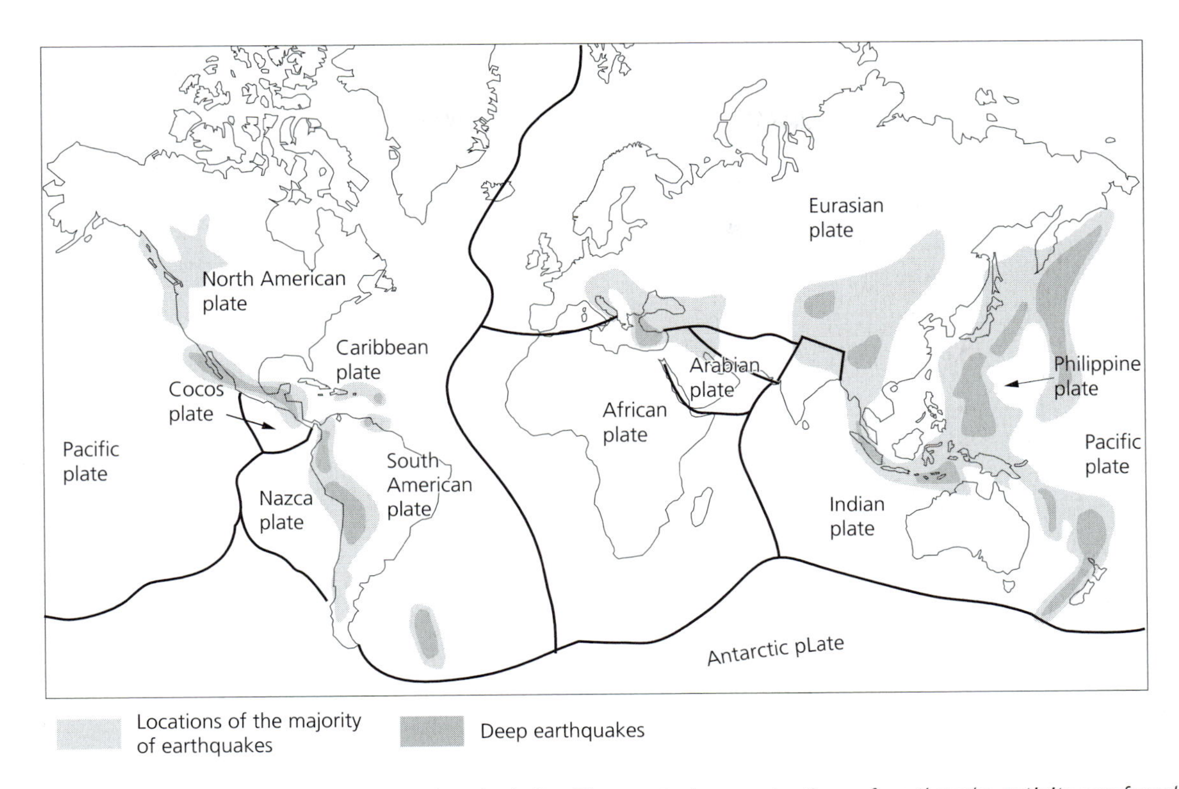

Figure 6.16 *Location of the major earthquake belts. The greatest concentrations of earthquake activity are found along or close to major crustal plate boundaries.*

Source: Figure 18.9 in Briggs, D. and P. Smithson (1985) Fundamentals of physical geography. Routledge, London.

Table 6.3 The Mercalli and Richter scales of earthquake size

*Mercalli number**	*Mercalli name*	*Characteristics*	*Richter scale***
I	Instrumental	Detected only by seismographs	<3.5
II	Feeble	Noticed only by some people at rest	3.5
III	Slight	Similar to vibrations from a passing truck	4.2
IV	Moderate	Felt generally indoors; parked cars rock	4.5
V	Rather strong	Felt generally; most sleepers wake	4.8
VI	Strong	Trees sway; furniture moves; some damage	5.4
VII	Very strong	General alarm; walls crack	6.1
VIII	Destructive	Weak structures damaged; walls fall	6.5
IX	Ruinous	Some houses collapse as ground cracks	6.9
X	Disastrous	Many buildings destroyed; rails bend	7.3
XI	Very disastrous	Few buildings survive; landslides	8.1
XII	Catastrophic	Total destruction; ground forms waves	>8.1

* earthquake intensity
** earthquake magnitude

communications into and around the damaged area, can be extremely costly to repair, and can cause additional risks through the fracture of gas pipes and water supply pipes (as happened in 1906 in San Francisco).

Earthquakes often trigger other geological hazards, too, including landslides and liquefaction (where underlying sediments are shaken so violently that they behave like liquids). Submarine earthquakes produce tsunamis (Box 6.13).

AMOUNT OF DAMAGE

The amount of damage and loss of life associated with earthquakes depends largely on four factors:

1. the number of people living close to the epicentre;
2. the kind of building they live in;
3. the density of buildings (closeness to each other);
4. the force of the shock itself.

The relative significance of these different factors can be appreciated by contrasting the experience of three large earthquakes that struck between 1993 and 1995. The most dramatic and widely publicised earthquake over the period occurred in the Kobe–Osaka region of Japan on 17 January 1995. The Kobe earthquake was magnitude 7.2, and it claimed more than 4,400 lives. It was slightly larger than the magnitude 6.6 earthquake that damaged parts of Los Angeles on 17 January 1994 but killed only 57 people. A major (magnitude 6.4) earthquake in Maharashtra in India, on 30 September 1993, killed an estimated 22,000 people.

Both Kobe and Los Angeles are in known earthquake zones. Houses were built to withstand quite powerful shocks, and local people had been educated about what to do in the event of an earthquake. Things were very different in central India, however, because there is no history of earthquakes in the area. Houses there were unstable and quickly fell down, and people had no idea what to do when the earthquake happened.

Earthquake damage is often most serious where buildings are not designed to withstand shaking or ground movement. Two very different groups of people were worst affected by the 1992 earthquake in northern Egypt. Many poor people in villages and the inner city slums of Cairo were killed or injured when their old, dilapidated adobe (mud-walled) houses collapsed. Many wealthy people were killed or injured when some modern high-rise concrete apartment blocks collapsed, some of which had been built without planning permission. Similar problems caused great loss of life and suffering in the 1988 earthquake in Armenia (Box 6.14). Recent excavations in Beirut, the war-torn capital of Lebanon, indicate that the city had been devastated by a major earthquake in AD 551 in which most buildings were levelled. This contradicts the popular belief that the damage was caused by a tsunami (see Box 6.13) triggered by earthquakes elsewhere in the eastern Mediterranean.

There is not necessarily a close correlation between the size of an earthquake and the damage it causes. Much depends on the area it affects, particularly the ability of

BOX 6.14 EARTHQUAKE IN ARMENIA, 1988

On December 1988, an earthquake hit Armenia in the former Soviet Union and caused widespread damage and destruction. Although it was only a moderate-sized event (magnitude 6.0), at least 25,000 people died, and many more were injured. Losses and suffering were particularly high in the cities of Spitak and Leninakan, where many buildings were badly damaged or collapsed into piles of rubble. Most resulted from the collapse of buildings that were badly designed and inadequately constructed. Ironically, the authorities in Armenia thought that they had learned lessons from major earthquakes that had badly damaged the area in 1926, and they claimed to have put up earthquake-proof buildings. Since the early 1970s, most apartment blocks throughout the former Soviet Union (including Armenia) were high-rise, with at least nine storeys. Almost all of these modern high-rise blocks collapsed in the area of the 1988 earthquake, causing massive suffering for the large number of residents in them at the time.

BOX 6.13 TSUNAMIS

Tsunamis are huge seismic waves (often incorrectly describes as tidal waves) that affect the entire water column. Tsunami waves are usually less than a metre high but can be several hundred kilometres long. They travel across the open oceans at more than 700 km an hour and can be detected long distances from the earthquake epicentre (sometimes on the other side of the world, literally). A tsunami rises to a great height when it reaches shallow water, such as off a beach or coastline, where it arrives with great destructive power and can cause vast amounts of damage over a very wide area.

buildings and structures to withstand violent shaking.

Disaster preparedness in Armenia (1988), Egypt (1992) and India (1993) was extremely limited, and the medical and rescue response was not as well planned or co-ordinated as it would have been in the USA, for example. Losses might have been reduced significantly if local people had been better warned about the likelihood of an earthquake happening and about what they should do to minimise their own personal risk. One lesson from such experiences of coping with earthquake hazards is that disaster relief could be greatly improved by strengthening self-reliance within the community.

Sometimes the damage caused by earthquakes has great cultural significance, and a number of important historical buildings and structures have been badly damaged by Earth movements. In September 1997, for example, an earthquake caused the roof of the thirteenth-century Basilica of St Francis in Assisi, Italy, to collapse. Priceless medieval frescoes by Giotto and Cimabue were damaged or destroyed, and restorers worked hard to piece together what they could of the shattered treasures. The restored building was reconsecrated in November 1999, but many people made homeless by the earthquake were still living in temporary shelters.

SAN FRANCISCO

One of the best-documented seismic events before 1980 was the earthquake that destroyed San Francisco in California in 1906 (Box 6.15). A great deal of archive material exists, including a striking set of aerial photographs of the city in ruins. San Francisco has been affected by a number of earthquakes since 1906, including one on 22 March 1957. This strong movement (magnitude 5.3) did not produce surface faulting, no lives were lost directly, and few buildings were damaged. San Francisco was badly affected by another earthquake on 17 October 1989, of magnitude 7.1 (Box 6.16).

Another major earthquake in the area is almost inevitable, but quite when it will happen is unknown. What is certain is that the 'Big One' is likely to produce massive damage, many deaths and much inconvenience, and it will have serious and long-lasting economic impacts.

OTHER MAJOR EARTHQUAKES

There is no shortage of examples of major earthquakes and the damage they can inflict on people and property. Two further examples will suffice.

One particularly spectacular example is the so-called Good Friday earthquake that rocked Anchorage in Alaska on 27 March 1964. This huge (magnitude 8.5) event released twice as much energy as the 1906 San Francisco earthquake, and it was felt over an area of nearly 1.3 million km^2. Over 130 people were killed, and an estimated US$500 million of damage was caused. It triggered large avalanches and landslides, which caused much damage. The earthquake also set in motion a series of tsunamis which travelled through the Pacific as far as California, Hawaii and Japan. Damage along the west coast of the USA was extensive.

The 1985 earthquake (magnitude 8.1) that badly damaged Mexico City was one of the most serious seismic disasters of the twentieth century. It caused massive and widespread damage and disruption. All aspects of the earthquake and how it was coped with were studied in great detail, and many important lessons were learned (if not always implemented). Mexico was again rocked by a powerful (magnitude 6.7) earthquake, with an epicentre

BOX 6.15 THE 1906 SAN FRANCISCO EARTHQUAKE

San Francisco is built over the San Andreas fault (see Box 6.10). This earthquake, the most serious one yet to affect the area (it was magnitude 8.3), struck at 5.12 am on 18 April 1906. It was short-lasting (a mere 5 seconds) but had a devastating impact, more or less flattening the city. The violent movements produced visible faulting, an upthrow of about 2 m on the east side of the fault and lateral displacement of nearly 4 m. The event was more than twenty-five times more powerful than the 1989 earthquake at Loma Prieta (see Box 6.16).

An estimated 700 people died, mostly through fires started by broken gas mains, which spread rapidly through San Francisco's wooden buildings. Attempts to dampen the raging fires were seriously hampered by ruptured water supply pipes. Many of the buildings not destroyed by fire were badly damaged and made unstable by subsidence, caused by liquefaction of the underlying clay (causing it to behave like a liquid) as it was violently shaken during the earthquake and aftershocks. Damage was estimated at about US$524 million.

BOX 6.16 THE 1989 LOMA PRIETA EARTHQUAKE, CALIFORNIA

This earthquake occurred on 17 October 1989 and was the most costly earthquake to hit the USA since 1906. It took place in the Santa Cruz Mountain segment of the San Andreas fault system, which is about 40 km long. The epicentre was located near Santa Cruz, about 80 km south of San Francisco. The south-west side moved nearly 2 m north-westwards and just over 1 m upwards relative to the north-eastern side. Most of the damage occurred in areas of unconsolidated sand and mud.

The earthquake was caused by underground rupture on a moderately dipping fault that lies alongside and is more or less parallel to the main San Andreas fault. The rupture occurred about 18 km below the surface but stopped at a depth of between 5 and 8 km (probably because the overlying material was more elastic). The Loma Prieta fault appears to be closer to the point of failure than the San Andreas fault over the same depth. The 1906 San Francisco earthquake and the 1989 Loma Prieta earthquake occurred on two different fault planes, although they are in the same area.

The 1989 earthquake lasted 15 seconds and caused an estimated US$6,000 million of damage. It struck at 5.04 pm, when many people were driving home from work. Sections of Bay Bridge, a two-level structure that carries interstate 880 across San Francisco Bay between Berkeley and San Francisco, collapsed and trapped people in vehicles. Some 60,000 people were evacuated from Candlestick Park football stadium and missed the end of their game! Oakland and San Francisco airports were closed for a time, after buildings and runways were damaged in the earthquake.

150 km south of Mexico City, on 14 June 1999. At least nineteen people were killed and hundreds injured, and many buildings (including nearly 300 churches) were at least partially damaged.

COPING WITH EARTHQUAKES

Many areas with a history of earthquakes have developed strategies for coping with the hazard. The object is normally to lessen the impact of earthquakes and thus save lives, buildings and money. Japan, which is sited on a particularly active part of the Earth's plate system, has a long history of earthquakes. Many steps are being taken there to minimise impacts, including prediction, building design, flood prevention and public information. Earthquake impacts in all areas with seismic risk can be reduced by better preparation and prediction.

PREPARATION

An estimated 1.4 million people lost their lives because of earthquakes during the twentieth century, and most of the deaths were caused by the collapse of unsuitable and poorly designed buildings. More than a third of the world's largest and fastest-growing cities are located in regions of high seismic risk, so the problems are likely to become worse in the future.

People cope with earthquake risks in a variety of ways. Prevention normally involves minimising the prospect of death, injury or damage by not building or constructing large structures in high-risk areas, rather than preventing an earthquake happening in the first place. In many places, warning systems have been introduced that advise people if an earthquake is about to occur and what to do when it actually happens.

Insurance schemes are another way of being prepared for damaging events, by sharing the costs between a wider group of people. Insurance companies face many problems in assessing real risks, which must take into account both scientific and statistical information.

BOX 6.17 CONTROLLING EARTHQUAKES

It would be very helpful if ways could be found to control the location, timing and size of earthquakes in order to minimise risk to people and property. As yet this is not practical, although at least in theory it might be possible to control some aspects of small earthquake events. Various strategies have been proposed, including altering the fluid pressure deep underground at the point of greatest stress in the fault line. This could trigger a series of small and less damaging earthquake events, thus releasing the energy that would otherwise build up to trigger a damaging large event. An alternative approach might be to use a controlled series of underground nuclear explosions, again with the object of releasing stress before it reaches critical levels.

BUILDING DESIGN

Land shortages in many areas, such as Japan, mean that even known high-risk areas are developed and built on. In these cases, structural protection can be adopted as an effective preparation for earthquake events. Buildings can be designed in such a way that they are better able to withstand the ground shaking that accompanies earthquakes and causes much damage. Single-storey buildings might be more suitable than multi-storey structures in particularly vulnerable areas, because this reduces population density (thus likely casualty levels) and the risk of collapse over roads and evacuation routes.

A number of recent earthquake disasters have been caused as much by inappropriate buildings as by geological forces, leading some geologists to stress that while it may never be possible to predict earthquakes, we can prevent disasters. An estimated 40,000 people died as a result of building collapse after a 7.2 magnitude earthquake devastated Izmit in Turkey in August 1999, for example. In Colombia, South America, a 6.0 magnitude earthquake in January 1999 caused widespread collapse of buildings, which trapped and killed up to 2,000 people in twenty towns spread across five provinces.

Some tall buildings are constructed with a 'soft storey' at the bottom (such as a car park raised on pillars). This is designed to collapse in an earthquake, so that the upper floors sink down on to it, and it cushions the impact. Basement isolation is also being widely adopted. This involves mounting the foundations of a building on rubber mounts, which allow the ground to move under the building, effectively isolating it from the tremors.

Building-reinforcement strategies are now used for large new structures in many seismically active cities, such as Los Angeles, Mexico City and Tokyo. These include building on foundations driven deep into underlying rocks (to reduce the impact on buildings of ground shaking) and constructing steel-framed buildings (designed to withstand shaking without collapsing).

Land-use planning is another important way of reducing earthquake risk. It is one way of making sure that high-density residential developments are located away from high-risk sites, and it helps to create access corridors to allow emergency teams to reach damaged areas in the event of an earthquake.

PREDICTION AND RISK ASSESSMENT

The second way of reducing earthquake damage is through prediction. A great deal of time, effort and resources have been invested in the search for reliable ways of predicting where and when earthquakes are likely to occur, and their likely size. As yet, the search has proved only partially successful.

The location of likely earthquakes is easiest to predict, because they are closely associated with the distribution of fault lines. Previous patterns and frequencies of earthquake events offer valuable clues to what is likely to happen in the future. Earthquake hazard risk maps can be helpful in identifying which areas are most at risk, although they inevitably rely heavily on historical patterns and assume that earthquakes will tend to recur in similar places.

Sometimes a long-dormant fault system can be reactivated, and when this happens an earthquake can occur in a totally unexpected location. Knowledge about where earthquakes might be expected, and the probability of an earthquake of a given size occurring, is absolutely vital in choosing risk-free sites for nuclear power stations and nuclear waste repositories (see p. 137).

The timing and size of likely earthquake events are much more difficult to predict than location. The most reliable approaches depend on the detection of geological and geophysical precursors, or early warning signs. Sometimes the build-up of stress along a fault line can be gauged by detailed measurement of small-scale ground surface changes, particularly where there is unusual uplift or subsidence.

Geologists now rely heavily on banks of instruments in continuous monitoring networks, recording small-scale variations in the properties of underlying rocks, including:

- ground tilt
- changes in rock stress
- micro-earthquake activity (clusters of small quakes)
- anomalies in the Earth's magnetic field
- changes in radon gas concentration
- changes in electrical resistivity of rocks.

Some areas with a history of earthquake activity are now extensively monitored in experiments designed to increase our understanding of how earthquakes work and to learn more about earthquake prediction. One particularly intensively studied site is Parkfield in California, on the San Andreas fault (Table 6.4). The Parkfield Earthquake Prediction Experiment (Box 6.18) is being watched with anticipation by geologists and emergency planners around the world.

Short-term earthquake predictions are difficult to make, because foreshocks are almost impossible to distinguish from background seismicity. But the hope is that with

Table 6.4 Monitoring of earthquake build-up in Parkfield, California

Instrument	*Purpose*
Seismometer	To record micro-earthquakes
Magnetometer	To record changes in the Earth's magnetic field
Near-surface seismometer	To record larger shocks
Vibreosis truck	To create shear waves to probe the earthquake zone
Strain meter	To monitor surface deformation
Sensors in wells	To monitor changes in groundwater levels
Satellite relay	To relay data to the US Geological Survey
Laser survey equipment	To measure surface movement

constant monitoring of many possible indicators it may be possible to interpret the signals of a forthcoming earthquake. Parkfield is heavily instrumented: strain meters measure deformation at a single point; two-colour laser geodimeters measure the slightest movement between tectonic plates; and magnetometers detect alterations in the Earth's magnetic field caused by stress changes in the crust.

Despite recent advances in monitoring technologies, methods of analysing seismic information and understanding how and why earthquakes happen, it is still not yet possible to give accurate long-term warnings about earthquake events. Yet the need for such advance warning is increasing as more development takes place in fault zones and high-seismic-risk areas, including the rapid growth of many towns and cities in developing countries.

EARTHQUAKES INDUCED BY HUMAN ACTIVITIES

Human activities can trigger earthquakes, or alter the magnitude and frequency of earthquakes, in a variety of ways. The three most widespread means are by increasing crustal loading, underground disposal of liquid wastes (Box 6.19) and underground nuclear testing and explosions (Box 6.20).

INCREASED CRUSTAL LOADING

Earthquakes can be triggered by imposing considerable pressures on previously stable land surfaces, such as through the weight of water stored behind large reservoirs. In 1935, for example, the Colorado River was dammed by the Hoover Dam to form Lake Mead. As the lake filled, and the underlying rocks adjusted to the load of over 40 km^3 of water, a series of long-dormant faults in the area were reactivated, resulting in over 6,000 minor earthquakes in the first ten years. Over 10,000 events had been recorded up to 1973, about 10 per cent of which were strong enough to be felt by residents. None caused damage.

Many other areas have experienced earthquakes within a few years of the completion and filling of large dams, including the area around Lake Nasser on the River Nile (see p. 63).

BOX 6.18 THE PARKFIELD EARTHQUAKE PREDICTION EXPERIMENT

Parkfield is a small town (population less than fifty) on the San Andreas fault in California. It proudly claims to be the 'earthquake capital of the world'. It is now home to a large-scale experiment to record seismic activity in the area using sophisticated networks of high-tech equipment. The object is to devise better means of predicting where and when an earthquake is likely to occur and how big it is likely to be. The site was chosen carefully, because in 1985 geologists predicted that a magnitude 6 earthquake was likely to occur in the area before 1993. It did not happen, and had not done so by mid-1996, but that does not mean that the problem has gone away. Quite the opposite!

The instrumentation has provided valuable information on the build-up and arrival of moderate-size earthquake events. One such earthquake, magnitude 4.7, hit the area around Parkfield on 20 October 1992. Geologists thought this might be a foreshock ahead of the magnitude 6 event, and a well-prepared contingency plan was activated. Public warnings were issued, emergency services were put on standby, and journalists and television news crews descended on the area.

The forecast was wrong, and the magnitude 6 event did not happen. But the incident was a good test of how well people in the area might cope with preparing for the real thing. The public warning scheme was judged a success. It was found, however, that people did not weigh up the probability of the earthquake risk but made black and white judgements about whether or not they were personally at risk and thus whether action was needed.

BOX 6.19 UNDERGROUND DISPOSAL OF LIQUID WASTE

Geologists were surprised at the seismic response to the underground disposal of waste water at Rocky Mountain Arsenal just outside Denver, Colorado, during the 1960s (Figure 6.17). Water was contaminated in the production of chemical warfare agents (such as mustard gas, white phosphorus and napalm), and the toxic wastes would be costly to transport off site for disposal elsewhere. The geology of the area beneath the site (fractured pre-Cambrian gneiss) looked quite suitable, so it was decided to pour it down a disposal well over 3,500 m deep at the arsenal.

Disposal began in March 1962. Soon afterwards, a series of minor earthquakes were detected in the area, which had no known history of seismic activity or instability. None of the minor earthquakes produced real damage, but they did give cause for concern. Between March 1962 and November 1965, over 700 minor earthquakes were monitored in the area.

It seems that the injection of the liquid waste into the bedrock of the area had inadvertently lubricated and reactivated a series of deep underground faults, which had been dormant (inactive) for a long time. The more waste water was put down the well, the larger the number of minor earthquakes. When the link was uncovered, geologists argued that disposal should cease for safety reasons. Disposal stopped late in 1965, and the well was filled in February 1966. Soon after closure, the number of minor earthquake events detected in the area fell sharply. In the mid-1980s, the US Army closed the site as a military facility, and it and the Shell Oil Company (which had been producing agricultural pesticides on site, under lease) embarked on an ambitious remediation (clean-up) programme. In 1996, a comprehensive management plan was approved for the site, which will turn it into a national wildlife refuge to protect the 300 or so species (including deer, coyotes and owls) that inhabit the site.

BOX 6.20 UNDERGROUND NUCLEAR TESTING

Underground nuclear testing has triggered earthquakes in a number of places. In April 1968, underground testing of a series of 1.2 kiloton bombs at the Nevada Test Site in the USA set off over thirty minor earthquakes in the area in the following three days. Since 1966, the Polynesian island of Moruroa (meaning literally 'place of a great secret') has been the site of over eighty underground nuclear explosion tests by France. More than 120,000 people live on the island. On 11 September 1966, a 120 kiloton nuclear device was triggered there, producing radioactive fallout that was measured over 3,000 km downwind and seriously rocking the whole area. More recently (in 1995) underground nuclear testing in the western Pacific was resumed by France, despite a massive show of public opposition and concern at the health and environmental risks of the operation.

VOLCANOES

The dynamic nature of the Earth's crust is rarely better displayed than in violent eruptions of volcanoes. More than 500 volcanoes have erupted over the past four centuries. Volcanic activity is often sudden, dramatic and spectacular, which is why the world's media pay so much attention to it. Such activity is far from uniform in time and space, so this type of geological hazard is highly variable and can be difficult to predict. The number of volcanoes reported as active appears to have risen progressively between about 1860 and 1980 (Figure 6.18).

Eruptions at the surface give clues to what is happening in the Earth's interior, particularly in the mantle beneath the crust, but our understanding of why volcanoes behave the way they do is still far from complete.

IMPACTS AND LOCATIONS

Volcanoes vary a great deal from place to place in appearance and activity. Some are totally inactive and pose no threat to people, whereas others are highly active and when they erupt can cause much damage to people and property. Volcanoes cause a great deal of death and destruction, and many of the problems associated with volcanic eruptions (Box 6.21) are outlined in the rest of this chapter. But volcanoes are not all bad news, and we must not overlook some of the benefits they offer.

BENEFITS OF VOLCANOES

Volcanic ash and dust are usually rich in minerals, so volcanic soils are generally very fertile and support high population densities. Paradoxically, the very thing that

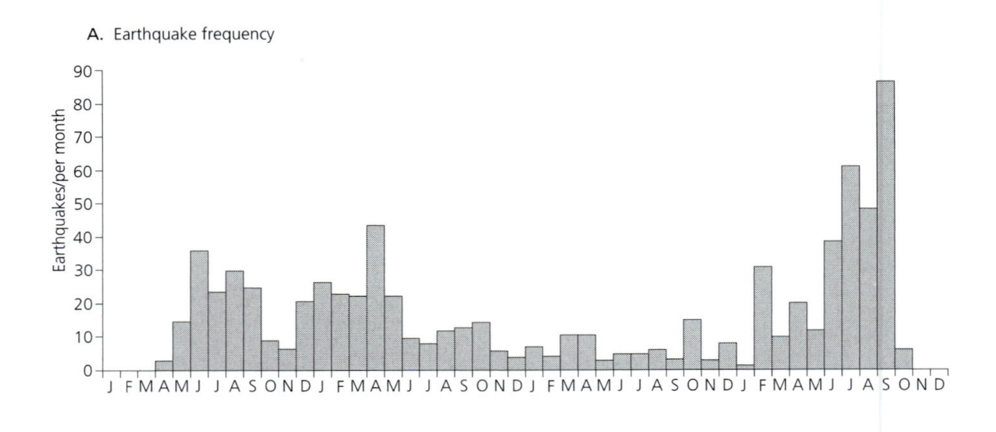

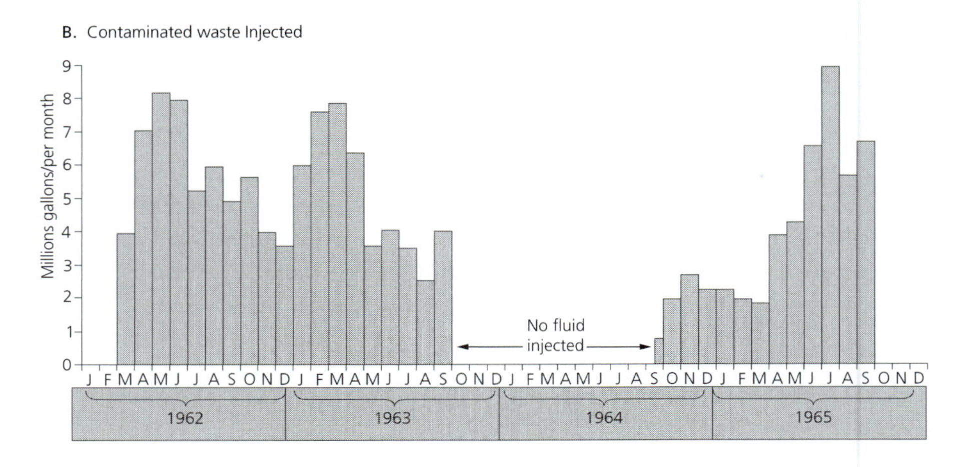

Figure 6.17 *Increasing earthquake frequency associated with underground liquid waste disposal, Rocky Mountain Arsenal, Colorado. There was a close association in timing between the disposal of waste water underground and the frequency of minor earthquakes in the surrounding area. See Box 6.19 for explanation.*

Source: Figure 6.25 in Goudie, A. (1993) The human impact on the natural environment. Blackwell, Oxford.

encourages people to farm and live in volcanic areas also puts them at risk. Life for many people without alternatives boils down to a game of Russian roulette with the volcanoes they depend upon.

Volcanoes also create new land and environments. A graphic example of this was the eruption in 1963 of the submarine volcano Surtsey, south of Iceland. It started with a violent explosive eruption, which poured out lava initially on to the sea bed. The lava cone quickly built up and eventually rose above the sea surface to create an entirely new island. As the lava cooled, it was slowly colonised by plants and animals (see p. 571). Kilauea, a volcano in Hawaii, was the world's most active volcano during the 1980s, and its quiet non-explosive eruptions are increasing the size of the island as the lava descends to the sea. The sudden appearance and rapid growth of Parícutin in Mexico (Box 6.22) is another graphic example.

Volcanoes also attract visitors and thus help local economies. Sometimes the attraction is simply the beauty of the volcanic peak, such as Fujiyama in Japan. The spectacular sight of a volcano erupting can attract visitors from around the world, as has happened in Hawaii in recent decades. The spectacle of volcanic eruptions has made Etna in Sicily a major tourist attraction, although from time to time tourist installations (such as cable car systems) are damaged by lava flows.

Some countries derive a number of benefits from volcanoes. Iceland is a good example. Before 1928 volcanoes

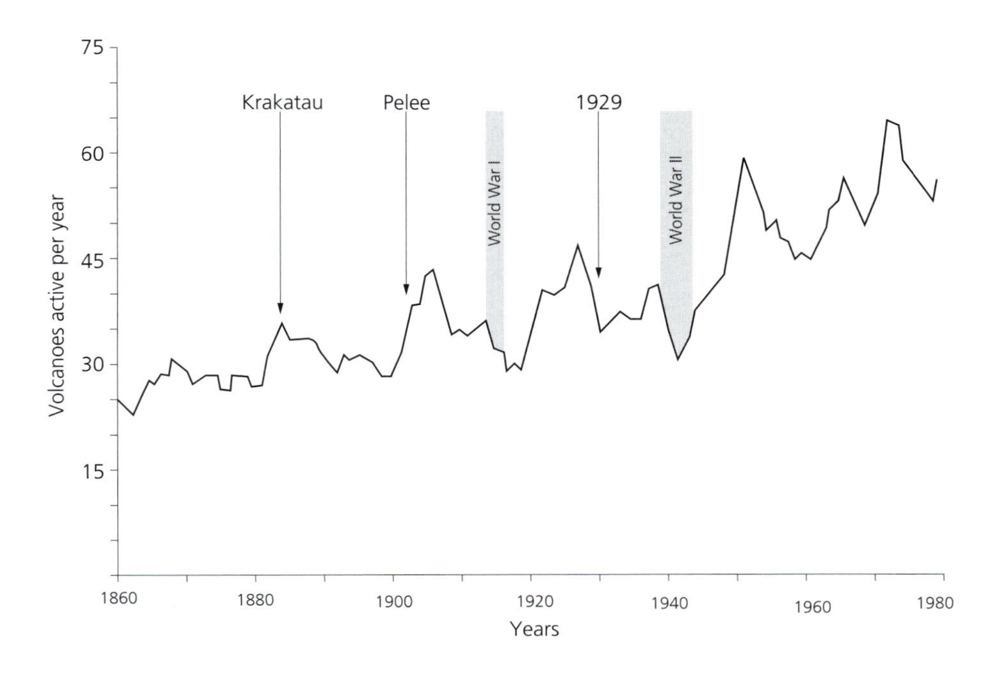

Figure 6.18 *The increasing frequency of active volcanoes, 1860–1980. The graph, based on the three-year running average number of active volcanoes, shows a great deal of short-term variation superimposed on an underlying increase through time.*

Source: Figure 2.2 in Smith, K. (1992) Environmental hazards. Routledge, London.

BOX 6.21 KRAKATAU (1883) – ANATOMY OF A VOLCANIC ERUPTION

Volcanic eruptions can be extremely spectacular, and rarely more so than the sudden explosion in 1883 of Krakatau. This small, steep, forested, uninhabited volcanic island in Indonesia, between Java and Sumatra, had been dormant for 200 years. It erupted with great force, and the explosions were heard nearly 5,000 km away in Australia. Two-thirds of the island was blown away in the explosions.

An estimated 17 km^3 of rock, ash and pumice was thrown high into the atmosphere. Much of the finer dust remained in suspension in the upper atmosphere (at heights of up to 27 km) for years and was blown around the world, creating spectacular bright red sunsets in many places for more than a year. The heavy load of dust in the atmosphere probably blotted out some incoming solar radiation, which would explain why average global temperatures were lower than normal for two years after the event.

The violent ground shaking triggered a catastrophic seismic wave (tsunami) (see Box 6.13), which flooded coastal towns and villages on the nearby coasts of Sumatra and Java and drowned 36,000 people.

posed problems to the people of Iceland but seemed to offer little in return. Since then, the hot water produced in volcanic systems has been used extensively, in both geothermal district heating schemes and geothermally generated electric power (see p. 140). Volcanic deposits (pumice and ash) are used as building materials, and new land created by volcanic eruptions is used for building and farming.

VOLCANO ZONES

Clues to the origin of volcanoes lie in their location:

- Active volcanoes are concentrated in parts of the world where earthquakes are most common.
- Volcanoes are often found in relatively young mountain belts (although some young mountain belts, such as the Himalayas, have no volcanoes).

BOX 6.22 PARÍCUTIN – BIRTH OF A VOLCANO

Explosive volcanoes grow very fast. Scientists could see this for themselves as they monitored the dramatic birth and growth of Parícutin in central Mexico. Over a two-week period early in 1943 the area around Parícutin had been shaken by earthquakes. On 20 February, the ground cracked and a vent opened up in a cornfield. First steam, then volcanic dust and then hot fragments started to pour out. Soon molten rock was pouring out, and it started to build up a cone around the central vent. The eruption continued for eight months, and the cone grew to a height of about 450 m. The accompanying lava flows buried the village of Parícutin and nearby settlements. By 1952, the volcano had gone quiet.

The distribution of volcanoes is strongly influenced by the major crustal plates, because most are clustered at or near the margins of the plates.

The area around the rim of the Pacific is particularly active and is referred to as the Pacific 'Ring of Fire'. It extends through the Andes in South America, into Central America and Mexico, along the west coast of the USA into Alaska, across the Aleutian Island chain, through the Kamchatka Peninsula, through Japan and the Philippines, into New Zealand and the margins of Antarctica. Other major volcanic regions include a zone running through the Mediterranean, areas along the rift zone in East Africa, and in the Lesser Antilles on the margins of the Caribbean.

Many volcanoes begin life underwater, on the sea floor. Surtsey started in this way, as did the Hawaiian Islands and many other volcanic islands in the Pacific. Etna and Vesuvius were originally submarine volcanoes.

Volcanoes form at convergent and divergent plate boundaries (see p. 150). At convergent boundaries, one plate is subducted beneath the other, and material in the subducted plate is dragged down into the mantle, where it melts. It rises to the surface again through vertical cracks and fissures and is released at the surface through a volcanic vent. The volcanic ranges of the Ring of Fire are thrust up at subduction zones, where the moving tectonic plates of the Pacific plunge beneath the confining plates to the east, north and west. Subduction of the Pacific plates provides the molten rock and pressure that drives the volcanoes. At divergent plate boundaries, the crust is stretched and rifted apart (as happens at the mid-ocean ridges). This forms a zone of weakness (spreading centre), which can be penetrated by the eruption of magma brought upwards by large convection currents within the mantle.

VOLCANIC MATERIALS

There are basically two types of volcanic material – magma and lava.

MAGMA

Magma is the molten rock beneath the Earth's surface. It is extremely hot and generally contains fairly large amounts of gas. Magma behaves like a plastic and can flow through weaknesses in the surrounding solid rock of the crust. It is the basic building material for all volcanoes and lava flows, and when it cools and solidifies it becomes igneous rock. Magma is not uniform in composition, and variations in the chemical make-up of magma cause the rock to behave differently, look different, weather differently on exposure and have different possible uses to humans.

Magma is produced within the upper 100 km of the Earth, by the partial melting of rock material within the asthenosphere. Melting is caused by pressure decrease and/or temperature increase.

Pressure decrease: hot material can be liquefied if the pressure decreases, even if the temperature remains the same. Pressure is reduced by the bending or cracking of rocks above, by removal (through weathering) of overlying rocks, and as tectonic plates move.

Temperature increase: crustal rock can be melted if it is heated by heat generated from the radioactive decay of rocks, by contact with hot plumes of rock from the mantle, or by the heat generated by plate movements and subduction.

LAVA

Lava is magma that has reached the surface. It is the molten material and gases that flow out of volcanic vents and fissures. The rock that solidifies from this material to produce hard igneous rocks is also called lava. Lavas range in temperature from 600 to 1200 °C. Magma temperatures below the surface must be much higher, because of the greater pressures. Different types of lava have very different appearances, and some have rather exotic names (Table 6.5).

Volcanic activity moves magma from inside the Earth towards the surface. If it spills out on to the surface it creates volcanoes and lava flows and is referred to as *extrusive vulcanism*. Sometimes the magma never reaches the surface but instead cools and solidifies underground. It is then referred to as *intrusive vulcanism*.

COMPOSITION AND CHARACTER OF VOLANIC ROCKS

The chemical composition of volcanic rocks (Table 6.6) is important, because it determines how the material flows when molten and significantly affects the speed and way it weathers after solidification. Silica (SiO_2) content is particularly important, because silica-rich magma and lava is viscous and sticky and does not flow far. Low-silica magma and lava, in contrast, can flow long distances. All volcanic rocks contain silica, although the proportion varies from about 45 to about 75 per cent. Viscosity (resistance to flow) is also affected by pressure (an increase in pressure tends to increase viscosity) and by temperature (a rise in temperature tends to decrease viscosity).

Geologists sub-divide igneous rocks into three groups (Box 6.23) – mafic, felsic and intermediate rocks – depending on their silica content. The radically different chemical composition of the mafic and felsic rocks indicates that they must originate in different parts of the Earth's interior (see p. 128). Mafic magma (including basalt) probably forms by the heating and melting of localised regions in the upper mantle. Most felsic magma (including granite) probably comes from the melting of the lower part of the continental crust.

INTRUSIVE VULCANISM

Volcanic activity moves magma from inside the Earth towards the surface, but sometimes the magma cools and solidifies underground. This is described as intrusive vulcanism, and it is commonly associated with a solid rock barrier beneath the surface, which prevents the rising magma reaching the surface.

An intrusion, to geologists, is a mass of igneous rock formed by the injection of molten magma into existing cracks beneath the Earth's surface. The magma may spread sideways, and it might eventually erupt into one or more volcanoes where there are cracks or weaknesses in the rock barrier.

Intrusive vulcanism produces a variety of different underground forms (Figure 6.19), which are collectively referred to as plutons. Some are too small to change the ground surface in any obvious way, but others can raise up large areas and affect surface drainage systems. Intrusions are described as concordant or discordant, depending on whether they cut across or lie parallel to the rocks into which they are emplaced.

Although the plutons are initially created underground, surface rocks can be removed by weathering to reveal the previously hidden forms. This is how the granite masses of Dartmoor and Bodmin Moor in south-west England were unroofed, for example. Once exposed at the surface, the intrusive forms are subjected to weathering, erosion and

Table 6.5 Some common types of lava

Name	*Appearance*
Aa (Hawaiian word pronounced 'ah-ah')	Lava that has a rough, jagged surface when it cools
Pahoehoe	Lava that solidifies into ropy or corded shapes with a smooth surface; the name is a Hawaiian word meaning satin-like
Pillow lava	Lava that has solidified under water and thus resembles heaps of pillows

Table 6.6 Some common types of igneous rock

Type	*Intrusive or extrusive*	*Comment*
Andesite	Extrusive	A fine-grained rock formed from the ejection of lava in continental not oceanic areas; named after the Andes in South America
Basalt	Extrusive	A fine-grained, heavy rock, which is usually dark grey or black in colour; the most common rock in lava flows
Gabbro	Intrusive	A coarse-grained rock, formed at great depth, with a similar chemical composition to basalt
Obsidian	Extrusive	A glassy rock, black or dark green in colour, formed from fast-cooling lava and with a similar chemical composition to granite
Pumice	Extrusive	A light grey, glassy rock; it contains masses of pores (holes sealed off from each other) formed by gases in the lava; the pores make it so light that it floats on water
Tuff	Extrusive	Rock formed from compressed volcanic ash and dust

BOX 6.23 THREE GROUPS OF IGNEOUS ROCKS

Mafic rocks: these igneous rocks are silica-poor and often contain relatively large amounts of magnesium, calcium and iron. The most common type of mafic rock is basalt, which is a dark, heavy, iron- and magnesium-rich rock with a low silica content. Many oceanic volcanoes and some continental volcanoes eject basalt, which cools in some interesting forms (including columnar jointing). Mafic rocks usually weather down into productive soils with mineral constituents and physical properties suitable for agriculture.

Felsic rocks: these igneous rocks are silica-rich and often contain relatively large amounts of aluminium and oxides of sodium and potassium. Ryolite, a relatively viscous extrusive type of granite, is a typical felsic rock – relatively light in weight, low in iron and magnesium, and rich in silica. Felsic rocks weather down quite readily, and they disintegrate without any strong soil structure. This produces soils that are poor for plant growth – coarse-textured, free-draining and lacking in essential minerals.

Intermediate rocks: these rocks have a silica content between the extremes of the mafic and felsic rocks and are typical of island arcs and young mountain ranges. Andesite is good example. Intermediate rocks weather down to produce quite fertile soils.

reshaping by surface processes. Intrusive igneous rocks are often more resistant than the surrounding rocks, so they may produce persistent upstanding landforms.

CONCORDANT FEATURES

Intrusive igneous features that lie parallel to pre-existing rock structures are described as concordant. The most common is the *sill*.

Sills: these are nearly horizontal sheets of igneous rock formed when molten magma that is intruded between sedimentary strata or between layers of volcanic ash or lava cools and solidifies. Buried sills rarely affect surface topography, but if overlying rocks are weathered and worn away the resistant sill can act as a protective cap rock and shield underlying rocks from erosion. Sills sometimes stand out along valley sides, rather like ribs.

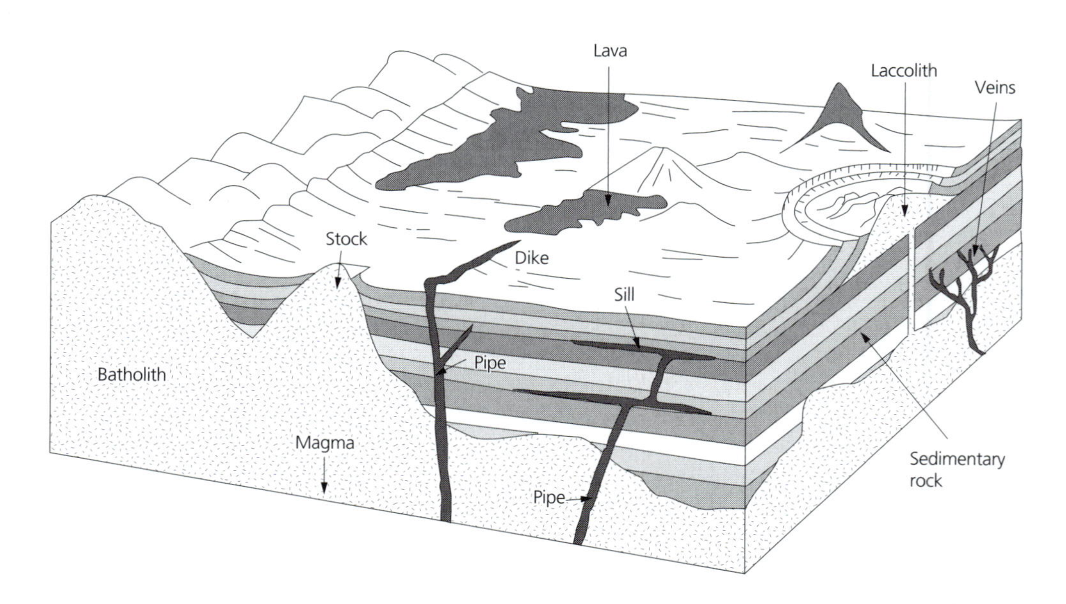

Figure 6.19 *The main features produced by intrusive vulcanism. The emplacement of volcanic material underground creates a number of characteristic features. See text for explanation.*

Source: Figure 10.33 in Doerr, A.H. (1990) Fundamentals of physical geography. Wm. C. Brown Publishers, Dubuque, Iowa.

DISCORDANT FEATURES

Intrusive igneous features that cut across existing sediments or layers of ash and lava are described as discordant. They come in various shapes and sizes (Box 6.24).

EXTRUSIVE VULCANISM

Magma is lighter than the rocks above it, so it tends to rise through overlying rocks if it finds an opening (such as a fracture). When it reaches the surface, the magma may pour out quietly as a fluid lava flow, or it may be thrown out in an explosive eruption.

VOLCANOES

Lava normally builds up into a cone (the volcano) with a bowl-shaped vent (crater) at the top (Figure 6.20). Craters are usually round, have steep sides and are much smaller features than calderas (Box 6.25). The crater is linked to the magma chamber underground by a vent (pipe), which is the supply line for the magma, gas and steam that pour or explode from the crater. Magma, gas and steam sometimes also escape from narrow openings (fissures) away from the crater.

Different types of magma produce different types of volcano (Figure 6.21). Some, including the Icelandic and Hawaiian types, are relatively low features created mainly by lava flows. Others, including the Vesuvian and Plinian types, are cone-shaped and are created by explosive eruptions.

A volcanic eruption is caused by the release of magma from the crater or fissures in a volcano, but eruptions also

BOX 6.25 CALDERAS

Large, steep-sided craters are found at the top of some long-dormant or extinct volcanoes, and these are called *calderas.* They are much larger than the original vent of the volcano and usually have relatively flat floors. Many are occupied by deep lakes (such as Crater Lake in Oregon), and some contain a family of small volcanoes produced by renewed volcanic activity.

Some calderas are formed by the collapse of the original volcanic peak into the magma chamber below, which has been emptied by repeated volcanic eruptions and can no longer support the weight of the volcano above it. Other calderas are the result of violent explosions that literally blow the top off erupting volcanoes. Krakatau in Indonesia (see Box 6.21) is an example of this explosive type.

BOX 6.24 COMMON DISCORDANT INTRUSIVE IGNEOUS FEATURES

Batholiths: these are the largest and most common type of pluton. They are at least 100 km^2 in size and can extend over thousands of square kilometres. Batholiths often form the core of major mountain ranges, such as parts of the Sierra Nevada and Rocky Mountains in the USA.

Laccoliths: laccoliths are similar in form but much smaller than batholiths. They resemble toadstools in cross-section and are formed by the emplacement of viscous magma, which cannot flow far. The ground surface sometimes bulges upwards above laccoliths. The Black Hills of South Dakota have a laccolithic core.

Stocks (bosses): stocks are smaller than batholiths, with no distinct shape. They are too small to cause surface uplift. Stocks typically occur at the outer edge of batholiths or laccoliths, or they develop separately from larger magma sources.

Dikes: dikes are vertical or steeply inclined pipes of magma which cool and solidify in cracks or weaker rock. They can extend for many kilometres and are often clustered in a radial pattern around a volcanic vent. Like other intrusive forms they are often more resistant to erosion than surrounding rocks, so they form radial ridges when exposed to surface weathering.

Veins: veins are small magma-filled cracks, which are abundant in some areas. Some volcanic zones have rich metallic mineral deposits, and these are usually concentrated in veins. Superheated water in contact with magma can dissolve minerals remaining in veins to produce metallic minerals (including precious metals). Carbon, subjected to great heat and pressure, can crystallise into diamonds, as has occurred in southern Africa.

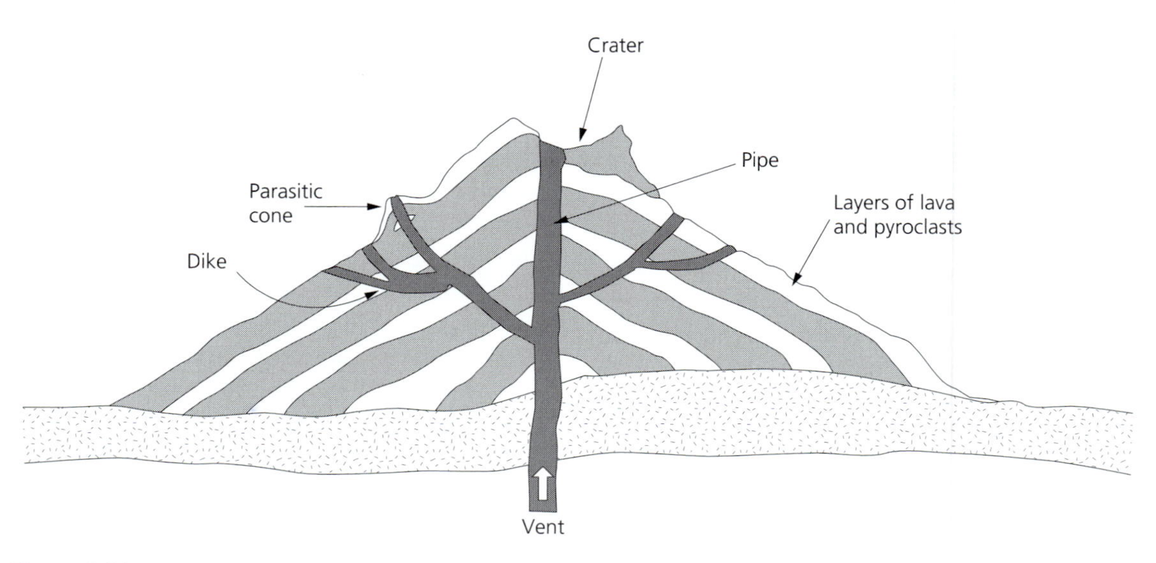

Figure 6.20 *The main features of a typical volcano.*

Plate 6.4 *Volcanic crater on Lanzarote, one of the Canary Islands in the Atlantic Ocean off the north-west coast of Africa. Note the symmetrical, relatively smooth-sided circular cone and the wide, deep crater.*
Photo: Philip Barker.

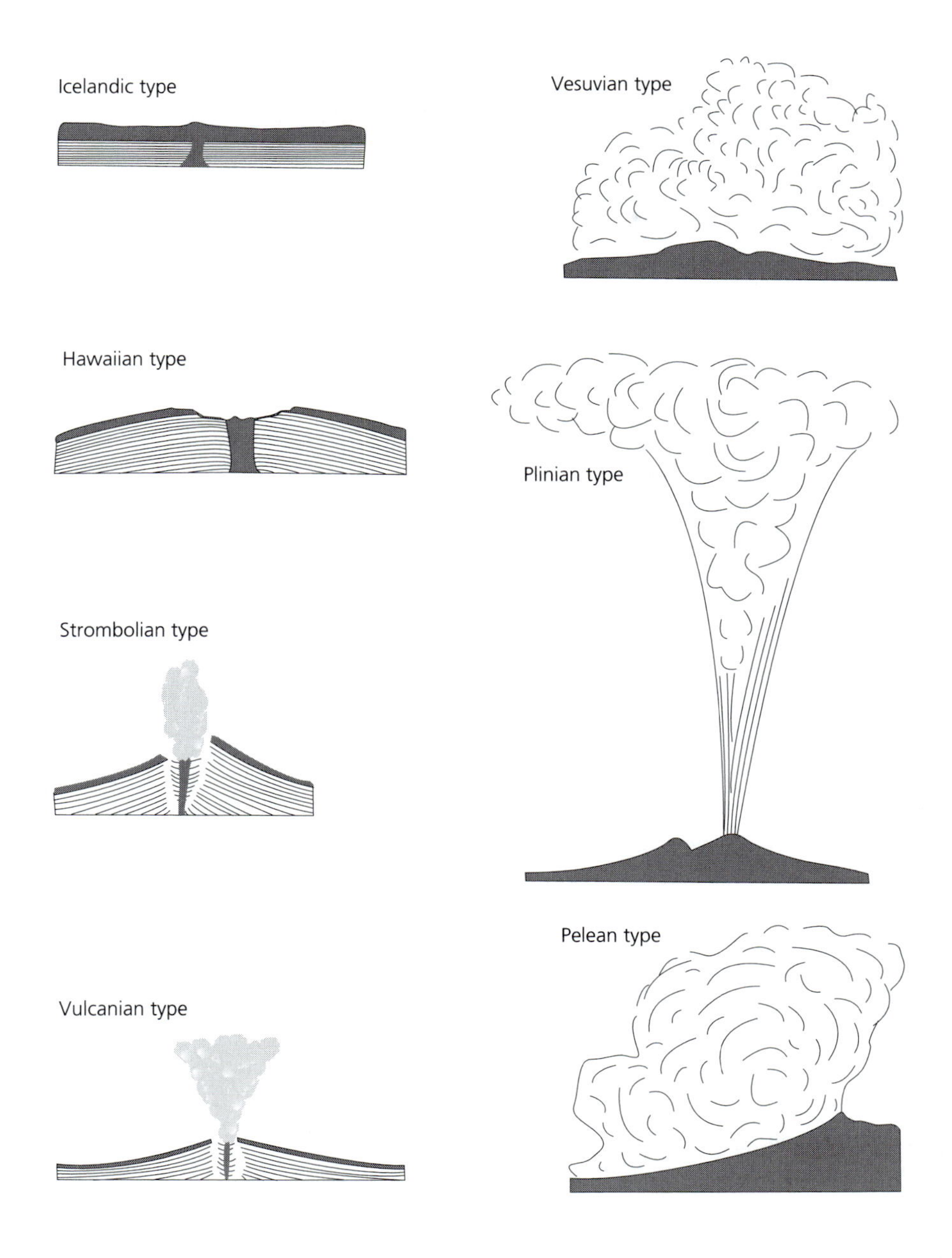

Figure 6.21 *The main types of volcanic eruption. The main types of volcanic eruption are shown here in increasing order of life-threatening hazard.*

Source: Figure 7.2 in Smith, K. (1992) Environmental hazards. Routledge, London.

release great amounts of gas and steam. Explosive eruptions also throw large amounts of ash, bombs and dust (Table 6.7) into the atmosphere, and the lighter material can be blown across a wide area by winds. Lava can produce some striking forms when it cools, depending on its constituent materials, the amount of trapped gases and the rate of cooling (see Table 6.5).

Some volcanoes no longer erupt lava and ash, but they do release steam and gases from small vents. Such a volcano is called a *solfatara*, after the name of a dormant volcano near Naples in Italy. The steam may be exploited for geothermal energy (see p. 139).

Table 6.7 Material released in an explosive volcanic eruption

Material	*Description*	*Comments*
Volcanic ash	Very fine fragments of magma	Formed when explosive gases shatter magma into fine fragments; forms a dark cloud when erupted
Volcanic bombs	Chunks of magma erupted into the air	Magma is erupted as molten rock, but the outside has usually cooled and hardened by the time it lands; volcanic bombs vary in size and appearance
Volcanic dust	Extremely fine volcanic ash	Clouds of volcanic dust can be carried great distances in the atmosphere by winds
Volcanic steam	Hot steam released in volcanic eruptions	Volcanic steam is also released in fumaroles and solfataras

LAVA FLOWS

Explosive eruptions normally occur when magma contains a high proportion (usually at least 75 per cent by volume) of gas. Magma with a lower gas content usually pours out from a vent or vertical fissure, and this produces a lava flow rather than a volcanic eruption. Lava flows of this sort have created thick sheets of basalt extending over hundreds of kilometres. Western North America contains several examples, including the great lava plain of the Snake River in Idaho. The Deccan Plateau in India and parts of East Africa are covered by lava flows. Smaller-scale lava flows have occurred in modern times in Iceland and Hawaii.

Under suitable climatic conditions, the basaltic rocks in lava flows can weather down into deep productive soils.

STATES OF VOLCANO ACTIVITY

Some volcanoes are much more active than others. Highly active volcanoes that erupt more or less continuously are rare, thankfully. One of the most constantly active volcanoes in recent times is Stromboli, in the Lipari Islands near Sicily. Other constantly active volcanoes are found around the Pacific Ring of Fire (see p. 150).

Most active volcanoes are more intermittent than these constantly active ones. They are moderately active for a period of time (months or years) and then become quiescent (dormant) for a period (months or years). Periods of activity and dormancy rarely match, and few active volcanoes spring into life in consistent cycles. Hazard risk is caused by the uncertainty about when and how big the next eruption will be. Quite often the eruption following a dormant phase is violent and causes widespread damage. This was certainly the case with the eruption of Mount St Helens in Washington, USA, in 1980 (Box 6.26) after a 123-year dormant phase. Mount Pinatubo in the Philippines (Box 6.28) exploded in June 1991 after six

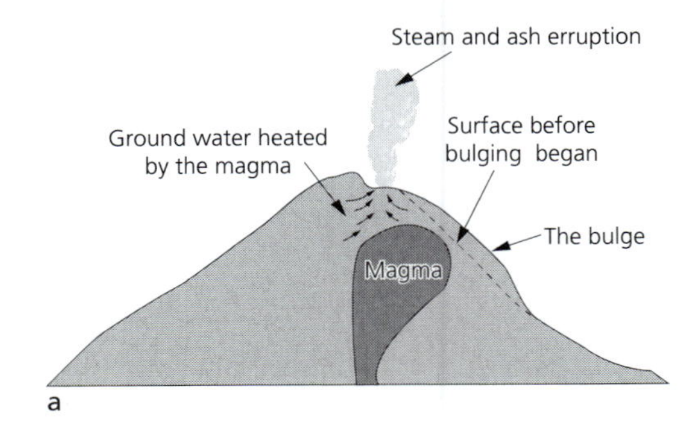

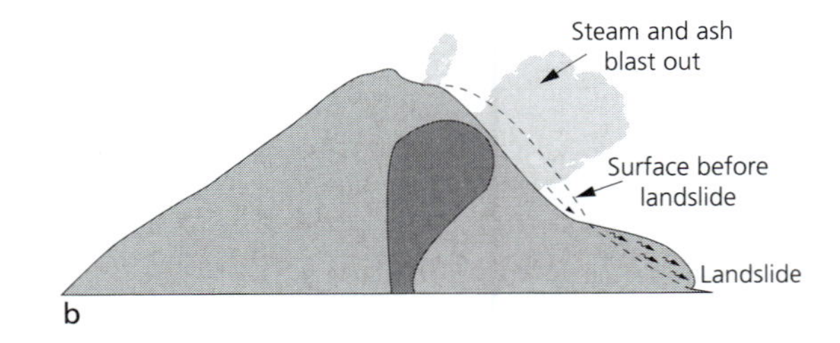

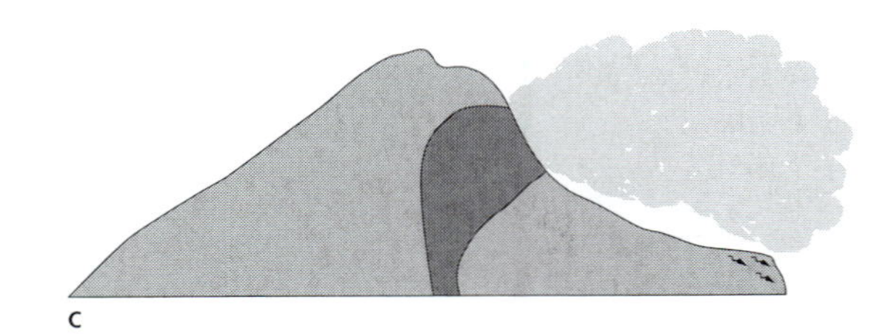

***Figure 6.22** Explosion of Mount St Helens on 18 May 1980. Before the eruption, magma was pushed upwards beneath Mount St Helens, causing the mountain to bulge and the ground surface to tilt (a). The explosive eruption released pressure on the magma, caused a massive landslide (b) and a powerful lateral blast (c), which deposited ash and blew down trees over a wide area.*

Source: Figure 10.17 in Doerr, A.H. (1990) Fundamentals of physical geography. Wm. C. Brown Publishers, Dubuque, Iowa.

BOX 6.26 THE 1980 ERUPTION OF MOUNT ST HELENS

Geologists from the US Geological Survey reported in 1978 that Mount St Helens, a little-known volcano in southern Washington state, would erupt again, perhaps within the next 20 years. This prediction was a surprise to many people, including locals, because the volcano had been dormant since 1857. But the evidence showed that Mount St Helens had a distinguished track record as the most active and most explosive volcano over the past 4,500 years of all volcanoes in the forty-eight contiguous states. The Cascades volcanic group, to which Mount St Helens belongs, is fed by magma that is melted as the small Juan de Fuca tectonic plate slowly bends downwards and is subducted beneath the Washington coast into the mantle.

Mount St Helens erupted violently on 18 May 1980, confirming the geologists' forecast. The eruption was explosive and destructive and followed a magnitude 5.1 earthquake. Seismic activity had been building up from 20 March, when a magnitude 4.2 earthquake occurred in the afternoon. Over the following seven days, up to 26 March, earthquake activity increased to as many as eight magnitude 4 events recorded each hour. Minor eruptions began the next day and continued up to the explosive eruption on 18 May. Earthquake frequency declined as eruption activity grew.

Damage from the May eruption was estimated at over US$2,000 million, and sixty-three people died. More lives would have been lost if warnings had not been issued, and if the eruption had not taken place on a quiet Sunday morning. Most people evacuated the danger zone when geologists warned that an eruption was imminent. The explosive eruption caused widespread damage to forests and wildlife in the area.

The explosion quite literally blew the top off Mount St Helens, changing its shape from a symmetrical cone 2,950 m high to a flat summit 400 m lower (Figure 6.22). It caused damage over a wide area, including mudflows and ash falls. Many trees were blown over in the enormous blast (Figure 6.23).

The 18 May 1980 eruption was followed by a series of other major eruptions of the volcano. Soon afterwards, steam began to pour from the top of Mount Baker, and unusual seismic activity was recorded near Mount Hood and Mount Shasta. Mount Bailey became more active than it had been over the previous 100 years.

The 1980 eruption of Mount St Helens is one of the most closely monitored and best-documented volcanic eruptions ever. Since then, geologists have carried out further monitoring in the Cascades Range in an attempt to further improve forecasting abilities. Continuous monitoring equipment has been installed on more active peaks to measure seismic activity (seismometers) and small-scale ground tilting and uplift (tiltmeters and laser surveys).

Since the 1980 event the volcano has continued to show signs of activity, including plumes of smoke, sporadic small ash eruptions and frequent earth tremors. Mount St Helens was made a national monument in 1982 in recognition of its scientific and tourism value.

centuries of quiescence. Both were dramatic events with significant impacts.

Long after a volcano stops erupting lava, it usually continues to release acid gases and vapour. This is described as the fumarolic stage of a volcano. After this, hot springs may arise from the volcano, producing features like the geysers (see p. 139) of Yellowstone National Park in Wyoming (see Box 18.21) and the hot springs of the North Island of New Zealand. As a volcano becomes more and more inactive, it ceases to release volcanic heat in any form. Crater Lake, Mount Shasta in California, Mount Hood in Oregon, and Mounts Rainier and Baker in the state of Washington are all examples of US volcanoes that have become inactive in recent geological times.

VOLCANIC ERUPTIONS

The violent eruption of a volcano is one of nature's most dramatic spectacles. Lumps of red-hot lava are thrown out from the vent with immense force, rise high into the atmosphere and fall to the ground over a wide area as bombs, cinders or ash, depending on their size and shape (Table 6.8). Lava is accompanied by steam and hot gases (including carbon dioxide, hydrogen, carbon monoxide and sulphur dioxide), which rise in a thick, dense cloud and often produce showers of rain around the volcano. Lightning is not uncommon in the volcanic cloud, particularly if it is heavily charged with dust particles. Lava rises in the vent of the volcano and often flows over the rim or oozes through a fissure in the side of the cone, and then flows downslope and buries almost everything that gets in its way.

BOX 6.27 CHANCES PEAK VOLCANIC ERUPTIONS, MONTSERRAT, 1995–97

Montserrat, in the eastern Caribbean – one of Britain's six remaining Caribbean colonies – has been described by long-haul tour operators as 'the emerald isle of the west'. The tropical island has long been regarded as a tranquil haven surrounded by warm seas and covered by bright skies, despite being badly battered by Hurricane Hugo (see Box 16.43) in 1989. Chances Peak, a dormant volcano in the Soufriere Hills on the southern part of the island that rises nearly 1,000 m above the sea, had not erupted for over 400 years. Yet the geological time-bomb was ticking for this island chain, which sits astride a subduction zone (see p. 153), and in July 1995 a strong earthquake rocked Montserrat and caused clouds of ash and steam to vent from Chances Peak. The same Earth movement triggered a volcanic eruption on the island of Guadeloupe, about 100 km to the north. Half the population of 10,000 who live on Montserrat evacuated the island, fearing a major eruption. Conditions for the remaining 5,000 became particularly difficult when the island chain was battered by strong winds from Hurricane Luis, the most destructive storm within the past 50 years, and a major tsunami (see Box 6.13), which between them destroyed 75 per cent of the buildings on Montserrat and adjacent Antigua. Another hurricane – Hurricane Marilyn – quickly followed, and ruined the ongoing rebuilding effort.

Disaster struck again in 1996, when Chances Peak erupted violently. Lava set fire to buildings in Montserrat's abandoned capital (Plymouth). A year later, on 25 June 1997, Chances Peak erupted again – this time claiming the lives of ten people, destroying ten houses and engulfing the runway of the island's airport. The British government began to consider the permanent relocation of all of Montserrat's citizens. In July, another unexpected eruption occurred and lasted two days uninterrupted. This caused avalanches of superheated rock and gas (pyroclastic flows; see table 6.8) to spew down the mountainside and engulf everything in their path. Two-thirds of the island was declared wasteland, and many more residents had to be evacuated. The lava flows ignited fires all over the island, which stretched fire-fighting crews to capacity. Two earthquakes on 5 August triggered more violent explosions from the volcano, which threw out grenades of pumice and covered the surrounding area in a thick layer of ash. Within 24 hours, ash flows had reached Plymouth and destroyed 80 per cent of the remaining buildings as it built up to more than 1 m in thickness. The eruptions of 5–6 August created superheated plumes, which reached a height of 12,000 m in the sky and spread over 40 km downwind, at least as far as Antigua.

Further eruptions continued between August and October 1997, and the British government announced a five-year plan to rebuild the northern part of the island and to compensate those people who wished to relocate off the troubled island. On 11 November, the volcanic dome began to collapse, releasing yet another pyroclastic flow, which carried rock and ash to the sea, where it created a new delta with an estimated volume of 2 million m^3. Explosions continued on and off until the dome collapsed totally on 26 December 1997.

Table 6.8 Some products of an explosive volcanic eruption

Name	*Description*
Fumarole	A hole in the ground on the side of a volcano from which steam and gases escape
Lahar	A landslide of volcanic debris, caused when melting snow, rain or water from volcanic steam or from a crater lake mix with ash or volcanic fragments on the sides of a volcano
Lapilli	Cinders or small lumps of lava that are thrown up from an erupting volcano
Nuée ardente	A 'glowing cloud' formed when a burning-hot cloud of ash, gas, steam and rock emerges from a volcano and spills downhill, burning all in its path
Pyroclasts	Broken fragments of rock ejected from volcanoes during eruptions; pyroclasts include lapilli, pumice, volcanic ash and volcanic bombs
Tephra	Volcanic ash thrown out in an explosive eruption

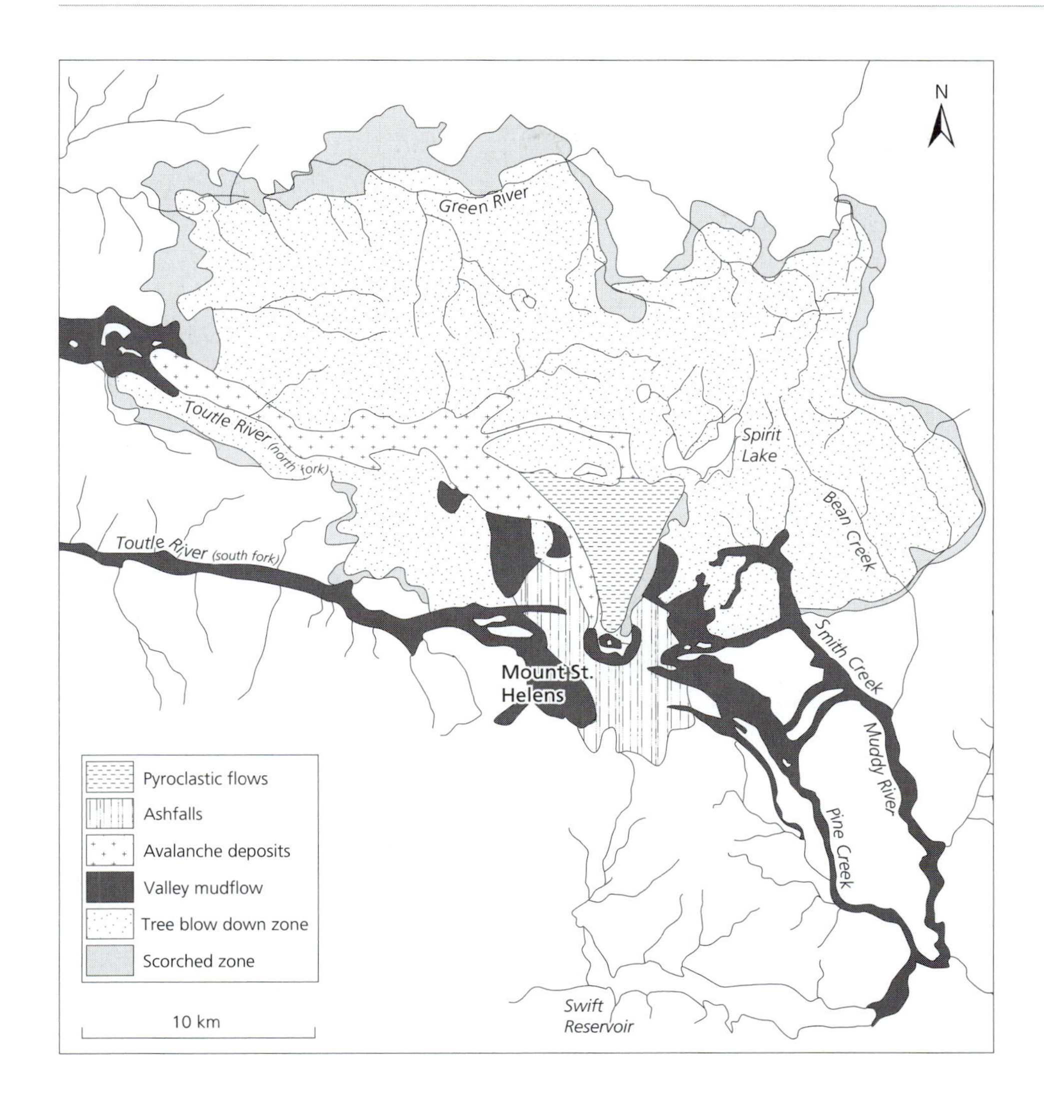

Figure 6.23 *Main environmental impacts of the Mount St Helens eruption. The explosive eruption (Figure 6.22) blew down many trees, spilled out lava and threw out vast amounts of fine volcanic ash.*
Source: Figure 7.7b in Smith, K. (1992) Environmental hazards. Routledge, London.

Probably the most explosive volcanic eruption ever, anywhere in the world, occurred in about 1450 BC on Santorini, an island in the eastern Mediterranean. The 1,500 m high mountain exploded, and about 130 km^2 of land on the island was completely destroyed. Most of the rest of the island was buried in up to 60 m of ash, and ash was spread over more than 200,000 km^2. The blast generated a tsunami that was up to 45 m high when it reached Crete, 112 km away.

TYPES OF VOLCANO

We tend to think of all volcanoes as explosive in character and conical in appearance, but volcanoes come in many different sizes and shapes (see Figure 6.21). The mode of formation and character of the lava involved largely determine the appearance of a volcano (Table 6.9). Basaltic lavas are fluid and tend to flow in a gentle and sluggish manner, building up low-slope shield volcanoes over a period of time. Granitic lavas, on the other hand, are viscous and tend to erupt explosively to build high-slope volcanoes.

Table 6.9 Some common types of volcano on land

Type	*Description*
Ash cone	A steep-sided volcano composed of fine ash erupted from the volcano
Cinder cone	Built from coarser material than ash cones, and usually steeper than them, e.g. Parícutin in Mexico
Plugdome volcano	Composite volcanoes that have a solidified intrusive plug at their core
Shield volcano	Built from repeated quiescent flows of lava, covering large areas and with relatively shallow slopes, e.g. many of the Hawaiian volcanoes, such as Kilauea, Mauna Loa and Mauna Kea

Many volcanoes are composite, having been built up over perhaps a million years from alternating layers of ash and cinders (from explosive eruptions) and lava (from lava flows). Composite volcanoes tend to be conical in profile and symmetrical in shape, rising from a wide base to a tall, thin peak. Fujiyama in Japan, with its much-photographed snow-capped summit, is a classic example. Other well-known composite volcanoes include Mount Rainier in the USA, Etna in Sicily and Vesuvius near Naples in Italy.

VOLCANO HAZARDS

Volcanoes affect people in a variety of ways, and at a variety of scales. The risks are real: more than 25,000 people around the world were killed by volcanoes during the 1980s alone. The US Geological Survey Volcano Hazards Program acknowledges the importance of hazards associated with lava flows, volcanic bombs and ash fallout, but it recognises that volcanoes also produce other hazards that may be a greater threat to people. These include avalanches, debris flows and river floods, which can affect areas many kilometres downstream from the volcano, and earthquakes and tsunamis. The experience of Montserrat between 1995 and 1997 (Box 6.27) demonstrates how complex and diverse the experience of coping with volcanic hazards can be.

Plate 6.5 *Volcanic ash deposits on Lanzarote, off the north-west coast of Africa. A barrier of dark ash separates a small saline (salt-water) lake from the sea.*
Photo: Philip Barker.

ASH FALLOUT

Some problems relate to the fallout of ash (tephra) after an explosive volcanic eruption, because the ash buries people, buildings and structures. The classic example is the eruption of Vesuvius in AD 79, which deposited ash over a wide area and entombed the town of Pompeii. Most of the people who died were killed by inhaling poisonous sulphur-rich gas or were suffocated under deposits of ash. Vesuvius has erupted on a number of occasions since then. Maps of ash fallout have helped in the reconstruction of the eruptive history of the volcano and may help to show areas at risk from future eruptions. Ash deposits also provide valuable marker horizons, which can be dated (using carbon-14 dating, for example) and used to piece together the environmental history of volcanic areas.

LAVA FLOWS

In some volcanoes, the main risk is from lava flows, involving the seepage of red-hot lava from the volcanic vent, which flows downhill (under the influence of gravity) and can smother anything that gets in its way. In sparsely settled areas, like around Mount St Helens, this changes landscapes but does not directly threaten many people. Sometimes lava flows create new land, as happened on Hawaii when lava flows from Kilauea between 1983 and 1991 added nearly 1.2 km^2 to the island. In populated areas, on the other hand, lava flows pose a real threat. When Etna erupted in 1983, for example, lava flows put a number of villages at risk. Local people were warned and evacuated, but geologists went one step further and tried to divert the lava away from the most vulnerable places. Lava diversion, using explosives and earth barriers, proved quite successful.

LANDSLIDES AND MUDFLOWS

People and property are not threatened only by molten rock and showers of ash and cinders when a volcano erupts. Sometimes the seismic disturbance sets off landslides and mudflows (lahars), which travel long distances at great speed, threatening towns and villages in their path. One such disaster followed the eruption on 13 November 1985 of Nevada del Ruiz in Colombia. The eruption melted snow and ice on the side of the high mountain, triggering an avalanche and mudflow that buried several villages in mud, water and ash and killed more than 25,000 people. Violent explosions during the eruption produced intense heating, strong convection currents and tall cumulus clouds; strong rain turned ash into streams of mud.

IMPACT ON CLIMATE

Explosive eruptions release enormous amounts of energy, and this is shown in the heights to which rock and ash are thrown. Fine ash from the 1883 explosion of Krakatau, for example, was carried by the uprush of gas and vapours to a height of 27 km. It was blown around the world and caused exotic sunsets and other climatic effects (Figure 6.24).

The vapour and dust clouds produced in volcanic eruptions can cause long-lasting atmospheric and climatic effects. For example, scientists think that the 1982–83 El Niño disturbance (see p. 482) might be at least partly related to an explosion in 1982 of the relatively small volcano El Chichón in Mexico, which threw dust clouds high into the atmosphere and dispersed them over a wide area. The 1991 eruption of Mount Pinatubo in the Philippines (Figure 6.25) has been blamed for widespread cooling during 1992 (Box 6.28).

Much more localised problems are sometimes posed by the release of toxic gases in volcanic eruptions. As noted earlier, magma often contains a large amount of gas as well as the molten rock. Many volcanic eruptions release carbon dioxide into the atmosphere. A curious example of hazards associated with carbon dioxide emission from volcanoes is the 1986 Lake Nyos gas explosion (Box 6.29).

VOLCANO PREVENTION AND PREDICTION

COPING WITH VOLCANOES

As yet it seems to be impossible to prevent or stop volcanic eruptions, so instead they have to be coped with. Some places – such as California, which has more than 500 volcanic vents, at least seventy-six of which have erupted (some repeatedly) during the last 10,000 years – have developed quite effective ways of coping with volcanoes.

A number of strategies can be adopted to limit damage from volcanoes. Hazard zonation maps, generally reflecting past experience, are useful for showing areas relatively likely to be affected by future eruptions. They can also be used to guide decisions regarding evacuation and other disaster response activities. Land-use planning is designed to avoid residential, commercial and industrial development in high-risk areas. Monitoring of active volcanoes is essential to provide early warning of likely eruptions. Other measures (including preparation of contingency plans) can

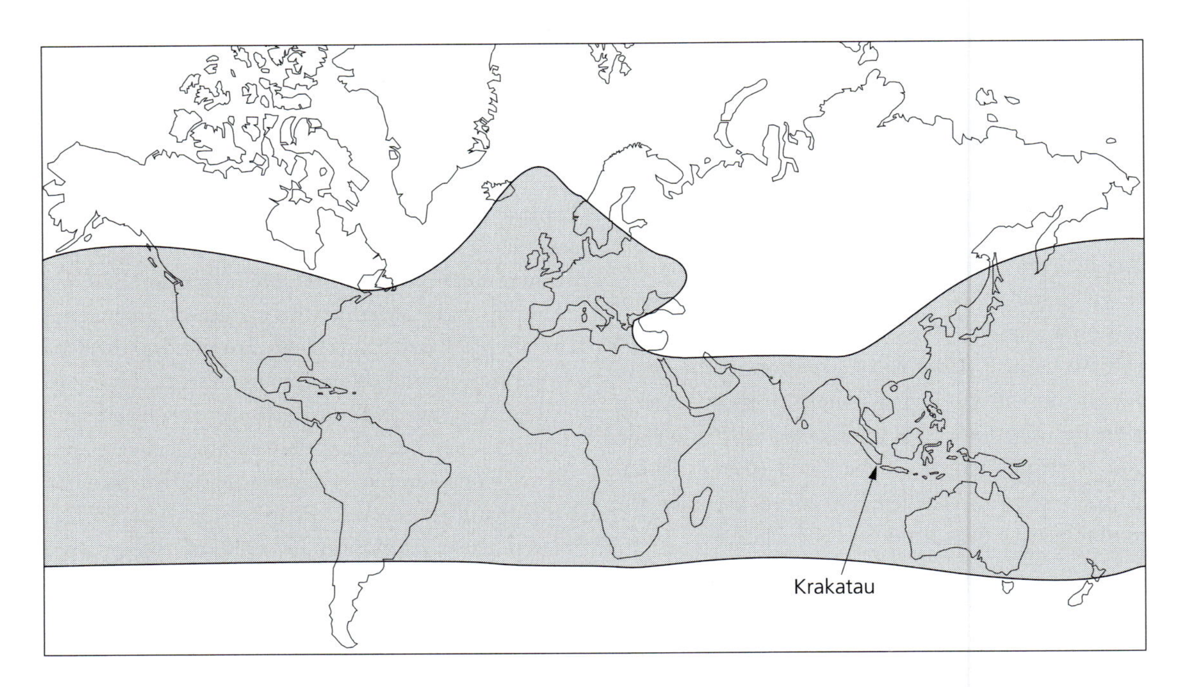

Figure 6.24 *Spread of volcanic ash from Krakatau around the world. The violent explosive eruption of Krakatau on 26 August 1883 sent a black cloud of ash high into the atmosphere, which was carried and dispersed by upper-atmosphere winds. Within two weeks, the dust had been blown right around the world. The dust scattered light, and this caused exotic sunsets in many places.*

Source: A map produced by the Royal Society in 1883 and summarised in the Guardian newspaper, 24 August 1990.

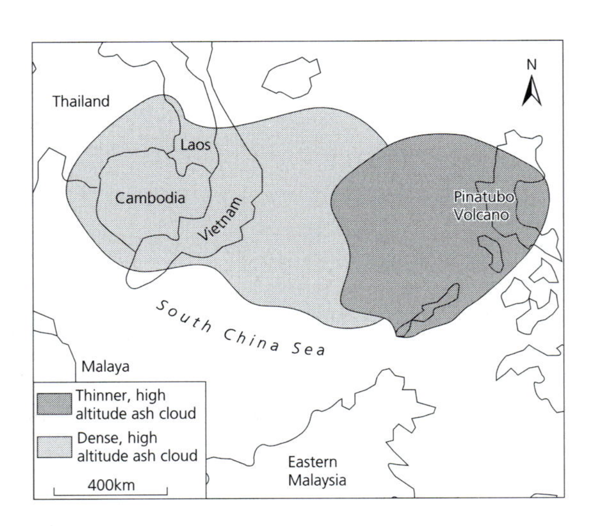

Figure 6.25 *Spread of the ash cloud from Mount Pinatubo, 1991. The ash thrown out by the June 1991 eruption of Mount Pinatubo in the Philippines was quickly spread over a wide area by upper-air winds.*

Source: Unknown.

be used to reduce the effects when vulnerable areas cannot be avoided.

Recent years have seen significant improvements in hazard assessment, volcano monitoring and eruption forecasting, particularly for less explosive eruptions, and lessons from disasters like Mount St Helens (1980) are helping to improve volcanic hazard management and reduce risk.

MONITORING AND PREDICTION

Techniques now exist that allow geologists to offer a fairly reliable prediction of an impending volcanic eruption, as the experience of Mount St Helens (see Box 6.26) shows. The most reliable forecasts depend on detailed monitoring of micro-earthquake activity in the vicinity of the volcanic cone, which indicates that magma is working its way upwards, filling the chamber(s) beneath the volcano. The side and summit of the volcano swell and are tilted by the emplacement of the magma, and the rocks within them are strained and may rupture, causing a cluster of small earthquakes. Seismographs on Hawaii sometimes register as many as 1,000 small quakes a day before a major eruption occurs.

One novel technique for predicting when an explosion might occur is to monitor the changing underground acoustics near a volcanic vent, rather like using a stethoscope to listen for irregularities in the heartbeat of a

BOX 6.28 THE 1991 ERUPTION OF MOUNT PINATUBO, PHILIPPINES

The 1991 eruption of Mount Pinatubo in the Philippines was the world's largest known volcanic eruption in more than a century. A series of small phreatic (groundwater) explosions on 2 April 1991 brought to a close more than four centuries of quiescence of Mount Pinatubo, and the major eruption occurred on 15 June. It released about ten times as much magma as the 1980 eruptions of Mount St Helens; some estimates suggest about 5 km^3 of dense magma.

Some 320 people died in the eruption, mostly due to the collapse of ash-covered roofs. Many more lives were saved because early warnings were issued, and at least 58,000 people were evacuated from the high-risk areas.

Management of the 1991 eruption seems to have been well co-ordinated and effective, and this probably saved many lives:

- State-of-the-art volcano monitoring techniques and instruments were applied.
- The eruption was accurately predicted.
- Hazard zonation maps were prepared and circulated a month before the violent explosions.
- An alert and warning system was designed and implemented.
- The disaster response machinery was mobilised on time.

The eruption threw up a vast cloud of volcanic dust, ash and gas, which was observed hourly by satellite. At 2.40 pm local time, the disc-shaped cloud covered 60,000 km^2 and within an hour it had spread to 120,000 km^2. Eyewitness accounts tell of a heavy ash fall just after 2.00 pm, intermittent falls of lapilli (lava bombs) starting at 2.20 pm, and continuous heavy falls of lapilli from about 3.00 pm onwards. Satellites were also used to track the movement of the ash clouds away from Mount Pinatubo over a period of months. Meteorologists were eager to provide warnings and advice, particularly to airline operators throughout Asia.

The Mount Pinatubo eruption gave scientists an opportunity to monitor the so-called volcanic radiative forcing. This is the effect of volcanic aerosols on radiation transfers within the atmosphere, which have a direct effect on temperatures via changes in albedo (reflectivity). NASA data show that short-wave forcing was stronger than long-wave forcing, the aerosols caused a cooling effect immediately, and the amount of cooling increased through to September. Cooling appears to have continued through much of 1992, and below-normal temperatures in much of the northern hemisphere between June and August 1992 have been attributed to the June 1991 Mount Pinatubo eruption.

BOX 6.29 THE 1986 LAKE NYOS VOLCANIC GAS EXPLOSION

On the night of 21 August 1986 a gas release occurred over a four-hour period at the lake in the volcanic crater Iwi at Nyos in Cameroon. More than 1,700 people and all animal life within 14 km of the lake were killed, but plant life was mostly unaffected. There were reports of a smell of bad eggs or gunpowder associated with the cloud of toxic gas, and survivors of the gas release suffered symptoms compatible with an asphyxiant gas like CO_2. Although the gas involved has never been definitely identified, it is believed to have been a toxic cloud of water and carbon dioxide.

The sudden gas release shook the lake so much that it triggered a small flood wave of water, which swept along valleys to the south. Some of the water thrown up in the gas release was turned into a fine mist, and this dense aerosol of water and carbon dioxide seeped rapidly along the valley to the north and through the three closest villages (Subum, Cha and Fang), leaving a trail of death and destruction in its wake.

Quite why this massive release of CO_2 occurred remains a mystery. There are two competing theories. One associates the gas release with a phreatic eruption of groundwater that was exceptionally rich in CO_2. The other explains it as a result of the overturn of water in the 220 m deep lake. Whichever explanation is true, the ultimate source of the CO_2 gas was earlier volcanic activity in the Nyos crater. An international inquiry was set up by British, French, Italian, Japanese and US scientists to determine what had caused the disaster. It concluded that natural gas releases like this are extremely rare and require such unusual circumstances that repeat events are highly unlikely.

person. Tests have shown that it is possible to detect subtle variations in sounds beneath the volcano, which show, for example, whether a river of fresh magma charged with explosive gases is surging up from the depths below.

Seismic monitoring is extremely valuable in volcano prediction, but it is virtually impossible to ground monitor all active volcanoes. Satellites offer the prospect of global coverage from space, and techniques are now being developed for remote sensing of the world's volcanically active regions. Weather satellites can be used to track the dispersion of eruption plumes around the world, infrared imaging can be used to monitor lava flows and volcanic craters, radar interferometry can be used to measure volumes of volcanic flows and cones, and ultraviolet monitoring can provide estimates of carbon dioxide releases.

The development of better volcanic surveillance programmes is one objective of the United Nations in the International Decade for Natural Disaster Reduction (IDNDR) (see p. 23).

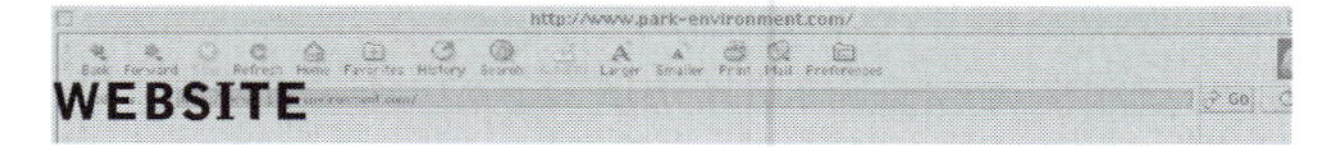

Links to relevant websites, a comprehensive bibliography, tools for teaching and learning, and downloadable images relevant to this chapter can be found at the website specially designed to accompany this book at http://www.park-environment.com

SUMMARY

This chapter focuses on how the dynamic crust of the Earth shapes surface features and creates a range of geological hazards. Our understanding of 'the big picture' relies heavily on plate tectonics, and we have explored how ideas in this field have changed through time and how new evidence – such as that relating to sea-floor spreading – has caused important changes in interpretation. There are still many unknowns (such as what drives the plate movements). We also examined some of the features created by crustal movements, including warping, folding and faulting. These large-scale geological features dominate the landscape in many places and often give rise to important geological hazards. One of the most dramatic of these is the earthquake, associated with fault movements and often causing widespread damage and destruction. Having looked at what causes earthquakes and how they are measured and described, we looked at a number of case studies that illustrate how earthquakes are coped with, what problems they pose and what prospects there are for minimising risk to people and property. We also considered ways in which human activities can trigger earthquakes or alter their magnitude and frequency. Volcanoes can be equally devastating, and they too are associated with sub-surface processes. We examined how and where volcanic activity occurs, the different types of volcanic materials and the differences between intrusive and extrusive vulcanism. Case studies allowed us to see how people cope with volcano hazards in different settings. The chapter ends with some reflections on the difficulties of preventing and predicting volcanic events.

FURTHER READING

Bennett, M.R. and P. Doyle (1997) *Environmental geology*. John Wiley & Sons, London. Examines the interplay between people and geological processes.

Bennison, G.M. and K.A. Moseley (1997) *Introduction to geological structures and maps*. Edward Arnold, London. Useful introduction to the principles and practice of interpreting geological information.

Bolin, R. and L. Stanford (1998) *The Northridge earthquake: vulnerability and disaster*. Routledge, London. Detailed case study of the 1994 Northridge earthquake in the USA, with a focus on vulnerability and post-disaster recovery strategies.

Bolt, B.A. (1993) *Earthquakes: a primer*. W.H. Freeman, Oxford. An authoritative review of the causes, occurrence and physical properties of seismic activity.

Bridges, E.M. (1990) *World geomorphology*. Cambridge University Press, Cambridge. A clear introduction to landform development and the role of plate tectonics in shaping the face of the Earth.

Brumbaugh, D.S. (1998) *Earthquakes: science and society*. Prentice Hall, Hemel Hempstead. Non-technical overview of the key processes and impacts of earthquakes, which outlines how the study of earthquakes has evolved over time.

Chester, D.K. (1993) *Volcanoes and society*. Edward Arnold, London. A comprehensive overview of how volcanoes work and how society responds to and makes use of areas with significant volcanic risk.

Decker, R. and B. Decker (1995) *Volcanoes*. W.H. Freeman, Oxford. A clear, non-technical introduction to volcanoes, how they work and what they do.

Donovan, S.K. and C.R.C. Paul (eds) (1998) *The adequacy of the fossil record*. John Wiley & Sons, London. Series of essays that explore the validity and completeness of the fossil record, and the implications of this for understanding how and why species have evolved over geological timescales.

Doyle, H. (1995) *Seismology*. John Wiley & Sons, London. An introduction to seismic waves, earthquakes and their effects.

Doyle, P. and M. Bennett (eds) (1998) *Unlocking the stratigraphic record: advances in modern stratigraphy*. John Wiley & Sons, London. Series of essays on the principles and practice of analysing and interpreting the stratigraphic record.

Hatcher, R.D. (1995) *Structural geology: principles, concepts and problems*. Prentice Hall, London. Useful overview of the basic principles and products of structural geology and geophysics, including stress and strain, deformation, seismic activity, Earth magnetism and gravity.

Keller, E.A. and N. Pinter (1996) *Active tectonics: earthquakes, uplift and landscape*. Prentice Hall, Hemel Hempstead. Exploration of the effects of earthquakes and active tectonic systems on people, landforms and landscapes.

Lomnitz, C. (1994) *Fundamentals of earthquake prediction*. John Wiley & Sons, London. A review of statistical and geological approaches to the important question of the reliable, practical prediction of earthquakes.

McClay, K. (1991) *The mapping of geological structures*. John Wiley & Sons, London. A useful guide to describing, measuring and recording geological structures in the field.

Moores, E.M. (ed.) (1990) *Shaping the Earth: tectonics of continents and oceans*. W.H. Freeman, Oxford. A collection of essays from *Scientific American* that review current understanding and evidence of plate tectonics.

Ollier, C. and C. Pain (2000) *The origin of mountains*. Routledge, London. Rejects the theory that most mountains were formed as a result of plate tectonics and concentrates on the evidence about the age of mountains, about landscape history, and about contemporary surface and subsurface processes.

Perkins, D. (1998) *Mineralogy*. Prentice Hall, Hemel Hempstead. Clear introduction to minerals, their properties and descriptions.

Skinner, B.J. and S.C. Porter (1995) *Dynamic Earth*. John Wiley & Sons, London. A clear introduction to physical geology, with an emphasis on environmental impacts of human activities.

Slaymaker, O. (ed.) (1996) *Geomorphic hazards*. John Wiley & Sons, London. A collection of essays that explore the assessment, perception, communication and management of landform hazards such as soil erosion, desertification, floods, seismicity and volcanic eruptions.

Smith, K. (1996) *Environmental hazards: assessing risk and reducing disaster*. Routledge, London. Clearly written, comprehensive assessment of environmental risk and the policy responses required to achieve a safer world.

Summerfield, M.A. (1991) *Global geomorphology: an introduction to the study of landforms*. Longman, Harlow. Detailed coverage of plate tectonics, the evolution of continents and large-scale landform development.

Summerfield, M.A. (ed.) (1999) *Geomorphology and global tectonics*. John Wiley & Sons, London. Detailed review of recent advances in understanding and studying the Earth's topography.

Windley, B.F. (1995) *The evolving continents*. John Wiley & Sons, London. Outlines a history of the Earth, tracing the causes and consequences of changes through time.

Chapter 7

Earth Materials

LEARNING OBJECTIVES

When you have finished studying this chapter, you should be able to

- *Appreciate the importance and describe the properties of minerals*
- *Distinguish between the three main types of rocks (igneous, sedimentary and metamorphic) and give examples of each type*
- *Outline the range of processes that give rise to sedimentary rocks*
- *Describe and explain the rock cycle*
- *Define the term 'weathering' and summarise the main processes and products of physical disintegration and chemical decomposition of rocks*
- *Define the term 'erosion' and outline the main processes of weathering*
- *Explain the main processes of mass movement, and show how they create landforms and affect landscapes*

As we saw in Chapter 5, the crust – that part of the solid Earth with which we are most familiar – constitutes less than 1 per cent of the planet's total mass, and it is made up of material that is much less dense than the material in the mantle and core (see Table 5.4). The continental and oceanic crust 'floats' on the mantle rocks below, and plate tectonic mechanisms make the continents drift and create mountains, folds, faults, earthquakes and volcanoes (Chapter 6).

The materials in the Earth's crust are important resources in their own right, and they are central ingredients in the global biogeochemical cycles (see p. 86). They form the basis of all soils (Chapter 19) and play vital roles in defining the structure and dynamics of ecosystems and biomes (Chapter 18).

In this chapter, we examine the rocks and minerals that make up this crustal material, focus on the main forms and products of weathering that alters the material, and explore the erosional processes that redistribute it and help to create and shape landforms.

ROCKS AND MINERALS

The Earth's crust is composed of different kinds of rocks, which are composed of one or more minerals in varying proportions. Rocks are divided into three major groups, depending on their origin – igneous, sedimentary and metamorphic. In this section, we look at what rocks are, how they are made and how they differ one from another.

MINERALS

Minerals are the basic building blocks from which the Earth's crust is made and the main ingredients of all rocks, deposits and soils (Box 7.1). Some of the common rock-forming minerals and their chemical composition are listed in Table 7.1.

An ore is any naturally occurring mineral or group of minerals from which economically important constituents (such as metals) can be extracted. Pyrite, for example, is a hard brittle yellow mineral consisting of iron sulphide (FeS_2), which occurs in all types of rock and is used in the manufacture of sulphuric acid and paper. The largest deposits of pyrite and associated minerals in Europe are at Rio Tinto in Spain, and these have been exploited since pre-Roman times for silver and copper.

Mining and processing of minerals is fairly widespread around the world, although the task of producing refined metals from raw mineral ores by smelting is often complex

BOX 7.1 MINERALS

A mineral is a naturally occurring, inorganic, homogeneous substance. There are many different rock-forming minerals, and each has:

- a definite chemical composition and crystal structure; and
- characteristic physical properties (such as colour, lustre, hardness and density).

and costly and generally creates pollution. New mineral-processing technologies are being developed that are cheaper, use less energy, produce less waste and release less greenhouse gases.

ELEMENTS

Minerals are made up of elements, which vary in number from one mineral to another. Some minerals are made from a single chemical element, but most are more complex compounds. Each mineral is always made up of the same or similar chemicals in relatively fixed proportions, and this is why it has fairly definite and stable properties. Thus quartz, for example, which makes up most sands (including sandstone rock) and gravels, is built from silicon (Si) and oxygen (O). Galena comprises lead (Pb) and sulphur (S) (Table 7.1).

Many minerals contain oxygen (O), which makes up about 47 per cent of exposed rocks, and/or silicon (Si), which makes up about 28 per cent (see Table 5.5). A further 24 per cent is made of aluminium, iron, potassium, calcium, sodium and magnesium (together). The most common minerals include the silicates, oxides and carbonates derived from combinations of silicon, oxygen and carbon with other elements.

Some elements, called native elements, occur by themselves in rocks as minerals. Gold (Au), the bright yellow precious metal, is a good example. It occurs naturally and is widely mined and has many uses, including currency, jewellery, printed circuits and semiconductors.

CRYSTALS

Most minerals have a crystalline structure. The size of the crystals varies a great deal from one mineral to another. In igneous rocks, the hard quartz crystals often look like bits of glass, and rock crystal (the purest form of quartz) is sometimes used in jewellery.

The individual crystals are joined together by physical and/or chemical bonds to make the minerals. Bonding can

Table 7.1 Some common rock-forming minerals

Chemical group	*Mineral*	*Chemical composition*
Carbonates	Calcite	$CaCO_3$
	Dolomite	$CaMg(CO_3)_2$
Halides	Fluorite	CaF_2
	Halite	$NaCl$
Oxides	Haematite	Fe_2O_3
	Magnetite	Fe_3O_4
	Quartz	SiO_2
Sulphates	Gypsum	$Ca(SO_4).2H_2O$
Sulphides	Galena	PbS
	Pyrite	FeS_2
Silicates	Olivine	$(Mg,Fe)_2SiO_4$
Amphiboles	Hornblende	$Ca_4Na_2(MgFe)_8(AlFe)_2(Al_4Si_{12}O_{44})(OHF)_4$
Feldspars	Orthoclase	$K(AlSi_3O_8)$
Micas	Biotite	$K(Mg,Fe)_3(AlSi_3O_{10})(OH)_2$
	Muscovite	$K_2Al_4(Si_6Al_2O_{20})(OH,F)_4$
Pyroxenes	Augite	$Ca(MgFeAl)(AlSi)_2O_6$

Table 7.2 Diagnostic properties of minerals

Property	*Comment*
Lustre	This refers to the way that light is reflected from the mineral surface. Some minerals have a glassy (vitreous) lustre, some have a metallic lustre (like polished metal), and some have a dull (earthy) lustre.
Streak	This is the colour of the powder produced by rubbing a mineral on an unglazed porcelain plate (a streak plate). The colour of streak for some minerals is constant and distinctive. Many minerals produce a white streak.
Specific gravity	This is the relative weight of a mineral compared with the weight of an equal volume of pure water (which has a specific gravity of 1). A mineral with a low density also has a low specific gravity.
Hardness	This is the degree of resistance to scratching and is usually measured on the Moh hardness scale (ranging from 1 to 10). Soft minerals cannot scratch a copper coin; medium-hardness minerals are harder than that but will not scratch ordinary glass; a hard mineral scratches glass.
Cleavage	Many minerals have a tendency to break (cleave) along smooth, flat planes. Different minerals have a different number and direction of cleavage planes. Mica, for example, has cleavage in only one direction, so it can be split into thin parallel sheets. Other minerals break along two or three cleavage planes.
Fracture	Most minerals do not break along smooth planar surfaces but have irregular or uneven fractures.

be simple or complex, weak or strong. The strength of the bonding determines rock hardness and thus its resistance to weathering and erosion. The type of bonding (physical or chemical) strongly influences rock resistance to different types of weathering (physical or chemical).

PROPERTIES

Each mineral has characteristic physical properties, and these help geologists to identify rock types. The most useful properties, certainly for identification purposes, are lustre, streak, specific gravity, hardness, cleavage and fracture (Table 7.2). Colour is often useful too, but it can vary a great deal and so is not very diagnostic. Some minerals always have the same colour, but others vary because their chemical composition varies or they contain impurities. Exposure to air or water can also alter colour.

Mineral hardness is usually measured on the Moh hardness scale (Table 7.3). This is based on analysis of what scratches a given mineral and compares it with ten standard minerals.

IGNEOUS ROCKS

Igneous rocks are formed when molten magma cools and solidifies, either underground or on the surface. They are often referred to as primary rocks because all rock is either igneous or ultimately derived from igneous rocks (Box 7.2). Igneous rocks make up the largest percentage of the Earth's crust, and they are often covered by sedimentary or metamorphic rocks.

Igneous rocks are constantly being destroyed and created. Once they are exposed at the surface (when overlying material is removed) they are weathered and break down to provide materials for secondary rocks, particularly sedimentary rocks. All types of crustal rock are carried downwards at subduction zones at the margins of some crustal plates (see p. 153), where they are remelted and might in the course of time be released again as new igneous rock.

INTRUSIVE AND EXTRUSIVE ROCKS

There are two main types of igneous rock, reflecting different modes of formation:

Table 7.3 The Moh hardness scale for minerals

Moh hardness	*Simple hardness test*	*Mineral*
1.0	Crushed by fingernail	Talc
2.0	Scratched by fingernail	Gypsum
3.0	Scratched by copper coin	Calcite
4.0	Scratched by glass	Fluorspar
5.0	Scratched by a penknife	Apatite
6.0	Scratched by quartz	Feldspar
7.0	Scratched by a steel file	Quartz
8.0	Scratched by corundum	Topaz
9.0	Scratched by diamond	Corundum
10.0		Diamond

BOX 7.2 PROPERTIES OF IGNEOUS ROCKS

Igneous rocks share the following properties:

- They contain no fossils.
- They usually look quite uniform in appearance.
- They are rarely layered (stratified).
- They have a crystalline structure, although the crystals in some igneous rocks may be microscopic in size.
- They have interlocking mineral structures (the individual minerals are tightly fitted together and fill all of the available space).

Intrusive igneous rocks: these rocks are emplaced within other rocks and solidify beneath the Earth's surface. Common intrusive rocks include granite, diorite, gabbro and diabase.

Extrusive igneous rocks: these rocks are pushed out and solidify on the Earth's surface. Common extrusive rocks include obsidian (volcanic glass), rhyolite (the extrusive equivalent of granite), andesite (the extrusive equivalent of diorite) and basalt (the extrusive equivalent of gabbro). Rocks thrown out in violent volcanic explosions, which generally involve the release of large amounts of gas as well as magma, are often porous and contain many cavities (scoria) when they cool and solidify. Volcanic ash often solidifies into light, porous, fine-layered pumice, which can develop as layers to form tuff when hardened.

The mode of production strongly influences the rate of cooling, and this strongly influences the rock's crystalline structure and hence resistance to erosion. Intrusive rocks cool more slowly than extrusive ones, so they have larger and better-developed crystal structures and are more resistant to erosion. Minerals in molten magma crystallise as they cool, and the crystal size largely determines the grain structure and overall character of the rock.

Plate 7.1 *The Giant's Causeway, Northern Ireland. The distinctive features of this coastline are the result of the cooling of basalt, an extrusive igneous rock, into an intricately interlocking series of hexagonal columns.*
Photo: Chris Park.

ACID AND BASIC ROCKS

As well as dividing igneous rocks into intrusive and extrusive types, we can also classify them according to their chemical composition. Acidic igneous rocks are rich in lightweight minerals (especially silica) and tend to be light-coloured, whereas basic igneous rocks are rich in the heavier ferromagnesian (iron and magnesium) compounds and tend to be darker. These are opposite ends of a spectrum, and some igneous rocks lie more towards the middle of the range.

Acidic rocks tend to disintegrate around the mineral crystals, and they often weather into coarse-textured and rather infertile soils. Many basic rocks decompose, and the crystalline structure breaks down into the main mineral constituents. Basic rocks weather down into more fertile and productive soils in which mineral elements (nutrients) are more freely available for plants.

There are many types of igneous rock, with different properties. Some of the most common types are mentioned in Table 7.4. Granite is a common, hard, coarse-grained intrusive igneous rock that varies in colour from pink to dark grey. It is usually formed in large underground masses and is exposed on the surface when overlying rocks are worn away. Nearly 15 per cent of the rocks exposed at the Earth's surface are granites. The most common extrusive igneous rock is basalt – a fine-grained, dark-coloured, basic rock.

Igneous rocks are used in a variety of ways. Resistant rocks, like granite, which is very hard, are widely quarried and used as building materials. Some exotic igneous rocks are mined for the chemical substances they contain, because they are valuable sources of ores (particularly metals) or precious gemstones.

SEDIMENTARY ROCKS

Sedimentary rocks are formed in very different ways to igneous rocks, and they are the product of surface processes rather than sub-surface (crustal) processes. Many are formed by the accumulation (build-up) and consolidation (compaction) of mineral and organic fragments that have been deposited by water, ice or wind. Others are formed from organic material, or from chemicals deposited from water (Box 7.3). Sedimentary rocks are sometimes referred to as secondary rocks, because they are derived from the weathering and deposition of pre-existing rocks.

Table 7.4 Some common igneous rocks

Grain size	*Light-coloured minerals dominant*	*Dark-coloured minerals dominant*
Coarse-grained	Granite	Gabbro
Fine-grained	Rhyolite	Basalt

BOX 7.3 FORMATION OF SEDIMENTARY ROCKS

Three types of material accumulate and consolidate to form sedimentary rocks:

1. cemented fragments of rocks worn from the land (clastic sedimentary rocks);
2. deposits of organic materials and remains of once-living things (organic sedimentary rocks);
3. chemicals deposited from water (chemical precipitates).

CLASTIC SEDIMENTARY ROCKS

Clastic means 'composed of fragments of pre-existing rocks', and clastic sedimentary rocks are formed by the deposition (laying down) and cementing together of weathered products of existing rocks to form new rocks.

Clastic sediments are composed of layers of sediment such as clay, sand, gravel or cobbles that are deposited by water, wind or ice. The size of the particles depends on:

- the source materials (fine-grained rocks produce smaller particles);
- the nature and power of the erosion process (glacier ice grinds rock into small particles, for example);
- the distance of the deposit from the source area (the largest particles tend to be deposited closest to the source).

Geologists usually express the size of individual particles in phi (ϕ) units (Figure 7.1). The phi scale is logarithmic, and it is used because it covers the whole range of particle sizes from the microscopic (clay-sized) to extremely large boulders.

Most clastic sedimentary rocks accumulate, layer by layer, in water. Deposition usually takes place in a depression or basin (such as the sea bed or base of a geosyncline). The sediments are initially loose and uncompacted, but over time they are gradually compressed and cemented together (Box 7.4). This process is called *lithification*.

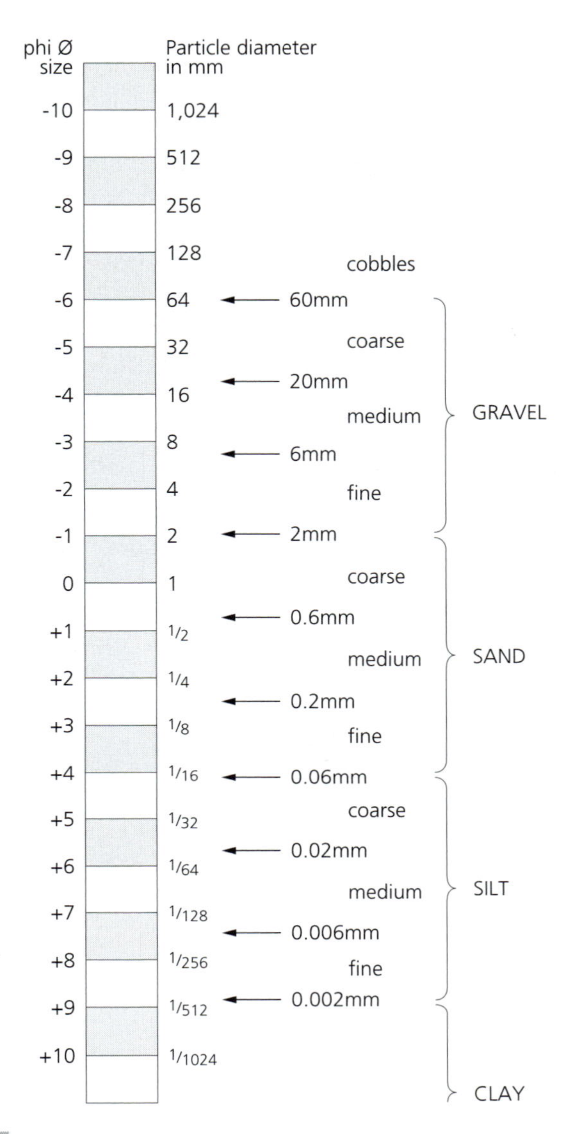

Figure 7.1 *The phi scale used to measure particles of sediment. The phi scale works by a factor of two, so each unit is half or double the size of the adjacent unit. Twenty phi units cover the range of sediment sizes from clay to large cobbles.*

The individual grains are not lost in lithification, and they rarely interlock completely to fill all available voids (spaces). Properties of the individual particles (such as size, shape and roundness) are preserved in the rock, and detailed analysis of the sedimentary particles can yield important clues about how the rock was formed and what the depositional environment was like. The character of a sedimentary rock is strongly influenced by the size of its particles (Table 7.5).

BOX 7.4 LITHIFICATION

The process by which sediment becomes a sedimentary rock is called *diagenesis* or *lithification*. It mainly involves:

- *Compaction*: of the individual grains, under the pressure of later deposits laid on top.
- *Cementing*: of the grains together, often by the growth of new minerals deposited by percolating groundwater, such as silica or lime.

Table 7.5 Common types of clastic sedimentary rock

Name	*Character*
Conglomerate	Large rounded particles (gravels, cobbles, boulders) cemented together in a matrix of fine materials
Breccia	Angular pieces of broken rock cemented together by a fine matrix
Sandstone	Sand-sized particles cemented together
Shale	Compacted and hardened silt and clay, with clear laminations (layers)
Siltstone	Compacted and hardened silt
Mudstone	Compacted and hardened clay, similar to shale but with less developed lamination

Shale is the most common sedimentary rock, accounting for about half (52 per cent) of the rocks exposed on the Earth's surface. Shales are grey or black in colour, are arranged in distinct layers (laminated) and are formed from layers of fine silt and clay. Sandstone, the most familiar and second most common sedimentary rock (15 per cent of the rocks exposed at the surface), forms about a third of the exposed sedimentary rocks. It varies in colour, depending on what mineral it contains and what material cements it together, and it consists of compacted and cemented grains of sand (mostly quartz – silica, SiO_2 – derived ultimately from the crystallisation of igneous rocks).

Geologists divide the clastic sedimentary rocks into three groups, depending on their particle sizes:

1. *argillaceous rocks*, such as siltstones and marls, which contain particles between 0.004 and 0.06 mm in size (although clays are even finer);
2. *arenaceous rocks*, such as sandstone, with particles ranging in size from 0.06 to 2 mm;

Plate 7.2 *Limestone scenery near Kuala Lumpur, Malaysia. This area is dominated by towering limestone cliffs, which are dissected by a series of caves and passageways.*
Photo: Chris Park.

BOX 7.5 STRATIFICATION

Sedimentary rocks are often layered or stratified, and individual beds or layers can be distinguished that have different properties (sometimes including fossil content) from the beds above and below. *Stratification* (the layering of sedimentary rocks, also called *lamination*) often reflects the seasonal inflow of sediment into lakes or the sea and variations in the erosional power of rivers or winds.

Well-defined bedding planes separate successive beds or strata. Stratigraphy is the branch of geology that deals with the study of the succession of strata in sedimentary rocks, which can reveal a great deal about how the sediments were formed and what conditions were like at the time. Geologists rely heavily on the principle of superposition of layers of sedimentary rocks, which states that older rocks are overlain by younger rocks (provided that the rock sequence has not been folded, faulted or otherwise disturbed). This helps to establish a relative chronology of deposition and to make geological associations from place to place.

Unconformities represent gaps in the sedimentary sequence caused by the erosion of existing strata before younger rocks are deposited on top. Gaps in the depositional record complicate the reconstruction of sedimentary histories.

Plate 7.3 *Sedimentary rocks exposed around the Grand Canyon escarpment, Grand Canyon, Arizona, USA. Differential resistance to erosion between strata within the sequence has produced the stepped valley side.*
Photo: Chris Park.

Plate 7.4 *Red Rocks amphitheatre, just outside Denver, Colorado, USA. The heavily eroded deep red sandstone outcrops that create this natural bowl-shaped area provide a dramatic backdrop for classical and rock concerts.*
Photo: Chris Park.

3. *rudaceous rocks*, such as breccias and conglomerates, which have particles larger than 2 mm across.

Some clastic sedimentary rocks are very hard, and many of them (particularly sandstones and limestones) are widely used as building materials.

ORGANIC SEDIMENTARY ROCKS

Some sedimentary rocks are composed of dead organic materials, the compacted and decayed remains of plants and animals. Limestone, derived from coral, is a good example, as is chalk (the purest form of limestone), which is composed of the remains of tiny sea creatures.

Peat, a soft fibrous material, is produced by the accumulation of organic remains in acidic waters. The acidic water slows down the rate of decay of organic material such as sphagnum moss, and this incomplete decomposition allows the organic material to accumulate to form peat bogs. Many countries, including Russia, Canada, Finland

and Ireland, have large peat deposits, which have long been dried and burned as a fuel for heating and cooking. Peat is also useful as a soil additive, increasing fertility by adding natural nutrients and improving soil structure.

Over time, peat can be compacted into *lignite* (a sedimentary rock with woody texture) as more partially decayed organic material accumulates on top. After further compaction and cementing (*lithification*) lignite is turned into *bituminous coal* – a soft, black coal that burns with a smoky yellow flame. *Anthracite* (a hard, black coal that burns slowly but gives off intense heat) is formed by the folding and hardening of sedimentary strata containing bituminous coal. As well as providing a valuable source of energy (Figure 7.2), coal preserves a record of environments in the geological past because it is the product of luxuriant vegetation growth in a hot, humid climate.

Oil shale is another organic sedimentary rock that can be used as an energy source. It is a fine-grained shale containing oil, which can be extracted by heating and crushing to yield petroleum.

BOX 7.7 COAL

There are much bigger reserves of coal and lignite (a low-grade coal with a relatively low heat value) in the world than there are of oil and gas. Roughly half (48 per cent) of the world's known coal reserves are in the former Soviet Union, Eastern Europe and China, and there are also sizeable reserves in Western Europe (9 per cent), Africa (6 per cent), North America (26 per cent) and Australia and Asia (9 per cent). Known reserves are thought to be large enough to meet projected world demand for up to the next 500 years, but this life expectancy will fall sharply if large amounts of coal are used to make synthetic liquid fuels as a substitute for oil. Coal mining has many environmental impacts, and large-scale opencast surface mining is often the cheapest option in the short term. Coal is the dirtiest of the fossil fuels when burned, and it contributes significant amounts of greenhouse gases and sulphur dioxide (an important ingredient in acid rain) to the atmosphere.

BOX 7.6 FOSSIL FUELS

Fossil fuels are probably the most valuable rock resource on the planet in terms of usefulness to people, because modern society relies so heavily on them as a major source of energy. They are non-renewable resources, because we use them much faster than they form naturally. They are among the world's most rapidly exploited non-renewable resources and will doubtless run out one day. Substitutes and alternative forms of energy are being actively sought and developed as a matter of urgency, before fossil fuel supplies run out completely or before they become so scarce that price rises make them non-viable as a major energy source.

The main fossil fuels are coal, oil and natural gas, which come from organic sedimentary rocks formed by the decomposition and compaction of the remains of plants that lived literally hundreds of millions of years ago. They are known as carbon or hydrocarbon fuels because they are based on organic compounds that contain the elements carbon and hydrogen. The exploitation and use of all types of fossil fuels create environmental impacts, some of them of major concern worldwide.

CHEMICAL PRECIPITATES

Some sedimentary rocks are formed by the precipitation or deposition of materials that were previously dissolved in water. Limestone and dolomite are good examples.

While some limestones are composed of organic material (such as the remains of dead corals), others are precipitated from seawater that contains calcium carbonate ($CaCO_3$). Oolitic limestone consists of small spherical grains of calcium carbonate that have formed around tiny nuclei. Limestones often contain nodules (small lumps) of quartz, called chert, deposited from solution. Flint is this type of deposit. Dolomite rock is a precipitate of the mineral dolomite, a calcium magnesium carbonate.

Evaporites are a distinctive type of precipitate, formed by the evaporation of former seas or salt-water lakes. These salt-rich rocks accumulate in enclosed water bodies that have a high dissolved mineral content and rapid evaporation. Rock salt (halite, an important source of table salt and other sodium compounds), gypsum (used in making plasters and cements) and anhydrite are typical examples. Evaporites form in arid and semi-arid environments, so their distribution is a useful indicator of climate change. In dry climates they are relatively resistant to erosion, but in humid conditions they break down rapidly by chemical weathering.

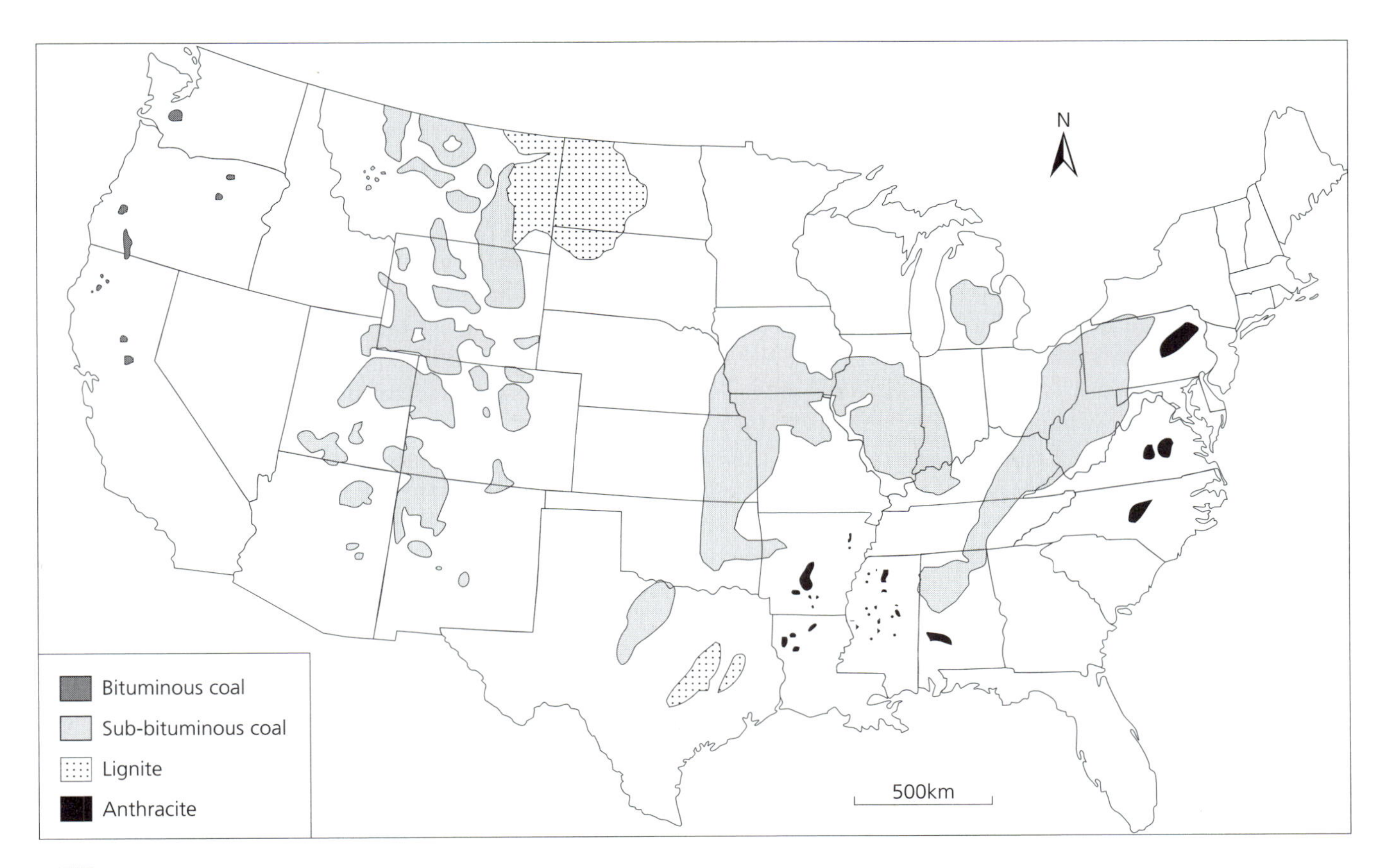

Figure 7.2 *Location of coal deposits in the USA. Bituminous coal and anthracite are found mainly in the east and midwest, and sub-bituminous coal and lignite are found mainly in the Rocky Mountain states and the High Plains.*

Source: Figure 18.10 in Cunningham, W.P. and B.W. Saigo (1992) Environmental science: a global concern. Wm. C. Brown Publishers, Dubuque, Iowa.

METAMORPHIC ROCKS

Many igneous and sedimentary rocks are altered by natural geological processes, which change their appearance, properties and character. The process of change is called *metamorphism* (Box 7.10), and the resulting rocks are called metamorphic rocks. Metamorphic rocks are described as tertiary rocks because they are derived from primary (igneous) and secondary (sedimentary) rock types. All metamorphic changes take place in solid rocks, because if the rock melts and then hardens again it becomes an igneous rock.

Metamorphic rocks often look like igneous rocks because their mineral structure also fills all available spaces, unlike sedimentary rocks, which contain voids (pore spaces). These metamorphic changes may cause the coarse minerals in a rock that contains more than one type of mineral to arrange themselves into wavy parallel bands (streaks) of alternating colour.

A good example of metamorphosis is the transformation of granite (an extrusive igneous rock) into gneiss by great heat and pressure, which radically alter its crystal structure, appearance and properties. Gneiss is banded (Table 7.6) and contains the same minerals as granite, but the minerals have been recrystallised by heat and pressure. If more heat and pressure is applied, banded gneiss can be further compacted to produce schist, in which the minerals are compressed into a plate-like arrangement. Many fine-grained sedimentary rocks metamorphose into schists.

Shale and mudstone (clastic sedimentary rocks) can be metamorphosed and recrystallised under great pressure into slate. Most slates are black or grey, but other colours occur too. Slate cleaves (splits) easily along distinct planes defined by the parallel alignment of the mineral grains into smooth, thin sheets. This property makes slate useful as a building and roofing material.

BOX 7.8 OIL

Oil is the principal world source of fossil energy resources at present, and known reserves are fast being depleted. Natural mineral oil, a thick greenish-brown flammable liquid, is found underground in permeable rocks. It is also called crude oil or petroleum, and it can be refined to produce a number of valuable products, including oil and petrol (gasoline). About two-thirds of the world's known recoverable reserves (some known reserves are too inaccessible to exploit profitably) are in the Middle East. Known world reserves of oil are expected to run out within 80 years using currently available technology, and the oil industry expects total world production to decline slowly after the year 2000. Since the 1970s, technology has allowed the pumping of oil from offshore fields (including the North Sea) and from the frozen north of Alaska.

The exploitation and use of oil resources has a number of environmental impacts. Burning of petroleum is a major cause of air pollution, and the transport of oil – particularly in large tankers at sea – has caused some major environmental disasters. Well-known examples include the 1967 oil spill from the *Torrey Canyon* off south-west England, the 1989 *Exxon Valdez* oil spill in Alaska and the 1992 *Braer* oil spill off northern Scotland. The largest onshore oil field in Western Europe was discovered under the Isle of Purbeck in Dorset, England, in 1973 and has since been exploited commercially. However, the area is also one of the most important nature conservation sites in the United Kingdom, and conservationists have argued that permission for drilling should never have been granted there.

BOX 7.9 NATURAL GAS

Natural gas is a mixture of flammable hydrocarbon gases (mainly methane) trapped beneath the ground, often found in association with oil reserves (Figure 7.3). It is now one of the world's three main fossil fuels (with coal and oil). Most is extracted from offshore wells (in the North Sea, for example) or from land-based wells in the USA and the Middle East. North America and the Middle East hold about 40 per cent of the world's known recoverable gas resources, and the former Soviet Union holds a similar amount. Natural gas is a very clean and convenient fuel, and it provides around a third of the energy used in the USA.

Rocks composed largely or entirely of one mineral are not banded by the intense heat and pressure of metamorphism. Limestone (a clastic or organic sedimentary rock) is metamorphosed into marble. Pure marble is white and much in demand by sculptors, but many marbles are discoloured by impurities. Sandstone (a clastic sedimentary rock) metamorphoses into quartzite, a resistant rock in which individual sand grains are cemented together by quartz.

THE ROCK CYCLE

Metamorphosis radically alters existing rocks, and it changes their character and properties. But it is rarely the

BOX 7.10 METAMORPHISM

Metamorphism involves three types of change, which might occur separately or together:

1. *Pressure*: rocks change when they are subjected to great pressure, as happens for example when tectonic plates are pushed together.
2. *Heat*: rocks are also changed when subjected to great heat, as happens along an active fault line when friction is generated by movement of the fault. Contact metamorphism is caused mainly by heat from pockets of magma, which bakes and hardens surrounding rocks.
3. *Chemical reactions*: rocks are also altered by chemical changes, as happens when gases and liquids are released into other rocks from pockets of magma.

Table 7.6 Common types of metamorphic rock

Grain size	*Other properties*	*Name*
Coarse-grained	Banded	Gneiss
Medium-grained	Uneven wavy surfaces	Schist
Fine-grained	Smooth flat surfaces	Slate

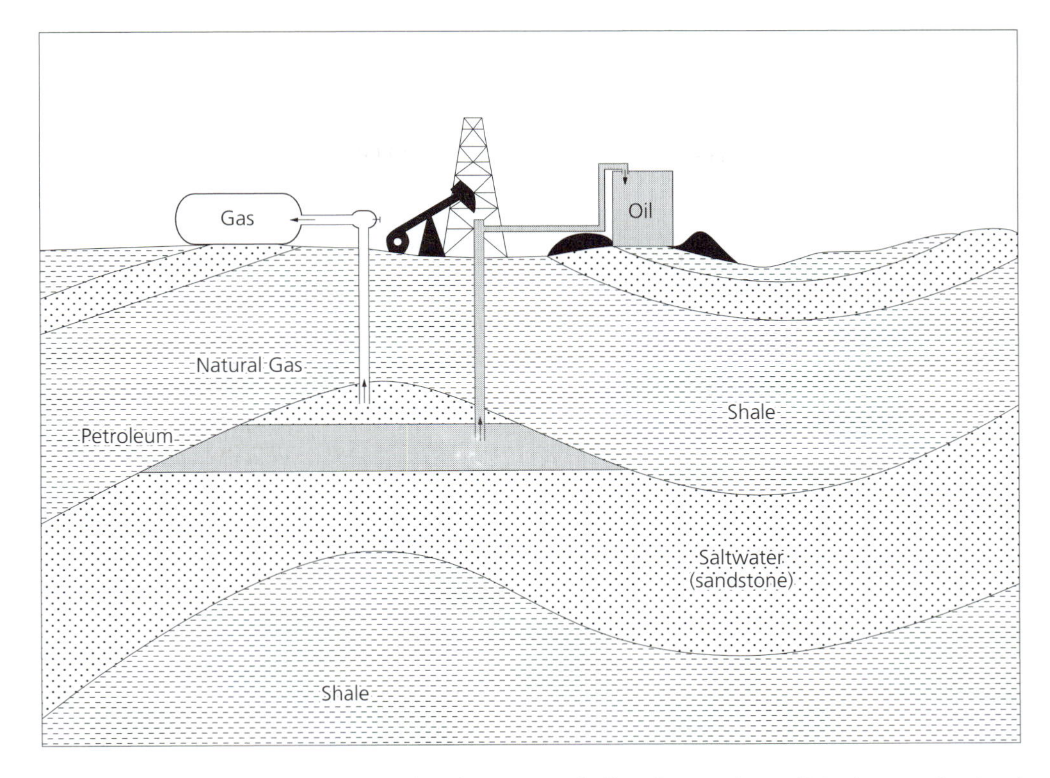

Figure 7.3 *Geological formations associated with reserves of oil and natural gas. Petroleum and natural gas are often found in sandstones.*

Source: Figure 7.7 in Marsh, W.M. and J.M. Grossa (1996) Environmental geography: science, land use and Earth systems. John Wiley & Sons, New York.

Plate 7.5 *China clay extraction at Lee Moor, on the south-western edge of Dartmoor, Devon, England. The china clay (kaolin) is the breakdown product of the metamorphic rock formed by the intense heating of existing rocks when the granite (an intrusive igneous rock) was forced into them as molten magma from below.*

Photo: Chris Park.

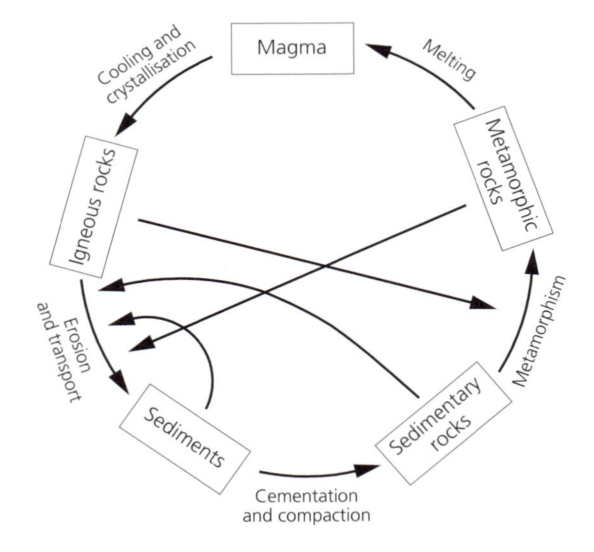

Figure 7.4 *The rock cycle. The rock cycle illustrates the interrelationships between the three main rock types (igneous, metamorphic and sedimentary), and the main processes that link them.*

Source: Figure 1.11 in Goudie, A. (1993) The nature of the environment. Blackwell, Oxford.

end of the story. Metamorphic rocks, like igneous and sedimentary rocks, can be weathered to produce the raw materials for new sedimentary rocks. In reality, all rocks are stages in a never-ending cycle (Figure 7.4), which continuously refashions, redistributes and recycles existing crustal material.

THE CYCLE

The ultimate source of all rocks on the Earth's surface is magma, hence the igneous rocks are described as primary rocks. Some igneous rocks are weathered and eventually buried and lithified into sedimentary (secondary) rocks. Other igneous rocks are transformed into metamorphic (tertiary) rocks by heat, pressure and chemical change. Sedimentary and metamorphic rocks at the edges of tectonic plates are subducted back into the mantle, where they are melted and can eventually be released as magma and form new igneous rocks. Sedimentary rocks can be transformed into metamorphic rocks by heat, pressure and chemical change. Metamorphic rocks can be weathered, buried and lithified into sedimentary rocks.

LINKS WITH OTHER ENVIRONMENTAL SYSTEMS

Surface processes (particularly weathering, erosion and deposition) play a vital role in creating sedimentary rocks. This links the rock cycle to the other major environmental systems, including:

- *The atmospheric system* (see p. 225): climate affects the processes, rates and patterns of weathering and erosion.
- *The hydrological system* (see p. 351): water is a major agent in rock weathering, erosion and deposition.
- *The biosphere* (see p. 519): weathering can be strongly influenced by plants and animals.

Sub-surface processes (particularly plate tectonics, folding, faulting and volcanic processes) play a vital role in creating igneous and metamorphic rocks. This links the rock cycle to the dynamics of the Earth's interior (see p. 128), which in turn are related to the evolution and cosmic setting of the planet (see p. 104).

The rock cycle operates in much the same way as the major biogeochemical cycles (see p. 86) and the global water cycle (see p. 352), but it operates much more slowly over the full span of geological time.

WEATHERING

DENUDATION

The Earth's surface is subject to constant change. Some of this is driven by sub-surface processes – particularly plate tectonics (see p. 145), folding (see p. 157), faulting (see p. 157) and volcanic activities (see p. 173) – which tend overall to raise the surface. Such processes, collectively, are referred to as *diastrophism* (building up). On the other hand, some of the change is driven by surface processes – particularly weathering and erosion – which tend overall to lower the surface. These processes, collectively, are referred to as *denudation* (wearing down). The rock cycle works non-stop building up and wearing down the Earth's surface materials.

BALANCE

Both sets of processes are at work everywhere, but not equally, so the balance between the two at any particular place determines whether it is being raised or lowered. The difference is not just important for topography and landscape – many of the major environmental systems can be significantly influenced by changes in relief. Changes in relief are associated with changes in weather and climate and changes in potential energy (thus sediment movement and landscape adjustment), and both of these are associated with changes in biogeochemical cycles, ecosystem dynamics and biological productivity (Chapter 18).

A feedback loop is built into these adjustments too, because relief influences the speed at which some important weathering and erosion processes operate. An increase in elevation (caused by uplift) increases gravitational forces and speeds up many erosion processes, and vice versa. Similarly, increased elevation often brings greater climatic variability and faster weathering of rocks, deposits and soils.

BOX 7.11 SURFACE LOWERING

Rates of surface lowering largely depend on:

- the nature of the materials being eroded;
- the amount of time they have been exposed on the Earth's surface;
- characteristics of the existing climate;
- the type of vegetation cover; and
- the average slope and height of existing surfaces.

PROCESSES

Denudation involves three main sets of processes to which all exposed rocks and landforms are subjected (Box 7.12). In essence, weathering prepares material for transport, and mass wasting and erosion move that material downhill towards some ultimate base level (normally the sea). Weathered rock products that are not removed by erosion become soils through time (see p. 600).

Weathering processes begin as soon as a new area of land is exposed at the Earth's surface. This occurs, for example, after a volcanic eruption deposits new lava or builds a new cone (see p. 179), after glacier ice has melted and retreated from an area (see p. 468), or after a fall in sea level exposes coastal rocks that were previously submerged (see p. 571). Weathering continues until the exposed surface has been worn down to base level.

Weathering and erosion take place in all environments all the time, but the processes involved and the speed at which they operate vary from place to place. They can also vary through time at a given place, driven particularly by environmental and climatic change (such as global warming; see p. 261).

Air pollution (see p. 241) can significantly increase rates of weathering of stone and other building materials. Statues and archaeological artefacts are disfigured and decay, building stone needs to be refaced or replaced, and in extreme situations air pollution control is required to prevent further damage. Tombstone weathering rates vary a great deal across the USA, being highest close to coal-fired power stations and lowest in dry areas, unpolluted wet areas and cold environments.

There are two main types of weathering – physical disintegration and chemical decomposition (Table 7.7). They usually occur together, but not necessarily at similar speeds. They produce different end-results and are controlled by different factors.

BOX 7.12 EARTH SURFACE PROCESSES

Weathering: weathering is the process by which exposed rocks are broken down *in situ* (in place). This is done mainly by the action of rain, frost and wind, but *biological processes are sometimes involved too.*

Mass wasting: mass wasting involves the downslope movement of rock materials, mainly by force of gravity but sometimes lubricated by water. It creates and shapes hill slopes and under extreme conditions can cause avalanches, landslides, earthflows and other hazardous processes.

Erosion: erosion is the movement of rock particles by active transporting agents such as running water, moving ice or blowing wind.

BOX 7.13 PRINCIPLES OF WEATHERING

There are a number of different weathering processes (Table 7.7), but some simple general principles apply to all rock weathering:

- All rocks are subjected to weathering.
- Some rocks weather faster than others (because of their mineral composition, bonding and particle cementing).
- The same rock type weathers differently and at different speeds in different climate regimes.
- Cracks and cavities in a rock enhance weathering processes by allowing them to occur deeper in the rock.

DISINTEGRATION

Disintegration, sometimes called physical weathering, involves the mechanical breakdown of pieces of rock into smaller pieces. It reduces the size of rocks and rock particles but does not affect their chemical composition. Neither does weathering itself move material to another place; this is done by erosion processes and gravity. Angular fragments of rock are often weathered into more rounded shapes.

A variety of processes, mainly physical ones, are involved in different climates. Physical disintegration rarely occurs on its own; it is usually accompanied by chemical decomposition. Four of the most common processes of disintegration are discussed below.

FROST WEATHERING

Frost weathering is a common and particularly effective form of physical weathering. Water seeps into cracks in the rock and freezes when the temperature drops. As it freezes the water expands, and expansion exerts pressure on the adjacent rock (Figure 7.5). The pressure is released when

Plate 7.6 *Limestone pavement near Ingleborough, Yorkshire Dales, England. The feature, typical of many karst landscapes, was formed by chemical weathering of the Carboniferous limestone rocks, probably under a soil cover that has since been eroded away.*
Photo: Chris Park.

Plate 7.7 *Exfoliation, a form of physical weathering, of basalt. The photograph was taken at the Giant's Causeway in Northern Ireland, and the camera case shown for scale is about 12 cm across.*
Photo: Chris Park.

BOX 7.14 CHEMICAL WEATHERING

Chemical weathering usually involves water in some form, and most chemical reactions are faster when heat is involved. As a result, decomposition is usually more rapid in hot, humid climates. There are a number of different types of chemical weathering, and their occurrence depends mainly on three factors:

1. rock type;
2. amount of available water;
3. the environment to which the rock is exposed (particularly climate).

Table 7.7 A classification of different weathering processes

Type of weathering	*Factor involved*	*Process*
Disintegration	Crystallisation	Salt weathering, frost weathering
	Temperature change	Insolation weathering (heating and cooling), fire, expansion of dirt in cracks
	Wetting and drying (especially of shales)	
	Pressure release by erosion of overlying material	
	Organic processes	Root-wedging
Decomposition		Hydration and hydrolysis oxidation and reduction solution and carbonation chelation
	Biological–chemical changes	Organic weathering

the ice melts. Repeated freeze–thaw cycles like this, over a long time, can split rocks into smaller pieces. The process is particularly effective in rocks with joints and weak beds. The process is also effective in cold climates, and periglacial areas with widespread frozen ground, typical of much of Britain during the Quaternary, often show signs of extensive frost weathering, including patterned ground, solifluction and talus slopes (see p. 444).

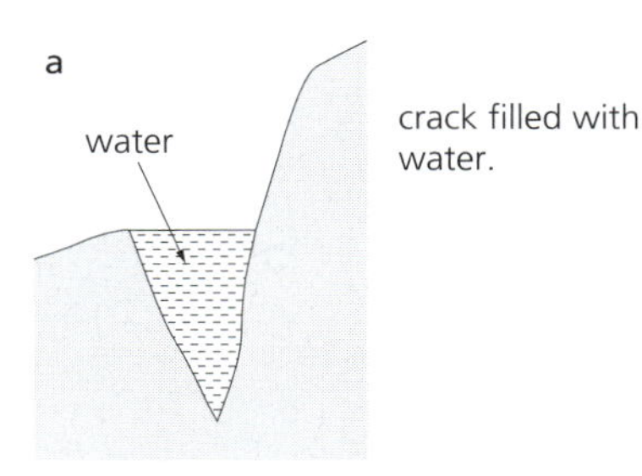
a
water
crack filled with water.

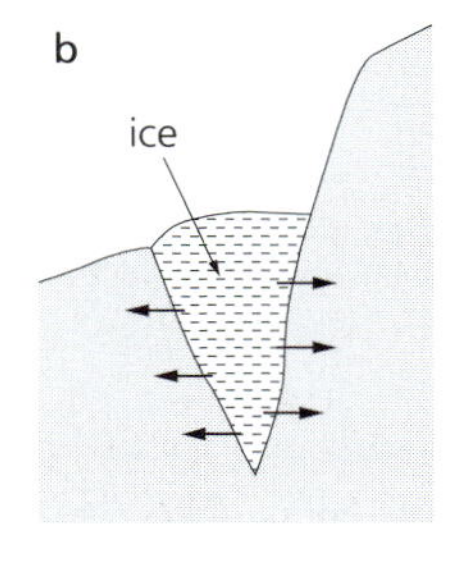
b
ice
ice extruded above top of crack on freezing, some pressure exerted

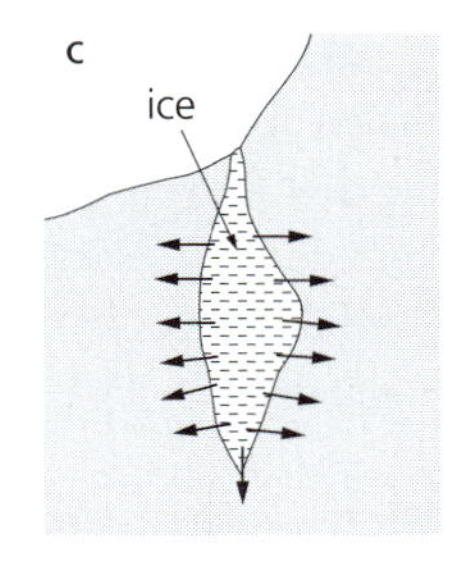
c
ice
ice more confined, restricted extrusion above top of crack on freezing, more pressure exerted

Figure 7.5 *Frost weathering of rocks.*

INSOLATION WEATHERING

Insolation weathering is similar in many ways to frost weathering, but it involves freeze–thaw cycles without frozen water. The rocks themselves heat up, perhaps during the day when they are exposed to strong sunlight, and then cool down, perhaps at night after sunset. This continues over many cycles, in this case over diurnal (daily) cycles. Repeated cycles of expansion on heating and contraction on cooling progressively break apart the grains of rock and cause granular disintegration. Over long periods, that can be a very effective agent of physical weathering. Insolation weathering appears to be responsible for the development of weathering pits on the Stone Mountain granite dome in Georgia, for example.

SALT WEATHERING

Salt weathering has some similarities with frost weathering too, because it involves the growth and expansion of material in cracks in rocks. In this case, it is salt crystals rather than ice that grow and exert pressure on the adjacent rock (Figure 7.6). The process is sometimes referred to as salt wedging and is most common in arid areas. Water in rocks generally contains dissolved salts, which are mainly derived

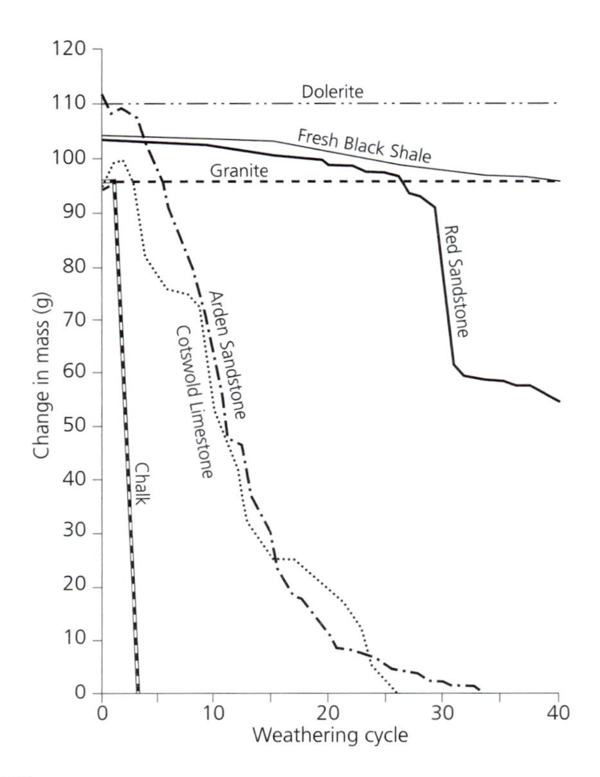

Figure 7.6 *Salt weathering of rocks. Samples of rock of different types were subjected to up to forty salt-weathering cycles in a solution of sodium sulphate, and the mass (weight) of each sample was measured after each cycle.*

Source: Figure 19.4 in Briggs, D. and P. Smithson (1985) Fundamentals of physical geography. Routledge, London.

from chemical weathering of soils and rocks but also originate in air pollution in many areas. The water is drawn through the interstices (small crevices or spaces) in rock, by capillary action, and as the water evaporates in the rock the salts (such as halite or gypsum) are deposited and crystallise. Over time the crystals grow bigger, and they start to exert pressure on the adjacent rock in exactly the same way as the growth of ice crystals. But, unlike the cyclical growth and melting of ice crystals, the salt crystals continue to grow.

Salt weathering occurs mainly in arid and semi-arid areas and is less widespread than frost weathering. Where it does occur, however, it affects concrete and natural stone buildings, structures and roads. Problems range from the visual (unsightly stains and blemishes on buildings) to the structural (serious building failure, which sometimes requires costly remedial works).

ORGANIC PROCESSES

Biological activity can also split rocks apart. One of the most common processes is root wedging. Plant roots (particularly from trees) grow in cracks in a rock, and they expand as they grow. This exerts pressure on the adjacent rock and can eventually split it. In some circumstances, burrowing animals might also expand existing rock cracks, particularly in relatively soft rock.

DECOMPOSITION

Decomposition, or chemical weathering, is very different from disintegration (physical weathering). It breaks down the chemical bonds that tie together the minerals and particles in the unweathered rock, chemically modifies the rock materials and produces new compounds. In some

a

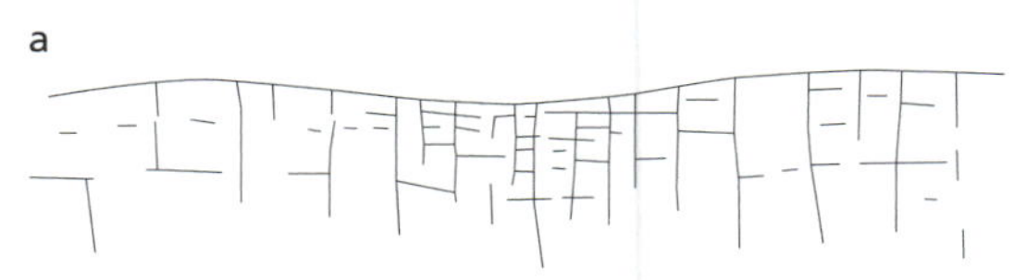

Original form. Variations in joint frequency due to inherent variations in granite.

b

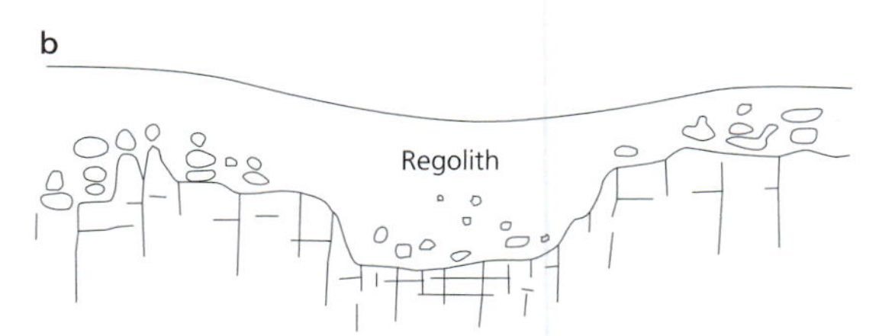

Deep weathering under a warm temperate climate; most intense in closely jointed granite.

c

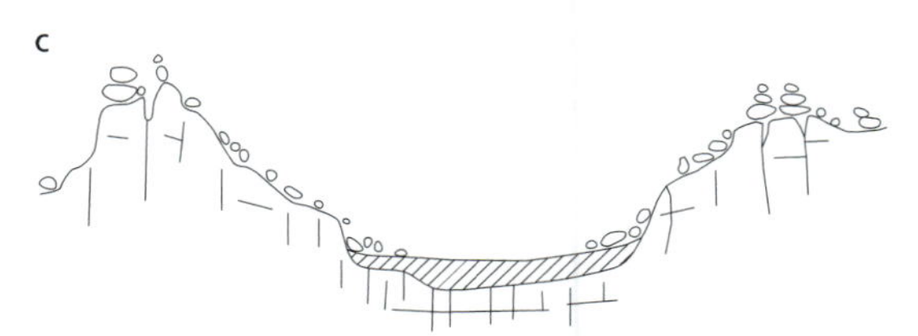

Removal of regolith by mass movement and river action to leave exposed tors.

Figure 7.7 *Development of tors on jointed granite. Linton explained the development of tors on granite in south-west England in terms of deep weathering of closely jointed rock.*

Source: Figure 19.8 in Briggs, D. and P. Smithson (1985) Fundamentals of physical geography. Routledge, London.

Plate 7.8 *Hay Tor on Dartmoor, Devon, England. The distinctive granite tors on Dartmoor (as on the nearby Bodmin Moor in Cornwall) are believed to have been formed by the chemical decomposition of granite under tropical climates during the Tertiary.*

Photo: Chris Park.

temperate areas (such as Dartmoor in south-west England), the development of tors on the surface of jointed granite (Figure 7.7) has been attributed to chemical weathering during the Tertiary (see Table 4.7), when climates were much warmer than today.

Decomposition always produces softer, less resistant rocks, which are more prone to physical disintegration. On the other hand, disintegration produces smaller fragments of rock with a larger surface area exposed to chemical decomposition. In these ways, both sets of weathering processes assist each other, and this is why they are usually found operating together.

The most common types of chemical weathering – oxidation, carbonation and hydrolysis – all involve oxygen and carbon dioxide in air and water. These same processes are important in the development of soils (see p. 600).

OXIDATION

Oxidation is a chemical reaction that involves the combination of rock materials (such as silicates or carbonates of iron or manganese) with oxygen and water. It produces new compounds that often have a reddish, yellowish or brown stain (often caused by the oxidation of iron material in the rock) and that are less resistant to further weathering and erosion. Many soils in hot or warm, humid environments are formed by oxidation of local rock, and many such soils are reddish or yellowish in colour because the rocks contain ferric (iron) oxides or hydroxides.

CARBONATION

Carbonation is a chemical reaction in which carbon dioxide (CO_2) and water (H_2O) combine to produce a weak solution of carbonic acid (H_2CO_3). The carbon dioxide comes from the decay of organic material (such as the strongly acidic leaf litter of a coniferous forest, or naturally acidic runoff from peat) or from air pollution (CO_2 is a greenhouse gas released into the atmosphere when fossil fuels are burned).

Carbonic acid combines with limestone or dolomite, which are composed of calcium carbonate ($CaCO_3$), to produce calcium bicarbonate, which dissolves quite readily in water. In this way, limestone and dolomite are readily weathered and broken down by acidified water.

HYDROLYSIS

In hydrolysis, the minerals in rocks react with the hydrogen (H^+) and hydroxyl (OH^-) ions in water. The reaction changes the character of the minerals concerned and produces material with a different volume and physical and chemical properties. Silicate minerals, which are common in many igneous rocks, are often involved. Thus, for example, granite weathers chemically, and the feldspar minerals in it break down into clay minerals (like kaolinite).

ORGANIC PROCESSES

Living organisms also contribute to chemical weathering processes in a variety of ways. The decomposition of organic material, particularly in soils, releases acids and carbon dioxide, which are important in chemical decomposition. Soil fauna (particularly insects and other invertebrates) move through soil and weathered rocks and deposits, mixing the material and assisting the penetration of air and water. Lichens grow on rock surfaces and absorb some mineral materials directly from the rocks. When lichens die they release organic substances, which dissolve in water to form weak acids, and these concentrate chemical weathering on the rock surface. Accelerated weathering of some granite rocks has been explained as the result of biological weathering by algae, compounded by repeated wetting/drying cycles.

EROSION

EROSION PROCESSES

All land surfaces are subjected to weathering, although the rates, processes and products of weathering vary from place to place. Weathering prepares material for transport, and erosion and mass movement transport that material downhill towards base level (normally sea level). Erosion is the movement of material, usually particles of sediment, by active transporting agents such as running water, moving ice or blowing wind. These agents are active in river systems (p. 364), coastal and ocean systems (p. 472), glacial and periglacial systems (pp. 456 and 444), and desert and semi-arid systems (p. 426). Something like 10 million tonnes of sediment is eroded from the world's continents each year, the vast majority (nearly 95 per cent) of which is eroded by rivers. Much smaller quantities are eroded by wind and ice.

Here we concentrate on the two main sets of processes that operate on hill slopes: erosion (by various forms of runoff, including overland flow and flow in small channels); and mass movement (gravity-driven processes including rockfalls, landslides and avalanches). Most of this material is ultimately transported to the oceans by river systems.

SLOPE EROSION

Particles of rocks and sediment are moved down slopes by two types of process. Slope erosion processes occur on all slopes, require water as a transporting agent and usually produce slow progressive changes in the form and appearance of the hill slope. Mass movement processes, on the other hand, tend to be concentrated on steep, unstable slopes, they do not always require water, and they often produce sudden, dramatic changes in hill slopes and can be a serious geological hazard. The three main erosion processes are rainsplash, wash and gully development.

Rainsplash: when raindrops land on the wetted surface of a soil or sediment (including weathered rock), they splash water and particles out in all directions. On a sloping surface, the downhill side of the splash moves further than the uphill side (Figure 7.8). Over a period of time, particles of sediment are progressively shifted downslope by repeated rainsplash movements. In this way, a hill slope tends to operate rather like a conveyor belt. Laboratory tests have shown that splash losses are controlled by rainfall intensity and duration, soil particle size and slope angle. Sometimes, particularly in fine materials such as loess, rainsplash causes a crust to build up on the surface of exposed soils and deposits, which strengthens the surface and significantly reduces further erosion.

Rainsplash is inhibited by a protective vegetation cover, such as grass, crops or forest. Conversely, removal of vegetation exposes more soil and deposits to rainsplash erosion, which is why soil erosion is often greatly accelerated by land-use change. Even temporary land-use change (for example during construction work, when topsoil is removed and piled up near the building work) can greatly increase soil erosion, change hill slopes, and lead to deposition and river channel changes downstream.

Wash: overland flow, in which water moves downhill across the slope surface (Figure 7.9), occurs when the intensity of rainfall exceeds the infiltration capacity of the

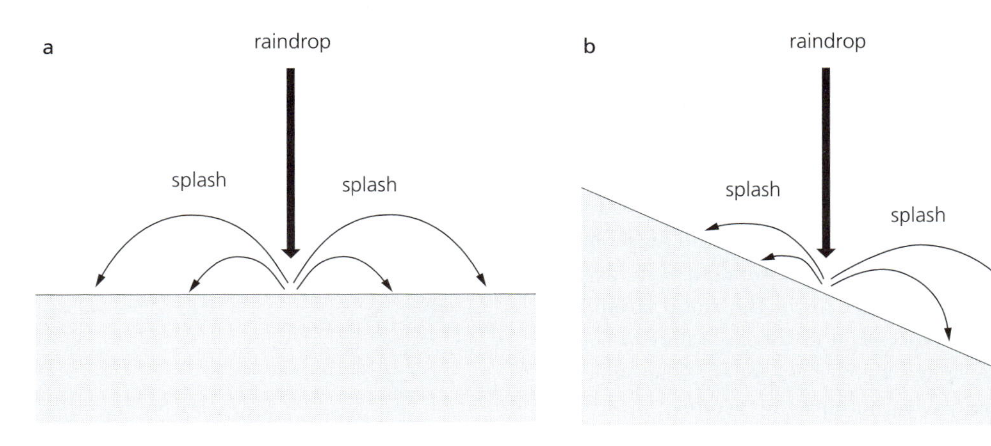

Figure 7.8 Rain splash erosion.

soil (this means that rain is falling faster than it can drain into the surface). Many hills have rounded or relatively flat tops, where overland flow is limited, but downslope from the crest of a hill, slopes are often steeper and gravitational pull is stronger. Overland flow increases in volume and speed until it is able to move small particles (usually sand-sized or smaller). This wash (rainwash) process can remove up to 4 cm or more of sediment from the surface of bare slopes in an average year. Rates of erosion are higher on steep slopes and where rainfall is higher and more intense. Monitoring has shown that rainwash removes an estimated 120 tonnes of soil a year from each hectare (0.01 km^2) of cultivated fields in Rwanda, for example. Vegetated slopes rarely experience significant wash or associated erosion.

Gully development: overland flow moves down a hill slope, and further down the slope the volume of overland flow increases. Wash dislodges and sets in motion individual particles of soil or sediment, and overland flow processes tend to concentrate in hollows and depressions on the slope surface. This concentrates the erosional processes in particular parts of a slope, where small rivulets (miniature rivers) form. These rivulets have more water than the adjacent slope surface, and erosion is concentrated within them. A positive feedback can be established, which encourages

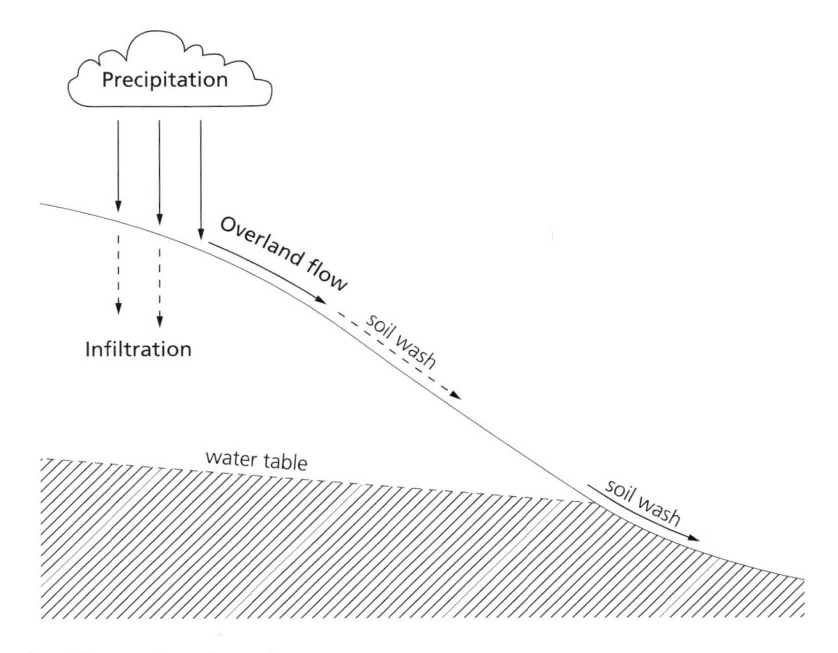

Figure 7.9 Overland flow and wash erosion.

more water and thus more erosion in the small linear depressions, which deepens them, further concentrating water and erosion within them, encouraging more erosion, further deepening, and so on. In this way rivulets grow, and downslope they coalesce to form ephemeral streams (which flow only during and immediately after rain), which can erode into gullies several metres deep. Gully development is also assisted by the collapse of natural pipes in soils.

Much of the sediment eroded from the slope is often deposited at the base of the slope to form an alluvial fan. Deposition is concentrated there because of the sudden drop in slope gradient, which reduces flow velocity in the gully.

Gully development is widespread in upland areas. In many parts of Scotland, for example, gully development is particularly common in areas with mineral soils and steep slopes. Parts of the loess plateau in China are suffering from severe erosion by gullies, and this is promoting the development of techniques for land evaluation and erosion risk assessment.

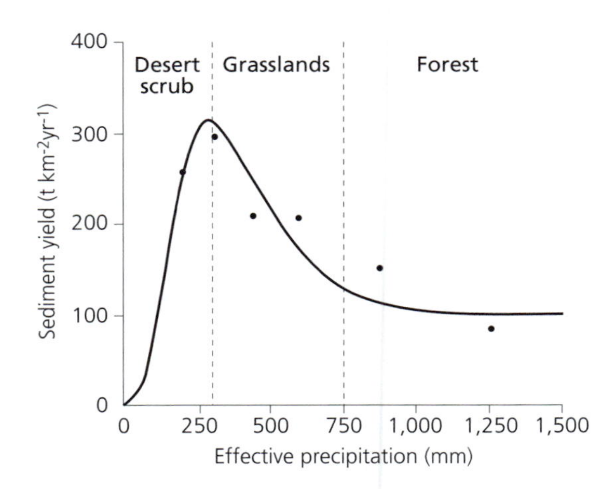

Figure 7.10 *The relationship between sediment yield and annual precipitation. Analysis of sediment yield under different climates in the USA shows peak yield under moderate rainfall, at the boundary between desert scrub and grasslands. Yield declines as rainfall increases past this point, as grassland and forest vegetation protect underlying soils.*

Source: Figure 4.3 in Walling, D.E. and B.W. Webb (1983) Patterns of sediment yield. In K.J. Gregory (ed.) Background to palaeohydrology. John Wiley & Sons, London, pp. 69–100.

EROSION AND CLIMATE

Rates of erosion vary a great deal from place to place, reflecting variations in key controlling factors such as:

- climate (particularly temperature and rainfall)
- vegetation cover
- changes in land use
- geology and soil type
- topography (particularly steepness and uniformity of the hill slope).

Measurements of rates of erosion in different climates show that erosion is relatively limited in arid and semi-arid areas, but it is high in semi-arid lands where precipitation averages around 300 mm a year (Figure 7.10). Semi-arid areas have limited vegetation cover because of the shortage of rainfall, but when rain does fall several times a year the often intense downpour falls on relatively bare slopes. This produces occasional but significant overland flow and erosion. Where rainfall is higher, vegetation grows more extensively and this protects slope surfaces and limits erosion. Where rainfall is lower, erosion is limited by the lack of overland flow.

Erosion rates also tend to be relatively high in strongly seasonal climates, which have clearly defined wet and dry seasons. Monsoon India is a typical case. Vegetation growth is limited during the dry season, and frequent intense rainstorms during the wet season produce high erosion losses from the bare hill slopes.

EROSION AND LAND USE

Land use affects erosion rates because it can significantly change the amount and character of protection offered to surface materials. Erosion rates tend to be quite high in humid climates where crop agriculture is widespread, such as large parts of the midwestern USA. Erosion is highest where cropping is seasonal and fields are left bare for several (up to six) months each year.

Human activities can significantly alter the pace and pattern of erosion. The United Nations Environment Program estimates that, as a result of human activities worldwide, 10,930 million km^2 of land have been seriously damaged by water erosion, 9.2 million km^2 by sheet and surface erosion, and 1.73 million km^2 by the development of rills and gullies. The main causes are clearance of natural vegetation and forest (43 per cent), over-grazing (29 per cent), poor farming practices such as cultivation of steep slopes (24 per cent) and over-exploitation of natural vegetation (4 per cent).

BOX 7.15 URBANISATION AND EROSION

Urbanisation alters the land surface and changes erosion rates and sediment yield. Erosion is normally greatest during construction, when vegetation is cleared and earth-moving machinery disturbs soils on site. When building work is complete, erosion rates tend to fall because much of the land surface is covered with impermeable materials (particularly roofs and roads) and planted vegetation (particularly lawns). Figure 7.11 shows the pattern of sediment changes in a small river in the USA since the early 1800s as natural forest was cleared and replaced by crops and grazing, then by suburban development.

BOX 7.16 FORCES ACTING ON A PARTICLE

Shear stress exists because gravity pulls objects downhill. Shear stress promotes downslope movement, and its strength is determined mainly by slope angle (steeper slopes have higher shear stress, all else being equal).

Shear strength is an opposing force that serves to resist movement. It is governed by a number of factors, including frictional resistance between particles, the cohesiveness of clays and the binding strength of plant roots.

MASS MOVEMENT

Mass movement moves weathered material downslope, but gravity is the main agent rather than a river, glacier or wind. Mass movement is also known as mass wasting or gravity transfer. A variety of processes are involved (Figure 7.12), some of which are extremely slow and operate more or less continuously through time, but others of which can be extremely fast and operate sporadically. Different amounts of material are moved, too, ranging from small groups of individual particles to landslides big enough to bury entire towns.

FORCES

Two sets of opposing forces operate on rock and soil particles on a slope – shear stress and shear strength (Box 7.16). The balance between these two forces determines whether the particles move or not. A slope is stable when its shear strength is greater than the shear stress on it. When the two forces are equally balanced, the slope is at a critical threshold; if either or both forces change, slope stability will be affected. A slope is unstable, and mass movement takes place on it, when shear stress is greater than shear strength.

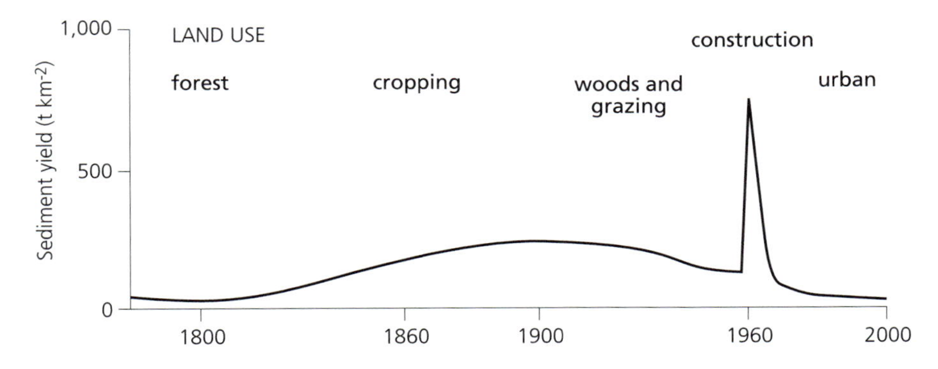

Figure 7.11 *Relationship between land-use change and sediment yield in the Piedmont region of Maryland. Sediment yield increased when forest was cleared and replaced by crops, and declined slightly when secondary woodland growth protected soils. A short phase of building activity in the drainage basin caused a sharp rise in erosion, followed by reduced yield in the urbanised basin.*

Source: Figure 24.23 in White, I.D., D.N. Mottershead and S.J. Harrison (1984) Environmental systems. George Allen & Unwin, London.

Plate 7.9 *Mass movement of surface materials on a hill slope on Plynlimon, near the headwaters of the River Severn in Mid-Wales. Photo: Chris Park.*

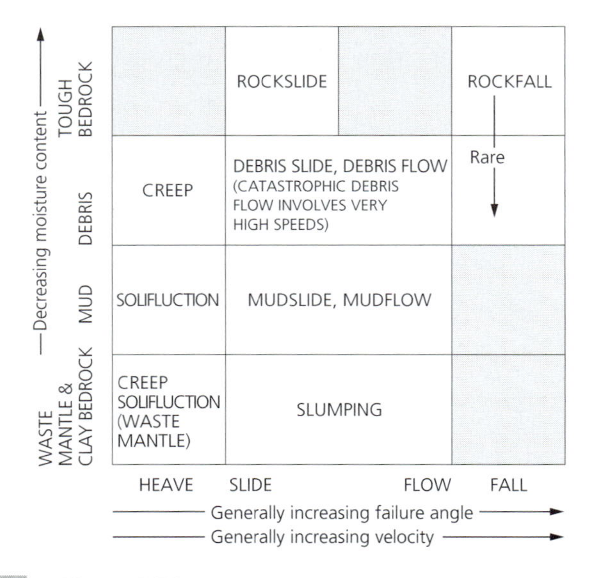

Figure 7.12 *A classification of mass-movement processes. This scheme classifies the most common mass-movement processes in terms of speed and wetness.*

Source: Figure 9.19 in Dury, G.H. (1981) Environmental systems. Heinemann, London.

MOVEMENT

Mass movement includes a range of processes, which operate in different ways and at different speeds (Figure 7.13). Some are dry processes, whereas others require water to lubricate the material (Figure 7.14). Most movement involves one or more of five types of motion – creep, flow, slide, slump and fall. The basic properties of each type are summarised in Table 7.8.

The type of motion that takes place on an unstable slope is related to three main factors:

1. the type of material involved;
2. the physical properties of the material; and
3. the moisture content of the material.

Brittle material – such as bedrock and sand, which is easily broken or cracked – ruptures when it fails and tends to move in slides, falls or slumps. Other material, like clay-rich soils, which behave more like plastics, can deform without rupturing, and they bend or creep downslope. Saturated materials with small grain sizes (sand, clay

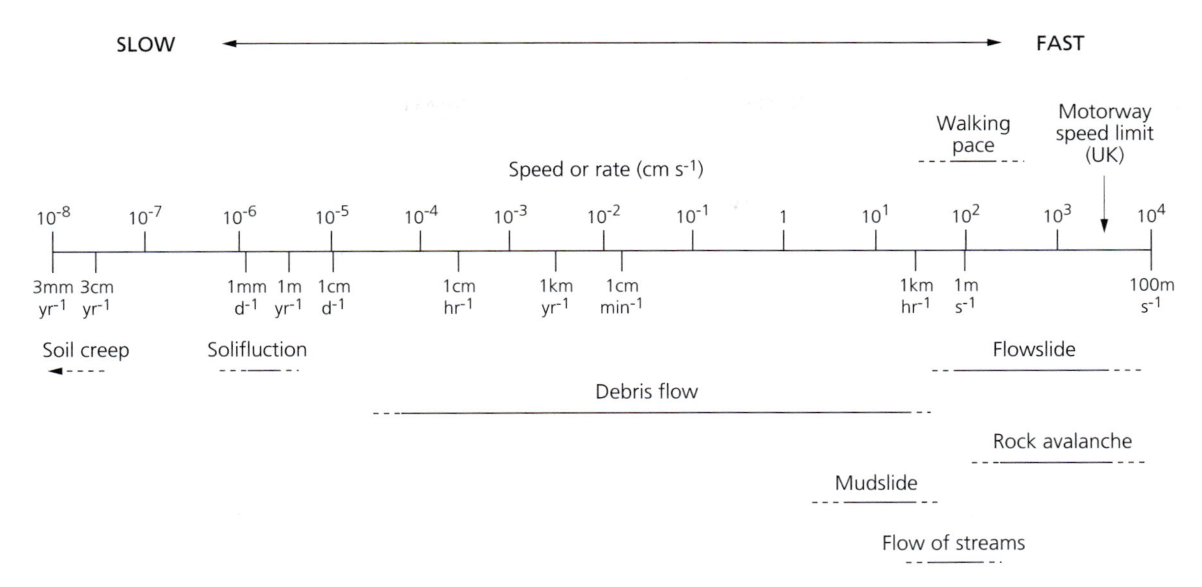

Figure 7.13 *Typical rates of operation of mass-movement processes. Mass-movement processes extend from the extremely slow (soil creep) to the extremely fast (rock avalanches). The speed scale shown is logarithmic, with each unit increasing or decreasing by a factor of ten compared with the adjacent unit.*

Source: Figure 11.3 in Goudie, A. (1993) The nature of the environment. Blackwell, Oxford.

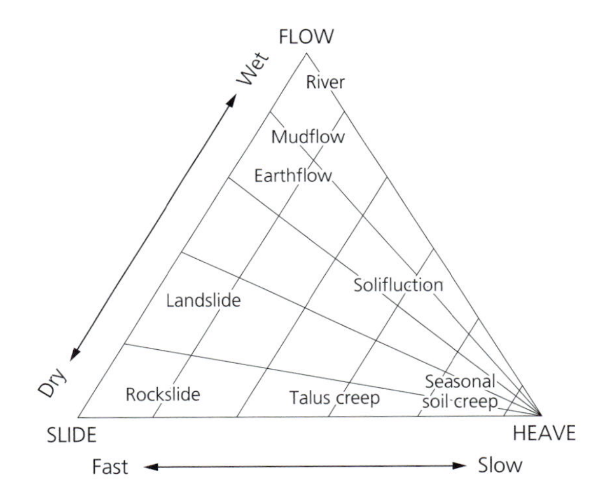

Figure 7.14 *Mass-movement processes defined in terms of wetness and speed. This way of classifying mass-movement processes is widely used.*

Source: Based on Figure 11.1 in Goudie, A. (1993) The nature of the environment. Blackwell, Oxford.

or silt), which behave more like liquids, literally flow downslope.

PROCESSES

Mass movement is really a family of processes, and different processes are usually dominant in different parts of a landscape (Figure 7.15). Common mass-movement processes include soil creep, mudflows and earthflows, rockfalls, landslides and avalanches.

Soil creep: creep is the very slow downslope movement of soil material, often associated with repeated freeze–thaw or wet–dry cycles. Inter-particle sliding, in which individual particles of soil and rock slide past and over each other, seems to be the main mechanism involved. Rates of soil creep appear to be directly proportional to slope angle, so that gentle slopes have slow creep and steep slopes have high rates of creep. Long-term average rates of soil creep in a mountain environment vary between about 3 mm a year on gentle slopes and up to 9 mm a year on steeper slopes. Rates of movement can be increased between two and three times by disturbance (including cultivation, tourism and winter sports). Telltale signs of creep include the tilting of vertical objects (such as monuments, gravestones and fences), the cracking of building foundations and curved tree trunks. Solifluction is a particular type of soil creep that occurs in areas underlain by permanently frozen ground (see p. 444).

Mudflow and earthflow: mudflows involve the downhill movement of fine-grained saturated material lubricated by water (Figure 7.16), perhaps after prolonged or unusually heavy rain. They occur particularly in areas with little or no protective vegetation cover, and they often flow down established drainage channels. Mudflows are particularly common in steep mountain areas like those in China, Norway and Colombia, and they can be triggered by intense rainstorms and earthquake shocks.

Earthflows are similar to mudflows, but they do not necessarily flow down existing drainage channels. Both occur most commonly on slopes with deep soils overlying weathered rock. Earthflows in the San Francisco Bay region of California are most common in old landslide deposits and where slopes are steeper than 15 degrees. Field studies in New Zealand have shown that earthflows move up to three times faster in grassed slopes compared with forested

Table 7.8 The five main types of mass movement

Type	*Features*
Creep	Absence of deformation in trees or soil surface; if soil is excavated, tree roots may reveal downslope bending
Flow	Lobes of deposited material form; trees are bent or upright; ground vegetation and soil are buried by the deposit
Slide	Concave scar on the slope, showing the back of the slide; the displaced mass of material comes to rest anywhere up to 1 km downslope
Slump	Bank or cliff exposure in soil or bedrock, showing the back of the slide; movement often has a rotational appearance; trees are uprooted and tipped back towards the slope
Fall	Cliff exposure with fractured bedrock; angular blocks and slabs of rock are piled up at the footslope; scarred and broken trees

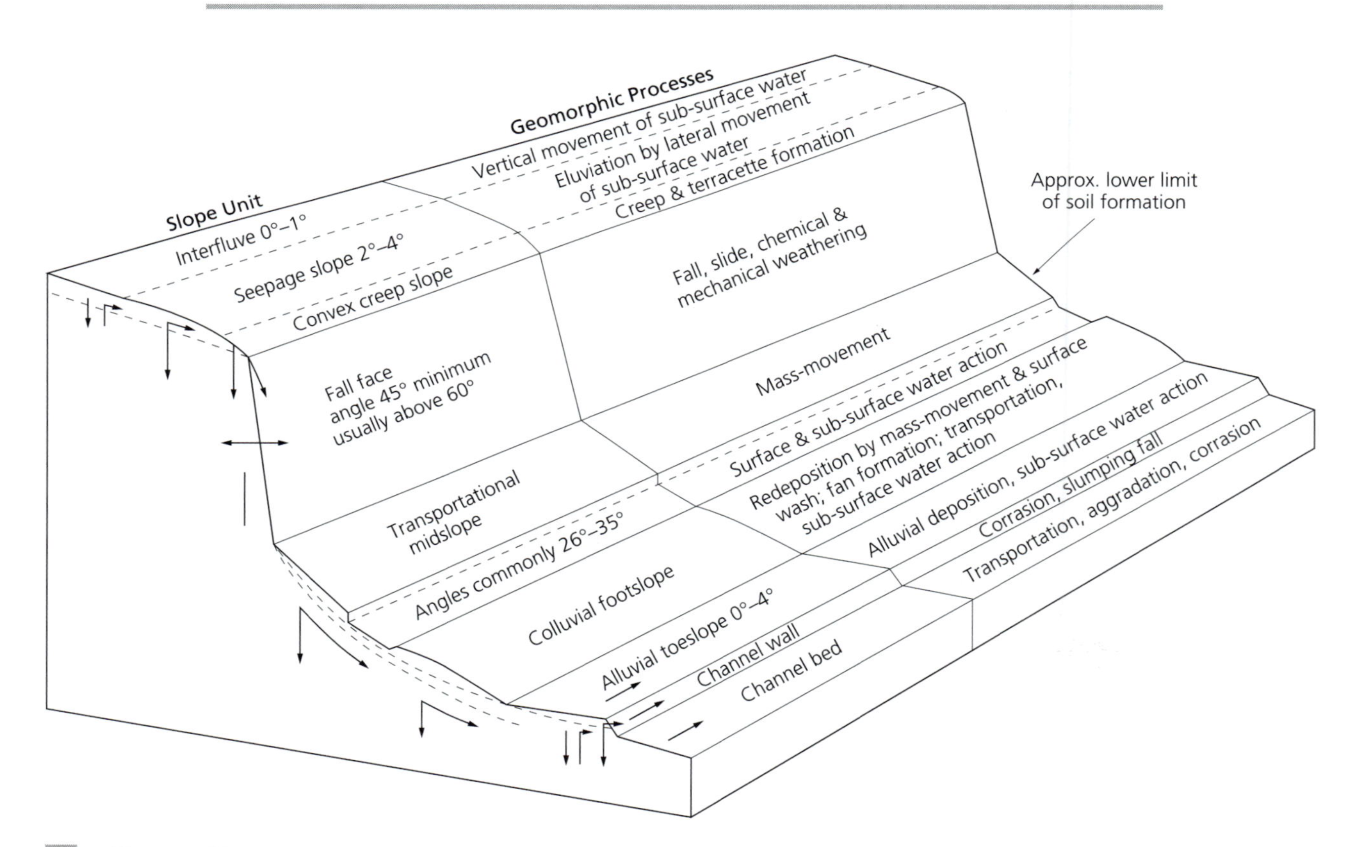

Figure 7.15 *Mass-movement processes on different parts of a slope system. Geologists identify nine different hill slope units, not all of which are present in any given hill slope. Each unit has a distinctive appearance and characteristic processes.*

Source: Figure 9.13 in Dury, G.H. (1981) Environmental systems. Heinemann, London.

Plate 7.10 *Rockfall from L'Eveque in the Pennine region of the Swiss Alps, resulting in a debris cone on the surface of the Upper Arolla Glacier.*
Photo: Alistair Kirkbride.

slopes. Mudflows and earthflows can create great problems for people and property, particularly on steep, bare slopes. Parts of southern California where vegetation is burned by natural fires during the hot, dry summers regularly suffer from mudflows in the wet winter months.

The collapse of the huge coal waste tip at Aberfan in South Wales on 21 October 1966 was an earthflow lubricated by a spring line. The collapse followed heavy rainfall, and it sent a flow of black sludge (with a volume estimated at 100,000 m^3) downhill into the village, almost without warning, killing 147 people (including 116 children in the local primary school).

Rockfall: this is a fall of weathered rocks down very steep slopes or cliffs, pulled down by gravity. The material that falls is usually deposited at the base of the slope or cliff as a pile of loose rocks called scree or talus (see Figure 7.16). Scree slopes (talus cones and aprons), composed of scattered boulders that have rolled beyond the base of the slope, are common in high mountain areas and in steep arid areas where vegetation is sparse. Rockfall hazard maps have been compiled for some places (such as Salt Lake County in Utah) to define threats to development and identify where control measures might be required. Slopes steeper than about 30 per cent seem particularly vulnerable to rockfalls, and in many places local building ordinances prohibit development of slopes of 30 per cent or steeper.

Landslide: a landslide is the sudden movement of a mass of soil and weathered rock down a steep slope (see Figure 7.16). It occurs when the slope becomes unstable, perhaps because of an earthquake or because an unstable slope is lubricated by water. The 1963 Vaiont Dam disaster in northern Italy (Box 7.17) was caused by a landslide lubricated by water. Landslides are also caused by fault line movements. Parts of South Wales are particularly susceptible to landslides, because thick, unstable deposits of glacial sediment overlie permeable rock strata in oversteepened valleys. Climate change, such as an increase in rainfall intensities since the mid-1920s, threatens to reactivate some old landslides and put valley-floor settlements at risk.

Landslides are among the most dramatic mass-movement processes because of their large size, great speed and sudden onset. A quarter of a million people were killed in China by landslides between 186 BC and AD 1987, and between 1951 and 1987 landslides in China caused at least 150 deaths and US$0.5 billion of damage each year. Landslides create problems in many areas, and the US National Landslide Information Center (NLIC) was established in 1990 to collect, analyse and distribute information on all aspects of the world landslide problem. Landslide hazard zone maps can help to determine which areas are most at risk, usually because of a mixture of natural factors and human activities. Remote sensing and geographical information systems (GIS) are proving very valuable in defining landslide sites and high-risk areas.

Avalanche: an avalanche is like a landslide, but it involves a mass of snow, ice and rock crashing down a mountainside under its own weight. Most avalanches occur in spring when the snow starts to melt, although some are triggered

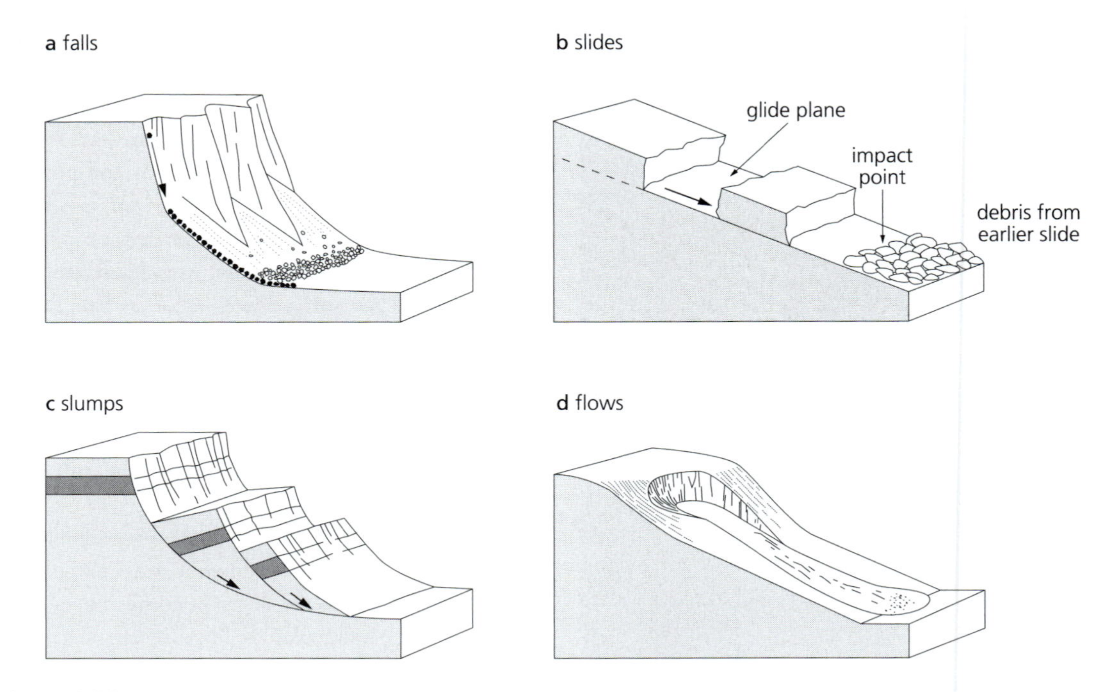

Figure 7.16 *Rapid mass-movement processes. Falls (a), slides (b), slumps (c) and flows (d) operate at different speeds and produce different landscapes.*

Source unknown.

BOX 7.17 THE 1963 VAIONT DAM DISASTER

The world's worst dam disaster occurred on 9 October 1963 in northern Italy. Vaiont Dam, a concrete arched dam, was completed in 1960. It was then the world's second highest dam (265 m high), with a reservoir storage capacity of 150 million m^3. Engineers had been concerned about the stability of slopes above the dam, and had recorded some small rock-slides and other mass movements as the reservoir filled up over a three-year period.

The 1963 disaster happened with remarkable speed. The engineers by then were expecting a slide, but certainly not on the scale of what happened (Figure 7.17). At 10.41 am, hill slope failure produced a slide of 140 million m^3 of material, mostly limestone rock. The debris slid within a minute down into the reservoir and was piled up 150 m above the height of the dam. This displaced water from the reservoir, and the water spilled over the crest of the dam in waves up to 100 m high. By 10.43 am, 2 minutes after the huge landslide, a wall of water up to 70 m high had passed down the steep-sided canyon below the dam and flattened everything in its path for many kilometres downstream. An estimated 2,600 people were drowned in the flood. By 10.55 am the flood waters had gone down, and a sombre silence fell over the devastated valley.

A number of factors contributed to the landslide at Vaiont, including:

- unsuitable geological conditions (the limestone in this area is relatively weak and has open fractures tilted towards the reservoir);
- very steep topography (the gravitational pull was particularly strong because of the steep slopes);
- increased water pressure in the underlying rocks because of the impoundment of water behind the reservoir;
- heavy rains immediately before the slope failure.

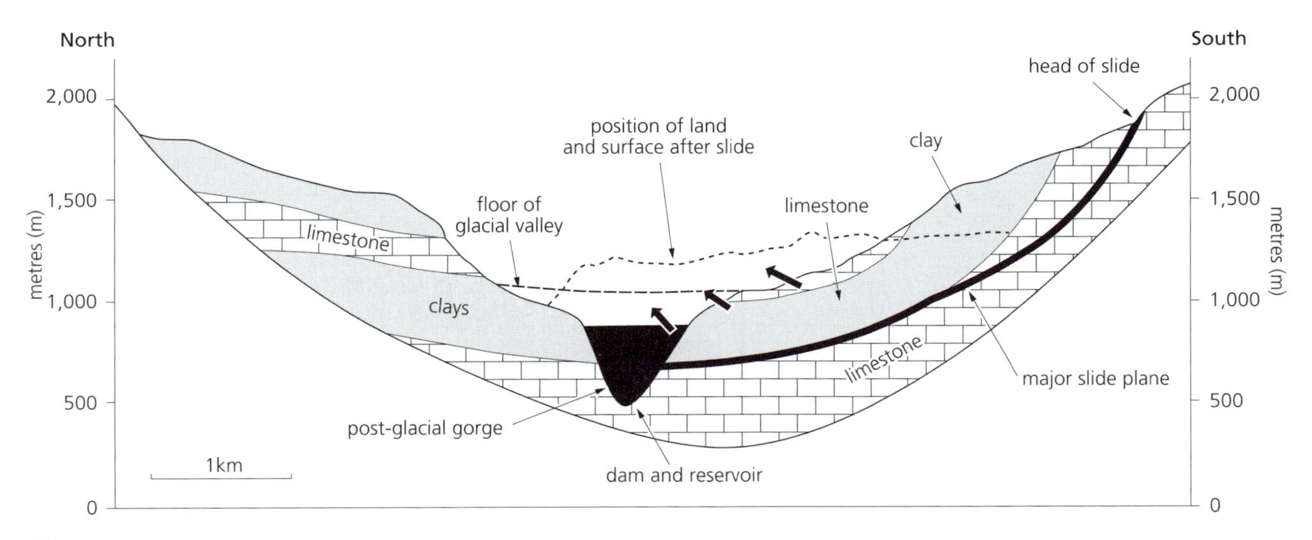

Figure 7.17 *The 1963 Vaiont Dam landslide. This famous landslide deposited a large amount of soil and rock in the reservoir behind the Vaiont Dam, causing water to spill over the dam and create a catastrophic flood.*

Source: Figure 3.15 in Waltham, B. (1978) Catastrophe – the violent Earth. George Allen & Unwin, London.

by earthquakes. Avalanches account for many deaths in some areas, such as the European Alps (particularly Switzerland) and the Rockies in North America. Around 150 people are killed by avalanches in the northern hemisphere in a typical year, and deaths occurring through recreational activities are on the increase (Figure 7.18).

Snow avalanches caused few deaths and little property damage in Britain until about 1950, since when winter mountaineering has become much more popular (particularly in places like Ben Nevis, the Cairngorms and Glen Coe) and many more people have been killed or injured by avalanches. Avalanche fatalities have also increased in the USA, from an average of 7.5 a year between 1950 and 1980 to nearly fourteen a year during the 1980s. As with many geological hazards – including earthquakes, volcanoes and landslides – avalanche hazard mapping is one way of reducing risk to people and property. Such strategies are being used increasingly in tourist areas, such as Vail in Colorado (Figure 7.19).

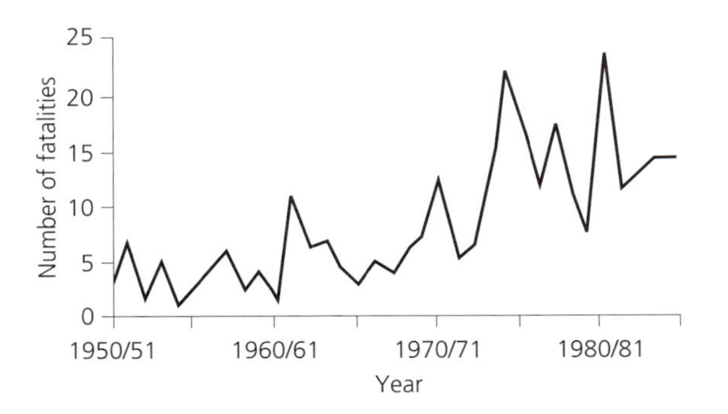

Figure 7.18 *Rise in the number of deaths in the USA associated with avalanches, 1950–1984. The number of deaths has increased in recent decades, partly in response to the development of tourism in mountain areas.*

Source: Figure 8.1 in Smith, K. (1992) Environmental hazards. Routledge, London.

INFLUENCES

Mass movement can be triggered by many different factors, including the following:

Soil water and groundwater: the shear strength and shear stress of slope material vary from season to season. Infiltration increases soil water during spring and winter. This fills voids between soil particles and increases shear stress, and it makes the material more like a liquid, which decreases its shear strength. Once the critical threshold is crossed, the material fails and moves downslope as a mudflow (if it is mainly clay and silt-sized material) or a debris flow (if larger particles are present). Landslides can also be caused by the addition of water to susceptible materials.

Ground frost: ground frost triggers several kinds of mass movement, including rockfalls produced by ice-wedging on cliffs and steep slopes. Repeated freeze–thaw cycles are important in rock weathering (see p. 206), but they also move particles downslope in much the same way as rain-

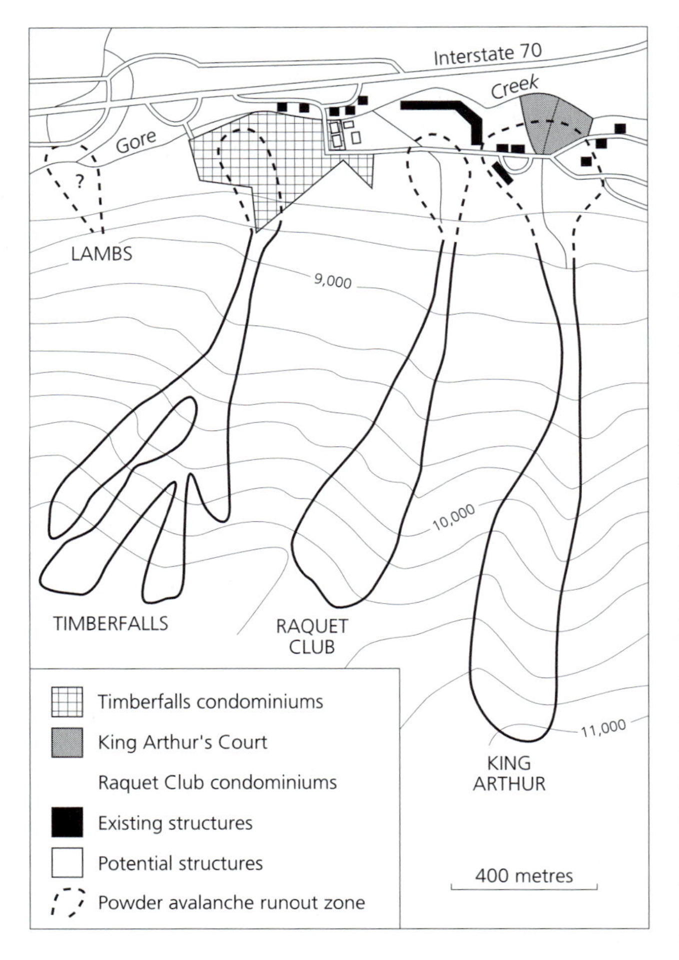

Figure 7.19 *Defining avalanche risk zones in Vail, Colorado. Hazard mapping helped to identify areas most at risk from avalanches, and this information was used in planning the development of King Arthur's Court.*

Source: Figure 8.5 in Smith, K. (1992) Environmental hazards. Routledge, London.

splash does, as a slow form of soil creep. Freeze–thaw in wet soils can give rise to solifluction, the movement of the active layer above permafrost (permanently frozen ground), which often deposits particles in the form of lobes (tongues) pointing downslope. Solifluction is very widespread in tundra environments (see p. 444).

Undercutting: many slopes fail and move (particularly in slides) because they are undercut by rivers, waves or other processes. Human activities such as terracing fields and cutting road and railway embankments also undercut slopes and can cause unexpected instability.

Earthquakes: some dramatic mass movements, particularly large landslides and avalanches, occur in steep mountainous areas and are triggered by earthquakes. As the seismic wave travels through the weathered rocks, deposits and soils, the bonding between particles is broken, the material's shear strength is greatly reduced, and it often fails instantaneously. Huge amounts of material are dislodged in a landslide or avalanche, and it can travel great distances at high speed because it can move over a layer of compressed air with little frictional resistance, much like a hovercraft. These mass movements can cause massive damage to towns, villages and structures that get in their way. Much of the damage associated with the 1964 Alaska earthquake (see p. 169) was caused by avalanches and landslides.

SUMMARY

The focus of this chapter is the materials that make up the Earth's crust and the processes that move them through the various stages of the rock cycle. We began by looking at minerals as the basic building blocks from which the crust is made, and the main ingredient of all rocks, deposits and soils – what they are composed of and how they are described. Next we considered the three types of rock (igneous, sedimentary and metamorphic), the key processes that create them and their main properties, before seeing how they form part of the global rock cycle. The second half of the chapter deals with surface processes of weathering and erosion. The two main types of *in situ* weathering (physical disintegration and chemical decomposition) were described, and typical processes and products were outlined. Erosion processes transport the products of weathering, and we looked at the main slope erosion processes and how they shape the land surface. Mass-movement processes tend to be more dramatic, and they create major changes in landscapes as well as posing serious risks to people and property.

WEBSITE

Links to relevant websites, a comprehensive bibliography, tools for teaching and learning, and downloadable images relevant to this chapter can be found at the website specially designed to accompany this book at
http://www.park-environment.com

FURTHER READING

Allison, R. (1997) *Rock slopes.* Blackwell, Oxford. A summary of recent research on how rock slopes develop and the main factors that influence their forms and processes.

Bland, W. and D. Rolls (1998) *Weathering.* Edward Arnold, London. A clear introduction to the scientific principles and the products of rock weathering.

Boggs, S. (1995) *Principles of sedimentology and stratigraphy.* Prentice Hall, Hemel Hempstead. Describes the processes that form sedimentary rocks and the important physical, chemical, biological and stratigraphic characteristics of these rocks.

Deer, W.A., R.A. Howie and J. Zussman (1992) *An introduction to rock forming minerals.* Longman, Harlow. Useful review of recent developments in mineralogy, including techniques for the study and description of minerals.

Dikau, R., L. Schrott, D. Brunsden and M.L. Ibsen (eds) (1996) *Landslide recognition.* John Wiley & Sons, London. Descriptions and comparative photographs of different types of landslide.

Doyle, P., M.R. Bennett and A.N. Baxter (1994) *The key to Earth history: an introduction to stratigraphy.* John Wiley & Sons, London. An overview of the principles of stratigraphy and patterns of Earth history.

Fry, N. (1991) *The field description of metamorphic rocks.* John Wiley & Sons, London. Handbook that gives clear details of how metamorphic rocks can be observed, recorded and mapped in the field.

Goldring, R. (1999) *Field palaeontology.* Longman, London. A comprehensive review of how to analyse and describe fossils and sediments, and of recent advances in palaeontology and the study of sedimentary rocks in general.

Goudie, A.S. and H.A. Viles (1997) *Salt weathering hazard.* John Wiley & Sons, London. Examines the causes, processes and consequences of salt weathering and explores why this environmental hazard is on the increase.

Hall, A. (1995) *Igneous petrology.* Longman, Harlow. Describes igneous rocks, their formation, character and composition.

Harper, D. and M. Benton (1996) *Basic palaeontology.* Longman, Harlow. A comprehensive introduction to all aspects of interpreting the fossil record.

Power, T.M. (1996) *Extraction and the environment: the economic battle to control our natural landscapes.* Island Press, New York. Argues that the quality of the natural landscape is a crucial economic resource that should not be degraded for short-term gain.

Ripley, E.A., R.E. Redman and A.A. Crowder (1995) *Environmental effects of mining.* St Lucie Press, New York. Detailed study of the environmental impacts of copper and nickel mining in the USA.

Robinson, D.A. and R.B.G. Williams (eds) (1994) *Rock weathering and landform evolution.* John Wiley & Sons, London. An overview of recent advances, approaches and interpretations.

Slaymaker, O. (1995) *Steepland geomorphology.* John Wiley & Sons, London. An overview of landforming processes on steep slopes and in tectonically active landscapes, with examples from around the world.

Spencer, E.W. (2000) *Geological maps: a practical guide to the interpretation and preparation of geological maps.* Prentice Hall, London. Clear guidance on how to prepare, read and interpret geological maps and features.

Thomas, M. (1994) *Geomorphology in the tropics: a study of weathering and denudation in low latitudes.* John Wiley & Sons, London. An overview of recent research on tropical weathering and the landforms it produces.

Thorpe, R. and G. Brown (1991) *The field description of igneous rocks.* John Wiley & Sons, London. Useful guide to the study and mapping of igneous rocks in the field.

Tucker, M. (1996) *Sedimentary rocks in the field.* John Wiley & Sons, London. Handbook that shows how sedimentary rocks can be studied, classified and mapped in the field.

Turk, J. and G.R. Thompson (1995) *Environmental geoscience.* Saunders College Publishing, Philadelphia. A broad overview of physical geology processes, with special emphasis on how they affect our everyday lives.

Yardley, B.W.D., W.S. Mackenzie and C. Guilford (1990) *Atlas of metamorphic rocks and their textures.* Longman, Harlow. An illustrated guide to the texture and composition of metamorphic rocks.

Part Three
The atmosphere

Chapter 8

The Atmosphere

LEARNING OBJECTIVES

When you have finished studying this chapter, you should be able to

- *Appreciate the importance of the atmosphere to life on Earth*
- *Describe and account for the composition of the atmosphere*
- *Outline the main sub-divisions of the vertical structure of the atmosphere and comment on the importance of each*
- *Summarise the properties of energy from the Sun and show the relevance of the electromagnetic spectrum*
- *Explain the main components of the Earth's energy budget*
- *Define the so-called 'greenhouse effect' and say why it is important for life on Earth*
- *Outline the main options for using solar energy as a renewable source of energy*
- *Distinguish between the main categories of air pollutants, and the main spatial scales at which air pollution poses problems*
- *Explain the relevance of ozone in the atmosphere, and the causes and impacts of ozone depletion*
- *Describe the causes and impacts of acid rain and suggest what can be done to reduce it*
- *Identify the main greenhouse gases, describe their properties and evaluate their contribution to global warming*
- *Outline the most likely environmental impacts of global warming and suggest why tackling global warming is such a difficult but pressing problem*

ATMOSPHERE AND LIFE

The atmosphere is the thin layer of air that surrounds the Earth. Our planet is literally wrapped in an envelope of gases, and conditions in this atmosphere make life possible. Humans can survive outside the Earth's atmosphere only by creating artificial atmospheres with similar properties (for example in a spaceship or in a moon suit).

IMPORTANCE OF THE ATMOSPHERE

People and atmosphere interact in many ways. Perhaps the most important aspect of the atmosphere is that it makes life on Earth possible in the first place.

The Earth is the only planet on which life is known to exist. Life is possible partly because the Earth is at a suitable distance from the Sun – it is neither too hot nor too cold (see Box 4.9). But that alone is not enough to create and sustain life. Life depends on the availability of energy and nutrients; the Sun provides the ultimate energy source, and nutrients are made available through the biogeochemical cycles (see p. 86). Water is also required, because it plays a vital role in many of the biogeochemical processes. The atmosphere provides the oxygen that humans breathe and the carbon dioxide that plants need to grow. It also receives the gases produced by human and animal respiration, and it acts as a reservoir for air pollutants and the products of natural fires.

The atmosphere provides both the setting in which these important processes and transfers take place and conditions in which all forms of life are created and sustained. The atmospheric system is finely balanced, dynamic and highly responsive to external change.

The atmosphere also produces weather (Chapter 10), which affects people both directly and indirectly. Through

BOX 8.1 A GAIAN VIEW OF THE ATMOSPHERE

The Earth's atmosphere is a product of physical, chemical and biological interactions on land and in the oceans over a very long period of time. According to the Gaia hypothesis (see p. 119), the entire Earth acts as a superorganism in which interactions between the atmosphere and the planet are crucial. These interactions create conditions for life, absorb many external changes and internal variations, maintain the Earth's temperature within a suitable range, and determine the balance and adjustment of the major biogeochemical cycles. Whether or not the hypothesis is correct, the importance of the atmosphere to life on Earth is beyond doubt.

weather it creates climate (Chapter 11), which in turn affects most other major environmental systems including weathering and erosion (see p. 204), soil development (see p. 600), and vegetation and biomes (Chapter 18).

THE CHANGING ATMOSPHERE

The balance of gases in the atmosphere is vital to the stability of life on Earth. This balance can be altered in a variety of ways, the two most important of which are:

- air pollution, particularly from the burning of fossil fuels, which releases carbon dioxide into the air;
- land-use change and vegetation clearance, particularly of tropical rainforest.

Throughout history, humans have altered local climate through activities like forest burning, land-use change and the smelting of metal ores. Increasingly through the twentieth century, these localised impacts have been overshadowed by more widespread changes, such as acid rain, ozone depletion and greenhouse gases, which threaten health and the environment and are becoming global environmental problems.

ATMOSPHERE – COMPOSITION AND STRUCTURE

COMPOSITION OF THE ATMOSPHERE

The atmosphere contains gases, liquids and solids. Because of its mobility – air is easily mixed and dispersed by winds – the detailed composition can vary a great deal. The variations are important for atmospheric stability, they can significantly affect people, and they take place in three dimensions:

BOX 8.2 ATMOSPHERE AND INSPIRATION

We often use the word 'atmosphere' to describe the mood or character of a place or a piece of creative work (such as a novel, painting or symphony), and this suggests a more subtle meaning of the atmosphere. While the atmosphere serves vital scientific functions, we also enjoy other benefits from it. Throughout history, the sky has inspired and excited people; it affects our moods and feelings. A calm, sunny sky triggers feelings of tranquillity, whereas a dark, cloudy, thundery sky brings feelings of insecurity, concern and oppression. The sky provides a setting in which we can view birds and planes, and it is an important part of scenery. Landscape painters, such as Turner and Constable, make effective use of shapes, patterns and colours in the sky. To many people, the sky also has spiritual significance; gazing at the clouds or the stars takes us beyond our day-to-day existence and can challenge us to ponder the bigger questions of life and meaning.

BOX 8.3 ATMOSPHERIC CHANGE

The atmosphere is very sensitive to change and disturbance from both natural and human causes. Natural changes are caused by various processes, including:

- *volcanic eruptions*: these produce gases such as sulphur dioxide and carbon dioxide (see p. 255);
- *dust*: particularly from desert storms (see p. 433), forest fires (see p. 575) and volcanic eruptions; this can reflect sunlight and reduce insolation (and thus lower temperatures), and it can increase rainfall by providing more condensation nuclei.

Rates of production of gases and dust in the atmosphere are highly variable through time and from place to place. This makes it extremely difficult to predict atmospheric changes and their likely impacts on weather and climate.

Atmospheric change is also promoted by human activities, particularly air pollution (see p. 241).

1. variations from place to place;
2. variations with height within the atmosphere; and
3. variations through time.

Some of the variability is caused by variations in natural processes, sources and sinks of material, such as volcanic eruptions (p. 173), natural forest fires (p. 583), ocean currents (p. 479) and global wind systems (p. 284). Increasingly, however, human activities (particularly pollution) are disturbing natural atmospheric equilibrium, sometimes with serious consequences.

GASES

The atmosphere is a mixture of gases, about ten of which are naturally and permanently present under all natural conditions. Two gases dominate, accounting for almost all of the atmosphere by volume (Table 8.1) – nitrogen (78 per cent by volume) and oxygen (nearly 21 per cent). The other 1 per cent of dry air (under clean conditions) is made up of small amounts of argon and extremely small amounts of carbon dioxide, neon, helium, ozone, hydrogen, methane, krypton and xenon. The amount of carbon dioxide naturally present in the atmosphere is very small indeed, but its importance far outweighs its size because plants use it for photosynthesis.

Nitrogen: nitrogen (N) is a colourless, tasteless, odourless and relatively unreactive gas that makes up nearly four-fifths of the atmosphere. It is an essential constituent of the proteins and nucleic acids that make up living matter, and it occurs in nature as nitrates and ammonium compounds. The nitrogen cycle is described in Chapter 3 (p. 89). Nitrogen oxides and ammonia also affect atmospheric chemistry through their involvement in acid rain and their reactivity with ozone in the troposphere (the lower atmosphere, where it is a pollutant) and the stratosphere (the middle atmosphere, where it is not a pollutant).

Table 8.1 Average composition of gases in the troposphere and lower stratosphere

Gas	Per cent by volume
Nitrogen	78.08
Oxygen	20.94
Argon	0.93
Carbon dioxide	0.035
Neon	0.0018
Helium	0.0005
Methane	0.00017
Ozone	0.00006
Hydrogen	0.00005
Krypton	trace
Xenon	trace

Oxygen: oxygen (O_2) is the colourless, odourless, tasteless and highly reactive gas that supports life and is essential for aerobic respiration. The Earth's early atmosphere probably contained little oxygen, but after the evolution of the first oxygen-producing plants (around 1,900 million years ago) the proportion of oxygen in the air increased steadily. It combines with most other elements to form oxides, such as water (H_2O), carbon dioxide (CO_2) and silicon dioxide or quartz (SiO_2). Oxygen is essential for combustion, including the burning of fossil fuels, which provide so much of our energy. Ozone (O_3) is an allotrope (one of the physical forms) of oxygen. Oxygen is also a chemical element – the most abundant element in the Earth's crust, of which it makes up nearly half (see Table 4.5).

Most gases in the atmosphere remain relatively stable in concentration to a great height because of turbulent mixing below the mesosphere (see p. 229). Water vapour is the main exception, because it decreases in concentration with altitude.

It is important to note that some of the main greenhouse gases – particularly carbon dioxide and methane (see pp. 256–8) – are present naturally in the atmosphere, but at much lower concentrations than are measured today with global air pollution. Other greenhouse gases – particularly oxides of nitrogen and CFCs (see p. 259) – are produced directly by the air pollution.

The gas composition of the lower atmosphere varies a great deal from place to place because it is a sink for many chemicals, particularly those released by air pollution. Concentrations of many pollutants tend to be much higher close to emission sources, even though winds disperse pollutants across wide areas. Dispersion also causes the pollutants to be diluted, mixing them with a larger volume of air. This is the basis of the saying that 'dilution is the solution to pollution'.

WATER VAPOUR

The atmosphere also contains water vapour, water in the gaseous state, which is an important part of the water cycle (see p. 352). When water vapour condenses it turns into a liquid state to form mist, fog, rain, sleet, hail and snow (see p. 291).

Natural concentrations of water vapour in the atmosphere are small but variable between nearly zero and about 5 per cent by volume. Concentrations are usually higher above oceans, seas and other large water bodies (including large lakes). Rather like carbon dioxide, the importance of water vapour is much greater than its quantity suggests; it plays a large role in determining weather conditions and weather changes (see p. 304). As already noted, the concentration of water vapour decreases with altitude, from about 4 per cent close to the ground to more or less zero above about 12 km.

SOLIDS

The atmosphere also contains a variable amount of fine solid material, held up in suspension by wind systems and convection currents. This material comes from a variety of sources and comes in various forms, including:

- dust – for example from volcanoes and desert sandstorms;
- pollen and mould spores – from natural vegetation;
- smoke – from natural forest fires, and from air pollution;
- salt spray – from the oceans.

This atmospheric loading of particles is mainly natural in origin, but air pollution is adding to it and changing spatial distributions within it. An estimated 725 million kg of dust falls from the sky to the ground each day, more than 180 million kg of which (25 per cent) is generated by human activities. A high loading of atmospheric solids close to the ground can decrease visibility, increase haze, decrease temperatures, increase cleaning costs, make buildings look dirty and promote breathing difficulties (including asthma) in people.

STRUCTURE OF THE ATMOSPHERE

Most people have direct experience only of the lowest part of the atmosphere, which is close to the ground. This is the part of the gaseous envelope in which we live and move, the part that creates and changes weather, and the part that determines what conditions on the ground are like outside our artificially heated or cooled buildings. Above this thin layer, we watch the clouds move overhead and the planes fly by, but we have no direct contact.

BOX 8.4 MONITORING THE ATMOSPHERE

Our understanding of what the atmosphere is like above the lowest layer is based largely on data sent back from instrumented balloons, planes and rockets that have travelled up and out towards the edge of the atmosphere. Over time the picture has become much clearer.

We now know much more about the structure and dynamics of the Earth's atmosphere, thanks to sophisticated atmospheric monitoring equipment, satellite observations, missile probes and manned orbiting observation vehicles. NASA's Upper Atmosphere Research Satellite (UARS) illustrates the new technologies being developed. The satellite was launched in September 1991 into a near-circular orbit above the Earth at an altitude of 585 km. It sends back invaluable measurements, including vertical profiles of temperature, changes in many trace gases, wind speeds and patterns, and solar energy inputs.

Until quite recently, it was assumed that above about the first 5 km the air became thinner and colder at a constant rate with increasing altitude. It was also assumed that the upper atmosphere had hardly any influence on weather conditions on the ground. Both assumptions are now being questioned on the basis of recent observations (Box 8.4).

VERTICAL ZONATION

There is no definite outer limit to the atmosphere. It simply becomes thinner with increasing height above the ground until it eventually becomes part of the solar atmosphere (at an altitude of about 80,000 km) and merges into space, where near-vacuum conditions exist. Weather is mostly confined to the lowest 16 km of the atmosphere, although it is now known that air movements in the lowest 30 km affect weather patterns and changes.

The atmosphere is composed of a number of layers or zones (Figure 8.1), each of which is quite clearly defined and has distinctive physical properties (particularly temperature and pressure). Abrupt changes in temperature and pressure mark the boundaries between successive zones. It is convenient to think in terms of three main layers in the atmosphere:

1. the lower atmosphere – the troposphere;
2. the middle atmosphere – the stratosphere;

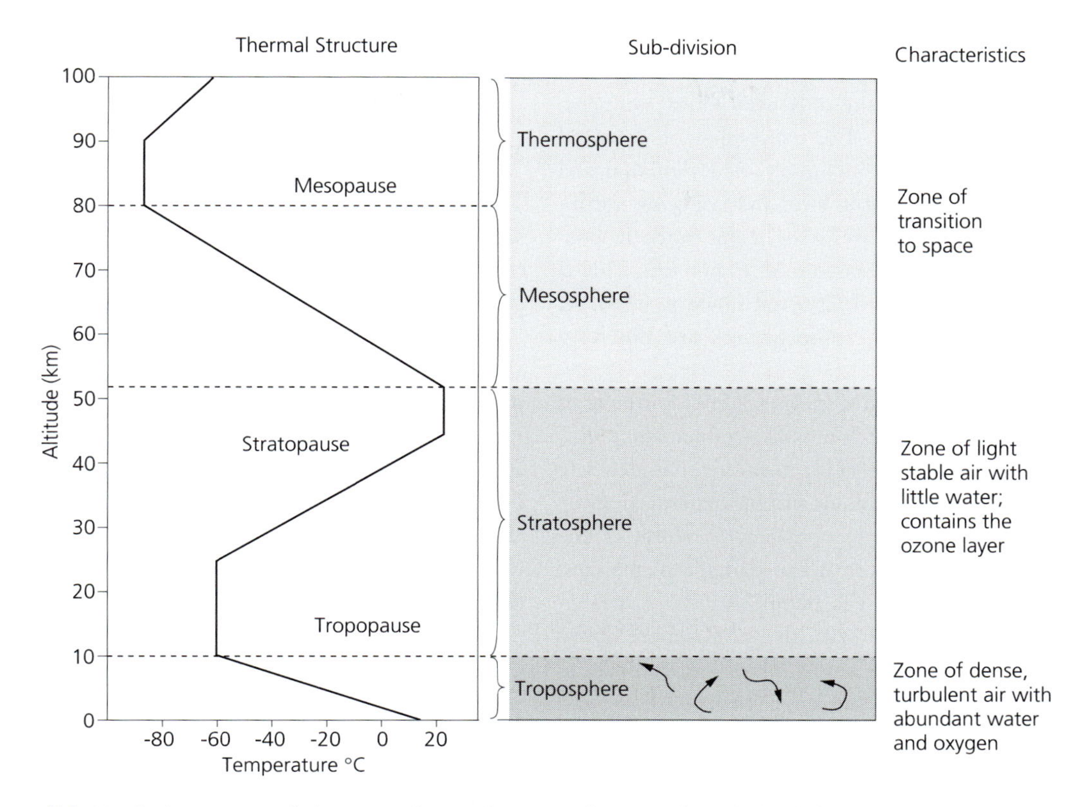

Figure 8.1 *Vertical structure of the atmosphere. The atmosphere can be sub-divided into three main vertical units, based on temperature. Most of the mass of the atmosphere is contained within the troposphere, in which temperature falls with height. See text for explanation.*

Source: Figure 8.2 in Marsh, W.M. and J.M. Grossa (1996) Environmental geography: science, land use and Earth systems. John Wiley & Sons, New York.

3. the upper atmosphere – the mesosphere, thermosphere and beyond.

TROPOSPHERE

The troposphere is the lowest layer in the atmosphere, closest to the ground. It extends up to about 8 km at the poles and 16 km at the equator, although the height varies through the seasons. Most of the troposphere lies within 10.5 km of the Earth's surface. Gravity pulls the atmosphere towards the Earth, and about 75 per cent of the total weight of the atmosphere is squashed into this lowest layer (which is why we have atmospheric pressure; see Box 9.6). As a result, air pressure is greatest closest to the ground, and it decreases with height. The troposphere also contains about 90 per cent of the moisture and dust within the atmosphere.

This is the most important part of the atmosphere for life on Earth, because weather patterns and weather changes originate in this zone, along with the wind systems

BOX 8.5 JET STREAMS

A series of strong winds, called jet streams, are found at the top of the troposphere. They follow large wave-like paths, which regularly shift location for largely unknown reasons and in largely unpredictable ways. The jet streams transfer heat from low to high latitudes, and shifts in their location are associated with major weather changes on the ground. They appear to cause movements in storm tracks, and to cause large air mass movements that allow very warm or very cold air to invade middle latitudes, producing cold spells or heatwaves accordingly. Heatwaves in the northern United States and southern Canada are often associated with a northward swing in the North American jet stream during the summer months, allowing warm tropical air to penetrate much further north across North America. Conversely, a southerly swing in the jet stream brings cold polar air and freezing conditions much further south.

and air currents that distribute heat, moisture and air pollutants from place to place (see p. 278). It is also the warmest part of the atmosphere, and the only part where temperatures are usually above 0 °C. The troposphere is warmed by the Earth, which absorbs incoming solar radiation and heats the overlying air. Except in local areas of temperature inversion, temperature decreases with height in the troposphere at an average rate of about 6.5 °C per km. This is called the normal or standard lapse rate (see p. 276). At the top of the troposphere, temperatures are down to around −60 °C.

Ozone occurs locally in the troposphere. Some of it comes from natural sources, but most is produced by pollution (particularly photochemical reactions involving vehicle exhaust fumes). It is a constituent of photochemical smog (see p. 247), it is probably responsible for some of the damage attributed to acid rain, it can damage plants, and it causes respiratory problems in people.

The troposphere interacts with the stratosphere above and with the land and oceans below, playing a vital role in linking together many important environmental flows and systems. It exchanges heat, energy, water, chemicals and aerosols with the Earth's surface. Atmospheric processes within the troposphere strongly influence movements and patterns of these important environmental factors across the Earth's surface.

The upper boundary of the troposphere is the tropopause, the level up to 18 km above the ground surface at which temperature stops decreasing with height and stabilises instead. Weather changes (Chapter 10) are largely confined to the troposphere below this zone of constant temperature, because rising air cannot continue to rise through it.

STRATOSPHERE

The stratosphere lies above the tropopause and extends up to the stratopause, about 60 km above the ground. Unlike the troposphere, the stratosphere is not uniform, and conditions change between the lower and upper parts of it.

The lower part of the stratosphere, up to about 35 km altitude, is a transitional zone above the tropopause in which air temperature is fairly constant in the region of −55 °C. Temperature rises with height in the upper stratosphere, between about 35 and 60 km, reaching about 0 °C at the stratopause. Because the air is very thin at this altitude, the conductivity of heat there is much lower than it is at the Earth's surface, so it would feel much colder than an equivalent temperature at ground level.

MESOSPHERE, THERMOSPHERE AND BEYOND

The outer part of the Earth's atmosphere contains two very different zones. Immediately beyond the stratopause lies the mesosphere, a zone stretching from about 50 to 80 km above the ground in which temperature decreases quite sharply with height down to about −90 °C.

The mesopause, at about 80 km, separates the mesosphere below from the thermosphere above. Temperatures rise again with height in the thermosphere. The thermosphere extends to about 350 km above the ground, where

BOX 8.6 THE STRATOSPHERE AND LIFE

The stratosphere is important to life on Earth in two critical ways. First, the most important global wind circulation systems are based in the stratosphere. This means that material (such as air pollutants or dust from explosive volcanic eruptions) that enters the stratosphere is likely to be distributed globally and to have much longer residence times than material that circulates within the troposphere. Much of the long-distance trans-frontier air pollution is driven by processes within the stratosphere.

Second, and more critically, the stratosphere contains the natural layer of ozone that screens the Earth's surface from biologically harmful ultraviolet rays from the Sun. Ozone is a minor constituent of the atmosphere overall, accounting for less than a millionth of the total volume (Table 8.1). But it is a highly reactive gas, and most of it is found in the stratosphere, particularly between about 20 and 26 km above the ground. Life on Earth would simply be impossible without this vital ozone shield. If the ozone filter was not there, or if it was not as effective in blocking the ultraviolet radiation, the Sun would burn our skin (causing skin cancer) and blind our eyes (causing cataracts). Many other species would also suffer, and their survival would be put at risk. This is why evidence of a decline in stratospheric ozone since the mid-1980s, associated with air pollution (particularly by CFCs), has been taken very seriously indeed (see p. 249).

BOX 8.7 THE IONOSPHERE

The outer parts of the atmosphere (between about 60 and 1,000 km above the Earth) were previously referred to as the ionosphere, because one of their most important characteristics is a high concentration of free electrons formed as a result of ionising radiation entering the atmosphere from space. Three regions were identified within the ionosphere:

1. *D region*: the lowest region of the ionosphere (60 to 90 km), which contains a low concentration of free electrons and reflects low-frequency radio waves.
2. *E region*: the middle region of the ionosphere (90 to about 150 km), which reflects radio waves of medium wavelength.
3. *F region*: the highest region of the ionosphere (about 150 to about 1,000 km), which contains the highest proportion of free electrons and is the most useful region for long-range radio transmission.

This part of the atmosphere is highly charged, and spectacular lights and colours (usually green, red or yellow) in the sky – called aurorae – can sometimes be seen within it, particularly in polar regions. These are the *aurora borealis* (northern lights) in the northern hemisphere and the *aurora australis* in the southern hemisphere. They are created when streams of charged particles from the Sun collide with molecules of air in the upper atmosphere and are deflected towards the magnetic poles. The collisions cause the particles to change their electric charge and glow, rather like particles in a neon tube.

temperature may be as high as 1,100 °C. The air is extremely thin at that altitude, so there is little conductivity of heat.

Beyond the thermosphere – in the exosphere – the air is extremely thin, but there are still traces of some gases (particularly hydrogen) as far out as 8,000 km. The exosphere has no clearly defined outer limit but instead fades off into the vacuum of space.

THE ENERGY SYSTEM

SUNLIGHT

All life on Earth is ultimately dependent on energy from the Sun, and solar energy is the basic source of power for all the Earth's environmental systems (see p. 4), including the atmospheric and ocean systems, and ecosystems. The quantity and distribution of energy from the Sun are vitally important for all species, including humans, and any changes (natural or human) that affect the receipt of solar radiation at the Earth's surface can alter the ability of the Earth to sustain life.

SUNSHINE AND PEOPLE

Sunlight affects people in many different ways. Once absorbed, it is used for plant growth and provides our daily bread. It drives the weather systems, which affect us both directly and indirectly.

Sunshine determines the character of a place. The sunniest place on Earth, the eastern Sahara Desert in North Africa, receives sunshine on more than 97 per cent of its daylight hours in a typical year. This contrasts sharply with the North Pole, the least sunny place, where no sun is recorded for winter periods up to 186 days long. The passage of day and night control the very rhythm of our daily lives. Lack of sunlight may even lead to problems such as vitamin shortage and seasonal affective disorder (winter depression). Ultraviolet radiation in sunlight is known to cause skin cancer and other health risks in humans. Diffuse solar radiation might affect the incidence of some diseases, such as multiple sclerosis, which attacks the central nervous system.

ENERGY

Sunlight can be converted into electricity, and solar heat can be used directly to warm buildings and water, desalinate water and cook food. Although we make use of sunlight both directly (for example in space heating through windows) and indirectly (for example through solar panels), it remains a largely untapped renewable source of energy.

The total amount of energy potentially available each year at the Earth's surface is estimated to be more than 20,000 million times greater than present energy consumption, and most of this is solar energy. This highlights the folly of depending so heavily on scarce non-renewable resources (such as oil, gas and coal) and explains why so much effort is being invested in developing new and more efficient methods of using solar radiation (see p. 239).

ENERGY FROM THE SUN

SOLAR CONSTANT

The Sun generates energy by nuclear fusion reactions that burn hydrogen into helium in its interior (see p. 106) and these reactions release vast amounts of energy into the solar system. Each square metre of the Sun's surface radiates an estimated 73.5 megawatts of energy. This energy is radiated out into space, and everything that lies in its path (including the planets) receives a proportion of it. Planets close to the Sun receive more energy than those further away. As a direct consequence, those planets that are closest to the Sun are too hot for life, and those furthest away are too cold. The Earth – 150 million km from the Sun – lies within the narrow zone in which solar energy receipt is large enough to sustain life but not so large as to endanger it.

THE ELECTROMAGNETIC SPECTRUM

We see sunlight as visible light, but in reality this is only a part of the solar radiation that reaches the Earth from the Sun. Much of the energy is radiant heat composed of wavelengths that cannot be detected (literally *seen*) by human eyes.

The Sun radiates energy across the whole electromagnetic spectrum, only a portion of which is visible. The electromagnetic spectrum (Figure 8.2) comprises waves of energy of different sizes or wavelengths (Table 8.2).

BOX 8.8 THE SOLAR CONSTANT

The outer edge of the Earth's atmosphere receives about 1,360 watts per square metre (W m^{-2}) of energy from the Sun. This is known as the 'solar constant', although it is not absolutely constant. The solar constant in fact varies a little over time, depending on:

- *Variations in solar output*: this is the total amount of energy released by the Sun, and it changes through the sunspot cycle (see Box 4.7).
- *Variations in the Earth's rotation around its axis*: these are cyclic and are associated with minor variations in climate (such as the Little Ice Age.
- *Variations in the Earth's orbit around the Sun*: these are cyclic and are associated with large-scale long-term climate change (see p. 335).

Different wavelengths have different properties and affect the Earth, its people and its environments in different ways. Long wavelengths are low-frequency and thus relatively low-energy, and as a result they are less damaging to people. Short wavelengths, on the other hand, are high-frequency, high-energy and damaging to people.

We see wavelengths between about 0.4 and 0.77 μm as visible light, and different wavelengths within that range appear as different colours of light. Short-wave radiation within the visible range appears as violet, and long-wave appears as red (Table 8.3). Rainbows display this colour sequence quite graphically.

The Sun radiates most intensely in the visible part of the spectrum, with a peak at around 0.5 μm in the green part

Table 8.2 The electromagnetic spectrum

Type of wave	*Wavelength or frequency*	*Typical applications*
Radio waves	3 to 30 GHz	Used to transmit radio and television
Microwaves	1 to 35 GHz	Used in radar (longer waves) and in cooking (shorter waves)
Infrared waves	1,000 to 0.77 μm	Heat waves, such as the energy radiated by the Sun
Light	0.39 to 0.77 μm	The visible portion of the spectrum
Ultraviolet light	0.39 to 0.01 μm	Highly energetic radiation visible to the naked eye
X-rays	0.01 to 0.0000003 μm	Even more energetic than ultraviolet light, and these can penetrate many substances, e.g. medical applications for photographing bones and detecting flaws in metal objects
Gamma rays	Down to 0.00000001 μm	Have very high energy levels and can pass through lead

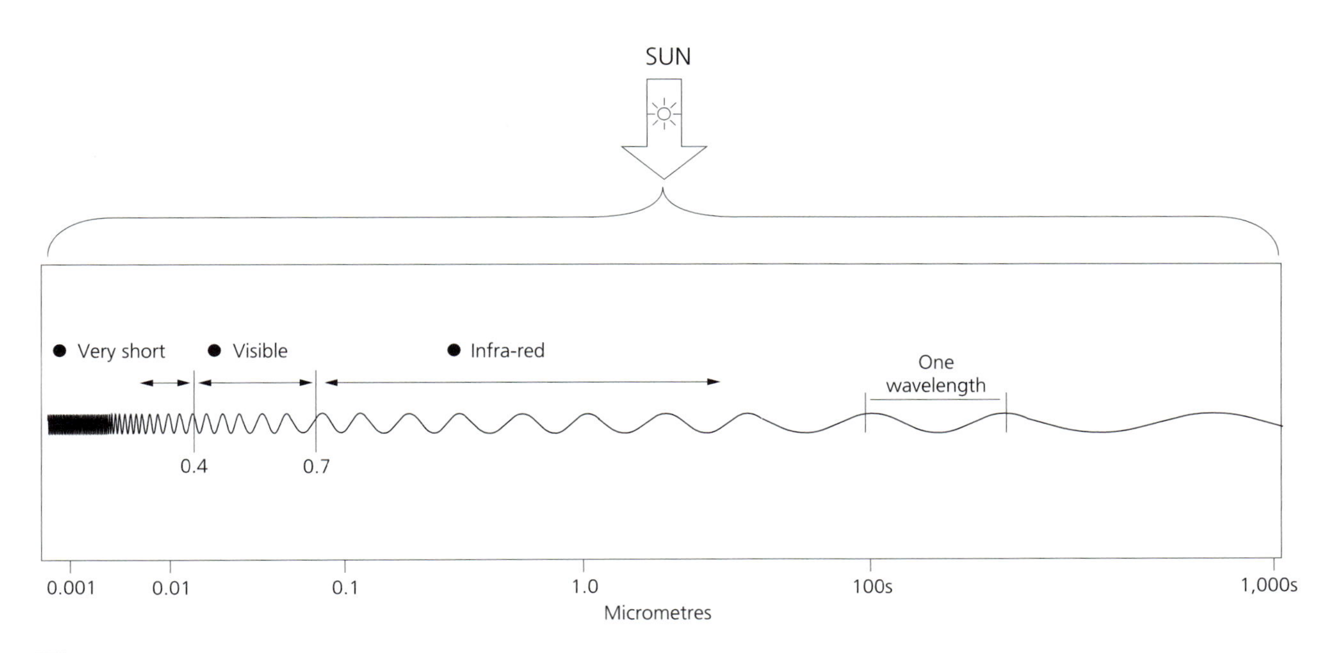

Figure 8.2 *The electromagnetic spectrum. There are three classes of radiant energy – very short wavelength (including ultraviolet, X-rays and gamma rays), visible (representing the colour spectrum from blue to red), and infrared (long wavelengths that are subject to absorption by carbon dioxide and water vapour).*

Source: Figure 8.6 in Marsh, W.M. and J.M. Grossa (1996) Environmental geography: science, land use and Earth systems. John Wiley & Sons, New York.

BOX 8.9 MEASURING WAVELENGTHS OF RADIATION

The wavelength of radiation is often measured in Ångstroms (Å), and there are 10,000 million in 1 metre (Å = 10^{-10} m). Scientists now prefer to express wavelengths in micrometres (μm), and there are 10,000 Å in 1 μm. Very long wavelengths are usually expressed in terms of frequency: the SI unit the hertz (Hz) describes the number of cycles per second. A gigahertz (GHz) is 1,000 million Hz.

Table 8.3 Composition of the visible light portion of the electromagnetic spectrum

Colour	*Wavelength (Å)*	*Wavelength (μm)*
Violet	3,900–4,550	0.390–0.455
Blue	4,550–4,920	0.455–0.492
Green	4,920–5,770	0.492–0.577
Yellow	5,770–5,970	0.577–0.597
Orange	5,970–6,220	0.597–0.622
Red	6,220–7,700	0.622–0.770

of the spectrum. Beyond visible red lie the infrared wavelengths (0.7 to 1,000 μm or 1 mm), and beyond that the long wavelengths represented by microwaves and radio waves.

ABSORPTION AND REFLECTION

The composition of the radiation that reaches the edge of the Earth's atmosphere is similar to that which left the Sun. As it passes through the Earth's atmosphere some wavelengths are filtered out or absorbed, so the composition of the radiation that reaches the Earth's surface differs from that at the edge of the atmosphere.

Very short-wave (high-energy) gamma and X-rays are absorbed in the upper atmosphere, and none reaches the surface. Much of the radiation in the ultraviolet (UV) range (0.2–0.4 μm) is filtered out by atmospheric gases, particularly at wavelengths below 0.29 μm, where most of the UV – which causes sunburn – is absorbed by stratospheric oxygen (O_2) and ozone (O_3). The atmosphere is transparent to wavelengths longer than 0.29 μm, but water vapour absorbs energy in a series of narrow bands between 0.9 and 2.1 μm.

As well as its composition, the amount of solar radiation also changes as it passes through the Earth's atmosphere. Less than half of the radiation that arrives at the edge

BOX 8.10 SCATTERING IN THE ATMOSPHERE

Scattering in the atmosphere, called Rayleigh scattering, reflects radiation in all directions and gives the sky its colour. On a clear day, when the Sun is high in the sky, violet light is scattered and absorbed high in the atmosphere and blue light is scattered and absorbed below it. Scattering diffuses the blue light, so the sky appears blue. Dust particles in the atmosphere scatter light of all wavelengths, and the sky then appears white. When the Sun is low in the sky (for example, towards sunset) dust particles scatter light in the orange and red wavelengths, so the sky appears orange or red.

of the atmosphere (the solar constant; see Box 8.8) eventually reaches the Earth's surface. Most of what is lost is reflected back into space and has no further impact on the Earth or its atmosphere. About 10 per cent of what is lost is absorbed or scattered (Box 8.10) by ozone, water vapour and particles of dust in the atmosphere.

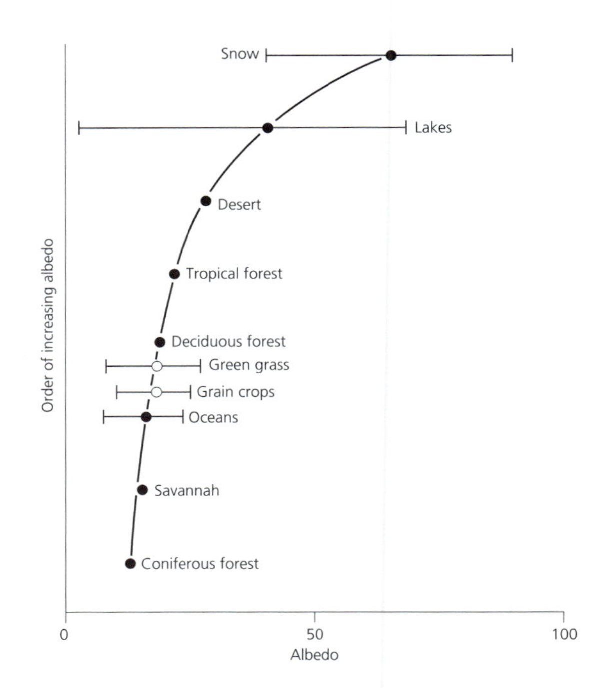

Figure 8.3 *Variations in albedo for different kinds of surface. Albedo means reflectivity, and surfaces with an albedo of 0 have no reflection, while those with an albedo of 100 have total reflection. Most natural surfaces have albedos of less than 50 per cent.*

Source: Figure 14.6 in Dury, G.H. (1981) Environmental systems. Heinemann, London.

ALBEDO

When we look at objects – such as trees, people or perhaps books – we see them as different colours, but some appear to be much brighter or duller than others. The colours we see are determined by the *wavelengths* of the light they reflect. Their brightness is determined by the *amount* of light they reflect. Fresh snow reflects up to 90 per cent of the light that falls on it, so it appears very bright compared with grass, for example, which reflects back only about 20 per cent of the light and thus appears relatively dark (Figure 8.3).

The proportion of light reflected by a surface is its *albedo* or reflection coefficient, and this is usually expressed as a percentage. Some typical albedo values are shown in Table 8.4.

Albedo is important not just from a visual point of view: it also determines the amount of radiant energy that an object absorbs or reflects, which in turn affects how much that body is heated. The Earth itself has an albedo of 31 per cent, partly because so much of the planet's surface is water and much of the rest is often covered by clouds (both of which are highly reflective). An understanding of albedo is useful in the design of solar heating systems (see p. 272).

Human activities can alter albedo by changing the character of the surface from which reflection takes place (Box 8.11). This, in turn, can promote climate change, which might further alter albedo, setting in motion a chain reaction involving feedback (see p. 79) that promotes or suppresses further change. The two main ways in which people alter albedo is via air pollution and land-use change.

Air pollution often leads to a change in cloud cover or cloud character, and this can directly alter albedo and thus atmospheric heating. Examples include the release of vast amounts of water vapour from power station cooling towers, the impact of cities on micro-climate (see p. 333), and the possible consequences of greenhouse gases on global circulation patterns (see p. 284).

Table 8.4 Some typical albedo values

Material	*Albedo (per cent)*
Fresh snow	80–90
Cumulonimbus cloud	70–90
Desert	25–30
Grass	18–25
Concrete	17–27
Tropical rainforest	7–15

BOX 8.11 LAND USE AND ALBEDO

Land-use changes alter albedo, sometimes in rather indirect but significant ways. The clearance of tropical rainforest to grow field crops, for example, might not reduce albedo much because both surfaces have albedos of between about 5 and 15 per cent. But a change from forest to pasture might double the albedo and set in motion a chain of adjustments involving negative feedback, which would ultimately stabilise conditions again:

1. the ground would then absorb less heat, because more light is reflected away;
2. the evaporation rate would fall;
3. there would be less cloud cover;
4. this in turn would reduce average cloud albedo;
5. this would increase the amount of radiation that reaches the surface;
6. the ground would then absorb more heat and be warmed;
7. the evaporation rate would rise; and
8. cloudiness would increase again.

Such feedback relationships should be taken into account when the climatic and environmental impacts of human activities are being assessed and predicted, but they can be extremely complicated.

THE EARTH'S ENERGY BUDGET

Energy drives environmental systems and flows through them, but it cannot be recycled (see p. 76). Strictly speaking, therefore, there is no energy cycle but rather a system. The Earth operates as an open system (see p. 70) because it receives a non-stop supply of energy from the Sun, which drives the planet's biological and environmental systems and warms its lower atmosphere enough to sustain life. In this section, we focus on the character of this natural energy system and on ways in which it can be altered by nature or people.

THE NATURAL ENERGY SYSTEM

We have already established some important characteristics of the Earth's energy system (Box 8.12), including the abundant provision of energy by the Sun. To illustrate that abundant provision, it is interesting to note the amount of energy from the Sun that reaches the outer edge of the Earth's atmosphere each year. Measured in kilowatt hours (kWh), which is defined as the energy expended when 1 kilowatt is available for 1 hour, the Earth receives an estimated 1.56 million million million kWh (1.56×10^{18}) a year. This sounds a lot, and indeed it is – it is equivalent to about 195 million million tonnes (1.95×10^{14}) of coal each year. For comparison, the total recoverable reserves of coal in the world in 1990 were estimated at just under 1 million million tonnes.

BOX 8.12 CHARACTERISTICS OF THE EARTH'S ENERGY SYSTEM

- The Earth receives a great deal more radiant energy from the Sun than it actually uses.
- The quantity and composition of that incoming solar radiation is changed a great deal as it passes through the Earth's atmosphere.
- The Earth's atmosphere acts as a vitally important filter that removes some of the more harmful wavelengths and thus makes life on the planet possible and sustainable.

The fate of all this renewable solar energy is known in general terms, but much research is being directed to collecting better measurements of energy flows within the atmosphere, often using remote sensing and satellites such as Nimbus. The object of such research is to build up a better understanding and construct more realistic simulation models of the Earth's energy budget and of how the atmosphere affects and is affected by energy flows and transfers. This is driven partly by scientific curiosity (the need to know how the world works), but its urgency is increased by the need to apply such understanding to solving global environmental problems such as the global warming associated with air pollution by greenhouse gases.

Table 8.5 The fate of incoming solar energy in the Earth's atmosphere

Fate	Per cent	Details	Per cent
Returned to space	32	Reflected back into space off clouds	21
		Reflected back into space from the ground	6
		Scattered by tiny particles in the air (diffuse radiation)	5
Reaches the ground	50	Direct radiation	30
		Diffuse radiation (scattered in the atmosphere)	20
Absorbed in the atmosphere	18	Absorbed by clouds; absorbed by dust and the	3
		ozone layer	15
Total	100		100

Roughly half of the incoming solar radiation reaches the Earth's surface (Table 8.5), is absorbed there and heats it (Figure 8.4). Some of this (30 per cent) arrives as direct radiation, and the rest (20 per cent) arrives as diffuse radiation having been scattered in the atmosphere on its way to the ground. Just under a third of the incoming solar radiation is reflected back into space and is not directly involved in heating the Earth or its atmosphere. Most is reflected off clouds, a much smaller amount is reflected off the ground, and a similar amount is scattered by tiny particles in the air and returns to space as diffuse radiation. The remaining incoming solar radiation (18 per cent) is absorbed within the atmosphere by clouds, dust and the ozone layer.

Plate 8.1 *Testing the efficiency of technologies to convert sunlight into usable energy, using photovoltaic cells. This particular experiment has been operating at the Centre for Alternative Technology at Machynlleth in West Wales.*
Photo: Chris Park.

BUDGET AND BALANCE

The precise details of this energy budget vary from place to place and over time, and the values quoted are long-term global averages. But the overall balance must be maintained, or the Earth would become progressively hotter or colder.

The amount of energy received from the Sun is controlled by various factors, including:

- length of day;
- the angle at which the Sun's rays strike the Earth (Figure 8.4);

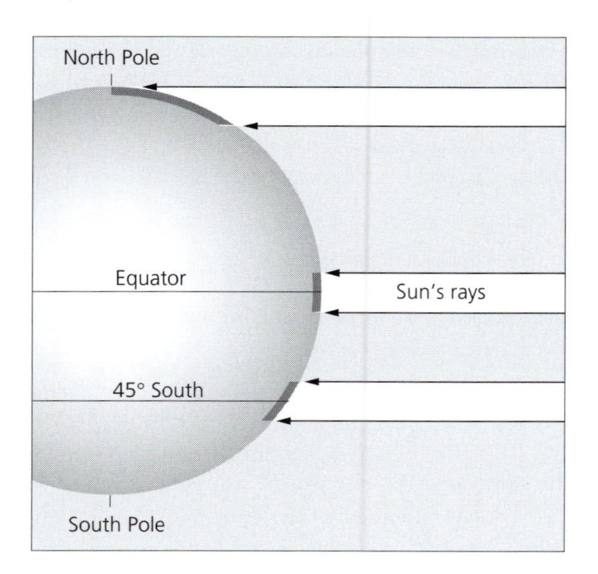

Figure 8.4 *Effectiveness of the Sun's rays in heating the atmosphere at different latitudes. At the equator, where the Sun is directly overhead, the Sun's rays strike the Earth's surface at a high angle (around 90°), so heating is concentrated. At higher latitudes, the Sun's rays strike at a lower angle, so they are more diffuse.*

■ conditions within the Earth's atmosphere.

The Earth's surface is warmed by the absorption of radiant energy from the Sun that strikes its surface. The Earth's atmosphere is warmed mainly by heat re-radiating from Earth back to space. Energy receipt is not even across the Earth's surface, and there are quite clear spatial patterns. Radiation is most intense at the equator, where

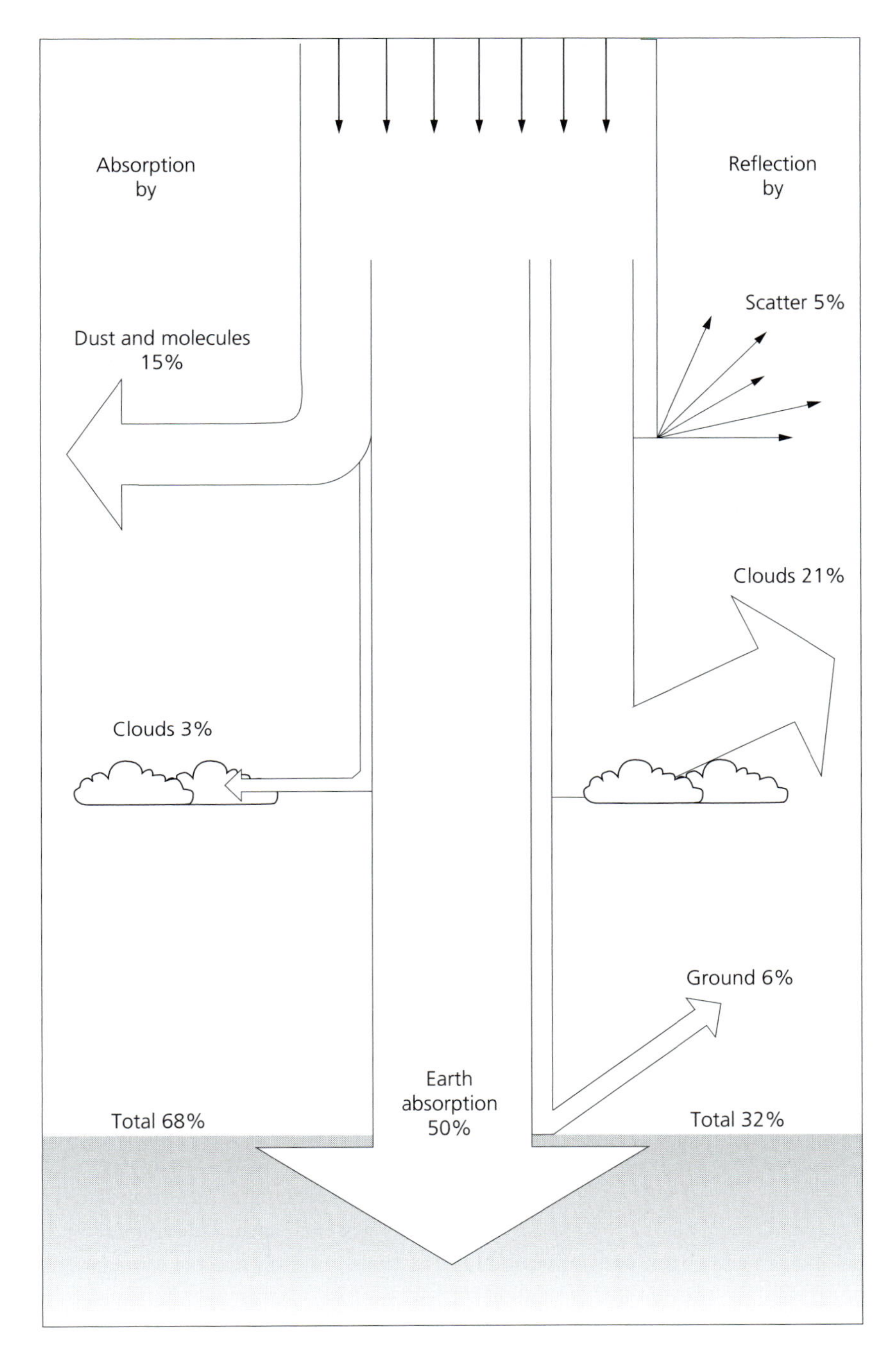

Figure 8.5 *The fate of incoming solar radiation. The figure shows the average effect of various factors that reflect and absorb incoming solar radiation. About half of the energy received at the Earth's surface is eventually released to the atmosphere and re-radiated back into space.*

Source: Figure 4.3 in Doerr, A.H. (1990) Fundamentals of physical geography. Wm. C. Brown Publishers, Dubuque, Iowa.

the Sun is more directly overhead. With increasing latitude, the Sun is lower in the sky at all seasons and its radiation is spread over a larger area.

The distribution of radiation is determined mainly by latitude, but cloudiness is also an important factor. Hence the equator – which is shaded by clouds (which reflect incoming sunlight) for much of the time – does not receive the most intense insolation. Tropical and subtropical deserts have much clearer skies, and they receive up to twice as much insolation as the equator. Dry continental interiors tend to be much sunnier than relatively cloudy maritime areas.

The flow and fate of solar energy in the Earth's atmosphere is without doubt the most important control over most of the world's environmental systems. Solar energy heats the atmosphere and oceans, particularly around the equator, and the redistribution of this energy to higher latitudes effectively drives the global circulation of the atmosphere (see p. 286) and oceans (see p. 480).

Predicting future climate variations is complicated because of the large number of possible external and internal forcing factors (Box 8.13) involved and the many different ways in which the atmospheric system might react to them.

BOX 8.13 EXTERNAL AND INTERNAL FORCING

The Earth's energy system can be altered by external factors (so-called external forcing) and internal factors (so-called internal forcing).

External forcing The most likely causes are:

- changes in the Earth's orbital characteristics (which alter the amount of solar radiation over geological timescales);
- changes in solar output (short-term changes are associated with sunspot cycles; longer changes may be associated with long-term changes in solar dynamics).

Internal forcing The most likely causes include:

- changing the albedo (reflectivity) of the Earth's surface, perhaps associated with urbanisation, land-use change or desertification. Natural causes of albedo change include melting or extension of the polar ice caps;
- changing cloudiness caused by natural climatic variation;
- changing levels of dust and aerosols in the atmosphere, perhaps because of explosive volcanic eruptions such as Krakatau in 1883 (see Box 6.21) or Mount Pinatubo in 1991 (see Box 6.28). Smoke and dust thrown up during the Gulf War in 1991 might also cause significant short-term atmospheric disruption and disturbance;
- air pollution by carbon dioxide and other greenhouse gases (see p. 255).

THE NATURAL GREENHOUSE EFFECT

BLACK BODY TEMPERATURE

The Earth as we experience it is a much warmer planet than it would be if it had no atmosphere. It is possible to calculate the so-called 'black body' temperature of the planet, using the Stefan–Boltzmann equation (Box 8.14). That black body temperature (the temperature that its surface would be if it were not warmed by its own atmosphere) is −23 °C. But the actual surface temperature of the Earth is about 15 °C. The difference (38 °C) is the amount by which the planet is warmed by the absorption of radiation within its atmosphere.

GREENHOUSE WARMING

This warming turns an otherwise frozen and inhospitable planet into our home, and it is the natural greenhouse effect. Without it life on Earth would be simply impossible – all water would be permanently frozen, temperatures would be too low for plants and animals to survive, and none of the natural biogeochemical cycles would operate.

The Earth is a much cooler body than the Sun, and – in accordance with Wien's law (Box 8.14) – it radiates energy of much longer wavelengths. Thus incoming solar radiation is short-wavelength, and re-radiated energy from the Earth is long-wavelength. This long-wave energy from the Earth is stored for a time in the atmosphere, trapped by gases that prevent it escaping back into space. The trapped long-wave energy heats the atmosphere, producing a rise in the Earth's temperature – the 'greenhouse effect'. Glass walls in a garden greenhouse trap heat in a similar way, hence the name, although the analogy is rather weak and much disputed.

BOX 8.14 PRINCIPLES OF ENERGY RADIATION

The Earth's energy system is much easier to explain if we have some understanding of two important principles:

The Stefan–Boltzmann equation The rate at which a body (such as a tree, a mountain or even an entire planet) emits radiation is a function of its temperature. The higher the temperature, the higher the rate of emission. The relationship is not linear but exponential. This means that the rate of emission increases at an increasing rate with temperature (as temperature increases, the rate of emission increases at a faster pace). Total emission is calculated by the Stefan–Boltzmann equation, which states that the energy (E) radiated from a body increases with the fourth power of its temperature (T) times a constant (σ);

$$E = T^4 \sigma$$

Wien's displacement law Most solar radiation has a relatively short wavelength, with a peak at about 0.5 μm, whereas most radiation from the Earth has a much longer wavelength, around 10 μm. The Earth receives energy as relatively short-wave radiation from the Sun, but it returns radiation as relatively long waves in the infrared part of the spectrum. Wien's law (its full name is Wien's displacement law) states that the wavelength of maximum-intensity radiation increases as the absolute temperature of the radiating body decreases. Thus a hot body like the Sun emits short-wavelength radiation, whereas a relatively cool body like the Earth emits longer-wavelength radiation.

GREENHOUSE GASES

Nitrogen and oxygen are the two most abundant gases in the atmosphere (see Table 8.1), and they are largely transparent to radiation at wavelengths longer than 0.29 μm. Some gases absorb some of the incoming (short-wavelength) solar radiation (Box 8.15).

A number of atmospheric gases absorb radiation at wavelengths longer than about 4 μm, and this is important because the Earth emits radiation at between 4 and 100 μm, with a peak at around 10 μm. About 94 per cent of this outgoing long-wave radiation is absorbed in the atmosphere by greenhouse gases – particularly carbon dioxide (CO_2), methane (CH_4) and water vapour (H_2O) – and it heats the atmosphere and the Earth's surface. The remaining 6 per cent of outgoing long-wave radiation escapes into space because it is of wavelengths between those at which it can be absorbed (the so-called 'atmospheric window' at about 8.5 μm and 13.0 μm).

BOX 8.15 GREENHOUSE GASES THAT ABSORB SHORT-WAVELENGTH RADIATION

- About 4 per cent of the total is absorbed by stratospheric ozone.
- About 20 per cent (in the infrared wavelengths) is absorbed by carbon dioxide.
- About 13 per cent is absorbed by water vapour (in three infrared wavebands of about 1.5 μm, 2.0 μm and 2.5–4.5 μm).
- About 6 per cent is absorbed by water droplets and dust.

SOLAR ENERGY RESOURCES

Solar energy is energy derived from the Sun's radiation, which (as we have seen) allows all life on Earth to exist. Nearly all the energy we use is ultimately solar, except perhaps for tidal energy (which is generated mainly by the gravitational pull of the Moon) (see p. 491) and nuclear energy (see Figure 1.5). Fossil fuels (see Box 7.6) represent solar energy captured by photosynthesis and stored over geological timescales in sedimentary deposits. Biomass fuels represent current growth cycles of plants that have captured solar energy by photosynthesis. Hydropower (see p. 394) and wind power (see p. 290) are generated from air and water circulations, which are ultimately driven by solar heating.

In this section, we examine the more direct use of solar energy as a renewable energy source. Solar energy has been exploited since the earliest development of settlement agriculture, but only in recent decades has much scientific and technical effort been invested in harnessing the power of the

Sun. Sunlight is widely used for industrial and domestic purposes, both for heating and for generating electricity.

Solar radiation arrives at the Earth's surface at a fairly low temperature, so sunshine is often thought of as low-grade energy that is not suitable for most uses. It can be made much more useful by concentrating the energy, using solar collectors.

SOLAR COLLECTORS

Solar collectors upgrade sunlight by concentrating it. One of the most spectacular examples of this technology is the huge solar furnace built at Odeillo in the French Pyrenees in 1970. It is based on a giant bowl-shaped mirror system, composed of thousands of mirrors that focus the Sun's rays to produce intense heat, which can be used for industrial, scientific or experimental purposes. It is a sustainable, renewable, non-polluting and relatively reliable energy source with a high capital cost of setting-up but small recurrent running costs.

More conventional solar heaters are much smaller, produce much less energy and serve individual buildings. Typical systems are based on heat-absorbing (usually black) panels containing pipes through which air or water is circulated, either by thermal convection or by a pump. The air or water is heated as it passes through the panel, and the heat is usually circulated around a building through pipes. Such collectors require a lot of sunshine to work efficiently, and are not really suited to high latitudes. The energy produced is relatively expensive because of the high capital cost of installing the heater systems.

SOLAR (PHOTOVOLTAIC) CELLS

Solar energy can also be harnessed indirectly using solar cells. These are made from panels of a semiconductor material (usually silicon), which generate electricity when illuminated by sunlight. Solar cells can be used at any latitude, and large installations are required because of the low efficiency of the technology (about 15 per cent efficient at present). The energy they produce is relatively expensive. With the exception of some exotic experimental solar-powered vehicles, their use is largely confined to space probes and artificial satellites.

SOLAR HEATING

Traditional approaches to heating buildings have relied heavily on passive solar heating, particularly in sunny places.

BOX 8.16 ACTIVE AND PASSIVE SOLAR HEATING

Passive solar heating uses solar energy to heat buildings directly, by trapping it within the structure of the building and then releasing it slowly (see Figure 9.1). It relies on natural air flow rather than pumps to distribute the heat within the building. Glass greenhouses and conservatories are good examples.

Active solar heating uses solar energy to heat buildings indirectly, by heating water, which is then circulated around a building by pumps and pipes (see Figure 9.2). Solar collectors are a form of active solar heating system.

More recently, technologies have been developed to pump heat around buildings, using active solar heating (Box 8.16).

Many solar energy systems, based on both active and passive heating, have been built into new and existing buildings in many countries. In some areas, whole neighbourhoods are using solar energy provided through centralised solar schemes. Nearly 2,000 houses are provided with electricity from a solar-powered boiler (which uses 1,800 mirrors) in the Mojave Desert in California. Even places that are not usually regarded as sunny can exploit solar power with the right technology. Such a system, which uses efficient solar collectors to heat water, which is then stored in an underground cavern, provides hot water and space heating for a community of 550 homes in northern Sweden.

Electricity generated from solar power is also being provided for some communities. The small, remote village of Long Dog in northern Ontario is a long way from the nearest mains electricity supply, and a local solar energy scheme is being tested there. An experimental solar village, called Lykovryssi, which incorporates many solar energy technologies, has been built near Athens.

Solar energy offers many benefits over energy generated from fossil fuels or nuclear sources, and its use may be expanded considerably in the future. Although costs are quite high and many technical problems remain, solar power does appear to have a bright future. Solar power is the most evenly distributed energy source, and one that can be harnessed with relatively few modifications in building construction and design. It is also the most sustainable and easily captured form of energy. One hour of sunlight could provide all the energy the world needs for a whole year!

The potential is there, but present technologies are simply not adequate to exploit it properly.

A great deal of effort and resources is being invested in research into 'solar houses' and 'solar architecture', involving a combination of approaches and technologies, including:

- solar panels
- photovoltaic systems
- conservatories, greenhouses and windows
- sensitive control mechanisms to regulate heat
- heat pumps and reservoirs
- good insulation.

BOX 8.17 SOLAR HEATING SYSTEMS – FUTURE POTENTIAL

There is great scope for expanding the use of solar heating systems because:

- The energy supply is inexhaustible.
- The technology is already fairly well developed, and it is likely to become cheaper and more efficient in the future.
- This type of energy use produces less pollution than burning fossil fuels or running nuclear reactors.
- Most places receive enough sunlight to make this approach possible, even if it supplements other forms of heating system, which can be switched on when there is insufficient sunlight.

Planners are starting to recognise the need to safeguard access to solar exposure around properties, because otherwise the Sun might be shaded out by adjacent buildings and structures. Questions are also being asked about the ways in which cities might develop in the future if more use is made of solar power. Some major changes would be required, including changes in transport systems and in patterns of working and living. But these might be better than the alternative, a continued reliance on dwindling, polluting and more costly fossil fuels.

AIR POLLUTION AND ENVIRONMENTAL CHANGE

AIR POLLUTION

NATURAL AND ANTHROPOGENIC POLLUTION

Air pollution is the contamination of the atmosphere with substances that, because of their nature or quantity, cannot be absorbed by natural environmental flows and cycles. Although there are a number of important natural sources of pollution (Box 8.18), the term 'pollution' is usually applied to substances released into the atmosphere as a result of human activities, which can be either:

- deliberate: such as the continual release of gases from factory chimneys; or
- accidental: such as the release of radioactive material from the damaged Chernobyl nuclear power station in 1986 (see Box 8.23), and the 1984 Bhopal explosion (see Box 8.24).

BOX 8.18 NATURAL AIR POLLUTANTS

A variety of natural sources release substances that can overload the atmospheric system locally, including dust, gases and aerosols from volcanic eruptions, forest fires and sea spray. While such materials are often referred to as natural pollutants, they are much less of a problem than pollution caused by humans (anthropogenic pollution) because:

- They do not release material that cannot be recycled in the biogeochemical cycles (see p. 86).
- They tend to be short-term problems (such as the dust released from the Mount Pinatubo volcanic eruption in 1991; see Box 6.28) and – unlike some other pollutants – do not built up progressively to major problems that can exceed the ability of environmental systems to cope.
- They tend to be released from particular places (point sources), whereas anthropogenic pollution is much more widespread and continuous.
- They are really part of nature and thus one cost of living on Earth, compared with human-caused pollution, which could – at least in theory – be eliminated completely.

For the rest of this chapter, 'pollution' is taken to refer specifically to anthropogenic pollution, i.e. that created by human activities.

HUMAN HEALTH

Some air pollution involves the release of toxic substances into the atmosphere (Table 8.6). Many toxic substances pose problems because they are:

- persistent (they survive a long time in the environment before being broken down)
- mobile (they can be transported long distances in air and water)
- harmful to people and wildlife.

Acutely toxic substances are sometimes released in accidents at chemical plants or stores, or while chemical materials are being transported by road or rail. At least 10 per cent of a sample of 317 urban areas in the USA experienced at least one toxic release a year during the 1980s.

Table 8.6 Types of toxic substance

Type	*Definition*
Toxin	Any of a group of poisonous substances produced by micro-organisms that stimulate the production of neutralising substances (antitoxins) in the body
Carcinogen	A substance that produces cancer
Teratogen	A substance that causes malformations in a foetus, including ionising radiation
Mutagen	A substance that can induce genetic mutation

Plate 8.2 *Photochemical smog in Mexico City. Air pollution, mainly from vehicle exhausts and industrial sources, is a serious problem in this rapidly growing city situated in a high basin surrounded by high mountains, which trap the polluted air and inhibit natural ventilation.*

Photo: Chris Park.

BOX 8.19 AIR POLLUTION AND HUMAN HEALTH

A number of serious and well-documented air pollution disasters – including those in London (1952), the Meuse valley (1930) and Donora, Pennsylvania (1948) – showed that extremely high levels of particulate-based smog could produce large increases in the daily mortality rate. Analysis of hospital records and death certificates in Philadelphia (1973–1980) showed strong associations between air pollution and the effect of respiratory problems on pneumonia, heart disease and strokes. Recent studies in different countries have also shown links between variations in particulate air pollution and daily death rates from lung problems, as well as increased symptoms and hospitalisation.

Not all air pollutants are toxic, but even those that are not can cause damage to people and wildlife because they are released or deposited in abnormally high concentrations. Many industrial cities around the world – particularly those with old heavy industries like iron and steel works, smelters and chemical plants, and those where coal burning is still a principal source of space heating and electricity generation – are surrounded by vegetation deserts (zones where natural vegetation has been badly damaged if not completely killed by air pollution). Air pollution is seriously affecting rates and patterns of weathering of buildings, structures and historic monuments in many places.

LOCAL AIR POLLUTION

In the past, most air pollution was concentrated in particular places, such as around large cities and industrial areas, and most ecological damage was centred on the areas of high fallout relatively close to the emission sources. Some cities are notorious for their poor air quality. Air pollution in Athens, for example, is damaging historic monuments and people's health, and emergency measures have been introduced to cut pollutant emissions, including reducing traffic volumes.

London, England, has long had a serious air pollution problem and is particularly well known for the December

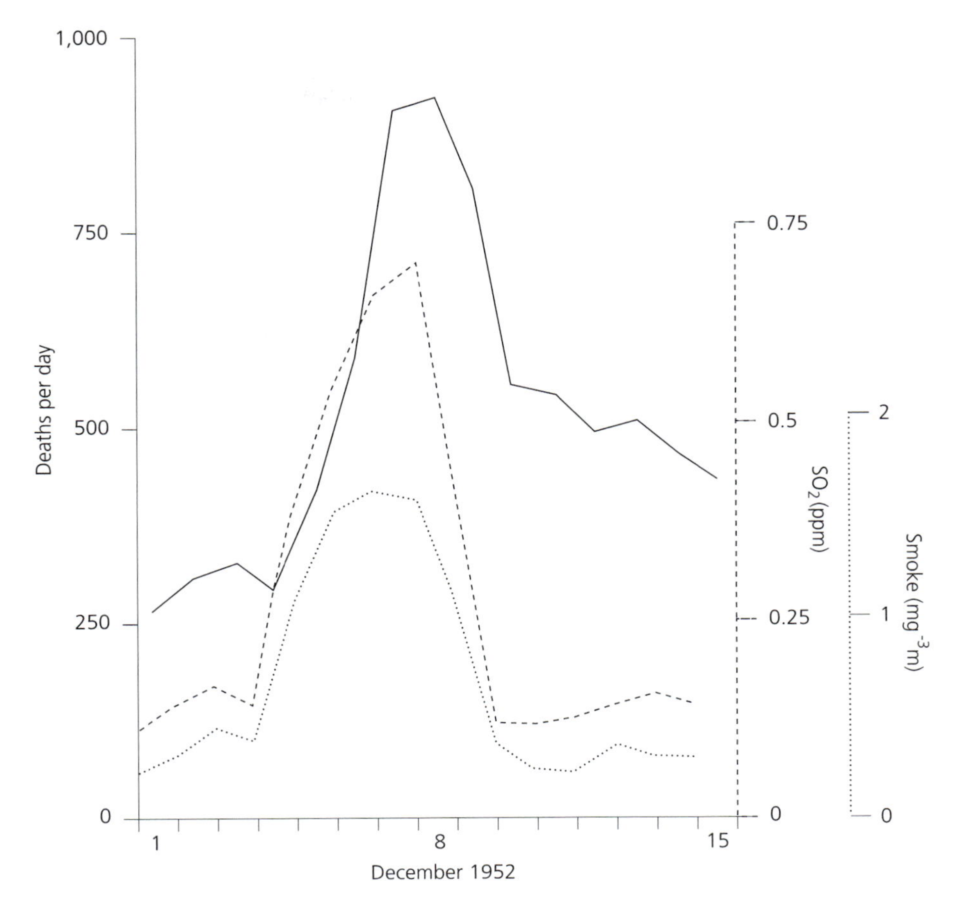

Figure 8.6 *The December 1952 London fog. The smog was caused by air pollution from vehicles and domestic chimneys, which caused a build-up of smoke (dotted line) and sulphur dioxide (broken line) under stable air conditions. Many people suffered breathing problems, and the death rate rose temporarily.*

Source: Figure 24.18 in Perry, A.H. (1981) Environmental hazards in the British Isles. George Allen & Unwin, London.

1952 smog incident (Figure 8.6), which damaged many people's health, put many in hospital and killed an estimated 4,000 people (from breathing problems). Air quality has been monitored in London since the early 1930s, and the records show a progressive improvement (declining SO_2 concentrations) since about 1963. Studies in other cities have shown that variations in the amount of SO_2 (sulphur dioxide), NO (nitric oxide) and dust in the air depend mostly on space heating, whereas NO_2 (nitrogen dioxide) depends more on traffic.

REGIONAL AIR POLLUTION

As some industrial areas and cities grew and spread out, air pollution became a regional environmental problem in some areas, such as around heavy industrial areas in Eastern and Western Europe, North America, Japan and China. The former Soviet Union was among the world's heaviest producers of air pollution, partly because of a continued reliance on coal burning, iron and steel production and other traditional heavy industries.

THE EMERGING GLOBAL PROBLEM

A major concern at the opening of the third millennium has been the growth of global air pollution (Box 8.20). Global air pollution requires detailed monitoring so that scientists can determine what is happening and where it is happening, how fast atmospheric equilibrium is being changed, where the problems exist and are likely to arise, and how close the critical thresholds important to environmental systems might be. Problems such as depletion of stratospheric ozone, development of photochemical smogs and acid rain involve complex chemical chain reactions in the

BOX 8.20 CONCERN ABOUT GLOBAL AIR POLLUTION

Increasingly, the most serious air pollution problems are becoming global in scale because:

- They involve the release of gases which can be blown across a wide area by wind systems.
- Many of the gases involved survive for days or longer in the environment.
- Many of the problems are cumulative, created by the progressive build-up of gas concentrations around the world over a long period of time.
- Without concerted international effort to reduce gas emissions, most countries have in recent decades released increasing amounts of the most important gaseous pollutants.

atmosphere, not all of which are fully understood and which therefore require more scientific study.

Global air pollution also requires careful management if many important environmental systems – on which life on Earth depends heavily – are not to be further damaged, and particularly if they are not to be damaged beyond the environment's in-built ability for self-repair and recovery (see p. 80). It also has many implications for natural resource planning, because the sustainable use and fair allocation of natural resources is difficult enough on its own, but it is made doubly difficult if resource managers are chasing the moving target of progressive long-term climate change too!

Because the atmosphere is so dynamic and mobile, material in it can be dispersed across a wide area very quickly, so places a long distance downwind can be seriously affected. This is how trans-frontier air pollution is created, because air movements respect no political or administrative boundaries in spreading pollution to countries downwind.

Air pollutants from Britain, for example, can be spread across Europe within a matter of days, and around the world in weeks. Such rapid dispersion means that some local processes can generate large-scale regional if not global impacts. A very vivid example was the fallout of radioactive pollutants from Chernobyl across Europe in 1986, within a week of the accident (see Figure 8.7 and Box 8.23). Some fallout from Chernobyl was traced across the Atlantic, in parts of the north-east USA. The trans-frontier movement of acid rain (see p. 251) is another example of the ease with which air pollutants can disperse over wide areas and the difficulties that this creates for being certain which countries have caused it and thus have responsibility for it.

COMPLEXITY

Monitoring the dispersion and distribution of pollutants is essential if scientists are to develop a better understanding of the processes and patterns involved. Analysis of the likely impacts and consequences of pollution is also vitally important, because it shows just how complex many of the interrelationships within and between environmental systems are in practice. Other problems have to be addressed too, including identifying links between cause and effect and trying to apportion responsibility for observed damage. All of these issues are made worse by the international nature of many environmental problems.

It is often difficult to establish which factor or factors cause particular pollution problems, because most such problems could involve more than one factor at a time. Sometimes the synergy (cumulative effect) of several factors operating together creates different problems, or familiar problems that mask underlying complexity. This is true in the case of forest dieback, for example (Box 8.22).

AIR POLLUTION AND AGENDA 21

A great deal of attention was focused on air pollution at the Rio Earth Summit (see p. 13). Chapter 9 of Agenda 21 describes how the atmosphere is under increasing pressure

BOX 8.21 MAJOR AIR POLLUTION PROBLEMS

The most important air pollution problems at the start of the new millennium are associated with changes in the composition of the atmosphere. They include:

- Sulphur dioxide from factory chimneys forms sulphuric acid in the atmosphere, which falls as acid rain (see p. 252).
- CFCs from aerosols and refrigerators drift into the stratosphere, where they break down the ozone layer (see p. 250).
- The build-up of carbon dioxide, water vapour and other gases keeps in the Earth's surface heat, producing the so-called greenhouse effect (see p. 258).

BOX 8.22 FOREST DEATH AND AIR POLLUTION

Forest death or tree dieback (*Waldsterben* in German) associated with air pollution is a good example of the difficulty of identifying basic causes of environmental damage. The first symptoms of damage to trees include the death of young shoots. Progressively larger branches are damaged and die, until eventually the whole tree dies. More than half of Germany's forests are affected, along with large areas in the Netherlands, Belgium, France, Britain, Switzerland, Austria, Eastern Europe, southern Scandinavia and parts of North America. Damage is widespread, particularly in industrial countries, and now appears to be occurring in many developing countries, including China. In the late 1970s, acid rain was believed to be the cause of most forest death in places like the Black Forest in Germany, but scientists now know that it is caused by a mixture of pollutants – including acid rain, ozone, sulphur dioxide and nitrogen oxides – and the chemical mix varies from place to place.

from greenhouse gases, which threaten to change the climate, and from holes in the ozone layer, which increase ultraviolet radiation received at ground level, which in turn causes cancer in humans and animals. It calls on governments to make more efficient use of existing energy sources and develop new, renewable energy sources such as solar, wind, hydro, ocean and human power.

Among other solutions to the growing global problem of air pollution, Agenda 21 calls for:

- promotion of energy standards;
- taxation of industries in ways that encourage the use of clean, safe technologies;
- development of improved substitutes for CFCs and other ozone-depleting substances;
- the transfer of all these technologies to poor developing countries;
- better control of trans-frontier acid rain air pollution (by regular exchanges of information, better training of experts and better application of international standards of pollution control).

MAJOR POLLUTION INCIDENTS

The most important global air pollution problems are progressive and cumulative, arising from the build-up to critical levels of substances that are normally naturally present in the atmosphere – such as greenhouse gases and sulphur dioxide. But from time to time serious pollution incidents occur that are:

- *Location-specific*: they happen in particular places, which might never have experienced such problems before.
- *Time-specific*: they happen at a particular point in time and are unlikely to happen again there.

The Lake Nyos natural gas release in Cameroon (see Box 6.29) was a natural pollution incident with devastating consequences. Most pollution incidents are associated with industrial accidents or similar events, including military activities.

INDUSTRIAL ACCIDENTS

Perhaps the most graphic illustration of an industrial accident that caused a major air pollution incident was the spread of radiation through the atmosphere from the nuclear accident at Chernobyl in the Ukraine in April 1986 (Box 8.23). Sadly, there are many other industrial accidents that provide detailed case studies of local and regional air pollution. One accident that has been studied in particular detail was the explosion at the Union Carbide pesticide plant in Bhopal, India, in 1984 (Box 8.24).

MILITARY ACTIVITIES

Military activities cause air pollution incidents in a variety of ways, involving both nuclear and non-nuclear releases.

Non-nuclear activities: fire is a traditional weapon of war, and through the ages advancing or retreating armies have set fire to their opponents' land, buildings and structures to prevent their subsequent reuse. Raging fires release great amounts of smoke and sometimes gases into the atmosphere, and this can cause local climatic disturbance.

One such case was the air pollution caused during the 1991 Gulf War, when commercial oil installations were deliberately set on fire by the retreating Iraqi army in an attempt to cripple the Kuwaiti economy. Skies in the region were blackened for several weeks by soot particles released from the blazing oil wells, causing commercial airlines to detour around the Middle East and temperatures downwind to be lowered because of the lack of sunlight.

BOX 8.23 THE 1986 CHERNOBYL NUCLEAR ACCIDENT

The incident happened on 26 April 1986 in the Ukraine, in the former Soviet Union, and is well documented. The accident, caused by human error and negligence, is on record as 'the world's worst nuclear accident' – *yet*. One of the four nuclear reactors at the plant was badly damaged by a chemical explosion after it seriously overheated and went out of control when an unauthorised experiment on the reactor went wrong. Highly radioactive and dangerous fission products poured out uncontrollably into the atmosphere over a period of ten days, while the struggle was under way to plug the leak. This material created an invisible radiation cloud, which blew over much of Western Europe within days, and many areas were contaminated by dry fallout (of particles and dust) when rain washed the material back to the ground.

The radiation cloud cast a long shadow over Europe (Figure 8.7). Reactions to the health risk differed from country to country as the cloud drifted first northwards over Scandinavia and later southwards across Europe and Britain. Many countries banned the import of foodstuffs (fruit, vegetables, milk, fresh meat, animals for slaughter, game and freshwater fish) from countries within 1,000 km of Chernobyl in order to reduce health risks to humans. Some countries (such as Britain) banned the sale of milk and dairy products from sheep and cattle that had grazed fresh, contaminated grass. Anxiety was heightened throughout Europe during the first week of May, but only in Poland was the radiation threat regarded as serious enough to issue stable iodine tablets and solutions as a prophylactic (preventive medicine) to an estimated 10 million children and youths.

The direct costs of Chernobyl include over 1,000 immediate injuries, thirty-one immediate deaths and the evacuation of 135,000 people from towns and villages around the reactor complex. Russian estimates put the total cost (including clean-up operations, damaged farmland and lost productivity) at around US$358,000 million. Most countries that received fallout from the accident were totally unprepared to cope with a radiation disaster, government responses were often very confused and poorly co-ordinated, and media coverage of the event was highly sensationalist and often very exaggerated.

In the order of 100 million people throughout Europe and the western (former) Soviet Union received some exposure to low levels of radiation (particularly iodine-131 and caesium-137) from Chernobyl. The whole world is believed to have received a radiation dose equivalent to all previous years' nuclear weapons fallout, and some individual doses in Britain were equivalent to one year of natural background radiation. Damage has been worst in the area around Chernobyl, where there has been a dramatic rise in the incidence of thyroid diseases, anaemia, cancer and symptoms of radiation sickness (including fatigue, loss of vision and appetite) among humans, and some grotesque birth defects among livestock. Many of the broader environmental impacts of the accident, including the possible long-term genetic consequences, are still not well understood.

There are many different estimates of the likely long-term human cost of the accident. One pessimistic estimate is that radiation from the accident weakened the immune systems and caused the premature deaths of between 35,000 and 40,000 people in the USA alone during the summer after the explosion. Other estimates of the worldwide number of deaths from radiation-induced cancers likely to be caused by Chernobyl up to the year 2030 vary between 1,000 and 500,000. The risk to individuals of death caused by cancer triggered by exposure to Chernobyl fallout is estimated at up to five in 10,000 in Europe and up to three in 100,000 across the northern hemisphere as a whole.

Nuclear activities: throughout the 1980s there was widespread concern about the possible consequences of fallout of radiation from military activities, both from nuclear testing and from nuclear warfare. The fallout from military nuclear explosions would include radioactive particles that give out harmful ionising radiation and are carried as dust on the wind and fall to the ground or are washed down in rain. Military nuclear testing is now banned under the Partial Test Ban Treaty, but before that came into effect in 1963 the USA had tested at least 103 nuclear bombs in the South Pacific. Only France continues to test nuclear bombs, on the two French Polynesian islands of Moruroa and Fangataufa, where it has tested over a hundred bombs since 1966, the most recent being in 1995. The testing is surrounded by great secrecy, and the French government has consistently refused to release information on the composition and spread of fallout, and on the disposal of radioactive wastes and accidents.

Concern about the possibility of a nuclear winter (Box 8.25) has died down since the end of the Cold War in the late 1980s, but the risk is ever-present in a world dominated by military superpowers with nuclear capabilities

BOX 8.24 THE 1984 EXPLOSION AT BHOPAL, INDIA

In December 1984, a major explosion occurred at the Union Carbide pesticide plant in Bhopal, India. Nearly 40 tonnes of methyl isocyanate (MIC) – a persistent and particularly toxic gas that attacks many different organs in the human body – was released in the explosion and blew over the residential area nearby. Nearly 200,000 people were exposed to the MIC, more than 3,000 people died within three days, and by 1993 over 4,000 deaths were directly attributable to the explosion. Many more have suffered long-term ill-health and disability because of exposure to the toxic cloud. Bhopal remains the world's worst industrial accident – certainly in terms of the death toll and human misery – and those who experienced it have since received little compensation, because of bureaucratic delays and inefficiencies.

(even when these are justified as deterrents). Scientific curiosity persists, even if the military prospects appear in the short term to be more optimistic.

LOW-LEVEL OZONE

Ozone (O_3) is a gas that is toxic to humans at a concentration as low as one part per million (1 ppm) in air. There are two very different forms of the ozone pollution problem, which have different causes and effects and take place in different parts of the atmosphere:

1. Depletion of ozone in the stratosphere (see p. 250).
2. Increasing ozone concentrations in the lower atmosphere.

Low-level ozone is a product of air pollution, mainly from industry and cars. They release nitrogen oxides (NO_x) and sulphur dioxide (SO_2) into the lower atmosphere, and these react with sunlight to create photo-oxidants, of which ozone is the most dangerous. Ozone is a major pollutant in hot summers, particularly in large cities and industrial areas where many people live and work.

Measurements of background levels of ozone, collected over 20 years at sites a long way from pollution sources (including the South Pole, Barrow in Alaska and Mauna Loa in Hawaii), show that ozone concentrations vary through the day and from season to season. Efforts are being made, through the International Geosphere–Biosphere Program (IGBP) (see p. 20), to collect more information on changing ozone levels in the atmosphere and on the ways that these might promote other changes in atmospheric chemistry.

IMPACTS

Some scientists predict that low-level ozone concentrations might double over the next 100 years. This could create a type of greenhouse effect – similar in some ways to the main carbon dioxide effect (see p. 255) – which absorbs radiation from the Sun, acts like a thermal blanket around the Earth and warms the Earth's surface. As well as changing temperatures and thus climate, low-level ozone pollution damages plants by stunting their growth. This might decrease world food supplies. Ozone also affects human health (Box 8.26) and corrodes certain materials.

PHOTOCHEMICAL SMOG

Near the ground, ozone pollution is a major ingredient in photochemical smog, a light haze that can develop during

BOX 8.25 NUCLEAR WINTER

One of *the* great environmental (and humanitarian) scares of the 1980s was the prospect of a nuclear winter, triggered by a major nuclear war, probably between the USA and the former Soviet Union. Various scenarios were proposed, and they have in common an expectation that massive clouds of smoke resulting from widespread fires would blot out sunlight from large areas for many weeks, creating a nuclear winter. Lack of sunlight would trigger atmospheric and climatic changes, which would almost inevitably result in a colder world climate.

Fears were expressed that a nuclear winter would bring long-term, irreversible and unwanted changes to the world's atmospheric, environmental and ecological systems, thus putting all life on Earth at risk. Worst-case scenarios predicted that nuclear war and nuclear winter would kill most plants and animals in the northern hemisphere and threaten the very survival of the human species. Some scientists argued that a nuclear winter was not an inevitable consequence of nuclear war, but few found it possible to dismiss the possibility altogether. Even if nuclear winter on a global scale seems unlikely, nuclear war would be most likely to have devastating effects on local and regional environments.

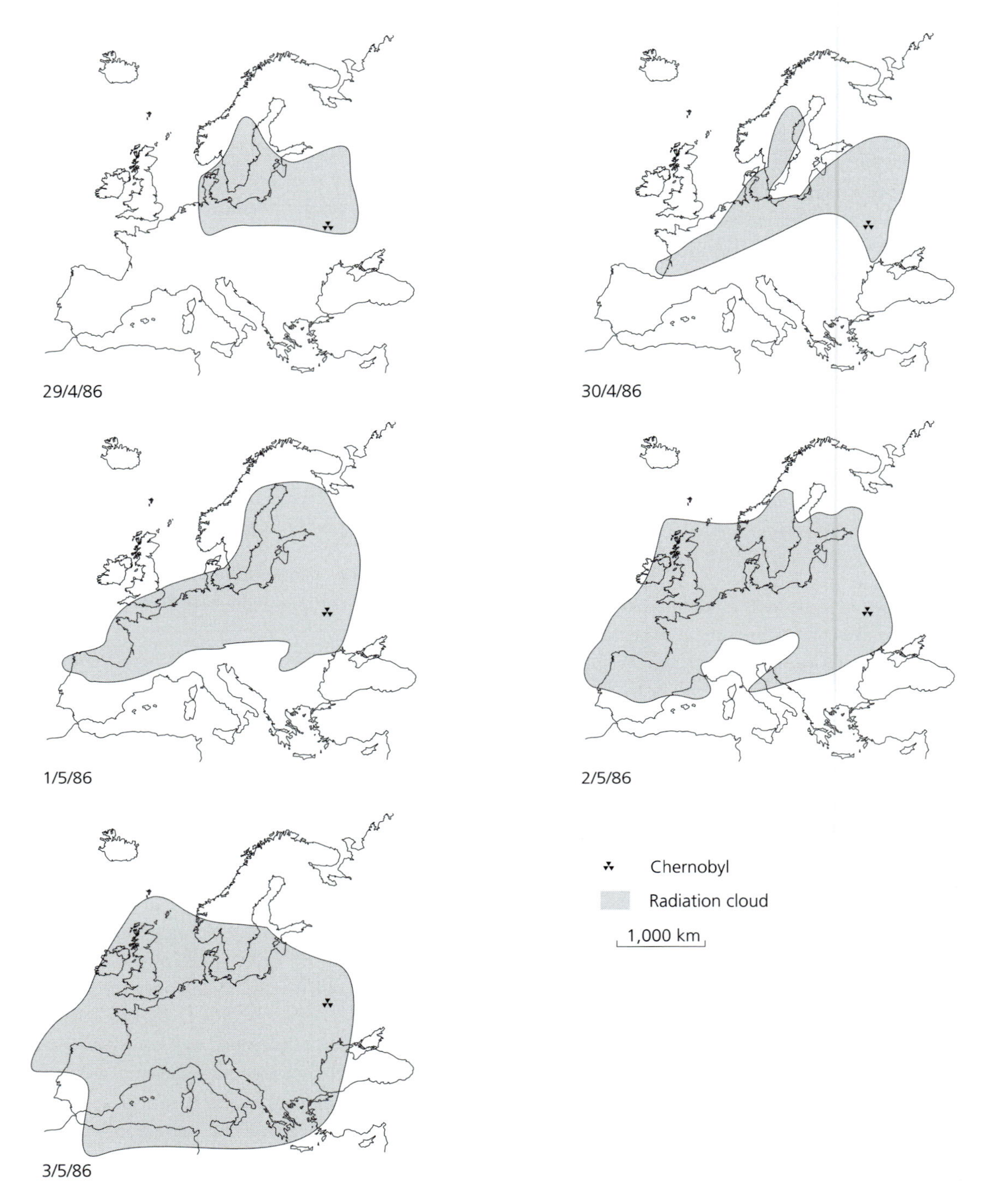

Figure 8.7 *The spread of fallout across Europe from Chernobyl, April–May 1986. Over the days following the accidental release of radioactive material from the Chernobyl nuclear power plant in the Ukraine, a dangerous radiation cloud spread across Western Europe and deposited radioactive particles and material across a wide area.*

Source: Figure 5.2 in Park, C.C. (1989) Chernobyl: the long shadow. Routledge, London.

BOX 8.26 OZONE AND HUMAN HEALTH

Exposure to high levels of ozone is dangerous to people, causing asthma attacks and other respiratory problems. Outdoor working practices (for example in the construction industry) might need to be changed to reduce long-term exposure to low-level ozone. The US Environment Protection Agency (EPA) recommends that people should not be exposed for more than one hour a day to ozone levels of 120 parts per billion (ppb), while the World Health Organisation recommends a lower limit of 76–100 ppb. The effects of chronic long-term exposure to ozone are not yet well understood, but it might include premature ageing of the lungs.

hot, dry weather when pollutants from vehicles and factories are trapped in still air. Such smogs were first observed in California in the 1940s, but they are becoming increasingly common in all heavily built-up areas during hot dry weather.

Many plants wilt and sometimes die when exposed to photochemical smogs, and people and animals suffer from irritation of the eyes and lungs. Ozone poisoning is now believed to contribute to the death of many trees and forests throughout Europe and North America (see Box 8.22).

One way of controlling ozone smogs is to cut down on the emissions of pollutant gases (particularly nitrogen oxide) from vehicle exhausts. During the 1990s, new vehicle exhaust emission controls were introduced in North America and in many European countries (except Britain) to try to tackle this problem. New catalytic converters on vehicle exhausts are designed to reduce exhaust gas emissions.

STRATOSPHERIC OZONE DEPLETION

THE OZONE LAYER

In the stratosphere, at a height of between 20 and 25 km, is the so-called 'ozone layer', where ozone accumulates naturally (see p. 229). Ozone is constantly being formed, broken down and reformed above about 40 km, and it sinks and accumulates at the 20–25 km level. Ozone is a highly reactive gas made up of three atoms of oxygen. It is formed when the molecule of the stable form of oxygen (O_2) is split by ultraviolet radiation or electrical discharge (as happens when lightning flashes through a cloud; see p. 314).

The ozone layer is not uniform around the world; it varies in density, being least dense over the equator and most dense over high latitudes (beyond 50° N and S). The ozone layer also changes through the year and is best developed over polar regions in early spring (because ozone is neither formed nor destroyed during the polar night, when ozone transported from lower latitudes is stored over the poles).

OZONE DEPLETERS

Chemicals that destroy the ozone in the stratosphere are described as ozone depleters. Most are chemically stable compounds containing chlorine or bromine, which remain unchanged long enough to drift up to the stratosphere. The best-known are chlorofluorocarbons (CFCs; Box 8.27), but there are many other ozone depleters, including:

- halons (used in some fire extinguishers);
- methyl chloroform and carbon tetrachloride (both solvents);
- some CFC substitutes;
- the pesticide methyl bromide.

Interest in possible damage to the ozone layer was first raised in the early 1970s, when atmospheric chemists expressed concern over the possible damaging effects of

BOX 8.27 IMPACTS OF OZONE DEPLETION

The ozone layer is a natural filter that prevents excessive amounts of harmful ultraviolet (UV) radiation from the Sun reaching the Earth's surface (see p. 229). Some scientists even think it unlikely that life would have developed on Earth without this ozone shield to 'filter out' harmful ultraviolet radiation.

If the ozone layer is damaged (thinned or punctured), more UV radiation would pass straight through the atmosphere. This would have a variety of possible impacts, depending on the extent and location of the damage. Agricultural crops would be scorched, and yields would fall. Marine plankton would be seriously affected. Human health would suffer – there would be more eye cataracts, more skin cancer, and more problems arising from damage to people's immune systems. The problems are likely to affect many if not all countries.

BOX 8.28 CFCS AND THE OZONE LAYER

CFCs are synthetic chemicals that are odourless, non-toxic, non-flammable and chemically inert. Until the early 1990s, they were widely used:

- as propellants in aerosol cans;
- as refrigerants in refrigerators and air conditioners; and
- in the manufacture of foam boxes for take-away food cartons.

They are believed to play a major part in destroying the ozone layer, and since the early 1990s most industrial countries (including the USA and United Kingdom) have agreed to phase out production of CFCs and a range of other ozone-depleting chemicals by the year 2000.

When CFCs are released into the atmosphere, they drift slowly upwards into the stratosphere, where, under the influence of ultraviolet radiation from the Sun, they break down into chlorine atoms, which destroy the ozone layer and allow harmful radiation from the Sun to reach the Earth's surface. CFCs can remain active in the atmosphere for more than 100 years.

CFCs in the atmosphere. The USA, Canada and Sweden took the issue seriously and unilaterally introduced bans on the use of CFC-propelled aerosols. These countries also sought major international co-operation on the ozone problem, and a 1977 United Nations Environment Program (UNEP) conference set goals for further research and consultation.

HOLE IN THE OZONE LAYER

Most countries refused to recognise ozone depletion as a serious problem until British scientists discovered a seasonal thinning (usually referred to as a hole), the size of the USA, in the ozone layer over the Antarctic in 1985. Field monitoring showed a 40 per cent thinning in the ozone layer. Their evidence suggested that the hole was growing bigger through time, and satellite surveillance confirmed the initial reports. The main cause of the thinning was believed to be a threefold increase in atmospheric CFCs within 10 years.

Annual measurements indicate that the ozone hole over the Antarctic appears to be growing bigger (Figure 8.8). More recently, reports that the ozone layer above the Arctic and parts of Western Europe is thinning have given rise to concern. While much of the thinning of the ozone layer can be blamed on air pollution by CFCs, some natural pollutants – including aerosols from the 1991 Mount Pinatubo volcanic eruption (see Box 6.28) – aggravate the problem.

INTERNATIONAL INITIATIVES

From the mid-1980s onwards, the issue of ozone depletion has been taken seriously. In 1985, the Vienna Convention for the Protection of the Ozone Layer – which sought to minimise human destruction of the ozone layer, mainly by reducing the production and emission into the atmosphere of CFCs – was initially signed by twenty-two countries. Not all countries agreed on the urgency of the problem, and three camps emerged:

1. The so-called Toronto group (including the USA, Canada, Sweden, Finland and Norway) agreed to ban or restrict the use of CFC-propelled aerosols.

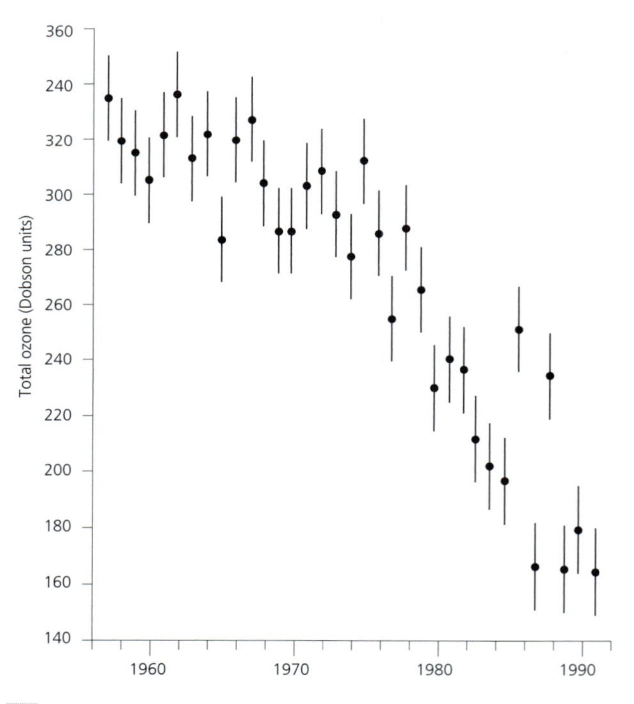

Figure 8.8 *Annual increases in ozone depletion over Halley Bay, Antarctica, 1957–91. The graph shows mean October levels of total ozone above Halley Bay in Antarctica based on measurements collected by the British Antarctic Survey.*

Source: Figure 6.5 in Elsom, D. (1992) Atmospheric pollution. Blackwell, Oxford.

2. The European Community (now the European Union) countries agreed on a reduction in the use of CFCs but took little action.
3. Japan and the former Soviet Union decided to wait for further scientific evidence.

Since 1987, the Montreal Protocol (Box 8.29) has been strengthened to work towards a faster phase-out of CFCs by 1996. By 1990, many European countries had made great progress in phasing out the use of CFCs. By the end of 1994, some industrial countries had already halted production of CFCs, and others were well on target to meet the 1996 deadline. Developing countries were given a 10-year extension.

The 1985 Vienna Convention and 1987 Montreal Protocol were at the time unique among international environmental initiatives, because it was probably the first time that serious attempts had been made to avert a major environmental problem *before* the more serious consequences and side-effects were obvious. Previously, governments would have insisted on waiting for conclusive scientific proof of cause–effect links before introducing appropriate preventive legislation and embarking on costly behaviour-changing strategies.

BOX 8.29 THE MONTREAL PROTOCOL

In 1987, thirty-nine nations signed the Montreal Protocol, under which they agreed to freeze CFC production at 1986 levels, and to decrease production by 20 per cent by 1993 and by half by 1999. The Montreal Protocol was widely welcomed, because it provided the international community with a mechanism for protecting the ozone layer that is:

- *effective* (the measures are likely to work, particularly as substitutes for CFCs are developed and make it easier to reach the emission reduction targets);
- *equitable* (the measures share responsibility for solving the problem among all the countries concerned); and
- *dynamic* (the measures get tougher through time and can be adapted further if new scientific evidence comes to light).

ACID RAIN

Acid rain has become one of the most widespread trans-frontier air pollution problems over the last decade, and it is regarded as one of the most serious environmental problems facing developed countries. Damage to forests, lakes and buildings has been reported downwind from most industrial areas of the world, but the worst damage to date appears to be in Scandinavia, parts of Western Europe, and eastern Canada and the USA. There are also some links between acid rain and human health – including increased incidence of respiratory problems such as bronchitis and lung cancer, and possibly also Alzheimer's disease (senile dementia) – but overall these are less common and less acute than the damage caused by ozone.

CHEMISTRY

Acid rain is measured using the pH scale (Figure 8.9 and Box 8.30). It is created by the transformation in the atmosphere of gaseous pollutants, principally sulphur dioxide (SO_2) and nitrogen oxides (NO_x). These are emitted mainly from coal-fired power stations and vehicle exhausts, respectively. These oxides interact with sunlight and other atmospheric contaminants such as ozone, photo-oxidants, hydrocarbons and moisture. The chemical chain reactions turn the primary pollutants (SO_2 and NO_x) into a variable cocktail of secondary pollutants.

BOX 8.30 THE PH SCALE

Acidity is measured on the pH (short for potential hydrogen) scale, which is logarithmic and runs from 0 (very strong acid) to 14 (very strong alkaline) (Figure 8.9). Normal rain has a pH of 5.6 or greater. This is more acidic than distilled water, which has a pH of 7, because even unpolluted rain is contaminated with natural products of volcanic eruptions, forest fires and spray from the oceans. Precipitation with a pH of below 5.6 is described as acidic or acidified. The lower the pH value, the greater the acidification. Because the scale is logarithmic, numbers can be deceptive – pH 5 is ten times as acidic as pH 6, pH 4 is 100 times as acidic, and so on. As a result, even small changes in pH values represent major changes in levels of acidity and thus potential damage.

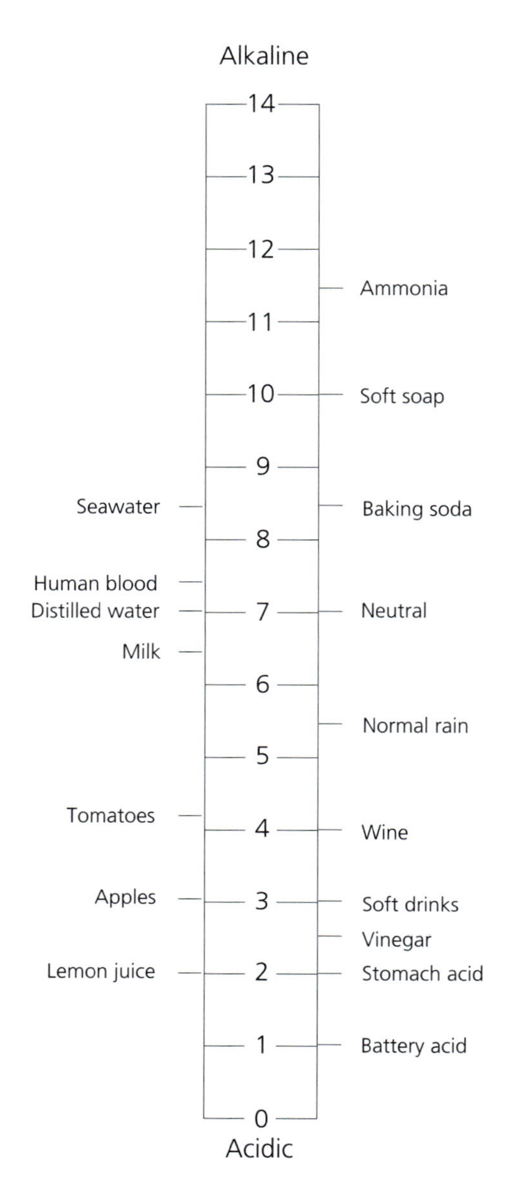

Figure 8.9 *The pH scale of acidity. The pH scale is a logarithmic scale, based on the negative logarithm of the hydrogen ion concentration in water. Low pH values indicate strong acidity, and high values indicate strongly alkaline material.*

Source: Figure 2.1 in Park, C.C. (1987) Acid rain: rhetoric and reality. Routledge, London.

The term 'acid rain' is shorthand for acid deposition, meaning the fallout of acidic material from the atmosphere (Figure 8.10). It can occur in dry and wet forms (Box 8.31). The oxides can remain active for at least seven days, during which time they can be blown thousands of kilometres by wind systems.

BOX 8.31 DRY AND WET DEPOSITION OF ACIDIC MATERIAL

Dry deposition involves the fallout of gases, aerosols and fine particles and usually occurs relatively close to the emission sources of the gases.

Wet deposition comes with precipitation (mainly in rain, but also in snow, mist, hail, and so on) and mostly occurs further downwind, after the complex chemical transformations have taken place.

GEOGRAPHY

There is often a distinct spatial zonation of dry and wet deposition downwind from a source of the gaseous oxides, with:

- a local 'ring of confidence' closest to the source, which receives little if any fallout;
- a zone of dry fallout; and
- a much wider zone of wet deposition further away.

Both wet and dry forms of acid deposition have been associated with environmental damage, but it is the wet deposition that has caused so much concern because it travels such vast distances (Box 8.32). Wind-blow of between 1,000 and 2,000 km in five days is not unusual, and this helps to explain why pollution and damage can be exported from donor countries to innocent countries downwind.

It has been argued that much of the SO_2 released from chimneys and power stations in the UK, France and Germany is eventually deposited over Scandinavia, causing great damage to freshwater lakes and rivers there. Swedish scientists estimate that up to nine-tenths of the sulphur deposited on Sweden is blown in from other countries. The USA is also accused of exporting acid rain from industrial centres in the north-east (particularly the Ohio valley), across the Canadian border into the province of Ontario (Figure 8.11).

Scientists are trying to develop a better understanding of how acid rain damages soils, trees and lakes, and one useful development during the 1990s was critical loads modelling. This allows some assessment of the likely consequences of a range of possible future emissions scenarios, based on identifying where rates of sulphur deposition might exceed critical environmental thresholds and thus cause much ecological damage.

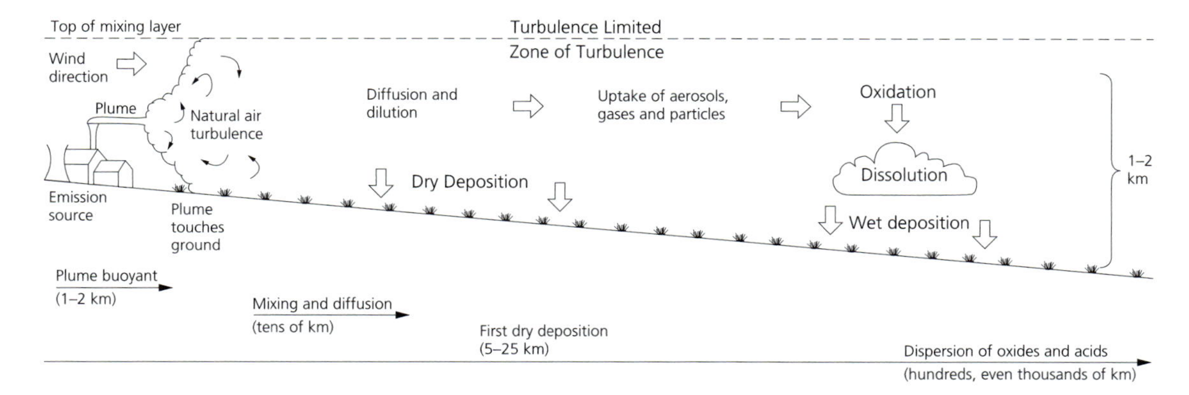

Figure 8.10 *Dispersion and deposition of acid substances. Sulphur dioxide and nitrogen oxides are released from point sources, then dispersed as winds blow the gases downwind. Some of the acidic material is deposited as gases, particles and aerosols quite close to the emission source, but most is blown long distances, converted and washed out of the atmosphere as wet deposition.*

Source: Figure 2.7 in Park, C.C. (1987) Acid rain: rhetoric and reality. Routledge, London.

BOX 8.32 ACID RAIN HOT-SPOTS

Despite notable advances over the last two decades, monitoring of rainfall chemistry is still rather patchy, and reliable data are not yet available from all parts of the globe. But the available evidence shows two key 'hot-spot' areas where acid rain is particularly problematic:

1. the north-east of North America (straddling the north-east USA and eastern Canada);
2. much of Western Europe (including Scandinavia and Britain).

Many parts of the heavily industrialised areas of Eastern Europe, including Poland and Czechoslovakia, are also believed to be badly affected by acidification.

The evidence from rainfall monitoring since the early 1970s shows that the problem areas are getting worse through time – precipitation pH in the hot-spots is falling (as levels of acidity rise), and the areas receiving acid rain are growing larger. New hot-spots are starting to emerge too, such as parts of the west coast of the USA.

HISTORY

While concern about acid rain has risen sharply since the early 1980s, the problem itself is far from new. Since the start of the Industrial Revolution, air pollution has been creating biological deserts around industrial centres, and in 1852 Robert Smith (Britain's first air pollution inspector) discovered links between sooty skies and acid fallout around industrial Manchester. Global emissions of SO_2 and NO_x have risen dramatically over the last 100 years, and slowly the problem has built up in many industrial areas.

From the early 1920s onwards, many lakes and rivers in southern Scandinavia have displayed signs of acidification, including large-scale fish deaths, declining fish populations and falling rates of reproduction. But only since the late 1960s have scientists beyond Scandinavia taken the acid rain problem seriously. In 1972, the Organisation for Economic Co-operation and Development (OECD) launched a major research programme designed to measure air quality at seventy sites in order to examine the long-range transport of air pollutants (LRTAP). By 1977, OECD studies had shown that the areas receiving acid rain were growing (in size and number) and that acid rain crosses national boundaries, so that some countries (such as Britain) are net exporters, while others (such as Sweden) are net importers. There is also mounting evidence of the adverse impacts of acid rain on forests and trees (see Box 8.22).

INTERNATIONAL INITIATIVES

The first international initiatives designed to deal with acid rain came in 1979, when a convention was drafted by the United Nations Economic Commission for Europe (ECE). This convention grew out of the LRTAP project, and it aimed to reduce acid rain by reducing and controlling emissions of SO_2 and NO_x using the 'best available control technology economically feasible'. Thirty-one countries signed initially, but the convention was not legally binding, and the commitment of some countries quickly evaporated.

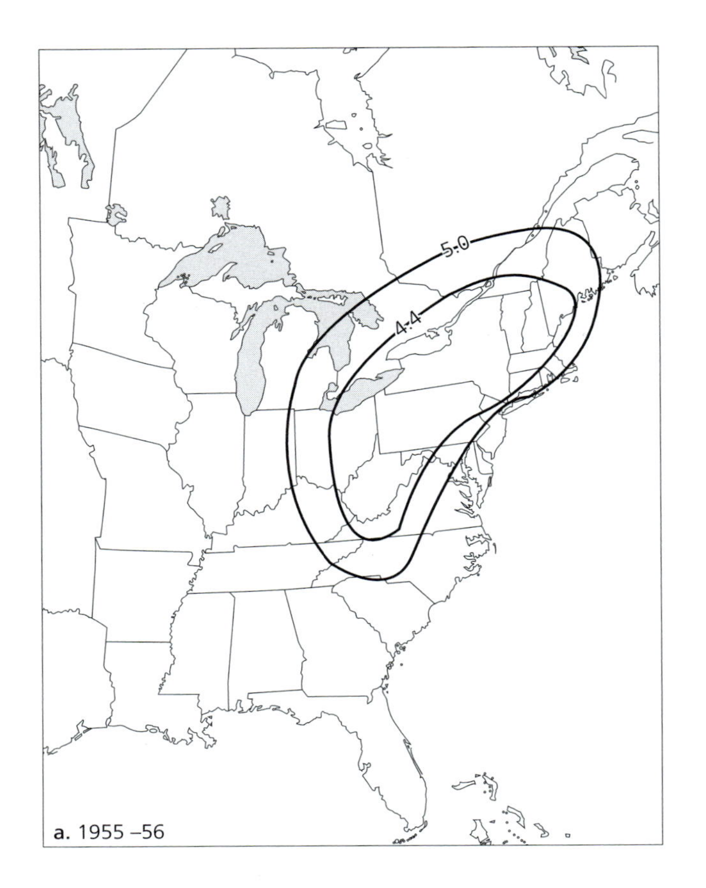

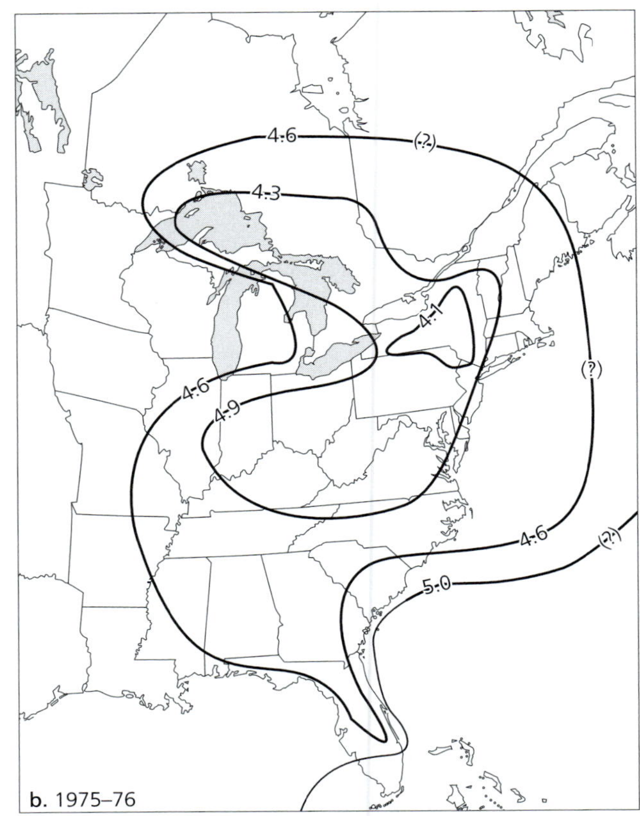

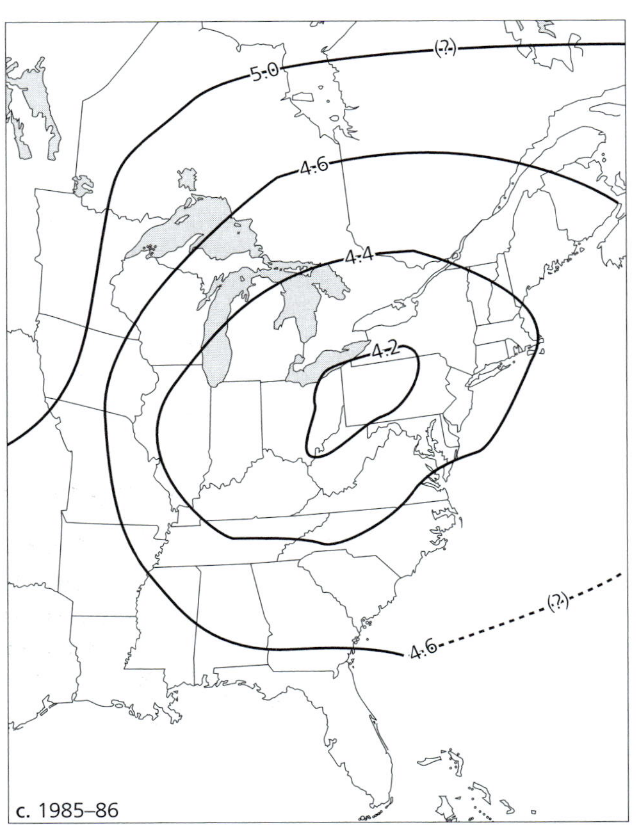

Figure 8.11 *The changing pattern of acid rain in eastern North America, 1955–85. The maps show increasing acidity (decreasing pH) and acid deposition over increasingly large areas over time.*

Source: Figure 9.9 in Marsh, W.M. and J.M. Grossa (1996) Environmental geography: science, land use and Earth systems. John Wiley & Sons, New York.

The Scandinavians were not satisfied with the 1979 convention, which they saw as inadequate to deal with the problem. They proposed that signatories should be responsible for reducing their emissions of SO_2 by 30 per cent between 1980 and 1993. As a result, a group of LRTAP countries met in Ottawa in 1984 to form what has since been called the '30 per cent club'. By July 1985, twenty-one of the LRTAP countries had joined the club by promising to cut their emissions by at least 30 per cent. The USA and UK initially refused to join the club and reduce emission levels, insisting that there was not yet enough firm evidence linking SO_2 emissions to observed damage to forests and lakes. The UK had a change of heart and joined in 1986.

Tackling acid rain will require action on a broad front, because there are a number of options for reducing the two main gases (SO_2 and NO_x) that create the problem (Table 8.7). A number of important scientific uncertainties (Box 8.33) also need to be addressed if politicians are to make further significant progress in tackling the problems of acid rain.

GREENHOUSE GASES

The Earth's atmosphere operates naturally like a greenhouse (see p. 258). Since the mid-1980s, concern has been rising over the ways in which some air pollutants – the so-called greenhouse gases – reinforce this natural greenhouse effect and appear to be causing global warming (Figure 8.12). The work of the Intergovernmental Panel on Climate Change (IPCC; Box 8.34) has been extremely important, not only in collecting together evidence of changing climate and in modelling possible future changes but also in shaping

Table 8.7 The main options for reducing emissions of SO_2 and NO_x

Objective	*Options*
Reducing SO_2 from power stations	■ burn less fuel
	■ use fuel with a lower sulphur content
	■ desulphurise fuel before burning it
	■ reduce the sulphur content of fuels at combustion – by fluidised bed technology (FBT) or flue-gas desulphurisation (FGD)
Reducing NO_x from vehicle exhausts	■ more and better public transport
	■ enforce stricter speed limits
	■ new forms of vehicle power (such as electric power)
	■ new standards of emission control (such as lean-burn technology)

BOX 8.33 UNCERTAINTIES SURROUNDING ACID RAIN

Many unresolved questions remain, both about the science of acid rain and about the most appropriate ways of tackling it and reducing the ecological and environmental damage associated with it. While scientific uncertainties persist, they offer an excuse for exporter countries to delay taking remedial action.

Even after three decades of research a number of critical problems remain, including:

- Establishing clear links between sources of gas emissions and areas of deposition. This is vital if importer and exporter countries are to be properly identified, and it is being assisted by plume-tracking experiments using heavily instrumented planes.
- Establishing clear links between acid deposition and ecological damage. Experimental studies under controlled field and laboratory conditions are proving helpful here.
- Establishing, with an acceptable degree of probability, the likelihood that reducing gas emissions will stop acid rain downwind. Detailed monitoring of atmospheric chemistry is helping to unravel some of the complex chain reactions that create acid rain.
- Establishing, with an acceptable degree of probability, that stopping acid rain will prevent further ecological damage. Some evidence suggests that the link between acid rain and damage is non-linear, meaning that a 50 per cent reduction in acid rain might not bring a 50 per cent reduction in damage. Evidence also suggests that there are time lags between periods of peak emissions and phases of worst damage, so that even if gas emissions were eliminated completely tomorrow, we should still expect to see damage occurring for some time to come.

BOX 8.34 INTERGOVERNMENTAL PANEL ON CLIMATE CHANGE (IPCC)

Concern about the possible impacts of greenhouse gases on global climate rose through the 1980s, and in the autumn of 1987 the United Nations General Assembly discussed a report prepared by the Brundtland Commission (WCED), which looked, among other things, at global climate change induced by human activities.

In 1988, the World Meteorological Organisation (WMO) and the United Nations Environmental Program (UNEP) established the Intergovernmental Panel on Climate Change (IPCC). The IPCC involves collaboration between hundreds of specialists from around the world, and it focuses on the likelihood and probable nature of induced climate change, based largely on forecasts from general circulation models (see Box 8.35).

The first and second IPCC assessment reports on climate change were published in 1990 and 1995, and the third assessment report is scheduled for publication in 2001. The IPCC has three working groups dedicated to particular aspects of climate change.

Working Group I: The science of climate change

This group works to assess available information on the science of climate change, and it is concerned with:

- developments in the scientific understanding of past and present climate, of climate variability, of climate predictability and of climate change modelling (including feedback from climate impacts);
- progress in the modelling and projection of global and regional climate and sea-level change;
- observations of climate, including past climates, and assessment of trends and anomalies;
- gaps and uncertainties in current knowledge.

Working Group II: Impacts, adaptation, vulnerability

This group's work focuses on assessing the scientific, technical, environmental, economic and social aspects of the vulnerability (sensitivity and adaptability) to climate change of, and the negative and positive consequences for, ecological systems, socio-economic sectors and human health, with an emphasis on regional sectoral and cross-sectoral issues.

Working Group III: Mitigation of climate change

This group is particularly concerned with the scientific, technical, environmental and economic and social aspects of mitigation of climate change.

In addition, the IPCC has a Task Force on Greenhouse Gas Inventories.

global policy responses as the search for ways of guaranteeing a more sustainable future gathers pace.

The main greenhouse gases (Box 8.36) are carbon dioxide (CO_2), methane (CH_4) and chlorofluorocarbons (CFCs). Their main properties are summarised in Table 8.8. Atmospheric concentrations of all the main greenhouse gases have risen dramatically since the Industrial Revolution – carbon dioxide concentration has risen by nearly 30 per cent, methane concentrations have more than doubled, and nitrous oxide concentrations have risen by about 15 per cent.

Carbon dioxide (CO_2): most of the CO_2 comes from burning fossil fuels and from forest fires (particularly associated with tropical deforestation). About 80 per cent of the CO_2 comes from energy burned to run cars and trucks, heat homes and businesses and power factories. An estimated 23,900 million tonnes of CO_2 were released into the atmosphere during 1996 alone from the burning of fossil fuels, cement manufacture and gas flaring. Roughly half of the CO_2 that is released is soon absorbed by the oceans or by increased plant photosynthesis, but the other half remains in the atmosphere for many decades. As a result, the average atmospheric concentration of CO_2 is increasing (Box 8.37). Industrial countries produce much more than their fair share of CO_2. The USA, for example, houses 5 per cent of the world's population and produces 25 per cent of the CO_2. Overall levels of CO_2 emission in the USA declined significantly between the early 1970s and mid-1990s (Figure 8.13), partly in response to changing

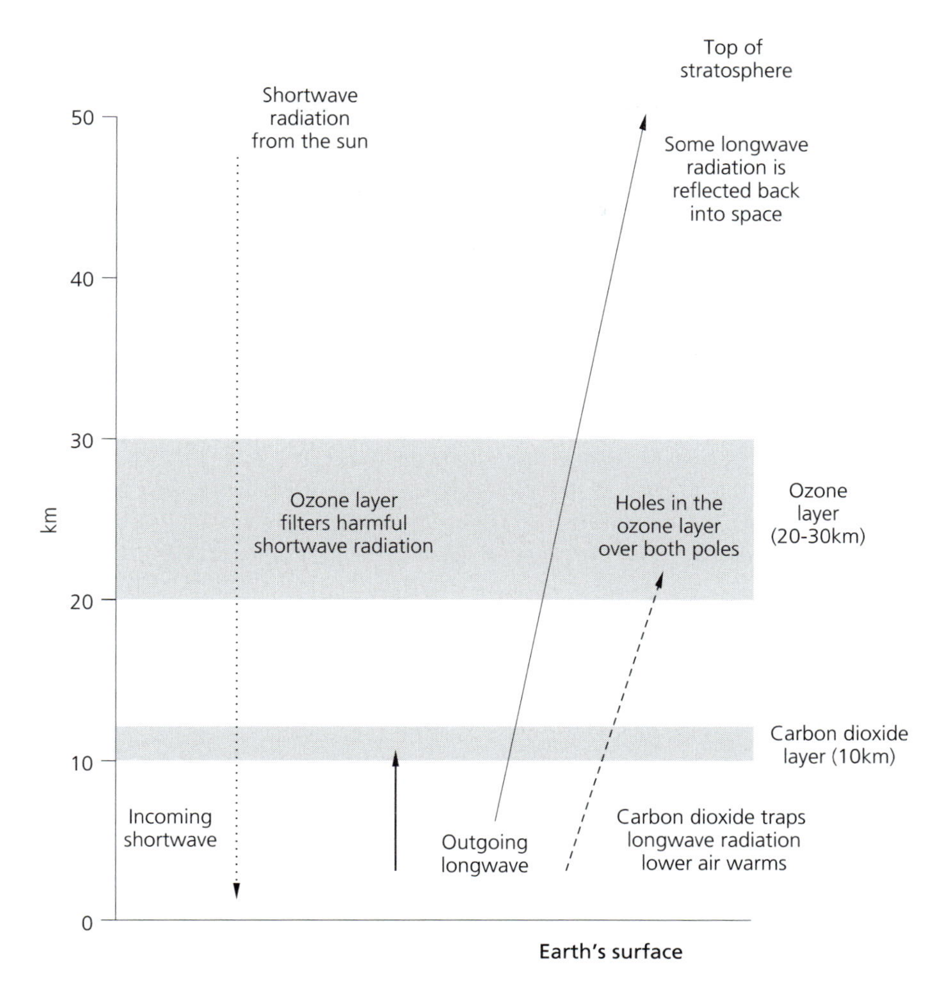

Figure 8.12 *The greenhouse effect. Short-wavelength incoming radiation passes through the Earth's atmosphere, interacts with the Earth's surface and is re-radiated as long-wavelength radiation. The carbon dioxide layer at the top of the troposphere traps long-wave radiation and thus the lower atmosphere is progressively warmed.*

Table 8.8 Greenhouse gases

Gas	*Rate of increase (per cent a year)*	*Sources*	*Lifetime (years)*
Carbon dioxide	0.2–0.7 average 0.4	Natural, combustion of fossil fuels	>100
Methane	1–2 average 1.3	Natural, food production	7–10
Nitrous oxide	0.2–0.3	Natural, combustion of fossil fuels, agriculture	100
CFC-11 and CFC-12	more than 5	Anthropogenic	50–100
Others:			
CFC-113	10	Anthropogenic	long
CFC-22	11		15
Methyl chloroform	5–7		6
Carbon tetrachloride	1–3		more than 50

BOX 8.35 GENERAL CIRCULATION MODELS

The IPCC uses sophisticated computer models to assess past variations in global climate and to project possible trends over the next century. Studies of climate change begin with an 'emissions scenario', which estimates likely future changes in the chemical composition of the atmosphere, based on an understanding of:

- the sources, sinks and transfers of greenhouse gases;
- how the sinks may respond to increased loading.

General circulation models (GCMs) are then built. These are complex computer models of atmospheric behaviour, based on a grid pattern, in which the evolution of atmospheric changes over a period of time is modelled, usually over a period of decades. GCMs are used to predict the speed and pattern of climate change.

While the IPCC models are regarded as sound and extremely useful, the inherent limitations of any type of model must not be forgotten. Any model is only as good as the assumptions on which it based, the data that are fed into it and the interpretations made of its output. The atmosphere is an extremely complex environmental system, and even the best GCMs are still not accurate enough to provide reliable forecasts of how climate might change. Different models can yield different – even contradictory – results.

BOX 8.36 GREENHOUSE GASES

Greenhouse gases are trace gases that alter the heating rates in the atmosphere by allowing incoming solar energy to pass through but trapping the heat radiated back by the Earth's surface. They share some important properties:

- they have strong radiative properties;
- they are increasing in concentration in the atmosphere; and
- they appear to be having significant cumulative impacts.

Simple solutions to the problems of global warming are difficult to find, because the greenhouse gases:

- come from various sources;
- are increasing at different rates;
- persist in the atmosphere for different lengths of time; and
- make different contributions to global warming.

patterns of energy use. Per capita emissions of CO_2 are much lower among developing countries, largely because per capita energy use is much lower.

While levels and concentrations of CO_2 are a major problem (Figure 8.15), the gas is much less effective as an absorber of infrared radiation than most of the other greenhouse gases.

Methane (CH_4): methane is a by-product of agriculture and comes mainly from cattle, termites, sheep and rice (Figure 8.16). The world cattle population has doubled in the last 40 years, and termites thrive on grasslands created by the clearance of forests for pasture. Little wonder, therefore, that atmospheric concentrations of methane have risen steadily since the 1950s. Concentrations are increasing more than twice as fast as CO_2. Analysis of ice cores reveals natural atmospheric concentrations of methane of about 0.3 ppmv at the peak of glaciation. Current levels are around 1.7 ppmv (compared with around 350 ppmv for CO_2). Methane is four times as effective as CO_2 as a greenhouse gas, so any further rise in atmospheric concentrations will add significantly to the greenhouse effect.

BOX 8.37 TRENDS IN ATMOSPHERIC CARBON DIOXIDE

Atmospheric concentrations of carbon dioxide have risen by about a quarter since the Industrial Revolution. Analysis of ice cores from Antarctica shows natural background levels of atmospheric CO_2 in the order of 200 parts per million by volume (ppmv) during glacial periods, rising to around 280 ppmv during warmer interglacials. In 1958, the atmospheric concentration was measured at 315 ppmv, and by the early 1990s it had risen to more than 350 ppmv (Figure 8.14). By 1997, the concentration stood at more than 360 ppmv, the highest level in 160,000 years. It is predicted to continue rising at a rate of 0.4 per cent a year, and to rise above 600 ppmv early in the twenty-first century.

Atmospheric concentrations of CO_2 are determined by the balance between production and consumption. Air pollution accounts for most production, but the consumption of CO_2 is decreased if and when carbon sinks (natural stores in the biogeochemical cycle; see p. 86) are removed. This is happening in various ways, including the burning of rainforests, the cultivation of soils, and the drainage and drying of wetlands, peats and tundra. Oceans and forests soak up carbon, so they need to be conserved if CO_2 concentrations are to be kept in check.

Nitrous oxide (N_2O): nitrous oxide is created naturally by microbial activity in soils, but it is increased by the use of nitrogen-based fertilisers and the burning of timber, crop residues and fossil fuels. Atmospheric concentrations are quite low, at around 0.31 ppmv, and they are rising much more slowly than methane (see Table 8.8).

Chlorofluorocarbons (CFCs): CFCs come entirely from industrial activities, and until recently have been widely used in refrigerators, air conditioners and fire extinguishers as aerosol propellants and solvents, and as foam-blowing agents for plastics. Atmospheric concentrations are very low – CFC-11 (0.00026 ppm) and CFC-12 (0.00044 ppm) – but the gases are long-lived and very efficient absorbers of infrared radiation.

Water vapour: water vapour is another greenhouse gas, in fact the most important of them all. Its concentration varies from place to place and over time, and it is affected much less by human activities than by normal atmospheric

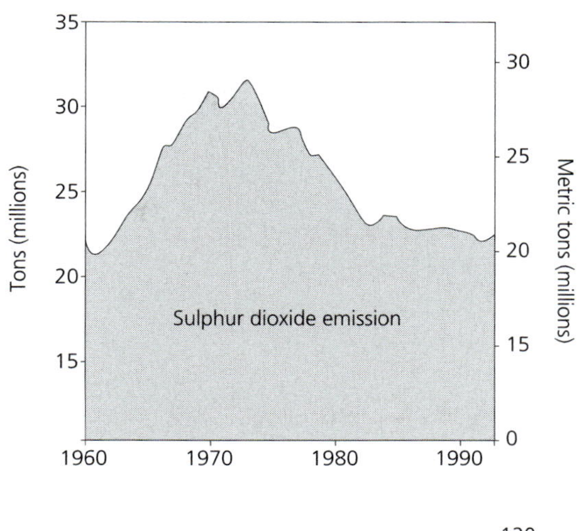

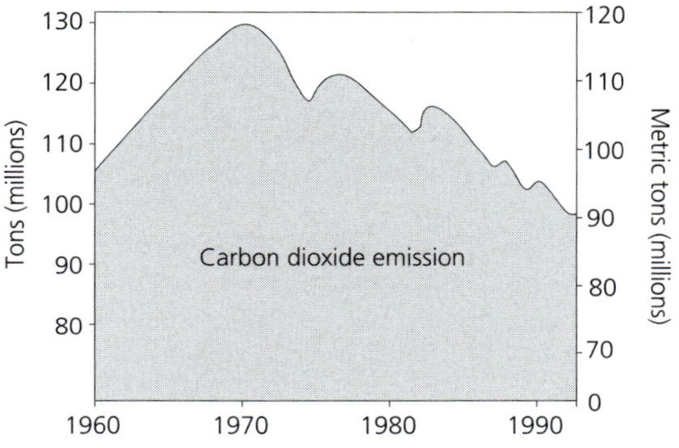

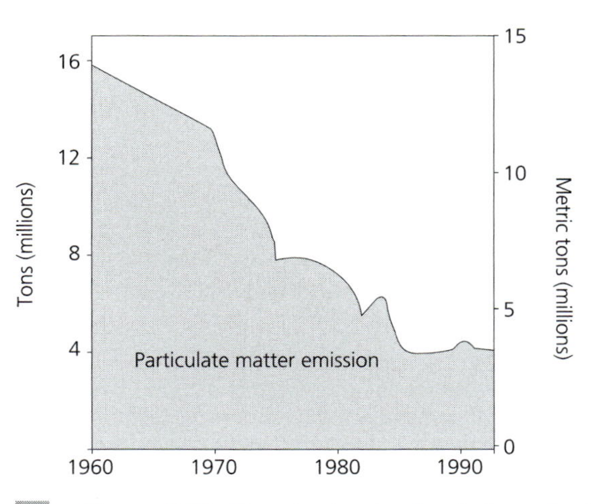

Figure 8.13 *Changing levels of emissions of sulphur dioxide, carbon dioxide and particulate matter in the USA, 1960–92. The skies over much of the USA have become much cleaner since the early 1970s, in particular, mainly in response to pollution control measures adopted by the public, industry and government.*

Source: Figure 9.14 in Marsh, W.M. and J.M. Grossa (1996) Environmental geography: science, land use and Earth systems. John Wiley & Sons, New York.

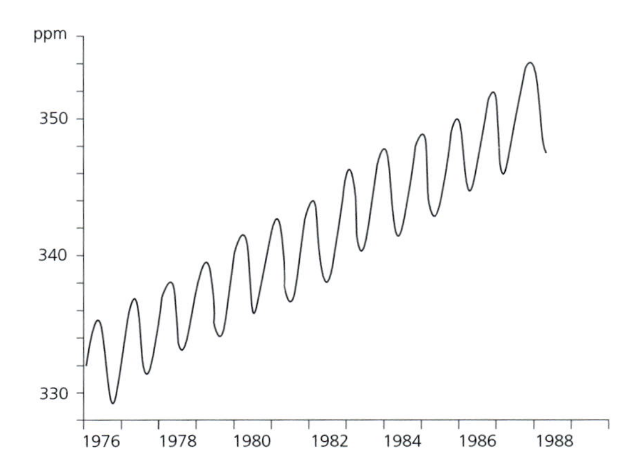

Figure 8.14 *Rising levels of carbon dioxide in the air over the Mauna Loa observatory, Hawaii, 1976–88. The graph shows a strong seasonal pattern superimposed on a steadily rising background level of carbon dioxide at this site, far removed from industrial sources of air pollution.*

Source: Natural Environment Research Council (1989) Oceans and the global carbon cycle. NERC, Swindon, p. 1.

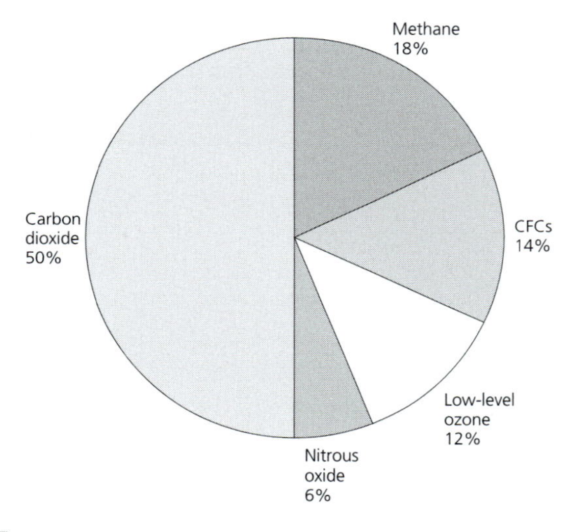

Figure 8.15 *The contribution of different greenhouse gases to global warming during the 1980s. Carbon dioxide and methane are the most important greenhouse gases, accounting for over two-thirds of the warming during the 1980s.*

Source: Natural Environment Research Council (1989) Oceans and the global carbon cycle. NERC, Swindon, p. 1.

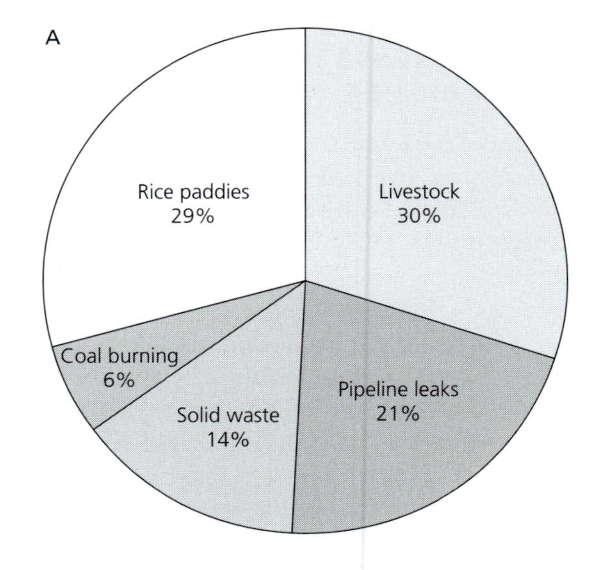

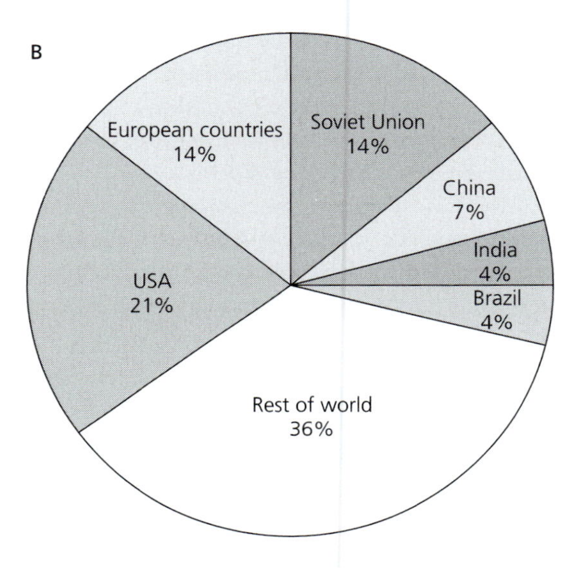

Figure 8.16 *Global sources of methane from human activities. The most important sources are livestock, rice paddies and pipeline leaks (a), and the USA contributes about a fifth of the global total (b).*

Source: Box Figure 17.5 in Cunningham, W.P. and B.W. Saigo (1992) Environmental science: a global concern. Wm. C. Brown Publishers, Dubuque, Iowa.

processes in the water cycle (see p. 352). Water vapour is involved in some important feedback links. A rise in temperature increases the capacity of the atmosphere to hold water vapour, and increased water vapour absorbs more energy, so this could further increase temperature. Increased atmospheric water vapour might also cause increased cloudiness, which (depending on the type of cloud formed) could affect the energy balance.

Many of the greenhouse gases are very long-lasting (see Table 8.8), surviving in the atmosphere for decades or longer. This has great implications for coping with global warming, because even if emission of all greenhouse gases stopped today there is a time lag built into the system, and the existing gases in the atmosphere would continue to have an effect for many years to come.

GLOBAL WARMING

Predicting likely future changes in atmospheric concentrations of the various greenhouse gases, and then using these figures to estimate likely atmospheric temperatures, is far from easy. Not only are emissions of the different greenhouse gases changing through time at different rates, and the different greenhouse gases have different life expectancies (see Table 8.8), but new evidence continues to emerge of new or previously overlooked sources of greenhouse gases (such as cement manufacture; see Box 1.2). What is more, each greenhouse gas contributes to the warming process in different ways and to different degrees. The so-called global warming potential (Box 8.38) is different for each greenhouse gas (Table 8.9), and this adds further complexity to what is already a very complex exercise.

BOX 8.38 GLOBAL WARMING POTENTIAL

IPCC expresses the effectiveness of different greenhouse gases in terms of 'global warming potentials' (GWPs). These indicate the contribution of each greenhouse gas to likely global warming, relative to carbon dioxide (Table 8.9). Thus, for example, methane is eleven times more effective than carbon dioxide, molecule for molecule. CFCs could clearly play a very significant role in global warming, so initiatives to cut down CFC emissions will help to reduce the greenhouse problem as well as helping to stop the attack on the ozone layer (see p. 249).

Estimates of future climate warming are based on scenarios that assume different rates of rise of carbon dioxide concentrations. This allows for the GWPs of all other relevant gases.

Table 8.9 Global warming potentials for the main greenhouse gases

Gas	*GWP*
Carbon dioxide	1
Methane	11
Nitrous oxide	270
CFC-11	3,400
CFC-12	7,100

BOX 8.39 IPCC FUTURE GREENHOUSE GAS EMISSION SCENARIOS

The most reliable forecasts of what the future might have in store are those based on a set of scenarios proposed by the IPCC:

Low-growth scenario
The lowest scenario assumes that world population will increase to 6.4 billion by 2100, the economy will grow by an average of 2 per cent a year by 2025 but slow down a great deal after that, and nuclear power costs will decline by 0.4 per cent a year. Under these assumptions, CO_2 emissions would increase from about 7.4 gigatonnes (Gt) in 2000 to 8.8 Gt in 2025 and then fall to 4.6 Gt by 2100. This would create an atmospheric concentration of CO_2 of 450 ppm by 2100, compared with 360 ppm in 2000.

High-growth scenario
The rapid-growth scenario assumes that world population grows to 11.3 billion people, the economy grows by 3 per cent a year over the next century, and nuclear power costs increase. Under these assumptions, CO_2 emissions would increase to 15.1 Gt by 2025 and to 35.8 Gt by 2100. This would create an atmospheric concentration of CO_2 of 900 ppm by 2100, compared with 360 ppm in 2000.

Middle-growth scenario
Under the middle scenario, annual emissions of CO_2 are expected to rise to 20.3 Gt by 2100, producing an atmospheric concentration of 700 ppm (double today's level) by 2100.

PREDICTED WARMING

IPCC scientists have predicted that if emissions of greenhouse gases continue at present rates, the Earth's temperatures will increase on average by about 2.5 °C by year 2050. Different scenarios (Box 8.39) put the likely increase as high as 4.5 °C or as low as 1.5 °C . This is very much a global average, and regional variations may be much higher than this, particularly in polar and near-polar regions.

Any rise of this magnitude would far exceed natural temperature changes over the last 8,000 years, which have

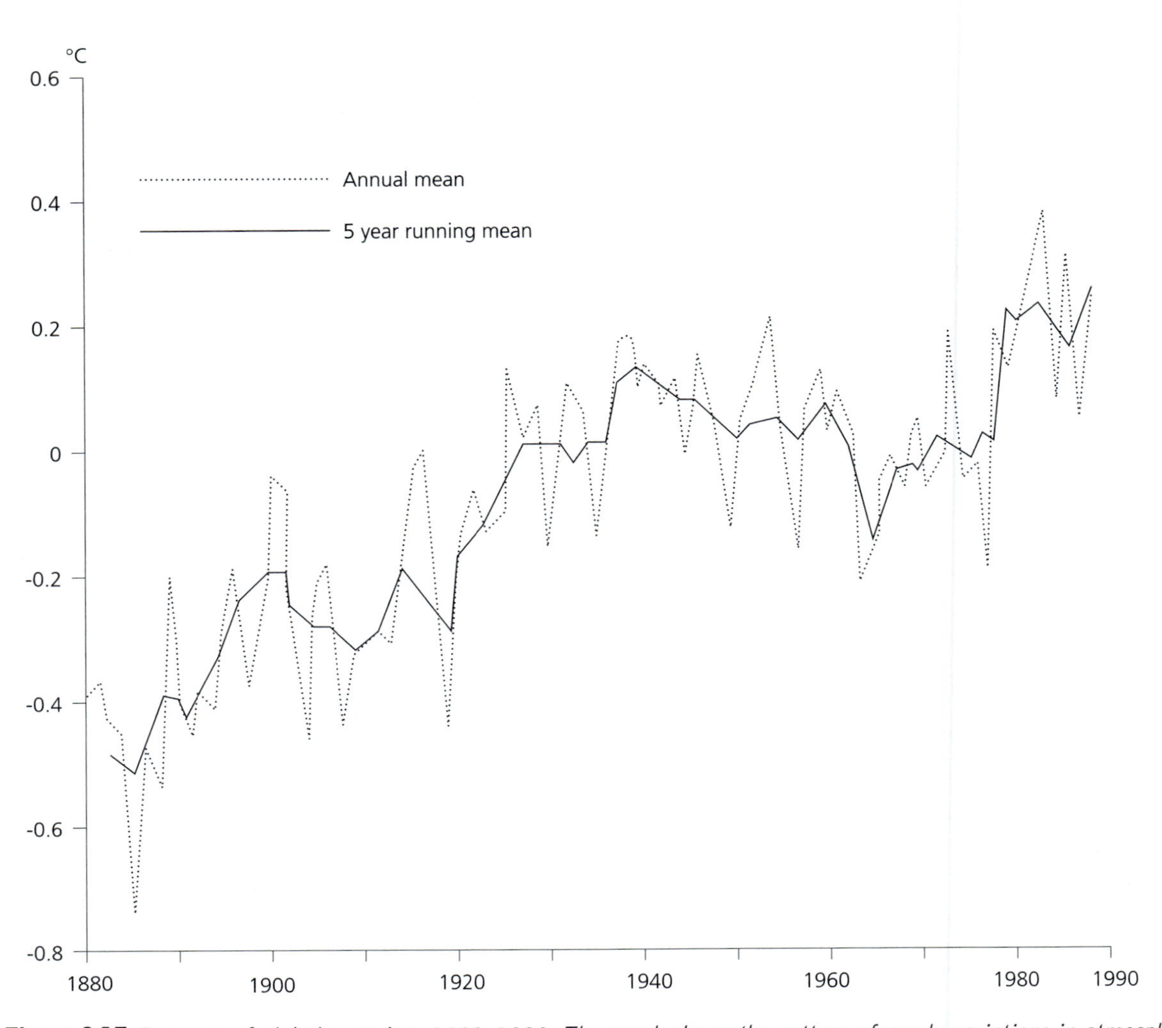

Figure 8.17 *Progress of global warming, 1880–1990. The graph shows the pattern of yearly variations in atmospheric temperature, with a smoothed five-year running mean superimposed to illustrate the general rise in temperature.*

Source: Figure 1.6 in Rees, J. (1991) Resources and the environment. In Bennett, R. and R. Estall (eds) Global change and challenge. Routledge, London, pp. 5–26.

been in the order of 1 °C. Since the Little Ice Age (between about 1450 and 1850), temperatures have risen by about 0.8 °C. Records show that over the past 100 years global mean temperature has risen by between 0.3 and 0.6 °C (Figure 8.17), so the next 100 years is likely to have some climatic surprises in store for us! Warming by as much as 5.0 °C would be greater than the difference between the depth of the last ice age (about 18,000 years ago) and today.

However, taking the Earth's temperature is not easy! The data used have been collected through a network of thermometers on ships, buoys and land-based weather stations, and are compiled by organisations such as the WMO (the World Meteorological Organisation), NASA and the NOAA (the US National Oceanic and Atmospheric Administration). These records go back to about 1860, and special care is taken to ensure that the records are reliable and consistent. Information on climate before 1860 must come from secondary sources, by examining biological indicators (such as tree rings and pollen records; see p. 341) and physical indicators (such as air bubbles locked in ancient ice; see p. 342).

Over the past 10 years, there has been some debate about whether global warming has started or whether it remains still just a possibility. The evidence of measurable changes is quite clear (Box 8.40), and research is uncovering some interesting variations in the response of environmental systems from place to place and over time (Box 8.41).

Most experts agree that a rise of between 1 and 2 °C above the average pre-industrial temperature is probably the most the Earth can cope with without massive and widespread environmental damage and disruption. Some fear that we may already be dangerously close to critical environmental thresholds (see p. 83), because the world is already at least 1 °C warmer than it was before the Industrial Revolution. If recent forecasts are correct, the warming will have reached 2 °C between 2040 and 2050.

If warming does occur as predicted, it will be relatively

BOX 8.40 HAS GLOBAL WARMING ALREADY STARTED?

Many scientists argue that it is too soon to know for certain whether global warming has begun. The IPCC insists that it will take at least 10 years of detailed monitoring (starting in the early 1990s) to detect for certain whether global warming has started, although it does not deny that it is likely. There are clear signs that the changes might already have started, however. For example:

- Global mean surface temperatures have risen by 0.45–0.6 °C (0.8–1.0 °F) since the nineteenth century.
- The surface of the ocean has also been warming at a similar rate.
- Studies that combine land and sea measurements have estimated that global temperatures have increased by 0.3–0.6 °C (0.5–1.0 °F) in the last century, and that about two-thirds of this warming took place between 1900 and 1940.
- Global temperatures have risen more rapidly since 1975 than in the period 1900 to 1940.
- Global temperatures are now higher than at any previous time during the period of instrumental records.
- The ten warmest years during the twentieth century all occurred in the last 15 years (1998 was the warmest year on record) (see Box 1.2).
- Snow cover in the northern hemisphere and floating ice in the Arctic Ocean have decreased since the start of the twentieth century; Europe's Alpine glaciers have lost half their volume since the 1850s.
- Worldwide precipitation has increased by about 1 per cent over the last 100 years.
- The frequency of extreme rainfall events has increased throughout much of the United States.
- Sea level has risen worldwide by 15–20 cm over the last century; roughly 2–5 cm of the rise has been caused by melting of mountain glaciers, and another 2–7 cm has resulted from the expansion of ocean waters.

BOX 8.41 VARIABILITY IN RESPONSE OF ENVIRONMENTAL SYSTEMS TO GLOBAL WARMING

- Surface temperatures are not rising uniformly: night-time low temperatures are rising on average about twice as quickly as daytime highs.
- The pace of change is not constant in all places: in the United States, temperatures over the last 50 years have cooled in the east and warmed in the west. The evidence suggests that regions downwind from major sources of sulphur dioxide emissions (see Box 8.32) have generally cooled.
- The pace of change is not uniform through the year: the winters in areas between 50 and 70° N (the latitude of Canada and Alaska) are warming relatively quickly, while summer temperatures show little trend.
- Urban areas are warming faster than rural areas, because of both changes in land cover and large-scale consumption of energy.
- Precipitation has increased by an average of 5 per cent over the last 100 years, but the increase in northern US states and southern Canada has been 10–15 per cent.
- Precipitation is increasing fastest in high-latitude areas, and it appears to have declined in some tropical areas.

rapid. Most forecasts are in the order of about 0.3 °C per decade, which is three times faster than ecosystems are believed to be capable of adjusting to. To make matters even worse, studies have shown that average global temperatures respond to elevated CO_2 concentrations via a process of lagged adjustment, so even cutting CO_2 emissions today might not be enough to prevent widespread environment disruption and damage.

DEBATE OVER CAUSES

While the records do show that global temperatures have been rising over the last century, not all scientists agree that observed changes have been caused by air pollution by greenhouse gases. One study of climate trends between 1870 and 1991 accounts for a slight decrease in global mean temperature between 1940 and 1975 as the influence of El Niño (see p. 483) and volcanic activity (see p. 173), and it dismisses as relatively insignificant the influence of sunspot activity and greenhouse gas emissions. Analysis of global and regional temperature variations recorded by NOAA satellites over the period 1979 to 1992 suggests that the Mount Pinatubo volcanic eruption in June 1991 (see Box 6.28) triggered a cooling trend that lasted at least a year, was strongest in the northern hemisphere and largely wiped out the warming from the 1991–92 El Niño.

Many human activities generate heat and release it directly into the atmosphere, and heat pollution must be taken into account when we are evaluating global warming. Heat pollution from anthropogenic sources may contribute between 5 and 10 per cent of the warming caused by greenhouse gases, but this additional heat loading may be significant, and it might push environmental systems closer to critical thresholds.

COMPLICATIONS

Scientists have generally agreed that a doubling of atmospheric CO_2 would increase the Earth's average surface temperature by about 1 °C. But the Earth's environmental systems are more complex than this simple prediction implies, because system response involves feedback which could either magnify or damp down warming. The major feedback loops involve natural processes such as the formation of ice, clouds, the circulation of the oceans and biological activity (Box 8.42).

Recent studies have suggested that the warming is likely to occur more rapidly over land than over the oceans. A further complication is that rises in temperature tend to lag behind increases in greenhouse gases (see p. 255). It is likely, therefore, that initially the cooler oceans will absorb much of the additional heat, which would serve to decrease the warming of the atmosphere. Only when the ocean comes into equilibrium with the higher level of CO_2 will the full atmospheric warming occur, after this time lag.

ENVIRONMENTAL CHANGE

Some startling scenarios have suggested what might happen if global warming were to proceed on the scale envisaged (Figure 8.18). It is impossible to predict precisely how climate will change, so there are many areas of uncertainty (see Table 3.9). The most important changes are likely to include:

BOX 8.42 MAJOR FEEDBACK PROCESSES IN THE GLOBAL WARMING SYSTEM

1. *Ice albedo feedback*: as the atmosphere warms ice caps will melt, so the proportion of the Earth's surface covered with ice will shrink as it is replaced by water or land. Ice is very efficient at reflecting solar radiation into space – it has a high albedo (see p. 234) – whereas land and water absorb more radiation (they have relatively low albedos). The Earth's surface will thus trap more heat and increase warming further. This is positive feedback (see p. 79).
2. *Water vapour feedback*: as the atmosphere warms, evaporation rates (see p. 291) will increase, and this will mean that there is more water vapour (see p. 292) in the atmosphere. Water vapour is a greenhouse gas (see p. 258) so it would contribute to the greenhouse effect by positive feedback.
3. *Cloud cover feedback*: greater evaporation means more water vapour in the atmosphere, and this in turn may increase cloud cover (see p. 296). It becomes very difficult to predict quite how this will affect global warming, because some types of cloud (particularly those that create a thick continuous cloud cover) produce negative feedback (damping warming by effectively blocking off sunlight). On the other hand, some clouds (high, light and discontinuous clouds, such as cirrus) may trap heat in the lower atmosphere and provide positive feedback.

BOX 8.43 UNCERTAINTIES SURROUNDING GLOBAL WARMING

There are a great many areas of uncertainty surrounding the issue of global warming, which makes the search for acceptable and sustainable solutions particularly difficult. Among the key uncertainties are:

- *Carbon sinks*: not all of the CO_2 emitted as air pollution stays in the atmosphere, because some is absorbed by vegetation on land (mostly by forests) and some is deposited in the oceans. If the rates at which these 'sinks' operate is altered, then the rate of build-up of the greenhouse gases in the atmosphere will also change. The net effect could be to either increase or decrease global warming.
- *Vegetation as a carbon sink*: continued felling of tropical rainforest (see p. 588) would reduce the carbon uptake by tropical trees and thus speed up global warming. Planting more trees would have the opposite effect and slow down global warming. The way in which vegetation responds to global warming may be critical, because higher temperatures and the fertilising effect of more CO_2 in the air will stimulate faster growth of trees and other vegetation. This would soak up more of the atmospheric CO_2 and thus reduce global warming. But this process of damping down global warming would only be possible provided that the water shortage does not constrain plant growth, and so long as the pace of climate change is not too fast for plants (particularly forests in northern latitudes; see p. 585) to adapt to it.
- *Oceans as a carbon sink*: another major carbon sink occurs in the oceans, and it begins with the 'conveyor belt' of the North Atlantic. The formation of ice in the North Atlantic makes the ocean water more saline (see p. 477) and thus denser, so it descends to the ocean floor. As it descends, it carries dissolved CO_2 with it. Some scientists expect that global warming might reduce the formation of ice, which would slow down the currents and/or make them carry less water. As a result, less CO_2 would be removed from the atmosphere. This would promote positive feedback, and global warming could happen even more quickly than predicted. Once it is dissolved in the water, much of the CO_2 is absorbed by plankton and other marine organisms. They turn it into organic compounds, most of which eventually fall to the ocean floor. So the strength and direction of this carbon sink depend a great deal on the biological productivity of the ocean, which itself is determined partly by water temperature. Some scientists have proposed adding iron dust to ocean water, because this would make the oceans more fertile and thus more effective carbon sinks.
- *Volcanic activity*: experience has shown that some major volcanic eruptions can affect atmospheric temperatures by ejecting particles that scatter sunlight. When Mount Pinatubo (see Box 6.28) erupted in June 1991, for example, it threw huge amounts of sulphate particles into the atmosphere, and for a short time afterwards the Earth's atmosphere was marginally cooler. It took up to two years for the volcanic material to settle out of the atmosphere and return temperatures to normal.
- *Sulphur pollution*: aerosols such as dust from soil erosion (see p. 609) and desert wind storms (see p. 433) can moderate atmospheric warming, and an interesting example of this is the sulphate particles emitted from the burning of fossil fuels, which fall as acid rain (see p. 252). These serve to reduce atmospheric warming, and in some heavily polluted industrial areas – such as parts of Central Europe and parts of China – high levels of sulphur pollution have been associated with atmospheric cooling. By a strange paradox, one form of air pollution can at least partially offset the impacts of another form!

Sea-level rise: as temperatures rise, the oceans warm and the water in them expands. Glaciers and possibly polar ice caps would also melt. The combined effect would be a rise in sea level. The best estimates are a rise of between 24 and 34 cm over the next 60 years; it is currently rising at about 2.4 mm a year. Sea-level rise associated with global warming is explored further in Chapter 16 (see p. 511).

Climate and weather changes: world climate could change in a number of ways. For example, cold seasons would become shorter, and warm ones would become longer. Northern latitudes would have wetter autumns and winters and drier springs and summers. There would be more rainfall in the tropics, and subtropical areas could become drier. Global warming is likely to cause shifts in the main climate belts around the world, and this will

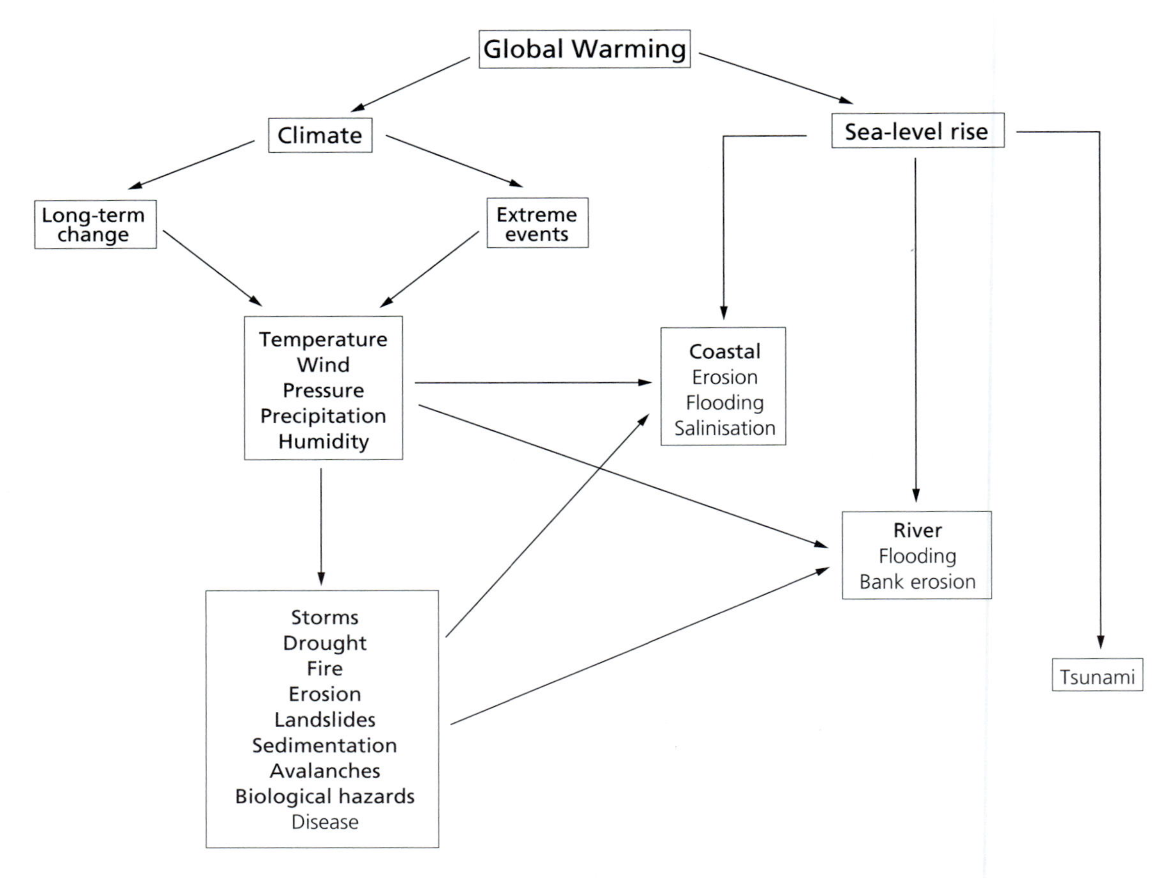

Figure 8.18 *Possible chain reaction of environmental impacts of global warming. Global warming might trigger a cascade of impacts, both directly (through the mechanism of climate change) and indirectly (via sea-level rise).*
Source: Figure 3.11 in Pickering, K.T. and L.A. Owen (1994) An introduction to global environmental issues. Routledge, London.

probably bring a rise in the frequency and intensity of floods, droughts, tropical cyclones and tornadoes in many areas. This would further aggravate the losses and hardships caused by sea-level rise.

Major disruptions to agriculture: there are likely to be winners and losers. If global warming causes a shift of climate zones towards higher latitudes, parts of the Sahara Desert and southern Russia would be wetter, and their agricultural output would rise accordingly. The grain lands of North America, southern Europe and the Commonwealth of Independent States would probably become warmer and drier, and suffer more heatwaves, droughts and shortages of irrigation water. Food production in developing countries might fall sharply, because increased flooding, drought, erosion and desertification would reduce yields and increase losses. Mass famine in some areas is not inconceivable.

Changes to other environmental systems: scientists expect global warming to trigger changes in many other environmental systems, including movements in the main vegetation belts (altitudinal and latitudinal), changes in runoff and soil moisture conditions in many areas, increased flooding of low-lying areas, changes in levels and patterns of precipitation, changes in glacier systems, melting of permafrost, and changes in the state of wetlands, deltas and coral reefs; other human activities may serve to compound the effects of global warming on geomorphological processes. Climate change might also promote changes in ocean circulations, which could further modify temperature patterns and the development of monsoon circulations and mid-latitude storms.

Mass extinctions of wildlife: wildlife would also be at risk from mass extinctions caused by rapid environmental change and lack of refuges (see p. 558). This could seriously affect habitats like the tropical rainforests, which contain over half (perhaps up to 90 per cent) of the world's species (see p. 527). Sea-level rise would also threaten wildlife dependent on habitats like salt marshes and coral reefs.

SOLUTIONS

While there are still many different opinions about the best way forward, by the mid-1990s there were signs of an emerging international consensus that a 'no regrets' policy must be adopted, which involves erring on the side of caution and planning well in advance. There is growing support for the 'precautionary principle', which states that if there is a risk of damage we should not wait until it can be confirmed scientifically before action is taken to minimise it.

The most obvious target in such a 'no regrets' policy is the greenhouse gases that contribute to the global warming. As we have seen, a number of important gases are involved, and they come from different sources, so there is no simple solution. Agenda 21, agreed at the 1992 Rio Earth Summit, proposed a wide variety of different ways of reducing greenhouse gas emissions (Table 8.10). The Kyoto Protocol (see Box 1.6) builds upon the Rio Climate Change Convention (see p. 15) and proposes a series of policy tools designed to encourage appropriate greenhouse gas emission reduction.

Global warming can probably best be slowed down by a combination of steps based on two broad strategies:

1. reducing emissions of greenhouse gases; and
2. enhancing the terrestrial sinks and stores for greenhouse gases (e.g. by planting forests to absorb carbon from the atmosphere).

The most realistic approach to tackling the global warming issue involves adopting both strategies together. This is clearly shown in a model of how predicted global warming might affect the United Kingdom. In a 'business as usual' scenario, the model predicts a temperature rise of 3.6 °C above pre-industrial levels by 2100, which would cause damage in the UK with a present value estimated at US$8.9 × 10^{18} (million million million). An optimal solution would require an immediate 12.7 per cent cut in emissions and the establishment of 371,000 km^2 of new forests, which would reduce temperature rise by 0.3 °C relative to 'business as usual'. New approaches to monitoring greenhouse gas emissions are required, such as emissions inventories (Box 8.44).

Table 8.10 Ways of reducing greenhouse gas emissions

The energy sector

- introduce a carbon tax on electricity generation
- introduce a higher carbon dioxide tax and retain the existing energy tax on non-energy-intensive industry
- introduce an energy tax on combined heat and power
- introduce measures to save electricity (including standards for domestic appliances)

The transport sector

- new rules on company cars and tax (to reduce long-distance travel)
- expand public transport systems
- set carbon dioxide emission limits on light vehicles
- further develop environmental classification systems for vehicles and fuels
- reduce average speeds on roads
- subject all transport plans and infrastructure investments to environmental impact assessments
- experiment with the introduction of electric vehicles

Other greenhouse gases

- reduce the agricultural use of nitrogen fertilisers
- expand methane extraction from waste tips
- reduce process emissions of fluorocarbons from aluminium smelters and ban their use as chemicals; reduce use of hydrofluorocarbons

BOX 8.44 EMISSIONS INVENTORIES

One of the important developments during the late 1990s was the creation of mechanisms for compiling emissions inventories. These are databases that allow the amount of greenhouse gases emitted from different countries and areas to be monitored and audited. The need for such inventories became obvious at the 1992 Rio Earth Summit (see p. 13) and during debate leading to the 1997 Kyoto Protocol (see Box 1.6).

An emissions inventory would usually include information on:

- the chemical or physical identity of the pollutants included;
- the geographical area covered;
- the institutional entities covered;
- the time period over which the emissions are estimated; and
- the types of activity that cause emissions.

Emission inventories are used in various ways. For example, they provide valuable inputs to air-quality models, allow the success of air-quality policies and standards to be evaluated, and enable regulatory agencies to establish compliance records with allowable emission rates.

Many of the proposed changes – particularly those based on reducing greenhouse gas emissions – would require radical changes in people's attitudes, values and behaviour. Time alone will tell whether (both individually and collectively) we take the risk of global warming seriously enough to act without delay. The stakes are high.

INTERNATIONAL INITIATIVES

The long-term and global character of the climate change problem requires an international long-term strategy based on internationally agreed principles such as sustainable development (see p. 26) and the precautionary principle (see p. 630).

Negotiations on an international climate change convention began in February 1991 under the auspices of the UN General Assembly. They were taken further in the Climate Change framework convention signed by heads of state at the 1992 Rio Earth Summit (see p. 13). The 153 countries (plus the European Union) that signed the Convention on Climate Change undertook to stabilise emissions of carbon dioxide by the year 2000 and then begin to reduce them. This will require radical changes in lifestyle and patterns of consumption for many people, as well as major changes in transport and energy systems.

BOX 8.45 PROBLEMS IN IMPLEMENTING INTERNATIONAL AGREEMENTS

Implementing agreed reductions in greenhouse gas emissions – such as those agreed under the Kyoto Protocol (see Box 1.6) – is not without real problems. This is partly because of the difficulties of measuring emission levels with acceptable levels of accuracy and reliability. Emissions from factories and power stations can be measured quite accurately, but other sources (such as land use) are much more difficult to monitor. Yet until such problems can be solved, it will not be possible to compile national inventories of greenhouse gas emissions, and without these it will be difficult to verify that targets are being reached. There are additional bureaucratic problems, particularly in the European Union, where many member states favour decision making based on national autonomy rather than central political authority.

The convention is an important milestone in the battle against global warming, but it is far from perfect. One particular weakness is that it is voluntary and thus contains no effective mechanism for enforcing the goal of reducing greenhouse gas emissions to 1990 levels by the year 2000. It also says nothing about goals after 2000.

The 1997 Kyoto Protocol attempts to create a framework within which both the spirit and the purpose of the Rio climate change agreements can be pushed forward, although time will tell whether more concessions will be needed to bring the major greenhouse-gas-emitting countries fully on board. Many scientists argue that there is no time for further delay and debate, because while the politicians stall for more time and quibble over fine points, greenhouse gas emissions continue to rise, atmospheric concentrations of greenhouse gases continue to mount, and global climate change continues to accelerate.

Developing countries stand to lose much more than developed countries. They are in a 'double bind' – reducing greenhouse gas emissions will seriously hinder their industrial and thus economic development, but they are also more vulnerable to global warming and its associated environmental changes. Most developing countries

SUMMARY

This chapter explores the structure and importance of the atmosphere, the relatively thin layer of air around the Earth that creates conditions suitable for life, within which weather systems operate, and which interacts with the other main environmental systems. After outlining the main ingredients of the atmosphere (gases, water vapour and solids), we looked at its vertical structure and saw the particular relevance of the troposphere and stratosphere. Energy from the Sun drives the atmospheric system and other environmental systems, and we considered the nature of that energy and the processes by which it interacts with the atmosphere and the Earth. The Earth's energy budget and the natural greenhouse effect (which make the Earth suitable for life) were explained next. Sunlight provides an abundant source of renewable energy, and we considered ways of making use of it (via solar collectors, solar cells and solar heating). Much of the chapter focuses on air pollution, its causes and impacts at various scales. A variety of types of air pollution were reviewed, including that created by industrial and military accidents, the problems of low-level ozone, and photochemical smog. The three most important air pollution problems today are depletion of stratospheric ozone, creation of acid rain and emission of greenhouse gases. The causes and impacts of and solutions to global warming caused by greenhouse gas emissions were examined in some detail, because we return to these themes in a number of the following chapters.

currently contribute relatively little to the global total output of greenhouse gases, but this is changing fast as population and energy consumption rise.

WEBSITE

Links to relevant websites, a comprehensive bibliography, tools for teaching and learning, and downloadable images relevant to this chapter can be found at the website specially designed to accompany this book at
http://www.park-environment.com

FURTHER READING

Adger, W.N. and K. Brown (1994) *Land use and the causes of global warming.* John Wiley & Sons, London. A detailed guide to the scientific and policy debate concerning the roles of agriculture, forestry and other activities leading to global warming.

Barry, R.G. and R. Chorley (1998) *Atmosphere, weather and climate.* Routledge, London. Useful and up-to-date seventh edition of the classic text on atmosphere and world climate.

Boehmer-Christiansen, S. and J. Skea (1991) *Acid politics.* John Wiley & Sons, London. Explores and tries to explain policy reactions in Britain and Germany to acid rain.

Brack, D. (1999) *International trade and climate change policies.* Earthscan, London. Explores the implications of the Kyoto Protocol for international trade, including energy pricing through taxation, setting of energy efficiency standards for industry and potential trade restrictions arising from enforcement of the protocol.

Brack, D., M. Grubb and C. Vrolijk (1999) *The Kyoto Protocol: a guide and assessment.* Earthscan, London. A detailed analysis of the agreement to limit greenhouse gas emissions and control climate change.

Bridgman, H. (1990) *Global air pollution: problems for the 1990s.* John Wiley & Sons, London. Detailed review of the scientific and social aspects of air pollution, including global warming, acid rain, ozone depletion and nuclear winters.

Cline, W.R. (1992) *The economics of global warming.* Longman, Harlow. A detailed study of the costs and benefits of global action to limit greenhouse warming.

Colls, J. (1996) *Air pollution.* E & FN Spon, London. Comprehensive survey of major air pollution issues.

Elsom, D. (1992) *Atmospheric pollution: a global problem.* Blackwell, Oxford. A comprehensive and balanced introduction to the causes, effects and controls of air pollution.

Elsom, D. (1996) *Smog alert: managing urban air quality.* Earthscan, London. A clear review of the problems of urban air pollution and the ways they can be tackled, drawing on case studies from around the world.

Graedel, T.E. and P.J. Crutzen (1993) *Atmospheric change: an earth science perspective.* W.H. Freeman, New York. A useful introduction to how the atmosphere works, recent natural changes and responses to human impacts.

Graedel, T.E. and P.J. Crutzen (1995) *Atmosphere, climate and change.* W.H. Freeman, Oxford. An accessible introduction to topics such as the greenhouse effect, the ozone layer, pollution and the effects of long-term climate changes.

Harvey, D. (2000) *Global warming.* Longman, Harlow. Detailed and clearly written overview of the science of global warming and the interrelationships between climate, weather and environmental change.

Hinrichs, R.A. (1996) *Energy*. Saunders College Publishing, Philadelphia. A clear exploration of the basic physical principles related to energy use and the environment, including global warming, radioactive waste management and the Rio Earth Summit.

Jackson, T., K. Begg and S. Parkinson (eds) (2000) *Flexibility in climate policy: making the Kyoto mechanisms work*. Earthscan, London. Examines the key economic, social and ethical issues and implications of the Kyoto Protocol and assesses the operational design of the flexibility mechanisms of joint implementation, including emissions trading.

Kemp, D.D. (1994) *Global environmental issues: a climatological approach*. Routledge, London. A clear introduction to atmospheric aspects of the environmental crisis, including global warming, ozone depletion, drought and acid rain.

Leggett, J. (2000) *Global warming and the end of the oil era*. Penguin Books, London. Interesting review of the political debate surrounding global warming during the 1990s, clearly and engagingly written by a former scientist with Greenpeace.

Mabey, N., S. Hall, C. Smith and S. Gupta (1997) *Argument in the greenhouse: the international economics of controlling global warming*. Routledge, London. Explores international issues and tensions in the search for effective solutions to the problem of global warming.

O'Riordan, T. and J. Jager (eds) (1995) *Politics of climate change: a European perspective*. Routledge, London. A critical analysis of the political, moral and legal response to climate change.

Park, C.C. (1987) *Acid rain: rhetoric and reality*. Methuen, London. Describes the nature of the problem, how it affects lakes, rivers and forests, and how it might be tackled.

Park, C.C. (1989) *Chernobyl: the long shadow*. Routledge, London. A detailed review of the causes and consequences of the world's worst nuclear accident to date.

Parry, M. and R. Duncan (eds) (1995) *The economic implications of climate change in Britain*. Earthscan, London. A detailed analysis that uses possible scenarios from the Intergovernmental Panel on Climate Change.

Paterson, M. (1996) *Global warming and global politics*. Routledge, London. Reviews the major theories within international relations and considers how these help to account for the emergence of global warming as a political issue.

United Nations Environment Program (1993) *Environmental data report 1993–94*. Blackwell, Oxford. A useful source of recent data on air quality and atmospheric change.

Wellburn, A. (1994) *Air pollution and climate change: the biological impact*. Longman, Harlow. A review of the ecological impacts of global warming, stratospheric ozone depletion and air pollution control.

Whyte, I. (1995) *Climatic change and human society*. Edward Arnold, London. Explores the interactions between climate change and society, drawing examples and case studies from history and the present day.

Chapter 9

Atmospheric Processes

LEARNING OBJECTIVES

When you have finished studying this chapter, you should be able to

- *Demonstrate the relevance of the main atmospheric processes that affect temperature, air movement and moisture in the atmosphere*
- *Summarise the factors that affect air temperature and cause its spatial and temporal variations*
- *Define what is meant by 'air stability' and show how it is defined*
- *Summarise the factors that affect air movement and generate winds*
- *Describe and account for the main features of global wind circulation*
- *Appreciate the principles of exploiting wind energy*
- *Explain what is meant by humidity in the atmosphere and describe the main phase changes of water*
- *Outline the main products of condensation and describe the main cloud types*
- *Describe how and why precipitation occurs*

The atmosphere is probably the most dynamic part of the global environmental system, and it responds quickly to external triggers (particularly variations in incoming solar radiation) and internal triggers (such as pressure differences). These responses are what create weather (Chapter 10) and climate (Chapter 11), and through this they have a strong influence over many different aspects of life on Earth, for all living organisms and not just humans.

In this chapter, we focus on the main processes that take place in the lower atmosphere and are responsible for the redistribution of heat, air and moisture from place to place across the Earth's surface.

ATMOSPHERIC TEMPERATURE

TEMPERATURE

Temperature is the degree of hotness, and levels and variations in temperature significantly affect people (Box 9.1) and environmental systems. People are particularly susceptible to extremes of temperature. Extreme heat puts great stress on the human body – we lose a lot of water and body nutrients through perspiration, have no energy and get tired easily, find it difficult to think properly, cannot perform routine tasks, and have short tempers. Heat stress is common in areas, like the Midwest USA, that regularly experience light winds and long, sunny days. Droughts and water shortages (see p. 419) often accompany prolonged hot, dry spells.

Heatwaves (spells of continuous abnormally hot weather) can occur in many places, and sometimes they are welcomed because they bring unusually warm summers to countries like Britain. Many areas with hot climates (such as Mediterranean Europe, Florida, the Caribbean islands and parts of tropical Africa) have well-developed and economically important tourism industries.

Systems have been developed for harnessing heat from the Sun's rays to warm buildings, both by passive collecting (Figure 9.1) and by active collecting (Figure 9.2) (see Box 8.16).

Extreme cold also affects people in different ways. On the plus side, persistent cold air (with temperatures mostly

BOX 9.1 TEMPERATURE AND PEOPLE

Air temperature is one of the elements of weather that we experience directly. Humans are very sensitive to levels and changes of temperature, because our bodies respond to preserve heat if it gets too cold or shed heat if it gets too hot. Clothes are designed partly with this purpose in mind, as are buildings. We develop technologies (central heating and air conditioning) to control air temperature artificially in buildings and keep conditions comfortable for living and working, and to create artificially cold conditions (refrigerators and freezers).

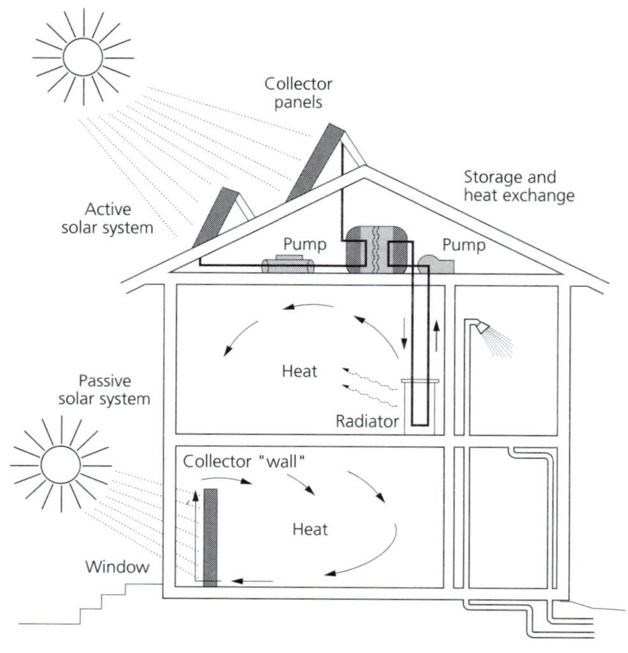

Figure 9.1 *Passive solar heating. Passive solar heating systems are common in houses and other buildings, in which short-wavelength incoming solar energy passes through a glass window, and the re-radiated longer-wavelength radiation cannot escape through the glass and thus heats the room or building.*

Source: Figure 7.16 in Marsh, W.M. and J.M. Grossa (1996) Environmental geography: science, land use and Earth systems. John Wiley & Sons, New York.

below freezing) allows snow cover to survive through much of the year in many high mountain areas, and this is exploited by the skiing industry. Antarctica, one of the world's great remaining wilderness areas (see p. 448), depends for its survival on extremely low temperature. Global warming associated with pollution by greenhouse gases (see p. 258) threatens the future of frozen places like Antarctica and the Arctic. On the minus side (literally, below freezing), the extreme cold in places like the Arctic severely restricts outdoor activities for visitors and permanent residents. Long polar winters have been known to make visitors to the Antarctic lethargic and sometimes even psychopathic (antisocial and violent, but without guilt), possibly because the severe climate induces transient hibernation and makes them less sensitive than normal.

Temperature affects people indirectly in many different ways. Probably the most important one is via food production, because the heat and light in sunlight are ultimately what power plant growth via photosynthesis. Plant growth in turn affects grazing animals and, through them, carnivores.

INFLUENCE ON ENVIRONMENTAL SYSTEMS

Many environmental processes and systems are also strongly influenced by temperature. Levels and variations of temperature are principal controls of the rates and patterns of rock weathering (see p. 205) and erosion (p. 207), for example. Temperature differences between places give rise to pressure gradients, which generate wind and air movements (p. 280). Indeed, the global air circulation system (p. 284) is driven by thermal gradients and the need to redistribute heat across the Earth's surface. Similarly, the major ocean current systems (p. 479) are strongly influenced by temperature variations.

MEASURING TEMPERATURE

Temperature is measured using a number of different scales:

Fahrenheit (F): this is the temperature scale most widely used in meteorology and climatology. On this relative scale water freezes at 32 °F and boils at 212 °F (at sea level). The Fahrenheit scale is widely used in the USA.

Celsius (Centigrade) (C): this temperature scale is also used widely. On this relative scale water freezes at 0 °C and boils at 100 °C (at sea level). The Celsius scale is widely used in the non-English speaking world and by scientists, and it is easy to convert temperatures between the Fahrenheit and Celsius scales (Box 9.2). Some common temperatures are listed in Table 9.1 for comparison, and record collectors might be interested in Table 9.2.

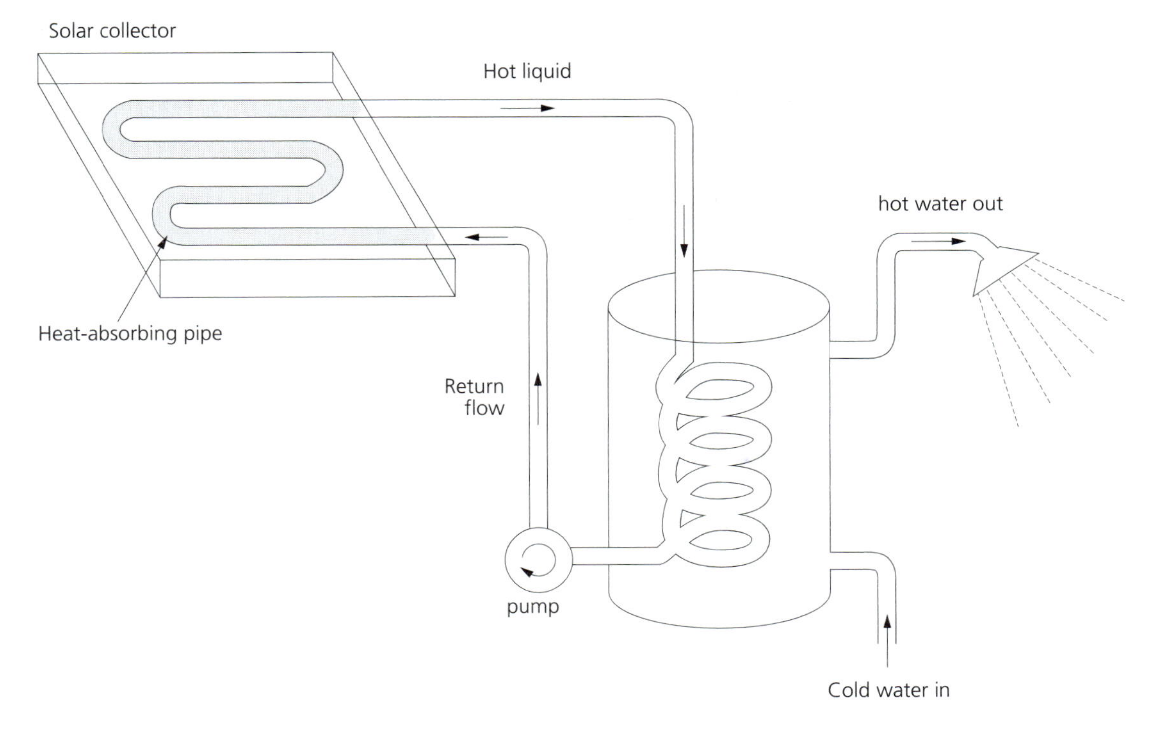

Figure 9.2 *Active solar heating. The figure illustrates the principles of an active, closed-cycle solar water-heating system, in which sunlight is used to heat a liquid in a pipe, so that cold water in contact with the pipe is itself heated.*

Source: Figure 20.7 in Cunningham, W.P. and B.W. Saigo (1992) Environmental science: a global concern. Wm. C. Brown Publishers, Dubuque, Iowa.

BOX 9.2 CONVERTING TEMPERATURES TO DIFFERENT UNITS

It is easy to convert temperatures from one scale to the other, using the formulae:

$$°C = 5/9\ (°F - 32) \qquad °F = 9/5\ (°C + 32)$$

To convert from Kelvin to Celsius, subtract 273.15 from the temperature (in K); to convert from Celsius to Kelvin, add 273.15 to the temperature (in °C).

Kelvin (K): this is an absolute temperature scale, on which there is an absolute zero temperature (0 K) and each degree is equal to that on the Celsius scale. Absolute zero on the Kelvin scale is −273.15 °C (−406 °F), and water freezes at 273.15 K. Note that degree symbols (°) are not used on the Kelvin scale, which is used at extremely low temperatures, mostly by physicists but rarely by climatologists or meteorologists.

Table 9.1 Some common temperatures for comparison

Situation	*Temperature (°C)*
Surface temperature of the Sun	5,330
Molten lava	1,730
Gold melts	1,064
Water boils	100
Some bacteria survive	70
Highest recorded shade temperature	58
Normal human body temperature	37
Comfortable room temperature	20
Pure water freezes	0
Arctic seawater	−1
Mercury freezes	−39
Temperature of outer space	−270

Table 9.2 Some record temperatures on Earth

Record	*Place*	*Temperature*	*Comment*
Hottest place on Earth	Dallol in Ethiopia	34.3°C	Annual average temperature
Highest temperature recorded on Earth	Al'Aziziyah in Libya	58°C	
Highest recorded temperature in North America	Death Valley, California	56.7°C	
Coldest place on Earth	Polus Nedostupnosti in Antarctica	−57.8°C	Annual average temperature
Lowest temperature recorded on Earth	Vostok in Antarctica	−88.3°C	

SOLAR ENERGY AND TEMPERATURE

The temperature of the air and ground varies from place to place and over time. These spatial and temporal variations are controlled mainly by the amount of solar energy received at different places on the Earth's surface after it has passed through space and then through the Earth's atmosphere.

ATMOSPHERIC HEATING

We have seen in earlier chapters that:

- The Sun radiates energy into space at a relatively uniform rate (see p. 107).
- A small amount of that solar energy (the solar constant) reaches the outer edge of the Earth's atmosphere (p. 232), and this amount varies through the Earth's orbit around the Sun (slightly less is received when the Earth is furthest from the Sun) (p. 118).
- The quantity and composition of the incoming solar radiation are changed as it passes through the atmosphere (p. 229).
- The atmosphere's natural greenhouse effect traps much of the outgoing long-wave re-radiation from the Earth, and this heats the atmosphere (p. 237).

These factors help to account for how and why the Earth's atmosphere is heated by sunlight, but they provide no explanation of why there are such pronounced differences in heating from place to place. The equator is hot, the poles are cold – but why?

SPATIAL VARIATIONS

Spatial variations in atmospheric heating reflect two main factors:

1. the angle at which the Sun's rays strike the Earth's surface (the angle of incidence); and
2. the length of day.

Angle of incidence: the angle at which the Sun's rays strike the Earth is important, because it affects the concentration or diffusion of heat across the Earth's surface. This angle varies with latitude. Where the Sun is more or less directly overhead and its rays shine directly down to the ground – as at the equator – the sunlight is concentrated and produces more heat. Towards the poles, at higher latitudes, the Sun's rays shine at a steep angle from the vertical, so they are more diffuse and less effective in heating the surface. With increasing latitude comes more diffusion and less direct heating. This helps to explain the spatial pattern of temperatures across the Earth's surface. Average temperatures are highest in low latitudes, close to the equator, where the Sun's rays are most nearly vertical for most of the year. Temperatures fall towards the poles, where the Sun's rays are more diffuse.

Length of day: the amount of solar heating also varies through the seasons, reflecting variations in length of day. Temperatures are higher when the Sun is high in the sky (during the summer), and cooler when the Sun is low in the sky. In middle and high latitudes, the length of daylight in summer is greater than the length of darkness (night), so daytime heating exceeds night-time cooling and the summer season becomes gradually warmer. The reverse applies in winter, when progressive cooling is caused by inefficient heating by the low-angle Sun coupled with the longer periods of darkness (and cooling).

Because of this progressive seasonal heating and warming, middle and high latitudes often experience a temperature lag or delay in heating. So, for example, the hottest month in the northern hemisphere is usually July, which comes after the summer solstice (21 June), and the coldest month is normally January, which follows the winter solstice (22 December) (see Table 4.8). There is a similar lag of cold and warm seasons in the southern hemisphere.

HEATING PROCESSES

The Earth's surface and atmosphere are heated by four main processes – radiation, convection, conduction and advection.

Radiation: radiation, in this sense, is the transfer of energy in the form of electromagnetic waves (see p. 233). This process allows heat to flow, because bodies that are warmer than their surroundings radiate heat energy to cooler areas around them. The Earth is warmed by shortwave radiant energy from the Sun (see p. 237), and it warms the overlying atmosphere by long-wave radiation (see p. 237). This heating process operates in much the same way as a domestic central heating radiator heats a room.

Convection: convection is the process of heat transfer through a gas or fluid. It explains why warm air rises and transfers heat vertically within the atmosphere. Air close to the ground is heated by re-radiation, and warm air naturally rises because it is less dense than the surrounding cooler air. When a gas is heated, it expands and becomes lighter (less dense). As the warm air rises it is replaced by cooler surrounding air, which is then heated by re-radiation, and so the process continues as long as heat is available at the ground to drive the process. Convectional uplift can be very powerful, and it plays an important role in cloud formation and the development of strong winds (see pp. 296 and 312). Convection currents are strong, particularly near the equator, where the hot ground heats overlying air, which rises and spreads out north and south to drive important thermal cells in the global wind system (see p. 286).

Conduction: conduction involves the transfer of heat through a medium from the warmer to the cooler parts of it. A rather obvious example is the conduction of heat along a metal poker left on an open fire – the tip in the fire is heated by radiation, then the heat is conducted along the poker towards the handle (hence the wisdom of a wooden handle, which does not conduct the heat so effectively!). Conduction warms the lowest part of the atmosphere, which is in direct contact with the relatively warm ground below. It also warms the upper layers of soil, rocks and deposits close to the ground surface.

Advection: advection is the horizontal or lateral transfer of heat in a moving stream of air. It redistributes heat across the Earth's surface in air mass movements and wind systems, which blow air from warmer or cooler areas (see p. 286). Air that moves across ground that is warmer or cooler will exchange heat with the ground surface, and air that comes into contact with surrounding air of a different temperature will exchange heat with it.

GLOBAL TEMPERATURE PATTERNS

World maps show temperature variations using *isotherms*, which are lines connecting places with the same temperature (just as contour lines show altitude). Values are reduced to sea-level equivalents to remove the distorting effects of altitude. The pattern of temperatures around the world reflects three main factors – latitude, relief and the distribution of land and water (Box 9.3).

SEASONAL VARIATIONS

The global pattern varies from season to season, reflecting variations in solar heating, the storage of heat in oceans and the continental land masses, and the redistribution of heat across the Earth by wind and ocean systems.

BOX 9.3 GENERAL PATTERN OF TEMPERATURE VARIATIONS

- Temperature decreases from the equator to the poles.
- Highlands are usually colder than surrounding lowlands.
- Temperature variations tend to be more extreme over land than over oceans and seas.

In January, temperatures are highest over the continents of the southern hemisphere (where it is summer), and the northern hemisphere continents (where it is winter) are much cooler. Land in the arctic and sub-arctic zones experiences extremely low temperatures in winter. Oceans bring warmer water further north in the northern hemisphere during winter, and this heats the overlying air. As a result, isotherms bend northwards over the oceans. Isotherms bend southwards over the continents because the land areas are much colder.

Patterns are very different in July. The northern hemisphere continents (now in summer) are warmer than those in the southern hemisphere (now in winter). Land heats more slowly but retains heat much longer than water, and during summer the continental areas warm up progressively. As a result, the isotherms in the northern hemisphere bend southwards over the oceans and northwards over the continents.

Temperatures in equatorial regions change relatively little from January to July. Beyond the tropics, however, continents experience a wider range of temperature variations from season to season than oceans do at similar latitudes. This has great significance for the detailed distribution of climate types around the world (see p. 335), and it makes the coastal zone – where maritime and continental climatic influences come into contact and interact – particularly important.

GLOBAL WARMING

Scientists have used present patterns of world climate to predict what is likely to happen in the future (Box 9.4) if greenhouse gas emissions are not reduced enough, or fast enough, to reduce the threat of global warming (see p. 261). Regional climates might change differently from the global mean, although there are many uncertainties in regional predictions. Temperature increases in southern Europe and central North America are predicted to be higher than the global mean, and these areas are also expected to face reductions in summer precipitation and soil moisture. Predictions for the tropics and the southern hemisphere are much less clear.

> **BOX 9.4 SCALE AND PATTERN OF GLOBAL WARMING**
>
> The IPCC 'business as usual' scenario (see p. 6) predicts a rate of increase of global mean temperature during the next century of about 0.3 °C per decade. If this happens, global mean temperatures will rise by about 1 °C between the mid-1990s and 2025 and by a further 2 °C between then and 2100. The land surface will warm more rapidly than the oceans, and high northern latitudes will warm more than the global mean in winter.

TEMPERATURE AND AIR STABILITY

Temperature decreases with height in a stationary parcel of air within the troposphere (see p. 229). The normal lapse rate is about 6.5 °C per km, although it varies from place to place and over time. Air temperature decreases from perhaps 20 °C on the ground to around −60 °C at the tropopause (Figure 9.3).

ADIABATIC LAPSE RATES

When a parcel of air rises it expands (because atmospheric pressure falls with altitude), and in expanding it uses energy. As a result it cools. The process is described as *adiabatic*, meaning that it takes place without loss or gain of heat. When the air rises and expands adiabatically, its

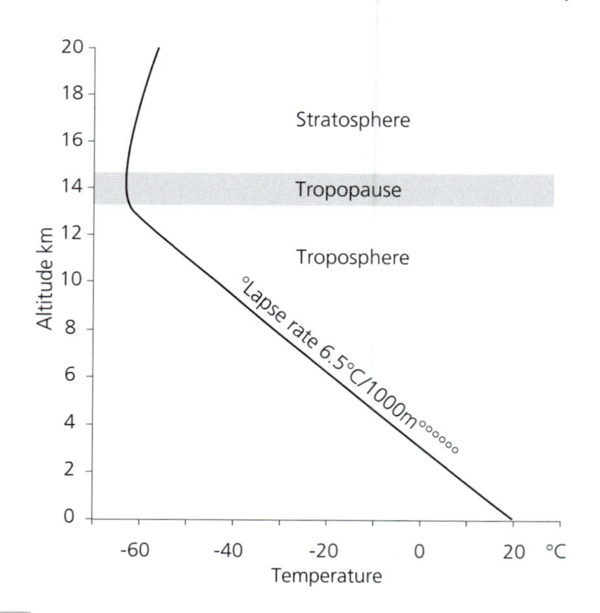

Figure 9.3 *Normal lapse rate in the troposphere. Air temperature decreases with altitude in the troposphere, and under normal conditions the lapse rate is 6.5°C per 1,000 m.*

Source: Figure 4.4 in Doerr, A.H. (1990) Fundamentals of physical geography. Wm. C. Brown Publishers, Dubuque, Iowa.

temperature falls as it expands to fill a larger volume. If a parcel of air is forced to sink in the atmosphere, for whatever reason, the reverse happens – it sinks, is compressed and becomes warmer. The change of temperature in a rising or sinking parcel of air is the *adiabatic lapse rate*.

Dry air cools faster than moist air when it rises because it expands faster and thus loses heat faster. Moist air cools more slowly on rising, because latent heat of condensation is released as the air expands. Rising dry air cools at the dry adiabatic lapse rate (DALR) (and sinking dry air warms at this rate), and rising moist air cools at the saturated adiabatic lapse rate (SALR) (sinking moist air warms at this rate). The DALR is higher than the SALR.

A typical DALR is about 10 °C per km of vertical rise. The SALR varies, depending on the moisture content of the parcel of air. Typical SALRs are between about 3 and 6 °C (the higher the moisture content, the lower the lapse rate).

BOX 9.5 SIGNIFICANCE OF ATMOSPHERIC STABILITY

The stability or instability of a parcel of air affects its vertical movements in the lower atmosphere, and it is also an important influence on cloud development (see p. 295) and the formation of precipitation (p. 296). It also determines the rates and patterns of dispersion of air pollutants released from point sources (such as sulphur dioxide from power station chimneys) (p. 252), and it can play a part in the build-up of photochemical smog (p. 243).

STABILITY AND INSTABILITY

The relationship between the normal lapse rate (in the surrounding air) and the adiabatic lapse rate (in the rising or sinking parcel of air) determines whether a parcel of air is stable or unstable, that is, whether it continues to rise or sink (Figure 9.4). Stability is important in a number of ways (Box 9.5). An unstable parcel of air continues to rise (if it is warmer than the surrounding air) or sink (if it is cooler than surrounding air). A stable parcel of air, on the other hand, will remain buoyant because it is the same temperature as the surrounding air.

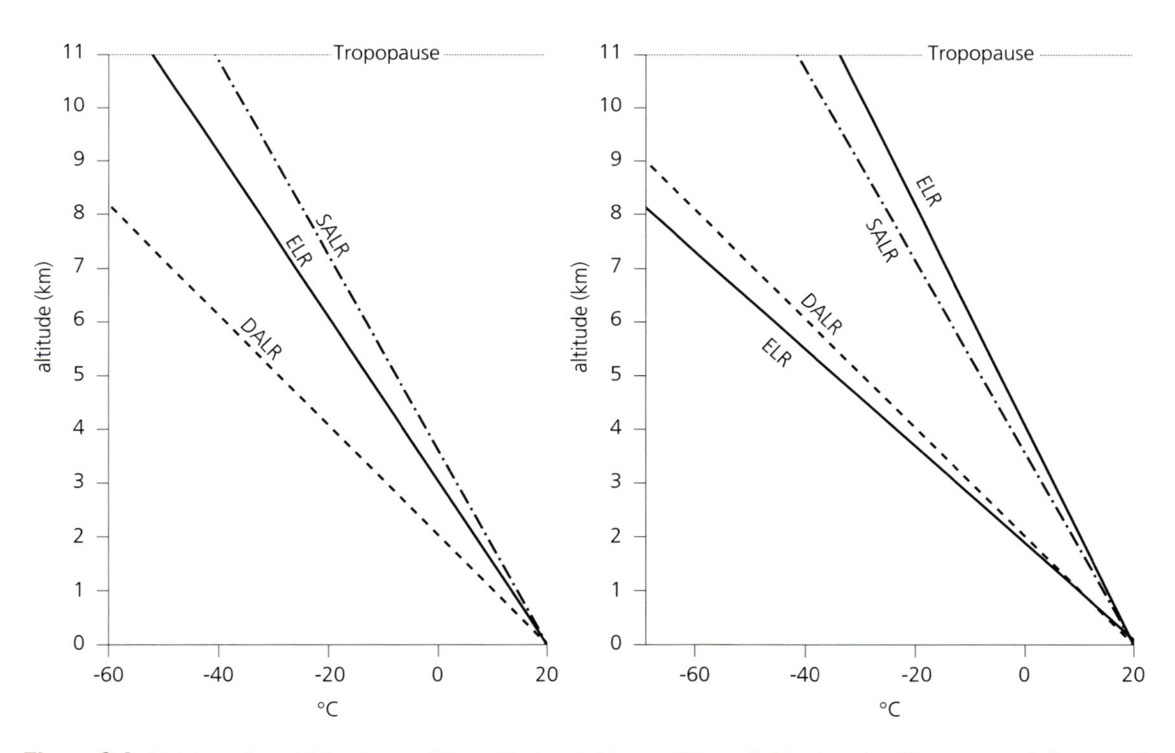

Figure 9.4 *Stable and unstable air conditions. Under stable conditions (left), dry air will not ascend, because the dry adiabatic lapse rate (DALR) is greater than the environmental (or normal) lapse rate (ELR). Under unstable conditions (right), the ELR is greater than the DALR, so ascending air will continue to rise.*

Source: Figure 16.1 in Doerr, A.H. (1990) Fundamentals of physical geography. Wm. C. Brown Publishers, Dubuque, Iowa.

Because the SALR is close to but sometimes less than the normal lapse rate, a parcel of air that contains a great deal of moisture cools more slowly than the surrounding air. As a result, it can be quite unstable and there can be substantial uplift, particularly in very moist air (with a low SALR). On the other hand, a parcel of dry air is often quite stable, because it cools faster than the surrounding air, and when the rising air and surrounding air have the same temperature there is no further uplift. Poor air quality in many large cities is generated partly by stable air trapping air pollutants, creating stationary air conditions with high levels of pollutants and irritants, which can significantly affect human health.

Conditional instability exists when a parcel of air rises, cools and condenses, so that whether or not it continues to rise depends on the relative humidity in the air (Figure 9.5).

ATMOSPHERIC PRESSURE AND WIND

ATMOSPHERIC PRESSURE

Air is a mixture of gases, and although the notion that air has weight might seem slightly strange, air in the atmosphere exists under pressure (Box 9.6). Atmospheric pressure is a physical pressure created by the interaction of the molecules that make up the gases in the air.

> **BOX 9.6 ATMOSPHERIC PRESSURE**
>
> It is estimated that the total weight of the atmosphere is about 5,000 million million tonnes, and half of this lies within the lowest 5 km (recall from p. 229 that the atmosphere extends out perhaps as far as 1,000 km). The lower layers are compressed, and average density decreases from about 1.2 kg m^{-3} at sea level to about 0.7 kg m^{-3} at about 5 km height

Atmospheric (air) pressure is relevant to humans in various ways. Pressure differences from place to place generate wind systems (see p. 283), and these diffuse pollutants, redistribute heat and bring weather changes. Variations in pressure at a particular place are closely associated with weather systems and weather changes – falling pressure brings the onset of poorer weather, and rising pressure brings brighter conditions.

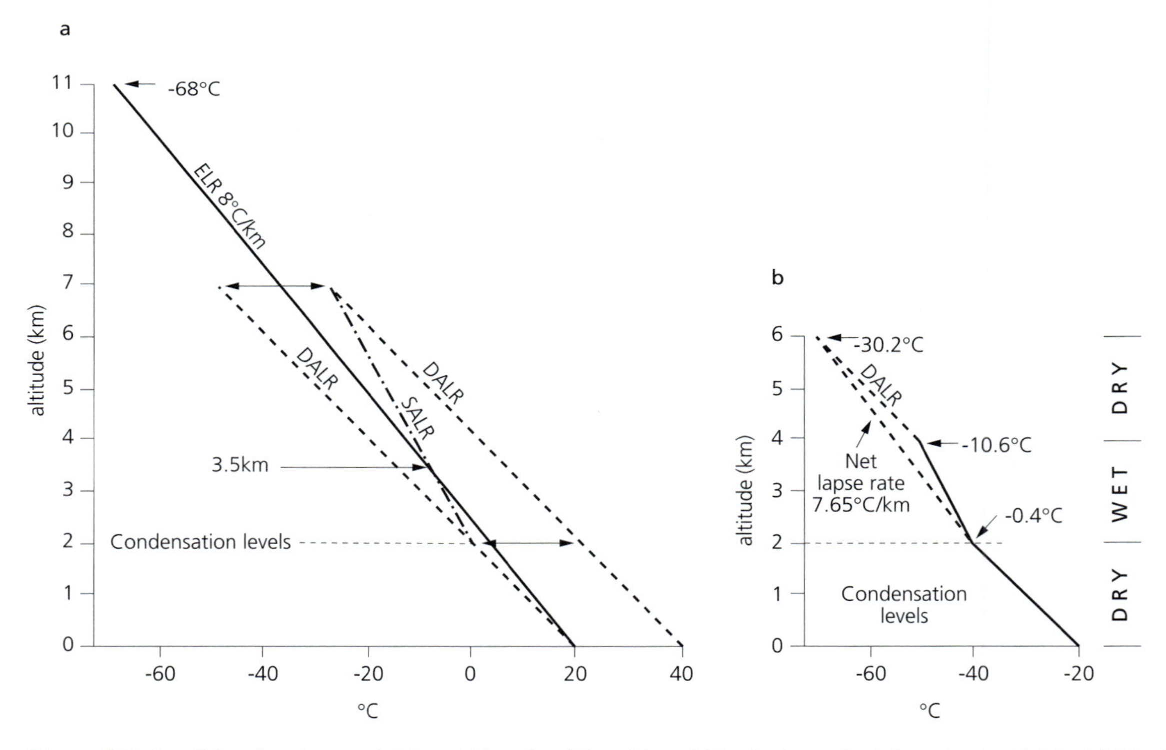

Figure 9.5 *Conditional and potential instability. Conditional instability is shown in (a), and potential instability is shown in (b). See text for explanation.*

Source: Figure 16.3 in Dury, G.H. (1981) Environmental systems. Heinemann, London.

MEASURING PRESSURE

Pressure is expressed in bars. One bar is the force required to lift a column of mercury a distance of 750.1 mm in a glass tube at 0 °C at 45° latitude. Scientists usually measure pressure in millibars (mb); there are 1,000 millibars in a bar (1 bar = 1,000 mb).

Pressure readings are normally adjusted to sea-level equivalents in order to eliminate the effects of altitude. Mean sea-level pressure on Earth is 1013.25 mb (this unit is defined as 1 normal atmosphere). High-pressure cells in the atmosphere have pressures up to about 1,060 mb (such as might occur in a Siberian winter anticyclone, for example). Low-pressure cells have pressures down to about 940 mb. While this range does not look particularly large (1,060 − 940 = 120 mb), it spans a wide variety of weather conditions (see p. 312).

Variations in pressure from place to place can be shown on weather maps by *isobars*. These are lines that join places with the same atmospheric pressure (usually measured in millibars), like contours on a relief map and isotherms on a temperature map.

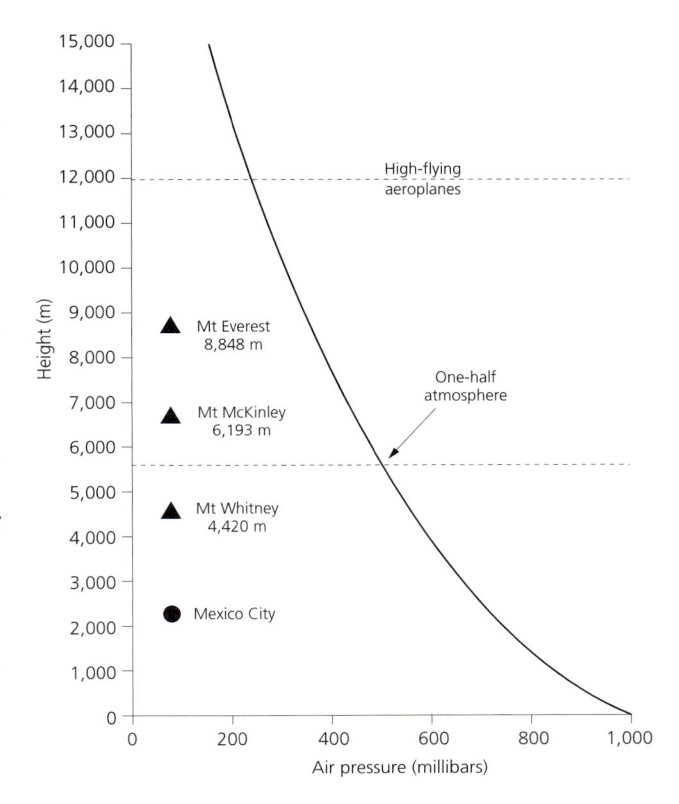

Figure 9.6 *Reduction in air pressure with height. Air pressure decreases with height and is down to one-half of sea-level pressure at about 5,500 m.*

Source: Figure 8.4 in Marsh, W.M. and J.M. Grossa (1996) Environmental geography: science, land use and Earth systems. John Wiley & Sons, New York.

VARIATIONS IN PRESSURE

Pressure is exerted in all directions, and the most important variations are altitude, place and time.

ALTITUDE

The lower part of the atmosphere, closest to the ground, is denser than the air above it because of the weight of overlying air. Pressure decreases with altitude, and at about 5.5 km it is about half of the normal pressure at sea level (Figure 9.6).

Temperature differences affect air movement (horizontally and vertically). High temperatures are associated with rising air and low pressure, and low temperatures with subsiding air and high pressure. Pressure varies with latitude (Box 9.7).

PLACE

Atmospheric pressure can vary a great deal from place to place, and these variations are reflected in the pattern of isobars on weather maps. High-pressure areas appear as ridges or domes, and low-pressure areas appear as troughs or depressions. The distance between isobars is an indication of the pressure gradient (or barometric gradient), which describes the steepness of the difference between high and low pressure. The pressure gradient affects wind speed and direction (see p. 281), and it is an important factor in weather forecasting because it strongly influences the development of weather systems (p. 307).

TIME

Pressure varies over a range of timescales, including the short-term (associated with weather changes; see p. 305) and the seasonal (associated with climate patterns around the world; see p. 332). Seasonal shifts of pressure are an important cause of the great monsoon circulations, with alternating wet and dry seasons. The tropics have relatively low pressure throughout the year, and cells of low pressure often extend south of the equator in January and north of the equator in July. In the northern hemisphere, in January, continental areas tend to be covered by high-pressure air, and low-pressure cells are common over the

BOX 9.7 VARIATIONS IN PRESSURE WITH LATITUDE

- There is a discontinuous belt of low pressure around the equator, caused by high temperatures in the tropics.
- Away from the tropics (at latitudes beyond 30° N and S) there is a zone of subsiding air, which produces high pressure.
- There is a second discontinuous belt of low pressure around the Earth at about 60° latitude.
- Closer to the poles is another belt of subsiding air, much colder than the one nearer the tropics, which also produces high pressure.

oceans. Patterns change towards July, when mid-latitude parts of the Atlantic and Pacific Oceans are overlain by high pressure, and South Asia and North America are low-pressure areas. Pressure belts and associated wind systems display more obvious zonal patterns in the southern hemisphere than in the northern hemisphere, because there are less pronounced differences between land and sea there.

WIND

Air moves in two directions, horizontally and vertically, as a result of differences in pressure and temperature:

1. *Air currents* are vertical movement of air.
2. *Wind* is the horizontal movement of air.

Winds are important for life on Earth in many different ways (Box 9.8). On a much smaller scale, pressure differentials in buildings can be exploited to create air movements as a form of passive cooling. This is a traditional element of architecture in hot countries, particularly the Middle East, and it holds great promise as a cheap and ecologically sound alternative to air conditioning.

MEASURING WIND

Wind speed is normally measured using the Beaufort scale (Table 9.3). This is a twelve-point scale in which high numbers indicate strong winds. It is a non-linear scale, so force 4 is not twice as fast as force 2, for example.

BOX 9.8 THE SIGNIFICANCE OF WIND

Winds are significant in many ways:

- They distribute heat across the Earth's surface, from low to high latitudes, and stop the Earth overheating in some places and becoming too cold in others.
- They control the invasion of colder air from high latitudes (so they influence weather changes).
- They transport moisture from oceans and seas and release it as precipitation over land and sea.
- They circulate and dilute air pollution from industrial areas and stop many areas becoming too badly polluted for life to survive.
- They transport soil and sand and so play a part in erosion.
- They cause environmental hazards such as sandstorms, strong winds and hurricanes.

At low wind speeds, the observed effects are very subtle, such as smoke patterns rising from chimneys and small movements on flags. At higher wind speeds (Beaufort force 4 and above) small branches start to sway on trees, and as winds become even stronger (force 8) twigs break off trees. Storm conditions (force 11) have winds stronger than 100 km h^{-1}, and hurricanes (force 12 upwards) have extremely high winds and cause widespread damage.

WIND MOVEMENT

There are two main components to wind movement – speed and direction. Both are controlled by the interaction of three main factors:

1. the pressure gradient
2. the Coriolis force
3. friction.

PRESSURE GRADIENT

The pressure gradient (see p. 279) affects wind direction and wind speed:

Wind direction: wind will always flow from high to low pressure, in the same way that water flows from high to

Table 9.3 The Beaufort wind scale

Beaufort Number	*Description*	*Speed ($km\ h^{-1}$)*	*Characteristics and observed effects*
0	Calm	<1	Smoke rises vertically
1	Light air	1–5	Direction shown by smoke
2	Light breeze	6–12	Wind vane moves; wind is felt on the face
3	Gentle breeze	13–20	Wind extends a light flag; leaves and twigs move
4	Moderate breeze	21–29	Raises dust and loose paper, moves small branches on trees
5	Fresh breeze	30–39	Small trees in leaf start to sway; flags ripple
6	Strong breeze	40–50	Large trees sway; flags beat; umbrellas used with difficulty
7	Moderate gale	51–61	Whole trees sway; walking into the wind is difficult
8	Fresh gale	62–74	Twigs break off trees; walking is hindered
9	Strong gale	75–87	Slight damage (e.g. to chimney pots and slates)
10	Whole gale	88–102	Trees uprooted; severe damage
11	Storm	103–120	Widespread damage
12–17	Hurricane	>120	Extremely violent; devastation

low ground. As a result, wind always blows into the centre of a low-pressure area (depression) and away from a high-pressure area (anticyclone).

Wind speed: the steeper the pressure gradient (i.e. the more closely spaced the isobars) the stronger the wind. This is like contours on a relief map, because steep slopes have closely spaced contours. Widely spaced isobars indicate a low pressure gradient and are associated with light variable winds; closely spaced isobars indicate a steep pressure gradient and thus strong winds.

CORIOLIS FORCE

If the Earth were stationary, wind direction would be controlled mainly by the pressure gradient. But the Earth rotates around its axis, from west to east (see p. 116), and this causes a deflection of the wind direction when viewed from the ground. This is called the Coriolis force (Box 9.9), although it is not a physical force (it simply explains the apparent deflection in terms of the motion of the Earth's surface beneath the air movement).

Air movement is triggered initially by the pressure gradient (air moves from high to low pressure), but the direction of movement is strongly influenced by the Coriolis force, which deflects it. The deflection is called the Coriolis effect, and it is to the right in the northern hemisphere and to the left in the southern hemisphere. The Coriolis force also causes ocean currents to deflect (see p. 480) in the same way and in the same direction as it affects winds. It affects aircraft in the same way too, and it must be taken into account in the planning of flight paths and the determination of fuel loads.

The Coriolis force varies with latitude (Figure 9.7). At the equator the force is zero, so there is no deflection; wind moves directly down the pressure gradient. The force is strongest at the poles, where deflection is consequently greatest.

BOX 9.9 THE CORIOLIS FORCE

One way of visualising how the Coriolis force deflects objects moving across a rotating surface is to imagine a disc spinning slowly around a central point (like a vinyl record on a turntable). If you draw a straight line with a pen and ruler from the centre towards the edge of the disc, that line would appear curved when you stop the disc rotating. This explains why wind movement triggered by the pressure gradient (the straight line) is deflected by the moving Earth below (the curved line). If you run the experiment again, rotating the disc faster, the line would appear to be even more curved. This explains why the Coriolis effect is stronger at the poles, which are smaller so they rotate faster than the equator.

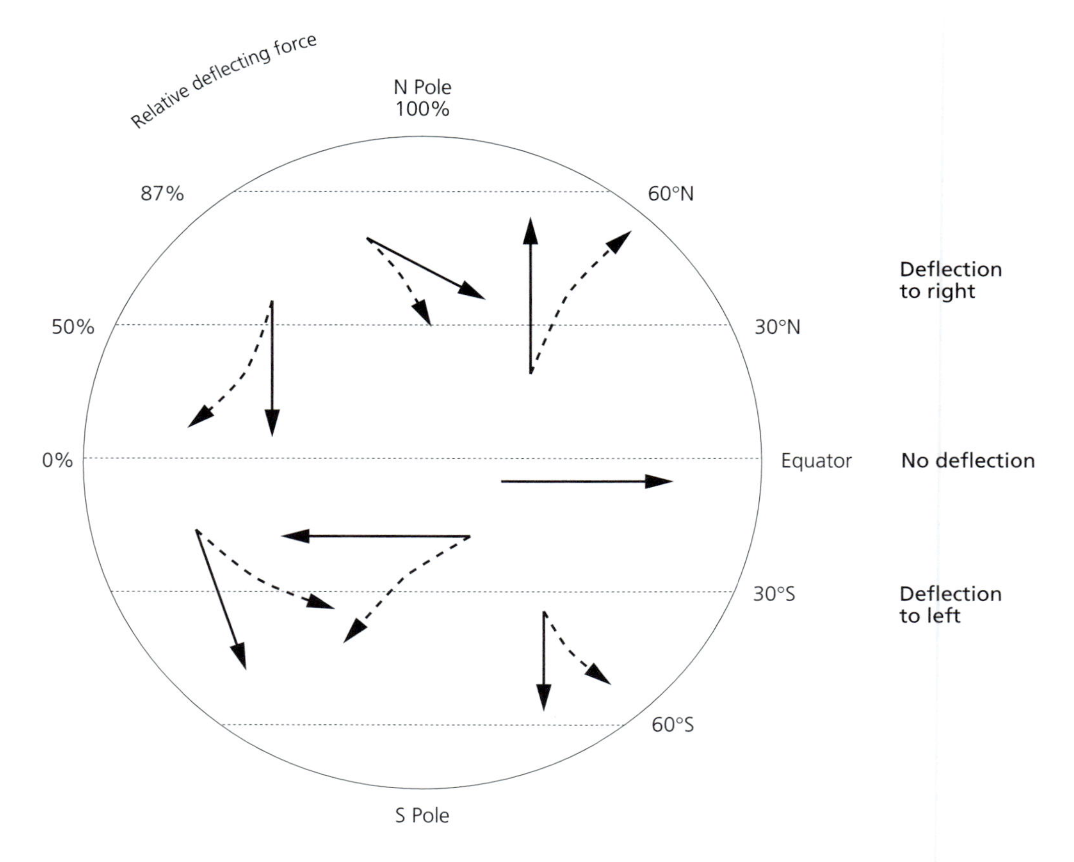

Figure 9.7 *Variations in the Coriolis force with latitude. The strength of the Coriolis force increases with latitude.*
Source: Figure 6.6 in Briggs, D. and P. Smithson (1985) Fundamentals of physical geography. Routledge, London.

GEOSTROPHIC WIND

Air movement reflects a state of balance between the pressure gradient and the Coriolis force. If either operated alone, air movement would be very different from what really happens. If the pressure gradient were the only control, winds would blow directly down the pressure gradient from high to low. If the Coriolis force were the only control, winds would (at least in theory) follow circular paths and simply blow around the world.

These two extremes do not happen, and air movement is controlled by a state of balance between the pressure gradient (forcing air to move from high to low pressure) and the Coriolis force (acting in the opposite direction). It is called the strophic balance, and it causes wind generally to blow at right angles to the pressure gradient – deflected to the right in the northern hemisphere and to the left in the southern hemisphere. This is the so-called *geostrophic wind*.

The geostrophic wind is an important component of the air movements associated with cells (areas) of high and low pressure, which in turn play important roles in the development of weather systems (see p. 304). In the northern hemisphere, air circulation is as follows (Figure 9.8):

- *Anticlockwise around a low-pressure cell*: air moves down the pressure gradient towards the centre of the low-pressure area, but is deflected to the right by the Coriolis force.
- *Clockwise around a high-pressure cell*: air moves down the pressure gradient away from the high-pressure cell (out from the centre) and is deflected to the right by the Coriolis force.

The pattern is reversed in the southern hemisphere – clockwise around a low and anticlockwise around a high. These patterns, determined by the interaction of the pressure gradient and the Coriolis force, are described by Buys-Ballot's law (Box 9.10).

Northern hemisphere

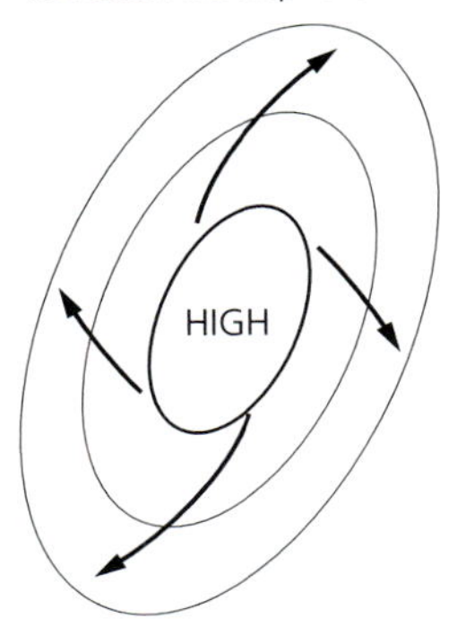

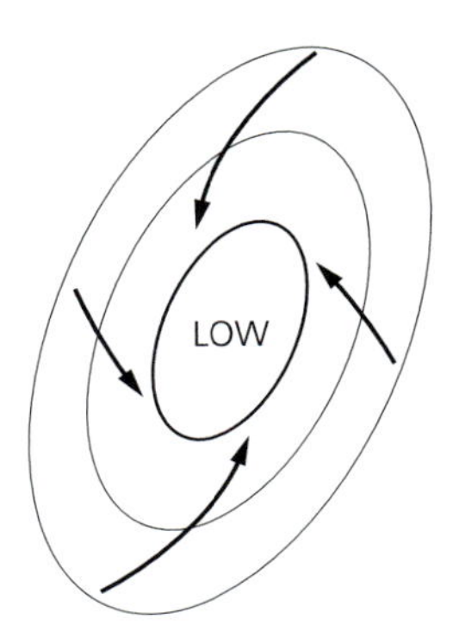

Southern hemisphere

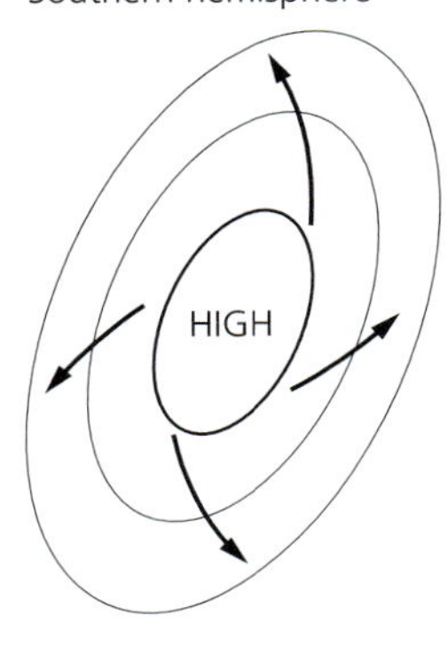

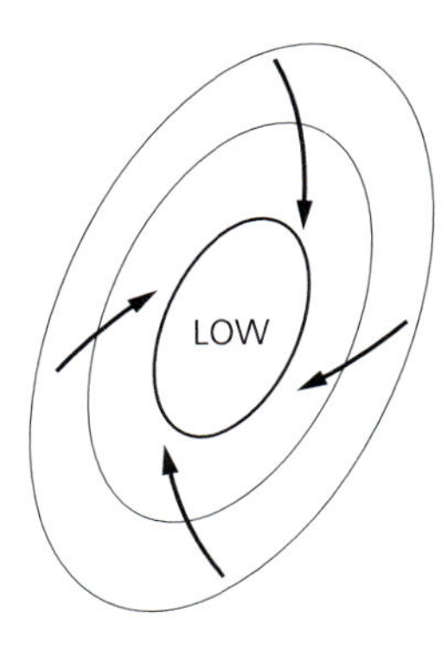

Figure 9.8 *Wind flow around high- and low-pressure cells. In the northern hemisphere, wind flows clockwise out from high-pressure cells and anticlockwise into low-pressure cells; the pattern is reversed in the southern hemisphere.*

Source: Figure 4.25 in Doerr, A.H. (1990) Fundamentals of physical geography. Wm. C. Brown Publishers, Dubuque, Iowa.

BOX 9.10 BUYS-BALLOT'S LAW

This law is named after a Dutch meteorologist, Buys-Ballot, who in 1857 defined the relationship between the pressure gradient and the Coriolis force. The simple rule of thumb is as follows:

- In the northern hemisphere, if you stand with the wind blowing into your back, low pressure is always to your left and high pressure is always to your right.
- In the southern hemisphere the pattern is reversed, and low pressure is on your right and high pressure on your left.

FRICTION

The strophic balance would determine wind speed and direction on a flat, featureless Earth with a smooth surface (like a marble). On this hypothetical planet, all winds would be geostrophic winds. But the real world is much less smooth than this, and its surface is both rough and variable. Mountains create large barriers to air movements, but wind flow is also heavily influenced by smaller features such as buildings, trees, rocks, rivers, small hills and structures such as fences, walls, roads and towers.

Anything that sticks up above the ground surface disturbs air flow in the lower atmosphere (below about 750 m), usually slowing it down. Any rough surface creates resistance or frictional drag, which reduces wind speed and changes the strophic balance. More friction means lower wind speed, and this reduces the Coriolis force and increases the relative effectiveness of the pressure gradient. As a result, the gradient effect is stronger and air flows more towards low pressure and is less deflected by the Coriolis effect. Under moderate wind speeds across an area of average roughness, surface winds are usually deflected by up to about 20° from the geostrophic wind (towards the direction of the pressure gradient).

This relationship between friction and the strophic balance has three important implications (Box 9.11).

BOX 9.11 IMPLICATIONS OF FRICTION ON AIR FLOW

- Air movements are likely to be different over smooth and rough surfaces. Over smooth surfaces, such as the oceans, surface winds closely match geostrophic winds.
- Wind speed influences wind direction because it determines how much the strophic balance is disturbed. In strong winds, friction effects are lower and air movement is determined largely by the geostrophic wind.
- Air movements are likely to change with altitude, as the friction effect is lower away from contact with the ground surface. The progressive decrease in frictional drag with altitude is displayed in the Ekman spiral (Figure 9.9). The thermal wind is the mean wind direction between geostrophic winds at the top and bottom of a column of air.

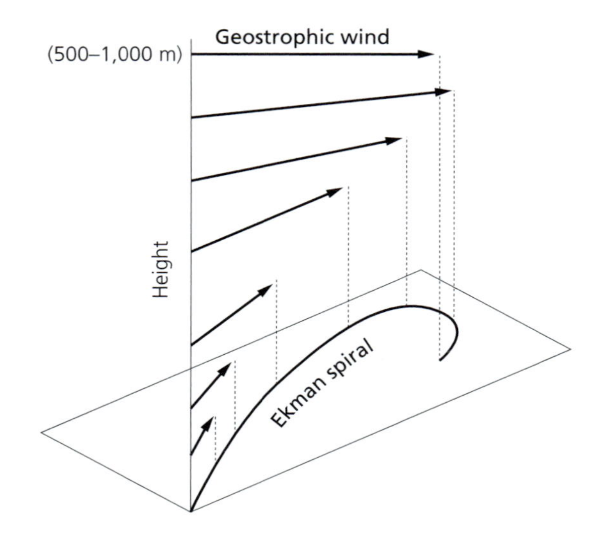

Figure 9.9 The Ekman spiral. In the northern hemisphere, wind direction veers increasingly to the right with increasing height above the ground (as shown) as the relative importance of frictional drag declines. In the southern hemisphere, it veers to the left.

Source: Figure 3.5 in Barry, R.G. and R.J. Chorley (1976) Atmosphere, weather and climate. Routledge, London.

UPPER-AIR WINDS

The atmosphere is three-dimensional and highly dynamic, so all parts of it can and do affect each other. Horizontally, this is clear through such controls on air movement as the pressure gradient and the Coriolis effect. Vertically, it is clear through the effects on air movement of friction, and through subsidence and uplift associated with convection processes (Figure 9.10) and adiabatic temperature change (see p. 277).

Relationships between the lower and upper parts of the atmosphere have a strong influence on air movements and wind systems. They operate at a range of scales and in a variety of ways:

- Wind speed increases and wind direction changes with height, because of reduced friction.
- Wind patterns in and around anticyclones and cyclones (depressions) are three-dimensional, involving both lateral movements and uplift/subsidence (Figure 9.11).
- Wind circulation around the Earth, driven by differences in temperature and pressure, involves both lateral and vertical movements.

Upper-air winds can travel much faster than winds closer to the ground, and they can affect weather conditions quite significantly. Particularly important are the jet streams (Box 9.12) and Rossby waves (Box 9.13).

GLOBAL WIND CIRCULATION

The global pattern of winds is dominated by four main belts that encircle the Earth. Wind direction and strength vary a great deal between successive wind belts, and the belts themselves shift slightly in position from season to season (reflecting variation in the length of day and night and variations in the angle of incidence of the Sun's rays).

Winds are named after the direction from which they blow, so a westerly wind blows from the west towards the east, and a north wind blows from north to south (Figure 9.13).

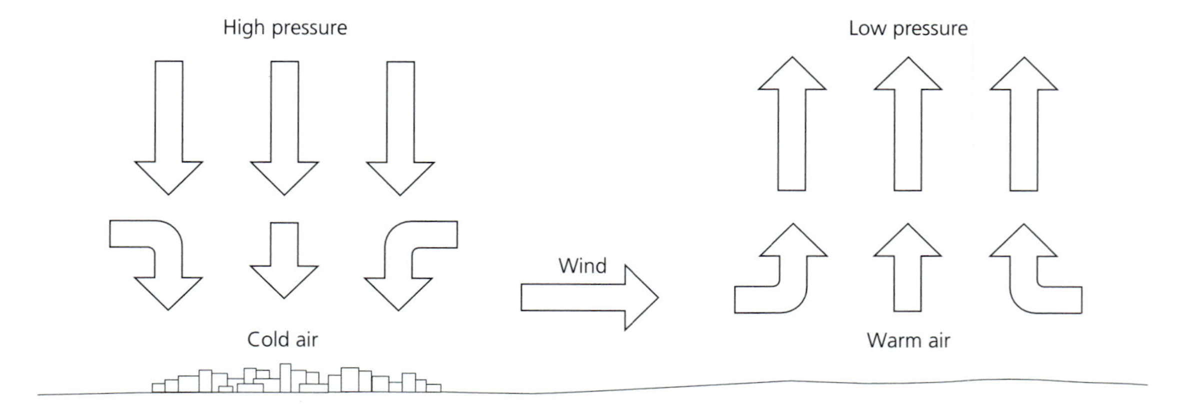

Figure 9.10 Subsidence and uplift of air. Air always blows from high pressure towards low pressure, down the pressure gradient. Warm air rises (causing uplift), and cold air sinks (causing subsidence); compensating air flows blow air from cold (high-pressure) to warm (low-pressure) areas.

Source: Figure 4.23 in Doerr, A.H. (1990) Fundamentals of physical geography. Wm. C. Brown Publishers, Dubuque, Iowa.

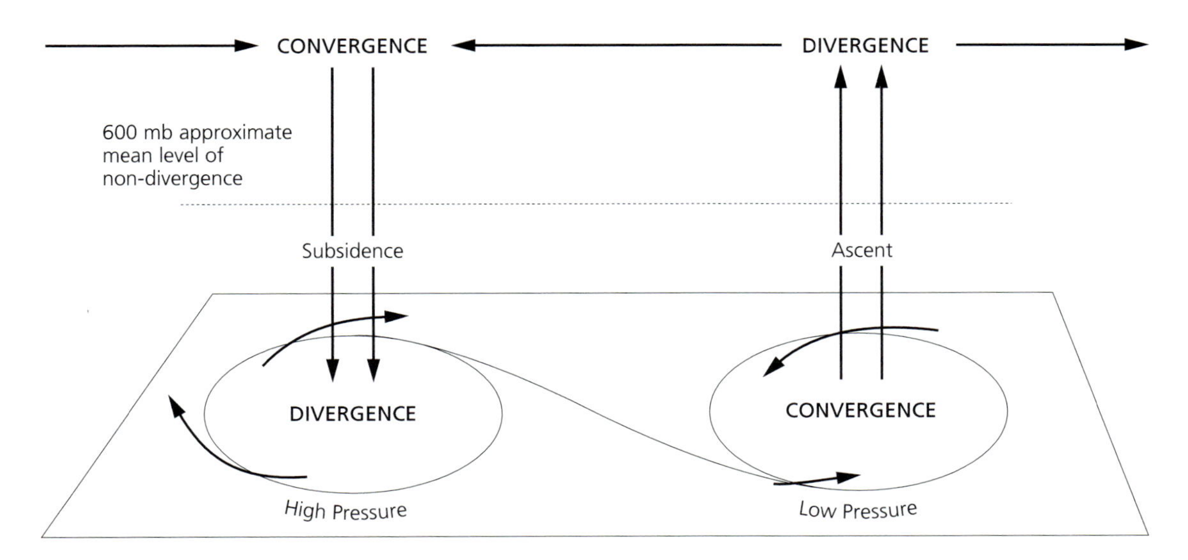

Figure 9.11 *Convergent and divergent air flow around high- and low-pressure cells. Air flow is divergent around high-pressure cells, from where it flows outwards towards the low-pressure cells, where it is convergent. The pattern is three-dimensional, because uplift over the low-pressure cell leads to divergence above, and subsidence into the high-pressure cell is fed by convergence above.*

Source: Figure 12.6 in Barry, R.G. and R.J. Chorley (1976) Atmosphere, weather and climate. Routledge, London.

BOX 9.12 THE JET STREAMS

Jet streams are narrow bands of very fast winds that blow at altitudes of 10–16 km in the upper troposphere or lower stratosphere (see Box 8.5). Speeds are mostly around 160 km per hour (km h^{-1}), but jet streams can sometimes reach up to 500 km h^{-1}. Aircraft pilots sometimes fly in jet streams to increase their speed and reduce fuel consumption. Jet streams are intense thermal winds, and they occur where temperature gradients are particularly strong. These conditions occur in two zones in each hemisphere (Figure 9.12):

1. the sub-tropical jet stream (around 30° latitude); and
2. the polar front jet stream (which shifts with the changing location of the polar front).

TEMPERATURE, PRESSURE AND CIRCULATION

The wind circulation system serves the vital role of redistributing heat around the Earth's surface, preventing overheating at the equator. This giant global heat engine starts with the heating of ground and air at the equator. The warm air expands and rises, creating a low-pressure belt around the equator. As it rises, the air cools adiabatically and it is blown by upper-air winds towards the poles.

This air sinks back to ground level at about 30° latitude, creating a belt of high pressure. Some of the descending air flows towards the equator (as the trade winds) and some flows towards the poles (as the westerlies). This circulation pattern is called the Hadley cell, and it effectively drives the global heat engine and wind system.

A dome of high pressure builds up at the poles from which air descends and flows towards the equator. It moves down the pressure gradient towards a low-pressure band at about 60° latitude, where it converges with the westerlies at the so-called polar front.

BOX 9.13 ROSSBY WAVES

Rossby waves are the large-scale meandering wave patterns in the upper atmosphere that geostrophic winds follow as they flow around the world. Often between three and six Rossby waves can be identified in each hemisphere. The waves change position over time, and this causes changes in air mass movements (see p. 307) and weather conditions on the ground.

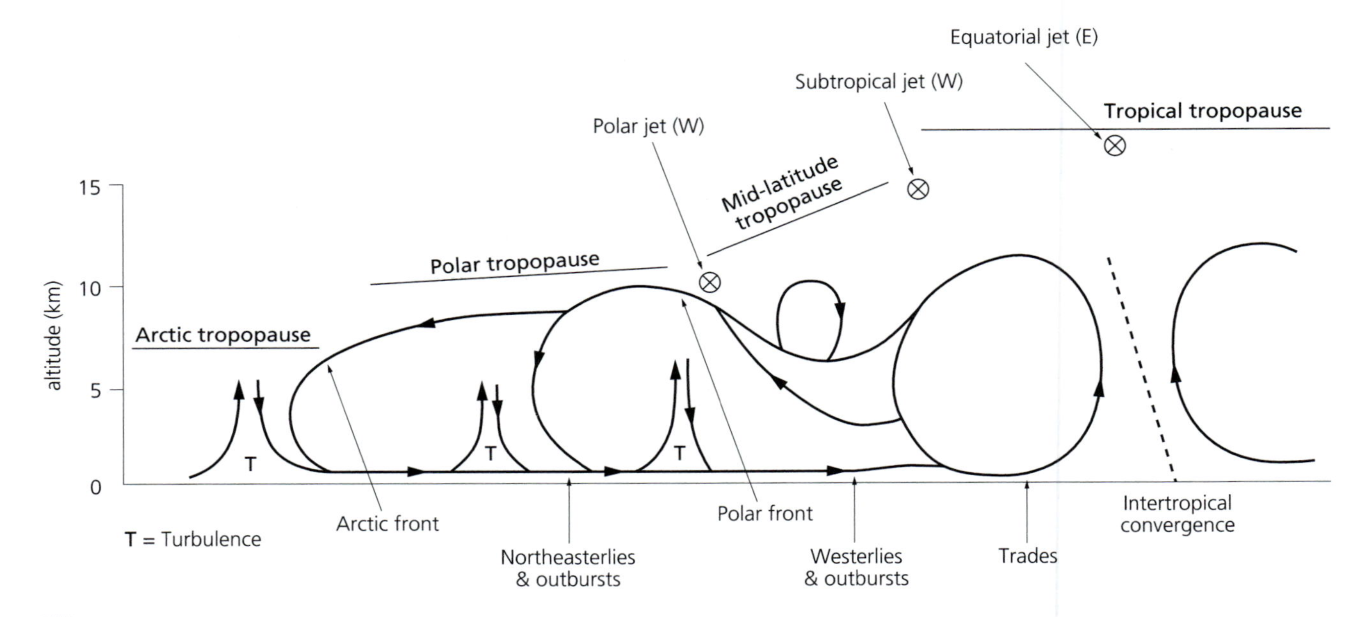

Figure 9.12 *Jet streams and global wind circulation. Three main jet streams (equatorial, sub-tropical and polar front) are associated with interactions between the tropopause and the major wind systems below. The pattern shown applies to the northern hemisphere.*

Source: Figure 14.14 in Dury, G.H. (1981) Environmental systems. Heinemann, London.

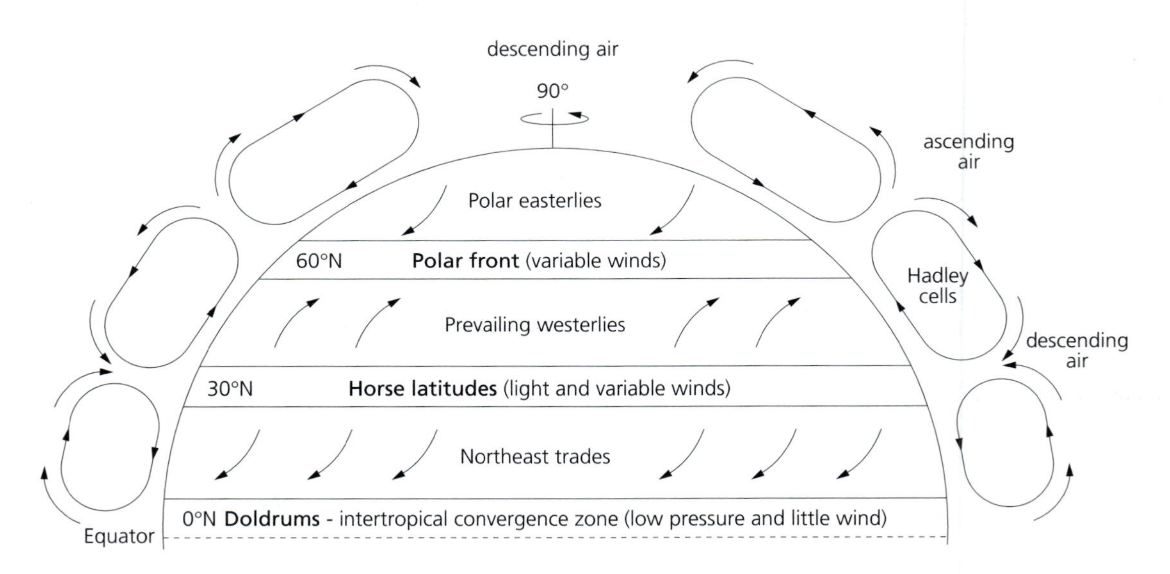

Figure 9.13 *General wind circulation pattern over the northern hemisphere. The generalised pattern shown is described in the text. The boundaries of the circulating Hadley cells and the local directions of the surface winds vary from day to day and season to season, and the circulation pattern is also complicated by variations in surface topography.*

Source: Figure 17.5 in Cunningham, W.P. and B.W. Saigo (1992) Environmental science: a global concern. Wm. C. Brown Publishers, Dubuque, Iowa.

At increasing latitude, from the equator to the poles, the main wind belts are as follows:

EQUATORIAL LOW (THE DOLDRUMS)

This wind belt is found throughout the area of low atmospheric pressure around the equator. It has light and variable winds, because pressure there varies little from place to place. Weather conditions are typically hot and humid, convectional uplift is strong, and this causes large amounts of precipitation.

To sailing ships – which rely on wind behind their sails as a form of power – light wind represents a real hazard (hence the origin of the expression 'in the doldrums', meaning a state of inactivity or boredom). Beyond the equatorial low, at higher latitudes, are the strong trade winds, which allowed sailing ships to cross the world's great oceans easily and quickly. The contrast with the becalming conditions in the doldrums was very marked!

The boundary between the equatorial low and the trade winds is called the inter-tropical convergence zone (ITCZ) (Box 9.14).

NORTH-EAST AND SOUTH-EAST TRADE WINDS

The trade wind belt lies beyond the equatorial zone up to about 30° latitude in each hemisphere. In this belt, the prevailing winds blow towards the equator, from the north-east in the northern hemisphere and from the south-east in the southern hemisphere. These winds are generated by the movement of air from higher latitudes towards the equator to compensate for the hot, rising air at the equator. The source area for the trade winds is the descending air in the subtropical highs (horse latitudes).

The pressure gradient in each hemisphere is straight down towards the equator from the subtropical belt of high pressure to the equatorial belt of low pressure. But winds are deflected by the Coriolis effect (see p. 282):

- to the right in the northern hemisphere, hence they flow from the north-east;
- to the left in the southern hemisphere, hence they flow from the south-east.

The light winds and unpredictable calms of the doldrums lie at the convergence of both sets of trade winds. Trade wind speeds are much higher than those in the doldrums.

The trade winds often carry a great deal of water vapour from the oceans across the continents, so they play an important role in the global water cycle (see p. 353). Some of this moisture can be dropped as orographic precipitation (see p. 330) on the windward slopes of mountain barriers (such as Hawaii, which has the greatest number of rainy days per year on Earth) that intercept the trades.

Trade winds are warmed progressively as they blow towards lower latitudes. If they blow over low-lying continental areas, this progressive heating can dry up the trade winds. Some areas in the path of the trades, such as the Caribbean islands, are relatively dry as a result.

WESTERLIES (PREVAILING WESTERLIES)

The third wind belt out from the equator, at about 30° latitude, is the westerlies (sometimes called prevailing westerlies). Winds in this belt flow towards the poles – northwards in the northern hemisphere and southwards in the southern hemisphere. They are generated by the same high-pressure zone of descending air (the sub-tropical highs) that gives rise to the trade winds at lower latitudes.

The pressure gradient in this belt is directly towards the poles, but the Coriolis effect deflects winds to the right in the northern hemisphere and to the left in the southern hemisphere. As a result, the dominant winds blow from the south-west (northern hemisphere) and the north-east (southern hemisphere). These are thus westerly winds in the northern hemisphere, and because the Coriolis force is stronger at these higher latitudes, the deflection is stronger.

BOX 9.14 THE INTER-TROPICAL CONVERGENCE ZONE

The boundary between the equatorial low and the trade winds moves north and south through the seasons, between about 7 and 15° latitude. Conditions in the boundary zone are changeable. During the summer months, when the Sun is high in the sky (and heating is particularly intense), rising air currents dominate the boundary zone with the trade winds. This zone is called the inter-tropical convergence zone (ITCZ) or inter-tropical front (ITF). Powerful convectional uplift in the ITCZ promotes strong vertical cloud development, intense thunderstorms and very unsettled weather. This produces a clearly defined wet season. At other times of the year, conditions in the boundary zone are dominated by the trade winds, which give rise to a dry season.

Winds in the horse latitudes are more variable and less persistent than the trade winds.

Descending air is warmed by compression (adiabatically), so it tends to be warm and dry. As a result, many areas in the horse latitudes have very dry climates, with light winds and many hot, bright, sunny days. It is no surprise, therefore, that some of the world's driest deserts – such as the Sahara in North Africa, the Mojave in the USA, the Atacama in Chile, and the Namib in southern Africa – are to be found in these latitudes (Chapter 14).

As the air masses are moved from west to east in the northern hemisphere, they redistribute pressure and weather (Box 9.15). Weather changes experienced on the ground reflect the arrival of successive air masses from different sources and with different properties (see p. 304). Low-pressure systems (depressions or cyclones) bring precipitation and changeable weather, whereas high-pressure systems (anticyclones) bring dry spells and bright, sunny weather. This mid-latitude zone tends to have very variable climates because of the regular passage across it of depressions and anticyclones.

SUB-POLAR LOWS AND POLAR HIGHS

The poles are centres of high pressure, where air subsides towards the ground from the upper atmosphere. When this subsiding air reaches the ground, it flows towards lower latitudes (towards the equator). This gives rise to a belt of winds that the pressure gradient forces to flow from north to south in the northern hemisphere (and from south to north in the southern hemisphere), and that are deflected by the Coriolis effect. Thus the winds blow from the north-east in the northern hemisphere and from the south-east in the southern hemisphere.

These easterlies are dry winds because they move over frozen ground for much of the time. Wind speeds can be very high in this belt, because of steep pressure gradients and low friction (from the smooth frozen surface).

The easterly winds (generated by the polar high) converge with the westerly winds (from the sub-tropical high) at the sub-polar low-pressure belt at about 60° latitude. Weather in this sub-polar-low zone is often very unsettled, with extensive cloud cover.

Precipitation across this whole zone is very limited, because the subsiding air at the polar high brings little water vapour. Polar and sub-polar environments (see p. 442) are in effect cold (frozen) deserts, with very limited precipitation.

BOX 9.15 SIGNIFICANCE OF THE WESTERLY WINDS

The westerlies play an important role in moving air masses and weather systems from west to east in the middle latitudes of the northern hemisphere. Many of the weather changes across much of North America and Europe are associated with the eastward movement of low-pressure (cyclonic) cells. While the westerlies move mainly from west to east, the main wind systems tend to follow meandering paths that reflect the influence of the Rossby waves (see Box 9.13) in the upper atmosphere. Rhythmic changes in the Rossby waves, which appear to swing northwards then southwards in cycles, help to explain why storm tracks appear to change position from time to time.

BOX 9.16 WIND TURBINES

The most common types of turbine have large propeller-type rotors mounted on a tall tower, which are connected to electricity generators. Modern wind generators are carefully designed to be efficient, using advanced aerodynamic design principles and techniques developed in aerospace and helicopter design. The amount of energy captured by a wind turbine depends on the size of its blades and on wind speed. The world's largest wind turbine is on Hawaii in the Pacific Ocean – it has two blades 50 m long on top of a tower twenty storeys high.

WIND ENERGY

Wind energy is energy extracted from the wind and, like solar power (see p. 236), it is derived from solar energy. Like solar power, it is also widely available and renewable, and it has been widely exploited throughout history as a source of power. Wind power is more variable over time than solar power, because it relies on air movements that are determined by pressure differences and weather systems (which are themselves ultimately driven by sunlight).

Many ways have been developed to exploit wind energy (Figure 9.14). The most common is the windmill (a mill with sails or vanes rotated by wind blowing against them), introduced into Europe from China in the twelfth century and widely used to drive machinery for grinding corn and pumping water. Electricity generators driven by the wind were first used in Denmark in the 1890s.

Plate 9.1 *Wind farm near Sellafield, Cumbria, England. Wind energy is exploited using a series of large rotary vanes, which generate electricity.*
Photo: Chris Park.

The 1980s saw a renaissance of wind-power technology, driven by the rising cost and declining reserves of fossil fuels and by growing awareness of the need to develop the sustainable use of non-polluting, renewable energy resources. Use of wind power around the world has risen rapidly since the early 1980s.

Electricity generation from wind power looks likely to increase further in the future, because it is cheap and environmentally clean, and it eases distribution problems in remote areas. Many countries have great untapped wind energy potential. This is certainly the case in Britain, for example, which is the windiest country in Europe thanks to the regular passage of low-pressure systems moving across the country from west to east. Hilly country in the north and west provides the exposed conditions necessary for high wind speed and consistent wind flow.

WIND TURBINES

The energy crisis of the 1970s and 1980s encouraged modern experiments with wind turbines, designed to use wind power on a large scale. They use the same basic principle as windmills – using wind to turn blades – but are much more sophisticated (Box 9.16). Studies show that on windy sites, currently available machines are among the cheapest generating options. Provided that suitable sites can be found, wind energy could provide between 20 and 50 per cent of total energy needs in Britain at economic prices.

WIND FARMS

Many wind turbines are small installations that provide electricity to isolated farms and houses. In remote places, wind power provides an ideal source of energy, but in built-up areas it is important to safeguard access to wind exposures around individual properties.

Modern turbines are usually grouped together in wind-power installations (wind farms). These are fairly large land users, because each turbine requires about 0.02 km^2 of land. Many large commercial wind farms were built in the USA (mostly in California) during the 1980s, partly because tax incentives were offered to encourage private investment in exploiting renewable energy sources. The Altamont wind farm in California – one of the largest and best-known wind farms in the world – has had a mixed response from the public. The USA has one of the world's largest resources of wind energy, and experience during the 1990s showed that wind power is a significant energy resource, that wind-generated electricity is not necessarily expensive and unreliable, and that existing technologies are both efficient and effective.

Large turbines and wind farms are also exploiting wind energy to generate electricity for national supply systems in France, the Netherlands, Denmark and parts of Britain (including south-west Wales, Devon, Lancashire, Orkney and Shetland).

While the energy source is freely available, renewable and non-polluting, the development of wind farms is not without controversy. Suitable sites are limited, because they have to be exposed, face dominant wind directions, be accessible to electricity supply networks, not be too close to existing or possible future buildings, and have a suitable rock base on which to construct the large turbines. There is growing public concern, particularly in the United Kingdom, about the development of wind farms and their impacts on the environment. Wind farm sites create a

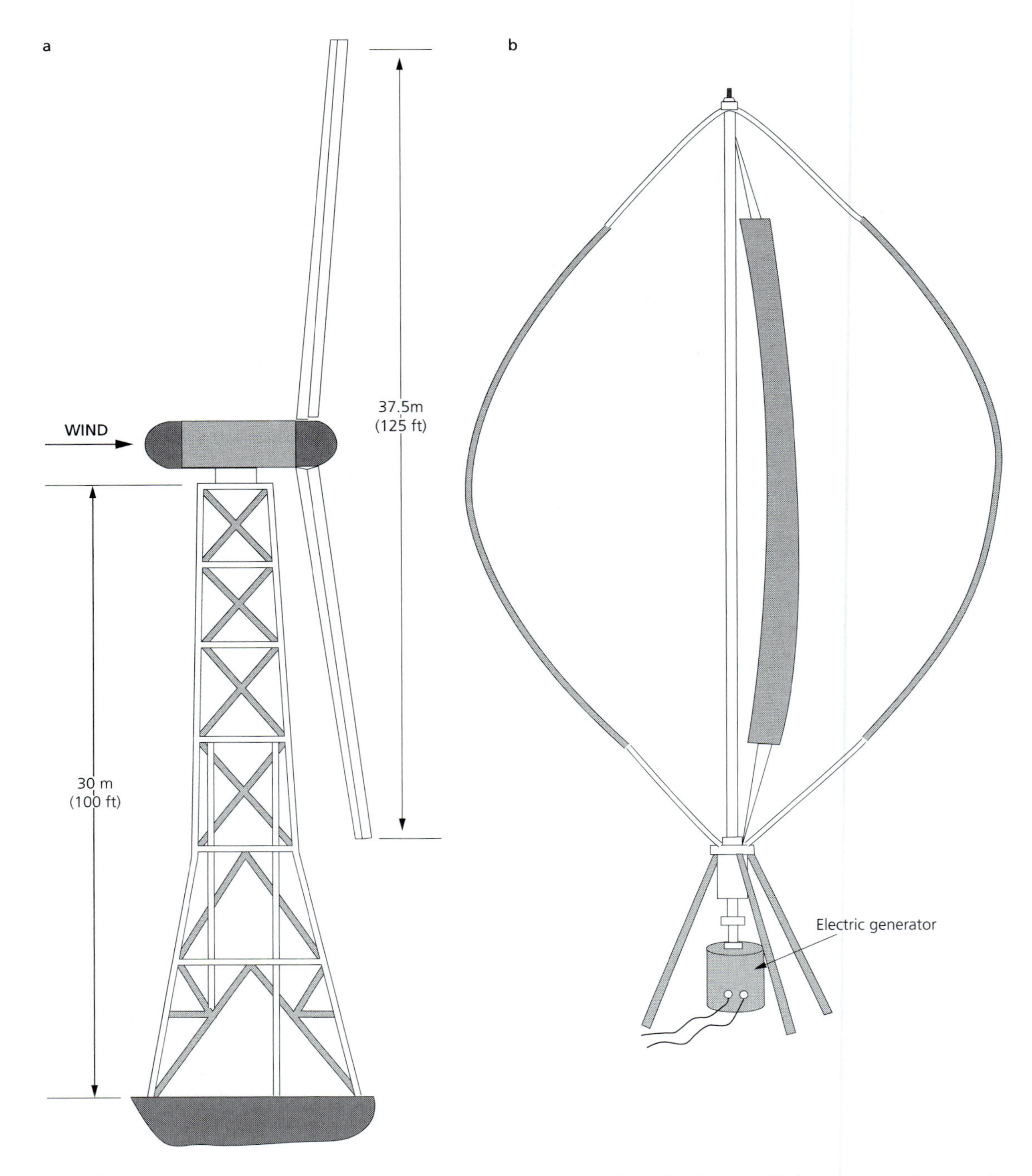

Figure 9.14 *Modern types of wind vane. The twin-bladed wind machine (a) is very efficient at extracting wind energy at high winds using its air-foil blades (which are similar in cross-section to an aeroplane propellor). Different forms of the Darrieus rotor (b), which is efficient at producing electricity at relatively low wind speeds, have been developed.*

Source: Figures 20.23 and 20.24 in Cunningham, W.P. and B.W. Saigo (1992) Environmental science: a global concern. Wm. C. Brown Publishers, Dubuque, Iowa.

BOX 9.17 PROBLEMS ASSOCIATED WITH WIND FARMS

- They can be extremely noisy and disturb local residents.
- They can be visually intrusive and are often visible from far away.
- They look unnatural and can spoil attractive landscapes.
- Some people are worried about electromagnetic interference generated by wind turbines.

number of problems (Box 9.17). Planners and operators believe that with careful planning and thoughtful design (including the use of buffer zones around wind farms) many of these problems can be minimised or resolved.

ATMOSPHERIC MOISTURE AND PRECIPITATION

WATER IN THE ATMOSPHERE

Water (H_2O), an oxide of hydrogen, normally exists as a liquid. It has no colour, taste or smell. Two-thirds of the Earth's surface is covered by water (in the great seas and oceans), and the abundance of moisture on this watery planet makes it unique in the solar system. No other planet has so much or such freely available water as exists on Earth. Water is the most abundant substance on Earth and is essential to all forms of life.

The global water cycle (see p. 353) involves stores and flows of water through all of the main environmental, ecological and biogeochemical cycles (see p. 86). It is driven ultimately by the global heat engine (see p. 275) and the global wind machine (see p. 286).

We know from experience that the atmosphere contains a great deal of moisture, and this is most obvious when it rains, when we see clouds in the sky and when we are surrounded by fog. But even under ordinary weather conditions, when the air seems quite clear and dry, it usually contains moisture that we are not directly aware of.

Water in the atmosphere exists mainly as an invisible gas, water vapour. Water vapour is an important ingredient of the atmosphere, accounting for up to 5 per cent of air by volume (see p. 256). It is also the most important of the greenhouse gases (see p. 258), so it plays a vital role in

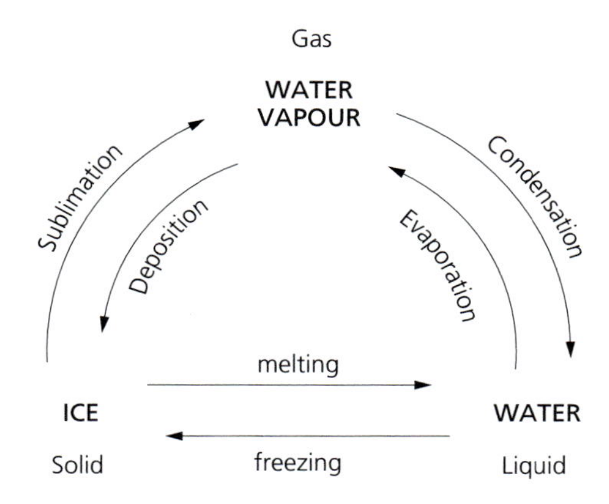

Figure 9.15 *Phase changes of water. Water exists in three physical states (phases) – as a gas (water vapour), as a solid (ice) and as a liquid (water). As it is converted from one to another, it undergoes phase changes.*

the greenhouse effect (natural and pollution-based) and in global warming.

Moisture exists throughout the atmosphere, but its form varies from place to place and over time (Box 9.18; Figure 9.15). Movements of water in the atmosphere are closely linked to temperature patterns and processes, especially stability and uplift (see p. 277). They are driven by the air

BOX 9.18 WATER – STATES AND PHASE CHANGES

Water vapour is one of the three states (phases) in which water exists in the atmosphere:

- solid: as tiny ice crystals
- liquid: as droplets of water
- gas: as water vapour.

Changes in temperature promote changes in which atmospheric moisture changes from one phase to another, and these help to determine weather patterns and changes. There are six different permutations between the three phases:

- solid to liquid, by melting
- liquid to solid, by freezing
- liquid to gas, by evaporation
- gas to liquid, by condensation
- solid to gas, by melting and evaporation
- gas to solid, by sublimation.

movements and wind systems (see p. 280) that redistribute air and the moisture it contains across the Earth's surface.

HUMIDITY

The amount of moisture in the atmosphere varies from place to place and over time. The amount that air can contain depends on the temperature – warm air can hold more moisture than cold air.

SATURATION

When air contains the maximum possible amount of water vapour at a given temperature, it is said to be saturated. The temperature at which this occurs in a given parcel of air is described as the *dew point*. The dew-point temperature varies because warm air can hold more water vapour than cold air.

It is possible for air to become super-saturated (when moisture continues to exist in gaseous form even after saturation temperatures are reached) for a short period of time under particular conditions.

HUMIDITY

The moisture content of a parcel of air (the amount of water vapour it contains) is referred to as its humidity, which can be expressed in absolute or relative terms (Box 9.19).

PHASE CHANGES

As noted above (p. 291), moisture exists in the atmosphere as a solid (tiny ice crystals), a liquid (droplets of water) and a gas (water vapour). These three states are described as phases, and moisture can undergo phase changes that convert it from one state to another (see Box 9.18).

Water can sometimes exist in all three forms (gas, liquid, solid) in the same parcel of air, if conditions (particularly temperature) vary from one part of it to another. This is common in large thunderclouds, for example. Most water in the atmosphere is in the gas phase, as water vapour. This is invisible, and it has the properties of a gas until condensation changes it into small water droplets, which are visible in clouds, fog and rain.

The two most important phase changes in the atmosphere – certainly as far as weather systems are concerned – are evaporation and condensation.

Evaporation: this is the phase change in which liquid moisture turns into water vapour without necessarily boiling. It occurs because fast-moving molecules escape from the surface of the liquid when heat evaporates water to form water vapour in the air. This phase change requires energy, and the energy used is termed the *latent heat of evaporation*. Evaporation is the opposite to condensation. The water vapour in the atmosphere is constantly replenished by evaporation from the oceans, seas, rivers, lakes and other water bodies, including reservoirs. Warm or hot and dry air is most effective in evaporating moisture. Evaporation rates are highest when warm, dry air moves over an open water body. The process is limited by cooler air (which holds less moisture) and by very humid air (which has a limited capacity to hold more moisture). Evaporation rates are also strongly influenced by air movements and wind systems, which bring new air into contact with the source of the moisture and thus promote continued evaporation.

BOX 9.19 ABSOLUTE AND RELATIVE HUMIDITY

Absolute humidity: this is the actual amount of water vapour in air, in g m^{-3} (number of grams of water per cubic metre of air). A different and less common way of expressing this is as specific humidity – the weight of water vapour in a given weight (rather than volume) of air. Another alternative is vapour pressure (the percentage of atmospheric pressure accounted for by water vapour).

Relative humidity: this is a measure of the amount of water vapour in the air compared with what the air could hold if it were saturated. The smaller the difference between the air temperature and the dew point, the higher the relative humidity. It is usually expressed as a percentage. A relative humidity of 50 per cent means that the air contains half as much water vapour as it could hold at that temperature; 100 per cent humidity means saturation. Relative humidity is reported regularly by weather forecasters because it has a strong influence on human comfort – high temperatures with high humidity mean hot, sticky conditions, whereas high temperatures with low relative humidity mean searing, dry heat.

Condensation: this is the phase change from a gaseous to a liquid state, and it occurs when water vapour condenses into water droplets in the air, around condensation nuclei (Box 9.20). This happens when the air temperature falls below the dew point (the temperature of air when it is saturated and has 100 per cent relative humidity), although air can be super-cooled (chilled well below the dew point) before condensation occurs, particularly in very still air. The condensation process releases energy (the energy taken up in evaporation). This energy is called the *latent heat of condensation*, and the heat is used to change the temperature of the surrounding air. It is a major source of heat transfer and provides much of the energy released in storms (see p. 312). Condensation is the opposite to evaporation. It can occur when atmospheric pressure increases as well as when temperatures are lowered. If temperatures fall even further once condensation starts to occur, moisture in the air can change from liquid (water) to solid (ice) by freezing.

Levels of water vapour in the atmosphere are also strongly influenced by *transpiration*, the process by which vegetation gives off large quantities of moisture to the overlying air. This increases concentrations of water vapour in the same way that evaporation from free water bodies does, and both processes are often measured and described together (as *evapotranspiration*).

BOX 9.20 CONDENSATION NUCLEI

The process of condensation requires small particles to be present in the air, around which the water droplets form. These so-called condensation nuclei or hygroscopic nuclei come from many sources. Natural sources include pollen, salt (from sea spray) and dust (for example from volcanic eruptions and soil erosion), and particulate air pollution (including soot and smoke) can increase the availability of such nuclei locally. Condensation is promoted by air turbulence, which increases mixing and brings water droplets into contact with each other. Droplets collide, coalesce and grow bigger until they are large and heavy enough to be pulled downwards by gravity and fall as precipitation.

PRODUCTS OF CONDENSATION

Moisture condenses in a number of different ways, and it can produce some striking effects in the atmosphere. Perhaps the most visible is the rainbow – an arc of coloured light formed in the sky when the Sun's rays are reflected and refracted (bent) by drops of water in rain or mist. Rainbows display the colours of the spectrum in sequence (see Table 8.3) – red, orange, yellow, green, blue, indigo and violet.

The two most obvious products of condensation are clouds (see p. 296) and precipitation (see p. 298). Here we consider dew, frost and fog.

DEW

Dew is formed when small drops of water (dewdrops) condense on a cool surface from water vapour in the overlying air. Moisture condenses when air is cooled to the dew point but temperatures remain above freezing. Ideal conditions include clear skies and a relatively calm atmosphere, and dew is commonly deposited at night on objects such as blades of grass and stones. The process by which objects become covered with small water droplets (formed by the coalescence of individual dewdrops) is known as *guttation*. Studies in Norfolk, England, show that dew and guttation can deposit around 0.1 mm of precipitation on grass surfaces in a typical night, most of which is evaporated the next day.

FROST

Frost consists of frozen moisture. It is a white deposit of ice particles that is often formed on objects out of doors at night. Frost forms under similar conditions to dew, but the dew point must be below freezing. There are a number of different types of frost:

Hoar frost (white frost): this is a deposit of needle-like, solid ice crystals formed on the ground or on window panes by direct condensation from the air at temperatures below freezing point.

Glazed frost (glaze ice, verglas, silver frost): this is a thin layer of clear ice that forms on surfaces that are below freezing point. It is caused by the freezing of rain or water droplets in the air when they come into contact with a cool surface or they refreeze after a thaw.

Rime: this is frost formed by the freezing of supercooled water droplets in fog on to solid objects. It appears as a mass of ice crystals on grass and other surfaces.

FOG

Fog consists of a mass of droplets of condensed water vapour suspended in the air. The droplets form as the air cools below dew point and water vapour condenses on particles of dust in the atmosphere. Fog is similar to a cloud and is formed in much the same way, but it occurs at a lower level and often covers the ground surface. Sometimes the moisture stored in fogs can be collected and used for water-supply purposes.

Few fogs persist for long periods of time. They tend to be short-lasting and are evaporated as air warms or are blown away as wind speed increases during the day.

Visibility is reduced by fog, often significantly. Indeed, fog is defined internationally as creating visibility of less than 1 km. Mist is a thin fog, with visibility of between 1 and 2 km. Fog varies in thickness a great deal from place to place and over time, depending largely on the concentration of water vapour and condensation nuclei present in the air. Reduced visibility from fogs often causes road vehicle accidents, and air and sea navigation problems.

Air pollution can create a particularly damaging type of fog called smog, in which natural fog is mixed with soot, smoke and gases from factory chimneys and other places where exhausts from fossil fuel combustion are released. Smog can seriously damage human health, as shown in numerous local air pollution incidents (see p. 242).

Two main sets of processes (radiation and advection) create fog, and other types of fogs are created by less common processes (Box 9.21).

CLOUDS

Clouds are one of the most visible components of weather, and they directly affect the weather we experience on the ground. Clouds shade out sunlight, reduce temperatures and often bring rain.

Most people make informed guesses about likely weather changes on the basis of the clouds overhead. Dark clouds suggest rain, thick thunderclouds suggest a storm, light clouds suggest perhaps changeable conditions, and no clouds suggests continued fine weather. These informed guesses merely reflect the fact that changes in cloud formations over time often follow a predictable pattern, and analysis of these changing patterns is an important part of weather forecasting and prediction.

BOX 9.21 THE MAIN TYPES OF FOG

Radiation fog: this is a common type of fog that occurs on clear nights when the lower atmosphere is cooled below the dew point by contact with the cooler ground beneath it. Radiation fogs tend to be short-lasting, and they develop best when the air is relatively calm and condensation nuclei are present. They often develop during the night and are evaporated next morning as solar radiation warms the ground and the air above it. Few survive past mid-morning.

Advection fog: this is caused by a different set of processes, is quite common and can be quite persistent, lasting longer and dispersing more slowly than radiation fogs. Advection fogs are formed when warm, moist air is cooled below the dew point as it moves over a cold surface (this could be either land or sea). This type of fog develops in the Midwestern USA during the winter months, when warm, moist air from the Gulf of Mexico blows northwards across the relatively cool ground. It is also common along the west coast of the USA. Such fogs regularly hang over western parts of the United Kingdom, when the warm waters of the North Atlantic Drift (see p. 480) move onshore over cooler land. Advection fogs also occur where cold ocean currents meet warm ones; warm, moist air above the warm current is cooled to the dew point as it mixes with the cooler air over the cold current. This type of advection fog is known as a sea fog or marine fog.

Mist fog: this less common type of fog is created when moist air overlying lakes or swamps is cooled by radiation cooling during the night. Mist fogs hang over the water surface during late evenings and early mornings but are quickly evaporated after sunrise.

Upslope fog: this is another unusual type of fog, which develops when slowly rising air in mountain areas is cooled adiabatically down to or below the dew point. It tends to concentrate at particular levels, where the dew point is reached. It is not unusual for the lower slopes in a mountain area to be fog-covered while upper slopes have clear skies and bright sunshine.

CLOUD FORMATION

Clouds are made up of masses of tiny water droplets or ice crystals that float in the atmosphere. The masses are visible, usually grey or white in colour, and can assume many different forms and appearances.

Clouds are formed when moist air cools, and water vapour in the air condenses around tiny dust particles and other condensation nuclei, in exactly the same way that fogs or mists develop (but above ground level). When the individual droplets or ice crystals coalesce, by being bumped together in the turbulent air within the cloud, they can grow big enough to allow rain or snow to fall beneath the cloud.

Cloud development is restricted to the troposphere (the lower part of atmosphere), because there are no rising air currents above this level that can contribute to uplift, cooling and condensation. There are three main types of cloud:

- *stratus*: 'layer' clouds, which often form a grey layer fairly low in the atmosphere;
- *cumulus*: billowing white or grey 'heap' clouds associated with rising air currents;
- *cirrus*: thin, wispy, fibrous clouds at high altitudes, composed of ice particles.

CLOUD FAMILIES

There are literally dozens of different types of cloud, and characteristic features of the most common ones are described in Table 9.4. Similar types of cloud are often found in groups or families because they have been formed in similar ways. Cloud families are defined on the basis of altitude (Box 9.22).

Clouds with vertical development: these clouds are formed in turbulent, rising air in which powerful convectional uplift creates very high clouds. The most common cloud types are cumulus and cumulonimbus (thunder clouds, see Figure 9.17). The cumulonimbus type in particular is often more than 500 m thick and can extend to more than 10,000 m in middle latitudes and up to

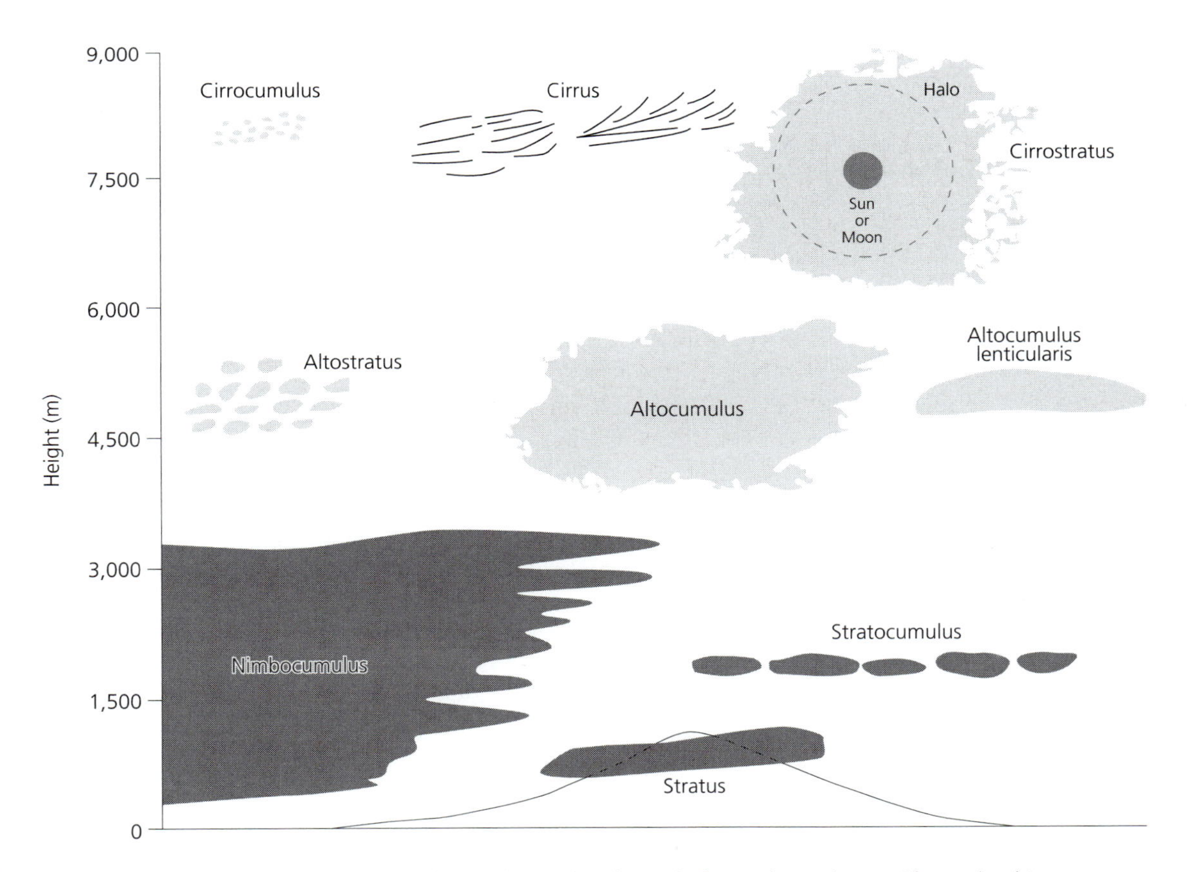

Figure 9.16 *Common cloud types. The figure shows, in schematic form, the main stratiform cloud types.*
Source: Figure 5.11 in Briggs, D. and P. Smithson (1985) Fundamentals of physical geography. Routledge, London.

Table 9.4 Common cloud types and their main characteristics

Altocumulus	These are middle clouds that appear as white or grey patches or layers. The patches generally look quite rounded and soft, and the effect is sometimes described as a 'mackerel sky' (because of the likeness to the mackerel fish, which has a greenish-blue body with wavy dark bands on the back). Altocumulus are associated with weather changes and often appear in mid-latitude cyclones
Altostratus	These are middle clouds with a fibrous appearance that form a dense grey cloud layer and often cover the whole sky. They often arrive ahead of a frontal system and signal weather changes about to occur
Cirrocumulus	These are thin high clouds that contain ice crystals and look like cotton wool. They often mark the leading edge of a frontal system and indicate that major weather changes are likely
Cirrostratus	These are high, thin, white clouds that appear in layers and look like a sheet. They contain ice crystals, which can refract moonlight and shine like a halo, and are often associated with weather changes
Cirrus	These are high-level, thin, wispy, feather-like clouds. They are composed mainly of ice crystals, which diffuse sunlight or moonlight and sometimes act like prisms and produce a halo effect. Cirrus clouds are associated with fair weather, but thick cirrus bring rapid weather changes
Cumulonimbus	These are particularly tall cumulus clouds (sometimes more than 15,000 m high), formed in strongly rising air. They are usually grey or dark grey, and they have anvil-shaped tops and dark flat bases. Cumulonimbus often bring heavy rain or hail (formed in the strong vertical uplift that gives the clouds their great towering height) and are sometimes associated with thunderstorms
Cumulus	These are dense, white, dome-shaped clouds with flat bases and rounded tops, which constantly change shape. Relatively thin cumulus clouds, which are common on warm summer afternoons in middle latitudes, indicate fair weather and relatively stable conditions
Nimbostratus	These are middle or low clouds. They form shapeless, thick, dark grey cloud layers and often bring rain or snow
Stratocumulus	These are soft, grey, rolling low clouds with some darker patches. They are thicker than stratus because rising air is lifted, cools and condenses to form thicker clouds, which usually have flat bases and rounded tops. As cloud cover thickens, weather changes and precipitation often follows
Stratus	These are low, uniform, grey, sheet-like clouds that shade out sunlight and moonlight, creating dull conditions on the ground below. They are commonly found during the winter in humid climates in middle and high latitudes, and they can persist for days or weeks

BOX 9.22 LOW, MIDDLE AND HIGH CLOUDS

Low clouds form from ground level up to about 2,000 m. The main clouds in this family are stratus, cumulus, cumulonimbus (thunderclouds), nimbostratus and strato-cumulus.

Middle clouds form between 2,500 and 6,000 m. The most common ones are altostratus and altocumulus.

High clouds form above 6,000 m up to about 10,600 m. The most common are cirrocumulus, cirrostratus and cirrus.

15,000 m in the tropics (where thermally induced uplift is particularly strong).

PRECIPITATION

'Precipitation' is a collective word used to describe the deposition on the Earth's surface of products formed by the condensation of water in the atmosphere. Rain, snow, sleet and dew are all forms of precipitation. Precipitation is important in various ways (Box 9.23).

DEVELOPMENT OF PRECIPITATION

Precipitation is created when some of the water vapour in the air condenses and clouds form. It requires three conditions:

1. The air temperature must fall below the dew point so that condensation can occur.
2. There must be enough condensation nuclei in the air for condensation to occur.

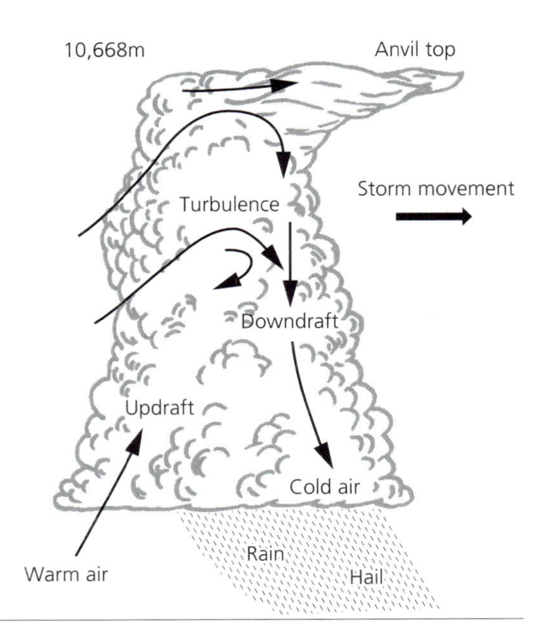

Figure 9.17 *Formation of a thundercloud. Towering thunderclouds are formed when warm air rises rapidly, and the strong updraught promotes condensation and the development of precipitation. Thunderclouds typically have flat, anvil-shaped tops.*

Source: Figure 4.27 in Doerr, A.H. (1990) Fundamentals of physical geography. Wm. C. Brown Publishers, Dubuque, Iowa.

3. There must be atmospheric mixing, normally as a result of turbulence, which makes tiny drops of rain or ice collide, coalesce and grow big enough to fall to the ground.

Moist air, with a high relative humidity, usually has to cool relatively little for temperatures to drop to the dew point, at which condensation occurs (see p. 292). Dry air requires much greater cooling to reach the dew point, as does warm air (which has a greater capacity to hold moisture than cold air has). This is usually caused by convectional uplift of a parcel of air that has been heated and is warmer than the surrounding air into which it rises.

Stability conditions in the atmosphere are critical, and they determine whether a parcel of air rises, and thus whether temperature can fall adiabatically to the dew point so that condensation occurs and precipitation forms. In stable conditions – when the dry adiabatic lapse rate is greater than the normal lapse rate (see p. 277) – a parcel of air that is caused to rise will tend to sink back to its original height without cooling to the dew point, so no condensation will occur. Stable air cannot create precipitation.

Instability exists when a rising parcel of air continues to rise because it is warmer (thus lighter) than the air around it. Precipitation can occur only in unstable air conditions, which allow a parcel of air to cool to the dew point, condensation to occur and water vapour to condense around the condensation nuclei to produce the initial tiny water droplets that eventually coalesce (through mixing of the air) into raindrops.

The strongest conditions for precipitation to develop occur where the normal lapse rate is much greater than the adiabatic lapse rate. This causes rapid uplift of air, rapid cooling, great turbulence and mixing, and it is typical of cumulonimbus clouds.

FORMS OF PRECIPITATION

Precipitation means the fallout of moisture from the atmosphere, and this can take a number of different forms, the most common being rain, snow, sleet, glaze and hail (Box 9.24). These are not mutually exclusive, and more than one type of precipitation can fall at the same time. Snow and rain, or sleet and rain, may fall at the same time if air

BOX 9.23 SIGNIFICANCE OF PRECIPITATION

Precipitation is a basic ingredient of weather systems, and when it falls it is the most obvious evidence of water vapour in the atmosphere. Differences in precipitation from place to place are important controls of climate regions, and some places have reputations based on precipitation (they are either particularly wet or particularly dry). Precipitation creates a number of natural hazards, particularly relating to extremes of shortage (drought) or over-abundance (floods), but more regular events (including thunderstorms) also put people and property at risk. Many air pollutants (gases, aerosols and particles) are washed back to the ground in precipitation – acid rain (see p. 251) is a graphic example.

Precipitation is a key component of the global water cycle (p. 353), and it provides the basic inputs to drainage basin systems (p. 355) and water resource management systems (p. 388). The moisture it brings plays important roles in rock weathering (p. 204), and the runoff it produces is an important agent of erosion (p. 210).

BOX 9.24 FORMS OF PRECIPITATION

Rain: rain is precipitation from clouds in the form of separate drops of liquid water that fall to the Earth's surface. Individual droplets of water condense around hygroscopic nuclei from water vapour in the air, and these droplets mix and collide. They coalesce and grow bigger until they eventually fall as raindrops. Raindrop sizes vary from very fine mist (drizzle) to large drops (typical of a thunderstorm).

Snow: snow is precipitation from clouds in the form of flakes of ice crystals formed in the upper atmosphere. It is formed when the dew point is below freezing and water vapour in the air condenses into hexagonal-shaped ice crystals, which float down to the ground. Snow can be dry and powdery or wet and heavy, depending on the temperature and moisture conditions at the time of snowfall.

Sleet: sleet is partly melted falling snow or hail, or (particularly in the USA) partly frozen rain droplets in the form of ice pellets. Sleet is formed when liquid raindrops are frozen as they fall towards the ground through the freezing atmosphere.

Glaze: glaze is created when rain falls on below-freezing surfaces and the liquid is frozen into a sheet of ice. Strictly speaking, this is not really a form of precipitation, because it is the rain that falls and then freezes on the ground.

Hail: hail is made up of small, clear pellets of ice (hailstones), which are formed when water droplets move repeatedly in and out of freezing zones in a cloud because of turbulence. Concentric layers of ice are added to the outside of the growing particle as it is repeatedly moved back into a zone of saturation. The size of hailstones reflects the degree of turbulence within the parcel of air (large hailstones suggest violent updraft and downdraft, causing many movements into and out of the freezing zones). The hailstones do not melt and become rain as they descend, because the air near the ground is cold and keeps them frozen. Hail is commonly produced in thunderstorms and falls from cumulonimbus clouds when there are very strong convectional rising air currents. On 25 March 1992, the Orlando area in Florida was battered by thunderstorms with hail (some hailstones were reported to be as big as grapefruit), which caused US$60 million of damage.

BOX 9.25 SPATIAL VARIATIONS IN PRECIPITATION

The global pattern of precipitation shows three key tendencies:

1. Precipitation tends to decrease from low to high latitudes (the tropics are wetter than middle latitudes, and these are wetter than polar regions).
2. Precipitation also tends to decrease from the margins of continents towards the interior (coastal areas are on the whole wetter than inland areas).
3. Windward slopes of mountain ranges close to the sea tend to be wetter than leeward slopes (because of the rain shadow effect; see p. 299).

temperatures are close to freezing. Rain and hail can fall from different parts of the same thundercloud if there are areas within it that have different turbulence.

SPATIAL VARIATIONS

Precipitation is quite variable from place to place, because it is formed in a number of different ways and precipitation-forming conditions can vary a great deal over time and from place to place. The general principles outlined in Box 9.25 do not apply in all places, and there are some notable exceptions. Some of the world's driest deserts (including the Karoo and Namib in southern Africa, the Atacama in Chile and the Mojave in the USA) are located along coastal areas (see p. 428), as a result of the interaction of cold ocean currents and high-pressure cells. The Amazon basin, in the interior of Brazil, is much wetter than the coast at the same latitude. Some mid- to high-latitude areas (including parts of North America) are wetter than some areas closer to the equator.

PRECIPITATION CHEMISTRY

Precipitation varies in quality as well as in quantity. Quality is defined by precipitation chemistry, and this reflects the varying chemical composition of rain, snow, sleet and hail. Some of the variations are caused by natural processes, such as volcanic eruptions and forest fires, but air pollution is becoming an increasingly important control.

Studies of precipitation chemistry, and how it varies from place to place and over time, provide valuable indicators of changing atmospheric composition, the dispersion of air pollutants, and processes involved in atmospheric portions of the biogeochemical cycles. Monitoring of the washout of organic pollutants in different parts of Switzerland, for example, has shown the importance of local atmospheric conditions and some of the dynamics of phase changes.

Long-term studies, based on repeated measurements at sample sites, can be very useful in establishing baselines of air quality in unpolluted areas and in pinpointing where and when air pollution is causing detectable changes in air quality. Analysis of snow chemistry at fourteen sites in the Sierra Nevada, over the period 1966 to 1985, has shown how acidity has increased slightly, probably as a result of regional increases in fossil fuel combustion.

RAINFALL

Rain is the most common form of precipitation, and it is generated in a number of different ways.

CONVECTION PRECIPITATION

Convectional rain occurs when heated air rises and cools, so that the air temperature drops to the dew point and condensation occurs. It is most common where the ground surface is strongly heated, promoting convectional uplift in unstable conditions. Hence it is a regular occurrence in middle latitudes during summer months, and it occurs throughout the year in the tropics.

Uplift promotes the growth of raindrops, so convectional rain often brings downpours that are:

- *localised*: weather conditions can be very variable over even short distances; it can be bright sunshine in one place while a heavy downpour of rain is falling a few kilometres away;
- *heavy*: rain can be torrential and might include violent thunderstorms.

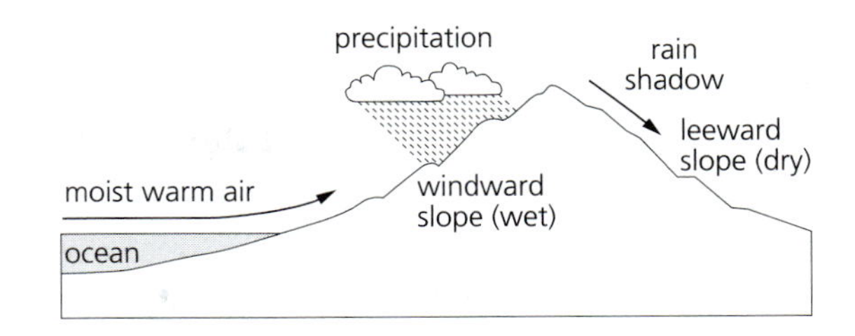

Figure 9.18 *Orographic precipitation. Warm, moist air rises up the windward slope of a mountain, water vapour condenses and precipitation falls.*

OROGRAPHIC PRECIPITATION

Orographic rain results from air currents rising over high ground, which intercepts moisture-laden winds (Figure 9.18). When warm, moist air is forced to rise over a mountain or hilly area, it cools adiabatically as it rises. If the temperature drops to the dew point, condensation occurs and rain falls on the windward (upwind) slope. Orographic uplift can be strong enough to generate thunderstorms.

The downwind (leeward) slope, beyond the barrier, is often dry and is described as a *rain shadow*. The descending air is compressed and heated adiabatically, and its temperature rises above the dew point. As a result, windward slopes are normally wetter than leeward slopes.

Orographic rain is usually lighter than convectional rain, but it occurs much more regularly and so total annual rainfall on windward slopes can be relatively high.

FRONTAL (CYCLONIC) PRECIPITATION

Cyclonic rain occurs in depressions when warm air rises above cold air. Such conditions occur in all latitudes, but they are most common in the middle latitudes, where westerly winds dominate, during the summer months. Here relatively dense air from higher latitudes comes into contact with warm, relatively light air from lower latitudes at the polar low zone (see p. 286). Warm air is forced to rise over the cold air at the front. It cools adiabatically, and if it reaches the dew point condensation occurs and rain falls. Rain usually falls across a wide area along the line of the frontal system, and it advances across an area as the front moves forward.

COMPOSITE PRECIPITATION

In many areas, most rainfall is of the frontal type, but sometimes more than one rain-forming process is at work at the

SUMMARY

Having explored the structure and importance of the atmosphere in Chapter 8, here we considered the atmospheric processes that control temperature, wind and moisture and give rise to weather and climate. We looked at the importance of temperature to humans, the four processes that heat the Earth's surface and atmosphere, the patterns and controls of temperature variations from place to place, and the relevance of temperature to air stability. Wind is caused by differences in air pressure, and we examined the nature and causes of pressure gradients, and the forces that determine wind speed and direction, at different heights in the atmosphere. The global pattern of wind systems was described, and the main options for exploiting wind energy were reviewed. Atmospheric moisture is important in many ways, and it is important to understand the different states that moisture exists in and the phase changes between them. Condensation creates a range of products, including frost, fog, clouds and precipitation, and we explored the processes that create clouds and precipitation, and the different forms of cloud and types of precipitation.

same time, giving rise to composite precipitation. For example, a frontal system that is moving fast across the ground, and is thus forcing warm air to rise rapidly above cold air, might generate frontal rain and convectional rain together. A depression that moves through a mountain area might bring frontal and orographic precipitation together.

Composite precipitation depends on the existence of conditions suitable for more than one rain-forming process at the same time, and these are not always predictable or regular occurrences. As a result, this multiple type of rain is relatively unusual, but it can produce heavy downpours and persist for some time.

WEBSITE

Links to relevant websites, a comprehensive bibliography, tools for teaching and learning, and downloadable images relevant to this chapter can be found at the website specially designed to accompany this book at
http://www.park-environment.com

FURTHER READING

Barry, R.G. and R. Chorley (1998) *Atmosphere, weather and climate*. Routledge, London. Useful and up-to-date seventh edition of the classic text on atmosphere and world climate.

Goudie, A. (1993) *The nature of the environment*. Blackwell, Oxford. A clear introduction.

Graedel, T.E. and P.J. Crutzen (1995) *Atmosphere, climate and change*. W.H. Freeman, Oxford. Good overview of the atmospheric system, with an emphasis on natural and induced changes.

Perry, A. and R. Thompson (1997) *Applied climatology: principles and practice*. Routledge, London. Comprehensive and wide-ranging review of how climatological knowledge can be applied to contemporary socio-economic systems and activities, with examples drawn from different areas and themes.

Robinson, P. and A. Henderson-Sellers (2000) *Contemporary climatology*. Longman, Harlow. A synthesis of contemporary scientific ideas about atmospheric circulation, its controls, impacts and significance.

Thompson, R.D. (1998) *Atmospheric processes and systems*. Routledge, London. Concise non-technical introduction to the atmosphere and to fundamentals of weather, well illustrated with case studies.

Chapter 10

Weather Systems

LEARNING OBJECTIVES

When you have finished studying this chapter, you should be able to

- *Distinguish between weather and climate*
- *Appreciate the importance of weather to human activities and human history*
- *Define what is meant by 'air mass' and describe and account for the main characteristics of the principal types of air mass*
- *Outline the relevance of fronts and frontal systems in the context of weather changes*
- *Summarise the main differences between warm and cold fronts, and ana-fronts and kata-fronts*
- *Describe and account for cyclogenesis*
- *Define the term 'anticyclone' and describe and account for the weather patterns commonly associated with anticyclones*
- *Explain the main types of storm and the weather associated with them*
- *Distinguish between tornadoes and hurricanes and account for their main features*

We experience the atmosphere (Chapter 8), and the operation and end-results of atmospheric processes (Chapter 9), through weather and climate. It is important to distinguish between the two, because they affect us and the world around us in different ways and are controlled by different sets of factors (Box 10.1).

In this chapter, we explore weather systems, how they work and what controls them. We examine climate, its influences and distribution, in Chapter 11.

IMPORTANCE OF WEATHER

WEATHER AND PEOPLE

Experiencing weather is probably the most direct, immediate and recurrent contact that most people have with the natural environment around them. Nomadic people, those who live in temporary shelters and those who live outside have non-stop contact with weather, and it determines just about everything they do, and how, when and where they do it. People who live in proper buildings can escape the worst excesses of weather – the searing heat, freezing cold, intense downpours – by retreating into their protective shelters and turning on the heating or air-conditioning systems to create a comfortable artificial climate that suits human needs (or, probably, human ideals).

WEATHER AND HISTORY

Weather patterns and changes have helped to shape human history. There are many examples, including Christopher Columbus's first voyage across the Atlantic in 1492 (Box 10.2). If that voyage had gone differently – the ship might have sunk, for example, or become becalmed and unable to continue – some aspects of world history might have turned out differently.

Maritime history includes many examples of weather control, because air conditions strongly influence ocean conditions (particularly waves), and ships at sea are very much at the mercy of the elements. This is well illustrated in the ill-fated maiden voyage of luxury British liner the *Titanic* in April 1912, which sank after hitting an iceberg during a storm at night, with the loss of 1,513 lives. The disaster might not have happened, or might have been

BOX 10.1 WEATHER AND CLIMATE

Weather is the day-to-day meteorological conditions experienced in a place or area. It is determined by factors that can vary a great deal over time – such as temperature, cloudiness and rainfall – as a result of natural atmospheric processes. The weather today might be hot, dry and sunny, whereas yesterday it might have been cooler and more overcast. Weather changes a great deal from place to place, even over relatively short distances. Weather can also change rapidly over a short period of time.

Climate is the long-term prevailing weather conditions in an area. These are determined by factors that are fixed, such as latitude, position relative to oceans or continents, and altitude. The climate in central Australia is hot and dry, whereas the climate in Britain is much cooler, wetter and more changeable. The climate in Antarctica is distinctly different, including freezing temperatures, strong winds and limited precipitation (in the form of snow). Climate changes from place to place, but much more slowly than weather because climate regions (areas that share similar climatic characteristics) are usually quite large. Climate also changes over time, but again much more slowly than weather – over decades rather than days.

BOX 10.2 THE WEATHER DURING COLUMBUS'S VOYAGE ACROSS THE ATLANTIC IN 1492

Analysis of the log of the voyage has suggested some interesting characteristics of weather conditions at the time, which can be compared with more recent weather patterns in the same part of the globe.

The Columbian pilots' descriptions of 'calms' related to travel slower than travel occurring during other portions of the voyage. That rate of travel compares favourably with calm winds and an ocean current of 0.4 knots (0.74 kilometres an hour), a value close to modern-day values.

The frequency of 'calm' events experienced by Christopher Columbus in 1492 is significantly higher than the most liberal estimates of calms in the North Atlantic over the last 100 years. The locations of the Columbian calms are generally in the same region currently experiencing the highest frequency of calms. Based on historical hurricane records from 1886 to 1989, the centre of a hurricane would have passed within 100 km of Columbus only once in the past 104 years.

handled differently, if there had been better visibility, smaller waves and a calmer sea at the time.

Weather patterns and changes also have to be taken into account when battle strategies and tactics are being planned. Reconstructions of conditions at the time of the Battle of Trafalgar (off south-west Spain) in 1805 have shown just how important weather changes were to sailing ships, which were heavily dependent on wind speed and direction. Battles in the skies are also heavily influenced by weather, because flight paths, safety and strategies inevitably depend heavily on atmospheric conditions (particularly visibility and cloud cover). Some important decisions in the 1990 Gulf War reflected the need for suitable weather conditions, and that conflict relied heavily on the availability of reliable weather-forecasting.

WEATHER AND HUMAN ACTIVITIES

Many human activities are affected by the weather. We clear snow from paths in the winter, look forward to sunny summer holidays and stay indoors if we can during raging storms. Adverse weather can reduce visibility, cause poor driving conditions on roads and delay transport systems (including trains and boats and planes). Some space and satellite missions have been disrupted because of delays in take-off or landing schedules, made necessary by poor weather.

Many outdoor activities are strongly influenced by the weather, and some have to be stopped for safety reasons during particularly bad weather. Building is a graphic example, and construction work is difficult and sometimes dangerous to continue during heavy wind and rain, or while temperatures are below freezing (when concrete will not set, for example). Prolonged bad weather can have significant economic impacts, because delays or disturbances in building projects can mean missed completion deadlines, financial penalties and serious knock-on inconvenience.

Weather influences recreation and tourism in a variety of ways, and it seriously affects many outdoor sports fixtures. Plans and schedules can be seriously disrupted by unusual or unexpected weather changes, as happened during the build-up to the 1992 Olympic Games in Barcelona (Spain), when snow fell out of season and at an unusually low altitude.

WEATHER AND HEALTH

Weather can affect human health both directly and indirectly. Many aspects of human happiness appear to be closely linked with daily and seasonal weather, which rules our moods and behaviour. The air we breathe has a profound influence on our well-being, and scientists are starting to discover some interesting patterns in bioclimatology, including the impact of electric radiation, climate rhythms and weather changes on health and disease. Death rates appear to increase once threshold temperatures are crossed, in both summer and winter, and particularly in areas where hot weather is uncommon. Mortality, particularly among the elderly, is also higher in cloudy, damp, snowy places.

Indirect effects of weather on health occur mainly through air pollution, because weather conditions can promote the build-up of serious air pollution incidents (see p. 245), which increase the incidence of respiratory problems and related illnesses.

WEATHER, IMAGE AND IMAGINATION

Weather is a regular topic of conversation in some countries, and images of some places are dominated by weather (for example, the North American view of England as a place where it is always raining or overcast). Weather also figures prominently in much creative art (Box 10.3).

BOX 10.3 THE WEATHER AND CREATIVE ART

Weather patterns and variations are a valuable source of inspiration for:

- *Writers*: many writers use weather as a backdrop and a context for dramatic events. This is well illustrated in Conan Doyle's misty nights and thick fogs in the Sherlock Holmes mysteries.
- *Poets*: Wordsworth immortalised clouds and daffodils in his poetry, and many other poets have painted pictures of the sky in words.
- *Painters*: landscape painters like Turner and Constable captured dramatic skies in oil on canvas, literally creating the atmosphere for particular scenes.

WEATHER EXTREMES

WEATHER DISASTERS

The variability of weather over time at particular places is a major source of environmental uncertainty and, particularly where climatic extremes are involved, it can be both a nuisance and a hazard. We shall look at some specific areas of risk later in this chapter.

So-called weather disasters are variable in time and space, and they can cause a great deal of damage. Weather disasters in the USA caused US$66.2 billion of damage (at 1991 prices) between 1950 and 1989, and this represents 76 per cent of the nation's insured loss over that period. Most such disasters occurred in the south, south-east, north-east, and central USA, and there were few in or west of the Rockies. Weather disasters were most common in the 1950s and 1980s, losses were highest in the 1950s, late 1960s and late 1980s, and the area affected was largest after 1975.

The USA is by no means the only country to have a long history of weather disasters. Records of severe weather incidents throughout Chinese history – including floods, droughts, plagues of locusts, hail, famines, unusually cold and warm spells, and harvest failures – help to provide a baseline against which to measure recent changes.

GLOBAL WARMING AND WEATHER CHANGES

Scientists predict that global warming caused by greenhouse gases (see p. 255) will probably be accompanied by significant regional weather changes. More common (and damaging) extreme weather events are likely in many places, and this will increase the frequency and intensity of floods, droughts, tornadoes and tropical cyclones. Some models forecast the spread of deserts and serious crop failures in the granaries of the USA, the Commonwealth of Independent States and China.

Scenarios for particular countries indicate some of the problems that might lie ahead. For example, climate models suggest that global warming could bring warmer, drier conditions to Mexico, with higher precipitation but an even greater increase in potential evaporation losses. This would cause serious water shortages and lasting problems for

BOX 10.4 WEATHER FORECASTING

Weather patterns and changes tend to be fairly predictable because they follow known physical laws and processes. This makes it possible to predict weather changes, at least over the short term, with reasonable levels of accuracy. Appropriate responses can then be selected and adopted. Traditional forms of prediction, often based on intuition and folklore, rely on repeated patterns of weather changes, and they can sometimes be quite useful. For example, the rule of thumb 'red sky at night, shepherd's (or sailor's) delight; red sky in the morning, shepherd's (sailor's) warning' has been used for many years.

More reliable predictions are based on observing weather changes (cloud sequences are often quite predictable) and on monitoring changes in air characteristics such as pressure (using a barometer) and humidity (using wet- and dry-bulb thermometers). The most detailed and reliable forecasts are based on sophisticated atmospheric modelling using high-powered supercomputers to analyse monitoring data from weather stations and satellites.

rain-fed and irrigated agriculture, urban and industrial water supplies, hydropower and ecosystems.

AIR MASSES AND FRONTS

AIR MASSES

An air mass is a large body of air in which the characteristics (temperature, moisture and pressure) are relatively uniform horizontally within the lower atmosphere. This relative uniformity means that most of the ground area beneath a particular air mass experiences similar weather conditions. Towards the edges of the air mass weather changes might be rather different, because there is generally a boundary zone (frontal system) between adjacent air masses.

PROPERTIES OF AIR MASSES

Air masses exist at a variety of sizes, from several km^2 to 100,000 km^2. They are usually classified in terms of temperature and moisture content:

- *Temperature*: air masses are designated as arctic (A), polar (P), tropical (T) or equatorial (E) depending on their source area and temperature. Tropical (T) and equatorial (E) air masses are warm, whereas arctic (A) and polar (P) air masses are cold.
- *Moisture*: air masses that originate over water are given the label m (maritime), while continental air masses are given the label c (continental). Continental air masses (c) are normally dry, and maritime air masses (m) are usually wet.

The composition and character of air masses are products of the areas in which they originate – their so-called source region. Source regions (Box 10.5) differ a great deal, which is why air masses vary so much. The main properties of the four most common types of air mass are summarised in Table 10.1.

AIR MASS MOVEMENTS

Air masses must remain stationary in a source region long enough for them to adopt their representative characteristics. But, after a period of time, most air masses are usually driven from their source regions by the prevailing wind systems. Although they often move great distances over the Earth's surface, most air masses can preserve their identity and characteristics (particularly temperature and humidity) for long periods. This allows them to redistribute heat and moisture from one place to another. Consequently, air mass movements play a critical role in the global heat system (see p. 275) and in the global water cycle (see p. 352). They also strongly influence weather conditions along their path (Box 10.6).

SECONDARY AIR MASSES

While air masses are internally quite homogeneous, particularly as they form in source regions, and thus have

BOX 10.5 PROPERTIES OF AIR MASS SOURCE REGIONS

Good source regions share two important properties:

1. limited air motion and calm conditions, which allow air masses to develop relatively uniform properties over time; and
2. relatively uniform surface conditions (such as oceans, deserts or large ice sheets), which allow air masses to develop relatively uniform properties over a wide area.

Table 10.1 Properties of common types of air mass

Type	*Source region*	*Properties*	*Effects on weather*
Tropical maritime (mT)	Tropical or sub-tropical oceans	High humidity and high temperatures	Often brings convectional rain in summer and drizzle in winter
Polar continental (cP)	Large land masses at high latitude	Usually dry and stable	Brings cool conditions in summer and cold conditions in winter
Polar maritime (mP)	Oceans in higher latitudes	Mild temperature and high humidity	Much cloud cover and dull conditions in winter; brings mild, fair weather in summer
Tropical continental (cT)	Continents in sub-tropical latitudes	Low humidity and high temperatures	Brings clear skies and low relative humidity

BOX 10.6 IMPACTS OF AIR-MASS MOVEMENTS ON THE WEATHER

Air masses often have characteristics that are different from those of the air into which they move, and this has two important consequences:

1. Weather, at a given location changes over time as one air mass is replaced by another, which is blown over the area by dominant wind systems.
2. The boundary zone between adjacent air masses – where the different bodies of air come into contact with each other – is usually very dynamic, and the frontal systems that develop there are the primary engines of localised weather patterns and changes.

distinctive characteristics that make them different from surrounding air masses and the air into which they move, they can change. When an air mass changes (Box 10.7) from its original character – it might become wetter or drier, warmer or colder, or any permutation – it is described as a secondary air mass.

Some regions, such as Britain, receive more secondary than primary (unmodified) air masses, and the weather in such places tends to be more variable and changeable as a result. An air mass can be warmer than the surface over which it travels (such as a warm tropical air mass that moves northwards across Africa towards Europe). It then cools progressively as it moves, and its lowers layers become more and more stable. Fog (see p. 294) and/or stratus clouds (see p. 296) are often associated with such conditions. Alternatively, an air mass can be cooler than the surface it moves across (such as cold polar air moving south towards Europe from the Arctic). It then warms and can become unstable, undergo rapid adiabatic uplift and generate thick, towering cumulus clouds (see p. 297).

GLOBAL AND REGIONAL PATTERNS

Polar/arctic and tropical air masses in each hemisphere migrate towards the middle latitudes, where they converge at the polar front and are blown eastwards by the westerly winds (see p. 286). Air flow above the polar front is often geostrophic (see p. 284), and cyclones (see p. 309) tend to move along the polar front. Little wonder, therefore, that the position and dynamics of the polar front play such an important part in determining weather conditions throughout the entire hemisphere.

The main air masses that affect the USA (Figure 10.1) are described in Box 10.8. Europe is affected by many of the same types of air mass as the USA. Continental polar air masses blow down from the Arctic ice sheet, Greenland and other high-latitude areas. Polar maritime (mP) air masses originate in the north Atlantic, and tropical maritime (mT) air masses come from the region between the North and South Atlantic, and from the Indian Ocean.

BOX 10.7 CAUSES OF CHANGE IN AIR MASSES

Change comes in two ways:

1. *Internal change*: air masses can be modified by internal processes, such as adiabatic change or subsidence associated with instability.
2. *External change*: air masses can be modified by contact with the ground surface over which they move.

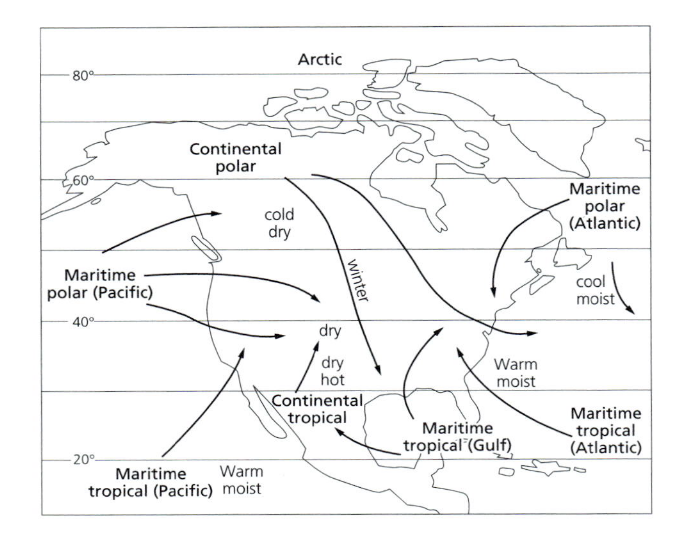

Figure 10.1 *Source regions of the main air masses that affect North America. Each source region generates air masses with distinctive properties (particularly temperature and humidity), and as these move across a region they can bring changes in weather.*

Source: Figure 4.28 in Doerr, A.H. (1990) Fundamentals of physical geography. Wm. C. Brown Publishers, Dubuque, Iowa.

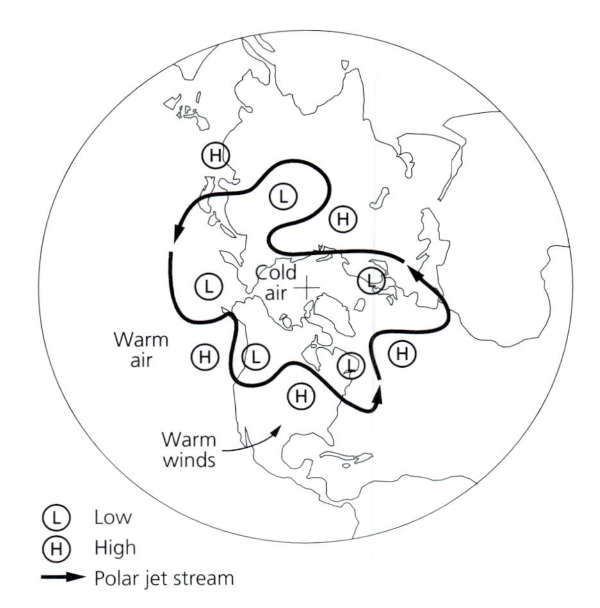

Figure 10.2 *Impact of the polar jet stream on air masses in the northern hemisphere. The figure shows a typical pattern of the arctic circumpolar vortex, a large, circulating mass of cold air that penetrates North America and Eurasia and is associated with the generation of storms. If the vortex stalls, weather becomes much more stable and some areas receive heavy rainfall, while others suffer droughts.*

Source: Figure 17.6 in Cunningham, W.P. and B.W. Saigo (1992) Environmental science: a global concern. Wm. C. Brown Publishers, Dubuque, Iowa.

Tropical continental (cT) air masses regularly arrive over Europe from source areas in North and Central Africa.

An important factor in determining which type of air mass is dominant over a mid-latitude area (including much of the USA) at a particular point in time is the location of the polar jet stream (see Box 9.12). This migrates northwards and southwards over time, partly in response to movements of the Rossby waves (see Box 9.13). When it moves south across North America it brings cold arctic air and low pressure; when it moves north it brings much warmer tropical air and high pressure (Figure 10.2).

BOX 10.8 THE MAIN AIR MASSES THAT AFFECT THE USA

The dominant air masses that affect the USA are mP (polar maritime), cP (polar continental) and mT (tropical maritime).

Most of the cold air masses are polar continental (cP) in origin, but from time to time arctic air masses cross the north-eastern USA. Polar continental air from source regions in northern Canada and Alaska regularly invade the Great Plains and the Midwest, sometimes stretching further south than the Gulf of Mexico. The polar maritime (mP) air masses in North America originate mostly over the North Atlantic and North Pacific; the latter, which are pushed eastwards by the westerly wind systems, have a stronger and wider impact on weather across North America.

Most of the tropical maritime (mT) air masses move northwards from the Gulf of Mexico and the Caribbean Sea, and they affect weather conditions east of the Rocky Mountains (particularly when the warm, moist Gulf air meets the cold, dry polar air coming down from Canada). Tropical continental (cT) air masses rarely pass over the USA, because the source area in Mexico and Central America is relatively small. When they do cross the area, they bring warm, dry air into the Great Basin and the south-west.

FRONTS AND FRONTAL SYSTEMS

We have seen (p. 304) that air masses have two important properties:

1. *Persistence*: air masses tend to retain their individual characteristics (particularly temperature and humidity) for a long period of time, even when they are moved to a contrasting area.
2. *Mobility*: once an air mass has formed and its character has stabilised (reflecting ground conditions in the source region), it can be moved by prevailing winds and can travel a long distance across contrasting ground surface conditions. Air masses move at different speeds, reflecting the interplay of forcing factors (pressure gradients, wind speeds), resisting factors (particularly ground surface roughness) and enabling factors (the nature and properties of the air mass itself).

It is inevitable, therefore, that air masses with different characteristics will come into contact with each other. When this happens, it can bring dramatic weather changes, particularly along the contact zone or boundary between adjacent air masses.

FRONTS

The boundary between air masses is marked by a front, and it is one of the most active weather zones where conditions change sharply. Frontal systems, defined by the invasion of air masses across an area, are major controllers of weather, particularly in middle and high latitudes. For example, cool, dry polar continental air often comes in contact with warmer, moister tropical maritime air masses over North America and Europe (see p. 306), and this can trigger predictable sequences of cloud and weather changes.

Scientific understanding of the formation and dynamics of fronts has advanced a great deal in recent decades, but the basic model proposed in 1917 by Vilhelm Bjerknes (a Norwegian meteorologist), which describes and explains weather changes at fronts, has largely stood the test of time. It remains an important tool in weather forecasting and helps to explain patterns of weather change over time and from place to place.

STRUCTURE OF FRONTS

A front is three-dimensional – it stretches along the ground the entire length of the boundary between adjacent air masses, and it normally extends up to the top of the troposphere (the tropopause) (see p. 227).

The contact zone (frontal zone) between air masses is not vertical, because a front always marks a boundary between air masses with different temperatures. The lighter, warmer air mass rides over the denser, cooler one, and the cool air forms a wedge under the warmer one. As a result, the frontal zone always slopes upwards when viewed in cross-section (Figure 10.3).

Diagrams often show the front as a steep line, but in reality most frontal zones slope very gently. A slope of about 1° is typical of a warm front, which means that a front stretching up to perhaps 8 km (the tropopause) might cover a zone 1,000 km wide on the ground. This low frontal slope means that an observer on the ground can watch cloud patterns change well in advance of the front, and this is useful in weather forecasting. Cold fronts are much steeper than warm fronts (frontal slope is often 2–3°), so the weather changes more quickly and covers a narrower zone (Box 10.9).

TYPES OF FRONT

Fronts are classified on the basis of temperature changes and air stability. In terms of air temperature there are two types, depending on which air mass is moving into and displacing the other:

1. *Warm fronts*: in a warm front, a warm air mass comes into contact with a cooler air mass and displaces it by pushing it forward.

BOX 10.9 WEATHER CHANGES AT A FRONT

Although fronts move across the ground relatively quickly (a warm front typically advances at about 50 km h^{-1}, for example), it can take many hours for the full sequence of cloud and weather changes to pass a particular location before the front itself arrives, because of the low frontal slope. The most marked weather changes are usually experienced close to and at the front, when cloud cover is lowest and there are obvious changes in air temperature, wind speed and direction, and humidity. Once the front has passed, the weather usually stabilises and conditions are determined by the character of the air mass behind the front.

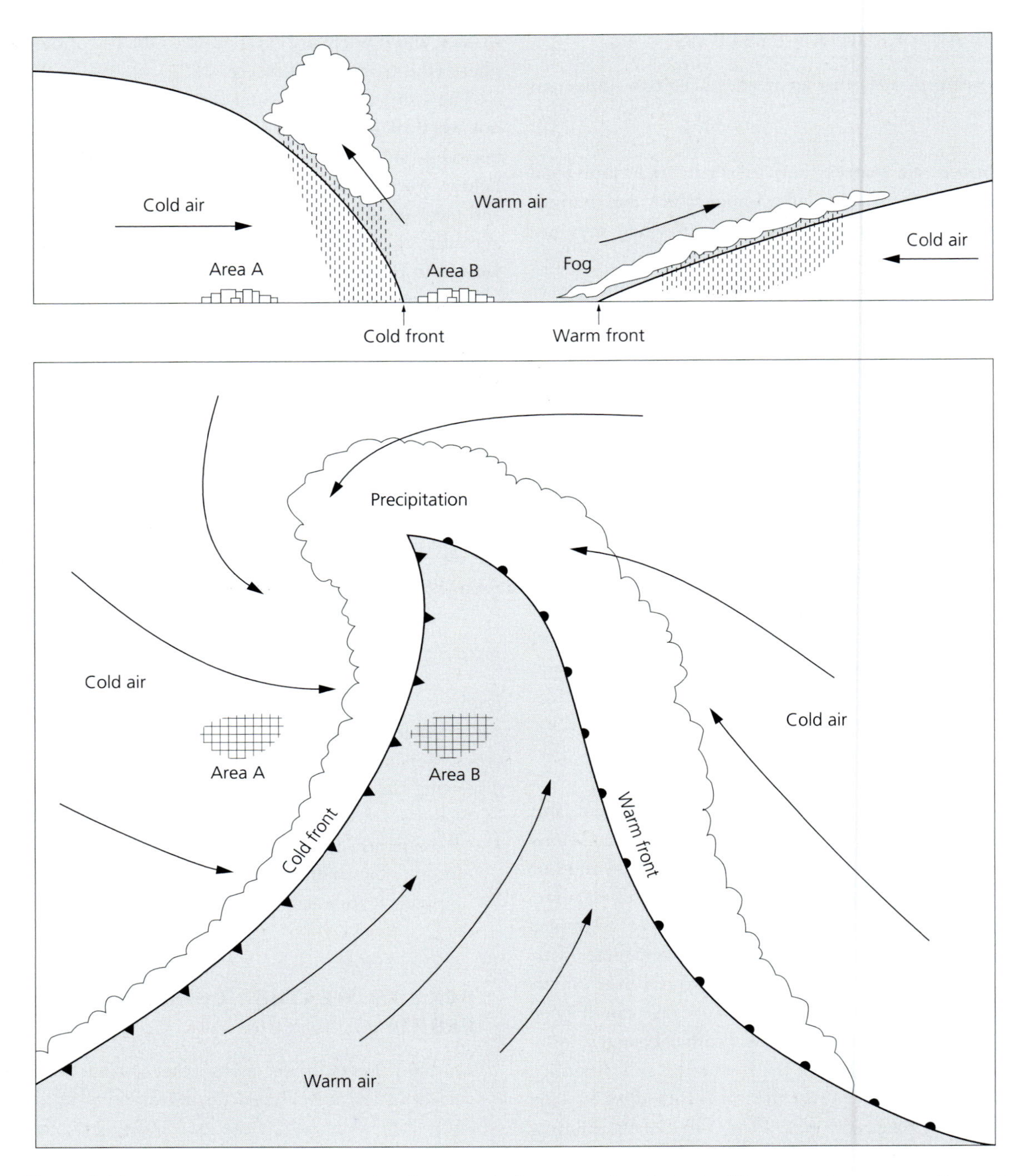

Figure 10.3 *Air-mass movements associated with typical warm and cold fronts. Warm air rises up over the cold air in front of it at a warm front, and cold air wedges in underneath the warm air in front of it at a cold front.*

Source: Figure 4.29 in Doerr, A.H. (1990) Fundamentals of physical geography. Wm. C. Brown Publishers, Dubuque, Iowa.

2. *Cold fronts*: in a cold front, a cold air mass comes into contact with a warmer air mass and displaces it by pushing it along.

There are two stability conditions at fronts, and these strongly influence the amount of activity associated with the frontal system (Box 10.10).

BOX 10.10 STABILITY CONDITIONS AT FRONTS

Ana-fronts: air at an ana-front is unstable and rises rapidly. Such fronts are therefore very active and are associated with uplift, condensation and pronounced weather changes.

Kata-fronts: air at a kata-front is stable, so it sinks. This limits condensation and suppresses weather changes.

WARM FRONTS

A warm front is marked by the advance of a relatively warm air mass into a cooler air mass. The warm air is lighter and rides over the cold air (see Figure 10.3). The frontal zone – which slopes up over the cold air, in advance of the warm front that pushes it – is usually quite shallow and broad. As a result, weather changes occur across a wide zone and persist for some time. The warm air might come from a tropical source region, and the cooler air might be polar or arctic.

Many of the warm fronts that cross North America and Britain are active, unstable ana-fronts. The warm air rises, it cools to the dew point and condensation occurs. Cloud cover is extensive, often thick and high, and cloud changes are visible well in advance of the arrival of the front. Precipitation usually starts well in advance of the front, and it continues until the warm air mass arrives. Kata-fronts are associated with more subdued weather changes, with smaller changes in temperature, pressure, and wind speed and direction. Thick, low-level stratocumulus cloud is common, bringing either no rain or light rain.

COLD FRONTS

A cold front is marked by the advance of a relatively cold air mass into a warmer air mass. The cold air is denser and pushed under the warm air, like a wedge (see Figure 10.3). The frontal zone is usually steep and relatively narrow, so weather changes more quickly over a narrower zone. The cooler air might be polar air moving southwards from the Arctic towards tropical air over central North America or Europe.

Unstable ana-fronts are once again the norm, with cloud patterns reversed from the warm ana-front sequence. Weather changes are more rapid, because the frontal zone is much steeper. Heavy rain from nimbus clouds often precedes the arrival of the front, which is followed by brighter weather, showers and scattered low clouds. Cold kata-fronts bring more limited cloud development and more gentle changes in temperature, pressure, and wind speed and direction.

STATIONARY FRONTS

A stationary front, as the name implies, does not move or at most moves forward very slowly. It can start to move forward again if either of the air masses is forced to move, perhaps because it is being pushed by prevailing winds. Stationary fronts often bring long spells of unsettled weather.

DEPRESSIONS

A depression is a low-pressure cell (Box 10.11) in the atmosphere. Depressions are also called lows, tropical cyclones (or just cyclones), hurricanes or typhoons, depending on where they occur (see Box 10.19). When they are well developed, they can produce violent thunderstorms with winds of force 12 or above on the Beaufort scale (see Table 9.3), which can cause widespread damage and destruction.

WEATHER

Both the warm and the cold fronts bring changes in cloud cover, changes in temperature and air pressure, and rain (see p. 298). Rain at both fronts can be heavy, particularly if the warm air mass is unstable and rises rapidly (causing cooling to the dew point). The passage of the warm front

BOX 10.11 LOW-PRESSURE CELLS

Depressions develop where two air masses with different properties converge into a central low-pressure zone. This commonly happens in middle latitudes in the northern hemisphere, where dense, dry, cold polar air from the north is brought into contact with lighter, damper, warmer air from the sub-tropics to the south. The contact zone is the polar front, and here the colder, denser air slides under the warmer, moister air and forces it to rise. These air movements – driven by warm air flowing into the polar front (the warm front) and cold air flowing in behind it (the cold front) – establish a rotating air mass with the lowest pressure at the centre.

is marked by a rise in temperature, a fall in air pressure, a shift in wind direction and (usually) a reduction in cloud cover. Precipitation occurs before and during passage of the cold front. Behind the cold front, stormy weather is replaced by clearer skies, air pressure rises, humidity declines and temperature drops as warm air is replaced by cold. Cloud cover becomes more patchy and lighter.

The violent storms and strong winds associated with depressions often follow similar paths and regularly pass over the same areas. It is possible to map cyclone tracks by comparing weather maps over time, and these tracks often determine weather conditions across an area. Lows are often associated with unsettled weather, and they bring rain to dry areas. They are important in the development of frontal systems.

CYCLOGENESIS

Cyclogenesis means the development or evolution of low-pressure cells and the frontal systems associated with them (Figure 10.4). Many frontal depressions in middle and high latitudes survive for perhaps four or five days before they disappear by occlusion.

Origin: the sequence begins, and is set in motion, with the arrival of two contrasting air masses on either side of the polar front, with cold polar air to the north and warm tropical air to the south in the northern hemisphere (the pattern is reversed in the southern hemisphere). The polar front will initially be relatively straight for a great distance, with contrasting air masses on either side of it.

Development: somewhere along the front a wedge of warm air starts to move in towards the cold air, and because it is lighter it starts to rise above it (with a gently sloping frontal zone) (see p. 309). This forms an indent, usually called a wave distortion, at the front. Warm air rushes into the wave distortion and rises above the cold air beneath it, encouraging rapid uplift, which draws more warm air in behind it. Pressure drops, and the centre of the wave becomes a low-pressure area. This sets in motion a chain reaction, with low pressure drawing in warm air, which rises, encouraging pressure to fall further.

Maturity: once the depression starts to develop it can quickly start to feed itself, so pressure drops lower and lower and the wave distortion grows bigger and bigger as a result. Two things happen that determine the fate of the depression. First, air in the warm sector of the growing depression is trapped between the cold air ahead (beyond the warm front) and the cold air behind (behind the cold front). It does not mix with the surrounding cold air but remains discretely trapped in its own sector within the growing wave distortion. Second, the depression moves eastwards in the northern hemisphere, driven along by the dominantly westerly winds at these latitudes.

The end-result is the eastward migration of the wave distortion, the warm and cold fronts and associated weather changes. The warm front moves ahead of the cold front, so an observer on the ground first experiences the weather changes associated with the warm front, followed some time later by weather changes associated with the cold front. Most depressions continue to deepen for some time, and this promotes more vigorous weather changes.

Decay: the cold front travels across the ground faster than the warm front in a depression, mainly because it has a steeper frontal slope. It is only a matter of time, therefore,

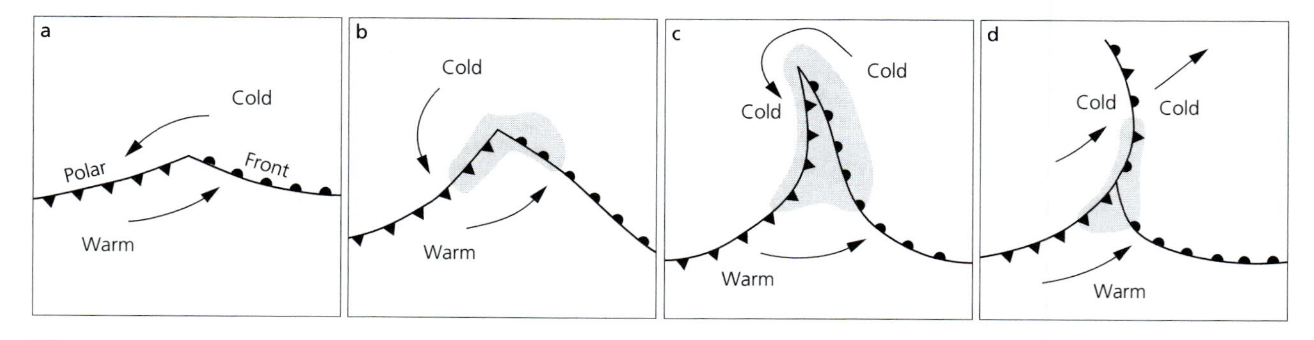

Figure 10.4 *Development of a low-pressure cell and a frontal system. In the northern hemisphere, a cyclonic cell starts to develop as a wave along the polar front (a), and cold air from the north rushes in as a cold front, which pushes the warm air towards and over the cold air ahead of it as a warm front (b). The cell intensifies (c) and the warm air is isolated; eventually, the cell closes in (d) and the storm dies as the polar front is re-established.*

Source: Figure 4.6 in Barry, R.G. and R.J. Chorley (1976) Atmosphere, weather and climate. Routledge, London.

before the cold overtakes the warm front. This process starts at the centre of the depression and works progressively outwards. The warm air is raised above the cold air, forming an occlusion, which is often accompanied by drizzle and unsettled weather (Figure 10.5). As the cold front overtakes the warm front, the warm air sector is lifted completely off the ground. This stops warm air flowing into the depression, so pressure stops falling. Over time, usually after several days, the depression fills in and disappears. The process of frontal decay is described as 'frontolysis'.

While scientists now have a fairly sound understanding of how frontal depressions develop once they form, the start of the sequence – how and why depressions form where and when they do – remains more of a mystery. It is known that surface depressions, which involve convergence of air flow, must be associated with divergence of air flow higher in the atmosphere. Acceleration and deceleration of air movement in the upper atmosphere as it flows around the giant meandering Rossby waves (see p. 286) seem to be a critical factor in generating depressions close to ground level.

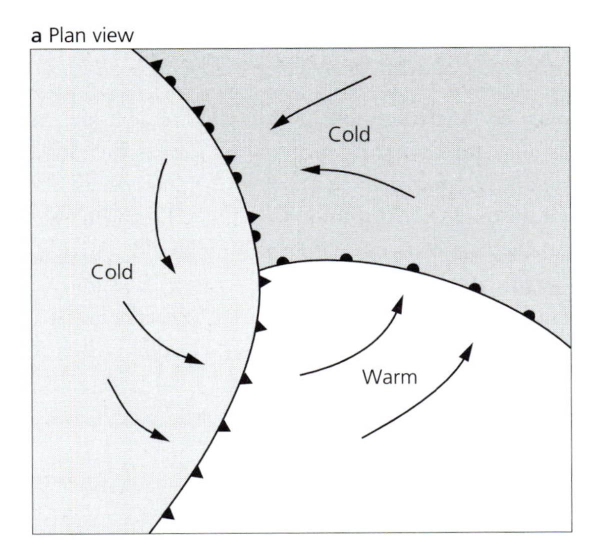

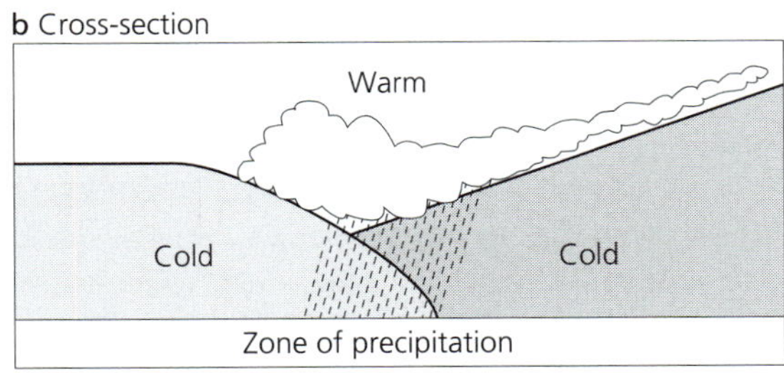

Figure 10.5 *Air flow and weather associated with an occlusion. An occluded front is created when a cold front begins to overtake the warm front ahead of it in a cyclonic cell (see Figure 10.4), forcing the warm air to be trapped and rise above the cold.*

Source: Figure 4.6 in Barry, R.G. and R.J. Chorley (1976) Atmosphere, weather and climate. Routledge, London.

ANTICYCLONES

An anticyclone – sometimes simply called a high – is a high-pressure cell, the opposite of a depression (low-pressure cell).

HIGH-PRESSURE CELLS

An anticyclone is a region of high air pressure in which pressure decreases from the centre (Box 10.12). Highs are generated when a mass of dense, cold air (generally from a polar source region) comes into contact with lighter, warmer air.

Anticyclones are shown on weather maps as domes of high pressure, with concentric isobars and a pressure gradient sloping outwards in all directions. The closer the isobars are spaced together, the steeper the pressure gradient and the more intense the anticyclone.

WEATHER

Weather conditions are relatively stable in anticyclones, which have no uplift, thus no condensation, cloud development or precipitation. Winds are usually calm or light at

BOX 10.12 AIR MOVEMENTS IN AN ANTICYCLONE

Air subsides (descends) in the high, and it moves out in all directions from the centre. This pattern of air movement determines the wind direction and weather associated with the anticyclone. Wind direction is:

- directed by the pressure gradient (which blows winds outwards from the centre of the anticyclone);
- modified by the Coriolis effect (which deflects to the right in the northern hemisphere and to the left in the southern hemisphere) (see p. 281).

As a result, winds circulate around the centre in a clockwise direction in the northern hemisphere and anti-clockwise in the southern hemisphere.

the centre of the high. Highs often bring cold spells in the winter and cooler, fresher spells in the summer. Precipitation can fall around the edge of the air mass at the centre of the high if cold air wedges below warm humid air, lifting it and cooling it to the dew point.

Weather patterns depend partly on the type of air mass involved in the anticyclone. Cold anticyclones, which typically develop in Siberia and Canada during the winter months, often bring long spells of stable, freezing conditions. Warm anticyclones, which are common in subtropical areas throughout the year, are warmer and have higher air pressures.

While anticyclones (highs) are the opposite of depressions (lows), there are some notable differences between the two (Box 10.13).

STORMS

STORM EVENTS

A storm is a violent weather episode that often includes extreme conditions (see Box 10.14), such as strong winds, rain, hail, thunder, lightning and snow. Many storms are associated with the passage across an area of mesoscale convective systems (MCSs) – weather systems up to 500 km across that persist for quite some time, involve significant convectional uplift and often bring severe weather phenomena such as hail, damaging winds, tornadoes, flooding and frequent lightning. MCSs are difficult to forecast because they are smaller than the synoptic (wide-area) weather systems on which weather forecasts are usually based, and their movements are difficult to predict because they have long durations.

BOX 10.13 DIFFERENCES BETWEEN ANTICYCLONES AND DEPRESSIONS

- Anticyclones are rarely as intense as depressions (the rise in air pressure in a high is rarely as big as the fall in pressure in a low).
- Anticyclones are usually larger features than depressions, so they affect weather over much larger areas on the ground.
- Anticyclones move more slowly across the ground than depressions, so weather changes are more gradual.
- Anticyclones tend to be more persistent than depressions, so they affect weather for a longer period of time.

BOX 10.14 STORMS AND WEATHER

Storms disturb normal weather conditions, and they vary a great deal from place to place in size, character and impacts. Some are small, local and brief. Many such storms produce no lasting damage and can in fact bring benefits – causing rain in dry areas, bringing cooling during hot spells and warming during cold spells, and cleaning air pollution over cities and industrial areas by increasing air mixing, dispersion and dilution. Other storms are intense and cause widespread damage and destruction, put lives at risk, and have significant impacts on local and national economies.

STORM ENVIRONMENTS

Some areas – such as middle latitudes in the northern hemisphere – regularly experience severe storms because they are regularly crossed by major depressions and associated frontal systems (see p. 306). Coping with storms is thus part of the cost of living within this highly dynamic part of the Earth's atmospheric system. Between 1950 and 1989, for example, there were more than 140 storms in the USA that caused losses of more than US$100 million each. They caused a total of US$66,200 million of damage, and most were located in the south, south-east, north-east and central USA.

Storms in dry areas pose particular problems and are often a major factor in erosion and landscape change. Dust storms, caused by strong winds in desert regions (see p. 433), throw vast amounts of dust into the air and make conditions very difficult for people and animals in the area at the time. Visibility can be reduced up to heights of 3,000 m or more during dust storms. Even more problematic, if somewhat more unusual, is the dust devil – a whirling column of rotating air and dust, which is usually only a few metres across but can cause damage to anything that crosses its path.

Storms also create serious problems in cold areas. Snowstorms (storms with heavy snow) and blizzards (strong, bitterly cold winds that are accompanied by widespread heavy snowfalls) can make life difficult and dangerous in the winter months in middle and high latitudes.

BOX 10.15 THUNDERSTORM CONDITIONS

Towering, anvil-shaped cumulonimbus clouds (see p. 295), thunder and lightning (p. 314), along with heavy rain or hail, usually accompany thunderstorms. These weather conditions create serious problems on the ground: strong winds can cause widespread damage; intense downpours of rain and hail can cause local flooding and direct damage; and lightning strikes can put people and property at risk. Thunderstorms also create problems in the air (of strong turbulence and powerful uplift and down-draughts, as well as the lightning), and are usually avoided by aircraft if at all possible.

THUNDERSTORMS

Thunderstorms are particularly intense storms created by strong rising currents of moist air (Box 10.15), caused by rapid heating of the ground surface. Conditions for generating violent thunderstorms are particularly common in tropical regions, where heating and thus convectional uplift are most intense.

FORMATION

Because they develop in warm, humid areas, where there is powerful convectional uplift, thunderstorms are common in the tropics and humid sub-tropical areas, particularly during the summer months, when the Sun is high in the sky.

Thunderstorms tend to occur in three different types of situation – where there is strong vertical uplift of air, at frontal zones, and where air is forced to rise over topographic barriers.

Air-mass thunderstorms: these are caused by convectional uplift associated with surface heating of an unstable air mass (see p. 277). They are accompanied by strong vertical (thunder) cloud development, powerful up-draughts and down-draughts of air, thunder and lightning.

Frontal thunderstorms: these are caused by the uplift of warm air along and in front of cold fronts, and by the uplift of cold air along and in front of warm fronts (see p. 307). Some active frontal systems (particularly fast-moving cold fronts) are preceded by a line of thunderstorms called a 'squall line'. Frontal thunderstorms can be particularly intense.

Orographic thunderstorms: these are created when a warm, moist air mass is forced to rise quickly over a mountain barrier or other major topographic obstruction (see p. 330).

Thunderstorms are quite common in the south-east USA, where there are strong regional variations in thunderstorm frequency and levels of activity that reflect a mosaic of climatic factors. They are also not unusual over Britain, particularly during the summer and late summer months, when convectional uplift is relatively strong. Such storms often bring quite heavy rainfall, strong gusty winds and lightning.

There are often quite marked seasonal and diurnal patterns in thunderstorm activity. In southern England, for example, June is often the most thundery month and November the least thundery. Activity often peaks in the late afternoon and is at a minimum just before mid-morning. Similar patterns have been reported from Greece, where 75 per cent of the annual total thunder occurs between May and August, and on a daily basis most thunder occurs between about 8 pm and 9 pm.

LIGHTNING

Lightning consists of a flash of light in the sky during a thunderstorm. We see lightning before we hear thunder because sound travels much more slowly than light. Lightning is caused by the discharge of electrical energy (Figure 10.6):

- in clouds
- between clouds
- between a cloud and the ground.

Although lightning frequency varies a great deal from place to place, it is a relatively common event in many environments. In a typical year during the early 1990s, for example, over 15 million cloud-to-ground lightning flashes were recorded in the USA alone. Scientists predict that global lightning activity is likely to increase by up to 30 per cent if global warming associated with greenhouse gases proceeds as anticipated (see p. 258).

For lightning to occur there must obviously be a store of electricity in a thundercloud, and this builds up by a combination of processes within the cloud. Electrical charges are created by frictional contact between water

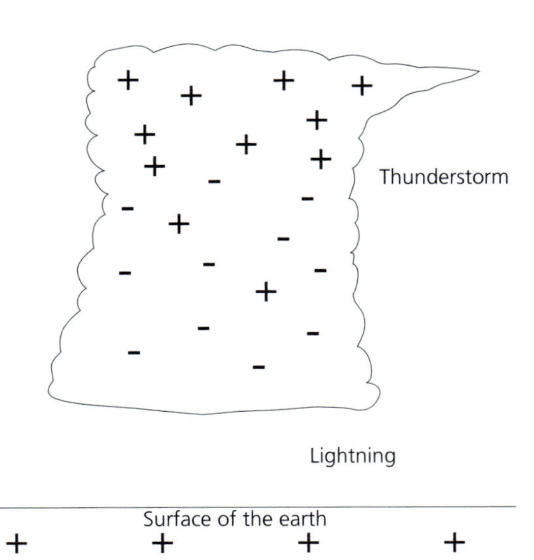

Figure 10.6 *Electrical charges and lightning in thunderstorms. See text for explanation.*

Source: Figure 5.2 in Doerr, A.H. (1990) Fundamentals of physical geography. Wm. C. Brown Publishers, Dubuque, Iowa.

droplets, ice crystals or suspended particles (of dust and ash). Charged particles are moved up towards the top of the cloud by convection currents, and oppositely charged particles interact in the up-draught to create so-called charge centres from which lightning is released. Suitable conditions are generally found in tall clouds, with powerful convectional uplift, and with a cloud base higher than about 1,800 m above the ground.

Cumulonimbus or thunderclouds (see p. 297) are electrically charged, with more positively charged particles at the top and more negatively charged particles at their base. Once the electrical charge has built up enough, some of the electrical energy is discharged as lightning (Box 10.16).

Some places experience a great deal more lightning than others. Lightning is more frequent in tropical latitudes than towards the poles, and the ratio of cloud–cloud to cloud–ground flashes appears to increase with latitude. Patterns can also be detected within continents. Analysis of lightning records in the USA, for example, has not surprisingly shown close associations with the pattern of thunderstorm activity, with most lightning activity concentrated in the south-east. Several particularly stormy areas were identified, including the Great Lakes, Florida, the Piedmont and Arizona. More than half of the lightning events studied were between clouds.

Lightning is not all bad news – it plays an important role in the global nitrogen cycle (see p. 90). The electrical energy in lightning causes nitrogen gas to react with oxygen gas in the atmosphere to form nitrate (NO_3^-). This form of nitrogen is then captured by falling raindrops and carried into the soil, from where it can be taken up and used by plants.

THUNDER

Thunder is the loud cracking or deep rumbling noise that is usually heard a short time after a lightning strike in a thunderstorm. It is a shock wave caused by the rapid and violent expansion of atmospheric gases that are suddenly heated (to perhaps 15,000°C) by lightning.

The time delay between lightning and thunder depends on how close the observer is to the storm. Close to the thunderstorm, the lightning flash is followed immediately by the crack of thunder, whereas further away there is a delay between the lightning flash (which is instantaneous) and the thunder. The delay is longer with increasing distance. A simple rule of thumb is that each 350 m of distance equates to roughly 1 second of delay, so a 3-second delay between a lightning strike and a thunderclap would mean that the storm was about 1,050 metres away.

WEATHER CHANGES

Thunderstorms are usually highly localised, but they move quickly across the ground because of the strong winds

> **BOX 10.16 LIGHTNING HAZARDS**
>
> Lightning poses a real danger for people exposed during a thunderstorm. Relatively few people have been killed by lightning in England and Wales over the last century, but more people are killed by lightning in the USA, often in job-related incidents, and death rates are highest among males and people aged between 10 and 19 years old. Most (86 per cent) lightning fatalities in Australia occur outdoors, are work-related (60 per cent) and occur along the populated south-eastern coast. Lightning also starts natural fires, which can destroy large areas of natural vegetation and anything else that crosses their paths. Many forest fires – in California and important wilderness areas in the USA, for example – are caused by lightning strikes on dry vegetation after prolonged drought or dry spells. Lightning can also damage wooden buildings and structures, as witnessed by the extensive fire damage to parts of York Minster in England in the mid-1980s.

involved. This means that they can affect fairly large areas over a relatively short period of time, and they develop quickly and often move on without delay. The intensity, unpredictability and high speeds of thunderstorms pose problems for people trying to avoid them or to minimise building damage associated with them.

Cloud patterns often give strong clues about the build-up of thunderstorm conditions. Strong convectional uplift encourages the development of tall cumulus clouds, particularly the anvil-topped cumulonimbus, which can rise as high as 18,000 m. The rapidly rising air cools adiabatically (see p. 277) and condenses when it reaches the dew point. If temperatures drop below freezing (0°C), water droplets and ice crystals form within the cloud, and these can fall to the ground as precipitation when they have grown big enough. When the thundercloud is well developed, there are within it areas of strong uplift and areas of strong down-draught, creating a great deal of air turbulence (which is dangerous to aeroplanes). The storm eventually dies out when precipitation and cool down-draughts deplete it of energy and damp down the strong convectional uplift.

Thunderstorms generally bring extreme weather conditions. Heavy precipitation (particularly rainfall) can often trigger localised but very damaging flash floods. Storms with powerful up-draughts can bring hail. Wind shear (differences in horizontal wind speed with altitude) causes strong turbulence for small aircraft and makes take-off and landing very difficult and dangerous.

TORNADOES

A tornado (also called a twister) is a particularly intense depression (see p. 312), which usually has a fairly short life span and some striking features (Box 10.17). Without doubt, the most striking feature of a tornado is the dark, funnel-shaped cloud that can be seen on radar images and satellite photographs, and seen in profile from the ground.

FORMATION

Tornadoes usually form when warm, humid air is sucked into a low-pressure cell. There it comes into contact with a cold front moving towards it from the opposite direction. The steep temperature gradient allows the tornado to develop along the squall line either in front of or along the cold front. In the USA, tornadoes are most common in the Midwest and along the east coast (Figure 10.7).

Nature's most violent storms are usually quite small and localised. They are generated, shaped and dominated by powerful winds that whirl around a small area of extremely low pressure, creating a revolving storm with the characteristic swirling, funnel-shaped clouds. Powerful up-draughts within a rising column of air give the tornado its strong vertical development, and the rotating circular form is created by strong winds that are drawn into the low-pressure centre. Tornadoes occur most commonly during late afternoon or early evening. They are most frequent during late spring or early summer.

BOX 10.17 COMMON FEATURES OF TORNADOES

Most continents experience tornadoes from time to time, but they are particularly common in the USA. Conditions are most suitable for tornado formation in the Great Plains, but they also occur in other parts, including the north-east (because of the movement of low-pressure systems across the Great Lakes). Tornadoes tend to recur in similar places, because these tornado zones regularly have suitable conditions. Identifying and mapping such risk zones helps to suggest appropriate land use, and this is useful in minimising loss of life and property.

CHARACTER

A tornado is created and driven by the very low pressure at its centre, which is often as much as 100 mb lower than in the surrounding air. This creates an extremely steep pressure gradient, which sucks in surrounding air and generates very high wind speeds. Wind speeds in excess of 300 km h^{-1} are not uncommon in tornadoes. Geostrophic winds blow clockwise around tornadoes in the northern hemisphere and anticlockwise in the southern hemisphere (see p. 283).

Most tornadoes have a diameter of less than a few hundred metres. A narrow rotating column of air that blows around a more or less vertical axis of low pressure and moves across the surface of the land is sometimes described as a 'whirlwind'. A dust devil is a strong miniature whirlwind that throws up dust, litter and leaves into the air.

Tornadoes form over dry land, but when the funnel-shaped vortex comes into contact with a lake or sea it sucks up particles of water and whirls them round in a spiral pattern as a 'waterspout'. Waterspouts have lower pressure gradients and weaker winds than tornadoes, and they are usually quite short-lived.

Figure 10.7 *Distribution of tornadoes in the USA, 1953–76. Tornadoes are most common in the Midwest and along the eastern coast.*

Source: Figure 6.12 in Marsh, W.M. (1987) Earthscape. John Wiley & Sons, New York.

DAMAGE

As well as the high wind speeds around the tornado, people and property are threatened by the great speed with which tornadoes move across the ground. Many tornadoes move at speeds of between 150 and nearly 500 km h^{-1}, destroying everything in their path. Fast-moving tornadoes cannot be outrun, and people caught in their path are generally advised to shelter or drive away at right angles to the narrow tornado track.

America's 'Tornado alley' is the scene of frequent twisters that bring great damage and destruction, but few periods see as much activity as on one night in mid-May 1999, when seventy-six twisters ripped through Oklahoma, Nebraska, Kansas and Texas. The tornadoes were the most destructive in decades, killing forty-seven people, destroying 2,000 homes and causing about US$500 million worth of damage.

Tornadoes are much less common in the United Kingdom, although the risks are perhaps higher than most people think; this means that simple precautionary steps that could limit tornado damage are not always taken in advance. Tornado frequency appears to be increasing in the USA and the United Kingdom, partly because improved monitoring is providing a much more reliable picture. The impact of tornadoes on buildings and structures also appears to be increasing, in part because bigger and more vulnerable structures – like large bridges – are being built.

In the USA, states with a high frequency of tornadoes have relatively low rates of tornado-related deaths, partly because of variations in population density, but the evidence shows a shift over time in the location of tornadoes, which is putting significant numbers of less well-prepared people and communities at risk. Death rates from tornado-related injuries in the USA are highest among people living in mobile homes, the elderly (over 60 years of age) and

BOX 10.18 TORNADO FORECASTING

It is very difficult to predict the arrival time, location or path of most tornadoes with any degree of accuracy, because they can and often do change direction quickly and repeatedly, generally without warning. This makes evacuation and evasion particularly difficult, because there is no certainty that the tornado will go in the direction you hope it will! They also have discontinuous contact with the ground and sometimes rise above the surface and cause no damage for a short distance before coming back to the ground and continuing their devastation. Forecasting is also made difficult by the short life span of most tornadoes and the tendency, when conditions are suitable, for them to develop in clusters.

In the USA, tornado watch advice is regularly broadcast on radio and television, encouraging people to be aware that conditions are suitable for possible tornado formation. If a tornado is sighted in an area, the next step is to issue a tornado warning, advising people to take cover.

people caught outside with no protection when the tornado passes by.

Tornadoes cause great damage (they often cause total destruction) where they touch the ground, because of the extremely strong winds and the powerful uplift within them. They often follow quite well-defined paths along the ground, and this is evident in the trail of damage they leave behind – including swathes cut through forests and narrow strips of buildings destroyed in residential areas. Some damage and losses could doubtless be reduced if it were easier to predict where and when tornadoes are likely to strike (Box 10.18).

HURRICANES

A hurricane is a severe storm, technically defined as a wind of force 12 or above on the Beaufort scale (see Table 9.3). Most occur in tropical and sub-tropical areas (Figure 10.8).

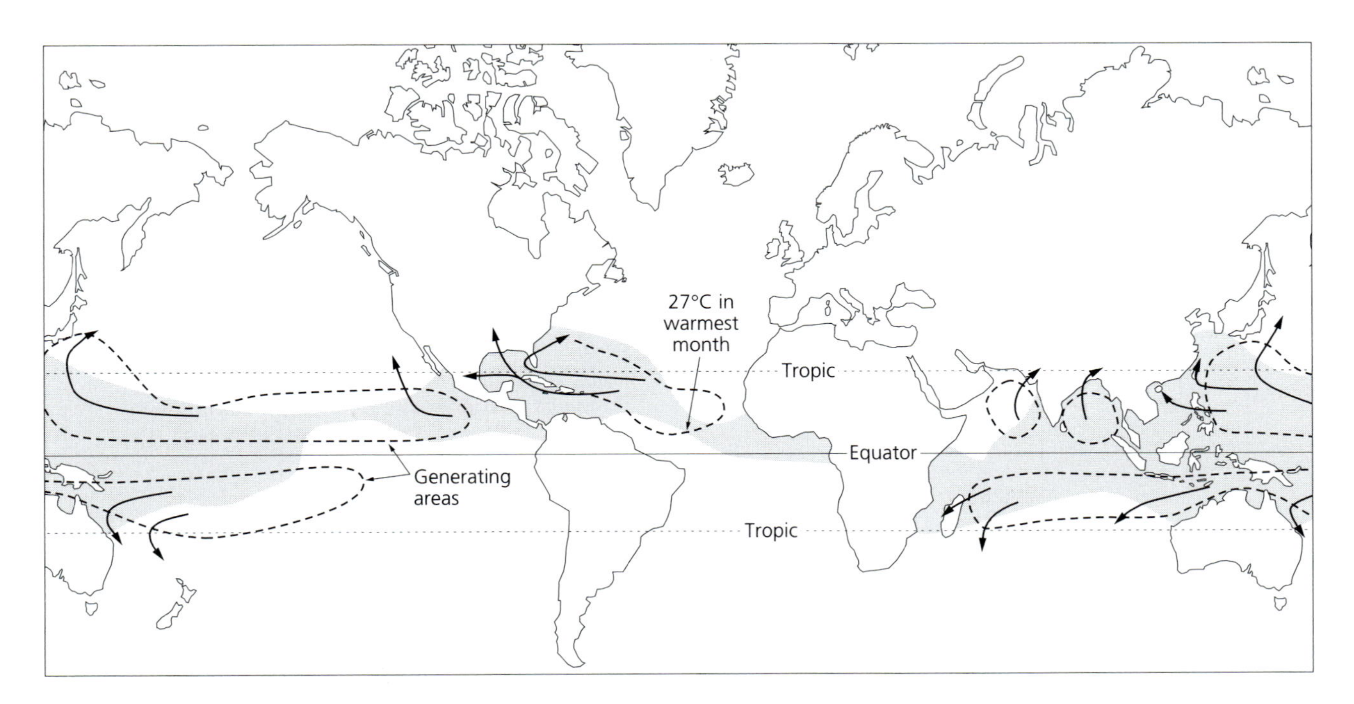

Figure 10.8 *Main hurricane source areas and tracks. Hurricanes are generated in the warm, moist air of the tropics, and they generally move from east to west.*

Source: Figure 16.14 in Dury, G.H. (1981) Environmental systems. Heinemann, London.

BOX 10.19 COMMON NAMES FOR SEVERE STORMS

Severe storms are known as:

- *hurricanes* throughout most of the western hemisphere;
- *typhoons* through most of Asia and across the Pacific;
- *cyclones* in the Indian Ocean and Australia;
- *willy-willies* in some South Pacific islands.

Such storms are called different things in different places (Box 10.19). The intensity of a hurricane is usually defined using a special scale (Table 10.2).

CHARACTERISTICS

Hurricanes generally cause much more damage than tornadoes because they are larger, last longer (several days is not unusual) and travel longer distances.

Most hurricanes are smaller than normal low-pressure cells typical of the mid-latitude zone (see p. 312), and they have different shapes. A typical hurricane has a diameter of between 150 and 1,500 km and is circular in shape, whereas lows are often more than 1,500 km in diameter and are usually oval in shape.

At the centre of a hurricane is a calm area (the eye) of descending air, with little wind and no clouds. Around the centre there is usually a zone of powerful up-draught (the eye wall), which has great turbulence, thick cloud cover and torrential rainfall, and it often has lightning too. Strong winds spiral inwards towards the centre of the storm (down the pressure gradient); the strophic wind is anticlockwise in the northern hemisphere and clockwise in the southern hemisphere. The strong winds can cause extensive damage.

North America is regularly affected by hurricanes that develop in the Atlantic Ocean. Between 1960 and 1989, there were 269 tropical hurricanes in the North Atlantic, seventy-six of which moved westwards across part of the USA and affected weather conditions there. The most intense and widespread impacts tend to be experienced along the east coast of the USA in late August and early September (Table 10.3). More than 80 per cent of the intense Atlantic hurricanes that affect the USA originate from African air masses, and the frequency of intense hurricanes has increased since the late 1970s.

Hurricane tracks often link weather conditions across a wide area. For example, a high frequency of cyclones in the Ohio valley, in the USA, appears to be associated with below-normal rainfall and snowfall and above-normal temperatures across much of the mid-Atlantic. A high frequency of storms along the south-east Atlantic coast is associated with above-normal rainfall and snowfall and below-normal temperatures in the region.

TROPICAL CYCLONES

Tropical cyclones (hurricanes, typhoons) usually originate over water between 5 and 20° latitude, in the warm months when the surface temperature of the ocean is above 27 °C.

Table 10.2 Hurricane severity scale

Category	*Wind speed (km h^{-1})*	*Storm surge (m)*	*Typical damage*
1	118–150	1.2–1.84	Minimal – minor damage to trees, low roads, piers
2	150–176	1.84–2.77	Moderate – trees and signs blown down. Marinas flooded. Coastal roads cut off by water. Some evacuations
3	176–208	2.77–4.0	Extensive – large trees and utility poles blown down. Mobile homes destroyed. Larger coastal structures hit by flying debris
4	208–248	4.0–5.54	Extreme – roofs, doors, windows destroyed. Major erosion. Complete evacuation needed within 3 km of shore
5	>248	>5.54	Catastrophic – buildings overturned. Evacuation needed within 8–16 km of shore

Table 10.3 The most damaging hurricanes of the twentieth century

Rank	*Area (name)*	*Category*	*Year*	*Deaths*	*Cost in millions of US$ (1996 prices)*
1	Florida	5	1935	408	less than 400
2	Mississippi (Camille)	5	1969	256	6,100
3	Florida (Andrew)	4	1992	26	30,500
4	Florida	4	1919	600–900	less than 400
5	Florida	4	1928	1,000–2,000	1,500
6	Florida (Donna)	4	1960	364	2,100
7	Texas, Galveston	4	1900	8,000–12,000	800

Tropical cyclone development sometimes reflects air–sea interactions and convection and rapid uplift.

Once formed, tropical cyclones are pushed westwards (in the northern hemisphere) by the trade wind systems (see p. 286), but they often move north-westwards, northwards or north-eastwards. Hurricanes usually develop during the warm summer months. Most hurricanes that affect the USA develop between June and October.

Tropical cyclones are intense low-pressure cells (see p. 312), with pressure at the centre up to 100 mb lower than in the surrounding air. This generates a steep pressure gradient and strong winds, with speeds of up to 320 km h^{-1}. These extremely high wind speeds occur in hurricanes; tropical storms have wind speeds up to 120 km h^{-1} and tropical depressions have much lower wind speeds (below about 60 km h^{-1}). Tropical cyclones usually last for several days, but they can persist for up to three weeks if conditions are suitable.

Hurricanes are driven along by prevailing wind systems, and they move forward at different speeds (mainly reflecting the intensity of the pressure gradient, the strength of the prevailing wind and the nature of the surface they are blowing across). They move much more slowly than tornadoes – typical hurricane speeds of 15–30 km h^{-1} contrast with tornado speeds of 150–500 km h^{-1}.

The most intense hurricane recorded during the twentieth century was Hurricane Gilbert (Box 10.20), which blew across the Caribbean in 1988 (Figure 10.9).

EXTRA-TROPICAL HURRICANES

The most damaging and frequent hurricanes occur in the tropics, but some storms beyond the tropics have winds high enough (greater than 120 km h^{-1}) to achieve hurricane status. Southern England experienced its strongest winds for three centuries in two storms with hurricane-force winds. The first occurred on the night of 15–16 October 1987 and caused extensive damage to buildings, disruptions to communications systems and problems for people. Electricity supplies were cut off in many areas when transmission lines were blown down or damaged. Just over two years later, on 25 January 1990, southern Britain was again ravaged by hurricane-force winds, which also caused great damage and inconvenience.

A completely different type of hurricane sometimes develops in arctic waters when warmer water is moved north from lower latitudes and heats the lower atmosphere. Arctic hurricanes tend to be relatively short-lived and mainly affect coastlines where few people live (so they cause less damage and attract less attention). They contrast with other hurricanes in a number of ways (Box 10.21).

DAMAGE

Hurricanes can be very destructive, and the strong winds can uproot trees, destroy buildings and flatten entire areas. The torrential downpours that accompany many hurricanes can cause flash floods. Some damage to walls, structures and buildings is caused by a combination of wind and rain.

Large hurricanes that move across built-up areas can cause extensive damage. In August 1992, for example, much of Dade County in southern Florida was devastated by Hurricane Andrew (see Table 10.3). Around 25,000 people moved away from the area over the following year, the clean-up operation was costly, and the future of farming in the area was put at risk. Similar problems were experienced with Hurricane Floyd, which hit the east coast of the United States in September 1999 (Box 10.22).

Most hurricane damage is usually confined to coastal areas, because hurricanes quickly lose their source of energy (warm, tropical water) when they move across land

BOX 10.20 HURRICANE GILBERT, SEPTEMBER 1988

The Caribbean has a long history of hurricanes, and hurricane frequency appears to vary over time – hurricanes are less common during El Niño conditions in the tropical Pacific (see p. 482), and they are more common close to major El Niño events, suggesting a link with rapid changes in ocean–atmosphere conditions. Historical records suggest high levels of cyclone activity in the whole or part of the Caribbean during the 1770s to 1780s, 1810s and 1930s to 1950s, while troughs in activity have been noted around the 1650s, 1740s and 1860s, and during the early twentieth century. From the mid-twentieth century onwards, hurricane tracks appear to have shifted noticeably towards the east.

The hurricane of September 1988 was the most intense tropical cyclone on record in the western hemisphere. Atmospheric pressure at the centre of the depression fell to 885 mb, and this was accompanied by sustained winds of up to 280 km h^{-1} and gusts of over 320 km h^{-1}. Jamaica and parts of Mexico were badly affected by the storm; about 200 people died in Monterrey.

Hurricane Gilbert has been labelled the 'storm of the century' because of the many meteorological records it set. These include size, straightness of track, atmospheric pressure, precipitation and total energy. After ravaging Jamaica as a category 3 storm, Gilbert made landfall in Yucatán as a category 5, one of only three hurricanes of such magnitude to do so in North America during the twentieth century.

Hurricane Gilbert made landfall on the Yucatán Peninsula on 14 September 1988, destroying valuable resort property in Cancun, Isla Mujeres and Cozumel. Salt flats (*salinas*) in northern Quintana Roo and Yucatán were flooded by the storm surge, and coastal marine ecosystems were devastated. Inland, the hurricane caused damage to houses, power lines and farm land. Twenty per cent of tropical rainforest suffered losses of canopy, and in the deciduous forests of north-central Yucatán populations of noxious insects increased. Hurricane Gilbert killed 318 people and caused property damage worth billions of dollars.

surfaces. Wind speeds are reduced by frictional drag with the rough surface conditions (including vegetation, landforms and buildings). After the strong winds die down a hurricane can still affect weather inland, particularly by causing heavy or prolonged rainfall, as happened with Hurricane Mitch in 1998 (Box 10.23), but the most damaging characteristics of the storm (low pressure and strong winds) have by then disappeared.

Vulnerability to hurricanes varies between developing and developed countries. Most of the heavy losses of life are concentrated in developing countries, and many developed countries have adopted land-use zoning and building design strategies to reduce exposure and risk. Loss of life from hurricanes has decreased in the USA in recent years, because of better monitoring and prediction of storm tracks and the adoption of more suitable emergency measures.

BOX 10.21 CONTRASTS BETWEEN ARCTIC AND TROPICAL HURRICANES

- Arctic hurricanes develop at much lower temperatures.
- They develop during the winter months (October to April, compared with June to September).
- They have much shorter duration (less than a day, compared with up to seven days).
- They tend to be smaller (less than 500 km in diameter, compared with up to 1,000 km).
- They have lower winds speeds (80–160 km h^{-1} compared with 120–200 km h^{-1}).
- They move forward much faster (up to 50 km h^{-1} compared with 16–32 km h^{-1}).
- They have higher air pressure at the centre (940 mb, compared with 900 mb).
- They have less vertical development (8,000 m, compared with up to 20,000 m).

STORM SURGES

In coastal areas, the main damage is usually caused by storm surges – high water levels that cause flooding and are created by high winds with a long fetch over open water (see p. 480). Cyclone Sally, which struck the Cook Islands in the Pacific in January 1987, caused a great deal of damage to coral reefs, the shoreline and low-lying buildings by a combination of strong winds and the storm surge.

Low-lying coastal areas are particularly at risk from storm surges. Most of Bangladesh lies close to sea level,

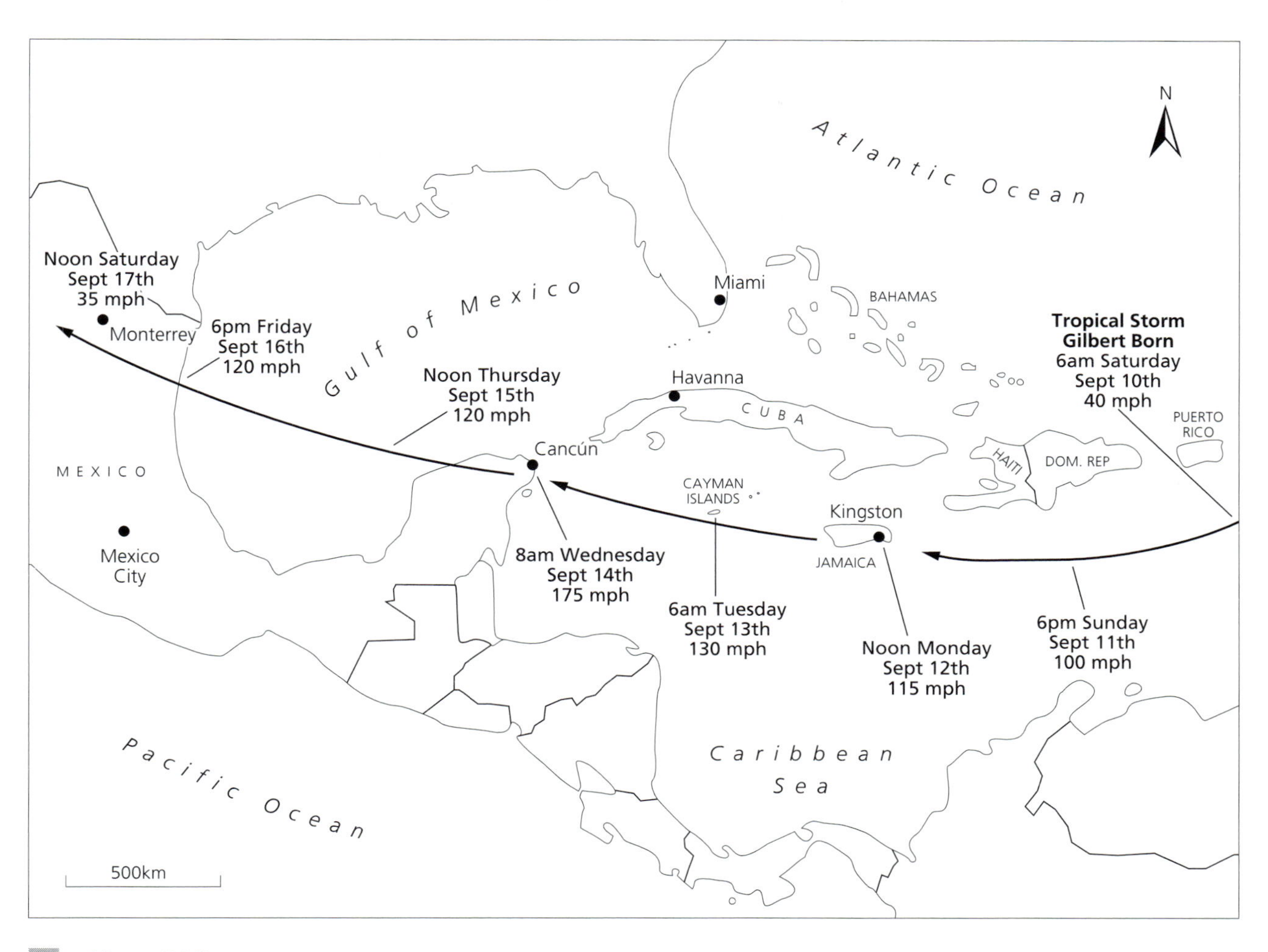

Figure 10.9 *The path of Hurricane Gilbert, September 1998. Hurricane Gilbert was formed in the eastern Caribbean Sea on 10 September and moved westwards across Jamaica, the Cayman Islands and the Yucatán Peninsula in Mexico before dying out near Monterrey (north Mexico) on 17 September.*

Source: A map published in The New York Times, 20 September 1988.

and throughout history the country has regularly been devastated by tropical cyclones and accompanying storm surges (Figure 10.10). One of the most damaging events occurred in 1991 (Box 10.24).

On 21–22 September 1989, a particularly intense tropical cyclone called Hurricane Hugo struck the southern USA (Box 10.25).

Hazard risk from storm surges appears to be increasing in many coastal areas, for a variety of reasons, including:

- higher population densities along coastal areas;
- increasing recreational use of coasts;
- a false sense of security offered by sea defence walls and retaining structures (which can be totally destroyed by storm surges).

Problems are likely to be compounded if global warming scenarios come true, which predict more frequent hurricanes and storm surges in most areas.

BOX 10.22 HURRICANE FLOYD, EAST COAST USA, SEPTEMBER 1999

Floyd – just short of the maximum category 5 storm (see Table 10.3) – was one of the strongest storms ever tracked in the Atlantic and the biggest to threaten America's east coast in decades. The storm developed in the warm, moist air over the western Atlantic early in September 1999. It moved westwards and hit the Bahamas on 14 September. Much damage was caused on the low-lying islands, which were battered by strong winds (up to 250 km h^{-1}), flooded by storm surges up to 6 m above normal and drowned by up to 25 cm of rain. Hurricane Floyd moved towards the south-eastern United States the next day, causing the evacuation of more than 2.5 million people between Florida and the Carolinas. Around 1.7 million people on Florida's Atlantic (east) coast were ordered to evacuate their homes in beach towns, barrier islands and trailer parks in what became the largest peacetime evacuation in US history. Nearly half a million people were told to evacuate coastal Georgia and parts of South Carolina.

The maximum sustained wind speed was around 220 km h^{-1}. Hurricane-force winds (over 120 km h^{-1}; see Table 9.3) extended outwards up to 200 km from the centre, and storm-force winds (103–120 km h^{-1}) extended outwards up to 450 km. Hurricane Floyd was three times the size of Hurricane Andrew, the 1992 storm that killed forty people and caused over US$30 billion of damage (Table 10.3) after it struck Florida near Miami.

The Kennedy Space Centre, where four US$2 billion US space shuttles were housed in structures built to withstand winds of up to 200 km h^{-1}, was evacuated except for a skeleton crew. There were fears that high winds and possible flooding could pose a serious risk to the US space programme. The world-famous theme parks around Orlando – Walt Disney World, Universal Studios and Sea World – were closed down, the first time they had ever been closed due to bad weather. Flooding was extensive in North Carolina.

As the storm moved northwards along the east coast, it weakened progressively. By 17 September, it had been downgraded to a tropical storm, but it still dumped 15 cm of rain on New York city and caused widespread flooding. The storm battered parts of Long Island with winds in excess of 100 km h^{-1}, leaving more than 280,000 people without power.

BOX 10. 23 HURRICANE MITCH, CENTRAL AMERICA, OCTOBER–NOVEMBER 1998

Hurricane Mitch was a powerful Atlantic storm that devastated most of Honduras and large parts of Nicaragua, El Salvador and Guatemala in late October and early November 1998. It caused massive destruction over a wide area and killed more people than any storm in the Western world for 200 years. Yet its impact was more the result of the deadly combination of local geography and poverty than of any record-breaking strength. The hurricane started to grow in the Caribbean on 22 October, and it quickly became the fourth-deepest depression recorded in the Atlantic region – atmospheric pressure at its centre was just 905 mb. Pressure at the centre of Hurricane Gilbert, which blasted the coast of Mexico in September 1988, was as low as 888 mb (see Box 10.20). By the time Mitch made landfall in Honduras on 29 October, it had been officially downgraded to a tropical storm, with a pressure of 994 mb and maximum wind speeds of around 97 km h^{-1}.

As the storm bumped into the mountains of Central America it was forced to rise over 2,000 m, and the moisture-laden air cooled adiabatically (see p. 277). Torrential rain followed – more than 600 mm fell on Honduras in one day, and within two days the area had received three-quarters of what would normally fall in a whole year. Many towns and villages were devastated by flash floods and landslides; more than 24,000 people died or went missing, and a million were declared homeless. Several towns in Honduras disappeared altogether. Damage was much worse than it might otherwise have been, mainly because land clearance for grazing and building had left many hill slopes prone to mass movement (see p. 215) and gully erosion (see p. 218). Many people were living in sub-standard housing in dangerous areas, and many towns had no storm drains. Communities had been allowed to build on river banks and unstable hill slopes. To make matters worse, three reservoirs in the hills failed or released unprecedented flows of water.

After Mitch has passed over El Salvador and Guatemala, it slowed down and lost intensity. It also changed direction and moved north-eastwards over the Yucatán Peninsula in Mexico and across the Gulf of Mexico towards the southern tip of Florida. By 6 November, the remains of the storm were heading off into the Atlantic.

BOX 10.24 TROPICAL CYCLONES – BANGLADESH, 1991

The low-lying country of Bangladesh is regularly damaged by floods, cyclones and storm surges and is one of the most disaster-prone areas in the world. Between 1797 and 1991, Bangladesh was hit by sixty severe cyclones, most of them accompanied by storm surges, with significant loss of life and damage to crops and properties.

One of the most serious recent cyclones hit Bangladesh on 29–30 April 1991. It was a severe cyclonic storm, with tidal surges up to 10 m high that battered coastal areas of Bangladesh for up to four hours. An estimated 140,000 people were killed, many of them women, children and the elderly. Most of the deaths were caused by drowning from the tidal wave that accompanied the storm. Few of the people most at risk had access to cyclone shelters – the 300 existing cyclone shelters had enough capacity for only 450,000 of the 5 million people affected by the cyclone. Cyclone shelters had been built after the 1971 disaster, and they did save lives, although the experience of the 1991 storm shows the importance of better early warning systems, evacuation advice and education.

Surveys of local residents in two coastal communities show that while most people had received cyclone warnings few had left their homes to seek shelter, either because they did not believe the warnings or because they feared that their property would be burgled and possessions stolen in their absence.

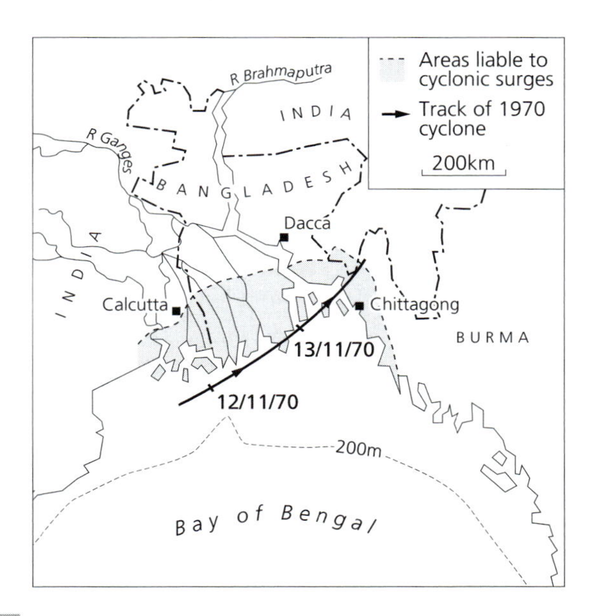

Figure 10.10 *Cyclone hazards in the Bay of Bengal. The low-lying Ganges–Brahmaputra Delta, at the north of the Bay of Bengal, has often been damaged by storm surges and floods. The figure shows the track of the 1970 cyclone, which killed an estimated 300,000 people.*

Source: Figure 275 in Carter, R. (1990) Coastal environments: an introduction to the physical, ecological and cultural systems of coastlines. Academic Press, London.

BOX 10.25 HURRICANE HUGO, SOUTH CAROLINA, SEPTEMBER 1989

Hurricane Hugo hit the southern United States on 21 September 1989. Air pressure at the centre fell as low as 918 mb, and Hugo brought winds speeds of up to 250 km h^{-1}. It badly affected the South Carolina coast, parts of which were flooded by storm tides of up to 6 m (see p. 320). Although the highest surges struck sparsely populated areas north of Charleston, South Carolina, damage was extensive, and forty-four people died. Eight of the deaths were due to drowning, and seven were the result of the collapse of mobile homes, houses or trees, but most were caused by electrocutions and fires. Wind damage was concentrated along the coast from 50 km south of Charleston to 160 km north, and it ranged from loss of roof and wall coverings to complete structural collapse. Many buildings collapsed because an inappropriate building code had been used in the area, it was difficult to enforce anyway, and residents had become complacent because of the unusually long interval since the last major hurricane to hit the area.

SUMMARY

We experience the atmosphere directly through weather, and in this chapter the focus has been on weather systems, how they develop and how they give rise to particular types of weather (including extreme weather conditions). After a broad overview of the relevance of weather to people, we examined the nature and importance of air masses and fronts – the two key drivers of weather changes. Frontal systems represent boundary zones between adjacent air masses, and weather changes most rapidly when fronts move overhead. The precise details of the weather changes depend on the type of front (warm or cold, ana- or kata-front). We examined the relevance of depressions (low-pressure cells) and of cyclogenesis, then moved on to consider the processes and weather associated with high-pressure anticyclones. Storms are perhaps the most dramatic extreme weather conditions, and we looked at the processes and impacts of thunderstorms, tornadoes and hurricanes.

WEBSITE

Links to relevant websites, a comprehensive bibliography, tools for teaching and learning, and downloadable images relevant to this chapter can be found at the website specially designed to accompany this book at
http://www.park-environment.com

FURTHER READING

Aguardo, E. and J.E. Burt (1999) *Understanding weather and climate.* Prentice Hall, London. Non-technical introduction that explains rather than simply describes the processes that produce the Earth's weather and climate.

Barry, R.G. and R. Chorley (1998) *Atmosphere, weather and climate.* Routledge, London. Useful and up-to-date seventh edition of the classic text on atmosphere and world climate.

Carlson, T. (1991) *Mid-latitude weather systems.* Routledge, London. A comprehensive introduction to the structure, dynamics and consequences of mid-latitude weather systems.

Djuric, D. (1994) *Weather analysis.* Prentice Hall, Hemel Hempstead. General introduction to weather analysis and how it is undertaken.

Holton, J.R. (1992) *An introduction to dynamic meteorology.* Academic Press, London. An up-to-date review of the fundamentals of meteorology and their relevance to weather and climate.

Linacre, E. and B. Geerts (1997) *Climates and weather explained.* Routledge, London. Broad introduction to the study of the atmosphere, which integrates climatology and meteorology. Covers basic principles, concepts and processes.

Morgan, M.D. and J.P. Moran (1997) *Weather and people.* Prentice Hall, Hemel Hempstead. Non-technical introduction to weather and its extremes, with particular emphasis on how weather affects human comfort and safety.

Peacock, W.G., B.H. Morrow and H. Gladwin (1997) *Hurricane Andrew: ethnicity, gender and the sociology of disasters.* Routledge, London. Detailed study of the conflict and competition inherent in the way that Miami prepared for, responded to and recovered from Hurricane Andrew.

Pielke, R.A. and R.A. Pielke (1997) *Hurricanes: their nature and impacts on society.* John Wiley & Sons, London. Examines the effects of hurricanes and their impacts on society, and explores how society responds to these storms.

Pielke, R.A. and R.A. Pielke (eds) (1999) *Storms.* Routledge, London. A collection of essays that explore the physical and social dimensions of storms in different countries and different contexts.

Robinson, P. and A. Henderson-Sellers (2000) *Contemporary climatology.* Longman, Harlow. A synthesis of contemporary scientific ideas about atmospheric circulation, its controls, impacts and significance.

Yarnal, B. (2000) *Local water supply and climate extremes.* Ashgate, London. Examines the impacts of weather and climate on local and regional water supplies, and the implications of these interrelationships for sustainable water management.

Chapter 11

Climate

LEARNING OBJECTIVES

When you have finished studying this chapter, you should be able to

- *Appreciate the different ways in which climate is important, for people and for environmental systems*
- *Outline the main factors that give rise to variations in climate from place to place*
- *Summarise the main ways in which human activities affect climate, both directly and indirectly*
- *Understand the need for climate classification schemes and the problems inherent in devising them*
- *Describe and account for the different types of tropical climate*
- *Describe and account for the different types of mid-latitude climate*
- *Describe and account for the different types of sub-polar and arctic climates*
- *Evaluate the main types of evidence of climate change over time*
- *Outline the principal features of climate change during the Pleistocene and Holocene, and their relevance*
- *Distinguish between the external and internal factors that cause climate change*

Climate reflects the long-term average weather conditions over a period of time for a given place, and weather reflects the state of the atmosphere at a given time in a particular place, as we saw in Chapter 10. We also use the word 'climate' to describe an area that has a particular type of climate.

Climate changes much more slowly than weather does over time and from place to place. It operates over longer timescales and broader spatial scales, yet to understand and explain climate and how it varies, we need to be familiar with the same atmospheric processes (Chapter 9) that create weather conditions. *Climatology* – the study of climates – seeks to explain how climate is developed, how it changes and what factors are involved.

In this chapter, we explore the factors that affect climate at different scales, describe how climate varies across the Earth's surface and examine the nature of climate change over time.

CLIMATE

IMPORTANCE OF CLIMATE

CLIMATE AND OTHER ENVIRONMENTAL SYSTEMS

Climate is very much a product of interactions between the atmosphere and the other major environmental systems (the lithosphere, hydrosphere and biosphere; see p. 4). In this sense, it is one of the most obvious and direct ways in which we become aware of environmental stability and change. Global warming and the climate changes associated with it (see p. 261) are perhaps the most significant illustration of this fact. These links between climate and environmental systems are two-way, and they are a core part of the Gaia hypothesis (see p. 120).

There are many links between climate and other environmental systems (Box 11.1). These interrelationships show something of the complexity of environmental systems, and they help to explain why human disturbance

BOX 11.1 LINKS BETWEEN CLIMATE AND OTHER ENVIRONMENTAL SYSTEMS

Climate exerts strong direct and indirect effects on such things as:

- the availability of water resources (p. 388)
- river flooding (p. 410)
- droughts (p. 419)
- desertification (p. 434)
- snow and ice cover (p. 442)
- the natural distribution of plant and animal species (p. 576)
- the development and distribution of soils (p. 604)
- patterns and processes of natural weathering (p. 204).

Climate affects ocean currents (p. 449) and sea level (p. 511), and these can significantly affect people via their impacts on:

- weather systems (p. 301)
- storm surges (p. 320)
- coastal erosion (p. 498)
- ocean fisheries (p. 486).

Some of the climatic effects are secondary impacts of other natural environmental processes, such as:

- variations in sunspot activity (p. 107);
- variations in atmospheric dust caused by volcanic activity (p. 189); and
- variations in tropical ocean currents associated with El Niño events (p. 482).

in some places can be translated into impacts in other places and impacts that affect other environmental systems. Sophisticated climate models are now being constructed that try to predict how climate is likely to change by taking into account physical, chemical and biological processes and the interactions between climate and oceans, soils, ecosystems, the rock cycle and the water cycle.

CLIMATE AND PEOPLE

Climate and people interact in various ways. This is a classic example of a symbiotic relationship, because each affects the other. People affect climate, because human activities – particularly land-use changes, and air pollution by greenhouse gases (see p. 258) and ozone depleters (see p. 249) – alter the atmospheric system and change the pace and/or pattern of natural atmospheric processes.

Climate affects people both directly and indirectly. Climate affects people indirectly in numerous ways, because (as we saw above) most environmental systems affect and are affected by climate. Direct effects are experienced through the opportunities and constraints that climate creates, which can strongly influence what people do, and where, when and how they do it. Many of the weather disasters we looked at in Chapter 10 (p. 301) affected people both directly and indirectly.

Some places (including most of the world's great deserts) receive too little rainfall and are too dry to sustain abundant life. Other areas (such as many tropical rainforests) receive vast amounts of rainfall and are threatened by serious flooding, mass movement and soil erosion. Temperatures in some places (such as Antarctica; see p. 448) are too low for people to survive without the shelter of artificial environments, and other areas (around the equator, for example) are simply too hot for outdoor work under the scorching midday sun.

Many climatic hazards (including heatwaves, cold spells and flooding) affect people in quite specific ways: they start off quite slowly and tend to build up over a period of time; they often affect some groups of people (particularly the young, the old and the poor) more than others; and they damage health.

BOX 11.2 CLIMATE AND THE MIGRATION OF PEOPLE

Climate has impacts on the distribution of people within and between areas. A favourable climate draws people to some places, and an unfavourable climate encourages people to avoid other places. These links are sometimes permanent – California is popular partly because it has a pleasant climate, while Alaska is altogether a more inhospitable and less popular place to live.

There are also strong seasonal links between climate and the movement of people in tourism. Skiing is a popular winter activity, so snow-capped places attract many visitors. Sunbathing, boating and swimming are popular summer activities, so many tourists (particularly from variable mid-latitude climates) fly off to hot, sunny coastal areas, despite the increasing risk of exposure to dangerously high levels of ultraviolet radiation (p. 233). Summer tourism poses some interesting paradoxes, including the question of why people (nationals and foreigners) prefer to visit Mediterranean parts of Greece during the climatically uncomfortable hot, dry summer months.

As well as affecting patterns of human activity, climate also affects human health. Health is directly affected in a number of ways, including our need for food and for protection against harmful ultraviolet radiation. Indirect effects of climate on health include the availability of food, the spread of insect-borne diseases, and respiratory complaints caused by poor air quality.

Climate can help or hinder the spread of infectious diseases. The spread of bubonic plague in Scotland is a graphic example. Bubonic plague, a highly infectious disease, swells a sufferer's glands and makes them delirious. It is caused by the bite of a rat flea infected with the bacterium *Pasteurella pestis*. Before the mid-fourteenth century, the climate across much of Scotland was probably too warm and dry for the fleas to survive. But in the north and west the disease spread fairly widely, possibly because of lower temperatures and high humidity there. During the Little Ice Age (see p. 341), ships continued to carry plague rats into ports, but low temperatures are believed to have inhibited the spread of rats and fleas, and hence of plague.

Early in the twentieth century, some scientists favoured the idea of climatic (or environmental) determinism, which argued that variations from place to place in climate (environment) are the most important control on human activities. The view is now regarded as too extreme and too simplistic, but there is no denying how important climate can be for life on Earth – it determines what we eat, how we dress, how we live, what we do, how we work and many other aspects of life.

CONTROLS OF CLIMATE

Climate varies from place to place, and this allows the surface of the Earth to be sub-divided into distinct climatic regions (see p. 335). Superimposed on this global pattern of variability are local and regional patterns, which reflect the interplay of smaller-scale factors and give rise to a detailed climatic mosaic.

Spatial patterns of climate are affected or controlled by a host of factors, many of them related to position on the Earth's surface. We shall examine the main controlling factors in this section. Climate changes over time are controlled by different factors, which are described later in this chapter.

SOLAR ENERGY

The atmospheric system, which controls weather and climate, is effectively a giant heat engine driven by solar

BOX 11.3 CLIMATE AND UNCERTAINTY

Climate is never constant, and people have had to develop coping mechanisms for dealing with climatic uncertainty. Much of this uncertainty comes from extreme climatic events – such as heatwaves, cold spells, strong winds, intense storms, heavy snowfalls and prolonged droughts – which are not always easy to predict and defend against. Rapid and large-scale climate change, whether natural or induced by human activities, poses particular problems for both people and the environment, and it increases climatic uncertainty.

Scientists and social scientists are particularly interested in the interactions between climate, environment and society because of the threat of global warming and large-scale, long-term climate changes, which could well have an impact on people around the world (p. 6). This is an important component of global environmental research (p. 17), although people and climate interact at all scales from the local to the global.

energy. Global wind systems, which are controlled mainly by variations in temperature and thus air pressure across the Earth's surface (see p. 286), serve to redistribute heat and moisture in air masses, and this gives rise to the major weather systems (Chapter 10).

Inevitably, therefore, climate is heavily influenced by variations in the amount of radiant energy from the Sun that is received at the Earth's surface. Insolation (*in*coming *sol*ar radi*ation*) varies across the Earth's surface with latitude (which determines the angle at which the Sun's rays strike the Earth; see p. 236) and season (which determines length of daylight; see p. 116). It varies over time with changes in solar output (through the sunspot cycle) (see Box 4.7) and changes in the Earth's distance from the Sun (through the Earth's annual elliptical orbit) (p. 118).

We saw in Chapter 8 (p. 237) how some of the incoming radiation is absorbed and reflected within the Earth's atmosphere, so that the Earth's energy budget is strongly influenced by greenhouse gases and the natural greenhouse effect. Air pollution by greenhouse gases is seriously disturbing natural atmospheric equilibrium. If recent trends continue unchecked, the IPCC (see Box 8.34) and other scientists predict that significant increases in global temperature and significant shifts in the main climate zones are likely within the next 100 years.

CONTINENTS AND OCEANS

If the Earth's surface had a uniform composition in all places, climate would have a broad zonal pattern determined largely by latitudinal variations in solar radiation receipt. But it does not, and the fact that about 70 per cent of the surface is covered by oceans and seas (see p. 126) is obvious from images of the Earth viewed from space.

Superimposed on the zonal pattern of climate are significant distortions created by the uneven distribution of land and sea within and between the northern and southern hemispheres. The distribution of water across the Earth's surface affects climate in two main ways – via its influence on the availability of water vapour and thus its impact on precipitation, and via its impact on air temperature (Box 11.4).

Ocean currents: currents also influence climate by redistributing heat, and this impact is superimposed on the temperature-stabilisation effect of large water bodies. The ocean

BOX 11.4 WATER, OCEANS AND CLIMATE

Water vapour: oceans and seas provide massive reservoirs of water, and they play important roles in the global water cycle (p. 353). Wind systems redistribute water vapour across the continents, but energy is used and released in the endless series of phase changes (particularly evaporation and condensation; p. 291), and this in turn affects temperature patterns.

Temperature: oceans and seas affect global patterns of temperature directly, because water bodies heat up and cool down more slowly than land. Several factors are involved:

- *Energy requirement*: water has a higher specific heat than land (this means that it takes more heat energy to heat a given volume of water than it does to heat the same volume of rock or soil).
- *Penetration*: light and heat penetrate water to some depth (unlike solid rock). This increases the thickness of material being warmed.
- *Reflection*: water surfaces reflect proportionately more solar energy than land surfaces do because they have a higher albedo (see Table 8.4).
- *Circulation*: water moves and circulates, so it transfers heat energy to other places.

In heating and cooling more slowly than land areas do, water bodies stabilise temperature. This significantly affects climate patterns, because:

- Air temperatures vary less through time (on a daily and seasonal basis) over oceans than they do over the continents.
- Coastal areas have much less pronounced temperature extremes than areas further inland.
- Continental interiors at similar latitudes tend to have much hotter summers and colder winters.

currents are driven by wind and pressure systems (see p. 480), and they move vast amounts of warm water from lower to higher latitudes and colder water from higher to lower latitudes to replace it. Ocean currents can have a strong impact on the temperature and humidity of adjacent air masses, and this in turn can influence climate and weather conditions along the path of the current (including over adjacent coastal and continental areas). The Gulf Stream, for example, transfers warm tropical water from the Gulf of Mexico north-eastwards towards the North Atlantic, and this brings milder weather and higher temperatures to parts of Western Europe, including Britain (see p. 480).

Teleconnections: oceans also influence climate through so-called 'teleconnections', the most important of which is the periodic El Niño/Southern Oscillation phenomenon (see p. 482). The climate in Peru, in South America, was significantly altered by an El Niño event between December 1982 and July 1983, which brought torrential rains and floods to the north of the country, landslides and flash floods in the central region, and severe drought further south. Such short-term variations can be highly variable from place to place and over time.

ALTITUDE

The distribution of land and sea and the ocean currents distorts the pattern of variations in climate from place to place around the world, but this two-dimensional spatial pattern is further altered by variations in altitude, which affect temperature (Box 11.5) and precipitation and can generate local and regional wind systems.

Precipitation: mountains form barriers over which air flows, rises, cools adiabatically to the dew point and condensation occurs, causing orographic precipitation (see p. 330) on the windward slope. Leeward slopes, in the rain shadow, are often relatively dry (Figure 11.1).

Wind: air that descends down the leeward side of mountain slopes is usually dry, so it warms at the dry adiabatic lapse rate of about 10 °C per km (see p. 276). This is much higher than the saturated adiabatic lapse rate (typically between 3 and 6 °C), so the descending air becomes much warmer at similar altitudes than the air that rose up the windward side of the slope. The longer the descent, the greater the warming. In this way, warm, dry winds are generated that can continue for several days at a time and bring extremely dry conditions. Many low-lying areas in the lee of mountains – including the Dead Sea area in Israel and Death Valley in California – are particularly dry for this reason. Dry winds like this (called *chinooks* locally) blow down the eastern side of the Rocky Mountains in North America, mainly in winter and spring. Similar winds (called *föhn winds* locally) blow along valleys in the northern Alps of Switzerland, having lost most of their moisture in rising over the southern Alps. Both winds melt snow in spring and uncover mountain pastures relatively early. Föhn winds are common around the world.

BOX 11.5 ALTITUDE AND TEMPERATURE

The atmosphere is heated mainly from the ground (p. 237), and air temperature decreases with altitude at the normal lapse rate of between 6 and 6.5 °C per km (p. 276). In stable conditions, therefore, if the air temperature at the foot of a 3,000 m mountain is 32 °C, the summit might be as cold as 14 °C or less. Little wonder that high-altitude climbers need warm protective clothing! Very high mountains have vertical zonations of climate that mirror spatial zonations and are reflected in changes in vegetation and soils with altitude. The base of Kilimanjaro (Africa's highest mountain, in north-eastern Tanzania) is tropical grassland, but the summit – more than 5,000 m higher and 30 °C cooler – is permanently snow-capped. In between are vegetation zones more typical of temperate and sub-arctic climates.

WIND SYSTEMS

Global wind systems, which are driven by thermally generated pressure patterns (see p. 275), play a vital role in redistributing air masses and weather systems across the surface of the Earth (p. 306). This inevitably has a strong impact on climate, by redistributing heat, moisture and air pressure. Wind systems also determine the paths of frontal systems (p. 307) and storms (p. 31).

The global wind systems are modified regionally by a range of different factors, including altitude and mountain barriers (the warming winds described above) and large-scale seasonal monsoon systems (Box 11.6). More local changes include land and sea breezes, mountain and valley breezes, and desert winds (Box 11.7).

Land and sea breezes: these are small-scale equivalents of the great seasonal monsoon circulations that occur in low and middle latitudes during the summer months. Land warms up and cools down faster than water. During the daytime, land heats faster than the sea, and this causes low pressure over the land (where the warm air rises by convection). Air flows down the pressure gradient, from sea (high pressure) to land (low pressure), causing a sea breeze (Figure 11.2). At night the opposite occurs – the land cools faster, the sea stores heat and becomes a low-pressure area, and air flows down the pressure gradient, which now runs from land (high pressure) to sea (low pressure), causing a land breeze. The stronger the temperature (thus pressure)

BOX 11.6 MONSOONS

Monsoons are seasonal winds that blow across some tropical areas (particularly South Asia, but also in Africa), blowing from the south-west in summer and from the north-east in winter. The so-called monsoon season is the rainy season, from about April to October, when the south-west monsoon blows moisture-laden air from over the sea, and this brings heavy rains. Monsoon circulations are associated with the movement of pressure cells and the inter-tropical front. When the inter-tropical convergence zone (ITCZ) moves much further north of the equator during the northern hemisphere summer, it pulls the south-east trades with it. These are deflected to the right by the Coriolis force (see Box 9.9) as they cross the equator, turning them into south-westerly winds that blow across the Indian sub-continent, drawn in by an intense thermally induced low-pressure area across north-western India and Pakistan (caused by rapid summer heating). Heavy rains during the monsoon often cause extensive and destructive flooding.

BOX 11.7 DESERT WINDS

Desert winds blow outwards from low-pressure cells over deserts into the more humid areas surrounding them. These winds are hot, dry and dusty (see p. 433), and they desiccate (dry) surrounding areas and wither plants. Two of the best-known desert winds are:

- *The sirocco*: a dry wind that blows from the Sahara in North Africa across the Mediterranean Sea (where it picks up moisture) towards Sicily and southern Italy, which experience it as a hot, humid wind.
- *The harmattan*: a dry wind that blows from the Sahara on to the coastal lands of West Africa between November and March.

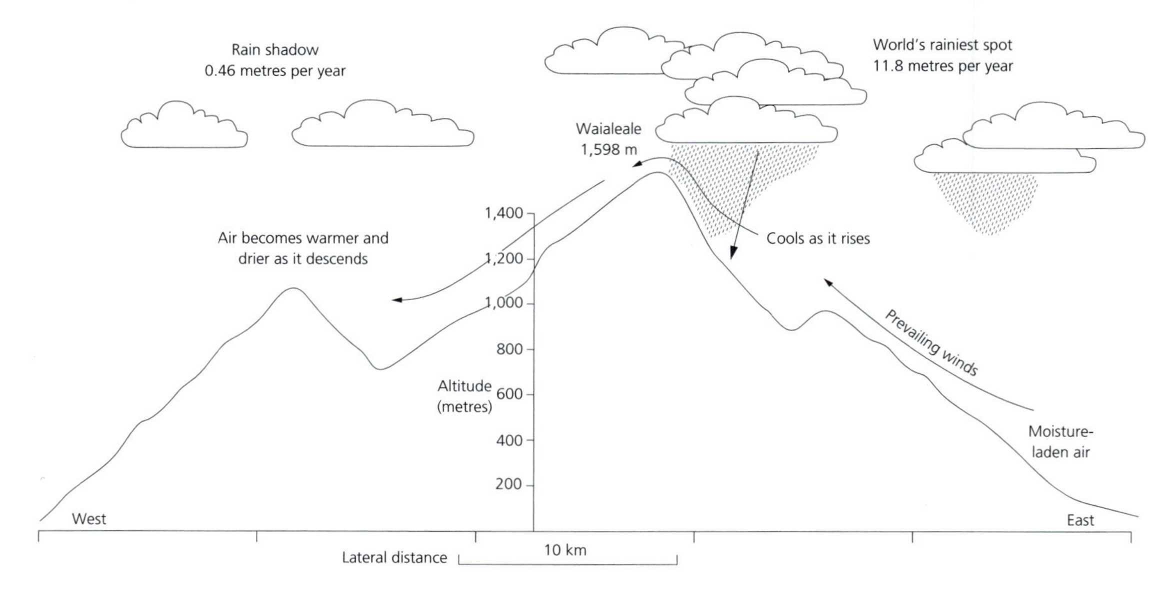

Figure 11.1 *Variations in rainfall across Mount Waialeale, Hawaii. The eastern side of the mountain, which is exposed to trade winds, which blow moist air onshore from the sea, receives more than twenty times as much rain as the west side, which is in a rain shadow.*

Source: Figure 16.5 in Cunningham, W.P. and B.W. Saigo (1992) Environmental science: a global concern. Wm. C. Brown Publishers, Dubuque, Iowa.

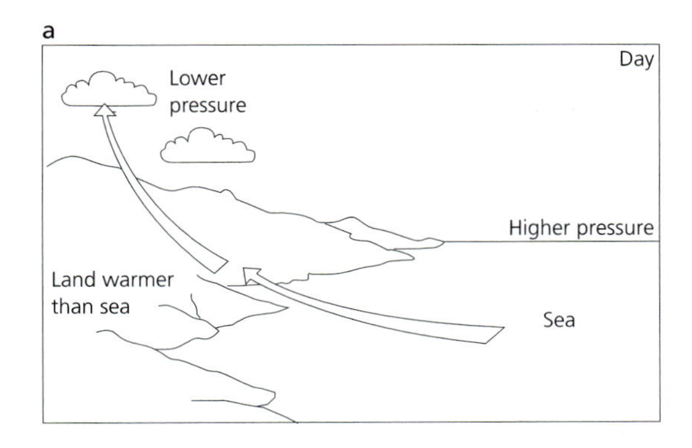

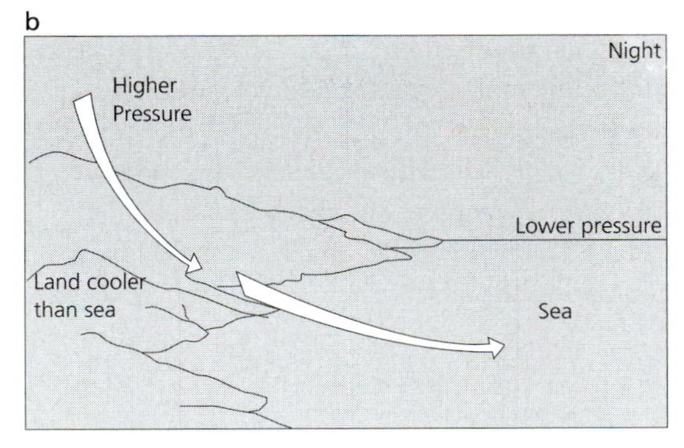

Figure 11.2 *Land and sea breezes. See text for explanation.*

Source: Figure 6.6 in Doerr, A.H. (1990) Fundamentals of physical geography. Wm. C. Brown Publishers, Dubuque, Iowa.

gradient between land and sea, the stronger the resultant wind. The upper air moves in the opposite direction to the land and sea breezes to compensate for the surface flows.

Mountain and valley breezes: these develop in similar ways to the land and sea breezes, and they reflect local topographic variations. Under stable, calm weather conditions in mountains, warm air rises up the mountain slopes and up-valley during the day as a valley breeze, and at night cold air sinks down the slope and down-valley as a mountain breeze (Figure 11.3). Both local winds are driven by the pressure gradient, and such winds are often very effective in mixing air and dispersing air pollutants in mountain areas. This contrasts quite sharply with the stable (trapped) air conditions and poor air quality that can cause pollution incidents (see p. 244) in cities like Mexico City and Los Angeles. Many mountain and valley breezes are very localised, but some persist for long distances and can be very strong, particularly when funnelled along narrow valleys. Such winds are called *katabatic winds* (or air drainage winds), and they often have local names such as:

- *the mistral* (a cold wind that blows mostly in winter in the north-western Mediterranean lands; it often blows out from high-pressure systems over the Massif Central and descends along the Rhône valley);
- *the bora* (a katabatic wind that descends from former Yugoslavia into the Adriatic Sea);
- *the taku* (a katabatic wind in south-eastern Alaska).

Similar winds are created as cold air flows out from frozen plateau areas, such as the interior of Antarctica and Greenland, and along deep, narrow valleys.

HUMAN IMPACTS

The influences on climate that have been considered so far are essentially natural ones, which reflect natural variations in environmental systems across the Earth's surface. If the Earth were uninhabited, they would explain most of the variability in climate from place to place. But human activities influence climate both directly and indirectly, particularly through air pollution and land-use changes. In

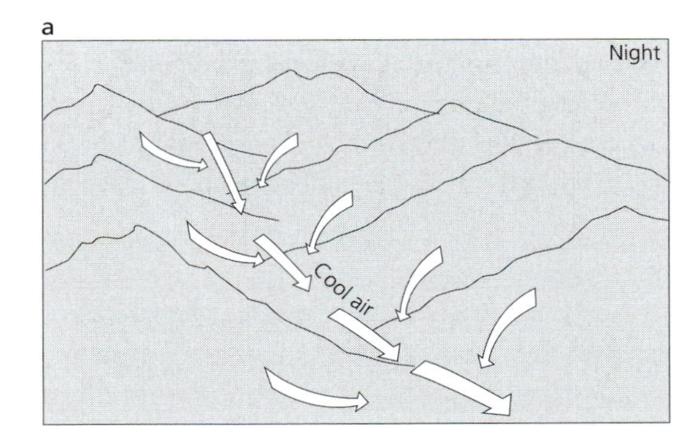

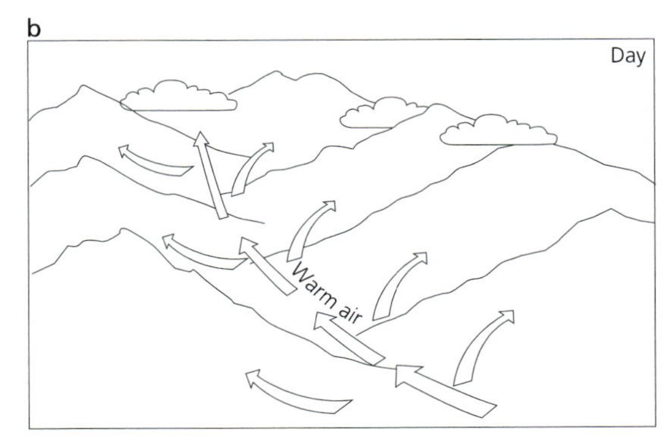

Figure 11.3 *Mountain and valley breezes. See text for explanation.*

Source: Figure 6.7 in Doerr, A.H. (1990) Fundamentals of physical geography. Wm. C. Brown Publishers, Dubuque, Iowa.

the past, this human impact has been concentrated on particular areas and regions where population was most heavily concentrated. Increasingly, however, human activities are affecting global climate.

Air pollution: much of the concern about the environmental impacts of air pollution centres on the damage it causes to wildlife, buildings and human health. But scientists are becoming increasingly aware of the significant ways in which air pollution can alter climate at a variety of scales (from the local, through the regional, to the global). At the local scale, these include the development of photochemical smogs and low-level ozone incidents (see p. 249). Globally, the most important changes appear to be those associated with depletion of stratospheric ozone (p. 249) and emission of greenhouse gases (p. 258), which both contribute to global warming and a redistribution of the major climate regions. This is without doubt the most pressing environmental problem that humanity faces at the close of the twentieth century.

Land use: superimposed on these often large-scale impacts of air pollution on climate are the many ways in which changes in land use affect local and regional climates. Natural climax vegetation (see p. 574) and soil patterns (p. 608) are heavily influenced by climate, and according to the Gaia hypothesis (p. 520) equilibrium conditions develop over long periods of time between vegetation, soils, the rock cycle, biogeochemical cycles, the water cycle and the atmospheric system. Land-use changes by humans normally disturb this equilibrium, and this can be reflected in local climate changes. Tropical deforestation (see p. 539), for example, can alter local and regional climates by changing the albedo (reflectivity) of the ground surface and changing local hydrology (particularly by changing evapotranspiration and rainfall). These changes tend to raise temperatures and lower rainfall, resulting in more regular and more serious droughts. Deforestation can also trigger global climatic change by altering atmospheric circulation patterns (via changes in atmospheric heating and heat exchange) and altering atmospheric chemistry (via changes in greenhouse gases). Perhaps the most dramatic form of land-use change is urban development, which replaces natural cover (vegetation and soil) with roads and buildings (Box 11.8). The main effects of urban areas on local climate are summarised in Table 11.1.

GLOBAL CLIMATE

Climate varies quite systematically across the surface of the Earth, because it is:

- created by the atmospheric system (Chapter 8);
- generated by atmospheric processes (Chapter 9);

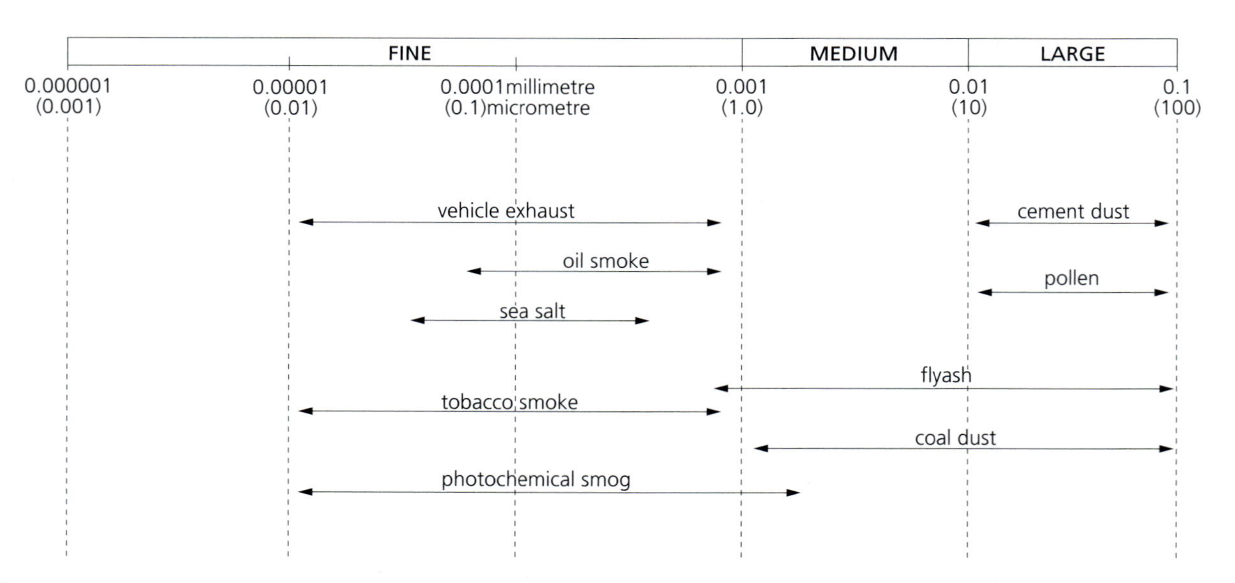

***Figure 11.4** Particulate air pollutants released over cities. Large amounts of particles of different sizes, released from different sources, are found in the air above urban areas.*

Source: Figure 9.5 in Marsh, W.M. and J.M. Grossa (1996) Environmental geography: science, land use and Earth systems. John Wiley & Sons, New York.

BOX 11.8 URBAN CLIMATES

An urban area modifies the local atmosphere directly and indirectly by:

- replacing natural soil and vegetation surfaces with impervious roofing and pavement materials (which alter the water budget and thermal properties of the ground);
- constructing buildings (which alter the wind and energy balances of the surface);
- direct modification of the atmosphere through artificial generation of heat and the release of air pollutants.

Urban areas are generally warmer than the surrounding countryside, particularly on calm, clear nights, mainly because buildings store heat. This creates an *urban heat island*, with the highest temperatures in the most heavily built-up areas. Relative humidity is usually lower over cities than over the surrounding countryside (because of the lack of evaporation from the urban surfaces), wind speeds are generally lower (because of friction from buildings), and rainfall is sometimes higher (because there is more convectional uplift in the warmer air, and because air pollution makes more condensation nuclei available) (Figure 11.4).

- influenced by a series of factors (see p. 327).

These all operate at a variety of scales to create a mosaic of climatic variability. In this section, we look at one of the most widely used schemes for describing different climates and for mapping their distributions.

CLIMATE CLASSIFICATION AND REGIONS

BASIS OF CLASSIFICATION

It is fairly obvious that the hottest places on Earth tend to be found around the equator, within the tropical zone, and that the coldest places are at the poles (the Arctic at the North Pole and the Antarctic at the South Pole) (Figure 11.5). A very simple classification would recognise three main types of climate, which we would encounter in sequence in moving from the equator to the poles:

- tropical (hot and wet)
- temperate (mild and variable)
- polar (cold and ice).

Most places could probably be fitted into a simple threefold scheme like this, but climate reflects more atmospheric factors than simply temperature and moisture. A more useful climate classification scheme would take into account all of the important factors, including temperature, humidity, precipitation, wind, pressure, sunshine and cloudiness.

CLASSIFICATION SCHEMES

Climatologists have proposed a number of different climate-classification schemes. The one we shall use here (Figure 11.6) was developed by Wladimir Köppen at the turn of the twentieth century. The Köppen scheme (like most other global climate schemes) is based primarily on

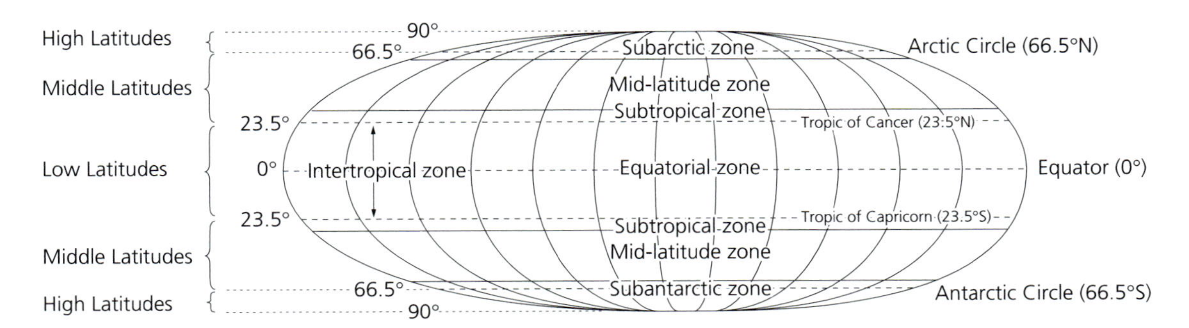

Figure 11.5 *Sub-division of the Earth into latitude.*

Source: Figure 1.4 in Marsh, W.M. and J.M. Grossa (1996) Environmental geography: science, land use and Earth systems. John Wiley & Sons, New York.

Table 11.1 The effects of urban areas on local climate

Climate element	*Comparison with rural climate*
Temperature	
Annual mean	0.5–0.8 °C higher
Winter minimum	1.0–1.5 °C higher
Relative humidity	
Annual mean	6 per cent lower
Winter	2 per cent lower
Summer	8 per cent lower
Cloudiness	
Clouds	5–10 per cent more
Fog in winter	100 per cent more
Fog in summer	30 per cent more
Radiation	
Total on horizontal surface	15–20 per cent less
Ultraviolet in winter	30 per cent less
Ultraviolet in summer	5 per cent less
Wind speed	
Annual mean	20–30 per cent lower
Extreme gusts	10–20 per cent lower
Calms	5–20 per cent more
Precipitation	
Amounts	5–10 per cent more
Days with less than 0.5 cm	10 per cent more
Dust particles	10 times more
Sulphur dioxide	5 times more
Carbon dioxide	10 times more
Carbon monoxide	15 times more

temperature and precipitation data, partly because these are most readily available but also because these two ingredients of climate either reflect or control many other climate variables. Like most other schemes, it is also based heavily on average conditions throughout the year, although some climate regions are defined on the basis of seasonal patterns or departures from the norm.

Other useful climate classification schemes that have been widely used include that proposed in 1948 by C. Warren Thornthwaite (an American climatologist) and a more recent global scheme suggested by Arthur Strahler (an American physical geographer). Where detailed climate records allow, quite sophisticated multivariate computer-based statistical schemes have been developed to sub-divide individual countries – like China – into discrete and meaningful climatic regions.

BOUNDARIES

All classification schemes seek to sub-divide the largest unit (usually the world) into smaller sub-units (in this case climate regions or zones) in such a way that variations between sub-units are larger and more important than variations within sub-units. It is important to be able to map the sub-divisions, so most schemes are based on defining measurable and meaningful boundaries between sub-units.

Boundaries are usually defined on the basis of threshold conditions, such as threshold temperatures or levels of precipitation. In reality, however, climate rarely changes so sharply from place to place that solid lines can be drawn around climatic regions. Transitions and gradual change are more normal than the abrupt boundaries suggested on climate maps.

The selection of boundary conditions – definition of what thresholds are important, and what critical values define those thresholds – is usually influenced by the purpose of the classification scheme. It is important to realise this, because Köppen was particularly interested in exploring how climate affects the distribution of vegetation around the world. He chose what he believed to be impor-

BOX 11.9 BENEFITS OF CLIMATE CLASSIFICATION

It is a fair question to ask why we should bother trying to develop rational ways of classifying climate at all, given all the complications involved. The answer is that without a sensible classification scheme it is impossible to:

- sub-divide the world into climatic regions;
- generalise about how climate changes from place to place;
- examine how climate affects other important environmental systems (such as vegetation, soils, rock weathering and sediment transport);
- compare different places that have similar or different climates; and
- have a baseline against which to measure how much, how fast and in what ways climate changes over time.

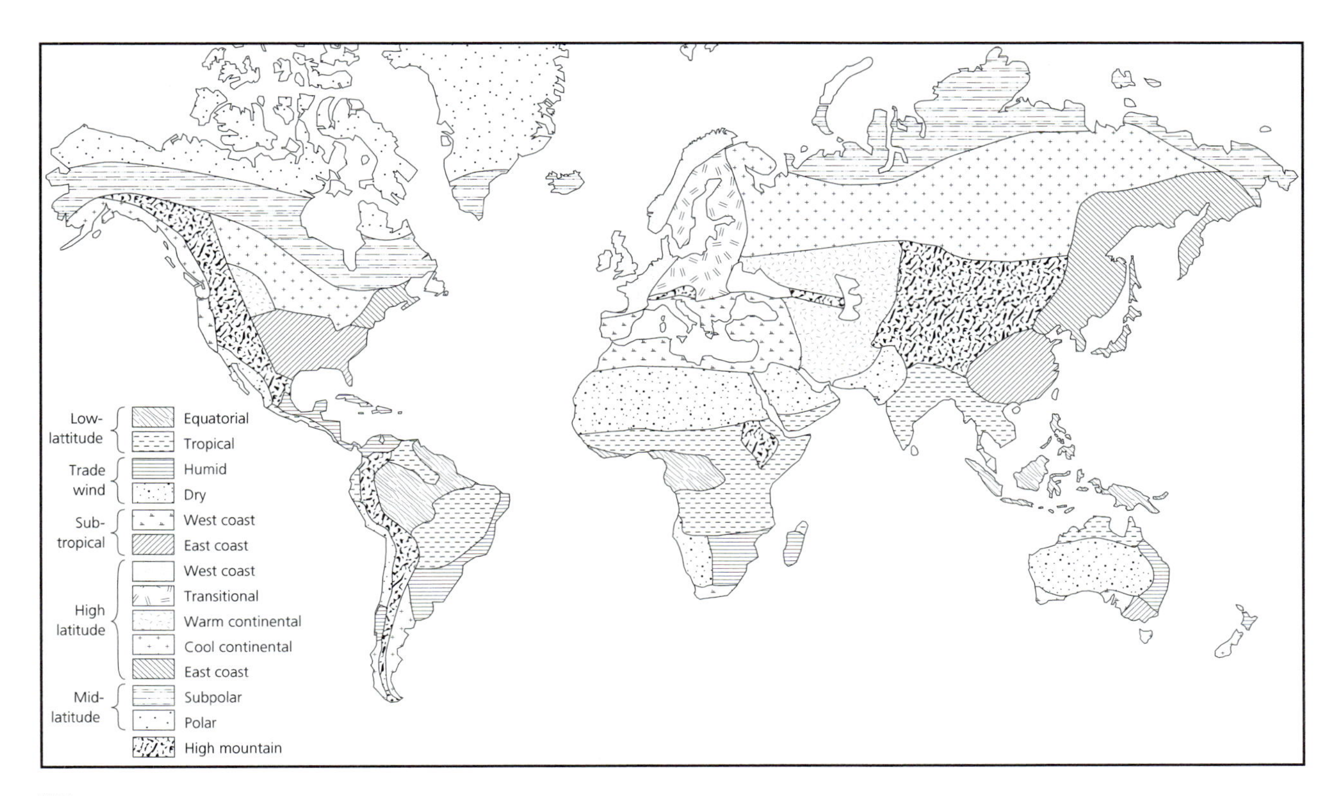

Figure 11.6 *Global distribution of climate zones. The map shows a simplified distribution of climate types.*
Source: Figure 15.2 in Dury, G.H. (1981) Environmental systems. Heinemann, London.

BOX 11.10 CLIMATE CHANGE

Climate regions shown on maps are time-specific, and they are liable to change over time if climate itself is undergoing change. Most published maps summarise fairly stable conditions in the recent past. This usually means the average over a 30-year period, although different countries have different periods, quantities and qualities of data available for defining climate. Inevitably, climate regions will change if global warming occurs as anticipated, and the world map will have to be redrawn.

tant ecological threshold conditions (of temperature and precipitation) for defining the boundaries between his climatic regions.

Some of Köppen's boundaries are now open to reinterpretation:

- We now have much more reliable and more extensive records than he had to work on.
- Köppen made some assumptions about links between vegetation and climate that ecologists would now question.

But the basic scheme has stood the test of time. While it has weaknesses, it:

- is as good as any other approach for most purposes;
- is easily understood;
- is still widely used around the world;
- has been adopted in many other classification schemes.

THE KÖPPEN SCHEME

The Köppen scheme recognises six major climatic groups (Table 11.2). Four groups are based on temperature, one (dry climates, type B) is defined largely by lack of rainfall, and one (highland climates, type H) is defined by elevation. The main climate regions are shown in Figure 11.6. We can now review the characteristics of each climate type, grouped into three broad latitude bands (tropical, mid-latitude and arctic).

Köppen included a highland zone (H) especially for mountainous areas, where climatic conditions can be

Table 11.2 The six major climate groups defined by Köppen

Primary group	*Type*	*Climate thresholds*
Tropical rainy	A	All mean monthly temperatures exceed 18 °C
Dry	B	Low rainfall, and evaporation exceeds precipitation
Humid mesothermal	C	At least one month less than 18 °C, coldest month greater than 0 °C, and the warmest month greater than 10 °C
Humid microthermal	D	Coldest month less than 0 °C, warmest month greater than 10 °C
Polar	E	Warmest month less than 10 °C
Highland	H	Mountain areas with great variations in climate

extremely variable because of variations in elevation, slope and exposure. Altitude is one of the major controls of climate, through its effects on temperature, precipitation and wind.

TROPICAL CLIMATES

Tropical climates (Köppen called them tropical rainy, type A) are typically hot and frost-free, with high rainfall either throughout the year or divided into distinct wet and dry seasons. Five different climate zones can be recognised within the tropical group.

EQUATORIAL WET

Weather and climate in the tropics are heavily influenced by the inter-tropical convergence zone (ITCZ) (see Box 9.14), which migrates northwards and southwards within about 15° from the equator (Figure 11.7). Temperatures in the equatorial wet zone are uniformly high and vary little through the year. Monthly mean temperatures are usually around 30 °C.

Precipitation is caused by convectional uplift in this continuously heated zone, which receives at least 60 mm of rainfall every month. Most days have at least some cloud cover. This zone is wet because easterly trade winds blow moist air masses from the oceans over coastal areas, and tropical cyclones regularly migrate into these coastal areas (particularly around the equinoxes). Periodic heavy downpours accompanied by thunder and lightning are common, and rainfall is usually greater than 1,500 mm a year.

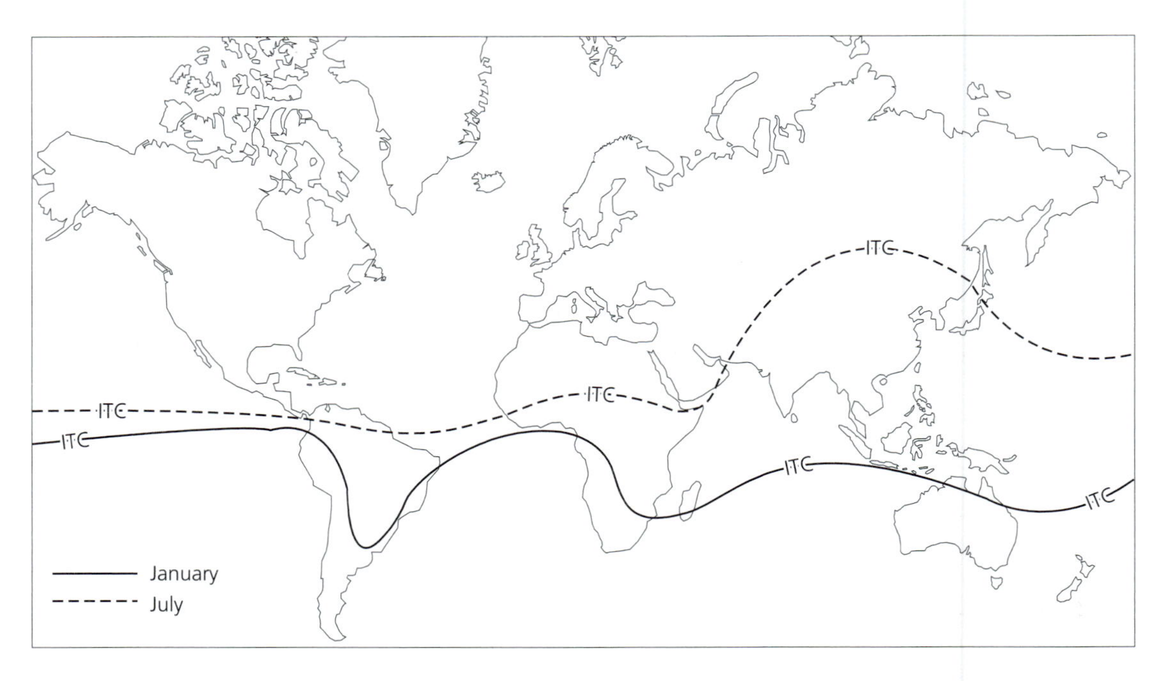

Figure 11.7 *Average position of the inter-tropical convergence zone in July and January. The ITCZ, which can strongly influence the strength and pattern of wind systems within the tropics, moves further south in January (the northern hemisphere winter) and migrates northwards again in July (the northern hemisphere summer).*

Source: Figure 4.27 in Doerr, A.H. (1990) Fundamentals of physical geography. Wm. C. Brown Publishers, Dubuque, Iowa.

Rainforest is the typical natural vegetation in this hot, wet climate zone, which is why Köppen called the zone tropical rainforest (Af). Major areas with an equatorial wet climate include the Amazon basin in South America, the Congo basin in Central Africa, and the mainland and islands of South-east Asia.

TROPICAL WET–DRY

Beyond the equatorial wet zone lies a more variable zone that is influenced by the ITCZ during the summer months and lies within the sub-tropical high-pressure cell during the winter. As a result, this zone has a summer wet season and a winter dry season. Annual precipitation is usually between 800 and 1,600 mm a year, and the rainy season (when up to 90 per cent of the rainfall occurs) lasts between five and eight months.

Places close to the equator have the longest and wettest wet seasons. Temperatures are high throughout the year. The ITCZ migrates to higher latitudes in some years, so this zone can experience quite variable weather from year to year. Savannah landscapes, dominated by scattered trees in extensive grassland, are the norm for this climate zone, which Köppen called the tropical savannah zone (Aw).

MONSOON

The monsoon system (see Box 11.6) is a winter–summer reversal of air flow in the tropics. It is caused by the movement of warm, moist air masses from the Indian Ocean over continental areas, where it is lifted by convection and orographic uplift as it passes over the foothills of the Himalayas (Figure 11.8). Precipitation is less than 60 mm during at least one month in the year, temperatures remain fairly high throughout the year, and there is pronounced seasonality (stormy, cloudy, wet summers and dry winters). The distinct wet–dry season associated with monsoon circulations is best developed in South Asia. Köppen defined the monsoon climate as type Am.

TROPICAL STEPPE AND DESERT

Beyond the tropical wet–dry climate zone, at higher latitudes, is a zone dominated by the relatively stable sub-tropical high-pressure cells throughout the year. Air descends in this low-pressure area and flows outwards from it. Air masses (containing moisture) can rarely penetrate the zone, and there is no convectional uplift (and thus no adiabatic cooling or condensation), so it remains exceedingly dry.

Many of the world's great deserts (see p. 426) – including the Sahara, the Sonoran and the Great Australian – are found in this climate zone. It is a hot, dry zone, with the world's highest amount of solar radiation at ground level and the least cloud cover. Temperatures are very high in summer, although winter nights may be quite cold. The semi-arid steppes (extensive grassy plains, usually without trees) mark the transition between the permanently dry desert climates and the tropical wet–dry areas (which are seasonally dry). Rainfall varies a great deal from year to year in the desert and steppe climate zones; variability here is the highest of all the climate zones.

WEST-COAST DESERT

There are relatively small areas of dry desert climate confined to narrow bands along the west coasts of all continents except Eurasia and Antarctica. The Atacama Desert in southern Peru and the Namib in south-west Africa are typical examples. In these tropical west-coast locations, the stable eastern sides of the sub-tropical high-pressure cells come into contact with cold ocean currents, which flow towards the equator, to create cool and very dry climates. Although rainfall is extremely limited, fog is frequently caused by cool oceanic (moist) air and cool nights.

MID-LATITUDE CLIMATES

The mid-latitude climates are more variable than the tropical climates because they lie within the broad zone dominated by westerly winds and seasonal interactions between the sub-tropical high-pressure cells and the polar front (Figure 11.9). During the summer months, the sub-tropical high-pressure cells intensify (because of heating) and expand polewards, and the polar front weakens and contracts polewards. These latitudes are also strongly affected by ocean currents and land/water differences. Precipitation is mainly cyclonic, and its distribution from place to place and variations over time largely reflect the movement of air masses and frontal systems.

MEDITERRANEAN

This zone is found on the west coast of each of the mid-latitude continents. It has warm, dry summers, caused by the poleward movement of the stable eastern sides of the sub-tropical high-pressure cells (which prevent frontal rainfall). Winter temperatures are much lower, and cyclonic storms (which pass closer to the tropics as the pressure

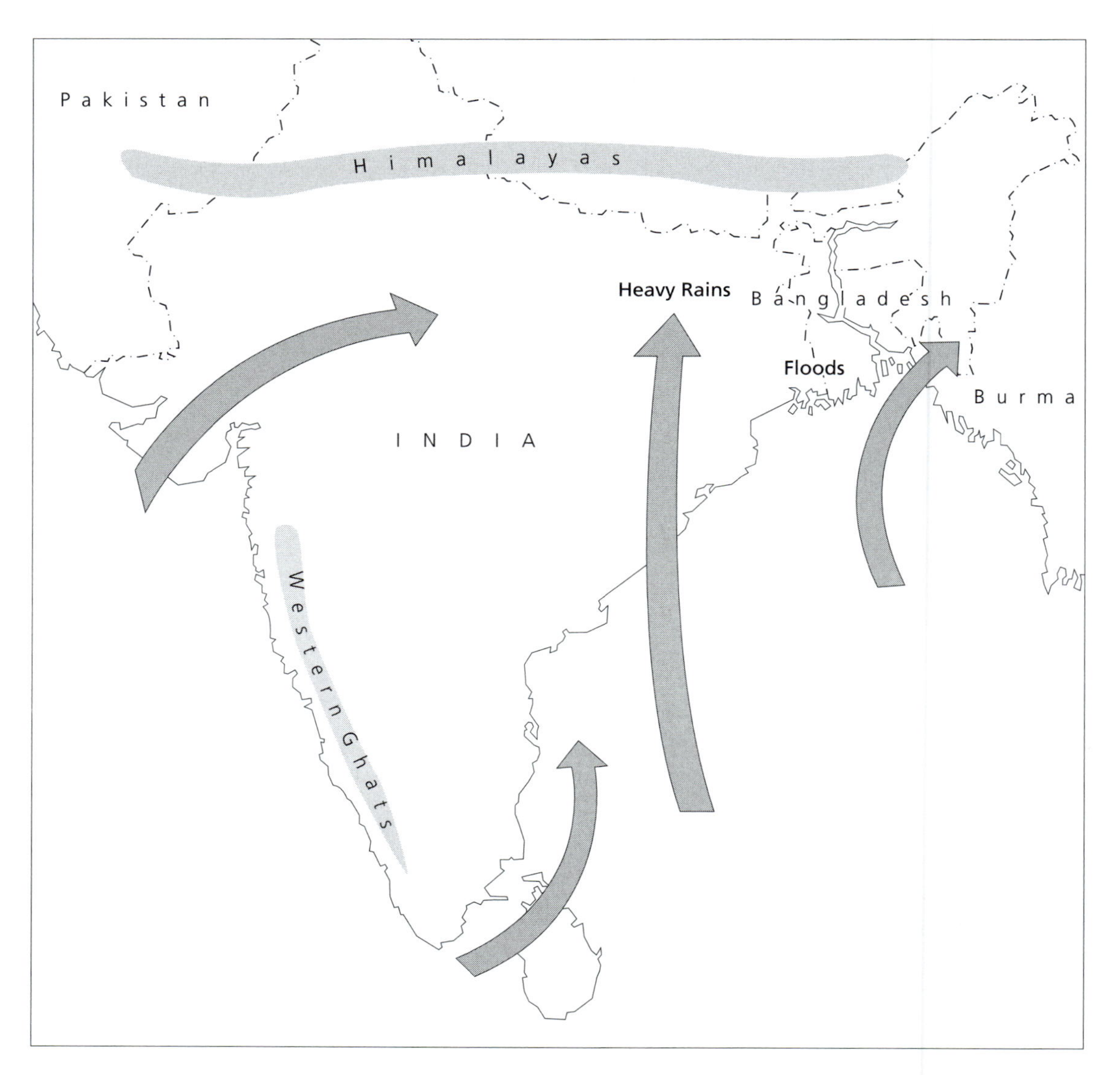

Figure 11.8 *Summer monsoon air flows over the Indian sub-continent. Warming air rises over the plains of central India in the summer, creating a low-pressure cell that draws in warm, wet maritime air. As this moist air rises over the Western Ghats or the Himalayas, it cools and heavy rain falls.*

Source: Figure 17.9 in Cunningham, W.P. and B.W. Saigo (1992) Environmental science: a global concern. Wm. C. Brown Publishers, Dubuque, Iowa.

zones migrate seasonally) bring precipitation across the zone. It is named after the Mediterranean area in southern Europe, but the climate zone is also found in California, Chile, South Africa and Western Australia. Scrubby grassland is a characteristic natural vegetation in the mediterranean zone.

MARINE WEST COAST

At higher latitudes than the mediterranean climate zone are the areas with marine west coast climates, which are warm and wet. Rainfall is often cyclonic, brought by the movement over the zone of cyclonic storms that originate over the oceans and are driven eastwards by the westerly wind systems. Orographic precipitation (see p. 330) can promote heavy rainfall along mountainous coastlines such as those along North and South America, where totals often exceed 500 cm a year. Forest cover often develops because of the heavy rainfall and moderate temperatures.

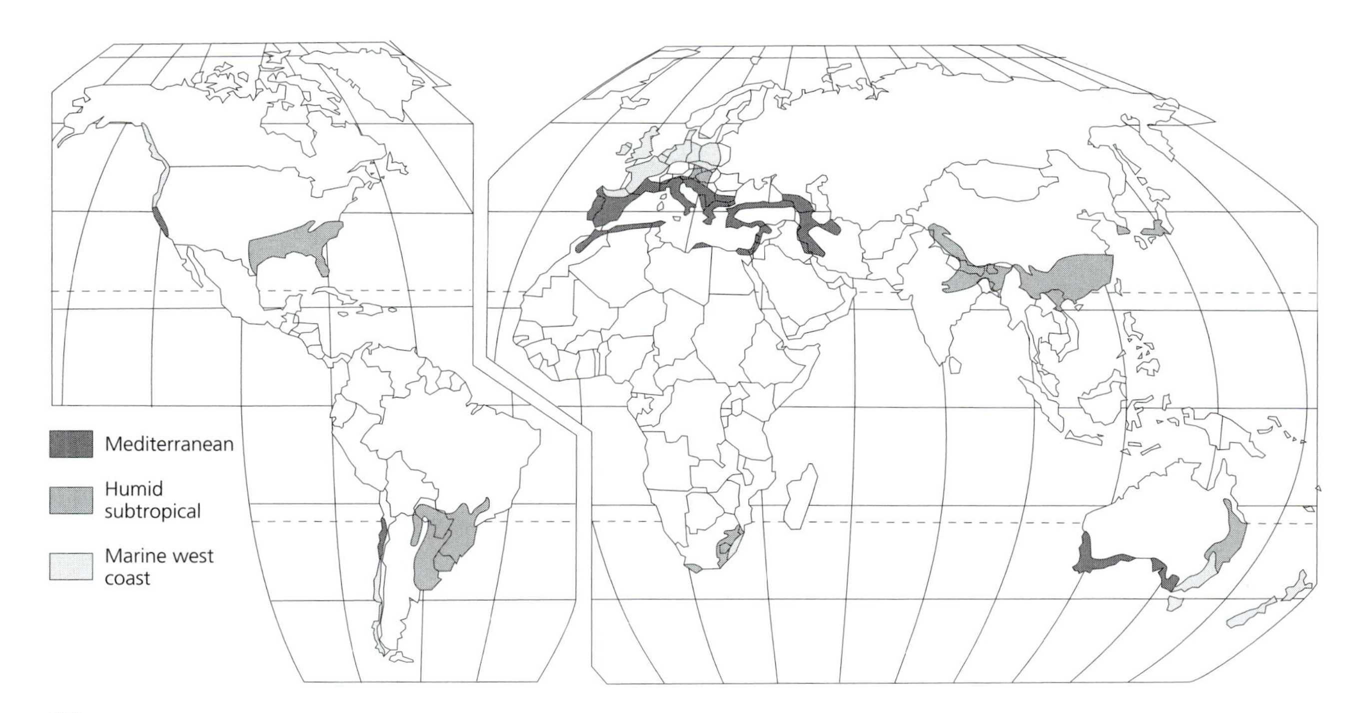

Figure 11.9 *Distribution of the main mid-latitude climate zones. The map shows the distribution of the humid mesothermal and humid continental climates.*

Source: Figure 7.13 in Doerr, A.H. (1990) Fundamentals of physical geography. Wm. C. Brown Publishers, Dubuque, Iowa.

HUMID SUB-TROPICAL

This zone is typically found on the south-eastern side of large continents, at relatively low latitudes (mostly between about 28 and 38°). Precipitation is relatively high (annual totals are usually in the range 80 to 160 cm), coming from both cyclonic and convectional storms. Mid-latitude cyclones dominate the zone in winter, but tropical cyclones cross coastal areas from time to time during spring and autumn (fall). Cold polar air masses regularly cross this zone in North America and Asia during the winter months, bringing cold and relatively wet conditions.

During the summer the polar front weakens, temperature rises, and convectional precipitation replaces frontal precipitation. Monthly precipitation rises during the summer months. In some coastal areas, particularly in China, monsoon air flow brings more rainfall. Seasonal variations in temperature tend to be much stronger in these east-coast locations than those in the west-coast mediterranean climate zone, because ocean circulations do not stabilise temperatures so much.

HUMID CONTINENTAL

This zone is typical of continental areas in middle latitudes, beyond the humid sub-tropical zone. It has harsh winters, with average temperatures below freezing (0 °C) in several months. Much of the precipitation that falls over the zone is brought in by mid-latitude cyclones, although during the summer months there is often a fair amount of convectional rainfall. Temperature varies a great deal, even over short periods of time, as different air masses (some tropical, others polar/arctic) cross the zone, particularly in spring and autumn. This zone is absent from the southern hemisphere because there are no large land masses in the middle latitudes.

MID-LATITUDE STEPPE AND DESERT

The continental interiors, further from the coast than the humid continental and humid sub-tropical climates, are often cut off from maritime air masses by mountain ranges. This makes such parts of North America, Eurasia and southern South America typically very dry, with steppe vegetation. Even further inland, steppe can be replaced by desert. Steppe and desert in mid-latitudes are less dry (particularly in winter) than they are in the tropics, and temperatures vary more throughout the year (temperatures below freezing are not uncommon in winter).

SUB-POLAR AND ARCTIC CLIMATES

CONTINENTAL SUB-ARCTIC

This zone is found at high latitudes in the continental interiors of the northern hemisphere (there are no large land areas at this latitude in the southern hemisphere). It determines the climate of northern parts of North America and Eurasia, where winters are very cold and relatively dry. High-pressure cells develop as a result of the winter cold, and this causes air to subside and blow outwards from this zone. As a result, the air is stable and dry. This zone is the source area of continental polar air masses, which are blown south and east into the humid continental zones. Permafrost (permanently frozen ground) up to a metre or more deep is found extensively through this climate zone. Summers are often relatively mild, with long hours of sunlight, but summers also tend to be quite short, with only two or three months without frost.

MARINE SUB-ARCTIC

Temperatures within the sub-arctic remain very low throughout the year, but conditions in west-coast areas are milder than those in the continental interiors because of the maritime influence – particularly the relatively warm ocean currents and heat storage in the oceans. Temperatures here remain around freezing throughout much of the winter, and this zone has some of the world's most cloudy and windy weather. Natural vegetation in this zone is either tundra (where it is coldest) or boreal forest (where it is milder).

ARCTIC AND ICE CAP

Permanent snow fields and ice caps are found where average air temperatures are below freezing (0 °C) every month of the year. The main areas of arctic climate are Greenland, Antarctica, and northern North America and Eurasia (above 70° latitude). Arctic conditions are found beyond this latitude in high mountain areas, even within the tropics. Beyond the arctic zone lies the tundra landscape (typically found in the marine sub-arctic zone), which is permanently underlain by permafrost (see p. 446) but has snow cover only during the winter.

CLIMATE CHANGE

Climate is the long-term prevailing weather conditions of an area (see p. 301), and both direct and indirect types of evidence indicate that climate changes over time as well as from place to place. In this section, we explore some of the main lines of evidence of climate change, examine the patterns and timing of these changes, and reflect on what might cause them.

EVIDENCE

Reconstruction of past climates relies heavily on the availability of evidence. Inevitably, the recent past is the easiest to reconstruct because the most reliable evidence is available for that time. The further back in time we go, the less complete the evidence and the more assumptions have to be made. Three main types of evidence are available – direct observation, historical evidence (Box 11.11), and physical and biological proxy indicators.

DIRECT OBSERVATIONS

By AD 1700, most of the standard meteorological instruments that are used to measure temperature, precipitation and other climate variables had been invented (they have since been greatly improved), although good-quality records going back centuries are available for only a few sites. Direct measurements of weather have been collected regularly only since about the 1850s, and many parts of the world (particularly in the tropics and southern hemisphere) have much shorter periods of records available.

Although it is extremely short relative to the full span of geological time, a century of climate records does offer a baseline against which to judge contemporary change, and detailed analysis of fluctuations over time throws some light on the pace, pattern and cause of change. Through the twentieth century, for example, annual precipitation increased in southern Canada by 13 per cent and in the contiguous USA by 4 per cent; between 1950 and 1990, northern Canada experienced up to a 20 per cent increase in annual snowfall and rainfall. The period 1970–1990 was dominated by extreme weather events worldwide, and this is generally regarded as a good indication that distinctive climatic change is under way at the present time.

PHYSICAL INDICATORS

A number of physical indicators are widely used (Figure 11.10), including the following:

BOX 11.11 HISTORICAL EVIDENCE OF CLIMATE CHANGE

Documentary records (including weather diaries, travel accounts and accounts of severe weather events or seasons such as floods, rivers freezing and snowstorms) provide evidence of weather conditions and extremes in the historical past. The direct evidence offered by documentary records is complemented by indirect evidence (including such things as harvest failures and phases of glacier advance and retreat).

Historical records from Britain going back as far as AD 1100 and weather observations from central England from the mid-seventeenth century onwards show a sequence of climate changes over the past 800 years. Highlights of this sequence include a warm, dry medieval phase followed by colder winters and wetter summers after about AD 1300, with a marked cooling particularly between 1550 and 1700 (when glaciers advanced in the Alps and in Scandinavia, and Arctic sea ice was much more extensive). In China, the documentary records (including evidence of the cultivation of citrus trees) show that the thirteenth century was the warmest period over the last 1,000 years, with temperatures up to 1 °C higher than at present.

Direct observations and historical sources allow climate reconstruction only back to medieval times at best in most countries. For longer timescales, scientists have to rely on proxy sources – physical or biological indicators of phenomena that are strongly influenced by climate.

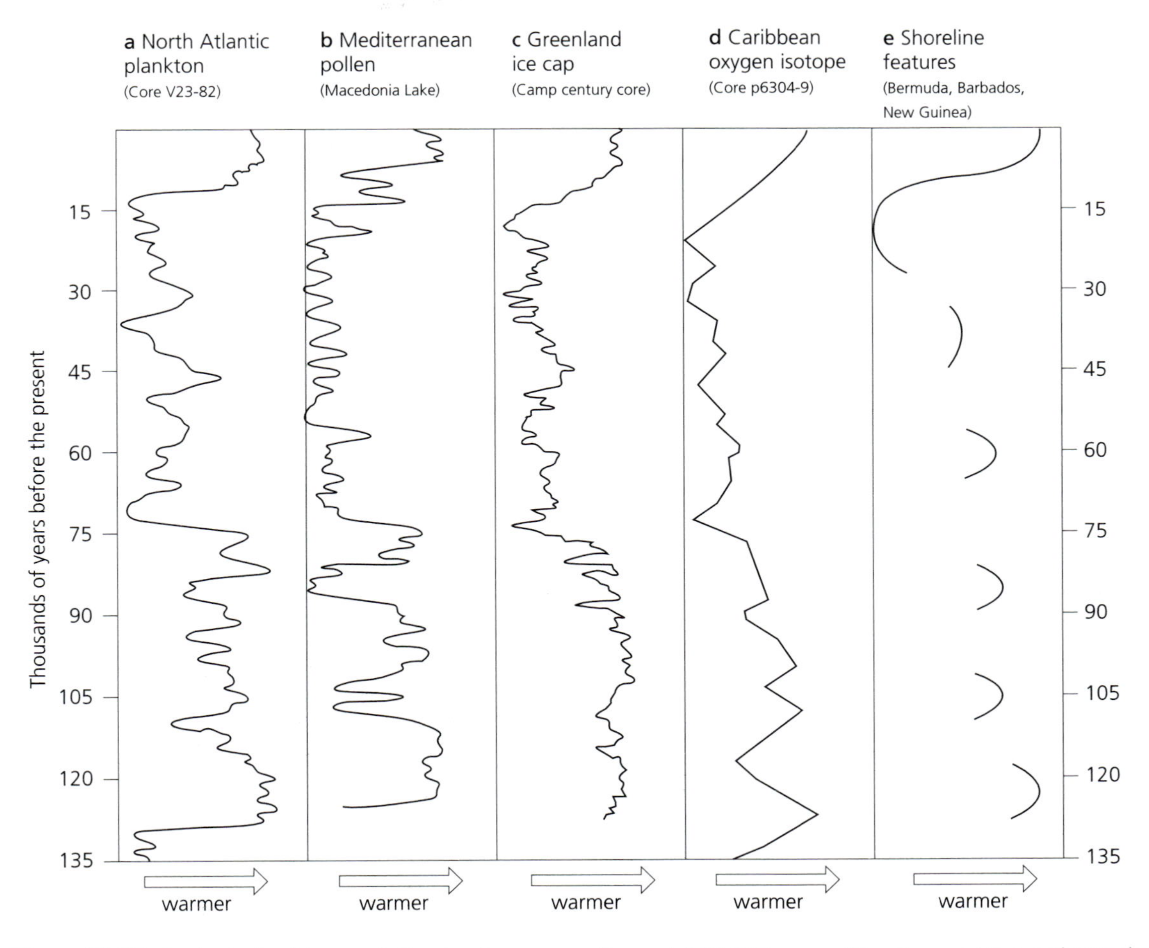

Figure 11.10 *Geological evidence of climate change over the past 135,000 years. Five time series are shown that indicate strong patterns of temperature changes over the past 135,000 years, based on (a) the proportions of different species of shell in a deep-sea core west of Ireland; (b) the changing proportions of tree pollen in sediment at the botton of a lake in former Yugoslavia; (c) a heavy oxygen record from a long ice core in Greenland; (d) a heavy oxygen record provided by shells in a deep-sea core from the Caribbean; and (e) changes in sea level recorded in various areas.*

Source: Figure 29 in Williamson, T. (1977) Exploring our changing climate. HMSO, London.

Ice cores: analysis can be made of the composition of ice that has accumulated over a period of thousands of years in parts of Greenland and Antarctica. Bubbles of air are trapped in the ice, and these preserved samples can be analysed to determine the composition of trace gases and aerosols (which are closely associated with climate) in the atmosphere at the time of deposition. The record goes back up to 125,000 years for Camp Century in Greenland.

Isotope analysis: past climates can also be reconstructed from detailed analysis of the oxygen isotope composition of ice cores, the remains of marine organisms and calcite deposits in caves.

Geological evidence: the stratigraphic (sedimentary rock) record (see p. 196) preserves a great deal of evidence (Box 11.12) of changes in climate and changing climate distributions over geological timescales. More recently, within the last 2 million years, there is extensive evidence of ice sheets covering much of Europe and North America during the Quaternary (see p. 451), and more extensive ice sheets than today in a number of places, including Siberia, South America and New Zealand.

BOX 11.12 GEOLOGICAL EVIDENCE OF CLIMATE CHANGE

The evidence includes:

- *glacial deposits*: tills and moraines (p. 465);
- *ancient soils*: palaeosols (p. 601) buried between two tills;
- *former dunes*: these record the dominant directions of past winds (p. 432);
- *former lake shorelines in present desert areas*: including the Sahara and the American south-west, which indicate much wetter conditions in the past (p. 437);
- *evidence of former periglacial conditions*: such as casts of former ice wedges and solifluction lobes (p. 446);
- *ocean sediment records*: these reveal rates and patterns of sediment deposition (p. 476), organic remains (including foraminifera and diatoms) and analysis of shell carbonate content.

BIOLOGICAL INDICATORS

Pollen evidence: pollen from each plant has a distinctive size and shape, and the frequency of each type reflects the frequency of that plant in the local vegetation at the time of deposition. The vegetation records can be converted into a climate record by knowing about relationships between climatic factors (e.g. summer temperature) and vegetation growth.

Dendroclimatology: this is the study of growth rings in trees, which reflect the climatic conditions of the growing season and can be counted to give absolute dates (one growth layer is normally added to the tree each year). Tree ring data from conifers in the Sierra Nevada in the USA have been used to reconstruct temperature and precipitation back to AD 800, including the Medieval Warm Period (from about 1100 to 1375) and the Little Ice Age (from about 1450 to 1850).

PATTERNS

GEOLOGICAL BACKGROUND

The geological record shows that in the past climate must have been very different from what it is today in many places. In Britain, for example, the rock record shows significant changes in temperature and precipitation, including evidence of forest swamps (in the Carboniferous Coal Measures, dating back 300–400 million years), desert dunes (in the Permo-Triassic sandstones, 200–300 million years old) and ice sheets (in the Pleistocene, within the last 2 million years).

Some of the changes up to the Tertiary Era (starting about 70 million years ago) were caused by plate tectonics and continental drift (see p. 150), but others were caused by changes in global climate. Between about 200 million and 70 million years ago, continental areas had wetter and warmer climates than they have today.

Global cooling began during the Tertiary (see Table 4.7), initially in Antarctica, where local mountain glaciers and sea ice formed about 37 million BP (before present). By the Miocene (14–10 million BP), there was a major Antarctic ice sheet. Glaciation of the Arctic and parts of South America began between 9 million and 3 million BP, and by 4–5 million BP the base of the ice sheet was below sea level in west Antarctica.

Over the last 3 million years or so, changes in the biosphere have reflected the broad pattern of climate change. The most recent 2 million years are the Quaternary, a period that includes the Pleistocene (ice age) and Holocene (post-glacial) epochs, and that continues today.

THE PLEISTOCENE

This period, which covers much of the last 2 million years, saw the most recent ice age (see Table 11.3). During the Pleistocene, climate switched between glacial and interglacial conditions through a number of cycles. Geological evidence from many locations throughout North America supports a four-phase model of glacial advances (called the Kansan, Nebraskan, Illinoian and Wisconsinian), although there is evidence that many parts of Europe experienced five glacial periods.

Glacial periods, when the ice advanced (Box 11.13), were separated by interglacial stages when the ice melted and temperatures were as high as if not higher than today. During the Ipswichian Interglacial (70,000 to 100,000 BP), for example, average summer temperatures in southern Britain were up to 3 °C higher than today. Conditions were so warm, in fact, that remains of elephants, hippopotamuses and rhinoceroses dating back to this time have been found in the London area. Glacial periods were also punctuated by relatively short periods of milder climates, called interstadials, which were cooler and shorter than the interglacials.

Table 11.3 Major ice ages in the geological past

Name	*Date (years ago)*
Pleistocene	1.7 million to 10,000
Permo-Carboniferous	330–250 million
Ordovician	440–430 million
Verangian	615–570 million
Sturtian	820–770 million
Gnesjo	940–880 million
Huronian	2,700–1,800 million

As ice sheets spread from high-latitude land areas in both hemispheres, non-polar climate regions were displaced towards lower latitudes. Vegetation zones migrated with them, and there was probably much less forest cover and more extensive deserts and tundra conditions. Arid conditions persisted in many tropical areas.

Sea level fell as more and more water was locked up in the growing ice sheets and ice caps. During the glacial maximum, when glaciers and ice sheets were at their most extensive, the sea level was up to 150 m lower than today. This exposed areas of land (called land bridges) in the Bering Strait (joining Alaska to Siberia), the Indonesian archipelago (joining New Guinea to Australia), the Caribbean and the North Sea, causing changes in ocean circulation patterns and creating opportunities for plant and animal species to expand their ranges and colonise new areas. Humans are believed to have reached most of the regions they now inhabit during the Pleistocene, thanks to these land bridges.

Analysis of deep ice cores from the Antarctic and Greenland shows that temperatures during the glacial phases were probably about 6 °C lower than today, and atmospheric concentrations of CO_2 and CH_4 were probably lower by factors of nearly 2 and 4, respectively. Levels of continental dust and sea-spray aerosols trapped in the ice were much higher during glacials than during interglacials, indicating that conditions were more favourable for their production and long-range transport.

BOX 11.13 THE PLEISTOCENE ICE AGE IN THE NORTHERN HEMISPHERE

Major ice sheets formed over Canada and Scandinavia, and the main source of moisture for this process was probably the North Atlantic. The Arctic Ocean has been largely ice-covered for at least the last 700,000 years. Continental glaciers in North America reached as far south as the Great Lakes, and an ice sheet spread over northern Europe as far south as Switzerland.

Glaciers expanded as the ice-age climate cooled, and this has left strong imprints on the landscapes of North America and Europe – including extensive glacial and periglacial landforms and deposits (p. 453), soil properties and distribution (p. 606), and plant and animal distributions (p. 579). When the ice caps and glaciers started to melt and retreat, they left behind bare surfaces that new plant and animal species could start to colonise so that secondary succession could begin (p. 573).

The most recent (last) glaciation is known as the Devensian in Britain and the Wisconsinian in North America. It began about 70,000 years ago and ended about 10,000 years ago. The last glacial maximum reached a peak about 18,000 years ago. At that time, summer temperatures were reduced by 8–15 °C over most of North America and Eurasia south of the ice sheets, and sea-surface temperatures were about 2–2.5 °C below the present.

BOX 11.14 GLACIAL CONDITIONS AND HUMANS

While scientists can reconstruct important aspects of how, where and when climate changed during the Pleistocene, according to current understanding it is still difficult to explain why ice sheets were built up as fast as they appear to have been. It has been speculated that early human interference, by burning and thus extending the savannah landscapes, particularly in Africa, might have contributed to the build-up of the Pleistocene ice age by destabilising climate, but this is a minority view.

On a geological timescale, we are currently experiencing an interglacial phase. Conditions as warm as we have today probably lasted for only about a tenth of each major glacial cycle. Before concern started to rise during the 1980s over the prospect of global warming (p. 261), scientists were predicting a return to glacial conditions some time in the future.

THE HOLOCENE

The Holocene (also called the Flandrian Interglacial) is the second and most recent epoch of the Quaternary period. It began about 10,000 years ago at the end of the Pleistocene, when temperatures and sea level began to rise again, and the ice started to retreat. The Scandinavian ice sheets disappeared about 9,500 BP, and the much larger North American ice sheet had gone by about 6,500 BP (see p. 453).

In the tropics, a very moist pluvial (wet) period occurred between about 9,000 and 8,000 BP. It caused high lake levels and brought a savannah-type vegetation to many areas in Africa, the Middle East and Australia that are now deserts. Modern arid environments (see p. 426) in these places were created by increasing aridity after about 4,500 BP. The climate outside the tropics became progressively warmer, and forest cover spread northwards.

Climatic Optimum: by about 5,500 years ago conditions had improved a great deal, and temperatures in mid-latitudes were about 2.5 °C higher than they are today. This period (between about 6,000 and 5,000 BP) is called the Climatic Optimum. Since then, much of North America and Europe has cooled and become wetter, and glacier records in many places indicate several phases of advance.

Climatic deterioration, particularly the cooling, began about 2,500 years ago, and evidence of it is widely preserved in sedimentary and biological deposits. Changes were extensive throughout the northern hemisphere, but they appear to have been mirrored in the southern hemisphere, as evidence from places such as southern Africa confirms. Cooling continued in many places up to about the early tenth century, sometimes punctuated by one or more shorter warm phases (which lasted centuries at most). It was followed by the Medieval Warm Period (Box 11.15).

Little Ice Age: the climate became much cooler after the Medieval Warm Period, and cold conditions persisted between about 1450 and 1850. This period is widely referred to as the Little Ice Age because many valley glaciers (particularly in mountainous parts of Europe) extended and there was renewed periglacial activity in many upland areas (Figure 11.11). The coldest conditions of the last 560 years were between 1570 and 1730 (the height of the Little Ice Age) and in the nineteenth century. The Little Ice Age seems to have been associated with longer, more severe

BOX 11.15 THE MEDIEVAL WARM PERIOD

There is evidence from many places that the climate became much milder, warmer and drier between about AD 900 and 1300. This phase is referred to as the Medieval Warm Period or Little Climatic Optimum. Much documentary evidence from this period survives, and it shows some interesting links between climate and history. During the eleventh to thirteenth centuries, for example, there were productive vineyards in southern England, which is now too cold for extensive wine growing. Icelandic sagas recall times around AD 1000 when the seas around Iceland and Greenland were more or less ice-free. This allowed the Vikings to sail to Greenland and settle there in colonies that were inaccessible from Europe in later centuries because of extensive sea ice.

During the height of the Medieval Warm Period (AD 1000–1100) Anasazi Pueblo Indians occupied sites in the northern Colorado Plateau region of the southwestern USA and practised dry farming there. Between 1100 and 1300, this became impossible with the onset of much drier, colder conditions, and the settlements were abandoned.

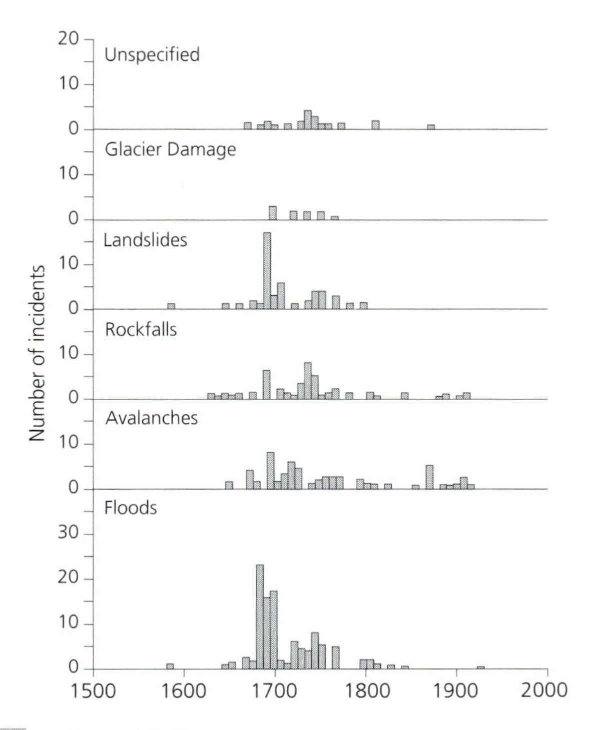

Figure 11.11 *Geological hazards during the Little Ice Age in Norway. The graph shows variations in the number of incidents of mass movements and other natural hazards during the Little Ice Age, based on an analysis of land rent records for Norwegian parishes.*

Source: Figure 4.10 in Goudie, A. (1983) Environmental change. Oxford University Press, Oxford.

BOX 11.16 TWENTIETH-CENTURY CLIMATE

Climate fluctuations over the last 100 years or so are easier to reconstruct than trends in earlier centuries, because they can be detected in instrumental records. Between about 1880 and 1950, summer and winter temperatures rose in many areas, including the Arctic, northern Scandinavia and the former Soviet Union, southern Canada and most of the USA. This warming led to increased sea temperatures in the northern Atlantic and in Arctic waters, with knock-on effects on deep-sea fishing and sea transport in northern latitudes. Agriculture in northern Finland and Canada benefited from the longer growing season. Unusually warm conditions have prevailed around the world since the 1920s, probably because of a relative absence of major explosive volcanic eruptions and higher levels of greenhouse gases.

winters. In the late seventeenth century, the River Thames completely froze over at London during some particularly cold winters, allowing 'frost fairs' to be held on the ice. Tree rings in the south-eastern USA preserve evidence of the Little Ice Age there too. Cold conditions did not exist only in the northern hemisphere. In southern South America, tree-ring evidence indicates a long, cold, moist period from AD 1270 to 1660, which was most pronounced between 1340 and 1640, and glaciers in many places advanced between the late seventeenth and early nineteenth centuries. The twentieth-century climate (Box 11.16) has been much warmer (Figure 11.12).

CAUSES

Many different factors are involved in climate change, and they fall into two categories – those external to the climate system (Box 11.17) and those internal to the climate system.

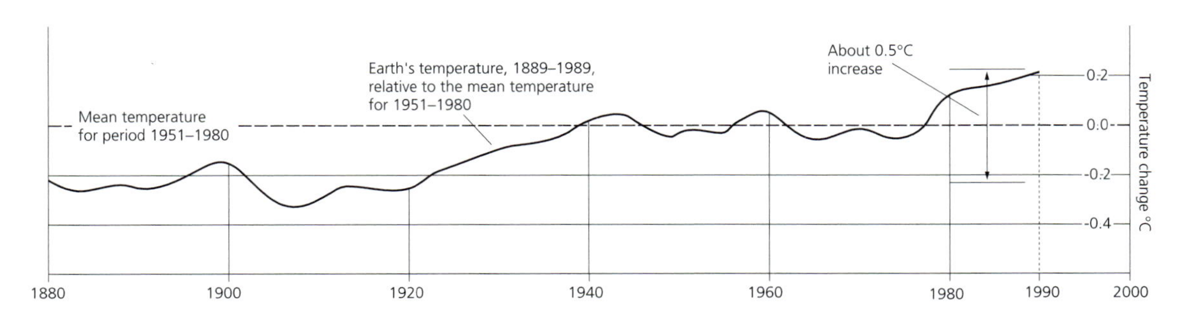

Figure 11.12 *Changes in mean atmospheric temperature of the Earth, 1880–1990. The graph shows changes in the mean temperature of the Earth relative to a standard mean temperature (for the period 1951–1980). Average temperature rose by about 0.5°C between 1880 and 1990.*

Source: Figure 8.11 in Marsh, W.M. and J.M. Grossa (1996) Environmental geography: science, land use and Earth systems. John Wiley & Sons, New York.

BOX 11.17 EXTERNAL FACTORS THAT PROMOTE CLIMATE CHANGE

Variations in solar output: there is little direct evidence that the Sun's output of energy is changing over the long term, but this is possible. Variations in solar output through the 11-year cycle of sunspot activity (see Box 4.7) are also important, and historical data suggest that colder conditions in Europe coincide with periods of reduced or no sunspot activity.

Variations in the Earth's orbit: the Earth orbits around the Sun in a complex manner (see p. 118). The Yugoslav astronomer M. Milankovitch worked out the basic underlying pattern (Figure 11.13), which is determined by three interacting effects:

1. variations in the shape of the elliptical orbit (over a 95,000-year cycle);
2. variations in the tilt of the Earth's axis of rotation (over a 42,000-year cycle); and
3. variations in the time of year when the Earth is closest to the Sun (perihelion) (over a 21,000-year cycle).

Analysis of ocean cores shows that all three cycles appear in the oxygen-isotope and faunal records, showing that they are associated with the timing of glacial and interglacial phases. The Earth's rate of rotation around its axis apparently speeded up from about AD 1000 to 1800.

Continental drift: over millions of years, the continents have been redistributed around the surface of the Earth by plate tectonic movements (see p. 150). This has created the major oceans and mountain belts that influence climate today (p. 128), but it also means that continental areas have been moved around and experienced radically different climates along the way. Around 450 million years ago, for example, what is now the Sahara Desert was located close to the South Pole, and it was glaciated. During the Permo-Triassic, what is now Britain was located much further south, in latitudes 20–30°, where trade winds created desert conditions that are now preserved in the New Red Sandstones.

INTERNAL FACTORS

Composition of the atmosphere: this strongly influences the amount of short-wave solar radiation that reaches the Earth's surface, and the amount of long-wave re-radiation from the surface that leaves the atmosphere and disappears back into space (see p. 237). While the composition of gases – particularly carbon dioxide – might have altered through time, scientists think that the most important factor has been variations in atmospheric dust particles (aerosols). Much of this fine dust is picked up by the wind as it blows over deserts or dry soils. Over land the dust scatters solar radiation and cools the atmosphere, but over the oceans (where albedo is relatively low) the dust increases the reflection of incoming particles and this heats the atmosphere. During ice ages, vast areas of bare ground were exposed around the margins of ice sheets in Europe and North America, much of it covered with extensive deposits of fine silt. Winds blowing across these plains would pick up huge amounts of fine particles and lift them into the atmosphere. Volcanic eruptions also throw large amounts of dust into the stratosphere (p. 180), which can reduce direct solar radiation at the Earth's surface but may at the same time increase scattered radiation. The atmosphere can show slight cooling (up to about 0.3 °C) for several years after a volcanic event (p. 189).

Ocean and ice cover feedback: there is a range of feedback relationships within the atmosphere that serve to reinforce climate change once it has begun. One involves snow and ice cover, which has a high albedo, reflects much of the incoming radiation and cools the air above. As a result, frontal systems are displaced to lower latitudes in autumn and winter, encouraging snow and ice cover to extend to lower latitudes, thus reinforcing the initial trigger. Other feedback involves large-scale changes in ocean currents and deep-water circulation patterns.

Many different factors play a part in promoting climate change, and once started it can be reinforced by feedback effects in the atmospheric, oceanic and hydrological systems. Changes in solar energy receipt are probably the most important trigger of climate change, but changes in heat storage and transport by the oceans, and changes in snow and ice cover, also play important roles.

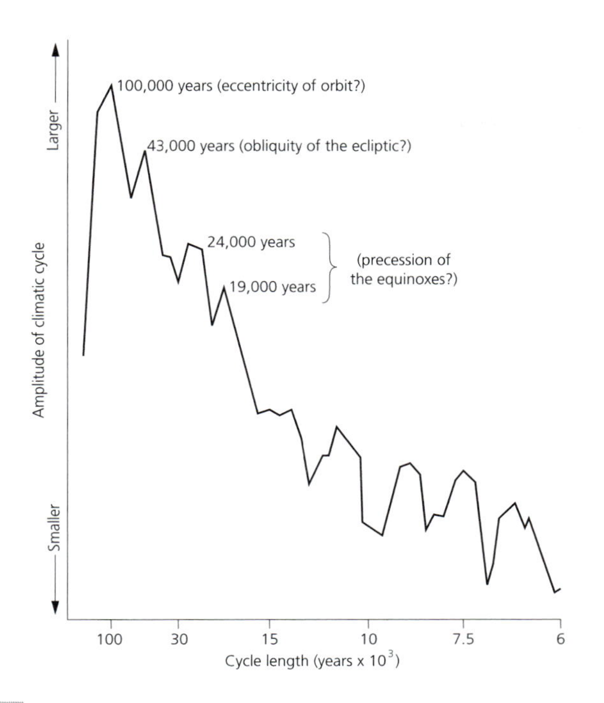

Figure 11.13 *Cycles of climate variation over the past 500,000 years. Climate has changed over cycles of varying lengths over the past half a million years, and these can be identified in the isotope record of sea-bed cores from the Indian Ocean. The cycles correspond quite closely to the pattern of variations in the Earth's orbit.*

Source: Figure 7.6 in Goudie, A. (1983) Environmental change. Oxford University Press, Oxford.

BOX 11.18 THE CHALLENGE OF GLOBAL WARMING

Global warming associated with the release of greenhouse gases threatens to set in motion more rapid climate change than anything the Earth has previously experienced, and many scientists fear that environmental systems will be unable to adjust to the pace and pattern of induced climate change. Quite what damage this will cause to the Earth, and how it will affect the ability of the planet to support life, remain unclear at this stage. Likely consequences could include movements in vegetation belts (altitudinal and latitudinal), changes in the intensity and frequency of tropical storms, transformations of runoff and soil-moisture conditions, the melting of ice bodies, and a rise of sea level and associated inundation and erosion. This remains a fundamental but critical area of environmental uncertainty.

NATURAL VERSUS INDUCED CHANGE

Until relatively recently, climate change has been caused by entirely natural processes. These have sought to maintain a balanced state between the atmosphere and other major environmental and biogeochemical cycles and systems, and they indicate Gaian self-regulation at work (see p. 119).

In recent centuries, however, and particularly since the Industrial Revolution in the eighteenth and nineteenth centuries, human activities have become an important trigger of climate change (see p. 261). Even in pre-industrial times, human activities (particularly forest clearance, vegetation burning and agriculture) probably had at least the potential to modify regional if not global climates. Detecting the contribution of human activity to observed climate changes is a difficult task, because induced changes are superimposed on natural variations and separating out the two sets of signals (trends) often requires quite sophisticated statistical analyses.

SUMMARY

Climate plays an important role in all the main environmental systems, and it dictates how comfortable we feel in different places. This chapter examined what climate is, how it is shaped, and how it varies from place to place and over time. It opened with an overview of the factors that control climate, from the global scale (for example the distribution of land and sea) to the local (for example mountain and valley breezes), and including human impacts. The main features of global climate were outlined next, including a description of the main climate zones. A major theme in the chapter is climate change, and we considered the main lines of evidence for climate change as well as the pattern of climate change at different timescales. We also looked at suggested causes, including both internal and external factors. Global warming associated with greenhouse gas emissions is the most serious environmental problem we confront today, and this chapter has sought to put recent changes into historical context.

WEBSITE

Links to relevant websites, a comprehensive bibliography, tools for teaching and learning, and downloadable images relevant to this chapter can be found at the website specially designed to accompany this book at http://www.park-environment.com

FURTHER READING

Barry, R.G. (1992) *Mountain weather and climate*. Routledge, London. A wide-ranging review of the relationships between mountains and weather, drawing on examples from around the world.

Barry, R.G. and R. Chorley (1998) *Atmosphere, weather and climate*. Routledge, London. Useful and up-to-date seventh edition of the classic text on atmosphere and world climate.

Bell, M. and M.J.C. Walker (1992) *Late Quaternary environmental change: physical and human perspectives*. Longman, Harlow. Explores the relationships between people and environment against a backdrop of climate change.

Bradley, R. and P. Jones (eds) (1994) *Climate since AD 1500*. Routledge, London. A collection of essays on the Little Ice Age and the climate of the twentieth century.

Downing, T. and R.S.J. Tol (eds) (1998) *Climate, change and risk*. Routledge, London. An overview of climatic hazards and climate change, with a focus on societal response, insurance and methodologies for analysis.

Hulme, M. and E. Barrow (eds) (1997) *Climates of the British Isles: present, past and future*. Routledge, London. A useful and well-supported description and explanation of the British climate, and its spatial and temporal variability.

Lamb, H.H. (1995) *Climate, history and the modern world*. Routledge, London. Examines what we know about climate, how the past record of climate can be reconstructed, the causes of climatic variation, and its impact on human affairs now and in the historical and prehistoric past.

Maddison, D. (2000) *The amenity value of global climate*. Earthscan, London. Examines the economic value of climate derived from its 'amenity value', which is the benefits that a particular climate provides to the people of a region or country. Includes case studies drawn from many different countries.

Roberts, N. (1998) *The Holocene: an environmental history*. Blackwell, Oxford. Reviews recent advances in understanding the causes, patterns and consequences of environmental change during the Holocene.

Tóth, F.L. (ed.) (1999) *Fair weather? Equity concerns in climate change*. Earthscan, London. A series of essays that explore the search for a fair solution to the problems of coping with climate change – how should the costs be measured, who should meet them, and how?

Wheeler, D.A. and J.C. Mayes (eds) (1997) *Regional climates of the British Isles*. Routledge, London. Outlines the pattern and causes of variations in climate across the British Isles.

Plate 1 *Many environmental problems arise directly from modern high-tech, consumer-oriented urban lifestyles, illustrated here by the high-rise and well-planned development of Singapore in South-east Asia.*
Photo: Chris Park.

Plate 2 *The Nile valley below the Aswan High Dam in Egypt. A green corridor of dense vegetation grows along the river edge, but it is hemmed in by the dry desert typical of this extremely arid climate.*
Photo: Philip Barker.

Plate 3 *Tower karst in Halong Bay, northern Vietnam. This scene reflects interaction and adjustment in natural environmental systems – the limestone tower karst has been flooded by rising sea level, but the unusual tidal regime of this coastal area (which has only one tidal cycle each day) reduces the amount of undercutting at the base of the cliffs, and this protects the towers from collapsing.* Photo: Andy Quin.

Plate 4 *Escarpment rift valley at Magadi in Kenya, part of the great African Rift Valley system. The distinct change of slope along the line of the fault is clearly visible, along with the signs of mass movement as material has slumped down towards the valley floor.* Photo: Philip Barker.

Plate 5 *Landscape heavily influenced by glacial erosion and deposition during the Pleistocene ice age, Great Langdale, Cumbria, England.* Photo: Chris Park.

Plate 6 *Gully erosion in Coverdale in the Yorkshire Dales, England. In this valley only the south-facing slopes are heavily dissected by gullies, and the continuous vegetation covers indicates that the erosional features have long been stable.* Photo: Chris Park.

Plate 7 *Mountain cloud cover in the Tyrolean Alps near Kitzbühel, Austria. The photograph was taken late one morning in early August, by when a more extensive early morning cloud cover had been burned off by the sun.* *Photo: Chris Park.*

Plate 8 *Causeway Bay typhoon shelter on Hong Kong island, East Asia. This area is regularly affected by intense tropical storms, and boats (including the characteristic Chinese sampans shown here) can take advantage of shelters in specially designated areas of the harbour.* *Photo Chris Park.*

Plate 9 The Grand Canyon at Grand Canyon National Park, Arizona. The Colorado River valley, up to 16 km wide at this point, is cut deep into the Colorado Plateau. Photo Chris Park.

Plate 10 Rice production under irrigation agriculture on alluvial flats in northern Vietnam. Photo: Andy Quin.

Plate 11 *Gypsum dunes at White Sands National Monument, New Mexico, USA. Small-scale ripples cover the surface of this dune ridge.* Photo: Andy Quin.

Plate 12 *Glaciated landscape in the Pennine Alps, Switzerland. The Matterhorn is on the left, and the Gornergletscher glacier is in the central foreground. The smooth surfaces of the glaciers and the exposed trim lines on the valley sides are evidence that the glaciers are retreating.* Photo: Alistair Kirkbride.

Plate 13 Construction of a sea-defence wall at Torcross, south Devon, England. The sea-defence works were built in 1980 to protect the village of Torcross from flooding and wave attack after it had suffered extensive damage during previous storms. Photo: Chris Park.

Plate 14 Thorn acacia scrub growing on basalt at the bottom of the rift valley at Magadi, Kenya, Africa. Photo: Philip Barker.

Plate 15 *Beach erosion on Sullivan's Island, Isle of Palms, near Charleston, South Carolina, USA. The protective beach system, previously several metres higher than shown here in 1999, was washed away during Hurricane Hugo in 1989 (see Box 16.43). Photo: Chris Park.*

Plate 16 *Salt precipitate (carbonate) at Salda Golü, South-west Turkey. Photo: Philip Barker.*

Part Four
The hydrosphere

Chapter 12

The Hydrological Cycle

LEARNING OBJECTIVES

When you have finished studying this chapter, you should be able to

- *Understand the relevance of the water cycle to other environmental systems and understand how the water cycle operates at a variety of scales*
- *Outline the main processes involved in the global water cycle*
- *Define what is meant by 'drainage basin' and outline the main processes involved in the drainage basin water cycle*
- *Describe and account for river flow variations over time*
- *Summarise the main features of river networks*
- *Describe the ways in which sediment is moved along rivers, and the landforms associated with river sediments*
- *Appreciate what studies of river channel geometry, pattern and slope indicate about channel equilibrium*
- *Explain the main causes and consequences of changes in channel equilibrium over short, medium and long timescales*
- *Show how and why river deltas form.*

Water is probably the most important natural resource on the planet, and in Part IV we examine the ways in which water flows through the environment and the consequences of these flows and stores for other environmental systems. Water is essential for life, and since the earliest times people have relied heavily on water to drink, obtain food from and use for irrigation. Places where water is too abundant or too scarce pose particular problems for human use, and throughout history there have been numerous conflicts over water rights and access to scarce water resources (the hydropolitics of the River Nile is a good example; see p. 58).

This chapter focuses on the water cycle, river systems and channel equilibrium. However, the movement of water is important in a number of environmental systems. Water resources are considered in Chapter 13, and we examine drylands (where water shortage gives rise to serious problems) in Chapter 14. Ice and snow create distinctive conditions in cold environments, and we look at glaciation and periglaciation in Chapter 15. Most of the water on Earth is stored in the great seas and oceans, which we consider in Chapter 16.

THE HYDROSPHERE

SIGNIFICANCE OF WATER

About 70 per cent of the Earth's surface is covered with water, as we have already seen (see p. 125). Water supports all forms of life on Earth. It occurs as standing water (in oceans and lakes), running water (in rivers and streams), and in the form of rain and water vapour in the atmosphere (Chapter 8). The study of water on Earth is known as *hydrology*.

The hydrological (water) cycle functions at a variety of scales, the two most important of which are the global cycle and the drainage basin cycle. It involves movements of moisture through the environment in its different phases (see p. 291) – as a liquid (water), a gas (water vapour) and a solid (snow and ice). These movements take place

BOX 12.1 IMPORTANCE OF THE HYDROSPHERE

The hydrosphere is the watery part of the Earth's surface, including the oceans, lakes, rivers and water vapour in the atmosphere. Because ecological processes play important roles in redistributing moisture within the environment, and because they are in turn strongly influenced by the availability of water, in practice it is almost impossible to draw a line between the hydrosphere and the biosphere.

Water vapour is an important component of the atmosphere, and it plays a key role in the energy cycle and in the natural greenhouse effect that heats the Earth's atmosphere (see p. 238). Most of the important biogeochemical cycles (p. 86) depend on water movements and stores within the hydrological cycle. Water also provides an important medium through which pollutants including acid rain (see p. 251) are dispersed through the atmosphere, through the oceans (Chapter 16) and along river systems across the continents (see p. 364). Many of the weathering processes require moisture in one form or another, and water movements strongly influence most of the erosion processes (see p. 210).

BOX 12.2 MAIN PROCESSES THAT DRIVE THE HYDROLOGICAL CYCLE

The cycle is driven by three main sets of processes:

1. evaporation and evapotranspiration (p. 291), which move water from the Earth's surface to the atmosphere;
2. precipitation (p. 298), which moves water from the atmosphere to the Earth's surface; and
3. air movement, including winds and weather systems (pp. 280 and 307), which redistributes water from place to place within the atmosphere.

between the atmosphere, lithosphere and biosphere, and in this way the water cycle integrates most of the other important environmental systems and strongly influences rates, patterns and processes in them. Earth is not the only planet to contain water (see p. 109), but conditions on Earth are particularly suitable for the continuous recycling of water, which in turn drives many other important environmental systems (Box 12.1).

THE GLOBAL WATER CYCLE

The hydrological cycle involves the continuous recycling of water between the atmosphere, land and oceans. Different processes redistribute water within and between these three realms. In the atmosphere, vertical and horizontal air movement and turbulent mixing transfer moisture from place to place, whereas large-scale currents transfer water in the oceans. Rivers and glaciers transfer water from land to the oceans. The cycle is driven by three main sets of processes (Box 12.2).

The basic structure of the global water cycle (Figure 12.1) is quite simple. Water is evaporated from the oceans, seas, lakes, rivers and vegetated land areas, and it becomes part of the atmospheric store of water vapour. Global and regional wind systems redistribute the water vapour across the Earth's surface. Condensation (see p. 291) creates clouds and precipitation, and the latter brings water back to the surface, where it enters the soil or flows directly into rivers or lakes. Rivers transport water from land surfaces into the oceans. Naturally, much of the precipitation falls directly over seas and oceans, effectively short-circuiting the land phase of the hydrological cycle.

While the basic structure of the global water cycle is clear, it has proved difficult to predict with any degree of certainty how the cycle is likely to respond to human impacts. Complex computer models are being developed to help with this task, but they require more and better observational data on many variables – including rainfall, evaporation, evapotranspiration, snow cover, snowmelt runoff and soil moisture – than are currently available.

The total volume of water in the global cycle is estimated at about 1,384 million km^3. The cycle is complex, with many different pathways, branches and stores. At any one point in time most of the water in the cycle is held in storage, mostly as water in seas and oceans (Figure 12.2; Table 12.1). Just over 2 per cent of the total water in the global cycle is fresh water, and most of that is locked up as polar ice caps and in glaciers. If all of the ice were to melt, it would release enough water to keep the world's rivers flowing at their normal rates for up to 1,000 years.

Rivers play a vital role in the global water cycle, because they drain water from the land into the seas and oceans. River flow rapidly carries water downhill to the sea, but rivers store hardly any water at the global scale (see Table 12.1). They are rapid conduits rather than stores.

Lakes store more water than rivers, and for longer. About two-thirds of all the fresh surface water on Earth is

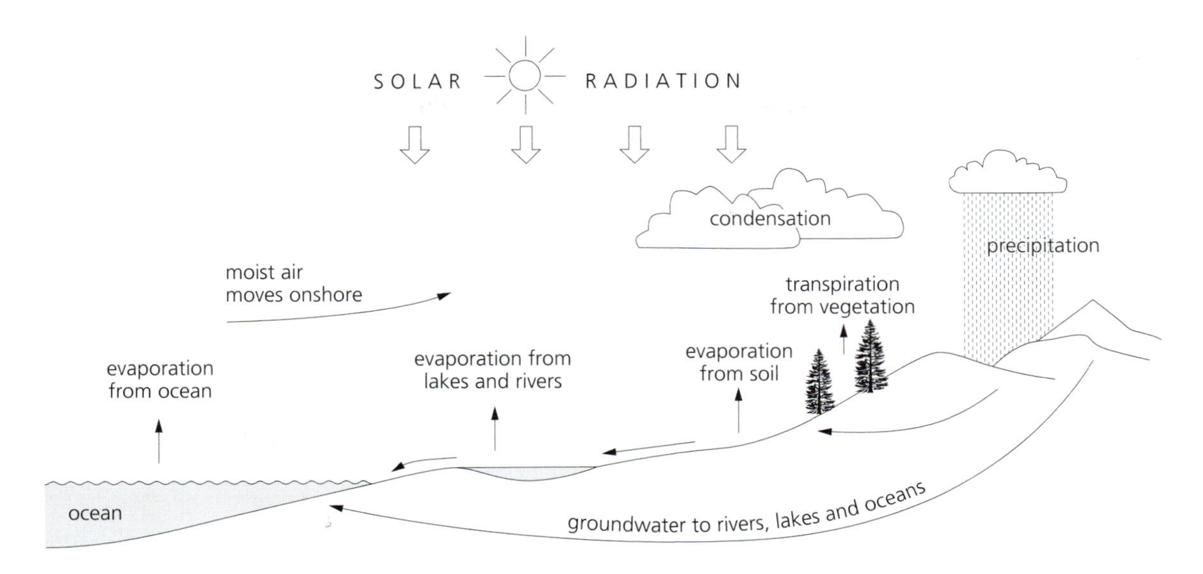

Figure 12.1 *The global hydrological cycle. The global cycle includes stores (in groundwater, surface water, the oceans and the atmosphere) and processes that transfer water between the stores.*

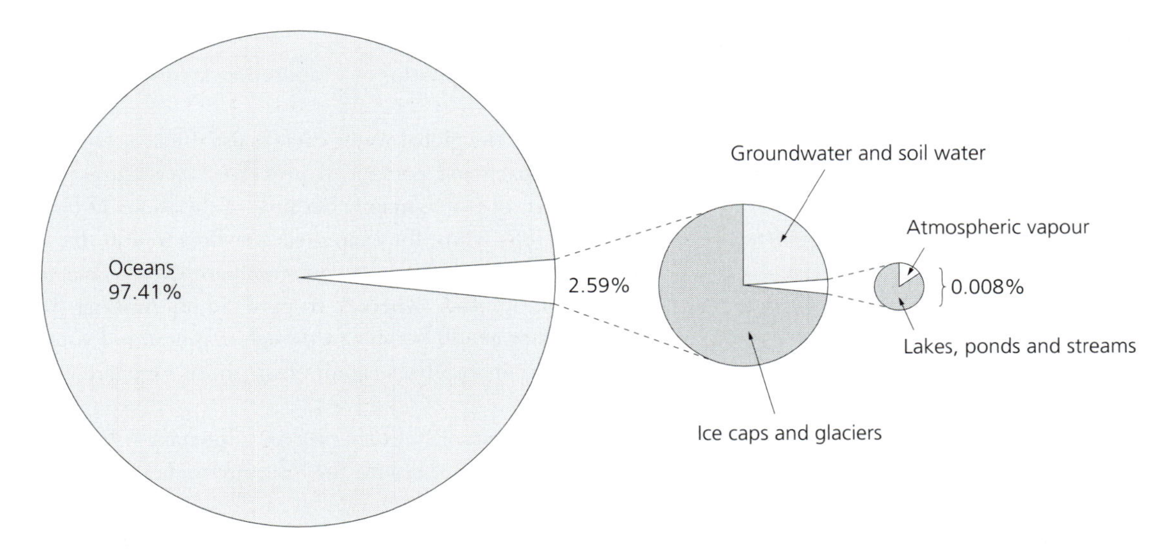

Figure 12.2 *Principal natural stores of water. The vast majority of the water on Earth is stored in the oceans, and most of the fresh water is stored in ice caps and glaciers.*

Source: Based on Table 10.1 in Marsh, W.M. and J.M. Grossa (1996) Environmental geography: science, land use and Earth systems. John Wiley & Sons, New York.

stored in about 250 large lakes, although these are coming under increasing pressure from over-use, pollution and drainage.

Water moves between the various stores, and these movements (fluxes) keep the hydrological cycle going. But different parts of the global cycle operate at different timescales, and water remains in some stores much longer than others (Table 12.2). Oceans, ice caps and rocks are long-term stores, whereas rivers and the atmosphere are short-term stores that play vital roles in recycling water. In effect, relatively small amounts of water are recycled very quickly, and most of the water is locked away in long-term stores.

Table 12.1 Major natural stores of water in the global hydrological cycle

Store	*Proportion of total (per cent)*
Oceans	97.41
Ice caps and glaciers	1.9
Groundwater	0.5
Soil moisture	0.01
Lakes and rivers	0.009
Atmosphere	0.0001

Table 12.2 The global water cycle – storage amounts and times

Store	*Size (km³)*	*Typical residence time*
Plants and animals	700	1 week
Atmosphere	13,000	8–10 days
Rivers	1,700	2 weeks
Soil	65,000	2 weeks – 1 year
Lakes, reservoirs, wetlands	125,000	years
Rock (groundwater)	7,000,000	days – thousands of years
Ice	26,000,000	thousands of years
Oceans	1,370,000,000	thousands of years

MASS BALANCE

The global water cycle is a closed system (see p. 70), because the Earth does not exchange moisture with the rest of the universe. In effect, therefore, the total amount of water in the global cycle remains constant over time. Over the short term, such as on a year-to-year basis, most parts of the global cycle appear to be reasonably stable. Global rates of precipitation and evaporation, for example, vary relatively little from one year to the next unless there has been a short-term disturbance to the global atmospheric system (such as a major volcanic eruption). Superimposed on this background stability are short-term and seasonal variations in the water cycle, which largely reflect regional weather changes.

This balance is brought about and maintained by variations from place to place in the detailed dynamics of the global cycle. Over the oceans, for example, evaporation exceeds precipitation, but the situation is reversed over most land areas (except from some deserts and drylands). Overall, global evaporation balances global precipitation.

BOX 12.3 ICE AGES AND WATER STORAGE

The distribution of water within the cycle can change over time if more or less water is stored in the different storage reservoirs in the system. This happens during an ice age, for example, when there is a significant increase in the amount of moisture stored in ice caps and glaciers. As the ice caps grow by accumulation, sea level falls as water is transferred into ice stores (see p. 454). Conversely, when the ice age ends and ice sheets melt, sea level rises as a direct result. Hence the concern about sea-level rise triggered by global warming (see p. 459).

If it did not, the amount of moisture in the atmospheric store would increase or decrease over time.

Over land, runoff along rivers balances precipitation (input) and evaporation (output), so the simple mass balance equation is:

$$\text{precipitation} = \text{evaporation} \pm \text{runoff}$$

While the global water cycle is usually in a state of balance, the pace and pattern of processes vary a great deal from place to place, mainly because of variations in climate (see p. 335). Thus, for example, river flow in hot, dry areas like Australia – where precipitation barely exceeds evaporation – is limited, whereas tropical South America has much higher runoff because, although evaporation losses are very high, it receives significantly more precipitation (Table 12.3).

The oceans clearly play an important role in the global water cycle, because they occupy such a large proportion

Table 12.3 Average water balance of land areas, by continent

Continent	*Annual precipitation (cm)*	*Annual evaporation (cm)*	*Annual runoff (cm)*
Africa	69	43	26
Asia	60	31	29
Australia	47	42	5
Europe	64	39	25
North America	66	32	34
South America	163	70	93
Total land areas	73	42	31

of the Earth's surface, store most of the water in the cycle and are the major source of evaporation and atmospheric water vapour.

INDUCED CHANGES

Human activities can alter the water cycle in many different ways, either by accident or by design. The movement of water vapour across oceans and continents can be altered by:

- air pollution, which can promote global warming and can significantly increase the availability and change the distribution of condensation nuclei; and
- weather modification programmes, such as cloud seeding with silver iodide crystals, to encourage rainfall over dry places.

Evaporation rates and patterns can be altered by changing ground surface conditions (such as through urban development or reservoir construction).

River runoff can be changed directly by increasing the length or density of the channel network (for example by installing drainage ditches, field drains and artificial channels) or by decreasing them (such as by direct building operations and channel removal or infilling). The size and hydraulic efficiency of river channels can be altered by channel clearance, realignment and diversion, and by direct engineering works. Infiltration can be influenced by changing ground surface materials, including:

- stabilising the bed and banks of river channels with concrete;
- changing local land use from natural vegetation to agriculture or buildings; and
- installing irrigation schemes and field drains.

Groundwater can be affected by pumping and abstraction works, which lower the local water table, and by increased inputs related to reservoir development and upstream drainage basin changes.

Flows within the water cycle can also be strongly influenced by changing vegetation – including vegetation removal and clearance, afforestation, cropping and grazing. Water resource management schemes (see p. 388) inevitably change the distribution of water from place to place and over time.

While the broad structure of the global water cycle is clear, and reliable estimates are available of the quantities stored in different parts of the cycle, of the residence times involved and of typical rates of movement between stores, many areas of uncertainty remain (Box 12.4).

DRAINAGE BASINS

THE DRAINAGE BASIN

OPEN SYSTEMS

A drainage basin (also called a river basin or a catchment) is the area drained by a river and its tributaries. The boundary of a basin is defined by the drainage divide (or *watershed*), and the area within the divide is referred to as the drainage area (or *catchment area*). Nearly all parts of the

BOX 12.4 UNCERTAINTIES IN UNDERSTANDING THE WATER CYCLE

An understanding of all the processes and mechanisms in the cycle is still far from complete, and detailed measurements have yet to be made of many of the fluxes involved. Particularly important gaps include the freshwater stores and their role in relation to the global energy budget, water supply and climate.

Uncertainty also surrounds the possible links between the water cycle and global warming. If the predicted changes in climate patterns (see p. 265) are realised, the water cycle will inevitably be affected, both directly and indirectly, via changes in albedo, cloud cover, atmospheric concentrations of water vapour, wind patterns, evaporation rates and condensation processes. Global warming also threatens to promote at least a partial melting of the major Arctic and Antarctic ice sheets (see p. 446), which would also change albedo as well as raising sea level and changing the global water balance. There are likely to be many feedback relationships between the water cycle and global climate change, but these are as yet not clearly understood.

Scientists are trying to plug some of these gaps in understanding the global water cycle by improved instrumentation, detailed monitoring and analysis of observed patterns and trends. Until we have a better understanding of how water cycles between the oceans, atmosphere and biosphere, it will be impossible to make more reliable predictions about how global warming is likely to affect different environments and the people who live there.

BOX 12.5 DRAINAGE DIVIDES

The area of higher ground between adjacent drainage basins (Figure 12.3) is known as a *drainage divide* or *watershed.* This can be defined both in the field and on maps, and it determines which way water will flow when it reaches the ground (and thus which river system it will become part of). In North America, the Continental Divide separates rivers that flow westwards to drain into the Pacific Ocean and those that flow eastwards and southwards into the Atlantic Ocean and the Gulf of Mexico.

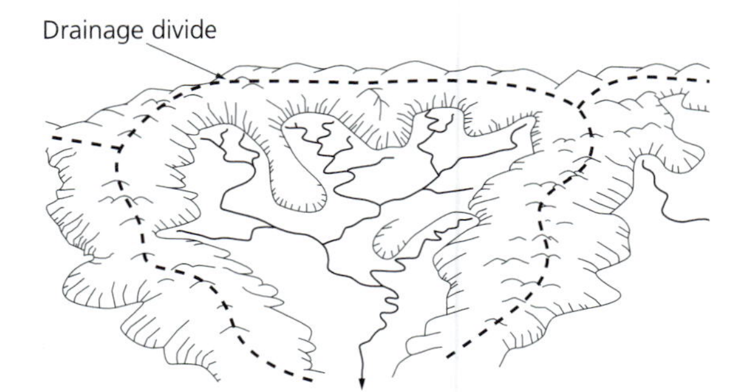

Figure 12.3 *The drainage divide. A drainage divide separates adjacent river systems, and it runs along the high ground (interfluves) between rivers.*

Plate 12.1 *The Continental Divide in the USA, which marks the boundary between river systems that drain into the Atlantic Ocean and those that drain into the Pacific Ocean. The divide is seen here at Milner Pass in the Rocky Mountains in Colorado, but it can be traced through from Alaska in the north to Cape Horn at the southernmost tip of South America.*
Photo: Chris Park.

Earth's surface belong to a river system, so drainage basins fit together spatially rather like the pieces of a jigsaw. Drainage basins are convenient units for sustainable development, land-use planning and natural resource allocation because they represent definable units on the ground that have real environmental relevance.

Drainage basins are open systems through which water and sediments are transported. Material and energy are input into the drainage basin system in the form of rainfall, weathered bedrock and sunlight, and they are output as river flow and sediments (particulate and in solution). Some of the material – such as floodplain deposits and lake water – is stored within the system, at least temporarily.

DRAINAGE BASIN MANAGEMENT

Drainage basins are convenient and logical units for managing water resource use, and for environmental planning in general, because they represent meaningful subdivisions of the land surface that are relevant to most environmental systems. In recent decades, attention has begun to focus on developing co-ordinated plans for drainage basins in order to optimise the use of available resources and minimise the environmental impacts of resource use.

Initially, the focus was on *integrated* river management, designed to reduce conflict between different users of drainage basins. The Connecticut River management programme, established in the early 1990s, illustrates a typical two-stage approach. Stage one involved compiling a rivers inventory to identify uses and characteristics of specific rivers, critical river issues and river segments of special significance. This created a state-wide database that could then be used in stage two to devise a river management strategy.

One of the best-known and earliest examples of integrated drainage basin management is the Tennessee Valley Authority (TVA) in the USA (Box 12.6).

Special water agreements have been adopted in attempts to reduce conflicts between different users of some large river systems. Such agreements are particularly useful in large rivers that cross national boundaries, such as the 1929 and 1959 Nile Water Agreements (see p. 63). Similar agreements have been introduced in large river systems in individual countries, with mixed success. Rational allocation of land, water and environmental resources in

BOX 12.6 THE TENNESSEE VALLEY AUTHORITY

The Tennessee Valley Authority (TVA) was created in 1933 to control and develop the Tennessee River and to improve socio-economic conditions for people living in the region. From the outset, the TVA sought an integrated approach to managing the drainage basin and its natural resources, with three primary objectives:

1. *Flood control*: this is being tackled by the co-ordinated management of a series of 47 dams and reservoirs in the basin.
2. *Improved navigation*: river management, including some realignment and dredging, is increasing the proportion of the 1,050 km river that can be navigated.
3. *Hydro-electric power generation*: the dams and reservoirs are used for power generation as well as flood control, and the system now has enough capacity to meet the needs of around 8 million people and nearly fifty very large industrial users.

The integrated development of the Tennessee valley region is widely judged to have been very successful. Not only have the primary objectives been met, but important natural resources in the drainage basin – including soils, forests, wildlife, fisheries and landscapes – have been conserved and enhanced at the same time as the local population has grown and industry has developed further in the region. Some important problems remain, most notably the increasing water pollution in the basin, which is partly a result of the increased number of lakes with little throughput of water.

The TVA has faced significant challenges in achieving this much, and in doing so it has gained valuable experience in adapting to changing circumstances and environmental uncertainty. This type of experience may prove very valuable in other areas confronted with the challenges of how best to adapt to potential climate change.

the Murray–Darling basin in Australia, for example, is essential to the development of the region, particularly because this dryland area is very susceptible to salinity problems. The 1914 River Murray Waters Agreement proved inadequate for solving water resource (particularly quality) problems within the basin, and in 1987 it was replaced by the Murray–Darling Basin Agreement, which sought to promote and co-ordinate the management of all natural resources within the basin.

BOX 12.7 SUSTAINABLE MANAGEMENT OF RIVER SYSTEMS

A high value is placed on:

- preserving the natural functions that water provides (for example in ecosystems and as an important flow pathway in the global biogeochemical cycles);
- taking into account the different uses of water by people;
- reducing demand for water and minimising wasteful uses;
- minimising the environmental impacts of water resource use;
- integrating water resource considerations into broader decisions about sustainable economic development.

Since the late 1980s, the focus has shifted towards the *sustainable* management of river systems (Box 12.7), reflecting the wider emergence of the sustainable development paradigm (see p. 26). This new emphasis is part of the strategy being adopted by the United Nations Environment Program and promoted since the 1992 Rio Earth Summit (see p. 13).

For basin-wide sustainable development to be successful in individual countries, it will require new institutional and legislative frameworks for decision making. It will also require more widespread adoption of such planning tools as environmental appraisal and strategic environmental assessment, and more widespread use of economic instruments. The ultimate objective is 'prevention rather than cure' – making sure that problems do not arise in the first place, rather than tackling them when they do.

THE DRAINAGE BASIN WATER CYCLE

The global water cycle is a good example of a nested hierarchy (see p. 71), because it can be studied at a variety of scales. From a human perspective the most important scale is that of the individual drainage basin, which is the most useful unit for water resource management (p. 355).

The fate of the precipitation that falls over land areas is determined largely by the ground surface conditions. Some precipitation falls directly into river channels, lakes and other water bodies. This is called *direct precipitation* and is rapidly transferred back to the oceans along rivers.

INTERACTION WITH VEGETATION

Precipitation that falls over vegetated ground interacts with the vegetation in a number of ways. A certain proportion will land on the leaves and branches of the plants, as *interception*. This can be evaporated back to the atmosphere as water vapour. If the weight of collected water exceeds the supporting capacity of the leaves, they bend downwards and the water drips off them on to the ground below. Water can also be knocked off leaves by the direct impact of other water droplets. Another variable amount will trickle down the stems of plants as *stemflow*, and although this slows down the water most of it will eventually reach the ground surface below the vegetation.

The rest of the precipitation will fall directly through the vegetation to reach the ground surface beneath, as *throughfall*. This fraction is clearly related to the type, density and structure of the vegetation it encounters. More precipitation will pass as throughfall through scattered woodland than through dense forest, for example.

INTERACTION WITH THE GROUND SURFACE

Precipitation reaches the ground, either directly or after interacting with vegetation, and it can then have several fates. On a sloping surface with saturated soil, or on a surface with an impermeable horizon in the underlying soil, much of the water will flow across the ground surface as *overland flow* (or *sheet flow*), its path of travel being directed towards the lowest parts of the ground surface. In this way, the water can reach the river channel and contribute to surface runoff. Not all parts of a basin produce overland flow, particularly if rainfall intensities are low and infiltration rates are high. Field studies have shown the importance of *contributing areas* within a basin (usually at the base of a slope, close to the channel) which can expand and contract during individual storms.

Much of the water that has not directly evaporated from the ground surface drains into the soil (Box 12.8).

BOX 12.8 WATER IN SOIL

Much of the water that has not directly evaporated from the ground surface drains into the soil by *infiltration*, and it becomes part of the soil water store. This soil water store has great ecological and agricultural importance, because plants take in water (and nutrients in solution) from the surrounding soil via root osmosis. If soil conditions are suitable, much of this water will continue to move downwards through the soil horizons by *percolation*, and it may eventually enter the underlying bedrock and contribute to the groundwater store (aquifer). Alternatively, and particularly on sloping surfaces, soil water will move downslope either through the pore spaces in the soil structure (*interflow*) or by flowing along impermeable horizons in the soil (*throughflow*). The water that moves downslope beneath the surface may appear in a seepage line or *percoline* (hollow) at the base of the slope.

GROUNDWATER

Water can percolate into and through rocks that are porous (they have pores) and permeable (water can pass through them), such as sandstone, where it accumulates as groundwater (Figure 12.4). The layer of rock through which the water percolates is known as an aquifer, and the water can be extracted for use by humans, usually by drilling wells. The water table is the boundary within the rock between the unsaturated zone and the saturated zone. Springs appear where an aquifer outcrops at the ground surface, and these can contribute runoff to streams and rivers.

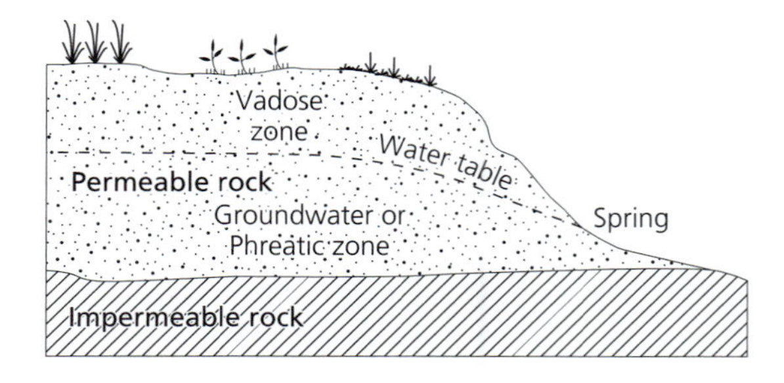

Figure 12.4 *The water table. The water table is the upper level of groundwater in permeable rocks beneath the ground surface, below which the rocks are saturated.*

Source: Figure 1.9 in Buckle, C. (1978) Landforms in Africa. Longman, Harlow.

Most groundwater near the surface moves slowly through the aquifer, while the water table stays in the same place. The depth of the water table reflects the balance between the rate of infiltration (recharge) and the rate of discharge at springs, rivers or wells. The water table usually follows surface contours, and it varies with rainfall.

Worldwide, between about 10 and 30 per cent of river flow is accounted for by groundwater, and groundwater provides about 6 per cent of the water and around 50 per cent of the dissolved sediment that rivers input into seas and oceans.

Groundwater levels can vary a great deal over time, and this can create problems for the stability of deep building foundations and flood tunnels. This seems to be happening in the London Basin, in England, where plans are being made to control the rising groundwater in the chalk aquifer by abstracting water from selected sites.

BOX 12.9 FLOW VELOCITY

Flow velocity is the speed at which water moves along a river channel, and it is usually measured in m s^{-1}. Velocity varies both over time and from place to place along a river, and it is controlled mainly by:

- *Slope*: water flows faster down steep slopes.
- *Discharge*: velocity increases as discharge increases.
- *Channel shape*: water flows faster through an efficient channel cross-section, and velocity is reduced by frictional resistance with rough channel beds and banks.
- *Size and shape of sediment in the channel bed and banks*: smooth channels allow water to flow rapidly, and rough channels slow down water movement by frictional resistance.

RIVER FLOW

The throughflow, interflow, percolation water and groundwater can be released into streams via springs, natural soil pipes and seepage through the bed and banks of the river channel. A number of streams join together downstream to form a river. Water flows down a river, ultimately to the sea. There are two important properties of river flow – velocity (Box 12.9) and discharge.

Discharge: this is the amount or volume of water that flows through a given cross-section of a river in a given unit

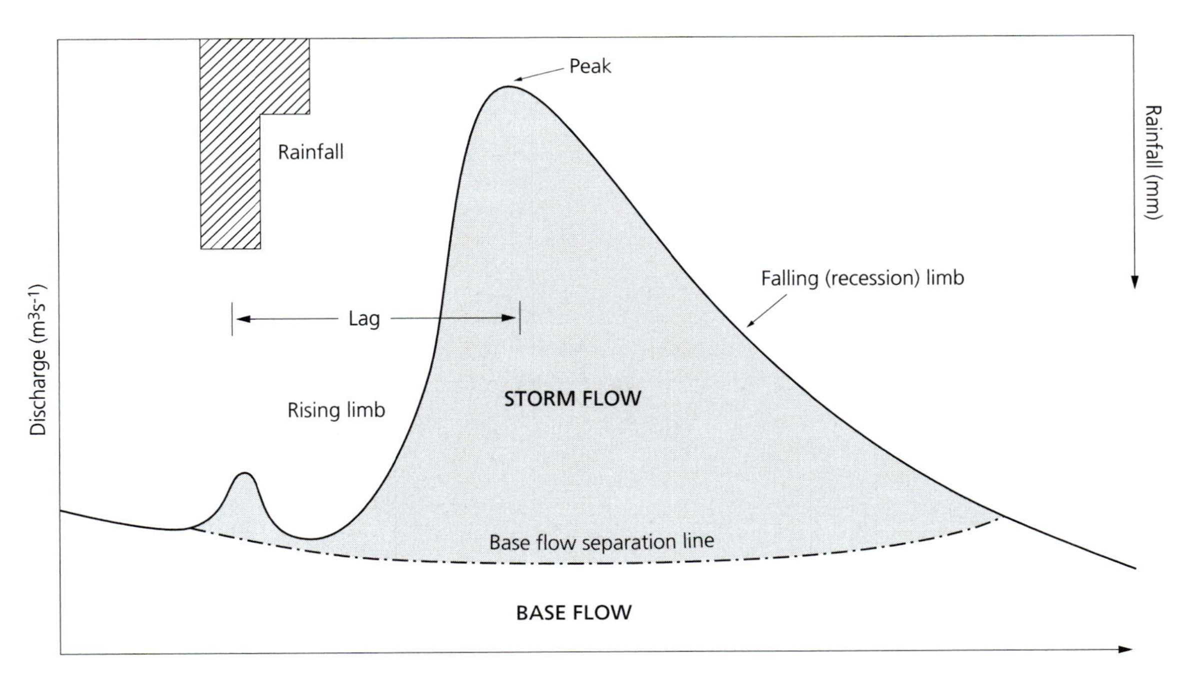

Figure 12.5 *The storm hydrograph. A storm hydrograph displays the variations in river discharge over time in response to the rainfall input by an individual storm event. Storm flow is the rapid runoff related directly to the storm input.*

BOX 12.10 FACTORS THAT CONTROL RIVER FLOW

River flow is controlled by many different factors, including:

- *Area of the drainage basin*: all else being equal, larger basins have larger average flows, but they respond more slowly to storm events; larger basins have flatter storm hydrographs (because water arrives from tributaries in a more staggered manner).
- *Shape of the drainage basin*: long thin basins tend to have flatter hydrographs than more circular basins, again because water arrives from tributaries in a more staggered manner.
- *Rainfall totals*: these inevitably affect mean and peak flows; rivers draining wetter areas have higher unit discharges (discharge per unit of drainage area, usually km^2) than those that drain dry areas.
- *Rainfall intensity*: when intensity is high the ground often cannot absorb all of the rain that falls, so a large proportion quickly drains off as overland flow. As a result, the river responds quickly and has a shorter response time.
- *Surface conditions in the drainage basin*: vegetation intercepts rainfall, and this slows down its movement into rivers and allows at least some to be evaporated before it ever reaches the river.

of time. It is usually expressed in $m^3\ s^{-1}$ (cumecs, short for cubic metres per second), and it can be estimated by multiplying the cross-sectional area of the channel (width × depth in m^2) by flow velocity through that cross-section (in $m\ s^{-1}$). Hydrologists usually refer to discharge as Q (or Q_w to denote water discharge, as distinct from Q_s, which is the sediment load).

Discharge varies over time in a river system, and these patterns can be plotted as hydrographs (Figure 12.5), which are plots of variations in discharge over time. Mid-latitude rivers usually have relatively low flows in summer and higher flows in winter, and this defines their annual flow regime or hydrograph. Superimposed on the annual cycles are the short-term responses of river flow to the rainfall associated with individual storms, and these can be plotted as storm hydrographs.

Continuous measurements are now collected from many large river systems. The analysis of these streamflow records is important in the assessment of hydrological hazards (flood and droughts), in evaluating water resources and in determining how climate change and land-use change are affecting river systems. Flow records for the River Nile show how the river has been affected by water management schemes (see p. 62), and detailed analysis of low flow sequences shows the influence on river flows of climate change. River flow is determined by a number of factors (Box 12.10).

Discharge increases downstream in most rivers because of increasing drainage area. As a result, the erosive power of rivers tends to increase downstream, they can carry much larger sediment loads, they flow faster, and they flood more frequently.

RIVER NETWORKS

Rivers serve many functions that are important to people. They supply water and have long been used for transporting people and goods. Many industries and activities (including agriculture) use water and produce liquid wastes, which are often released directly into rivers. Rivers also serve as international boundaries, although natural river channel changes sometimes complicate international diplomacy!

In this section, we explore how river networks develop, how they are affected by geological controls and how they can be described and compared.

RILLS AND GULLIES

Running water is by far the most efficient agent of erosion (see p. 211), and overland flow running down a hill slope can dislodge particles of soil and sediment, start to erode simple small-scale channel-like features and transport sediment and deposit it downslope or down-valley. Vegetation cover can protect underlying material from erosion, and bare areas are much more susceptible to erosion by running water.

Rills are small channels formed by soil erosion on hill slopes and gently sloping surfaces. They commonly develop on bare spoil heaps and in recently ploughed fields, often at the end of summer, when there is no vegetation cover to protect the soil surface from erosion. Rills often form very dense networks, and they usually disappear in winter (either because the soil is ploughed or soil creep and frost destroy the rills).

Once running water is confined within a channel, even a very small one like a rill, it becomes much more effective and more efficient at eroding and transporting sediment and at shaping landforms. If rills survive long enough, they

BOX 12.11 GROWTH OF CHANNEL SYSTEMS BY POSITIVE FEEDBACK

River channels grow from small rills by a series of processes involving positive feedback (see p. 79) in which:

- Water flows towards the channel (because its bed is lower than the surrounding surface).
- The depth of water flowing in the channel increases (because it contains more water).
- Flow velocity increases (because it is deeper, so frictional resistance decreases).
- Erosion by flowing water is concentrated within the channel (because of the higher velocity).
- The base of the channel is lowered, and it grows wider (as a result of erosion).
- More water is encouraged to flow into the evolving channel, and the chain reaction continues.

promote further erosion and more and more water concentrates into the linear depression to form gullies. Gullies, in turn, can grow into river channels, and if river channels continue eroding long enough they can create river valleys. A positive feedback loop is created (Box 12.11) that leads to further growth of the channel system.

STREAMS AND RIVERS

Flowing water usually concentrates into distinct watercourses or channels. Small channels tend to be called streams, and larger ones are called rivers. There is no particular size threshold to distinguish a stream from a river. Streams and rivers carry water, but not always continuously (Box 12.12).

RIVER NETWORK PATTERNS

Rivers form distinct networks, like the branches of a tree. Small streams (tributaries) join the main river and as rivers develop they generate interlocking networks of tributaries within a drainage basin. Major river systems (Table 12.4) drain large areas of land, have many tributaries, usually have large discharges and carry large sediment loads.

The most common drainage pattern is the dendritic network, which from above resembles the trunk and branches of a tree (Figure 12.6). Tributaries flow into the main rivers at quite sharp angles. Over time the network grows by headward extension, in which the outermost tributaries extend by erosion into the hill slopes of the drainage divides in the network and between it and adjacent networks.

BOX 12.12 TYPES OF RIVER FLOW

Three types of river flow are common, and rivers tend to be given the same descriptive names:

1. *perennial flow (or river)*: these flow continuously, although discharge can vary a great deal over time. Discharge usually increases downstream. Most rivers, particularly in middle latitudes, are perennial and rarely run dry. Perennial rivers can dry up after prolonged drought, but they start to flow again when regular precipitation is restored.
2. *ephemeral flow (or river)*: these flow only at certain times of the year or during and immediately after a heavy downpour of rain. Many rivers in semi-arid areas are ephemeral.
3. *intermittent or episodic flow (or river)*: these flow discontinuously from time to time, and sometimes from place to place within a river. Desert channels (*wadis*) are episodic.

BOX 12.13 RIVER NETWORKS AND GEOLOGICAL CONTROL

Rivers tend to exploit lines of weakness, such as areas of softer rocks, and this is often preserved in the network pattern long after surrounding rocks have been worn away. This creates two distinctive types of river network:

1. *Antecedent rivers*: in these, an initial network pattern, established in the geological past, has been preserved while the area has been uplifted. Rates of river down-cutting must have been higher than rates of uplift for this to occur.
2. *Superimposed rivers*: in these, an initial network pattern, determined by original rock structures and types, has been preserved as the river has cut down through underlying rocks, thus superimposing the original pattern on underlying structures and rock types.

Table 12.4 The world's ten longest rivers

River	*Continent*	*Length (km)*
Nile	Africa	6,648
Amazon	South America	6,275
Mississippi–Missouri–Red Rock	North America	6,210
Ob–Irtysh	Asia	5,569
Yangtze	Asia	5,519
Hwang Ho	Asia	4,670
Congo (Zaire)	Africa	4,666
Amur	Asia	4,508
Lena	Asia	4,269
Mackenzie	North America	4,240

Geology often controls the overall shape of a river network, because it can strongly influence rates and patterns of erosion. Dendritic patterns develop on uniform rock types where there is no strong structural control (such as fault lines or folds), but most other patterns (Table 12.5) reflect geological influences (Figure 12.6).

RIVER NETWORK GEOMETRY

Since the 1950s, geologists have viewed drainage basins as open systems and sought ways of describing, comparing and explaining basins and the river networks in them. These studies focused on equilibrium and adjustment in river systems, and they were accompanied by other studies looking at equilibrium in channel form (see p. 369).

One of the most valuable approaches has been the stream-ordering system introduced originally in 1945 by engineer Robert Horton. This established a way of classifying rivers of different sizes and of examining geometric properties of their networks. It also allowed comparisons within and between individual rivers, and comparisons with other natural networks (such as the veins on a leaf). The smallest streams in a river network, which have no tributaries, are designated first-order streams (Figure 12.7). When two first-order streams meet, they form a second-order stream; the junction of two second-order streams defines a third-order stream, and so on. Every river system can be ordered in this way.

Analysis of the geometry of a large number of river networks, based mainly on maps, has suggested a series of so-called 'laws of drainage basin composition' (Box 12.14).

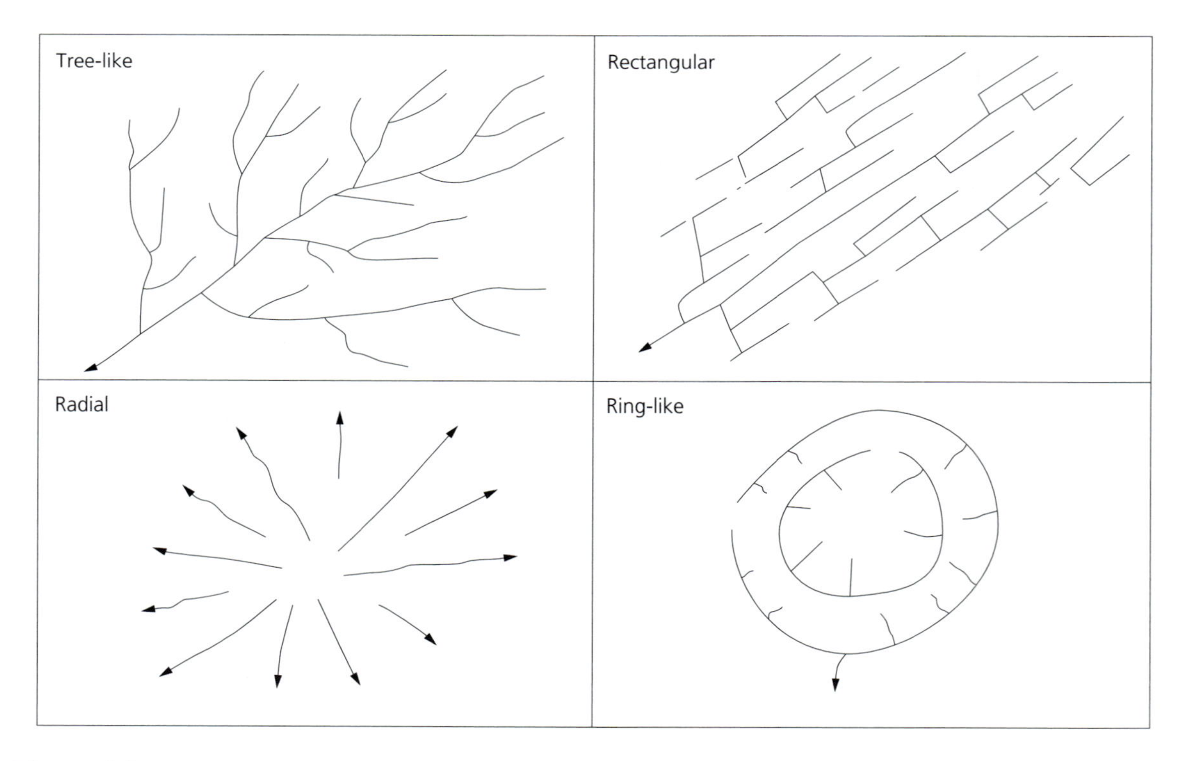

Figure 12.6 *Some common river network patterns that are controlled by geology.*
Source: Figure 1.9 in Dury, G.H. (1981) Environmental systems. Heinemann, London.

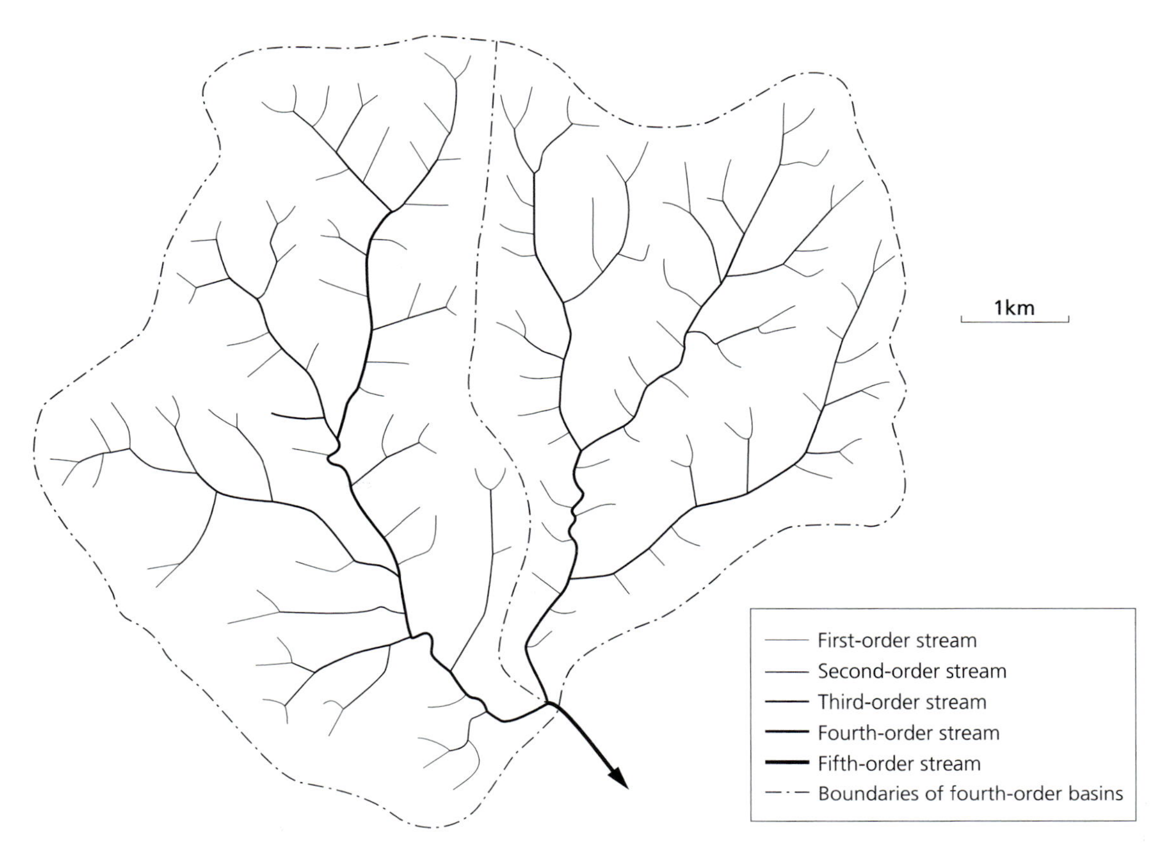

Figure 12.7 *Stream orders. This system of ordering river networks was developed by Robert Horton in 1945. See text for explanation.*

Source: Figure 7.3 in Dury, G.H. (1981) Environmental systems. Heinemann, London.

Table 12.5 Major types of river network pattern controlled by geology

Type	*Description*
Trellis	Common in areas where there are rocks of different strengths (thus resistance to erosion) and in areas with regular series of folds (anticlines and synclines). Tributaries join at high angles, often approaching right angles
Rectangular	Similar to the trellis pattern, but here the pattern is largely controlled by faults or joints that intersect at high angles. Rock dip and strike are relatively unimportant
Annular	This pattern typically develops on a dome, where concentric outcrops of rocks of different resistance to erosion are exploited by river erosion
Radial	Rivers radiate outwards from a central high point, which might be the dome of a uniform rock outcrop or perhaps the centre of a volcanic cone
Centripetal	This is the opposite of a radial pattern, and here the rivers drain in towards the centre of a basin, like the spokes of a wheel
Parallel	Small rivers sometimes flow parallel to one another over a sloping surface of uniform rock resistance
Deranged	There is no obvious pattern in a deranged river network, which is sometimes found in areas of low relief, low slope and large sediment loads

BOX 12.14 THE LAWS OF DRAINAGE BASIN COMPOSITION

These regular patterns (not physical laws) are suggested by the analysis of river network geometry:

- There will be more first-order streams in a drainage basin than all other stream orders combined.
- Stream length increases with increasing stream order.
- Size of drainage area increases with increasing stream order.
- Average gradient decreases with increasing stream order.
- River discharge increases with increasing stream order.

They tell us, for example, that as we progress downstream in a large river system, we will encounter fewer tributaries of higher order, we have to travel increasingly far to move up from one order to the next, the overall drainage area grows, slope decreases and discharge increases. Studies of this type are important, because network structure determines how flood waters collect in the channels, which in turn determines the magnitude and frequency of river flooding.

The nature of a river network varies a great deal from place to place (Box 12.15). This mainly reflects differences in:

- precipitation (amount, intensity and distribution)
- surface material
- slope
- vegetation cover.

BOX 12.15 DRAINAGE DENSITY AND STREAM FREQUENCY

One useful measure of network structure is drainage density, which is the length of river channel per unit of drainage area (km km^{-2}). High values of drainage density are associated with very dense networks, such as those often found in semi-arid and tropical environments (up to 100 km km^{-2}). Low values indicate less dense networks, such as those in temperate areas like Britain (where typical values are around 3 km km^{-2}).

Stream frequency is a different measure of network geometry, based on the number of streams per unit of drainage area (no. km^{-2}). Dense networks tend to have high stream frequencies.

Both measures reflect the degree of dissection of a particular landscape. Fine-grained landscapes, with high drainage densities and high stream frequencies, tend to develop in areas with erodible soils and rocks and high rainfall. Badland topography, with dense networks of interlocking gullies and ridges, is a classic example.

Coarse-grained landscapes, with low drainage densities and low stream frequencies, tend to develop in areas with resistant rocks and low rainfall.

RIVER SYSTEMS

RIVER PROCESSES

Rivers are the most important agents of erosion and landscape change in temperate mid-latitude areas, and it is important to have some understanding of how they operate. This helps us to appreciate a number of important things about rivers, including:

- how they form and develop;
- how they transport water and sediment from the continents to the oceans;
- how and why they appear to have remarkably similar appearances from place to place (even in strongly contrasting environments);
- how river equilibrium is established;
- how rivers adjust to external change.

TYPES OF FLOW

Water flows downhill across the ground surface and in a channel. The speed of flow depends mainly on three factors (Box 12.16).

Moving water can flow in two very different ways, by laminar and by turbulent flow. *Laminar flow* involves the movement of layers of water, one over the other, with no mixing between them. If a coloured dye is injected into a laminar flow, it flows straight and is not dispersed. Laminar flow is relatively rare in nature, and it occurs when flow velocity and water depth are low. Laminar flow cannot carry or set in motion particles of sediment, because of the lack of mixing and uplift within the moving water.

Turbulent flow is more common, both on hill slopes and in channels. It involves mixing within the moving water,

BOX 12.16 CONTROLS ON SPEED OF WATER FLOW

There are three main controls:

1. *Gravity*: which pulls it downhill and in this sense depends largely on slope. All else being equal, water flows much faster down steep slopes.
2. *Friction*: this is a resistance to flow that is created by contact between the moving water and the stationary material it is flowing on or in (such as the bed and banks of a channel). All else being equal, water flows much faster across a smooth surface.
3. *Properties of the water*: particularly viscosity (effectively its ability to flow smoothly), which depends mainly on temperature.

BOX 12.17 RELATIONSHIP BETWEEN FLOW VELOCITY AND SEDIMENT SIZE AND TRANSPORT IN RIVERS

- The smallest and largest particles require higher velocities for entrainment than particles between about 0.1 and 1.0 mm diameter (because the greater cohesion of very small particles causes them to resist entrainment).
- Higher velocities are required for entrainment, for a given size of particle, than to keep sediment in motion once entrained.
- When velocity falls below the critical level for a given particle size, those particles are deposited and sedimentation (deposition) occurs.

which is caused by turbulent eddies. Turbulent flow can carry and entrain (set in motion) sediment particles.

SEDIMENT ENTRAINMENT

Entrainment is simply the process of picking up and starting to move particles of sediment on a slope or from the bed and banks of a channel. The process is critical in determining where and when erosion will occur, and thus in promoting landscape change. It can happen only in unconsolidated materials – rocks and deposits that have been broken into smaller fragments by some earlier processes. Entrainment cannot occur in solid rock unless erosion processes have first reduced the consolidation of the material.

Running water can be highly effective at entraining particles of sediment. Once entrained, the material can be carried along in the water flow by a variety of sediment transport processes, including sliding, rolling, jumping and suspension.

For entrainment to occur, the moving water must be able to exert a force on the particle that is larger than that particle's force of resistance (which prevents it from moving). The balance of these forces is determined largely by flow velocity, and for particles of a given size there is a critical velocity at which entrainment occurs. The critical velocity differs for particles of different sizes, and geologist F. Hjulstrom constructed a graph that shows how velocity, particle size and sediment behaviour are interrelated (Figure 12.8). The Hjulstrom graph shows three important characteristics of sediment behaviour (Box 12.17).

SEDIMENT TRANSPORT BY RIVERS

River channels are shaped and maintained by the movement through them of water and sediment. The fluvial (river) sediment also produces features within channels, such as point bars and shoals. Erosion of the bed and banks of a river by the running water, which entrains particles, adds to the sediment load supplied to the river from tributary slopes and streams. Rivers transport sediment in three main ways (Box 12.18).

Most rivers transport sediment in all three ways most of the time, although the proportion carried by different processes varies a great deal from river to river and over time within a given river. Rivers that flow across highly soluble rocks, such as limestone, usually have high solution loads. Rivers that flow through fine materials tend to carry large loads of suspended sediment.

On the flat bed of a channel, particles of sediment are entrained and grains start to move when the flow velocity reaches the critical threshold for the size of material present (see Figure 12.8). Flow might initially be laminar, but if the velocity increases it becomes turbulent and particles are suspended in the moving water.

The quantity of particles in motion increases with increasing velocity, up to a limit defined by the *capacity* of

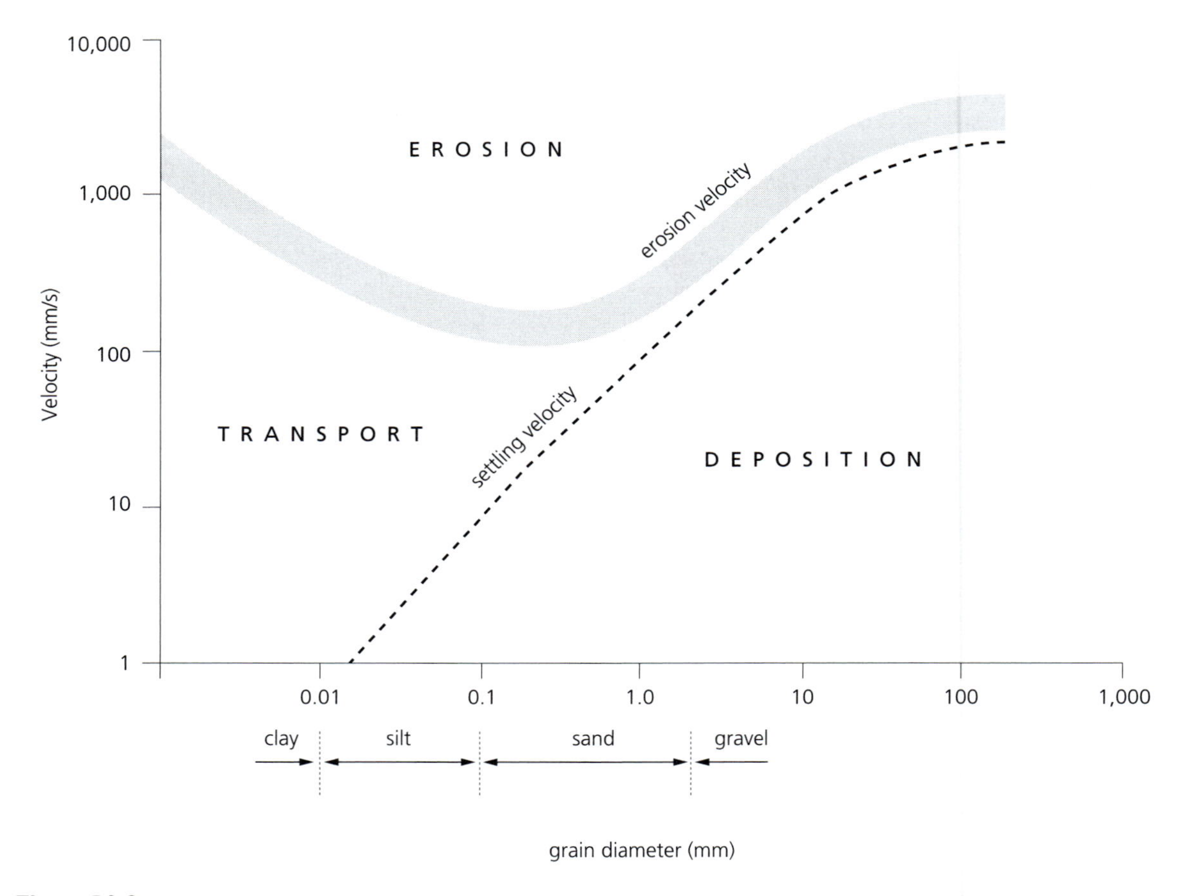

Figure 12.8 *Relationship between flow velocity and sediment size and transport. This widely used graph, based on the work of Hjulstrom, indicates the critical threshold velocities for the entrainment and deposition of sediment particles of different sizes.*

Source: Figure 4.20 in Buckle, C. (1978) Landforms in Africa. Longman, Harlow.

BOX 12.18 SEDIMENT TRANSPORT PROCESSES IN RIVERS

Rivers transport sediment in three main ways (Figure 12.9):

1. *Solution*: this involves the transport of dissolved material in the flowing water, and it is described as the dissolved solids or solute load.
2. *Suspension*: this is the movement of small particles (such as silt and clay), which are held in suspension by the upward thrust of turbulent eddies and carried along by the water movement.
3. *Bedload*: this generally involves the larger particles of sediment, which are too heavy to be suspended in the flowing water and which remain at or near the river bed. Bedload is usually coarser than suspended load, but the size carried in suspension depends largely on flow velocity. Bedload is moved in two main ways. Particles in the coarse sand to gravel size ranges are normally rolled or slide along the bed of the river, by *traction*. Only small particles are moved at low velocities, but quite large particles can be moved at higher velocities (Figure 12.8), particularly over relatively smooth beds. Smaller particles, generally within the sand size range, hop or bounce along the stream bed in the turbulent water flow, by *saltation*.

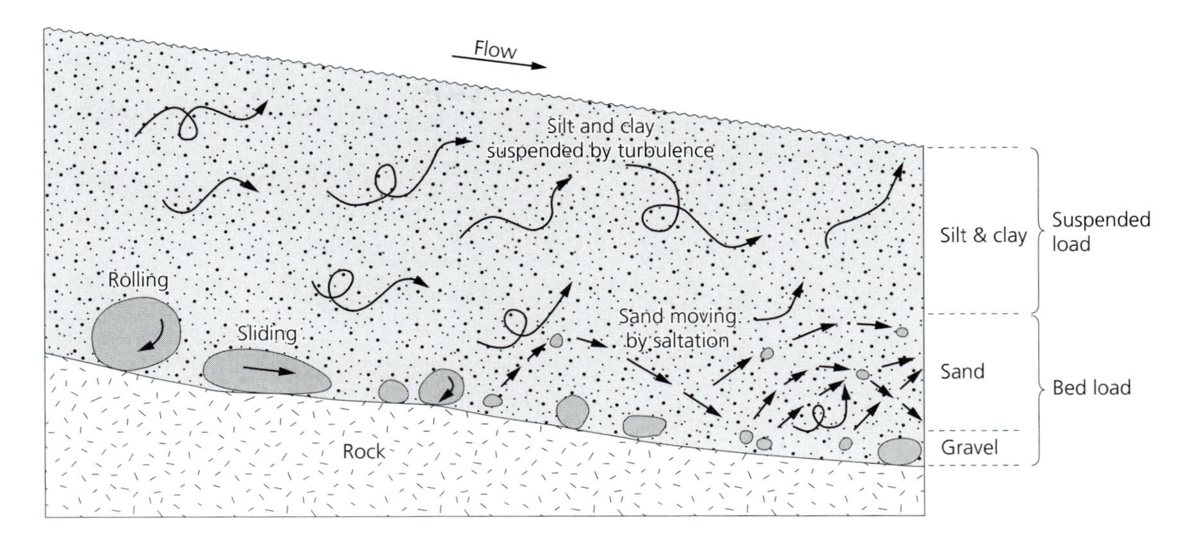

Figure 12.9 *Modes of sediment transport in rivers. Rivers carry chemicals in solution (dissolved), and they carry particles in suspension and as bedload.*

Source: Figure 12.6 in Doerr, A.H. (1990) Fundamentals of physical geography. Wm. C. Brown Publishers, Dubuque, Iowa.

BOX 12.19 THE PHI SCALE OF SEDIMENT SIZES

Sediment size is often expressed using the phi (ϕ) scale, which is based on the logarithm of size rather than size in millimetres (Table 12.6). This scale (see Figure 7.1) is used because it creates a bell-shaped (normal) distribution of sediment, and it includes very large particle sizes. The scale sets zero in the fine sand range, with positive values for finer material and negative values for coarser material.

Table 12.6 Size classification of sediment particles

Sediment class	*Size (mm)*	*Size (ϕ units)*
Boulders	> 256	< −8
Cobbles	64–256	−6.0 to −8.0
Gravel	2–64	−1.0 to −6.0
Sand	0.06–2	4.0 to −1.0
Silt	0.002–0.06	9.0 to 4.0
Clay	< 0.002	> 9.0

the river. The largest size of material that a given river can transport is referred to as its *competence*.

CHANNEL BEDFORMS

Channels cut into fine material, or which transport fine material, display a sequence of different bedforms created by sediment movement as flow increases (Figure 12.10).

At low velocities, near the threshold of grain movement, the bed is usually smooth and flat. If the velocity increases by even a small amount, grains are entrained once the threshold velocity has been reached, and then the bed material is moved around. Initially a series of small ripples develops. If the velocity continues to increase, the ripples grow bigger into bedforms called dunes, which move slowly downstream (much more slowly than river flow velocity). At higher velocities (around 1 m s^{-1}) grains are further redistributed around the channel bed, and the bed sometimes becomes smooth and flat again. Whether it does or does not become flat again, a further increase in velocity causes the development of antidunes, which migrate upstream (as sediment is scoured from the downstream face and deposited on the upstream side) and are accompanied by standing waves. Chutes and pools are formed at the highest velocities.

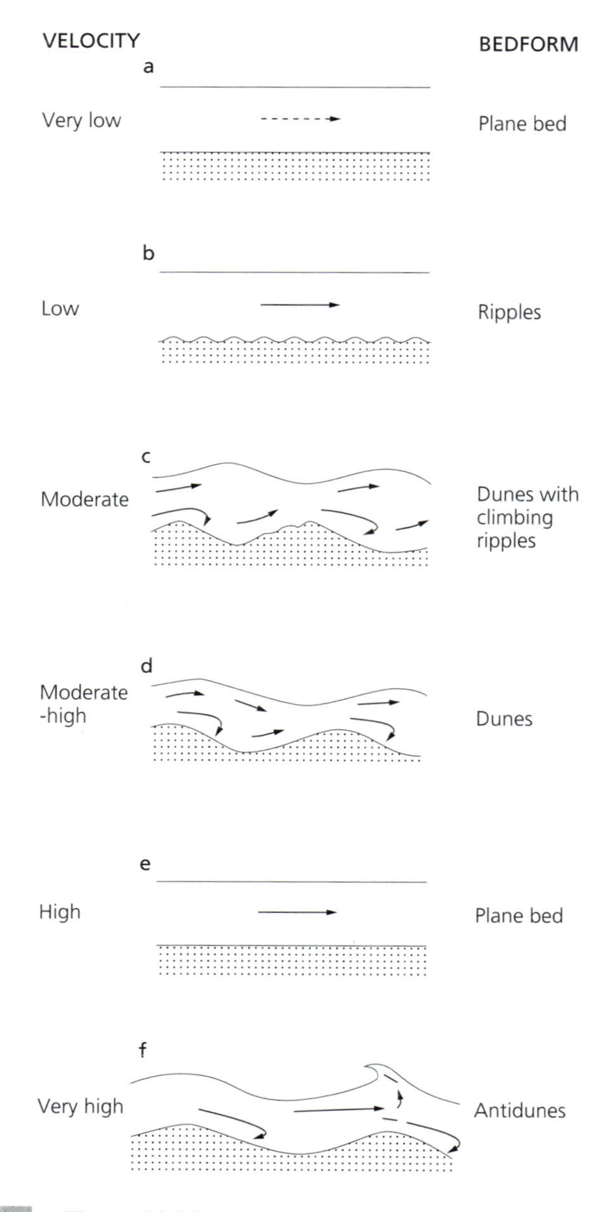

Figure 12.10 *Relationship between flow velocity, bedform and sediment movement. See text for explanation.*

Source: Figure 22.8 in Briggs, D. and P. Smithson (1985) Fundamentals of physical geography. Routledge, London.

BOX 12.20 DEPOSITION OF RIVER SEDIMENT

Deposition of river sediment often happens:

- when floodwaters recede (on the falling limb of a storm hydrograph);
- where stream slope is suddenly reduced (for example where a steep tributary stream flows into the main river valley);
- where a river flows into a stationary water body, such as a lake, a reservoir or the sea.

The sediment deposited by a river in river beds, on flood-plains and in estuaries is fine-grained (consisting of mud, silt and sand). Sediment deposited by a river is known as *alluvium*, in contrast to *colluvium*, which is deposited by hillslope processes. Alluvium is usually very fertile because it contains dissolved nutrients, but alluvial areas like floodplains and deltas often face the risk of flooding.

DEPOSITION

Particles of sediment are dropped and deposited (Box 12.20) when flow velocity decreases below the threshold required to keep particles of that size in motion (see Figure 12.8). There are three main depositional processes in rivers – lateral accretion, over-bank deposition and growth of channel bars.

Lateral accretion: this deposition occurs on meanders (channel bends) as a result of inward flow near the bend, which leads to sediment being dropped on the inside of the bend.

Over-bank deposition: this deposition occurs in rivers that have floodplains (large flat areas on each bank separating the channel from the valley slope). When discharge increases in response to a storm, water level rises in the channel. If discharge increases enough, water spills over the top of the banks and flows out on to the adjacent flood-plain, sometimes causing scouring through entrainment. Low-lying areas (called flood basins) are filled with water, and when these are overtopped the water can eventually flow down the floodplain. Flow velocity usually decreases away from the main channel (because of friction and reduced flow depth), and particles are deposited when the velocity falls below the critical level for a given particle size (see Figure 12.8). Levees are formed where fine alluvial material is deposited on the floodplain by the channel.

Growth of channel bars: coarse gravel bars are deposited on the beds of rivers with steep slopes where the velocity decreases (usually where there is a sharp decrease in slope, such as where a steep tributary stream flows into the main river).

Attempts have been made to establish sediment budgets for individual rivers, and for specific reaches (sections) in rivers. Special tracers – material whose movement through the river system can be monitored – are useful in working out where and for how long sediment is being stored in a river system, and this helps to clarify the processes and dynamics involved. Studies of the movement of radionuclides, from Chernobyl fallout, downstream along the River Severn in Wales and England have shown that about 23 per cent of the river's suspended sediment is stored for some time within the floodplain, and 2 per cent is stored in the channel.

RIVER CHANNEL GEOMETRY

River channels often display evidence of dynamic equilibrium (see p. 80), reflecting a balance between erosion and deposition. This is evident in the three dimensions of channel geometry – notably in cross-sections, plan (meanders) and slope – and at a variety of spatial scales. Over relatively short reaches, perhaps a few hundred metres long, channel width and depth often show systematic variations related to the nature of the channel bed and banks. Over longer reaches, particularly those that include tributary junctions, channel geometry shows adjustment to increasing discharge – width and depth increase, meanders grow bigger, and slope decreases.

BOX 12.21 APPROACHES TO THE STUDY OF RIVER CHANNEL GEOMETRY

Channel cross-sections can be studied in two particular ways – morphometric analysis and hydraulic geometry. Morphometric analysis is based on the analysis of changing channel size, shape and geometry in a downstream direction, regarding the channel as a three-dimensional component of the landscape. Plots of channel width, depth, cross-sectional area, width–depth ratio and similar indicators of channel size and shape, measured at different points downstream along a river, often show systematic changes that can be interpreted as channel adjustments to the discharge of water and the transport of sediment.

Since the early 1950s, geologists have shown great interest in channel equilibrium expressed through the hydraulic geometry of natural river channels. The term 'hydraulic geometry' was proposed in 1953 by Luna Leopold and Thomas Maddock, who showed how width, depth and velocity (the three components of discharge) vary with discharge, both from place to place and over time. Their approach has since been used widely.

AT-A-STATION HYDRAULIC GEOMETRY

'At a station' simply means at a particular cross-section on a river. A cross-section is chosen, and over a period of time a number of measurements are made of flow depth and discharge at that site. As discharge increases, flow depth usually increases (the water gets deeper), and the two variables normally show a fairly close association, which can be graphed (Figure 12.11). Such graphs are conventionally plotted on logarithmic graph paper (or on normal graph paper if both variables are transformed to log values), and this produces a straight-line relationship known as a power curve.

The relationship can be described by an equation, known as a power function, which in this case has the general form:

$$d = cQ^f$$

where d is depth, Q is discharge, c is the intercept (the value of d when Q is zero) and f is the exponent (gradient of the line).

If depth increases rapidly with discharge, the line will be quite steep (so f will have a large value). Conversely, if depth increases slowly with discharge, then f will be small. Studies in many rivers of different sizes and in different geological and climatic settings show that most river channels have a value of f of about 0.4 (Table 12.7). This indicates that most channels have a similar shape, even if the size varies a great deal.

The graph (Figure 12.11) illustrates the relationship between depth and discharge at a specific point along the river channel, and it is called the *at-a-station relationship*.

Similar hydraulic geometry graphs can be drawn to show how water surface width and flow velocity change as discharge varies at a station. The three relationships have the same general form:

$$w = aQ^b$$
$$d = cQ^f$$
$$v = kQ^m$$

where Q is discharge ($m^3\ s^{-1}$), w is width (m), d is depth (m), v is velocity ($m\ s^{-1}$); b, f and m are the exponent values, and a, c and k are intercept values.

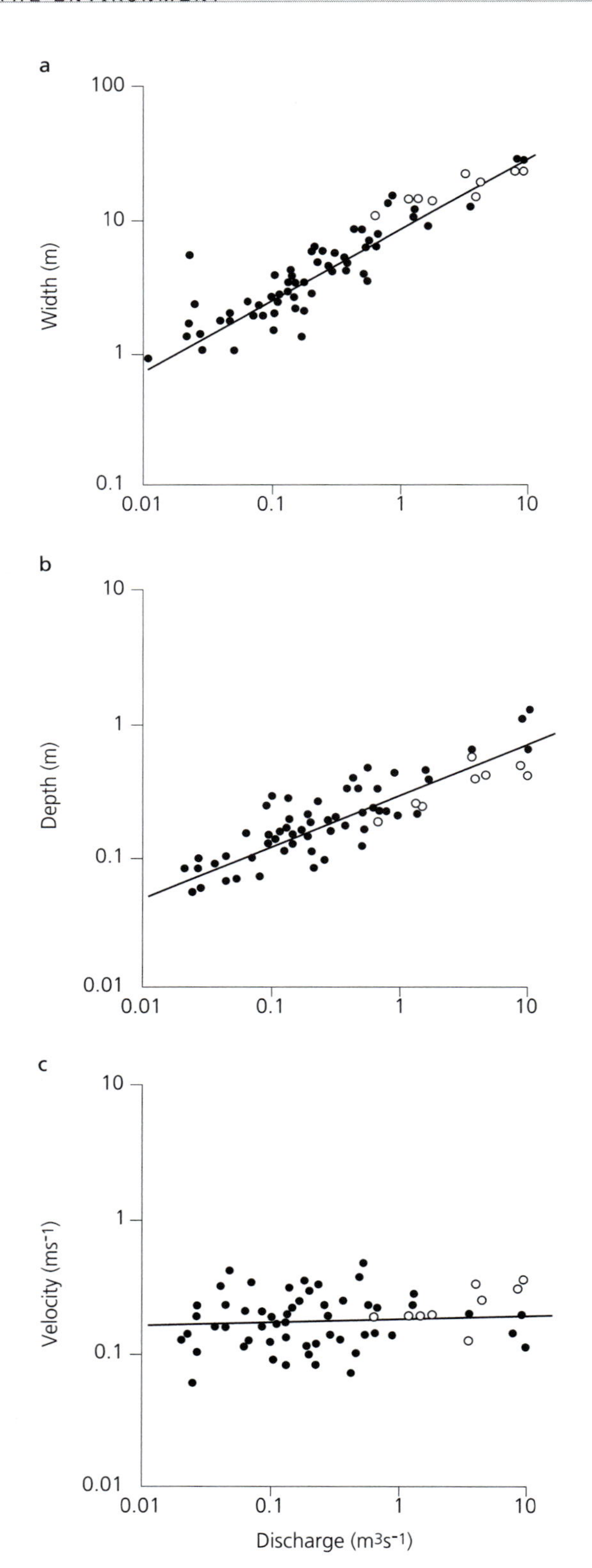

Figure 12.11 *At-a-station hydraulic geometry. The graph shows the relationship between discharge and channel width (a), depth (b) and velocity (c) as discharge varies over time at a particular cross-section.*

Source: A figure in Thornes, J. (1979) River channels. Macmillan Education, London.

BOX 12.22 HYDRAULIC GEOMETRY AND CHANNEL EQUILIBRIUM

Because discharge is the product of width, depth and velocity ($Q = w\ d\ v$), the laws of mathematics dictate that:

- the sum of the exponents *must* equal 1 ($b + f + m = 1$);
- the product of the intercepts *must* equal 1 ($a\ c\ k = 1$).

To maintain continuity and abide by the mathematical laws, if one of the three exponent values changes, one or both of the other two *must* change to keep the sum equal to 1. So, if for some reason the relationship between depth and discharge at a cross-section changes, adjustments *must* follow in one or both of the other relationships (width and velocity against discharge). In this way, channel cross-sectional equilibrium is maintained at a station.

Studies have shown that the hydraulic geometry exponent values appear to vary relatively little from one river to another (Table 12.7), even between different rock types and different climatic zones. Geologists interpret this as evidence that rivers have some in-built tendency to behave in quite a regular manner, implying some kind of equilibrium.

While the evidence indicates the existence of equilibrium, it is not possible to predict for certain how a river channel will change in response to a change in external factors (such as climate change or land-use change). This is called indeterminacy, and it arises because a river channel can adjust in many different ways.

Table 12.7 Average values of hydraulic geometry exponents from rivers in the Midwest USA

	b	*f*	*m*
At a station	0.26	0.40	0.34
Downstream	0.50	0.40	0.10

DOWNSTREAM HYDRAULIC GEOMETRY

Discharge increases downstream in most rivers, because the drainage area (which contributes runoff) increases. Most river channels also grow bigger downstream as a response to increasing discharge. If width, depth and velocity measurements are made at a series of sites along a river channel, downstream relationships can be plotted in the same way as the at-a-station graphs. The downstream hydraulic geometry relationships have the same general form as the at-a-station ones, notably:

$$w = aQ^b$$
$$d = cQ^f$$
$$v = kQ^m$$

where Q is discharge ($m^3\ s^{-1}$), w is width (m), d is depth (m), v is velocity ($m\ s^{-1}$); b, f and m are the exponent values, and a, c and k are intercept values. Downstream relationships have different exponent and intercept values from the at-a-station relationships, even in the same river system.

The downstream hydraulic geometry relationships must be based on comparable situations at the different sites. This is usually done by plotting data relating to events of the same frequency (such as the bankfull discharge or the average annual discharge).

Width tends to increase downstream faster than depth (the downstream value of b is usually much larger than the value of f), indicating a downstream change in channel shape. Flow velocity usually increases slightly downstream, despite the popular misconception that mountain streams are fast and large rivers flow very slowly.

RIVER CHANNEL PATTERNS

Straight channels are relatively rare, except on unusually steep slopes. In fact, rivers are rarely straight for lengths of more than ten times their average width. A meandering river flows around a series of bends, but braided rivers flow through multiple channels with islands in between them (Figure 12.12). All three channel types – straight, meandering and braided – can occur along a single river, depending on variations from place to place in slope, discharge and sediment size (Figure 12.13).

Meandering is the most common channel pattern (Box 12.23), and to understand how and why meanders develop we need to consider how bed material is moved along and shapes the bed of a channel.

BOX 12.23 CAUSES OF MEANDERING

It is generally believed that meanders are caused by large turbulent eddies (flow distortions) in the river flow. The length of these eddies is related to channel width: the wider the channel, the longer the eddy that can develop. Alternating erosion and deposition of bed material in response to these turbulent eddies may cause the pool-and-riffle patterns.

However the meander bend starts, once there is a deflection of flow within the channel this sets in motion a concentration of erosion (by undercutting) on the outside of the bend and compensating deposition (lateral accretion) on the inside of the bend. Flow velocity is greatest on the outside of the bend, so erosion tends to be concentrated on that bank. Velocity is much lower on the inside of the bend, and this causes deposition of sediment and the build-up of point bars.

BEDFORMS AND MEANDER DEVELOPMENT

Straight channels rarely have smooth, regular beds and perfectly straight banks. They usually have narrow, deep sections with fine sediment on the bed, alternating with wider, shallower sections with gravel beds. This is called a pool-and-riffle sequence, and it appears to play an important role in the development of meanders (Figure 12.14). The spacing of pools and riffles in straight reaches appears to be similar to the spacing of pools and riffles in a meandering channel, which suggests that even straight channels have a tendency to meander.

MEANDERS AND SLOPE

For meandering to develop, a river must have enough power to overcome the resistance of its banks. Power increases with slope, and as a result meandering tends to occur when slope increases. Field studies in many areas have shown that if slope is increased further, with no increase in discharge, the single meandering channel pattern might be replaced by a wider braided channel.

For a given discharge, there appears to be a critical (threshold; see p. 83) slope below which meandering is the norm and above which rivers tend to be braided (see Figure 12.13). At small discharges, meanders can survive on fairly steep slopes, and as discharge increases braiding

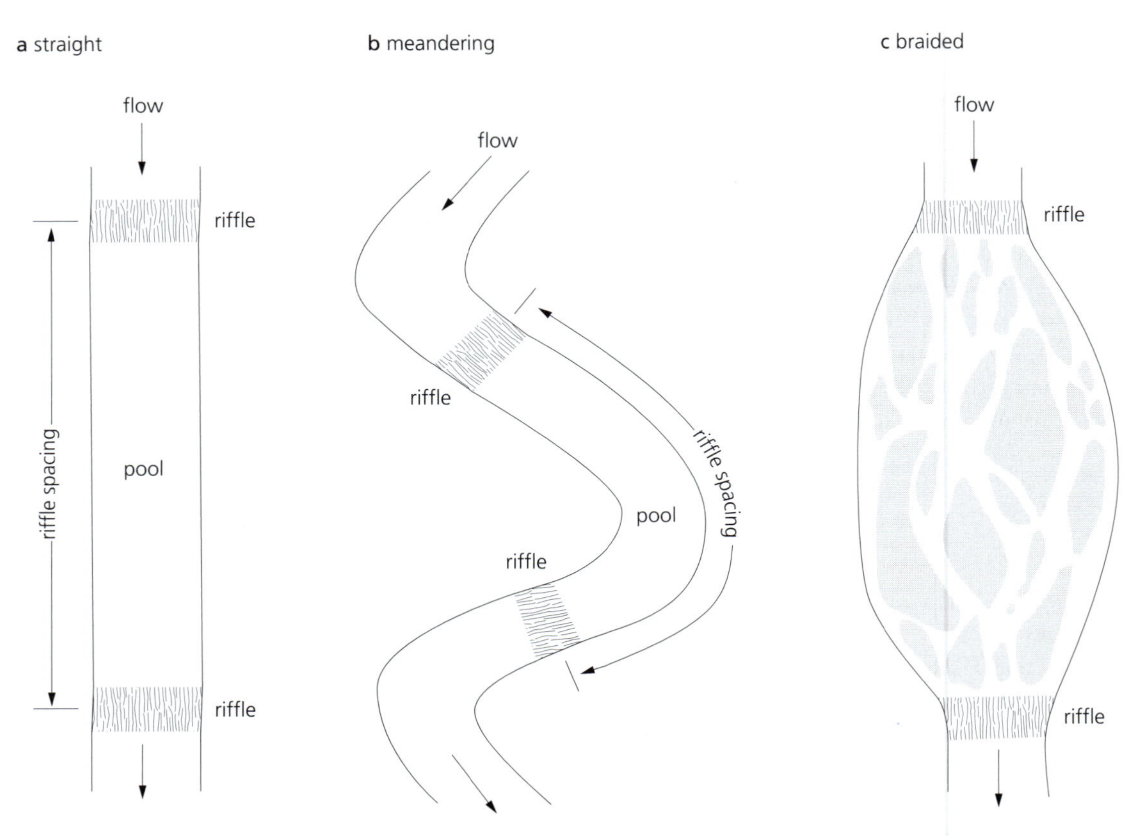

Figure 12.12 *Main types of channel pattern. There are three main types of channel pattern – straight (a), meandering (b) and braided (c). Meandering is the most common.*

Source: Figure 15 in Thornes, J. (1990) River channels. Macmillan Education, London.

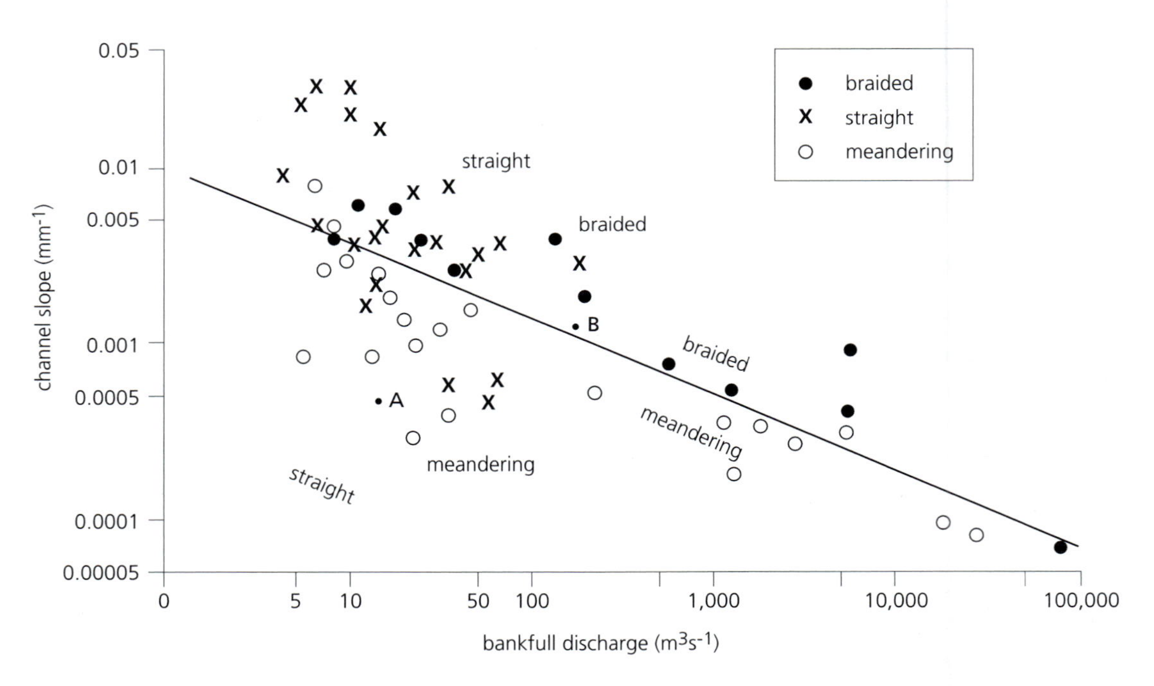

Figure 12.13 *Channel pattern in relation to discharge and slope. Field data collected from different rivers in the USA show that, for a given discharge, meandering rivers tend to have lower slopes than straight or braided channels.*

Source: A figure in Thornes, J. (1990) River channels. Macmillan Education, London.

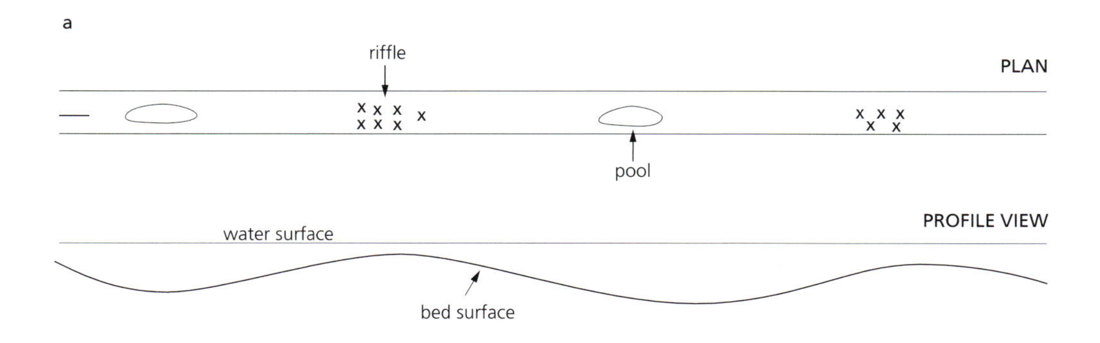

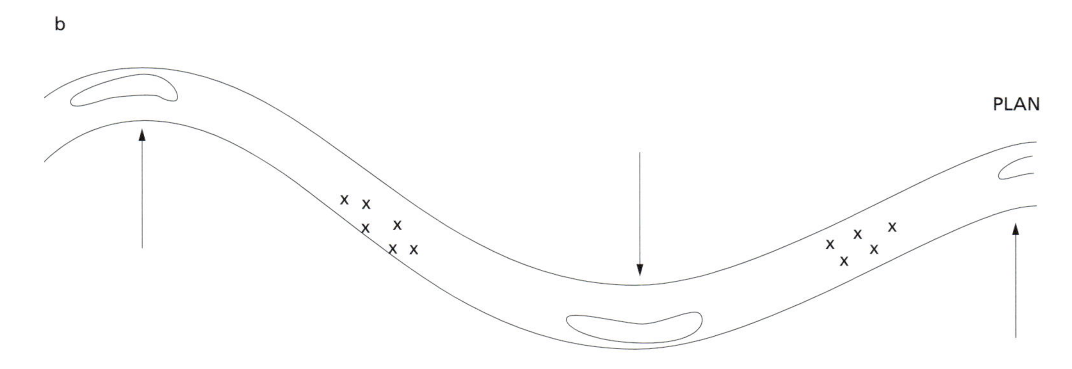

Figure 12.14 *Pool-and-riffle sequences. The figure shows the relationship of pool-and-riffle sequence to channel plan in a straight channel (a) and in a meandering channel (b).*

Source: Figure 8.11 in Dury, G.H. (1981) Environmental systems. Heinemann, London.

Plate 12.2 *Meandering meltwater channel on the surface of the Lower Arolla Glacier in the Val d'Herens, Switzerland. Even streams that do not flow in sediment, or carry heavy loads of particulate sediment, display sinuous meandering patterns typical of alluvial river systems. The ice axe, shown for size, is about 1 m long.*

Photo: Chris Park.

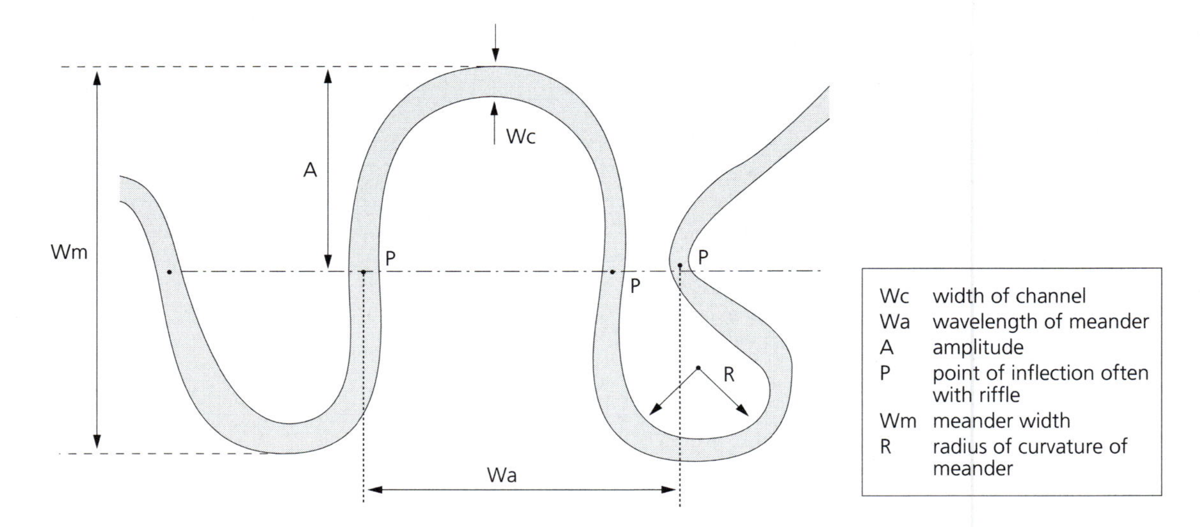

Figure 12.15 *Meander geometry. A range of different measures of meander size and geometry have been developed, which can be used to study relationships between meander properties, and between these and possible causal factors (such as slope or discharge).*

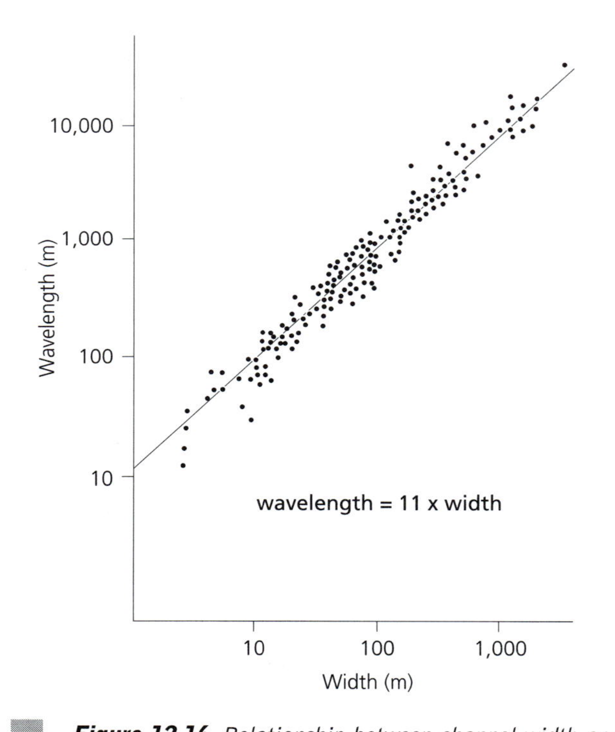

Figure 12.16 *Relationship between channel width and meander wavelength. The relationship shows a regular increase in meander wavelength as channel width increases over a wide range of channel sizes.*

Source: Figure 8.9 in Dury, G.H. (1981) Environmental systems. Heinemann, London.

can occur on progressively gentler slopes. Rivers that are close to the threshold slope can quickly change from meandering to braided with even a relatively small change in discharge. Such channels are unstable and liable to change almost without warning. Rivers that are well away from the threshold slope are much more stable, and they will preserve their channel pattern unless there is a radical change in discharge and/or slope.

MEANDER GEOMETRY AND MOVEMENT

Meanders often display remarkably similar appearances, even in rivers of radically different size and between rivers in very different geological and climatic environments. The most widely used, and easily measured, geometric property of meanders is the wavelength (Figure 12.15).

A commonly used measure of the degree of meandering within a particular river is the *sinuosity*, which is the ratio of stream length to valley length. Tightly meandering rivers travel much further over a given length of valley, and they have high sinuosity. This depends partly on the dominant mode of sediment transport in a river, and rivers with high suspended sediment loads tend to meander most and to have the highest sinuosities (Table 12.8). Rivers that carry coarse sediments (sand and gravel) tend to have braided patterns.

Some meander bends can be stable for long periods of time, but old maps and repeated photography, coupled with field evidence of floodplain stratigraphy (sediment

BOX 12.24 GEOMETRIC REGULARITY OF RIVER MEANDERS

Strong relationships have been found in many different rivers between:

- meander wavelength and channel width (Figure 12.16) (wider channels have larger meanders);
- riffle spacing and channel width (wider channels have more widely spaced riffles);
- meander wavelength and discharge (higher discharges promote longer meander wavelengths).

These suggest that meander development is an important aspect of channel equilibrium.

sequences) and topography (surface expression), often show that meanders move across and down floodplains over time. Lateral (side-to-side) migration of meanders tends to redistribute floodplain sediments, and repeated cycles of lateral planation over a long period of time can produce very flat, featureless floodplains.

When channels meander across a floodplain, they sometimes leave evidence of former channel courses. An *ox-bow lake* or cut-off, formed when a river cuts through the narrow neck of a large meander (Figure 12.17), is a typical example.

CHANNEL SLOPE

Slope or gradient strongly influences flow velocity and thus sediment transport in a river, and this dimension of channel geometry appears to be much more stable and persistent

Table 12.8 A classification of river types

Type	*Width-to-depth ratio*	*Sinuosity*	*Slope*	*Bedload as a percentage of total load*
Bedload channels	> 40	< 1.3	steep	> 11
Mixed-load channels	10–40	1.3–2.0	moderate	3–11
Suspended load channels	< 10	> 2.0	gentle	< 3

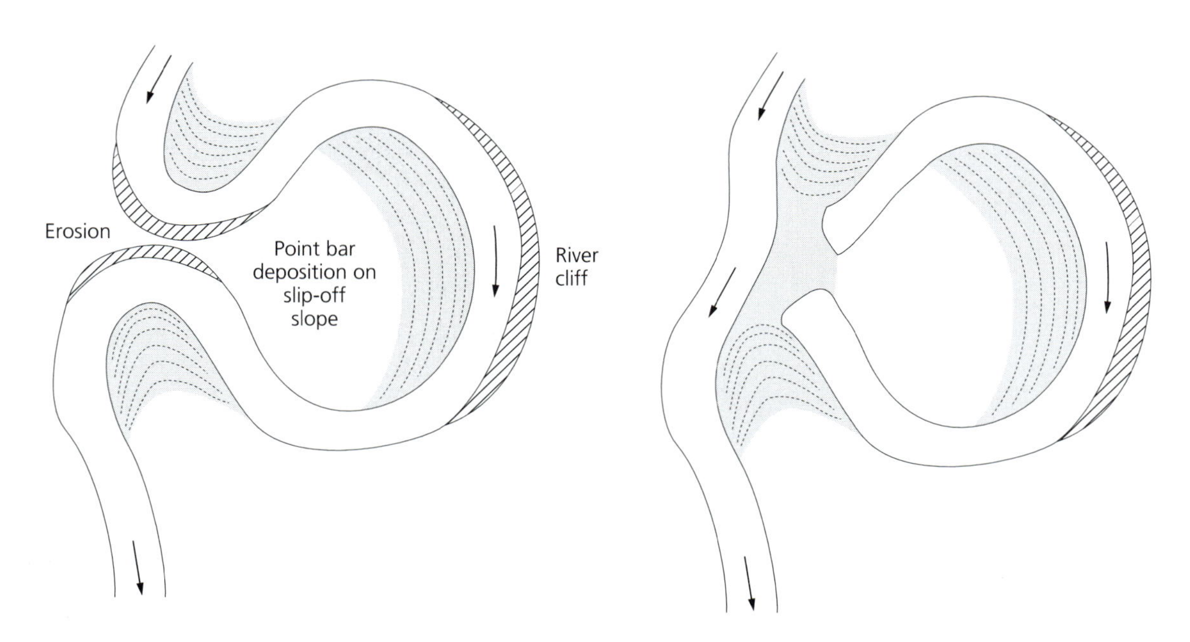

Figure 12.17 *Development of an ox-bow lake. The ox-bow lake is formed by the erosion of a tight meander bend, leaving the former meander effectively separated from the main channel.*

Source: Figure 5.6 in Buckle, C. (1978) Landforms in Africa. Longman, Harlow.

BOX 12.25 SLOPE AND VELOCITY

Rivers with a steep slope tend to flow more rapidly than rivers flowing down shallower slopes, all else being equal. On this basis, we should expect flow velocity to decrease downstream because slope usually decreases downstream, but studies have shown that flow velocity increases downstream in many rivers (see p. 371) because channels become smoother (so frictional resistance is reduced) and have more efficient cross-sectional shapes in a downstream direction.

Plate 12.3 *Fast-flowing, steep mountain river in Rocky Mountain National Park, Colorado, USA.* *Photo: Chris Park.*

than cross-sections and meanders. Slope is normally expressed as fall (drop in height) per unit distance along the valley, in m km^{-1}.

LONG PROFILE

The long profile is a section through the channel from its headwater to its mouth at the sea (Figure 12.18), which shows how slope changes downstream. Many rivers have long profiles that are concave upwards, because their headwaters are steep, and slope decreases progressively in a downstream direction. Rivers tend to down-cut and lower their valleys over time, so the tendency is for slope to decrease over time. The theoretical lower limit for erosion, the so-called base level, is sea level, but many rivers have more local base levels, determined by outcrops of resistant rock (Figure 12.19) or by artificial controls such as dams.

SLOPE AND EQUILIBRIUM

Slope is the most stable and persistent aspect of channel geometry. River cross-sections generally respond to short-term changes (such as the passage of a flood wave), and river patterns tend to be adjusted to the flow conditions and type of material being supplied over timescales of decades or centuries. Slope appears to adjust to long-term controls that operate over at least thousands of years. The most likely distribution of slope in a river system lies between two extremes – one that performs the least amount of work (sediment transport) overall, and one in which the distribution of work done is uniform along the river (Figure 12.20).

CHANNEL EQUILIBRIUM AND CHANGE

Rivers are open systems that can adjust their morphology (form) to the throughput of water and sediment. In regulated rivers, some of the adjustments are to external control, such as the altered river flows associated with large dams and reservoirs. Rivers can adjust over a variety of

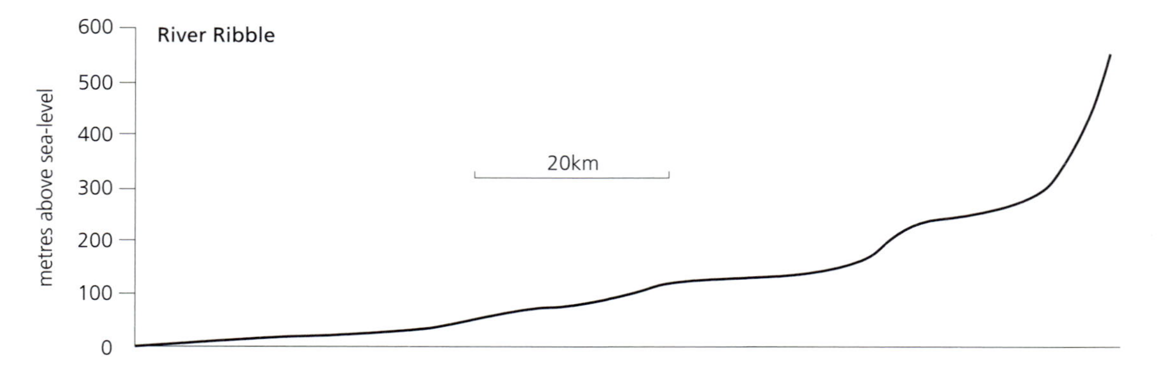

Figure 12.18 *Long profile of the River Ribble, north-west England.*

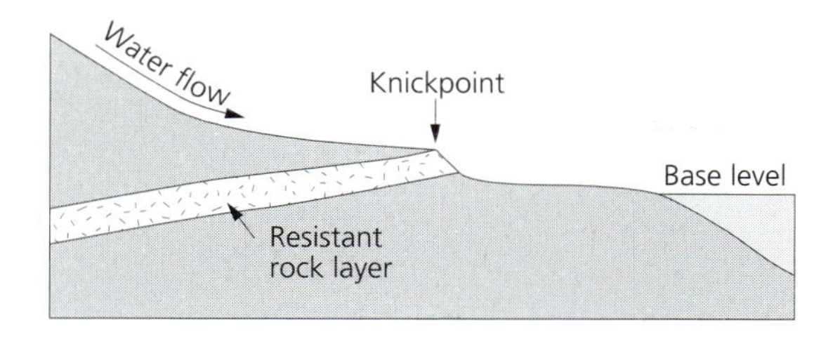

Figure 12.19 *Development of a knick-point along a river. Knick-points usually develop where river down-cutting is inhibited by an outcrop of resistant rock, causing a step in the long profile that in time will be removed by erosion.*

Source: Figure 12.18 in Doerr, A.H. (1990) Fundamentals of physical geography. Wm. C. Brown Publishers, Dubuque, Iowa.

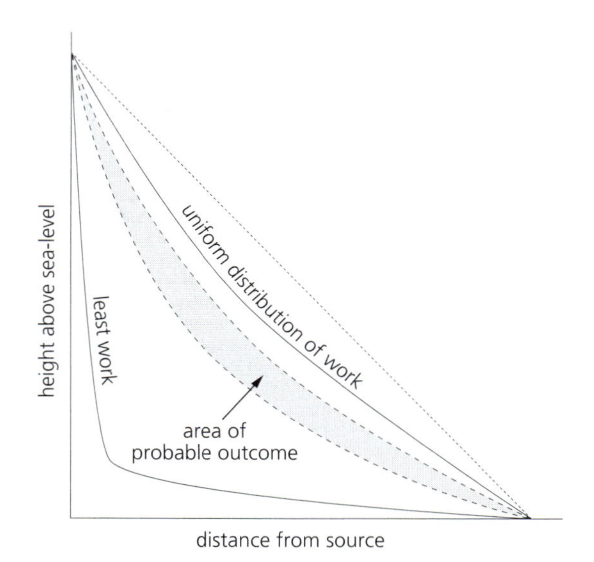

Figure 12.20 *The ideal river long profile, defined by thermodynamics. Most river long profiles lie between the positions defined by two of the principles of thermodynamics – the principle of least work and the principle of uniform distribution of work.*

Source: Figure 5.2 in Clowes, A. and P. Comfort (1982) Process and landform. Oliver & Boyd, Edinburgh.

BOX 12.26 WATERFALLS

In theory, rivers tend to develop smooth long profiles, but they often contain irregular sections where slope is affected by local geological factors (such as resistant hard rock outcrops). Rapids form where the slope of a river increases and flow velocity increases as a result. The increased velocity promotes greater river erosion, which will tend to wear down the obstruction over a long period of time and ultimately remove it. Waterfalls often occur where relatively soft rocks (which are easily eroded by river flow) are overlain by resistant rocks (which take much longer to wear away), and this interrupts river flow.

The highest waterfalls in the world include Angel Falls in Venezuela (979 m), Tugela Falls in South Africa (948 m) and Yosemite Falls in California (739 m). Niagara Falls, on the border between Canada and the USA, carries one of the largest water flows (average flow is around 6,000 $m^3 s^{-1}$) and attracts large numbers of tourists. Recent research suggests that water first flowed over Niagara about 12,000 years ago, when melting glacier ice poured out of Lake Erie. The falls may have dried up between 10,000 and 5,000 years ago, as the lake level dropped too low, but surface rebound (see p. 156) – after ice left the area – gradually raised the land and made the falls active once again.

timescales, and for convenience we can focus on short-, medium- and long-term changes, although these are obviously part of a continuous spectrum of change over time.

EQUILIBRIUM

Field evidence from many different environments shows that most natural rivers establish and maintain a state of equilibrium in which sediment supply, erosion and deposition are in some sort of balance in the drainage basin (Box 12.27), depending largely on slope, climate and land use. Studies in the Yosemite National Park in California have shown that, between 1919 and 1989, many natural river channels had reaches where channel width increased and others where it decreased, but the net increase over the period was only 4 per cent.

RIVER REGULATION

Throughout history, rivers have been regulated and directly modified, for a variety of different reasons. For thousands of years, the Nile (see p. 63) has been regulated to irrigate large areas of its floodplain and delta, for example. In China, attempts to regulate flooding on the Hwang Ho

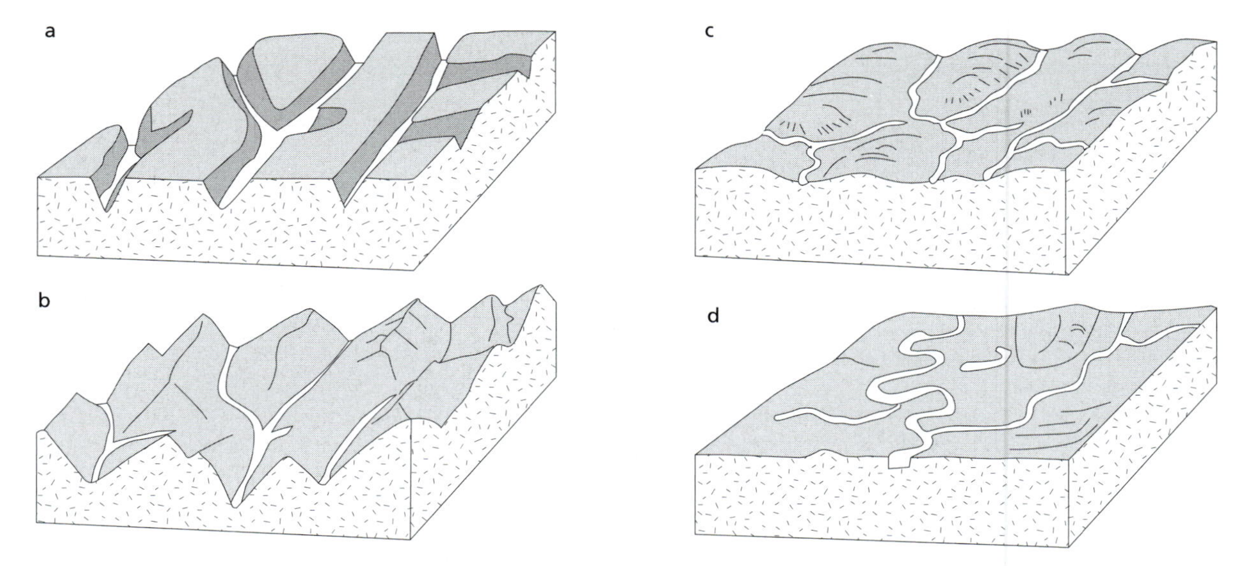

Figure 12.21 *The cycle of erosion. William Morris Davis defined a progressive sequence of landform changes through the cycle of erosion, starting from an uplifted surface dissected by some rivers (a), through a youthful landscape with steep valley sides and steep long profiles (b), to a mature landscape dominated by low hills and low river slopes (c), to the gentle undulating landscape of old age (d).*

Source: Figure 12.19 in Doerr, A.H. (1990) Fundamentals of physical geography. Wm. C. Brown Publishers, Dubuque, Iowa.

BOX 12.27 MODELS OF LONG-TERM RIVER DEVELOPMENT

William Morris Davis, an American geologist, proposed a *cycle of erosion* in 1899 (see Figure 12.21) based on the assumption that slopes (both in the river and on the divides) decline progressively through time. The cycle appeared to explain how temperate landscapes, in which river activity is the dominant mode of landform development, change over geological timescales (see p. 115). It was widely accepted until the 1950s, when hydraulic geometry studies (p. 369) established that flow velocity increases downstream in most rivers and does not decrease (because of decreasing slope) as the cycle model assumed. In the 1960s, John T. Hack proposed a steady-state type of equilibrium in slopes, similar to that in cross-sections and patterns, in which form remains constant while water and sediment move through it. This implies that a concave upwards channel slope is formed and persists through the long term as a steady-state equilibrium.

BOX 12.28 EQUILIBRIUM AND ADJUSTMENT

Rivers have an ability to adjust their hydraulic geometry, pattern and slope to maintain a balance of energy moving through the system. Equilibrium exists when the energy provided to do work (that is, to transport the sediment and shape the river landscape) is just enough to perform the work that has to be done. Field and map surveys show that channels in the River Thames in England, for example, appear to be adjusted to rock type, slope, land use and channel management.

This is a type of dynamic equilibrium (see p. 80), and the balanced state can be altered if any of the controlling factors change past critical thresholds. Most such change is promoted by factors that are external to the river system, such as:

- rainfall change, associated with climate change;
- sediment load changes, perhaps associated with land-use changes;
- slope (thus velocity) changes, perhaps associated with tectonic uplift or fault line movements.

Plate 12.4 *Karst cliff undercut by a river in northern Vietnam. Higher-level undercuts can be seen on the cliff face, indicating local base-level change.*
Photo: Andy Quin.

Plate 12.5 *Modern-day river system flowing through the glacial valley of Borrowdale in the Lake District, Cumbria, England. The river has reworked and transported Pleistocene glacial deposits, and the river morphology reflects the interaction of contemporary fluvial processes and inherited glacial deposits and valley features.*
Photo: Chris Park.

(Yellow River) by building artificial levees have promoted increased channel changes.

The history of river regulation in Britain illustrates the changing pace and pattern of change. Rivers in Britain have been directly modified and controlled since at least the first century AD, when the Romans undertook land-drainage schemes and river navigation improvements. Many small rivers were affected by flow-control structures and water extraction to power watermills (more than 5,000 of which were recorded in the Domesday Book of 1086), and local regulation was common until the seventeenth century. From the end of the eighteenth century, the scale of regulation increased a great deal as large drainage schemes were implemented and many small water supply dams were built. During the twentieth century, the pace of change grew faster, with the construction of large dams and large inter-basin water transfers and the adoption of large multi-purpose river management schemes.

By the close of the twentieth century, virtually all major rivers in the UK were regulated directly or indirectly by large dams, inter-basin transfers, pumped storage reservoirs or groundwater abstractions. Flows on 60 per cent of all river gauging stations were by then being significantly modified by various forms of regulation.

BOX 12.29 REGULATION OF THE RIVER SEVERN

The River Severn, which has a source in central Wales and flows into the sea at the Bristol Channel, is Britain's longest and best-studied regulated river. It has long been regulated for water resource purposes and has a long history of flood problems. Demand for water continued to rise during the twentieth century, and an ambitious drainage basin management scheme was developed to augment low river flows, increase the amount of water that could be abstracted for supply purposes and decrease the flood risk. Clywedog Reservoir, a storage reservoir on the headwaters of the Severn, is an important component of the scheme. Water is stored at Clywedog under high flow conditions and released from it under low flow conditions. Since it was completed in the 1960s, low flows along the Severn have been increased by nearly a quarter, normal flows reduced by about half and the mean annual flood reduced by nearly a third. Detailed analysis of these flow changes, however, reveals that climate change has probably been a more important control than river regulation.

SHORT-TERM CHANGES

Short-term channel changes occur over timescales of years or less. They are caused mainly by variations in discharge, such as those associated with the passage of a storm across the drainage basin.

Most flows in a typical river are small to medium in size (magnitude), and they remain well within the bankfull channel (Box 12.29). However, experience shows that the most dramatic channel changes are caused by less frequent large flows, including floods that overtop the banks and spill on to the adjacent floodplain.

Extremely large flows can cause significant channel changes, but they are relatively rare. Catastrophic events of low frequency (particularly very large floods) perform only a small proportion of the total work of a river. For example, the 1952 flood at Lynmouth in south-west England, with an estimated return period (see p. 415) of about 150 years, caused extensive damage by destroying bridges and

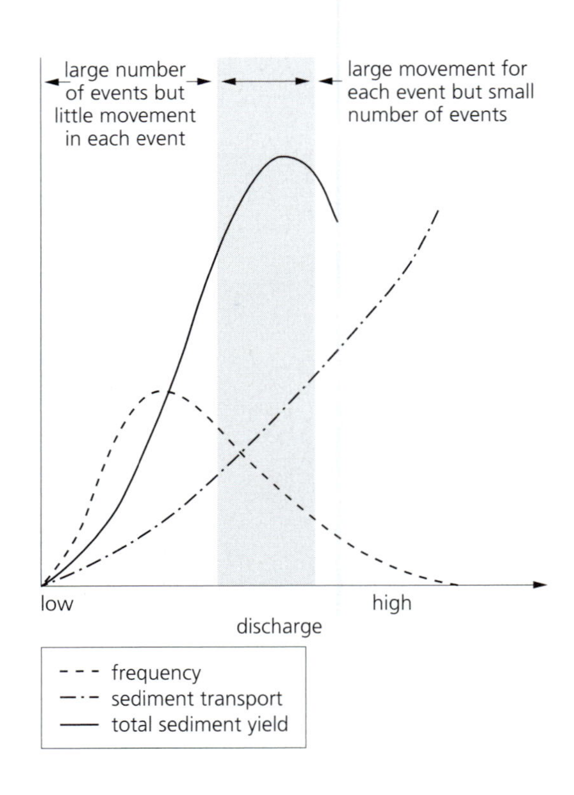

Figure 12.22 *Magnitude and frequency relationships. See Box 12.30 for explanation.*

Source: Figure 17 in Thornes, J. (1990) River channels. Macmillan Education, London.

BOX 12.30 MAGNITUDE AND FREQUENCY CONCEPTS

The amount of erosional work done in a river is related to both the magnitude of the flow and the frequency with which that flow occurs (Figure 12.22). In simple terms, work = magnitude × frequency. Studies show that most of the work done by rivers is actually done by relatively frequent flow events of moderate size. Most of the suspended and dissolved load is carried by such flows, for example, and most of the channel changes appear to be related to such flows.

buildings, cutting new channels on the peat moorlands on Exmoor, depositing large boulders on the stream bed, diverting some channels and cutting off some meander bends. Since 1952, the River Lyn appears to have established a new equilibrium, and many of the features caused by the 1952 flood are left as residual or persistent elements in the landscape. A major flood in 1990 on the River Tay in Scotland, the third-largest flood since 1800, caused widespread channel adjustments, damaged some artificial flood embankments, deposited some large gravel bars on the river bed and changed the location of many meanders.

MEDIUM-TERM CHANGES

Over the longer term, perhaps measured in thousands of years, channel changes appear to be controlled mainly by climatic change. Concern has been expressed about the implications of contemporary global warming for river management, although scenarios are difficult to establish because of the complexity of river response and limitations in current understanding of river equilibrium and adjustment.

Two of the most widespread indicators of medium-term channel changes are terraces and misfit streams, both of which illustrate channel adjustment to changing throughputs of water and sediment.

A river terrace is a flat platform of land that lies above the level of the valley floor (Figure 12.23). Terraces are often remnants of an earlier valley floor higher than the present one, which was cut into when the river's erosive power increased (Box 12.31). The terrace survives as a remnant of the abandoned floodplain.

Misfit (or underfit) streams are relatively small river meanders – created by the present-day river and in equilibrium with the water and sediment load it carries – that lie within much larger valley meanders (Figure 12.24). The valley meanders were cut by much large rivers, with higher discharges, in the past. Misfit streams indicate adjustment of channel pattern to changing climatic or hydrological conditions, such as a decrease in river flow since the end of the Pleistocene (see p. 452) less than 10,000 years ago.

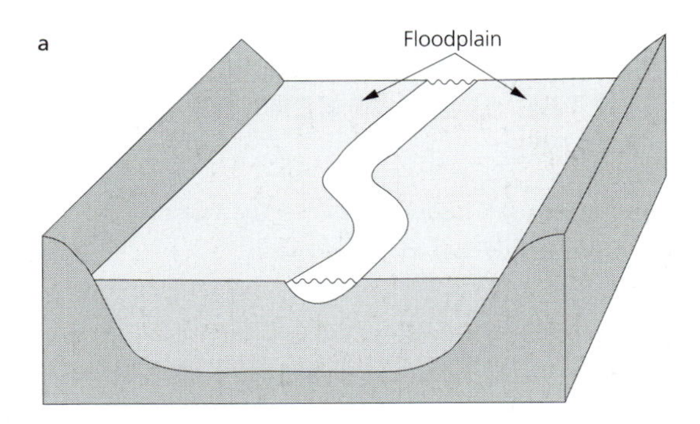

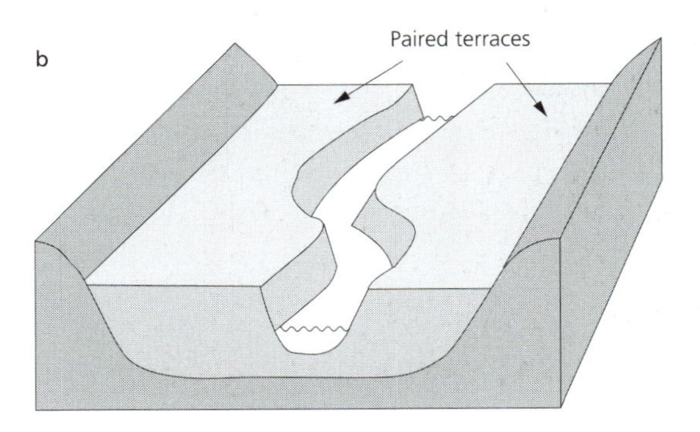

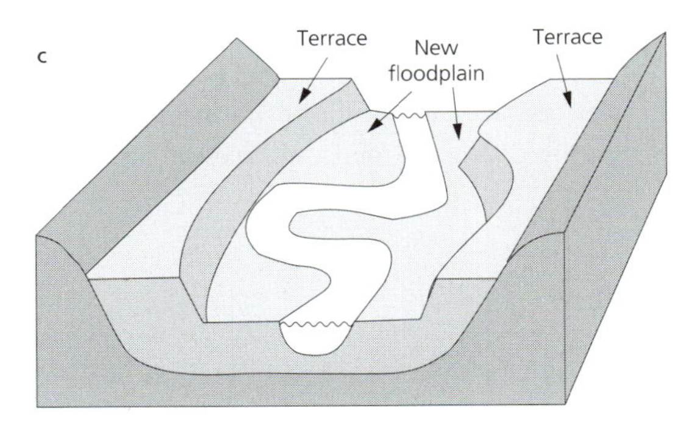

Figure 12.23 *Development of a river terrace. Terraces are formed by incision (lowering by erosion) of a river into the floodplain, leaving remnants of the former floodplain that are now higher than the active floodplain.*

Source: A figure in Doerr, A.H. (1990) Fundamentals of physical geography. Wm. C. Brown Publishers, Dubuque, Iowa.

BOX 12.31 FACTORS ASSOCIATED WITH THE DEVELOPMENT OF RIVER TERRACES

Terraces represent adjustments of the river to changing circumstances, which might include:

- sea-level fall, which creates a lower base level;
- uplift of the land, which increases river slope and thus velocity;
- decreased sediment load, perhaps associated with upstream impoundment of water and sediment behind a dam;
- increased river flow, perhaps associated with the melting of upstream glaciers or with land-use changes within the drainage basin.

LONG-TERM CHANGES

Many river systems preserve evidence of their long-term evolution over geological timescales. Examples include:

- Erosion surfaces that indicate remnants of the long-term evolution of the landscape and can be interpreted in terms of the denudation chronology (the erosional history) of the area, reflecting the history of uplift and wearing down of the regional landscape.
- Signs of rejuvenation (renewed down-cutting), particularly in river long profiles – which sometimes display a step-like structure, with a series of concave-upwards segments indicating former base levels – related to falling sea levels.
- River capture or piracy, in which an actively eroding river might capture the headwaters of an adjacent river. The topography of the western shore of the Chesapeake Bay in Maryland shows that five rivers that now flow eastwards into the bay are the pirated headwaters of rivers that previously flowed westward from a vanished Pliocene upland now occupied by the central Chesapeake.

DELTAS

Many rivers eventually flow into the sea or a lake, where they deposit sediment when velocity falls below that required to keep particles of that size in motion (see Figure 12.8). This sediment often builds up into a delta composed of fine-grained deposits (Figure 12.25). The word 'delta' is derived from the Greek letter Δ (delta), which is shaped like a triangle, and the large delta at the mouth of the River

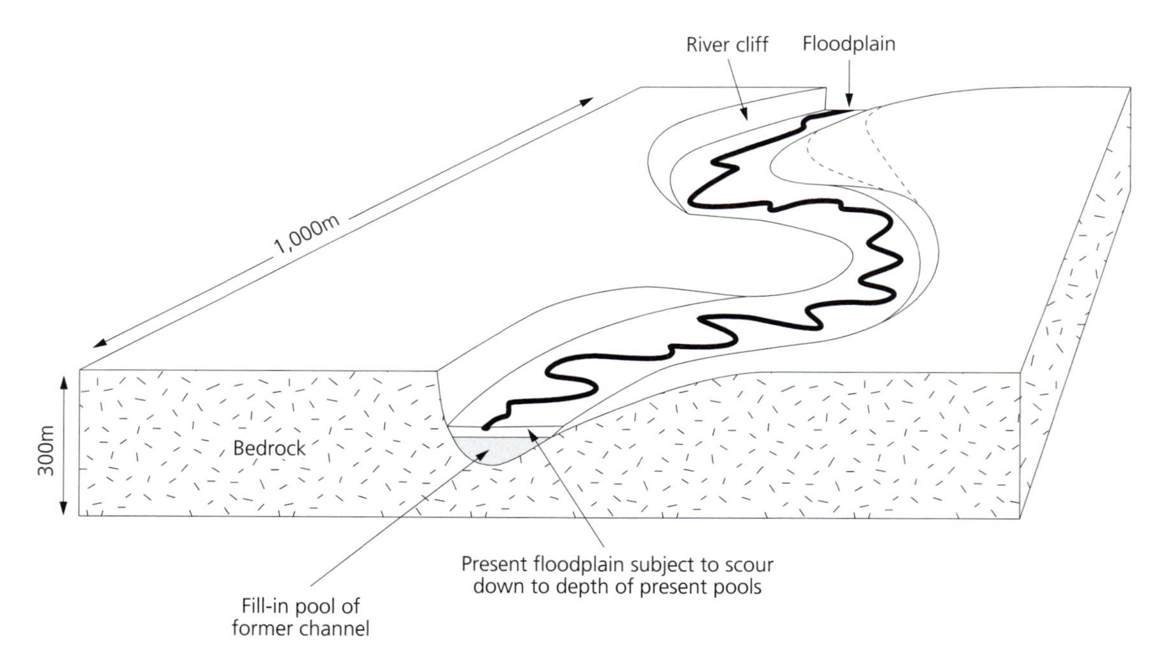

Figure 12.24 *A misfit stream. A misfit (or underfit) stream is a small meandering stream within a larger meandering valley, usually created as a response by the river to a significant reduction in discharge.*

Source: Figure 18.15 in Dury, G.H. (1981) Environmental systems. Heinemann, London.

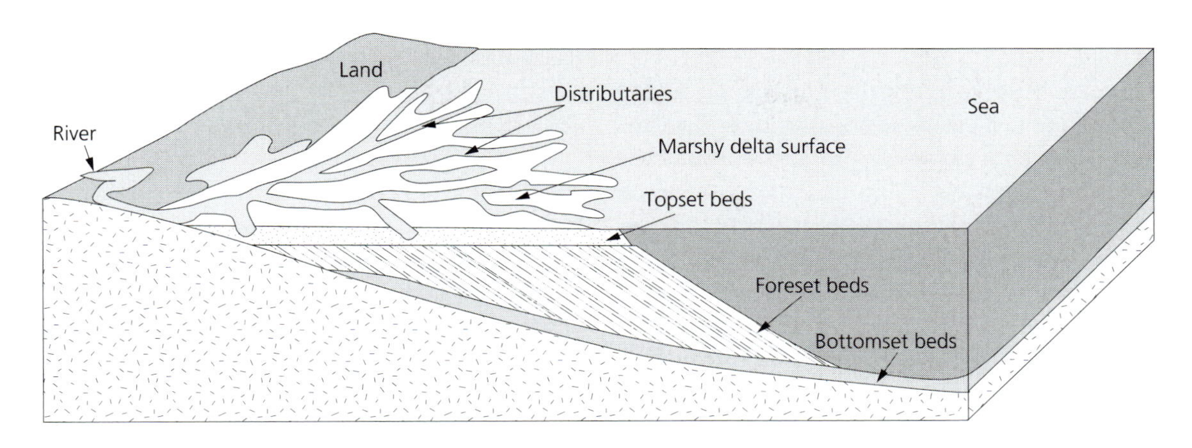

Figure 12.25 *Structure of a delta. The river deposits sediment at its mouth, which builds up as a delta. Note the relationship between the topset, foreset and bottomset beds.*

Source: Figure 12.31 in Doerr, A.H. (1990) Fundamentals of physical geography. Wm. C. Brown Publishers, Dubuque, Iowa.

Plate 12.6 *Delta deposits in Cartref Reservoir in the Brecon Beacons, South Wales. These sediments were deposited by the tributary rivers that drain into the reservoir and were exposed during the prolonged British drought of 1975–76, when the reservoir was downdrawn.*

Photo: Chris Park.

Table 12.9 A typology of delta forms

Form	*Example*	*Comments*
Arcuate (fan-shaped)	Nile delta	A triangular delta across which the main stream and its distributaries (side-branching channels) flow *en route* to the sea
Bird's foot	Mississippi delta, in the USA	This type develops when major distributaries build deposits adjacent to them, with open water in between. It is encouraged by frequent dredging of the main channel
Compound	The deltas at the mouths of the Ganges River in India and the Brahmaputra River in Bangladesh	These form where two or more large rivers flow into the sea or lake in the same area and have distributaries that are linked together
Estuarine	The delta at the mouth of the Mobile River in southern Alabama, which flows into the Gulf of Mexico	These grow where former drowned river mouths are progressively filled in by a river and its distributaries, which deposit sediments and fill the drowned area
Cuspate	Rhône delta in France	These form as a result of wave action (longshore currents), which modifies the delta form along its seaward side

Nile in the southern Mediterranean Sea (see p. 66) is a classic example.

The particles of the sediment deposited on the bed of the sea or lake accumulate as *bottomset* beds. As the delta grows, the river continues to deposit material as *foreset* beds. Very fine material, including the finest particles of silt and clay, settle last as the *topset* beds. These different beds can often be seen in stratigraphic sequences from old delta deposits and in trenches cut through contemporary deltas.

Deltas grow over time as sediment is deposited. They can also change over time in response to changes in the delivery of water and sediment from the drainage basin upstream. Such changes can be promoted by climate change, by land-use change and by river management schemes. The delta at the mouth of the Hwang Ho in China has altered a great deal over historical time because of drainage basin changes.

While all deltas accumulate where rivers flow into lakes or the sea, there are a variety of different delta forms (Table 12.9), which indicate different modes of formation.

Deltas are usually very fertile areas and are extensively used for agriculture. They contain good soils, have abundant water supplies available for irrigation and – in natural rivers that are not controlled upstream – are frequently flooded, which brings regular inputs of nutrients and fertile silt. The Nile delta (see p. 66), for example, has been a centre of settlement and civilisation for many thousands of years.

Many deltas are being developed for human use, including reclamation by drainage to create new farmland and flood protection by construction of coastal engineering schemes. Deltas also provide important wetland habitats (see p. 401), which are particularly important for migrant bird populations. They are among the world's most productive natural habitats (see p. 564).

WEBSITE

Links to relevant websites, a comprehensive bibliography, tools for teaching and learning, and downloadable images relevant to this chapter can be found at the website specially designed to accompany this book at
http://www.park-environment.com

FURTHER READING

Acreman, M. (ed.) (1999) *The hydrology of the UK*. Routledge, London. A detailed assessment of the changing hydrology of the UK, with thematic chapters contributed by experts in the field.

Benito, G., V.R. Baker and K.J. Gregory (eds) (1998) *Palaeohydrology and environmental change*. John Wiley & Sons, London. Useful collection of essays which explore characteristics of past hydrological cycles and their role in past environments.

Brown, A.G. (ed.) (1995) *Geomorphology and groundwater*. John Wiley & Sons, London. A series of essays dealing with themes such as river–groundwater interactions, the location and evolution of aquifers, and karst and landform evolution.

De Wall, L.C., A. Large and M. Wade (eds) (1998) *Rehabilitation of rivers: principles and implementation*. John Wiley & Sons, London. A collection of essays that explore the principles and practice of river rehabilitation, drawing on case studies from different countries.

SUMMARY

This chapter has focused on the water cycle, and the main processes involved in it, and the ways in which water creates rivers and fluvial landscapes. It opened with an overview of the global water cycle, its main processes, stores and transfers, and the ways in which human activities can alter these. At the drainage basin scale, we looked at the processes involved and the way in which these interact with vegetation, soils and geology to generate river flow. River networks are the conduits along which rivers carry water and sediment, and we explored how studies of networks have progressed from large-scale geological control to scale-independent studies of network geometry. Next, we examined the processes and products of sediment transport in rivers, noting the importance of deposition processes and the depositional environment. Studies of river channel geometry – in cross-section, plan and slope – since the 1950s have revealed important features of channel equilibrium and change, which help us to understand how channels work and what the main controls are. This allowed us to reflect on issues of equilibrium and change, noting the importance of river regulation and river change over different timescales. Finally, we looked at deltas and what causes them.

Gordon, N., T. McMahon and B. Findlayson (1994) *Stream hydrology: an introduction for ecologists*. John Wiley & Sons, London. A readable introduction to measuring, understanding and explaining river processes.

Graf, W.H. (1998) *Fluvial hydraulics*. John Wiley & Sons, London. Explains the basic concepts of hydrodynamics and illustrates their application to real-world problems.

Gregory, K.J., L. Starkel and V.R. Baker (eds) (1995) *Global continental palaeohydrology*. John Wiley & Sons, London. A collection of essays that look at recent developments in reconstructing past river forms and behaviour based on an understanding of how present-day rivers operate.

Gurnell, A. and G. Petts (eds) (1995) *Changing river channels*. John Wiley & Sons, London. A collection of essays that review recent research on river channel forms, processes and adjustment.

Hickin, E.J. (ed.) (1995) *River geomorphology*. John Wiley & Sons, London. A collection of essays that outline recent developments in river sediment transport, channel form and change, deposition and human impacts.

Knighton, D. (1997) *Fluvial forms and processes: a new perspective*. Edward Arnold, London. A fresh look at the theory and practice of river studies.

Macklin, M. (1997) *Holocene river environments*. Blackwell, Oxford. Looks at how modern-day river systems have developed and how they operate.

Manning, J.C. (1997) *Applied principles of hydrology*. Prentice Hall, Hemel Hempstead. Non-technical treatment of all the major components of the water cycle, including ocean water evaporation, precipitation, infiltration, groundwater and streamflow.

Newson, M. (1997) *Land, water and development: sustainable management of river basin systems*. Routledge, London. Clearly written and abundantly illustrated overview of the theory and practice of catchment management planning in both developed and developing countries.

Thorne, C.R. and N. Newson (eds) (1997) *Applied fluvial geomorphology*. John Wiley & Sons, London. Detailed review of methods of studying, analysing and interpreting rivers.

Chapter 13

Water Resources

LEARNING OBJECTIVES

When you have finished studying this chapter, you should be able to

- *Appreciate the need for sustainable use and fair allocation of water resources*
- *Identify the main sources of water resources and comment on the potential for increasing them*
- *Describe and account for the use of water as a source of energy and in irrigation*
- *Outline the problems and benefits of large dams*
- *Understand the need to conserve wetlands*
- *Summarise the main environmental problems associated with salinisation*
- *Outline the main sources of freshwater pollution and the problems they cause*
- *Define the term 'eutrophication' and say how it is caused and how it can be managed*
- *Explain how floodplains form and what role they play in the drainage basin water cycle*
- *Describe the steps involved in carrying out a flood-frequency analysis*
- *Outline the main approaches to managing the river flood hazard.*

Water covers about three-quarters of the Earth's surface (see p. 125), and it is essential to life (Box 13.1).

In this chapter, we explore some of the critical issues surrounding the use of water resources. The first section deals with water resource use, and our primary focus is on demand for water and some of the key ways in which water is used, including the generation of hydroelectricity and irrigation. Large dam construction schemes have attracted mounting controversy during the latter part of the twentieth century, and we explore that debate. We also reflect on wetlands, which are critical habitats for many important wildlife species and serve important environmental functions. The second section concentrates on some of the more widespread problems that arise from water resource use, including land drainage and waterlogging, salinisation and desalination, water pollution, and eutrophication (nutrient enrichment).

BOX 13.1 THE IMPORTANCE OF WATER RESOURCES

People rely heavily on fresh water, and throughout history the sustainable development and management of available water resources has been of paramount importance in all cultures. Equally importantly, many human activities disrupt or disturb the natural water cycle, and this affects the major biogeochemical cycles as well as the availability of water resources.

Water remains one of our most important natural resources, and the allocation of supplies between competing users in ways that are fair, sustainable and produce acceptable levels of environmental damage remains a rather elusive goal in most countries. The United Nations Environment Program has concluded that there is no single best approach to sustainable water management and that decisions are best made by each country.

Plate 13.1 *Local villagers make extensive use of river water for potable water supplies in Abo Mkpang village, Cross River State, Nigeria.*

Photo: Uwem Ite.

BOX 13.2 INCREASING GLOBAL DEMAND FOR WATER

Demands on global water resources are increasing because of the growth of the world's population, rising standards of living and the expansion of agriculture and industry. Consumption of fresh water is believed to have increased by a factor of 35 since about 1700, and during the early 1990s global water use was growing at rates of between 4 and 8 per cent a year. Growth rates have been highest in the developing countries. By the year 2000, total world water demand was expected to be around 4,350 km^3 each year.

Plate 13.2 *Meltwater from the Arolla Glacier is used for hydroelectricity generation in the Grande-Dixence HEP scheme, southern Switzerland. Some of the meltwater is diverted from the main river through these abstraction culverts and carried along extensive underground pipes to a nearby storage reservoir, from where it is released through turbines to generate electricity.*

Photo: Chris Park.

Table 13.1 Freshwater withdrawals, by country and sector

Country	Year	Total freshwater withdrawal (km^3 per year)	Estimated per capita withdrawal in 2000 (m^3 per person per year)	Domestic use (per cent)	Industrial use (per cent)	Agricultural use (per cent)
USA	1995	469.00	1,688	12	46	42
Canada	1990	43.89	1,431	11	80	8
Ethiopia and Eritrea	1987	2.20	31	11	3	86
Japan	1990	90.80	718	17	33	50
China	1980	460.00	360	6	7	87
United Kingdom	1994	11.75	201	20	77	3
Russian Federation	1994	77.10	527	19	62	20

WATER RESOURCE USE

Water is a renewable resource. After it has been used it returns to the water cycle and in time it will be used again. In theory, there is plenty of water in the global cycle (see p. 352) to meet all present and expected human needs, but water resources are used unevenly around the world, and much water use is wasteful and inefficient. Problems of maintaining reliable supplies of usable water are not confined to drylands.

DEMAND FOR WATER

Water is the most basic and most important of all natural resources. Human bodies are largely water (by weight), and most food contains a high proportion of water. We require regular supplies of water, but it must be fresh water. Most comes from rivers, lakes and underground aquifers. Sea water can be used, but the salts dissolved in it must first be removed (by desalination; see p. 401) before it is suitable for people to drink or use directly. Demand for water has increased markedly in recent centuries (Box 13.2).

WATER USE

Water is used in many different ways, including:

- *Domestic uses*: such as water used for showering, washing clothes, and watering lawns and gardens.
- *Industrial uses*: water used for processing, washing and cooling in facilities that manufacture products.
- *Agricultural uses*: this includes water used in irrigation systems.
- *Thermo-electric uses*: including water used for cooling to condense the steam that drives turbines in the generation of electric power with fossil fuels and nuclear or geothermal energy.
- *In-stream uses*: water used for hydroelectric power generation, navigation, recreation and ecosystems.

Worldwide, irrigation agriculture consumes over 70 per cent of all the water used by people, industry accounts for a further 20 per cent, and domestic and municipal uses account for most of the remainder. The amount used, and what it is used for, varies a great deal from country to country (Table 13.1).

Definite improvements were made in global water provision during the 1980s – the International Drinking Water Supply and Sanitation Decade. For example:

- The proportion of rural poor without access to water was reduced from 80 per cent in the 1960s to less than 60 per cent by 1989.
- Clean drinking water was made available to 1.3 billion people.
- Sanitation services were offered to an additional 748 million people.
- Overall, the cost of water use was reduced and efficiency was increased through a combination of improved management practices, the use of appropriate, low-cost technologies, and the integration of supply and waste management systems

BOX 13.3 EFFICIENCY OF WATER USE

Most water resource use involves waste and inefficiencies. This arises partly because water is piped directly into most homes in developed countries, and it is widely regarded as a 'free good' that is there to be used. The true cost of water supplies, in terms of environmental costs – such as the damage caused by large dam schemes (see p. 396) – engineering costs and opportunity costs, is generally overlooked. Many water supply distribution networks also lose a great deal of water through leaks. Up to half of the water supplied to Western cities is lost and thus wasted by leaks in the supply systems.

It is estimated that nearly one-third of the world's inhabitants live in countries with severe water problems. By the early 1990s, a number of countries were using almost all of their available renewable water supplies. Belgium, Malta, Israel and Egypt use more than 70 per cent of their available resources. Libya, Qatar, Saudi Arabia, United Arab Emirates, the former Yemen Arab Republic and the former People's Democratic Republic of Yemen use more than their annual supply of renewable water resources, supplementing it with groundwater (from non-renewable aquifers) and desalination plants.

INEQUALITY AND INEFFICIENCY

Less than a tenth of the planet's freshwater resources are currently used by people, but those resources are not evenly distributed around the world, and neither is demand. As a result, some countries are threatened with

Plate 13.3 *Traditional irrigation agriculture in the rice fields of northern Vietnam, involving extensive use of manual labour assisted by water buffalo.*
Photo: Andy Quin.

critical water shortages either now or in the foreseeable future. The problem is made worse by the fact that water supplies in many areas are decreasing in quality because of pollution by chemicals, radioactive materials and suspended sediment. Extensive use is made of water from the Hwang Ho (Yellow River) in China, for example, but pollution – paradoxically, often caused by the very industries and activities that water supplies have encouraged to develop along the river banks – is a growing threat.

Water is a constantly renewable resource, and if it is used wisely there is enough water in the global water cycle to meet human demands for agricultural, industrial and domestic use. But there are real imbalances in the amount of water used in different countries, so available supplies are not evenly or fairly shared out. A typical person in the West uses up to 300 litres of water a day, nearly a third of which is flushed down toilets. A typical person in India uses about 25 litres a day. It is estimated that to maintain adequate health, people need a minimum of about 100 litres of water a day for drinking, cooking and washing.

Although global demand for water continues to rise, the development of water resources in some countries has been restricted by regional water shortages and environmental damage associated with existing water resource schemes.

WATER RESOURCE MANAGEMENT

Complex systems have been developed to try to match water demand and supply, over time and from place to place. Many problems arise from the fact that water resources are unevenly distributed, in both space and time, and it is not always of adequate quality for human use (Box 13.4).

As a direct result of these types of problem, water resource management requires co-ordinated long-term planning and long-term investment in engineering schemes (including storage reservoirs, water treatment plants, pipelines, aqueducts and sewerage systems).

Sustainable water resource management is now an objective in many countries, but it creates particularly difficult challenges in large countries that extend across a number of climatic zones. The former Soviet Union illustrates the problems well. Economic development of the former superpower was heavily dependent on large-scale integrated water resource development schemes designed to improve inland transportation, increase flood control, produce hydroelectricity and extend agricultural irrigation. Some huge public works were built into water management

BOX 13.4 WATER RESOURCE PROBLEMS

There are many different problem areas in water resource management, particularly:

- *Problems in space*: many sources of water are some distance from major areas of demand, so storage and transfer schemes are needed to redistribute it to where it is needed.
- *Problems in time*: rain hardly ever occurs when the water is most needed, and storage is essential so that it can be released when needed.
- *Problems of quality*: water often has to be treated to make it suitable for human use.

schemes, such as the ambitious plan to reverse flows in some major rivers, which was started in 1984. Such schemes are likely to create many environmental problems, including irreversible damage to estuaries, forests, groundwater and floodplain soils.

Water management strategies require constant monitoring and updating, because the level and type of demand change regularly over time. This is particularly true in the context of climate change (Box 13.5), which threatens to alter radically the availability and distribution of usable water supplies. Problems of water shortage are likely to be particularly serious in tropical countries, although many other countries are also highly susceptible to the hydrological consequences of climate fluctuations.

Water looks likely to be the focus of major international and inter-regional resource conflicts in the future (Figure 13.1). This will happen between countries as they debate who should have access to available water supplies in trans-frontier river systems. The long history of controversy over the allocation of water along the River Nile (see p. 58) is a good example.

Conflict is also likely within countries, between different users. In many countries, there is already lively debate about the best allocation of scarce water resources between agriculture, industry and domestic uses, and this will inevitably intensify as demand continues to grow. In California, for example, farmers and city planners have long been in conflict over water resources, and a great deal of effort has been invested in the search for a rational water policy that serves everyone and protects the environment at the same time.

BOX 13.5 WATER RESOURCES AND GLOBAL WARMING

According to a 1997 report by Resources for the Future:

- The timing and regional patterns of precipitation will change, and more intense precipitation days are likely.
- General circulation models (see Box 8.35) used to predict climate change suggest that a 1.5 to 4.5 °C rise in global mean temperature would increase global mean precipitation by about 3 to 15 per cent.
- Although the regional distribution is uncertain, precipitation is expected to increase in higher latitudes, particularly in winter. This conclusion extends to the mid-latitudes in most GCM results.
- Potential evapotranspiration (ET) (see p. 292) rises with air temperature. Consequently, even in areas with increased precipitation, higher ET rates may lead to reduced runoff, implying a possible reduction in renewable water supplies.
- More annual runoff caused by increased precipitation is likely in high latitudes. In contrast, some lower-latitude rivers may experience large reductions in runoff and increased water shortages as a result of a combination of increased evaporation and decreased precipitation.
- Flood frequencies are likely to increase in many areas, although the amount of increase for any given climate scenario is uncertain, and impacts will vary between rivers. Floods may become less frequent in some areas.
- The frequency and severity of droughts could increase in some areas as a result of a decrease in total rainfall, more frequent dry spells and higher ET.
- The hydrology of arid and semi-arid areas is particularly sensitive to climate variations. Relatively small changes in temperature and precipitation in these areas could result in large percentage changes in runoff, increasing the likelihood and severity of droughts and/or floods.
- Seasonal disruptions might occur in the water supplies of mountainous areas if more precipitation falls as rain than snow and if the length of the snow storage season is reduced.
- Water quality problems may increase where there is less flow to dilute contaminants introduced from natural and human sources.

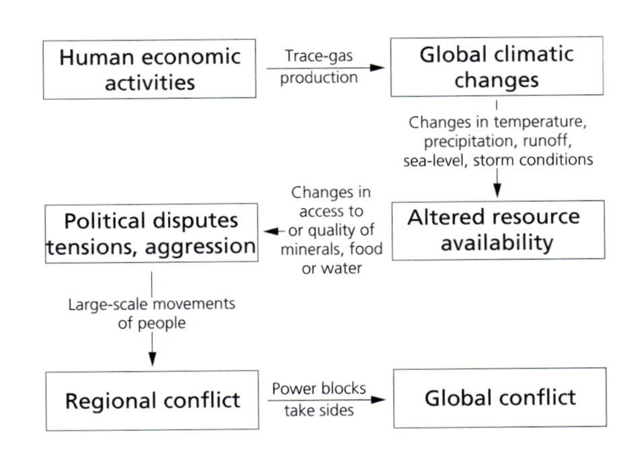

Figure 13.1 *Generation of international conflict over water resources as a result of global warming. Global warming is likely to change the availability of natural resources (particularly water), and this could promote political tensions between countries, which could escalate to regional or even global dimensions.*

Source: Figure 6.16 in Newson, M.D. (1995) Land, water and development. Routledge, London.

The need for clean water is escalating because of rapid population growth and industrialisation, coupled with rising standards of living. Three issues are crucial and must be addressed, dealing with availability (Box 13.6), access (Box 13.7) and quality (Box 13.8).

GROUNDWATER ABSTRACTION

Groundwater pumped from underground aquifers (see p. 393) is a major source of water supplies in many areas. In many aquifers, the groundwater has to be pumped out through boreholes or wells. As water is abstracted, the water table is lowered around the borehole.

Pumping is not necessary to exploit groundwater in confined aquifers, where a porous layer of water-bearing rock is sandwiched between two impermeable layers of different rock or rocks. In low-lying areas, the groundwater in the confined aquifer is under hydrostatic pressure, which forces it to flow out of an artesian well (Figure 13.2) so long as the aquifer is constantly recharged.

BOX 13.6 PROBLEMS OF WATER AVAILABILITY

- Only 3 per cent of the world's water is fresh, and 99 per cent of that is locked in ice caps and glaciers or flows or is stored underground. As a result, only 1 per cent is readily available for use.
- Water shortage is a major and recurrent problem in many regions, including most of Africa, the Middle East, much of South Asia, a large proportion of the western United States and north-west Mexico, parts of South America, and nearly all of Australia.
- In 1999, twenty-six countries were experiencing water scarcity; by 2025, it is likely that sixty-five countries will (including India, Korea, Nigeria, Peru and Poland).
- Poverty-stricken countries tend to be in the climate zones most susceptible to drought and other water problems, so they are least able to afford to import water from other sources.
- Regions with scarce water tend to have the highest rates of population growth. The result is that more than 300 million people live in regions affected by water shortages. This figure is expected to rise to over 3 billion people by 2025.

BOX 13.7 PROBLEMS OF ACCESS TO WATER RESOURCES

- Less than 20 per cent of the population in many developing countries has access to clean drinking water.
- In 1990, 243 million urban and 988 million rural dwellers were without access to potable (drinking) water, and this figure was expected to rise by an extra 2.1 billion people (813 million in cities and 1.3 billion people in the countryside) by 2000.
- In 1990, 377 million urban and 1.4 billion rural inhabitants were without adequate sanitation facilities. This was expected to rise by an extra 2.1 billion people (947 million in cities and 1.7 billion in the countryside) by 2000.
- Water supplies to an area can be increased in various ways (including building dams, canals and pipelines, and desalination plants), but these engineering solutions are expensive.
- Access to and shared use of water are major issues for many neighbouring countries, and hydropolitical issues can lead to major confrontations between countries.

BOX 13.8 PROBLEMS OF WATER QUALITY

- Untreated or partially treated sewage, agricultural chemicals and industrial effluents are major contaminants, with nitrates, metals and pesticides the main problems. As a result, many water supplies are being damaged by pollution, with declining water quality.
- Human and animal wastes introduce pathogens that cause serious diseases (including typhoid, cholera, amoebic infections, dysentery and diarrhoea). This accounts for over three-quarters of all disease in developing countries. Improvements in water supplies and sanitation could reduce child mortality by more than a half.

If rates of abstraction exceed rates of groundwater recharge in an aquifer, the water table can fall across a wide area. Borehole yields decline when this happens, and wells begin to dry up. Sometimes it is necessary to sink boreholes deeper into the aquifer, which lowers the water table further. Groundwater depletion follows, and water supplies are threatened if not reduced. Water shortages caused by over-extraction of groundwater commonly affect agriculture in dry environments, such as the Great Plains, California and southern Arizona in the USA.

Water table downdraw also creates problems for water quality (chemistry), particularly in coastal areas, where salt water enters the aquifer to recharge it (see p. 404). In some rock formations, toxic mineral salts dissolve in groundwater, and they increase in concentration as the volume of water is reduced.

Rising water tables pose a threat in some areas. The most common cause is a change in water use, such as the decline of heavy industries, which use large amounts of water and

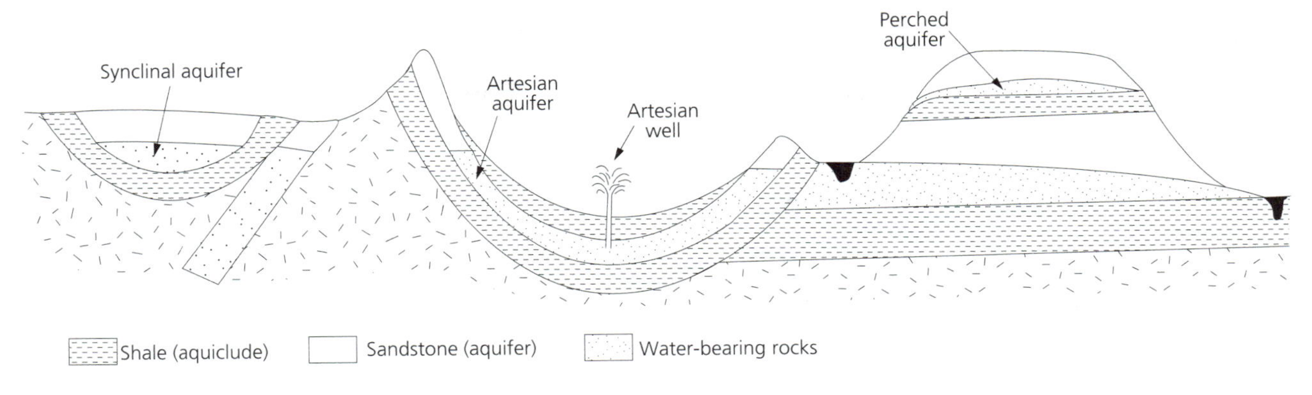

Figure 13.2 *Types of aquifer.*
Source: Figure 14.17 in Briggs, D. and P. Smithson (1985) Fundamentals of physical geography. Routledge, London.

BOX 13.9 SUBSIDENCE AND GROUNDWATER ABSTRACTION

Lowering the water table can cause ground subsidence (see p. 156), as the volume of material in the saturated zone decreases on drying. Between 1865 and 1931, for example, the ground surface in London (England) was lowered by between 0.91 and 1.21 mm a year, producing overall subsidence of up to 0.08 m. In a similar way, Mexico City has been sinking at a rate of up to 300 mm a year, and Tokyo subsided by nearly 4 m between 1892 and 1972 (at annual rates of up to 500 mm).

which previously obtained their water supplies from groundwater. London, England, is a good example (Box 3.10).

HYDROPOWER

Hydropower, or water power, is a free and renewable source of energy provided by falling water. Compared with the burning of fossil fuels, hydropower is an environmentally friendly way of producing power because it emits no carbon dioxide and little pollution. It is not trouble-free, however, because – among other things – it can displace people, damage land, destroy habitats and cause soil erosion and silting of rivers.

TECHNOLOGIES

Water has been used as a source of power for many centuries. Watermills make direct use of the force of water

BOX 3.10 RISING WATER TABLE UNDER LONDON

London's water table has been rising steadily in recent decades, threatening the stability of many important buildings and structures. The decline of heavy water-using industries such as brewing and engineering has caused the water table to rise by as much as 3 m a year. The water table was about 100 m below Trafalgar Square in 1905, but by the 1950s it was 80 m below. By 1995, it had risen to 50 m below the surface, and in 2000 it was about 40 m. This has put the water table within reach of some of the deepest parts of the Underground network and deep foundations in the City of London, some of which are only 20 m below the ground. Recovery of the water table continues, and it is expected to return to its normal level of 0–20 m below street level in parts of Westminster and the City by 2010. Experts fear that unless urgent action is taken, buildings with deep foundations and basements, some underground car parks, telecommunications and electricity cables, and parts of the London Underground system may become unstable by as early as 2005. Plans have been developed to drill a network of fifty boreholes to siphon off up to 70 million litres a day by 2004, at a cost of around £10 million. Some of the water is suitable for drinking, but close to the centre of London the groundwater is older and has a high salt content, so it is likely to be used for other purposes (such as park watering). Other cities threatened by rising water tables include Birmingham and Manchester in England, Paris (France), and Milan (Italy).

BOX 13.11 PUMPED STORAGE SCHEMES

Some large-scale modern HEP schemes (such as the Dinorwig scheme in North Wales) use pumped storage. Here the water that flows though the turbines is recycled by pumping it back to the upper storage reservoir during off-peak periods, where it can be used again to produce electricity during peak periods. Tidal power stations (see p. 492) are HEP plants that exploit the rise and fall of tides.

directed against the paddles of a waterwheel to make the wheel turn. This motion is transferred by simple gears to turn millstones for grinding grain.

It has only been in recent decades that efficient hydroelectric power (HEP) generation technologies have been developed and widely adopted. Traditional HEP schemes make use of natural slopes and topography to create a head of water. In a typical scheme, water is stored in a reservoir (often created by damming a river), from where it drops – under the influence of gravity – down pipes into water turbines, which are coupled to electricity generators. Steep mountain rivers with high discharges, such as glacial meltwater streams in Switzerland, provide ideal sites for hydropower developments. Not all HEP schemes are based on large rivers: by the mid-1990s there were more than 80,000 small 'mini-hydro' projects in China alone.

CONSTRAINTS

While the power that falling water produces is renewable, and almost any site on a river with a sloping bed could at least in theory be used, there are constraints that limit the number of sites that would be commercially viable (Box 13.12). Installations are capital-intensive, and initial set-up costs can be restrictively high.

Some proposed hydropower schemes – such as the Gordon-below-Franklin dam in Tasmania – have been so controversial and attracted such widespread public opposition that they have not been built. Other controversial schemes are going ahead, but in the face of massive international concern and opposition. Recent examples include the Three Gorges Dam in China (see Box 13.16), which will displace more than a million people, and the Narmada River dams in India, which will force the movement of up to 10 million people.

BOX 13.12 SITE REQUIREMENTS FOR HYDROPOWER PLANTS

Most hydropower sites require dams and embankments to store and provide a head (drop) of water, and these are costly to build. Silt is often deposited in the storage ponds and reservoirs, so over time their storage capacity decreases. Dams and reservoirs have other environmental impacts (see p. 396), including loss of agricultural land, loss of natural habitat, displacement of local people, spread of water-borne diseases and rising local water tables. These should all be taken into account in the evaluation of the cost-effectiveness and suitability of HEP schemes, but they rarely are.

BOX 13.13 PUBLIC ATTITUDES TOWARDS HYDROPOWER

Many rivers with the greatest water power potential are already being exploited. Public opposition to large hydropower installations is mounting, and this is promoting a re-evaluation of sustainable energy options in many developing countries. The mighty Amazon River in Brazil has, at least in theory, vast potential for HEP, but other alternatives (including biomass) are now being explored as well. Small hydropower plants, using small tributaries rather than the main river and designed to serve local towns, have many advantages over large centralised HEP schemes because they have much smaller environmental impacts, supply energy continuously, require little maintenance and do not need long transmission lines running through the Amazon rainforest.

ENERGY PRODUCTION

HEP provides more than 20 per cent of the world's electricity, and it is now the largest renewable source of electricity around the world. Capacity has increased fourteen-fold since 1950, but the spread between countries is very uneven. Hydropower provides more than two-thirds of total electricity production in thirty-five countries, and many African countries rely particularly heavily on hydropower (Table 13.2). Developing countries generate 37 per cent of the world HEP total. The USA and Canada

Table 13.2 Countries that were heavily reliant on hydroelectricity production in 1996

Country	*Percentage of electricity generated by hydropower*	*Hydroelectric production (GWh a^{-1})*
Democratic Republic of Congo (formerly Zaire)	99.9	5,550
Zambia	99.8	8,102
Malawi	99.5	850
Congo	99.5	352
Cameroon	98.5	2,778
Rwanda	98	230
Uganda	98	940
Mozambique	92.2	336
Namibia	90.1	1,134
Ethiopia	87	2,000

produce more hydroelectricity than any other countries (about 13 per cent of the world total each). Norway gets 95 per cent of its electricity from falling water, and Sweden gets roughly half.

In other countries, HEP provides relatively little power. Less than about 2 per cent of the electricity produced in the United Kingdom, for example, comes from hydropower, and most of that is concentrated on steep upland rivers in Scotland.

Hydropower is exploited in many countries, but most plants are small and produce only limited amounts of electricity. In the USA, for example, by the late 1990s there were only about five HEP plants with generating capacities larger than 1,000 MW. Some plants have great capacity. The Grand Coulee plant, in Washington State (USA), for example, has a power output of about 10,000 MW, while the Itaipu power station on the Paraná river (Brazil/Paraguay) has a potential capacity of 12,000 MW.

It is estimated that about a quarter of the world's HEP potential had been exploited by the mid-1990s. Large-scale growth of hydropower is unlikely to continue significantly in the future, partly because of the lack of suitable sites and partly because of the environmental and social impacts of such schemes. The high capital cost per unit of electricity produced is a further constraint. Small-scale local hydropower schemes appear to have much greater potential, particularly in developing countries, because they are cheaper, quicker to install, easier to maintain and cause less damage.

IRRIGATION

Most plants need water for healthy growth, and water shortages decrease plant yields and increase plant stress and susceptibility. Many countries suffer from water shortages, and complex irrigation schemes have been developed to supplement water supplies to dry agricultural areas by means of dams and artificial channels. Irrigation agriculture is a traditional form of agriculture along the Nile valley (see p. 58), for example.

Irrigation of croplands is by far the biggest agricultural use of fresh water, and it grows bigger over time as the area under irrigation continues to expand. By 1990, about 15 per cent of all the farmland in the world was irrigated, although the proportions varied from 6 per cent in Africa to 31 per cent in Asia. The total area under irrigation increased by more than a third between 1970 and 1990, with most of the increase concentrated in developing countries.

Output from the irrigated land is more than double that from the land not irrigated, and one-fifth of the world's food is grown on irrigated land. Food production, particularly in many developing countries, can often be increased only by improving or rehabilitating existing irrigation systems and by building new irrigation projects.

The continued expansion of irrigation agriculture is causing mounting concern about the sustainability of water supplies in many areas. So much water is now abstracted from some major rivers – including the Nile, Euphrates and Jordan (in the Middle East) and the Colorado (USA) – for irrigation and other uses that flows are seriously reduced in dry years. Some areas are facing serious environmental problems because of the over-extraction of water for irrigation.

ENVIRONMENTAL PROBLEMS

There is also mounting concern over the environmental problems associated with extensive irrigation. Irrigation generally brings significant benefits by allowing an expansion of the area under agriculture and thus an expansion of food production. But many irrigation schemes, particularly poorly designed ones, give rise to some important environmental problems. Two widespread problems are waterlogging (p. 399) and salinisation (p. 403) of soils, often as a result of excessive inputs of water into soil systems with

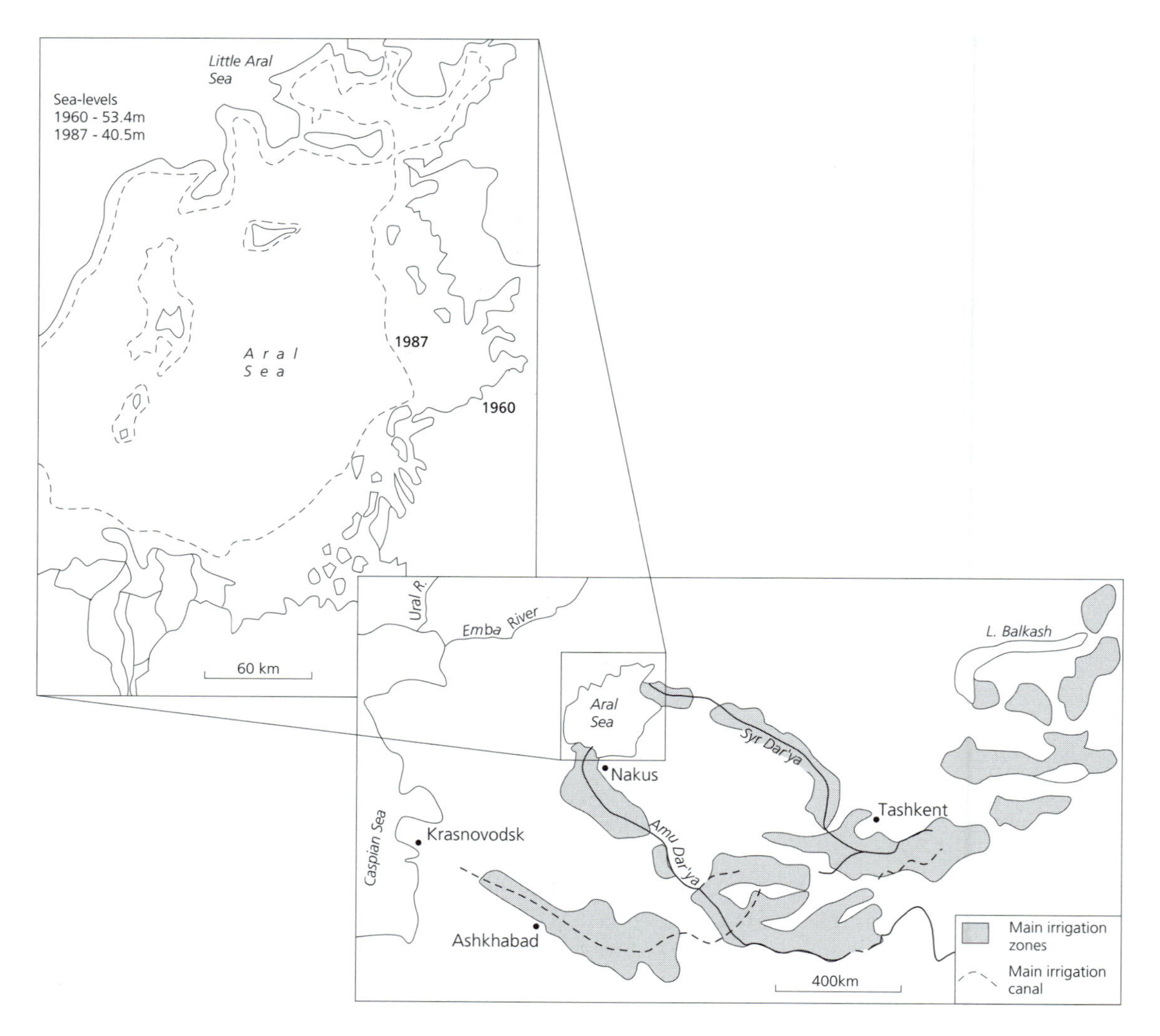

Figure 13.3 *Changes in the extent of the Aral Sea between 1960 and 1987. See Box 13.14 for explanation.*
Source: Figure 6.13 in Marsh, W.M. and J.M. Grossa (1996) Environmental geography: science, land use and Earth systems. John Wiley & Sons, New York.

inadequate drainage. Crop productivity and yields generally decline because of rising water tables (waterlogging) and the build-up of mineral salts in the waterlogged soils.

One of the most dramatic examples of damage associated with irrigation is the desiccation of the Aral Sea (Figure 13.3) in Central Asia (Box 13.14). Among the worst-affected areas are irrigated croplands in parts of Egypt, India, Iraq, Mexico, Pakistan, the former Soviet Union and the USA. Up to 3,000 km^2 of irrigated cropland around the world is believed to be seriously affected by salinity problems.

Many large irrigation schemes, particularly those built during the 1950s and 1960s, include large reservoirs that create social and environmental damage. A number of them, including the Aswan High Dam in Egypt (see p. 62), have turned out to be financially questionable as well as environmentally damaging. River silt, which usually contains high concentrations of nutrients, is often deposited behind irrigation storage dams rather than being made available to land and fisheries downstream.

LARGE DAMS

Many water resource schemes rely on dams to impound river flow and allow it to be redistributed from place to place, and over time, so that it is made available when and where it is needed. We have already seen many of the important environmental impacts of large dams in the

BOX 13.14 DESICCATION OF THE ARAL SEA

The Aral Sea is really a very large freshwater lake – formerly the world's fourth largest lake – on the border between Kazakhstan and Uzbekistan in Central Asia. In the past it supported a thriving fishing industry, but since the 1960s the lake and surrounding area have been seriously damaged as a consequence of unwise and unsustainable water resource management. This part of the former Soviet Union was earmarked by the Soviet authorities for major agricultural development, to be made possible by a large-scale expansion of irrigation. The objective was to make this area self-sufficient in cotton and increase rice production. The strategy involved diverting water from the two main rivers – the Amu Dar'ya and the Syr Dar'ya – that flow into the Aral Sea and provide 90 per cent of its water.

The scheme went ahead, large quantities of water were diverted, and the irrigated area was more than doubled from about 0.03 million km^2 in 1950 to nearly 0.07 million km^2 by the late 1980s. Almost half of the water that previously drained into the Aral Sea was diverted into irrigation systems, and as a result the lake is drying up. Since 1960 its surface area has shrunk by nearly half, and its volume has decreased by two-thirds. The former shoreline is now abandoned, often long distances from the current shore of the shrunken lake. By the early 1990s, the late had separated into two relatively independent water bodies, and most scientists concluded that total restoration would be impossible.

The lake is drying up because of a lack of input from river flow, but the problem is made worse by the high evaporation rates in this hot, dry climate. As water is evaporated from the lake, the salts dissolved in it are left behind, and salt concentrations in the lake water are increasing. Salinity has increased threefold since 1960, which makes the previously fresh water in the Aral Sea now almost as salty as water from the open ocean. Most native freshwater fish species have disappeared from the lake because of rising salinity. This has had a significant knock-on effect on the local fishing industry, which in the early 1960s caught nearly 40,000 tonnes of fish but is now almost extinct. Fish species that can tolerate high salinity have been introduced, but competition between the native and introduced species has led to population declines among both.

As the lake level falls, vast areas of what was previously fine-grained lake bed deposits are exposed to drying heat. The surrounding area is rapidly turning into a wasteland of salty bogs and desert. Dust storms are becoming more frequent, and strong winds blow the sediments and salts (including sulphates and chlorides, which are poisonous to plants) across a wide area, spreading the environmental damage across the entire region.

In the past, the large water body had a mediating effect on local climate, which was milder and less extreme than the surrounding area. As the lake shrinks, this climatic influence is declining, which also spreads the impacts of the irrigation scheme across a wide area. Local climate change has set in motion a feedback loop that increases desertification, through higher temperatures (summer temperatures reach 45 °C) and greater aridity (humidity has fallen by 28 per cent).

Water that drains off the irrigated fields usually contains relatively high levels of pollutants (from toxic pesticides, defoliants and fertilisers used on the cotton and rice fields) and dissolved mineral salts (washed out of soils). Much of this water finds its way back into the two main rivers (Amu Dar'ya and Syr Dar'ya) and contaminates local groundwater water supplies. As a result, human health has been affected throughout the area, with increased incidence of cancer, liver and kidney disease, and child mortality.

It is clear that not enough thought was given in advance to predicting the likely environmental impacts of expanding irrigation in the area. Some Soviet scientists warned that the scheme would be disastrous if it went ahead, but their warnings were largely ignored. With the benefit of hindsight, it has become obvious that not enough attention was paid to the likely environmental and health effects of the scheme when it was originally conceived.

Various strategies for solving the problems of the Aral Sea have been proposed, including better conservation of existing water resources in the area and restoration of the sea itself. This would require a total revision and reorientation of the development strategy for the region and would require extensive modernisation of the irrigation system. Planting of forests might help to stabilise the desert expansion and modify local climate. There are even ambitious proposals to divert water from western Siberia into the Aral Sea region to try to restore the hydrology there.

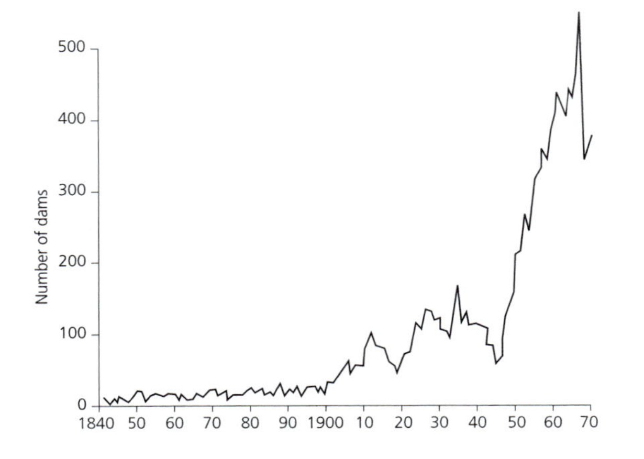

Figure 13.4 *History of large dam construction, 1840–1970. A number of large dams were built during the nineteenth century, but during the twentieth century dam building increased at an almost exponential rate, at least until the late 1960s.*

Source: Figure 1 in Beaumont, P. (1978) Man's impact on river systems – a world wide view. Area 10: 38–41.

BOX 13.15 HISTORY OF LARGE DAMS

Dams have been built on many river systems over the last 5,000 years. The first dams were small constructions designed to control floods and to supply water for irrigation and domestic use. Dams were later used to drive waterwheels and more recently to generate hydroelectric power (see p. 394). Since the 1930s, many large dams have been built, particularly between the early 1950s and early 1970s (Figure 13.4). By 1986 there were more than 36,000 large dams (over 15 m high) in the world, nearly two-thirds of which were in Asia (half of the world total was in China).

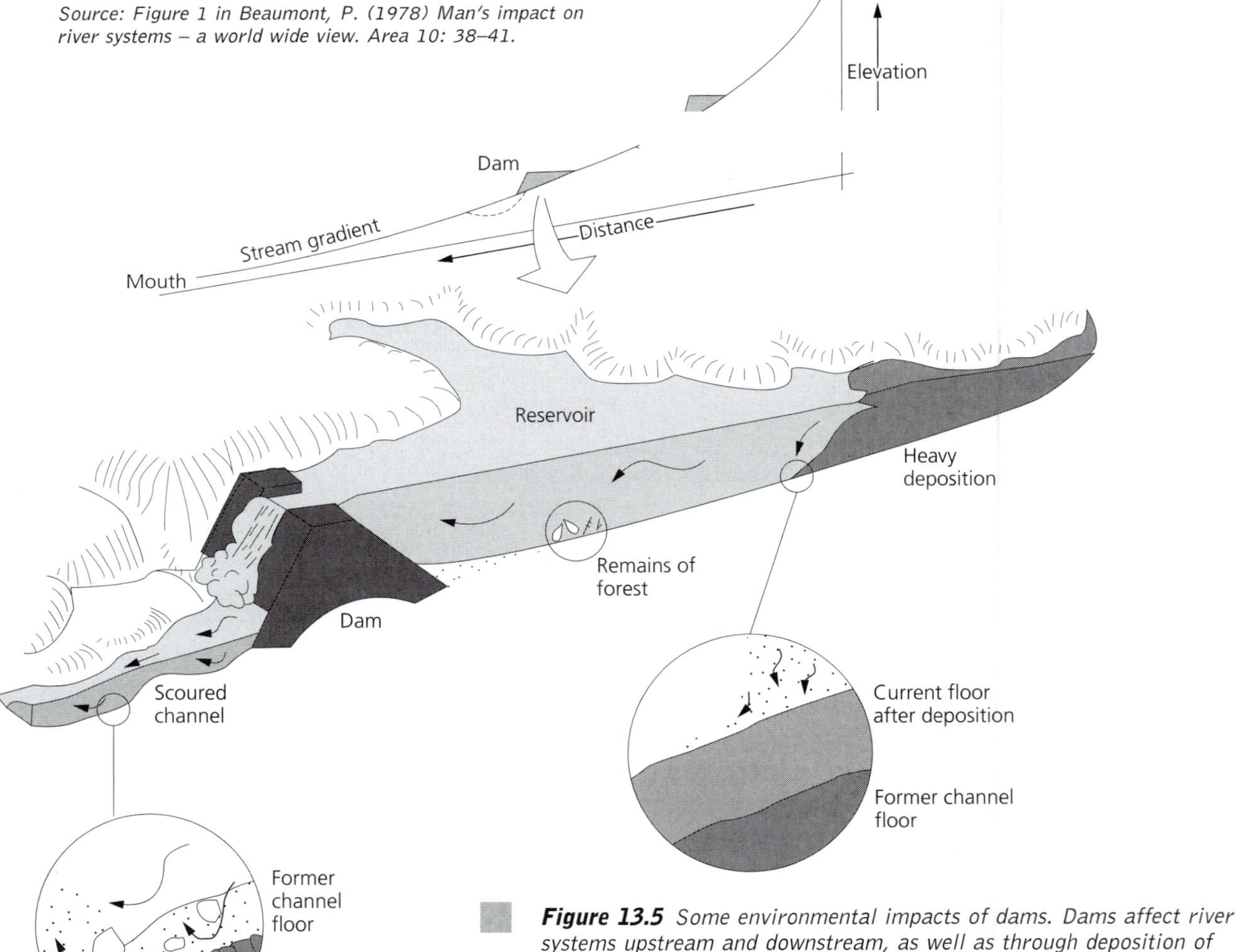

Figure 13.5 *Some environmental impacts of dams. Dams affect river systems upstream and downstream, as well as through deposition of sediment on the reservoir floor.*

Source: Figure 10.17 in Marsh, W.M. and J.M. Grossa (1996) Environmental geography: science, land use and Earth systems. John Wiley & Sons, New York.

Plate 13.4 *Hoover Dam on the Colorado River on the border between Arizona and California, USA. Lake Mead, behind the dam, is a storage reservoir that drowned part of the Grand Canyon. The hydroelectricity plant is clearly visible below the arch-shaped concrete dam.*
Photo: Chris Park.

River Nile case study (see p. 68). Many large dams produce environmental impacts upstream, downstream and in the vicinity of the reservoir (Figure 13.5).

The Balbina Dam in Brazil illustrates many of the controversies surrounding large dam developments. The dam, in the state of Amazonas, floods a relatively flat area and it covers 2,360 km^2 of what was previously natural rainforest. It receives runoff from a small drainage basin, so river flow is limited and the scheme generates only about 112 MW of electricity. The dam was completed late in 1987, and it was designed to supply much of the electricity needs of the city of Manaus. But it took so long to build, and the population in the city rose so fast during construction, that additional alternative energy sources were soon required. Among the more serious short-term consequences of dam construction were loss of valuable rainforest and displacement of about a third of the surviving members of the Waimiri Atroari tribe.

Few dam schemes are so controversial and have attracted such widespread attention and condemnation on environmental grounds as the Three Gorges Dam on the Yangtze River in China (Box 13.16).

WATER RESOURCE PROBLEMS

WETLANDS AND DRAINAGE

WATERLOGGING AND DRAINAGE

Waterlogged soils are soils that are saturated with water. This typically happens when drainage is poor and more water is added to the soil than can be lost from it by evaporation and transpiration. It often accompanies irrigation and can be tackled by installation of suitable drainage and reduction in the supply of irrigation water. The process of waterlogging is usually a cumulative one, building up over a period of time.

Waterlogging makes soils less suitable for plant growth, and agricultural yields can decline significantly when it occurs. As the soil water content builds up, soil around plant roots becomes saturated, even if no water is visible at the surface. Saturation reduces plant productivity and growth and sustained waterlogging can kill the plants and turn the area into a marshy wetland.

Excess water can be removed by installing appropriate drainage systems, which are often an integral part of large-scale irrigation schemes. Traditional farming practices include land drainage on sloping ground by digging ditches aligned downslope, to drain water into nearby streams and rivers. Underground drainage systems are now widely used as well, particularly on flat ground. There are many different types of underground drainage system, including perforated pipes buried in trenches and mole drains (which are formed by dragging a metal cylinder beneath the ground at a pre-defined depth).

WETLANDS

A wetland is an area of swampy or marshy ground. While many important wetlands are found on or near the coast (see p. 497) – particularly along undrained coastlines – there are also many different types of freshwater wetland. These include swamps, marshes, fens, wet meadows,

BOX 13.16 THE THREE GORGES DAM SCHEME ON THE YANGTZE RIVER IN CHINA

Throughout the 1980s, the Chinese authorities developed plans to build a massive dam across the Yangtze River in the Three Gorges (Sanxia) area. It was to be a large multipurpose water resource project designed to achieve four main goals:

1. *Flood control*: the primary objective of the scheme was to prevent flooding along the densely populated middle and lower reaches of the Yangtze.
2. *Energy generation*: the project was also intended to generate the large amounts of electricity needed for industrial expansion in south-central and eastern China.
3. *Water supply*: the scheme should significantly increase the availability of water for agricultural, industrial and domestic users in the region.
4. *Navigation improvement*: the dam was also designed to improve the navigation of modern boats between Chongqing and Yichang, which in the past has been severely restricted by a series of rapids, shoals and shifting shallows.

Environmentalists argued that the dam could be a disaster because of its huge costs and uncertain cost-effectiveness, its possible environmental and social impacts, and the possibility that it would fail to meet all or even most of its main objectives. Controversy over the Three Gorges project began well before construction work started and was fuelled by a polarisation of views between those who supported the scheme and those who opposed it. In the late 1980s, for example, a group of Canadian engineering companies carried out a study of the proposed scheme which concluded that it should go ahead. At the same time, a group called Probe International examined the proposals and criticised them heavily. The pro-dam engineers and anti-dam protesters had little in common, and project planning was heavily influenced by the engineering view.

Although the ambitious scheme has gone ahead, criticism of it has not died down. Many engineers doubt that the dam will provide effective flood protection over the whole area it is intended to. Some fear that it could actually increase the risk of a high-magnitude, low-frequency flood (with a recurrence interval of perhaps 1,000 years). Critics also point out that the dam could become a strategic military target, which would be hard to defend and catastrophic if it were badly damaged. Geological studies indicate that a reservoir of the size of Three Gorges, in that setting, could possibly trigger strong earthquakes of magnitudes between about 4.7 and 6.

There is also concern over the wider possible impacts of the scheme. It has already displaced up to 1.2 million people from that portion of the Yangtze valley, from more than 320 villages and 140 towns. This has set in motion large-scale population migration, which will affect the areas that receive the involuntary refugees. Much of the land that is being flooded in the scheme has been intensively farmed, and the land loss will inevitably increase pressure on remaining habitats and environments adjacent to the river. Fears have also been expressed that the dam scheme will alter regional hydrology so much that there will be significant increases in salinisation and waterlogging of local soils, further reducing agricultural output.

Early in 2000, China announced plans to build another large dam on the Qingjiang River (a tributary of the Yangtze) near the Three Gorges Dam. The new dam would cost around US$1.5 billion, be completed in 2009 and would regulate flooding in the lower Yangtze basin. Critics say that this is a costly attempt to address some of the problems that the Three Gorges scheme should have been designed to tackle.

The dam is scheduled to be complete by 2009, and it will have been the biggest, most expensive (at US$25 billion) and perhaps most hazardous hydroelectric project in world history. The dam – 2 km long and over 200 m high – will create a reservoir the size of Lake Superior, stretching over 600 km through the canyons of the valley between the dam site near Yichang upstream to the large city of Chongqing.

Table 13.3 Types of freshwater wetland

Type	*Description*
Bog	Wet spongy ground consisting of decomposing vegetation, which ultimately forms peat
Fen	Low-lying flat land that is marshy or artificially drained
Lake	An expanse of water entirely surrounded by land
Marsh	Low, poorly drained land that is sometimes flooded and often lies at the edge of lakes and streams
Meadow	A low-lying piece of grassland, often boggy and near a river
Pond	A pool of still water, often artificially created
Swamp	Permanently waterlogged ground that is usually overgrown and sometimes partly forested

ponds, lakes and both lowland and upland bogs (Table 13.3).

Attitudes towards the world's remaining wetlands have changed a great deal since about the 1960s. Before that, wetlands were usually regarded as wastelands that would be of value only after they had been drained, reclaimed or filled in. Many wetland sites were used for dumping rubbish and sewage. Since the 1960s, as more and more wetlands were destroyed and understanding of the many different values of wetlands grew, there has been growing concern to protect them much better.

LOSS AND PROTECTION OF WETLANDS

Appreciation of the values of wetlands has increased as the total area of wetlands has shrunk. It is estimated that by 1985 1.6 million km^2 of wetlands around the world had been drained, and the destruction continues. The most important causes of loss and conversion are drainage for agriculture, plantation forestry and urban expansion. The total area of wetland in the USA fell by 1.1 per cent between 1982 and 1987: 48 per cent of the loss was attributed to conversion to urban and built-up land, and 37 per cent to agricultural development.

As appreciation of the importance of wetlands continues to grow, so too does concern to protect what remains. In some countries, new wetland areas are being created on nature reserves, and in many countries important existing wetlands are protected in nature reserves and by appropriate local land-use planning. The Somerset Levels in south-west England cover an area of nearly 600 km^2 and are one of Britain's largest surviving lowland wet meadow habitats and home to a rich variety of plants and animals. Agricultural improvement of the wetlands, particularly through extensive grant-aided land drainage between the early 1970s and mid-1980s, significantly reduced the area and ecological value of the habitat. In 1986, the Somerset Levels were designated an 'Environmentally Sensitive Area', and landowners are now given financial incentives to maintain agreed land management practices that benefit wetland conservation.

In the Netherlands, where wetland habitats make up a large proportion of the land area, attempts are being made to develop conservation strategies and protect the wetland areas (Box 13.18).

BOX 13.17 WETLAND ECOSYSTEMS

All wetlands are important ecosystems that provide habitats for a wide variety of plants, animals and birds, but wetlands are used by people in many different ways, and they offer a range of important benefits and services. Many wetlands are centres of great conflict between different groups of users, who each seek different things from the same resource. Competition between users often leads to over-use or even destruction of the resource.

The lowland wetlands of the Norfolk Broads in eastern England illustrate the diversity of interests represented. The Broadlands are extremely popular for recreation, and to the tourist the wetlands are synonymous with summer boating holidays. To the conservationist they represent a unique, internationally important ecosystem. To local residents they are an appealing place to live, work and relax. To farmers they are a place to earn a living from agriculture. Unless more sustainable ways can be found of reconciling and reducing these often conflicting demands, the survival of wetland areas like the Broads will be threatened.

SALINITY

Most water in the land phase of the global water cycle is fresh water, while water in the oceans and seas, which contains dissolved material (particularly metallic salts), is saline. The composition of ocean water appears to vary relatively little from place to place (Table 13.4), although the overall concentrations of dissolved salts do vary a great deal. Fresh water in lakes, rivers, soils and groundwater

BOX 13.18 CONSERVATION OF WETLANDS IN THE NETHERLANDS

Most of the Netherlands is low-lying, close to sea level, and much of the present land area has been reclaimed and drained since the seventeenth century (Figure 13.6). Indeed, more than half of the land area is made up of reclaimed wetlands, and 16 per cent of the land is internationally important wetland. There are four main types of wetland:

1. coastal ecosystems;
2. large riverine systems;
3. base-rich freshwater systems;
4. nutrient-poor freshwater systems.

Survival of many of the country's wetland areas is threatened over the short term by eutrophication (see p. 409), toxic pollution and hydrological changes, and over the longer term by climate change (particularly by increases in temperature and ultraviolet radiation). High population density and lack of alternative sites pose particular problems. Great efforts are being made to conserve the remaining wetland sites and to rehabilitate and restore degraded and damaged sites. The ecologically most important wetland sites are registered for international protection under the Ramsar Convention – the Convention on Wetlands of International Importance – especially as waterfowl habitats. Management strategies are being developed in local and national policy plans to improve the sustainability of wetland use throughout the Netherlands.

Table 13.4 The composition of salts in seawater

Compound	Per cent
Sodium chloride	77.8
Magnesium chloride	9.7
Magnesium sulphate	5.7
Calcium sulphate	3.7
Potassium chloride	1.7
Calcium carbonate	0.3
Other	1.1

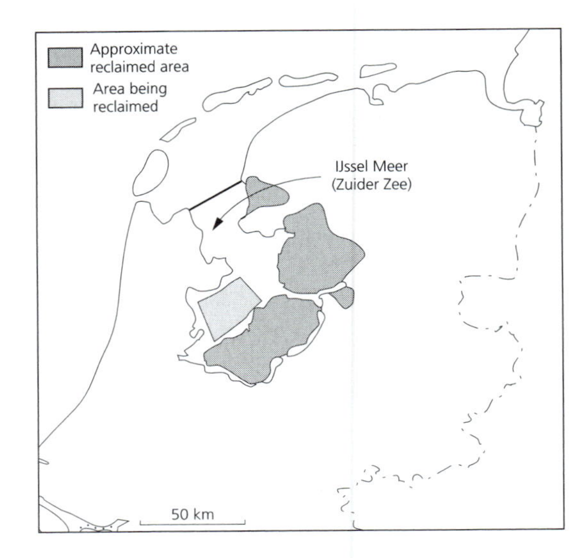

Figure 13.6 *Reclaimed land in the Netherlands. The map shows the approximate area reclaimed from the Zuider Zee to produce low-lying but fertile farming land.*

Source: Figure 19.14 in Doerr, A.H. (1990) Fundamentals of physical geography. Wm. C. Brown Publishers, Dubuque, Iowa.

does contain variable amounts of dissolved material, including salts, but at nowhere near the concentrations of seawater. The material dissolved in river flow is the solution load (see p. 367).

Two critical problems are associated with water that has high salinity:

1. It is damaging to and can kill plants, so it is unsuitable for irrigation agriculture and damages natural vegetation and forests.
2. It cannot be drunk by people, so it is unsuitable for domestic water supply.

Salinity is not a fixed property of water, however, because it can be increased (usually inadvertently, through salinisation) or decreased (on purpose, by desalination).

SALINISATION

Salinisation is the process by which the salinity of soils and groundwater increases. It occurs naturally in two main ways:

1. by the intrusion of seawater into a coastal aquifer, which replaces fresh water with saline water (Box 13.20; Figure 13.7); and

BOX 13.19 MEASURING SALINITY

Salinity is a measure of the concentration of dissolved salts in a body of water, usually expressed in parts per million (ppm) by volume. Seawater usually has a salinity of around 35,000 ppm, about 30,000 ppm of which is sodium chloride (NaCl, common salt). Some typical salinity values are listed in Table 13.5. Brackish water is slightly salty, but 'potable' – a word used widely in the water industry that simply means drinkable.

Table 13.5 Some typical salinity values

Source of water	*Salinity (ppm)*
Typical seawater	35,000
Arabian Gulf	44,000
Baltic Sea	2,000–3,000
Brackish well water	1,500–6,000
Approximate limit for irrigation	1,000
Potable well water	1,300–500

BOX 13.20 SALINISATION OF COASTAL GROUNDWATER

Groundwater intrusion is a principal cause of salinisation in many coastal areas. It occurs mainly where large quantities of fresh groundwater are abstracted for supply purposes (usually by pumping it out) and more is taken out than naturally flows in to replenish it. Along the coastal zone there is an ever-present risk that salt water, which is permanently stored in sediments beneath the sea bed, will percolate inland beneath the freshwater aquifer. This is most likely to happen when the aquifer is depleted by over-extraction and the boundary between the salt and fresh water moves further inland and closer to the surface. If the process continues, salt will eventually penetrate the soil, where it damages plants.

2. by the evaporation of moisture from soils in hot dry climates, which leaves a salty residue on the soil surface.

Salt water intrusion in coastal aquifers has affected many areas, particularly in arid climates. The United Nations Environment Program estimates that by the early 1990s it had already affected 70,000 km^2 in China, 200,000 km^2 in India, 32,000 km^2 in Pakistan and the Near East, and 52,000 km^2 in the USA. It also affects parts of southern Europe.

Salinisation also occurs away from the coast, in hot, dry areas. Water is lifted in soils from the aquifer below by capillary attraction (see p. 358). It is taken into plant roots by osmosis and helps the plants to grow. But, particularly in hot, dry climates, as the water rises in the soil some of it evaporates, and the salts that are naturally dissolved in it are precipitated (deposited) on and near the soil surface. The process partly drives itself, because as more water is evaporated, capillary attraction is increased, which forces more water to rise from below to replace what has been lost. Precipitation of dissolved salts is thus a non-stop process, and salts accumulate on and near the surface. Over time, salinity in the upper soil increases and plants start to show symptoms of stress. This type of salinisation is widely associated with irrigation agriculture in drylands (Table 13.6), particularly in Australia, China, Egypt, India, Pakistan, the former Soviet Union and the USA.

History records many examples of salinisation associated with irrigation schemes. Mesopotamia, in the Middle East, provides a graphic example. Here inadequate supervision of irrigation canals led to increased salinity in soils, which caused decreased agricultural yields and helped to bring about the collapse of eleven empires in the period between 2400 and 1700 BC.

Table 13.6 Extent of salinisation of irrigated cropland in some dryland countries

Country	*Percentage of irrigated land affected by salinisation*
Australia	15–20
China	15
Egypt	30–40
Iraq	50
Pakistan	35
Spain	10–15
Syria	30–35
USA	20–25

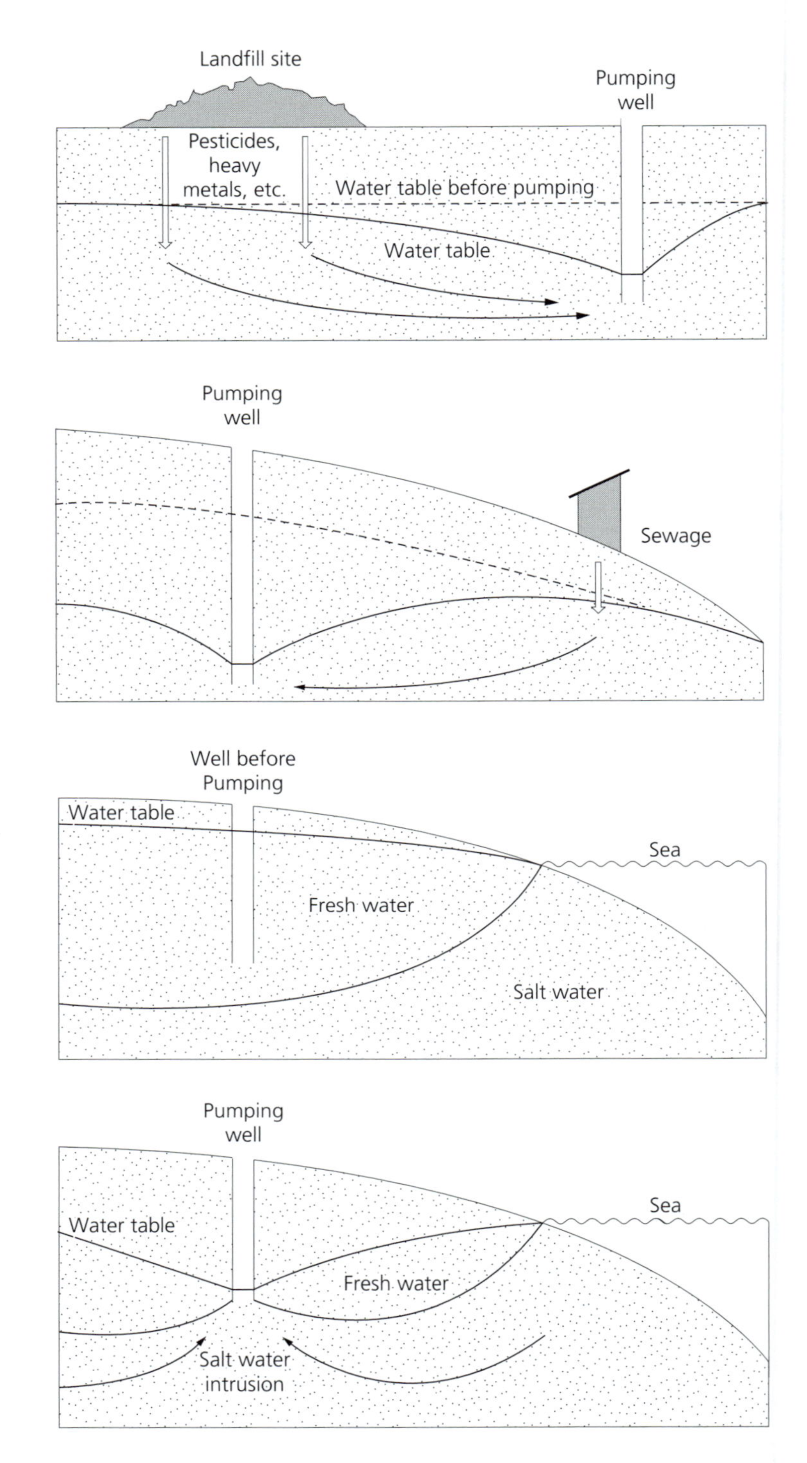

Figure 13.7 *Intrusion of seawater into coastal aquifers. The lower two figures show how saline seawater can seep into coastal aquifers and alter the suitability of well water for drinking and other uses. The upper two figures show similar effects of toxic waste materials and sewage.*

Source: Figure 14.8 in Doerr, A.H. (1990) Fundamentals of physical geography. Wm. C. Brown Publishers, Dubuque, Iowa.

Salinisation also occurs naturally, particularly in dry areas where irrigation agriculture is not practised. So-called secondary salinisation, induced by human activities, accounts for only about 10 per cent of Africa's saline soils, although it does affect half the area of irrigated land.

DESALINATION

Desalination is the process of removing dissolved salts, particularly from seawater, so that it can be used for domestic, agricultural and industrial purposes. The two most common uses of desalinated water are for drinking and for irrigation. There are a number of ways of achieving this (Box 13.21).

Desalination is widely used in the Middle East and the USA (Table 13.7). All of the different processes are expensive (particularly in energy terms). One problem associated with desalination is what to do with the concentrated brine (salty solution) that is left. Some of the salt – including sodium chloride (common salt), metallic magnesium, magnesium compounds and bromine – can be extracted and sold. But there are few economic uses for much of the saline waste, for which suitable and cost-effective methods of disposal have yet to be found. Brine wastes are usually released from desalination plants back into the sea, where they increase water temperature, salinity and turbidity, and harm marine wildlife.

Table 13.7 Desalination capacity for the ten countries with the largest capacity, 1996

Country	*Capacity (m^3 per day)*
Saudi Arabia	5,006,194
United States	2,799,000
United Arab Emirates	2,134,233
Kuwait	1,284,327
Libya	638,377
Japan	637,900
Qatar	560,764
Spain	492,824
Italy	483,688
Iran	423,427

BOX 13.21 TECHNOLOGIES FOR DESALINATION OF WATER

A number of approaches are available, including:

- *electrodialysis*: this involves passing an electric current through the saline water in a container. The current causes salts to be concentrated in compartments in the container, and salinity declines in the rest of the water. It is an expensive and energy-intensive process.
- *reverse osmosis*: this involves passing the saline water through a permeable membrane (film), normally under pressure. Relatively pure water passes through the membrane, leaving much of the dissolved salt behind.
- *freezing*: when a solution of salty water is frozen, crystals of pure water grow, which can be separated from the salty concentrate left behind.
- *distillation*: this involves boiling or evaporating seawater to produce water vapour, which can then be condensed. The water that condenses is relatively salt-free.

Despite the high costs, energy requirements and waste disposal problems, it seems highly likely that desalination will be more widely adopted in the future as demand for fresh water continues to rise and water shortages – real or threatened – become more common. Global warming and associated shifts in the main climate zones will doubtless accelerate this trend.

An alternative approach, rather than desalinating water

Plate 13.5 *Desalination plant on the coast of the Canary Island of Lanzarote, off the north-west coast of Africa.*

Photo: Philip Barker.

for use in agriculture, is to make more use of the salty water that is available in such large amounts. Techniques are being developed for seawater-based agriculture, in which seawater and nutrients from the sea are brought on to the land for aquatic animal production and halophyte (plants that grow in very salty soil) farms.

WATER QUALITY AND POLLUTION

Water pollution is one of the most widespread problems within water resource management. While great efforts have been made to protect and improve water quality in many countries, water pollution continues to become more common and cause more damage, particularly in many developing countries. Water quality poses many challenges in the search for more sustainable approaches to development (see Box 13.8).

POLLUTION AND POLLUTANTS

Pollution means the contamination of fresh water, which decreases its purity and often makes it unsuitable for water resource use if not dangerous to human health. Pollution also damages aquatic habitats and the plants and animals that live in them. We tend to think of pollutants as being poisonous, and while this is certainly true of many types of water pollutant, many others are harmful rather than poisonous. Some pollutants are highly toxic and thus damaging at any concentration, whereas many are only a problem when they accumulate to critical concentrations. Many organic pollutants cause oxygen depletion in lakes and rivers, and when this happens the aquatic ecosystem is altered for some distance downstream (Figure 13.8).

Some pollutants are entirely synthetic, whereas many – indeed, most – are natural materials that are simply released in unacceptably high concentrations at particular places. The synthetic pollutants create particular problems because no natural biogeochemical cycles exist for them (see p. 96), so they cannot be readily broken down in the environment by natural processes. This is why many of them are persistent (see Box 3.27) and do not disappear, even after fairly lengthy periods of time.

There are far too many different types of water pollutant to deal with them all here. Two of the most widespread water pollution problems during the 1990s were eutrophication (see p. 409) and acidification. Freshwater resources in most industrial areas around the world are showing signs of acidification, associated with the deposition of the gaseous air pollutants sulphur dioxide (SO_2) and nitrogen oxides (NO_x) (see p. 255). Acidification damages aquatic ecosystems, and recent studies show that ecological damage begins soon after acidity starts to increase (when pH reaches about 6.5; see Figure 8.9). For sensitive organisms, the chemical threshold of acid water is about pH 5.5; this marks the end-point rather than the onset of acidification.

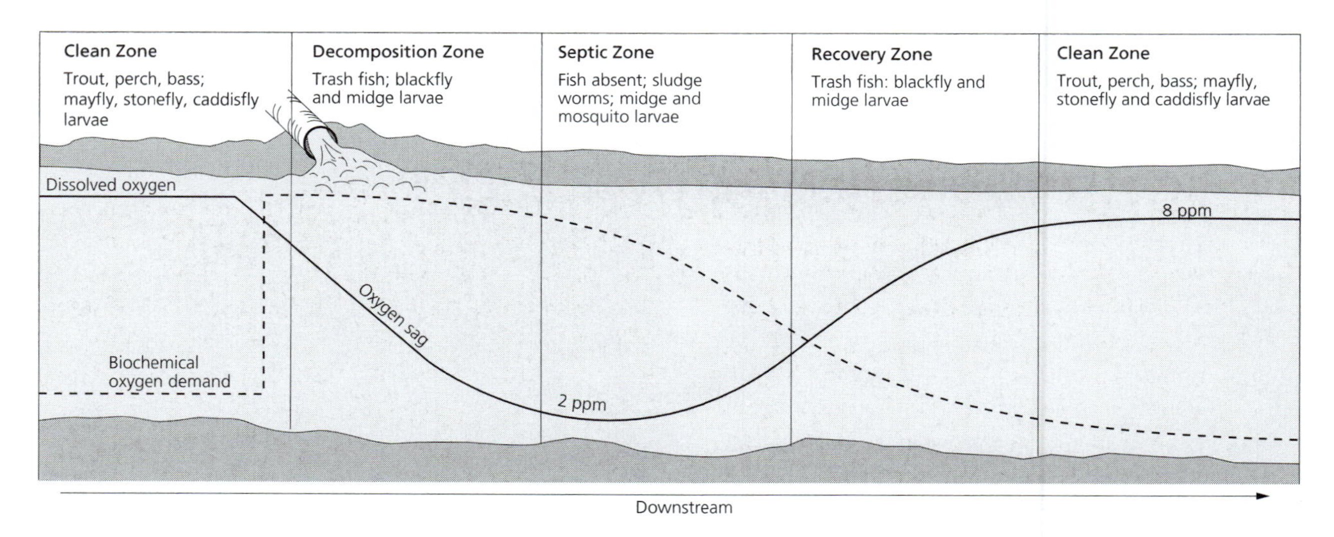

Figure 13.8 *Downstream ecological impact of oxygen depletion in a river. Oxygen depletion caused by increased biological activity downstream from a pollution source can radically alter the species composition of the river system.*

Source: Figure 22.5 in Cunningham, W.P. and B.W. Saigo (1992) Environmental science: a global concern. Wm. C. Brown Publishers, Dubuque, Iowa.

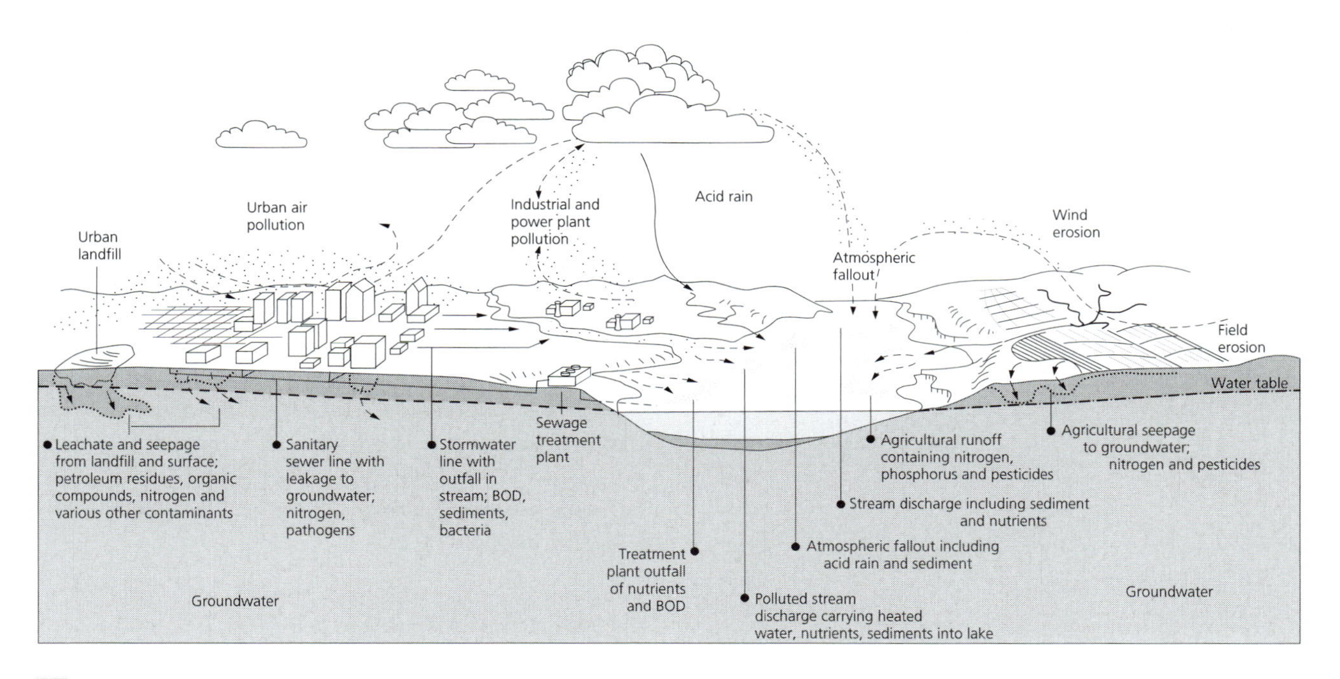

Figure 13.9 *Multiple sources of water pollution. Water pollution is generated from many different sources and by many different processes, and land-use changes can significantly alter the amount and type of pollution in a river system.*

Source: Figure 11.3 in Marsh, W.M. and J.M. Grossa (1996) Environmental geography: science, land use and Earth systems. John Wiley & Sons, New York.

BOX 13.22 WATER POLLUTION SOURCES

Studies in the USA in the late 1980s suggest that up to 55 per cent of total river length was polluted by agriculture, about 16 per cent by towns and cities, 13 per cent by mining activities and a further 13 per cent by habitat modification. Other sources of pollution – each accounting for less than 10 per cent of total river length – included storm drains, forestry, industrial activities, construction work and waste disposal.

In industrial areas, there may be many different sources of water pollution. Along the Yangtze River in China, for example, there are twenty-two major cities and more than 300 sources of industrial pollution. In 1982, of a total pollution load estimated at around 4,000 million m^3, the chemical industry contributed 41 per cent, iron and steel industries 21 per cent and light industries 18 per cent.

SOURCES OF POLLUTION

There are many different sources of freshwater pollution (Figure 13.9), and they vary from place to place and over time. Some are point sources – specific places such as factories and mines that release contaminated water. Much pollution comes from diffuse or non-point sources, of which agriculture is the most widespread. Agriculture causes water pollution in a variety of ways, including the leaching of nitrates (from fertilisers), eutrophication, and the accumulation of pesticides and other contaminants.

Most water pollution problems arise from the continuous or intermittent release of contaminants into streams, lakes, rivers and groundwater. From time to time there are major pollution incidents in particular places, which create problems because they are usually unexpected but can spread pollution across wide areas. Good examples include the spread of radioactive fallout across Europe after the 1986 nuclear accident at Chernobyl (see Box 8.23), and the more local accident that seriously polluted the Camelford area in south-west England in 1988 (Box 13.23).

BOX 13.23 THE CAMELFORD WATER POLLUTION INCIDENT IN SOUTH-WEST ENGLAND, 1988

This pollution incident was caused by the accidental release of 20 tonnes of an 8 per cent solution of aluminium sulphate into the water supply of the Camelford area of north Cornwall on 6 July 1988. About 20,000 people in the area depend on that water supply and were exposed to levels estimated at 2,000 times the European Union acceptable limit. Soon after the accident some people fell seriously ill, and around 60,000 fish died in the Rivers Camel and Allen. Further problems were caused by the chemicals that were added to the water supply in an attempt to neutralise the aluminium sulphate. These chemicals reacted with water pipes, and this caused high levels of some toxic heavy metals (copper, lead and zinc) to be released into the drinking water. Many local people suffered health problems after the incident – including skin trouble, arthritis, nausea and kidney problems – but no firm links were established with the aluminium or heavy metals.

GROUNDWATER POLLUTION

Groundwater supplies are often susceptible to water pollution because they receive water from many different sources and pathways (Figure 13.10), and as water is stored in aquifers for long periods of time, pollutants can accumulate to critical levels. The intrusion of seawater into coastal aquifers (see p. 403) is a form of pollution because it limits the uses to which the groundwater can be put. One of the chief concerns about long-term underground storage of high-level nuclear waste (see p. 137) is the possibility that leaks or seepage of radioactive material might contaminate local and regional groundwater supplies and put people's health and lives at risk.

Even other types of pollution incidents can lead to groundwater pollution. There is evidence from different parts of Europe that radionuclide concentrations in groundwater up to 120 m below the surface are up to twice what they were before the Chernobyl accident in 1986 (see Box 8.23), for example.

MANAGING WATER POLLUTION

Considerable investment is now being made in attempts to restore rivers, lakes and aquifers that are suffering from

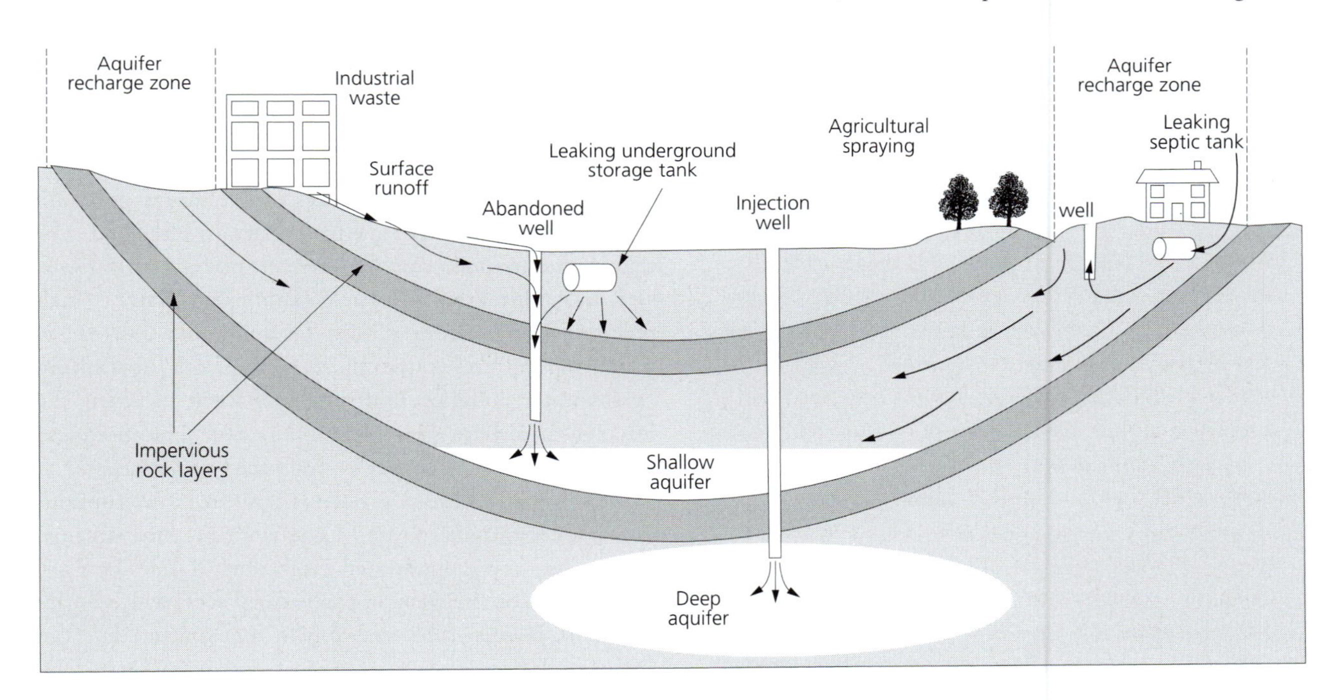

Figure 13.10 *Sources of groundwater pollution. There are many different sources of pollution of groundwater, including septic systems, landfills and industrial activities on aquifer recharge zones.*

Source: Figure 17.2 in Cunningham, W.P. and B.W. Saigo (1992) Environmental science: a global concern. Wm. C. Brown Publishers, Dubuque, Iowa.

BOX 13.24 INTERNATIONAL WATER POLLUTION CONTROL

While the drainage basin is the most obvious and relevant unit around which to devise water quality management strategies, water pollution problems are often regional or even multinational in scale because of the mobility of groundwater and river flows. Levels and types of water pollution vary a great deal between rivers in Europe, for example. Cadmium concentrations appear to be particularly high in some Eastern European and Greek rivers, and many European rivers are suffering from organic pollution and eutrophication. Although the levels and character of water pollution vary from river to river across Europe, member states have in the past had their own ways of tackling the problems of water quality. In recent decades, great efforts have been made in the European Union to:

- harmonise water quality standards between member states;
- introduce appropriate legislation to protect the aquatic environment;
- evaluate the effectiveness of different economic instruments in support of that legislation; and
- encourage the introduction of more environmentally acceptable products and processes (for example, by the use of eco-labelling and environmental auditing).

water pollution. Since the 1970s, water pollution has become one of the most important environmental problems in China, although the success of pollution management and control measures there has been hindered by a dramatic increase in new sources of pollution and by the persistence of poor waste treatment facilities in thousands of small towns.

Waste water purification systems are one important and potentially fruitful way of reducing water pollution, and they are being adopted in many countries. The strategy adopted in Denmark, for example, includes biological treatment of waste water and the removal of phosphorus from waste water in most towns, along with significant reductions in the output of organic material and phosphorus from fishponds and farms. It is hoped that this strategy will preserve water quality in major lakes and rivers and reduce the input of nutrients into estuaries and coastal waters.

Integrated water quality management schemes, designed to meet multiple objectives, are being introduced in some large drainage basins (see p. 357). The Mersey basin campaign, in England, is a case in point. This basin-wide initiative was launched in 1984 in an attempt to revitalise the River Mersey and the waterfront along the Mersey estuary. The core of the campaign is major investment in improved sewerage and sewage disposal systems in order to improve water quality in the river and estuary.

EUTROPHICATION

NUTRIENT ENRICHMENT

Eutrophication is a process by which particular pollutants are washed into a water body – a lake, river, wetland or shallow sea – and overload it with organic and mineral nutrients. Two main materials are involved:

1. nitrate fertilisers washed from the soil by rain; and
2. phosphates from fertilisers and detergents in municipal sewage.

Quite large amounts of nitrate are often leached from the soil during those times of the year when arable fields are bare, which often coincide with times of heavy rainfall. It

BOX 13.25 BENEFITS OF FERTILISERS

Nitrates and phosphates both exist naturally in the environment, and there are well-developed biogeochemical cycles (see p. 86) for each. Both are essential for plant growth, and both are highly soluble in water and therefore very mobile within environmental systems. Although eutrophication is widely viewed as an environmental problem, when the process first begins it has some advantages, particularly by supplying nutrients that are important for plants and animals to nutrient-poor freshwater bodies. This was certainly the case in the Netherlands during the first half of the twentieth century, when the inwash of fertilisers into many freshwater wetlands initially improved their natural nutrient-poor status. Continued inwash has turned the initial benefit into a longer-term problem.

is washed into lakes and rivers and enriches them. Agricultural policies in the European Union have strongly encouraged the widespread use of nitrate fertiliser and have thus contributed to the problem of freshwater eutrophication in Europe. This is a good illustration of compartmentalised environmental decision making, in which policies in one area (agriculture) conflict with policies in another area (protection of water quality).

IMPACTS

Fast-flowing rivers generally dilute, disperse and effectively remove the soluble compounds (including nitrates and phosphates) that are washed in naturally from soils and the atmosphere. But these materials can accumulate in still or slow-moving water, and this gives rise to the eutrophication.

The pollution enriches the water body into which it flows. In reality, it over-enriches or overloads it. This encourages the rapid growth of aquatic plants – particularly algae and bacteria – which can form an unsightly layer on the surface of a stationary water body such as a lake. This is sometimes described as an 'algal bloom'. Throughout the twentieth century such blooms were an increasingly regular occurrence in the most heavily polluted parts of the Great Lakes in North America, but they have declined as water quality has improved as a result of improved pollution control and water quality management strategies.

As the algae and bacteria grow rapidly, they use up much of the available oxygen supply dissolved in the water. Oxygen depletion then makes the water body unsuitable for other organisms (including fish) to survive in, so individuals and ultimately entire populations of fish, insects and other wildlife disappear from a eutrophic lake or water body. Some of the algae and bacteria produce relatively large amounts of toxins, which further disrupt the aquatic ecosystem.

When the process is well advanced, a water body can appear to be ecologically dead. Water turbidity increases (because it has more organic matter suspended in it), and rates of sediment deposition increase.

MANAGEMENT IMPLICATIONS

Eutrophic lakes are much less useful than unpolluted lakes for activities such as water supply, recreation, navigation or sport fishing. As the eutrophication process continues in a particular water body, water quality declines and sustainable use of the water body becomes more problematic.

BOX 13.26 TACKLING EUTROPHICATION

A variety of different approaches to reducing eutrophication have been tried. Prevention is always better than cure, and the most logical objective here is to control the release of nutrients into water bodies. Reducing the nutrient content of wastes and/or disposing of wastes in ways that prevent them from being released into lakes and rivers are two obvious approaches. Prevention is a more difficult, expensive and long-term approach than cure (corrective action), and most attempts to control eutrophication have as yet been curative. A number of different techniques have been tried, ranging from expensive dredging of nutrient-rich lake sediments to short-term and quickly effective chemical dosing. Dilution is another possibility, which would involve recharging a eutrophic lake with oligotrophic water (which has a much lower nutrient concentration).

Lake Muggelsee is the main drinking-water reservoir for Berlin in Germany, but between about 1900 and 1970 it was showing signs of increasing eutrophication. Efforts are under way to reduce nutrient inputs into the lake, mainly by introducing control measures designed to reduce the inwash of phosphorus into the lake and its drainage basin.

FLOODS AND FLOODPLAINS

FLOODPLAINS

A floodplain is the flat lower part of a valley floor, adjacent to the river channel, that is inundated by flood flows which spill over the channel bank when discharge periodically exceeds the bankfull. It is formed by the deposition of layers of alluvial sediment by the river during floods (Figure 13.11), and the stratigraphy of floodplains often preserves evidence of channel changes and large past floods.

The floodplain effectively separates the river from the adjacent hill slopes. Upland portions of rivers usually have close associations between the channel and hill slopes, whereas floodplains make such links much less strong on lowland reaches.

Rates and patterns of deposition on floodplains can vary over time and from place to place. Many floodplains have altered a great deal in historical times because of land-use

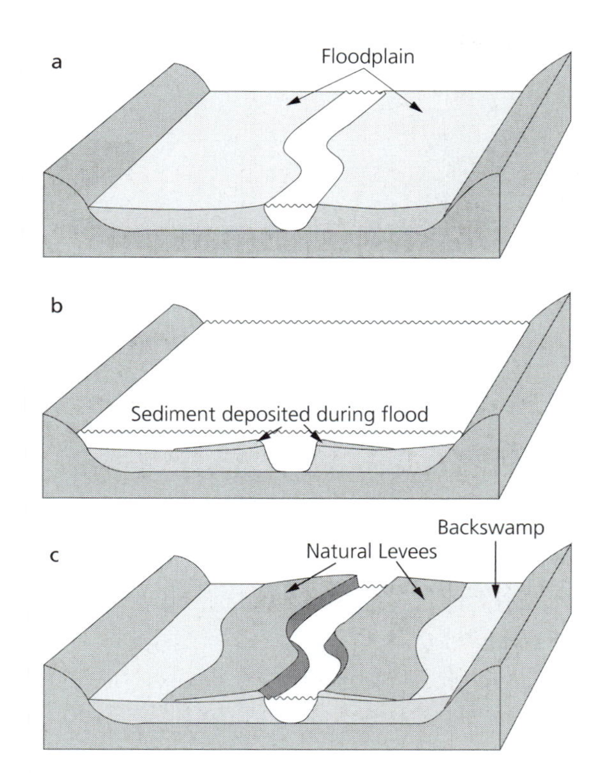

Figure 13.11 *Development of a floodplain and levees. A river normally flows within its channel along the floodplain (a), but during floods it deposits sediment adjacent to the channel (b), which builds up to form natural levees (c).*

Source: Figure 12.28 in Doerr, A.H. (1990) Fundamentals of physical geography. Wm. C. Brown Publishers, Dubuque, Iowa.

changes in the upstream drainage basin. Forest clearance, for example, usually increases river sediment loads and rates of floodplain deposition. Agriculture is often associated with quite high rates of sediment transport and deposition. Floodplain surfaces usually contain fluvial features indicative of present and past river activity, including meanders (see p. 372), ox-bow lakes (p. 375) and levees.

LEVEES

A levee is a small ridge or embankment on the floodplain that lies beside and runs parallel to the river channel. Natural levees form when a river overflows its banks during floods and deposits material along the edge of the floodplain (Figure 13.11). Artificial levees are built to prevent flooding, but if they are overtopped or breached during large floods they can trap flood waters behind them and prolong flooding.

BOX 13.27 FLOODPLAIN DEVELOPMENT

Floodplains provide valuable land for development, including urbanisation, because they are flat, have supplies of water and gravel available and are accessible (floodplains have traditionally been important transport corridors). Many early settlements grew up on floodplains, and continued growth over time has led to the establishment of many large cities adjacent to rivers. When the river floods and the floodplain is inundated, extensive areas of housing, offices and other uses can be affected, and the cost of damage and inconvenience can be extremely high.

Natural levees can also restrict the drainage of flood waters, so many floodplains lack good natural drainage after floods. For this reason, many floodplains contain swampy areas (back swamps), which are a nuisance to agriculture and a barrier to access. Levees can also make it difficult for tributary streams to enter the main river channel, so sometimes the tributary stream flows alongside the main channel for some distance (as a so-called 'yazoo tributary').

FLOODS

A flood is a discharge that exceeds the channel capacity of the river (it is bigger than the bankfull discharge), so it inundates the adjacent floodplain. When this happens, the

BOX 13.28 CAUSES OF RIVER FLOODS

Floods are caused by many different factors, but the most important ones include:

- climatic extremes, particularly heavy or prolonged rainfall
- melting of snow and ice
- collapse of dams
- landslides.

Flooding tends to be most frequent during the wet and/or the melt seasons. In some parts of the world, including Britain, intense convectional storms can produce flooding during the summer.

Plate 13.6 *Flood deposits around the village of Inja in the Altay region of southern Siberia. This region is dominated by catastrophic flood deposits of unknown age that resulted from the drainage of large lakes dammed by glacier tongues. The terrace on which this village is built and the large bar upslope are both deposits from single floods, and they are up to 280 m thick.*

Photo: Alistair Kirkbride.

channel and the floodplain together allow the passage of flood waters. Floods are part of the normal range of river flow conditions. Much of the Midwest of the USA was flooded by the Mississippi in 1993 (Figure 13.12), for example. Although the flood caused widespread damage and inconvenience, most environmental systems regained equilibrium relatively quickly afterwards. The massive flood in Mozambique in 2000 caused much more serious human suffering and loss (Box 13.29).

The flood (or storm) hydrograph records the passage of flood water over time at a given point in the river system (see Figure 12.5). The detailed timing and shape of flood hydrographs reflects relationships between climatic and drainage basin controls, and these can vary from basin to basin and over time within a basin.

A range of factors serves to intensify floods when they happen or to increase the probability that flooding will occur. Such flood-intensifying conditions can be either permanent or transient:

- Permanent factors include characteristics of the drainage basin and the river network (such as drainage density, basin size, slope and network pattern).
- Transient factors include the duration, intensity and path of the storm, evaporation rates and infiltration rates.

Frozen ground is a common flood-intensifying factor in high latitudes.

BOX 13.29 THE MOZAMBIQUE FLOOD, 2000

Early in 2000 much of southern Africa – centred on Mozambique but spanning South Africa, Zambia, Zimbabwe, Botswana, Swaziland and Madagascar – was badly affected by floods. The most disastrous flooding was centred on low-lying areas along the Limpopo River and the Save River in southern Mozambique. Heavy rain started to fall there on 4 February, and the situation was made worse by Cyclone Eline, which swept over the Mozambique coast near Beria on 23 February. Eline brought further heavy rain to central and southern provinces in the country and into south-eastern Zimbabwe. Flood waters swept through the entire valley of the Limpopo, engulfing the cities of Xai Xai and Chikwe. Over a thousand people died, and hundreds of thousands were displaced. Media around the world showed dramatic pictures of people stranded on rooftops, in trees and on telegraph poles; many were rescued by helicopters, but many others drowned.

During the flood, and in the immediate aftermath, the spread of disease was a major concern. Many communities (even large towns) and groups of stranded people were completely isolated with no clean water, no food, no sanitation, no medicine and poor communications. After a delay of days, the UK, USA and other countries sent emergency supplies of transport, food, medicine and other essential resources to the most badly affected areas. The threat of disease was widespread as decomposing human bodies and cows contaminated the surrounding flood waters. Malaria and diarrhoea were rampant, especially among the children.

External assistance came in two phases. During the first phase, the priority was to provide immediate emergency relief to people in the form of food, medicines and basic survival kits (tarpaulins, tents, jerry cans, salt, powdered milk, blankets, soap, kitchen utensils and buckets). Phase two is designed to provide displaced families with materials to rebuild their houses and agricultural seeds and tools to help them to start farming again as quickly as possible. It took several weeks for the flood water to recede, and many of the displaced people were unable to return to their homes and land for many months.

The 2000 floods were considered the worst in more than forty years. The UN World Food Programme estimated that Mozambique lost at least a third of the staple maize crop and 80 per cent of its cattle. A quarter of the country's agriculture was damaged.

FLOOD-FREQUENCY ANALYSIS

Records of flood hydrographs can be analysed to reveal important properties about the magnitude and frequency (see Box 13.31) of river floods. This involves flood-frequency analysis, which determines the recurrence interval (or return period) for floods of a given size at a particular location in a river system. The recurrence interval is the average period of time between two successive floods of that particular magnitude (discharge). It is not a prediction, in the sense that it allows hydrologists to predict when the flood will occur; it is a long-term average. Two floods of the same size could well occur within a short period of time, and there might be a long gap before the next event of that magnitude.

Flood-frequency analysis is based on analysis of long-term hydrograph records from instrumented cross-sections, usually gauging stations (Box 13.30). Whichever data series is adopted, the recurrence interval is calculated in the same way. First, all the peak discharges are ranked

BOX 13.30 DATA USED IN FLOOD-FREQUENCY ANALYSIS

Hydrologists use one of two types of data in flood-frequency analysis:

1. *Annual maximum series*: this includes the highest peak discharge each year in the series of records, so only one flood each year (the largest flood) is included in the analysis.
2. *Partial duration series*: this includes all peak flows with a discharge greater than a chosen threshold (which might be the bankfull discharge). For some years there may be a number of floods that were greater than the threshold discharge, while for others there may be one or none.

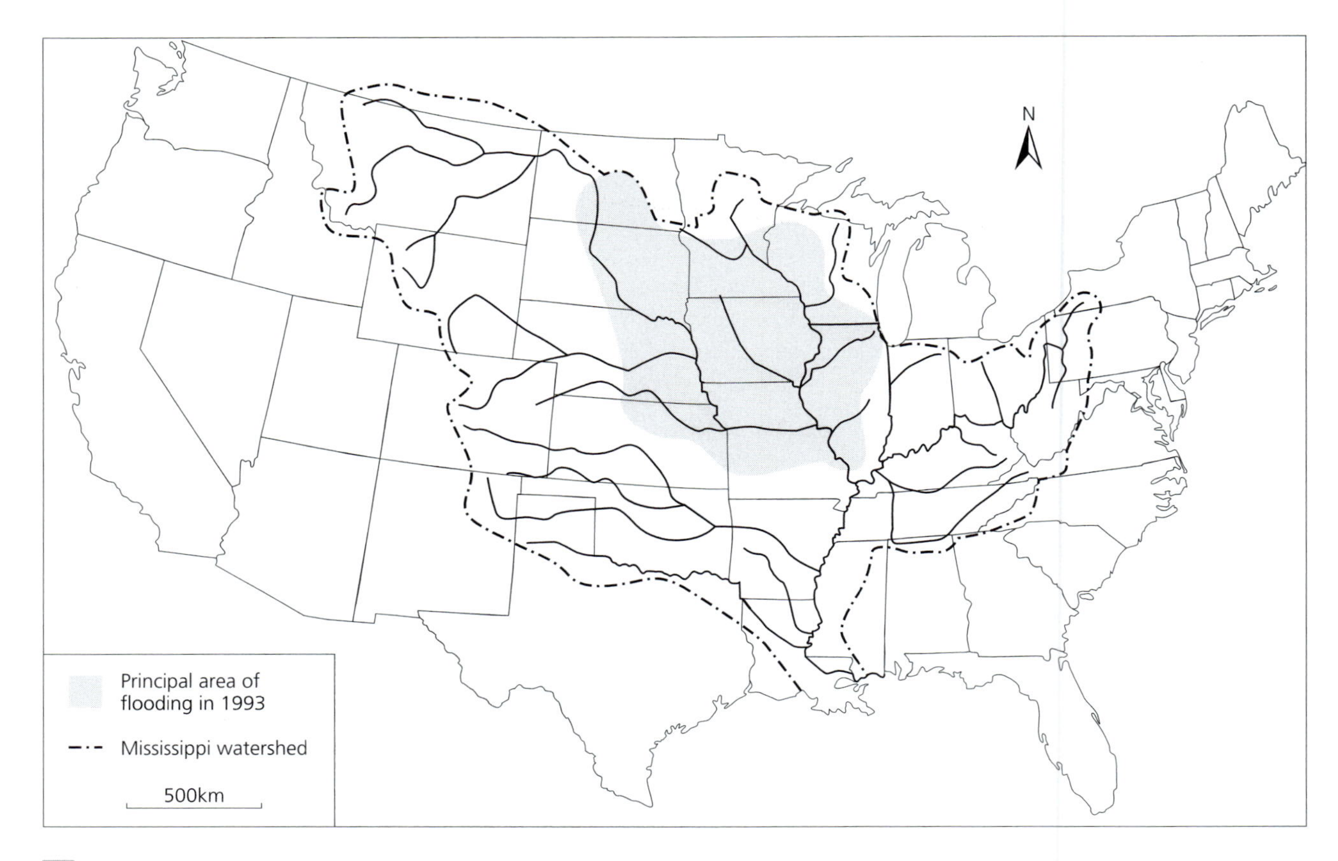

Figure 13.12 *Extent of the 1993 Mississippi flood. Unusually prolonged high rainfalls over much of the central part of the Mississippi drainage basin during 1993 caused widespread flooding, damage and inconvenience.*

Source: Figure 17.2 in Marsh, W.M. and J.M. Grossa (1996) Environmental geography: science, land use and Earth systems. John Wiley & Sons, New York.

in descending order of magnitude. Second, a recurrence interval is calculated for each peak discharge in the ranked series, as follows:

$$T = (n + 1)/m$$

where T is the recurrence interval (in years) of that peak discharge, n is the number of peak discharges included in the analysis, and m is the rank order of that peak discharge in the ranked series. A simple worked example is given in Table 13.8.

Once the recurrence interval has been calculated for each peak discharge in the series, the data can be plotted as a flood-frequency curve (Figure 13.13). This plots discharge against recurrence interval, and it shows that large floods with a high discharge are relatively rare (they have a long recurrence interval) and smaller flows are much

Table 13.8 Worked example of how to calculate flood recurrence interval

Discharge (m^3 s^{-1})	*Rank*	*Equation*	*Recurrence interval (years)*
25	1	$T = (5 + 1)/1$	6
20	2	$T = (5 + 1)/2$	3
18	3	$T = (5 + 1)/3$	2
14	4	$T = (5 + 1)/4$	1.5
3	5	$T = (5 + 1)/5$	1.2

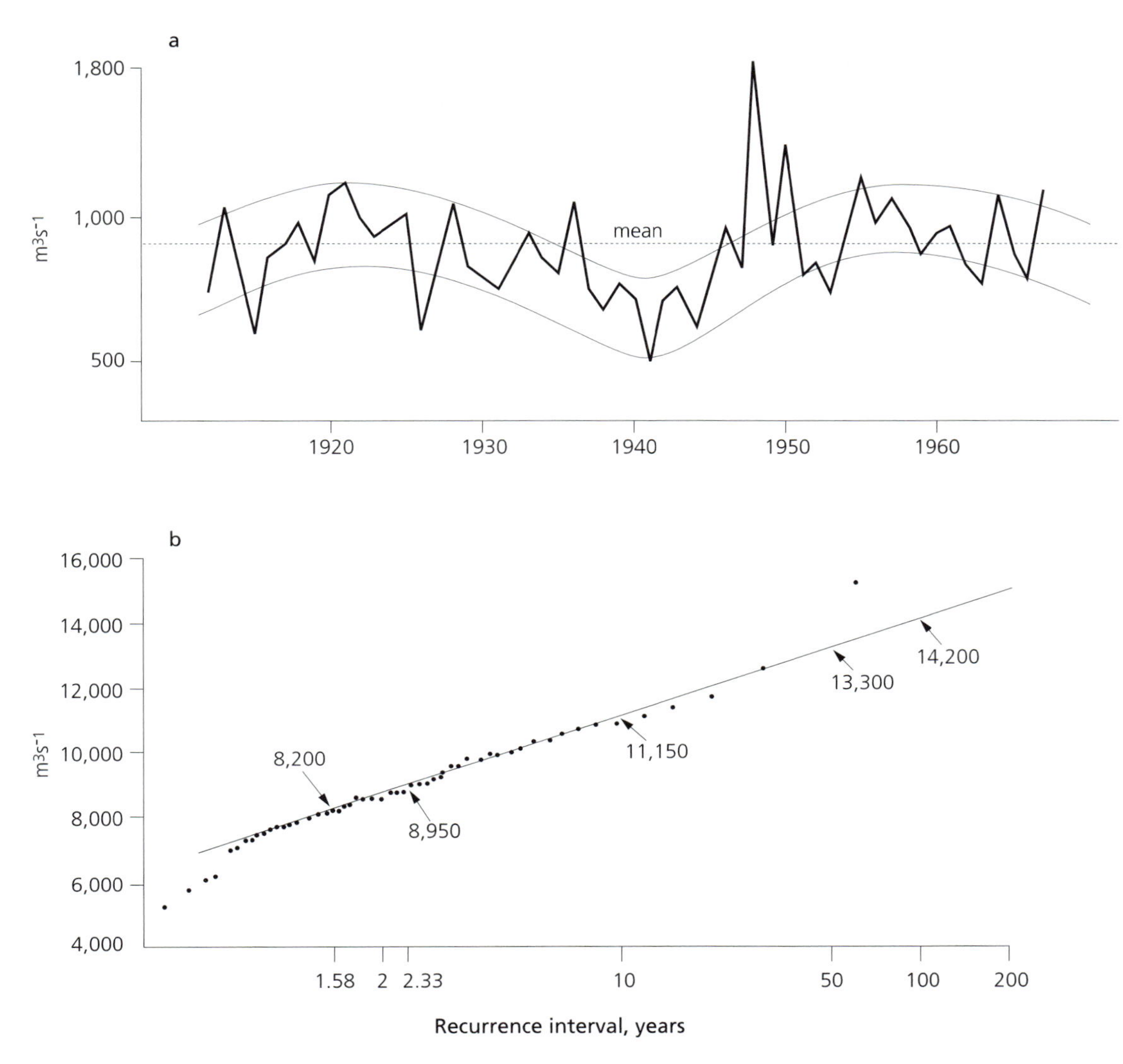

Figure 13.13 *Flood-frequency curve. The annual peak discharge on the Fraser River at Hope, British Columbia, varies a great deal from year to year, and between the 1920s and the 1960s the variations seemed to follow a cyclical pattern (a). Flood-frequency analysis (see text for explanation) indicates that the floods conform to a normal pattern, which plots as a straight line on probability graph paper.*

Source: Figures 6.10 and 6.11 in Dury, G.H. (1981) Environmental systems. Heinemann, London.

more common. This is the basis of magnitude–frequency analysis.

Studies on a large number of rivers in different climatic and geological settings have shown that rivers flood with remarkable regularity. Most natural rivers have a recurrence interval of between 1.5 and 2.3 years for the bankfull discharge, above which flow spills out on to the floodplain and causes flooding. This seems to hold true for all sizes of river and in all climatic zones. The explanation lies in river channel adjustment, because channel cross-sectional size adjusts to carry the dominant discharges in a given river system. Rivers in areas with high rainfall and/or areas with relatively large amounts of surface runoff (perhaps because of geological factors) tend to have large channels that are adjusted to transport high discharges.

MANAGING THE FLOOD HAZARD

We must not lose sight of the fact that floods and flooding are natural phenomena that occur in all river systems. Flood disasters and flood damage are caused by humans, in the sense that people and property are at risk on floodplains,

BOX 13.31 APPLICATION OF FLOOD-FREQUENCY ANALYSIS

A flood-frequency curve can be interpreted in two ways:

1. What is the discharge with a given recurrence interval at that site on the river? For example, what discharge might we expect to occur on average every ten years?
2. What is the recurrence interval of a particular discharge at that site on the river? For example, how often would we expect discharge to exceed bankfull?

Such information has a number of important applications, including:

- *floodplain management*: for example in deciding what types of land uses are most appropriate in different parts of a floodplain;
- *river management*: for example in deciding whether to invest in engineering schemes to control flooding;
- *engineering design*: such as deciding how big to build artificial levees and control structures intended to reduce flood risk, or how big to build a water storage dam.

BOX 13.32 CHANGING FLOOD FREQUENCY

A change in any of the factors that cause or intensify floods is likely to promote changes in flood frequency. This is particularly true with climate change. Analysis of a 7,000-year record of flood deposits in the upper Mississippi in the USA has shown that between 3,300 and 5,000 years ago the climate was warmer and drier and large floods were rare. After about 3,300 years ago the climate became cooler and wetter, and floods became much larger. Even larger floods occurred between about AD 1250 and 1450, during the transition from the Medieval Warm Period to the cooler Little Ice Age (see p. 344).

Land-use change can also radically alter the flood frequency on river systems. The most common changes include intensive agriculture, forest clearance and urbanisation, all of which can increase flood magnitude and frequency and thus increase the flood risk along floodplains. River management, including channel dredging and flood prevention schemes, can decrease flood levels and frequency, as appears to have happened on the River Thames in England since about 1940.

and land-use changes in upstream drainage basins can significantly alter the magnitude and frequency of flooding in river systems.

Floodplain invasion is the primary cause of flood hazard in most developed countries. In the USA, for example, more than 2,000 towns and cities are located on floodplains, and nearly 60 per cent of the total floodplain area is occupied by urban development. The problem is often made worse by ineffective remedial measures, which can give a false sense of security to floodplain occupants. To eliminate the flood risk entirely would require the complete evacuation of floodplains, but this is not viable socially or economically.

Thus we usually have two options – to control how we use floodplains or control the floods. The former can involve various approaches, including:

- flood-proofing structures
- regulating land use on floodplains by zoning

BOX 13.33 FLOOD CONTROL ON THE YELLOW RIVER IN CHINA

Attempts to control flooding along China's Yellow River illustrate some of the complexity of trying to tame nature. The Yellow River has a long history of disastrous flooding, and flood control is largely based on dams, storage basins and artificial levees. The flood control system is designed to cope with floods with a recurrence interval of about 60 years, but such events are becoming more common because of upstream land-use changes and rapid sedimentation along the river. One possible solution is to breach the levees at predefined points, allow the floods to happen, issue advance warnings and evacuate people from the areas most at risk.

- insurance
- co-ordinated emergency action (when floods occur).

Flood control can be tackled in various ways. Preventive flood control might involve trying to reduce the build-up of floods by upstream measures, including building flood storage basins, land-use changes (including planting forests, replacing crops with grass) and channel management. Corrective flood control might involve downstream protection of important or sensitive areas of floodplain.

Flood protection is the more common approach, and it usually involves structural (engineering) measures. One option is to construct embankments (levees and dikes) and flood walls, and this is often done using alluvial deposits excavated on site. Embankments along the lower Nile are of this type. Another option is channel improvements, including:

- decreasing channel roughness (perhaps by lining the channel with smooth concrete bed and banks);
- widening and/or deepening the channel (by dredging);
- shortening the channel by straightening it and cutting off meander loops.

A third approach is to build river diversion schemes, which are designed to redirect flood flows away from built-up areas, normally by constructing artificial channels.

Many countries are now investing heavily in developing better abilities to forecast when floods are likely to occur so that flood warnings can be issued and people advised to evacuate high-risk areas. Such a scheme, based on analysis of meteorological and hydrological data in real time, has been introduced to forecast floods on the River Arno in Italy and to provide flood warnings up to 12 hours ahead in order to evacuate the city of Florence.

Despite such advances, experience shows that many countries have simply not learned the lessons from past floods. In Britain, for example, although the total flood loss potential is rising, there is no indication that local authorities and the emergency services are responding to the increased risk by adopting appropriate land-use zoning and development measures.

WEBSITE

Links to relevant websites, a comprehensive bibliography, tools for teaching and learning, and downloadable images relevant to this chapter can be found at the website specially designed to accompany this book at
http://www.park-environment.com

FURTHER READING

Anderson, M.G., D.E. Walling and P. Bates (eds) (1996) *Floodplain processes.* John Wiley & Sons, London. Summary of recent advances in understanding how floodplains develop, function and respond to change, with implications for floodplain management and risk assessment.

Baird, A.J. and R.L. Wilby (eds) (1999) *Eco-hydrology.* Routledge, London. Examines the water relations of plants in drylands, wetlands, temperate and tropical rainforests, streams and rivers, and lakes.

Best, G., E. Niemirycz and T. Bogacka (1997) *International river water quality.* E & FN Spon, London. Reviews the key issues, challenges and possible solutions to problems of freshwater pollution and restoration.

Cook, H.F. (1998) *The protection and conservation of water resources: a British perspective.* John Wiley & Sons, London. Explores the problems and realities of water resource management in Britain, with a special emphasis on recent policy developments.

SUMMARY

Building on Chapter 12, we looked in this chapter at the importance of water as a renewable natural resource. We began with an overview of how and why water is regarded as a resource, where the main demand comes from and where supplies are drawn from. Water also provides a valuable form of renewable energy and is used extensively in irrigation systems. We examined some of the environmental problems associated with large dam schemes. Other environmental problems arising from water resource use were examined, including wetlands and drainage, salinisation and desalination, water quality and pollution, and eutrophication. The final section in the chapter dealt with floods, flooding and floodplains. Here we considered how floods are generated, how we can analyse their magnitude and frequency, and how the flood hazard changes over time.

Cosgrove, W.J. and F.R. Rijsberman (2000) *World water vision: making water everybody's business*. Earthscan, London. Comprehensive analysis of the world's water resources – retrospect and prospect.

Drever, J.I. (1997) *The geochemistry of natural waters*. Prentice Hall, Hemel Hempstead. Examines the theory and practice of natural water geochemistry, with an emphasis on water pollution and natural processes.

Frederick, K. (1997) *Water resources and climate change*. Resources for the Future, Washington. An analysis of the likely impacts of climate change on water resources, based on IPCC forecasts.

Gleick, P.H. (2000) *The world's water 2000–2001*. Island Press, Washington. A major biennial report on the world's freshwater resources, the threats they face and challenges they pose, and the need for sustainable water management.

Gray, N.F. (1994) *Drinking water quality*. John Wiley & Sons, London. A detailed exploration of the structure, regulation and operation of the water supply industry within the framework of British, European, US and WHO standards.

Haslam, S.M. (1992) *River pollution: an ecological perspective*. John Wiley & Sons, London. A detailed introduction to the causes and impacts of natural and cultural pollution on freshwater rivers, with an emphasis on processes and monitoring techniques.

Kay, B. (1998) *Water resources: health, environment and development*. E & FN Spon, London. Broad exploration of the health aspects of the environmental impacts of water resource development.

Jobin, W. (1999) *Dams and diseases: ecological design and health impact of large dams and irrigation schemes*. E & FN Spon, London. Case studies from twenty-five countries that focus on assessing, predicting and preventing major water-associated diseases.

Jones, J.A.A. (1996) *Global hydrology: process and management implications*. Longman, Harlow. A clear introduction to hydrological systems and processes, water resource management, and the potential impacts of hydropolitics and global warming.

Manning, J.C. (1997) *Applied principles of hydrology*. Prentice Hall, Hemel Hempstead. Non-technical treatment of all the major components of the water cycle, including ocean water evaporation, precipitation, infiltration, groundwater and streamflow.

Mason, C.F. (1996) *Biology of freshwater pollution*. Longman, Harlow. A comprehensive introduction to water pollution, which includes case studies of the Great Lakes and the River Rhine.

Merrett, S. (1997) *Introduction to the economics of water resources*. UCL Press, London. Comprehensive overview of international issues relating to the supply and use of water resources.

Newson, M. (1997) *Land, water and development: sustainable management of river basin systems*. Routledge, London. Clearly written and abundantly illustrated overview of the theory and practice of catchment management planning in both developed and developing countries.

Parker, D. and J. Handmer (eds) (2000) *Floods*. Routledge, London. A collection of essays that explore the physical and social dimensions of floods in different countries and different contexts.

Pereira, L.S. and J. Gowing (1998) *Water and the environment: innovation issues in irrigation and drainage*. E & FN Spon, London. Outlines the challenges and explores the options.

Rijsberman, F.R. (ed.) (2000) *World water scenarios: analysing global water resources and use*. Earthscan, London. Explores the factors that influence demand for and supply of water, illustrates the importance of more efficient and productive water use, and speculates about what the future might have in store if current trends are not altered.

Schwab, G., D. Fangmeier and W. Elliot (1996) *Soil and water management systems*. John Wiley & Sons, London. A detailed introduction to hydrology, erosion control, water supply, drainage and irrigation.

Smith, K. and R. Ward (1998) *Floods: physical processes and human impacts*. John Wiley & Sons, London. Explores the causes and consequences of river and coastal floods and discusses possible responses to the flood hazard.

Twort, A.C., D.D. Ratnayaka and M.J. Brandt (1999) *Water supply*. Edward Arnold, London. Fifth edition of a classic textbook on the engineering dimensions of water supply systems.

Williams, M. (ed.) (1993) *Wetlands: a threatened landscape*. Blackwell, Oxford. A series of case studies of wetlands, the pressures they are under, and some approaches to the management and conservation of wetland resources.

Yarnal, B. (2000) *Local water supply and climate extremes*. Ashgate, London. Examines the impacts of weather and climate on local and regional water supplies, and the implications of these interrelationships for sustainable water management.

Chapter 14

Drylands

LEARNING OBJECTIVES

When you have finished studying this chapter, you should be able to

- *Appreciate the problems caused by droughts and desertification*
- *Distinguish between the four main types of drought*
- *Describe the major natural processes that give rise to drought and the human factors that cause, intensify or prolong it*
- *Show how and why responses to drought differ between developed and developing countries*
- *Explain what is meant by 'desert' and account for the location of the five main desert provinces*
- *Outline the main features of a typical desert landscape and the processes involved*
- *Define 'desertification' and say why it is such an important environmental problem today*
- *Describe and account for the main processes that cause desertification, including human factors.*

Water plays a central role in many environmental systems, and the availability of adequate supplies of water (of the right quality, in the right place, at the right time) is vitally important, as we saw in Chapter 13. Water shortage can create serious problems for people and put critical environmental systems under increased stress. Many places suffer from periodic water shortages associated with droughts, and in the first section of this chapter we explore the meaning of drought and look at its impacts and how it is coped with.

There is abundant evidence to show that drylands are expanding, because of a combination of natural processes (climate change) and human impacts. Desertification – the focus of the third section – is already a serious environmental problem in many arid and semi-arid areas, and it is likely to become worse because of rising population pressure and global warming.

DROUGHT

Drought causes problems because it disturbs natural water flows through the regional hydrological cycle (see p. 351). The most significant impact of this is serious water shortage. People suffer as water supply wells and reservoirs dry up, crops are damaged and yields fall, drinking water is limited and heavy water users (including industry) face restrictions on supplies.

DEFINING DROUGHT

DROUGHT – THE CONCEPT

Drought can be defined as a prolonged period of unusually dry weather (i.e. little rainfall) in a region where some rain might normally be expected. As such a drought differs from a dry climate, which is usually associated with a region that is normally or at least seasonally dry. A drought is also drier and lasts much longer than a dry spell, which is usually defined as more than 14 days with no significant precipitation that causes a relatively short period of moisture deficiency. Droughts often last for years!

Drought differs from aridity (climatic dryness) because it is temporary, whereas aridity is a permanent feature of the climate of areas with low rainfall. Drought is a natural

BOX 14.1 WATER SHORTAGE

Serious and longer-lasting problems of water shortage are concentrated in the world's main dryland environments. Drylands are areas with a moisture deficiency, where evaporation losses exceed precipitation inputs. These areas lie mostly within the arid and semi-arid climates that cover about a third of the Earth's land surface. They contrast with the humid climates at higher latitudes, where moisture is usually much more abundant. The dryland areas usually support small populations of people, most of whom either remain close to available water sources or have developed a nomadic lifestyle that makes sustainable use of the natural resources of deserts. Some drylands – such as the Nile valley in Egypt; see p. 58) – are able to support high population densities because sophisticated water management systems have been developed to exploit groundwater and river flows. Deserts occur in extremely arid climates, and we examine the desert environment in the second section of this chapter.

feature of climate, and – particularly in some climates – it is neither a rare nor a random occurrence. Long-term climate and historical records indicate that major droughts have occurred with remarkable regularity (roughly every 22 years) in the USA, for example, particularly in the prairie and Midwestern states. The serious drought of 1933–35, in which large areas of the Great Plains (Figure 14.1) became the Dust Bowl (Box 14.2), is part of this ongoing trend.

Drought is in some ways a relative thing, because it reflects water shortage below normal conditions in a particular area. Thus, for example, drought occurs in mid-latitude countries like Britain when rainfall is considerably below the long-term average. But this might still involve significantly greater rainfall and water availability than under normal non-drought conditions in many naturally arid areas (such as North Africa).

Operational definitions of drought are useful because they help water resource managers to define the start, end and severity of a particular period of water shortage. Most operational definitions adopt a critical threshold of rainfall reduction, usually comparing the current situation with the long-term average. The onset of drought might, for example, be defined as 75 per cent of average rainfall over a stated time period.

BOX 14.2 THE DUST BOWL IN NORTH AMERICA DURING THE 1930S

The Dust Bowl is a large area in the southern part of the Great Plains region of the USA that suffered serious drought and soil erosion during the 1930s. The area – which included parts of Kansas, Oklahoma, Texas, New Mexico and Colorado – has a long history of persistent droughts and had a natural vegetation of grass that protected the fine-grained soils from wind erosion. Between about 1885 and 1915, many people moved into the area to settle, many new farms were established, and large areas were given over to wheat and crops and to cattle grazing. The marginal land was unsuited to intensive agricultural use, and serious soil erosion began in the early 1930s when a period of severe drought affected the area. Up to 10 cm of valuable topsoil (including organic matter, clay and silt) was blown from some places, removing the very resource on which farming was dependent, and it drifted against fences and buildings in other areas. Thousands of families migrated westwards, abandoning their farms, and up to a third of the families who stayed behind survived only with the aid of government relief. Farm incomes were seriously depressed during the 1930s drought, and they also suffered badly during a drought in the 1950s (Figure 14.2).

From 1942 onwards, concerted efforts were made by federal and state governments to develop programmes for soil conservation and for rehabilitation of the Dust Bowl region. These included seeding large areas with grass, introducing three-year crop rotations, use of contour ploughing, terracing, strip planting and shelter belts. While these measures have protected much of the remaining soils in the region, droughts in the 1950s, 1960s and late 1970s caused further wind erosion and reinforced the need for more sustainable resource use in this marginal area.

There are various forces at work that threaten to increase further the drought risk in the Dust Bowl area, including global warming, continued expansion of irrigation agriculture and large-scale groundwater abstraction from the Ogallala aquifer.

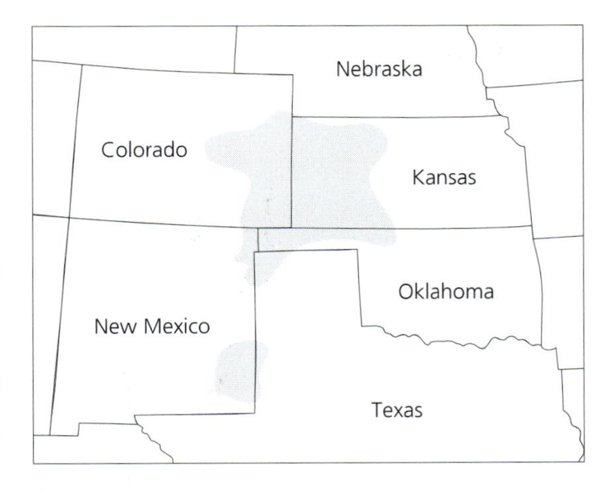

Figure 14.1 *The Dust Bowl in the Midwest USA.*

Source: a map on page 128 in Marsh, W.M. and J.M. Grossa (1996) Environmental geography: science, land use and Earth systems. John Wiley & Sons, New York.

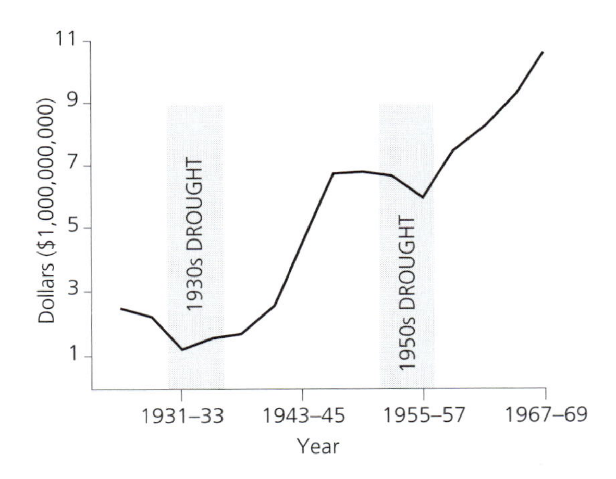

Figure 14.2 *Gross farm incomes in the Great Plains region, 1925–69. Over this period, gross farm incomes rose in the Great Plains, but the trend was interrupted by the very severe drought of the 1930s and another less severe drought during the mid-1950s.*

Source: Figure 4.4 in Hewitt, K. (ed.) (1983) Interpretations of calamity. George Allen & Unwin, London.

SEVERITY

Some droughts are much more serious than others. A mild drought in mid-latitude countries, with well-developed water resource management systems, normally causes minor disturbance – for example, the use of hosepipes in gardens is banned temporarily, and car wash machines are closed down. A severe drought in an already arid area can cause many deaths through famine and starvation.

Severity of drought is usually gauged by a number of factors, the most important of which are:

- degree of moisture deficiency (the greater the deficiency, the worse the drought);
- duration of deficiency (the longer it lasts, the worse the drought);
- size of area affected (the bigger the area, the worse the drought);
- number of people affected (the greater the number, the worse the drought).

Shortage of rainfall is the most common cause of droughts, but it is not the only cause, and rainfall deficit alone does not always produce visible impacts.

CAUSES OF DROUGHT

Many factors are involved in creating droughts and the famines that are often associated with them in arid areas. Quite often, droughts are the combined result of climate and human activities in the area.

CLIMATE

Droughts can occur in any area, but some areas are more susceptible to severe drought than others. The most frequent and most serious droughts tend to occur between latitudes 15 and 20° in areas adjacent to the permanently arid continental regions. These areas are permanently arid because warm tropical air masses descend to the Earth's surface and become hotter and drier (see p. 335). If and when the prevailing westerly wind system shifts polewards, as it does from time to time, this brings the high-pressure

BOX 14.3 TYPES OF DROUGHT

There are four main types of drought:

1. *meteorological drought*: usually defined by rainfall deficit or the degree of dryness compared with some 'normal' or average amount;
2. *hydrological drought*: usually defined by river flow deficit;
3. *agricultural drought*: usually defined by soil moisture deficit;
4. *famine drought*: usually defined by food deficit.

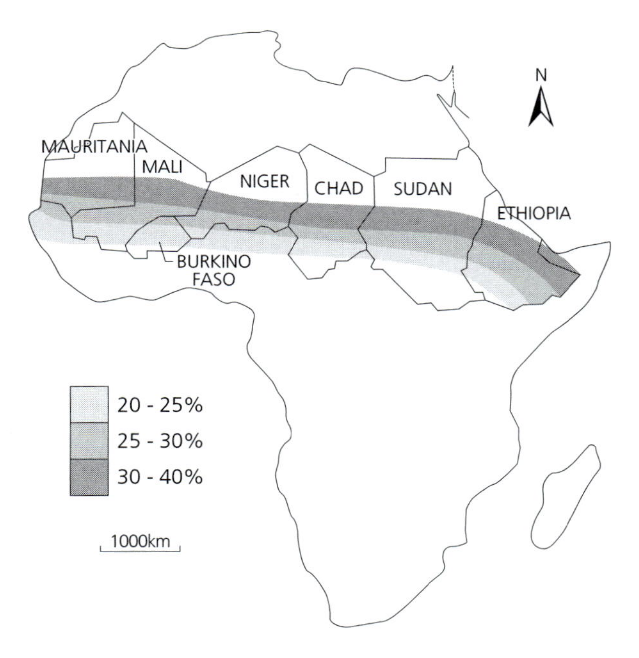

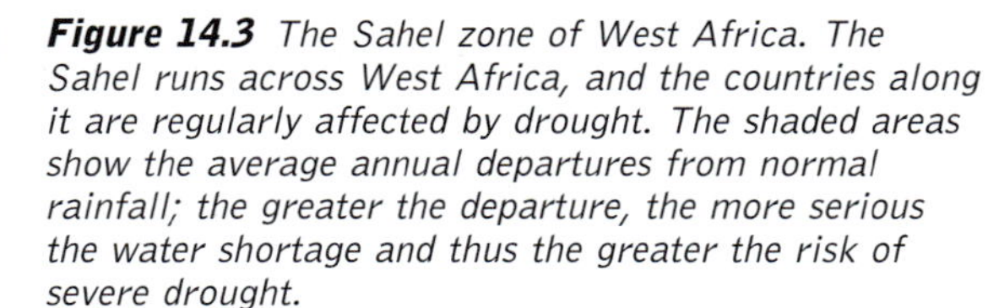
Figure 14.3 *The Sahel zone of West Africa. The Sahel runs across West Africa, and the countries along it are regularly affected by drought. The shaded areas show the average annual departures from normal rainfall; the greater the departure, the more serious the water shortage and thus the greater the risk of severe drought.*

Source: Figure 12.4 in Smith, K. (1992) Environmental hazards. Routledge, London.

Plate 14.1 *Edge of the Sahara Desert, Tunisia, North Africa.*

Photo: Philip Barker.

BOX 14.4 THE SAHEL DROUGHT

The Sahel is a dry region in West Africa, in the arid zone between the desert to the north and the wet tropics to the south. It is a belt about 500 km wide along the southern edge of the Sahara Desert and is one of the most drought-prone areas in the world. Annual rainfall is normally between 100 and 200 mm, and most of the rain falls between June and September. The natural vegetation is mainly sparse savannah grassland and shrubland, and traditional forms of agriculture include nomadic herding and (in some places) the cultivation of peanuts and millet.

The area has always been marginal because of the limited rainfall and naturally arid conditions, but matters were made much worse when an extended drought set in during the late 1960s and lasted intermittently until the early 1980s. Water is always in short supply in the Sahel because of the low total rainfall, the concentration of rain into the wet season and the persistently high temperatures (thus rapid rates of evaporation). But water shortage became particularly serious between 1968 and 1974, when rainfall in the Sahel was significantly below the long-term average. This triggered the worst drought in West Africa in about 150 years and was probably associated with an increasingly arid regional climate. Between 1969 and 1990, rainfall in the Sahel was below the long-term average in every year.

The tragedy of drought is well illustrated in the Sahel. During the drought of 1973–74, for example, an estimated 100,000 people died, mostly through starvation. Millions of cattle died too. The situation grew worse in the 1980s as the continued lack of rainfall caused widespread famine. Initially, many people died through lack of drinking water, but as the drought continued, lack of water caused crop failures and livestock had nothing to graze. The land became completely barren, and the soil was baked by the sun to form a hard surface that was impossible to plough.

anticyclonic conditions of the permanently arid regions to areas where the climate is usually dominated by seasonally wet low-pressure weather systems, and drought follows.

A southward shift in the westerlies indirectly caused the long drought that affected the Sahel region of West Africa (Figure 14.3) between 1968 and 1980 (Box 14.4) by preventing the northward movement of moist equatorial air from the West African coast. Climate records throughout the twentieth century show interesting associations between rainfall in West Africa and the frequency of intense Atlantic hurricane activity – when drought breaks in the Sahel region, there appears to be an increase in the frequency of intense hurricanes that reach the US east coast and the Caribbean basin.

Analysis of lake sediments from Lake Naivasha in Kenya's Rift Valley has been used to reconstruct past changes in climate in East Africa over the last thousand years. Results show a long dry period between 1000 and 1270, and three shorter but intense droughts in 1380–1420, 1560–1620 and 1760–1840. Each coincides with an era of famine, political unrest and mass migration, according to oral history. The droughts were all associated with variations in the radiative output of the Sun (see p. 107).

HUMAN FACTORS

The effects of water shortage and drought are often complicated and intensified by human mismanagement. For example, the impacts of the Dust Bowl drought in the USA (see Box 14.2) were certainly aggravated by over-cropping, over-population and lack of effective relief measures.

Sometimes human activities increase vulnerability to droughts rather than directly causing them. This has been the case in some parts of Mexico since the 1970s, where new agricultural technologies (such as irrigation, fertiliser and improved seeds) and land tenure structures (particularly the increasing number of large private farms) have increased the frequency and severity of drought losses.

FEEDBACK AND CHANGE

Once desertification starts in an arid area, the problem is amplified initially by a series of positive feedback mechanisms (see p. 79). As the desert area increases, the regional climate becomes drier and this further increases the risk of drought. Increasing aridity and expansion of desert margins reduce the area of land suitable for growing food (crops and grazing), so more and more people are forced to migrate to these shrinking areas in search of food and water. This further increases pressure on available resources and makes famine much more likely when drought occurs. Negative feedback mechanisms (see p. 79) eventually restore some form of equilibrium, as population declines (as people die or move away) and thus pressure on resources drops.

EXPERIENCING DROUGHT

ONSET OF DROUGHT

Drought is a progressive hazard because it builds up over a period of time and its impacts change over time. When drought begins, the first impacts are usually experienced by the agricultural sector, which relies heavily on stored soil water, which can be depleted rapidly during long dry spells. If the dry spell continues, with little or no rainfall, the impacts are felt more widely. When this happens, other water users, who rely on surface water supplies (including lakes and reservoirs) and on sub-surface supplies (from groundwater), are affected by shortages. This happened in England and Wales during the drought of 1975–76, for example (Figure 14.4).

BOX 14.5 RESOURCE USE AND DROUGHT

While climate (particularly shortage of rainfall) is usually the primary trigger of drought, the situation is often made much worse by the way people use resources in the arid zone. Human impacts are evident in various ways, including causing deserts to spread (desertification) by felling trees for firewood and building houses, and by poor farming techniques (particularly overgrazing). The Sahel drought in West Africa (see Box 14.4) was aggravated by non-climatic factors such as over-cropping, military conflicts and political unrest. In Ethiopia during the mid-1980s, for example, the government systematically redirected funds sent for drought relief and used them for political and military purposes. In a wider sense, some observers argue that most serious droughts in developing countries are more a function of modern development policies than climate – caused by deforestation, over-exploitation of land and water resources, and loss of traditional knowledge of natural resource management in dryland environments.

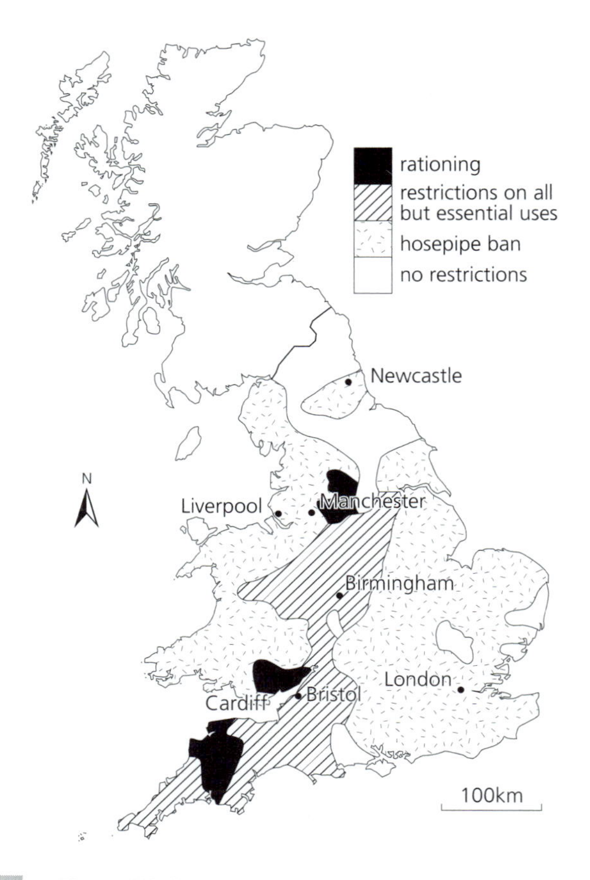

Figure 14.4 *Water restrictions in England and Wales during the 1975–76 drought. Below-average rainfall across much of England and Wales caused serious water shortages, so the use of hosepipes (for example for watering gardens) was widely banned and some areas faced restrictions on water use. Water was strictly rationed in a few particularly dry areas.*

Source: Gregory, K.J. and J.C. Doornkamp (eds) (1978) Atlas of drought. Institute of British Geographers, London.

Some water resource management systems are capable of maintaining water supplies during short droughts (which last up to about six months), particularly if they include large-scale water storage (in reservoirs) and groundwater abstraction. Sometimes water supply systems are unable to cope with a prolonged or particularly severe drought, and water rationing and shortages are inevitable.

IMPACTS OF DROUGHT

Drought produces a series of impacts that usually extend far beyond the area that is experiencing the actual water shortage. These impacts are both direct (Box 14.7) and indirect.

BOX 14.6 ARIDITY–STRESS–RESPONSE CYCLE

Traditionally, the cycle of aridity–stress–response has kept populations in many drylands within the long-term carrying capacity of the environment. This situation is punctuated by periodic instability but has tended to be sustainable over the long term. However, the traditional balance has been upset in recent decades by two opposing trends:

1. Carrying capacity is declining as more drylands are over-used and are unable to restore long-term stability between successive droughts and the environmental stresses associated with them.
2. Population levels are rising as better medicine, food and shelter become available, reducing the death rates, illness and out-migration.

The net effect is increased stress on available resources – including soil quality, natural vegetation, grazing land, water resources and land – from deforestation, overgrazing and over-cultivation, each of which is directly related to population.

BOX 14.7 DIRECT IMPACTS OF DROUGHT

Direct impacts include:

- decreased biological productivity (and thus lower yields) in cropland, rangeland and forest land;
- increased fire hazard;
- reduced water levels;
- increased mortality rates among livestock and wildlife;
- damage to fish and wildlife habitats.

Indirect impacts arise via the direct impacts. Lower yields, for example, mean reduced income for farmers and forestry owners, increased prices for food and timber, increased unemployment, and decreased tax revenues. Some of the important economic, environmental and social impacts of drought are summarised in Table 14.1.

Table 14.1 Some important impacts of drought

Type of impact	Examples
Economic	Loss from crop production
	Loss from dairy and livestock production
	Loss from timber production
	Loss from fishery production
	Loss from recreational businesses
	Loss of hydroelectric power
	Loss to industries directly dependent on agricultural production
	Increased food prices
	Loss of navigability on rivers and canals
	Cost of new water resource development
Environmental	Damage to animal species
	Wind and water erosion of soils
	Damage to fish species
	Damage to plant species
	Water quality effects (e.g. salinisation)
	Air quality effects (e.g. dust blow)
Social	Food shortages
	Loss of human life
	Conflicts between water users
	Health-related low-flow problems (e.g. increased pollutant concentrations)
	Population migration

One of the most serious impacts of prolonged drought in arid areas is enforced mass migration – the involuntary movement of large numbers of people from place to place (see p. 39). This happens particularly in arid developing countries, and it involves the wholesale movement of many people from rural areas, who are unable to obtain drinking water or food. They move to areas where they hope to find abundant supplies of both. Mass migration in West Africa – such as that associated with the Sahel drought (see Box 14.4) – has generally involved movement southwards to the wetter regions. The areas into which the migrants flock are often already seriously overcrowded, and they often find it very hard to cope with the sudden arrival of vast numbers of dependent newcomers. In this way the impacts of drought are spread over a larger area. Large population movements were also triggered by the failure of farming in the American Dust Bowl during the 1930s (see Box 14.2).

COPING WITH DROUGHT

In many industrial countries, droughts can cause water rationing and the temporary closure of heavy water-using industries, but it is possible to take precautions in drought-prone areas. These include:

- constructing reservoirs to store emergency water supplies;
- education to avoid over-cropping and over-grazing;
- programmes to limit settlement in drought-prone areas.

Drought is a much more serious problem in arid developing countries, where a range of strategies have been adopted, including emergency aid (Box 14.8), small-scale projects and large-scale projects.

BOX 14.8 EMERGENCY AID

In times of severe drought and famine, relief operations involving international agencies such as the United Nations and the Red Cross are required to bring emergency supplies of food and medicines to the worst-affected areas, and to co-ordinate the setting up and running of refugee camps and manage resource distribution. Food aid is often sent in a somewhat haphazard, *ad hoc* and opportunistic manner, and the management of food aid during famine crises has not always gone smoothly. Food aid is often badly targeted too, so it does not always get to those who are most in need.

Regional co-operation is often vitally important in making sure that international aid and practical assistance reaches places where it is needed at the appropriate time. This was the case in southern Africa in 1993, when an area roughly the size of Australia was parched by the century's worst drought and 20 million people were at risk of hunger and disease. All the southern African nations worked together on a major relief effort, helped by the international community, and disaster was averted. Relief operations are short-term and disaster-specific. Longer-term solutions are required to tackle the root causes of the problem rather than just the most immediate symptoms.

SMALL-SCALE PROJECTS

Many different types of small-scale project can significantly decrease the risk of prolonged water shortage in drought-prone areas, including:

- drilling of deep water wells;
- installation of water pumps;
- building simple irrigation channels and systems.

Such schemes are often related to the needs of individual villages. They are usually cheaper and quicker to implement, and easier for locals to maintain and manage, than large projects (like dams and reservoirs). Local aquifers can be exploited by sinking boreholes and wells, providing water supplies to local vegetable gardens, farms and villages. Problems often arise with local wells because extraction can cause the water table to drop (see p. 358), so wells often have to be dug deeper and deeper over time.

A range of different water-harvesting techniques can also be used to make best use of available rainwater. They are usually small-scale and suit local needs. One common strategy is to build a series of small stone walls across a hillside to catch rainwater as it runs downslope. The small barriers interrupt the natural downslope flow of the water and allow more of it to infiltrate the soil.

Another type of small-scale project is the tree-growing programme. This is generally designed to stop the spread of a desert margin by providing shade, which reduces the rapid evaporation of soil moisture. Peasant farmers can also adapt their farming techniques when confronted with prolonged drought. In Sudan during the 1980s, for example, many farmers switched from growing traditional low-yield late-maturing crops to growing new higher-yield quick-maturing crops.

LARGE-SCALE PROJECTS

Large-scale projects often involve constructing dams and reservoirs and irrigation canal systems. Such schemes are designed to increase total quantities and the reliability of water supplies, and they require massive financial investment (often with international assistance) and long-term planning. Large-scale projects can produce unwanted environmental side-effects (see p. 68), and they are usually difficult for local people to maintain and operate. This increases dependency on engineering specialists and international funding bodies, neither of which might be in the best long-term interests of local populations. Large schemes can provide valuable water resources for a variety of purposes, including water and hydropower (see p. 394).

DESERTS

THE NATURE OF DESERTS

Hot deserts make up about 20 per cent of the Earth's land surface in two belts between about 20 and 30° north and south of the equator (Figure 14.5). These are surrounded by semi-arid belts. Together they form a third of the Earth's land area and are home to nearly a billion (1,000 million) people.

In hot deserts, evaporation rates exceed precipitation and temperatures can be very high. Daytime shade temperatures of up to 55 °C are not uncommon. Desert surfaces heat up rapidly during the day in the scorching sun and lose heat rapidly at night by radiation back to the overlying air. Night-time temperatures can fall close to freezing.

ARIDITY

Water shortage, associated with extreme aridity, is a hallmark of hot deserts. Aridity is best defined in terms of water balance (see p. 392), which is the relationship in a given area between:

- precipitation input (P)
- evapotranspiration loss (Et)
- changes in storage (in soil moisture, groundwater, and so on) (s).

An arid region has a deficit in the water balance over a year, and the degree of aridity is defined by the size of the water

BOX 14.9 DEFINITION OF DESERT

Deserts are defined as areas that receive less than 25 cm of rainfall a year. There are two types of desert:

1. Cold deserts, at high latitudes near the poles, are always covered by snow and ice. Antarctica is a cold desert.
2. Hot deserts, closer to the equator, are hot and dry. The Sahara in North Africa is a hot desert.

This section is concerned with hot deserts; cold deserts are covered in Chapter 15.

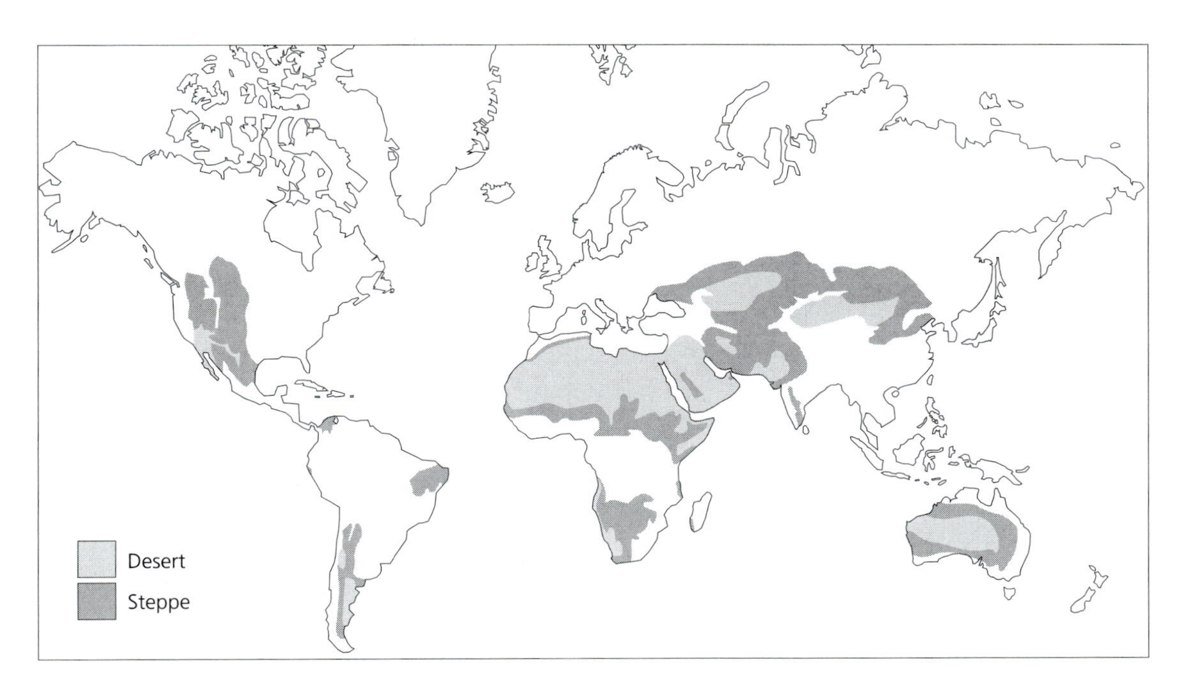

Figure 14.5 *Distribution of desert and steppe climates.*
Source: Figure 8.2 in Doerr, A.H. (1990) Fundamentals of physical geography. Wm. C. Brown Publishers, Dubuque, Iowa.

BOX 14.10 DRY CLIMATES

Dry climates share some important characteristics:

- low precipitation amounts;
- uneven distribution of precipitation from month to month and from year to year;
- high evaporation rates, which usually exceed precipitation;
- precipitation is often in the form of intense, short-duration showers;
- precipitation reliability decreases as aridity increases.

Plate 14.2 *Giant cactus plants in the Mojave Desert, California, USA. The cactus in the foreground is about 5 m tall.*
Photo: Chris Park.

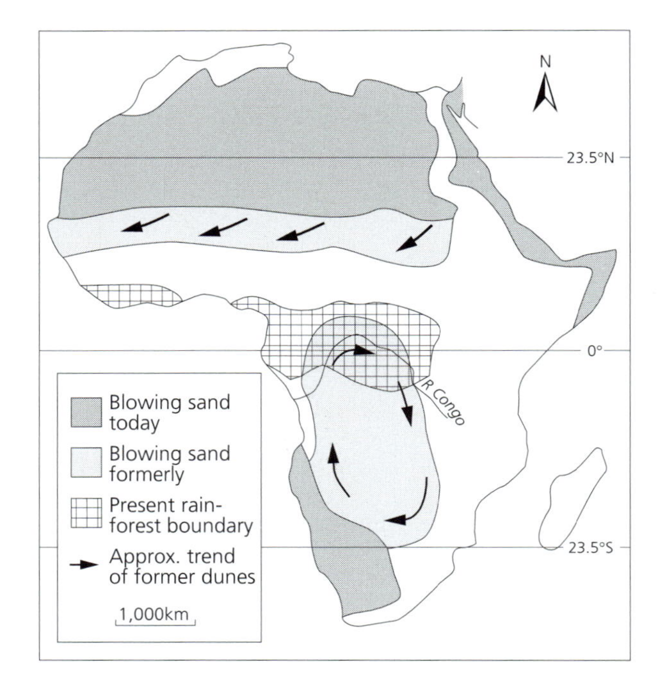

Figure 14.6 *Former active desert dune fields in Africa. Desert dunes extended over a much broader area in tropical Africa in the past than they do today, and the orientation of former dunes gives important clues to the dominant wind direction at the time they were formed.*

Source: Figure 18.13 in Dury, G.H. (1981) Environmental systems. Heinemann, London.

deficit. A sub-humid area has a slight surplus in the water balance over a year, whereas a semi-arid area has a slight deficit. As the deficit in the water balance increases, the environment becomes increasingly arid. Extreme aridity occurs in places that have no regular seasonal pattern of rainfall, when no rainfall at all is recorded over a period of at least twelve consecutive months.

About 4 per cent of the Earth's land surface is extremely arid, and a further 30 per cent is arid (15 per cent) and semi-arid (14.6 per cent). In the past, the area covered by active deserts was much larger. This is borne out, for example, by the distribution of former sand dunes in Africa (Figure 14.6).

Deserts are characterised by water deficit, and in some years they receive no rain at all. Water shortage coupled with high temperatures give rise to distinctive desert environments, defined in terms of soils, vegetation, animals, landforms and human activities. Many deserts contains *wadis* – dry watercourses that contain water only immediately after a storm.

Problems of water resource management are particularly acute in and around hot deserts. Human settlements in desert environments are usually concentrated around oases, where relatively reliable and accessible supplies of water are available. Some oases are small springs, whereas others are long valleys such as the Nile valley (see p. 66).

LOCATION

Dry climates occupy large areas in low latitudes from the interior to the western coasts of continents (e.g. the Sahara in North Africa). In middle latitudes, they occupy areas to

Table 14.2 The ten largest hot deserts in the world

Desert	*Location*	*Approximate area (km²)*
Sahara	North Africa	8,600,000
Arabian	South-west Asia	2,330,000
Gobi	Mongolia and north-east China	1,166,000
Patagonian	Argentina	673,000
Great Victoria	South Australia	647,000
Great Basin	South-west USA	492,000
Chihuahuan	Mexico	450,000
Great Sandy	North-west Australia	400,000
Sonoran	South-west USA	310,000
Kyzyl Kum	Kazakhstan–Uzbekistan	300,000

BOX 14.11 DESERT PROVINCES

There are five main desert provinces (Figure 14.7):

1. the Sahara, the largest province and an associated series of deserts stretching eastwards through Arabia into Central Asia;
2. the southern Africa province, which includes the coastal Namib Desert and the inland deserts of the Karoo and Kalahari;
3. the South American desert province has two coastal deserts, the Atacama Desert along the west coast and the Patagonian Desert along the east coast;
4. the North American desert province, which covers much of Mexico and the south-western USA, including the Mojave and Sonoran Deserts;
5. the Australian province.

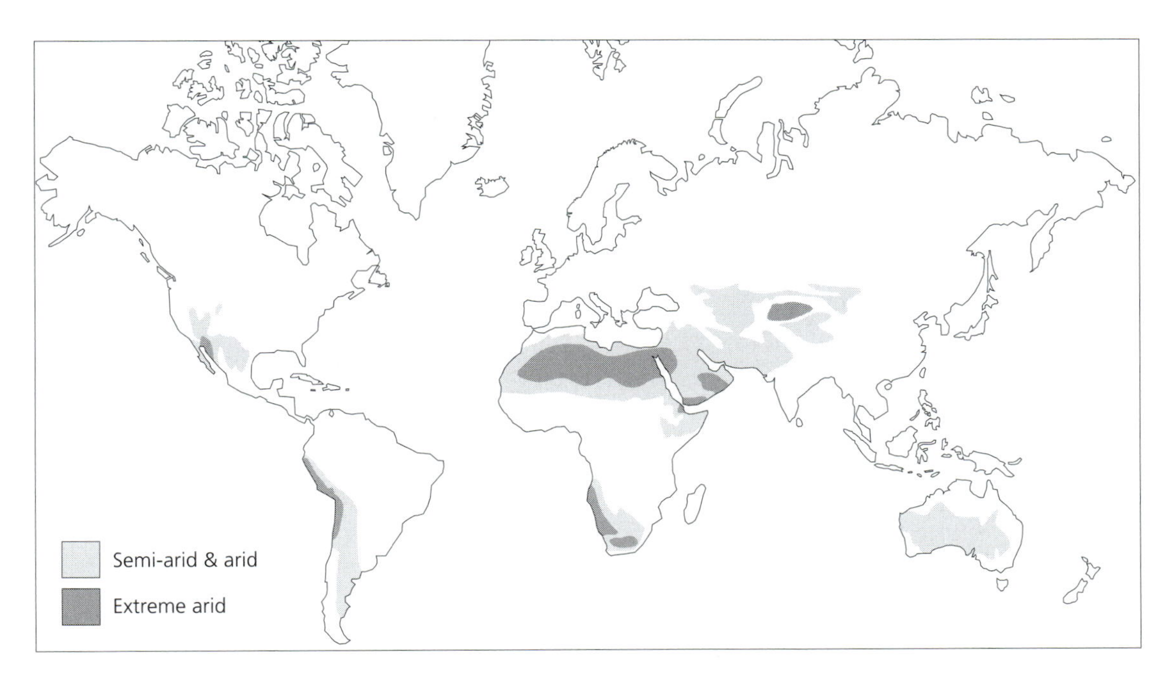

Figure 14.7 *Location of the major desert provinces. Most of the major deserts are located within the tropical climatic zone, but some (such as the Gobi Desert in Asia) are located in the dry continental interiors.*
Source: Figure 10.2 in Goudie, A. (1993) The nature of the environment. Blackwell, Oxford.

the east of high mountain barriers and in the interiors of continents. The Sahara is by far the largest desert in the world (Table 14.2), covering more than 8 million km^2. It is nearly four times as big as the Arabian Desert, which is the second largest.

BOX 14.12 CHARACTERISTICS OF THE DESERT CLIMATE

- wide temperature variations (especially between daily maxima and minima);
- low relative humidity;
- limited precipitation;
- more or less continuous sunshine throughout the year;
- little cloud cover and clear skies;
- occasional intense local convectional showers.

CLIMATE

Most dry deserts are located between 15 and 30° north and south of the equator. This desert zone is often dominated by subsiding air associated with the sub-tropical highs, which is warmed by compression and brings dry conditions. It is also sometimes affected by trade winds, which blow towards warmer latitudes and pick up moisture from these already dry areas.

Like all climate zones (see p. 335), deserts are far from uniform, and there are many variations from place to place. But the desert zone – type Bw in the Köppen climate classification (see Table 11.2) – has a distinctive climate (Box 14.12).

The low-latitude deserts that develop in the zone are hot and dry, and this climate supports minimum vegetation cover. Summer shade temperatures often exceed 38 °C. This climate is extremely inhospitable to humans, who face the risk of sunstroke, skin cancer and eye damage in the scorching sun. The extreme day-to-night temperature changes – daytime highs of up to 38 °C followed by night-time lows of around 20 °C – are highly uncomfortable for people without adequate shelter or clothing. This significant diurnal change also promotes mechanical weathering of rocks in the desert (see p. 207). People are also appreciably affected by the persistent aridity of low-latitude deserts, which makes them extremely uncomfortable but has the benefit of preserving organic remains (including mummies from ancient Egypt) for long periods of time.

Plate 14.3 *Wadi (dry watercourse) in the desert of Tunisia, North Africa.*
Photo: Philip Barker.

BOX 14.13 VISIBILITY OVER THE DESERT

Sky conditions over deserts are often ideal for flying because of the lack of cloud cover, the regular sunshine and the reliability of clear weather. NASA takes advantage of such conditions and uses the Edwards Air Force Base, in the California desert, as the main landing site for the space shuttle. While good visibility and lack of air pollution are typical of desert regions, they can be disturbed by human habitation. A 10-year study of the changing meteorology and air chemistry of the desert surrounding a rapidly growing small city in the Mojave valley, on the Nevada/Arizona border, has shown significant human impacts. These include increased temperature, relative humidity, particulate pollution and nitrogen oxide concentrations.

BOX 14.14 TYPES OF DESERT SURFACE

There are three main types:

1. *erg*: a desert covered by sand dunes;
2. *hammada*: a desert landscape consisting of bare rock, with no sand or gravel;
3. *reg*: a desert landscape covered by loose stones and gravel, but with no sand.

Mid-latitude deserts are cooler than the low-latitude deserts, and they receive low and unreliable rainfall. Most are located in the dry rain shadows on the lee side of topographic barriers, or in the interior of large land masses far from maritime sources of moisture. Such deserts have much less extreme climates than the low-latitude deserts and are much less inhospitable to people.

DESERT LANDSCAPES

We tend to think of deserts as vast seas of shifting sand, like the Sahara. While this is certainly true of many deserts, and it makes dramatic photographs, it is not always the case or even necessarily the norm. Much of the character of the desert is determined by the surface material (Box 14.14), and there are many different types of hot desert, one classification of which is summarised in Table 14.3. Typical desert landscapes are quite stark, and the landforms are shaped by wind and water.

WIND

Dry areas often experience strong and quite persistent winds, partly because of the sparse vegetation, which reduces frictional drag. Winds play two important roles in shaping desert landforms:

1. They sand-blast rocks; some deserts contain rocks called *yardangs*, which have exotic shapes determined by the exploitation of layers of weakness by the wind.
2. They build up sand dunes.

Table 14.3 Some common types of hot desert

Type of desert area	*Character*
Desert uplands	In this type of desert, geological controls of relief are important, bedrock is exposed, relief is high
Desert piedmonts	These are zones of transition separated from the uplands by a change of slope, but which receive runoff and sediments from the uplands. These areas have erosional forms (such as pediments) and depositional forms (such as alluvial fans)
Stony deserts	These consist of stony plains and structural plateaux, and they may have a cover of stone pavement
Desert river and floodplains	These are features of desert lowlands
Desert lake basins	These are depressions that are often salty and receive drainage from inflowing streams
Sand deserts	These are often covered by sand dunes and dominated by wind action

Unconsolidated surface materials are often easily moved by the wind. Grains of sand (blown from alluvial deposits, lake shores, sea shores and weathered rocks) can be transported across the surface of a desert. The sediment is usually sorted during transport and is often deposited as large fields of regularly formed sand dunes (*ergs*; Box 14.14). Up to a quarter of the world's deserts are covered by wind-blown sand.

Sand dunes are mounds of sand, sometimes higher than 200 m, that are built up by prevailing winds (Figure 14.8). They are typical features in many hot deserts, including the Sahara and Arabian Deserts, although in some deserts (such as the North American deserts) sand dunes may cover only 1–2 per cent of the total area. Loose sand is blown and bounced along by the wind, up the windward side of a dune. The sand particles then fall to rest on the lee side, while more are blown up from the windward side. In this way, the dune builds up and moves gradually downwind. Dune sands pose particular problems for engineers, because they create environmental hazards during strong

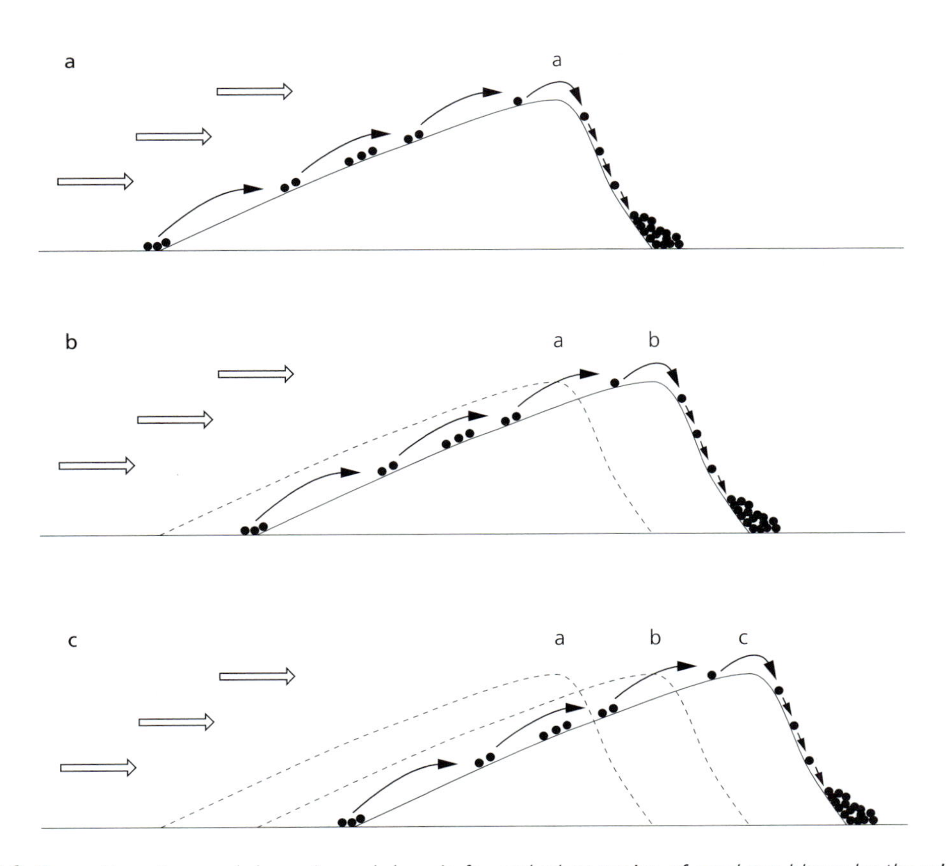

Figure 14.8 *Formation of a sand dune. A sand dune is formed when grains of sand are blown by the wind; the dune migrates forward as the individual sand grains are pushed downwind.*

Source: Figure 15.19 in Doerr, A.H. (1990) Fundamentals of physical geography. Wm. C. Brown Publishers, Dubuque, Iowa.

winds and sand storms, and they are not always suitable for construction purposes.

Desert dunes can be extremely mobile, and their speed, direction and pattern of movements largely depend on the availability of a sand supply and on the dominant local wind direction (Box 14.15). Remote-sensing imagery can be used to monitor sand dune changes over time. Studies in the Rajasthan Thar Desert in India have shown remarkable stability in dune boundaries between 1973 and 1986, despite significant land-use changes and progressive desertification. Ancient dune systems often preserve evidence of climate change over various timescales. Reconstruction of ancient dune systems in parts of southern Africa reveals three distinct phases of increased aridity, when dunes developed and were active, separated by milder phases when dune activity ceased.

BOX 14.15 DUNE FORMS

Dunes assume different shapes depending on wind speed and direction and the availability of sand (Figure 14.9). Crescent-shaped dunes, called *barchans*, are common in deserts with strong prevailing winds and relatively little sand (such as the coastal deserts of Peru). The horns of barchan dunes point downwind, and the dunes themselves move continuously downwind across the desert floor. *Seif* dunes are long ridges of wind-blown sand aligned parallel to the dominant wind direction. Star-shaped dunes are formed in areas where the wind blows from all directions.

Plate 14.4 *Sand dunes in Death Valley, California, USA. These dunes are highly mobile and appear to be encroaching on to the surrounding vegetated area.*

Photo: Chris Park.

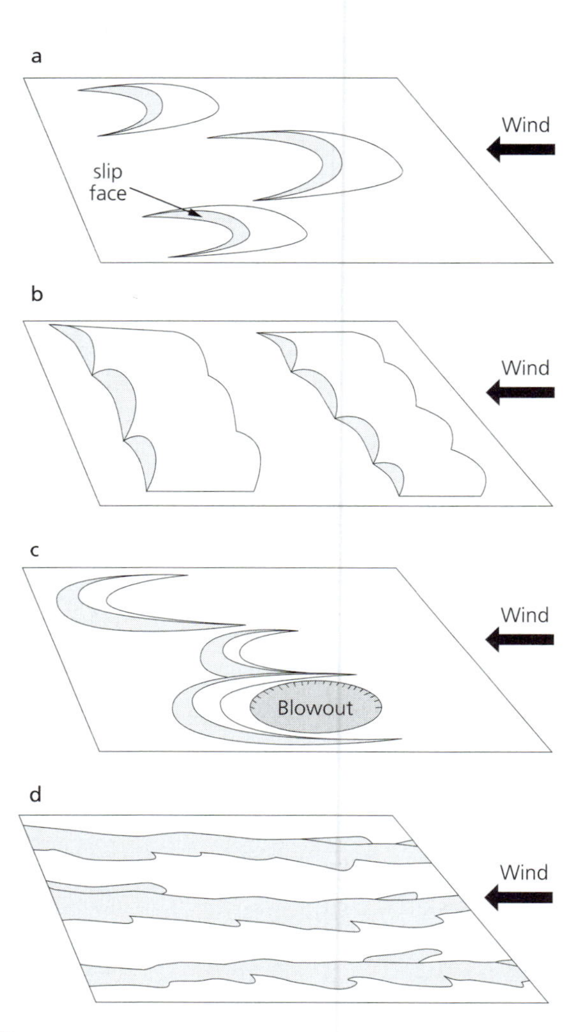

Figure 14.9 *Dune shapes and movement. Desert dunes assume a variety of different shapes and forms, the most common of which are the barchan (a), the transverse dune (b), the parabolic dune (c) and the longitudinal dune (d).*

Source: Figure 15.20 in Doerr, A.H. (1990) Fundamentals of physical geography. Wm. C. Brown Publishers, Dubuque, Iowa.

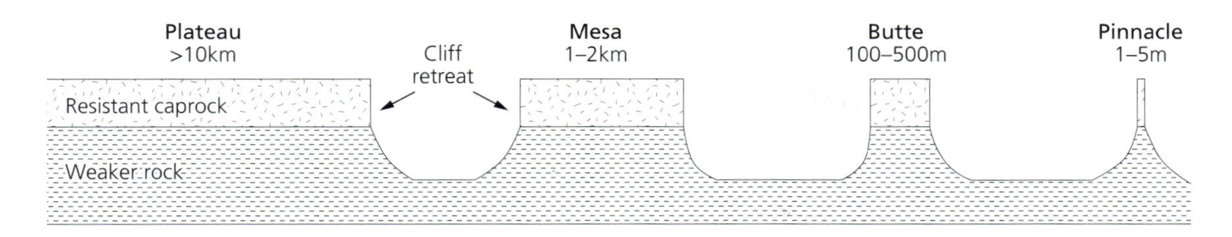

Figure 14.10 *Mesas, buttes and pinnacles. Plateau areas, often composed of a protective caprock overlying a more easily erodible rock, are eroded in the semi-arid environment, and cliff retreat progressively wears back the remnants to form mesas, buttes and eventually towering pinnacles.*

Source: Figure 10.2 in Dury, G.H. (1981) Environmental systems. Heinemann, London.

WATER

While rain is rare, when it does fall it can have a significant effect on the desert surface, which is unprotected by vegetation and erodes easily. Where water flows across unvegetated or sparsely vegetated sand it erodes gullies or canyons, which are often called *arroyos*. Resistant rock often rises above the desert floor as angular peaks (Figure 14.10), and as this material is eroded it is generally deposited as a series of alluvial fans (called *bajadas*) sloping away from the sediment source (Figure 14.11).

The depositional slopes often grade together in a relatively flat-floored basin called a *playa* (Figure 14.12). Playas fill with water when infrequent but often intense rain falls over the desert. Once the rainwater has evaporated, dissolved salts are left as a crusty layer on the desert floor, giving it a powdery white appearance. Salinisation

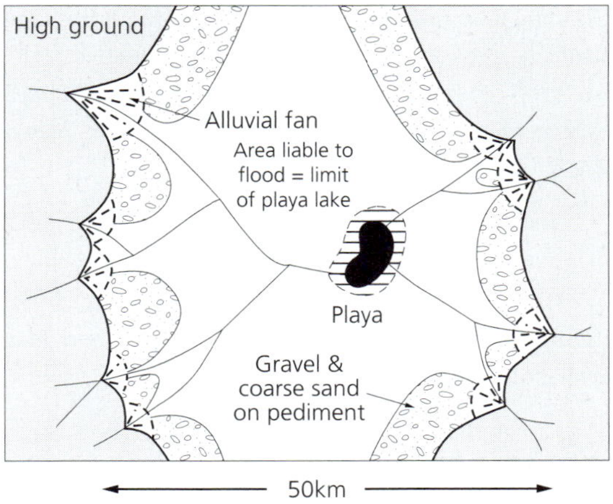

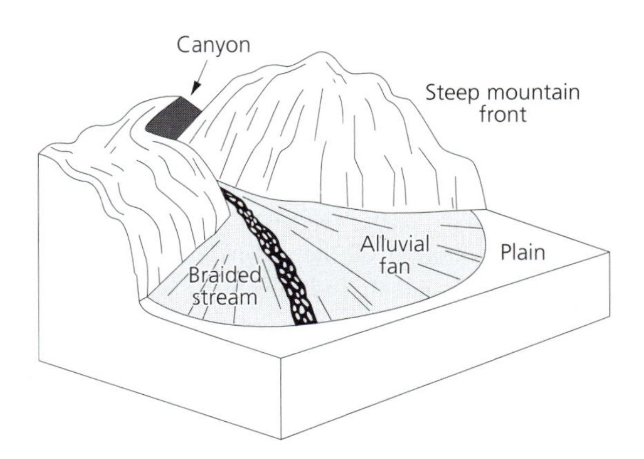

Figure 14.11 *Alluvial fans. Alluvial fans accumulate in desert areas where streams flow out from steep-sloped mountain areas carrying large loads of particulate sediment, much of which is deposited where the stream flows on to the flatter adjacent plain.*

Source: Figure 15.14 in Doerr, A.H. (1990) Fundamentals of physical geography. Wm. C. Brown Publishers, Dubuque, Iowa.

High ground
Alluvial fan
Area liable to flood = limit of playa lake
Playa
Gravel & coarse sand on pediment
50km

Figure 14.12 *Playas. A playa is a flat area at or near the lowest point in a desert basin, and many playas flood from time to time.*

Source: Figure 11.4 in Dury, G.H. (1981) Environmental systems. Heinemann, London.

BOX 14.16 DUST STORMS

Strong wind blowing across a desert surface also creates dust and sand storms, which can significantly reduce visibility, affect communications and pose a major threat to agriculture. The frequency of dust storms in the Sahel region appears to be increasing by a combination of drought and human mismanagement of the sensitive dryland environment. Changes in wind direction can distribute desert dust across wide areas. With strong southerly winds, for example, dust can easily be blown from the Sahara Desert across much of mainland Europe.

BOX 14.17 DEFINING DESERTIFICATION

The United Nations Conference on Desertification (1977) defined desertification as 'the diminution or destruction of the biological potential of the land . . . leading ultimately to desert-like conditions'. An alternative way of defining desertification is 'natural and cultural processes leading to an encroachment or intensification of desert conditions in arid lands and their marginal zones'.

Both definitions embody the idea that the desert is advancing or encroaching on previously productive land, causing a decrease in the agricultural potential of the land. This is caused by both natural and cultural processes and is a highly complex and variable phenomenon.

(see p. 402) can be a major problem in many drylands, threatening agricultural land, water supplies and historic and archaeological sites.

DESERTIFICATION

DEFINITION AND AWARENESS

Desertification is a process by which fertile land, particularly in the tropical steppe (see p. 578), is turned into barren land or desert. The term was originally used to mean the spread of desert conditions over adjacent areas as a result of climatic factors and/or human resource use, but it is now used more broadly to describe a range of processes occurring in many arid and semi-arid areas (Box 14.17). It has proved difficult to find a precise definition that covers all situations. Drought can quite readily be defined as a deficiency of rainfall, but there have been numerous attempts to define desertification.

An estimated 135 million people are directly affected by desertification, mainly in Africa, the Indian sub-continent and South America, but the problems are not confined to the particular places where desert expansion and soil degradation are concentrated. At the global scale, the major impacts include stress on food-producing capacity, reduction in biodiversity and modification of climate.

BOX 14.18 AWARENESS OF THE DESERTIFICATION PROBLEM

The worldwide scale of desertification first came to light in 1972 at the Stockholm United Nations Conference on the Human Environment. Within a year, the Sahel drought, which had started in 1968, was well established and widely publicised. But the international community failed to recognise how important the desertification problem was, and what the risks were of repeated drought disasters and associated famines, even after the United Nations held a Conference on Desertification (UNCOD) in Nairobi in 1977.

Between 1981 and 1985 a major famine occurred in the Sahel (see Box 14.4), particularly within Ethiopia and Sudan, and while development agencies warned of the seriousness of the problem, little international aid was given. But by the mid-1980s there was growing concern about the extent of desertification. In 1984, for example, Mostafa Tolba (director of the United Nations Environment Program) concluded a United Nations review of progress in combating desert expansion by noting that 'currently, 35 per cent of the world's surface is at risk . . . each year 21 million ha [210,000 km^2] is reduced to near or complete uselessness'.

AWARENESS

Drought and desertification have been linked together in Africa throughout the twentieth century, particularly since the major famine in West Africa in 1913, but the severity of the potential problem was not realised for many decades. Even by the 1990s there was ongoing debate about the nature and seriousness of the desertification 'problem'. Some scientists argued that the problem was being overstated and that 'desertification' implies a once-and-for-all process on a massive scale, whereas what seems to happen on the ground is the development of areas of land degradation, in some cases quite severe, but which with care and two or three good rainy seasons might return to their former status. Others argued that the problem was being neglected because of its complexity and the poverty and marginality of the arid lands.

EXTENT AND DISTRIBUTION

Desertification is not restricted to developing countries, and neither is it an entirely new phenomenon, but developing countries are facing the most serious desertification problems, and the problem appears to be worsening.

In 1977, the United Nations Conference on Desertification (UNCOD) concluded that desertification was then one of the most serious environmental problems in the world. UNCOD data suggest that about 6,000 km^2 of land is turned into desert each year, and that the productive capacity of a further 210,000 km^2 is ruined each year (making it unprofitable for farmers to use). More recent data suggest that over 3.1 million km^2 of the world's rangelands (80 per cent of the total), 3.35 million km^2 of rainfed cropland (60 per cent of the total) and 0.4 million km^2 of irrigated drylands (30 per cent of the total) are threatened by moderate to severe desertification.

THE PROCESS OF DESERTIFICATION

The word 'desertification' suggests sand dunes rolling across farms and villages, turning everything into desert, but in reality deserts expand and contract with the weather over a variety of timescales. The real problem is soil degradation (see p. 608), which refers to the way in which good land slowly becomes less and less fertile.

Desertification is concentrated mainly in arid and semi-arid areas (Figure 14.13), and it usually begins with the partial or complete removal of natural vegetation (the possible reasons for which are outlined below). The onset of the process is therefore often marked by a decline in the diversity of plant and animal communities, a decrease in overall cover and an increase in the proportion of bare ground. This loss of vegetation in turn causes some critical changes in soils – reduction of organic matter, decline in soil structure, loss of water-retention capacity – which lower soil fertility and encourage further loss of vegetation. The spiral of decline accelerates when the bare surface is attacked by wind and water erosion. Barren desert-like conditions – extensive bare areas of infertile and badly eroded soils – are often the inevitable end-result, and these can be persistent and extremely difficult to reverse.

Desertification tends to feed on itself, promoting further change through positive feedback. Climate is not always the main trigger, but once the chain reaction has started the resulting changes to the character of the land surface (particularly its albedo and energy budget; see p. 234) can promote changes to the local climate that are enough to deepen the problem in the source area and to spread it

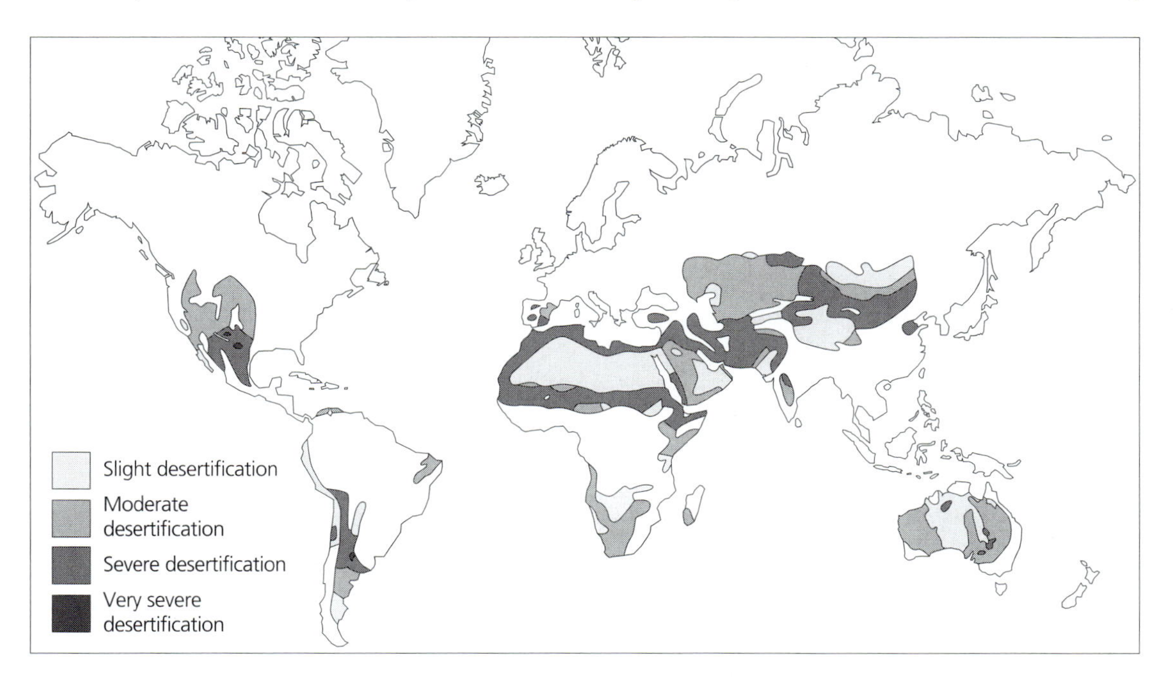

Figure 14.13 *Distribution of desertification. Desertification is affecting large areas both in and beyond the tropics, although the threat is not constant from place to place. See text for explanation.*

Source: Figure 12.12 in Cunningham, W.P. and B.W. Saigo (1992) Environmental science: a global concern. Wm. C. Brown Publishers, Dubuque, Iowa.

across a wider area. Human-induced albedo changes appear to be particularly dramatic along desert margins.

CAUSES OF DESERTIFICATION

Many different factors cause or contribute to the spread of desertification in marginal areas around the world. Climate is inevitably the dominant natural factor, and the main cultural factors (which are usually promoted by the pressures of expanding populations) are over-grazing, over-cultivation and deforestation. Like most other major environmental problems, desertification is surrounded by uncertainty and complexity (Box 14.21), and projected global warming might make an already difficult situation much worse in many areas (Box 14.22).

BOX 14.19 MONITORING DESERTIFICATION

As the problem has grown, more effort has been invested in monitoring and assessing what is happening and where it is happening. Increasing use is being made of remote-sensing technologies (see p. 21) and geographical information systems (GIS) (see p. 18) to establish the global extent and pattern of desertification and to understand better why severity varies so much from place to place. There is now a *World atlas of desertification* based on data held in the UNEP Global Resources Information Database (GRID).

Problems of defining precisely what desertification is and what it means hinder the quest for a complete understanding of the nature and extent of the problem. Natural variations in dryland ecosystems caused by inherent environmental instability need to be distinguished from degradation of dryland soils caused by human activities, for example, otherwise studies are not comparing like with like.

BOX 14.20 EARLY ACTION TO CONTROL DESERTIFICATION

The rapid spread of desertification is one reason why it is vital to try to halt it at the early stages, before the chain reactions and feedback relationships start to take hold. Appropriate treatment at an early stage (perhaps by changing land-management strategies) often has a good chance of success, but it requires:

- good monitoring (to detect the onset of desertification);
- agreement about what is happening, and what is causing it;
- an understanding of appropriate ways of tackling the problem;
- an appreciation that early intervention, before the problem becomes serious but also before conclusive symptoms appear, is without doubt the best solution;
- a willingness to act.

BOX 14.21 DESERTIFICATION AND COMPLEXITY

Evidence is being collected about contemporary desertification in different places. It shows that desertification is a complex process that involves multiple causes and proceeds at varying rates in different climates. It may intensify a general climatic trend towards greater aridity, or it may initiate a change in local climate. The history of desertification in southern Europe, for example, reflects complex interactions over time between natural factors – including climate (particularly prolonged droughts) and topography (particularly steep, unstable slopes) – and human impacts (particularly land-use change, agriculture and tourist developments).

Desertification has occurred in parts of the Great Plains in the USA over the past 1,000 years, although geological and archaeological evidence suggests that variations in temperatures and precipitation have been relatively slight. It appears that this area has long been close to a critical threshold in the relationship between climate and the behaviour of environmental systems; even minor variations in climate can promote significant environmental responses.

The evidence also shows that it does not occur in linear patterns that are easy to map – deserts advance erratically, forming patches on their margins. Areas currently far away from natural deserts can quickly degrade to barren soil, rock or sand through poor land management. The presence of a nearby desert has no direct relationship to desertification.

BOX 14.22 GLOBAL WARMING AND DESERTIFICATION

The impacts of the global climate change associated with global warming on desertification are as yet unclear, but it is highly likely that desert conditions will become more widespread as the climate belts migrate to higher latitudes. The long-term implications of these shifts for world food supplies could be very serious.

A projected temperature rise of up to 3 °C by 2050 is expected to have a number of impacts in the Mediterranean region, including:

- an increase in evaporation of about 200 mm a year;
- a sea-level rise of 20 to 60 cm;
- abandonment of most cereal growing in northern areas;
- increased growing of plant species that are sensitive to cold;
- a sharp increase in the spread of desertification.

BOX 14.23 ORIGIN OF THE SAHARA DESERT

Just 6,000 years ago the Sahara was covered with grasses and shrubs, and climate modelling of why it suddenly turned into one of the driest regions on Earth shows that the transition from grass to sand could have occurred in just 300 years. The model starts with a slow reduction in solar heating of the atmosphere (to simulate the effect of changes in the Earth's tilt and orbit; see p. 118), which gradually weakens the monsoons (see p. 338) over India and North Africa, thinning the vegetation cover. After a few thousand years of gradual change, the impoverished vegetation could no longer preserve soil moisture and maintain the cycle of evaporation, atmospheric circulation and precipitation that drove the African monsoon. It is suggested that this triggered an abrupt switch to a desert climate about 5,400 years ago, producing the initial Sahara Desert, which has since changed in response to climate change and human disturbance.

CLIMATE

Climatic change, over a variety of timescales, can be an important influence on the distribution and spread of desertification. Climate anomalies, involving natural changes in precipitation and/or temperature and lasting up to decades, could start the process of desertification, and cultural factors could amplify the changes and make them more long-lasting. The Sahel drought in West Africa (see Box 14.4) is probably such an anomaly. Rainfall throughout Africa was abnormally low during the 1970s and 1980s compared with historical and geological times. Superimposed on this general aridity has been a significant reduction in rainfall, which was discontinuous over time and from place to place.

Over the longer term, climate change lasting for centuries or millennia could radically alter the distribution of the main climate zones, which would have direct and lasting effects on where and when desert conditions prevailed. The Medieval Warm Period (see p. 344) was such a phenomenon. Over a timescale of thousands of years, the margins of the Sahara have expanded and contracted in harmony with global and regional temperature changes (see Figure 14.6).

OVER-GRAZING

Traditional farming practices in arid lands rely heavily on grazing animals, and the tendency has been to increase the number of animals grazing a particular area. Various factors promote this increase (Box 14.24). Until the latter half of the twentieth century, relatively little attention was paid to the carrying capacity of different grazing lands, so stocking levels have often been increased beyond what local land resources can sustain. Pressure on the remaining grazing land has intensified, too, as desertification has ruined much grazing land and as nomadic activity has declined in many drylands (particularly in Africa and the Middle East).

Once over-grazing starts, the species composition of the grazing land changes, usually for the worse. Palatable and nutritious plants disappear because they are over-grazed, and they are replaced by more hardy but unpalatable plants. Grazing quality declines, which increases the pressures on remaining pasture and encourages movement to less over-grazed areas (which in turn become over-grazed). As over-grazing continues, the amount of bare ground – occupied by deflation patches, sand sheets and sand dunes – increases. Large areas of bare ground begin to appear, often first around villages and watering holes. Sheep, goats and cows tend to graze within 5–6 km of accessible water, so

BOX 14.24 FACTORS THAT PROMOTE OVER-GRAZING

- *Status*: wealth is widely measured by the number of animals that an individual or tribe owns.
- *Food security*: having more animals would decrease the risk of starvation during drought.
- *Food supply*: more animals are required for food supply as the local population continues to increase.
- *Rise of export agriculture*: many of the animals are now exported, and this has encouraged farmers in some dryland areas to switch from horticulture to cattle rearing.
- *Veterinary care*: the number of animals that died because of diseases has declined sharply since about 1950 because of widespread vaccination and animal health programmes.

Plate 14.5 *Dryland vegetation near Marbella, southern Spain.*
Photo: Chris Park.

the over-grazing is concentrated initially close to water sources. Over time, as over-grazing decreases the quality of grazing land and ultimately destroys it, the animals have to graze over a wider area, moving increasingly far from water. It sets in motion a circle of decline, which becomes more widespread and serious over time.

BOX 14.25 POPULATION GROWTH AND DESERTIFICATION

It is widely argued that population growth – particularly among the rural poor – is a fundamental cause of land degradation and desertification, but this view can be challenged. The evidence from many developing countries shows that the poor need not and generally do not destroy their own resources in order to survive. Many are displaced from access to land and livestock and become increasingly dependent on low, unstable, off-farm incomes and more vulnerable to food scarcity.

OVER-CULTIVATION

Desertification is also associated with over-cultivation of marginal arid lands, particularly those where population levels and densities have increased sharply during the twentieth century. Over-cultivation is not such a widespread cause as over-grazing, but where it does occur it creates particular problems. It arises partly through the introduction and use of mechanised machinery, such as tractors and disc ploughs, which can destroy native perennial vegetation and encourage soil degradation and desertification. Once the protective soil cover is removed (even temporarily) by large-scale ploughing, the soil surface is exposed to wind erosion. When this happens, the valuable topsoil can be blown away, leaving infertile, dry soil behind.

Irrigation schemes have been widely introduced to exploit the agricultural potential of drylands, but, as we have seen in Chapter 13, irrigation often causes salinisation (see p. 402) and waterlogging (p. 399). While these are in effect problems associated with excess water, not water shortage, they can lead to desertification if the irrigated areas are abandoned as unworkable or unprofitable.

DEFORESTATION

Deforestation is another factor that encourages desertification in arid and semi-arid areas. Wood is used for a variety of purposes, including making furniture, housing and other items. Wood collection for use as a fuel is a primary cause of deforestation throughout much of Africa, where it is widely used for cooking (oil, gas and kerosene are too expensive). As trees are removed, the land is exposed to weather and climate, and weathering and soil erosion often increase dramatically.

Deforestation also triggers desertification quite large distances downwind by altering regional water cycles. Clearance of tropical rainforest (see p. 588) is particularly important in this respect, because rainforest 'short-circuits' the water cycle by rapidly releasing back into atmospheric storage up to 90 per cent of the precipitation it receives by means of evapotranspiration. Under normal conditions, prevailing winds then redistribute downwind the moisture-laden air masses that develop over rainforest areas. With deforestation, the rapid evapotranspiration cycling is lost, so rainfall declines downwind. Some of the desertification around the Sahel in West Africa (see Box 14.4) is believed to be associated with clearance of tropical rainforest upwind along the west coast of Africa in Cameroon and the Ivory Coast.

COPING WITH DESERTIFICATION

As we saw above (p. 434), desertification appears to involve a series of positive feedback mechanisms capable of setting in motion an almost irreversible chain reaction that can turn previously productive land into barren desert. Inevitably, given the large area already affected, and the rapid growth and spread of desertification, there is growing concern to find ways of halting and ideally reversing the process. Concern currently seems to be outpacing progress.

BOX 14.26 SUSTAINABLE RESOURCE USE IN DESERTS

Not all scientists agree that people make deserts, and there is plenty of evidence to show how people who live in a dryland environment understand it, make good use of it and are well capable of looking after it and using it sustainably. In Central Asia, for example, the Uygur have developed a simple and effective means of stabilising sand dunes by using square mats of straw, like a chequerboard. The mats break the force of the wind and restrict sand movement to within empty squares. Over many generations, the nomadic Bedouin of the Arabian Desert have evolved sustainable ways of grazing animals on marginal drylands, making efficient use of scarce water resources and grazing opportunities. Adjustments like these illustrate how important it is to study how past cultures have adapted to climatic fluctuations, because this could hold important clues about sustainable ways of using drylands in the future.

BOX 14.27 DESERTIFICATION IN CHINA

North-central China has a serious desertification problem, and efforts are under way to halt desert encroachment on surrounding productive land and to reclaim desertified land. A range of techniques is being used there, including:

- windbreaks (to reduce wind erosion of bare soil);
- irrigating with silt-laden river water (to restore soil in badly eroded areas);
- dune stabilisation using straw chequerboards and planted xerophytes (plants that can withstand prolonged water shortage);
- land enclosure (to reduce wind erosion);
- redistribution of material from palaeosols (to restore soils);
- chemical treatment (to restore soil fertility).

A variety of different strategies can be adopted to combat desertification. Some seek control in indirect ways, by raising awareness of the problem and trying to change attitudes and behaviour thought likely to promote desertification. Such measures include:

- education
- research
- training
- publicity and increased awareness
- monitoring and assessment.

While such measures are essential, they need to be carefully targeted and co-ordinated, and they must be accompanied by field control, without which dramatic improvements are highly unlikely. Field control measures include:

- stabilisation of dunes and desert sands (by special planting of marram grass and trees);
- increasing the water-retaining capacity of desert soils (perhaps using water-absorbent plastic polymers) to enable crops to grow;
- reduction of salinisation in irrigated lands.

Many scientists call for much greater investment in field projects coupled with a better understanding of the

Table 14.4 Some important ways of tackling desertification recognised in Agenda 21

- increase knowledge of mountain and desert ecosystems by developing a world information centre and identifying areas most at risk from floods, soil erosion, etc.
- provide better environmental education for farmers
- prevent desertification by not polluting soil, by using land soundly and by planting trees that retain water and soil quality
- pass laws to protect endangered areas
- make plans to ensure that potential drought victims survive

socio-economic factors underlying desertification. Without this, they argue, real progress in implementing the UN Plan of Action to Combat Desertification (agreed in the late 1970s) will be extremely slow.

Desertification figured on the agenda at the 1992 Rio Earth Summit (see p. 13), where concern was expressed over the poverty and starvation associated with it. Chapter 12 of Agenda 21 recognised that 'poverty is a major factor in soil degradation. We need to restore fragile lands and find new jobs for farmers thrown out of work'. While Agenda 21 spelled out some important ways of tackling the problem of desertification (Table 14.4), it proposed no firm solutions and concluded that much better understanding of the desertification process is required.

Many of the Agenda 21 suggestions are incorporated in a proposed International Convention on Desertification, which has been the subject of much debate and negotiation at the United Nations. Developed countries are reluctant to support the convention, partly because they would be obliged to provide most of the money needed to make it work. Many developing countries, on the other hand, welcome it because it would help to tackle one of their most serious environmental problems. Nearly 37 per cent of the world's desert land is in Africa, and African delegates at Rio felt that their interests and concerns were largely overlooked.

WEBSITE

Links to relevant websites, a comprehensive bibliography, tools for teaching and learning, and downloadable images relevant to this chapter can be found at the website specially designed to accompany this book at http://www.park-environment.com

FURTHER READING

Abrahams, A. and A. Parsons (eds) (1993) *Geomorphology of desert environments.* Chapman & Hall, London. A collection of essays that review present understanding of the landforms and processes of the desert environment.

Agnew, C. and E. Anderson (1992) *Water resources in the arid realm.* Routledge, London. An introduction to the need for and problems of water resource management in drylands.

Beaumont, P. (1993) *Drylands: environmental management and development.* Routledge, London. A detailed and up-to-date introduction to the world's drylands and their problems, with an emphasis on managing and developing this fragile environment.

SUMMARY

Drylands and droughts occur in many parts of the world, and this chapter explores how water shortage occurs, how it affects landforms and landscapes, what problems it poses for people, and how it is coped with. It begins with a review of drought, which distinguishes different types of drought and identifies a number of causal factors (including human factors). The experience of coping with drought varies a great deal from place to place, depending on drought severity and human preparedness; the Sahel drought during the 1970s was particularly difficult and damaging. Deserts are widely distributed around the world, and we considered what controls their location and character, what processes operate in them, and how these processes create and shape landforms. Desertification is a major theme in the chapter because it is already a major problem in many places and a growing problem in many other places. We looked at how awareness of the problem has changed over time, what processes are involved, and how the interplay between natural causes and human mismanagement often intensifies the desertification problem. Global warming caused by air pollution is likely to make desertification an even more difficult problem to cope with in the future.

Goudie, A.S. and H.A. Viles (1997) *Salt weathering hazard.* John Wiley & Sons, London. Examines the causes, processes and consequences of salt weathering and explores why this environmental hazard is on the increase.

Kliot, N. (1993) *Water resources and conflict in the Middle East.* Routledge, London. A series of interesting case studies of hydropolitics in the Euphrates, Tigris, Nile and Jordan basins.

Lancaster, N. (1995) *Geomorphology of desert dunes.* Routledge, London. A detailed introduction to how desert dunes are formed, how they change, and their environmental importance.

Lewis, L.A. and D.L. Johnson (1994) *Land degradation.* Blackwell, Oxford. An overview of land degradation, defined as a decline in natural, biological productivity that is either irreversible or that may not be recovered for at least one human generation.

Livingstone, I. and A. Warren (1996) *Aeolian geomorphology: an introduction.* Longman, Harlow. A concise introduction to the work of wind in shaping the physical landscape in deserts, coasts and periglacial areas.

Mairota, P., J.B. Thornes and N. Geeson (eds) (1997) *Atlas of Mediterranean environments in Europe: the desertification context.* John Wiley & Sons, London. Detailed study of desertification in this dryland environment, which draws on recent research and sets the problem into context.

Middleton, N. (1999) *The global casino: an introduction to environmental issues.* Edward Arnold, London. Clear introduction to major environmental issues, with an emphasis on underlying causes, human factors that have contributed to the problems, and possible solutions.

Middleton, N. and D. Thomas (1992) *World atlas of desertification.* Edward Arnold, London. An up-to-date analysis of the scale of desertification and its various forms.

Millington, A. and K. Pye (eds) (1994) *Environmental change in drylands.* John Wiley & Sons, London. A series of essays that explore geomorphological and ecological responses to climate and culturally induced changes in drylands around the world.

Mortimore, M.J. and W.M. Adams (1999) *Working the Sahel.* Routledge, London. Detailed study of how people in the semi-arid conditions of the Sahel cope with their harsh environment, with a particular focus on how they organise their labour to manage fields, crops and other resources.

Slaymaker, O. and T. Spencer (1998) *Physical geography and global environmental change.* Longman, Harlow. Examines the principles and concepts of environmental systems, and the implications of these for society.

Thomas, D.S.G. (ed.) (1997) *Arid zone geomorphology.* John Wiley & Sons, London. Wide-ranging exploration of geomorphological processes that operate in the arid zone.

Thomas, D.S.G. and N.J. Middleton (1994) *Desertification: exploding the myth.* John Wiley & Sons, London. Explores how the 'desertification myth' started, and the political and institutional factors that shaped it, sustain it and protect it against scientific criticism.

Webb, P. and J. Von Braun (1994) *Famine and food security in Ethiopia: lessons for Africa.* John Wiley & Sons, London. Explores the reasons why famine, once a universal problem, is now largely restricted to Africa.

Wilhite, D.A. (ed.) (1999) *Drought: a global assessment.* Routledge, London. A collection of essays that explore the physical and social dimensions of drought in different countries and different contexts.

Williams, M.A. and R.C. Balling (1996) *Interactions of desertification and climate.* Edward Arnold, London. A summary of current knowledge on the interactions of desertification and climate in dryland areas, which includes recommendations for future dryland management strategy.

Yarnal, B. (2000) *Local water supply and climate extremes.* Ashgate, London. Examines the impacts of weather and climate on local and regional water supplies, and the implications of these interrelationships for sustainable water management.

Chapter 15

Cold and Ice

LEARNING OBJECTIVES

When you have finished studying this chapter, you should be able to

- *Appreciate the problems and challenges posed by snow and ice and understand the relevance of glaciation to many contemporary landscapes*
- *Describe the climatic conditions that give rise to tundra, permafrost and ice caps*
- *Outline the main features of the Arctic and Antarctic environment and the reasons for preserving these frozen wildernesses*
- *Define what is meant by 'glaciation' and summarise the main causes of it*
- *Show how ice forms and how glaciers accumulate and move*
- *Describe the main processes of glacial erosion and their impacts on landscape*
- *Describe the main processes of glacial deposition and their impacts on landscape*
- *Account for the main processes and products of glacial meltwater*

Our concern with cold climates is not confined to the present day, because many mid-latitude areas that currently enjoy temperate climates have been significantly affected by ice and meltwater activity during the Quaternary sub-era (see Table 4.7). It is difficult if not impossible to understand how the landscapes and environments of these areas have been formed, and thus why they are the way they are, without an awareness and appreciation of the relevance and impacts of cold climate processes in the past. In this chapter, we focus on cold climates at the present time and then take a look at glaciation and its impacts, past and present.

COLD ENVIRONMENTS

SNOW AND ICE

Snow and ice are both associated with cold conditions, where temperature remains below freezing (0 °C) for some time. Ice is water in the solid state, formed by the freezing of liquid water. It can form anywhere that has liquid water and freezing temperatures, and it persists as long as temperatures remain below freezing.

We saw in Chapter 9 (p. 298) that snow is formed when the dew point is below freezing and water vapour in the air condenses into hexagonal-shaped ice crystals, which float down to the ground. If temperatures remain below freezing, snow accumulates on the ground, and it can be blown by the wind and deposited in thick snowdrifts, which cover fences, buildings and other structures. The snow melts when temperatures rise above freezing.

SNOWFALL AND SNOW COVER

In the next section, we look at areas that have permanently cold climates, but we must not overlook the importance of short-term cold weather conditions in other areas. These are often important climatic hazards that cause great disruption to normal life (Box 15.1).

Snow cover hinders outdoor movements and activities, and – at least in theory – areas where snow cover is more common should have more closely spaced buildings and thus higher densities of buildings and populations. Detailed

BOX 15.1 THE RELEVANCE OF COLD AND ICE

Many environmental systems, and their interactions with people, are strongly influenced by cold and ice. For example, cold and ice play important roles in rock weathering (see p. 204) and sediment transport (see p. 210), so they can exert powerful control over landform development and stability. Cold and ice also determine important properties of ecosystems and habitats, such as the availability of water and nutrients, rates and patterns of growth and lifestyle of plants and animals, and the availability of food supplies.

Many aspects of human life and activities can also be heavily influenced by cold weather, particularly snow and ice, which affect all outdoor activities and are reflected in human adjustments such as central heating systems, building insulation and thermal clothing. Extreme cold also affects people's health, both directly and indirectly (see p. 272). Permanent ice cover and freezing temperatures are the norm in some environments, most notably Antarctica and the Arctic, where special adaptations to the cold are essential.

analysis of urban areas in the USA – which showed a strong correlation between average annual snowfall and population density, even after taking into account factors such as topography, city age, population size, income and race – confirms this. Transport systems (particularly roads, railways and airports) are often highly vulnerable to cold weather and icy conditions, which are not always easy to forecast with adequate reliability far enough ahead to avoid serious disruption and cost.

Snow cover is a big enough problem in itself, but when it is accompanied by driving winds it creates blizzards and severe snow storms, which put lives and property at risk. In exposed upland areas such as much of Scotland, coping with blizzards puts major strains on the emergency services such as the police and medical systems, and it creates great logistical problems such as search and rescue and clearing snowdrifts from roads littered with abandoned vehicles.

Countries that regularly have to contend with extensive and persistent snow cover tend to cope with the snow hazard better than countries where snowfall is less regular, often because they have developed more sophisticated ways of predicting where and when snow events are likely.

BOX 15.2 THE SNOW LINE

In middle latitudes, snow and ice are common ingredients of winter weather, but at most other times of the year temperatures are usually too high. Snow and ice cover tend to be permanent features of high latitudes, where temperatures are much lower throughout the year (Figure 15.1), but even at low latitudes, in the tropics, snow and ice can form on high mountain tops because air temperature naturally decreases with altitude (see p. 275). The summit of Kilimanjaro, in northern Tanzania, is permanently snow-covered because temperatures at that altitude (5,895 m above sea level) remain below freezing all year long. The altitude above which snow remains on the ground throughout the year is called the snow line. Around the equator it is between 5,200 and 5,480 m above sea level, but it decreases with increasing latitude so that near the poles it lies at sea level.

BOX 15.3 SNOW RISKS AND RISK ASSESSMENT

Risk assessment is often based heavily on past experience, and it helps in planning how best to cope with the snow hazard. Risk assessment of snow hazards relies heavily on analysis of climatic records, although even in developed countries the database tends to be patchy and incomplete. For example, in the late 1980s accurate and complete information on snowfall and snow cover was being collected at only just over half of the official climate recording stations in the USA where regional snows occur each year.

Analysis of yearly variations in snow cover in North America since the early 1970s has shown that higher temperatures during the 1980s were accompanied by a significant reduction in winter snow cover and snowfall. Most of the change is accounted for by a reduced frequency in snow cover in areas that have traditionally had persistent snow on the ground. Whether this is a direct result of global warming and whether the trend continues remain to be seen.

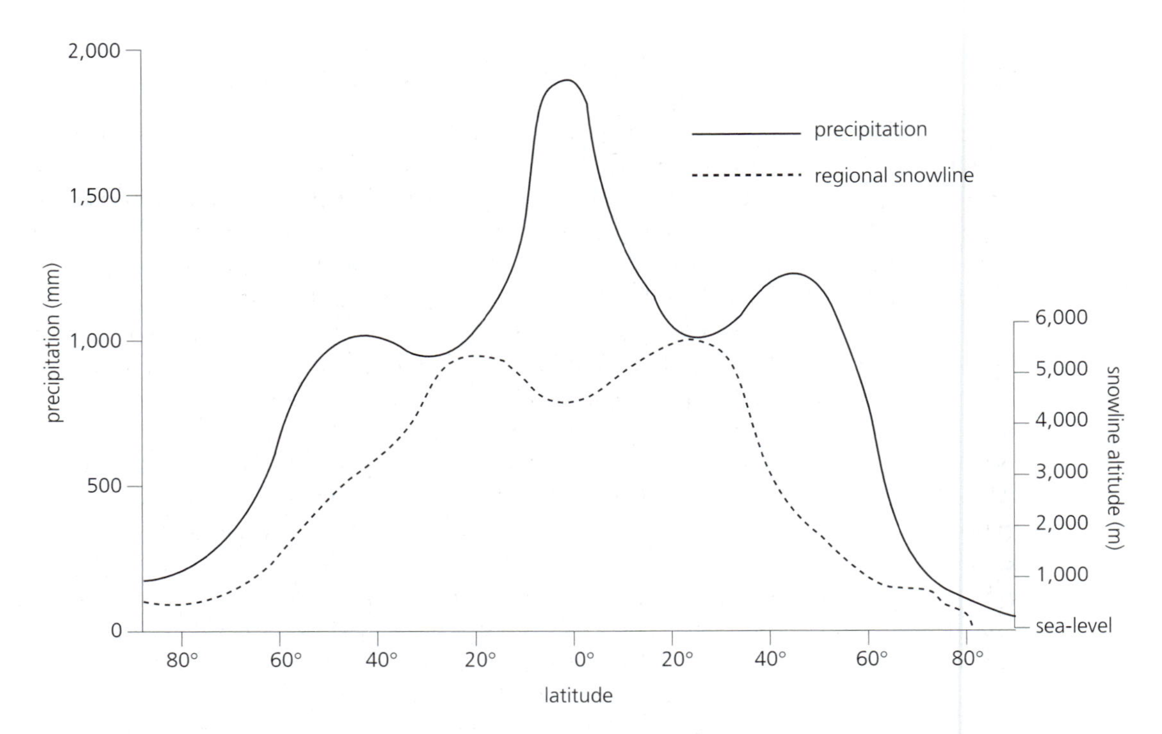

Figure 15.1 *Variations in height of the snow line with latitude. The snow line lies relatively close to sea-level at high latitudes, and it rises progressively towards the warmer climates of the tropics. Near the equator the snow line is below 5,000 m, which explains why some of the high mountains of equatorial Africa are snow-capped.*

Source: A figure in Sugden, D. and B. John (1976) Glaciers and landscapes. Edward Arnold, London.

In China, for example, snow hazard maps have been compiled that divide the entire country into snow hazard regions and sub-regions.

BENEFITS OF SNOW

While it often causes serious problems, snow cover is not always bad news. Many mountain areas rely heavily on good snow cover and regular supplies of fresh snow to support their alpine sports and tourism industries, which are often the backbone of the local economy. Skiing is a growth industry in Scotland, for example, but it requires a relatively large number of winter days with snow cover, and this is not always guaranteed. Much depends on the source of the air masses (see p. 304) that pass over Scotland. Continental air masses from the east bring prolonged cold spells, which promote extensive snow cover, whereas maritime air masses from the west bring less snow.

COLD CLIMATES

The areas that have persistently cold climates (Figure 15.2) are at high latitudes and include the Arctic Ocean and the Antarctic continent. Köppen called these the polar climates, type E (see Table 11.2), and they are defined as areas with temperatures of less than 10 °C in the warmest month. There are two main types of polar climate (Box 15.4) – tundra and polar ice caps.

TUNDRA AND PERMAFROST

The tundra zone is found at high latitudes in the northern hemisphere. There is no large land mass at equivalent latitudes in the southern hemisphere. It lies to the north of the sub-arctic climatic zone (see p. 340) and the boreal forest belt (p. 578), along the margins of the Arctic Ocean. This zone was much larger, and extended much further south, during the Pleistocene.

Köppen defined the southerly limit of this zone by the 10 °C warm-month isotherm (line of equal temperature during the warmest month), which he believed to be the climatic limit to tree growth. Precipitation is limited by the cold air masses, and most falls during the summer months. Because of the low temperatures, much of the precipitation, particularly during the winter months, is in the form of snow, which accumulates on the ground.

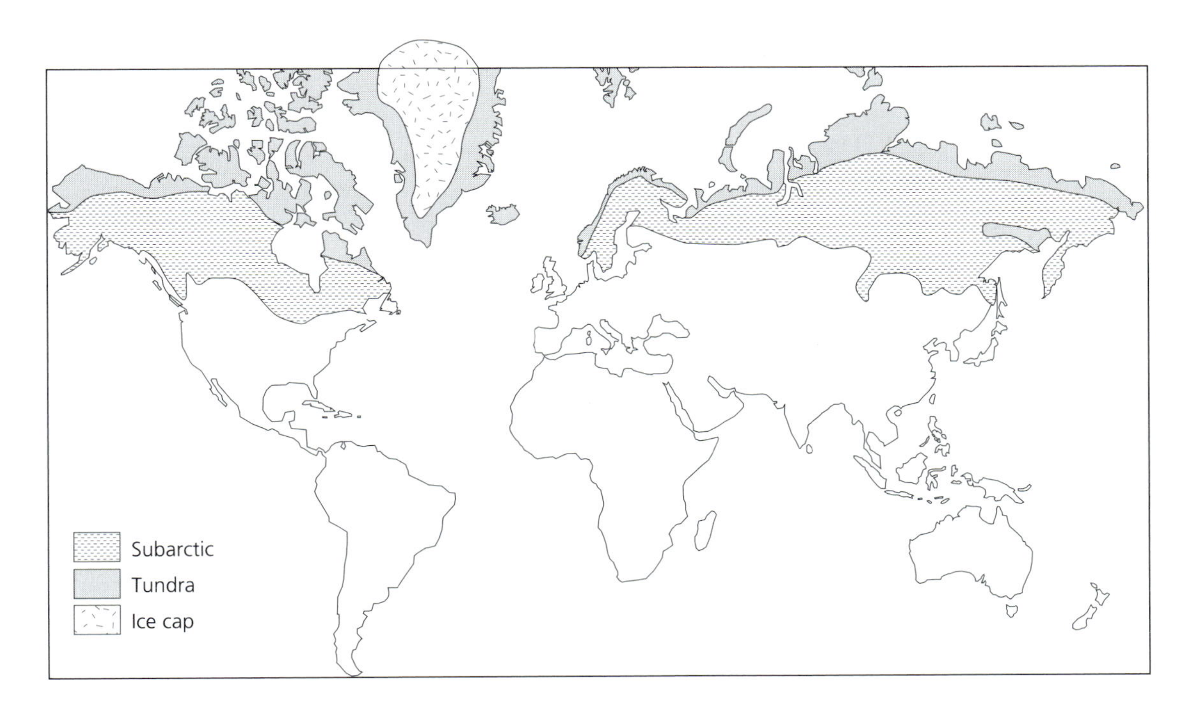

Figure 15.2 *Location of arctic and sub-arctic climates. As climate becomes progressively colder with latitude in the northern hemisphere (there is little land at high latitudes in the southern hemisphere), the sub-arctic zone gives way to the tundra zone.*

Source: Figure 7.18 in Doerr, A.H. (1990) Fundamentals of physical geography. Wm. C. Brown Publishers, Dubuque, Iowa.

BOX 15.4 POLAR CLIMATES

The polar climates have no warm season, receive limited precipitation and experience penetrating cold. This makes them extremely harsh environments and goes a long way towards explaining why so few people live in them. The scattered settlements in such areas are mostly populated by indigenous peoples, who over many generations have adjusted to the harsh environment and base their subsistence economy on exploiting its few resources. Thus, for example, the Inuit (Eskimos of North America and Greenland) have traditionally survived mainly by fishing and hunting. Polar areas have been home to fur trappers (for example, in Canada during the nineteenth century) as well as the scattered groups of indigenous peoples, and more recently to growing numbers of personnel visiting scientific and military bases (for example, in Antarctica; see p. 448).

BOX 15.5 CLIMATE IN THE TUNDRA ZONE

The climate is strongly influenced by the cold air masses that dominate this high latitude:

- Summers are short and cool.
- Winters are long, cold and harsh.
- Annual precipitation is limited (usually less than 250 mm a year).
- Mean annual temperatures are below freezing.

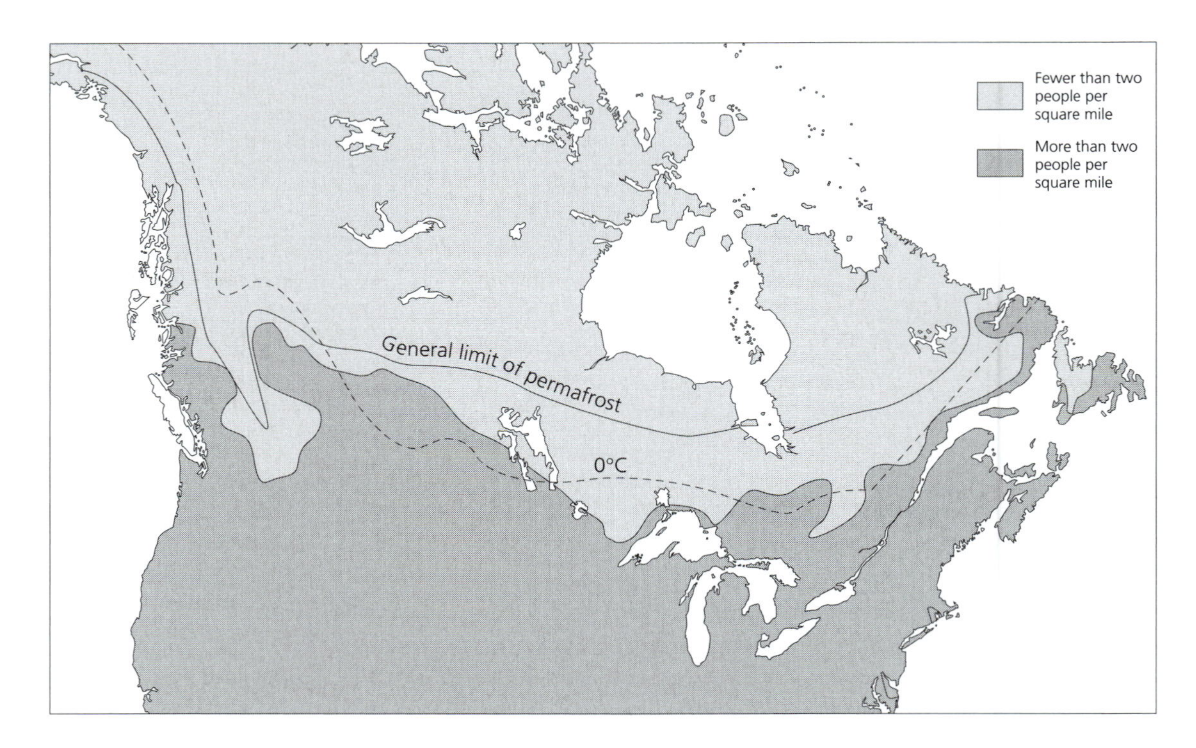

Figure 15.3 *Distribution of permafrost in North America. Much of northern Canada is underlain by permafrost, and the 0 °C mean annual temperature isotherm passes further south across much of the continent. This cold zone has low population densities.*

Source: Figure 2.9 in Marsh, W.M. and J.M. Grossa (1996) Environmental geography: science, land use and Earth systems. John Wiley & Sons, New York.

Most of the ground surface in the tundra is *permafrost* (permanently frozen ground). This inhibits the percolation of water into the soils and deposits, and there is a great deal of surface water. Evaporation rates are low because of the low temperatures. Swampy surface conditions are common, particularly during the summer melt months. This restricts the uses that can be made of many permafrost areas, making access to and travel within them difficult and creating serious problems for constructing buildings. Population densities in the permafrost zone are extremely small (Figure 15.3).

Permafrost presents many difficulties for building and engineering because of the instability of ground conditions and their susceptibility to mass movement. Special techniques have been developed to allow building in such an environment. These include maintaining natural soil temperatures by building on stilts and conducting energy, water and soil pipes above ground on insulated structures called utilidors.

BOX 15.6 WEATHERING AND EROSION PROCESSES IN COLD CLIMATES

Freeze–thaw processes dominate weathering, sediment transport and landform development. Patterned ground, often in the form of polygons, develops in relatively well-drained sites as periodic freezing and thawing of the ground redistribute rock particles. On sloping sites, the patterned ground can be oriented downslope to produce stone stripes (Figure 15.4). Solifluction terraces can be formed by the seasonal thawing of permafrost-covered slopes during summer melt, as the lubricated surface material creeps and slumps (see p. 218) downslope under the influence of gravity. Hummocks develop on poorly drained sites, where particle sorting is less pronounced.

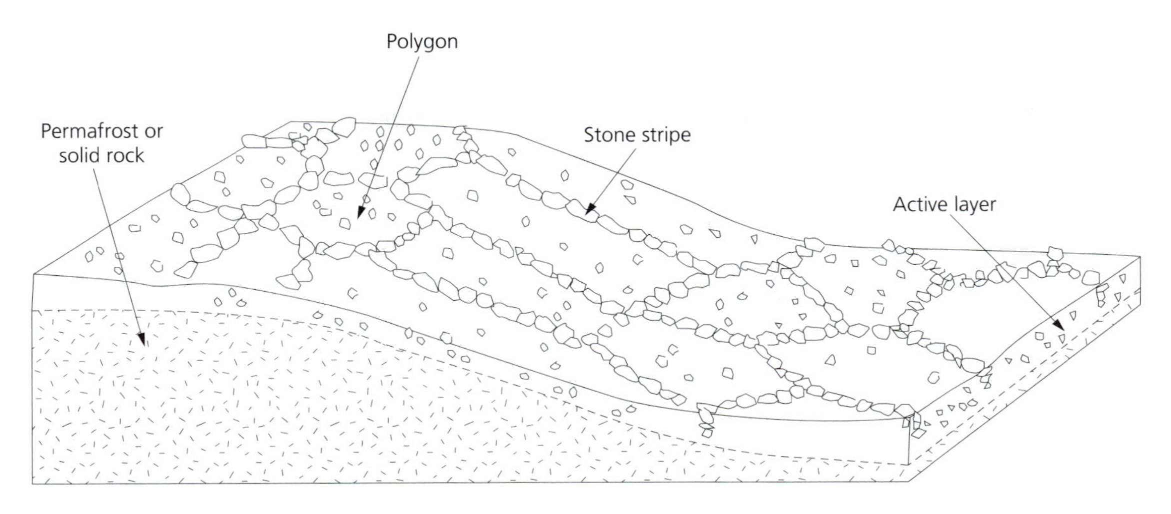

Figure 15.4 *Sorted ground and permafrost. Freeze–thaw processes sort surface materials, giving rise to some characteristic features, including sorted stone polygons (on low-sloping ground) and stone stripes (on higher slopes).*
Source: Figure 8.4 in Clowes, A. and P. Comfort (1982) Process and landform. Oliver & Boyd, Edinburgh.

POLAR ICE CAPS

Beyond the tundra, in the coldest and harshest high latitudes, lie the permanent ice caps. Antarctica and the interior of Greenland are the only large land masses in this zone. The total amount of precipitation that falls on this zone is small (usually less than 250 mm a year), but the average temperature in each month is below 0 °C (freezing point), so a permanent cover of snow and ice accumulates and persists because in a typical year none of the ice melts. Regional ice sheets often have mountain ridges sticking up through them, which are called *nunataks*. These are usually angular with sharply defined ridges, products of intense freeze–thaw weathering processes (see p. 207).

The seas in the vicinity of the polar ice caps also display evidence of the permanently cold conditions in the form of pack ice (seawater that is frozen into thick blocks) and icebergs (floating blocks of ice that have generally broken off glaciers or ice sheets that flow into the sea).

Conditions in the polar ice cap zone are extreme, with freezing temperatures throughout the year and frequent strong winds, which accentuate the wind chill factor. These areas are extremely inhospitable to humans, and they support only a few small, scattered permanent settlements.

BOX 15.7 DEFINING THE ARCTIC REGION

The Arctic is the large cold region around the North Pole. Unlike Antarctica, there is no obvious boundary to the Arctic, which can be defined in a number of ways:

- the area north of the Arctic Circle (latitude 66° 30″ north);
- the area north of the 10 °C summer isotherm; or
- the area north of the tree line (which connects places where trees will not grow because of the cold climate).

THE ARCTIC AND ANTARCTICA

THE ARCTIC

The Arctic region is dominated by the Arctic Ocean, but it also includes large land areas in Canada, Russia, Greenland, Scandinavia, Iceland and Alaska. Some of the land areas, including most of Greenland, are permanently ice-covered, and pack ice is common throughout the Arctic Ocean.

Icebergs break off glaciers that reach the sea and can be carried into lower latitudes by ocean currents. They can rise over 100 m above the surface of the sea and have nine-tenths of their mass hidden beneath the surface; North Atlantic icebergs have been mapped over 3,000 km away from the Greenland ice sheet. The British luxury liner the *Titanic* was sunk by such an iceberg during its maiden voyage from Liverpool to New York in April 1912, with the loss of more than 1,500 lives.

CLIMATE

Climate in this high-latitude zone is extreme, with short, cool summers and long, cold winters. Temperatures are particularly low in the continental interiors – average mid-winter temperatures on the Greenland ice cap are around −33 °C, for example. Precipitation is generally low (less than 250 mm a year on average) and is well distributed throughout the zone. Large river and lake systems are rare, because of the low precipitation, but shallow lakes, ponds and marshes are common in areas underlain by permafrost. Many polar areas are badly affected by air pollution, often reflected in hazy conditions, although they are far from industrial sources of pollutants.

WILDLIFE

Although the Arctic has the appearance of a vast, frozen desert, it does support wildlife. Signs of life are difficult to find during the cold, dark winter months, but some species of mammals and birds carry extra insulation (such as fat) to survive the winter. The Arctic appears to wake up in spring. More than 400 species of flowering plant grow in the Arctic, and most of the land that is not ice-covered is tundra, with a natural vegetation of low, creeping shrubs, grasses, thick growths of lichens and mosses, and herbs and sedges. The Arctic region is home to a wide variety of birds and fish, and mammals include the polar bear, arctic fox, arctic wolf, walrus, seal, caribou, reindeer, the notorious lemming and many species of whale.

BOX 15.8 HUMANS IN THE ARCTIC

The small, scattered populations of indigenous peoples in the Arctic have adapted to the harsh climate and limited natural resources of their environment, depending entirely on hunting and fishing and using natural materials for clothing, tools, homes and vehicles. Climate is mostly too cold for agriculture, although reindeer herding has long been important in northern Scandinavia and Russia, and sheep are raised in south-west Greenland and Iceland. The Arctic Ocean is among the world's most important fishing grounds (see p. 486), and many countries send fishing boats to it. Large amounts of cod and shrimp are caught off western Greenland.

BOX 15.9 DEFINING ANTARCTICA

The outer limits of Antarctica are defined by the Antarctic convergence, which marks a sharp change in the physical characteristics of the Atlantic, Indian and Pacific Oceans between latitudes 48 and 60° S. Warmer waters moving south from the mid-latitude oceans meet and mix with colder waters moving north from Antarctica at the convergence, where there is a steep temperature gradient in the ocean water. The area within the Antarctic convergence is known as the Southern Ocean.

ANTARCTICA

What is now Antarctica began life as part of the former super-continent Gondwanaland (see p. 147). It drifted from the tropical zone to its present polar position about 100 million years ago, when the super-continent broke apart and the main continental land masses went their separate ways.

Antarctica (Figure 15.5) is the fifth largest of the seven continents (see Table 5.1). It is centred on the South Pole and located mostly within the Antarctic Circle (south of latitude 66° 30″ in the southern hemisphere). Great stretches of sea ice form around the margins of Antarctica during the winter, nearly doubling its size.

ENVIRONMENT

Antarctica is by far the coldest continent. It is in effect a cold desert, with an average annual precipitation in the interior of about 50 mm. Climate in the interior is dominated by extreme cold and light snowfall, and temperatures are milder and precipitation is much higher (up to about 380 mm a year) around the coastal fringe. The lowest temperature ever recorded anywhere on Earth (−88.3 °C) was on 21 July 1983, at Vostok Station. The continent is also swept by strong winds, with speeds of up to 300 km h^{-1}, which cause raging blizzards and sizeable snowdrifts.

Conditions in Antarctica today reveal a great deal about what conditions must have been like in many high-latitude areas during the Pleistocene ice age. Over 95 per cent of the land surface is now covered by ice, and the ice cover is variable in thickness but is sometimes thicker than about 2,000 m. This thick ice cover makes Antarctica the highest

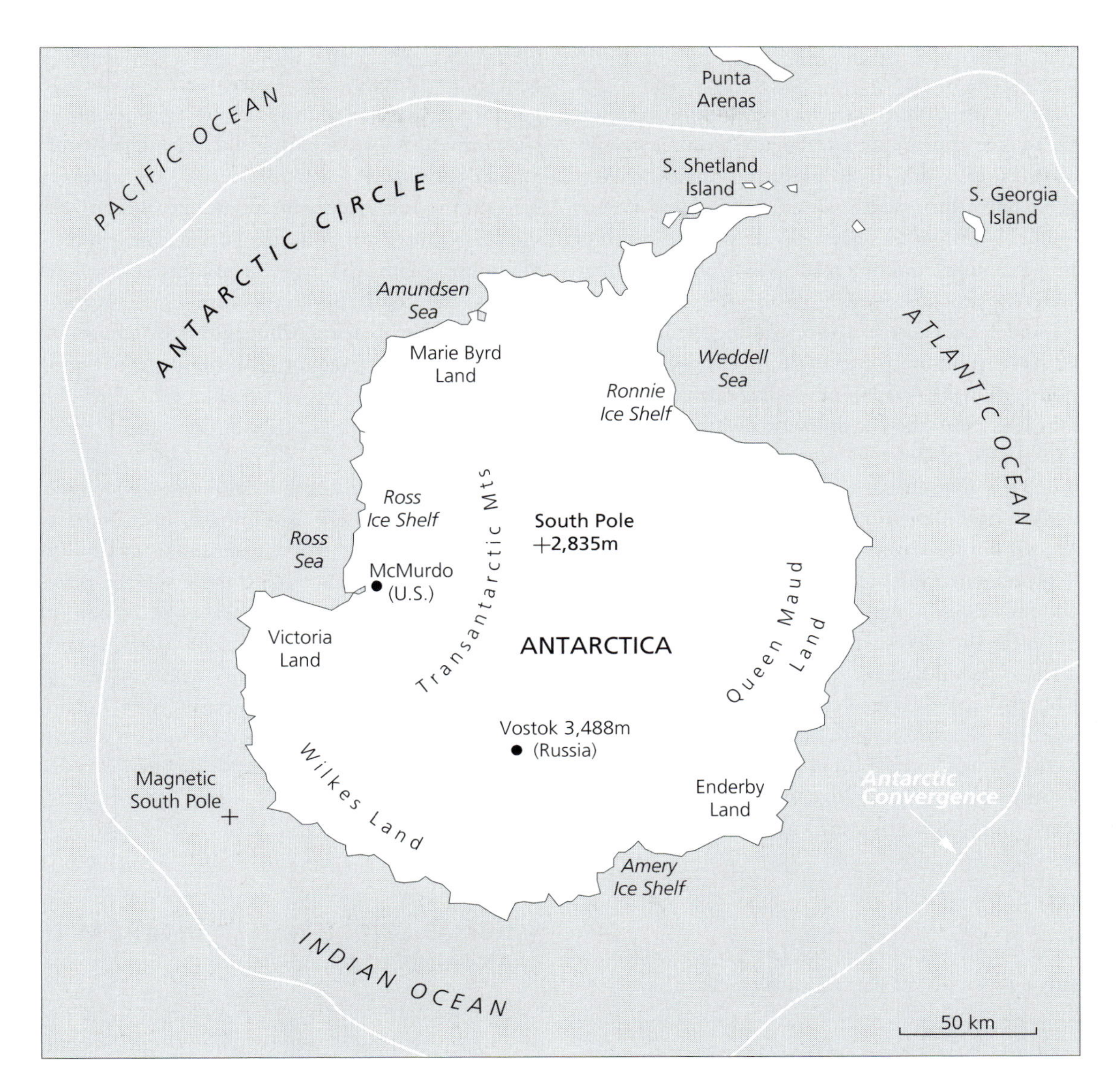

Figure 15.5 *Antarctica. The frozen continent of Antarctica is surrounded by cold waters within the Antarctic convergence.*

BOX 15.10 WATER RESOURCES AND ANTARCTICA

The giant Antarctic ice cube is vitally important to the global water cycle (see p. 352), and thus to global water resources (see p. 388), because it contains about 90 per cent of the world's fresh water. This would obviously change if global warming continues and the major ice caps continue to melt, contributing to sea-level rise and causing adjustments to the global water balance and the spatial distribution of storage within the system. Ambitious schemes to use the water resource potential of Antarctica have been considered, including exploring the possibility of towing icebergs to arid regions (including Australia and Saudi Arabia) for fresh water.

continent overall, with an average elevation of about 2,300 m above sea level.

Conditions on Antarctica are dynamic, and the evidence shows that the ice appears to have been retreating steadily over thousands of years. It is predicted that the West Antarctic Ice Sheet may vanish within 7,000 years with or without global warming. Research has shown that much of the melting is natural and not related to global warming. West Antarctica contains enough ice to raise sea level by between 5 and 6 m. There is also mounting evidence that Antarctica has shrunk more rapidly in recent decades than was previously thought. Analysis of whaling records from the 1920s to the late 1980s shows a dramatic shrinkage in the ice sheet over a period of about 15 years, beginning in the late 1950s. It is likely that changes in ocean circulation caused that particular phase of melting, rather than global warming.

Wildlife on the ice-covered areas is extremely limited; the most prominent examples are the penguins (particularly Adélie and emperor penguins), which breed on ice and live on ice and in the surrounding oceans. The sea has more abundant life, including six species of seal and large numbers of whales, which feed on krill (small, shrimp-like crustaceans that swarm in dense shoals and feed on tiny diatoms). The human population of Antarctica is extremely small, because there is no native population, and most residents are short-term scientific visitors.

THREATS

Antarctica is widely regarded as the last great wilderness. Until relatively recently, its natural environments have been preserved more by lack of exploitation (because the continent is so remote, inaccessible and uninviting) than by purposeful action. But this is changing as geological exploration reveals more details of the large deposits of valuable mineral resources (particularly coal, oil and natural gas) beneath the cold continent and its surrounding continental shelf. Pressures on Antarctic marine resources are also mounting, with extensive commercial exploitation of whales and krill. Seven countries – Norway, France, Australia, New Zealand, Chile, Great Britain and Argentina – have claimed sovereign rights over parts of the continent.

BOX 15.11 CONSERVATION THREATS TO ANTARCTICA

Conservationists and scientists are anxious to protect the Antarctic wilderness from further exploitation and development, because:

- it represents one of the few surviving areas of natural environment anywhere in the world;
- it is a unique wildlife habitat and feeding ground for millions of whales, seals, penguins and seabirds;
- it plays a key role in regulating global climate and ocean ecosystems;
- it provides an important facility for scientific research, particularly research relating to global climate change.

SCIENTIFIC INTEREST

Remoteness from major industrial areas, which are concentrated in the northern hemisphere, has served to isolate Antarctica from the worst excesses of air pollution. The snow and ice that accumulate there are the purest in the world. But they still contain traces of contaminants and pollutants, which reflect ambient (background) global concentrations of many gases, aerosols and particles. Analysis of the chemical composition of bubbles of air trapped in different layers of ice provides a valuable means of reconstructing changes in air quality over time (see p. 342).

BOX 15.12 SCIENTIFIC INTEREST IN ANTARCTICA

Long-term scientific study of Antarctica and its environment has been ongoing since the International Geophysical Year (IGY) 1957–58. More than sixty scientific stations were established in Antarctica during IGY by twelve countries, and many of these have continued scientific monitoring, exploration and research there ever since. Continuous weather and climate records have been collected in Antarctica since the early 1960s, and these provide an extremely useful baseline of changes over time against which to compare trends in more industrialised areas. They are also invaluable in the study of climate change and global warming (see p. 261). The so-called 'hole' in the ozone layer, attributed at least in part to air pollution by CFCs (see p. 249), was first detected over the South Pole and reported by British scientists in 1985. Research is under way into the biological effects of decreased ozone on the Antarctic environment.

PRESERVATION

An important step towards preserving the unique environment of Antarctica was taken in 1959 with the signing, by the seven countries that claim sovereign rights over the continent, and twelve other countries, of the Antarctic Treaty. This important early international environmental initiative recognised that 'it is in the interest of all [hu]mankind that Antarctica shall continue forever to be used exclusively for peaceful purposes and shall not become the scene or object of international discord'. It dedicated the whole continent to peaceful scientific investigations, and all existing territorial claims were suspended when it came into effect in 1961.

Since the treaty was signed in 1959, international concern has grown over the likelihood of renewed mineral prospecting in and around Antarctica, and there have been repeated calls for the continent to be designated a World Park and protected for ever against development. Among the supporters for such a designation is the Antarctic and Southern Ocean Coalition, a partnership of environmental organisations set up in 1978. In 1991, twenty-four countries approved a protocol to the treaty (called the Madrid Protocol) that would ban oil and other mineral exploration for at least 50 years. An important component of the protocol is the gathering, analysis, storage and exchange of scientific information, and the co-ordination of environmental assessment and monitoring procedures, using geographical information systems and other approaches.

GLACIATION

ICE AGES

Ice ages are periods in the Earth's history when ice sheets spread over areas that now have warm or temperate climates. The geological record suggests that for about 10 per cent of the last 4,000 million years the Earth has experienced ice age conditions, and it preserves evidence of a number of major ice ages lasting different lengths of time over the last 2,000 million years (see p. 115).

Ice ages tend to occur roughly every 150 million years, and they typically last a few million years. One particularly extensive ice age, which covered southern Africa, India and parts of South America and Australia, began in the late Carboniferous period and continued into the Permian period.

BOX 15.13 THE EXTENT OF PLEISTOCENE GLACIATION

At its maximum, Pleistocene ice covered most of Canada, about half of Alaska, much of the northern USA, all of the Scandinavian countries, much of Poland, northern Germany, the United Kingdom and about a third of the former Soviet Union. Only high-altitude areas (above the snow line) beyond the ice sheets were directly affected by glaciation, including much of the Alpine region in Switzerland, Austria and northern Italy. A wide zone beyond the ice sheets was indirectly affected by permafrost and periglacial conditions, which significantly influenced the distribution of plants and animals, the development of soils and deposits, and the formation of distinctive landforms by ice and meltwater.

There are few large land masses in the southern hemisphere, and Pleistocene glaciation had less of an impact there, except for Antarctica, which has retained its ice cover. Much of the area previously covered by Pleistocene ice, particularly in the northern hemisphere, has since risen because of isostatic adjustment (see p. 156) in response to unloading of the great thicknesses of ice.

THE PLEISTOCENE

The most recent ice age – the Pleistocene (see p. 343) within the Quaternary – has had a significant and lasting impact on landscapes and environments in many mid- and high-latitude areas. We know much more about the Pleistocene glaciations than about earlier ones because it is the most recent one, and most of the evidence for it (in deposits and landforms) survives, mainly because weathering and sediment transport processes have had relatively little time to remove or refashion them.

The Pleistocene began about 1.5 million years ago and lasted until about 10,000 years ago. Ice covered much of the northern hemisphere (Box 15.13; Figures 15.6 and 15.7). It also largely removed all traces of earlier landscapes across much of Europe and North America and replaced them with distinctive glacial landforms, including the Great Lakes. Moraines (see p. 466) indicate the outer limits of ice advance, which correlates closely with the distribution of glacial deposits and landforms.

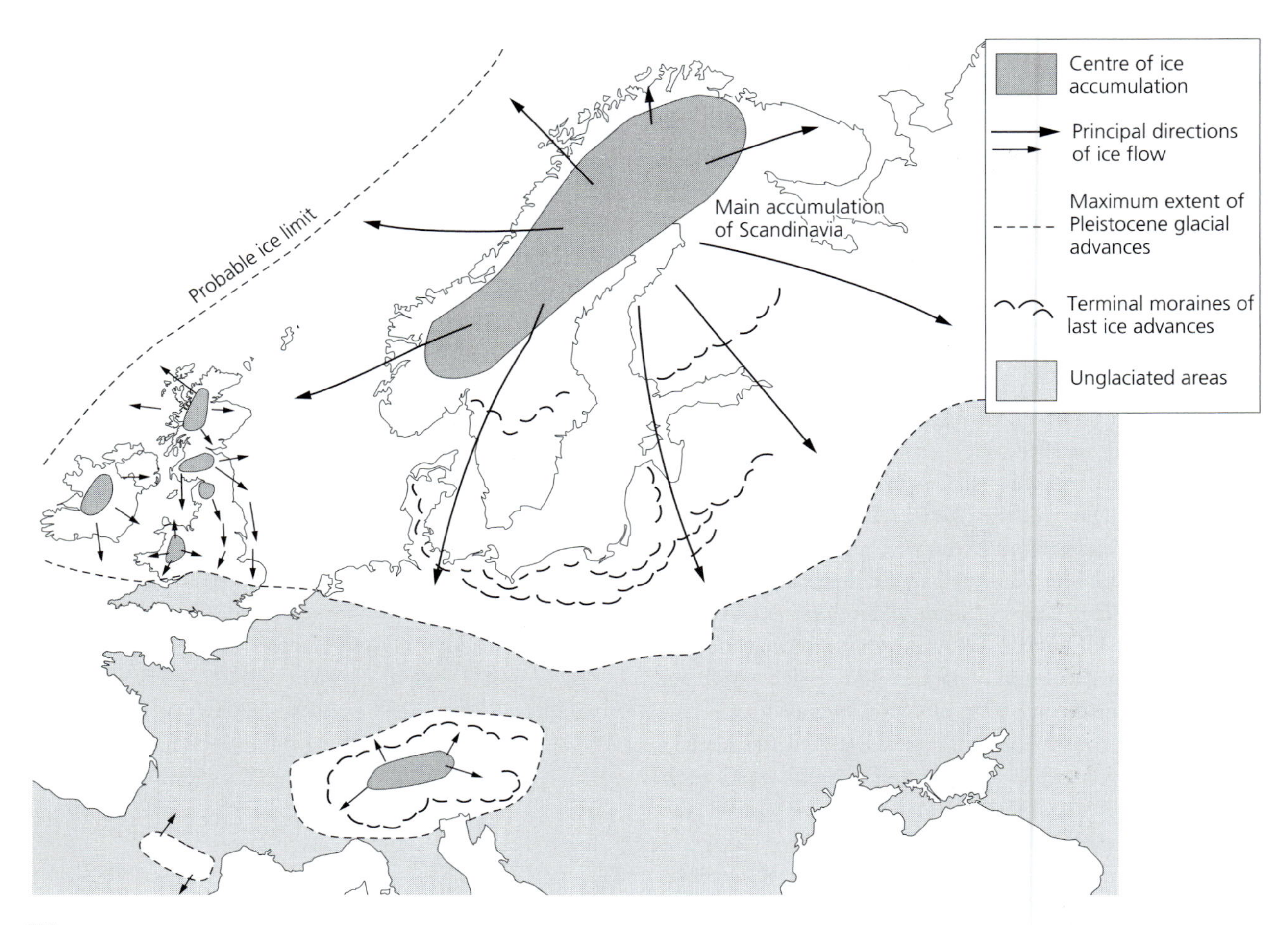

Figure 15.6 *Pleistocene glaciation in Europe. Much of north-west Europe was covered by ice during the Pleistocene ice age, with ice radiating outwards in all directions from a number of high-altitude source areas.*
Source: Figure 14.18 in Clowes, A. and P. Comfort (1982) Process and landform. Oliver & Boyd, Edinburgh.

GLACIALS AND INTERGLACIALS

The Pleistocene involved a number of phases of glacial stages when climate was particularly cold and ice accumulated and advanced, separated by interglacial stages when the climate warmed and the ice melted and most of it retreated. Many areas experienced multiple phases of glaciation, and this has produced complex landscapes that often make it difficult to reconstruct the glacial history of the area.

There is some debate about exactly how many glacial and interglacial stages there were, and whether everywhere in the northern hemisphere experienced each stage and at the same time. Most evidence supports four glacials separated by three interglacials, and the geological record suggests that we are currently living through a fourth interglacial. This implies that a fifth glacial period will follow

Plate 15.1 *The Lower Arolla Glacier in the Val d'Herens, southern Switzerland.*
Photo: Chris Park.

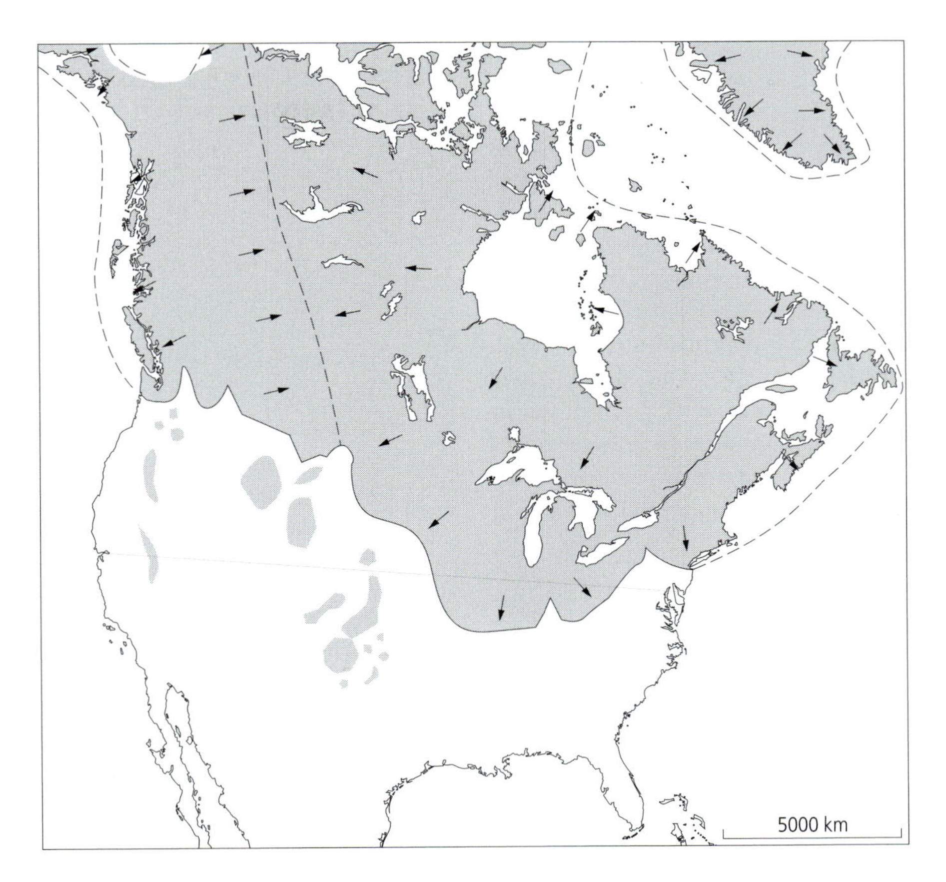

Figure 15.7 *Pleistocene glaciation in North America. At its maximum, Pleistocene ice extended further south than the Great Lakes and was fed from source areas in the Rocky Mountains and the northern Canadian Shield.*

Source: Figure 13.1 in Doerr, A.H. (1990) Fundamentals of physical geography. Wm. C. Brown Publishers, Dubuque, Iowa.

BOX 15.14 IMPACTS OF GLACIATION

The impacts of Pleistocene and present-day glaciation include:

- shaping landforms and landscapes, particularly in mountain areas such as the Alps in Europe, the Himalaya in Asia, the Rockies in North America and the Southern Alps in New Zealand;
- determining rates and patterns of sediment entrainment, transport and deposition across much of North America and Europe;
- strongly influencing the nature of surface deposits and soils across much of North America and Europe;
- creating habitats and refuges for alpine plants and animals;
- storing vast amounts of fresh water in glaciers and ice sheets;
- affecting regional and global climate. Extensive ice caps such as Antarctica often have stable air conditions and are part of the global circulation system.

some time in the future, but this would inevitably be postponed (possibly indefinitely) if induced global warming continues.

Most of the Pleistocene ice had melted by the start of the Holocene (post-glacial period), about 10,000 years ago, but valley glaciers remain in some high-altitude mountain areas in both hemispheres, and ice caps survive in Greenland and other parts of the Arctic Circle, and in Antarctica.

Recent glaciation has strong imprints on today's landscapes, particularly at high altitudes. One direct consequence of this has been the creation of popular tourist resources, such as the Alps in Europe, which are endowed with striking scenery and year-round snow and ice and are now popular for alpine walking, climbing and skiing.

GLACIER HAZARDS

Glacier dynamics and changes also create environmental hazards. At the global scale, warming associated with rising concentrations of greenhouse gases (see p. 255) threatens to increase rates of ice melting. Fears have been voiced about thermally induced thinning of the Antarctic ice cap in particular, which could trigger or at least contribute to significant sea-level rise (see p. 511).

Glacier hazards are not confined to the global scale, and many alpine areas today suffer from a range of hazards including slow and progressive glacier advance, sudden avalanches (see p. 215), meltwater floods and snowstorms. Efforts are being made to assess the risks from potential glacier hazards that threaten the safety of settlements and other fixed installations in high mountain areas. In some areas, maps of potentially dangerous zones are compiled to help to plan appropriate safety measures. Hazard risk changes over time, particularly in response to climate change. Progressive warming through the nineteenth and twentieth centuries has caused many North American glaciers to shrink, and in some areas this has increased the incidence of serious flooding (caused by the collapse of moraine dams), mass movement, river deposition and rock and debris avalanches (caused by the collapse of unstable over-steepened valley walls).

CAUSES OF GLACIATION

MULTIPLE CAUSES AND THRESHOLDS

Scientists cannot say for certain what causes an ice age to set in and glaciers to advance. Quite obviously a drop in temperature is required, but how big must that be? And what might cause it? Answers to these questions are important, not just to satisfy academic curiosity but because a better understanding of what causes glaciation could prove invaluable in the future for coping with glacier advance and the growth of ice sheets. An awareness of what causes glaciation should also reduce the risk of people unsuspectingly using environmental resources in ways that might increase the glacier hazard.

BOX 15.15 WHAT CAUSES GLACIATION?

Two important things are clear from recent scientific research:

1. There is probably not one single cause of glaciation; it is much more likely that multiple factors are involved, that they reinforce each other and that different factors act as triggers in different glaciations.
2. Glacial activity can be triggered by relatively small reductions in global temperature; the Earth's climate and environmental systems are generally in equilibrium but often close to critical thresholds, so even relatively small changes (usually in external factors) can promote major internal change and adjustment.

FEEDBACK AND AMPLIFICATION

The geological and climatic evidence suggests that a period of glacial activity could probably be triggered by only slightly larger changes in surface temperatures than have occurred naturally over the past few thousand years. Early views of glaciation were based on significant temperature reductions over long periods. These have now been replaced by a better appreciation of the significance of small temperature reductions over even relatively short periods, which can be amplified by positive feedback in the Earth's environmental systems. Feedback (see p. 79) involving changes in cloudiness, reflectivity and albedo seems particularly important.

For most of the past 4,000 million years, the average surface temperature of the Earth has been about 20 °C. Present understanding is that a fall in average summer temperatures of around 2–3 °C, over a period as short as decades, would be enough to promote the progressive accumulation of snow and ice, which can set in motion widespread glaciation. When summers are cooler than usual, some snow survives throughout the year without melting. Over a period of time, so long as temperatures remain relatively low, this snow can accumulate and eventually change into ice. As the extent of ice cover increases it changes surface albedo, which can have feedback impacts on atmospheric energy balance and cloud cover, which in turn can promote further temperature changes.

PRINCIPAL FACTORS

Three factors appear to play significant roles in causing glaciation – changes in solar radiation, changes in the albedo (reflectivity) of the Earth and its atmosphere, and cyclical variations in the Earth's orbit.

Changes in solar radiation: the amount of solar radiation that reaches the outer edge of the Earth's atmosphere from the Sun (the solar constant) is relatively stable over time (see p. 232). But it does vary with changes in solar output, particularly in response to variations in sunspot activity (see Box 4.7), which might be associated with changes in the strength and pattern of thermonuclear reactions in the Sun. Analysis of long-term climate records shows that cold weather conditions on Earth often coincide with phases of maximum sunspot activity.

Changes in albedo: temperatures in the lower atmosphere can be strongly influenced by the albedo of the Earth's surface and the atmosphere (see p. 235). Snow and ice have much higher albedos than rock or vegetation (see Table 8.4), so that once ice starts to accumulate (for whatever reason) it can begin to alter the energy balance of the lower atmosphere and thus affect temperatures. The process involves positive feedback (see p. 79), which amplifies the initial change. Clouds also have relatively high albedos (Table 8.4), and increased cloudiness might help to set the albedo–cooling–ice feedback in motion. Direct cause-and-effect relationships are very difficult to isolate, because reflectivity in the atmosphere can also be strongly affected globally by volcanic dust clouds (p. 181) and locally by processes such as desert wind storms (p. 433), forest fires (p. 575) and particulate air pollution (p. 241).

The Earth's orbit: while the timing and duration of glacials and interglacials within an ice age can vary a great deal, there is strong evidence to indicate that the underlying variations are often cyclical, with cycles lasting in the order of 100,000 years. Much of the variability appears to be related to cyclical variations in three key aspects of the Earth's orbit (Box 15.16), which were first identified by Milutin Milankovitch (see p. 346).

Sediment cores taken from ocean-floor deposits preserve (mainly biological) evidence of fairly regular cycles of climate changes over the past 300,000 years. Interestingly, the cycles have periods of about 100,000 years, 42,000 years and 23,000 years, which coincide remarkably well with the orbital changes that Milankovitch identified.

BOX 15.16 GLACIATION AND VARIATIONS IN THE EARTH'S ORBIT

The three variations identified by Milankovitch that affect glaciations are in:

1. *Eccentricity of the orbit*: these are variations from an almost circular path around the Sun, and take place over a cycle of 93,408 years. When the Earth and Moon are further from the Sun, the Earth–Moon system spins more slowly and the Earth's magnetic field is weaker. This reduces the number of high-energy particles that reach the Earth from the Sun and lowers temperatures.
2. *Tilt of the Earth's axis of rotation*: these are variations in the tilt of the axis around which the Earth rotates, from about 22 to 25° from the vertical. These variations occur over a cycle of roughly 42,000 years and may explain up to a quarter of the temperature differences between glacials and interglacials.
3. *Cycle of precession*: these are variations in the orientation of the Earth's axis, over a cycle of about 21,000 years, rather like the wobble of a spinning top. This determines which hemisphere is closer to the Sun in winter, which in turn influences how mild or severe summers and winters are. Currently the northern hemisphere has relatively mild summers and winters, but about 11,000 years ago summers were warmer and winters were colder. Winter ice survived longer into the summer at high latitudes, and there were droughts in sub-tropical areas.

GLACIERS AND ICE FORMATION

A glacier is a large mass of ice that forms above the snow line (see Box 15.2) in high mountains or in high latitudes. Here the rate of snowfall is greater than the rate at which snow melts, and the ice is formed from compressed snow. Most glaciers move slowly downhill, either along valleys or across the land surface, under the force of gravity. As they move downhill they are replenished by new ice that forms at their source. In this sense, glaciers are effectively slow-moving rivers of ice.

There are many different types of glacier (Box 15.17), reflecting variations in surface topography and regional climate.

BOX 15.17 GLACIER TYPES

The four most common types of glacier are:

1. *Alpine glaciers*: these are the valley glaciers that are seen today in high mountain areas such as the Alps in Europe, the Rockies in North America and the Himalaya in Asia. Most of the rest of this section relates to alpine glaciers.
2. *Piedmont glaciers*: these are extensive glacier sheets that are formed when a number of alpine glaciers converge and coalesce at the foot of a range of mountains. They are particularly common in Alaska and include the Malaspina Glacier there, which covers an area of about 3,900 km^2.
3. *Ice-cap glaciers*: these are a combination of the alpine and continental types, with a central ice sheet overlying a high plateau, surrounded by a series of alpine glaciers, which flow down steep slopes. The Svalbard Islands, off Norway in the Arctic Ocean, are of this type.
4. *Continental glaciers*: these are extensive blankets of thick ice that cover entire land surfaces. Antarctica is covered by such a continental glacier nearly 13,000,000 km^2 in size, and most of Greenland is covered by one that stretches over 1,800,000 km^2. During the Pleistocene, continental glaciers covered much of North America and Europe.

ACCUMULATION AND COMPACTION

It stretches the imagination to associate flurries of tiny hexagonal snowflakes, which are often so light that they are carried in the air and take some time to land on the ground, with the striking and often large-scale landforms produced by glaciers. Ice – particularly thick ice flowing downslope

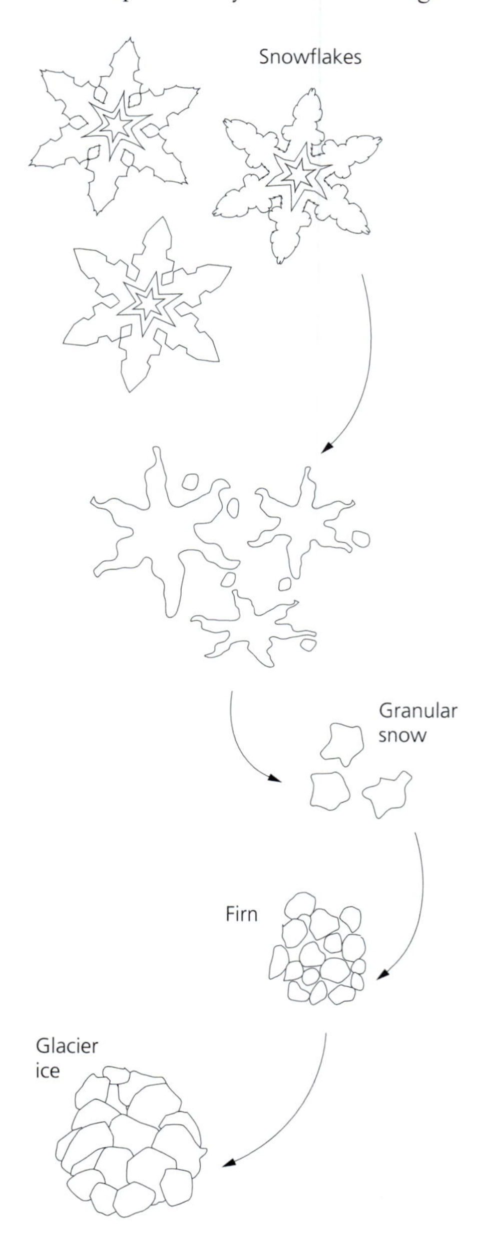

Figure 15.8 *Progression from snow to glacier ice. Snowflakes are progressively compacted into granular snow, then firn and eventually into glacier ice.*

Source: Figure 13.2 in Doerr, A.H. (1990) Fundamentals of physical geography. Wm. C. Brown Publishers, Dubuque, Iowa.

– has great power to erode, transport and deposit particles of rock of different sizes. But how is it formed?

Two key sets of processes are involved in turning snow into ice (Figure 15.8) – accumulation and compaction (Box 15.18). As a result of these processes, a thick column of ice in a glacier usually contains material at all stages along the spectrum between fresh snow and solid ice:

1. a thin layer of freshly fallen snow on the top, with a low density (less than 0.1);
2. a layer beneath that where snowflakes have been compressed to form granular snow (with a density of around 0.3) and lower down *firn* (with a density of 0.5);
3. at the base a layer of dense (density up to 0.8) clear ice, which has the ability to flow like a thick, viscous fluid.

Large ice sheets can be extremely thick – up to 1,600 m in North America during the Pleistocene, for example, and over 2,500 m in Greenland today. Such thicknesses of ice are well capable of depressing the land surface beneath it, sometimes to levels much below sea level.

BOX 15.18 ACCUMULATION AND COMPACTION OF SNOW

Snow turns into ice by the combined effect of these two processes:

1. *Accumulation*: the snow that falls in areas above the snow line can accumulate to considerable depths because it melts much more slowly than it falls because of the low temperatures, particularly during the winter months. When the individual snowflakes land on the ground, they pack together under their own weight to form granular snow (called *firn* or *névé*) in that the individual grains are joined together. Over time, the firn builds up and becomes thicker and thicker.
2. *Compaction and compression*: as the firn accumulates it is compacted and compressed by the weight of overlying snow and firn to produce glacier ice, which has a granular structure. Air bubbles trapped in the snow and firn are squashed out, and the individual grains of ice are squeezed tightly together during compression, so the ice becomes much more dense. In areas where temperatures remain mostly below 0 °C – the freezing point of water – ice is formed by sublimation (a phase change directly from solid to gas) and recrystallisation.

BOX 15.19 ICE MOVEMENT

A mass of ice moves in two ways (Figure 15.10):

1. The lower part slides over the underlying surface (often bedrock) on a film of meltwater, which is created by friction and overlying pressure.
2. The main body of the ice moves by laminar flow, which involves plastic deformation within the dense, compacted ice.

MOVEMENT

Once snow and ice have accumulated to a depth of about 30 m, the mass of material can usually begin to move slowly downslope under the influence of gravity, either down-valley in a glacier or outwards in all directions (in a continental glacier). Glacier flow speeds vary a great deal, depending on factors such as ice volume, ground slope, nature of the surface they are flowing over and air temperature. Most glaciers flow at speeds of less than about 1 m a day. Flow continues as long as new ice accumulates in the main supply area.

Ice movement not only extends the ice over a wider area but also plays a critical role in glacial erosion and thus landform development. Meltwater, in particular, is important because it helps to lubricate the base of a glacier and is a major factor in freeze–thaw processes.

As ice moves down-valley, its ease and speed of flow are regularly disturbed by irregularities on the valley floor, such as bands of resistant bedrock. As it flows, it erodes the underlying surface, which tends to smooth out such irregularities over a period of time. But many glaciers flow down valleys that have alternating relatively high-slope and relatively low-slope sections, and these can have a pronounced effect on glacier flow. Flow along steep sections often involves straining and stretching of the ice, which is reflected in the production of crevasses (cracks in the ice, which are often filled with sediment) (Figure 15.9). Jagged pinnacles of ice, called *seracs*, appear on the glacier surface. Flow along lower slopes is often described as compressional, because the ice is compressed as it is pushed along

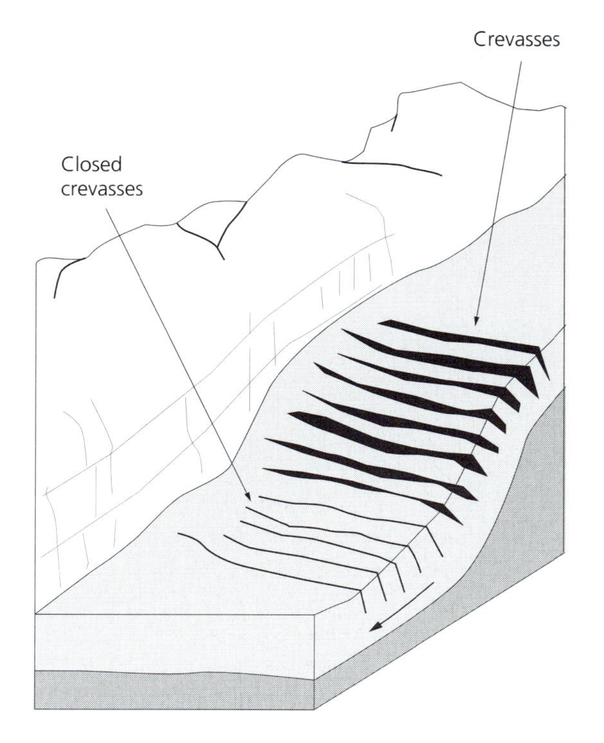

Figure 15.9 *Crevasses in a glacier. Crevasses occur where glacier ice flows over a change in slope in the underlying valley floor – crevasses by extension where slope increases and closed crevasses by compression where it decreases.*

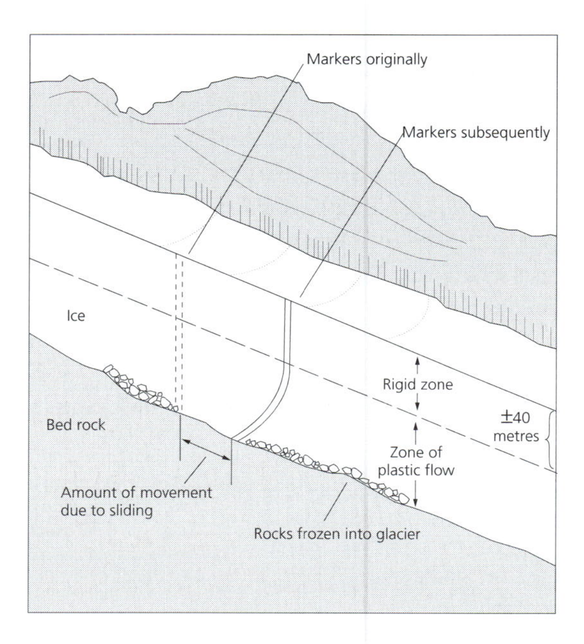

Figure 15.10 *Movement of glacier ice. Most of the ice movement in a glacier occurs by plastic flow within the ice mass itself, and relatively little occurs by sliding along the valley.*

Source: Figure 13.3 in Doerr, A.H. (1990) Fundamentals of physical geography. Wm. C. Brown Publishers, Dubuque, Iowa.

Plate 15.2 *Crevasse at the side of the Tsijore Nouve Glacier in the Val d'Herens, southern Switzerland. The author is standing on a lateral moraine beside the ice margin.*

Photo: Chris Park.

by the weight of ice further up the glacier. Surface cracks are relatively rare in compressional flow.

MASS BALANCE

As with all open environmental systems that have inputs and outputs of material and energy (see p. 70), it is possible to construct the mass balance of a glacier or ice sheet, linking inputs and outputs of ice. The mass balance of a glacier reflects two main sets of processes:

1. precipitation input, radiation and temperature conditions in the accumulation zone;
2. speed of output (outflow of ice from the area).

In the upper part of a valley glacier ice accumulation exceeds ablation (melting), and in the lower part ablation usually exceeds accumulation. The so-called firn line (where the surface of the glacier is firn rather than snow) separates the two sections of the glacier (Figure 15.11).

When the mass balance is positive, inputs exceed outputs and the glacier grows. The larger the positive balance, the faster the glacier grows. Glaciers and ice caps are fed with ice when snow accumulates faster than it melts. This happens fastest when precipitation is high and temperatures are low. Conversely, when the mass balance is negative, outputs exceed inputs and the glacier melts and shrinks. The larger the negative balance, the faster the glacier shrinks.

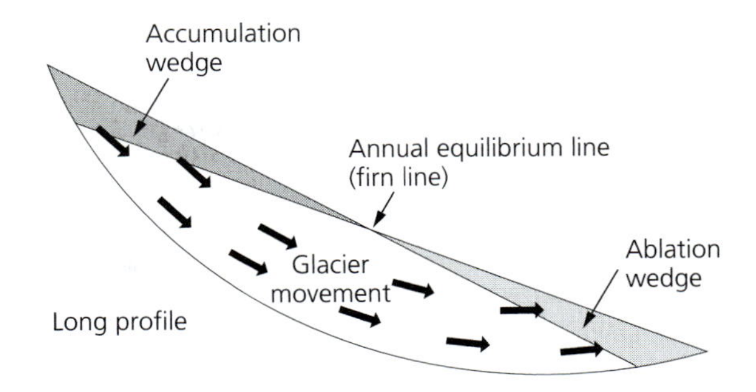

Figure 15.11 *Basis of glacier mass balance. In a glacier that is in equilibrium the ablation wedge equals the accumulation zone. The accumulation zone exceeds the ablation wedge in an expanding glacier and is less than in a retreating glacier.*

Source: Figure 6.5 in Clowes, A. and P. Comfort (1982) Process and landform. Oliver & Boyd, Edinburgh.

BOX 15.20 GLACIER MASS BALANCE AND GLOBAL WARMING

Global warming threatens to alter the mass balance of glaciers and ice sheets in most places, and this in turn might promote further regional changes in climate (via changes in surface albedo and atmospheric energy flows). Predicting exactly how mass balances might change is not easy, because ablation (melting) will doubtless increase, but this may well be offset at least partially by increased snowfall. Even small-scale changes to the mass balance of an ice sheet the size of Antarctica, which contains an estimated 24–29 million km^3 of ice, could have important global consequences. Models of how the ice sheet might respond to global warming suggest that a 2 °C rise in mean annual surface temperature over 40 years could increase ablation in the Antarctic Peninsula region enough to cause a 1 mm rise in global sea level. That in itself is not a great threat, but if it continued over a long period or if the warming was greater, it could pose problems for many low-lying coastal areas.

There are already clear signs that global warming is causing ice melting. Most of the Alpine glaciers in Europe have lost up to half their volume since 1850 in response to warming, and the US government predicts that no more glaciers will be left in Montana's Glacier National Park by 2030. Most of the glaciers in the Himalaya are expected to vanish by 2035 if recent trends continue; for example, the Gangorti Glacier, at the head of the River Ganges, is retreating at a rate of 30 m a year.

Comparison of satellite data between 1978 and 1988 reveals that the Arctic ice cap is shrinking at a rate of about 37,000 km^2 a year – an annual loss greater than the area of Wales. Recent studies of the Arctic suggest that the loss of sea ice has begun to accelerate, and some parts of the Arctic ice sheet are up to 40 per cent thinner than they were 20–40 years ago. It is feared that between 10 and 20 per cent of Arctic ice may melt in less than 50 years unless urgent action is taken to curb global warming.

GLACIER ADVANCE AND RETREAT

Glacier dynamics are heavily dependent upon climate, and climate change can promote significant adjustments in valley glaciers and ice caps:

- Glaciers expand and ice advances when temperatures decrease and/or precipitation increases.
- Glaciers shrink and ice retreats when temperatures increase and/or precipitation decreases.

Expansion and advance usually accompany cooler periods. This was certainly the main cause of the major phases of glaciation during the Pleistocene, but the trend has continued during the Holocene. Dated deposits in the Southern Alps of New Zealand, for example, indicate at least three distinct phases of glacier advance within the last 14,000 years.

Shrinkage and decay – by backwasting (retreat) and downwasting (thinning) – generally accompany warmer periods. This happened during the main Pleistocene interglacials, but it also happens over much shorter timescales in response to changing climate. There is abundant documentary and photographic evidence of the retreat of alpine glaciers in the European Alps since the nineteenth century, for example. More recently, remote-sensing imagery, from such sources as the Landsat Thematic Mapper (TM), have clearly shown significant glacier recession within less than a decade during the 1980s.

GLACIAL EROSION

Glaciers and ice sheets are powerful agents of erosion that radically alter landscapes and carve distinctive landforms. Landscapes of glacial erosion tend to be very angular, often with relatively steep slopes, increased relief and sharp-edged mountains. Alpine scenery in Europe and elsewhere bears the striking imprint of prolonged glacial erosion.

RATES OF EROSION

Rates and patterns of erosion depend largely on the volume of ice involved and the speed at which it moves across the underlying surface. Erosion is generally fastest in large glaciers and in fast-moving ones. For these reasons, most erosion is concentrated in the upper and middle sections of long valley glaciers; the lower portions tend to be dominated more by deposition (see p. 463). Erosion rates also reflect rock hardness and strength, and – all else being equal – resistant bedrock slows down erosion.

EROSION PROCESSES

The two most effective processes of erosion by glaciers and ice sheets are quarrying and abrasion (Box 15.21). The combined effect of quarrying and abrasion can be extremely effective in producing characteristic ice scour features, including:

- carving out deep basins at the head of valley glaciers, where the ice originally collects;
- removing rock barriers on the floor and sides of glacial valleys;
- deepening the valley;
- straightening the valley and trimming off minor bends and intervening spurs (hills);
- creating extensive lowland rock surfaces (known as ice scour planes), such as the Laurentian Shield in Canada, which often contain glacial lakes and deposits.

The most striking alpine landforms are created by scour, which creates the sharp edges and angular form of many glaciated landscapes.

BOX 15.21 QUARRYING AND ABRASION PROCESSES

- *Quarrying (plucking)*: ice at the base of a glacier or ice cap is melted by friction and pressure, and this often refreezes in cracks in the rock that it comes into contact with (Figure 15.12). Freeze/thaw weathering weakens the rock, and fragments of it are plucked out by the ice that flows over it. This is the process of quarrying, and the fragments are incorporated into the bed of the moving ice, where they contribute to the process of abrasion.
- *Abrasion*: frozen ice that contains fragments of rock (often as a result of quarrying) can be very effective at grinding the surface it moves over, particularly under great pressure from the weight of ice above. This grinding process has two main consequences – rough surfaces can be polished quite smooth (particularly when polished by very fine particles of rock called glacial flour), and smooth surfaces can be scratched by abrasion to create striations (or striae) aligned in the direction of ice movement. Striations are very useful in reconstructing directions of ice movement in the past.

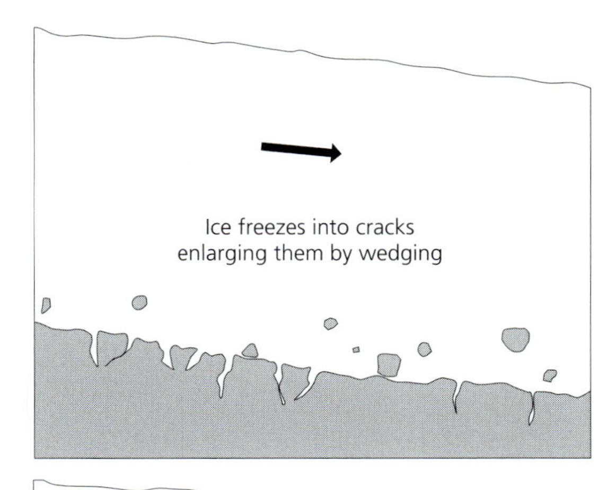

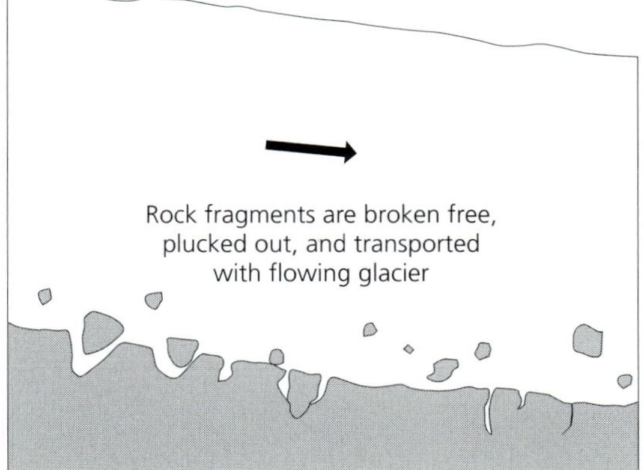

CIRQUES AND ARÊTES

Bowl-shaped depressions, called *cirques*, are eroded by quarrying processes at the head of valleys in which glaciers form (Figure 15.13). There are many regional names for cirques, including *corries* (Scotland) and *cwms* (Wales). A typical cirque is semicircular in shape when viewed from above, with a steep headwall (at the back) and sidewalls, and a much flatter floor. Many cirques have been over-deepened by erosion, and after the ice has melted they contain lakes (*tarns*), which are often ponded behind mounds of glacial deposits. When cirques are being formed and ice is present in them, a wide, deep gap often develops between the ice and the headwall (called a *bergschrund*) as the ice is pulled away from the headwall and moves downslope.

Mountain ridges are often eroded on more than one side by ice, and this can create some distinctive landforms indicative of glacial scour. When glaciers erode adjacent cirques on the opposite sides of a mountain ridge, this can create a long, sharp, knife-edged ridge with a serrated top, which is called an *arête* (Figure 15.14). Such features produce dramatic alpine scenery and provide high-level pathways for hillwalkers. Striding Edge in the English Lake District is a typical arête.

Notches or natural passes cut through such mountain ridges, called *cols*, are often the result of further back-to-

Figure 15.12 *Glacial quarrying. See Box 15.21 for explanation.*
Source: Figure 13.5 in Doerr, A.H. (1990) Fundamentals of physical geography. Wm. C. Brown Publishers, Dubuque, Iowa.

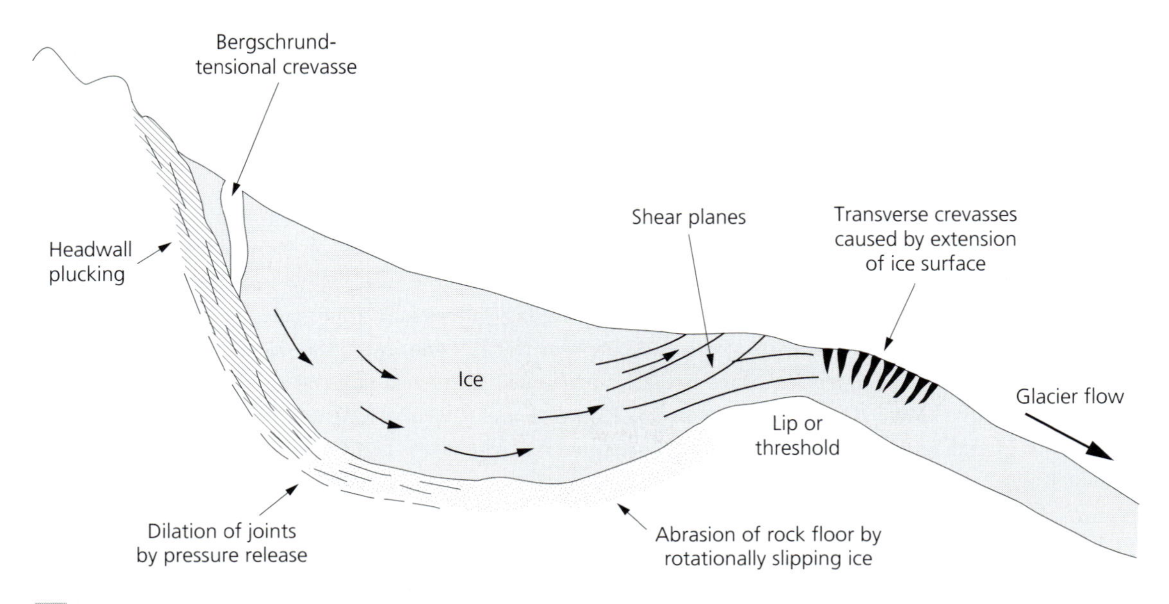

Figure 15.13 *Formation of a cirque. See text for explanation.*
Source: Figure 7.4 in Clowes, A. and P. Comfort (1982) Process and landform. Oliver & Boyd, Edinburgh.

Plate 15.3 *Goat's Water, a corrie tarn (lake) near Coniston, Cumbria, England. This post-glacial lake sits in a glacially excavated headwater basin, with the headwall to the left and the lip of the corrie to the right.*

Photo: Chris Park.

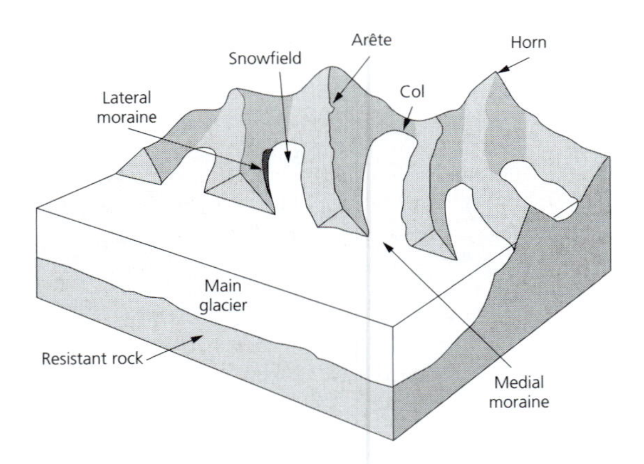

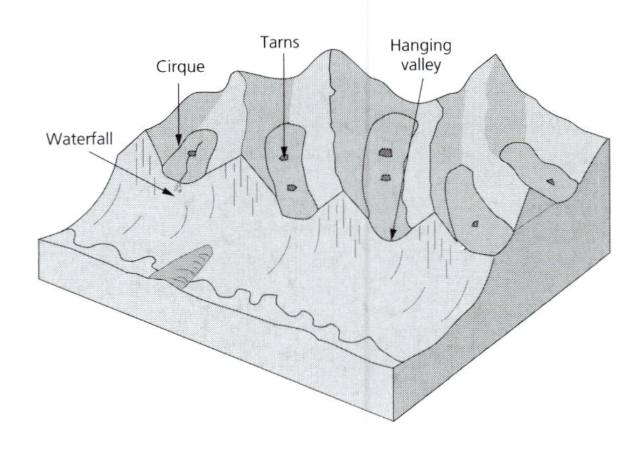

Figure 15.14 *Landforms created by ice scour. Characteristic features of ice scour in glaciated mountains and valleys include cirques, hanging valleys, cols, horns and arêtes. See text for explanation.*

Source: Figure 13.17 in Doerr, A.H. (1990) Fundamentals of physical geography. Wm. C. Brown Publishers, Dubuque, Iowa.

back cirque erosion. When cirques erode simultaneously on more than two sides of a mountain, the sharp sidewalls of the cirques can intersect to produce a sharp-edged mountain peak called a horn. The Matterhorn in the Swiss Alps (see p. 158) is a classic example.

U-SHAPED VALLEYS

One of the most characteristic features of glacial erosion is the distinctive U-shaped valley or trough carved from an original V-shaped valley that was probably eroded by a pre-glacial river. U-shaped valleys typically have parallel sides with steep slopes, and they are usually either straight or gently meandering. They are formed by glacial scour as a glacier extends and advances down-valley. Quarrying and abrasion processes deepen the valley by scouring its floor, widen it by scouring the sides and straighten it by removing small obstructions and hills.

While many glacial valleys have a U-shaped form, with steep sides and relatively flat floors, the bedrock surface is not always so smooth and regular. Many U-shaped valleys contain thick deposits of glacial and fluvioglacial (meltwater) sediments, which mask the more irregular bedrock surface beneath them.

The long profile of glacial valleys is often stepped rather than smooth, with steeper sections reflecting the location of resistant bedrock or the increased erosional power where two glaciers converge. The flatter sections often contain wide river floodplains or series of small lakes (called *paternoster lakes*) after the ice has melted.

U-shaped valleys are often excavated quite deeply by glaciers, so that the mouths of tributary valleys are left hanging high above the new valley floor as hanging valleys. Many hanging valleys have long, steep waterfalls cascading down the valley side into the main valley floor (the Lauterbrünnen in Austria is a classic example).

In some areas, over-deepened glacial valleys are now occupied by large lakes. The Great Lakes in North America are perhaps the most obvious example, but others include many of the large lakes in Scandinavia and the distinctive radial pattern of the English Lake District. *Fjords* (deep, steep-sided inlets) are formed where such valleys flow into cliffed coastlines, as commonly occurs in Greenland, the South Island of New Zealand and Norway.

OTHER FORMS CREATED BY GLACIAL SCOUR

While cirques, arêtes and U-shaped valleys are probably the most characteristic landforms created by glacial scour, they are by no means the only ones. Landscapes in many previously glaciated areas preserve evidence of glacial erosion, and careful interpretation of these can yield vital clues to reconstructing the history and pattern of glacial activity in the area.

Two of the most widespread other forms created by glacial scour are *rôches moutonnées* (Figure 15.15) and *crag-and-tail topography* (Figure 15.16; Box 15.22).

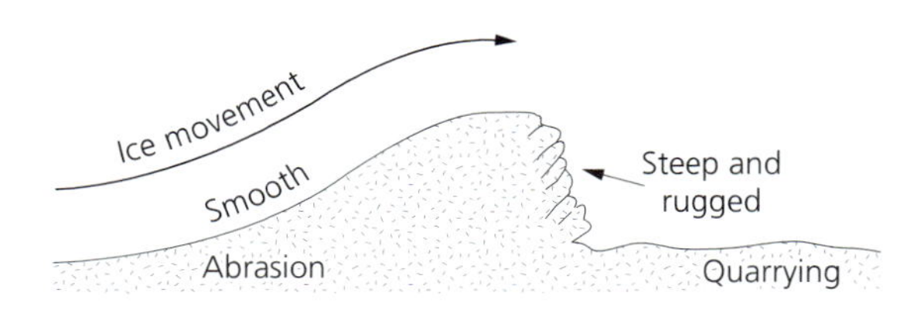

Figure 15.15 *Rôche moutonnée. See Box 15.22 for explanation.*

Source: Figure 9.2 in Buckle, C. (1978) Landforms in Africa. Longman, Harlow.

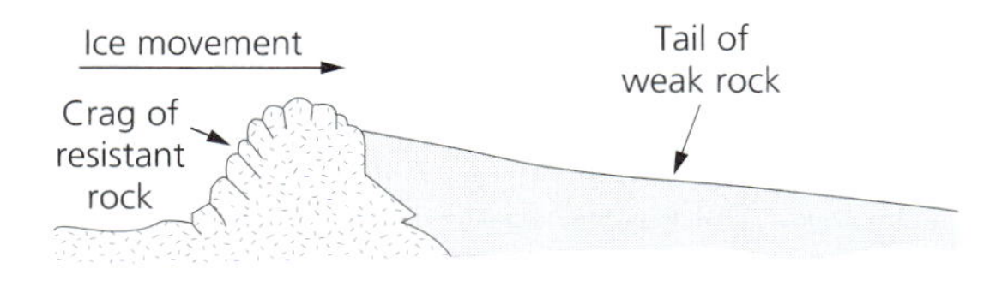

Figure 15.16 *Crag-and-tail. See Box 15.22 for explanation.*

Source: Figure 9.3 in Buckle, C. (1978) Landforms in Africa. Longman, Harlow.

BOX 15.22 RÔCHES MOUTONNÉES AND CRAG-AND-TAILS

- *Roches moutonnées*: these are outcrops of bare, resistant rock that have been shaped as glacier ice moved over them. The upstream side (called the stoss), shaped by abrasion, is usually smoothed, rounded, streamlined and often striated. The downstream side (the lee) is usually steep and angular because it is formed by quarrying processes.
- *Crag-and-tail topography*: a crag-and-tail can also be formed around resistant rock underneath moving ice. Quarrying gives a rocky, angular appearance to the upstream side (the crag), while the downstream side (the tail) has a gentle, streamlined slope that is aligned in the direction of ice movement and is covered by glacial deposits.

GLACIAL DEPOSITION

SEDIMENT TRANSPORT

Glaciers are very effective at transporting sediment from one place to another (Box 15.23) and depositing it. Much of the sediment load is derived from quarrying processes (see Box 15.21) underneath and at the sides of a moving glacier, but some can also come from rock falls and other mass-movement processes on the valley sides above the glacier. In an area that has already been directly affected by glaciation in the recent past, it is also quite likely that much sediment is reworked – previous deposits are picked up by the new glacier advance and transported, sorted and ultimately redeposited further down the valley.

The sediment carried by a glacier is also highly variable in size and character, depending on factors such as the geology, mineral composition and hardness of the original bedrock, the mode of transport, and the distance over which it is transported.

BOX 15.23 GLACIER MOVEMENT OF SEDIMENT

Glaciers move sediment in a number of different ways:

- Some is literally frozen to the base of the moving ice (where it acts as an abrasive and plays an important role in ice scouring).
- Some is frozen inside the ice.
- Some is transported at the edge of the ice.
- Some is transported on the surface of the ice.
- Some can be pushed ahead of the ice as it moves down-valley (rather like a bulldozer).

DEPOSITION

Glaciers deposit particles of sediment across a broad area, and the nature of the glacial deposit reflects the character of the sediment being transported and the particular environment in which it is deposited. Sediment that is deposited directly by melting ice, for example, usually contains particles of a wide variety of sizes, which are angular in shape because they have been weathered by frost shattering. Beyond the down-valley limit of the glacier (the *snout*), sediment that is deposited by meltwater tends to have much rounder and smaller particles, and it is often stratified (layered).

TILL

The sediment that is deposited directly by a glacier or ice sheet, without being reworked by meltwater, is called *boulder clay* or till. Till varies a great deal in particle size from place to place, and it usually includes a mixture of rocks, sand and clay. A layer of till of variable thickness is often deposited over irregular bedrock surfaces (such as the floor of U-shaped valleys), creating a gently undulating ground surface that largely masks the sub-surface variability.

While till sheets often produce a relatively flat ground surface, depressions (called *kettle holes*) are formed where blocks of ice buried in the till melt and the overlying

BOX 15.24 TILL AND ERRATICS

Extensive till deposits (till sheets) cover many high-latitude areas in the northern hemisphere that were glaciated during the Pleistocene. Such areas often have poor drainage and are susceptible to flooding because the till often contains a large amount of impermeable clay. The distribution of till reveals a great deal about the pattern and history of glacial advance and retreat in an area.

Other valuable information about the glacial history of an area is provided by erratics – large particles of rock that are of a different geology to the area in which they are deposited. Erratics of exotic or unusual rocks, for which there are few possible sources, are particularly useful. Detailed analysis of the erratic content of a particular till sheet can reveal a great deal about the source area (provenance) of the rock, and this in turn indicates the directions and amounts of ice movement.

Plate 15.4 *Norber erratics in the Yorkshire Dales, England. These large blocks of Silurian Grit were lifted from the valley floor by ice during the Pleistocene and deposited on the limestone surface of adjacent hilltops. Post-glacial weathering of the surrounding limestone surface has left the erratics perched on pedestals, the height of which provides a measure of the amount and rate of erosion.* Photo: Chris Park.

material slumps down to form a hollow after the glacier or ice sheet has retreated. Lines of kettle holes commonly form along lines of stagnant ice. Kettle holes are often filled with water to form lakes, and many former kettle lakes are now swampy or infilled with vegetation because of succession (see p. 571).

DRUMLINS

A third type of directional information is the orientation of streamlined hills of glacial till called *drumlins*, which are long, low features (usually up to several kilometres long and up to 50 m high) that are symmetrical and oval-shaped (Figure 15.17).

The upstream (stoss) slope is sometimes much steeper than the downstream (lee) slope. They are deposited beneath an advancing glacier, often on a wide plain rather than in a confined valley, and they are aligned parallel to the main direction of ice movement. It is not uncommon to find large groups of drumlins together, all aligned more or less parallel to one another, in a drumlin field (or swarm). Such groups are sometimes referred to as 'basket of eggs' topography, because from the air the surface resembles eggs in a basket. Drumlin fields are relatively common in the north-east USA and Canada, and throughout northern Europe.

MORAINES

Moraines are deposits of glacial till in the form of hummocky hills or hill-like features. There are a number of different types of moraine, which are deposited at different parts of a glacier. The history and pattern of glaciation in an area can be reconstructed if these different types of moraine can be identified and mapped after the ice has retreated.

BOX 15.25 TYPES OF MORAINE

The main types of moraine (Figures 15.18 and 15.21) are:

- *Lateral moraines*: these are deposited at the side of a valley glacier and are often composed of rock particles that have fallen off the sidewalls of the valley as a result of frost-wedging (see p. 207).
- *Ground moraine*: this is till.
- *Medial moraines*: these form in the centre of a glacier downstream from the confluence of neighbouring valleys as adjacent lateral moraines join together.
- *Terminal moraines*: these are deposited at the lower end of a melting glacier and are composed of sediments that have been carried in, on and under the ice as it advanced. As a valley glacier retreats, it rarely retreats fully in one go; rather, it retreats then halts, often a number of times in succession. Each halt can be marked by a terminal moraine across the valley, so these are valuable in reconstructing the glacial history of an area.

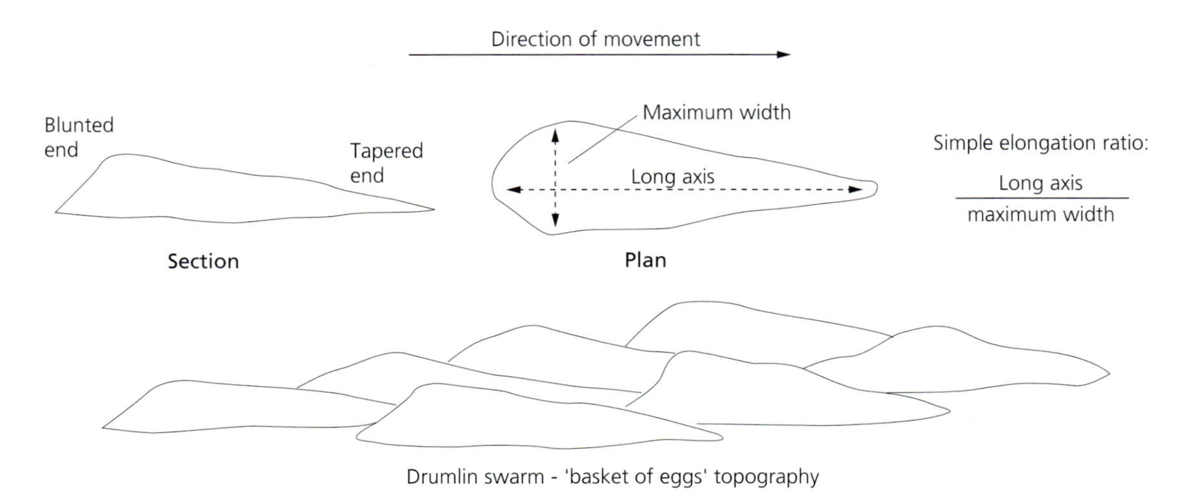

Figure 15.17 *Drumlin shape and a drumlin swarm. See text for explanation.*
Source: Figure 7.7 in Clowes, A. and P. Comfort (1982) Process and landform. Oliver & Boyd, Edinburgh.

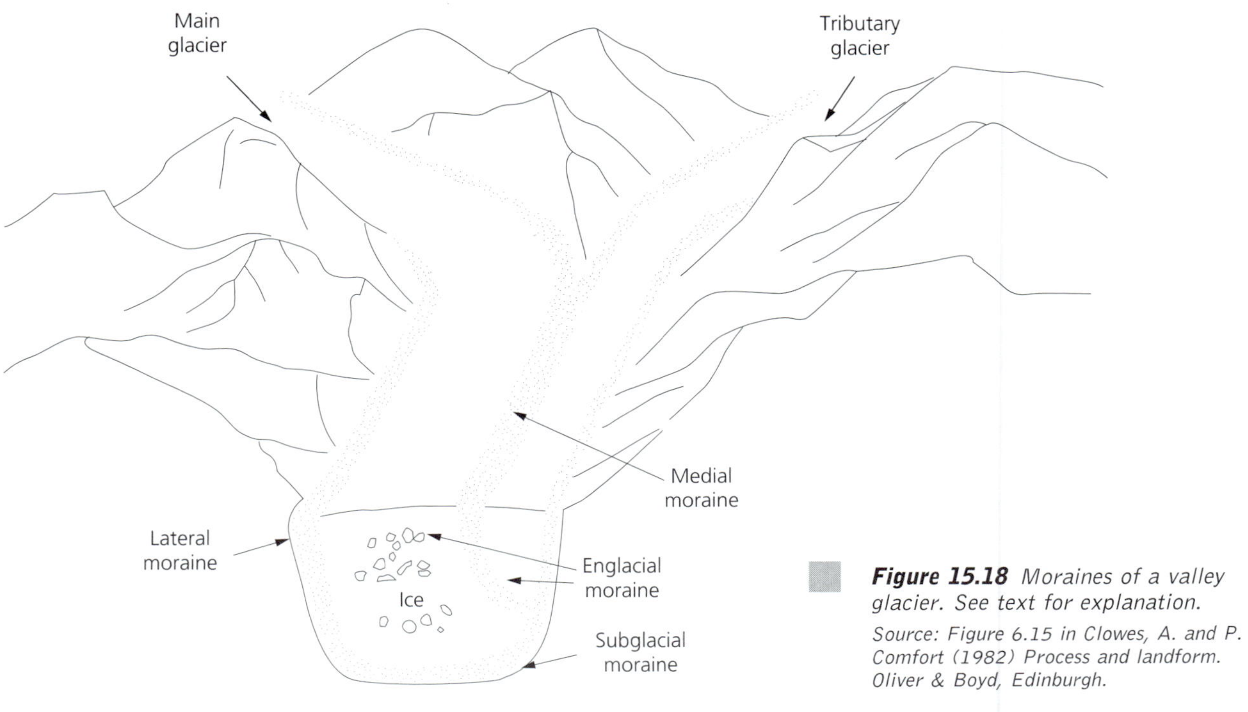

Figure 15.18 *Moraines of a valley glacier. See text for explanation.*

Source: Figure 6.15 in Clowes, A. and P. Comfort (1982) Process and landform. Oliver & Boyd, Edinburgh.

Plate 15.5 *Snout of the Ferpecle Glacier in the Val d'Herens, southern Switzerland. This ice-melt environment is characterised by abundant meltwater and extensive deposits of unsorted and unstratified glacial debris.*

MELTWATER

Although most of the erosion and deposition in glaciated environments is done by the moving ice, meltwater also plays an important role in creating and reshaping landforms and landscapes. As a glacier or ice sheet melts, it stops eroding and deposits the sediment it has been carrying. A glacier retreats (becomes shorter by melting progressively upstream) and grows thinner by melting. The sequence of retreat can often be reconstructed from the detailed analysis and dating of terminal moraines (Figure 15.19).

Under ice-age conditions, when temperatures remain below freezing for most if not all of the time, meltwater is created underneath the ice, particularly at the contact with bedrock below, by friction and pressure. In high-altitude and high-latitude areas where valley glaciers survive today, all three ablation processes contribute to the production of meltwater. In some heavily glaciated areas and areas with permanent ice sheets (such as west Greenland), meltwater appears to offer useful potential for generating hydroelectricity (see p. 394).

Wherever the meltwater is created, much of it eventually drains out in front of the glacier or ice sheet and flows down-valley as meltwater streams. These generally carry very high loads of dissolved, suspended and bedload sediment (see p. 366), which are provided by the many sources of supply in, on, beside and under the ice. Many meltwater streams have such high concentrations of fine suspended sediment (derived mainly from the rock flour) that they appear milky. Much of the sediment load is deposited downstream from the ice margin, often relatively close to it, as so-called *fluvioglacial deposits*.

Fluvioglacial deposits tend to be stratified (see p. 196) because they are laid down by sequences of high flows, each of which sorts particles into different sizes and usually deposits the largest particles first. Glacial deposits, in contrast, are not stratified, because the ice deposits particles of all sizes.

Deposition of fluvioglacial sediments by meltwater streams creates a series of distinctive features including eskers and kames, outwash plains, pro-glacial lakes and meltwater channels.

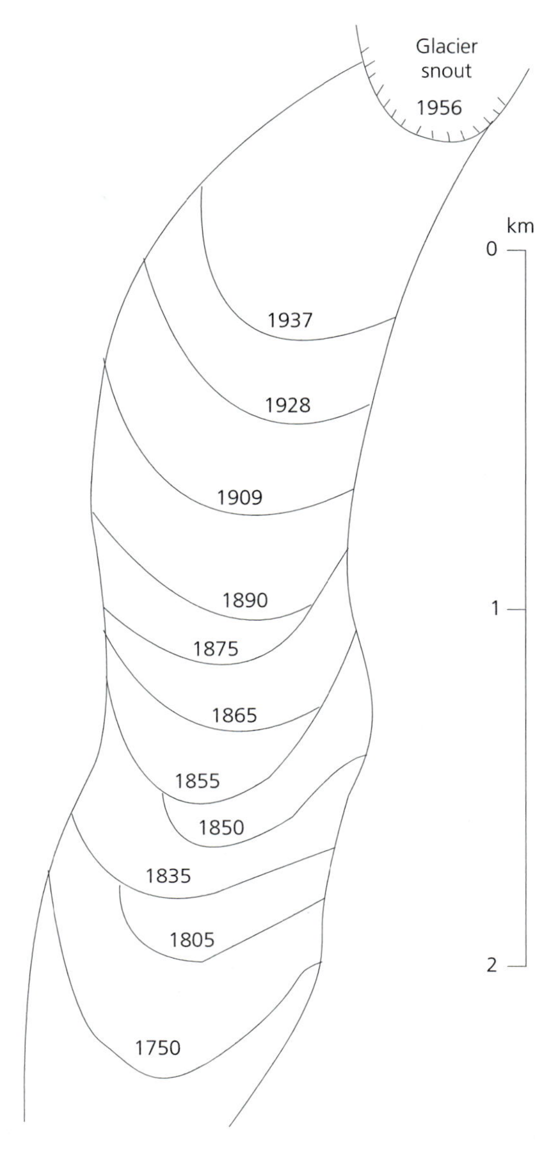

Figure 15.19 *200 years of recession on the Austerdalen Glacier, Norway. The pattern of ice retreat in this Norwegian glacier was reconstructed by detailed mapping and analysis of a series of terminal moraines.*

Source: Figure 13.11 in Dury, G.H. (1981) Environmental systems. Heinemann, London.

BOX 15.26 ICE MELTING

The process of ice melting is called *ablation*, and it occurs for three main reasons:

1. frictional drag over the underlying rock, which creates meltwater at the base of the ice;
2. pressure of overlying ice, which creates water in and under the ice;
3. temperature rise, which creates water, particularly on the surface and at the sides of the ice.

BOX 15.27 GLACIER DRAINAGE NETWORKS

Meltwater generally collects into stream channels that flow in, on and under the ice in a glacier or ice sheet, as:

- *supraglacial streams,* which flow over the ice surface;
- *englacial streams,* which flow within the body of the ice;
- *subglacial streams,* which flow at the base of the ice, across the underlying material.

Many glaciers have integrated drainage networks that link together these three different types of glacial stream and make meltwater channels in glaciers effectively three-dimensional networks.

ESKERS AND KAMES

Eskers and *kames* are two particularly distinctive fluvioglacial features that are formed by meltwater below or adjacent to a glacier or ice sheet (Figure 15.20).

Eskers are long, narrow, sinuous ridges formed from fluvioglacial sediments deposited by meltwater streams flowing in or beneath a glacier or ice sheet during stagnation. Their length and sinuous pattern reflect their fluvial

Plate 15.6 *Extraction of sand and gravel from a large glacial esker at Carnish on the Hebridean island of Harris, off the west coast of Scotland.*
Photo: Chris Park.

a Glacial landscape

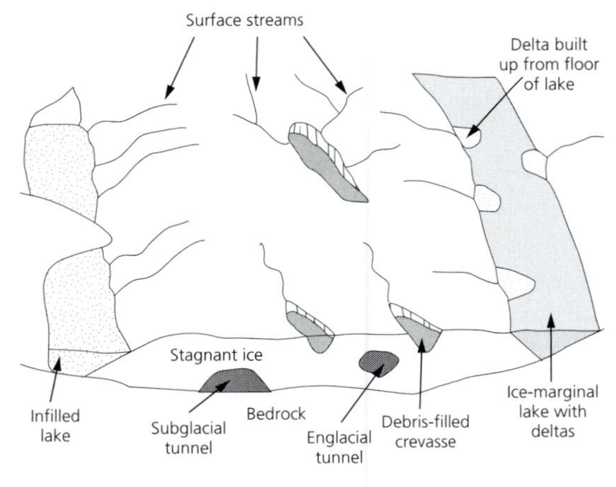

b Post-glacial landscape

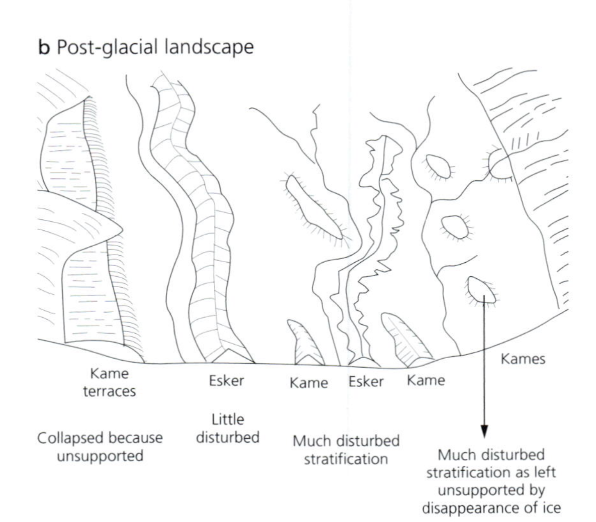

Figure 15.20 *Ice-contact depositional features. Characteristic ice-contact features include kames and eskers (see text for explanation).*
Source: Figure 7.12 in Clowes, A. and P. Comfort (1982) Process and landform. Oliver & Boyd, Edinburgh.

origin, rather like superimposed drainage patterns in ordinary river systems (see p. 362). After the ice has melted, the deposits remain as ridges, which generally have relatively flat tops but steep sides. Some eskers appear to defy gravity because they meander over hills and up slopes. This is because they were formed under the ice, under hydrostatic pressure, which enables water to flow uphill if there is sufficient pressure from the water behind it.

Kames are also long ridges of fluvioglacial sediments, but they were deposited along valley slopes at the margins of the glacier by supraglacial streams, which flow over the surface of the glacier. The deposit remains along the valley side after the glacier ice has melted, providing a valuable indicator of the relative height of the ice margin. Some kames are continuous features stretching along much of a valley side, and these are generally described as kame terraces. Other kames are short, discontinuous and often conical in shape. Kettle holes are often dotted along kames, creating a distinctive hummocky kame-and-kettle topography.

OUTWASH PLAINS AND PRO-GLACIAL LAKES

An *outwash plain* is an area in front of a retreating glacier or ice sheet that is covered by fluvioglacial deposits laid down by a meltwater stream draining from the ice (Figure 15.21). Most outwash plains are wide and extensive, but some (called valley trains) are confined within a narrow valley.

Meltwater draining from the terminal moraine of a melting glacier is often ponded behind an earlier terminal moraine or other glacial barrier down-valley to create a *pro-glacial lake*. Fluvioglacial sediment is deposited in the lake, including large quantities of fine glacial flour and suspended sediment. Lake Arsine, a pro-glacial lake in the French Alps, first appeared in about 1960, and by 1985 it was up to 40 m deep and covered an area of 0.06 km^2. The lake bed had risen by 8 m because of deposition, and the height of the barrier had fallen by 10 m because of the melting of dead ice trapped in it. This greatly increased the risk of overflow and flooding downstream.

Pro-glacial lake beds are preserved long after ice has finally left the area, as flat and often relatively featureless plains that can occupy the full width of a valley and extend several kilometres downstream. Post-glacial rivers generally meander across this inherited floodplain, reworking the fluvioglacial sediment, which was itself reworked by the meltwater stream from initially glacier-scoured material.

BOX 15.28 MELTWATER FLOOD IN CENTRAL CANADA

Geologists have reported evidence of a great flood in North America more than 8,000 years ago, when two vast freshwater lakes burst in central Canada. The lakes formed at the end of the last ice age, when meltwater from the great North American ice sheet was trapped by ice in Hudson Bay to the north. Eventually, more than 2,000 years after the thaw began, the ice gave way and floodwater roared through the Hudson Strait and into the North Atlantic. An estimated 200,000 km^3 of water (three times the amount of water in the North Sea today) rushed into the ocean, raising sea level worldwide by 30 cm. The flood of fresh water disrupted the ocean system and may have caused a drop in temperature in Europe of around 3 °C.

MELTWATER CHANNELS

When a large valley glacier or ice cap melts, particularly if this happens quickly because of rapid climate change, it produces vast amounts of meltwater, which flows down-valley. The high-discharge meltwater flows, with a high particulate sediment load, can cause extensive erosion, and this is reflected in the large overflow channels left behind after the melting has finished.

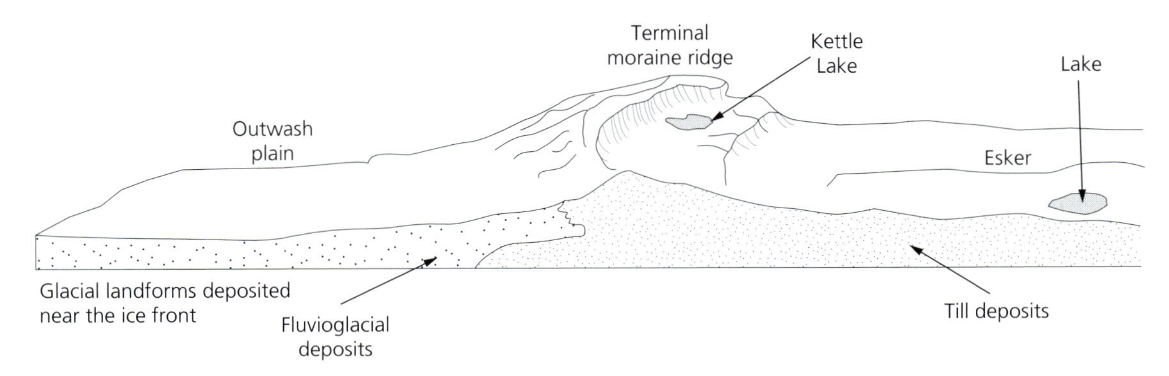

Figure 15.21 *Glacial outwash plain. A typical assemblage of landforms near the ice front includes an outwash plain, the terminal moraine and ice-contact features behind the moraine.*

Source: Figure 9.5 in Buckle, C. (1978) Landforms in Africa. Longman, Harlow.

BOX 15.29 MELTWATER FLOOD TRIGGERED BY VOLCANIC ERUPTION, ICELAND, 1996

One of the largest volcanic eruptions on Iceland during the twentieth century occurred in October 1996 beneath the Vatnajökull Glacier. While the volcano itself posed no real threat to people and property, it gave rise to a powerful flood which presented a significant risk to the south-east area of the island. Barbarbunga, one of two volcanoes that lie beneath Vatnajökull, started to emit magma on 29 September, and an eruption occurred the next day. This volcanic activity melted ice, and one initial response of the glacier was a significant reduction in ice thickness in the vicinity of the activity. By 2 October, the activity had melted through the ice cover, spreading ash and gas over a wide area. Activity continued until 14 October. Throughout, the Icelandic authorities were anxious about the risk of a huge meltwater flood, and steps were taken to control flooding and minimise likely flood damage downstream. Few people were living in the area, so a mass evacuation was not necessary. Scientists monitored water flows at the snout of the glacier, hoping to predict when the flood would happen. It eventually started on 5 November, over four weeks after the initial eruption. Water rushed out from the front of the glacier at a rate of around 45 $m^3 s^{-1}$, carrying with it up to 100 million tons of volcanic material, which was deposited in the sea. The flood caused an estimated US$15 million worth of damage to power lines, communication systems, bridges and other structures.

BOX 15.30 BRAIDED CHANNELS

When they first form, outwash plains are dominated by braided channels (see p. 372), which flow around a series of islands and bars within the channel. These islands tend to be covered by relatively fine-grained sediment, and the active channels generally have well-rounded coarse gravel beds. Braided channels change position regularly, and over time the plain builds up but preserves evidence of its evolution in the sub-surface stratigraphy. A typical mature outwash plain, long after ice has left the area, has a relatively flat surface.

Some overflow channels survive as dry valleys, which are often relatively flat-floored and steep-sided, shaped like large sweeping meanders, but with no present-day river. Many overflow channels do contain a river, but one adjusted to the much smaller discharges produced by today's climate. These misfit streams (see p. 382), with relatively narrow channels and small meanders, flow within much wider valleys with large meanders, which are the meltwater channels.

WEBSITE

Links to relevant websites, a comprehensive bibliography, tools for teaching and learning, and downloadable images relevant to this chapter can be found at the website specially designed to accompany this book at
http://www.park-environment.com

SUMMARY

Cold and ice dominate many environments, and glacial and periglacial processes have left strong impressions on many landscapes – particularly in high latitudes. This chapter focuses on the processes and products of glacial environments, and explores the relevance of these to people. It opens with an introduction to cold environments, which outlines the nature, causes and distribution of cold climates, and some characteristic features. The Arctic and Antarctica represent the world's largest expanses of ice today, and we examined how these environments were formed, why they matter, what threats they face, and what is being done to preserve them. The theme of glaciation is a major one within the chapter, and the nature and causes of ice ages and of glaciation were outlined. Next we considered how glaciers form and how they move, before examining the processes and products of glacial erosion and deposition. Meltwater processes, associated with the melting of glaciers and ice sheets, also create distinctive landforms and pose serious flood hazards for people. Many glaciers are thinning and shrinking in response to global warming, and this trend is set to continue a long way into the future.

FURTHER READING

Ballantyne, C.K. and C. Harris (1994) *The periglaciation of Great Britain.* Cambridge University Press, Cambridge. A detailed overview of frost action in Britain during the Quaternary, and the landforms and deposits that resulted from this.

Benn, D.I. and D.J.A. Evans (1997) *Glaciers and glaciation.* Edward Arnold, London. Outlines the nature, origin and behaviour of glacier systems, and the geological and geomorphological evidence for their former existence.

Bennett, M.R. and N.F. Glasser (1996) *Glacial geology: ice sheets and landforms.* John Wiley & Sons, London. Useful examination of methods and evidence for reconstructing glacial landforms and sediments.

Dawson, A. (1991) *Ice age Earth.* Routledge, London. A clear introduction to the glacial environment and its processes and landforms.

Ehlers, J. (1996) *Quaternary and glacial geology.* John Wiley & Sons, London. Explains the principles and practice of reconstructing Quaternary environmental changes in Europe and North America.

Evans, D.J.A. (ed.) (1994) *Cold climate landforms.* John Wiley & Sons, London. A collection of important papers published in languages other than English between 1909 and 1992.

Evans, I.S. (1996) *The geography of glaciers.* Edward Arnold, London. A comprehensive examination of the past and present distribution of glaciers, which emphasises spatial differences and interactions, environmental relationships and effects on the landscape.

French, H. (1996) *The periglacial environment.* Longman, Harlow. Outlines the geomorphological processes and landforms typical of high-latitude environments and shows how periglacial features can be identified in what are now temperate areas (including North America and Europe).

Hanson, J. and J. Gordon (1998) *Antarctic environments and resources.* Longman, Harlow. Charts the history of human use of the area, the development of conflicts between different resource users (particularly fishing, tourism and science) and the emerging framework for future environmental management developed under the Antarctic Treaty.

Hooke, R. LeB. (1998) *Principles of glacier mechanics.* Prentice Hall, Hemel Hempstead. Clear explanations of the physical processes by which glacial landforms are developed.

Lowe, J.L. and M.J.C. Walker (1996) *Reconstructing Quaternary environments.* Longman, Harlow. Examines the different types of evidence used to establish the scale and history of environmental changes during the Quaternary.

Roberts, N. (1998) *The Holocene: an environmental history.* Blackwell, Oxford. Reviews recent advances in understanding the causes, patterns and consequences of environmental change during the Holocene.

Simpson-Housley, P. (1992) *Antarctica: exploration, perception and metaphor.* Routledge, London. Traces the ways that images and perceptions of Antarctica have changed over time.

Slaymaker, O. and T. Spencer (1998) *Physical geography and global environmental change.* Longman, Harlow. Examines the principles and concepts of environmental systems, and the implications of these for society.

Williams, M.A.J., D.L. Dunkerley, P. De Deckker, A.P. Kershaw and T. Stokes (1993) *Quaternary environments.* Edward Arnold, London. A study of global environments during the Quaternary period, with an emphasis on geological, biological and hydrological processes.

Wilson, R.C.L., S.A. Drury and J.L. Chapman (1999) *The Great Ice Age.* Routledge, London. Wide-ranging examination of the natural climate changes that have occurred over the past 2.6 million years, along with the environmental changes associated with them.

Chapter 16

Oceans and Coasts

LEARNING OBJECTIVES

When you have finished studying this chapter, you should be able to

- *Appreciate the importance of the oceans as a source of resources and an important environmental system that interacts with the other major environmental systems on Earth*
- *Describe the main structures of the ocean floor*
- *Summarise the properties of ocean water and explain the relevance of salinity*
- *Explain how waves and tides are developed, and why and how ocean currents operate*
- *Define the El Niño phenomenon and explain how it affects many other environmental systems*
- *Outline the main ways in which oceans provide resources (fishing, minerals, energy) and comment on the problems of using these resources in sustainable ways*
- *List the most common forms of ocean pollution and describe the problems they give rise to*
- *Describe the main processes of coastal erosion and deposition and their impacts on landscape*
- *Outline the main environmental problems associated with estuaries*
- *Appreciate the need for coastal management, and the main approaches used*
- *Understand the relevance of sea-level change, how it is caused and how it can be coped with*

The oceans are the vast expanses of salt water that cover more than 70 per cent of the Earth's surface. Throughout history, people have relied heavily on the oceans and seas as a means of communication and as a source of food and raw materials. Coastal and island communities have often looked to the sea to offer protection from invasion or attack by their enemies, yet this same sea has regularly attacked them itself with storms and floods.

Oceans and seas are also interesting environmental systems in their own right, and in the first section of this chapter we look at the structure and operation of ocean systems. Coasts, which represent the interface between land and sea, are among the world's most dynamic environments because they are the scene of endless interactions between terrestrial and maritime processes. In the final sections of this chapter, we explore the coastal system and how it functions.

THE OCEANS

Oceanography is the study of the oceans, their origin, composition, history and ecology. It examines water movements (via currents, waves and tides) and the chemical and physical properties of seawater. As well as studying ocean dynamics, oceanography also seeks to describe and account for the origin and topography of the ocean floor. This includes the role of plate tectonics in forming ocean trenches and ocean ridges (see p. 149), and the role of sea-level change (see p. 511) in creating coastlines and continental shelves.

IMPORTANCE OF THE OCEANS

Largely because of their vast size (see Table 5.2), most of the major oceans remain relatively unexplored. In many

BOX 16.1 OCEANS AND ENVIRONMENTAL SYSTEMS

Oceans and seas cover roughly two-thirds of the Earth's surface (see p. 125), and they interact with or are critically important components of most of the important environmental systems. They exert a strong influence on climate (p. 327), play a significant role in the major biogeochemical cycles – particularly the carbon, nitrogen, phosphorus and sulphur cycles (p. 86) – and the global water cycle (p. 352). They are also important ecosystems (p. 551) that offer habitats for wildlife. As repositories for the sediment washed down river systems (p. 364), oceans and seas are also important depositional environments, and they play key roles in the sedimentary parts of the rock cycle (p. 202).

ways the oceans – particularly their remote parts, far from direct human interference – represent one of the world's last remaining wilderness resources.

PROSPECTS AND UNCERTAINTIES

Scientists believe that the oceans are likely to prove immense storehouses of valuable natural resources (including sea-bed minerals), if cost-effective technologies can be developed to explore and exploit this potential. Some have called for some of the vast amount of resources devoted to space research to be redirected towards ocean research, so great are the potential rewards. There are still many areas of scientific uncertainty concerning some potentially very important aspects of ocean systems (Box 16.2).

BOX 16.2 AREAS OF UNCERTAINTY CONCERNING THE OCEANS

Among the key areas of uncertainty, where greater scientific understanding is essential, are:

- the role of the oceans in creating some important environmental hazards (such as tsunamis; see p. 168);
- the role of the oceans in causing some important environmental changes (such as the El Niño phenomenon; see p. 482);
- the dynamics of ocean ecosystems (particularly fish populations: see p. 486);
- the role of ocean stores and processes in the major biogeochemical cycles (see p. 86), particularly as a source of sulphur and its compounds through submarine volcanic activity and the production of dimethyl sulphide (DMS) from plankton, as a sink for carbon dioxide and for phosphorus;
- the significance of the oceans as thermostats and regulators in the Gaia hypothesis (p. 121);
- the influence of oceans on regional climate (p. 328);
- the likely impacts of global warming (p. 512) on sea level and ocean and coastal systems.

PEOPLE AND PRESSURE

According to figures presented at the 1992 Rio Earth Summit, 3,000 million people – 60 per cent of the world's population – now live in coastal areas, within 100 km of the shoreline. Many large cities have developed along the coastal zone, and some of them are growing very rapidly. By the year 2000, global population is expected to reach 8,000 million, 70 per cent (5,600 million) of whom will live in coastal regions.

As population levels and densities in the coastal zone continue to rise, pressure increases further on coastal ecosystems. These are among the most productive ecosystems in the world (see p. 564), and they play a significant role in many important food chains. At least 2,000 million people already depend entirely on coastal waters for their food supplies, coastal waters supply 95 per cent of total world fish catches, and over half of the intake of animal protein in tropical coastal developing countries comes directly or indirectly from the sea.

EXPLOITATION AND DAMAGE

The natural resources of the oceans have long been exploited, often in very sustainable ways that do no long-term damage to the resource base. Increasingly, however, particularly in heavily populated developed countries, ocean resource use is becoming unsustainable largely through over-use or because of conflicts between incompatible uses. Key pressures include the development of the offshore oil and gas industry, changes in the marine trading system, over-fishing and waste dumping in the oceans (Box 16.3). China, for example, relies heavily on the adjacent oceans for fishing, shipping, offshore oil and gas, marine

BOX 16.3 OCEANS AS DUMPS

Despite the great significance of the oceans to many of the world's most important environmental systems, people have traditionally treated them as dumping grounds for all sorts of unwanted material. All rivers ultimately flow into the sea, discharging material from towns and cities, farms and forests – including refuse, bacteria, pesticides, oil products, heavy metals, PCBs, nutrients, eroded soil, manure and artificial fertilisers – into the ocean stores.

The seas and oceans have long been looked upon as limitless dumps, largely because of their great size (coupled with the 'out of sight, out of mind' mentality of many people). Yet this is simply not true, and many oceans, seas and coastlines around the world are now facing serious problems of pollution, instability and environmental damage. Many coastal areas have a long history of local pollution, but many problems have now become regional and threaten large areas. Damage can spread through oceans and seas, like most environmental systems, because of the mobility of the water in them. Damage can also be transferred to other important environmental systems – particularly the biosphere, atmosphere and hydrosphere – with which the oceans and seas interact.

minerals, coastal land for reclamation, and marine conservation.

The ocean system is now being pushed to its limit in many places. It is likely that some critical environmental thresholds (see p. 83) are being approached or exceeded. If and when this happens, complex responses and chain-reaction adjustments are likely to occur, possibly causing irreversible change and irreparable damage and instability.

THE EARTH SUMMIT AND THE OCEANS

The problem of managing the world's oceans in sustainable ways was high on the agenda at the 1992 United Nations Conference on Environment and Development at Rio (see p. 13). The debate about living marine resources was one of the most controversial debates in the whole conference, and conflicts about conservation and management of high-seas fisheries proved impossible to resolve. As a compromise, Agenda 21 recommended a new international conference, under the auspices of the United Nations, to seek improvement in the provisions of the 1982 Law of the Sea Convention, which deals with stocks of highly migratory fish.

Most of the delegates at Rio recognised the merits of effective international ocean management, even if they could not always agree about the best ways of achieving it. Agenda 21 contains some useful policy proposals designed to tackle the problems of ocean pollution, over-fishing and degradation and to maintain the life-support capacity of the world's oceans (Table 16.1).

Table 16.1 Agenda 21 proposals concerning oceans

- Protect and check environmental damage to coastal areas nationally and internationally
- Polluters should pay for the damage they cause, and those using cleaner methods should be rewarded
- Protect marine life by controlling what materials may be removed from ships at sea and by banning removal of hazardous waste
- Nations should share new technologies
- Set limits on how many fish may be caught
- Encourage fishing by skilled local people
- Stop fishing for species at risk until they recover to normal numbers
- Ban destructive fishing practices (dynamiting, poisoning and others) and develop new practices to replace them

STRUCTURE OF THE OCEANS

GEOGRAPHY

Although there is in effect one world ocean that spans the globe, it is convenient to sub-divide it into three major ocean basins – the Atlantic Ocean, the Pacific Ocean and the Indian Ocean (see Figure 5.1). The boundaries of these basins are defined by the continental land masses or by ocean ridges or currents. In the southern hemisphere, the three oceans converge below 40° S in the Antarctic Ocean, which is defined by the Antarctic Circumpolar Current or West Wind Drift. In the northern hemisphere, the Arctic Ocean is nearly circular and is mostly defined by the polar coastlines of North America, Europe and Asia.

OCEAN FLOOR TOPOGRAPHY

The average depth of the ocean floor is more than 3,650 m below sea level, and the total volume of the world's oceans is estimated at around 1,370 million km^3 (see Table 12.2).

Not only is more of the Earth's surface covered by water than by land, but the ocean floor overall lies further below sea level than the land area overall rises above it (see Figure 5.2). But just as topography varies a great deal from place to place across the continents, so also the topography of the ocean floor is far from uniform.

The following topographic units can be identified in ocean basins (Figure 16.1):

- *Continental shelf*: this is the gently sloping sea bed that extends out to sea beyond the shore, sloping downwards at a gradient of between about 1:500 and 1:1,000. The continental shelf extends out to sea an average distance of about 75 km, although it varies in width from nearly zero to 1,500 km. The ocean is generally about 200 m deep at its outer edge. Although the shelf is covered by shallow sea, it is effectively part

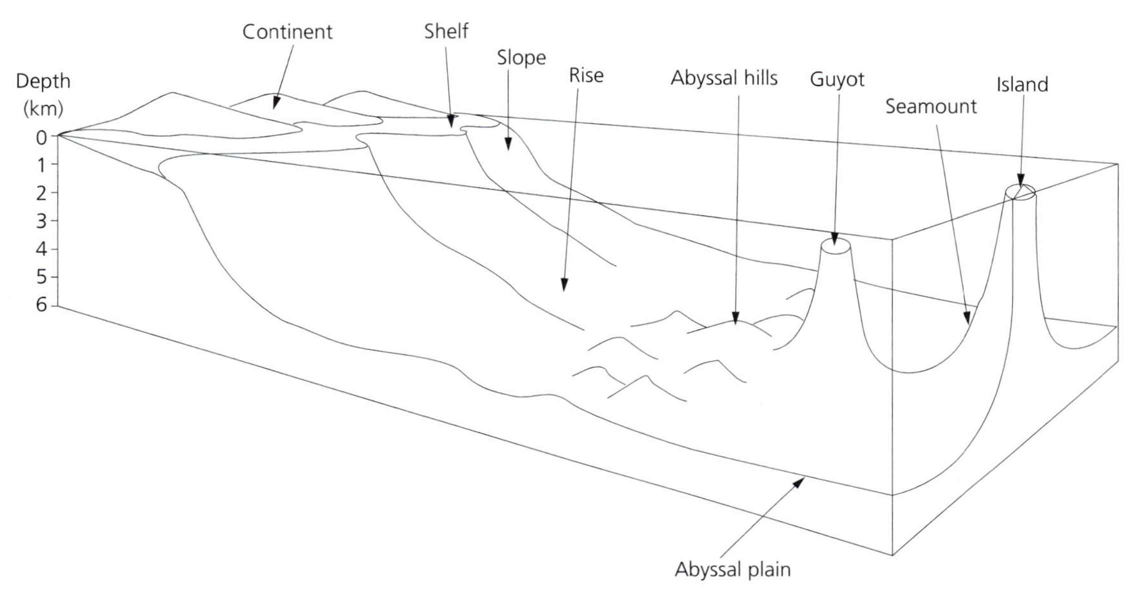

Figure 16.1 *Topographic features of the oceans. Beyond the continental shelf the ocean floor is far from flat or featureless. The seamount and guyot are greatly exaggerated in size.*

Source: Figure 16.3 in Doerr, A.H. (1990) Fundamentals of physical geography. Wm. C. Brown Publishers, Dubuque, Iowa.

BOX 16.4 TOPOGRAPHY OF THE OCEAN FLOOR

For a long time the ocean floor was thought to be flat, but it is now known to contain volcanic mountains, long high ridges and deep trenches, which are all associated with plate tectonics. The volcanic mountains are found along the major destructive boundaries between crustal plates (see Box 6.5), sometimes as island arcs or archipelagos. The Pacific Ring of Fire (see p. 150) is composed of such mountains. Submarine mountains can be as high as their terrestrial counterparts (see Box 5.2) – one, off the Tonga Trench, is 8,687 m tall, and its summit lies only 365 m below sea level.

Ocean trenches are created by the subduction of one plate beneath another at a destructive plate boundary (see Box 6.5). The deepest known ocean trench is the Marianas Trench in the Pacific Ocean, which is 11,033 m below sea level (Table 16.2) – more than 2,000 m deeper than Mount Everest is high!

Mid-ocean ridges, extensive mountain chains that can tower thousands of metres above the surrounding ocean floor, are created by the rise of magma to the ocean surface along submarine constructive plate boundaries (see Box 6.3). The ridges are part of a continuous system that runs for 60,000 km through all the oceans. For example, the Mid-Atlantic Ridge (Figure 16.2) stretches from the Norwegian Sea through the volcanic islands of Iceland and the Azores to the South Atlantic, where it lies roughly halfway between Africa and South America. It continues into the Indian Ocean, cuts between Australia and Antarctica and extends into the eastern South Pacific. The East Pacific Rise extends north to the Gulf of California and includes the volcanic islands of Easter Island and the Galapagos.

of the continent and would be uncovered by a large-scale drop in sea level (during an ice age, for example).

- *Continental slope*: the outer edge of the continental shelf is often marked by a sharp increase in gradient, at a depth of about 200 m. Beyond this lies the continental slope, which extends at an average gradient of about 1:700 down to a depth of about 2,000 m.
- *Continental rise*: this marks the outer edge of the continental slope, where it comes into contact with the relatively flat ocean floor. It represents the real boundary between the continents and the oceans. Sediment washed down from the shelf and slope accumulates in the rise, which is considered to be part of the ocean bottom. The rise extends about 600 km from the base of the continental slope to the deep-ocean floor.
- *Ocean floor*: this zone, also known as the abyss, is the main ocean bed beyond the continental shelf, slope and rise. Most of the area occupied by the major oceans (see Table 5.2) is accounted for by ocean floor topography, which is often far from flat (Box 16.4).

The sea bed at depths greater than about 2,000 m is largely covered by fine-textured sediment (ooze), which is made up mostly of organic matter. Across the oceans as a whole the average thickness of sediment is about 0.5 km, but it can be up to 7 km thick (for example in the Argentine basin in the South Atlantic). New submarine surfaces, such as newly formed parts of the mid-ocean ridges, are largely sediment-free.

Variations from place to place in ocean topography are important because they determine water depth. This in turn exerts a strong influence over conditions in the ocean water – particularly in terms of temperature, availability of sunlight and nutrients, and rate and patterns of vertical mixing in the water body. Plant and animal life in the oceans responds to conditions in the water.

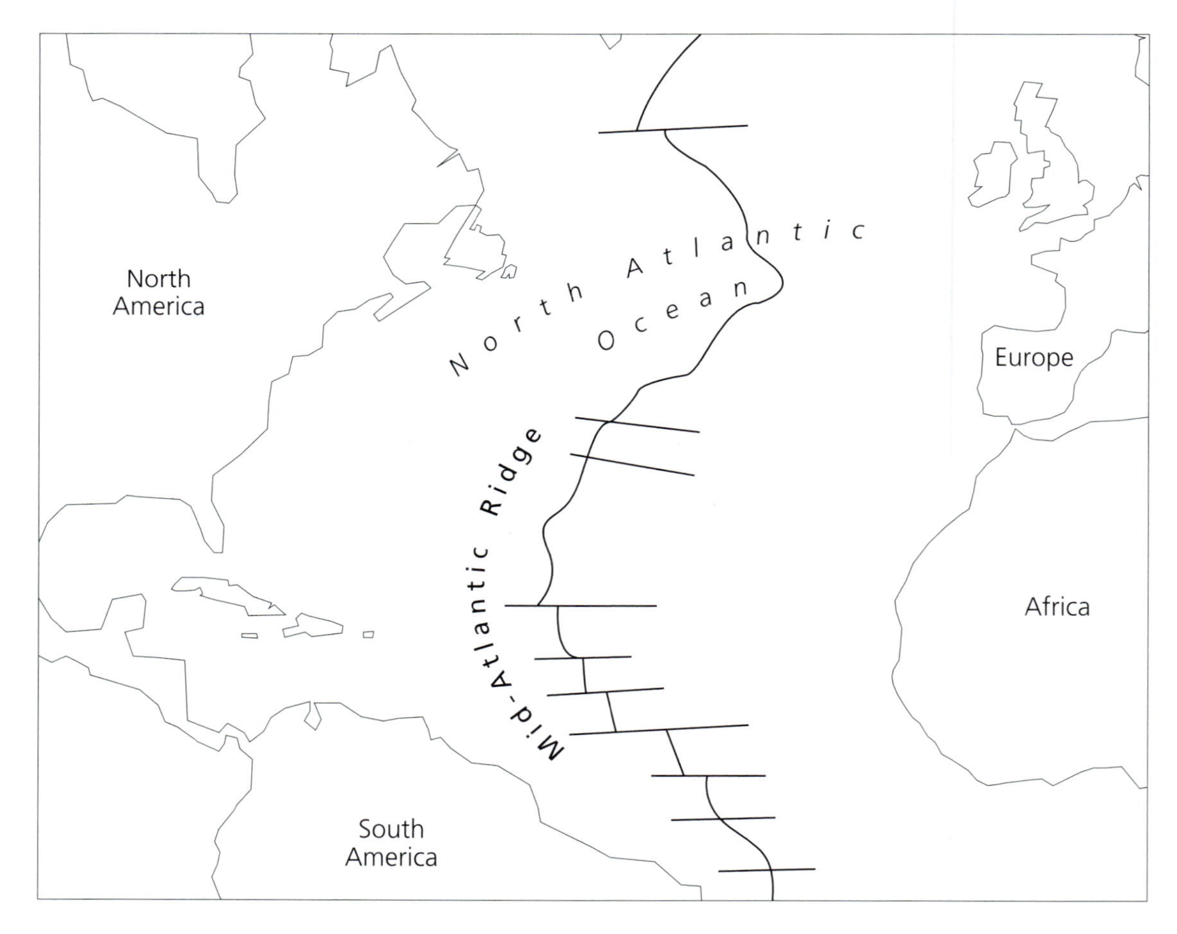

Figure 16.2 *The Mid-Atlantic Ridge. The Mid-Atlantic Ridge is one of the important sea-floor spreading zones, which drive the lateral movement of crustal plates.*

Source: Figure 16.2 in Doerr, A.H. (1990) Fundamentals of physical geography. Wm. C. Brown Publishers, Dubuque, Iowa.

Table 16.2 The world's five deepest ocean trenches

Trench	Location	Depth (m)
Marianas	Pacific Ocean	11,033
Puerto Rico	Atlantic Ocean	8,381
Diamantina	Indian Ocean	8,047
Cayman	Caribbean Sea	7,010
Kei	Malay Sea	6,505

BOX 16.5 ORGANISMS IN THE OCEANS

Two main groups of organisms live in the oceans:

1. *Benthic organisms*: these inhabit the sea floor and account for nearly 98 per cent of all marine animals.
2. *Pelagic species*: these can move vertically upwards and downwards within the ocean water body between the surface and the sea floor. Some pelagic species – particularly the plankton, comprising zooplankton (small animals) and phytoplankton (single-celled plants) – have no motor power of their own, and they drift around with the tides and ocean currents. Other pelagic species, such as fish and mammals, can swim about freely in search of food.

BOX 16.6 ECOLOGICAL ZONES IN THE OCEANS

Four ecological zones can be identified in the oceans:

1. *Neritic zone*: this corresponds to the level above the continental shelf down to a depth of about 200 m.
2. *Bathyal zone*: this is the upper part of the ocean, above the continental slope, at a depth of between 200 and 2,000 m.
3. *Abyssal zone*: this zone includes the continental rise and the ocean floor, and it accounts for three-quarters of the deep ocean floors. Conditions are cold (temperatures remain steady at about 4 °C) and dark in this zone, which lies between 2,000 and 6,000 m below the ocean surface.
4. *Hadal zone*: this is the deepest part of the ocean, confined to the lower levels of ocean trenches.

PROPERTIES OF OCEAN WATER

The two most important properties of ocean water, particularly from the point of view of ocean ecosystems and suitability for human use, are salinity and temperature. Waves are also important from the point of view of coastal erosion and deposition (see p. 498).

SALINITY

Seawater is saline in that it contains dissolved salts, which are derived from weathering and erosion of rocks on the continents, as well as material derived from natural forest fires, volcanic eruptions and air pollution. Salinity is expressed as concentration of dissolved salts in parts per thousand parts of water, or parts per million (ppm). Seawater in the open ocean away from obvious sources of pollution has an average salinity of about 35 parts per thousand (35,000 ppm), varying between about 34 and 36 parts per thousand. Runoff in rivers normally has a salinity of more or less zero, whereas salinity in the Great Salt Lake in the USA is more than 150 parts per thousand (150,000 ppm) because of the high evaporation loss in the hot, dry climate. Some comparative salinity values are given in Table 13.4. The salinity of the oceans varies slightly around the world, being generally highest in the tropics (Figure 16.3).

The bulk of the dissolved material in seawater is sodium, with lesser amounts of magnesium and calcium, and trace amounts of potassium (see Table 13.4). The chemical composition of unpolluted ocean water varies relatively little from place to place, but there are some broad-scale variations in the total concentration of dissolved material (that is, the salinity).

Some or most of the dissolved salts can be removed from seawater by desalination (see p. 405), but the process tends to be costly, requires a great deal of energy and cannot remove all of the dissolved material. Desalination technologies are developing fast, however, and as well as offering the promise of increased freshwater supplies they are opening up the prospect of commercial extraction of commodities such as halite, magnesium and bromine from seawater.

TEMPERATURE

The temperature of ocean water varies with depth, through the seasons and from place to place. Surface temperatures

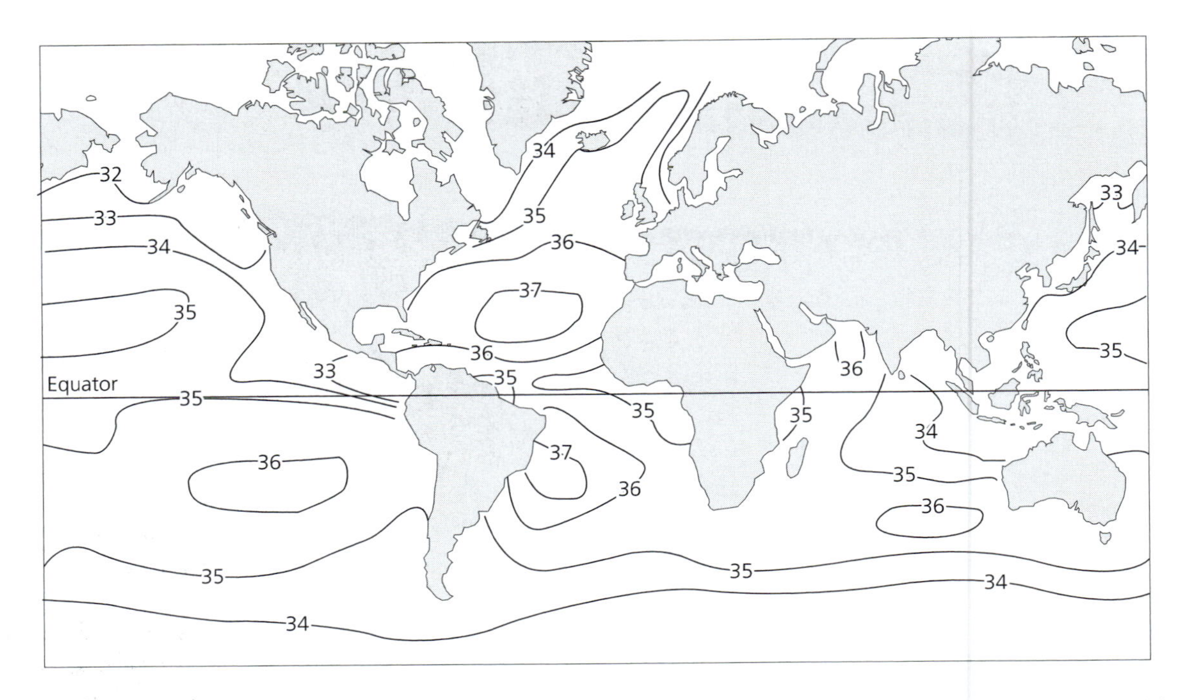

Figure 16.3 *Variations in ocean salinity around the world. The salinity of the ocean surface varies a great deal from place to place, being highest around the tropics and generally declining towards high latitudes.*
Source: Figure 17.6 in Dury, G.H. (1981) Environmental systems. Heinemann, London.

BOX 16.7 VARIATIONS IN OCEAN TEMPERATURE WITH DEPTH

There is a pronounced vertical change of temperature in the oceans. Water in the upper 100 m or so usually has temperatures similar to those at the surface, although the uppermost layer, which is in contact with the atmosphere above, heats and cools more quickly and thus displays more short-term variability. Below that upper zone is a region known as the *thermocline*, which extends from about 100 m down to about 1,000 m, and the temperature drops sharply to about 5 °C. At even greater depths, below about 1,000 m, the temperature falls to just above freezing.

tend to decrease with increasing latitude, from an average of about 26 °C in the tropics to −1.4 °C (the freezing point of seawater) at the North Pole. Temperatures in the oceans vary much less from season to season than they do on land (see p. 275).

WAVES

Waves are rhythmic movements on the surface of the ocean, produced by wind as it blows over the water. Ocean waves are mechanical because they transmit energy through the water without a mass movement of the water itself (Figure 16.4). Periodic oscillations (indicated by the wave peaks and troughs) reflect a displacement of energy along the path in which the energy moves. Molecules of water oscillate around their own equilibrium position, and only the energy moves continuously in one direction.

Waves do not move sideways, they move up and down. Because of this, in the open ocean, objects floating on the surface of the sea bob up and down with the waves but are carried along by currents or blown along by the wind. Thus, for example, boats with engines that fail and without sails can float helplessly for long periods of time. Similarly, objects such as plastic bottles and other material that floats can remain in the same position, even in a sea with high waves; they are moved from place to place only by currents and the wind.

Wave characteristics are strongly influenced by wind. Wave direction in the open ocean depends on wind direction, while sea-bed topography can significantly affect wave direction in shallower water.

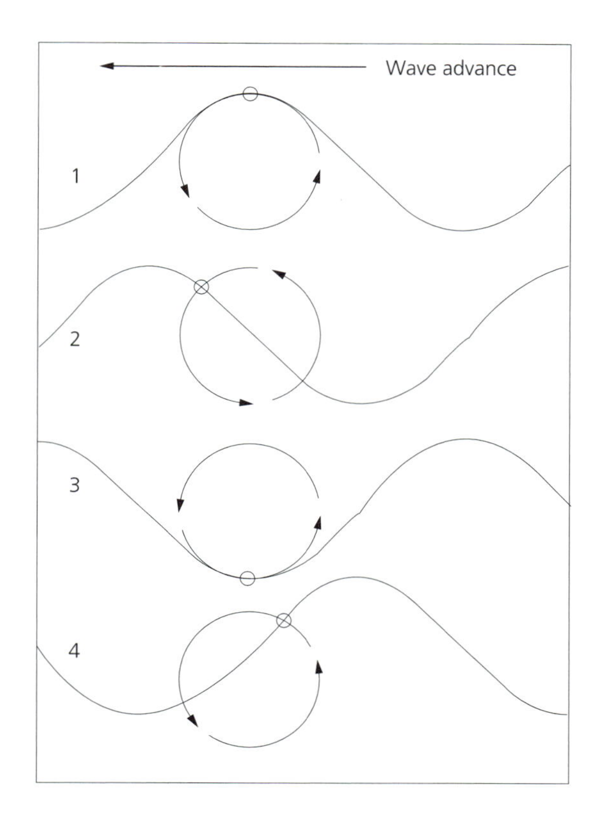

Figure 16.4 *Orbital movement of water within a wave. Although waves are blown across the surface of the open ocean by wind, the water in them actually follows an orbital path, bobbing up and down without necessarily any strong lateral movement.*

Source: Figure 12.8 in Dury, G.H. (1981) Environmental systems. Heinemann, London.

BOX 16.8 WAVE SIZE

Wave size depends mainly on:

- wind speed (strong winds tend to produce large waves);
- the length of time during which the wind blows (short gusts tend to produce small waves);
- fetch, which is the distance over water that the wind blows (a long fetch tends to produce long waves);
- interactions with waves generated elsewhere.

Wave size can be expressed in a number of ways (Figure 16.5), including:

- *wave height*: the distance between the crest and the trough of the waves;
- *wavelength*: the distance between adjacent wave crests;
- *wave frequency*: the number of waves per second (which reflects wavelength).

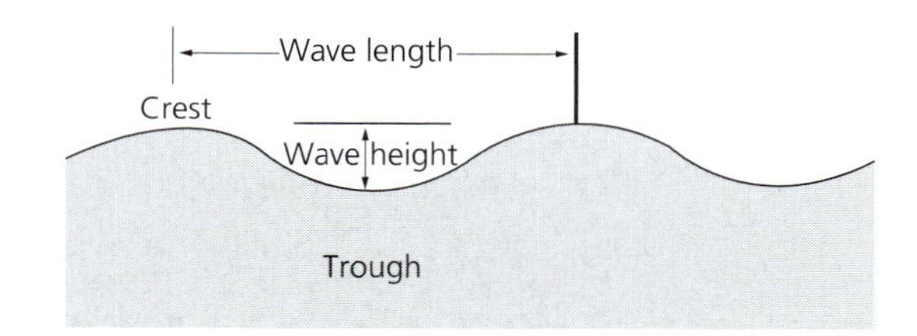

Figure 16.5 *Wave height and wavelength. Wave geometry is defined between adjacent wave crests: wave height is the vertical distance between the crest (high point) and trough (low point) of the wave, and wave length is the horizontal distance between two succeeding wave crests or troughs.*

Source: Figure 16.4 in Doerr, A.H. (1990) Fundamentals of physical geography. Wm. C. Brown Publishers, Dubuque, Iowa.

Most waves in the open ocean have heights in the order of several metres, but the highest instrumentally measured ocean wave is 26 m. Oceanographers calculate that the highest ocean wave possible is about 34 m. Tsunamis (see Box 6.13), huge seismic waves caused by submarine earthquakes or volcanic activity, have been recorded as high as 67 m.

OCEAN CURRENTS

A current is a distinct flow of a body of water that moves in a definite direction within the oceans. Major current systems flow through all the oceans, redistributing ocean water (and its properties, particularly temperature) across the Earth's surface, transferring heat from low to high latitudes and modifying regional climate (see p. 335). There are similarities with the way in which major wind circulation patterns (p. 286) redistribute air and air masses (p. 308) and give rise to weather systems.

TYPES OF CURRENT

Another similarity between ocean currents and global wind circulation is that both are three-dimensional, and flows at different heights in the atmosphere and different depths in the oceans can vary a great deal in speed and direction. Surface ocean currents are mainly caused by atmospheric pressure systems and winds or by density differences in the water. Sub-surface currents flow well beneath the surface,

BOX 16.9 TYPES OF CURRENT

There are three basic types of oceanic current:

1. *drift currents,* which are broad and slow-moving;
2. *stream currents,* which are narrow and fast-moving;
3. *upwelling currents,* which bring cold, nutrient-rich water from the ocean bottom.

Ocean currents can also be classified into warm and cold currents, relative to the temperature of the surrounding water:

- Warm currents generally flow from lower (warmer) to higher (cooler) latitudes.
- Cold currents generally flow from higher (cooler) to lower (warmer) latitudes.

often in the opposite direction to the surface currents, to compensate for the surface flows – in exactly the same way as convergent and divergent air flows compensate for each other in the atmosphere (see p. 285).

Areas where warm and cold currents meet tend to have regular foggy conditions, as the overlying warm and cold air come into contact with each other. They also tend to have high biological productivity, because plankton growth is encouraged by the mixing of warm and cold waters. Some of the world's most productive ocean fishing zones are located where warm and cold currents converge.

GYRES

Surface currents in the ocean are dominated by large *gyres* (meaning literally 'swirls') (Figure 16.6). These are large-scale permanent currents that are driven mainly by prevailing winds, but as water is moved to or from the equator its motion is influenced by the rotation of the Earth

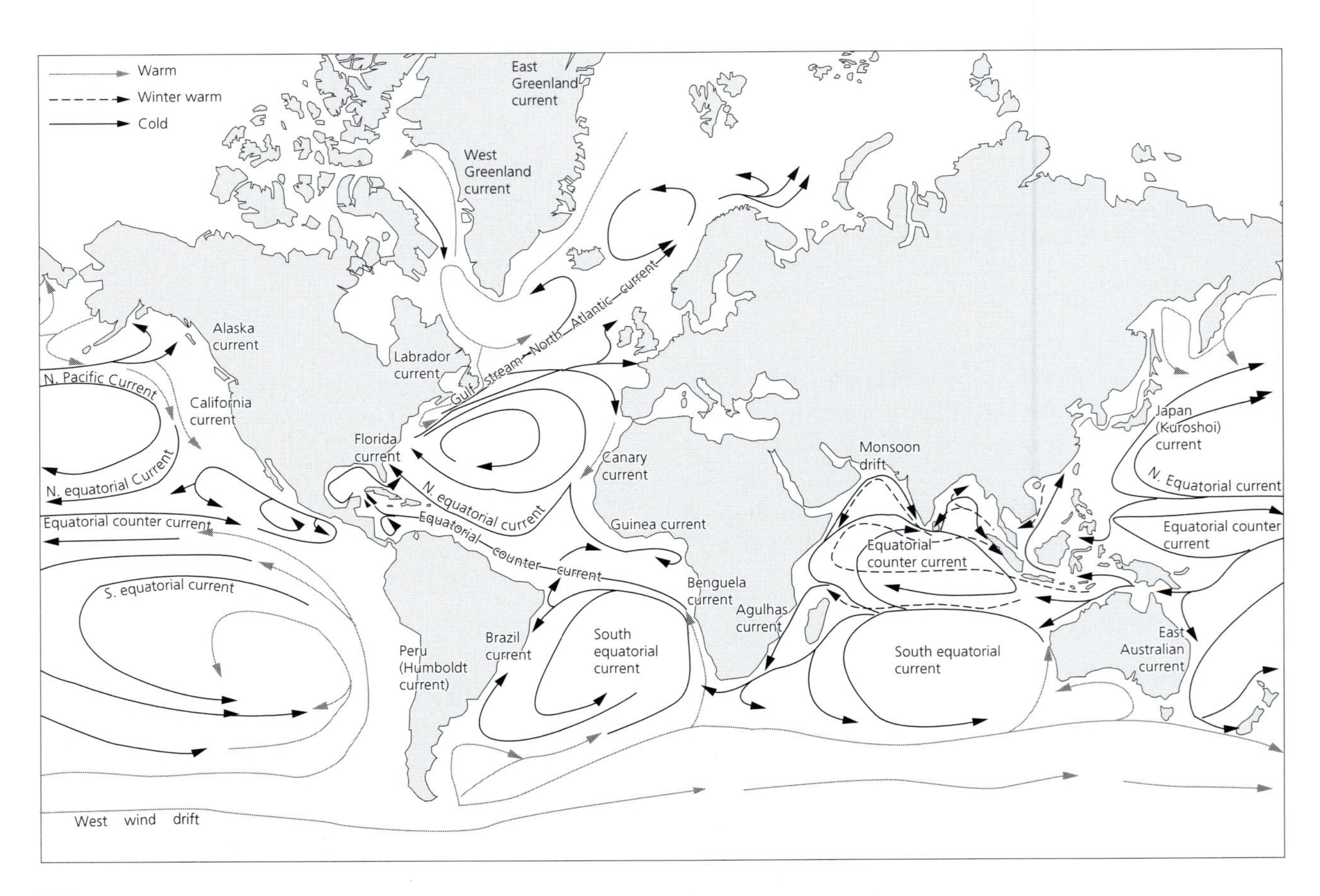

Figure 16.6 *Distribution of the main surface ocean currents. See text for explanation.*
Source: Figure 6.1 in Doerr, A.H. (1990) Fundamentals of physical geography. Wm. C. Brown Publishers, Dubuque, Iowa.

BOX 16.10 IMPORTANCE OF GYRES

Gyres play a vitally important role in redistributing heat across the Earth's surface (see p. 339). Warm equatorial water is driven westwards by easterly winds around the equator (p. 286) to create the Equatorial Current, and it flows towards the poles along the eastern coasts of the continents in both hemispheres, warming them. Relatively cold water from high latitudes is driven eastwards by westerly winds as it flows towards the equator, and it cools the western coasts of the continents as it passes them. The western margins of the continents are further cooled by the upwelling of cold water raised to the ocean surface by persistent trade winds.

This redistribution of heat by the gyres in turn affects surface temperatures across the continents and can affect patterns and processes of precipitation, particularly in terms of fog (see p. 294) and rainfall (p. 298).

BOX 16.11 UPWELLING

In some areas – such as the west coast of Mexico and along the coasts of Peru and Chile – prevailing winds blow offshore, and this forces ocean surface waters to flow offshore too. This sets in motion a three-dimensional pattern of water circulation in which the surface waters moving offshore are replaced by colder water drawn up from within the ocean.

This process of upwelling brings to the surface nutrient-rich water from up to 300 m below the surface, where decomposition of organic matter is greater than production because lack of light at these depths inhibits photosynthesis. Because of the significantly increased availability of nutrients, areas where upwelling occurs usually have high biological productivity. Many of them support fishing industries, of which the Peruvian anchovy industry has long been a good example.

(Coriolis force) in exactly the same way as wind patterns are deflected from the pressure gradient by the Coriolis force (see p. 281). This deflection gives rise to major currents, which follow roughly circular paths within the North and South Pacific, the North and South Atlantic, and the Indian Ocean. The gyres flow in a clockwise direction in the northern hemisphere and anticlockwise in the southern hemisphere, rotating around centres located in the sub-tropical high-pressure centres (see p. 311) about 30° north and south of the equator.

GULF STREAM AND NORTH ATLANTIC DRIFT

Perhaps the best known of the current systems is the Gulf Stream in the North Atlantic. The Gulf Stream is a narrow, fast-flowing stream current, which flows at speeds of up to 2.5 metres a second (9 km h^{-1}). It starts in the Gulf of Mexico and – driven by the westerly winds and deflected by the Coriolis force – it flows north-eastwards across the Atlantic and warms the climate of the east coast of the USA. The current splits around the latitude of Spain, and part of it flows south (heating the coast of south-west Europe and eventually rejoining the Equatorial Current). Part of the Gulf Stream continues to flow towards the north-east, as the North Atlantic Drift, and this brings relatively mild climatic conditions to Britain and north-west Europe.

Plate 16.1 *Waves breaking on arrival at the shallower water near the coast. The picture shows the Atlantic coast of south-west Spain.*
Photo: Chris Park.

Short-term variations in the climate of this whole region appear to be strongly influenced by shifts in the strength and location of the Gulf Stream. North-west Europe usually experiences fairly cool summers and cold winters when the Gulf Stream is relatively weak or moves further south than normal. Mild winters and particularly warm summers occur in north-west Europe when the Gulf Stream is strong or flows further north.

It is difficult to predict how global warming might affect the strength and location of the Gulf Stream, but this will doubtless have consequences for climate change across north-west Europe.

EL NIÑO

The main ocean currents usually follow fairly clearly defined and stable paths, and this helps to create relatively stable global climatic patterns. But because the currents are driven by forces that can themselves change through time (particularly atmospheric pressure systems and wind circulations), they can be subject to periodic or occasional shifts in their strength and/or direction. We noted above, for example, the possible impacts on weather in north-west Europe of shifts in the North Atlantic Drift.

CYCLES

The best example of periodic shifts in the ocean currents is the so-called El Niño, a change in the ocean current in the Pacific off the coasts of Peru and Ecuador that flows southwards. El Niño events normally occur every three to five years, lasting six to eighteen months, and peak around Christmas time, which is why Peruvian fishermen called the phenomenon the 'boy child' (which is El Niño in Spanish). Recent El Niño disturbances occurred in 1953, 1957–58, 1972–73, 1976, 1982–83 and 1997–98.

The 1997–98 El Niño was one of the strongest on record, developing more quickly and with the highest temperatures ever recorded. The episode developed rapidly throughout the central and eastern tropical Pacific Ocean in April–May 1997. During the second half of the year, it became more intense than the major El Niño of 1982–83, with sea-surface temperature anomalies across the central and eastern Pacific of 2–3 °C above normal. The 1982–83 event was estimated to have been responsible for 2,000 deaths and about US$13,000 million worth of damage worldwide.

CAUSES

Scientists have a fairly good understanding of what happens when the current changes occur, but the processes involved and the initial causes are not clearly understood. Some interesting correlations have been established between the timing of El Niño events and the timing of other environmental phenomena – including submarine seismic activity along the East Pacific Rise, one of the Earth's most rapidly spreading mid-ocean ridge systems – but associations do not prove causality.

In the years between ENSO events, the ITCZ remains in its more normal position much further north, the trade winds are stronger, the South Equatorial Current is stronger, and the ocean surface waters off South America remain relatively cool. The southward migration of the ITCZ appears to be cyclical in nature, although the exact cause is not known. Sometimes it is a response to atmospheric disturbances associated with volcanic eruptions in the equatorial zone, as happened after the 1982 eruption of El Chichón in Mexico.

> **BOX 16.12 CAUSES OF EL NIÑO**
>
> El Niño is linked to a reversal of the dominant equatorial wind pattern. The cycle of changes appears to start with the distribution of atmospheric pressure over the Indian and Pacific Oceans around South-east Asia. This causes the inter-tropical convergence zone (ITCZ) (see p. 287) to move further south than normal between December and February. The southward migration of the ITCZ is known as the Southern Oscillation, and the incidents that accompany El Niño current changes are usually referred to as El Niño–Southern Oscillation events (ENSO).
>
> The Southern Oscillation causes the southern hemisphere trade winds to become much weaker than normal, and they can even change from their predominantly south-easterly direction. As a result of the wind changes, the South Equatorial Current in the Pacific Ocean weakens or even reverses, and warm water starts to accumulate off South America. During the 1983 El Niño, for example, sea-surface temperatures were about 1 °C higher than normal.

CONSEQUENCES

El Niño regularly brings heavy rain to the coast of Peru, because air movement across the warm ocean surface increases evaporation and atmospheric water vapour, and the moisture is condensed and precipitated when it reaches the coast (particularly by orographic precipitation; see p. 330). Torrential rains across Peru are not uncommon during El Niño, and they can trigger landslides and flash floods in the Andes. The main impacts of the 1997–98 El Niño event are summarised in Box 16.13.

BOX 16.13 IMPACTS OF THE 1997–98 EL NIÑO

North America

- Unusual jet stream patterns over North America led to severe storms over the eastern North Pacific and the west coast of the United States.

South America

- Guyana, severely affected by drought, began water conservation measures.
- The coasts of Ecuador and northern Peru received 350–775 mm of rain in December 1997 and January 1998, compared with the normal 20–60 mm.
- Torrential rains soaked southern Brazil, south-east Paraguay, most of Uruguay and adjacent parts of north-east Argentina.
- Rain on Colombia's Pacific coast increased the threat of landslides, while inland forest fires destroyed about 1,500 km^2.
- The sea level in the Colombian Pacific rose by 20 cm.

Africa

- Unusually warm weather was reported in most of South Africa, southern Mozambique and the central and southern portions of Madagascar.
- Heavy rains fell across central and southern Mozambique, the northern half of Zimbabwe and parts of Zambia, causing flash floods in places.
- Kenya was particularly hard-hit by flooding, many villages being cut off and the main Nairobi–Mombasa road becoming impassable.

Asia and the Pacific

- Long-term dryness persisted over Indonesia and the Philippines.
- Tropical storms Les and Katrina caused heavy rain in northern Australia.
- There were torrential rains in southern China.

Farmers benefit from El Niño because it waters their grazing land, but it causes great hardship for the fishing people of Peru because it disturbs and often eliminates the normal pattern of upwelling, which sustains the anchovy and fishing industries by increasing nutrient supplies. Fish catches decline dramatically during ENSO events, with consequent impacts on food supplies and regional economies.

When the El Niño flow pattern changes, the ENSO events can have widespread impacts on climatic and environmental systems. These impacts are not confined to the South Pacific, because they can be transferred around the world via adjustments to global climatic and environmental systems via so-called 'teleconnections'. The significant heating of the eastern equatorial Pacific that occurs in strong El Niño events appears to be associated with significant climatic anomalies and abnormal weather in many parts of the world. Historical records of large-scale droughts and floods in India, Africa, Indonesia and parts of China show close correlations with the chronology of El Niño events in South America at least as far back as 1750.

TIDES

The surface of all ocean waters (including the open sea, gulfs and estuaries) rises and falls periodically as a result of the gravitational attraction of the Moon and Sun on the Earth and its waters. The results of these two sources of gravitational pull are called lunar tides and solar tides accordingly (Figure 16.7). Lunar tides are stronger than solar tides because the Moon is much closer to the Earth than the Sun is and so exerts a stronger influence over tides on Earth. There are usually two high tides and two low tides each day.

LUNAR TIDES

All bodies in the universe (including the Earth and its moon) have gravity, which is the force of attraction that tends to pull things towards its centre. The Earth's gravity stops us from simply floating off into space! Gravitational attraction is the force of attraction that bodies exert on one another as a result of their mass.

While the Moon is only a quarter of the size of the Earth and has a gravity one-sixth that of the Earth (see p. 108), it still exerts gravitational attraction on the surface of the Earth. As a result, when the Moon is directly over a particular point on the Earth's surface, it exerts a pull on the water. The pull is powerful enough to raise the water above

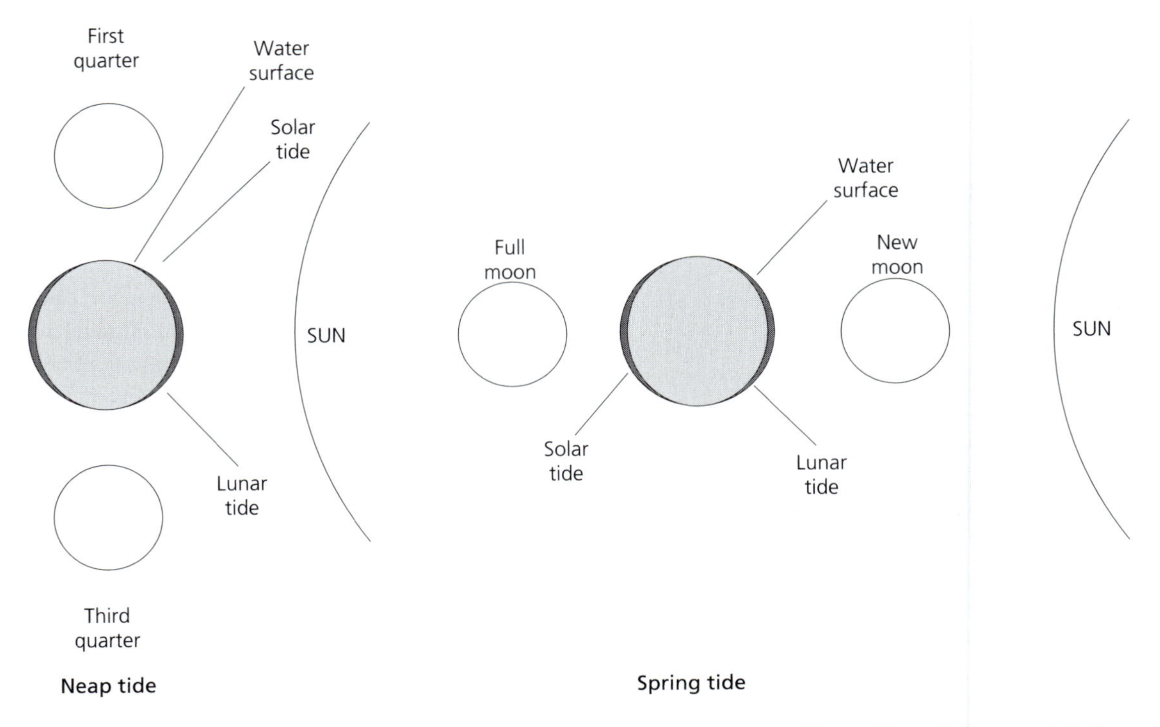

Figure 16.7 *The formation of tides. Tidal ranges are greater during spring tides and less than normal during neap tides.*

Source: Figure 16.6 in Doerr, A.H. (1990) Fundamentals of physical geography. Wm. C. Brown Publishers, Dubuque, Iowa.

its normal level, creating a high tide (high water). This is called a direct tide. Water on the opposite side of the Earth, furthest away from the Moon, is also subject to the gravitational pull. This creates a second high tide, called an opposite tide. As the water is raised to create these two high tides, it is lowered around the circumference of the Earth at right angles to the tidal axis to create phases of low tide (low water).

SOLAR TIDES

Like the Moon, the Sun also exerts a gravitational attraction on the Earth and its waters. This also gives rise to two high tides on opposite sides of the Earth, but the Sun's tidal pull is only about half as strong as the Moon's, so the solar tide is superimposed on the more dominant lunar tide.

BOX 16.14 TIDE CYCLES

High tides and low tides alternate in a continuous cycle, being effectively giant waves (see p. 478) that span the globe. Tidal range is the difference in height between successive high and low tides at a particular place, and this varies over time because of the solar tides (see below).

Most coastal sites around the world experience two high tides (one direct and one opposite) and two low tides each lunar day, with successive high and low tides generally roughly the same height. A lunar day lasts 24 hours 50 minutes and 28 seconds, and this means that high and low tides are experienced nearly an hour later each day. Tide tables are based on detailed computation of when high and low tides will occur at named places over the year.

Tidal cycles affect many coastal, estuarine and near-shore environmental systems and ecosystems that are alternately flooded (at high tide) and exposed (at low tide). They also affect the ease with which boats can enter and leave harbours and navigate through shallow waters, and they can be exploited in tidal power stations (see p. 491).

BOX 16.15 SPRING AND NEAP TIDES

Tidal condition is closely related to the relative position of the Sun and Moon (see p. 118):

- *Spring tides*: high water is higher and low water is lower than normal during spring tides, which occur during periods of new and full moon. This is because the Sun, Moon and Earth are directly in line with each other, and the solar and lunar tides coincide.
- *Neap tides*: high water is lower and low water is higher than normal during neap tides, which occur during the first and third quarters of the Moon. Then the Moon is at right angles to the Sun relative to the Earth, so the tidal pull of the Moon and the Sun oppose one another.

BOX 16.16 TSUNAMIS

Tsunamis are giant waves (not tidal waves) that are triggered by submarine earthquakes or volcanic eruptions (see Box 6.13) and move across the ocean at speeds of more than 700 km h^{-1}. The word 'tsunami' is Japanese for 'harbour wave'. In the open ocean, these giant waves have long wavelengths and short heights (perhaps about a metre), but when they arrive in shallow water they slow down and build up. This creates large, powerful waves that can sweep inland and cause widespread damage and destruction, with great loss of life.

A tsunami was generated off the west coast of North America on 27 March 1964 by the Good Friday Alaskan earthquake. It swept down the west coast of Canada from the Gulf of Alaska and caused serious damage to Port Alberni on Vancouver Island. Despite an elaborate tsunami warning system now being in place, many coastal towns along the west coast of Canada are still at risk because many residents do not regard tsunamis as a serious threat and do not always follow the advice given.

The forces exerted by the Sun and the Moon sometimes complement one another, and sometimes they partly counteract one another, because the relative positions of the Sun, Moon and Earth vary through the year (see p. 118). As a result, prevailing tidal conditions at a particular location reflect the relative positions of the Sun and the Moon at the time. This gives rise to two distinctive tidal conditions – *spring* and *neap tides* (Box 16.15).

TIDAL CURRENTS

Tidal currents are movements of water in confined coastal areas, such as estuaries and inlets, where water usually flows towards the shore as high tide approaches and away from the shore towards low water. Each tidal current lasts for about 6 hours 12 minutes, corresponding to the two phases of high water and low water in each lunar day.

The tidal current that flows with a rising tide is called the *flood current* and that which flows with a falling tide is called the *ebb current*. The reversal between successive flood and ebb currents, when the water is relatively calm, is described as *slack water*.

OCEAN RESOURCES

WET COMMONS

The oceans are one of the world's most important remaining common resources, and like all commons (see p. 11) – including Antarctica (p. 448) – they are subjected to over-exploitation and unsustainable use. As global population continues to rise, and because the oceans have traditionally been used as a source of materials and energy, the threat to the ocean commons is likely to become more serious. Some activities that use ocean resources conflict with one another, such as oil extraction and fishing. Many ocean areas, particularly in the northern hemisphere near industrial countries, already suffer from serious over-use. Increasing use of the oceans for a range of activities, including mariculture, waste disposal, recreation and transportation, is highly likely over the next two decades.

Many countries are realising just how important the oceans are to their economies and future prosperity, and this is promoting greater efforts to ensure the rational management and allocation of ocean resources. The USA, for example, has developed a plan called US Ocean Resources 2000. This is designed to reduce conflicting uses (such as between waste disposal at sea, marine transportation and environmental protection) and to develop potential resources (including fisheries, mariculture, biotechnology, oil, gas, minerals, mining and alternative energy resources) in sustainable ways. Britain, being an island, has always been heavily dependent on the sea, and in the post-Rio era serious thought is being given to the protection, management and development of ocean resources.

FISHING

Fishing is a traditional form of harvesting natural resources, and it has long been an important source of income and employment in coastal areas, and of food for local, national and global markets. Nearly 1,000 million people depend on fish for their primary source of protein, and demand for food fish is projected to increase from about 75 million tonnes in 1995–96 to 110–120 million tonnes in 2010. It is argued that, with careful management, the marine catch could be increased sustainably by about 10 million tonnes a year.

PRODUCTIVITY AND HARVESTING

Some parts of the oceans are among the most productive ecosystems on Earth (see p. 564), and this encourages the

BOX 16.17 GLOBAL FISH CATCHES

World fish catches increased by nearly 50 per cent between the early 1970s and the mid-1990s, but concern over this aspect of ocean resource use is rising because:

- World fish catch is based on a very small number of species (ten species account for over a quarter of the total catch).
- Population levels among many species of fish in the oceans have declined a great deal in recent decades, largely as a result of over-fishing but also in response to increasing ocean pollution.
- Pressures on marine fish are not evenly distributed, either by location (fish stocks in some popular and accessible fishing grounds are now more or less exhausted) or by source (two-thirds of total world catch in 1992 was caught by twelve leading fishing countries).

Plate 16.2 *Fishing boats tied up at low tide in the harbour at Newlyn, Cornwall, England.*
Photo: Chris Park.

Table 16.3 Global marine fish catch, 1975–95

	Marine fish catch (Million tonnes per year)				
	1975	*1980*	*1985*	*1990*	*1995*
Africa	6.75	7.44	9.17	10.96	10.28
Asia and the Pacific	22.94	25.18	29.74	34.72	40.94
Europe and Central Asia	12.11	12.41	12.89	20.91	18.50
Latin America and the Caribbean	6.29	9.18	13.27	15.71	21.11
North America	3.76	4.90	6.16	7.27	6.22
West Asia	0.36	0.29	0.32	0.36	0.43
Total	52.21	59.40	71.55	89.93	97.35

view that ocean fishing is likely to continue to be a major source of food in the future. Production is the amount of organic matter fixed (changed into stable compounds) by photosynthetic organisms in a given unit of time. Total world ocean production of organic matter is estimated at around 130,000 million t a^{-1} (metric tonnes per year).

Biological productivity in the oceans depends heavily on phytoplankton, the primary producers at the base of the marine food web (see p. 555). Phytoplankton are tiny aquatic plants that can convert inorganic carbon into organic matter by photosynthesis. The process requires sunlight, so primary productivity is much greater in the upper parts of the ocean; lack of sunlight inhibits photosynthesis at greater depths. Fish and zooplankton (tiny aquatic animals) feed on the phytoplankton, and they in turn are eaten by predators.

The world catch of fish, shellfish and other aquatic organisms reached an all-time record of nearly 90 million tonnes in 1986, according to estimates by the UN Food and Agriculture Organisation (FAO). Between 1980 and 1986, world fisheries catch rose by 25 per cent, most of the increase being accounted for by developing countries (particularly in Latin America).

Fish catches have risen progressively since the 1970s (Table 16.3) in response to growing demand and the introduction of new technology, including high-tech fishing gear, sonar fish tracking systems, and on-board processing and refrigeration, which allows boats to stay at sea for many weeks. While the global marine fish catch rose from around 50 million tonnes in 1975 to more than 97 million tonnes in 1995, the picture is complicated because new species of fish and new fishing grounds have been exploited and depleted.

Since the 1950s, the world's fishing fleets have caught fewer large predatory fish – whose populations have declined through over-fishing – and more of the smaller fish further down the food chain (see p. 555). This partly explains why total fish catches have remained relatively steady despite over-fishing of cod and other important species. Around 60 per cent of the world's ocean fisheries are now at or near the point at which yields decline.

SUSTAINABLE YIELD

The quantity of marine biological resources that is harvested over a given period of time (usually a year) is known as the yield, and it varies from place to place and over time.

Yields can be increased when demand for marine biological products increases and/or when new or improved fishing technologies are adopted, but there is an upper limit beyond which harvesting becomes unsustainable. This upper limit is the maximum sustainable yield (MSY), which is the maximum yield that can be obtained from a given crop or species if it is to maintain equilibrium. MSY reflects carrying capacity.

BOX 16.18 TRADITIONAL FISHING METHODS

Traditional fishing methods are still used by many countries. These include:

- baited hooked lines;
- trawl nets, which are towed through the water;
- ring (seine) nets, which are like curtains that are drawn out to enclose a shoal of fish;
- drift nets, which are like series of curtains that hang vertically.

In practice, the MSY of a fish population is the largest number of fish that can be harvested in a given year without altering the size of the fish population. This is a good example of steady-state equilibrium in open systems (see p. 72). It can be calculated for a given area (such as the North Atlantic Ocean) using a simple mass balance, which depends mainly on:

- the number of individuals born in the area;
- the number of individuals that migrate in and out of the area.

The maximum sustainable yield of the world's oceans is estimated at between 150 and 200 million t y^{-1}.

Recruitment (by birth and migration) can vary a great deal from year to year, reflecting variations in factors such as weather and climate and the availability of food supplies. In the light of this inherent variability, harvesting the maximum sustainable yield every year would be extremely unwise and doubtless very risky, because it would inevitably promote an overall decline in the fish population concerned.

A more realistic approach is based on optimum sustainable yield (OSY), which is usually calculated as half of the carrying capacity. Estimates of optimum sustainable yield for different fish populations and different parts of the global oceans are used as the basis for calculating yield quotas for commercial fisheries.

BOX 16.19 FACTORS THAT AFFECT FISH YIELD

Yield depends on a number of factors, including:

- *Technology*: modern fishing boats and techniques can collect much larger catches from across a much bigger area than can more traditional boats and techniques.
- *Taste*: when fish is a popular food the market for fish increases, and this encourages larger catches and promotes the development of better fishing technologies.
- *Sustainability*: catches also depend heavily on the ability of the marine ecosystems to sustain continued harvesting, and they decline when over-fishing leads to reduced stocks.

Global warming is another important factor, because rising temperatures in the North Sea are damaging cod stocks, which appear to be particularly sensitive to temperature rises. Since the 1960s, sea-surface temperatures in the North Sea have risen on average from about 7.75 to over 8 °C during the 1990s. Cod populations have declined as temperature has risen, from around 300,000 tonnes a year in the early 1980s to around 100,000 tonnes in 2000. Fisheries experts have called for a significant reduction in quotas, if not a complete moratorium.

BOX 16.20 FACTORS THAT PROMOTE OVER-FISHING

A number of factors promote over-fishing, including:

- difficulties of establishing and getting international agreement on optimum sustainable yields for different fish populations;
- reluctance by the fishing industry (and many individual countries) to limit catches voluntarily;
- problems of monitoring and enforcing catch quotas, where these have been agreed;
- difficulties of negotiating international marine conservation legislation;
- introduction of new non-specific intensive fishing techniques (including some that use equipment like a vacuum-cleaner, which literally sucks all marine life out of the sea);
- significant growth in the use of fish as animal feed.

OVER-FISHING

When fish are harvested at rates that exceed the optimum sustainable yield, over-fishing occurs, which leads to a net decline in the fish population. Once this happens, a spiral of decline can quickly set in unless fish catches are quickly reduced well below the OSY level and the populations are given time to recover and stabilise naturally.

There is no shortage of examples of over-fishing in recent decades. The North Atlantic, for example, has traditionally been a popular fishing ground for many countries and – because the open oceans have been viewed as common resources and thus available for all to exploit – extraction of marine biological resources has been exten-

BOX 16.21 CONSERVATION OF FISH STOCKS

Some form of active intervention is generally required to conserve declining stocks of marine biological resources. This takes various forms, which are often backed up by international legislation. Examples include action on:

- *Catch size*: for example, quotas on all commercially important species have been agreed between member states of the European Union, which fix the total permissible catch and share it among fishing nations.
- *Appropriate equipment*: there are now European Union regulations on the size of drift nets (which cannot be longer than 2.5 km), with minimum mesh sizes prescribed (to avoid catching young individuals).
- *Protected species*: fishing of a number of species (including several types of whale) whose populations have declined so much that their survival is at risk has been prohibited.
- *Reducing the size of fishing fleets*: one proposal to halt the decline in the British fishing industry has been to introduce a decommissioning scheme to pay fishermen to scrap their boats.
- *Extending territorial rights and exclusion zones*: fish stocks can be protected by preventing competition between different countries for available yields. In the 1980s, for example, Canada extended its territorial waters out into the Newfoundland Grand Banks to protect the Atlantic cod industry there from large-scale international competition.

sive. Cod and haddock populations have been seriously depleted in recent years, and the herring population is on the verge of extinction as a result of over-fishing. Fishing on Canada's Grand Banks, previously one of the world's richest cod fisheries, collapsed in the early 1990s. At the global scale, the decline of the whaling industry in recent years is clear evidence of the short-sightedness of not keeping extraction within sustainable limits.

CONSERVING STOCKS

Economists point out that fishing any particular species to extinction is highly unlikely, because as catches decline that species becomes uneconomic to fish, so attention switches to other species. As a result, they argue, populations that have been damaged and depleted by over-fishing are allowed to recover naturally. In practice, such recovery can be an extremely long process, particularly if the over-fishing continued so long that the surviving population became very small and scattered.

One way of helping to conserve natural wildlife in the oceans is to develop mariculture (the cultivation of marine plants and animals in their natural environment). As capture fishing approaches the limits of sustainable yield in many areas (including the USA), attention is starting to switch towards the farming of marine finfish, shellfish, crustaceans, molluscs and seaweed to meet the growing demand for seafood. China now has a well-developed mariculture industry, which is based heavily on fish and shrimp but also includes mussels, seaweed and molluscs.

OCEAN MINERAL RESOURCES

Until quite recently, little attention was paid to exploring the mineral resources of the sea, but interest is growing in extracting minerals from seawater and in mining minerals

BOX 16.22 MINERAL POTENTIAL OF THE OCEANS

Seawater clearly contains great potential for mineral extraction, and this is likely to be exploited much more heavily in the future if:

- Supplies of minerals and metals from other sources start to run out (because of over-extraction).
- Mineral and metal prices on the world market change (making it cost-effective to extract minerals and metals from seawater).
- New techniques and technologies are developed (which make it cost-effective to extract minerals and metals at lower concentrations than at present).

At the present time, however, more investment is being directed towards developing cost-effective methods of mining material from the sea bed.

BOX 16.23 NATIONAL SOVEREIGNTY AND OCEAN ACCESS

Expansion of ocean-floor mining has implications for ownership and rights of access to the wet commons (see p. 11), particularly for landlocked nations, which have no coastal areas of their own. Such problems were recognised in the 1982 United Nations Convention on the Law of the Sea (UNCLOS), which introduced the concept of 200-mile (322 km) exclusive economic zones (EEZs). These give coastal states sovereign rights to explore, exploit, conserve and manage natural resources.

Countries outside the EEZs rightly argue that they are being denied access to valuable resources, in much the same way as some countries have claimed rights and others have been denied rights of access to mineral resources in Antarctica (see p. 450).

from the ocean floor (Box 16.22). This raises important questions about who owns the oceans and how access to ocean resources should be shared out (Box 16.23).

SEAWATER

Seawater contains a large amount of dissolved materials, which are washed in from the continents by river systems. Some minerals are present in high enough concentrations in seawater to make extraction cost-effective; the most important ones are magnesium, bromine and sodium chloride (salt) (see p. 403).

The sea also contains valuable metals. These are present in very low concentrations dispersed over a huge area, which makes it uneconomic if not impossible to extract and recover them. The world's oceans contain an estimated 10,000 million tonnes of gold, for example, at extremely low concentrations.

MINING THE OCEAN FLOOR

The ocean floor is a vast storehouse of material that can be used if prices are right and the technology is available to collect it and bring it ashore. Sand and gravel have traditionally been extracted from the sea bed along the coast and near-shore environments for use by the construction industry.

Deeper parts of the ocean floor also contain many useful mineral resources, few of which have yet been extracted or exploited on a large scale. One such mineral, which exists on the ocean floor and has great potential as an agricultural fertiliser but has yet to be extracted, is phosphorite (a mineral that consists mainly of calcium phosphate). The most valuable ocean-floor minerals, given present technology and potential uses, are manganese nodules. These are spherical lumps containing about 20 per cent manganese, 10 per cent iron, 0.3 per cent copper, 0.3 per cent nickel and 0.3 per cent cobalt, all of which can be extracted and used. The US and Japanese governments, in particular, have invested heavily in developing the technology required to mine the mineral wealth of the ocean floor.

While sea-bed mining is still in its infancy, there is great potential for it to expand. Planners will have to exercise caution in encouraging its growth, because it could seriously affect coastal environments and communities. Federal initiatives designed to promote ocean mining in the USA recognise the potential conflicts of interest while at the same time recognising that it is likely to grow along the US coastal zone.

OCEAN ENERGY RESOURCES

As well as offering considerable potential as a source of minerals and metals, the oceans are a valuable source of renewable energy. Despite some public opposition, based mainly on the likely environmental impacts of large-scale marine engineering systems, the use of the oceans for power generation is becoming increasingly attractive and realistic. Technologies are now available for harnessing ocean energy in a variety of ways, and considerable investment is being made to improve them and increase their cost-effectiveness.

By the early 1990s, ocean power (discounting the French tidal power plant) was providing less than 100 MW, but the

BOX 16.24 OCEAN ENERGY OPTIONS

At present, the most important marine sources of renewable energy include ocean thermal energy, tidal power and wave energy. Other options that are being actively explored include the construction of offshore wind power stations, exploitation of submarine geothermal energy (p. 140) and marine biogas energy, and energy generated by exploiting salinity gradients (p. 403) within the oceans.

projected total energy availability from the oceans was greater than global energy consumption. The oceans remain a largely untapped source of non-polluting, inexhaustible energy.

THERMAL ENERGY

One way of tapping some of the vast amount of energy in the oceans is ocean thermal energy conversion (OTEC). This exploits the temperature difference between warm surface water and cold bottom water to generate electricity. A number of different approaches are being used and developed, and OTEC plants are already operational in Hawaii, Japan, the USA and Tahiti. Most OTEC plants to date have been designed to generate electricity, but new applications are now being explored that make direct use of the deep, cold, nutrient-rich seawater. Applications with considerable potential include mariculture (which would make use of the nutrient-rich water), freshwater production and air conditioning.

If technologies can be developed to exploit such applications commercially, the future could well see many more OTEC plants and the continued search for new applications. This is true particularly in the many islands of the Pacific Ocean, which have few alternative renewable energy sources with such vast potential.

TIDAL ENERGY

Perhaps the simplest way of generating power from the oceans is by using the rise (flow) and fall (ebb) of the tides. The most common way of exploiting the tides is by so-called 'tidal-range energy', or simply 'tidal energy', which exploits the difference between high and low water to power turbines or a waterwheel (Figure 16.8). This is a traditional form of power generation, used in ancient Egypt and throughout Europe until the middle of the twentieth century. It is sustainable and non-polluting, and in theory it could provide around 20 per cent of present global energy production.

Large tidal range schemes have been operating in France, China and the former Soviet Union for some years, and they have proved that this approach is both sustainable and cost-effective. The best-known example, at the mouth of the River Rance at St Malo in north-west France, was built in 1966. It uses both the inflowing and outflowing tides of the English Channel, and the high (8.5 m) tidal range in the Rance estuary is capable of producing up to 240 MW (240,000 kW) of electricity. The Bay of Fundy in Canada has great (as yet untapped) potential for producing tidal power, because it has the highest tidal range (18 m) in the world.

Most tidal power schemes are based on exploiting tidal-range energy, but research is under way into ways of exploiting 'tidal-flow energy', based on the passage of tidal currents through narrow channels.

BOX 16.25 TIDAL ENERGY SYSTEMS

Most tidal energy schemes are based on barrages, usually across natural inlets. The reservoir is filled with water as the tide rises, and the sluice gates are shut at high tide. As the tide falls, water drains out from the reservoir. In traditional systems, the water rushing out turned a waterwheel to provide direct mechanical energy for grindstones and pulleys. In modern systems, the water turns turbines to generate electricity. Some tidal energy schemes use only the outflowing tide to produce energy, and they operate for about 30 per cent of the time. Others use both inflowing and outflowing tides to generate energy about 60 per cent of the time.

WAVE ENERGY

The main alternative to producing energy from the tides is to harness the energy in sea waves to generate electricity. As the waves move across the surface of the sea (see p. 479) they transmit energy, which can be converted into usable forms. Records of wave size and frequency, collected using a wave recorder, can be used to compute wave power potential in a given area.

Wave power currently contributes very little to global energy supplies, but it has great potential – up to 1 million MW (1 TW) according to some estimates. It also has minimal environmental impacts. Many different types of wave power plant have been proposed and evaluated. After an initial introductory phase, which would require heavy subsidies, wave energy is expected to become economically competitive. The cost-effectiveness of wave energy schemes is likely to increase as technical improvements and new designs are introduced.

Renewed interest in wave power in the UK was shown at the close of the 1990s, after sixteen years of neglect, when the British government seriously considering backing a range of new and improved technologies for turning wave energy into electricity after being advised that they are now

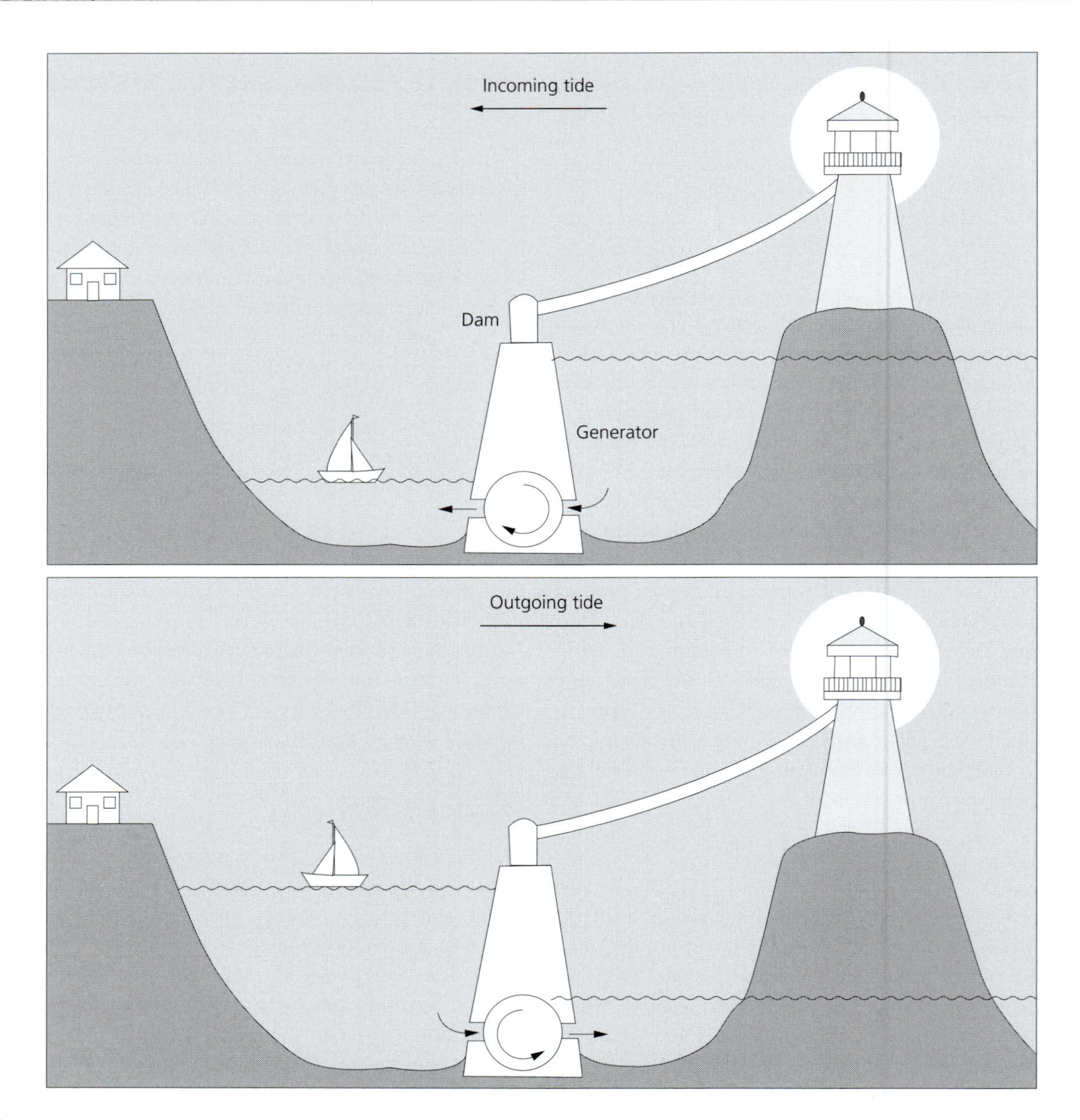

Figure 16.8 *Principles of a tidal power station. A tidal power station operates by holding back incoming and outgoing tides behind a dam. The difference in water levels generates electricity in both directions as water is passed through reversible turbogenerators.*

Source: Figure 20.26 in Cunningham, W.P. and B.W. Saigo (1992) Environmental science: a global concern. Wm. C. Brown Publishers, Dubuque, Iowa.

economically viable. The estimated price of wave energy (in pence per kilowatt hour, 1998) was 2.6, compared with gas (2.5), wind (3.0), coal (4.0) and nuclear (4.5). It is estimated that waves in British waters could produce up to 100,000 MW of electricity, more than 20 per cent of the country's needs. Waves in waters off the Irish Republic could generate nearly 75 per cent of that country's electricity.

By 1998 at least fifteen wave generators were planned around the world – nine in Europe, four in the Far East, one in the USA and one in India. Eight of these were intended to be producing electricity in 1999.

> **BOX 16.26 WAVE ENERGY TECHNOLOGIES**
>
> The basic idea of wave energy systems is to use a floating platform that rises and falls with the tide and has some wave-powered system of driving turbines for power generation. Various approaches have been tested since the early 1970s, including:
>
> - transforming the motion of the water surface into a flow of air, which can then drive a turbine;
> - transforming the motion of the water surface using floating devices that rise and fall at different rates. One well-known example is the Salter 'bobbing duck' (a floating boom whose segments nod up and down with the waves);
> - using underwater devices that are activated by sub-surface water movements;
> - allowing waves to break over a barrier into a storage lagoon, which provides the head of water to turn a turbine and so produce electricity.

OCEAN POLLUTION

According to the United Nations Environment Program, the open ocean is still relatively clean. Most marine pollution monitoring is concentrated around the coastal zone, where the impacts of human activities are most obvious and can affect the use or exploitation of seawater and its resources.

Most of the pollution that affects the oceans originates on land, and it reaches the sea at or near the coast from rivers, pipelines or underground seepage. Population increase, the spread of urbanisation, industrial development, extensive deforestation and land-use change, and intensive agriculture have together produced a dramatic rise in the discharge into the sea of sediment and nutrients washed down by river systems.

Pollutants discharged directly into the sea (such as sewage disposal) or washed down by rivers (domestic, industrial and agricultural wastes) are dispersed and diluted by waves, tides and currents. Dilution reduces the concentration of the pollutant, and thus decreases the risk of serious ecological damage, but dispersion spreads the pollution over a much wider area and thus exposes more of the environment to the risk.

There are two other main sources of marine pollution. Some is discharged directly into the sea from marine accidents, such as oil spills. Some is deposited directly into the sea from the atmosphere, such as nuclear fallout from the 1986 Chernobyl reactor accident (see Box 8.23).

Many different types of pollution affect the world's ocean resource base, including oil and chemical pollution, and damage caused by algal blooms as a result of eutrophication (see p. 410). Pollution tends to aggravate the ecological problems caused by over-fishing because it normally leads to population decline among affected species. Such problems are likely to become even worse in the future as the impacts of global warming on the marine food chain become clearer.

One particularly difficult class of marine pollution to deal with effectively is the introduction of materials such as artificial radionuclides and plastics, which have long residence times and have known impacts on aquatic ecosystems.

OIL POLLUTION

One of the most visible forms of ocean pollution is the oil spill – a release of oil that is dispersed across the ocean surface by wind, waves and currents as an oil slick. In one three-year period alone (1975–78), visual observations of nearly 100,000 oil slicks were reported around the world, and doubtless many more smaller ones went unreported.

Many oil slicks are caused by accidental spillage from tanks, tankers and pipelines, but sometimes the oil is discharged deliberately (as happens, for example, when oily water ballast is discharged from a ship into the sea). Most of the serious oil pollution at sea is concentrated along the major tanker lanes and shipping routes, but oil pollution of the oceans now seems to be a truly global problem, because analysis of seawater quality in even remote parts of the ocean has detected measurable amounts of dissolved and dispersed petroleum residues up to 1 m below the surface.

Oil pollution causes extensive ecological and environmental damage. At sea, toxic chemicals leach out from the oil and contaminate the surrounding water, poisoning sea life around and below the spill. Light oils can float freely on the ocean surface and be dispersed over a wide area. Heavy oils sometimes form globules (small spherical lumps), which sink to the sea bed and poison plants and animals that live there. When an oil slick drifts or is blown towards the coast it can seriously pollute the shore, fouling beaches and killing all shore life. The oil clogs up the feathers of birds, making it impossible for them to fly or

BOX 16.27 THE *EXXON VALDEZ* OIL SPILL, 1989

One of the worst oil spills in recent years occurred when the *Exxon Valdez* oil tanker ran aground off the coast of south Alaska on 24 March 1989. More than 41 million litres of crude oil was discharged into Prince William Sound, creating an oil slick that covered 12,400 km^2 and polluted at least 1,100 km of coastline. Damage was extensive, partly because the area is ecologically sensitive and thus not able to withstand great stress. The problem was made worse by a lack of disaster preparation and an inability to use containment methods, so large amounts of damaging detergents were used to try to disperse the oil. The impacts on wildlife were devastating – at least 34,000 seabirds were killed, along with around 10,000 otters and up to sixteen whales. The experience of coping with the *Exxon Valdez* incident taught many lessons about the strategic management of oil spills at the global scale and showed the importance of appropriate prevention and remediation (post-event restoration) policies.

BOX 16.28 CONTAINING OIL SPILLS

Various procedures are used to control and remove oil slicks, including:

- *Containment*: floating booms are used to contain the oil slick, and the oil can then be transferred into tanks by suction.
- *Absorption*: oil can be absorbed on to special materials (such as a special nylon 'fur'), from which it can be squeezed out for reuse or disposal.
- *Dispersal*: chemical detergents can be used to break up the slick. This method has been widely used in the past but is problematic because it does not remove the oil, and the detergents used contain toxic ingredients that are harmful to birds and marine life (particularly shellfish).

even swim. Most other coastal creatures are suffocated by the oil.

Two recent oil spills in which tanker accidents caused ecological damage were the 1989 *Exxon Valdez* spill in Alaska (Box 16.27) and the 1993 *Braer* oil spill off the Shetland Islands to the north of Scotland.

Many geological structures under the sea floor are reservoirs for petroleum, and offshore oil and gas wells currently supply about 17 per cent of world petroleum production. Most offshore oil wells are located in the shallow waters of the continental shelves, but exploration is under way in the deeper waters beyond. Oil was discovered beneath the North Sea in the 1970s, and since then it has been extracted from a number of submarine oil fields off north-east Scotland and pumped ashore from the oil platforms along sea-bed pipelines. This clearly creates a risk of major oil pollution incidents in the event of an accident or leakage from the rig or pipeline. Such prospects have been taken into account in assessments of the possible environmental impacts of developing the hydrocarbon reserves of the North Sea. Careful monitoring of the physical, chemical and biological environment offshore and onshore is designed to give early warning of any unwanted side-effects of the oil extraction.

MARINE EUTROPHICATION

Eutrophication, caused by nutrient enrichment, is a common problem in freshwater lakes and rivers (see p. 410), but it can also affect marine ecosystems. Increases in the amount of nutrients washed into the sea from land-based sources can promote blooms of phytoplankton (particularly dinoflagellates and diatoms), which can cause short-term health problems in humans around the worst-affected coasts.

Marine eutrophication problems are most acute where dilution and dispersion are limited, particularly in enclosed or semi-enclosed seas such as the Black Sea and the Baltic Sea. The Black Sea is virtually enclosed, and it receives large inputs of nutrients from rivers in Turkey. Extensive monitoring has established that serious ecological damage is occurring in the Black Sea. Most is attributed to eutrophication, coupled with pollution from pathogenic microbes and toxic chemicals.

The Baltic Sea is partially enclosed, and it too is suffering from problems of nutrient enrichment. Nutrient inputs from river systems have increased dramatically in recent years; inputs of nitrogen and phosphorus to the Baltic Sea rose by up to six times during the 1980s. Recent eutrophication has increased primary production in the Baltic Sea by up to 70 per cent, biomass and production of benthos have increased in the warmer upper layers, and fish catches have increased more than tenfold.

BOX 16.29 RED TIDES

Severe algal blooms often cause a change in the colour of seawater, which creates what are commonly known as 'red tides' (although the water may be turned red, green, yellow or brown). Red tides have appeared regularly throughout South-east Asia since at least the early 1980s. They also seem to be becoming more common around the world and more regular where they have appeared in the past. Whether this represents a real increase in incidents of red tide, or better awareness of a stable situation (particularly via satellite-based remote sensing and monitoring), is not yet clear.

INCINERATION OF TOXIC WASTES

A small amount of ocean pollution, concentrated in a few particular areas, is created by the practice of burning toxic wastes at sea on board specially adapted ships. Toxic wastes are difficult to dispose of safely on land, because they can contaminate groundwater (see p. 358) if buried underground or in landfill sites, and long-term safe storage is costly. Some countries prefer to incinerate the toxic wastes at sea, to reduce the volume (by up to 85 per cent) and weight (by up to 75 per cent) of material for disposal, and to keep toxic air pollutants well away from centres of population.

Air pollution from incineration at sea is rapidly dispersed by the wind, and it probably causes no serious biological or environmental problems. The relatively small amounts of ash left after incineration are virtually sterile, and they can be dumped at sea or brought back to land for disposal. Health and environmental risks appear to be based largely on the prospect of spillage or accidental discharge, although concern has been expressed about the lack of monitoring of incineration emissions.

More than a million tonnes of chlorinated organic chemical wastes from industries throughout Western Europe was burned aboard ships in the North Sea during the 1970s and 1980s. Supporters of incineration at sea claim that the European experience proves that the technology is efficient and safe, although the practice was phased out on safety and environmental grounds. Incineration of hazardous wastes in the North Sea ended in 1994.

BOX 16.30 CONTROLLING OCEAN POLLUTION

Tackling ocean pollution at the global scale will require at least three steps, which are politically difficult but ecologically essential:

1. *Reduction at source*: it is important to reduce and restrict the use of damaging chemicals in all aspects of industry and land use.
2. *Effective waste management*: significant improvements are necessary in the infrastructure needed to handle sewage and waste, particularly from large cities, and these will require great financial commitment from industrial countries and private companies.
3. *Co-operation*: ongoing commitment to manage the global oceans jointly is required at all levels of community and government.

POLLUTION CONTROL

While pollution is not an environmental concern across the oceans as a whole, it is clearly a serious problem in certain particular places and as such it needs to be controlled. Moreover, the fact that the oceans are a common resource puts a responsibility on all users of the oceans, and on all countries that discharge material into the sea (knowingly or accidentally), to protect the resource base and use it in sustainable ways. Because there are so many point sources and diffuse sources of ocean pollutants, it is extremely difficult to control effectively, even if the goodwill existed to do so.

OCEAN MANAGEMENT

Given so many different pressures on the world's oceans, it is not surprising that efforts are being made to manage them and use them in more sustainable ways. A central problem is the overall lack of responsibility that most countries feel towards the open seas – the wet commons belong to no particular nation, but to the whole world. Individual countries make great efforts to protect the marine areas within their territorial limits and express great concern when pollution is washed in from beyond their limits or when ocean resources within their limits are exploited by outsiders. But until relatively recently there was little sense

BOX 16.31 COMPREHENSIVE OCEAN MANAGEMENT

A comprehensive ocean management regime would be ideal, because it would seek to conserve all marine resources (no matter where they are located) and to use them in the most sustainable ways. But many obstacles have to be removed before significant progress can be made, including the availability and sharing of information between countries, the significant costs that would have to be shared, traditional attitudes towards single uses of the oceans (rather than multiple use), and the short-term timescales over which many political decisions tend to be made. Comprehensive ocean management might best be tackled by an ecosystem approach (see p. 570), based on multiple-use criteria applied at a regional scale.

of shared responsibility towards the truly global ocean commons.

An ecosystem approach is important because it focuses on interrelationships and flows within the ocean system, seeking to move beyond sectoral interest and embrace the reality that the oceans are dynamic open systems that cannot be managed piecemeal. The 1982 United Nations Convention on the Law of the Sea (UNCLOS) emphasised that large marine ecosystems could become a tool for managing fisheries, because they are convenient units in which to determine sustainable yields and allowable catches.

Marine ecosystems also provide useful conservation units, and some countries are putting a great deal of effort into developing sustainable ocean ecosystem management strategies. An example that is proving effective in practice is the Galapagos Marine Resources Reserve established by the government of Ecuador in May 1986. The Galapagos Islands were originally made famous by Charles Darwin's descriptions of inter-island variations in bird species (see p. 530), and more recently they have attracted attention because of the abundance and variety of wildlife that lives there (including endangered giant tortoises). As a result, the area – which is ecologically very important but also extremely fragile – attracts many visitors, particularly scientists and tourists. A zoning policy has been drawn up to regulate and manage human activities in the area, with the aim of conserving wildlife, making sustainable use of natural resources and reconciling the needs of local fishing, tourism, local recreation and scientific research.

The regional scale is important because it brings together all countries with coasts around the ocean in question, and it seeks to encourage co-operation between countries, harmonisation of environmental standards and criteria, and the search for agreement about shared objectives and goals. A regional approach to ocean management was favoured at the Rio Earth Summit, and its merits were recognised by UNCLOS. The Mediterranean Action Plan is a good example of an effective regional approach to ocean management; it is designed to restore the environmental balance in the region, to protect the coastal environment and to highlight the need for appropriate environmental management on land in all countries tributary to the Mediterranean.

COASTS

IMPORTANCE OF THE COAST

The coast is the zone where the land meets the sea, and as such it marks the interface between terrestrial and marine environmental systems (Figure 16.9). Coasts tend to be highly dynamic open systems (see p. 72), where physical and ecological processes interact with cultural influences and controls.

Coasts are also the stage on which many environmental battles take place. Some of these are natural and relate to hazards such as erosion, storm damage, flooding and coastal instability. Other environmental battles are cultural and include pollution, resource use, development and unsustainable exploitation. Many countries recognise the environmental, ecological and cultural importance of their coasts, and efforts are now being made to conserve and protect them. China, for example, has recognised the need to develop and manage its coastal zone in sustainable ways and has adopted a three-stage strategy based on:

1. *Inventory*: a comprehensive study of the country's coastal resources will help in the formulation of a long-term programme for development.
2. *Selective development*: selected locations will be identified for investment within the long-term development programme. The investment will be aimed at improving reclaimed salty soil, strengthening marine fishing measures, developing mariculture, establishing forest protection systems, preventing coastal erosion, regulating coastal transport and establishing environmental protection zones.
3. *Legislation*: stage three will be the drafting of legislation and management measures for the coastal zone.

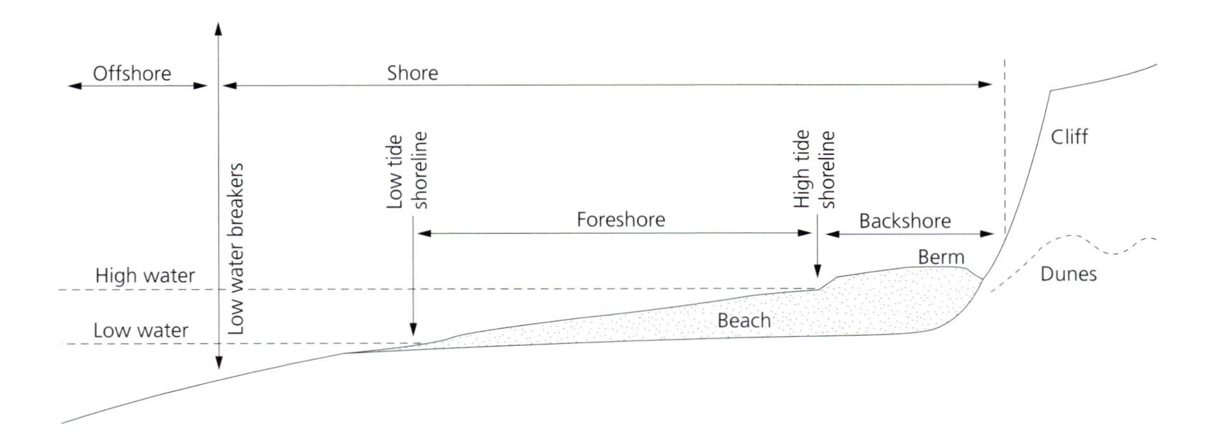

Figure 16.9 *Zonation of the coast. The coast can be sub-divided into a number of distinct zones on the basis of high- and low-water levels.*

Source: Figure 9.2 in Clowes, A. and P. Comfort (1982) Process and landform. Oliver & Boyd, Edinburgh.

BOX 16.32 COASTS AND PEOPLE

As well as being important environmental systems in their own right, coasts can be important administrative and political boundaries. This creates problems particularly if natural or cultural coastal processes cause changes in the position of the coastline – moving seawards by deposition to create new land, or moving shorewards by erosion to reduce the land area. Sudden coastal changes, such as those associated with storms, do not normally affect political and administrative boundaries. Coasts also define important property rights for natural resource management. In England and Wales, for example, the law defines property rights based on *terra firma* (solid ground), non-tidal watercourses, tidal watercourses, the foreshore and sea bed.

The coastal zone is also a focus for human settlements because it has traditionally offered ease of access, various forms of building material, foodstuffs and natural resources. About half of the world's population lives within 80 km from the sea, and much of the expected increase in the future is likely to be concentrated on that relatively narrow strip. This means increasing population density, greater competition for space and access to resources, more pollution and wastes to be disposed of, and more environmental change caused by or associated with human activities.

COASTAL FORMS AND PROCESSES

CLASSIFICATION

Many distinctive landforms are found along the coast, but they fall into two main groups:

1. beach or cliff elements that are adjusted more or less closely to present-day processes;
2. other, generally more resistant elements, which have been only slightly modified by the sea in the few thousand years that the sea has been at its present level.

Geology can strongly control the form of a coastline, particularly via rock type and erodibility. So-called upland or steep coasts, characterised by alternating cliffs and bays and often backed by quite gentle relief, are common in areas with alternating hard (cliffs) and soft (bays) rocks. Geology, both in terms of rock type and geological structures (such as fault lines and folds; see Chapter 6), usually determines the location, size and erodibility of the resistant elements, which might not all be in equilibrium with present-day processes.

Cliffs are eroded largely by mechanical weathering (physical disintegration) processes (see p. 207). Waves crash against cliff faces with great force, and this can directly erode the cliff face in soft or heavily jointed rocks. Slightly different processes are at work in more resistant rocks. Each incoming wave compresses air in cracks in the rock, and the air expands again when the wave recedes. This cycle, repeated many times over, puts great pressure on the

BOX 16.33 EROSIONAL AND DEPOSITIONAL COASTS

One useful way of classifying coastlines is into two types, depending on whether erosion or deposition processes are dominant:

1. *Erosional coastlines*: where erosional processes dominate, and so the sea is progressively extending further inland. Typical landforms include steep cliffs, stacks (isolated pillars of rock in the sea), bays and inlets.
2. *Depositional coastlines*: where deposition processes predominate, so the land is progressively extending further out to sea. Typical landforms include sandy beaches and dunes, and depositional forms such as bars and tombolos.

rock and weakens it. The process involves some positive feedback, because as the rock is weakened more cracks are created in it, thus intensifying the erosional impact. Eventually blocks and pieces of rock are so weakened that they fall from the cliff face directly into the sea below. Over time, cliffs are eroded back and much of the sediment is deposited nearby to fill in bays, ultimately producing a relatively straight coastline with a beach and cliffs behind it (Figure 16.10).

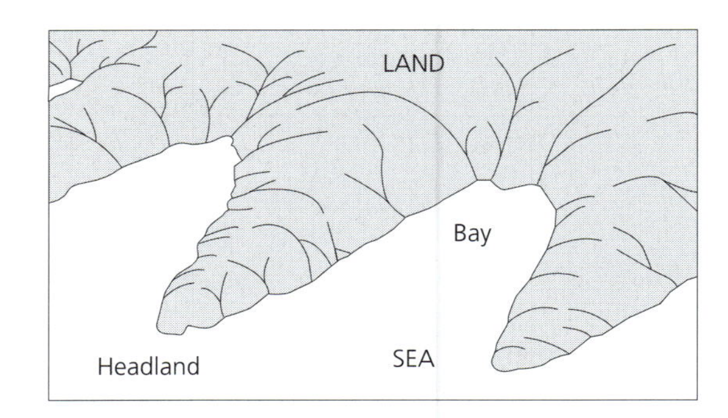

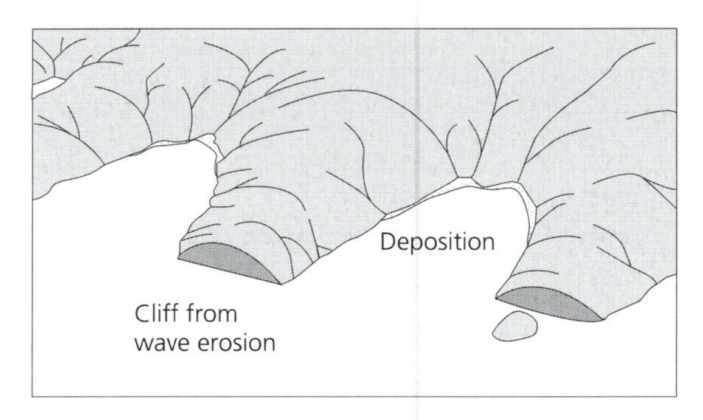

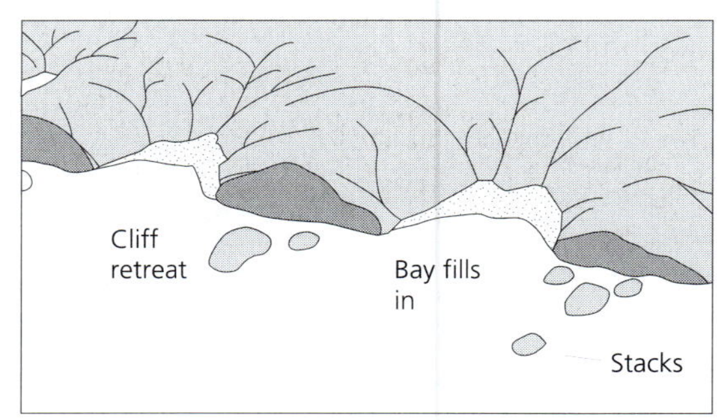

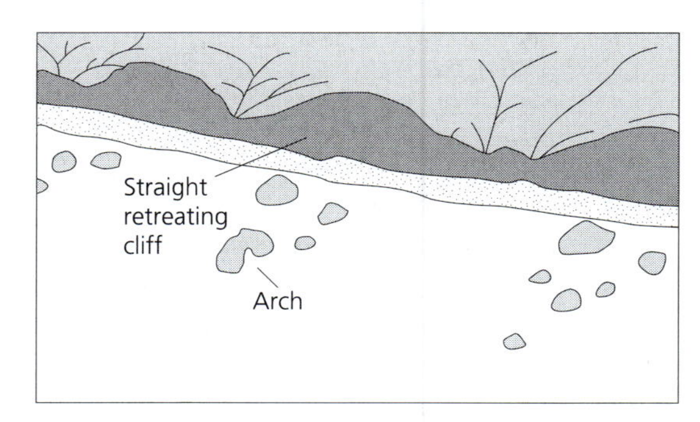

Figure 16.10 *Cliff attack and coastal development. Waves attack headlands, which stick out into the sea, and sediment derived from wave erosion is deposited in bays. Over time, cliffs are worn back and bays fill in, leaving arches and stacks as remnants of the former headlands.*

Source: Figure 16.9 in Doerr, A.H. (1990) Fundamentals of physical geography. Wm. C. Brown Publishers, Dubuque, Iowa.

DEPOSITIONAL LANDFORMS

From a human point of view, beaches are perhaps the most important coastal landforms. A beach is a strip of land that borders the sea. It is a depositional form, usually composed of sand and mud on sheltered coasts and of boulders and pebbles on more exposed coasts. Most beaches are created and sustained by longshore drift (see below). The upper and lower limits of a beach are usually defined by the high-water mark and the low-water mark, respectively (Figure 16.9). Below the low-water mark the beach material is always submerged; above the high-water mark beach material is transported by wind processes rather than by waves and tides.

Beaches and other coastal depositional landforms are usually composed of sand – grains of quartz derived from the weathering of igneous and other rocks inland and transported downstream to the sea by rivers. Much of this

Plate 16.3 *Cliff and beach forms along the coast of Caxton Bay near Scarborough, Yorkshire, England.*
Photo: Peter French.

material is initially deposited at the mouth of a river estuary, from where it is subsequently set in motion again and transported by tides and currents.

Beaches (including bars, spits and tombolos) are built by waves. So-called 'spilling breakers' move up a beach before breaking, and because they push sand particles upslope they tend to be constructional waves. So-called 'plunging breakers', on the other hand, collapse when they first encounter the beach, so they break early. They produce a strong backwash that pulls sand particles back from the beach towards the sea, and as a result they erode more than they build up a beach.

LONGSHORE DRIFT

Longshore drift plays an important role in redistributing coastal sediment and creating depositional landforms. It occurs because – unlike waves in the open sea, which move in directions determined by the prevailing wind direction – waves close to the shore move in directions that are largely determined by the angle between the shoreline and the dominant water movement. This is called wave refraction (Figure 16.11). If waves arrive at an angle of about 90° they will flow more or less directly onshore, with no lateral movement of particles of sediment along the beach. If the angle is less than 90°, the waves will curve in parallel to the shore as they move onshore, carrying particles of sediment along the beach with them and thus creating longshore drift (Figure 16.12).

CORAL REEFS

Coral reefs (low ridges of coral) are among the most interesting features of some coastlines, from both geological and biological points of view. Coral reefs and islands are organic sedimentary rocks (see p. 196), composed of the remains

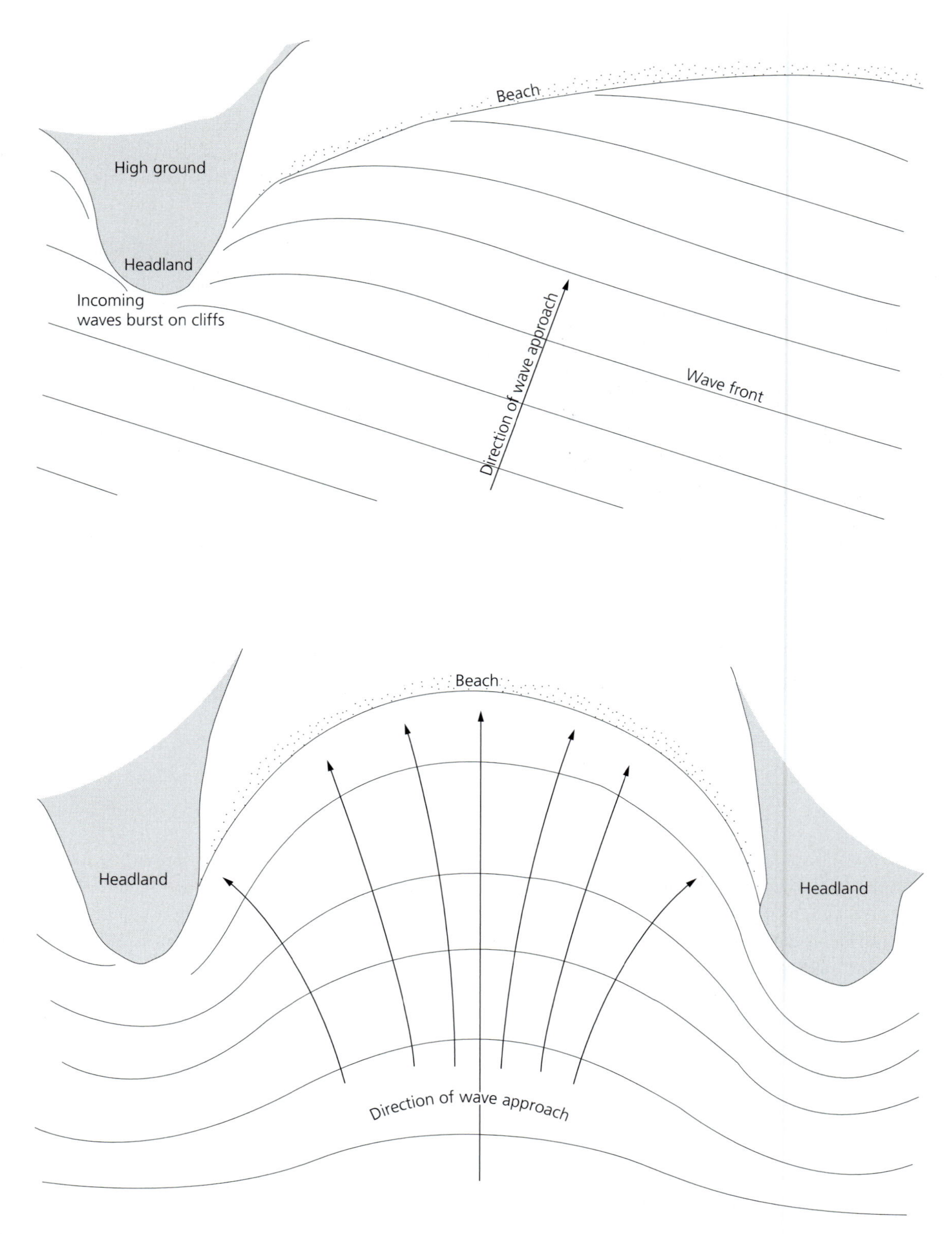

Figure 16.11 *Wave refraction. Waves are refracted as they arrive onshore. Waves that approach a long beach at a shallow angle are refracted in the direction of the beach, and waves that approach a beach confined between two headlands are refracted in a circular arc.*

Source: Figure 12.13 in Dury, G.H. (1981) Environmental systems. Heinemann, London.

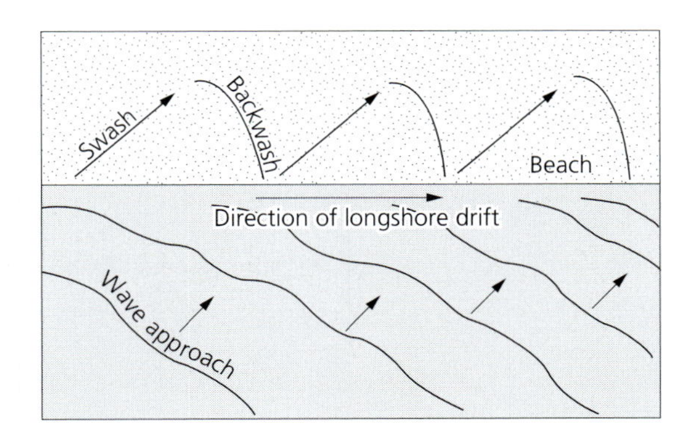

Figure 16.12 *Longshore drift. See text for explanation.*

Source: Figure 10.2 in Buckle, C. (1978) Landforms in Africa. Longman, Harlow.

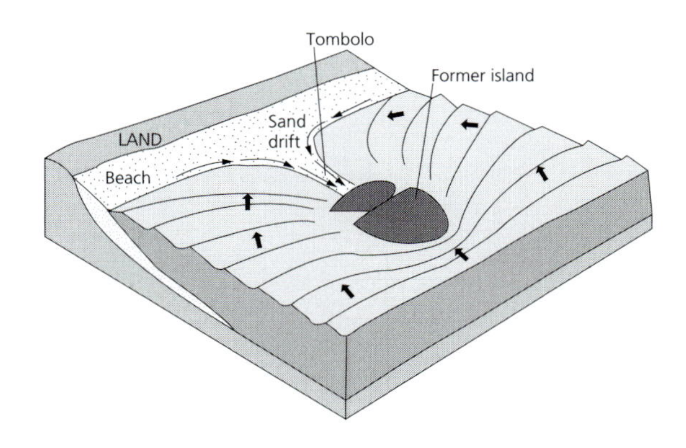

Figure 16.14 *Development of a tombolo. A tombolo forms where a former rock or offshore island is connected to the coast by the deposition of a bar.*

Source: Figure 16.16 in Doerr, A.H. (1990) Fundamentals of physical geography. Wm. C. Brown Publishers, Dubuque, Iowa.

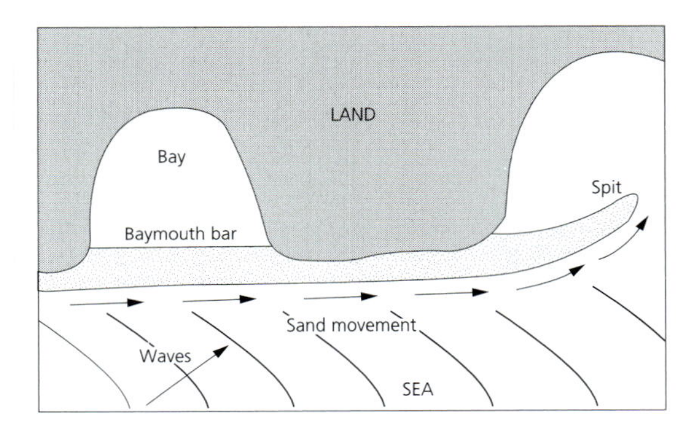

Figure 16.13 *Development of bars and spits. Bars and spits are formed by the deposition of sediment moved along the coast by longshore drift.*

Source: Figure 16.15 in Doerr, A.H. (1990) Fundamentals of physical geography. Wm. C. Brown Publishers, Dubuque, Iowa.

BOX 16.34 LONGSHORE DRIFT AND COASTAL EQUILIBRIUM

Longshore drift accounts for the development of long, sandy beaches and linear depositional forms such as tombolos and bars along coastlines with abundant supplies of sand-sized particles. In this sense, many coastal depositional landforms are open systems in dynamic equilibrium, in which the internal structure of the system (beach form) is adjusted to the balance of inputs and outputs of energy (waves) and matter (sediment particles). This dynamic equilibrium can readily be disrupted, however, by a change in either or both of the inputs and outputs. Thus, for example, beach erosion often accompanies storms with high tides and large, crashing waves. A sudden increase in sediment supply, on the other hand – perhaps associated with land-use changes in the drainage basin of one or more major rivers tributary to that particular coastline – would often trigger deposition and the build-up of beaches and bars.

Beach erosion typically occurs if the sediment supply from further along the shore is reduced or cut off. Sometimes this happens naturally, by a cliff, for example, which prevents the normal longshore drift of sediment along a coastline. Many forms of coastal management also cut off upstream sediment supply – either deliberately (groynes are built to prevent beach erosion, but in doing so they encourage local deposition but downstream erosion) or incidentally (sea defence walls are built to protect cliffs or property from erosion, but this then transfers the erosion problem along the coast by reducing sediment supply downstream).

BOX 16.35 COASTAL DEPOSITIONAL LANDFORMS

As well as beaches, coastal depositional forms created by longshore drift include:

- *bar*: an offshore ridge of sand, gravel and pebbles, often parallel to the coastline, sometimes across the mouth of a river, bay or harbour (Figure 16.13);
- *spit*: a low ridge of sand and gravel projecting from the shore out into the sea;
- *tombolo*: a narrow sand or shingle bar linking a small island to another island or to the mainland (Figure 16.14).

Bars and spits often have shallow lagoons (or *haffs*) behind them.

of coral polyps – small, jelly-like creatures that live in warm, clear seas. The polyps build hard, cup-like external skeletons around themselves by secreting calcium carbonate ($CaCO_3$), and the hard shells accumulate to form coral reefs and islands.

Australia's Great Barrier Reef (Box 16.36) is the largest and best-known coral reef in the world. Many coral reefs are found close to the shore, on the continental shelf. In the open sea coral builds up to create *atolls* – low-lying coral islands, which are usually circular or horseshoe-shaped, with a lagoon in the middle. The tops of coral reefs and atolls lie close to the surface of the sea, so they have traditionally been a major hazard to shipping. Many boats have been sunk or damaged by hitting them.

The survival of many coral reefs is threatened by a combination of natural and human factors. A build-up of carbon dioxide in the atmosphere can lead to increased acidity of the surface waters in the ocean. This, together with increased penetration of ultraviolet light into the ocean, can reduce phytoplankton productivity. It can also change the carbonate content in surface waters, which could impair the growth of corals. Extensive coral bleaching has recently been associated with the warming of surface waters (Box 16.37).

The resources of many coral reefs around the world are heavily exploited for subsistence and local market use because many objects are made from coral, with its exotic colourings, including jewellery. Cruise ships also cause extensive damage to coral reefs, and studies in the Grand Cayman Islands in the West Indies suggest that it might take up to 50 years for a damaged reef to recover properly. Many isolated coral atolls also provide valuable land that is intensively used for tree crops, root crops, fisheries, mariculture and cottage industries, while some atolls have experimented with commercial fisheries and tourism.

Plate 16.4 *Wave refraction along the Atlantic north coast of Northern Ireland, near Mussendon Temple, County Londonderry.*
Photo: Chris Park.

Another source of stress for coral is pollution, and recent studies of coral reefs off the Florida Keys in the USA have shown dramatic damage associated with human disease pathogens (including cryptosporidium) and coral infections. In 1996, researchers found unhealthy reefs in twenty-four places and nine species of coral infected; by 1998, there were unhealthy reefs in 131 places, and thirty-one species were infected.

BOX 16.36 THE GREAT BARRIER REEF

The Great Barrier Reef is in fact a chain of coral reefs nearly 2,010 km long in the Coral Sea, off the north-east coast of Australia. This unique environment is the world's largest coral reef system, and it appears on the World Heritage List. The reef is economically important to the state of Queensland and the country of Australia, because it supports significant tourism and fishing industries. The reef is separated from the mainland of Australia by a channel that varies in width from about 16 km in the north to about 240 km in the south. The channel is relatively shallow, and it contains many coral atolls, which pose a hazard to boats.

The coral in the Great Barrier Reef is composed of the external skeletons of many different species of polyp, cemented together by a limestone substance produced by a type of algae to form exotic shapes and colours. It provides a habitat for many larger water animals, including over 1,000 species of fish.

Since the early 1960s, the reef has suffered extensive damage. Much of the damage has been natural, arising from the invasion of parts of the reef by the crown-of-thorns starfish, which feed on and destroy the coral. Detailed monitoring is also establishing many of the subtle ways in which human activities affect the reef, including:

- intensive recreational use (particularly in the outer reef);
- shipping accidents;
- runoff from the mainland, which increases turbidity, reduces salinity and introduces some pollutants (particularly on near-shore reefs);
- global climate change (via sea-level rise, ocean warming and increased air temperatures).

The Australian government has introduced legislation designed to limit destruction of the coral and preserve this unique ecological resource. The Great Barrier Reef Marine Park has been established to preserve the reef in perpetuity, and a comprehensive programme of water quality management within the park is designed to protect the reef and the industries that depend on it.

BOX 16.37 GLOBAL WARMING AND CORAL BLEACHING

Coral bleaching was first described in the 1920s. It occurs when corals expel the symbiotic algae that live in their tissues as a response to environmental stress. Stresses include particularly high sea temperatures, high solar radiation, varying salinity levels, extremely low tides, or a combination of these factors. Coral bleaching began to occur on a large scale in the mid-1980s. In 1998, coral bleaching was more severe than ever before, and it was reported from more than sixty countries. Although the exact cause is still open to debate, many scientists argue that only global warming could have triggered such widespread bleaching over such a short period of time.

The world's largest coral atoll system – the Chagos Islands, a remote archipelago of coral reefs uninhabited except for the US–British naval air base of Diego Garcia – has been severely damaged by the warm waters that washed through the Indian Ocean during the 1998 El Niño. The islands were home to 220 coral species, 380 molluscs and more than 700 species of fish, but by early 1999 most of the coral on the seaward reefs was dead, and up to half of the fish were dead.

Although reefs can recover from occasional bleaching events, the records show that widespread bleaching has struck tropical coral reefs seven times over the last 20 years. Computer models have predicted that reefs in South-east Asia and the Caribbean could experience annual bleaching by 2010, reefs off Tahiti are likely to be affected on an annual basis by 2030, and Australia's Great Barrier Reef could be affected between 2020 and 2030.

ESTUARIES

The mouth of a river, through which it flows into the sea, is known as an estuary. Estuaries are effectively arms of the sea that extend inland. Conditions in an estuary are heavily dependent upon the sea, because fresh water flowing down the river mixes with saline water from the sea, and tidal effects are felt. An interesting feature of many funnel-shaped estuaries is the bore – a tidal flood that surges upstream in a wall-like wave and occurs during high tides (particularly during spring tides) (see p. 484).

LANDFORMS

Many estuaries are drowned river valleys (*rias*) that were flooded when sea level rose in the past relative to the land level. This could be caused by a rise in sea level (eustatic change; see p. 511), a fall in land level (isostatic change; see p. 511), or both.

Sand bars form at the mouths of some estuaries, where fresh water and seawater meet. Most of the sediment is transported downstream along the river, but some can also be supplied by tides, ocean currents and longshore drift. Bars are also deposited within many estuaries, and they often shift location in response to changes in sediment input into the estuary system, causing problems for navigation and coastal stability.

ECOSYSTEMS

Estuary ecosystems are heavily influenced by the typical increase in salinity towards the mouth of an estuary; most estuary organisms are marine, so species diversify generally down the estuary towards the mouth. Biological productivity in most estuaries is extremely high (see p. 564), mainly because nutrients are trapped in the estuary by the interaction of river flow and tides. Many of the nutrients are stored temporarily in mudflats and salt marshes.

Estuaries normally support a wide variety of different species – particularly birds, amphibians and fish – because of the diversity of habitats (including sheltered inlets) and abundant nutrient supplies.

THREATS

The survival of many estuaries and the salt marshes that fringe them is threatened because of the damage caused by pollution, oil spills, dredging and reclamation. Recent attempts to manage the environmental resources of Chesapeake Bay (Box 16.38) illustrate some of the major difficulties also experienced elsewhere.

The success of the Chesapeake Bay Program is being eagerly watched by scientists and coastal zone managers elsewhere, because it is one of the pioneer elements in the US National Estuary Program (NEP). The NEP is designed to develop effective strategies for adaptive estuary management throughout the USA, in which:

- problems are identified at an early stage before they build up to critical proportions;
- remedial action is introduced to address the problems; and
- the solutions adopted are appropriate to each individual estuary.

In some ways estuaries are victims of their own success, in that they have traditionally been attractive sites for industry because of the ease with which industrial effluents can be disposed of in the sea, where they can be dispersed and thus diluted – once again illustrating the maxim that 'dilution is the solution to pollution', as with air pollution (see p. 241). Traditional approaches to estuary pollution control in Britain have relied heavily on the concept of assimilative capacity (estimating the maximum amount of pollution that can be dispersed within a given estuary system without causing long-term environmental damage). In recent years, this has been replaced by an environmental protection philosophy based on the precautionary principle (that pollution should be reduced at source rather than dealt with downstream).

COASTAL MANAGEMENT

Coasts are among the most dynamic and complicated zones within the environment because they represent the interface between marine and terrestrial environmental systems. We have already noted the heavy and growing concentration of people in the coastal zone, and the importance of natural coastal resources. As a result, coastal systems are very important but are often highly changeable, and this makes it vitally important to manage coasts effectively and sustainably. A sound understanding of coastal processes, landforms and adjustments is essential if this goal is to be achieved, and a great deal of research has been invested in coastal systems in recent years.

BOX 16.38 THE CHESAPEAKE BAY PROGRAM

Chesapeake Bay is a large inlet of the Atlantic Ocean on the east coast of the USA, between Maryland and Virginia. It is about 320 km long, between 6 and 64 km wide, and navigable by deep-water vessels along its entire length. Many rivers drain into the bay through estuaries, including the James, York, Rappahannock, Potomac, Patuxent and Susquehanna rivers. The bay has long been an important source of seafood (including oysters and crabs), as well as a valuable wildlife resource and recreational amenity.

Water quality has declined significantly in Chesapeake Bay in recent years as a result of pollution from upstream. An ambitious, expensive, wide-ranging, long-term clean-up programme involving many federal, state and local government agencies has been designed to tackle three key problems:

1. nutrient over-enrichment (eutrophication) (see p. 410);
2. contamination by toxic substances;
3. decline and loss of submerged aquatic vegetation.

Reconstructions of the environmental history of the bay, based on analysis of the pollen, diatom and chemical composition of sediment cores, show that eutrophication and declining water quality have increased a great deal in Chesapeake Bay since the time of European settlement. Since about 1760, sedimentation rates have increased tenfold, the input of total organic carbon has increased by a factor of 35, and the diversity of diatom species has declined steadily.

Clean-up will not be an easy task, because of the great variety of sources of pollutants into the bay and its estuaries. Much of the pollution (particularly by nitrogen) comes from non-point sources, which are notoriously difficult to regulate compared with identifiable point sources (such as factories and water treatment sites). Roughly a quarter of the anthropogenic nitrogen entering the bay, and about 14 per cent of the ammonium, comes via air pollution, so the programme must extend well beyond water resource management. It must also extend outwards beyond the bay itself, because much of the sediment input into the bay appears to be derived from coastal erosion. Coastal erosion contributes up to twelve times as much sediment as the major tributary rivers, most of which deposit their particulate load in marshes and on the bay bed within 1 km of the river mouth.

There are other complications too, including the fact that some of the pollution processes in the bay are not well understood, so detailed monitoring and research are required. New analytical approaches are being developed that incorporate remote sensing, geographical information systems and real-time monitoring of variations in temperature, salinity, dissolved oxygen concentration, turbidity, toxin levels, chlorophyll and nutrient levels.

SEA-LEVEL RISE

Without doubt the greatest threat at present comes from the prospect of sea-level rise associated with greenhouse gases and global warming (see p. 261). Global warming will inevitably promote thermal expansion of seawater (warm water occupies a greater volume than cooler water), as well as melting of the major polar ice caps – both of which will contribute to a progressive rise in sea level for the foreseeable future.

Induced sea-level rise will inundate many low-lying coastal areas, but it will also greatly increase the threat of coastal erosion. Coping with these problems will require major decisions, without delay, because massive investment in coastal protection schemes and long-term planning may be needed to ensure optimum protection. One particular problem in coastal protection is to ensure that an entire coastline is protected, not just key sites or sensitive locations. It would be easy to increase the protection offered to some parts of a coastline (for example by building solid sea defence walls) and at the same time transfer the erosion problem further along the coast by reducing sediment supply and interrupting the natural flow of longshore drift (see p. 501).

HARD AND SOFT PROTECTION

Many countries with extensive coastlines, and where major centres of population and industry are concentrated along

BOX 16.39 COASTAL STABILITY

The most pressing management problems in the coastal zone centre upon the management of shoreline change, particularly erosion. This requires a good understanding of coastal sediment transport processes, and of rates and patterns of coastal change at a variety of timescales.

Coastal stability is always difficult to guarantee, because it depends on the interactions between land and sea, and these are highly variable over time. Over the long timescale, sea level would fall again from its recent relatively high level if global temperatures fell far enough to promote the advance of ice sheets, the growth of valley glaciers and the development of a new ice age (see p. 343). On the other hand, a local rise in sea level would occur if the land were to sink relative to the sea – such as through subsidence associated with extensive groundwater abstraction (p. 393) or crustal warping associated with glaciation (p. 156).

the coast, have invested a great deal of resources and effort in coastal protection in recent decades. This is true of Japan, for example, where shore protection works are designed to reduce erosion and coastal flooding while preserving coastal scenery, maintaining access to waterfronts and enabling the use of beaches for recreation.

Engineers have traditionally favoured hard protection strategies, particularly where the survival of important structures, settlements or roads is threatened. But, increasingly, the practical and economic values of soft protection strategies, or – even better – comprehensive strategies involving both hard and soft protection, are being recognised.

Much of the coastline of south-east Florida is low-lying and at risk from erosion and flooding, and hard protection measures have been used extensively to protect the interstate highway that runs along the back of the beach parallel to the shoreline. Rock rubble – piles of large boulders dumped in front of the road, towards the back of the beach – has been used a great deal in the past, but more recent defence works have used interlocking concrete revetments (retaining walls). Both types of engineering structure have failed in places, and since the early 1970s a more comprehensive approach has been adopted, including beach replenishment (Box 16.41). This does not offer cost-effective protection, but it creates a recreational beach, which becomes a major economic asset.

BOX 16.40 APPROACHES TO COASTAL PROTECTION

There are many different ways of increasing coastal protection, but they fall into two main categories:

1. *hard protection*: by large-scale engineering schemes such as sea defence walls, which offer protection by replacing natural shore materials with more solid structures designed to withstand infrequent large storms and reduce coastal erosion under more normal wave and tide conditions;
2. *soft protection*: by using natural coastal materials (such as sand bars and beaches) and processes in such a way as to absorb much of the energy input from large storms and normal waves and tides.

BOX 16.41 BEACH REPLENISHMENT

One way in which the dynamic equilibrium of depositional coasts can be disturbed is by beach replenishment (beach nourishment). This involves the deliberate addition of sand to beaches where erosion is a major problem in order to stabilise losses and restore badly eroded (and sometimes unsightly) beaches. Sometimes a beach is replenished with sand mined further along the same coast (to reduce the cost of transporting the sand over long distances), often by dredging in the near-shore environment, which effectively short-circuits natural longshore drift processes and replaces wave action with trucks and diggers! Beach replenishment is widely used to protect beaches in north-west Europe, and in recent years cost-effective and environmentally sustainable working practices have been developed that could be adopted elsewhere.

Elsewhere along the eastern seaboard of the USA – in South Carolina, for example – recent experience suggests that soft protection is preferable to hard protection, and that coastal communities should designate zones behind the shoreline where development should be restricted to low-risk uses (a so-called setback policy). Similar strategies are being adopted in Britain, including along the coast of Norfolk in eastern England, which is faced with an increased threat of coastal erosion and flooding associated

Plate 16.5 *Palm Beach, Florida, USA. The beach acts as an important natural shock absorber to reduce the damaging impacts of storm surges associated with tropical hurricanes, and the sand is regularly replenished with sand brought in from the more sheltered west coast of southern Florida.*
Photo: Chris Park.

with global warming. The strategy there includes extensive use of soft protection, introduction of a setback policy, and planning regulations that require developments in areas of flood risk to be built with floors above potential flood levels.

BEACH MANAGEMENT

Natural beach systems often demonstrate dynamic equilibrium because they are able to adjust to changes in inputs and outputs over a variety of timescales. This is why beaches that have been badly damaged by storms can quickly recover stability, for example. But human activities can radically alter beach dynamics, either deliberately or accidentally, through such activities as sand mining, beach replenishment and dune stabilisation.

There is an added dimension to human interest in beach systems, because beaches provide important natural means of protecting coastlines, acting effectively as shock absorbers that reduce the damaging impacts of tides, waves and storms on coastal areas. Increasing attention is being paid to the creation and management of beaches for coastal defence purposes – the soft protection strategy – often as part of integrated coastal management strategies.

SAND MINING

One way in which the dynamic equilibrium of coastal systems can be disturbed is by sand mining, the deliberate extraction of sand from beaches and other depositional landforms for use elsewhere (including the construction industry). In some places, large quantities of sand have been taken away, and this can be a valuable and sustainable form of natural resource use if the quantities extracted are replenished by natural coastal processes (including longshore drift). But beach erosion and coastal instability quickly follow if rates of extraction exceed rates of replenishment. Predicting the threshold level that defines sustainable use is often difficult, and sometimes it has to be learned by trial and error, which can be a costly and damaging way to learn.

Coastal sand mining is common along some of the major beaches in New Zealand, where the Resource Management Act (1991) is designed to promote the sustainable management of natural and physical resources. Under the Act permission to mine sand from a coastal sand body is granted only after the likely environmental impacts of such operations have been determined. But critics argue that such assessments do not establish the sustainability of sand-mining operations, because they do not define the dimensions of the active sediment system, quantify the volume of the related resource, state the period within which sustainability is achievable, or consider the long-term cumulative effects of the extraction.

DUNE STABILISATION

Some beach systems have extensive sand dunes, formed by the accumulation of wind-blown sand in much the same way as desert sand dunes are formed (see p. 432). Dunes often provide natural sea defences, which protect low-lying coastal areas (like much of the Netherlands) from flooding. Beach dunes often have a natural vegetation cover that stabilises the dune and prevents excessive wind-blow of sand over the surrounding area. Species such as marram grass, which has an extensive, tightly knit underground root system and can survive in the relatively salty, nutrient-poor free-draining sand, are particularly common on beach dunes.

Dunes are also popular for recreation, partly because they offer a variety of places and often some privacy for people. Yet the vegetation cover is often fragile and can be easily disturbed, particularly when over-used (Box 16.42).

Dune erosion often starts with one or more blowouts. These are concentrated areas of sand erosion, which

BOX 16.42 DUNE EROSION

Once the protective vegetation cover has been broken, dune sand is exposed to onshore breezes and winds, and erosion quickly follows. This causes two types of damage:

1. loss of sand from the dune (thus instability of the landform);
2. problems where the wind-blown sand is deposited behind the dune (where natural soil and vegetation can be buried under a new sand cover).

Plate 16.6 *Stabilisation of sand dunes at the Hook of Holland, the Netherlands, by fencing (to keep walkers out) and planting marram grass (which has extensive networks of lateral roots to bind the dune sand).*
Photo: Chris Park.

become bigger over time as more sand is eroded and the surrounding vegetation is damaged. Positive feedback is involved in the growth of the blowouts – destruction of the vegetation cover causes erosion, which exposes more sand to erosion so the blowout grows bigger, destroying more vegetation, and so on.

One way of tackling the problems of dune erosion is to plant species that help to stabilise the sand surface. Dune stabilisation has been carried out since 1845 in the Cape Province of South Africa, for example, using Australian *Acacia* species. Initially a foredune was formed at the source of the drift sand by constructing barriers with wooden poles or marram grass; then the area behind this was thatched with brushwood or seeded with grass before alien woody species were introduced. By the late 1940s, it was becoming obvious that the introduced species were more hardy and competitive than most indigenous species, and this policy was radically altering the natural vegetation of the dune systems. The use of introduced species was stopped in 1974. The present policy is to stabilise areas only when absolutely necessary, using indigenous species.

COASTAL HAZARDS

Coastal areas are subjected to a wide variety of natural hazards, but among the most problematic are storm surges and coastal flooding.

STORM SURGES

In many coastal areas the most damaging hazards are storm surges, which have high water levels accompanied by strong winds and are often associated with the passage over the coast of a hurricane (see p. 318). Storm surges are extremely powerful, and the combination of high water and strong wind can cause extensive damage over a wide area. Some are powerful enough to flatten everything that lies in their path, including buildings, structures, trees and electricity lines. The storm surge associated with Hurricane Gilbert (see Box 10.20), the most intense hurricane of the twentieth century, caused damage along the barrier coast of north Yucatán in Mexico in 1988. Damage to beachfront structures, highways, boats and salt ponds was extensive along this coastline, which has been the focus of beachfront vacation home construction since the 1950s.

Many storm surges also cause rapid changes and adjustments to coastal landforms, particularly to beaches and dune systems, which can be severely eroded and take a long time to regain equilibrium after the storm surge has passed. The storm surge associated with Hurricane Hugo, which caused extensive damage along the South Carolina coast in 1989 (Box 16.43), was typical of such events.

Storm surges are created by high winds with a long fetch (path without obstructions) over open water. They are not tidal waves. The surface of the sea rises as a response to falling atmospheric pressure in the cyclone; water level rises about 2.5 cm for each 2.5 mb drop in pressure. This raises the water level overall, but the problem is made worse by the strong winds, which generate large waves. A 1 m rise in sea level during a violent storm might produce a storm surge 6 metres high, which crashes against the shore and can flood a large area along low-lying coastlines.

BOX 16.43 THE STORM SURGE ASSOCIATED WITH HURRICANE HUGO, 1989

Hurricane Hugo, which developed in the western Atlantic Ocean around the Caribbean Sea in September 1989, was one of the strongest hurricanes to hit the western hemisphere during the twentieth century. It caused an estimated fifty deaths and more than US$4,000 million in damage. In comparison, Hurricane Andrew, which affected the same area in 1992 (see p. 319), killed more than fifty people, caused an estimated US$12,000 million in damage and left many thousands of people homeless.

Hugo caused a great deal of damage along the coast of South Carolina when its arrival there coincided with high tide on 22 September 1989. Wind speeds near the eye of the storm were as high as 217 k h^{-1}, and the storm surge near the eye raised water level by nearly 6 m. Natural storm damage was extensive along the entire coastline, but the most serious damage was concentrated along the developed shorelines. In fact, the human changes to natural coastal topography made matters much worse than they might have been:

- Beach scour below sea defence walls was much higher than elsewhere.
- Major roads running at right angles to the shore allowed wash from the storm surge to extend much further inland (and cause damage over a much wider area) than elsewhere.
- Closely spaced buildings created tight channels, which concentrated the storm surge ebb and focused the damage on the buildings.

Much of the shoreline along the most badly affected parts of the South Carolina coast was covered by sand dunes, some of which were high enough not to be overtopped. Many small dunes were eliminated by the storm surge, but tall ones at least 30 m wide survived relatively intact. About half of the buildings that were completely destroyed or badly damaged were located behind narrow beaches (less than 3 m wide) and dune fields (less than 15 m wide).

FLOODING

Many coastal areas are low-lying, with natural sea defence systems (including beaches and dunes) that can readily be overtopped during storm surges, tsunamis or unusually high tides.

The evidence from many areas shows that coastal flooding is becoming a more serious hazard as the number of people living in the coastal zone continues to increase, at the same time as sea level is undergoing a progressive long-term rise and global warming threatens increasing hurricane and storm surge activity.

Many coastal defence schemes – particularly hard defence schemes – are constructed to prevent coastal flooding, but all such engineering solutions are designed to withstand events of a particular magnitude and frequency (the so-called 'design flood'). Many existing coastal defence schemes are unlikely to offer adequate protection if sea level continues to rise as expected (see p. 512). As a result, large areas – often heavily populated, and containing strategically important communication and commercial facilities – are at risk of flooding in the future. Venice in northern Italy has a longstanding flood problem, and recent attempts to find a long-term solution (Box 16.45) have considered both 'hard' and 'soft' engineering options.

Low-lying coastal areas face particularly difficult decisions, because of:

- the vast investment in coastal defence schemes required to increase protection to adequate levels;
- the long lead time required to make such important decisions; and
- the uncertainties surrounding projections of sea-level rise.

A case in point is the Netherlands, most of which lies below 50 m above sea level. Much of the western part of the country in particular is below sea level, and it is protected from regular flooding by an extensive network of barriers, sea defence walls and dikes. Much of the low-lying land is polders – areas reclaimed from the sea by being artificially drained and protected from flooding by building dikes (see Figure 13.6). Continuous pumping is required to keep the drained areas dry. The sea defence systems, in which the

BOX 16.44 COASTAL DEFENCE IN BRITAIN

Much of coastal Britain is also low-lying, and the country's sea defence system relies heavily on hard protection, particularly sea walls and embankments. These too would need to be raised to cope with possible sea-level rise if important industrial, residential, commercial and agricultural areas are to be protected from recurrent flooding. The Thames Barrier, a movable structure across the Thames just downstream from London, is intended to prevent much of central London from being flooded by the sea. It was designed (in the late 1960s and early 1970s) to cope with known variations in river levels in the Thames, and with known tidal currents and rates and patterns of silting in the Thames Estuary. There is mounting concern that the barrier and associated sea defence wall will not be high enough to withstand unusual combinations of high tides and storm surges running down the North Sea, particularly if the general sea level continues to rise as a result of global warming.

In 2000, the UK government estimated that it would cost at least £1.2 billion over the next 50 years to improve sea and river defences in England and Wales to cope with global warming. It was also accepted that some coastal areas (especially in eastern England) will increasingly be abandoned to the sea because they have become uneconomic to defend.

BOX 16.45 SOLVING THE FLOOD PROBLEM IN VENICE

Venice is a popular coastal holiday resort on the north-east coast of Italy, with a rich historical and artistic heritage and a thriving tourism industry. But it also has a recurrent flood problem, and particularly since the city was seriously flooded in 1966 great thought has been invested in the search for a lasting solution to the problem. Venice is built at the north end of a small cluster of low-lying islands set in a large lagoon of tidal mud flats, measuring 50 by 10 km, and separated from the Adriatic Sea by a natural barrier of low islands.

Engineers have favoured construction of a series of giant barriers to form a movable shield against the waters of the Adriatic. The scheme, which would cost an estimated US$2.6 billion to construct and US$2 million a year to maintain, is called the Moses Project. It is based on installing seventy-nine huge hinged flaps (each one 30 m long and weighing up to 350 tonnes) at the three main entrances to the lagoon, each anchored to the sea bed. During major floods (which happen about six times a year and which flood up to 10 per cent of the city), the hollow flaps would be filled with compressed air and would rise up to block the entrances. The scheme envisaged that the barriers would be raised for 4–5 hours roughly twelve times a year, taking 30 minutes to erect and 15 minutes to deflate.

Early in 1999, the Italian government abandoned the ambitious barrier scheme in favour of a more ecologically acceptable alternative (a soft engineering solution rather than a hard engineering one; see Box 16.40). The aim of the new scheme is still to isolate the lagoon from the sea, but this time using silt rather than concrete. Stage one involves many small simple measures to help to protect the city, including flood-proofing buildings, raising banks and walkways, fitting drains with valves to stop them back-washing, barricading the most vulnerable islands, and keeping the canals between the islands free from debris. Stage two is more long-term, and it involves restoring the Venetian lagoon landscape closer to its original nature before extensive engineering schemes were carried out over the past 600 years. Two key steps in stage two, designed to reduce the sea's scouring action and prevent high tides pouring into the lagoon, will be to narrow the wide entrances to the lagoon (which were dredged out 100 years ago), and to stop dredging the shipping channels so that they refill to half their current depth of 20 m. However the problem is tackled, it remains imperative that steps be taken to protect Venice and its treasures from further flooding. In 1900, St Mark's Square was flooded on average six times a year, for a day or so each time. In 1996, it was under water for 101 days, and it was flooded eighty times in 1998. By 2050 – if nothing is done, current trends continue unchecked, and sea level continues to rise as projected by the IPCC – the square could well be inundated at every high tide.

dikes play a crucial role, have been substantially expanded and extended since 1953, when the area was badly flooded. But expensive further strengthening of coastal protection (particularly raising the heights of the dikes) will be necessary to cope with a projected rise in sea level in the order of 1 m over the next 100 years.

SEA-LEVEL CHANGE

Sea level – the mean level of the surface of the sea – is important because it determines the location of the boundary between land and sea. The evidence from most coastlines indicates that sea level is rarely constant (even allowing for daily and seasonal patterns of the tides), and that the present-day sea level in many places is higher than it has been in the past. Large-scale landform and geological evidence also indicates that sea level has been a great deal higher in the geological past.

Sea level is taken to be the mean level of the surface of the sea between high and low tide, and the universal datum is based on the tidal observatory at Newlyn in Cornwall. Sea level is used as a standard base for measuring heights and depths on the Earth, and it is the baseline for plotting the global hypsometric curve (see Figure 5.2) and for calculating the heights of mountains (see Table 5.3) and the depths of ocean trenches (see Table 16.2). Atmospheric pressure readings are usually expressed as sea-level equivalents in order to eliminate the effects of altitude.

BOX 16.46 RELEVANCE OF SEA LEVEL

Sea level influences such factors as:

- ease of coastal navigation and access to harbours, canals and waterways;
- exposure or inundation of beach systems and their associated dunes;
- rate and pattern of coastal erosional and depositional processes;
- suitability of estuary and coastal sites for industrial and residential development;
- the amount of land that is permanently flooded;
- areas suitable for reclamation by building embankments and dikes;
- the location and extent of areas of the coastline that are subject to storm surges and periodic flooding;
- the height and extent of coastal defence schemes;
- choice of locations best suited to tidal power generation.

BOX 16.47 RELATIVE SEA LEVEL

Sea level is a relative phenomenon that can change for one or both of two reasons:

1. The sea level changes (because of changes in the volume of water stored in the oceans); such changes are described as eustatic. Rapid eustatic falls in sea level are caused mainly by the melting of ice sheets (see p. 354), but more gradual eustatic changes can be associated with movements of the ocean floor and with sedimentation on the ocean floor.
2. The land level changes (because it is moved upwards or downwards by tectonic forces); such changes are described as isostatic (see p. 156).

CAUSES OF CHANGE

A global rise or fall in sea level must, by definition, be eustatic (Box 16.47), and the geological evidence indicates that most such changes are associated with the formation and melting of ice sheets and with the thermal expansion and contraction of ocean water.

Ice sheets and the oceans are major stores in the global water cycle (see p. 352), which contains a finite amount of water. Together they account for nearly 99 per cent of all of the water on Earth (see Table 12.1). Climate changes that trigger adjustments in the location and size of the major ice sheets also trigger changes in sea level, both directly (by the thermal expansion and contraction of seawater as air temperature and thus sea temperature rise and fall) and indirectly (sea level is high when ice sheets melt, and it is low when ice sheets form).

Ignoring the thermal expansion of seawater, to keep the calculation simple, a 1 m rise in sea level around the world would be the equivalent of about 360,000 km^3 of water. This is the surface area of the oceans (360 million km^2) multiplied by 1 m depth of water. There is an estimated 26 million km^3 of water stored as ice in the global water cycle (see Table 12.2), which means that if – at least in theory – all the ice were to melt, sea level would rise by about 72 m above where it is today. This would be high enough to flood many of the major cities in the world, and the world land map would look very different from what it is today.

NATURAL CHANGE

As noted above, most of the natural long-term variations in sea level are associated with climate change, particularly the magnitude, duration and timing of ice ages. At the close of the Pleistocene ice age, when the Holocene began about 11,000 years ago, global sea level was about 54 m lower than it is today. Nearly half of the present US continental shelf was then a coastal plain with vegetation ranging in nature from subarctic tundra to coniferous woodland. As temperatures rose so did sea level, during the Flandrian transgression, but the rise was not always at the same rate. The transgression reached the present coastline about 6,000 years ago, drowning river valleys to create estuaries and embayments. During the late Holocene (see p. 344), global sea level has fluctuated a little in response to a series of warm stages (when storms were more frequent)

BOX 16.48 EVIDENCE OF SEA-LEVEL CHANGE

While sea level has varied a great deal over geological time, the evidence suggests that:

- Average sea level over all the changes of the last few glaciations and interglacials has been about 50–60 m below its present level.
- During the height of the last glaciation (about 18,000 years ago), it was between 110 m and 140 m lower than it is today, and large areas of the continental shelves were dry land.
- Sea level has risen through the post-glacial period (this is called the Flandrian transgression).
- The post-glacial sea-level rise was fast initially, but it levelled off about 5,000 to 6,000 years ago.
- Sea level has remained relatively close to its present position over the last 5,000 to 6,000 years.

Plate 16.7 *Roman harbour in Tunisia, North Africa, which is now partly inundated as a result of sea-level rise over the last 2,000 years.*
Photo: Philip Barker.

punctuated by cooler intervals (when valley glaciers advanced).

Warming since the Little Ice Age (see p. 344) has resulted in continued sea-level rise, which is creating difficulties for many countries because, among other problems, it causes coastal erosion and the loss of tidal wetland habitats. Analysis of tide data from around the world indicates that eustatic sea level has risen by between 12 and 15 cm over the past century.

The slow, progressive rise in sea level continues, and this is likely to increase further the risk of coastal flooding in many areas. The problem is most serious along some delta coasts, such as Louisiana in the USA, where land subsidence has increased relative sea-level rise by almost 10 times the present world average. Relative sea-level changes in China over the last 80 years also vary from place to place because of local crustal deformations and subsidence associated with groundwater extraction.

Superimposed on these largely natural changes in sea level are the observed and possible effects of sea-level changes associated with global warming.

INDUCED CHANGE

A key component of the global warming debate (see p. 6) is the likelihood of an accelerated rate of sea-level rise during the next century as a result of culturally induced global climate change. A great deal of scientific research – often in collaborative international programmes such as the Intergovernmental Panel on Climate Change (IPCC), the Regional Seas Program of the United Nations Environmental Program (UNEP/RSP), and the 1992 United Nations Conference on Environment and Development (UNCED) – is now being devoted to the problem of sea-level change associated with global warming. It is clearly a global problem, which requires a global solution.

Debate continues about the likely speed, overall level and timetable of sea-level rise, but many scientists now expect a rise of up to 60 cm by 2050, up to 1 m by 2100 and up to several metres by 2200. Environmental change on such a scale would be without precedent, and the impacts would be serious (Box 16.50). Coping with such changes will require vast investment (of the order of billions of dollars) in large coastal protection schemes, wetland conservation measures, flood protection and water supply for coastal cities.

A 1999 report from Britain's Hadley Centre predicted that – whatever action governments take to halt global warming – sea levels are set to rise by at least 2 m over the next few hundred years (because of thermal expansion of the ocean as heat penetrates deeper and deeper into the ocean).

Scientists accept that some global increase in sea level is now inevitable, but they stress that the rate and extent of change depend on what action is taken by society today. Land-use policies can determine whether an area will have developed 100 years hence, and such decisions must take into account the real possibility of substantial sea-level rise

BOX 16.50 POSSIBLE IMPACTS OF PROJECTED SEA-LEVEL RISE

There are many possible impacts, including:

- damage to many important coastal ecosystems, including deltas, coral atolls and reefs;
- flooding of many densely populated areas (with a total population of up to 1,000 million people);
- damage to port facilities and coastal structures;
- severe coastal erosion in many countries, including loss of beaches and dunes;
- salinisation of many important groundwater resources through salt water intrusion (see p. 403);
- decline or loss of production in up to one-third of the world's croplands.

BOX 16.49 OBJECTIVES OF OCEAN RESEARCH

Key objectives of ocean research include:

- monitoring current rates and patterns of environmental change;
- identifying critical thresholds in the relevant environmental systems;
- establishing a better understanding of how the different environmental systems fit together and interact;
- modelling the global atmospheric system and the interactions between it and the ocean system;
- forecasting likely rates and patterns of sea-level rise;
- evaluating a range of coping strategies for dealing with the problem.

over that timescale. Important decisions will also have to be made to determine which coastal areas should be protected with dikes and embankments, like the Thames Barrier (see Box 16.44).

Among the places that stand to lose most if projected sea-level rises actually occur are the small island states. Most of these developing countries are low-lying, so even a moderate sea-level rise could flood much if not all of their land area. In 1987, a special Oceans and Coastal Areas Program Activity Centre was established under the auspices of the United Nations Environment Program to identify which countries were most at risk. Already, rising sea level is causing serious ecological problems in some parts of the Pacific Ocean. The Carteret Islands, among the most densely populated coral atolls in the Pacific, lie to the north-east of Papua New Guinea and they are experiencing problems that prevent the inhabitants from supporting themselves from their own resources. Resettlement on the east coast of Bougainville began in 1984 but has not been entirely successful, and since 1989 the number of people emigrating from the islands has increased greatly. The entire population will have to be relocated at some time in the future if recent problems continue.

One of the six island states that could disappear completely when sea level rises is Tuvalu (formerly the Ellis Islands), which lies roughly 1,000 km north of Fiji in the South Pacific Ocean. Tuvalu consists of nine coral atolls with a total area of 26 km^2, which are largely covered by coconut palms. The 9,000 or so people of Tuvalu would have little option but to abandon their homeland and subsistence agriculture to its maritime fate if sea level rises as high as some scientists expect, and to resettle elsewhere on higher land. Their culture, history and archaeology could be lost to the sea for ever. Another island state whose future looks bleak is Kiribati in the South Pacific, which is spread over 3.6 million km^2 (larger than Western Europe), 99.9 per cent of which is ocean. The land area is a mere 700 km^2 on thirty-three islands. Kiribati was the first place on Earth to see in the third millennium because of its location astride the International Date Line (see p. 117).

WEBSITE

Links to relevant websites, a comprehensive bibliography, tools for teaching and learning, and downloadable images relevant to this chapter can be found at the website specially designed to accompany this book at
http://www.park-environment.com

SUMMARY

Oceans cover more than two-thirds of the Earth's surface, and in this chapter we explore their relevance to people and their interactions with other major environmental systems. The chapter opens with an overview of the importance and structure of the oceans and the properties of seawater. Ocean currents play a vital role in distributing heat around the Earth, and we looked at gyres in general and the North Atlantic Drift in particular. Special attention was given to El Niño events, their causes and impacts, because they affect many other environmental systems and cause widespread damage and disturbance. We also examined tides and what influences them. The oceans offer a range of important resources, including fish for various uses, and we considered factors causing changes in fishing behaviour and yields. The oceans also hold potential for extracting valuable minerals and metals, and as a renewable source of power (thermal, tidal and wave-based). Many ocean resources are being badly damaged by pollution, and we highlighted oil pollution and eutrophication as the worst problems overall. The final section in the chapter deals with coasts, as the point of contact between land and sea. Here we examined the major erosional and depositional processes that create and shape coasts, and we considered the threats to the world's coral reefs and estuaries. We also looked at the challenges of coastal management, including beach management, coastal hazards, and sea-level change (natural and induced by humans). Global warming is already causing changes to many parts of the ocean system, and governments are having to adopt costly measures to protect coasts, preserve ocean resources and cope with the wider consequences of changing ocean currents and processes.

FURTHER READING

Bird, E.C.F. (1996) *Beach management*. John Wiley & Sons, London. Outlines the origin of beaches, the processes at work on them, the causes of beach erosion and deposition, and different approaches to beach management.

Briggs, D.J., P. Smithson, K. Addison and K. Atkinson (1997) *Fundamentals of the physical environment*. Routledge, London. Comprehensive introduction to the complex interactions of the natural environment.

Carter, R.W.G. (1990) *Coastal environments: an introduction to the physical, ecological and cultural systems of coastlines*. Academic Press, London. A comprehensive review of the physical and biological resources of coastlines, and how they are exploited and used.

Coull, J.R. (1993) *World fisheries resources*. Routledge, London. A broad introduction to ocean fishing.

Davis, R.A. (1993) *The evolving coast*. W.H. Freeman, Oxford. Introduction to the evolution, nature and dynamics of coastal systems, with special emphasis on the role of tectonic plate movements and sea-level change.

Deacon, M., T. Rice and C. Summerhays (eds) (1999) *Understanding the oceans*. UCL Press, London. A series of essays that chart the development of new ideas, theories and approaches over the last century.

Dyer, K. (1997) *Estuaries: a physical introduction*. John Wiley & Sons, London. Introductory textbook that describes the basic physical processes in estuaries, and examines their controls and impacts.

French, P. (1997) *Coastal and estuarine management*. Routledge, London. Useful and clearly structured review of the problems and prospects of coastal management.

Gren, I.M., R.K. Turner and F. Wulff (2000) *Managing a sea: the ecological economics of the Baltic*. Earthscan, London. Interdisciplinary examination of the causes, consequences and importance of pollution in the Baltic Sea.

Iudicello, S., M. Weber and R. Wieland (1999) *Fish, markets and fishermen: the economics of over-fishing*. Earthscan, London. Looks at fisheries under different management regimes and concludes that the reasons for over-fishing are mainly economic.

Jones, E.J.W. (1999) *Marine geophysics*. John Wiley & Sons, London. Reviews the techniques for studying, and geophysical evidence of, the structure and dynamics of the ocean floor.

Kay, R. and J. Alder (1998) *Coastal planning and management*. E & FN Spon, London. Comprehensive review of practical options and management solutions.

Komar, P.D. (1998) *Beach processes and sedimentation*. Prentice Hall, Hemel Hempstead. Introduction to beach processes from an engineering perspective.

Lonsdale, P. (1997) *Deep sea geomorphology*. Blackwell, Oxford. An up-to-date review of a largely neglected field of study – the development and significance of landforms on the ocean floor.

Nordstrom, K.F. and C.T. Roman (eds) (1996) *Estuarine shores: evolution, environments and human alterations*. John Wiley & Sons, London. Series of essays that explore the interrelationships between different parts of estuary systems.

Nunn, P. (1994) *Oceanic islands*. Blackwell, Oxford. Explores the climate, geology, evolution, environments and development of oceanic islands around the world.

Pilson, M.E.Q. (1998) *Introduction to the chemistry of the sea*. Prentice Hall, Hemel Hempstead. Stresses the interrelationships between the major marine processes, emphasising the chemistry of seawater as a common link.

Pirazzoli, P.A. (1996) *Sea-level changes: the last 20,000 years*. John Wiley & Sons, London. Describes and accounts for the last 20,000 years of sea-level change, spanning the last glacial maximum and recent global warming.

Rice, T. and P. Owen (1999) *Decommissioning the* Brent Spar. E & FN Spon, London. Outlines the history of the *Brent Spar* saga and charts how the environmental problem was handled and what lessons can be drawn from the whole experience.

Ross, D.A. (1995) *Introduction to oceanography*. Longman, Harlow. Useful introduction to the physical, chemical and biological aspects of ocean systems.

Simmonds, M. and J. Hutchinson (eds) (1996) *The conservation of whales and dolphins*. John Wiley & Sons, London. Analyses the complex issues surrounding the conservation of whales, dolphins and other cetaceans, with an emphasis on threats and responses.

Summerfield, M.A. (ed.) (1999) *Geomorphology and global tectonics*. John Wiley & Sons, London. Detailed review of recent advances in understanding and studying the Earth's topography.

Thurman, H.V. and A.P. Trujillo (1999) *Essentials of oceanography*. Prentice Hall, London. Useful broad overview of the scientific principles needed to understand ocean systems.

Upton, D. (1996) *Waves of fortune: past, present and future of the UK continental shelf*. John Wiley & Sons, London. A detailed examination of the North Sea, its problems and prospects.

Viles, H. and T. Spencer (1995) *Coastal problems: geomorphology, ecology and society at the coast*. Edward Arnold, London. A wide-ranging overview of the pressures on coastal systems and of scientifically informed management strategies, which draws on examples from around the world.

Wells, N. (1997) *The atmosphere and ocean*. John Wiley & Sons, London. Explores the interrelationships between atmospheric and ocean systems, including recent research on climate change.

Part Five
The biosphere

Chapter 17

The Biosphere

LEARNING OBJECTIVES

When you have finished studying this chapter, you should be able to

- *Appreciate the relevance of the biosphere and of biodiversity, and of the threats they both face*
- *Define what is meant by 'biosphere' and 'biodiversity' and outline how they have evolved*
- *Show how organisms are classified and suggest why this matters*
- *Summarise the difficulties of establishing how many species there are and say why conserving them is important*
- *Describe the principles and applications of evolution and outline how species have evolved over geological time*
- *Explain how extinction occurs and why it matters*
- *Understand the need for conservation and the main approaches to it*
- *Outline the main categories of protected area and explain how they fit into a global conservation strategy*

In this chapter, we examine the nature of the biosphere and explore the meaning and significance of biodiversity, which reflects the number and variety of species in the biosphere. Both are a product of the evolution and extinction of species, and both require conservation if they are to survive the stresses and impacts of modern human activities.

LIFE ON EARTH

Although the Earth is relatively large, certainly compared with the largest landforms and features on its surface (see p. 126), life on the planet exists within a relatively small zone around it.

DEFINITION

The biosphere is the part of the Earth that is occupied by living organisms (*biota*), and this includes parts of the atmosphere, hydrosphere and lithosphere. The concept of the biosphere as the Earth's integrated living and life-support system was first proposed in the 1920s, but only in recent decades has it been widely adopted and used. Originally, the concept was applied just to the Earth's surface where plants and animals obviously make their home. But it has more recently been extended by the Gaia hypothesis (Box 17.1) to include parts of the atmosphere and sub-surface geology, which were previously thought of as *abiotic* (non-living).

BOUNDARIES

Conditions throughout most of the Earth's system – certainly below the outermost part of the crust (see p. 128) and above the lower layers in the atmosphere (see p. 226) – are largely unsuitable for life, because they are either too hot, too cold, under too much pressure or too little pressure. Most of the living organisms on Earth are found on or near the surface of land or water, so the biosphere is a very thin layer around the Earth (Figure 17.1). Above the surface, in the atmosphere, only a few bird and insect species survive above about 6.5 km. The distribution of life in the atmosphere is constrained mostly by air density (which affects ease of flight), availability of oxygen for respiration, and temperature.

BOX 17.1 GAIA AND THE BIOSPHERE

The biosphere is made up of the living parts of the Earth's system, including plants, animals and humans. Until quite recently, scientists explained the evolution of plant and animal species in terms of adaptation to the chemical and physical properties of the environment, as a cause–effect relationship. The Gaia hypothesis (see p. 119), in contrast, sees the living (biotic) and non-living (abiotic) parts of the Earth system as mutually interdependent, each having been influenced by the other, and with the survival of each dependent on the survival of the other. This new perspective has many implications for the future of life on Earth, and it raises many questions about why and how nature should be conserved.

BOX 17.2 THE BIOSPHERE AND ENVIRONMENTAL SYSTEMS

The biosphere integrates the other main environmental systems. It:

- links the lower atmosphere (troposphere) with the lithosphere;
- provides a vehicle for the transfer of chemicals via the biogeochemical cycles;
- plays important roles in the hydrological cycle;
- affects rates and patterns of weathering within the lithosphere; and
- contributes to the global energy system.

BOX 17.3 LIMITS TO THE BIOSPHERE

The biosphere extends from less than 11 km below sea level to the tropopause, which is less that 17 km above sea level. This gives it a maximum thickness of 38 km – roughly 0.5 per cent of the radius of the Earth (6,371 km). The Earth's thin living skin is much more important than its size suggests!

BOX 17.4 EVOLUTION OF THE BIOSPHERE ACCORDING TO THE GAIA HYPOTHESIS

According to the Gaia hypothesis (see p. 119), as life evolved on Earth it both created and was created by biological processes that have radically altered the chemistry of the planet and its environment in ways that promoted further biological evolution of plants and animals on land and in the sea. This planetary positive feedback (p. 119) gave rise to large-scale dynamic equilibrium between life and the planet, which we would be foolish to ignore or disturb.

A key part of this complex feedback mechanism was the creation by early life forms (by photosynthesis) of an oxygen-rich atmosphere. The evolution of a soil cover was also highly important, because it links the evolution of life forms to the evolution of landscapes and the initiation of soil formation. Feedback also involved interrelationships between the evolution of plants and animals, the development of weathering processes, and the stability of the lithosphere.

Below the surface, the distribution of organic life is also constrained by environmental factors. Most soil organisms are found in the top few centimetres, although organisms do exist at much greater depths. For most of the twentieth century, scientists believed that bacteria lived down to about 1.5 m in the topsoil or ocean mud, then quickly disappeared. However, with the development of better techniques for detecting bacteria and more extensive subsurface exploration by coring, new evidence is emerging that there is life – single-celled bacteria – at least 500 m below the surface on land and at least 750 m beneath the ocean floor. It seems likely that microbes are involved in many subterranean geochemical processes, such as diagenesis and weathering, and in the oxidation or reduction reactions of metals, carbon, nitrogen and sulphur.

In the oceans, organic life is concentrated near the surface, constrained by the penetration of light (for photosynthesis). The biologically productive euphotic zone (see p. 477) is generally less than 150 m deep, below which light levels are poor. A small number of organisms live on the deep ocean floor, relying on organic matter that sinks down from above and on dissolved minerals released from hydrothermal vents associated with ocean trenches.

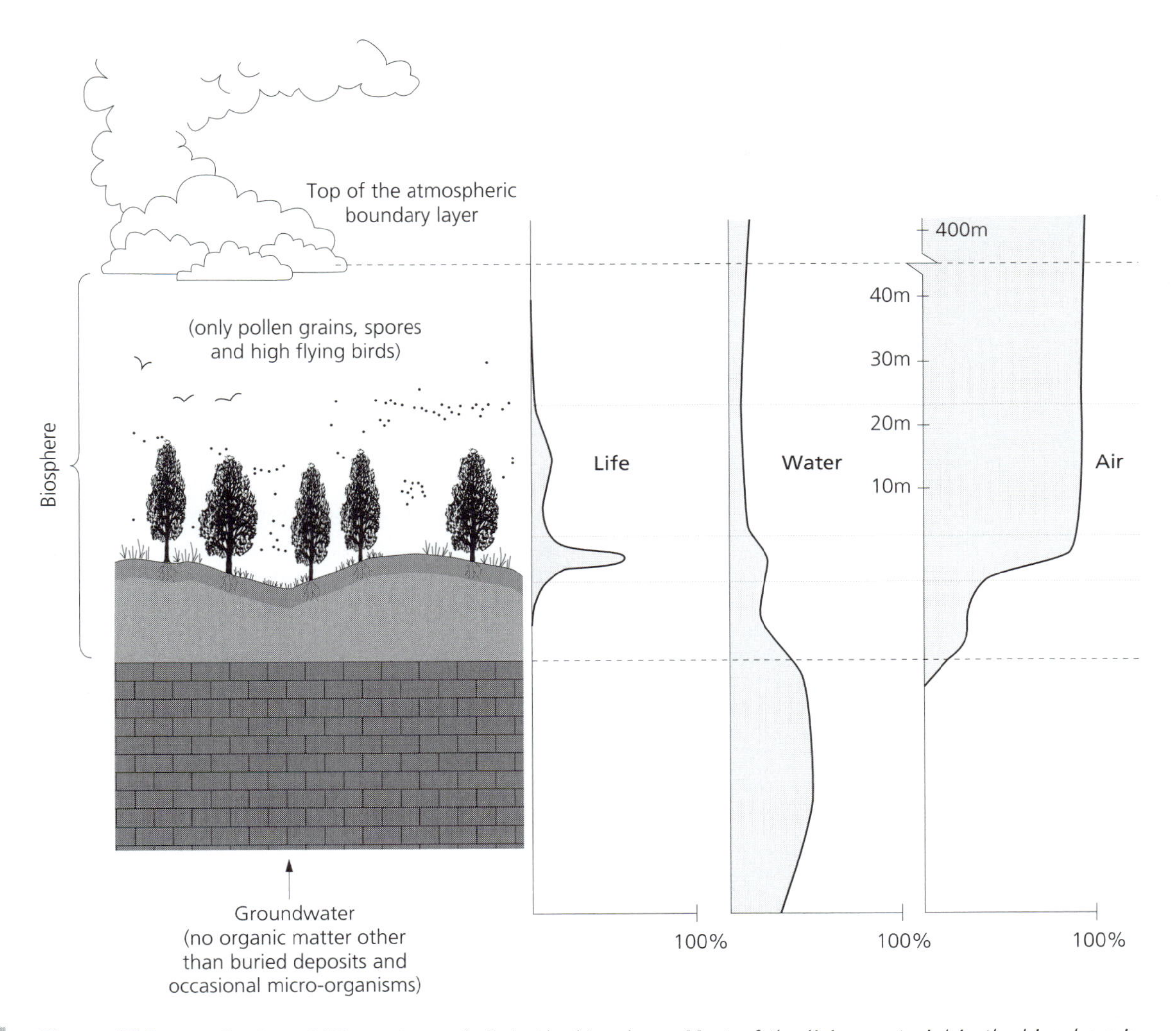

Figure 17.1 *Distribution of life, water and air in the biosphere. Most of the living material in the biosphere is concentrated within the relatively thin layer comprising the soil and the lowest 30 m of the atmosphere.*

Source: Figure 3.3 in Marsh, W.M. and J.M. Grossa (1996) Environmental geography: science, land use and Earth systems. John Wiley & Sons, New York.

EVOLUTION OF THE BIOSPHERE

The Earth is probably one of the few planets in the entire universe where conditions exist that are suitable for life (see p. 109), at least in forms we would recognise. The ultimate origin of this living material remains shrouded in mystery. The basic building blocks were probably derived from organic compounds found in interstellar clouds and in old parts of the solar system such as comets and asteroids, and the first cells probably emerged during the Archaean (see Table 4.7), around 3,500 million years ago.

Detectable forms of life have existed on the Earth for three-quarters of its history – 3,500 million years out of the 4,600 million years since the planet was formed (see p. 113). Over that time life has evolved, and as it evolved it diversified into more and more different species, giving rise to the abundant biodiversity we can witness today.

FUTURE PROSPECTS

If the origin and evolution of the biosphere are uncertain, there is little more certainty surrounding its future prospects. Without the intervention of human activities the biosphere would probably have evolved to be richer, more varied, more stable and more resilient to change than it is today.

Given the many ways in which human activities affect the biosphere (Figure 17.2), generally for the worse, it is

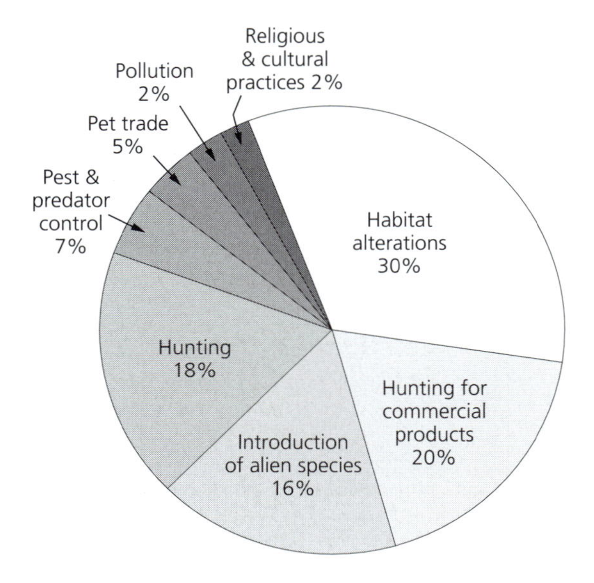

Figure 17.2 *Human activities that cause species extinction. Many different factors give rise to the extinction of species, among the most important of which are habitat alterations, hunting for commercial products, hunting and the introduction of alien species. More than one activity may be involved in causing the extinction of a particular species.*

Source: Figure 14.11 in Cunningham, W.P. and B.W. Saigo (1992) Environmental science: a global concern. Wm. C. Brown Publishers, Dubuque, Iowa.

BOX 17.5 HUMAN IMPACTS ON THE BIOSPHERE

There are many different ways in which humans can affect the biosphere, including:

- environmental pollution;
- habitat removal or damage;
- changing the structure and distribution of vegetation and soils;
- over-exploitation of renewable resources;
- altering the species composition of an area (such as through agriculture or forestry);
- introducing exotic organisms (including genetically engineered ones) into new places or environments;
- replacing natural biological variety with monocultures (single-species farming and forestry);
- elimination of biological pests (which can disturb natural feeding relationships);
- inappropriate disposal of wastes;
- disturbing the equilibrium of important environmental systems.

currently depleted and vulnerable and its long-term future looks questionable. Computer modelling of the long-term consequences of the greenhouse effect (see p. 258) suggests that the biosphere could survive for at least another 900 million to 1,500 million years, depending on whether CO_2 or temperature, respectively, is the limiting factor. Less than 1,000 million years after that, the Earth may lose its water to space to become like Venus, its sister planet (see p. 109).

HUMAN IMPACTS

People are important parts of the biosphere, and the evolution of human forms and human intelligence cannot be separated from the evolution of the biosphere at large. Yet we are the only species that consciously alters the environment and the ecosystems that depend on it, for our own ends. In doing so, we are altering our own life-support system, often in ways that are unsustainable. Quite when we will reach the most critical environmental thresholds – when irreversible change sets in that threatens the very survival of people and the planet – is a matter of debate among scientists. Many hope that we find out before it is too late and we have crossed them!

In the past, most human impacts were local and regional, so that while some areas showed signs of ecological stress or damage, the biosphere at large was relatively stable and environmental self-repair (see p. 79) was possible. As the human population has continued to grow, spread across the globe, develop new technologies and intensify its use of natural resources (see p. 10), many of these problems are now becoming global in scale and significance (Box 17.5).

Loss of biodiversity (see Box 17.9) as a result of human activity is now a major concern. Many different types of human activity, including agriculture, forestry and urbanisation, have tended to reduce biodiversity. As a result, they have led to the extinction of particular species, which in turn has reduced the gene pool from which new species might have evolved and which provide natural variability (a source of adaptability and resilience) in existing species.

One positive way in which humans affect the biosphere is via conservation schemes designed to protect species, habitats and ecosystems. Ironically, these are usually required to repair the damage that people caused in the first place!

BOX 17.6 BIOSPHERE II

Biosphere II was a major ecological test project, described by some scientists as a 'planet in a bottle', designed to simulate the way the real world (Biosphere I) works. The experiment began in the Arizona desert in May 1991. It involved artificially recreating a range of habitats under a geodesic glass dome, using nearly 4,000 natural species representing tropical rainforest, salt marsh, desert, coral reef, savannah and intensive agriculture. Eight humans were part of the experiment, and they lived on a self-sufficient basis inside the sealed dome for two years, effectively isolated from the outside world.

The aim of Biosphere II was to see how effectively the recycling of air, water and waste could work in an enclosed environment, and to evaluate whether a stable ecosystem could be created under the experimental conditions. It was, to all intents and purposes, the equivalent of a prototype space colony. The experimental dome was self-contained and self-sufficient, except for some electricity fed in from outside. The people in the experiment also maintained contact with the world outside via a computer link with a team of scientists, who continuously monitored the changing environment within the dome.

The project cost an estimated US$100 million, some of which was recouped by charging visitors to watch the experiment through the glass dome.

BOX 17.7 SCALES OF INTEREST IN ECOLOGY

Ecology stresses unity and interrelationships. It operate at a variety of scales, including:

- *individual organisms*: where the main interest might lie in behavioural ecology or feeding strategies;
- *populations*: perhaps with an emphasis on population dynamics; and
- *the community*: where the focus might be on competition between species for access to resources.

Many ecologists favour the ecosystem – short for ecological system – as the fundamental unit of study, because it represents the entire community of organisms and the habitat in which they live.

BOX 17.8 CORE THEMES IN ECOLOGY

Core themes in modern ecology, which define it as a science and as a (global) context within which to make political and management decisions, include:

- the relationship between habitat stability, rate and direction of evolution and biodiversity;
- the relationship between environmental transformations caused by human activities and by extinction;
- the importance of diversity as an attribute of ecosystems and of the biosphere, and the significance of losses of diversity;
- carrying capacity and the extent to which 'useful productivity' can be enhanced and sustained;
- resilience: the extent to which nature can restore itself.

ECOLOGY

Ecologists were among the first scientists to express particular concern about the state of the environment, and they have played a leading part in trying to find solutions to the environmental crisis and ways of implementing sustainable development (see p. 26) ever since. This reflects their interest in the interconnectedness of nature, which includes humans and their actions.

DEFINITION

Ecology is the study of the relationship of plants and animals to their surroundings and to each other. The emphasis is on relationships between organisms and the environments in which they live, including all living (biotic) and non-living (abiotic) components. The term was first used in 1866 by the biologist Ernst Haeckel, but more recently it is generally taken to mean the study of the structure and function of nature.

APPLICATIONS

A knowledge of ecology is essential for the conservation of wildlife around the world, but ecology is increasingly being looked upon to provide answers to broader questions, such as how the Earth system operates and how it is responding to global change (particularly global warming).

Some critics argue that ecology has limited use because of the difficulties associated with defining foundational ecological concepts and with formulating regularities about allegedly unique phenomena. Supporters stress that the emergence of ecology has encouraged the development of large-scale inter- and multidisciplinary research efforts, which complement the more traditional single-discipline, single-investigator approach in science. Certainly, without ecology the broader field of environmental management would be much poorer and less able to propose viable approaches to sustainable development of renewable resources. Most of the rest of this chapter and Chapter 18 rely heavily on the theories, principles and practices of modern ecology.

BIODIVERSITY

There is huge variety in nature. It encompasses a wide range of sizes of individuals, from the smallest plankton to giant whales and from tiny lichens to giant redwood trees. This richness and variety is an important part of nature – created by evolutionary processes (see p. 531), expressed in biodiversity (Box 17.9) and worthy of conservation (p. 540).

BOX 17.9 THE MEANING OF BIODIVERSITY

Biodiversity – short for biological diversity – is a measure of variation, the number of different varieties, among living things. It can be expressed in a variety of ways, including the number of genetic strains (differences) within species and the number of different ecosystems in an area. The most common expression of biodiversity is the number of different species in either a particular area (local biodiversity), a specific habitat (habitat biodiversity) or the world (global biodiversity)

Biodiversity is not static. It continually changes as evolution gives rise to new species and some existing species disappear with the arrival of new ecological conditions.

CLASSIFICATION

The richness of nature can be described in different ways, but they all require some form of classification of organisms into groups and sub-groups. The most obvious division within nature is between plants and animals (Table 17.1), but each can be further sub-divided into groups and sub-groups that share important properties and are distinguished from one another by important differences.

Modern biology has devoted a great deal of attention to taxonomy – the study of the classification of living organisms. This is because the basis of a chosen classification can have a strong influence on the groups that are defined, and on the interpretation and explanation of differences between groups.

CLASSIFICATION OF PLANTS AND ANIMALS

The largest grouping used in the classification of living organisms is the kingdom. The simplest division is between the plant and animal kingdoms.

Table 17.1 Similarities and differences between plants and animals

Similarities

- both are composed of cells, with nuclei, chromosomes, enzymes, and so on
- both need nutrition for growth and health
- both digest food, excrete wastes, grow and reproduce
- simple animals are difficult to distinguish from simple plants

Differences

Plants

- most have stiff cell walls
- they usually manufacture their own food by photosynthesis (exceptions are some parasitic plants)
- they do not move from place to place (they are described as *sessile* organisms)

Animals

- most have flexible cell walls
- they cannot manufacture their own food but take food by eating other organisms (energy is obtained from the oxidation of glucose)
- most animals are free to move about (they are described as *motile* organisms)
- advanced animals have complicated nervous and muscular systems

BOX 17.10 SUB-DIVISIONS OF THE PLANT AND ANIMAL KINGDOMS

The plant kingdom is sub-divided into several divisions, and the animal kingdom is sub-divided into phyla (singular is phylum). Divisions and phyla are, in turn, divided into classes, which are further sub-divided into orders. Orders are divided into families, which are sub-divided into genera (singular is genus), and genera into species. Thus the normal classification scheme is hierarchical, going from the kingdom (highest level) to the species (lowest level):

kingdom → division/phylum → class → order → family → genus → species

The individual species is the smallest grouping used in the classification of living organisms.

The system for naming species, which is used around the world, was developed by Swedish botanist Carolus Linnaeus (1707–1778). His hierarchical system of classification groups similar-looking species into genera, genera into families, families into orders, and so on. Linnaeus also established the binomial (two names) system of naming species, in which each species is given a scientific name consisting of two Latinised names, followed by the authority or organisation responsible for the name. For example, the scientific name for the giant panda is *Ailuropoda melanoleuca*: *Ailuropoda* is the name of the genus, and *melanoleuca* indicates the particular species. A sub-species is denoted by a third Latinised name after the species name.

Simple organisms such as protozoa are sometimes placed in a third kingdom, called the Protista. A protist is a tiny organism, and most are too small to be seen without a microscope. About 120,000 species in the kingdom Protista have been named, about half of which are fossil kinds. All animals, plants and fungi are believed to have evolved from protists.

Members of the same species can breed with each other and produce fertile offspring, whereas members of different species are not normally able to breed successfully. This preserves the identity of the individual species over time, and it raises questions about how new species can evolve from existing ones (see p. 529).

HOW MANY SPECIES ARE THERE?

TOTAL NUMBER

It is impossible to say exactly how many species there are on Earth, because many of them have yet to be found, described, named and classified. It is easy to count the mammals, birds and flowering plants, because they are relatively large and obvious, but most life forms are very small (less than 3 mm long) and they hide in the mud, the sea, in lagoons, swamps and forests. Many species are insects or tiny nematode worms or molluscs or fungi.

Estimates are based on what is known to exist and predictions about how many as yet unknown species there are likely to be. Published estimates vary a great deal, between about 5 million and 80 million different species, but most scientists think that a realistic figure is about 30 million species. The counting and naming game continues, because biologists have only recently discovered the amazing ability of life to adapt to inhospitable environments such as soda lakes, pools of acid, volcanic vents, ice sheets and the dark ocean floor.

Up to the present, about 1.5 million species have been described and named (Table 17.2). The list includes 950,000 species of insect, 45,000 vertebrate species and

Table 17.2 Estimated number of species in the world

	Number described (thousand)	*Estimated total number (thousand)*
Viruses	5	500
Bacteria	5	400
Fungi	7	1,000
Protozoa	40	200
Algae	40	200
Nematodes	15	500
Molluscs	70	150
Crustaceans	40	100
Arachnids	75	600
Insects	950	4,000
Vertebrates	45	50
Higher plants	250	300
Total	1,542	8,000

Plate 17.1 *Zebras grazing near Nairobi, Kenya, Africa.*
Photo: Philip Barker.

250,000 different species of plant. But the real total is likely to be much higher than 1.6 million, because the list excludes some invertebrate groups such as worms, sponges, hydras and jellyfish, as well as plant groups such as ferns and bryophytes (including mosses).

The species that have been found represent only a fraction of the total number, and of the total list of known species less than 100,000 have been studied in detail. This means that we know very little about global biodiversity at the present time. Nonetheless, difficult decisions have to be made about which species are most in need of conservation, even though most of them have yet to be discovered and identified.

Many different theories have been put forward to explain why so many species appear in the tropics (Box 17.12), ranging from the geometry of the Earth to the impacts of ice ages. One obvious answer is that there is more habitable space at low latitudes than there is near the poles – at the poles, 1 degree of latitude includes less than 40,000 km^2 of land, whereas a 1-degree belt around the

BOX 17.11 UNCERTAINTY AND DISCOVERY

Recent research suggests that there are vast numbers of invertebrates, fungi, algae and other micro-organisms waiting to be discovered and named. New species are being discovered all the time. For example, between 1978 and 1987 five new species of bird were identified, as well as twenty-six species of mammal, 231 species of fish and more than 7,000 species of insect. Fears are mounting that many species are likely to become extinct before we ever realise that they exist! The implication is that valuable natural resources will be wasted and lost for ever, and that we will never know what has been lost or how important it might have been.

BOX 17.12 VARIATIONS IN BIODIVERSITY

Biodiversity, however measured, is far from uniform around the world. It varies with a number of factors, including:

- *Latitude*: species diversity may be up to six times higher in the warm tropics than in cooler northern Europe and North America. Many explanations have been offered for this latitudinal gradient. It might be due to greater 'effective' evolutionary time (evolutionary speed) in the tropics, probably as the result of shorter generation times, faster mutation rates and faster selection at high temperatures.
- *Habitat*: certain habitats – including tropical seas, coral reefs, rainforests and wetlands – are biologically much more varied than most others.
- *Human impact*: species diversity varies across a region depending on landscape diversity and the nature and extent of human interference (primarily by land-use changes, pollution and natural resource management).

equator covers more than a hundred time as much. A recent study of the habits of almost 4,000 species of marine snail living off the coasts of North America has shown that distribution patterns are closely dependent on variations in energy input and stability of food supply, both of which are strongest in the tropics.

Scientists recognise that current understanding of the scale and pattern of biodiversity is limited, and that much more research is urgently required to establish what exists, where it is located, how fast it is changing and how much of it is currently protected. So-called 'gap analysis' is designed to identify which components of biodiversity are unprotected, and this is normally done using geographical information systems (GIS) to overlay maps of biodiversity, land ownership and land management.

A recent survey by the Nature Conservancy found that America contains twice as many species of plant and animal as previously thought, but it cautions that up to a third of these are under threat because their natural habitats are being destroyed. The total number of US species is expected to continue to rise as new ones are discovered – every year, about thirty new species of flowering plant are named in the United States. The study found that one-third

BOX 17.13 IN DEFENCE OF CONSERVATION

Conservation of biodiversity is important for a range of practical reasons, including:

- maintaining essential ecological processes and life-support systems;
- preserving genetic variety; and
- ensuring the sustainable use of species and ecosystems.

of the 15,320 flowering plants are endangered, and 202 of the 292 types of freshwater mussel are considered to be at risk. Many new native species have been found, and thousands more are expected to be identified, while 500 named species became extinct or were lost during the twentieth century.

SIGNIFICANCE OF BIODIVERSITY

There has been much discussion in recent years about preserving biodiversity, and *Caring for the Earth* argued that biodiversity must be conserved as a matter of principle, as a matter of survival and as a matter of economic benefit. Producing the Biodiversity Convention was one of the successes of the 1992 Rio Earth Summit (see p. 13). But why does biodiversity matter, and why should it be conserved?

SIGNIFICANCE AND VALUES

The ultimate reason for preserving biodiversity is that it exists, and it has a right to continue in existence! This essentially ethical argument also underpins the wider issue of conserving nature, landscapes and wilderness (see p. 539).

Some scientists put forward utilitarian reasons for conserving biodiversity, arguing that nature provides a wide variety of goods and services, which we make use of in a variety of different ways. Perhaps the most utilitarian argument is that nature should be conserved because it offers a rich storehouse of genetic material that can be used in agriculture, medicine and industry (Box 17.14). Much of this genetic material is already being exploited using traditional techniques (such as harvesting fruit and other products from tropical rainforests), but developments in biotechnology are likely to make much more of it accessible and valuable.

BOX 17.14 REASONS FOR CONSERVING THE GENE POOL

Preservation of the gene pool in an unspoiled state will keep open options for the future, for example in the improvement of existing agricultural species and the deliberate creation of new or altered species, perhaps by genetic engineering (see p. 532). Ten species of bird and wild animal provide the basic genetic material on which 98 per cent of all livestock production around the world is based, and wild species are useful for improving the characteristics of domesticated ones. There are an estimated 75,000 edible plants in the world, but only twenty of them are widely used as a source of food. About 90 per cent of all grain production is accounted for by wheat, rice, maize and barley, and all four of these cereals have been improved with the help of genes from closely related wild species.

It is also important to preserve the genetic variation of wild species because an unknown number of them might be raw materials for the development of new medicines. In recent years, for example, the rosy periwinkle (a tropical forest plant originating in Madagascar) has been found to contain active substances that significantly increase the chances of survival of a child suffering from leukaemia.

BOX 17.15 HUMAN ACTIVITIES AND BIODIVERSITY

Human activities cause loss of biodiversity in a number of ways, including:

- destruction and loss of habitat; the most threatened ecosystems (those with the smallest proportion remaining in a nearly natural condition) are fresh water, wetlands, coral reefs, oceanic islands, mediterranean climate areas, temperate rainforests, temperate grasslands, tropical dry forests and tropical moist forests;
- pollution of air, water and land resources;
- hunting or exploitation of commercially important species.

There are other reasons for preserving nature and the biodiversity it contains that spring from the fact that it enriches our culture and our lives. Nature is not just utilitarian; it inspires ordinary people as well as artists, writers, poets and philosophers. Nature uplifts, and it points us to ethical and religious questions beyond ourselves and our day-to-day world.

PRESSURES AND LOSSES

The emergence, expansion, decline and ultimate disappearance of species are all basic facts of life on Earth. Throughout geological time species have come and gone, largely in response to changing environments and to competition between species (see p. 530). Against this background of non-stop long-term change in the size, composition and diversity of the world's plant and animal populations, there are fears that a quarter of the world's species could be endangered and threatened with extinction within the next 30 years. The loss of species and rates of extinction are now much faster than at any time in the history of the Earth, as the impacts of human activities accelerate the decline and extinction of species and alter the conditions within which new species can evolve.

The need to protect biodiversity grows stronger as more and more species are faced with extinction. As with many environmental problems, intervention and remedial action are required without further delay if irreversible environmental change is to be avoided and the future of people and plants is not to be put at risk.

The argument about inter-generational equity, borrowed from the sustainable development debate (see p. 26), applies well to biodiversity. By allowing species to become extinct we assume (on their behalf) that future generations will be able and happy to survive with far fewer species than we have access to today. This is questionable, to say the least.

SPECIES AT RISK

While both species and habitats face the same pressures, and the survival of both is at risk, species are generally regarded as endangered rather than habitats. Until quite recently, more effort was invested in protecting particular endangered species than in protecting habitats. Species, the smallest sub-division in plant and animal classifications (see Box 17.10), have also tended to be the basic units of conservation.

Detailed information about the conservation status of endangered species is collected and collated by the World

BOX 17.16 CLASSIFICATION OF SPECIES AT RISK

Most conservationists use the IUCN classification, which defines species in terms of the degree of threat they face. The main categories are:

- *extinct;*
- *possibly extinct;*
- *endangered*: species likely to become extinct if present threats continue;
- *vulnerable*: species likely to become endangered if present threats continue;
- *rare*: species that are uncommon but not necessarily at risk;
- *no longer threatened*: species from which threats have declined;
- status unknown.

Conservation Monitoring Centre (WCMC). The information is published by the International Union for Conservation of Nature (IUCN) in the form of special reports (the IUCN Red Data Books), which provide an inventory of what survives and where it is to be found. The IUCN classification of species at risk (Box 17.16) is used around the world.

Endangered species are those most likely to disappear if they are not offered adequate protection. Identifying and protecting these endangered species are matters of great significance and urgency, and a growing number of countries are introducing appropriate legislation. In the USA, for example, an Endangered Species Act was first introduced in 1973. Critics argue that the very strict provisions of the Act sparked off controversies between species preservation and economic concerns. Analysis of the plants and animals proposed as additions to the endangered species list between 1985 and 1991 has shown that about 80 per cent were full species (rather than localised populations), and that for plants listed the median population size remaining was less than 120 individuals, compared with about 1,000 for animal species.

Earlier listing of declining species could significantly improve the likelihood of successful recovery, but to do so would require much more detailed monitoring of plant and animal populations.

EVOLUTION AND EXTINCTION

The world today houses a rich variety of plants and animals, but the situation is far from static. We know from the fossil record that there were many different species in the geological past, from the history of domestication of plants and animals that particular characteristics can be enhanced or eliminated in populations, and from the current environmental crisis that some species have become extinct within the very recent past. The situation is one of constant ebb and change, with the emergence of new species (evolution) and the disappearance of existing species (extinction).

EVOLUTION

Evolution is not a process but a theory or a way of explaining things. It explains how all organisms change over time, based on a gradual change in the characteristics of a population of animals or plants over successive generations. This allows us to account for the origin of existing species from ancestors that are unlike them and to explain how new species can be formed from pre-existing species.

To most modern biologists, evolution is the most logical way of explaining how higher forms of life (including humans) have developed from more primitive forms through a series of very slow, gradual changes stretching over a long period of time. The scientific theory of evolution is usually traced to Charles Darwin's (1859) *On the origin of species by means of natural selection*. Darwin argued

BOX 17.17 IMPLICATIONS OF CO-EVOLUTION

Evolution creates a never-ending sequence of changes in biodiversity, species composition and the character of individual species, which is of central interest to ecologists. But it also has much wider significance because, according to the Gaia hypothesis (see p. 119), the long-term evolution of plants and animals has made the planet fit for the forms of life we know today. This co-evolution (the biotic and abiotic parts of the Earth evolving together) has shaped the world we know and care for. It is the basis of the planet's life-support systems, and it plays a fundamental role in maintaining the chemical balance of the Earth via biogeochemical cycles (see p. 86), stabilising climate (p. 326), protecting watersheds from erosion (p. 357) and renewing soils (p. 600).

that varieties of plant and animal life evolve gradually from each other over long periods of time, adapting themselves to their environment in the process.

NATURAL VARIABILITY

The basic building block in evolution is the species. As we have seen (Box 17.10), the species is the lowest level of biological classification, and members of a species can breed with each other to produce fertile offspring.

While all members of the same species display all of the characteristics of that species, there are natural variations between individuals. Some variations are visible and obvious – such as differences in size, behaviour or colouring. Other variations (such as in biochemistry) may be almost unnoticeable. Some variations are environmental, caused by interactions between the individual and its environment. Much of the natural variability is inherited in the genetic material of which the individual is constructed, which comes from both parents and sometimes also includes mutations (changes in the genes of cells).

BOX 17.18 ADAPTATION TO ENVIRONMENT

Adaptation to an environment takes place naturally over many generations by natural selection, and it allows each species to adapt to its environment so that it can survive there. This helps to explain why the global pattern of biomes (see p. 579) is heavily influenced by the main climate zones (p. 335). It also explains why many species do not survive if they are transplanted into other environments to which they are not adapted.

Sometimes adaptation improves the individual's ability to perform a particular function, so that some species of bird can fly much further or longer than others, for example. But the main benefit of adaptation is that it improves a species' ability to thrive, and thus its chances to survive and reproduce effectively, in a particular environment.

ADAPTATION

There are very few parts of the Earth's surface – on land or in the sea – where life is impossible, and even the great ice caps of Antarctica and Greenland are home to single-celled species and algae. Yet there are huge variations in the character of the environment around the world, and each part of the global system supports its own plants and animals.

This shows that plants and animals can survive in even the most hostile environments (such as parched deserts and freezing mountain tops), because they adjust to particular environments. Many plants and animals are well adapted to particular climates. Thus, for example, cacti are adapted to the prolonged drought and intense heat of deserts because they can store water inside their bodies and they lose relatively little water through transpiration. Many plants are adapted to survive in very windy environments by their short size, compact form and thick skin. Organisms that live on the deep ocean floor are adapted to the lack of light by not containing chlorophyll (green pigment) and not deriving their energy supply from photosynthesis (which requires sunlight).

NATURAL SELECTION

Scientists believe that adaptation occurs initially as a result of random variations in the genetic make-up of organisms, some of which are caused by mutation. Once an adaptation occurs that benefits the species and improves its ability to thrive in a particular environment, that adaptation is reinforced in the species by natural selection. This process drives evolution.

The creation of new species from existing ones by the inbreeding of adaptations through successive generations is known as speciation. It divides one species into two, which

BOX 17.19 FINCHES ON THE GALAPAGOS ISLANDS

Darwin based important parts of his theory of evolution on a detailed study of birds in the Galapagos Islands, off the coast of Ecuador in South America. He recorded fourteen separate species of finch on different islands there, and he concluded that they were probably all descended from an initial single species that migrated to the islands from the mainland. But each of the finch species looked different and had different feeding habits; some had long beaks and others short beaks; some ate nectar, while others ate seeds. Darwin concluded that the finches had adapted quite quickly (over a relatively small number of generations) to their own island environments, which dictated what food sources were available and determined how the birds adapted.

> **BOX 17.20 DARWIN'S IDEAS ON EVOLUTION**
>
> Darwin's basic ideas about evolution, based on natural selection, can be summarised as follows:
>
> - Within any population of a species, individuals are not identical in every respect.
> - Populations generally produce more offspring than are required to replace their parents.
> - On average, population numbers remain stable over time; no population can increase in numbers indefinitely.
> - Competition for breeding opportunities, food and other resources must occur among offspring, only some of which survive long enough to breed.
> - Survivors are those best adapted to the environment (i.e. the fittest).

at least in theory are unable to interbreed. Rapid speciation, like the Galapagos finches (Box 17.19), is described as adaptive radiation.

Darwin explained the evolution of species in terms of natural selection, based on adaptedness – the degree to which a species is suited to the environment it lives in. In natural selection, the organisms that are best fitted for survival (by being best adapted to their environment) breed and pass on their features to the next generation. This is the essence of the 'survival of the fittest' argument. A species evolves slowly over a long period, generation by generation.

APPLICATIONS OF EVOLUTION

ADAPTABILITY AND EXTINCTION

Natural selection raises the question of what happens when prevailing environmental conditions, to which many present-day species are adapted, change rapidly. Global environmental change is occurring much faster than any natural change previously experienced during the history of the Earth, so the question is important! In effect, when environmental change is rapid, species chase a moving target. Because adaptation and natural selection are long-term processes that usually occur over many generations, many existing species are left disadvantaged by not being optimally adapted to the new environment. This places them at risk from sudden population decline (e.g. by pest attack, climate extreme or lack of food resources) and thus extinction. Many ecologists are concerned about the threat to the survival of many species of plant and animal posed by global warming and associated environmental changes.

INDUSTRIAL MELANISM

Critics of the theory of evolution point out that rarely, if ever, are individual plants or animals found that obviously display the transition from one species to another, either in the fossil record or in the world today. This, they argue, discredits Darwin's ideas and undermines the notion of natural selection. But some evidence in support of this form of biological evolution comes from *industrial melanism*. This explains the occurrence of dark varieties of animals (particularly moths) in smoke-blackened industrial areas, in which they are well camouflaged, in terms of adaptation and natural selection (Box 17.21).

SELECTIVE BREEDING

The process of natural selection has been exploited deliberately in the selective breeding of domesticated animals such as dogs, cats and farm animals. This involves so-called artificial selection, the selective breeding of individuals that exhibit particular characteristics. In plants, this has promoted the breeding of particular plants that are resistant to disease, high-yielding or attractive in appearance.

> **BOX 17.21 INDUSTRIAL MELANISM AND THE PEPPERED MOTH**
>
> A good example is the peppered moth (*Biston betularia*), which evolved melanic (adapted) forms in Manchester, England, during the nineteenth century. The moth exists in several forms (some pale, some dark), and it rests on tree trunks and wooden fences. It is hunted by birds, which seek their prey visually. Between 1848 and 1895, as sooty air pollution from industries darkened the skies, trees, fences and buildings in Manchester, darker moths became much more common than lighter moths, because they were much better hidden from predators. Later, as air pollution decreased and trees and fences became lighter again, the darker moths (now more visible) largely disappeared and the lighter ones (now better camouflaged) became dominant.

Artificial selection in animals includes the development of particular breeds of cattle for improved meat or milk production.

GENETIC ENGINEERING

Genetic engineering – also known as genetic modification, genetic manipulation or biotechnology – is the deliberate manipulation of genetic material in living organisms. Biochemical techniques are used to alter the structure of the chromosomes, which are the basic building blocks of genes within the nucleus of individual cells, by introducing genetic material previously alien to them. Such manipulation is sometimes done for pure research purposes, to discover more about genetics and the way it works. But, increasingly, the commercial benefits of genetic engineering are being exploited in the breeding of plants, animals and bacteria for particular purposes that benefit humans (so-called 'designer breeding'). Such organisms with an alien gene added are referred to as transgenic.

There are many potential applications of biotechnology, including hazardous-waste disposal and the recovery and recycling of natural resources. Many potential and existing applications – including pollution control, metal recycling, production of food and biomass fuels – lead to more efficient resource use and enhanced environmental quality. Most success to date has been achieved using genetic engineering in agriculture, particularly by improving productivity by increasing the resistance of plants and animals to pests and disease.

Some uses of biotechnology offer great promise to sustainable agriculture by replacing synthetic products (such as chemical fertilisers and insecticides) with genetically engineered natural products. A case in point is the natural insecticide *Bacillus thuringiensis*, which was first identified in 1911. Cultures of it were developed by genetic engineering and first sold commercially in France in 1938. It was first introduced to tobacco in 1987, to rice in 1993 and to potatoes in 1995.

Manipulation of human genetic material is as yet in its infancy and is surrounded by complex ethical questions. Some developments have clear health benefits, including the production of human insulin, human growth hormone and a number of other bone-marrow-stimulating hormones. There are widespread hopes that biotechnology will play important roles in developing effective drugs for cancer and AIDS.

While genetic engineering doubtless offers many advantages and has great potential, it also presents a series of major disadvantages. Some scientists fear that transgenic organisms could do more damage than good. There is also widespread public concern over the risk that new and harmful strains might be produced (accidentally or on purpose) when genes are transplanted between different types of bacteria. The release of such material into the environment would create a nightmare scenario. Genetically modified plants might breed with wild species and thus spread their genes far and wide. For example, if a gene for herbicide resistance were to find its way into a weed, it could create a super-weed that might be very difficult to control and might thus dominate an ecosystem.

Cultural problems surrounding biotechnology have to be addressed too, including technology transfer between the developed and developing worlds. There is also a clear need for internationally accepted mechanisms to ensure that all humankind will benefit.

Most of the raw materials for the booming bioengineering industry come from rainforests, and many scientists are concerned about the hunt for genetic riches in the developing world. There is a widespread belief that the cures for AIDS and other killer diseases are to be found somewhere in the rainforest, and the race is on to be first to find the important gene or cell and then patent the

BOX 17.22 NEW APPLICATIONS OF GENETIC ENGINEERING

New developments since 1998 include:

- testing of crops genetically modified to make them more pest-resistant, which would reduce the need to spray chemical pesticides, herbicides and insecticides on crops and would thus benefit wildlife;
- experiments to genetically modify rice as a cheap way of producing useful proteins (such as those used to stop blood products from clotting) and enzymes (such as those used in washing powders);
- experiments with the genetic modification of garden plants to control when and how they bloom, or to change the colour range of flowering plants (both of which would be of enormous value to the horticulture industry);
- the development of salt-tolerant crop species (including tomatoes, melons and barley), which could be grown in salty soils in arid areas affected by salinisation (see p. 403)

BOX 17.23 ETHICAL QUESTIONS SURROUNDING GENETIC ENGINEERING

Application of the new genetic engineering technologies raises a number of ethical issues, including who owns the genes and who is going to benefit most from their exploitation. The UN Convention on Biodiversity (see p. 541) touches on some of these issues by declaring that biological resources (such as plants, cells and genes) belong to their country of origin rather than being the common property of humanity. Before the Biodiversity Convention, it was possible for biotechnology companies to acquire soil and plant samples from anywhere in the world, and if they discovered anything that could be exploited commercially, the rights and profits belonged to the company. Now companies must enter into formal agreements with governments before collecting any samples, and some of the profits made from commercial exploitation must be ploughed back into the source country.

Ethical issues are even more problematic with the Human Genome Project, which was set up in 1990 and is studying differences in the genetic make-up of ethnic populations and sampling DNA from populations around the world. The aim of the project is to identify the 60,000 to 80,000 genes carried by humans, which when analysed alongside data on the incidence of disease may point to genetic causes and possible treatments. Native American groups in the USA objected to their genes being studied, concerned that the information would be used to exploit them or discriminate against them.

These examples highlight the significance of the ethical questions about who owns the genetic information and thus the right to exploit it commercially. At heart, the question is whether the genes themselves might be patentable (even those that have no known function). Even though patents are usually granted on inventions rather than on discoveries, patents have been applied for by the companies that are investing vast sums of money in bioengineering research and development. These have often been granted.

'invention'. Profits may be slow to start with but will be massive once pharmaceutical or agricultural products reach the market after trials and safety testing. Supporters of the genetic 'gold rush' call it 'bioprospecting' – transforming relatively common or cheap raw materials through detailed scientific research and high technology into refined commodities and much-needed medicines. Critics call it 'biopiracy', and they question who owns the biodiversity in developing countries and who should be able to profit from exploiting it.

EVOLUTION OF LIFE ON EARTH

PHYLOGENY

Scientists have long been interested in ideas of how species have evolved from common ancestors, and a great deal of research has been directed to tracing the family trees of individual species and fitting that into the long-term evolution of life on Earth. This focus is known as *phylogeny* (the historical sequence of changes that occur in a given species during the course of its evolution), and it contrasts with *ontogeny* (the process of development of a living organism through its life-cycle).

BOX 17.24 PUNCTUATED EQUILIBRIUM

Evidence from modern studies of the fossil record suggests that the appearance of new species is a separate process from the gradual evolution of adaptive changes within a species. It also suggests that rather than one long continuous flow of evolutionary development of new species (which scientists call *phyletic gradualism*), the history of life on Earth is characterised by long periods of relative stability interrupted by relatively short phases of rapid evolutionary change. This so-called punctuated equilibria model helps to explain some important discontinuities in the fossil record, including:

- mass extinctions (in which many different species appear to have died out roughly at the same time); and
- the lack of intermediate forms (fossil evidence of the existence of transitional forms between species, which would indicate stages in the long-term evolution of new species).

BOX 17.25 HUMAN EVOLUTION

The evidence suggests that human evolution – at least in a biological sense – has been driven by essentially the same processes as the evolution of other forms of life, even though it appears that the rate of evolution has slowed down in humans. Being part of the overall stream of evolution, humans are by definition part of the complex web that binds all life on Earth together. This implies that we have a responsibility (as well as a need) to find ways of coexisting and co-operating with the rest of nature, which is effectively our extended family – rather than damaging it, over-exploiting it and manipulating it for our own short-term ends. Exercising such a responsibility is particularly important given recent fears about the rate and direction of global environmental change (such as global warming and ozone depletion) and uncertainties about the ability of many existing species to adapt.

Evidence from modern genetics suggests that a number of factors have been at work in phylogeny, including random mutations, molecular changes within organisms and the acquisition of inherited genomes (chromosomes with a single nucleus).

IMPLICATIONS

A number of implications flow from a modern understanding of phylogeny and how it appears to operate. One is that rates of evolution clearly vary over time and from species to species; there is no underlying pace at which species evolve. Another implication is that adaptive change in species depends heavily on the availability of genetic diversity. This underlines the need to preserve biodiversity and the genetic reservoir within it if plants and animals are to survive in a rapidly changing world. The Gaia idea of planet Earth evolving as a super-organism (see p. 119) also has interesting implications about the long-term evolutionary history of the biosphere and the global environment.

EMERGENCE OF LIFE

The evidence suggests that single-cell plants and animals developed from inert materials in the Earth's crust. As a result, the first primitive organisms appeared about 2,500 million years ago, roughly 1,100 million years after the birth of the planet (see p. 114). These early organisms helped to create an atmosphere suitable for the development of other (higher) orders of life. Since then, many different life forms have evolved (Table 17.3) by natural selection.

We know relatively little about what the earliest forms of life on Earth were like, because there are very few fossils in the pre-Cambrian rocks, which account for more than 85 per cent of the Earth's history. Early life forms must have had soft bodies, which decayed quickly after they died, leaving few traces in the rock record. Proterozoic rocks contain traces of simple life, such as bacteria and algae, but little else.

Evidence of higher forms of life is available only for the most recent geological eon, the Phanerozoic – which means 'interval of well-displayed life' and covers the last 590 million years. Geologists divide this vast stretch of time into three major phases – the Palaeozoic, Mesozoic and Cenozoic eras.

PALAEOZOIC ERA

Palaeozoic means 'ancient life', and this era extended from about 590 million years ago to about 248 million years ago. The climate was generally relatively warm, but the mildness was interrupted by a series of short ice ages (see Table 11.3). Continents were very different from the present ones, but towards the end of the era all were joined together as a single world continent called Pangaea (see p. 147). Palaeozoic rocks contain many types of fossil, which indicate the early evolution of life in the sea and the progressive emergence of life on land. The first marine animals with hard outer parts (*exoskeletons*) and internal skeletons, which survive as fossil remains, emerged during the Cambrian period.

The end of the Palaeozoic was marked by an abrupt change in marine life during the Permian period, when many corals and trilobites (marine invertebrates with exoskeletons divided into three segments) became extinct. On the land, deserts were widespread, amphibians and small mammal-like reptiles flourished, and cone-bearing plants (*gymnosperms*) evolved and spread rapidly.

MESOZOIC ERA

The second phase of the Phanerozoic, the Mesozoic era, lasted from about 248 million years ago to about 65 million years ago. It saw equally wide-ranging changes in life forms

Table 17.3 Emergence and evolution of life on Earth

Era	*Period*	*Epoch*	*Began (Ma)*	*Life forms*
Priscoan			4,600	
Archaean			4,000	Earliest known rocks on Earth
Proterozoic			2,500	Earliest living things
Phanerozoic				
	Cambrian		570	First fossils
	Ordovician		500	First vertebrates (armoured fish) appear in the sea
	Silurian		430	First land plants; first jawed fish
Palaeozoic	Devonian		395	First forests and land animals; first amphibians; age of fishes
	Carboniferous		345	First reptiles and insects; forests formed coal
	Permian		280	Expansion of reptiles
	Triassic		225	First mammals and small dinosaurs
Mesozoic	Jurassic		190	Giant dinosaurs; first birds
	Cretaceous		136	End of dinosaurs; spread of flowering plants
	Tertiary			
		Palaeocene	65	Early mammals flourish
		Eocene	53	First horses and elephants
		Oligocene	37	Modern mammals
Cenozoic		Miocene	26	Whales and apes
		Pliocene	12	Mammals spread
	Quaternary			
		Pleistocene	2	Mammoths; woolly rhinoceroses; appearance of modern humans
		Holocene	0.01	End of Ice Age

BOX 17.26 EVOLUTION DURING THE PALAEOZOIC

Fairly early on in the Palaeozoic life in the sea was abundant, but land areas were bare. Life was to evolve quite rapidly over the following 200 million years:

- Invertebrate animals (without backbones) appeared during the Cambrian, and marine algae were widespread.
- The first fishes (including armoured fish – the first animals with backbones) and reef-building algae evolved in the Ordovician period.
- The earliest land plants evolved in the Silurian period.
- Corals were abundant in the seas, land plants flourished and insects evolved on land during the Devonian; the first amphibians (which live on land but breed in water) evolved from air-breathing fish and crawled out of the seas in the late Devonian.
- Many land areas had extensive forest cover (producing coal deposits), amphibians were abundant and the first reptiles evolved in the Carboniferous.

BOX 17.27 EVOLUTION DURING THE MESOZOIC

Geologists divide the era into three periods:

1. The first mammals evolved during the Triassic, and Triassic sediments contain remains of early dinosaurs and other reptiles that are now extinct.
2. Great reptiles (dinosaurs) dominated the land and the first birds appeared during the Jurassic.
3. Angiosperm (seed-bearing) plants evolved and dinosaurs and other reptiles reached their peak and were dominant during the Cretaceous.

and species. Mesozoic means 'middle life', and this era is best known as the 'age of reptiles' because during it reptiles (including dinosaurs) roamed around and dominated the Earth.

By the close of the Mesozoic, climate around the world was warm, the continents had begun to assume their present positions, flowering plants were dominant, and many of the large reptiles and marine fauna were becoming extinct. The end of the Cretaceous period – the so-called K–T boundary – around 65 million years ago was marked by the mysterious extinction of the dinosaurs and other great reptiles, which had dominated the Earth throughout the Mesozoic era. The mass extinctions may have been caused by a giant meteorite impact (see Box 4.3).

CENOZOIC ERA

The most recent geological era, the Cenozoic, spans the last 65 million years. Most of this time is accounted for by the Tertiary sub-era (2–65 million years ago), which is often described as the 'age of mammals'.

The Quaternary, the most recent sub-era, occupies the most recent 2 million years. During most of this time, through the Pleistocene, ice-age conditions were widespread (see p. 434). Modern humans are believed to have evolved in the Quaternary. The last 10,000 or so years (the Holocene; see p. 444) has seen the spread of post-glacial conditions and the formation of the landscapes that we are familiar with today. Final details have been added to the evolutionary family tree by such processes as industrial melanism, selective breeding and genetic engineering.

EXTINCTION

The number of species at any one point in time – biodiversity – reflects a balance between speciation and extinction. Extinction is the complete disappearance of a species of plant or animal, with no survivors left to reproduce. Extinction is for ever, because once a species has become extinct its genetic composition is lost and cannot be replaced naturally. A valuable resource and its future potential are lost to present and future generations. This is the very essence of the extinction problem and the principal reason why conservation of species is of paramount importance.

NATURAL EXTINCTION

Species do not persist indefinitely. This a simple fact of life; extinction is natural, inevitable and a key part of the circle of life. Extinction has occurred throughout Earth's history for entirely natural reasons, either because of major catastrophes (such as meteorite impacts) or because species were unable to adapt quickly enough to natural environmental change.

Remains or evidence of the former existence of animals and plants is preserved in the fossil record of sedimentary

BOX 17.28 EVOLUTION DURING THE TERTIARY

During the Tertiary:

- Mammals took over all the ecological niches left vacant by the extinction of the dinosaurs.
- Animals diversified and developed rapidly; the variety of mammal species reached its peak in the Miocene epoch, and the number of species has decreased since then.
- Animal life began to assume forms similar to those we see today; ancestors of some modern mammals first began to evolve during the Eocene epoch, such as the first member of the horse family (*Eohippus*), and many modern kinds of mammal evolved during the Oligocene epoch.
- The continents assumed their present positions; Gondwanaland started to break up during the Palaeocene epoch.
- Climatic zones (see p. 335) and vegetation zones (see p. 579) as we know them became established.

BOX 17.29 MASS EXTINCTIONS

The fossil record shows that from time to time in the past the general pace of extinction has greatly accelerated and then slowed down again. There have been at least five of these so-called mass extinctions, during which more than half (and up to 90 per cent) of all existing species disappeared in a relatively short period of time. A good example is the wholesale extinction of the dinosaurs, other big reptiles and many marine invertebrates at the K–T boundary (see p. 115).

In the short term, mass extinctions have a catastrophic impact on biodiversity, but the long-term impact is generally limited because mass extinctions have usually been followed by phases of rapid speciation. Surviving species generally evolve rapidly, by adaptive radiation (see p. 530), to occupy the ecological niches vacated by the species that became extinct. Extinction of the dinosaurs, for example, was quickly followed by rapid speciation among mammals.

BOX 17.30 CULTURAL EXTINCTION

Many types of human activity cause or accelerate species extinction, including:

- over-hunting of animals; classic examples include two flightless birds, the dodo of Mauritius and the moa of New Zealand;
- over-collection of plants, eggs and other prized specimens;
- introduction of non-native species, which can create intense competition as a result of which native species disappear;
- removal of habitat, which also removes food supply and forces species to move into environments for which they are not as well adapted (so that natural selection works against them);
- environmental change, for example by pollution.

rocks. This charts the rise and fall of a great many different species throughout geological history (see Table 17.3), which has been punctuated by five major phases of extinction (Table 17.4). Biodiversity has often remained relatively steady over the long term, as species extinction has been more or less matched by speciation.

HUMAN IMPACT

While extinction is a natural process and has occurred throughout geological history, most recent extinctions are the result of human impacts. Many different types of human activity cause or accelerate extinction, which leads to loss of biodiversity (Box 17.30).

The human impact often increases the threat of extinction for species that are already facing serious environmental stress. This was the case with an endangered local type of fritillary butterfly in the San Juan Mountains of south-west Colorado, which declined significantly during the 1980s, putting its survival at risk. The species was already facing problems associated with the limited genetic variability of the local population (inhibiting adaptation and natural selection) and persistent drought, and over-collecting by humans pushed the local population to the very edge of extinction.

The shrinking range of the North American bison from the turn of the nineteenth century (Figure 17.3) is a direct result of over-hunting. At the beginning of the nineteenth century more than 60 million bison (commonly known as buffalo) roamed freely across the Great Plains, but the

Table 17.4 The 'big five' extinctions

Time	*Years ago*	*Impact*
End of Ordovician	440 million	About 85 per cent of species were wiped out
Late Devonian	*c.* 365 million	Two waves of extinction, a million years apart; marine species were particularly hard-hit
End of Permian	*c.* 251 million	The largest mass extinction of all – 96 per cent of all species
End of Triassic	205 million	An estimated 76 per cent of species were lost, mainly marine species
End of Cretaceous	65 million	Probably 75–80 per cent of all species went extinct. This is the most famous mass extinction because it signalled the end of the dinosaurs

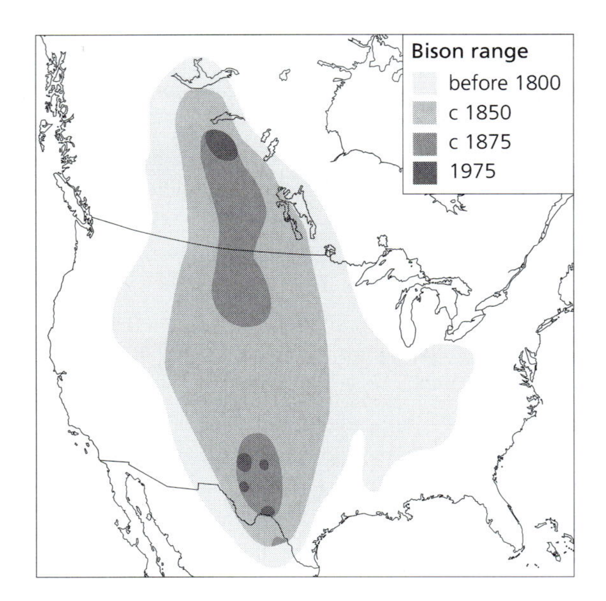

Figure 17.3 *Declining range of the North American bison since the early nineteenth century. Hunting reduced local bison populations, particularly during the late nineteenth century, leading to a progressive shrinkage in the geographical range of the species. By the 1970s, relatively few areas supported natural bison populations.*

Source: Figure 14.4 in Marsh, W.M. and J.M. Grossa (1996) Environmental geography: science, land use and Earth systems. John Wiley & Sons, New York.

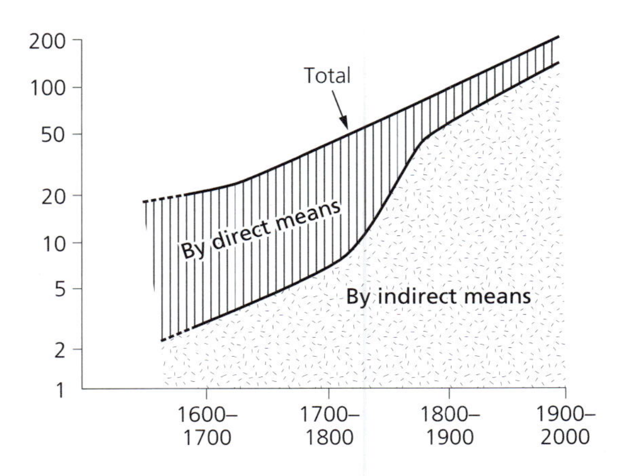

Figure 17.4 *Rates of extinction of mammals since the early seventeenth century. Rates of extinction of mammal species have risen progressively since the early seventeenth century, particularly by indirect means.*

Source: Figure 23.4 in Dury, G.H. (1981) Environmental systems. Heinemann, London.

Union Pacific railway brought in hunters, who shot the animals in their hundreds from the comfort of luxury carriages. The population was decimated, but during the 1990s buffalo ranchers have reintroduced the bison in private herds. By 1999 there were around 250,000 bison across America, and the population was growing at a rate of around 20 per cent a year.

It is known that 485 species of animal and 584 species of plant have become extinct since 1600, largely if not entirely as a result of human activities. For mammals, for example, direct pressure from humans (such as hunting) accounted for much of the extinction until the early nineteenth century, but since then indirect pressure (particularly through habitat clearance) has become much more important (Figure 17.4).

More than a tenth of the world's plant species are heading towards extinction, according to the first fully comprehensive study. The IUCN *Red list of threatened plants* (published in 1998) includes 33,798 species, 380 of which are extinct in the wild, 371 may be extinct, 6,522 are endangered and the rest are vulnerable or rare. Around 90 per cent of the species listed are endemic to just one

BOX 17.31 CONTEMPORARY SPECIES DECLINE AND EXTINCTION IN BRITAIN

The 1998 WWF report indicates that:

- 154 species of plant and animal became extinct in Britain during the twentieth century.
- The short-haired bumble bee (*Bombus subterraneus*) is now extinct in Britain.
- The water vole, which has declined by 94 per cent since 1900, could be extinct by 2003.
- The high brown fritillary butterfly has also declined by 94 per cent since 1900, and it might also be extinct by 2003.
- The skylark, which has declined by 54 per cent since 1975, could be extinct by 2009.
- The song thrush, whose population has fallen by 73 per cent in 25 years, is likely to die out in 2004.
- The grey partridge, whose numbers have halved in 25 years, looks likely to have disappeared by 2011.
- The march fritillary could be extinct by 2020.
- 95 per cent of Britain's peatland has been lost or degraded during the twentieth century.
- Half of the ancient semi-natural woodland has disappeared since 1950.
- Ancient meadows have declined by 97 per cent since 1945.

country, and those growing on isolated islands are especially vulnerable because they can be displaced by plants and animals introduced by humans.

A 1998 report by the World Wide Fund for Nature reported that the current rate of extinction in Britain is three species every two years. It warned that the rate of extinction could accelerate over the next 20 years because of global warming, increased development and intensive agriculture. The report provided many indicators of the serious rate of change (Box 17.31).

Recent scientific estimates suggest that between a quarter and half of all living species may ultimately be lost. Ecologists fear that human activities could trigger another of the mass extinctions that have occurred several times during the Earth's history. If this happened the consequences could be catastrophic, because environmental change is too rapid and time too short to allow natural rapid speciation to come to our rescue.

BOX 17.32 LOSS OF RAINFOREST SPECIES

Recent evidence indicates that the tropical rainforest – which contains more than half of the world's known species (see p. 525) – has the dubious distinction of being the world's most threatened biome because so many of its species are facing extinction. Tropical deforestation accelerated during the 1980s, putting the survival of many tropical species at risk.

Best estimates of likely future losses are alarming. Between 4 and 8 per cent of closed tropical forest species are likely to become extinct over the next 25 years, and by 2040 – if forest clearance continues to accelerate – between 17 and 35 per cent could be committed to extinction. The pace of extinction is driven by some unfortunate positive feedback links, because as more forest is cleared increasing numbers of species have fragmented populations, and this generally causes a serious decline in their genetic diversity. This decreases their chances of being able to adapt and survive by natural selection and increases the likelihood of much higher rates of extinction in the future. The loss of rainforest species is not simply an ecological problem, because the extinctions also mean the permanent depletion of the gene pool and loss of genetic variety, which could be of real benefit in upgrading existing crops and developing new ones.

CONSERVATION

The implications of species extinction and loss of biodiversity are clear – species and habitats need to be protected to increase their likelihood of survival. In this section, we explore why and how this is happening and set nature conservation into its wider context within the environmental crisis (see p. 4).

CONSERVATION THEORY

CONSERVATION OR PRESERVATION?

The ultimate object of nature conservation is to prevent the extinction of species and thus preserve genetic diversity, but there are many different ways of achieving this. The debate hinges partly on differences in perspectives and proposed solutions between supporters of two different approaches – preservation and conservation (Box 17.33).

The difference is not just linguistic, because preservation is more static than conservation and usually involves keeping things the way they are. Such an approach seems entirely proper for buildings of historical or architectural importance and for particular sites with archaeological or scientific interest, but less so for natural habitats or environments. Preservation ignores the dynamic equilibrium that underpins all open environmental systems (see p. 80), including the biosphere.

Conservation of nature is now widely preferred to the preservation of nature (Box 17.34), and it has two fundamental implications:

1. an awareness that people must work within ecosystems rather than from outside them;
2. an appreciation that nature conservation is as much for the sake of what is being conserved as it is for the human enjoyment of it.

Since the 1960s, the word 'conservation' has been used more widely to embrace the rational use of all types of natural resource (biotic and abiotic) (Box 17.35). This is a core belief of modern ecology and a central tenet of sustainable development (see p. 26).

No longer is the protection of wildlife simply the domain of the scientist or natural historian, because conservationists are often locked into heated debates with those who favour development of natural environments. This was the case in Alaska in the late 1980s, for example, when the question was raised of whether or not to open up the

BOX 17.33 CONSERVATION VERSUS PRESERVATION

Preservation involves maintaining areas unchanged from the conditions they were in when their importance was first recognised.

Conservation recognises that natural communities of plants and animals are not static, and it involves preventing any development that would alter or destroy natural habitat but not interfering unduly with ecological changes that occur naturally.

coastal plain of the Arctic National Wildlife Refuge to oil exploration. The area in question lies close to the enormous reserves of petroleum and natural gas in Prudhoe Bay, but it is also the primary calving ground for the porcupine caribou herd of around 180,000 animals, which migrates each year between Alaska and Canada. Ecologists saw the debate as a test case of whether public opinion would support nature conservation in preference to development of this important area.

WHY CONSERVE WILDLIFE?

The most fundamental reason why nature should be conserved is the same reason that climbers risk their lives to ascend Everest – because it is there! There are strong ethical arguments that humans have no right to destroy any form of life on Earth, knowingly or inadvertently, and no matter whether or not it has any obvious or potential benefits to people. Existence alone should justify the right to continued existence. But there are a range of other powerful arguments in favour of conserving wildlife (see Box 17.34).

BOX 17.34 ARGUMENTS IN FAVOUR OF CONSERVING WILDLIFE

There are many different arguments, including:

- *Economic or utilitarian arguments*: plant and animal resources provide the material basis for human life, and the genetic variation in plant and animal species provides the requisite materials for sustaining and improving farm production, forestry, animal husbandry and fisheries.
- *Moral or ethical arguments*: we owe it to future generations to leave an environmental heritage or estate for them; all species have a right to live.
- *Aesthetic arguments*: wildlife is attractive and interesting, and it makes the world pleasurable, e.g. watching birds and butterflies, hearing birdsong, seeing wild flowers and natural forest.
- *Environmental arguments*: natural wildlife is important to environmental systems (e.g. biogeochemical cycling) on which all life on Earth depends.

CONSERVATION AND CLIMATE CHANGE

The rapid pace and global nature of environmental change today make it even more important and necessary to develop workable strategies for conserving nature, given recent rates of species extinction and the associated loss of biodiversity. This is particularly true in the face of global warming triggered by greenhouse gas emissions (see p. 255).

Although the long-term pattern and consequences of global warming are as yet far from clear, most ecologists agree that species abundance and distributions will inevitably change, and careful monitoring and management of the changes at conservation sites will be particularly important. A great deal of careful thought will be required to formulate management plans for conservation sites that include protection against climatic impacts. On some sites

BOX 17.35 THE WORLD CONSERVATION STRATEGY – CONSERVATION OBJECTIVES

The World Conservation Strategy, a major international initiative launched in 1980 (see p. 26), defined three key objectives of nature conservation:

1. to maintain essential ecological processes and life-support systems;
2. to preserve genetic diversity, which is being dangerously impoverished;
3. to ensure the sustainable use by us and our children of species and ecosystems.

minimal intervention might be the best strategy, using the logic that 'nature knows best' how to adapt to prevailing environmental change.

Many surviving habitats are likely to become fragmented – separated from one another by fairly large distances. Some ecologists have proposed that a global network of 'greenways' (corridors of natural vegetation) be established to allow the free movement of species between conservation areas. Coastal habitats, in particular, face the real threat of significant changes caused by global warming and sea-level rise. Ensuring their conservation and survival will doubtless require very careful consideration.

Special plans will also have to be drawn up to protect forests and woodlands during the transition towards a warmer world over the next 50 to 75 years. Some scientists believe that, with appropriate management, many forests might not be too badly affected by potential climatic changes.

CONSERVATION PRACTICE

In this section, we explore a range of practical approaches to the conservation of species and habitats that are in common use today. Not all approaches are used in every country, and there is great variety from country to country in approach and practice.

CONSERVATION IN CONTEXT

Turning the general objectives of conservation into workable and effective practical tasks is never easy, and there is no single approach or method that is guaranteed to be successful in every situation. Much depends on the context of the decision making. Sometimes quite simple solutions can be adopted, whereas at other times difficult compromises have to be reached between multiple and possibly conflicting objectives.

Conservation must be seen as part of a wider environmental agenda, in which the practical protection of species is one – admittedly a significant one – of a series of agreed objectives.

In situ conservation involves conserving species in their natural habitat. *Ex situ* conservation, in contrast, involves taking steps before populations become critically low that would allow the conservation of species outside their natural habitat. Common measures include captive breeding and propagation programmes aimed at conserving the species in the wild and providing specimens for reintroduction where appropriate. Agenda 21, the action programme agreed at the Rio Earth Summit (see p. 13), proposed an equally wide-ranging set of priorities for saving biodiversity (Table 17.5).

BOX 17.36 THE UNCED BIODIVERSITY CONVENTION

A major breakthrough in international commitment to conserve wildlife was the Convention on Biodiversity adopted at the 1992 Rio Earth Summit (see p. 13). The convention calls on countries to identify endangered species and conserve the places where they live. It was signed by 142 countries, including the United Kingdom but initially excluding the USA (which saw it as a threat to free use of private property). The convention will become law when at least thirty countries ratify it (by making it a law in their own country); by August 1993, twenty-six countries had ratified it, including the USA, which signed it in April 1993.

Under the convention, each nation agreed to prepare a national action plan or biodiversity strategy by December 1993. The strategies should be specific, measurable, achievable and realistic. Such requirements pose significant challenges to the scientists responsible for ensuring that targets are precise and realistic and that progress is monitored.

BOX 17.37 EXTINCTION HOT-SPOTS

A report published by Conservation International in 2000 highlights the scale of the global nature conservation problem – nearly half the world's plant, animal and bird species are concentrated in twenty-five areas, which they call 'hot-spots' and which account for a mere 1.4 per cent of the Earth's surface. The twenty-five sites contain the last remaining habitats of 133,149 plant species (44 per cent of all the known plant species worldwide) and 9,645 (35 per cent) of known reptile, amphibian, bird and mammal species. About 38 per cent of the 'hot-spots' are already in protected areas or national parks, but many of them do not have adequate habitat conservation schemes. The study did not consider lower life forms, such as insects, or the marine environment.

Table 17.5 Biodiversity and Agenda 21

Agenda 21 proposals suggest various ways of saving biodiversity, including:

- create a world information resource for biodiversity
- protect biodiversity; this should be a part of all government plans on environment and development
- offer indigenous peoples the chance to contribute to biodiversity conservation
- make sure that poor countries share equally in the commercial exploitation of their products and experience
- protect and repair damaged habitats
- conserve endangered species
- assess every big project (dams, roads, etc.) for its environmental impact

BOX 17.38 DEBT-FOR-NATURE SWAPS

An emerging approach to the conservation of wildlife is the debt-for-nature swap. This involves conservation organisations (usually from developed countries) acquiring part of the international debt of a particular developing country at an agreed discount price. The organisation then redeems the debt in local currency and uses it to fund conservation activities such as the setting up and running of nature reserves and national parks.

Supporters of such schemes claim that there are no losers and few financial risks in this innovative form of international financial transaction, and for this reason it is an ideal way of raising the financial resources required for major conservation initiatives. The debt-for-nature swap mechanism also helps to break down some of the persistent barriers that inhibit collaborative policies in a world of sovereign states – including lack of co-operation, different agendas and objectives, short-term perspectives, and inward-looking and territorially defined concerns.

UNDISTURBED ENVIRONMENTS AND HABITATS

One of the most important requirements of any nature conservation strategy is to preserve the largest possible amount of undisturbed environment so that species and habitats can continue in as near natural a state as possible. All other approaches to conservation must be seen as secondary to this primary objective.

To achieve this object will require concerted efforts between and within countries. It will require extensive pollution control, along with strict control over the development and alteration of natural landscapes and existing ecosystems. For it to be effective, nature conservation will have to assume priority over resource development, which may require significant changes in attitudes and values.

Conflict often surrounds situations where environmental disturbance threatens the survival or well-being of plant and animal communities, and the forces of destruction often vastly outweigh the forces of protection. In the Amazon rainforest in Brazil, for example, native peoples have a direct interest in preserving wildlife resources, whereas outside ranchers, loggers and gold miners stand to gain a great deal from exploiting the resources. Even in less contentious surroundings, the conservation–development conflict can be significant. The survival of wildlife on the Hawaiian volcanic mountains of Mauna Kea and Mauna Loa, for example, is threatened by the rapid pace of lowland (coastal) urbanisation and population growth. Traditional agricultural and extractive forest land uses tend to respect natural habitats, whereas new pressures are being created by tourist-related coastal economic development, which is encouraging more recreational access to the mountains, second-home development, golf courses and forest watershed protection.

PROTECTED AREAS

One of the most effective strategies for conserving species and habitats is to designate particular areas for special protection, which in practice usually means preventing or carefully guiding development and change within them. There are many different types of protected area, ranging from national parks (where conservation is one of a number of priorities) to nature reserves (where it is the main object). The system in Russia (Box 17.39) is perhaps more diverse than in most other countries.

In 1990, according to UNEP data, the world's 6,930 protected sites covered 6.5 million km^2 (4.4 per cent of the Earth's land surface). Clearly this is a relatively small area overall, but most of these sites were chosen because they represent particularly valuable, vulnerable or interesting habitats and landscapes. In an ideal world the proportion of the land area that is protected for conservation purposes would be much higher, but it is certainly expanding as more and more countries realise the importance of conserving their biodiversity and natural landscapes. The

BOX 17.39 PROTECTED AREAS IN RUSSIA

The protected nature area system in Russia is well developed in general, although not as well in the Arctic. On 1 January 1994, the total area of all types of Arctic reserve covered about 197,000 km^2 (comprising about 10.2 per cent of the area of the Russian Arctic). There are five categories of protected nature area:

1. strict nature reserves (*zapovedniki*)
2. national nature parks (*natsional'nyye parki*)
3. nature monuments (*pamyatniki prirody*)
4. special-purpose reserves (*zakazniki*)
5. nature-ethnic parks (prirodno-etnicheskiye parki)

The system of the *zapovednik* is unique. Because of the unstable economic and political situation in Russia, the nature protection system is difficult to manage effectively. The legal structure that defines the purpose of and responsibility for these areas is sometimes not completely clear, and a great deal is dependent on presidential decrees that, over time, have limited validity.

Global Environment Facility (see p. 16) has helped to fund the setting up of protected areas in many developing countries around the globe, particularly in tropical forests.

In the past, most areas set aside for conservation purposes were selected on pragmatic grounds – because they were available, their ecological value was known, they were largely unspoiled by human activities, and they could be managed with minimum effort. But, increasingly, more sophisticated decisions are required if the overall network of protected areas is to have some underlying logic, and if it is to safeguard the widest possible range of species and habitats and be manageable. Guided by the spirit and purpose of Agenda 21, many countries are now approaching the issue of set-aside protected areas from a broader perspective, seeking to optimise their network of sites on the basis of global and national criteria.

In Florida, for example, the Game and Fresh Water Fish Commission has developed a Comprehensive Statewide Wildlife Habitat System. Landsat imagery and a geographical information system (GIS) are being used to establish a biological basis for acquiring lands that should be included in the system, which is designed to provide adequate living space for all of the state's game and non-game wildlife.

Plate 17.2 *Boundary demarcation for a tropical forest/gorilla conservation project in the Cross River National Park, Nigeria.*
Photo: Uwem Ite.

Set-aside is not enough by itself, and conservationists recognise the importance of detailed monitoring and surveillance of protected sites to check how things are going. Still in Florida, NASA is using GIS technologies to monitor protected wildlife at the Kennedy Space Center. Monitoring is designed to provide regular updates on the fate of threatened and endangered species at the centre, and

BOX 17.40 MANAGEMENT OF PROTECTED AREAS

All protected areas are managed to some extent, although the form and extent of management that is regarded as appropriate varies a great deal between the different types of site. Common management tasks in many protected areas include:

- culling species that become too numerous, to keep animal populations in balance with available food supplies;
- clearing waterways of plants that choke them and deplete the dissolved oxygen, which is required by aquatic species;
- allowing natural fires to burn without interference. In some areas, carefully controlled deliberate burning is used, for example on some heather moorlands to encourage new shoot growth.

BOX 17.41 OPTIMAL DESIGN FOR NATURE RESERVE NETWORKS

In a number of countries there has been a shift in emphasis away from simply designating as many sites as possible for protection, towards creating an optimum network of protected sites. Such a network takes into account such factors as:

- the need to conserve as wide a range of species and habitats as possible;
- the need to manage protected sites in sustainable ways, using available resources;
- the need to have a balance between special sites (with some unique properties) and ordinary sites (which protect representative or characteristic species and habitats that may not be endangered themselves).

to measure the local effects of exhaust deposition from space shuttle launches.

NATURE RESERVES

Nature reserves have long been used to provide protection for particular species and habitats. These are tracts of land set aside and managed for the conservation of wild plants and animals, where a particular habitat is protected from human activity. They are usually much smaller than national parks. Many endangered species of plant and animal are protected inside nature reserves.

Human activity other than scientific research, and sometimes controlled educational use, is prohibited in most nature reserves in order to minimise the disruption of natural ecological and environmental processes. Public access to most reserves is controlled or denied, because they exist solely for conservation purposes. Reserves can be privately owned or in public ownership.

Set-aside as nature reserves is clearly an important option for conserving wildlife in threatened habitats. This is true particularly in the tropical rainforests of America, Africa and Asia, where few forest areas have yet been set aside specifically to protect the gene pool. Some 'extractive reserves' have been established in the state of Acre, in the Brazilian Amazon, for those wishing to live sustainably off the forest's resources, but the effectiveness of these reserves in safeguarding genetic variety is as yet unproved.

BOX 17.42 ISLAND BIOGEOGRAPHY AND NATURE RESERVES

Reserves vary in size, but many – particularly in crowded countries like Britain, where sites are mostly acquired as and when opportunities arise – are small and isolated. Ecologists have established clear links between habitat fragmentation and species extinction, which means that larger reserves are usually much more valuable for conserving species than small ones. A great deal of research is being directed towards the design of effective reserves, often based on the principles of island biogeography. Important decisions have to be made about the optimum size, shape, pattern and geographical coverage of reserves. There is also growing awareness of the need to include conservation corridors (like hedges or strips of natural vegetation between reserves) to allow migration and movement of species between the isolated islands of protection.

BOX 17.43 WORLD HERITAGE SITES

A growing number of internationally important nature reserves are being designated (under the international Convention Concerning the Protection of World Cultural and Natural Heritage) as world heritage sites. At the beginning of 1992, a total of ninety-five world heritage sites had been designated around the world – twenty-eight in Africa, twenty-two in North America, nine in South America, thirteen in Asia, eleven in Europe and twelve in Oceania. These are a special kind of nature reserve that protects areas of special importance, either because of their plant or animal populations or because of their geology. The choice of sites for inclusion on the world heritage site list often creates controversy, because this designation is regarded as the premier league but also because such designation brings significant restrictions on how the area can be used. In Tasmania, for example, where 20 per cent of the land area has been given world heritage status, major political tensions have arisen between those who favour recreational use of the land and those who favour world heritage listing.

NATIONAL PARKS

National parks are much larger than nature reserves, and they have broader objectives. Most national parks are areas of land that have great natural beauty, which are set aside and protected for the conservation of scenery as well as plants and animals. People are allowed into national parks to enjoy the scenery and wildlife, but visitor management is often required to reduce conflicts between recreational and conservation objectives.

The world's first national park was Yellowstone in the USA (see Box 18.21). It was designated in 1872, occupies almost 9,000 km^2, and its striking scenery and unspoiled landscapes attract millions of visitors each year.

A 1975 IUCN definition of national parks – as large areas of land that have not been altered materially by human activities and are of scientific, aesthetic, educational or recreational importance – has been widely adopted around the world (Box 17.44). Parks vary widely in size.

National parks are largely natural and unchanged by human activities, but many of them already had existing human impacts (such as those arising from mining, forestry, agriculture and recreation) before they were designated for protection. Many such uses have been allowed to continue, often in a restricted and heavily managed form, after designation.

Many parks are under constant threat from human activities, including pollution, mining, power projects and over-use by visitors.

BOX 17.44 IUCN DEFINITION OF NATIONAL PARKS

In 1975, the World Conservation Union (IUCN) defined a national park as

> a relatively large area
>
> (a) where one or several ecosystems are not materially altered by human exploitation and occupation, where plant and animal species, geomorphological sites and habitats are of special scientific, educative and recreational interest or which contain a natural landscape of great beauty, and
> (b) where the highest competent authority of the country has taken steps to prevent or eliminate as soon as possible exploitation or occupation in the whole area and to enforce effectively the respect of ecological, geomorphological or aesthetic features which have led to its establishment, and
> (c) where visitors are allowed to enter, under special conditions, for inspirational, cultural and recreative purposes.

BOX 17.45 ZONING IN BIOSPHERE RESERVES

Each biosphere reserve consists of three zones:

1. the core, which is legally protected and where only the minimum amount of human activity is allowed, to enable plants and animals to thrive without disturbance from people;
2. the buffer zone, which is also legally protected, but a wider range of resource-using activities are permitted there, including some hunting and careful tree felling. The buffer zone helps to protect the inner core zone;
3. the transition zone, the outer zone where reserve management is broadened to make it compatible with local socio-economic development.

BIOSPHERE RESERVES

Biosphere reserves are a special kind of protected area that grew out of the UNESCO Man and the Biosphere (MAB) Program in the 1970s. Between 1976 and 1992, 300 biosphere reserves had been designated in seventy-six countries, and the list continues to grow.

The Maya Biosphere Reserve, established in Guatemala in 1990, illustrates the pattern of zones in biosphere reserves (Box 17.45). The reserve protects 16,000 km^2 of sub-tropical forest and its rich natural and cultural resources. Its core is made up of four national parks and three protected biotopes (small areas that support their own distinctive community of plants and animals). A multiple-use zone of roughly equal size surrounds the core.

While biosphere reserves are designed to protect unique areas and their wildlife, they are also used for research, monitoring, training and demonstration of best conservation practice. They were developed as a means of integrating conservation with regional development objectives, thereby including protected areas within the resource

Plate 17.3 *Giant redwood trees in Muir Woods, San Francisco, California, USA. These giant trees grow as high as 70 m and are up to 5 m in diameter.*
Photo: Chris Park.

management of inhabited areas. As such, they form an integral part of the sustainable development framework (see p. 26).

The pursuit of multiple objectives is well illustrated in the Cévennes National Park and Biosphere Reserve in the southern part of the French Massif Central. Management of the reserve is designed to preserve the historical, cultural and natural landscape and to protect the traditional agricultural practices associated with that landscape. Eco-development, an environment-friendly approach to sustainable development, is being pursued through a series of projects dealing with energy requirements, agricultural practices, vernacular buildings and building materials, educational programmes, and the development of an eco-museum.

Sometimes pre-existing land uses in the core area of biosphere reserves are allowed to continue, provided that they do not conflict with the primary conservation objectives of the reserve. For example, the Mapimi Biosphere Reserve near Durango, Mexico, was established to conserve the ecology of the Chihuahuan Desert. It has proved highly successful in protecting an endangered species of tortoise, even though extensive cattle grazing is allowed in the core zone.

WILDERNESS

The largest tracts of land set aside and managed for conservation purposes are wilderness areas. In the USA, wilderness areas are specially designated by Congress and protected by federal agencies (Box 17.46).

A wilderness is any wild, uninhabited and uncultivated area, usually located some distance from towns and cities. This definition includes sandy deserts, areas of snow and ice (such as Antarctica) and undeveloped portions of tropical rainforest – indeed, it includes any land that has never been permanently occupied by humans or intensively used by them. Many scientists now regard Antarctica (see p. 448) as the last great wilderness area in the world. Although it faces the prospect of mineral extraction and increased exploitation of its important marine resources, the wild continent is currently fairly well protected by the Antarctic Treaty System.

Natural areas large and unspoiled enough to classify as wilderness are relatively rare in Europe but are more common in the USA. Even in the USA, however, only about 13 per cent of the land area can now be described as wild, compared with 100 per cent three centuries ago.

Many wilderness areas have primary forest as the main habitat, largely because this is the natural climax vegetation (see p. 574) in the temperate zone.

BOX 17.46 US DEFINITION OF WILDERNESS

According to the 1964 Wilderness Act, which established the National Wilderness Preservation System in the USA:

> A wilderness . . . is an area where the earth and its community of life are untrammelled by man [*sic*], where man himself is a visitor who does not remain . . . an area of undeveloped federal land retaining its primeval character and influence, without permanent improvement or human habitation.

Wilderness areas are carefully managed to preserve their natural state as much as possible. Protecting the wildlife and landscape is greatly assisted by prohibitions on the construction of roads into or through wilderness areas and tightly regulated controls on the number of people who are permitted (often by licence) to visit them at any one time.

PROTECTED SPECIES

A great deal of effort has been invested in designating and managing protected areas, and this is entirely logical given the need to protect unspoiled habitats and landscapes for the benefit of plants and animals. But sometimes particular species are at risk, and other strategies have to be adopted to conserve them. Protection of species and protection of areas must not be seen as separate pursuits or conflicting objectives, however, because both are aimed at exactly the same thing – the conservation of biodiversity.

TRADE IN ENDANGERED SPECIES

Many endangered species are highly prized by collectors, as rarity increases value. As a result, there has long been a buoyant trade in endangered species and their products (including eggs and skins), both within and between countries. This has significantly increased the tendency for poachers to find ways of hunting, capturing and selling their quarry illegally without being detected. In turn, this has greatly increased the threat of extinction for species whose survival is already questionable. The logical solution is to ban the exploitation of certain high-risk species, and to find ways of making the bans effective.

There has long been an international trade in live plants and animals, and in wildlife products such as skins, horns, shells and feathers. The trade has made some species more or less extinct, including many of the largest whales, crocodiles, marine turtles and some wild cats. Populations of black rhinos and African elephants have disappeared from some areas because of illegal hunting for their tusks (which are made of valuable ivory). The future of a number of species of seabird in Jamaica – including the sooty tern (*Sterna fuscata*) and brown noddy (*Anous stolidus*) – is currently at risk because of the collection of eggs for the local luxury food market. But the problem does not stop with the loss of the target species, because sometimes whole ecosystems (such as coral reefs) are threatened by the wildlife trade.

Ultimately, the trade in endangered species will be controlled by a combination of demand management and supply management. Demand management involves trying to dissuade people from wanting to buy endangered wildlife products. Supply management involves controlling which wildlife products are available in the marketplace. Extinction, which will inevitably follow if trade in endangered species is not curbed, is the ultimate form of supply management.

BOX 17.47 CITES

Wildlife trade is to some extent regulated by the Convention on International Trade in Endangered Species (CITES), which was launched under the auspices of the IUCN and has been signed by more than ninety countries since 1973. The objective of CITES is to regulate the trade in endangered species of animals and plants. It lists eighty-seven species of plant and nearly 400 species of animal as 'threatened with extinction' and prohibits trade in those species. Although it has certainly not solved the problem completely, CITES has made it much more difficult to continue the international trade in endangered species, and it is regarded as an important step in the right direction.

CAPTIVE BREEDING AND REINTRODUCTION

For the most seriously endangered species there are few viable alternatives to captive breeding programmes, which are designed to try to save what survives, ideally for subsequent release back into the wild.

Many botanical gardens were initially set up with the purpose of growing, studying and exhibiting plants, particularly exotic species, but they are increasingly becoming repositories of endangered plants, where at least small breeding populations can be conserved in as near-natural a state as possible. These populations can often be used to stock replanting programmes, designed to reintroduce locally extinct species into their native habitat.

Zoological gardens, or more simply zoos, serve a similar purpose for animals. Most were built originally as places where animals – particularly exotic or unusual species – could be kept, studied, bred and exhibited to the public. Critics have long argued that zoos are like prisons, and similarly lacking in dignity, freedom and quality of life for their animal inmates. But here, too, things are changing as many zoos refocus their resources and objectives towards the conservation of species. Many zoos now have well-founded captive breeding programmes, particularly for endangered species. The hope is that viable breeding populations of a number of endangered animal species can be created by releasing captive-bred individuals into the wild, into their natural habitats.

Aquaria – pools or buildings in which aquatic animals and plants are kept for pleasure, study or exhibition – have a similar history to that of zoos. Many aquaria, too, are switching towards conservation and education objectives.

GENE BANKS

Another strategy for conserving endangered species, literally the last straw, is to store genetic material for possible future use. In this way, even if the species in question does become extinct, at least its genetic make-up is not lost for ever.

A gene bank is a collection of genetic material. This material can take many different forms, including seeds, tubers, spores, bacterial cultures, yeast cultures, live animals and plants, frozen sperms and eggs, or frozen embryos. These are stored for possible future use in agriculture, in plant and animal breeding, or in medicine or genetic engineering.

The stored genetic material can also be used in the breeding of replacement plants and animals for restocking wild habitats where natural species have become extinct. Many rare or endangered plants are now preserved in seed banks, where stored seeds remain viable for up to 20 years. It seems highly likely that gene banks will become more common and more important as the depletion of biodiversity continues.

WEBSITE

Links to relevant websites, a comprehensive bibliography, tools for teaching and learning, and downloadable images relevant to this chapter can be found at the website specially designed to accompany this book at
http://www.park-environment.com

FURTHER READING

Adams, W.M. (1996) *Future nature: a vision for conservation.* Earthscan, London. A radical reassessment of nature conservation that embraces recent thinking about biodiversity and sustainability.

Allen, K.C. and D.E.G. Briggs (1991) *Evolution and the fossil record.* John Wiley & Sons, London. A clear introduction to the study of how species are believed to have evolved over geological timescales, and the interpretation of fossil evidence.

Barbier, E.B., J.C. Burgess and C. Folke (1994) *Paradise lost? The ecological economics of biodiversity.* Earthscan, London. Explores the relationships between economics and ecology, with a particular focus on threats to and preservation of biodiversity.

SUMMARY

This chapter has explored the meaning, structure and composition of the biosphere and commented on its relevance both to people and to other environmental systems. We began by looking at how the biosphere evolved and how it is influenced by human activities, and we recognised the need to understand some basic principles of ecology. Biodiversity refers to the number and variety of different species there are, and we saw some of the difficulties in estimating the numbers involved. We then examined the pressure on biodiversity. After outlining the principles of the theory of evolution, we looked at some important applications including industrial melanism, selective breeding and genetic engineering. The sequence of evolutionary changes that have happened throughout geological history was also outlined. Natural extinction is a fact of life in the biosphere, but recent times have seen dramatic increases in extinction rates because of human impacts. It is abundantly clear that without greater efforts to conserve what remains, biodiversity will shrink even further. We explored the basics of conservation theory and practice and examined the most important types of protected area. A final theme, which reflects the globalisation of nature conservation efforts, was the international trade in endangered species and approaches such as captive breeding and gene banks. It is ironic that humans – an integral part of the biosphere – are putting at risk the very stability and survival of the biosphere through pollution, removal of habitat, induced environmental change, unsustainable use of natural resources and lack of adequate conservation.

Bradbury, I.K. (1998) *The biosphere*. John Wiley & Sons, London. Comprehensive introduction to functional, historical and geographical aspects of the biosphere.

Bruce, D. and A. Bruce (eds) (1999) *Engineering Genesis: the ethics of genetic engineering in non-human species*. Earthscan, London. Explores the key issues and challenges in the current genetic engineering debate – is genetic engineering playing God?

Brum, G., G. Karp and L. McKane (1993) *Biology*. John Wiley & Sons, London. A clear introduction to the principles and applications of biology.

Colinvaux, P. (1993) *Ecology*. John Wiley & Sons, London. A broad overview of general ecology, with an emphasis on changes over time and space.

Cunningham, A.B. (2000) *Applied ethnobotany: people, wild plant use and conservation*. WWF/UNESCO/Earthscan, London. Explains how local people can learn to assess the pressures on plant resources and what steps to take to ensure their continued availability.

Gardner, E.J., M. Simmons and D. Snustad (1991) *Principles of genetics*. John Wiley & Sons, London. An up-to-date introduction to the basic principles of genetics, including the new field of genetic engineering.

Goldsmith, F.B. and A. Warren (eds) (1993) *Conservation in progress*. John Wiley & Sons, London. A useful set of essays that focus on different aspects of nature conservation in Western Europe in the 1990s.

Hardman, D., S. McEldowney and S. Waite (1995) *Pollution: ecology and biotreatment*. Longman, Harlow. An introduction to the principles of pollution and pollution control, with special emphasis on biotreatment.

Huggett, R. (1998) *Fundamentals of biogeography*. Routledge, London. Comprehensive introduction to the theory and practice of biogeography, including a useful overview of the principles of ecology.

Hutton, J. and B. Dickson (1999) *Endangered species, threatened convention: the past, present and future of CITES*. Earthscan, London. Examines the record of CITES, its controversies, successes and future direction.

Jeffries, M. (1997) *Biodiversity and conservation*. Routledge, London. Defines and explains biodiversity, drawing on examples from different environments, and emphasises the need to protect habitats and species.

Kate, K.T. and S.A. Laird (1999) *The commercial use of biodiversity*. Earthscan, London. Explores key issues relating to access to genetic resources and sharing the benefits of this access.

Kellert, S.R. (1996) *The value of life: biological diversity and human society*. Island Press, New York. Explores the impacts of loss of biodiversity on the physical, emotional and intellectual well-being of people.

Laird, S.A. (ed.) (2000) *Biodiversity and traditional knowledge: equitable partnerships in practice*. WWF/UNESCO/ Earthscan, London. Draws on case studies to show how to arrive at equitable and successful arrangements over access to and the commercial development of genetic resources.

Lappé, M. and B. Bailey (1999) *Against the grain: the genetic transformation of global agriculture*. Earthscan, London. A critical evaluation of the wisdom of the current rapid introduction of genetic engineering, and an analysis of its challenges and implications.

Lohnert, B. and H. Geist (eds) (2000) *Coping with changing environments: social dimensions of endangered ecosystems in the developing world*. Ashgate, London. A series of essays that explore endangered ecosystems, their perception and coping strategies for their conservation and management, based on case studies from Asia, Africa and Latin America.

McGinnis, M.V. (ed.) (1998) *Bioregionalism*. Routledge, New York. A series of essays that explore the meaning, relevance, opportunities and constraints of bioregionalism from an interdisciplinary perspective.

McNamara, K. and J. Long (1998) *The evolution revolution*. John Wiley & Sons, London. Outlines new and emerging ideas about the evolution of life based on the fossil record.

Mooney, H.A. and R.J. Hobbs (eds) (2000) *Invasive species in a changing world*. Island Press, Washington. Explores how and why introduced species take hold in new habitats, using examples from freshwater, marine and terrestrial ecosystems.

Scragg, A. (1999) *Environmental biotechnology*. Longman, Harlow. Explores the use of biotechnology in bioremediation of contaminated sites, treatment of waste from domestic, industrial and agricultural sources, and pollution prevention.

Secretariat of the Convention on Biological Diversity (2000) *The handbook of the Convention on Biological Diversity*. Earthscan, London. The official handbook of the convention, which presents a guide to the decision adopted at the 1992 Rio Earth Summit.

Snape, W.J. (ed.) (1996) *Biodiversity and the law*. Island Press, New York. A collection of essays that argue that protection of biodiversity should be recognised as the main tenet of environmental law and policy.

Snustad, D.P., M.J. Simmons and J.B. Jenkins (1999) *Principles of genetics*. John Wiley & Sons, London. A review of recent developments in genetics, including the Human Genome Project.

Spellerberg, I.F. (ed.) (1996) *Conservation biology*. Longman, Harlow. Explores how we view, value and protect nature, wildlife and biological diversity.

Spellerberg, I.F. *et al*. (eds) (1991) *The scientific management of temperate communities for conservation*. Blackwell Scientific, Oxford. Explores the theory and practice of nature conservation designed to preserve species and habitats.

Tacconi, L. (2000) *Biodiversity and ecological economics: participatory approaches to resource management.* Earthscan, London. An accessible introduction to the subject of applied ecological economics, based on case studies of biodiversity conservation.

Tivy, J. (1993) *Biogeography: a study of plants in the ecosphere.* Longman, Harlow. A clear introduction to the biosphere, ecosystems and biotic resources, and ecological interactions in environmental systems.

Ward, P.D. (1991) *On Methuselah's trail: living fossils and the great extinctions.* W.H. Freeman, Oxford. Describes modern animals and their ancient relatives and unravels a fascinating history of palaeontology.

Whitmore, T.C. and J.A. Sayer (eds) (1991) *Tropical deforestation and species extinction.* Chapman & Hall, London. A collection of essays that focus on rates, patterns and causes of extinction of ecologically important rainforest species.

Wood, A., P. Stedman-Edwards and J. Mang (eds) (2000) *The root causes of biodiversity loss.* WWF/Earthscan, London. A series of case studies and a proposed new interdisciplinary framework that go beyond pollution and habitat change and embrace the socio-economic factors that drive people to degrade their environment.

Chapter 18

Ecosystems, Succession and Biomes

LEARNING OBJECTIVES

When you have finished studying this chapter, you should be able to

- *Appreciate the role and significance of ecosystems within the biosphere*
- *Understand the principles and relevance of food webs, photosynthesis, trophic relationships, habitat and niche*
- *Outline the processes by which energy and nutrients flow through ecosystems and the factors that affect primary productivity*
- *Summarise what controls ecosystem equilibrium and what disturbs it*
- *Show how ecosystem principles can be used in developing restoration and sustainable management strategies for particular areas*
- *Define what is meant by 'succession' and outline the five main stages in succession*
- *Explain how and why vegetation changes through succession and explain the different possible end-points of successional development*
- *Define what is meant by 'biome' and list the major biomes*
- *Outline the factors that affect the distribution of biomes*
- *Describe and account for the main characteristics of the main biomes (cold climates; temperate grassland; temperate forest; hot climates; tropical forests)*

There is a great deal of order in nature, and in this chapter we look at three particular ways in which this order is established and reflected. Many ecologists argue that the ecosystem is the most important unit for studying plants and animals, and the fundamental building block of the biosphere. In the first section, we explore the meaning and significance of ecosystems, with a particular emphasis on how they are structured and how they function.

New land areas and recently exposed land surfaces rarely survive for long without some form of vegetation developing on them. Over time, vegetation develops through a series of stages from patchy towards an ultimate end-point defined largely by climate. This is the notion of an ecological succession, which is the focus of the second section of this chapter. The global pattern of vegetation, explored in the final section, shows striking associations with the global patterns of climate (see p. 335) and soils (p. 606), and it reveals a great deal about the factors that control the development and character of the major ecosystems.

Throughout this chapter, the emphasis is on studying patterns in nature and how they vary over time and from place to place. The *flora* (plant populations) and *fauna* (animal populations) of an area often exert a strong influence over the whole environment, so the interaction between biotic (living) and abiotic (non-living) aspects of the environment is another important integrating theme.

ECOSYSTEMS

THE ECOSYSTEM CONCEPT

The ecosystem is a convenient scale (Table 18.1) at which to consider plants and animals and their interactions because it is more localised and thus more specific than the whole biosphere (see p. 519), and yet it includes a sufficiently wide range of individual organisms to make regional generalisations feasible and valuable.

Table 18.1 Five levels of ecological organisation

Level	*Unit*	*Definition*
1	Organism	An individual plant or animal
2	Population	A group of individuals of one species
3	Community	The sum of the different populations of species within a given area
4	Ecosystem	The sum of the communities and the abiotic environment in an area
5	Biosphere	The sum of all ecosystems

BOX 18.1 ECOSYSTEM – A DEFINITION

An ecosystem is an ecological system in which organisms interact with each other and with the environment. The term was first used by British ecologist Arthur Tansley in 1935, who visualised ecosystems as being composed of two parts:

1. *the biome*: the entire complex of organisms (both plants and animals) that live together naturally in harmony;
2. *the habitat*: the physical environment within which the biome exists.

In Tansley's view:

> all parts of such an ecosystem – organic and inorganic, biome and habitat – may be regarded as interacting factors which, in a mature ecosystem, are in approximate equilibrium; it is through their interactions that the whole system is maintained.

INTERACTION

An ecosystem is the sum of all natural organisms and substances in an area, and it is a good example of an open system. Like all open systems, an ecosystem has a series of major inputs and outputs, and these effectively 'drive' the internal dynamics of the system (see p. 78). In a woodland ecosystem, for example, the inputs include solar energy, rainwater, dust from the atmosphere, minerals in the soil (derived from rock weathering and decomposition) and soil water; outputs include the timber and logs that are taken away and used, woodland animals that migrate elsewhere, water losses (drainage through the soil, evaporation and transpiration losses from the trees) and heat loss (from animal respiration and movement).

Many ecologists regard ecosystems as the basic units of ecology (see Box 17.8) because they are complex, interdependent and highly organised systems, and because they are the basic building blocks of the biosphere.

ECOSYSTEM STRUCTURE

ORDER IN NATURE

The biosphere is highly ordered, and it is possible to examine regularity within nature in a variety of ways. The most obvious distinction is between the biotic (living) and abiotic (non-living) aspects of the environment – plants, animals and people are the main biotic elements, and weather, climate, energy flows, the lithosphere and the water cycle are among the more important abiotic elements. Within the biotic elements, the main distinction is between plants and animals. While most higher plants and animals can obviously be distinguished apart – a polar bear and an oak tree have little in common, for example – there are some single-celled organisms that do not fit readily into either group.

FOOD CHAINS AND WEBS

The distinction between producers and consumers offers a basis for a hierarchical structure of organisms, depending on feeding habit (Figure 18.1; Box 18.2).

The green plant is the most important component, and the lowest level in the hierarchy. Plants manufacture their own foods by photosynthesis (Box 18.3) from:

- carbon dioxide, which is taken in through leaf walls from the atmosphere; and
- inorganic salts (such as phosphorus and nitrates) and water, which are taken in from the soil via plant roots, by the process of osmosis.

Animals do not contain chlorophyll, so they cannot photosynthesise and produce their own food. As a result, they are fully dependent on plants and other animals for food – hence they are consumers. Animals lie above plants in the hierarchy of organisms (Figure 18.3) because they eat them! Thus, for example:

- Sheep and cattle (consumers) eat growing grass (primary producer).

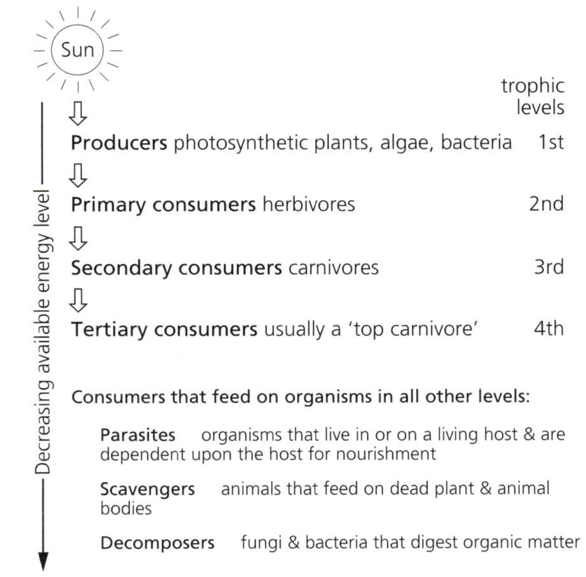

Sun

trophic levels

Producers photosynthetic plants, algae, bacteria 1st

Primary consumers herbivores 2nd

Secondary consumers carnivores 3rd

Tertiary consumers usually a 'top carnivore' 4th

Consumers that feed on organisms in all other levels:

Parasites organisms that live in or on a living host & are dependent upon the host for nourishment

Scavengers animals that feed on dead plant & animal bodies

Decomposers fungi & bacteria that digest organic matter

Decreasing available energy level

Figure 18.1 *Hierarchical structure of organisms. Organisms in an ecosystem can be classified on the basis of what they eat or by consumer level in a feeding hierarchy. See text for explanation.*

Source: Figure 3.5 in Cunningham, W.P. and B.W. Saigo (1992) Environmental science: a global concern. Wm. C. Brown Publishers, Dubuque, Iowa.

BOX 18.2 FEEDING HABITS

A convenient and meaningful basis on which to classify living organisms is by feeding habit. The simplest division (Table 18.2) is between:

- organisms that can produce their own food from simple inorganic materials; these are called *autotrophs* (which means, literally, 'self-feeders') or primary producers;
- organisms that cannot produce their own food and have to obtain their energy by eating autotrophs or other organisms; these are called *heterotrophs* ('other feeders') or consumers.

The autotrophs can be further sub-divided according to the source of energy necessary to convert inorganic into organic forms. The heterotrophs can also be sub-divided according to the way in which the organism actually takes in the organic matter from the autotrophs.

Table 18.2 Simple classification of the main types of organism on the basis of feeding habit

Autotrophs		These contain chlorophyll-like substances, make organic matter from inorganic ingredients and need energy (light and chemical)
	Phototrophs	These derive energy from light by photosynthesis. Examples include mosses, ferns, conifers, flowering plants, algae and photosynthetic bacteria
	Chemotrophs	These derive energy from inorganic substances by oxidation. Examples include certain bacteria and blue-green algae
Heterotrophs		These feed on the organic matter produced by autotrophs. This is available in three forms – living plants and animals, semi-decomposed plants and animals, and organic compounds in solution
	Saprophytes	These feed on soluble organic compounds from dead plants and animals. Some absorb compounds that are already dissolved, others (including some fungi and many bacteria) break down undissolved foods by secreting digestive enzymes on to them
	Parasites	These are organisms that at some time or times in their lives make a connection with the living tissues of another species on which they rely for food. Food may be in the form of soluble or insoluble compounds. There are many animal parasites, many parasitic fungi and a few parasitic plants (including mistletoe)
	Holozoic organisms	These eat by mouth, and they can absorb large particles of undissolved food and soluble compounds. All higher animals (including rabbits, cows and humans) are holozoic

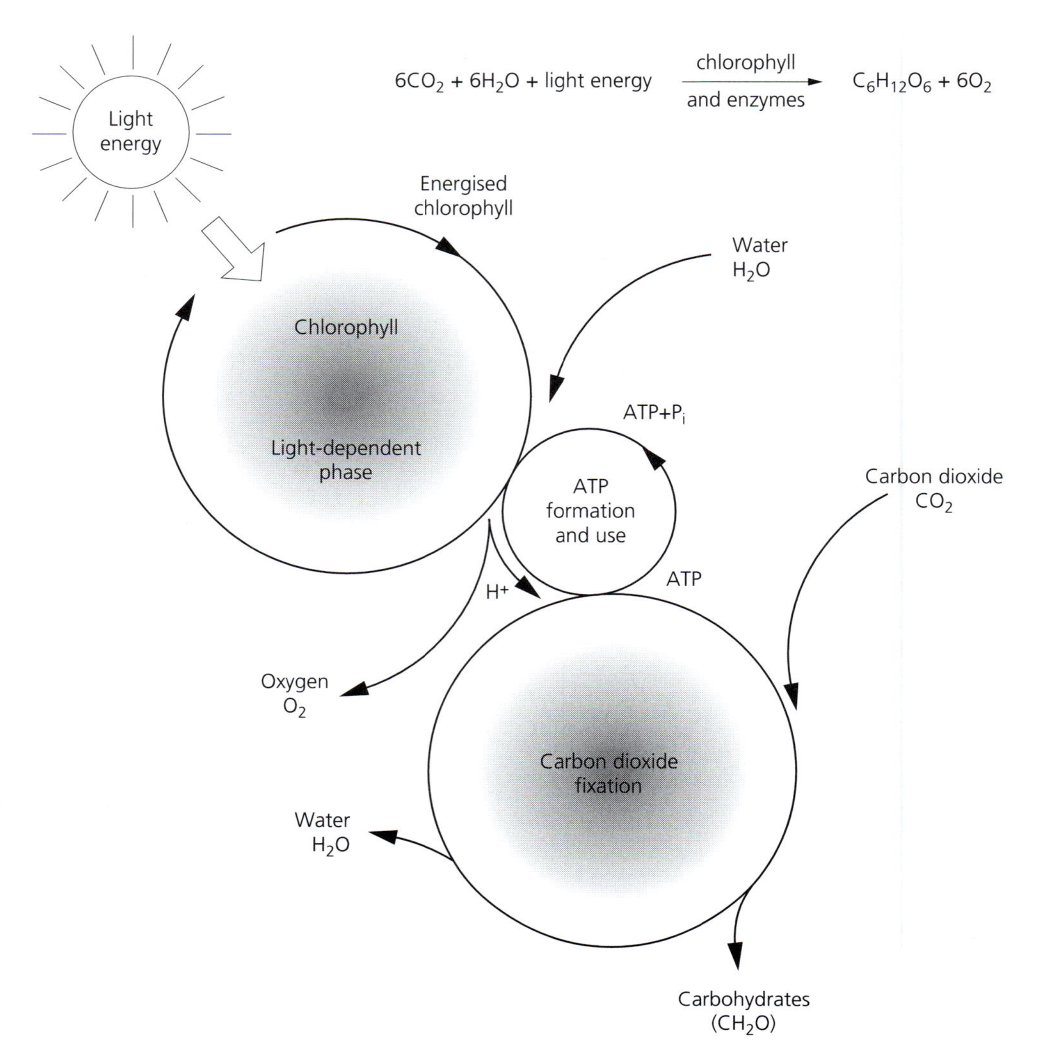

Figure 18.2 *Photosynthesis. Photosynthesis involves the conversion of carbon dioxide and water into forms of energy that can be used directly by plants. The figure shows the flow of energy through photosynthesis, from light units to food units in the form of carbohydrate molecules.*

Source: Figure 3.2 in Cunningham, W.P. and B.W. Saigo (1992) Environmental science: a global concern. Wm. C. Brown Publishers, Dubuque, Iowa.

BOX 18.3 PHOTOSYNTHESIS

The manufacturing process involves the conversion of water and carbon dioxide into starch and sugar in the presence of sunlight (Figure 18.2). The process is called *photosynthesis* ('building by light'), and it requires the presence of *chlorophyll*, a green, energy-trapping pigment in the leaves of the plant. All organisms that contain chlorophyll are technically classified as plants regardless of whether or not they have roots (in land plants) or swim around in water (as free cells as with green algae, or in cell clusters).

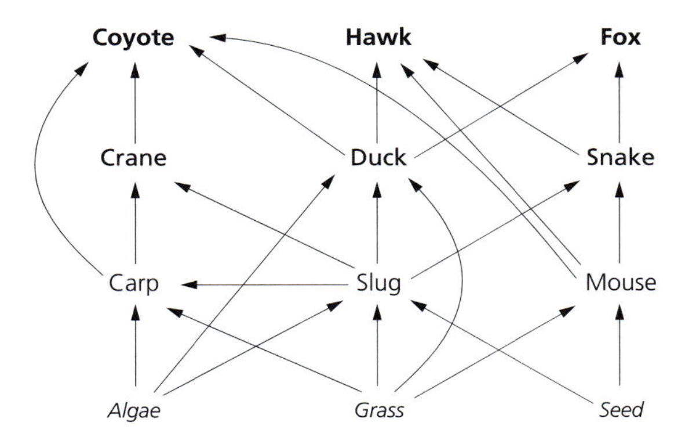

Tertiary consumer
Secondary consumer
Primary consumer
Producer

Figure 18.3 *Simple food web. The food web involves three overlapping food chains, each of which has producers at the bottom, which are consumed by primary consumers. These in turn are consumed by secondary consumers, and ultimately by tertiary consumers.*

Source: Figure 4.1 in Marsh, W.M. and J.M. Grossa (1996) Environmental geography: science, land use and Earth systems. John Wiley & Sons, New York.

- Goats and giraffes eat the leaves and young shoots of trees and bushes.
- Insect larvae eat stems, shoots and leaves.

Level 3 in the hierarchy will be occupied by animals that eat other animals, that is, by carnivores. A fourth level might also exist in the hierarchy, occupied by so-called top carnivores, which can eat almost anything! Humans, for example, can use the plant energy produced at the first level (for example by eating fruit, nuts, grains and other vegetable products), and they can also use the energy produced by the herbivores and carnivores at levels 2 and 3 (by drinking milk and eating meat, eggs and other animal products).

Although the food chain (Box 18.4) is a convenient way of summarising feeding relationships between organisms, the relationships are rarely simple and linear. Most animals vary their diet, for example, and they are preyed upon by a variety of different consumers. Food chains generally overlap to form a food web. Although a large number of species may be present in a lateral sense within the web, webs are rarely longer than four vertical links, because of the progressive loss of energy at each of the levels (see p. 560).

BOX 18.4 THE FOOD CHAIN

All organisms can be seen as part of a chain of feeding habits – a food chain. A simple marine food chain would be a diatom (autotroph), which is eaten by a crustacean (herbivorous consumer), which in turn is eaten by a herring (carnivorous consumer), which might eventually be eaten by a human (top carnivore). The simple food chain would then be:

diatom → crustacean → herring → human

BOX 18.5 BIOACCUMULATION

The concept of food chains is useful in tracing the movement of chemically stable products that are consumed by primary producers and primary consumers and accumulate (often in high concentrations) higher up the food chain. This is described as bioaccumulation, and it explains how pollutants can have significant impacts on many species within an ecosystem. In 1962, Rachel Carson wrote an influential book, *Silent spring*, which drew attention to the knock-on ecological effects of chemical insecticides, which are used extensively in modern farming. An example is the insecticide DDT, which was used in the 1950s to control insect pests. The DDT was ingested by earthworms, which in turn were consumed by robins. The bioaccumulation of DDT in robins eventually exceeded a threshold level and killed the birds. More recently, concern has been expressed about the mobility of toxic heavy metals (such as cadmium) through food webs.

There are still problems surrounding the idea of food webs, because it is often difficult to know exactly where to draw the line between overlapping webs. Many animals can alter their feeding habits and food sources in times of food shortage, so the web would usually be constructed to show normal feeding relationships. It is also unclear whether food chains are controlled from the top down or from the bottom up.

BOX 18.6 DECOMPOSITION

Decomposition is a fundamental part of the ecological system that works in a number of ways, including:

- Plants die and their structures are broken down by decomposers, which releases energy and chemicals in the plant organs and tissues.
- Animals produce waste material and they eventually die; these both produce organic matter, which can be decomposed and broken down.
- Humans produce wastes and also die; this material can be decomposed slowly after burial or instantly on cremation.

BOX 18.7 TROPHIC PRINCIPLES

There are a number of important generalisations about the relationships between trophic levels in all ecosystems:

- Species at tropic level 2 are entirely dependent on trophic level 1 as a source of energy.
- Species at higher trophic levels (3 and above) are usually generalists rather than specialists in terms of feeding habits; they can often obtain energy by eating from any of the lower trophic levels, and predators are usually less specific in their food preferences than their prey are.
- The relative loss of energy due to respiration grows progressively larger at higher trophic levels.
- Species at progressively higher trophic levels become more efficient in using their available food supply.
- Higher trophic levels tend to be less discrete than lower ones, because of the generalist species there.
- Food chains tend to be relatively short (usually up to four levels at most) because of the progressive loss of available food energy at each successive level.

DECOMPOSITION

The hierarchy of organisms also includes bacteria, fungi and other micro-organisms, which play a vital role in breaking down dead tissue from plants and animals into its constituent elements, which are normally simpler chemical compounds. The processes of decomposition, and the elements they recycle within ecosystems, play vital roles in the major biogeochemical cycles (see p. 86) and in the development of soils (p. 600).

The decomposer part of the system is composed of bacteria and fungi in grassland, fungi in woodland, soil organisms such as earthworms and wood lice, and aquatic organisms such as worms, bacteria and molluscs.

TROPHIC LEVELS

The different levels in the feeding hierarchy are called *trophic levels* (trophic means relating to nutrition). Because all organisms belong to and play a part in food webs, it follows that all organisms can be grouped into a series of more or less discrete trophic levels, with each level depending on the one beneath it for its energy supply. The ultimate source of energy for all ecosystems (and thus for all organisms) is solar radiation, which is trapped by the autotrophic green plants and converted into chemical energy in the plants themselves by the process of photosynthesis (see Box 18.3).

Table 18.3 Trophic levels in an ecosystem

Trophic level	*Feeding habit*	*Example*
$\wedge_1$	Primary producers (autotrophs)	Green plants on land, phytoplankton in oceans
$\wedge_2$	Primary consumers (grazers)	Sheep and cattle
$\wedge_3$	Secondary consumers (carnivores)	■ on land – lions eat wildebeest, hawks eat field mice ■ in the soil – soil organisms decompose herbivore remains ■ in oceans – herrings eat copepods
$\wedge_4$	Top carnivore	These are generally not eaten while they are alive; a good example is the lion

Each trophic level in a food web can be given a number, which indicates its feeding level (in a vertical sense) relative to the primary producers at trophic level 1 ($\wedge_1$). Thus $\wedge_1$ are the primary producers, $\wedge_2$ are the primary consumers, and so on (Table 18.3). The decomposer part of the system can be in trophic levels 2, 3 or 4 depending on the source of food, because it receives dead material from levels 1, 2 and 3 and waste materials from levels 2 and 3.

While the notion of trophic level is very useful, some complications remain. For some species, the trophic position changes through the life-cycle of the individual. Thus, for example, the feeding relationships of the tadpole are different from those of the frog. Many species can also change their feeding habits, depending on the availability of food supplies.

BOX 18.8 PYRAMID OF NUMBERS

A typical pyramid for a food web where primary producers are small in size but need to be large in number to support the smaller number of consumers that feed on them has a wide base and a narrow top. An entire field of grass, for example, is needed to feed a small flock of sheep. In some food webs, where the primary producer is a large individual (such as a tree), each individual plant can support a large number of herbivores, so the pyramid has a narrow base and wide trophic level 2. Inverted pyramids, with a narrow base and a wide top, are produced in a parasitic feeding relationship, where a large number of parasites ($\wedge_2$) and hyper-parasites ($\wedge_3$) are supported by a single primary producer ($\wedge_1$).

ECOLOGICAL PYRAMIDS

In most ecosystems, there are a relatively large number of primary consumers but far fewer carnivores. The relationship between the number of individuals and their trophic level in a food web can be plotted (Figure 18.4) as a 'pyramid of numbers' (Box 18.8).

The pyramid of numbers takes no account of the size of the individual plant or animal; a blade of grass and an oak tree are both primary producers ($\wedge_1$), for example. As a result, ecologists generally prefer to compare ecosystems on the basis of the biomass (total weight of organic matter, in g m^{-2}) of each trophic level, rather than simply the number of individuals present. The 'pyramid of biomass' provides an ecologically more meaningful indication of food web structure, and the normal pyramid is wide at the base and narrow at the top. The sum total of the biomass at each trophic level (i.e. the total weight of organic matter) represents the standing crop of the ecosystem at that point in time.

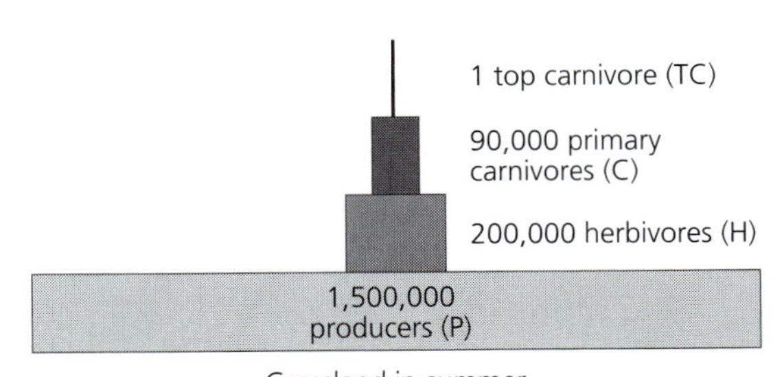

Figure 18.4 *Pyramid of numbers for a grassland food web. The pyramid of numbers shown here, representing the number of individuals per 1,000 m² of typical grassland in the summer, has over 1.5 million producers supporting one top carnivore.*

Source: Figure 3.8 in Cunningham, W.P. and B.W. Saigo (1992) Environmental science: a global concern. Wm. C. Brown Publishers, Dubuque, Iowa.

The pyramid of numbers and the pyramid of biomass tell us nothing about the biological productivity of either the individual trophic levels or the entire food web. These are important dimensions of ecosystem structure and function, so a 'pyramid of energy' is sometimes used to display the total amount of energy used at each trophic level. Energy (in calories) is expressed per unit of area (m^2) over a given period of time (usually a year), so the units used are cal m^{-2} a^{-1}. The pyramid of energy allows comparisons to be drawn between the productivities of different types of ecosystem, such as deserts and tropical rainforests.

HABITAT AND NICHE

Habitats (Box 18.9) vary in size according to the species that occupy them. Carnivores usually occupy and require much larger habitats than herbivores, because each carnivore (primary or secondary consumer) consumes many herbivores and thus derives its food from a wider area. For a species to survive, its habitat must support a population large enough to sustain itself by breeding. Many animal species lack the ability to breed once their numbers fall below a critical threshold, so their populations may collapse and even become extinct. This is the basis of much of the concern about habitat clearance and loss of biodiversity (see p. 539).

'Habitat' refers to where an organism lives, feeds and breeds. If habitat is an organism's or species' address, its

BOX 18.9 HABITAT

Habitat describes the local surroundings in which a plant or animal lives. Each habitat has a particular environment, and habitats are usually defined on the basis of geology, vegetation and location (Figure 18.5). Thus, for example, there are wetland habitats, woodland habitats, heathland habitats, grassland habitats, mountain habitats, coastal habitats, freshwater lake habitats, and so on. Areas within a habitat that are occupied by a particular species or community are described as *microhabitats*. Habitats are mosaics of microhabitats.

BOX 18.10 NICHE

Niche is a functional concept that describes the role played by a particular species within the ecosystem. That role is defined in terms of such things as what the species does, how it feeds, what and when it eats, when and for how long it is active, how it reproduces, how it behaves, what particular part of the habitat it uses, and how it responds to temperature and moisture.

profession is defined by its *ecological niche* (Box 18.10). In most ecosystems, a number of species could perform that role, but at any one point in time only one species actually does. In grassland ecosystems, for example, bison (in North America), antelopes and zebras (in Africa) and kangaroos (in Australia) play more or less the same role in different places. No two species can occupy exactly the same niche, because they would be in direct competition for the same resources at every stage of their life-cycle.

INTRODUCED SPECIES

Sometimes an exotic (non-native) species of plant or animal is introduced into an area, either accidentally or more usually deliberately. If the introduced species is not well adapted to the climate and environment of its new location, it is unlikely to survive. In the eighteenth and nineteenth centuries, for example, British explorers brought back a wide range of exotic plant, animal and bird species from distant lands, although the climate of Britain proved too wet and/or cold for many of them to survive and reproduce outdoors.

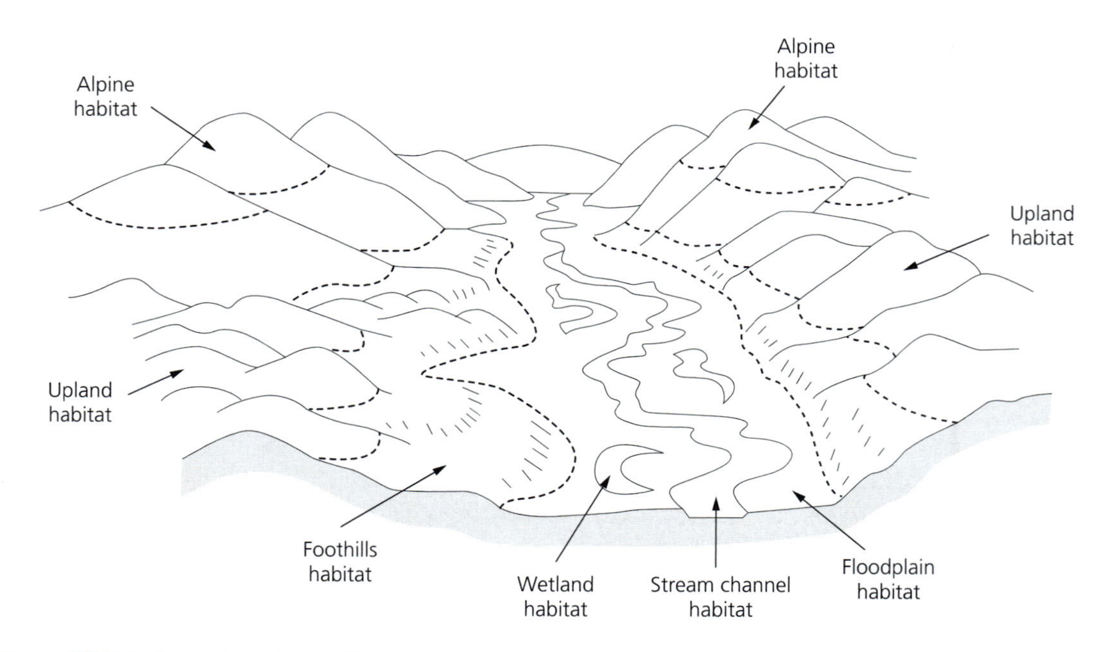

Figure 18.5 *Habitats in a river valley. Most river valleys and surrounding uplands comprise a number of different types of habitat for plants and animals.*

Source: Figure 14.8 in Marsh, W.M. and J.M. Grossa (1996) Environmental geography: science, land use and Earth systems. John Wiley & Sons, New York.

BOX 18.11 IMPLICATIONS ABOUT NICHE

There are four important implications of ecological niches:

1. Many species have overlapping niches, but species with identical niches cannot coexist in a single ecosystem. This so-called *competitive exclusion principle* is explained by assuming that dominant species are specialists that have evolved adaptations to a particular aspect of the environment.
2. The greater the diversity of niches in an ecosystem, the greater the variety of energy flows and thus the more stable the ecosystem. Fluctuations of species populations in the system will be less extensive than in systems with less diverse niches, because predators have a wider choice of prey.
3. The introduction of exotic species into an ecosystem, either deliberately or by accident, can create problems because they sometimes compete with native species for ecological niches and food supplies.
4. The removal of a species (or group of closely related species) can leave a niche empty, and this can soon lead to reduced energy flow through the entire system, which can radically affect the structure and stability of the overall ecosystem. Species can be removed from ecosystems in a variety of ways, including pollution, over-exploitation, loss of habitat or environmental change (see p. 529). Maintenance of a diverse range of species, habitats and niches is of fundamental concern in conservation (see p. 539).

Many introduced species are well adapted to their new environments, and their populations often increase rapidly, particularly if there are few or no predators and parasites to attack them. This was certainly the case with the house mouse (*Mus musculus*), a native of the Middle East that was introduced into Britain by people in pre-Roman times. It is also true of the European house sparrow (*Passer domesticus*), which was deliberately introduced into North America in the early 1850s and is now widespread.

Ecologists usually oppose the introduction of species to new areas because of the competition and disturbance that they can cause to existing species.

ECOSYSTEM DYNAMICS

Ecosystems are typical open systems. As such they are driven by the continuous throughput of energy and materials that are input from outside the system boundary and that in turn are output either back to the environment (in the case of gases and waste heat) or to adjacent ecosystems.

The two main movements through ecosystems are energy (which flows) and nutrients (which are cycled). Energy derived ultimately from solar radiation (see p. 237) passes through the food web and becomes an output from the system via respiration from successive trophic levels (Figure 18.6). Nutrients, on the other hand, are derived ultimately from rock weathering (see p. 203), soil formation (p. 602) and biological decay (p. 556), and they cycle through ecosystems.

ENERGY

The movement of energy through an ecosystem is closely governed by the first two laws of thermodynamics (see p. 76). The ultimate source of energy for ecosystems is the Sun, but only about 43 per cent of the solar radiation that reaches the Earth's surface is of suitable wavelengths (see p. 233) to be useful in the process of photosynthesis.

Photosynthesis by plants consumes carbon dioxide and releases oxygen; respiration by animals consumes oxygen and releases carbon dioxide. Both processes are essential to the maintenance of an atmosphere on Earth that is suitable for life, including human life. Much of the concern over extensive clearance of the tropical rainforest centres on the disruption that this causes to global atmospheric chemistry, particularly because vegetation helps to filter out naturally rising levels of CO_2 caused by air pollution.

Solar energy enters the plants through leaf walls, and in the presence of chlorophyll in the leaves the energy is transformed into a usable form by photosynthesis (see Box 18.3). Some of the energy converted in this way is lost by plant transpiration, and the rest is used in plant growth, where it is stored in a chemical form in plant tissues. When the plants are eaten by primary consumers (herbivores), the energy stored in the plants enters trophic level 2, from which some is lost by animal respiration and the rest is used for animal growth and movement. This transfer process continues along the food chain (Figure 18.7).

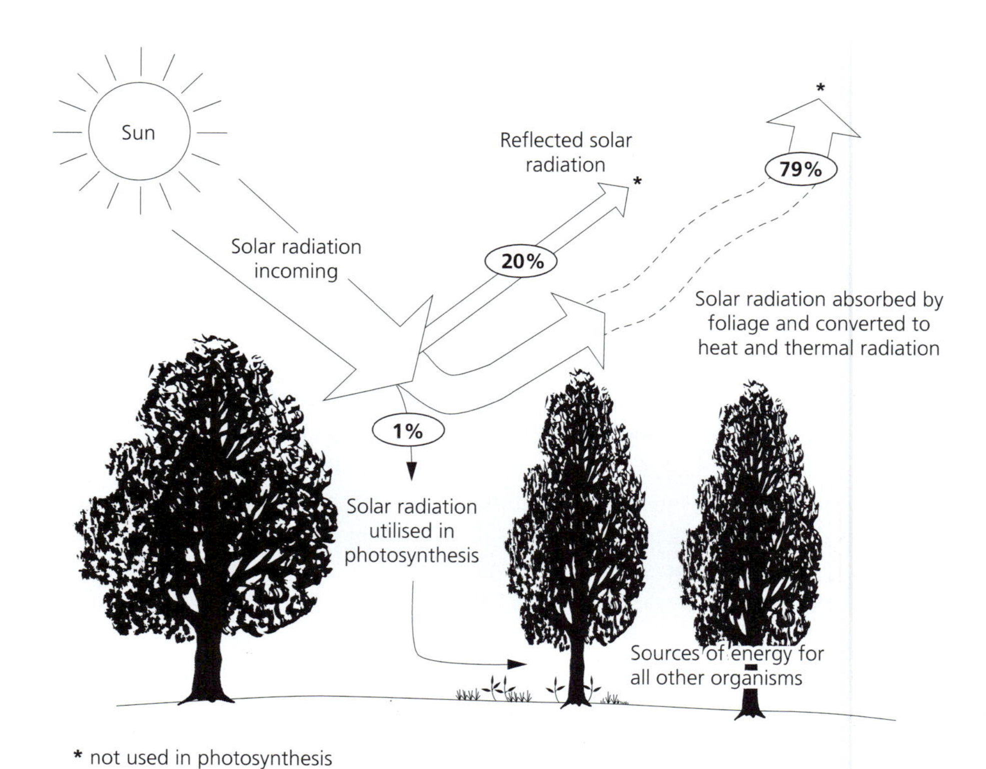

Figure 18.6 *Input and output of energy in a woodland ecosystem. Solar energy is broken down into heat, radiation and organic compounds.*

Source: Figure 4.3 in Marsh, W.M. and J.M. Grossa (1996) Environmental geography: science, land use and Earth systems. John Wiley & Sons, New York.

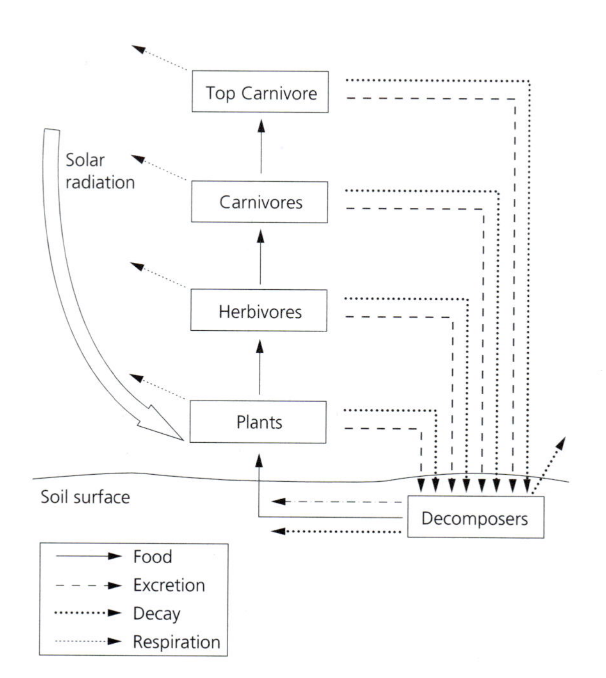

Figure 18.7 *Principal energy flows in an ecosystem. Energy in ecosystems is ultimately derived from solar radiation, but it passes through the ecosystem along the food webs. Decomposition is an important part of this process.*

When plants and animals die, if their bodies fall to the ground the decomposer organisms (see p. 556) inherit the remaining energy, which can then be released back to the atmosphere by respiration or used in the growth and movement of the decomposers.

Inefficiency in the use of energy in ecosystems (Box 18.12) explains why there are rarely more than about four vertical links in a normal food web, because so little energy would be available above trophic level 4 to support and sustain omnivorous species (Figure 18.8). Results from a classic study of energy flows through Silver Springs, Florida (Figure 18.9), show that just over 5 per cent of the energy produced by the primary producers becomes available to the top carnivores.

A useful way of expressing the efficiency with which different parts of an ecosystem use available energy is in

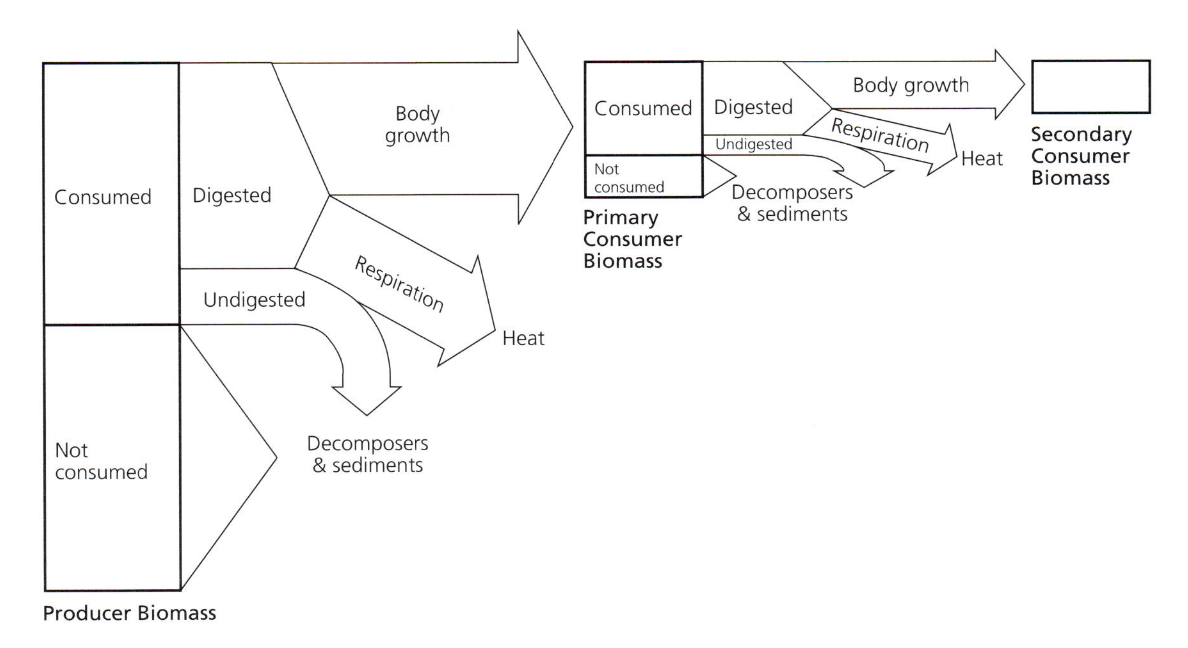

Figure 18.8 *Distribution of energy flows along a food web. The figure is schematic, and it shows how only part of the relatively large amount of energy available in the biomass of producer organisms (plants) becomes available to the primary consumers, and only part of that energy in turn becomes available to secondary consumers.*

Source: Figure 3.7 in Cunningham, W.P. and B.W. Saigo (1992) Environmental science: a global concern. Wm. C. Brown Publishers, Dubuque, Iowa.

BOX 18.12 ENERGY EFFICIENCY IN ECOSYSTEMS

In general, ecosystems are very inefficient in their use of available energy, for three main reasons:

1. Green plants (on which the rest of the trophic system heavily depends) make only about 7 per cent of the solar radiation available to the primary consumers.
2. Only a small but variable fraction of the energy available at any particular trophic level is passed on to the next; the rest is lost in respiration.
3. Because the animals at trophic levels 2 and above are mobile, they respire proportionally more than the plants at trophic level 1. As a result, the relative loss of energy increases at successively higher trophic levels.

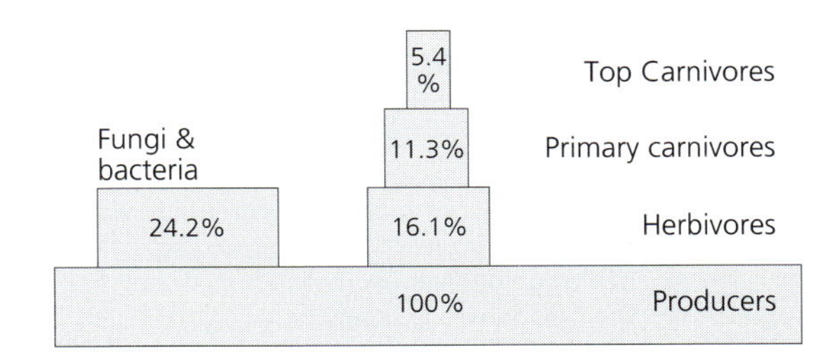

Figure 18.9 *Energy flow through Silver Springs, Florida. This pyramid of energy is based on a classic study of an aquatic ecosystem at Silver Springs, Florida, in which the producer energy base was calculated to be 20,810 kcal m^{-2} a^{-1}. At each step along the food web energy is lost through respiration, so relatively little of the original energy is ultimately available to the top carnivores.*

Source: Figure 3.6 in Cunningham, W.P. and B.W. Saigo (1992) Environmental science: a global concern. Wm. C. Brown Publishers, Dubuque, Iowa.

terms of the ecological efficiency – the ratio between the productivities at successive trophic levels. Productivity (the amount of energy passed per unit of time from one trophic level to the next) is measured in cals cm $^{-1}$ unit time^{-1}. Ecological efficiency is expressed as a percentage. It must always be less than 100 per cent, obviously, and in practice is often between 5 and 20 per cent.

NUTRIENTS

Unlike energy (which flows through, is used by, then lost from ecosystems), nutrients flow through ecosystems, are used by them and can then be recycled and reused. The flow of nutrients through ecosystems is an important component of the biogeochemical cycles (see p. 86).

The basic source of many of the chemical elements is the physical and chemical weathering of bedrock and deposits (see p. 204), although human inputs in the form of fertiliser applications and chemical pollution can be important in many areas.

The chemicals are withdrawn from the underlying material by green plants by the process of root osmosis, in solution in the water. Once taken in by the plants, these elements are subsequently passed up through the trophic structure along the food web. The dead organisms from each trophic level, and waste material from trophic level 2 upwards, are broken down by decomposer organisms, which release the elements back into the cycle.

Equilibrium in the cycle can be disrupted in a number of ways, including soil erosion, the leaching of elements deep into soil profiles beyond the reach of plant roots, water flow into streams and rivers, and extraction for resource use (such as guano). In these cases, the output of minerals from the ecosystem exceeds the input, and this often leads to ecological imbalance and soil depletion.

BOX 18.13 MAJOR ELEMENTS AND TRACE ELEMENTS

Many different chemical elements cycle through ecosystems, including major elements – such as nitrogen (N), phosphorus (P), potassium (K), calcium (Ca), magnesium (Mg), sulphur (S) and iron (Fe) – which are required in relatively large amounts. There are also a large number of trace elements that are required in smaller amounts, but are nonetheless essential for plants and animals. Each chemical element meets a different need (Table 18.4), and all are required in stable amounts.

Table 18.4 Ecological significance of particular chemical elements

Element	*Significance*
Oxygen	Very active chemically; it combines with most elements in nature and the process of oxidation speeds up metabolic processes. Oxygen forms an average of 70 per cent of the atoms in living matter and is a basic building block of carbohydrates, fats and proteins
Hydrogen	This is also an important ingredient of living matter. On average it constitutes about 10.5 per cent of the atoms
Carbon	Another important constituent (about 18 per cent on average) of living matter
Nitrogen	An important constituent of protein molecules
Sulphur	An important, though small, constituent of some proteins
Phosphorus	A constituent of protein molecules, including ATP (the chemical energy-storing compound that is important in photosynthesis). This is the element that normally limits the growth of ecosystems

BOX 18.14 PRINCIPLES OF NUTRIENT CYCLING

There are a number of important principles of nutrient cycling in ecosystems:

- The total amount of mineral nutrients in an ecosystem depends on the rate of movement of nutrients into and out of the system.
- The amount of nutrients in the living biomass, litter and soil in the ecosystem is a function of the transfer rates of those nutrients between the components.
- Over time, an ecosystem tends towards an equilibrium condition in which the quantity of nutrients in the system and in each compartment remains relatively stable.

BOX 18.15 CONTROLS ON PRODUCTIVITY

The process of building organic matter in an ecosystem depends closely on the availability and movement of energy through the system, and the movement is ultimately driven by solar energy. Two main factors determine the ecological productivity of an ecosystem:

1. the amount of solar radiation that is available to the primary producers for photosynthesis;
2. the efficiency with which the autotrophs convert the solar energy into usable forms by photosynthesis.

BOX 18.16 NET PRIMARY PRODUCTIVITY

Net primary productivity varies a great deal from ecosystem to ecosystem (Table 18.5). The most productive ecosystems are reefs and estuaries, with natural forests and fresh water a close second. The lowest productivities are found in the open oceans and in desert environments. While mean net productivity levels on land are more than four times as high as those in the oceans, when these productivities are weighted according to the total area involved the oceans far exceed land in area, so total net primary productivity on land is only twice that in the oceans. Recent estimates suggest that over 70 per cent of terrestrial net production is concentrated between 30° N and 30° S latitude, which underlines the ecological importance of the tropical environment.

PRODUCTIVITY

One of the most important characteristics of an ecosystem is its productivity. This reveals a great deal about the condition of a particular ecosystem, and different systems can be compared directly on the basis of productivity. Analysis of variations in productivity allows the significance of various limiting factors to be evaluated. Techniques are now being developed that make it possible to estimate primary productivity using remote sensing, which is opening up the prospect of better monitoring of changes and analysis of causes of variation.

All of the energy used by plants is converted into chemical energy, so, at least in theory, measurement of the amount of sugar produced in plants should provide an index of their energy uptake, and thus a measure of productivity. The total amount of energy produced at trophic level 1 in an ecosystem is the *gross primary production*.

Some of the energy produced by photosynthesis is lost by respiration, and this fraction varies from species to species and within a given species depending on local environmental conditions such as weather. As a result, gross primary production data reveal little about the amounts of energy available to higher trophic levels, and yet from the point of view of cropping or extraction this is the key to ecological productivity.

A more useful measure is gross primary production minus respiration losses at trophic level 1. This is the *net primary productivity* (Box 18.16), and it represents the total amount of usable organic material produced per unit of time at trophic level 1. It is conventionally expressed in g m^{-2} a^{-1} (grams per square metre per year), and it varies from ecosystem to ecosystem (Figure 18.10).

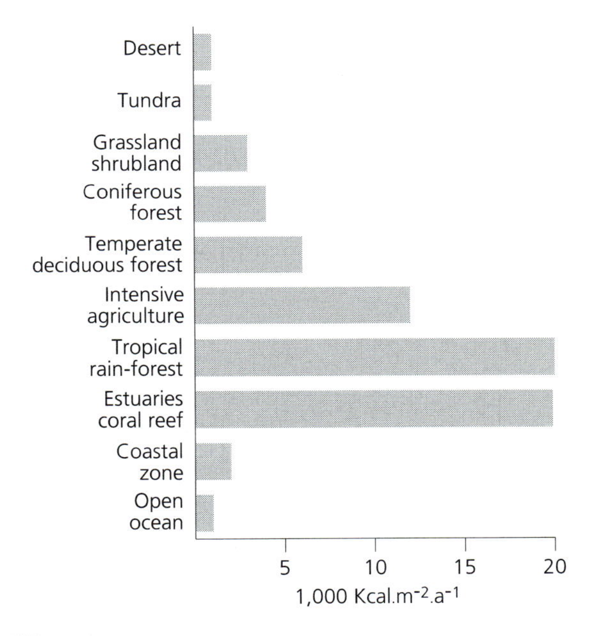

Figure 18.10 *Variations in productivity between ecosystems. Gross primary productivity varies a great deal from one type of ecosystem to another. Tropical rainforest, estuaries and coral reefs are the most productive natural ecosystems in the world.*

Source: Figure 4.1 in Cunningham, W.P. and B.W. Saigo (1992) Environmental science: a global concern. Wm. C. Brown Publishers, Dubuque, Iowa.

Table 18.5 Estimated net primary productivity for major world biomes

Biome	Mean productivity ($g\ m^{-2}\ a^{-1}$)	Total for area ($10^9\ t\ a^{-1}$)
Reefs and estuaries	2,000	4.0
Forest	1,290	64.5
Fresh water	1,250	5.0
Cultivated land	650	9.1
Woodland	600	4.2
Grassland	600	15.0
Upwelling zones	500	0.2
Continental shelf	350	9.3
Tundra	140	1.1
Open ocean	125	41.5
Desert scrub	70	1.3
Desert	3	–
Total continental	669	100.2
Total oceanic	155	55.0
World total	303	155.2

The productivity of cultivated land overall is 650 $g\ m^{-2}\ a^{-1}$, which is not substantially higher than that of natural grassland (600 $g\ m^{-2}\ a^{-1}$). This suggests that perhaps cultivation is not a particularly efficient form of land use (in an ecological sense) when the amount of time, effort and resources devoted to it are taken into account. Some particular crops – such as wheat, rice and potatoes – have still lower productivities (Table 18.6).

Table 18.6 Estimated net primary productivity of cultivated ecosystems

Ecosystem	Productivity ($g_m^{-2}\ a^{-1}$)
Mass algal culture, outdoors	4,526
Sugar cane (world average)	1,726
Rice (world average)	496
Potatoes (world average)	402
Wheat (world average)	34

BOX 18.17 GLOBAL WARMING AND PRODUCTIVITY

Scientists are concerned about the possible long-term impacts of global warming (see p. 261) on biological productivity around the world. If productivity declines, particularly in the main food-producing zones, there could be serious risk of wholesale food shortages and widespread famines. Computer modelling of the possible effects of continued rise in atmospheric CO_2 and associated climate changes has shown that productivity responses in tropical and dry temperate ecosystems are likely to be dominated by CO_2, but those in northern and moist temperate ecosystems are more likely to be caused by the effects of temperature on nitrogen availability.

LIMITING FACTORS

The productivity of ecosystems is clearly dependent on the availability of solar radiation in suitable amounts and at suitable wavelengths. But this is not the only important factor that influences productivity; others include temperature, nutrient supply and water. Ecosystems with each of these factors present in suitable forms and quantities often show relatively high levels of productivity.

Quite often one or more of the important factors is in short supply, and this can inhibit efficient and productive ecological development. Such a factor is called a *limiting factor* (Figure 18.11). For example, in many desert areas there are plentiful supplies of nutrients, sunlight and temperature to support a reasonably healthy vegetation, but vegetation is generally sparse because shortage of water is a limiting factor.

The concept of limiting factors can be traced back to the 1840s, when a German agricultural chemist, von Leibig, found that the yield of a crop could be increased only by supplying the plants with more of the nutrient that was present in the smallest quantities. This suggested to von Leibig a law of the minimum, which states that the productivity, growth and reproduction of organisms is constrained if one or more limiting factors is present in less than a minimum threshold quantity, and that yields can increase in direct proportion to the amount of that limiting factor which is added.

There should also be a law of the maximum, because productivity and growth can also be constrained by an excess of critical factors. Between 1950 and 1990, for example, the average temperature of surface water in the Pacific Ocean off California increased by 1.2–1.6 °C, and this warming is believed to account for an observed 80 per cent reduction in zooplankton in the ocean water. Many plants on land are badly affected by air pollutants, particularly sulphur and nitrogen oxides, and ozone.

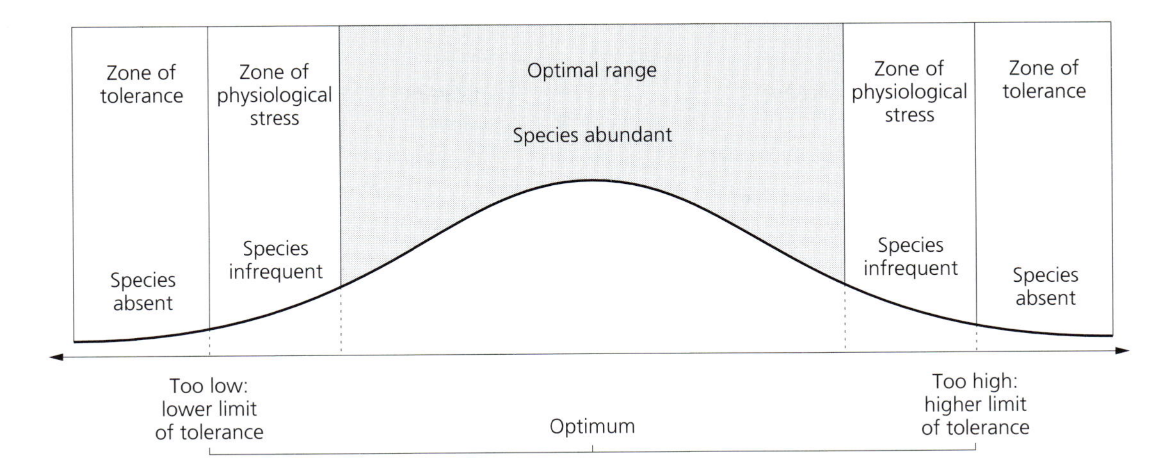

Figure 18.11 *Limiting factors and tolerance range. The presence of a species is closely related to its ability to survive within a specified range of variation for each factor of the environment. See text for explanation.*

Source: Figure 4.4 in Cunningham, W.P. and B.W. Saigo (1992) Environmental science: a global concern. Wm. C. Brown Publishers, Dubuque, Iowa.

BOX 18.18 LIMITING FACTORS

There is a spectrum of significance of any particular limiting factor, with upper and lower critical threshold conditions and an optimum. In theory, therefore, many limiting factors have three critical values:

1. a minimum condition, below which the organism dies;
2. an optimum condition, where the organism thrives;
3. a maximum condition, above which the organism dies.

BOX 18.19 DISTURBANCE AND RECOVERY

All ecosystems can withstand a certain amount of disturbance but still maintain their in-built ability for self-repair. But for each ecosystem there is a critical level of disturbance beyond which natural recovery is very difficult. The disturbance can take many different forms, including recreational pressure in which over-use by walkers can seriously damage natural vegetation, cause extensive soil erosion and radically alter local landscapes.

Critical thresholds are not established for all ecosystems. They often become apparent only when they have been reached or exceeded, and when irreparable damage is done or irreversible changes are set in motion. Present-day environmental change associated with global warming (see p. 261) may well prove to be too fast for most ecosystems to adapt to.

If these critical conditions are scaled up from the individual organism to the whole population of a particular species, it suggests that a population can exist only over a portion of the total range of variation of a given limiting factor (this is the tolerance range for that factor), and that the population is better adapted to certain parts of the range (the optimum) than to others. This notion of adaptedness to environment lies at the heart of Darwin's theory of evolution by natural selection (see p. 531).

EQUILIBRIUM

Ecosystems are characterised by a complex web of relationships between individual organisms (created by flows of energy and nutrients) and by a mosaic of feedback loops whereby a change in one part of the system may lead to adjustments in many other parts of the same system and in adjoining systems. Ecosystems that have evolved over long periods of time can reach a steady-state condition characteristic of open systems in general, in which the impact of local disturbances from the environment can be stabilised by ecosystem constraints and by the process of self-adjustment.

The adjustment occurs over a spectrum of timescales. Adjustment to natural environmental change is required to cope with three main types of pressure:

1. gradual change, such as changes in climate and sea level;
2. disturbances, such as hurricanes, freezes and fires;
3. natural periodicities, such as the cycles of dry and wet seasons.

EVALUATION AND APPLICATION

There is no doubt that the ecosystem concept is one of the most useful ideas of ecology; it allows us to examine how nature works in an objective way. Ecosystems are dynamic, self-regulating open systems that link together the main environmental systems (the hydrosphere, lithosphere, atmosphere and biosphere) and are vital to the existence and survival of life on Earth.

PROBLEMS

A number of difficulties surround the theory and practice of ecosystem studies, including:

- *Scale*: an ecosystem can be of any size from a goldfish bowl to the global ocean, or from a single tree to the entire biosphere, yet the nature of an ecosystem clearly depends on the scale it is examined at. In practice, the size of system adopted for measurement and analysis often depends on the particular focus of the study, or on convenience.
- *Boundary definition* (Box 18.20).
- *Measurement*: measuring and describing ecosystem forms and processes is often much harder in practice than it is in theory, because of the large number of organisms and flow pathways involved.
- *Human modification*: relatively few ecosystems are entirely natural, because even those that have been least affected by direct human activities (including remote portions of the tropical rainforest and some remote ocean areas) have probably been affected indirectly, either by global air and water pollution or by induced environmental change (including global warming; see p. 261).

BOX 18.20 ECOSYSTEM BOUNDARIES

Nature is a continuum, and with few exceptions sharply defined boundaries are rare. But for ecological analysis, measurement, modelling and application, boundaries have to be defined for a study ecosystem. Sometimes a biome boundary will suffice (such as the edge of a woodland or the outer margin of a desert), and often a landscape boundary is convenient (such as the watershed of a drainage basin; see p. 356) or logical (such as the high-water mark for coastal ecosystems; see p. 497). But material and energy flow freely through the environment, and defining the end of one ecosystem and the start of another is often very difficult. Ideally, the boundary would be defined to include all functioning processes (such as nutrient cycling) in the study system.

BENEFITS

Despite these problems, the ecosystem approach has some significant benefits. These include:

- *Convenient scale*: the ecosystem is a convenient scale at which to consider plants and animals, and their interactions with each other and with their environment.
- *Unifying framework*: ecosystems provide a convenient framework in which to view the interactions between the biotic and abiotic aspects of the environment.
- *Fundamental ecological unit*: the ecosystem is a convenient unit on which to base environmental management strategies.
- *Applied analysis*: ecosystems are also valuable for analysing the effects of human activities on the environment.

APPLICATIONS

By the 1960s, the theory of ecosystems was quite well developed but was not widely applied. This is perhaps surprising, given the emphasis in ecosystem studies on a comprehensive, integrated approach. This approach is the real strength of adopting an ecosystem perspective, because it stresses unity, interaction and synergy (the whole is greater than the sum of the parts).

Natural resource management up to the 1960s relied heavily on managing parts of ecosystems as more or less independent units. Thus, for example, forest management, wildlife management and water and soil resource management were widely practised, but integrated ecosystem management was not. One important development since

BOX 18.21 THE GREATER YELLOWSTONE ECOSYSTEM PROJECT

Yellowstone National Park covers an area of 8,983 km^2 in the central Rocky Mountains (mostly in Wyoming) (Figure 18.12). It was established in 1872 and is the oldest and largest national park in the USA. It has striking scenery, dominated by spectacular geysers and hot springs (p. 139) – the park has more than 3,000 geysers and hot springs, the greatest concentration anywhere in the world – along with high waterfalls and deep canyons. Most of the national park is covered by coniferous forest, which provides an undisturbed wilderness habitat for wildlife. It is home to a striking diversity of birds (more than 200 species) and animals (including grizzly bears, elk, moose and bison). Perhaps inevitably, the very qualities that need conserving also attract vast numbers of visitors, who are causing damage by over-use. Other activities in the national park are also causing problems, such as land-use changes, resource extraction and pollution.

In 1987, as a result of these increasing pressures, the US Forest Service and the National Park Service launched a joint co-ordination and planning process for the Greater Yellowstone region, commonly called the 'Vision' exercise. A report was published in 1991, and it recommended much better co-ordination of the Greater Yellowstone Ecosystem (GYE) in order to protect its integrity and interrelationships. The report attracted widespread public opposition. Not surprisingly, the traditional resource-extracting industries opposed it because they felt it restricted their operations and future potential. Many local residents believed that ecosystem management would negatively affect their communities and lifestyles. Perhaps most surprisingly, most conservationists also failed to support it, dismissing it as too vague and lacking clear objectives.

Considerable effort is being invested in trying to preserve the naturalness of the Yellowstone landscape and restore the natural ecosystems in the area. One initiative raises interesting questions about the rights of wildlife – the park's thriving population of wolves, which was reintroduced by the US Fish and Wildlife Service between 1994 and 1997, has been declared illegal. The 'experimental population', which might compete with native wolves that occasionally stray into the area from further north, is not covered by the US Endangered Species Act. This means that ranchers whose herds are threatened by the reintroduced wolves can shoot the animals. Another native animal is also under attack in Yellowstone – the bison. In 1902, only twenty-five bison were left in Yellowstone National Park, but by the late 1990s the population had grown to around 2,200. Half of them carry the infection *Brucella abortus,* but few are seriously affected by it. However, the bison are believed to pass on the disease to cattle, where it causes spontaneous abortion, infertility and slow growth. Any bison that leave the park boundary to graze in surrounding pastures – where cattle also graze – are now trapped, and any that test positive for the disease are killed.

the early 1970s has been the widespread adoption of an integrated ecosystems framework and perspective in natural resource management, for example for planning and managing protected areas such as national parks and wilderness areas. The Greater Yellowstone Ecosystem (GYE) project (Box 18.21) is an interesting illustration of how difficult it often is to balance ecological and socio-economic factors in making conservation decisions.

This broadening of approaches in natural resource management to embrace an ecosystems framework is part of a wider shift in environmental management. Scientists are increasingly being called upon to formulate simple and clear strategies for effective management of environmental problems, and an ecosystem approach might be highly relevant.

One area where important concepts and approaches have been borrowed from ecosystem analysis and applied in environmental planning is landscape ecology. This seeks to manage the countryside in a balanced, sustainable and integrated way, and it is recognised that ecosystem units and principles are logical starting places. Landscape ecology has grown rapidly in recent years, partly because it holds great promise in sustainable development but also because it has been stimulated by access to the new technologies for handling spatial information, particularly remote sensing and geographical information systems.

One of the most ambitious applications of ecosystem ideas is in the study of urban areas and how they function. A city contains all the ingredients of a natural ecosystem – structure, functional relationships, flows of materials and energy, population dynamics, and so on – and so it should be possible to adopt an ecosystem framework for the comprehensive analysis of urban systems. One such study – the Baltimore Ecosystem Study in the north-east United States – is currently under way. It is costing around US$1 million a year, is expected to last for several decades and is the first time that an entire city has been looked at as an ecosystem. The goal of the study is to understand the city's hydrology, microclimate, nutrient cycles and flows of energy, and to work out predator–prey relationships and the competition between species in that urban habitat.

Ecosystem principles are also proving extremely useful in restoration ecology, where the object is to restore damaged plant and animal communities, habitats and ecosystems as part of a broader nature conservation strategy. This often requires great investment, long-term commitment and an appreciation of the wider values of nature and natural habitats. A good example is the long-term programme to restore the natural wetlands of the Everglades in Florida (Box 18.22).

The ecosystem approach is an extremely useful and effective way of taking account of the interrelationships between water, land, air and wildlife. For this reason, it has been used as a basis for developing remedial action plans (RAPs) for restoring degraded areas of the Great Lakes in North America (Box 18.23).

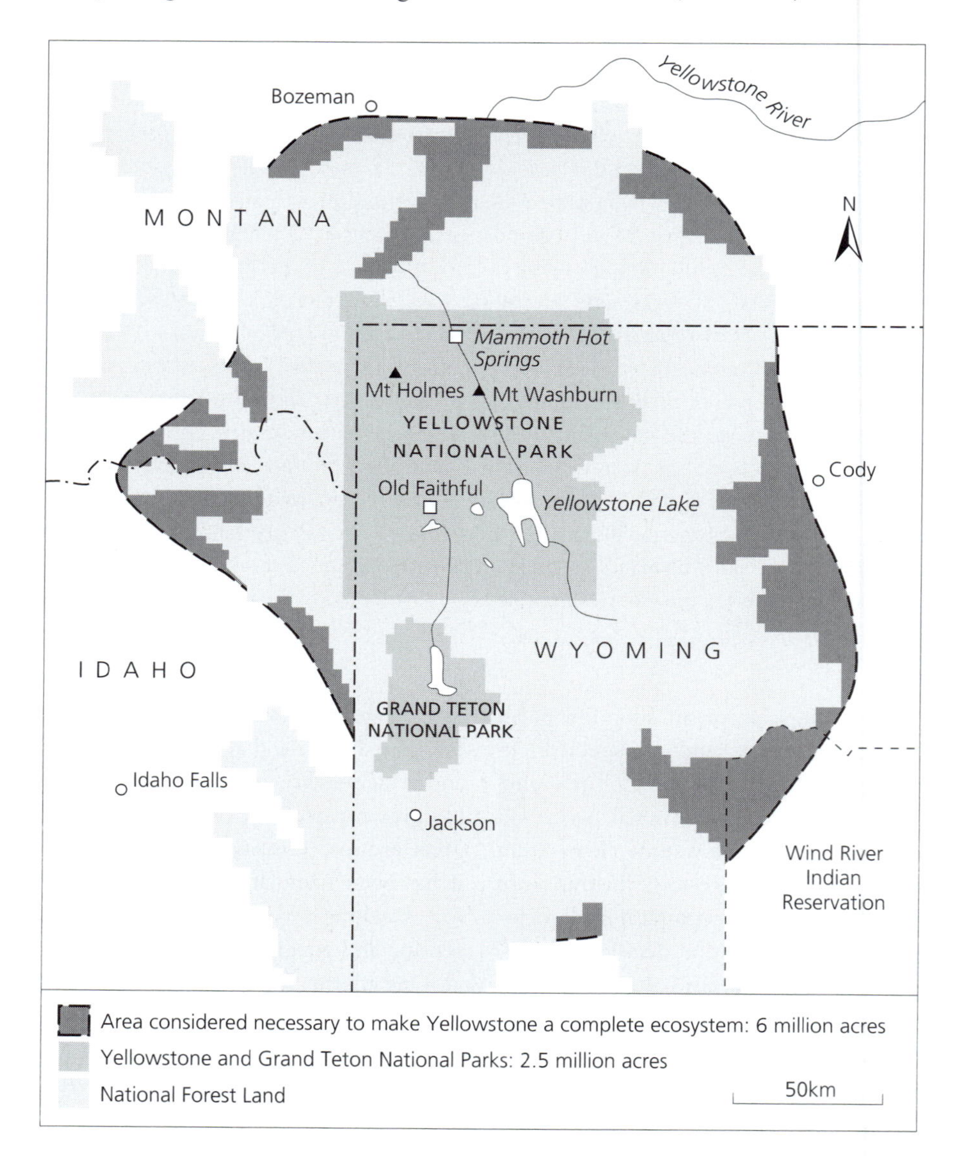

Figure 18.12 *The Greater Yellowstone ecosystem. The Yellowstone ecosystem complex (or biogeographical region) extends well beyond the boundaries of the national park. Ecologists believe that it is necessary to manage the whole ecosystem complex if the national park is to remain viable.*

Source: Figure 13.20 in Cunningham, W.P. and B.W. Saigo (1992) Environmental science: a global concern. Wm. C. Brown Publishers, Dubuque, Iowa.

BOX 18.22 RESTORATION OF THE FLORIDA EVERGLADES

The Everglades is a large natural freshwater marsh that covers nearly 13,000 km^2 of southern Florida. The wetland was created by regular overflowing from Lake Okeechobee during the wet season, which flooded the Everglades up to 1 m deep in water. Most of the area is a wilderness of swamp, savannah and primary forest, and it contains a rich diversity of wildlife, much of which is protected within the Everglades National Park (which covers 6,100 km^2). The marsh is covered with dense saw grass, which rises up to 3 m above the water level, making access difficult in many places. Many small islands covered with thick bushes and trees (including cypress, mangrove and palms) are scattered throughout the marsh, all of which is below 2 m above sea level.

The resident human population of the Everglades is confined to several hundred Seminole Indians, but the area has nonetheless been altered significantly, mainly by a series of large and ambitious drainage programmes in the late nineteenth and early twentieth centuries designed to reclaim up to half of the original marshland for agriculture (mainly fruit, vegetables and sugar cane). Further changes came in the 1960s, when the US Corps of Engineers started to build a 1,600 km network of channels with levees to remove flood water more quickly and store some of it in 'water conservation areas'.

The various drainage schemes have seriously altered the Everglades wetlands:

- Marshes have dried and become more saline (by intrusion of seawater into the groundwater).
- The area of wetland habitat has declined by half, adding 68 species to the threatened or endangered list.
- Populations of wading birds and other vertebrates have declined by up to 90 per cent.
- Fish populations have been threatened.
- Aquifers have been depleted, creating water supply problems for Florida's cities.

In 1999, a programme of large-scale restoration work began, which will cost an estimated US$7.8 billion over 20 years and return the Everglades to their former glory. The plan covers an area of 28,000 km^2, stretching from Orlando to the Florida Keys and including Lake Okeechobee (the third-largest freshwater lake wholly within the USA) and the Kissimmee River. Levees will be lowered, meanders will be restored on straightened channels and 800 km of canals will be closed. Over time, the objective is to return the whole drained area to natural wetland. A special US$13.85 million Everglades Nutrient Removal Project has been started alongside the drainage restoration programme to restore and protect the Everglades from over-enrichment (see p. 409) caused by runoff of phosphates from agricultural areas.

One particularly interesting feature of the long-term restoration programme is that it is based on adaptive management (see Box 20.16). This involves focusing all restoration work on one area at a time and then adapting the programme on the basis of experience so as to optimise the benefits.

A great deal of research is undertaken on sensitive ecosystems, and this raises serious questions about what types and levels of intervention and disturbance should be allowed. A framework for scientists working on sensitive ecosystems, developed initially in Australia for research on the Great Barrier Reef (see Box 16.36) but equally applicable elsewhere, includes the following ethical rules:

- Researchers should not work in sensitive areas if alternatives are available.
- Research that can be undertaken without harming or killing endangered or threatened species is preferable in almost all circumstances, even if the alternatives are more costly and/or time-consuming.
- Before work begins, researchers will be asked questions such as whether the research and its outcomes are compatible with the management objectives of the area.
- Scientists should involve indigenous people in the planning of their project where this is appropriate.
- Any biological specimens collected during research should be made available to other researchers so that they do not have to gather further specimens unnecessarily.

Plate 18.1 *The Everglades, Florida, USA. Photo: Chris Park.*

BOX 18.23 AN ECOSYSTEM APPROACH TO RESTORING THE GREAT LAKES

Various factors favour the development of an ecosystem approach, including:

- The natural resource is shared and highly valued.
- Many of the pollutants in the lakes are long-lasting.
- The lakes are important water resources, providing supplies to nearly 23 million people.
- The integrity of the lakes is threatened by a number of induced changes, including water pollution and water diversion.
- Special institutional arrangements are required for the effective management of nationally shared resources.

Table 18.7 Clements' five stages in succession

1. *Nudation*: the initial creation of an area of bare land
2. *Migration*: the arrival of the first plant seeds
3. *Ecesis*: establishment of the seeds and growth of the plants
4. *Reaction*: competition between the established plants, and their effects on the local environment
5. *Stabilisation*: populations of species eventually reach an equilibrium condition, in balance with the surrounding environment, which is sustainable

BOX 18.24 TYPES OF SUCCESSION

Different communities of plants develop on different types of habitat, so the seral stages can be identified by appropriate prefixes. Thus, for example:

- *Hydroseres* occur on wet sites such as swamps and lake shores.
- *Xeroseres* are found on bare rock, where little moisture is freely available.
- *Psammoseres* occur on moving sand such as sand dunes.

Succession is heavily controlled by environment, and there are three main types of environment – wet (*hygric*), dry (*xeric*) and intermediate (*mesic*). Thus hydroseres are hygric successions and psammoseres are xeric successions. Figure 18.13 shows how different successions can develop from slightly different starting points.

SUCCESSION

One striking feature of vegetation development in most environments is the fact that groups of plants (which are not necessarily genetically linked to each other) often grow together. Groups of plants that are adapted to local environmental conditions are referred to as plant associations or communities. In this way, vegetation often has an apparent orderliness and continuity.

CONCEPT OF SUCCESSION

Studies of vegetation development over time have been made in many areas, and these generally show two important factors:

1. Development is a slow process involving a large number of small sequential changes.
2. Ultimately, a relatively stable and self-perpetuating plant community is often established.

Succession describes the way in which ecological associations succeed each other in a particular area (Table 18.7). The ways in which vegetation changes over time on the new surface tend to follow a similar pattern from one area to another. The sequence of steps is similar, but the precise species present and the timescale of change can vary a great deal from one environment to another.

The term 'succession' was coined by the American ecologist Frederick Clements in 1916. He saw vegetation development as an orderly and predictable sequence that develops along definite pathways towards predictable end situations. Clements identified five phases in the development of a succession (Table 18.7).

There are three main prerequisites for succession to occur:

1. A new surface is created.
2. Environmental conditions on that surface must be within the tolerance range of existing species.
3. Coloniser species must have access to that new surface.

DYNAMICS OF SUCCESSION

COLONISATION

Few bare land surfaces have no organic life at all on them, and in a relatively short period of time some types of plant will start to grow on most new surfaces (Figure 18.13).

Plate 18.2 *Hydrosere in the Tyrolean Alps, Austria. This small lake is infilling with sediment, vegetation is growing around the lake margin, and over time successional development of vegetation can be expected to radically alter the size and character of the lake.*
Photo: Chris Park.

The first plants to establish themselves are usually ones with light seeds, which can be blown by the wind. Colonisation occurs fastest where seeds have to travel only short distances; barriers such as mountains or seas slow down the process. The natural revegetation of Mount St Helens after the 1980 volcanic eruption has provided valuable opportunities for ecologists to study what happens in this interesting natural laboratory (Box 18.25).

The first species to colonise newly exposed ground are opportunist pioneers that can tolerate strong direct sunlight (in the absence of shade), grow in thin soils with few nutrients, and survive temporary drought or waterlogging. Most such plants are annuals which have some important features in common:

- They grow rapidly.
- They complete their life-cycle within one year.
- They produce many more seeds than are required simply to establish the species on the new site.
- They are unable to survive competition from other species.

SERAL STAGES

Most pioneer plants are herbs, but over time they are replaced by woody plants, which grow more slowly but grow taller than the herbs, in time shading them out. This

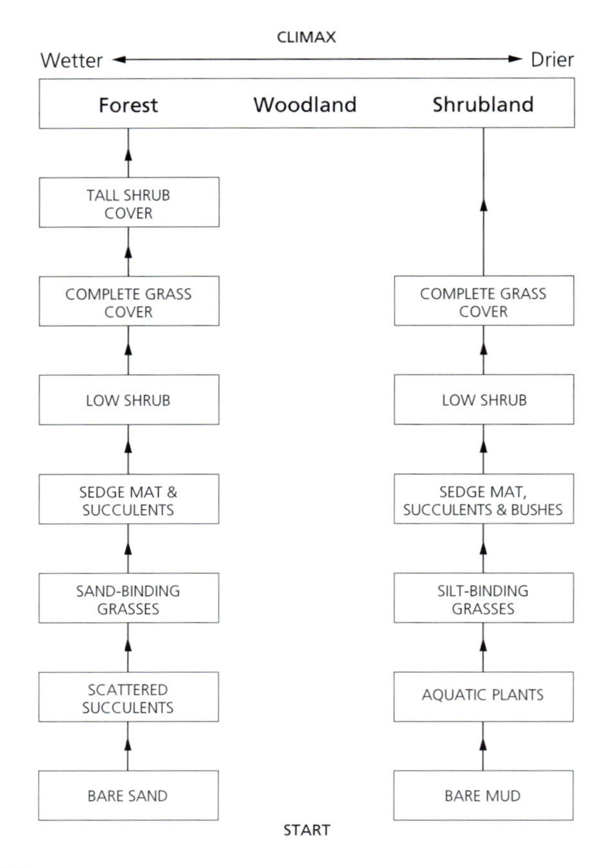

Figure 18.13 *Progress of succession from bare sand and bare mud. Two vegetation successions typical of a region with a sub-tropical east coast climate are shown, reflecting differences in the initial conditions.*

Source: Figure 21.8 in Dury, G.H. (1981) Environmental systems. Heinemann, London.

BOX 18.26 PLANTS AND ENVIRONMENT DURING SUCCESSION

There is a close association between the biotic and abiotic components of an ecosystem as the plant community progresses through the seres in a succession, because the plants and animals change their local environment in ways that encourage and facilitate colonisation by other species. Environmental changes are numerous, and they include decomposition of organic materials, which releases mineral nutrients, development of a soil profile (see p. 601) by the interaction between plants and the parent material, and alteration of local microclimate by providing shade and changing the land-surface albedo (see p. 601).

A progressive chain reaction is set in motion in which plants and animals change the environment, which becomes more suitable for other plants and animals, which in turn further change the environment, and so on. Successful colonisation by pioneer species prepares the way, literally, for colonisation by other species further along the chain of succession. In a strangely altruistic way, species tend to 'put themselves out of business' by changing the environment in such a way as to encourage competition from other species and thus eliminate themselves from the succession as it progresses to the next stage.

BOX 18.25 SUCCESSION ON MOUNT ST HELENS

The violent volcanic explosion that occurred on Mount St Helens in 1980 (see Box 6.26) radically altered the landscape of the area, but it also created a great deal of bare ground, which over time has been recolonised and revegetated by succession. Ecologists have monitored the vegetation changes in detail, and they reveal some interesting facets of the successional process. For example, plants quickly started to colonise a debris avalanche on the side of the mountain, so that within three years there were seventy-six (mainly wind-dispersed) species present with an average density of almost two plants per square metre. Total recovery was very slow, however, because within the first four years only 30 per cent of the species present before the eruption had re-established themselves, and average plant cover was less than 1 per cent. Within nine years plant cover averaged 18 per cent. While deciduous and coniferous trees had returned within three years, and within nine years willows (*Salix*) were common and red alder (*Alnus rubra*) was on the increase, the evidence suggested that coniferous forest typical of the Pacific north-west is very unlikely to be fully re-established on Mount St Helens for many decades.

Primary succession on Mount St Helens has been very slow because most habitats are isolated and physically stressful. Well-dispersed species lack the ability to establish until physical processes improve the site. Plant species that are capable of establishment lack suitable dispersal abilities.

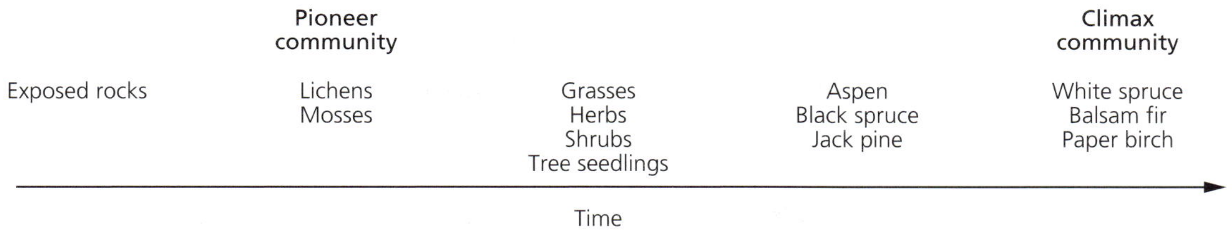

	Pioneer community			Climax community
Exposed rocks	Lichens Mosses	Grasses Herbs Shrubs Tree seedlings	Aspen Black spruce Jack pine	White spruce Balsam fir Paper birch

Time →

Figure 18.14 *Typical primary succession from rock to forest. The primary succession usually begins with a new exposed surface (such as bare rocks), which is colonised by a pioneer community of lichens and mosses. Over time the succession evolves towards a climax community, often dominated by tall trees.*

Source: Figure 4.14 in Cunningham, W.P. and B.W. Saigo (1992) Environmental science: a global concern. Wm. C. Brown Publishers, Dubuque, Iowa.

encourages further changes in species composition as shade-tolerant species start to appear.

Vegetation change is sequential and continuous throughout a succession, which comprises a series of transitional plant communities heading towards the final equilibrium state. The transitional stages are described as *seres*, and the whole sequence of seral communities is a *prisere*. The final equilibrium is the *climax community* (Figure 18.14).

BOX 18.27 TYPICAL SUCCESSION IN A TEMPERATE AREA

A typical succession in a temperate area might take up to 150 years from open ground to mature woodland. Each sere would have its typical plants and animals:

- Initially, there is bare ground.
- This is colonised by annual herbs, which cover some or most of the ground, but at least some bare ground remains.
- The pioneering herbs are succeeded by perennial plants, which establish a complete ground cover so that eventually no bare ground remains.
- The perennials are followed in turn by small trees and shrubs.
- Young woodland trees progressively replace the small trees and shrubs.
- Eventually, a mature woodland is established.

The woodland is likely to remain relatively stable in terms of species composition so long as there is no environmental change (such as climate change and global warming) or human impacts (such as woodland clearance, management or exploitation).

Studies of the progress of succession on a relatively recent earth flow in Colorado display many of the properties of progressive change. The early colonisers were plants with good dispersal and stress tolerance abilities, but they were replaced over time by more generalist species. Species diversity increased a great deal during the first 20 years but then stabilised, while the actual species present continued to change.

PRIMARY AND SECONDARY SUCCESSION

Ecologists recognise two basic forms of succession – primary and secondary. Primary succession involves the

BOX 18.28 SUCCESSION ON KRAKATAU

The development of vegetation on the island of Krakatau after the violent explosion in 1883 (see Box 6.21) provides a classic example of primary succession. The result of the explosive eruption was to bury the whole island under up to 30 m of ash and pumice, which completely destroyed and removed all vestiges of the original luxuriant tropical forest vegetation. Natural succession began with initial recolonisation, and an extensive plant cover quickly developed (Table 18.8). Within 14 years, the centre of the island was covered with tall grass several metres high, and within 30 years a distinctive shore vegetation and inland plant community had developed. By the early 1950s, within 70 years of the explosion, the succession was heading towards re-establishment of climax vegetation. The rapid succession on the island is partly attributable to the volcanic ash soils, which are rich in plant nutrients, pervious to water and penetrable by plant roots.

Table 18.8 Primary succession on Krakatau after the 1883 volcanic eruption

Year	*Total number of plant species*	*Vegetation on the coast*	*Vegetation on lower slopes*	*Vegetation on upper slopes*
1883		Volcanic eruption kills all life on the island		
1884			No life survives	
1886	26	Nine species of flowering plant	Ferns and scattered flowering plants, blue-green algae beneath them on the ash surface	
1897	64	Coastal woodland develops	Dense grasses	Dense grasses with shrubs interspersed
1908	115	Wider belt of woodland with more species, shrubs	Dense grasses up to 3 m high, woodland in the larger gullies and coconut palms	
1919			Scattered trees in grassland, single or in groups with shade species beneath. Thicket development in large gullies	
1928	214			
1934	271		Mixed woodland largely taken over from savannah	Woodland with smaller trees, fewer species taking over
early 1950s		Coastal woodland climax	Lowland rainforest climax	Sub-montaine forest climax

development of vegetation on newly exposed land areas that have never in the past had any vegetation cover – such as newly emerged volcanic islands (p. 174), melt zones in front of retreating valley glaciers (p. 466) and ice caps, sand dune systems that are still developing along a shoreline (p. 506), recently exposed river sediments (such as the point bars of a braided channel; p. 470), and completely artificial habitats such as spoil heaps and motorway embankments.

Secondary succession involves the redevelopment of vegetation on land areas that have had a vegetation cover at some time in the past. The original vegetation might have been destroyed by natural processes (including lightning fires, severe storms and volcanic eruptions) or by human interference (including forest clearance, building construction and fires). Secondary succession is a very important natural means of self-repair in ecosystems.

PRODUCTS OF SUCCESSION

CLIMAX VEGETATION

The association of species in the succession changes over time, in a directional sequence. The changes all tend to work towards a stable vegetation that is in balance with the environment – particularly soil, climate and nutrients – of the area. Given a long enough time without environmental change or human interference, a final equilibrium state is reached, which is known as the climax vegetation (or, more correctly, climatic climax vegetation because climate is usually the dominant controlling factor). In general, the closer to the climax the vegetation of a region is, the more slowly it changes and the longer it takes to regenerate when destroyed.

BOX 18.29 PROPERTIES OF CLIMAX VEGETATION

Climax vegetation has a number of important properties:

- It is the final end-point of successional development.
- Its species composition is stable over time so long as its environment remains unchanged, because the growth of other species is inhibited by competition for food, energy, nutrients and space.
- It is self-sustaining by encouraging the continued growth of the characteristic species that are associated with the climax.
- There is equilibrium between vegetation, soil and climate.
- It would normally take a long time to re-establish if destroyed.

BOX 18.30 SUB-CLIMAX AND PLAGIOCLIMAX

Vegetation in many areas is clearly affected by natural or human factors, which inhibit the normal sequence of successional stages and lead to arrested development. This is reflected in two main ways:

- *Sub-climax* exists where a succession has been retarded or inhibited by natural arresting factors, such as natural fires, storms or lack of long-term environmental stability.
- *Plagioclimax* exists where human activities arrest the succession. This can take many forms, including pollution, habitat change/clearance and over-grazing.

Different environments have different climax vegetation, reflecting the same adaptation within species that drives evolution by natural selection (see p. 530). Woodland is the normal climax vegetation in temperate areas (see Figure 18.14).

ARRESTED DEVELOPMENT

The climax vegetation is the natural end-point of successional development, but there is some debate among ecologists about whether or not it is a permanent state. Some argue that it is a real end-point, reflected in the stability of many major biomes (see p. 579) in different environments. Others argue that it is a theoretical end-point, which in practice is rarely reached because the environment is usually not stable long enough for dynamic equilibrium to be achieved.

Vegetation in the Kosciusko National Park in New South Wales, Australia, displays many signs of human disturbance, which makes it a plagioclimax community. Reconstruction of vegetation changes there over the last 1,000 years has

BOX 18.31 FIRE MANAGEMENT IN YELLOWSTONE NATIONAL PARK

Like all natural areas, Yellowstone (Figure 18.12) has been seriously affected in the past by natural fires, usually caused by lightning, which have damaged or destroyed large areas of natural vegetation, disturbed ecosystems and at least temporarily decreased biodiversity. When Yellowstone was established as the USA's first national park in 1872 (see Box 18.21), one of the goals of park managers was complete protection from fire. Vegetation fires were extinguished as quickly as possible, and land management was designed partly to hinder the spread of fire across the area. This policy was maintained until the early 1960s, when ecologists started to debate how much and what forms of human intervention in natural ecosystem processes are acceptable. In 1968, Sequoia and Kings Canyon National Parks became the first North American parks to create a 'natural fire zone', within which natural fires would be allowed to blaze unchecked. The strategy seemed to be effective and beneficial, and as a result the US National Park Service changed its fire policy from fire control to fire management. Under the new policy, new fires may be attacked and extinguished, held to a specified area or allowed to burn, as long as they meet certain criteria.

The new fire management policy was adopted in Yellowstone as in other parks, and although it looks good in theory it has proved difficult to implement in practice. During the 1988 fire season, in particular, extensive areas in the Yellowstone National Park were seriously damaged by fire. Nearly 45 per cent of the total park area (4,000 km^2) was burned, and about 80 km^2 was almost destroyed. That experience called into question the wisdom of the fire management strategy and triggered widespread public and political debate on the subject. Ecologists opposed to the strategy pointed out that without some form of active intervention in the natural fire regime of this type of environment (such as small controlled fire), litter accumulates on the ground surface and eventually burns, setting off a much larger fire.

Ecologists have been surprised at the fast rate of recovery of wildlife after the 1988 fire. The fires opened the forest canopy and allowed sunlight in; they also enriched the soil with nutrients from dead trees. With ten years, tens of millions of trees – particularly lodgepole pines – had started to grow in the ashes. Over 150 detailed studies have monitored how wildlife (particularly insects, plants and animals) has recovered after the fires in different habitats, including swamps, streams, meadows and forests.

shown minimal prehistoric disturbance (mainly by fire), followed by significant vegetation changes associated with changes in the erosion and fire regimes and the introduction of new grazing animals and exotic plant species since the arrival of European settlers.

FIRE AS AN ARRESTING FACTOR

Fire ravages dry vegetation, burning everything in its way. Under prolonged dry conditions with even moderate winds fire can spread across a wide area very quickly, stretching the emergency services and testing their ability to contain and manage the spread and to minimise risk and damage to people and property.

Natural fires (perhaps sparked initially by lightning) have traditionally caused extensive damage to grassland, woodland and forest. Yet these ecosystems have often adapted to the natural fire regime of their environment, and the complete removal of existing vegetation by fire provides an opportunity for secondary succession to revegetate the burned area. Indeed, in some situations (particularly in natural forests), this can actually increase biodiversity by creating a wider range of habitats at different stages along the successional continuum from bare ground to climax vegetation. This so-called intermediate disturbance hypothesis has been used to explain the species richness of some tropical rainforests, where indigenous people have traditionally used controlled burning to clear small patches of forest for cultivation. The patches are abandoned, usually after between five and seven years (when soil productivity declines), to be healed naturally by secondary succession while the tribe continues its shifting cultivation elsewhere in the forest.

Fire management is widely used to limit the damage that fires (however started) cause to protected areas and habitats. The history of fire management in the Yellowstone National Park (Box 18.31) shows interesting changes in attitude and approach, which reflect the constant tension between those who favour *preservation* (which would seek to eliminate natural fire damage) and those who favour *conservation* (which would allow natural fire damage) (see Box 17.33).

GLOBAL PATTERNS OF VEGETATION

Vegetation is far from uniform around the world, and the biosphere displays interesting spatial patterns at a variety of scales. Local variations reflect differences in soils, landforms and geology, and they give rise to differences in habitat within ecosystems. At the regional scale, differences in vegetation often reflect human impacts, particularly in areas with a long history of natural resource exploitation and land-use change. At the global scale, the pattern of variations in vegetation is strongly influenced by the major climate zones (see p. 335), and soil classification schemes (see p. 606) are often based on climate and vegetation.

BIOMES

Vegetation units can be identified at a variety of scales, from the small patch (as small as a few square metres in area) to the *formation*, which is a community of plants – such as a tropical rainforest – extending over a very large area (up to hundreds of square kilometres or larger). A *biome* is of a similar size to a plant community, but it includes animals as well as plants.

DEFINITION

Biomes are strongly influenced by climate, and their distributions often coincide with climate regions. Many other factors also influence the distribution of plants and animals (Box 18.32). As climate changes, it promotes corresponding adjustments in the nature and distribution of the major biomes (Figure 18.15). Climatic reconstructions

BOX 18.32 BIOMES – CONTROLLING FACTORS

The size, location and character of a biome reflect the interplay of a wide variety of environmental factors, including:

- length of daylight and darkness (which affects when and for how long photosynthesis can occur)
- mean and extreme temperatures
- length of the growing season (which is partly a function of latitude)
- precipitation (including the total amount, variations over time and intensity)
- wind (speed, direction, duration, frequency)
- soil
- slope
- exposure
- drainage
- other plant and animal populations.

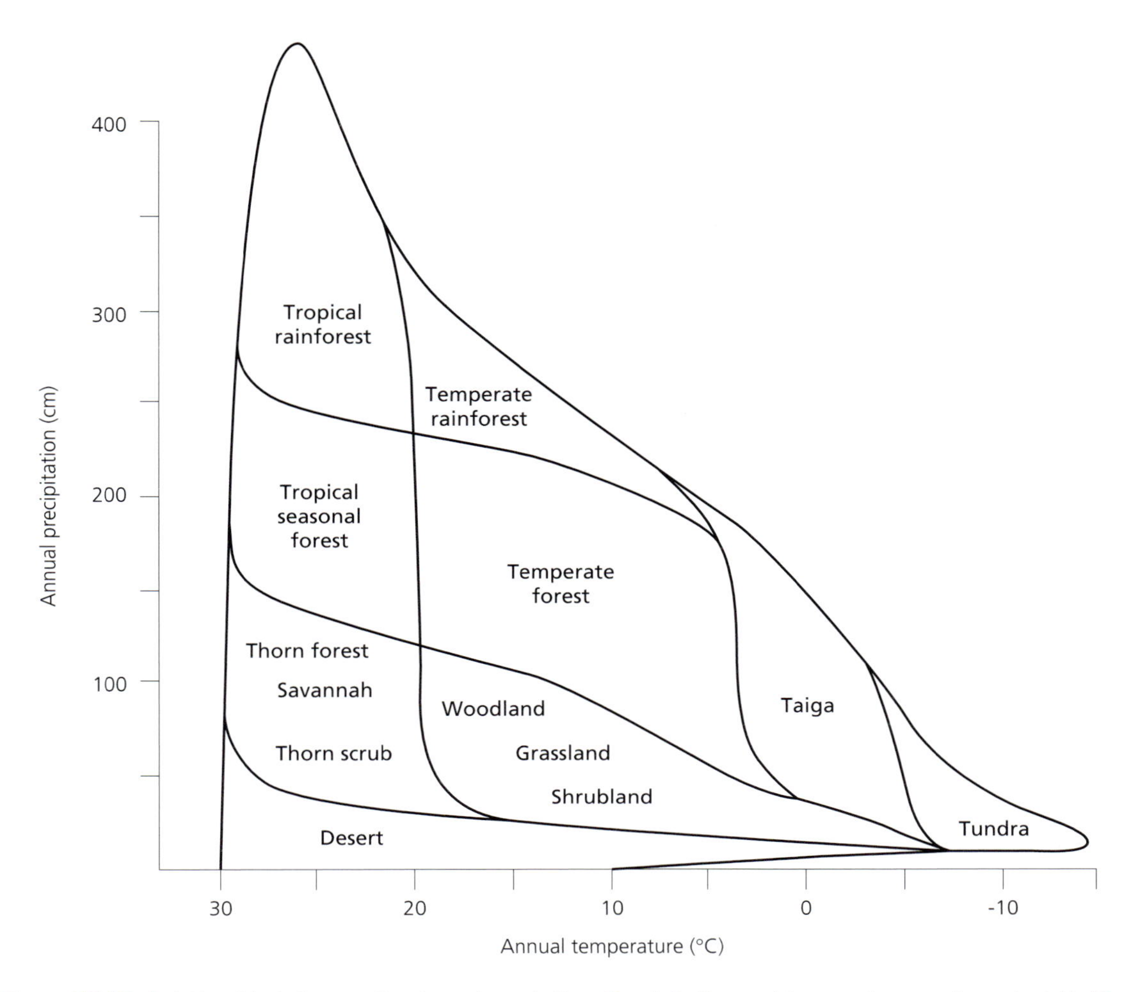

Figure 18.15 *Relationship between climate and vegetation. Precipitation and temperature are important limiting factors for most biomes, and this graph shows the critical values of each factor that determines large-scale patterns of vegetation.*

Source: Figure 5.2 in Cunningham, W.P. and B.W. Saigo (1992) Environmental science: a global concern. Wm. C. Brown Publishers, Dubuque, Iowa.

indicate that during the Pleistocene ice age much of North America, Europe and Siberia would have been covered by tundra and taiga (high-latitude biomes), whereas today those areas are dominated by temperate grassland and woodland.

Biomes are characterised by a dominant vegetation and defined by the species in them. Thus, for example, the desert biome contains radically different species from the tropical rainforest biome or the savannah grassland biome. Species composition can vary from place to place within a biome (because of local differences in soils, drainage, topography, microclimate and other factors). It can also vary within and between continents for a given type of biome, reflecting adaptation to environment and local speciation (see p. 530) in widely separated populations. Different trees dominate temperate forests on different continents, for example.

In each biome there are usually strong relationships between the species and their environment, and in this sense the biomes are effectively large ecosystems. Many biomes have also been affected by human activities, at least in places. Natural plant communities are often affected by such things as fires, clearance, introduced species, over-grazing and pollution. Most biomes are thus mosaics, with much local variability, rather than extensive blankets of uniform vegetation.

The major biomes are forests, grasslands, shrub/scrub, deserts and tundra, and their distribution is strongly

BOX 18.33 ECOTONES

Biomes rarely have sharp boundaries. For example, natural forests often have scattered trees towards their margins rather than an abrupt edge. The boundary between adjacent biomes is a transitional border zone, called an ecotone, which contains species from both ecological communities.

Ecotones are far from fixed, and they can change location as the species in adjacent biomes adjust to environmental change. Ecologists believe that many of the ecotones we see today are still in the process of adjusting to post-glacial climate change. It is possible, for example, that the northern edge of the taiga is still advancing northwards as climate warms from the Pleistocene ice age (see p. 343). Human factors can also cause shifts in the boundaries between plant communities. Over the last two centuries, for example, the boundary between short grass (in the east) and tall grass (in the west) in the central plains of North America has moved progressively westwards, partly in response to grazing and other human pressures.

BOX 18.34 ZONAL PATTERN OF BIOMES

The distribution of the major biomes is strongly influenced by the main climate zones:

- deep snow in the polar regions;
- tundra at high latitudes: a cold climate zone dominated by lichens, mosses, sedges, flowering herbs, and dwarf shrubs and trees;
- taiga or boreal forest, dominated by coniferous trees;
- the continental interiors support temperate grassland (such as the prairie of North America, pampas of South America, veld of South Africa and steppe of Eurasia) or tropical savannah that is partly wooded in places;
- mediterranean climates support chaparral (grassland with drought-resistant scrub);
- deserts are common in the sub-tropics;
- the humid equatorial regions support rainforest.

affected by the major climate zones (Figure 18.16). Thus, moving from the poles towards the equator, in a typical continent we would expect to find a pattern that mirrored the main climate zones (Box 18.34).

ALTITUDINAL PATTERN

Just as climate varies with altitude in a way that mirrors the latitudinal pattern of the climate regions (see p. 335), so the major biomes have a vertical zonation too. This is displayed most graphically in low-latitude mountains, where the latitudinal zonation of biomes is mirrored in the way vegetation changes with altitude. High mountains in Africa close to the equator, for example, have rainforest at their base, which gives way to temperate forest further up. This in turn is replaced by tundra, and sometimes even permanent snow at the summit.

COLD CLIMATES

The two biomes associated with cold climates are the *tundra* and the *taiga*. Polar regions are mostly ice-covered (see p. 445), so they offer few opportunities for plants and animals. They do support some animal populations, such as polar bears in the Arctic and a range of penguin species in Antarctica, but biodiversity is greater in the oceans surrounding the ice caps, which support a number of varieties of fish, walruses, seals and whales.

TUNDRA

The high-latitude climate is cold and harsh, and it supports a distinctive vegetation dominated by sedges, shrubby bushes and lichens. The tundra biome is controlled by a short growing season, low temperatures, limited precipitation and permafrost (see p. 445), and soils are frozen to a great depth, which inhibits the growth of plant roots.

Few plant species can survive this harsh environment, so biodiversity is low. Tundra species are well adapted to the short, frost-free summer growing season – most are short and compact and are able to sprout, grow, flower and produce seeds within a few weeks. Conditions are not suitable for tree growth, except for occasional willows and birches along river banks. Plants tend to get shorter, more dispersed and fewer in number, and lichens and mosses increase as conditions become progressively colder at high latitudes.

Most tundra areas support a rich range of animals, including predatory birds, foxes, bears and wolves. The habitat also provides grazing for large herbivores such as the caribou and musk ox. The tundra comes alive during the summer thaw, when flowering plants support large populations of mosquitoes and flies, which in turn provide food for large numbers of migratory waterfowl that summer in the tundra.

TAIGA

The boundary between the tundra and the taiga, in the sub-polar zone, is marked by a decrease in permafrost and a corresponding increase in larger plants, particularly trees. At lower latitudes, beyond the tundra, permafrost disappears and this allows the proper development of tree roots, which can extract nutrients and water from the soil. Trees are smaller and more widely spaced near the boundary with the tundra, where the cold and lack of moisture are important limiting factors (see p. 445).

Taiga is associated with humid microthermal climates (see Table 11.2), and it covers large areas of Alaska, Canada, Scandinavia and Russia. The characteristic vegetation is coniferous forest, often dominated by spruce, fir and pine. Some areas have deciduous trees such as birch, poplar, willow and ash. There is a wide variety of habitats, which often create a mosaic of plant communities and ecosystems within the taiga. Animal life is fairly rich and varied. It includes large grazing animals such as deer, moose, elk and caribou, carnivores such as bears and wolves, and smaller predators such as otters, mink, ermine and foxes which have long been hunted and trapped for their fur.

Although tree cover is extensive in the taiga, trees grow relatively slowly because of the short growing season and relatively low precipitation. This slow rate of regrowth is important, because it means that human impacts on taiga forests can be extremely long-lasting. Extraction of wood for pulp and timber is now damaging many taiga forests, particularly in Russia, and there are fears for the survival of the forest in some areas. Other threats include resource extraction (particularly petroleum, natural gas and gold), road construction and expansion of settlements.

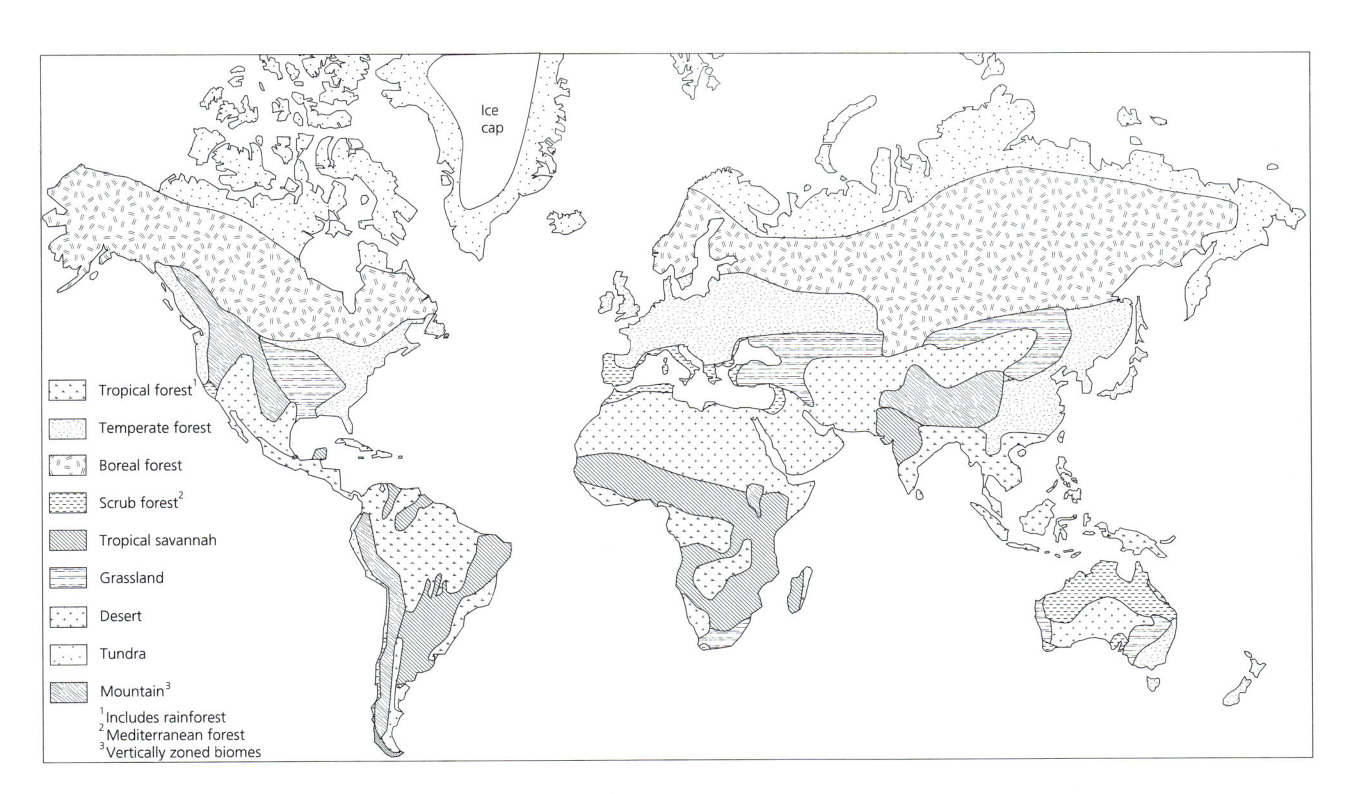

Figure 18.16 *Distribution of the major biomes. This simplified map shows the global distribution of terrestrial biomes, which closely reflects the distribution of the world's climatic zones.*

Source: Figure 4.8 in Marsh, W.M. and J.M. Grossa (1996) Environmental geography: science, land use and Earth systems. John Wiley & Sons, New York.

BOX 18.35 HUMAN IMPACTS ON TUNDRA

Ecologists are concerned about the impact of human activities on the tundra, because the vegetation takes a long time to recover after disturbance. Oil exploration and extraction is threatening important areas of Arctic tundra, particularly around Alaska's Prudhoe Bay oil field and the nearby coastal plain of the Arctic National Wildlife Refuge. There are fears too about the possible consequences of global warming on tundra ecosystems, which depend on permafrost and low temperatures.

Plate 18.3 *Alpine plants in the Rocky Mountains National Park, Colorado, USA. Typical alpine vegetation is scattered, low-growing and capable of withstanding long cold periods, including frost.*
Photo: Chris Park.

BOX 18.36 TAIGA AND THE CARBON CYCLE

Scientists recognise that the taiga biome plays an important role in the global carbon cycle (see p. 92) as a major carbon sink. The coniferous forest takes some of the carbon out of the biogeochemical cycle and puts it into store, in biomass. This probably helps to limit some of the worst consequences of air pollution by greenhouse gases (see p. 255), but the effect might be relatively short-lasting. Even without global warming, it looks likely that the carbon sink will change into a carbon source as the taiga vegetation completes its adaptation to the warmer climate since the Little Ice Age. The implication is that the forest resources of the taiga need to be carefully managed to optimise their benefits in terms of global warming. A major international research effort – the Boreal Ecosystem–Atmosphere Study (BOREAS) – is under way (1990–96) to improve understanding of the structure, controls and dynamics of the boreal forest biome.

TEMPERATE GRASSLAND

The coniferous forest of the taiga gives way to extensive grassland, particularly in the interiors of the continents, at lower latitudes as the climate becomes progressively milder. Trees are adapted to cold, wet conditions, whereas grassland is better suited to lower and more variable precipitation (particularly during the growing season) and higher rates of evaporation. Trees also require deeper soils than grassland. The two main types of temperate grassland are the *prairies* and the *steppes*.

PRAIRIE

A prairie is a treeless grassy plain found in middle latitudes. Most occur along the drier margins of the sub-humid and semi-arid climates in the continental interiors, away from the generally wetter coastal zone, which often has forest cover. Prairies are scattered in distribution and relatively limited in area compared with most other biomes. The prairies of the central USA and southern Canada and the pampas of temperate South America (particularly Argentina) represent the two largest units that are defined by climate. There are other, smaller, prairies – such as the prairies of east Texas – which are probably the result of particular soil conditions and/or repeated burning of natural woodland.

Prairie grassland is dominated by tall grasses (up to 2 m tall) with deep roots, and it grows best on soils with a good mineral content and a large amount of organic matter. The natural prairie biome provided extensive grazing for herds of large herbivores, such as the bison in North America, particularly before they were hunted to the edge of extinction during the nineteenth century (see Figure 17.3).

BOX 18.37 HUMAN USE OF THE PRAIRIES

Most temperate grassland has been converted to grazing or crops, particularly during the twentieth century. Early European settlers in the USA tended to clear forests first to create farming land, under the false impression that forest soils would be more fertile than the thicker prairie soils. More recently, the prairies have been turned into major grain-producing areas, which has made them susceptible to extensive wind erosion during times of drought, such as the Dust Bowl experience (see Box 14.2) during the 1930s.

BOX 18.38 IMPORTANCE OF FORESTS

Forests are important in a variety of ways:

- They are productive ecosystems in their own right (see p. 564).
- They provide many different habitats for other species to exploit (p. 558).
- They help in soil formation, by creating fertile humus (p. 600).
- They protect soils from erosion (p. 611).
- They play important roles in the biogeochemical cycles for carbon, nitrogen and oxygen (p. 86).
- They influence global energy flows, via their albedo (reflectivity) (p. 234).
- They influence climate, via evapotranspiration, frictional resistance to winds, shelter effects, and so on.

STEPPE

Steppe is an extensive grassy plain, usually without trees but often with scattered shrubs, found in the semi-arid zone at middle latitudes. Much of Eurasia, particularly Russia, is covered by steppe grassland (called the Steppes). Steppe grasses differ from prairie grasses in two main ways:

- They are shorter.
- They are different species and form different associations.

The steppe biome stretches over a fairly wide range of climatic conditions, and this is reflected in some major differences from place to place in vegetation and habitat. In the wetter, more humid areas of steppe, grass cover is more or less continuous and is dominated by short bunch grasses. At the drier margins of the biome, the grass becomes less continuous and is progressively replaced by desert. Like the prairie grassland, there is little surviving natural steppe because most of it has been ploughed to create land for growing grains, or heavily grazed. Some steppe damaged by over-grazing has been invaded by scrub vegetation such as sage or mesquite.

TEMPERATE FOREST

Forest is the natural climax vegetation of much of the temperate zone because it is self-adapted to the climate,

Plate 18.4 *Ancient oak woodland in Wistman's Wood Nature Reserve in the Dartmoor National Park, Devon, England.*
Photo: Chris Park.

BOX 18.39 THREATS TO THE FOREST

Many forests are facing serious stress from both natural and human causes. One of the more important natural stresses is wind damage from hurricanes and other violent storms (see p. 318), which can flatten or significantly damage extensive tracts of forest. Disease also attacks many tree species, so that in the USA, for example, the forest chestnut has declined sharply as a result of chestnut blight and the elm has declined because of Dutch elm disease. There are four main human causes of stress:

1. forest clearance to create land for agriculture and urban development;
2. use of hardwood tree species in construction and for making furniture;
3. replacement of broadleaf trees by fast-growing conifers, which yield a commercial tree crop much more quickly;
4. air pollution.

Air pollution, particularly by acid rain (see p. 251), is causing widespread forest decline and tree death in parts of Germany, Britain and the USA. It is not clear whether the damage is direct (the direct effect on leaves of gaseous air pollutants and acid mist) or indirect (via nutrients and water taken in from acidified soils).

soils and environment there. Trees need adequate moisture, fairly deep soils and a relatively long growing season, and for much of the Holocene (see p. 344) tree cover has developed naturally over much of North America, Europe and other areas in response to rising temperatures and natural succession.

Forests play a vital role in suppressing the worst impacts of global warming associated with greenhouse gases by acting as a major sink for carbon dioxide (an important greenhouse gas). Like all vegetation, trees take in carbon dioxide from the atmosphere, which they require for the process of photosynthesis (see p. 554), and they give out oxygen to the atmosphere (by the process of respiration). Forests are more effective carbon sinks than most other types of vegetation, so it is vitally important that they be conserved as part of a broader strategy of coping with global warming.

Scientists are concerned about the real possibility of a positive feedback link, in which climate change promotes a major short-term effect on tree distribution, which decreases the effectiveness of this particular carbon sink and in turn promotes further climate change. This makes it even more important to conserve the world's remaining forests and to adopt forest management strategies designed to meet a number of objectives – including conserving biodiversity, optimising the natural carbon sink, controlling global warming and protecting natural landscapes. A great deal of discussion at the 1992 Rio Earth Summit (see p. 13) centred on forests, their values and conservation, and Agenda 21 includes some wide-ranging proposals designed to protect the interests of indigenous people, future generations and the global environment (Table 18.9).

There are a number of different types of forest in the temperate zone, the two most important of which are humid continental mixed forest and marine west coast forest.

HUMID CONTINENTAL MIXED FOREST

This forest biome is found in middle latitudes where precipitation is higher than about 750 mm a year. Trees grow fastest where there is abundant precipitation and where droughts and drying winds are rare. Grassland tends to replace forest towards the drier margins of the biome.

The natural climax vegetation of this zone is deciduous broadleaf trees, which appear in a number of associations reflecting variations from place to place in soil and climate.

Table 18.9 Agenda 21 forest proposals

- plant new forests
- increase practical knowledge of the state of forests (planners often need even basic information on the size and types of trees in forests)
- more research into forest products such as wood, fruits, nuts, dyes, medicines, gums, etc.
- replant damaged areas of woodland
- breed trees that are more resistant to environmental pressures
- encourage local business people to set up small forest enterprises
- limit and aim to stop slash-and-burn farming methods (the basis of subsistence-level shifting agriculture)
- keep wood waste to a minimum; find ways of using trees that have been burned or thrown out
- increase tree planting in urban areas.

Oak, maple, beech, hickory and elm are common tree associations in the USA. Most trees are shorter than about 20 m.

MARINE WEST-COAST FORESTS

The other main type of temperate natural forest biome is the coniferous forest concentrated along the marine west coast of some continents. Good examples include the forests of north-west Europe and southern Chile, and the natural forest cover of New Zealand is of a similar type.

The grandest marine west-coast forests are to be found in North America, particularly in the Pacific north-west. Common species include spruce, hemlock, redwood, Douglas fir and sequoia. Individual trees taller than 30 m, with diameters greater than 2 m, are not uncommon here, and many trees (particularly in the sequoia and redwood forests of northern California) are well over 1,000 years old. Some tracts of this great forest have been set aside as forest and game sanctuaries for nature conservation purposes.

HOT CLIMATES

A number of different biomes are established in areas with hot climates, including mediterranean shrub and scrub vegetation, tropical savannah grassland, humid sub-tropical forest and true desert.

MEDITERRANEAN SHRUB AND SCRUB

Mediterranean climates, which have limited precipitation through the year and largely rainless summers, are mainly found on the west coast of each of the mid-latitude continents (see p. 335). Vegetation composition varies depending on aridity and the history of human interference.

Woody scrub and sclerophyll forest (which has widely spaced trees, most of which are less than 6 m high) is the dominant natural vegetation in the dry zone. Little of the original forest remains around the Mediterranean Sea in Europe after a long history of clearance and land-use change. Elsewhere, forest clearance for timber and firewood has also been extensive, although some native tree species (such as the cork oak and the olive) have been cultivated for use.

Vegetation is more scattered, more scrubby and generally lower in other parts of the mediterranean climate zone, which are dominated by isolated groups of shrub-like, drought-resistant plants. There are many different local names for this particular biome – in the south-western USA it is called *chaparral*, and in France it is *maquis*, for example. Plants have various adaptations to cope with extreme aridity. Some have extensive root systems for exploiting sub-surface water sources; many have thick bark and small leathery leaves, which limit water loss from evapotranspiration.

BOX 18.40 IMPACTS OF FIRE ON SHRUB AND SCRUB

A wide variety of bird and animal species were supported by the scrub and shrub biome when it was more extensive and natural, but large areas have been substantially modified by human activities, including the expansion of settlements and cultivation of olives and other cash crops. Fire, both accidental and deliberate, is common in the mediterranean zone during the long, dry summers, and it spreads quickly through the plants (many of which contain relatively large amounts of resins and oils), often fanned by strong mistral winds. Although fire is a natural part of this climatic regime, it does pose serious problems, particularly in terms of loss of life and damage to property.

TROPICAL SAVANNAH

Savannah (or savanna) is open grassland, usually with scattered bushes or trees, that is found mainly in tropical Africa between about 5 and 15° latitude (Figure 18.17), beyond the tropical rainforest. This zone has distinct rainy and dry

BOX 18.41 SAVANNAH – NATURAL OR CULTURAL?

Savannah grass is coarse, and it can grow to a height of about 2 m. However, in many areas it is much lower, either because it has been grazed or because it is recovering from deliberate burning (which is designed to encourage the growth of new shoots and ensure that the grassland survives). Ecologists have long debated whether savannah grassland is the natural climax vegetation for this climate zone or whether it is more the result of clearing or burning of tropical rainforest and monsoon forest by humans over many generations.

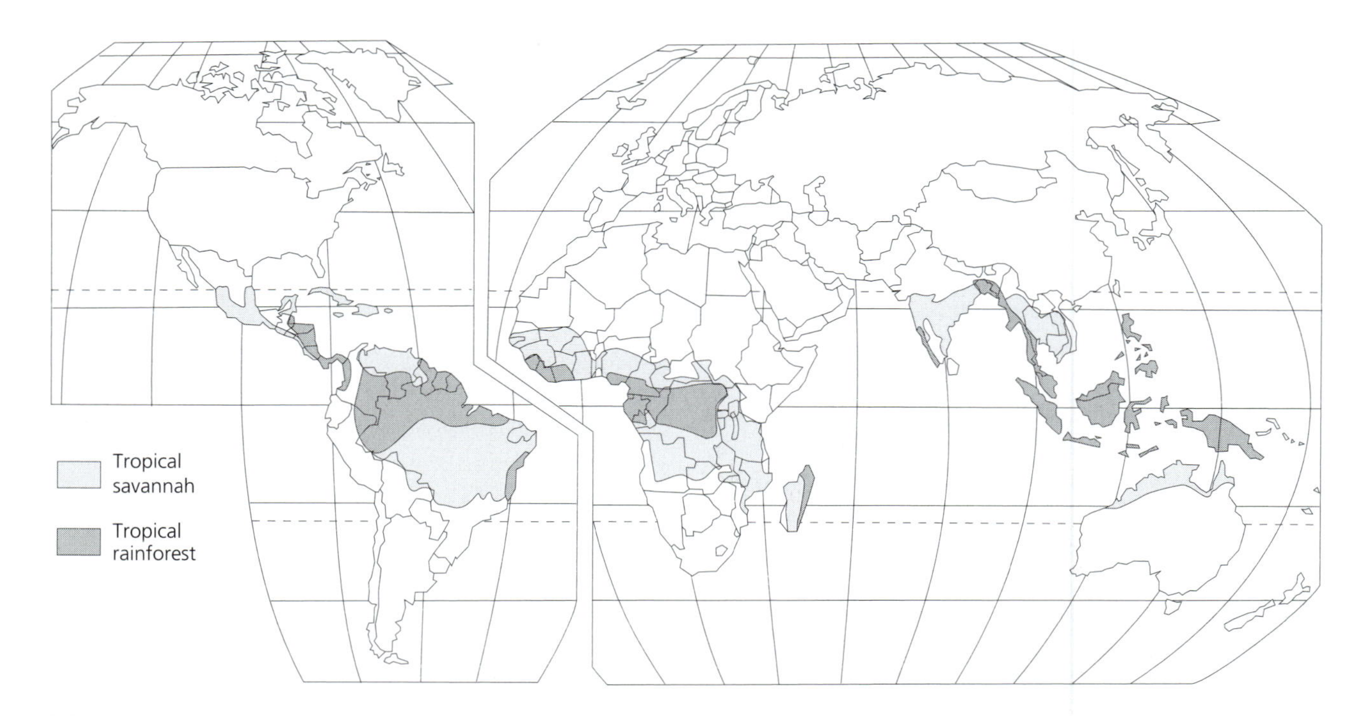

Figure 18.17 *Location of savannah and rainforest biomes.*
Source: Figure 7.2 in Doerr, A.H. (1990) Fundamentals of physical geography. Wm. C. Brown Publishers, Dubuque, Iowa.

seasons and high evaporation rates, and it is drier than the rainforest zone. The rainy season becomes shorter and rainfall lower and less reliable with increasing latitude.

Vegetation changes at the margins of the biome. At high latitudes, where it is much drier, the tall tropical grassland is replaced by short grass and widely scattered scrub. The boundary between this and the adjacent steppe is very indistinct. At low latitudes, where it is much wetter, scattered trees start to grow in the grassland, and it is eventually replaced by monsoon forest or tropical rainforest.

The animal population of savannah grassland has some well-known members, including much of the African big game beloved of hunters and documentary film makers. Herbivores include zebras, antelopes, rhinoceroses, elephants and hippopotamuses. The first two, and other smaller herbivores, are hunted by large carnivores such as lions and hyenas. There are also many scavengers, such as vultures and safari ants.

Many of the large savannah animals are endangered, mainly by poaching and hunting. A number of them fall under the protection of CITES (see Box 17.47) and are being conserved in protected areas, including national parks (see p. 542).

HUMID SUB-TROPICAL FORESTS

The humid sub-tropical climate zone in North America, Asia, South America and Australia has its own distinctive forest biome. It is dominated by coniferous trees, particularly pine in the south-eastern USA. A typical humid sub-tropical forest has a dense undergrowth and many climbing plants, and its habitats support a wide variety of animals (particularly reptiles, mammals and birds) and insects.

Where environmental conditions (particularly soils, slope and climate) are suitable, the biome also includes broadleaf evergreen and deciduous trees. Common deciduous trees include oak, magnolia and hickory.

DESERT

Conditions in hot deserts are very inhospitable for plants and animals, particularly the extreme aridity and high daytime temperatures (see p. 429). Because evaporation exceeds precipitation, there are very limited supplies of moisture, and plant and animal species have had to adapt to survive.

Most desert plants are *xerophytes* (adapted for growing and living in dry surroundings). Such plants have developed a range of evolutionary strategies to withstand serious and

Plate 18.5 *Cedar–oak–pine forests in the middle Atlas Mountains, Tunisia, North Africa.*
Photo: Philip Barker.

BOX 18.42 FORESTS AS CARBON SINKS

Each year CO_2 emissions from human activity add just over 6 billion tonnes of carbon to the atmosphere, and around a third of this is absorbed by the world's forests. Some scientists have proposed planting new forests to soak up the greenhouse gas CO_2 as one alternative to cutting emissions of the gas from power stations and vehicle exhausts. The concept of forests as carbon sinks was widely welcomed in discussions leading to the Kyoto Protocol (see Box 1.6) and was eagerly seized upon by countries (including the USA) that are reluctant to reduce their greenhouse gas emissions. However, a report published by the IPCC in 1999 warns that this is not a viable solution, because while new forests do initially function as carbon sinks, they quickly become saturated with carbon and then begin to return it to the atmosphere, accelerating rather than slowing down global warming. Shortage of CO_2 often limits photosynthesis in plants and organic breakdown in soils, and when atmospheric levels of CO_2 are raised (as through greenhouse gases), rates of photosynthesis and organic breakdown increase, releasing more CO_2 into the atmosphere. Critics of the Kyoto Protocol, which gives the planting of forest sinks equal value to emissions cuts as a way of meeting greenhouse gas targets, point out that the carbon sink concept gives false hope and may delay real progress in cutting emissions.

BOX 18.43 PLANT ADAPTATIONS TO THE DESERT

Plants have adapted in many different ways, such as:

- *Ephemeral lifestyles*: some desert plants live for only a very short period of time, normally a few days at most. Their seeds remain dormant in the desert sand for long periods of time (often years), and they quickly germinate and bloom when the infrequent rain falls.
- *Root systems*: many woody desert plants have developed root systems that are capable of reaching whatever moisture is available in the sand. Some have very long root systems that can exploit deep water sources (the mesquite from the south-western USA, for example, has tap roots up to 10 m long). Others (such as the creosote bush) have shallow spreading roots that can exploit dew and occasional rain.
- *Leaves*: most desert plants have small leaves, and this adaptation conserves water by limiting the surface area from which transpiration can occur. Some desert plants shed their leaves during dry periods, and plant stems then take over photosynthesis.
- *Water storage*: some desert plants are succulents, which store water in their leaves, stems and roots. Cacti such as the rounded prickly pear and the tall, straight saguaro – which have fleshy bodies, stems or leaves that act like sponges – are classic examples.

BOX 18.44 ANIMAL ADAPTATIONS TO THE DESERT

Examples include:

- Some amphibian species (vertebrates that usually live on land but breed in water) remain dormant for long periods of time when it is dry and then quickly mature, mate and lay eggs when rain falls.
- Many birds and rodent species reproduce only during or after winter rain, which encourages plant growth.
- Some desert rodents, such as the African gerbil, feed on dry seeds and have metabolisms (body processes) that can conserve and recycle water.
- Some desert mammals, most notably camels, can survive prolonged dehydration by storing water in the fatty parts of their bodies (the humps in the case of camels).
- Most desert mammals and reptiles are nocturnal and remain in the shade during the day.

prolonged water shortage (Box 18.43). Most xerophytic areas have widely scattered plants surrounded by much bare ground. Some deserts, such as the Atacama Desert (in South America) and the interior of the Sahara Desert (North Africa), are so dry that they support no large plants at all.

A number of animals thrive in deserts, particularly reptiles, rodents, insects and birds. Like plants, many of them have evolved adaptations (Box 18.44) in order to survive in the hot, dry desert environment.

The desert biome grows only very slowly, so it takes an extremely long time for it to recover after it has been damaged (for example, by over-grazing or over-collection of wood for fuel).

TROPICAL FORESTS

There are two main types of tropical forest biome – the rainforest and monsoon forest.

Plate 18.6 *Tropical forest on Centosa Island, Singapore.*
Photo: Chris Park.

RAINFOREST

Tropical rainforest is perhaps the most important biome in the world, but it is also the most seriously threatened one. Home to an unrivalled diversity of species, the rainforest is under threat from a wide variety of human impacts, and some ecologists argue that without significant international co-operation and some well-conceived conservation strategies, within a few decades there will be hardly any natural rainforest left to conserve.

Rainforest is the climax vegetation for the equatorial wet climatic zone (see p. 335), which extends in a band around the equator, mostly between latitudes 5 and 10°. In this humid zone (Figure 18.17), temperatures remain uniformly high throughout the year, there are high diurnal variations in temperature, each month receives at least 60 mm of rainfall (so there is no dry season), annual rainfall usually exceeds 1,500 mm, and there are periodic torrential downpours accompanied by thunder and lightning.

The rainforest environment is hot, wet and humid. Ecologists also believe that it has been remarkably stable over exceedingly long periods of time. This has encouraged plants and animals to adapt to their environment (see p. 530), which in turn has promoted speciation and the evolution of unparalleled biodiversity.

More than half of the remaining rainforests are in Central and South America (particularly in Amazonia), and the rest are in South-east Asia (particularly Sumatra, Borneo and Papua New Guinea) and Africa (particularly the Congo basin).

Animal life in the rainforest is rich and varied. It includes forest elephants (whose ivory tusks are highly prized by poachers), many different reptiles (including snakes, alligators, crocodiles, turtles and lizards), a range of primates (including monkeys and apes), some predatory big cats and a huge variety of birds and insects.

Contrary to popular belief (because they look so abundant and productive), rainforests grow on very poor soils. Most of the nutrients within the rainforest ecosystem are stored in the biomass (living vegetation) rather than in the soils, and the decomposer part of the food web (which includes fungi, bacteria and insects) rapidly breaks down organic material, including dead trees, broken limbs and branches, leaf litter and dead animals. The nutrients are quickly recycled through the forest soil into tree roots, so

Figure 18.18 *Stratification in the rainforest. Typical tropical rainforest has a strongly stratified appearance, with distinctive forms of vegetation at each layer. See text for explanation.*

Source: Figure 21.12 in Dury, G.H. (1981) Environmental systems. Heinemann, London.

BOX 18.45 CHARACTER OF RAINFOREST VEGETATION

Tropical rainforest, which is also known as *selva*, has a distinctive character. The vegetation is dominated by tall broadleaved evergreen trees, epiphytes (plants such as mosses that grow on other plants but are not parasitic on them), lianas (climbing vines) and giant ferns. Rainforest trees are closely spaced, and they are often tied together by lianas and epiphytes. Most are tall and straight, have very shallow root systems and are anchored to the ground by a series of buttresses. Many different tree species are present in even quite small areas of rainforest.

Vegetation is stratified (layered), with different habitats and environmental conditions in each layer (Figure 18.18). Most trees grow to the so-called intermediate layer about 30 m above the ground, where their canopies join to give more or less complete cover. Scattered emergent trees rise through the canopy, sometimes to a height of 60 m or so. Beneath the intermediate layer is the understorey on the forest floor. Relatively few species of plant can grow in this dark zone, heavily shaded by the canopy above. Where light reaches the forest floor – such as at the edge of the forest, on steep slopes or along rivers – a dense brush can grow, which contrasts with the relatively bare floor of the forest interior and makes access to the interior difficult.

forest soils have little organic matter and only a very thin cover of leaf litter.

Because the forest soils contain so few nutrients, they are not suitable for continuous cultivation after the forest has been cleared. Rapid leaching caused by heavy tropical rainfall makes the problem even worse. Soil productivity declines quickly (after between three and six years), and cultivated areas are then abandoned. Secondary succession leads to the growth of scrubby vegetation and then to savannah (if the cleared area is large) or secondary rainforest (if it is a relatively small patch that can revegetate and self-repair).

BOX 18.46 CAUSES OF TROPICAL DEFORESTATION

Clearance is promoted by many different factors, including:

- over-intensive shifting cultivation;
- timber extraction;
- over-collection of fuelwood for cooking and heating, and for making charcoal;
- encroachment and clearance by landless peasant farmers;
- clearance for pasture or crops, promoted by cheap land and government tax and financial incentives to encourage international investment;
- road and highway construction.

Ecologists are concerned at the pace and pattern of clearance of tropical rainforest, which once covered about a tenth of the Earth's surface. More than 40 per cent of the rainforest has been cleared since the 1940s, and clearance continues at a rate of about 200,000 km^2 a year. Less than 1 per cent of the Brazilian Amazon had been cleared before 1975, but between 1975 and 1987 the rate of clearance increased exponentially.

Overall, about a quarter of the loss is due to forestry, but the balance of factors varies from place to place. Much of the wood felled from rainforests is destined for the international timber market, where tropical hardwood commands a high price. This promotes extensive deforestation in places like Borneo and the Malay Peninsula. It is a classic resource depletion problem, because as tropical timber supplies in South-east Asia start to decline, major importers such as Japan are searching for new sources, particularly from Brazil and other South American countries.

Ecologists argue that if clearance continues at recent rates, all of the world's primary (undisturbed) rainforest is likely to have disappeared or be damaged by 2020. This would mean the loss of an irreplaceable biological asset – rainforests contain about half of all the wood growing on the Earth and at least 40 per cent of all known species of plant and animal. They are among the most diverse and complex ecosystems on the planet. As well as a significant loss of biodiversity, clearance of rainforests causes the loss of valuable natural resources, including hardwoods (such as mahogany, rosewood and teak) and tree products (such as quinine, vegetable gums and rubber).

Plate 18.7 *Clearing and burning secondary forest (fallow) areas for farming, near Bokalum village in Cross River state, Nigeria.*
Photo: Uwem Ite.

BOX 18.47 INDONESIAN FOREST AND PEAT FIRES IN 1997

For more than two months in the summer of 1997, fires raged across Indonesia and for more than four weeks caused poor air quality and low visibility over much of South-east Asia. Many forest areas were burned out completely, and the cost in lost timber stocks was estimated at more than US$5 billion. While a significant number of the fires were set deliberately during land disputes, the biggest culprit was government-sponsored land clearance projects. Government targets required the clearance of around 400 km^2 of forest in 1997 to create space for large-scale agricultural and forestry plantations, and burning was the only practical method. Many of the fires quickly went out of control in the drought-stricken climate, and few serious attempts were made to dampen the flames. Fire is a fairly common hazard in this region – during 1982–83 fires swept through 36,000 km^2 of Borneo – and in most years the fires would have been out by late September. But seasonal rains were delayed in 1997 by El Niño (see Boxes 16.12 and 16.13), which regularly causes a long drought season in the western Pacific. Ironically, the heavy smoke caused by the fires will have suppressed rainfall – because the clouds contain more but smaller water droplets than normal air, and hot air would increase evaporation within the clouds – and thus caused a devastating positive feedback (see p. 79).

Concern has also been expressed over the burning of peat bogs in Indonesia beneath the forests. One study estimated that a fifth of the estimated 600 billion tonnes of carbon stored in the world's peat bogs is in Indonesia, and it predicted that if the forested peat bogs continue to burn for six months they would release 1 billion tonnes of carbon. For comparison, Western Europe emits just under 900 million tonnes of CO_2 into the atmosphere each year. Some scientists fear that the 1997 Indonesian fires could have a significant impact on global warming.

Plate 18.8 *Mangrove roots in tidal silt, Mai Po Marshes, New Territories, Hong Kong.*
Photo: Chris Park.

But the cost is not just ecological: rainforests also provide invaluable environmental services by helping to regulate:

- global weather patterns
- soil erosion in the tropics
- river flooding in the tropics.

Tropical deforestation may also contribute to the greenhouse effect and global warming by removing an important carbon sink.

Various attempts have been made to conserve the remaining rainforests. For example, some governments (including Costa Rica, Panama, Brazil and Zaire) have designated forest reserves to conserve representative tracts of this important biome, and some endangered tropical hardwood species have been listed in the Convention on International Trade in Endangered Species (CITES; see Box 17.47).

MANGROVE

Mangrove is a tropical evergreen tree or shrub that grows in coastal areas, particularly on tidal flats and in estuaries. Many deltas at the mouths of large rivers in South-east Asia have mangrove swamps on them. The mangrove tree has stilt-like roots, which knit together to form dense thickets. These often trap coastal sediment and encourage coastal accretion (build-up), which allows the mangrove to extend further out from the shore. Mangrove swamps provide important habitats for aquatic and amphibious species.

BOX 18.48 MONSOON FOREST CLEARANCE

There are fewer tree species in the monsoon forest than in the rainforest, and this makes commercial timber extraction easier and more cost-effective. Some of the rich teak forests in South-east Asia are extensively worked and are in danger of disappearing altogether. Other monsoon forests (particularly in India) are being cleared to create land for agriculture and settlements, leaving small patches in remote areas.

Many areas of mangrove are supporting local economies. In Hong Kong, for example, mangrove covers an area of about 60 km^2 and supports local mariculture and fish farms. In Thailand, the remaining 1,960 km^2 of mangrove supports a range of economic activities from charcoal to methanol production and from prawns to pharmaceuticals. In other areas – such as south Sulawesi (in Indonesia) – large tracts of mangrove forest have been cleared for timber, fuelwood, and conversion to *tambak* (brackish ponds).

MONSOON FORESTS

Monsoon forests grow in tropical places that have a monsoon climate (see p. 335) with distinct wet and dry seasons. Most are in South-east Asia, including north-eastern India, Burma, Thailand, Cambodia, Laos and Vietnam.

The monsoon forests grow adjacent to the tropical rainforests. Many species are found in both, but in the monsoon forest they have adapted by natural selection (see p. 530) to cope with drought during the dry season. Adaptations include a period of dormancy, shedding of leaves and wider spacing between trees (which compete for water). Monsoon forests also have a much denser ground vegetation of shrubs, because the wider spacing of trees allows more sunlight to reach the forest floor.

GLOBAL WARMING AND BIOME CHANGES

A great deal of interest has been shown in predicting how successfully species and habitats will be able to migrate and adapt to global warming. The evidence suggests that tropical grasslands – which support grazing and agriculture in areas of high population – are likely to come off worst.

Table 18.10 Maximum rate of movement of some European tree species during the past 100,000 years

Tree species	*Maximum rate of movement (m per year)*
hazel	1,500
elm	1,000
ash	500
oak	500
lime	500
beech	300

Tropical forests are also predicted to decline (above and beyond the ongoing decline associated with clearance, logging and other land-use changes; see p. 582) and suffer further declines in biodiversity. The tropical forest changes will be particularly important, because tropical forests provide very effective carbon sinks that absorb huge amounts of carbon dioxide from the atmosphere – a decline in this biome could further amplify global warming, by positive feedback (see p. 79). Temperate and boreal forests are likely to increase in size with global warming, but they are nowhere near as species-rich as the tropical rainforest, which will decline.

Changes in the size and distribution of tundra (see p. 578) will affect rates of global warming, because as the moist tundra melts more methane will be released from the rotting vegetation. The IPCC estimates that 16 per cent of permafrost areas will have melted by 2050.

The great boreal forests of North America and Eurasia, which cover 27 per cent of the Earth's land surface, are less effective as carbon sinks than tropical rainforests. Many tree species have narrow temperature niches, and a rise of 1 °C would create real problems for them because they migrate only very slowly (Table 18.10). Scots pines in Britain take a long time to mature and to establish themselves – they 'move' at only 4 km a century, compared with some trees that can 'move' 200 km a century. The IPCC predicts that 65 per cent of boreal forests will undergo major changes in vegetation types over the next 50 years.

In many areas, global warming will cause a mixture of positive and negative changes. The Highlands of Scotland, for example, are likely to become wetter (annual rainfall is likely to increase by 14 per cent by 2100), windier (mean annual wind speed is likely to rise by 25 per cent) and duller (solar radiation is likely to decrease by 6 per cent). A 1999 report by Scottish Natural Heritage concludes that this will threaten the survival of large tracts of ancient Caledonian pine forest, along with coastal meadows, mountain birds, flowers and insects. Rare communities of alpine plants, mosses, liverworts and lichens will shrink, and high-altitude stoneflies and spiders will also be at risk. On the other hand, many plants and animals will thrive with the higher temperatures. It is expected that lowland flowers will do better, over-wintering birds will become more numerous, bat populations will flourish, and butterflies will spread further north. Peat bogs are also likely to expand in Scotland.

WEBSITE

Links to relevant websites, a comprehensive bibliography, tools for teaching and learning, and downloadable images

SUMMARY

Ecosystems are the basic functional units in the biosphere, which we examined in Chapter 17. In this chapter, we focus on the role and relevance of ecosystems, the development of vegetation over time and the global distribution of vegetation. After outlining important ecosystem properties such as food webs, energy dynamics, trophic structure, habitat and niche, we looked at key aspects of ecosystem dynamics (particularly flows of energy and nutrients). The importance of productivity and the role of limiting factors were also touched upon, and we noted the value of ecosystem approaches in managing protected areas and restoring altered environments. Succession describes the way in which vegetation and soil change over time at a particular place, and we looked at how and why these changes occur, what stages are involved and what factors influence the outcomes. The global distribution of vegetation can be described in terms of biomes, and we considered the basis of biome definition before taking a quick trip around the world to see what biomes in different climates look like, how they operate and what threats they face. Global warming is likely to promote major changes in the distribution and composition of biomes.

relevant to this chapter can be found at the website specially designed to accompany this book at
http://www.park-environment.com

FURTHER READING

Alexander, R. and A. Millington (eds) (1998) *Vegetation mapping*. John Wiley & Sons, London. Series of essays that evaluate and illustrate contemporary techniques in vegetation mapping at a variety of spatial scales.

Barraclough, S.L. and K.B. Ghimire (2000) *Agricultural expansion and tropical deforestation: poverty, international trade and land use*. Earthscan, London. Draws together case studies from different tropical countries to illustrate the importance of factors such as the misguided policies of national and regional authorities, and issues such as systems of land tenure, forces of trade and globalisation.

Dickinson, G. and K. Murphy (1998) *Ecosystems*. Routledge, London. Outlines the basic concepts and processes in ecosystems and explores the implications for environmental decision making.

Doornbos, M., A. Saith and B. White (2000) *Forests: nature, people, power*. Blackwell, Oxford. Contemporary perspective on forests as highly contested spaces in which both trees and forest dwellers often find themselves on the losing side, which calls into question many of the received wisdoms and highlights complexity and uncertainty.

Fairhead, J. and M. Leach (1998) *Reframing deforestation: global analysis and local realities – studies in West Africa*. Routledge, London. Suggests that the scale of destruction brought by West African farmers during the twentieth century has been greatly exaggerated and that global analyses have unfairly characterised them.

Huggett, R. (1997) *Environmental change: the evolving ecosphere*. Routledge, London. Explores past, present and future environmental change, stressing complexity and interdependence within environmental systems.

Hviding, E. and T. Bayliss-Smith (2000) *Islands of rainforest: agroforestry, logging and eco-tourism in the Solomon Islands*. Ashgate, London. Detailed case study of modern initiatives in the tropical rainforest from the perspective of indigenous peoples.

Jepma, C.J. (1995) *Tropical deforestation: a socio-economic approach*. Earthscan, London. Analysis of the underlying socio-economic causes of deforestation that concludes that measures are needed to deal with population and land-use policies.

Kellman, M. and R. Tackaberry (1997) *Tropical environments: the functioning and management of tropical ecosystems*. Routledge, London. Comprehensive introduction to the complex environmental systems of the tropics that integrates biophysical and human management issues.

Mannion, A.M. (1995) *Agriculture and environmental change*. John Wiley & Sons, London. A detailed review of how agriculture affects and has affected the environment in different cultural settings.

Mather, A. (1992) *Global forest resources*. John Wiley & Sons, London. A summary of the history, distribution, character and dynamics of forests around the world.

Milliken, W. and J. Ratter (1998) *Maráca: the biodiversity and environment of an Amazonian rainforest*. John Wiley & Sons, London. Detailed account of a scientific survey by the Royal Geographical Society that outlines the results and their implications.

Mistry, J. (2000) *World savannas*. Prentice Hall, London. Outlines the distribution, characteristics and dynamics of savannahs, and sets the importance of this tree and grass biome into historical context.

Morley, R.J. (1999) *Origin and evolution of tropical rain forests*. John Wiley & Sons, London. Detailed review of the history of tropical rainforests over the past 100 million years, with commentary on twentieth-century change and the need for preservation of this important species-rich ecosystem.

Oliver, C. and B.C. Larson (1996) *Forest stand dynamics*. John Wiley & Sons, London. An in-depth analysis of forest growth patterns in different environments.

Park, C.C. (1992) *Tropical rainforests*. Routledge, London. An introduction to the tropical rainforest environment, its history, dynamics and relevance, with particular emphasis on rates, causes and patterns of clearance and options for conservation and management.

Reading, A.J., R.D. Thompson and A.C. Millington (1995) *Humid tropical environments*. Blackwell, Oxford. Reviews the world's humid tropical areas – which contain some of the richest, most diverse, most important and most threatened environments – and the pressures they face.

Wadsworth, R. and J. Treweek (1999) *GIS for ecology*. Longman, Harlow. Clear introductory guide to GIS and how they can be applied in ecology and environmental sciences, with an emphasis on data acquisition, handling and analysis.

Chapter 19

Soils

LEARNING OBJECTIVES

When you have finished studying this chapter, you should be able to

- *Appreciate the importance of soil as a resource and the need to use it in sustainable ways*
- *List the main properties of soil and comment on their relevance*
- *Distinguish between soil texture and soil structure and outline the classification systems for each*
- *Describe and account for the main features of a soil profile*
- *Summarise the main soil-forming factors and processes*
- *Outline the main factors that influence soil distributions and summarise the main soil classification schemes*
- *Understand the principles and purpose of land evaluation*
- *Be familiar with the principles, processes and products of soil erosion*

Soil is the thin layer of disintegrated rock particles, organic matter, water and air that covers most of the land surface, to an average depth of only about 20 cm. Its significance is out of all proportion to its depth – it is without doubt one of the most precious natural resources on the planet, and the key to the survival of the human species. We could live in a world with fewer species and few attractive landscapes, where synthetic materials had largely replaced natural ones (although our quality of life would be much the poorer for it), but soil is a central part of food production without which our very survival would be at risk.

Soil is essential to life on Earth, and its very existence reflects continuous two-way relationships between biotic (living) and abiotic (non-living) parts of the environment – soil supports life, but it also requires biological activity to form. It is formed by natural processes operating over long periods of time, and there are close links between the development of a soil cover on Earth and the development of the biosphere.

In many ways, soil is a typical open system (see p. 72) with inputs, throughputs and outputs of matter and energy, and it is intimately linked to other environmental systems, particularly climate and vegetation (Figure 19.1). Through weathering and erosion, soils are linked to the lithosphere system and have a place (albeit a short-term one) within the rock cycle.

Throughout history, people have been aware of the need to use soil resources wisely and conscious that failure to do so could seriously affect food production and thus human survival. A number of great civilisations from the past have suffered badly, or even declined fully, through lack of stewardship of their soil resources. Soil erosion – mainly caused by over-use or inappropriate use – is causing serious problems in many areas today, but scientists generally consider that the quantity and quality of global soil resources are capable of supporting the world population until at least well into the twenty-first century.

In this chapter, we focus on three key themes – the nature and development of soils, the classification and distribution of soils, and the management of soil resources. These are central themes in *pedology* (soil science), which is the study of the formation, characteristics and distribution of soils. This field of study is becoming more applied and more important as the real value of soils in the biosphere is better understood.

BOX 19.1 IMPORTANCE OF SOIL

The soil system plays a central role in many of the biogeochemical cycles (see p. 86), providing a short-term store for many chemical elements and the main interface between abiotic processes (such as rock weathering) (p. 204) and biotic processes within ecosystems (p. 551). Soil plays many important roles in the biosphere: it is a natural body in its own right, a medium for plant growth, a structural mantle or cover for rock, a medium through which water is transmitted and redistributed, and a component of ecosystems. Without soil, few species of plant other than lichens and mosses (which can survive on bare rock) can colonise an area. Grasses, shrubs and trees need soil for support and sustenance.

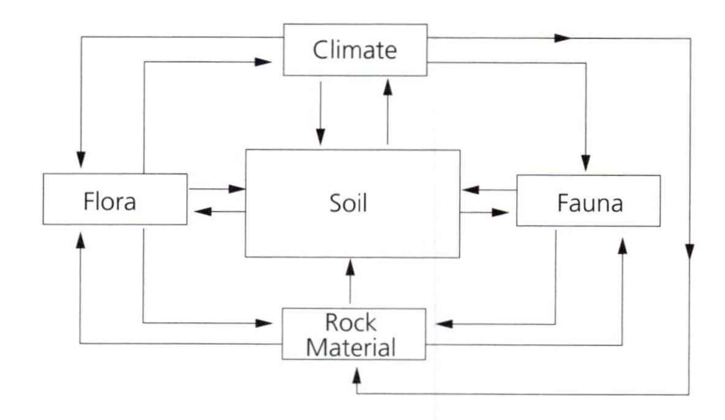

Figure 19.1 *Soil as an open system. Soil is an open system that reflects interactions with rock materials, climate and organisms (plants and animals).*

Source: Figure 20.5 in Dury, G.H. (1981) Environmental systems. Heinemann, London.

NATURE AND DEVELOPMENT OF SOILS

Soils vary a great deal from place to place, and they develop over time, making them highly dynamic. The way in which they change and evolve depends on a number of important factors, including parent material, climate and vegetation.

SOIL COMPOSITION

Soil is ultimately a collection of particles of disintegrated rock, so variations in soil from place to place partly reflect the pattern of underlying rocks and deposits. But there is a great deal more to soil than this, because it is very much a product of interaction between living organisms (plants and animals), climate (particularly temperature and moisture), geological materials (rock type) and geological processes (particularly weathering and erosion).

There are five main ingredients in all soils – minerals, organic compounds, living organisms, air and water. They are present in different combinations, and this determines what a particular soil is like. Some soils are fertile and support luxuriant vegetation, for example, whereas others are infertile and support only a few hardy plants. Thus the combination of ingredients strongly influences how useful particular soils are in agriculture, and how productive or susceptible to erosion different soils are.

MINERALS

Minerals are the core ingredients of soils; without these basic building blocks there would simply be solid rock

BOX 19.2 PARENT MATERIAL

Parent material is the substance from which a soil has developed. Most soils develop from the weathering of solid rock, volcanic ash (see p. 180) or deposits left behind by retreating glaciers (p. 467), migrating rivers (p. 375), stabilised dunes and desert materials (p. 431), and newly exposed coastlines (p. 511). Inevitably, the mineral composition of the resulting soil is strongly influenced by the mineralogy of the parent material, although this can be modified by weathering processes and biological activity as the soil develops. Thus, for example, soils that develop on limestone ($CaCO_3$) will contain relatively large amounts of calcium and carbon, whereas soils that develop on sandstone (SiO_2) will contain more silicon.

This close link between soil minerals and parent material has three main implications:

1. The mineral content of soils that develop over single rock types tends to be relatively uniform over a large area.
2. Mineral content varies between soils that have developed on different rock types.
3. There can be sharp changes in the mineral content of soils across a boundary between contrasting geologies.

BOX 19.3 MINERALS AND PLANT GROWTH

Plants require different amounts of different minerals for healthy growth. Some minerals are required in relatively large amounts, as nutrients. Others – the so-called trace elements such as copper, zinc and boron – are required in relatively small concentrations. Shortage of any of these minerals can inhibit plant growth and productivity, reduce crop yields and increase the susceptibility of plants to other environmental stresses (such as frost, drought or disease).

covered by the decaying remains of plants and animals. A mineral is a naturally occurring solid inorganic substance that has a characteristic crystalline form and a homogeneous chemical composition. These are the solid particles that form the very fabric of a soil.

There are great variations from place to place in the type and character of minerals in the soil. This strongly influences the nature and properties of the soil, and it provides a basis for classifying soils into different types (see p. 605). The variations reflect differences in parent material and in the weathering and soil-forming processes involved.

As a result of the close link with rock type, soils tend to form mosaics across areas of variable geology. The mineral content of soil can significantly affect its suitability for different types of plant, so underlying geological patterns can also be reflected in vegetation mosaics.

All rocks weather down to produce a range of minerals (see p. 193), the most common of which are clay minerals and quartz. Both play vital roles in the development of productive soils:

- Clay minerals help to hold water and nutrients in the soil, which are attracted to the surface of individual soil particles by the negative electrical charge that most clay minerals develop. This is essential to the growth of plants in the soil.
- Quartz grains are much larger than the clay minerals, and they help to create a proper soil structure that aids the circulation of air and the percolation of water through the soil. This is also vital for plant growth.

Some minerals that are naturally present in soils are used by growing plants. In natural vegetation this rarely creates a serious problem, because the nutrients are recycled by decomposition within the ecosystem (see p. 556). Cultivation often depletes the soil nutrient reservoir, particularly of nitrogen (N), phosphorus (P) and potassium (K), because nutrients are exported from the ecosystem in the crops and produce. Mineral or organic fertilisers and manures are widely used to restore the depleted minerals in agricultural soils.

Not all minerals in a soil are beneficial for plant growth. Some naturally occurring substances (such as salt) are harmful to plants, particularly in large concentrations. Consequently they limit a soil's suitability for agriculture without appropriate management and control.

ORGANIC MATTER

Soil also contains organic matter derived from living plants and animals, which contains carbon. This is concentrated mainly on the soil surface and in the uppermost layers of the soil. Plant debris such as leaves and twigs falls off living and dead plants, and it accumulates on the soil surface as litter. It protects new plants and provides a habitat for ground-dwelling insects and other small animals.

The litter is broken down by soil organisms to become humus in the soil. This partly decomposed organic matter increases fertility and water retention in the soil, and it improves soil texture by making it more workable. As it decays it releases organic compounds in the form of colloids (mixtures between a solution and a suspension), which provide some of the mineral nutrients required for plant growth. The decomposing humus is usually dark-coloured (dark brown or black), so the uppermost layer (horizon) of the soil where it accumulates is generally darker than the soil below.

There is a positive feedback association between the fertility and the organic content of a soil, linked with plant growth. Soils with a high organic content tend to be more fertile than those without, and this promotes plant growth, which in turn provides the litter and humus that maintain or improve the organic content. Success breeds success, literally, in this part of the soil system!

SOIL ORGANISMS

The third major ingredient of soils is living organisms. Some of these play a vital role in the decomposition of organic matter, some help to aerate soils by burrowing through them, and all contribute to the organic content of the soil. Most of the soil organisms are concentrated in the upper

BOX 19.4 SOIL ORGANISMS

Macro-organisms include the larger animals such as earthworms, ants and beetles. These affect the physical structure and texture of soils by burrowing, which creates a network through which air and water can spread through the soil. But they also influence the organic content of the soil when they excrete bodily waste (faeces) and when their remains are decomposed. Some macro-organisms, such as earthworms, can have a considerable effect on the structure and composition of the upper levels of the soil.

Micro-organisms are much smaller than macro-organisms. Indeed, many of these minute animals, plants and fungi are invisible to the naked eye. The micro-organisms play vital roles in the chemical decomposition of organic material and the physical disintegration of mineral particles, both of which release nutrients to plants and are essential for soil development.

BOX 19.5 POROSITY AND PERMEABILITY

Porosity is a measure of the amount of pore space in a soil, which determines the capacity of that soil for holding water. It is expressed as the amount of pore space as a percentage of the total volume of soil.

Permeability is a measure of the ease with which water moves through the soil and is also known as *hydraulic conductivity*. It is the rate of flow of water through a given area of soil over a stated period of time (usually expressed in $m^3\ s^{-1}$).

Both porosity and permeability depend largely on the texture and structure of the soil, which vary a great deal from place to place. Porosity is heavily affected by the size, shape and packing of individual mineral particles in the soil – large, rounded particles usually have larger voids than small, flat particles, which pack closer together.

part of the soil, mainly because they feed off the organic matter that is also concentrated there. Relatively few organisms exist deeper than about 1 m below the surface.

There is a two-way relationship between soil organisms and soil organic content – each promotes the other. Soil organisms depend on the colloids and organic materials as a source of nutrients, as do the plants that grow in the soil.

All soils contain living organisms, but some soils support much larger populations than others. The latter tend (all else being equal) to be the more productive soils used in agriculture. Ecologists distinguish between larger animals (*macro-organisms*) and much smaller organisms (*micro-organisms*).

The speed at which soil organisms break down leaf litter and other organic matter depends partly on temperature. Decay is relatively slow in cold environments (so high-latitude biomes usually have quite thick litter layers on the forest floor) and much faster in hot and humid climates. Thus, for example, nutrients are quickly recycled in the tropical rainforest, where there is little partly decayed organic material on the soil surface.

WATER AND AIR

Plants and animals require a regular supply of water, which the plants take in from the soil through their roots by osmosis. Air is also a vital ingredient of soil because it allows the macro-organisms and micro-organisms to survive underground, and it is necessary for the organic decay that recycles nutrients and improves soil structure.

Most water enters the soil via precipitation on to the ground surface followed by infiltration into the soil. Most soils have voids (open spaces or pores) between individual particles of mineral matter, which allow water and air to move into and through the soil.

The two most important aspects of soils, from the point of view of water retention, are porosity and permeability (Box 19.5). A soil that is porous and permeable can take in water when it rains, and it can dry out rapidly afterwards. Compacted soils generally have low porosity (because the individual soil particles are packed close together) and low permeability, and as a result little water can get into the soil, so most of it remains as pools on the surface. Conversely, very free-draining soils (such as desert sands) have extremely high porosity and permeability, and water drains so freely through them that relatively little is stored and accessible to plants. The most suitable soils for cultivation have porosities and permeabilities between the two extremes.

BOX 19.6 SOIL WATER

Soil water is held in three forms:

1. *Gravitational water*: this water is pulled down into the soil by gravity, moves between the soil particles and fills the pore spaces between them. The higher the permeability, the more and the faster gravitational water moves through the soil. It drains quickly into the soil when it rains and drains quickly out afterwards (as throughflow and interflow within the soil, and into the aquifer below; see p. 393).
2. *Capillary water*: this water is held by surface tension in and on soil particles and in parts of the pore spaces between particles. Capillary water can move in any direction in the soil, pulled by capillary tension. Thus, for example, when warm, dry air and light winds promote evaporation of water from the surface of a soil, capillary action pulls capillary water up towards the soil surface. This supplies more water for evaporation, it progressively dries the soil out, and it lifts dissolved salts in the soil to the surface (and can lead to salinisation; see p. 403). Plants use capillary water, and capillary uplift in the soil helps to maintain healthy plant growth between periods of rainfall (particularly during long dry periods). Capillary water survives much longer in the soil than gravitational water and disappears only after prolonged dry conditions.
3. *Hygroscopic water*: this water is held as a very thin layer around each individual particle of soil. It is not very accessible to plants, and at least some of it remains in the soil even after extreme drought.

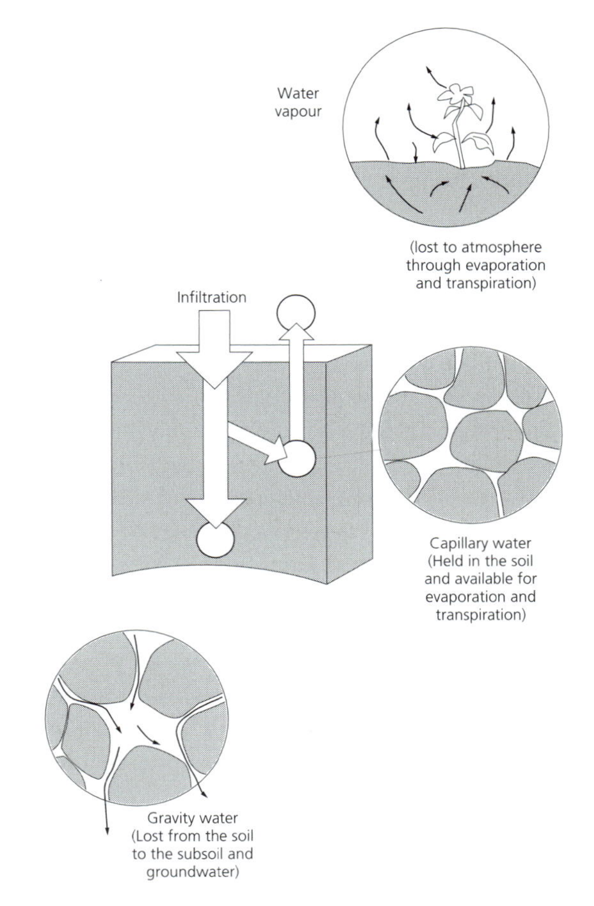

Figure 19.2 *Water in soil. Water infiltrates soil and then moves through it as capillary water and gravity water. See text for explanation.*

Source: Figure 13.9 in Marsh, W.M. and J.M. Grossa (1996) Environmental geography: science, land use and Earth systems. John Wiley & Sons, New York.

SOIL PROPERTIES

Soils vary in character from place to place, reflecting the interactions between parent material, environment and soil-forming processes. The soil produced by these never-ending interactions is dynamic, and it can change over time in response to changes in the environment (particularly climate) and soil-forming processes.

Soils develop naturally over time, so some of the variability in soil character in an area might be caused by differences in the age of the soils. This might happen if there has been a major natural change in vegetation in one part of the area, for example, or if a new surface has been created (perhaps by mass movement, volcanic activity or river deposition) on which soils can develop.

There can be significant variations in the character of soils across even relatively short distances, particularly if these embrace major changes in landscape. Different soils develop on floodplains, low slopes, high slopes and flat summits, so the mosaic of landscape units in an area is generally mirrored in the mosaic of soils (and, in turn, the mosaic of vegetation).

Soils vary in appearance because they have different textures, structures and colours, and they also vary in depth. These characteristics strongly influence soil stability and fertility, and they reflect the interaction of climate and parent material. They are very useful in classifying soils (see p. 604) and in land evaluation, which is designed to assess how suitable soils are for different types of use.

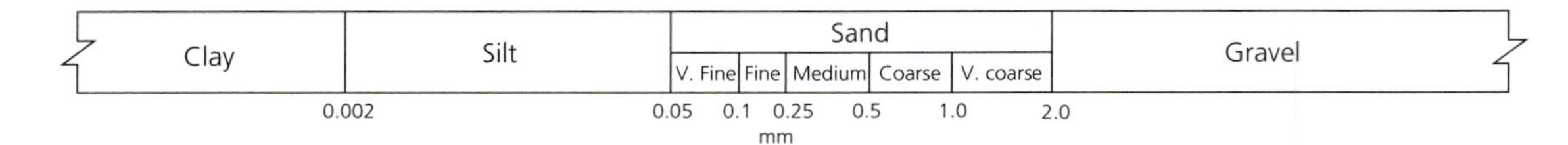

Figure 19.3 *Size classification of soil particles. This classification of particle sizes is used around the world.*

SOIL TEXTURE AND STRUCTURE

Mineral grains produced by the weathering of rocks are inert fragments that provide the basic particles of which all soils are composed. These particles can vary a great deal in size, shape and mineralogy (depending largely on the nature of the parent material). Particle sizes range from very fine clays (less than 0.002 mm in diameter) and silts (between 0.002 and 0.05 mm), to coarse sands (range from 0.05 to 2.0 mm) and gravels (larger than 2 mm) (Figure 19.3).

Soil texture: texture is defined by the mixture of particle sizes in a soil, which can vary from place to place, even over relatively short distances. A standard basis for defining soil texture classes (Figure 19.4), based on the proportion of particles in different size classes, was proposed by the US Department of Agriculture in the early 1950s and has been widely used since.

Soil structure: structure reflects the way individual soil particles stick together as clumps or aggregates. Soil aggregates are remarkably strong and durable. They rarely break down into individual particles naturally but will do so if deliberately crushed under great force.

Structure varies from soil to soil, depending on climate (particularly moisture and temperature), parent material and particle sizes. Organic and mineral materials play important roles in binding the particles together, and these also vary from soil to soil and from place to place.

Soil structure affects important conditions in the soil, such as water retention and nutrient availability. Like texture, soil structure is an important determinant of productive capacity, environmental quality and agricultural sustainability.

BOX 19.7 IMPORTANCE OF SOIL TEXTURE

Texture strongly affects the ability of a soil to retain and circulate air and water, which in turn affects the supply and availability of nutrients in the soil – and hence plant growth, productivity and yield. Texture also influences root penetration in a soil and is a major determinant of soil tilth (workability) in agriculture. Most plants grow best in a loam soil, which contains 20 per cent or less of clay, 30–50 per cent of silt and 30–50 per cent of sand. Such a texture has good air and water retention, good nutrient availability and good root penetration.

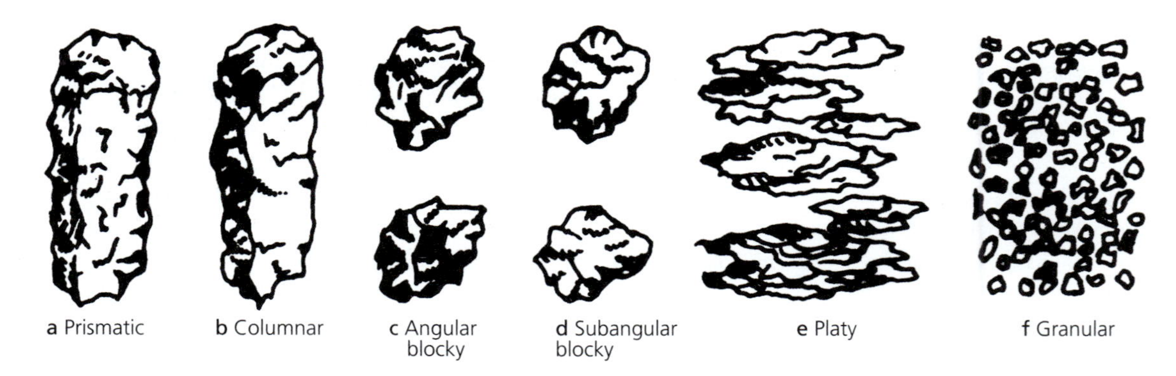

Figure 19.4 *Soil texture classes. This widely used sixfold classification of soil textures was developed by the US Department of Agriculture in the early 1950s.*

Source: Figure 17.5 in Doerr, A.H. (1990) Fundamentals of physical geography. Wm. C. Brown Publishers, Dubuque, Iowa.

BOX 19.8 SOIL AGGREGATES

Particles stick together in different ways, creating different aggregates, which create different soil textures. Minute clay particles stick tightly together, for example, and the aggregate behaves differently under different conditions – when wet it behaves like a plastic, but when dry it can be hard and brittle. The US Department of Agriculture classification of aggregates (Figure 19.5), which defines six major groups of soil structures, is widely used around the world. It defines broad differences in texture. Thus, for example, a soil with a granular structure is composed of relatively small, rounded aggregates, whereas a soil with a blocky structure is made up of aggregates with sharp corners and irregular

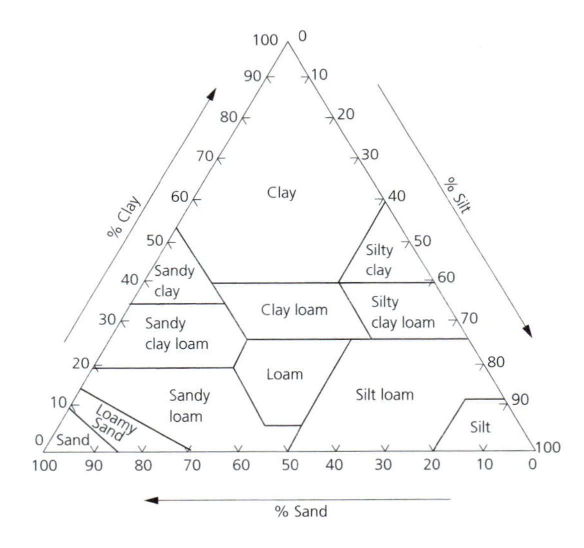

Figure 19.5 *Soil structure classes. Soil texture can be defined using this classification based on the proportion of clay-, silt- and sand-sized particles in a particular soil.*

Source: Figure 20.1 in Dury, G.H. (1981) Environmental systems. Heinemann, London.

SOIL COLOUR

One of the most obvious differences between soils is in their colour – some are grey or black, others are shades of brown and some soils are even bright yellow or red. These variations in colour are caused by chemical reactions in the soil (see p. 602), which are heavily influenced by its mineral content.

Like most other soil properties, colour reflects the interaction of parent material, climate and soil-forming processes. In turn, colour can influence other characteristics of a soil, such as temperature (and thus suitability for soil organisms and speed of chemical reactions). Darker soils have a lower albedo (reflectivity) than lighter soils, so they absorb more heat. This can affect plant growth and productivity, with a knock-on effect on organic content. This feedback link indicates that soil colour can be in equilibrium with the prevailing environmental conditions and as such is also liable and able to adjust to environmental change if necessary.

As a general rule, soils that contain more organic material (and are more fertile) tend to be darker. Within a particular soil, the upper layers, which contain the most humus, are usually darker than the lower layers. There is a clear zonal pattern of variations in soil colour, which mainly reflects differences in organic content. Thus, for example, high-latitude soils that lack organic matter are mostly greyish, while many mid-latitude soils with a high organic content range from brown to black.

BOX 19.9 CONTROLS ON SOIL DEPTH

Soil depth is controlled by a range of factors, including:

- *Age of the soil*: soil depth usually increases over time as bedrock weathers and organic material builds up.
- *Climate*: climate (particularly temperature and humidity) influences rates and patterns of rock weathering, and this directly affects soil depth. Thus weathering is slow and soil is thin in arid areas, and weathering is fast and soils are correspondingly deeper throughout much of the humid tropics.
- *Topography*: soils tend to be deeper on flat and gently sloping ground, whereas steep slopes usually have thin soils because much of the surface material is regularly removed by surface runoff (see p. 211) and mass-movement processes (see p. 215).
- *Human impacts*: soil depth is affected by many different types of human activity, including compaction (which directly decreases depth) and cultivation and over-grazing (which both increase the risk of soil erosion and thus loss of topsoil). Soil conservation measures (see p. 612) are designed to combat these problems and preserve soil depth, structure and fertility.

Superimposed on this broad zonal pattern are variations that reflect differences in mineral content (created by a combination of parent material, climate and soil-forming processes). Soils with a relatively high content of iron and aluminium oxides (which are common in humid tropical and sub-tropical environments) are often red or yellow, whereas soils containing little iron (such as poorly drained soils) are often grey or blue.

SOIL DEPTH

Soils also vary in depth, and this can exert a strong influence on plant growth (because soil depth affects root penetration and water and nutrient storage) and suitability for agriculture. Depth variations are associated with a range of factors (Box 19.9).

SOIL FORMATION AND PROFILES

Natural vegetation and food production rely heavily on the topsoil – the upper layer, which normally varies in depth between about 10 and 45 cm. But soils are much more than just topsoil, and as they develop they typically display a series of horizontal layers in which different processes operate.

SOIL FORMATION

The basic source of all soils is rock weathering (see p. 205), which breaks down the solid rock into smaller mineral particles, which can be eroded and transported away by a range of processes (including wind, rivers, ice and the sea). The particles are affected by a range of natural chemical processes (p. 206), and over a long period of time these processes – coupled with the decomposition of organic material and the activities of soil organisms – turn the inert mineral material into living soil that is dynamic and adaptable to environmental change.

BOX 19.10 IMPORTANCE OF TOPSOIL

Topsoil is important in three particular ways:

1. It contains the organic matter (decayed remains of vegetation) that plants need for active growth.
2. It contains a range of soil organisms (including earthworms) that assist the decay processes.
3. It provides a firm foundation for plants to grow in.

BOX 19.11 PEDOGENESIS AND SUCCESSION

Pedogenesis (the origin and development of soil) is much like ecological succession (see p. 570) in the sense that once the process is started it tends to be self-perpetuating. Both aspects are also progressive and sequential, with soils and vegetation becoming more stable and mature over time. Both also reflect adjustments between biotic and abiotic aspects of ecosystems, and they can be significantly affected by rapid environmental change. It is also possible to monitor both sets of processes at work, either by comparing sites of different ages or by monitoring changes over time at particular sites. Studies have been made of the development of soils on the volcanic ash deposited by the 1980 explosion of Mount St Helens (see Box 6.26), for example, and they show that some soil changes (particularly in the availability of nutrients) could be detected within as little as two years.

The speed at which soil develops depends on a number of factors. Climate is particularly important, because it largely determines which processes operate and how effective they are. Soil development is much faster in moist climates than in dry climates, for example. Living organisms are also essential, because they increase the organic content and help to release chemical compounds in soluble forms accessible to plants.

Soil formation (*pedogenesis*) effectively begins when organisms interact with the products of rock weathering to produce a mixture of mineral and organic material. Over time, plants colonise the young soil, and a progressive sequence of vegetation changes (natural succession; see p. 570) further assists in the development of a mature soil (Box 19.11). Plants are important in various ways – their litter decays and releases nutrients into the soil, and their roots penetrate the soil and help the sub-surface movement of water, air and nutrients.

Another important part of pedogenesis is the redistribution of mineral material within the soil by natural processes (p. 602). The most common process is the leaching (washing) of soluble minerals from the upper part of the soil and the deposition of this material lower down in the soil. Once this starts the soil tends to develop two layers – an upper one depleted of soluble minerals and a lower one with a high concentration of minerals.

Differences in soil properties between the topsoil and the subsoil tend to increase over time, leading to the development of soil profiles.

SOIL PROFILES

As most soils develop, distinct layers can be identified in them that are usually the product of natural soil-forming processes. These layers are called *soil horizons,* and they are most distinct in mature soils that have developed over a long period of time. Soil horizons form the basis of most soil classification schemes (see p. 604) because they strongly affect soil properties and preserve evidence of soil development. Within a given soil, each horizon can have different physical and chemical characteristics, even though the entire thickness of soil has evolved from the same parent material.

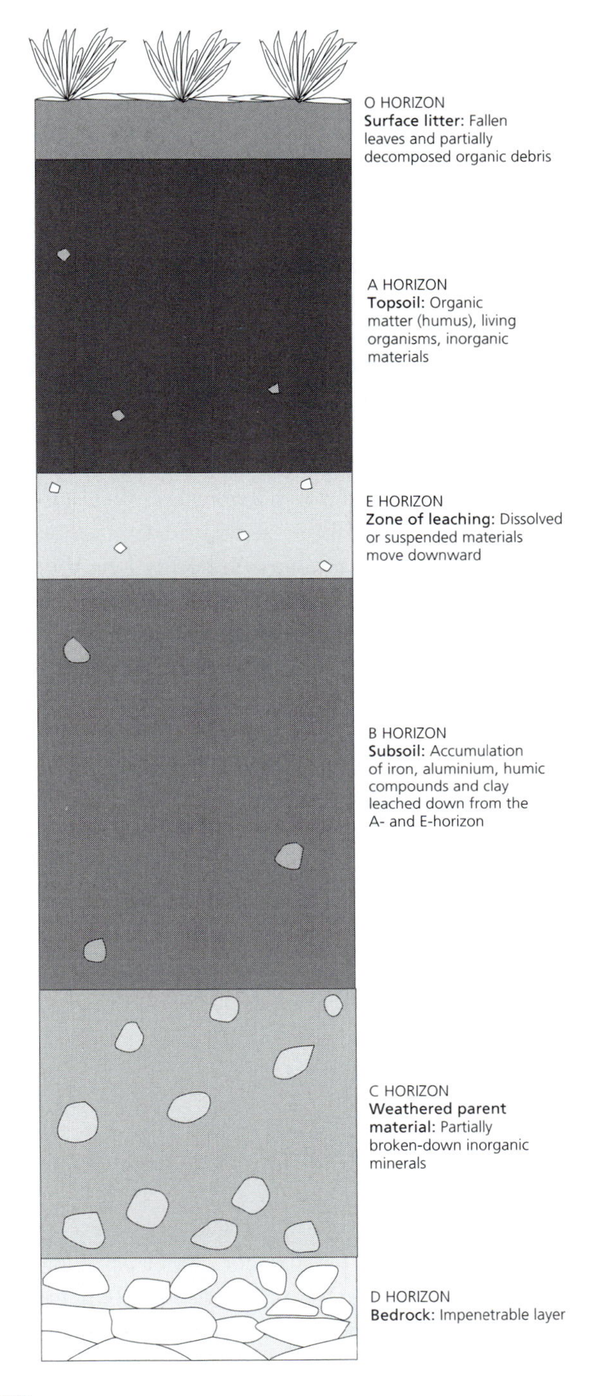

Figure 19.6 *Characteristics of a typical soil profile. Most soil profiles display evidence of a number of different horizons, although the number, thickness and character of each horizon varies between different soil types.*

Source: Figure 11.5 in Cunningham, W.P. and B.W. Saigo (1992) Environmental science: a global concern. Wm. C. Brown Publishers, Dubuque, Iowa.

BOX 19.12 SOIL HORIZONS

Most profiles contain four distinct horizons:

1. *A horizon*: this is the fertile topsoil or surface layer of a soil and is dominated by two processes – the accumulation of organic matter (from the decomposition of litter) and the leaching (washing downwards to the B horizon below) of soluble material and the eluviation (removal in suspension) of fine materials.
2. *B horizon*: this horizon is dominated by the deposition (illuviation) of material leached from the A horizon, so it is a zone of accumulation and enrichment. It contains little organic material and is usually more compacted than the A horizon because clay washed down from the A horizon fills many of the voids between particles.
3. *C horizon*: this layer is the part of the soil profile that is the least weathered, and it receives little or no additions of material from above. Over time this material will develop into the B horizon by continued weathering and soil-forming processes. In areas where rock weathering is particularly fast, such as the wet tropics, the C horizon is likely to be very thin.
4. *D horizon*: this is the unweathered bedrock beneath the soil and is now widely referred to as the R horizon (rock).

A cross-section through the horizons in a well-developed soil is known as a *soil profile* (Figure 19.6). Different types of soil formed under different conditions (particularly parent material, climate and age) have different profiles. Most profiles contain four distinct horizons (Box 19.12).

The A horizon and B horizon together form the proper soil (*solum*), where organisms live and organic material is concentrated; beneath them is the largely inert parent material. Over time, as a soil develops, the horizons above the weathered bedrock (D horizon) grow progressively thicker, and the overall soil profile grows deeper.

Soil erosion (see p. 611) has the opposite effect on profile depth by removing much if not all of the topsoil (A horizon). The B, C and D horizons are often not seriously affected by soil erosion in the short term, but re-establishment of a proper A horizon takes a long time and is unlikely to occur while the erosion continues. If soil erosion is particularly serious or long-lasting, it can remove horizons below the topsoil, ultimately stripping away everything down to the solid bedrock.

Some soils, called *palaeosols*, have been buried by material that is deposited on top of them. Burial might be by a number of different processes, including lava flows, avalanches and migrating sand dunes. If the overlying material is subsequently eroded, the palaeosol is exposed once again and its development can continue by adaptation to its new environment.

SOIL-FORMING PROCESSES

SOIL-FORMING FACTORS

Soil is formed from weathered bedrock (*regolith*), but the type of soil that is developed depends on the interaction of a wide variety of factors (Box 19.13).

Human activities are also important, because they can radically alter the pace and pattern of soil development, either accidentally or on purpose. Accidental effects include the ecological consequences of air pollution or the unplanned impacts of land-use changes such as forest clearance. Deliberate changes include the addition of chemical fertilisers to agricultural soils, land drainage (to prevent the waterlogging of soils), land reclamation and restoration, and the creation of new land surfaces that have soil added to them that is transported from elsewhere (as with some landscape gardening for example).

Many different processes operate in soils. There are three main physical processes (Box 19.14) and four main chemical processes (Box 19.15). These physical and chemical processes, in various combinations, give rise to the five main soil-forming processes: laterisation, podsolisation, calcification, acidification and salinisation.

BOX 19.13 SOIL-FORMING FACTORS

The most important natural soil-forming factors are:

- parent material
- climate
- natural vegetation
- living organisms
- slope
- time.

All factors are important, but one or more of them may dominate in a particular place or time.

BOX 19.14 PHYSICAL SOIL-FORMING PROCESSES

The three main physical processes are:

1. *leaching*: the removal of soluble substances dissolved in percolating water, usually from the A horizon;
2. *eluviation*: the removal of fine particles of material in suspension, usually from the A horizon (Figure 19.7);
3. *illuviation*: the deposition, usually in the B horizon, of material (including colloids and mineral salts) that has been washed down from above (usually the A horizon).

LATERISATION

Laterisation involves the deposition of a hard layer of metallic oxides (called *laterite*) in the A horizon. It most commonly occurs in humid tropical and humid sub-tropical areas, where precipitation is high. It is formed after natural vegetation has been removed and most of the soluble materials are leached down through the soil. This leaves a residue of insoluble oxides of iron and aluminium

BOX 19.15 CHEMICAL SOIL-FORMING PROCESSES

The four main chemical processes are:

1. *hydrolysis*: a chemical reaction in which an insoluble compound reacts with water to produce other compounds that are more soluble;
2. *hydration*: a chemical reaction in which a compound breaks down into water and another compound. The compound combines with water but does not react chemically with it, so the process is reversible;
3. *oxidation*: a chemical process in which an element reacts with oxygen to form an oxide (a compound of oxygen and the element);
4. *reduction*: a chemical process in which an element reacts with hydrogen to form a hydride (a compound of hydrogen and the element).

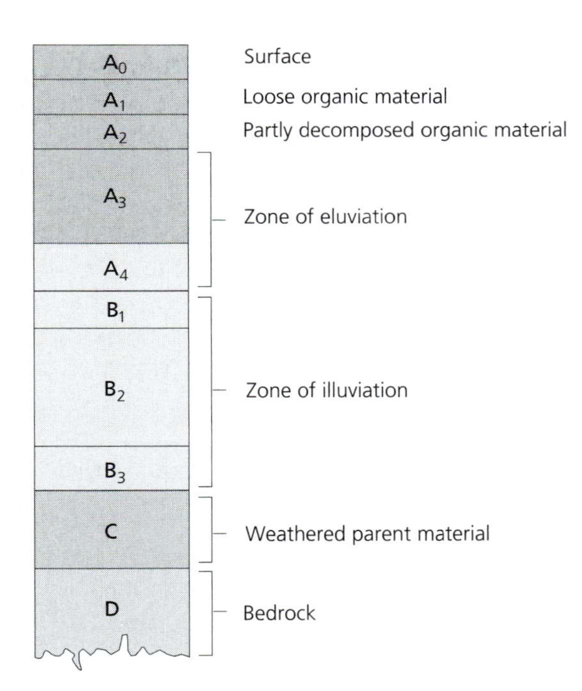

Figure 19.7 *Zones of eluviation and illuviation in a soil profile. This is a more detailed way of describing the horizons in soil profiles (see Figure 19.6) that reflects the dominant soil-forming processes.*

Source: Figure 17.9 in Doerr, A.H. (1990) Fundamentals of physical geography. Wm. C. Brown Publishers, Dubuque, Iowa.

in the A horizon, which is also coarse-textured because most of the fine particles have been washed down into the subsoil. For the same reason, the A horizon contains relatively little organic material.

Lateritic soils (*latosols*) are often very deep, because of rapid weathering in the hot, wet conditions, but they have a relatively thin A horizon. The presence of the iron and aluminium compounds gives lateritic soils a distinctive reddish or yellowish colour. The laterite deposit is usually impermeable, which prevents water infiltrating the soil and increases the likelihood of surface runoff, which can seriously erode the topsoil. The hard layer also limits the agricultural use of these soils, because it is difficult to plough or break up. Lack of organic matter further reduces their suitability for agriculture.

PODSOLISATION

Podsolisation is the process by which the upper layer of a soil becomes acidic through the leaching of bases, which are deposited in the lower horizons. It typically occurs under forests (see p. 581) in areas with cool, moist climates and quite severe winters. Decomposition of organic material is inhibited by the low temperatures and long frozen spells, so a peaty mat of acidic organic material builds up on the soil surface. There are relatively few organisms in these soils, so mixing of organic matter and mineral matter within the soil is very limited.

Under these conditions, the topsoil is often heavily leached and eluviated, leaving an A horizon with a sandy or granular texture. Soluble bases and iron and aluminium compounds are leached into the B horizon, where they accumulate. Tiny clay particles are also illuviated in the B horizon, which has a high clay content and is often quite well compacted as a result.

Podsols are shallow, greyish-white acidic soils with relatively little organic content or soil organisms. Many are infertile and unsuitable for agriculture, particularly in colder climates.

CALCIFICATION

Calcification is the process by which the upper layers of a soil are hardened by the deposition of calcium salts. It occurs mainly under grass or xerophytic shrub (see p. 583) in sub-humid, semi-arid and arid climates. Leaching and eluviation of material from the A horizon down to the B horizon is minimal, because precipitation is light. This allows calcium and other bases to accumulate in the upper soil horizons, being constantly supplied from dead grass and shrubs.

The depth within the soil at which calcium accumulates depends mainly on humidity. In very dry areas it occurs more or less on the soil surface, but in humid areas it can be as much as 1 m below the surface. Calcification produces neutral or basic soils.

ACIDIFICATION

Like all other properties of soil, acidity varies a great deal from one soil to another, and it depends on interactions between parent material, living organisms and soil-forming processes. Many plants will tolerate a wide range of acidity conditions, and different plants will grow along the acidity spectrum between a pH of about 4 (highly acidic) and about 10 (highly alkaline). Laterisation and podsolisation produce acid soils, which have a pH of less than 7.2 (see Box 8.30).

Acidification occurs naturally in some soils, such as those beneath coniferous forest, which sheds acidic needles and creates acidic litter. Scientists are concerned about induced acidification associated with acid rain (see p. 251), which is caused by the burning of fossil fuels and the release of sulphur dioxide (SO_2) and nitrogen oxides (NO_x). Studies have shown that acid deposition can seriously damage forests, vegetation and crops via its effects on soils, and that acidified soils can in turn lower the pH of some lakes and cause ecological damage there too.

Acidic soils, however caused, can be treated to improve their productivity. Treatment normally involves adding mineral fertilisers or neutralising materials (such as crushed limestone or dolomite), which do reduce acidity and improve fertility. But the process generally needs to be repeated regularly because over time the additives are leached out of the soil.

SALINISATION

Salinisation is the process by which unusually high concentrations of dissolved salts accumulate in soils. This is often the result of large-scale irrigation schemes in semi-arid areas (see p. 61), where much of the soil water is evaporated off, leaving behind the salt residue. Salts can accumulate over time to threshold concentrations at which they become toxic and inhibit commercial agriculture, as has already happened in parts of California. The salinisation risk appears to be increasing in parallel with the continued rises in global population, demand for food and irrigation agriculture in arid areas.

BOX 19.16 PEDON AND CATENA

The smallest three-dimensional unit of soil, which defines a block with relatively uniform properties, is called a *pedon*. The pattern of soil variations across a landscape is made up of many different pedons, and it reflects the interactions between parent material, organic activity and soil-forming processes. Drainage exerts a powerful influence on this pattern, and a *transect* (cross-section) from the top of a hill down to the bottom often shows a clear sequence of profile changes – known as a *catena* – which indicates the importance of soil moisture.

CLASSIFICATION AND DISTRIBUTION

SOIL MOSAICS

Soils vary a great deal from place to place because of variations in the main soil-forming processes. This creates a mosaic of soil variations, which can be studied at a variety of scales from the local to the global. At the local scale, soils vary with topography and land use (Box 19.16). At a broader regional scale, the impact of climate on the pattern of soil variations is more obvious, particularly in areas that are close to or on major boundaries between climatic zones.

Larger still, at the global scale, the pattern of soil variations closely reflects the influence of climate and vegetation, and there are many similarities between the world maps of climate, biomes and soils. This is no surprise, given the importance of climate and organic activity to soil development, but it is reinforced by the fact that the choice of where to define the boundaries in global climate classification schemes is usually based partly on ecological criteria.

SOIL CLASSIFICATION

MATURITY AND ZONALITY

Most schemes for classifying soils are based on the characteristics of mature soils. This is logical, given that soil properties change as the soil develops, and to compare soils at different stages in their development would not be comparing like with like. The rationale is the same as classifying biomes on the basis of natural climax vegetation (p. 574), which appears to be stable over time and in dynamic equilibrium with the prevailing environment.

BOX 19.17 ZONAL, INTRAZONAL AND AZONAL SOILS

The three types of soil are:

1. *Zonal soils*: which are typical of the climatic zones in which they occur. They have well-developed characteristics, are widely distributed within those climatic zones and have well-developed profiles.
2. *Intrazonal soils*: which are untypical of their climatic zone because they have been strongly affected by more local factors, such as parent material. These soils transcend the major climate and vegetation zones and appear as separate islands within the surrounding sea of zonal soils.
3. *Azonal soils*: these are poorly developed soils with young and evolving horizons. Alluvium (river deposits) and loess (wind-blown deposits) are typical azonal soils. Over time these types of soil are likely to mature and develop into one of the other two types.

Most soil classification schemes also recognise that all soils can be assigned to one of three broad categories (Box 19.17), depending on the extent to which they are in equilibrium with their environment.

THE RUSSIAN–AMERICAN SOIL CLASSIFICATION SCHEME

This classification scheme, originally developed by Russian soil scientists and revised in the late 1930s by the US Department of Agriculture, was widely used around the world until the 1980s. It is based on the relationships between soil type, vegetation and climate, and it divides zonal soils into two main sub-orders – *pedalfers* and *pedocals* (Box 19.18).

BOX 19.18 PEDALFERS AND PEDOCALS

Pedalfers are soils in which aluminium and iron accumulate. They are produced by laterisation and podsolisation (see pp. 602–603) and occur mainly in humid climates.

Pedocals are soils in which calcium accumulates. They are produced by calcification (p. 603) and occur mainly in sub-humid, semi-arid and arid climates.

Each of the sub-orders is then sub-divided into great soil groups, defined to maximise the between-soil variations and minimise the within-soil variations. The great soil groups within the pedalfers are the lateritic soils, red and yellow podsolic soils, grey-brown podsols, podsols, tundra soils and prairie soils. Within the pedocals the main great soil groups are the chernozem (black earth) soils, chestnut and brown soils, sierozems and desert soils.

Many soil scientists felt that the Russian–American Soil Classification Scheme was too coarse for the detailed study and description of local soils, and that it was based too heavily on assumptions about how different soils have developed. It was eventually replaced by the American-based Comprehensive Soil Classification System.

THE COMPREHENSIVE SOIL CLASSIFICATION SYSTEM

This new approach to the classification of soils was developed by the US Department of Agriculture. It is a more sophisticated scheme than the earlier one it replaced, and it is based on six levels of classification. All soils are divided into eleven soil orders, which in turn are sub-divided into forty-seven sub-orders, 185 great soil groups, many sub-groups, and a great many soil families and series. The main characteristics of the different soil orders are summarised in Table 19.1, and Figure 19.8 shows the global distribution of the main soil orders.

Inevitably, there are close associations between the two classification systems. Thus, for example, the oxisol order covers soils that in the earlier scheme would have been classed as latosols and lateritic soils, the mollisol order is similar to the earlier chernozem group, and the aridisol order is much the same as the earlier desert soil group.

SOILS, VEGETATION AND CLIMATE

Soils within each of the orders have similar structures and characters because this was an important factor in devising the classification. Thus it should be possible to draw important clues about soil development from a map of soil distributions and a knowledge of how the soil orders were defined.

The aridisols, for example, develop under desert conditions. With little if any vegetation growing on the soil

Table 19.1 The Comprehensive Soil Classification System

Order	*Main characteristics of typical soils*
Alfisols	These soils develop in areas with 510–1,270 mm of rainfall each year; most develop under forests; clay accumulates in the B horizon. Figure 19.9 shows a typical profile
Andisols	These are volcanic soils that are deep and have a light texture. They contain iron and aluminium compounds
Aridisols	These are desert soils with little or no organic content but significant amounts of calcium. They are often affected by salinisation. Figure 19.10 shows a typical profile
Entisols	These are soils with little or no horizon development, which are often found in recent floodplains, under recent volcanic ash and as wind-blown sand
Histosols	These are organic soils, found in bogs, swamps and wetlands
Inceptisols	These are young soils in which the horizons are starting to develop. Figure 19.11 shows a typical profile
Mollisols	These soils form mainly under grassland. They are dark-coloured, with upper horizons rich in organic matter
Oxisols	These are infertile, acidic, deeply weathered soils that contain clays of iron and aluminium oxides
Spodosols	These are sandy soils that develop under forests, particularly coniferous forest. They are acidic and have accumulations of organic matter and iron and aluminium oxides in the B horizon
Ultisols	These are acidic, deeply weathered tropical and sub-tropical soils with clay accumulations in the B horizon. Figure 19.12 shows a typical profile
Vertisols	These are clay soils, which expand when wet and crack when dry. They develop in climates with marked wet and dry seasons.

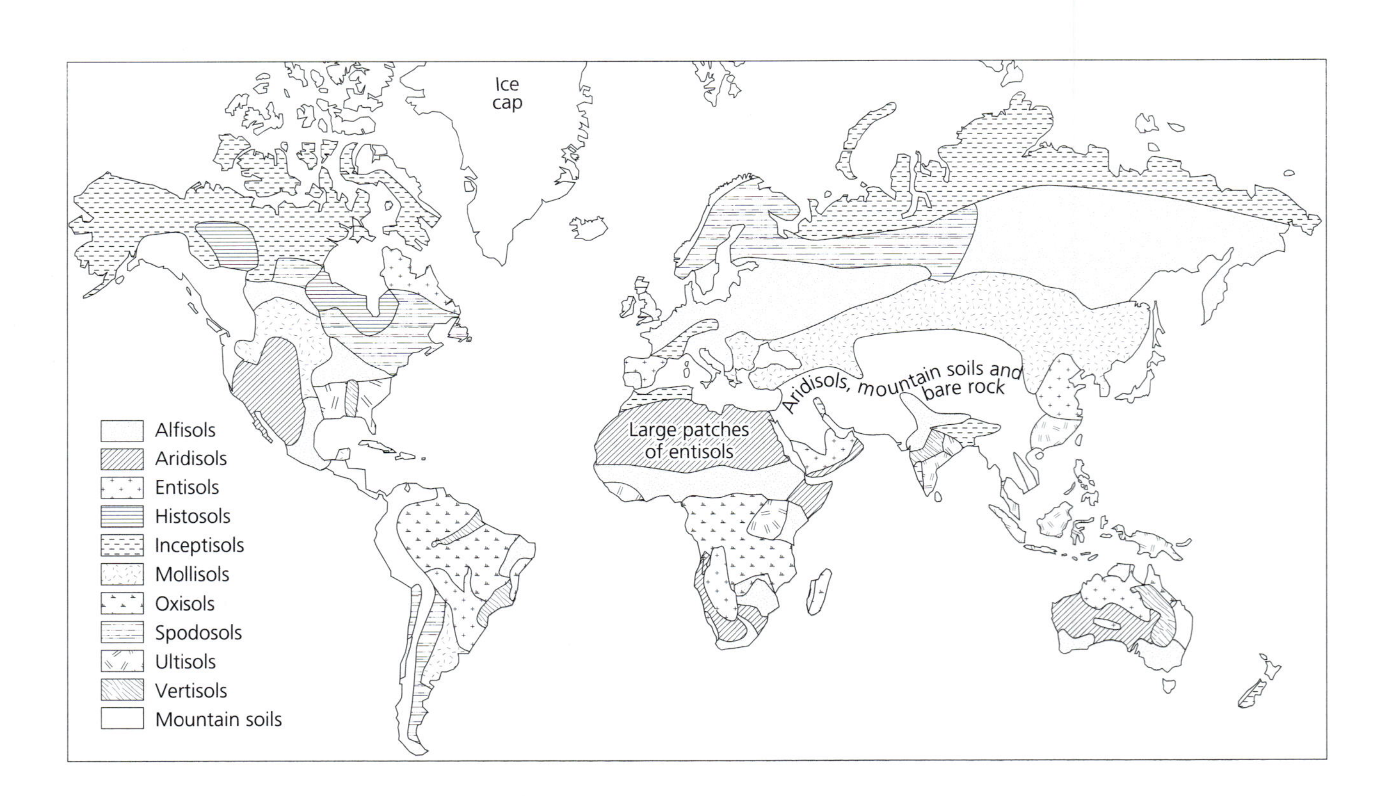

Figure 19.8 *Global distribution of the main soil orders. This generalised soil map shows the global distribution of the ten soil orders.*

Source: Figure 20.8 in Dury, G.H. (1981) Environmental systems. Heinemann, London.

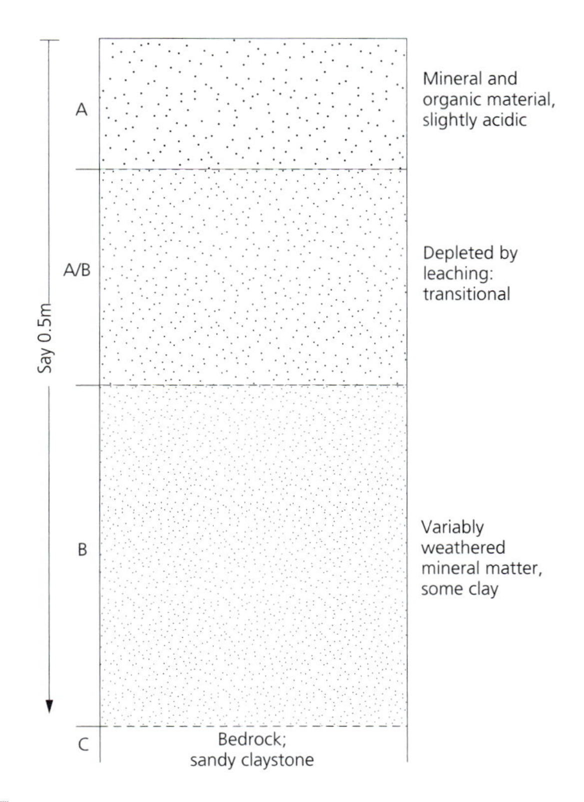

Figure 19.9 *Profile of an alfisol. This type of soil is typical of a mid-latitude west coast climate.*

Source: Figure 20.14 in Dury, G.H. (1981) Environmental systems. Heinemann, London.

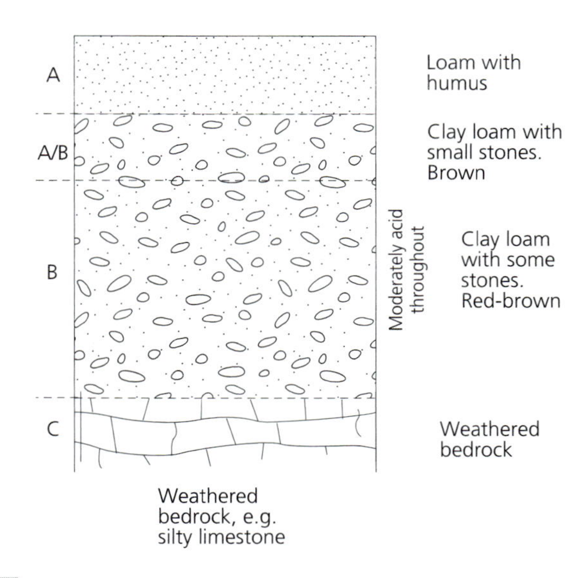

Figure 19.11 *Profile of an inceptisol. This type of soil is typical of a mid-latitude west coast climate.*

Source: Figure 20.9 in Dury, G.H. (1981) Environmental systems. Heinemann, London.

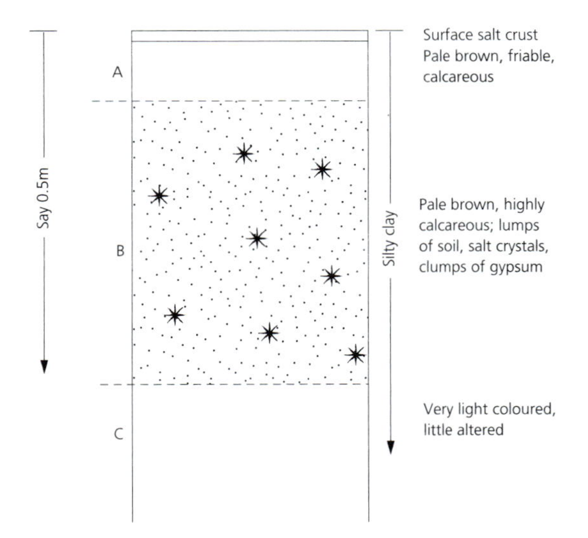

Figure 19.10 *Profile of an aridisol. This type of soil is typical of a warm, dry continental climate.*

Source: Figure 20.12 in Dury, G.H. (1981) Environmental systems. Heinemann, London.

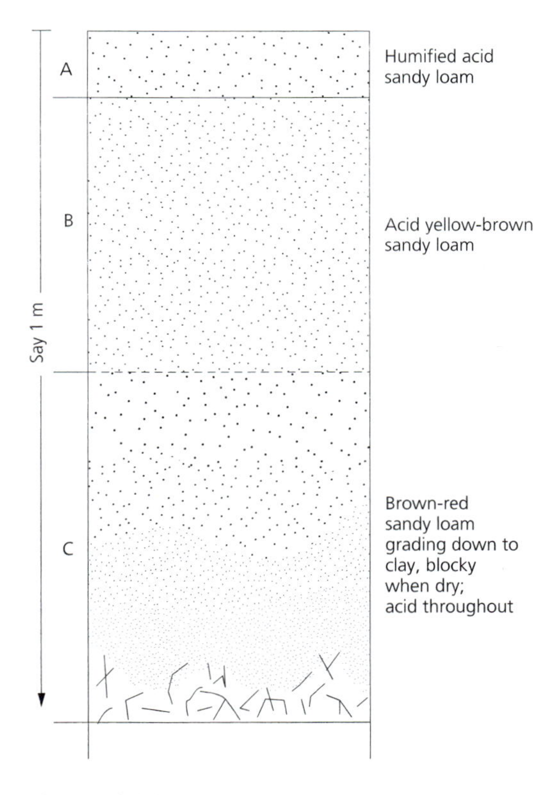

Figure 19.12 *Profile of an ultisol. This type of soil is typical of a sub-tropical east coast climate.*

Source: Figure 20.15 in Dury, G.H. (1981) Environmental systems. Heinemann, London.

BOX 19.19 SOILS AND CLIMATE

The global distribution of soil orders closely mirrors the major climate zones (see Figure 11.6) and biomes (see Figure 18.16). Thus, for example:

- *oxisols* are found in the humid tropics;
- *alfisols* are found in temperate regions;
- *mollisols* are found in the prairies, pampas and steppes;
- *spodosols* are found in the temperate zone around North America and Eurasia.

BOX 19.20 HUMAN ACTIVITIES AND SOILS

The stability and future of many soils is under threat from a wide variety of human activities, including overgrazing, poor agricultural practices, land-use change and forest clearance. Many human activities degrade soils and alter their structure, fertility and usefulness. Many activities also cause or promote soil erosion, in which the valuable topsoil is washed or blown away. Soil erosion creates problems where it happens (the most fertile soil is lost) and problems where the eroded material is deposited (it fills lakes and reservoirs, blankets floodplains, makes river channels unstable, is blown by wind against structures and buildings, and buries other soils and vegetation). Moreover, while soil is a renewable resource, erosion usually strips it away much faster than natural soil-forming processes can replace it, so erosion sometimes means that it has gone for ever. Even if the erosion were to cease, most soils would take centuries to be fully re-formed.

surface, there is no surface litter, no humus and no dark organic layer in the A horizon. There is a relatively deep B horizon, where soluble compounds accumulate that have been leached down from above. Alfisols generally develop beneath broadleaved forest, and they have a much thicker A horizon (partly because leaves and other litter decompose quickly during the mild, wet winter) and a thick B horizon, where plant nutrients accumulate and are accessible to tree roots.

MANAGING SOIL RESOURCES

SOILS UNDER THREAT

Soils are one of the most important natural resources we have, because they play a central role in food production and in the major biogeochemical cycles (see p. 86). They are the main interface between the atmosphere, hydrosphere, lithosphere and biosphere, so any changes in any of these environmental systems are likely to have knock-on effects on soils and soil development.

Problems of degradation and erosion (see p. 609) are becoming more serious as the human population continues to grow. Population growth further increases the demand for food and promotes more intensive agriculture, increasingly in environments and on soils for which that is not a sustainable type of land use. At the global scale, soil erosion is the main threat to soil resources, and poor land management is making it worse. But many soils are also threatened by pollution, including the fallout of air pollutants (such as acid rain), deposition of water pollutants, particularly in times of flood, leaching of material from waste dumps and sanitary landfill sites, and contamination with toxic heavy metals and other pollutants from local industries.

It is not all bad news, however, because soils also offer great potential as carbon sinks (Box 19.21), which might prove very helpful as a means of tackling the problems of greenhouse gas emissions.

BOX 19.21 SOILS AS CARBON SINKS

Soils provide natural sinks for carbon dioxide, and some scientists think they could soak up as much CO_2 as forests in the battle to control global warming (see p. 585). Organic carbon in soils accounts for an estimated three-quarters of the carbon stored in ecosystems on land – much more than that in forest timber. It is suggested that modern techniques of soil conservation could soak up enough carbon to meet a sixth of the American and Canadian targets for reducing emissions of greenhouse gases over the next 20 years, mainly by changing ploughing regimes and making more use of pesticides and fertilisers. There is growing pressure to include soil conservation among projects recognised under the Kyoto Protocol (see Box 1.6) that aim to expand natural carbon sinks.

BOX 19.22 SCALES OF LAND EVALUATION

Land evaluation can be carried out at a variety of scales, and an appropriate unit for analysis must be selected. Logically, this unit should have some meaningful environmental basis – such as a soil pedon (see Box 19.16), a stand of vegetation or a full ecosystem – if the evaluation is to guide sustainable development of natural resources based on ecological principles. One proposal is to use an 'environmental unit' (an arbitrary ecosystem) defined on the basis of landform, geology, soils and vegetation. Within each unit, relevant factors such as climate, water, wildlife and human activities would be examined and evaluated as a basis for deciding on the land capability for that particular unit. Ultimately, a map of land capability variations can be drawn, which should guide land-use planning and decisions about the suitability of each unit for different types of use.

An alternative to the area-based approach is to use a grid-based approach, based on a standard unit of analysis such as a 1 × 1 km grid square. This allows soil information to be considered alongside other environmental and climatic variables, often derived from maps, so that overlapping patterns and possible causal associations can be examined using geographical information systems (GIS) techniques. Such an approach has been used to evaluate crop growth potential across the countries of the European Union and to identify which areas of land are most suitable for set-aside, designed to control agricultural production and increase nature conservation.

LAND EVALUATION

One way in which soil scientists are responding to the need to find more sustainable approaches to the use of valuable soil resources is by developing ways of evaluating how suitable different soils are for particular uses. This is known as *land evaluation*, and various approaches have been explored (Box 19.22).

An important consideration when deciding on the appropriate type and level of use of soil resources is the carrying capacity. This is the maximum level of use that a particular soil can sustain without suffering from depletion, exhaustion, erosion or other unwanted side-effects. The notion of carrying capacity (see p. 48) also applies to individual ecosystems and human populations at a variety of scales, in all of which it reflects the balance between population density and available resources.

The main objective of land evaluation is to guide decisions about land-use change by identifying which soils are best suited to particular uses. Present use and optimum use are often rather different, because current land use reflects the interplay of many more factors than simply environmental suitability (including history and continuity of use, personal preferences among landowners, and commercial and market forces). Physical land evaluation has established that about 65 per cent of the land in the Netherlands is suitable for growing potatoes, for example, while the area actually under potatoes is a great deal smaller.

One useful application of land evaluation is in evaluating different scenarios about possible future land-use changes. There may be powerful socio-economic pressure to increase a particular land use in an area, for example, but environmental constraints (particularly inherent limitations in the soil) clearly need to be taken into account before big decisions are made that could well have long-term consequences. In this sense, land evaluation has been used to evaluate three scenarios for a region in Greece based on forestry, agriculture and soil protection. This was part of a larger comprehensive approach to regional land-use planning based on land capability (its physical ability to produce vegetation) and sensitivity to soil erosion.

SOIL EROSION AND CONSERVATION

Sustainable development requires the careful management of soil resources and appropriate farming procedures and practices. Many of the world's more fertile soils are already extensively used in agriculture and forestry, and soil erosion is an increasing problem in many areas.

SOIL DEGRADATION

Degradation can often involve loss of soil fertility without necessarily physical removal. It is a particular problem in the drylands of the tropics and sub-tropics, where it is promoted by high precipitation intensities and high rates of decomposition of organic matter. Human factors that contribute to the problem include over-exploitation of the scattered dryland vegetation, over-grazing, over-cultivation

BOX 19.23 SOIL EROSION AND DEGRADATION

The expression 'soil erosion' is widely used in two senses, to mean either the degradation of soil (which reduces its quality and usefulness) or the physical removal of soil. The two have much in common, because similar factors are responsible for both changes, both cause serious problems for humans who rely on the eroded soils, and both cause major problems for ecosystem stability and continuity.

and collection of fuelwood. If soil degradation in the tropics continues over a period of time, it can lead to desertification (see p. 434) and the complete loss of livelihood for many pastoral peoples.

Many agricultural activities promote the degradation of soils. Nearly 40 per cent of the soils in the Canary Islands (off North Africa) are showing signs of degradation, caused by a combination of wind and water erosion and salinisation. The area's arid climate does not help, but natural damage to the soils is being aggravated by intensive agricultural practices, which include monoculture and the indiscriminate use of agrochemicals.

Remote-sensing technologies, including the Landsat satellite monitoring system, are proving extremely useful in pinpointing where soil degradation is occurring and in monitoring changes in the severity and extent of the degradation over time. Soils over nearly 52 per cent of a sample area of alluvial plain in Rajasthan, India, were found to be degraded by soil stripping, sheet erosion and gully development (see p. 218), and a further 8 per cent was degraded by salinisation (p. 403).

SOIL EROSION

Soil erosion involves the removal of topsoil. All soils are liable to be affected by natural erosion processes that operate on the ground surface. But under natural conditions, most soils are protected from serious erosion by the overlying vegetation. Rapid erosion normally occurs only when an ecosystem and its vegetation have been disturbed or damaged, perhaps by a major storm (see p. 312) or by natural fires (p. 583).

Two of the most widespread processes of soil erosion are rainsplash and sheet erosion (see p. 211). Rainsplash is particularly effective on bare, sloping surfaces. Average-sized raindrops fall at a speed of about 9 m s^{-1}, which is enough to detach soil particles and splash fine sands and silt up to a height of 60 cm and a distance of up to 1.5 m. Strictly speaking, splash redistributes soil particles rather than directly removing them, but on slopes this leads to a net downslope transfer, which can effectively strip the topsoil over a period of time. Sheet erosion is caused by surface runoff (overland flow), which occurs where rainwater is unable to drain into the soil (p. 211). This happens

BOX 19.24 SOILS AND NUCLEAR CONTAMINATION

Some soils are better able than others to reduce the medium-term problems arising from the fallout of nuclear material, such as happened after the 1986 Chernobyl explosion (see Box 8.23). Much depends on the efficiency with which a soil passes on contamination, which depends on how well it binds caesium. Particles of caesium can become trapped within the structure of clay minerals (such as illites), which make it difficult if not impossible for them to be taken up by plants and thus enter the food chain. Peaty soils contain fewer of these minerals than other soils, so caesium is more readily available on peat. In the UK, the evidence shows that peat bogs transfer radioactive caesium to plants, and then on to sheep and cows, far more efficiently than sandy and clay soils. Most of the peat soils are in Scotland and other upland areas, whereas sandy and clay soils are much more common in England and most of Europe. By late 1998, more than 47,000 sheep on twenty-three farms in Scotland were still subject to restrictions on their slaughter and movement because of contamination from Chernobyl, 12 years earlier.

Caesium-137, which has a half-life of 30 years, is the most important radionuclide left from the Chernobyl explosion, and around the nuclear plant an area roughly the size of Luxemburg has been declared off-limits for habitation or cultivation because of radioactive contamination. Tests have shown that mulching the contaminated soils helps to prevent the caesium getting into crops planted in the area. Full recovery of the soils is a long-term aim, but initial progress has been extremely slow.

BOX 19.25 SOIL FORMATION AND EROSION

The expression 'soil erosion' is usually used to describe the removal of topsoil as a result of human activities, particularly if the protective vegetation has been removed. This is effectively accelerated erosion, in which soil is removed at a much faster rate than natural processes of weathering and soil formation can replace it.

The natural rate of soil formation is probably about 8 mm per century, although soil formation under agriculture might be as high as 80 mm per century, because ploughing aerates the soil and increases the rate of leaching. As a result, soil develops on cultivated land at a rate of around 0.8 mm each year. Net soil erosion is the loss of topsoil above this rate of production. When this happens, the land declines in productivity and eventually becomes infertile.

either because rain is falling faster than infiltration can occur or because the soil surface has a hard, impenetrable crust formed by packing together of the soil particles.

Heavy rains can wash away the topsoil from bare surfaces, particularly where soils are sandy or loamy. Dense networks of rills are created by sheetflow across the soil surface, and these can grow into deep gullies as erosion continues (see p. 216). Particles of soil and rock that are removed from eroded areas are blown by the wind and transported along rivers, and are ultimately deposited elsewhere (burying crops, roads and other structures).

Many farmers in the Midwest USA lost their farms and livelihoods during the Dust Bowl of the 1930s (see Box 14.2), when vast expanses of topsoil – exposed by ploughing and reduced by aridity to a fine dust – were blown away. Much of the badly eroded land was returned to grass (the area was originally prairie grassland) after the drought ended.

Soil erosion rates are highest in areas with fine-grained soils, steep slopes and periodic intense rainfall. Probably the worst soil erosion problems in the world are encountered on the loess plateau in China, where soil is being blown and washed away at a record rate of more than 33,000 tonnes per square kilometre a year. Such high rates of erosion are causing serious problems of silting and flooding along the lower reaches of the Yellow River and

BOX 19.26 ESTIMATING RATES OF SOIL EROSION

Some soils are much more susceptible to erosion than others; differences reflect variations in basic soil-forming factors such as climate, topography and parent material. An estimate of the amount of soil that is likely to be lost each year can be made using the so-called universal soil loss equation. This predicts soil erosion from a number of contributory factors and has the general form:

$$A = R \times K \times L \times S \times C \times P$$

where:

- A is the amount of soil lost from a given field each year.
- R is the erosivity of the rainfall (based on amount and type of rainfall).
- K is the erosivity of the soil (liability of the particular soil to erosion).
- L is the length factor of the field (ratio of the length of the field to a standard field of 22.6 m).
- S is the slope factor (ratio of the soil lost to the amount lost from a field with a 9 per cent gradient).
- C is the crop management factor (ratio of soil loss to that from a field under cultivated bare fallow).
- P is the soil conservation practice factor (ratio of soil loss to that from a field where no care is taken to prevent erosion).

BOX 19.27 FACTORS AFFECTING SOIL EROSION

A number of factors promote soil erosion, including:

- deforestation
- over-grazing
- compaction (which destroys soil structure)
- farming on steep slopes without terraces
- the practice of leaving ground bare for long periods in winter.

BOX 19.28 SOIL EROSION IN INDIA

Surveys have established that in 1997 around 57 per cent of the land in India was affected by soil degradation, and that the area of critically eroded land had doubled since 1980. The total cost of soil erosion and depletion in India is estimated at between US$2.5 billion and US$6.5 billion a year (equivalent to 1–2 per cent of the country's gross domestic product). Various factors have contributed to the problem, including:

- *deforestation*: trees that protect the soil are regularly cleared for firewood or to create new farmland;
- *intensive farming*: which has rapidly depleted many soils of nutrients;
- *extensive irrigation*: which has raised the water table in many areas and increased soil salinity.

are inhibiting the agricultural and industrial development of the region.

While rates of soil erosion are highest in arid and semi-arid environments, soil erosion is nonetheless an emerging problem in most areas where modern agriculture is practised. In Britain, for example, problems of soil compaction and erosion have not yet been widespread, but concern is growing over the increasing incidence of soil erosion. It has been estimated that around 20 per cent of the world's cultivated topsoil was lost between 1950 and 1990, and this trend appears to be continuing. The worst-affected areas are steep hillsides, particularly in mountain regions where forests have been cleared.

SOIL CONSERVATION

While soil is an organic resource and thus at least theoretically renewable, in practice it takes so long to form that it should be treated as a non-renewable resource. As a result, rational approaches to the conservation of soil resources are required in areas where soil erosion is heavy or advanced (to prevent further losses) but also as a preventive strategy in areas where soil erosion is not currently taking place. Co-ordinated soil conservation programmes were first introduced in the USA during the Dust Bowl years of the 1930s (see p. 420), and since then they have become integral aspects of natural resource management in many countries.

The most logical approach to soil conservation is to maintain a protective ground cover of vegetation (either natural or cultivated) and minimise the amount of disturbance to the soil surface (for example by avoiding soil compaction from the use of farm machinery). If these are not possible, or fully effective, the best strategy is to try to control the two main processes involved in soil erosion – detachment of particles (by rainsplash or wind) and transport away (by wind or water).

One promising approach to the conservation of soil resources is to switch land use in many areas away from

BOX 19.29 CONTROLLING SOIL EROSION BY WATER

Soil erosion by running water can be tackled in a number of ways, including:

- *Maintaining a protective plant cover on vulnerable soils*: perhaps by planting crops in alternate strips (grass and grains).
- *Reducing the amount of time that bare soil is exposed*: by using mulch (such as leaving straw stubble lying on the surface after harvesting to cover the ground) or reintroducing crop rotations.
- *Contour ploughing*: a method of ploughing that follows the contour of the land, producing parallel furrows at right angles to the slope. It produces much less soil erosion than ploughing downslope, which creates channels along which surface water can collect, encouraging rill and gully development.
- *Terraces*: these intercept and stop the downslope movement of soil. This is a traditional and highly effective approach to soil conservation.
- *Netting*: a temporary form of erosion control by placing netting over bare soil surfaces after they have been seeded. The netting holds the soil in place until the plants grow and offer better protection. This technique is widely used to stabilise roadside verges and embankments.

BOX 19.30 CONTROLLING SOIL EROSION BY WIND

Widely adopted measures designed to tackle soil erosion by wind include:

- maintaining a plant cover to reduce wind speed (thus energy);
- contour ploughing;
- planting strips of row crops in intervals across a field of cereals;
- using trees and shrubs as shelter belts or windbreaks.

intensive cultivation and let natural processes of secondary succession (see p. 572) re-establish a protective plant cover. Such a strategy has been adopted in the USA under a programme called 'Breakthrough on soil erosion', which was designed to convert around 160,000 km^2 of eroding cropland (11 per cent of total cropland) to grassland or woodland between 1988 and 1991. Individual states – such as Minnesota – have implemented schemes to retire marginal land from crop production and set it aside for nature conservation purposes.

SUMMARY

Soils are the product of interactions between the main environmental systems, and they are vitally important to people for growing food in. The three main themes in this chapter are the nature and development of soils, soil classification and distribution, and managing soil resources. We saw how soils develop from parent material, and how their composition and properties reflect the interplay between minerals, climate and vegetation. Soil profiles reveal a great deal about the nature and effectiveness of soil-forming processes, and they also influence the way in which different soils can be used. A number of ways of classifying soils have been proposed, and the US Comprehensive Soil Classification System is widely used. Climate and vegetation exert powerful influences on the geographical distribution of soils. The principal resource management themes we encountered were approaches to land evaluation, processes and controls of soil erosion, and the problems and prospects of soil conservation. Soil is much more than dirt – it is the mainstay of most agricultural systems, a vital natural resource, and it is under threat from soil erosion and land management practices that are not sustainable.

WEBSITE

Links to relevant websites, a comprehensive bibliography, tools for teaching and learning, and downloadable images relevant to this chapter can be found at the website specially designed to accompany this book at
http://www.park-environment.com

FURTHER READING

Coleman, D.C. and D.A. Crossley (eds) (1996) *Fundamentals of soil ecology*. Academic Press, London. Summary of recent advances in understanding the physics, chemistry and biology of soil ecosystems.

Collins, D. (2000) *Contaminated land; managing legal liabilities*. Earthscan, London. Explores the practical, financial and legal issues surrounding the clean-up of contaminated land, with particular reference to the UK.

Davidson, D.A. (1992) *The evaluation of land resources*. Longman, Harlow. Describes the use of geographical information systems and modelling techniques in land evaluation in order to determine the most effective use of natural resources.

Ellis, S. and T. Mellor (1996) *Soils and environment*. Routledge, London. Study of the properties, processes and classification of soils, with an emphasis on environmental history and soil–human interactions.

Foth, H. (1991) *Fundamentals of soil science*. John Wiley & Sons, London. Describes and accounts for the major biological, chemical and mineralogical properties of soil.

Gerrard, J.G. (1992) *Soil geomorphology*. Chapman & Hall, London. A detailed exploration of the associations between soils and landforms.

Gerrard, J. (2000) *Fundamentals of soil.* Routledge, London. A comprehensive introduction to soils and the workings of soil systems.

Morgan, R.P.C. (1995) *Soil erosion and conservation.* Longman, Harlow. An up-to-date review of the principles and processes of soil erosion, and the theory and practice of soil conservation.

Ollier, C. and C. Pain (1995) *Regolith, soils and landforms.* John Wiley & Sons, London. A review of links between the formation and destruction of soils and the evolution of landforms.

Retallack, G.J. (1997) *A colour guide to paleosols.* John Wiley & Sons, London. An introduction to the study of palaeosols and an exploration of how burial changes soils.

Roth, H. (1991) *Fundamentals of soil science.* John Wiley & Sons, London. A comprehensive textbook on soils, their properties, formation and classification.

Rowell, D.L. (1994) *Soil science: methods and applications.* Longman, Harlow. A practical introduction to methods of assessing the properties of soil.

Schwab, G., D. Fangmeier and W. Elliott (1995) *Soil and water management systems.* John Wiley & Sons, London. Outlines the principles and practice of managing soil and water resources, including controlling wind and water erosion.

Part Six
Reflections

Chapter 20

Retrospect and Prospect

LEARNING OBJECTIVES

When you have finished studying this chapter, you should be able to

- *Appreciate how our current understanding of the environment is shaped by our values and attitudes and by experience of coping with environmental problems, and how concerns for the future of the environment are influenced by uncertainty and complexity*
- *Outline how the 'environmental debate' has evolved over the last 20 years*
- *Distinguish between the views of the 'optimists' and the 'pessimists' and explain why each group claims to be right*
- *Define the environmental priorities that need to be tackled as a matter of some urgency*
- *Explain how new ideas about scientific knowledge and expertise are challenging conventional approaches and perspectives*
- *Describe and account for new approaches such as risk assessment and adaptive management*
- *Understand how and why never enough is known about environmental problems to be absolutely certain about the best ways of tackling them, but also why waiting for further research to be completed is not always a sensible or realistic option*

Lester Brown, writing in the Worldwatch Institute *State of the world 2000* report, stressed that:

> caught up in the growth of the Internet, we seem to have lost sight of the Earth's deteriorating health. It would be a mistake to confuse the vibrancy of the virtual world with the increasingly troubled state of the real world.

This final chapter is a convenient point at which to reflect on where we stand at the opening of a new millennium, charged with the task of looking after the planet and its environmental systems and services. It is also a good vantage point from which to look forward, to see what the future might hold in store. In doing so, we can bring together many of the threads that run through previous chapters and offer an overview of the key themes (Box 20.1).

BACK TO THE FUTURE

Human culture and society have developed in amazing ways in recent decades, and these changes must be appreciated in their historical context. Three thousand years ago, there were less than 100 million people on the planet, populations were scattered, technology was limited, and the overall environmental impacts of human activities were limited and isolated. The situation is very different today, with more than 6 billion people, much wider distribution of population and heavy concentrations in large cities, much more advanced technologies, and significantly more and worse environmental impacts. Advances in technology have allowed people to exploit more and more environmental resources, and in doing so generate wastes and pollutants that damage environmental systems. The pace of change has accelerated a great deal since the Industrial Revolution, and the scale and complexity of human impacts show no sign of slowing down. Little wonder, therefore, that many experts are voicing concerns about mounting environmental damage, shrinking environmental resources,

BOX 20.1 CRITICAL ASPECTS OF THE RELATIONSHIP BETWEEN PEOPLE AND THE ENVIRONMENT

- the interdependence of major environmental and socio-economic systems;
- the needs and rights of future generations;
- the values of diversity;
- the existence and importance of limits to growth;
- the need for precautionary approaches.

declining environmental quality and the likelihood of irreversible environmental change.

CONTINUITY AND CHANGE

One theme that emerges throughout this book is the issue of continuity – many environmental problems are not new, they have been developing for a long time. Many examples illustrate this theme, including land-use changes such as forest clearance (see p. 582) and urban development (p. 47), species extinction (p. 536) and adaptation (p. 529), and pollution of air (p. 241) and water (p. 388) resources.

CHANGE

While many problems are not in themselves new, the pace of change has quickened, and the pattern of change has spread for many environmental issues since the start of the twentieth century. This is certainly true for population, most forms of pollution and loss of biodiversity (see p. 524), and problems such as soil erosion (p. 609), mass movement (p. 213) and exploitation of ocean resources (p. 485). The problem of global warming caused by greenhouse gases (p. 255) fits into this category too. Alongside this, new environmental problems have emerged as a result of the introduction of new technologies – such as declining ozone concentrations in the stratosphere (related to the release of CFCs; p. 249), increased use of genetically modified crops and organisms (p. 532), and new forms of pollution such as synthetic pesticides and nuclear waste products (p. 137).

PROGRESS

A further important dimension of change is the increasing effort being devoted to tackling environmental problems and trying to bring them under control. At the start of the twenty-first century, for example, great efforts are being made to tackle environmental issues such as freshwater pollution (see p. 392), air pollution (p. 241), over-use of the oceans (p. 485), loss of biodiversity (p. 524) and urban expansion (p. 47). The biggest challenges centre on how best to cope with global warming associated with greenhouse gas emissions (p. 255), because if this mega-problem cannot be brought under control within a reasonable time, there are likely to be much greater and more complex environmental problems ahead for all countries.

Identifying environmental problems is one thing, but tackling them effectively is another. Perhaps inevitably, with increased awareness of the nature, scale and distribution of environmental problems has come a better awareness of how serious many of them are (Box 20.2). The *State of the world 2000* report, for example, shows that:

BOX 20.2 HUMAN IMPACTS ON ENVIRONMENTAL SYSTEMS: THE SITUATION IN 2000

- Up to half of the land surface has already been transformed or degraded by land-use changes, including agriculture and urban development.
- More than half of the accessible surface fresh water is already being exploited.
- Atmospheric concentrations of CO_2 have increased by 30 per cent as a result of air pollution.
- More than half of terrestrial nitrogen fixation is caused by human activity (including the use of nitrogen fertilisers, the planting of nitrogen-fixing crops and the release of reactive nitrogen from fossil fuels into the atmosphere.
- About 20 per cent of the plant species found on continents have been introduced to their present locations by humans; on many islands, the figure is as high as 50 per cent.
- About 20 per cent of bird species have become extinct over the last 200 years, mostly because of human activity.
- About 8 per cent of the primary productivity of the oceans is used for food.
- Over 20 per cent of marine fisheries are over-exploited or depleted, and an additional 40 per cent are at the limit of exploitation.

- Every person alive today has in their bodies around 500 synthetic chemicals that had not been invented in 1950.
- Half of the world's people are medically malnourished, suffering from either obesity or from diets with inadequate calories, vitamins or minerals.
- The 3 parts per million increase in the atmospheric concentration of carbon dioxide (CO_2) in 1998 was the largest ever recorded.
- Spreading water shortages threaten to reduce the global food supply by more than 10 per cent.

CAUSES AND COMPLICATIONS

There is little hope of finding sustainable ways forward if we do not understand what causes particular environmental problems. It is comforting that a great deal of effort is being invested in scientific research designed to establish cause–effect links (see p. 78), clarify the nature and importance of critical environmental thresholds (p. 83) and lags (p. 83), and identify how and why adjustments and responses (p. 79) occur. However, despite real progress in recent years, many uncertainties remain, and there are still a great many important gaps in understanding.

A great deal of research is also seeking to provide a better understanding of underlying socio-economic factors, including:

- levels and patterns of consumption (which continue to be unsustainable in many countries);
- demographic and socio-economic factors that promote rapid population growth (which in many places puts available environmental resources under constant pressure);
- questions about why environmental issues are often regarded as a reasonably low priority by national and regional governments;
- the political factors that give rise to international disputes and wars, which also cause wholesale environmental stress and damage.

Again, there are many more questions than answers, but at least at the start of the new millennium there is a much clearer understanding of the need to identify and tackle the root causes of environmental problems, not just the symptoms. Many of these root causes reflect human values, attitudes and behaviour, so it is right that people come under scrutiny as well as the environment.

PEOPLE–ENVIRONMENT RELATIONSHIP

A number of attempts have been made to measure the relationships between population, consumption and environmental impacts. One widely discussed approach was proposed in the 1970s by ecologist Paul Ehrlich, who defined the population–environment relationship in terms of a formula:

$$I = P \times A \times T$$

in which I is the environmental impact (for example, pollution), P is population size, A is affluence (usually expressed as average gross domestic product (GDP – see Box 20.3) per capita) and T is technology (a measure of efficiency, for example of energy use).

Although the $I = PAT$ formula is a useful way of studying the relationship between factors that govern environmental change, critics have pointed to two important weaknesses in this sort of approach:

1. The factors contributing to any particular impact can vary a great deal, depending on the environmental impact in question. For example, different factors contribute to depletion of the ozone layer and to loss of biodiversity.
2. The equation suggests that the three factors (P, A and T) operate independently, while in reality they may well interact with each other.

PERSPECTIVES AND PROGNOSES

There has been an ongoing debate since the 1960s over whether economic growth and human activity are causing irreversible damage to the natural environment. There are strongly held views on both sides of the debate, occupied by the 'pessimists' and the 'optimists'.

PESSIMISTS

The pessimist view is founded on the writings of the economist Thomas Malthus, who in 1798 argued that the potential population size is limited by the amount of cropland (and thus food) available for human consumption. He assumed – based on his observations of eighteenth-century English society – that if population growth continues unchecked, then population would outstrip the food available and this would cause widespread famine and death. He also described a feedback mechanism – when the population became too large for the available food supply,

BOX 20.3 PERFORMANCE INDICATORS OF GROWTH AND PROGRESS

Indicators provide useful means of measuring performance or achievement, although until recently the most widely used performance indicators of 'growth' have been essentially economic indicators. The conventional method of measuring economic growth is gross domestic product (GDP).

Gross domestic product: This index measures level of economic activity but without indicating whether that activity increases or decreases human well-being. GDP, by default, assumes that all economic activity is positive. Thus, for example, forest clearance, strip mining, over-fishing and air pollution are counted as positive indicators because they generate increases in GDP.

Economic indicators like GDP usually drive national policies – if GDP is rising, most governments take that to be a good sign, so they promote more of the same. But such policies are doomed to fail if the indicator does not properly measure what is going on. Other performance indicators have been proposed that place more emphasis on people and environment. They include:

Genuine progress indicator: This measure of economic and social well-being measures overall economic activity, but it deducts for resource depletion and declines in social health. Thus, for example, it adds for housework, child rearing and volunteer work, but it deducts for loss of leisure time, family breakdown and pollution.

Index of social health: Although this measure does not take into account environmental impacts, it does consider sixteen social indicators – including infant mortality, child abuse, poverty among children and the elderly, drug abuse, health insurance coverage, unemployment, average weekly earnings, and the gap between rich and poor.

International human index of suffering (HIS): This index can be used to show genuine changes in human well-being. It incorporates a number of key variables, including life expectancy, daily calorie intake, clean drinking water, infant immunisation, secondary school enrolment, GDP per capita, rate of inflation, communications technology, political freedom and civil rights.

increased mortality would reduce the population to the level that could be sustained by the amount of food produced.

Malthus's ideas shaped the so-called *neo-Malthusian* view of the relationship between population, economic growth and resources. This viewpoint grew in popularity, particularly between the 1940s and the 1960s, when population growth and economic development were particularly strong in many countries. Many experts concluded that rapid population growth would eventually be checked by some absolute limit on resources (such as food, energy or water). There was also mounting evidence that continued population growth and the environmental stresses associated with economic development could cause irreversible damage to the environmental systems that support life. This school of thinking was widely promoted through books such as *The population bomb* by Paul Ehrlich (1968) and *The limits to growth* by Donella Meadows and colleagues (1972) (see Figures 20.1 and 20.2). Pessimists believe that essential stocks of renewable resources are already in decline, and that technology alone cannot reverse these declines. This group focuses on sustainability.

Lester Brown (president of the Worldwatch Institute) is a leading promoter of the anti-growth lobby (Box 20.4). He argues that in many parts of the world forests, groundwater, pasture land and fisheries are being used faster than they can be restocked; nearly a quarter of the world's population live in extreme poverty, despite half a century of relentless economic growth; and – despite the Green Revolution – the security of global food supplies is worryingly low.

OPTIMISTS

The neo-Malthusian view has not been universally accepted, and the 'optimists' (sometimes called the 'cornu-

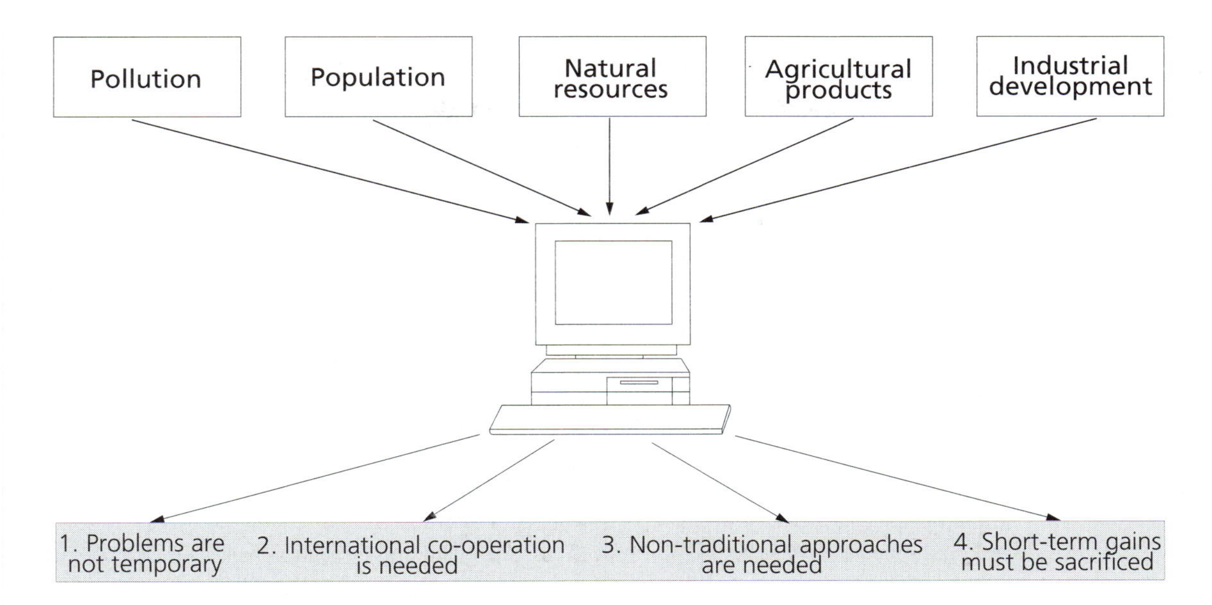

Figure 20.1 *Basis of the Limits to Growth computer simulation model. The dynamic computer model simulated the complex interrelationships between a range of important variables (shown in the upper boxes). Successful management of these relationships must recognise some important constraints (shown in the lower boxes).*

Source: Figure 1.7 in Cunningham, W.P. and B.W. Saigo (1992) Environmental science: a global concern. Wm. C. Brown Publishers, Dubuque, Iowa.

copians') have rejected it flatly. They argue that population growth is a positive influence on economic development, and they propose that human ingenuity and creativity – particularly through the development of technology – would overcome any potential environmental constraints on development. One early leading 'optimist' was Ester Boserup, who in the 1960s and 1970s argued that the need for more food can stimulate adaptive responses such as the adoption of better farming techniques or the sharing of high-yield plant varieties. Optimists believe that natural resources are not finite in an economic sense; they argue that any scarcity is temporary until price increases encourage efforts to discover new sources of raw materials and new technologies to extract and process them. This group is pro-growth and favours free trade.

Julian Simon (former economist and writer) was a leading optimist and promoter of the pro-growth lobby (Box 20.5). He argued that consumption of resources is good because it stimulates their production. His solution to the problem of carrying capacity is thus to concentrate on economic development, production and consumption rather than to try to stabilise population or lower consumption. Simon used indicators such as increased life expectancies, global economic growth and lower price trends for energy to support his argument.

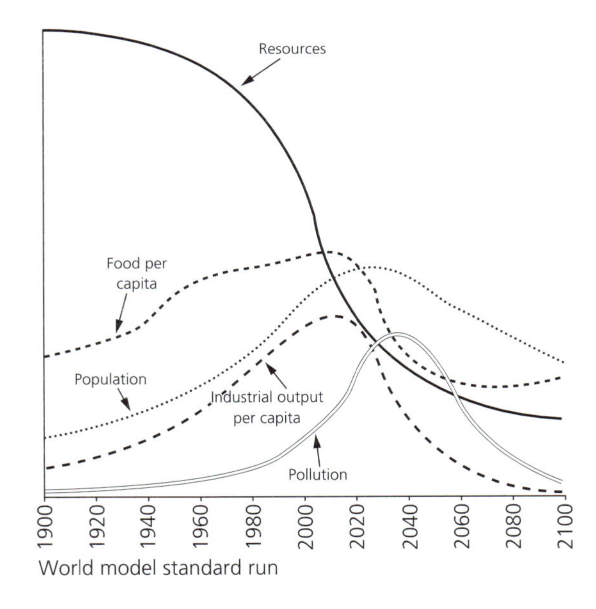

Figure 20.2 *Limits to Growth predictions of global population, resources and pollution. A number of different simulations were made by the MIT team using the dynamic computer model (Figure 20.1), based on different assumptions. The one summarised here, referred to as the standard run, assumed that declining resources and increasing degradation of the environment would eventually result in declines in human aspects of the model.*

Source: Figure 1.8a in Cunningham, W.P. and B.W. Saigo (1992) Environmental science: a global concern. Wm. C. Brown Publishers, Dubuque, Iowa.

BOX 20.4 LESTER BROWN'S PESSIMIST'S VIEW OF THE ENVIRONMENTAL CRISIS

Brown argues that traditional market-based solutions to environmental problems are inappropriate, because:

- Low energy costs do not include 'externalities' (external side-effects such as pollution or environmental damage).
- Commodity prices are kept artificially low by government subsidies for large energy, forestry, mining and agribusiness interests.

In the 1998 *State of the world* report, Brown wrote:

> While economic indicators such as investment, production and trade are consistently positive, the key environmental indicators are increasingly negative. Forests are shrinking, water tables are falling, soils are eroding, wetlands are disappearing, fisheries are collapsing, rangelands are deteriorating, rivers are running dry, temperatures are rising, coral reefs are dying, and plant and animal species are disappearing. The global economy as now structured cannot continue to expand much longer if the ecosystem on which it depends continues to deteriorate at the current rate.

BOX 20.5 JULIAN SIMON'S OPTIMIST'S VIEW OF THE ENVIRONMENTAL CRISIS

In Simon's view, spelled out in his 1977 book *The ultimate resource* (and *The ultimate resource 2*):

- Per capita energy costs have fallen over the past 50 years because increasing demand has promoted greater production.
- Resources are not really in decline, otherwise more commodity prices would rise more than we have seen.
- Concerns about rapid rates of species extinction are greatly over-stated.
- Global warming and ozone depletion are less of a problem than many people think they are.
- Population growth is a benefit, because more people means that more human intelligence and creativity can be focused on solving problems.

DEBATE

Both sides – the pessimists and the optimists – agree on many of the basic facts. For example, neither side disputes that since 1950 global

- population has more than doubled;
- economic output has increased almost sixfold;
- water use and demand for grain, firewood, beef and mutton have tripled; and
- use of fossil fuels has increased nearly fourfold.

Where the two sides disagree is over the causes and consequences of these trends, and thus on how best to respond to them.

Since the early 1990s, the debate between 'optimists' and 'pessimists' has lost intensity and changed direction due to the emergence of the sustainable development debate (see p. 26). This has sought common ground in the middle of the debate and has explored practical ways forward. It holds great promise for the future.

IMPROVEMENTS AND PROGRESS

Despite the rhetoric about environmental crises and the need for radical changes in human values and aspirations, there is much to celebrate. Recent years have seen continued success in many areas, and many current initiatives promise to improve things even further.

Major successes identified in the UNEP report *Global environment outlook 2000* include:

- increasing public concern over environmental issues;
- voluntary action taken by many of the world's major industries to reduce resource use and eliminate waste;
- major successes by governments in developed regions in reducing levels of air pollution in many major cities;
- the halt and reversal of deforestation in parts of Europe and North America;
- local Agenda 21 (see p. 16) initiatives, which promote sustainable development policies;
- the success of the Montreal Protocol (see p. 251) in preventing the ozone depletion problem from getting worse;

- the adoption of steps to tackle the issue of global climate change under the UN Framework Convention on Climate Change (1992) and its Kyoto Protocol (see p. 15).

Governments are at last seeking to meet the challenges posed by the Rio Earth Summit and the rise of sustainable development thinking, for example by adopting more and better policy responses to environmental problems. Initiatives at all levels in society – from the local, through regional, to national and international – are also starting to bear fruit. It is still early days, and there are many more and bigger challenges ahead. But it is a start, and a useful one at that. Tackling today's environmental problems is like trying to turn a huge supertanker at sea – it has to be done slowly but deliberately, with forward planning and a clear resolve, and a good understanding of the benefits to be enjoyed in the future by acting now.

Industry and technology are also changing in response to environmental challenges. For example, the energy industry introduced many improvements during the 1990s, including cleaner and more sustainable methods of producing energy (see, for example, p. 288), and more efficient ways of using energy.

FUTURE PROSPECTS

NEW CHALLENGES

A great deal of progress in protecting the natural environment has been made since the 1970s, yet the issue remains of major concern both globally and nationally. Many traditional environmental problems (such as common forms of industrial pollution) have been tackled in places, but few have been successfully tackled at the global scale, and new problems continue to emerge.

At the Rio Earth Summit in 1992, world leaders agreed that current patterns of economic and social development are not sustainable and that major changes are required. Throughout the 1990s, many initiatives were developed and introduced – by governments, business and other agencies – to try to make development more sustainable, and some are already producing positive results.

However, in tandem with these searches for sustainable ways forward, there has also been growing awareness that environmental protection is a bigger and more complex challenge than was previously thought. There are various reasons for this (Box 20.6).

BOX 20.6 SOURCES OF COMPLEXITY IN SUSTAINABLE DEVELOPMENT DECISION MAKING

Decision making within sustainable development is inherently complex for various reasons, including:

- Fundamental change: achieving sustainable development will require significant changes in the way economic activities are organised and lifestyles are led. Both producers and consumers play key roles in reducing this environmental pressure.
- Integration: environmental policies will only be effective if they are fully embedded within all appropriate policy areas. They cannot be regarded as free-standing and separate from broader social and economic policies. Environmental policies will succeed only if they also promise and deliver social benefits (for example, reducing poverty) and if they are compatible with increased business competitiveness.
- Spatial scale: globalisation and increasing international co-operation mean that political decision making has to take into account issues and interests at a variety of spatial scales, including the global scale.
- Interests: decision making is increasingly having to include consideration of wider sets of interests than purely political ones, including business and society.
- Uncertainty: uncertainty is central to environmental problem solving. Decisions must embrace conflicting opinions about environmental problems, as well as different notions of risk posed to different sectors of society, the need to build and maintain public trust and confidence, and the engagement of people in solutions (ownership).

PRIORITY AREAS

While the successes are to be welcomed, hope for the future is tempered by awareness that many existing environmental problems continue to get worse, and new environmental problems continue to emerge, and there is an appreciation of the complexities involved in tackling them all (Box 20.6).

The signs are that we are winning individual environmental battles but not the overall environmental war.

BOX 20.7 EXISTING ENVIRONMENTAL EMERGENCIES, ACCORDING TO UNEP'S GLOBAL ENVIRONMENT OUTLOOK 2000

- The world water cycle seems unlikely to be able to cope with the demands that will be made of it in the coming decades, and widespread serious water shortages are expected.
- Land degradation has reduced fertility and agricultural potential, and soil erosion has more than offset advances made through expanding agricultural areas and increasing productivity.
- Tropical forest destruction has gone far too far to prevent irreversible damage.
- Many species have already been lost or condemned to extinction.
- Many ocean fisheries have been seriously over-exploited, and their recovery will be slow.
- More than half the world's coral reefs are threatened by human activities.
- Urban air pollution problems are reaching crisis dimensions in many mega-cities in developing countries.
- It may be too late to prevent global warming as a result of increased greenhouse gas emissions.

UNEP's *Global environment outlook 2000* points out that time is running out in the search for sustainable solutions to many of the most serious environmental problems, and that full-scale emergencies exist on a number of issues (Box 20.7).

The UNEP report highlights four key areas that need to be tackled as a matter of priority (Box 20.8). As the report puts it:

> the challenge for policy-makers in the next century will be to devise approaches that encourage a more efficient, fair and responsible use of natural resources by the production sectors of the economy, that encourage consumers to support and demand such changes, and that will lead to a more equitable use of resources by the entire world population.

Lester Brown, writing in the Worldwatch Institute *State of the world 2000* report, stresses that:

> Two big challenges in this new century are to stabilise climate and population. If we cannot stabilise both, there is not an ecosystem on Earth that we can save. Everything will change. If we can stabilise population and climate, other environmental problems will be much more manageable. . . . The scale and urgency of the challenges facing us in this century are unprecedented. We cannot overestimate the urgency of stabilising the relationship between ourselves, now 6 billion in number, and the natural systems on which we depend. If we continue the irreversible destruction of these systems, our grandchildren will never forgive us.

BOX 20.8 PRIORITY AREAS TO BE TACKLED, ACCORDING TO UNEP'S GLOBAL ENVIRONMENT OUTLOOK 2000

Filling the knowledge gaps: There are many gaps in current understanding of the interactions and impacts of global and regional processes, and in monitoring the impacts of existing environmental policies and expenditure.

Tackling the root causes: Many root causes are relatively unaffected by strictly environmental policies, which tackle symptoms. For example, resource use is a major factor behind much environmental change, but tackling resource use will require changes on many fronts, including measures to reduce population growth, change patterns of consumption, increase efficiency of resource use and make structural changes to the economy.

Adopting an integrated approach: This will take various forms, including integrating environmental thinking into mainstream decision making, broadening the framework of decision making away from single-interest issues and single-sector issues towards broader social goals and objectives, and better integration of international initiatives and activities.

Mobilising action: The need to increase co-operation between all of the groups involved in environmental decision making (including individuals, NGOs, industry, local and national governments, and international organisations) is widely recognised.

DEALING WITH UNCERTAINTY

Recent research on environmental decision making has revealed that there are rarely easy choices to be made, because options and risks are usually surrounded by uncertainty. This new and emerging awareness of environmental uncertainty has shown the importance of three key themes:

1. how people attach values to the environment and its services and functions;
2. how we deal with uncertainty and complexity in environmental systems;
3. what role and credibility are given to scientific knowledge and expertise in environmental decision making.

VALUES

Sound decision making should be rational, and traditional approaches have attempted to ensure this by adopting quantitative approaches. Cost–benefit analysis (Box 20.9), which attaches monetary values to things, is a classic example. In recent years, questions have been raised about how best to include issues such as natural resources, which have no specific price tag, in such analyses. How much is clean air worth, for example? Economists have tried to answer such questions by proposing exchange values for environmental goods and services, but these are just proxy (surrogate) values because environmental goods and services are not traded in the marketplace, and this is how monetary value is usually determined.

Since the early 1990s, numerous attempts have been made to devise ways of expressing the value of ecological and environmental factors, which would be helpful in making decisions about rational and sustainable resource use. Such initiatives are as yet in their infancy, but they will inevitably become more important in the future as decision makers seek to evaluate what are often hidden costs and benefits of environmental factors in policy decisions. Recognition of the need to include the value of natural resources in national accounting systems provides another reason for making progress in this field.

Awareness of the benefits of devising more inclusive valuation techniques is growing. It is now widely recognised that, while economic valuation can be an important tool in the policy-making process, those who use it must be aware of its limitations and accept that it does not and cannot establish 'true' and 'objective' prices.

BOX 20.9 COST–BENEFIT ANALYSIS

Cost–benefit analysis (CBA) is a technique that is widely used to evaluate proposed projects in terms of social outcomes as well as the more normal economic outcomes (profit and loss). A CBA study begins with an assessment of likely costs, benefits and drawbacks of the proposed project, including externalities (such as generation of pollution). Financial values are assigned to each cost and benefit, and the decision whether or not to proceed with the project is made based on the comparison of anticipated costs and benefits.

Critics of the approach point out that CBA cannot properly assign values to intangible things (such as aesthetics), and they question the wisdom of evaluating everything in financial terms only. The CBA approach assumes that everything has a monetary value, and that this value is more important than any ecological, health or environmental issues involved, which have not been costed. It also has to take into account future conditions, which it normally does by discounting future value (for example, by assuming that values fall by 5 per cent a year over the next 20 years of a project). Critics also stress that it cannot include political judgements about what is acceptable or not at a particular point in time, or moral judgements about what is right and wrong (and who should decide these issues). CBA as a technique does not distinguish between who may or should benefit from a proposed scheme, and who may or should pay the costs.

UNCERTAINTY AND COMPLEXITY

A central theme of this book is that many gaps and uncertainties remain in our understanding of how natural environmental processes operate, and of cause-and-effect links within and between environmental systems. The importance of these uncertainties, and the need to recognise them and to reduce them as much and as quickly as possible, becomes particularly clear when we try to tackle complex issues such as global warming (see p. 261) and the impacts of new technologies such as genetic engineering (see p. 532).

Uncertainty often reflects lack of research – we have simply not studied a particular process or phenomenon enough to fully understand how it works or what it does. There are many other sources of uncertainty too, including indeterminacy (in essence, the fact that some things can never be established, if only because the very act of trying to measure them changes what is being measured).

Unpredictability is a consequence of uncertainty – if we do not know how something works, we cannot predict how it will or may respond to other factors. While we now have a much better understanding of how environmental systems operate than we did several decades ago, we also know that many of them can and do include unpredictable changes. These arise when thresholds are crossed (see p. 83), when different forms of adjustment are involved (see p. 79) and when time lags occur (see p. 83). Unpredictability also reflects complexity, in which there are far more interactions between variables than we can untangle. Lovelock's Gaia model of the Earth as a super-organism (see p. 119) illustrates this point clearly, as does the problem of fully understanding the El Niño phenomenon (see p. 482).

Without doubt the most critical area where uncertainty is still too high is global climate change (see p. 261), because the stakes are so high. Despite huge investment in research since the early 1990s, big improvements in data gathering and analysis, and significant advances in understanding the climate system, there are still important gaps in our knowledge and understanding. For example, we still have quite a limited understanding of the interactions between factors that create and change climate, such as atmosphere–ocean interactions, and the role of atmospheric chemistry (particularly air pollutants). Given these uncertainties, it is difficult to project how a future change in greenhouse gas emissions might affect atmospheric conditions. The further ahead we try to predict or project, the greater the uncertainty.

SCIENTIFIC KNOWLEDGE AND EXPERTISE

One hallmark of environmental studies since the early 1990s is a growing awareness, drawn from research in the social sciences, that knowledge is both conditional and situated. It is conditional in the sense that understanding grows over time, so new breakthroughs in understanding build on previous understanding and are often inspired or driven by the need to explain things that earlier or existing models cannot provide satisfactory explanations for. It is situated in the sense that all knowledge and understanding are rooted in particular times and particular places, and usually in the work of particular people.

Since the Renaissance, science has been seen as a rational and value-free area of enquiry. The view is that – in scientific research – data are gathered and analysed, and interpretations made, with few likely sources of error or ambiguity, in a process that is balanced and objective. From this perspective, scientists are seen as neutral experts whose views and interpretations are accepted without question by the general public. The Intergovernmental Panel on Climate Change (see p. 256), for example, is made up of scientists drawn from different countries and different disciplines, who offer expert views and feed scientific knowledge into global negotiations. Their views and interpretations are based on an agreed consensus about research findings and research methodology (they regard the best way forward as including climate change models and economic cost–benefit analysis, for example).

But the view that scientists are neutral experts is now being challenged, for various reasons (Box 20.10). The emergence of participatory approaches in environmental decision making has promoted the adoption of approaches that can combine lay (public) and specialist (scientific) knowledge and expertise. This is driven partly by a desire to build and maintain public trust in decision making, but it is also part of a much bigger debate about who knows best and who is best placed to make good decisions about the environment.

It is now becoming apparent that there are different forms of 'expertise' in addition to 'expert' science, and that these have much to offer. In developing countries, these are known as 'indigenous knowledges' (plural because there is no single way of looking at things or interpreting things), and in developed countries they are known as 'lay expertise' and 'citizen science'. Such local knowledges are embedded in cultural assumptions and socio-political values that can be very different from 'expert' science, and it is important to recognise them and include them in decision making. Recent policy initiatives on tropical rainforests (see p. 588), for example, take into account the interests and understanding of local forest farmers and forest product users as well as the expert judgements of scientists.

NEW APPROACHES

The evidence – summarised in reports such as UNEP's *Global environment outlook 2000* and the Worldwatch Institute's *State of the world 2000* report – shows that the global system of environmental management is moving in the right direction, but much too slowly and in much too patchy a way. Many countries have introduced environmental laws, set up environmental institutions and adopted direct regulation (by so-called command-and-control policies). Few have yet adopted multimedia strategies, designed

BOX 20.10 CHALLENGES TO THE TRADITIONAL VIEW OF SCIENTIFIC EXPERTISE

The traditional view of science as objective and rational is being challenged on various grounds, including:

- The question of how to take into account broader issues such as ethical questions, social interests and concerns, and political factors.
- Recent accidents and incidents which have called into question the dominance of technical expertise. The 'received wisdom' that the specialist scientist knows best has been challenged repeatedly in recent years (a good example is the ongoing public debate about the wisdom of growing genetically modified crops).
- Mounting evidence that the general public has many different views and understandings about the environment, and these are not always fully sympathetic to the view of so-called experts.
- Increasing understanding that people's knowledge and understanding of complex issues like the environment is based on their direct experience and is embedded in their social and political relationships. Understanding is thus described as 'socially contingent' and socially constructed.
- Awareness that scientific approaches and results often reflect value judgements and underlying assumptions about society, and that these assumptions reflect the particular positions, locations and interests of expert institutions. Examples include value judgements over what levels of health risk can be regarded as acceptable, or over the relative weight that should be attached to local or global effects of air pollution.
- Experience that shows the value in environmental decision making of listening to public concern and of taking into account lay understanding, which can help to 'frame' environmental issues and define the context in which they are treated.
- Growing awareness of the need to build and maintain public trust and confidence in government decision making and policy formulation, which in turn requires trust and confidence in the scientific advice given to government and in the scientists who give that advice.
- Realisation that sound and socially acceptable environmental decisions can be made on the basis of (traditional) formal science coupled with 'citizen science' or 'indigenous technical knowledge'.

BOX 20.11 REQUIREMENTS OF AN INTEGRATED APPROACH TO THE MANAGEMENT OF LAND AND WATER RESOURCES

An integrated approach would require:

- making full use of economic instruments that treat land and water as scarce economic resources that are part of the Earth's natural capital;
- co-ordinating the management of land and water resources as closely as possible;
- establishing secure land and water property rights where these do not exist;
- reorganising land and water management policies at a river basin level;
- introducing the concept of shared and equitable water use to resource allocation strategies;
- reformulating regional and national agricultural and food security strategies to bring them into line with the principles of sustainable development;
- providing people with alternatives to the use of marginal land;
- reducing water wastage in urban areas.

BOX 20.12 ENVIRONMENTAL IMPACT ASSESSMENT

Environmental impact assessment (EIA) refers to a range of evaluation techniques that assess the possible and likely environmental impacts of particular projects, with an emphasis on minimising irreversible damage to environmental systems and resources. EIA was first introduced in the USA in 1969 and has since been widely applied. It is being adopted in one form or another in an increasing number of countries as a basis for making informed and rational judgements about what sorts of development are environmentally acceptable. EIA is also being adopted in some of the Eastern European countries with damaged environmental legacies. Developments of basic EIA techniques include social project appraisal and strategic environmental assessment (SEA). The evidence from using SEA in a number of countries suggests that it might provide a framework for implementing the objectives of sustainability.

to tackle environmental damage (such as pollution) to air, water and land resources simultaneously; in most countries, environmental policies are still based on sectoral approaches (for example by tackling air pollution separately from land or water pollution).

Two important areas in which many countries are making progress are in adopting more comprehensive approaches and in moving away from heavy reliance on direct regulation.

COMPREHENSIVE APPROACHES

With better understanding of the nature and causes of environmental problems comes a sounder awareness of the need for comprehensive, integrated policy making. Environmental issues are all multidimensional and complex, and recent initiatives have reflected this complexity by seeking to move beyond the piecemeal approaches of the past and seek comprehensive solutions that tackle root causes as well as environmental symptoms, and that tackle issues together rather than separately. For example, issues such as deforestation and land degradation have usually

BOX 2.13 SOME PRINCIPLES AND APPROACHES COMMONLY ADOPTED IN PRO-ACTIVE ENVIRONMENTAL DECISION MAKING

Polluter pays principle: based on the idea that the polluter is primarily responsible for controlling pollution and for cleaning up environmental damage caused by that pollution.

Integrated approach: an approach in which the solution to environmental problems is tackled in a properly planned and integrated way, not in a piecemeal way.

Precautionary principle: given that most environmental problems are complex and involve a large amount of uncertainty, policies should reflect this and adopt a precautionary approach, which always errs on the side of safety and minimises risk taking. For example, the principle implies that the burden of proof of safety should be on those who wish to introduce a new chemical into the environment, not on those who claim to have been injured by it.

Pollution prevention: potential problems should be anticipated in environmental policies, and preventive measures should be taken before problems occur.

Proximity principle: the requirement to treat wastes close to where they arise (ideally within the boundary of the plant or community in which they are generated), to avoid exporting them to another place.

Best available techniques not entailing excessive costs (BATNEEC): an approach to pollution control (previously adopted in the UK) based on adopting the most effective techniques for an operation at the appropriate scale that are commercially available, and where the benefits gained are greater than the costs of obtaining them.

Best practicable environmental option (BPEO): an approach to pollution control (adopted in the UK) that seeks to establish the option that causes the least damage to the environment at an acceptable cost, taking into account the total pollution from a process and the technical possibilities for dealing with it.

been dealt with separately in the past, but new approaches seek to deal with them together while also trying to meet the needs and aspirations of the people most directly concerned.

The UNEP report *Global environment outlook 2000* emphasises the fundamental links between land issues and water management, and it suggests a holistic approach to the management of water and food (Box 20.11).

Comprehensive approaches to environmental management tend to be more pro-active (anticipating problems before they arise and taking steps to minimise impacts on human health and environmental systems). More traditional approaches have tended to be reactive (responding only when evidence of a problem emerges and the need to do something about it becomes apparent). Pro-active responses are greatly aided by strategies such as environmental planning and techniques such as environmental impact assessment (Box 20.12). Box 20.13 summarises some of the principles and approaches that are commonly used in pro-active environmental decision making.

DEREGULATION

Although most countries still rely heavily on institutions and regulations, a growing number are moving away from direct regulation towards deregulation. This can take various forms, including:

- making more use of economic instruments and subsidy reform;
- relying more on voluntary action by the private sector;
- inviting more participation in decision making from the public and NGOs (non-governmental organisations).

The 'people' side of the people–environment relationship is now being taken more seriously than ever before. Many of the more fruitful initiatives and developments of the 1990s placed an emphasis on including and empowering people and on listening to their views and ideas, taking their interests and concerns seriously, and seeking to meet their needs. This trend is manifest in various ways (Box 20.14).

DECISION MAKING

One of the most important new developments since the early 1990s has been the search for new and more appropriate ways of making decisions regarding the environment. A range of formalised decision-making tools and approaches – including risk assessment (Box 20.15) and cost–benefit

BOX 20.14 EMPOWERING AND INCLUDING PEOPLE IN ENVIRONMENTAL DECISION MAKING

New approaches to including people other than specialists and professionals in environmental decision making include:

Consultation: involving all stakeholders (people and groups with direct and indirect interests in the issue) in environmental decision-making processes, through consultation, liaison, working partnerships and other means.

Indigenous knowledge: listening to what native people think and believe, and incorporating their views into environmental decision making.

Community involvement: allowing all stakeholders to have access to reliable and up-to-date information, to be able to exercise their right to have their voices heard in environmental debates, and to be able to protest against decisions they are unhappy with and to have their concerns heard and taken seriously. This is an important step towards making environmental decision making more democratic and inclusive.

Environmental education: introducing schemes to educate people better about environmental issues, not just through formal education (in schools, colleges and universities) but as part of a lifelong learning strategy designed to produce more environmentally aware citizens.

BOX 20.15 RISK ASSESSMENT

Risk is the likelihood of possible outcomes as a result of a particular action or reaction, and environmental risks are usually very difficult to estimate. Risk is usually assessed by means of probability – the likelihood of an event occurring, expressed on a scale of 0 (absolute impossibility) to 1 (absolute certainty). It can also be expressed as a percentage.

BOX 20.16 DECISION-MAKING APPROACHES IN NATURAL RESOURCE MANAGEMENT

The political/social approach: in which the main concerns are public and political response to a decision. Common outcomes of this form of decision making include the decision to do nothing or to delay action until more data are available, and many outcomes are designed to keep options open for the future and/or satisfy particular powerful interests.

The conventional-wisdom approach: in which managers use a historical method or rule of thumb that has been applied to similar situations in the past. Managers rely on historical knowledge of the situation and the resource involved and assume that the response to management will be similar to that experienced previously. Examples include decisions about how many fish to stock, or what level of pollutant to allow.

The best-current-data approach: which uses current data collected through new or existing sampling programmes. Managers analyse the data using the latest techniques, assess their management options and then choose the one best option to implement. Examples include habitat enhancement projects and management for various forms of optimal sustained yield.

The monitor-and-modify approach: here a policy decision is usually made using the conventional-wisdom or best-current-data methods, and then the policy is implemented along with a monitoring plan. Monitoring data are then used to evaluate and periodically modify the policy relative to a specific goal, and the policy can be 'fine-tuned' in the light of experience. Examples include management that involves annual resource assessments, such as marine fish stocks.

The adaptive management approach: which usually begins by bringing together interested parties (stakeholders) in workshops to discuss the management problem and the available data, and then to develop computer models that express participants' collective understanding of how the system operates. The models are then used to assess the significance of data gaps and uncertainties and to predict the effects of alternative management actions. The stakeholders then develop a management plan that will help to meet management goals and will also generate new information to reduce critical gaps and uncertainties. The management plan is then implemented, along with a monitoring plan. As monitoring proceeds, new data are analysed, and the management plans are revised as understanding of how the system works is increased.

analysis (see Box 20.9) – have been widely used since the 1960s. While they are much better than simple guesswork or intuitive decision making, there are drawbacks in using them in isolation and in placing too much weight on them to the exclusion of other relevant issues.

Traditional decision-making approaches

- are based on rational analysis, which relies heavily on 'facts' that are regarded as objective and which can be quantified and used in mathematical models;
- are much less useful when the problem is complex, when there is a great deal of uncertainty (for example when there is not enough 'hard evidence' to base the analysis on);
- are also unsuitable for dealing with situations that involve differing attitudes and values, because these are often subjective and can only really be handled qualitatively. This is true, for example, of ethical, social and political issues.

The search for more appropriate ways of making environmental decisions also reflects a much better understanding of how different types of decision maker use and interpret information. Most people assume, without question, that quantitative information (such as rates of forest clearance, emission levels of CO_2) are more robust, reliable and meaningful than qualitative information (such as what a particular group of people think is right and wrong behaviour). This is not necessarily true in all cases.

Nearly all decision making has to cope with incomplete information and disagreement between experts over how best to analyse and interpret the available information. This is largely unavoidable. Given that fact, decision makers have a responsibility to identify and acknowledge areas of high uncertainty and not to give the impression that there are no serious gaps in knowledge and understanding and no major differences of opinion between experts in the field.

ADAPTIVE MANAGEMENT

The unsuitability of 'traditional' decision making in the face of environmental uncertainty is well illustrated in the emergence of so-called 'soft disasters'. These are environmental and political crises that emerge very slowly but cause serious social concern when they do. Recent examples include the BSE crisis and the genetically modified (GM) food debate in the UK.

Because such problems are very complex and have yet to be fully studied, the public quite rightly asks whether scientists really understand the issues, and it is left wondering who it can trust for advice and guidance. Public confidence in government decision makers can be eroded quickly when complicated problems like these arise. The issue is more than just political, because people then start to question and doubt the legitimacy and relevance of 'scientific' advice.

One way in which complex issues like these can be approached is via an adaptive management approach (Box 20.16), which goes beyond the linear decision making of more conventional approaches and allows decisions to be steered by experience of policies in action. Adaptive management reflects more flexible and responsive decision making, such as the range of initiatives and policies that the UK government has introduced since 1997 as means of reducing greenhouse gas emissions (Box 20.17).

BOX 20.17 NEW POLICIES AND STRATEGIES INTRODUCED BY THE UK GOVERNMENT SINCE 1997 TO TACKLE GREENHOUSE GAS EMISSIONS

- a new climate change levy (tax) to encourage business to use energy more efficiently;
- a new target to double energy generation through combined heat and power schemes of at least 10,000 MW by 2010;
- an increase in the amount of money available for energy efficiency and fuel poverty programmes in the domestic sector, which will reduce CO_2 emissions;
- a new integrated transport policy that will deliver a better-quality transport system with lower CO_2 emissions;
- changes in transport taxation to stimulate fuel efficiency;
- measures at the European Union level, such as an agreement with car manufacturers that will improve the fuel efficiency of new cars by 25 per cent by 2008;
- active consultation with businesses, local government, the general public, associations, trade unions and other key stakeholders as the government develops its new climate change programme.

NEW AND EMERGING PROBLEMS

Optimists (see p. 624) argue that much of the rhetoric in the environmental debate is misplaced scaremongering, and that things are not as bad as the pessimists would have us believe. The jury is still out on the question of which side is right and which one is wrong, but in many ways the debate since the mid-1990s has moved away from seeking simple answers and has moved forwards to embrace the principles and practice of sustainable development.

ENVIRONMENTAL THREATS

Evidence of continued environmental change is emerging, and new environmental threats are being recognised all the time. Since 1995, for example, a number of new problems have been identified, and some existing problems have become worse (Box 20.18).

Tackling existing environmental problems is a major problem in its own right, but the challenge is made much bigger by the difficulties of anticipating what lies ahead. Many of the most critical environmental problems – such as global warming, loss of biodiversity, soil erosion and decline in ocean resources – are large-scale, long-term and complex. They cannot all be tackled at the same time, because of constraints of time, expertise, resources and political commitment. Establishing priorities is vitally important, and as we move forwards from the present to the future it becomes increasingly difficult to evaluate one problem against another and draw up an objective list of priorities.

BOX 20.18 NEW AND WORSENED ENVIRONMENTAL PROBLEMS SINCE 1995, ACCORDING TO UNEP'S GLOBAL ENVIRONMENT OUTLOOK 2000

Problems that have emerged or worsened since 1995 include:

- a global nitrogen problem, leading to widespread eutrophication (see p. 410);
- more and larger forest fires (see p. 589), caused by a combination of unfavourable weather conditions and land uses that made susceptible areas more likely to burn;
- increased frequency and severity of natural disasters, which kill and injure many millions of people every year and cause serious economic losses;
- global warming: 1998 was the warmest year on record (see p. 6), and climate change problems coupled with the most severe El Niño to date (1997–98) (see p. 483) have caused many deaths and many economic problems;
- the economic and ecological importance of species invasion (see p. 531), an inevitable sign of increasing globalisation, which appears to have become more significant;
- declines in some countries in the quality of governance, which have weakened capabilities to solve national and regional problems and to manage the environment;
- declining government and media focus on urgent environmental issues (see p. 27) as attention has been diverted by political and economic upheavals;
- new wars, which, like all wars, threaten both the environment of the war zone and that of neighbouring states (see p. 247);
- the environmental importance of refugees (see p. 39), who place many demands on natural resources in the receiving area.

PRIORITIES

In 1999, the Scientific Committee on Problems of the Environment (SCOPE) carried out a survey among 200 scientists in fifty countries asking what they thought would be the environmental issues that would require attention in the twenty-first century (Table 20.1). Most respondents expect that the major environmental problems will stem from the continuation and worsening of existing problems that currently do not receive enough policy attention, such as climate change, availability of water resources and water pollution. Other major issues included deforestation, desertification and problems arising from poor management at national and international levels.

Priority issues varied from region to region. Only a few were mentioned regularly in all regions: scarcity of fresh water, environmental pollution (mainly chemical), invasive species, reduction in human immunity and resistance to disease, the collapse of fisheries, and food insecurity. Key environmental issues mentioned in Africa and West Asia were air pollution (particularly urban air pollution), industrial emissions and contamination from waste disposal. Experts in North America voiced greater concern about biodiversity, changes in ocean systems, emerging diseases, sea-level rise and space debris. The SCOPE survey suggests that many of the major environmental problems expected in the next century are problems that exist now but which are not yet receiving enough policy attention.

UNEP's *Global environment outlook 2000* outlines a series of global and regional trends which are likely to get worse during the twenty-first century (Box 20.19). The report expects three sets of environmental issues to become priorities in the twenty-first century:

Unforeseen events and scientific discoveries: doubtless many current trends (Box 20.19) will continue into the future, but the situation – already complex – will become worse because the future will also contain surprises. Recent decades illustrate clearly the ways in which previously undetected environmental issues come to the fore. Good examples include the discovery of depletion of the ozone layer in the stratosphere, which was first discovered in 1974 and only became a major issue in 1985 when the presence of the so-called 'ozone hole' above Antarctica was discovered (see p. 249). Acid rain became an issue in the 1970s, but it had been causing damage and

Table 20.1 Major emerging environmental issues identified in a 1999 survey by the ICSU Scientific Committee on Problems of the Environment

Issue	*Percentage of respondents who mentioned it*
Climate change	51
Scarcity of fresh water	29
Deforestation/desertification	28
Freshwater pollution	28
Poor governance	27
Loss of biodiversity	23
Population growth and movements	21
Changing social values	20
Waste disposal	20
Air pollution	20
Soil deterioration	18
Ecosystem functioning	17
Chemical pollution	16
Urbanisation	16
Ozone depletion	15
Energy consumption	15
Emerging diseases	14
Natural resource depletion	11
Food insecurity	11
Biogeochemical cycle disruption	11
Industrial emissions	10
Poverty	9
Information technologies	7
War and conflict	7
Reduced resistance to disease	7
Natural disasters	7
Invasive species	6
Genetic engineering	6
Marine pollution	6
Collapse of fisheries	5
Ocean circulation	5
Degradation of the coastal zone	5
Space debris	4
Persistent bioaccumulative toxins	4
El Niño effects	3
Sea-level rise	3

problems for many decades before that (see p. 252). Many people are concerned about the prospects of genetically modified (GM) organisms being accidentally or intentionally released into the environment, with potentially enormous consequences for gene pools and natural systems.

Sudden, unexpected transformation of old issues: the emergence of 'new' (as in 'newly discovered') environmental problems will be only part of the problem, because many existing problems are likely to worsen. Some might even change dramatically, in scale or nature, if critical environmental thresholds (see p. 83) are crossed. For example, acid materials – derived ultimately from acid rain (see p. 251) – might accumulate in a lake until the buffer capacity of the lake (its ability to neutralise the acid input) is exhausted, and then the pH of the lake water may fall sharply. At this point, toxic heavy metals are likely to be mobilised and released, causing major damage to aquatic life.

Neglected issues: a third area of concern is the likelihood that issues that are already known about, but largely ignored, will become much worse in the future. Examples include the collapse of ocean fisheries in many areas as a result of continued over-fishing (see p. 486), and the progressive build-up of greenhouse gases in the atmosphere as a result of air pollution (see p. 241). Early detection and evaluation of environmental problems and potential problems, coupled with a resolve to act quickly to bring problems under control rather than wait until the environmental impacts are serious, are clearly crucial.

FINAL REFLECTIONS

It is abundantly clear at the start of the twenty-first century that we need to appreciate and take seriously some critical aspects of the relationship between people and the environment. These include the interdependence of major environmental and socio-economic systems, the needs and rights of future generations, the values of diversity, the existence and importance of limits to growth, and the need for precautionary approaches. There is still much to be discovered about how the environment works, how resilient it is, how it changes in response to human pressure, and what its limits and critical threshold are.

Throughout this book, we have tried to identify some of the current and emerging themes within environmental management, and the enormous size and complexity of many of the problems should by now be self-evident.

BOX 20.19 GLOBAL AND REGIONAL TRENDS LIKELY TO WORSEN DURING THE TWENTY-FIRST CENTURY, ACCORDING TO UNEP'S GLOBAL ENVIRONMENT OUTLOOK 2000

- *Nitrogen overload*: many environmental systems are being fertilised as a result of human activities, and little is yet known about the possible impacts of such large-scale disruption of the nitrogen cycle. Similarly, little is known about possible synergistic effects between major biogeochemical cycles (for example, between the nitrogen and carbon cycles).
- *Environment-related disasters*: both natural disasters and those associated with human activities are becoming more frequent and more severe, bringing greater human loss and suffering and causing ever-higher economic losses.
- *Degradation of coastal areas and their resources*: many environmental problems are concentrated in coastal areas – fisheries have been badly mismanaged, coastal land is badly affected by poorly planned and regulated land-use changes (including urbanisation, industrialisation, aquaculture, tourism, port development and flood control), and many nearshore coastal waters continue to deteriorate.
- *Chemicals*: there is rising concern over chemical compounds that persist in the environment and that can affect the health and reproduction of organisms at the molecular or reproductive level.
- *Species invasions*: increasing introductions of non-indigenous species, either deliberately or accidentally, are putting the survival of many native species at risk.
- *Climate extremes*: 1998 was the warmest year on record globally, and it was the twentieth consecutive year with an above-normal global surface air temperature. The 1997–98 El Niño was the most powerful on record.
- *Growing global water crisis*: a major water crisis looks highly likely, which would increase water stress, particularly in low-income countries.
- *Land degradation*: vulnerability of land to erosion by water, especially where marginal land conversion is widespread, has increased.
- *Urbanisation*: half of the world's population will soon live in cities; where urban growth is uncontrolled or badly managed, it can give rise to many environmental problems (including waste disposal, air pollution and human health).
- *Environmental importance of refugees*: the influx of large numbers of refugees into an area can seriously stretch the ability of environmental systems and services to cope.
- *Vulnerability of small island developing states*: small remote islands are particularly vulnerable to pressures such as global warming, natural hazards, shortage of fresh water, coastal erosion and uncertain energy supplies.

SUMMARY

This chapter has ranged over a wide variety of themes because it seeks to look back over the main implications of the earlier chapters, and look forward to what might lie in store for us in the future. A number of important themes recur throughout the chapter, including issues of complexity and uncertainty, difficulties of forecasting and predicting the future, problems of reconciling different viewpoints and sets of values, and the non-stop way in which everything appears to change through time. Another recurrent theme is awareness of what we do not know, and sensitivity to the limitations of current models, information, interpretations and understanding. The chapter opened with some reflections on where things stand today in this so-called 'environmental crisis', in terms of continuity and change, causes and complications, perspectives and prognoses, and improvements and progress. This section was designed to highlight just how multi-faceted, multi-dimensional and complex the environmental debate actually is. Moving on from past to future, we looked at the new challenges which are emerging, particularly in terms of how to deal with uncertainty, and the new approaches being adopted to make more informed decisions about the environment. We also paused to reflect on new and emerging problems, and the challenges they confront us with.

In terms of progress within this field, we have come a long way over the last decade, but the Holy Grail of having sufficient understanding to solve all environmental problems remains if anything more elusive now than ever before. The more questions we ask, and the more answers we seek, the more we become aware of just how little we really know – an allegory of life itself!

WEBSITE

Links to relevant websites, a comprehensive bibliography, tools for teaching and learning, and downloadable images relevant to this chapter can be found at the website specially designed to accompany this book at
http://www.park-environment.com

FURTHER READING

Baker, S., M. Kousis, D. Richardson and S. Young (eds) (1996) *The politics of sustainable development.* Routledge, London. A series of essays that debate the meaning of sustainable development and explore how sustainable development has been interpreted in different countries.

Barry, J. (1999) *Environment and social theory.* Routledge, London. Accessible introduction to how people value, use and think about the environment, which explores the relationship between the environment and social theory.

Bromley, D. (ed.) (1995) *The handbook of environmental economics.* Blackwell, Oxford. An up-to-date review of recent thinking in environmental and natural resource economics.

Brown, L.R. (ed.) (2000) *State of the world 2000.* Earthscan, London. Up-to-date review of the state of the global environment, the factors leading to environmental change and responses to change.

Brown, L.R., M. Renner and B. Halweil (eds) (2000) *Vital signs 2000–2001: the environmental trends that are shaping our future.* Worldwatch Institute, Washington. Excellent source of up-to-date information on the state of the environment and sustainable development.

Carley, M. and I. Christie (2000) *Managing sustainable development.* Earthscan, London. Explores the organisational dimension of sustainable development, with case studies and examples from different contexts.

Chatterjee, P. and M. Finger (1994) *The Earth brokers.* Routledge, London. A provocative look at the main initiatives undertaken by national governments to deal with the global threat of environmental degradation.

Connelly, J. and G. Smith (1999) *Politics and the environment.* Routledge, London. Comprehensive introduction to environmental politics, which explains key concepts and issues and draws on case studies from different countries.

Cooper, D.E. and J.A. Palmer (eds) (1998) *Spirit of the environment: religion, value and environmental concern.* Routledge, London. A series of essays that explore different aspects of the relationship between people and the natural world.

Cutter, S.L. and W.H. Renwick (1998) *Exploitation, conservation, preservation.* John Wiley & Sons, London. A clear overview of the physical, economic, social and political dimensions of natural resource management.

Edwards, M. (2000) *Future positive: international co-operation in the twenty-first century.* Earthscan, London. A non-technical review of how international systems operate, what pressures they face and the changes they must undergo.

Elliott, J. (1993) *An introduction to sustainable development.* Routledge, London. A clear introduction to the origins, expression, application and problems of sustainable development.

Engel, J.R. and J.G. Engel (eds) (1992) *Ethics of environment and development: global challenge and international response.* John Wiley & Sons, London. A set of essays focusing on environmental ethics, exploring how we decide what is right and wrong, and what is appropriate behaviour towards the environment and what is not.

Gore, A. (2000) *Earth in the balance: forging a new common purpose.* Earthscan, London. Reissue of a best-selling popular book by US Vice-President Al Gore, which calls for a strengthening of environmental protection across the board, from global warming to stabilising world population.

Holdgate, M. (1995) *From care to action: making a sustainable world.* Earthscan, London. A call to action by one of the world's leading environmentalists, who urges people to work for change at every level in society.

Howitt, R. (2000) *Rethinking resource management.* Routledge, London. Offers a new perspective on resource management, based on a deeper understanding of relationships between resource projects and indigenous peoples.

Hussein, A.M. (1999) *Principles of environmental economics.* Routledge, London. Useful introductory textbook that reviews the economic and ecological principles essential for a clear understanding of contemporary environmental and natural resource issues and policy considerations.

Huxham, M. and D. Sumner (2000) *Science and environmental decision making.* Prentice Hall, London. Explores the roles and significance of scientific uncertainty and ignorance, and of politics and ethics, in making decisions about the environment.

Mabey, N., S. Hall, C. Smith and S. Gupta (1997) *Argument in the greenhouse: the international economics of controlling global warming.* Routledge, London. Explores international issues and tensions in the search for effective solutions to the problem of global warming.

Moss, N. (2000) *Managing the planet: the politics of the new millennium*. Earthscan, London. Explores what politics might look like in the future if the central focus becomes sustainable development.

O'Riordan, T. (1999) *Environmental science for environmental management*. Longman, Harlow. Explores issues and challenges in environmental management, set against a backdrop of changing environmentalism and emerging views about science.

Pavlinek, P. and J. Pickles (2000) *Environmental transitions: transformation and ecological defense in Central and Eastern Europe*. Routledge, London. Reviews the pace and pattern of environmental change in Central and Eastern Europe, both under state socialism and during the period of transition to capitalism.

Pearce, D. and E.B. Barbier (2000) *Blueprint for a sustainable economy*. Earthscan, London. Shows how ideas about sustainable development have evolved and altered, and explores how these ideas can be applied to countries and companies.

Pritchard, P. (2000) *Environmental risk management*. Earthscan, London. Outlines recent developments in environmental risk management as they relate to commercial organisations, including consideration of contaminated land and fire-related risks.

Rao, P.K. (2000) *Sustainable development: economics and policy*. Blackwell, Oxford. Interdisciplinary treatment of new and emerging concepts and approaches, including environmental accounting, questions about scientific knowledge and indicators of sustainability.

Rietbergen-McCracken, J. and H. Abaza (eds) (2000) *Economic instruments for environmental management*. UNEP/Earthscan, London. A series of case studies drawn from around the world that explore the power and importance of economic instruments in managing the environment.

Sarewitz, D., R.A. Pielke and R. Byerly (eds) (2000) *Prediction: science, decision-making and the future of nature*. Island Press, Washington. Ten detailed case studies that explore efforts to generate reliable scientific information about complex natural systems, and to use that information in making sound policy decisions.

Simon, J. (1995) *The state of humanity*. Blackwell, Oxford. A collection of essays that review the state of the Earth and its inhabitants at the close of the twentieth century. Subjects include deforestation, water pollution and air pollution.

Smith, M. (ed.) (1999) *Thinking through the environment: classic and contemporary readings*. Routledge, London. A useful reader of essays and articles that embraces ecological thought, environmental policy, environmental philosophy, social and political thought, historical sociology and cultural studies.

Trudgill, S.T. (1990) *Barriers to a better environment: what stops us solving environmental problems?* John Wiley & Sons, London. Argues that the nature and interaction of technological, economic, social and political barriers stop us!

Turner, R.K. (ed.) (1993) *Sustainable environmental economics and management*. John Wiley & Sons, London. A wide-ranging overview of the principles of sustainable approaches to environmental problems, with case studies drawn from many different countries.

United Nations Environment Program (1999) *Global environment outlook 2000*. UNEP/Earthscan, London. Comprehensive and up-to-date review and analysis of the state of the environment, at global and regional scales.

Vig, N.J. and R.S. Axelrod (eds) (1999) *The global environment: institutions, law and policy*. Earthscan, London. Examines the way in which disagreements are created and resolved, including debate over the meaning of sustainable development, conflict over the exclusion of developing countries from the Kyoto Protocol and the role of international financial interests in promoting incompatible forms of development.

Bibliography

Abrahams, A. and A. Parsons (eds) (1993) *Geomorphology of desert environments.* Chapman & Hall, London.

Acreman, M. (ed.) (1999) *The hydrology of the UK.* Routledge, London.

Adams, W.M. (1992) *Green development: environment and sustainability in the third world.* Routledge, London.

Adams, W.M. (1996) *Future nature: a vision for conservation.* Earthscan, London.

Adger, W.N. and K. Brown (1994) *Land use and the causes of global warming.* John Wiley & Sons, London.

Agnew, C. and E. Anderson (1992) *Water resources in the arid realm.* Routledge, London.

Aguardo, E. and J.E. Burt (1999) *Understanding weather and climate.* Prentice Hall, London.

Alexander, R. and A. Millington (eds) (1998) *Vegetation mapping.* John Wiley & Sons, London.

Allen, K.C. and D.E.G. Briggs (1991) *Evolution and the fossil record.* John Wiley & Sons, London.

Allison, R. (1997) *Rock slopes.* Blackwell, Oxford.

Anderson, M.G., D.E. Walling and P. Bates (eds) (1996) *Floodplain processes.* John Wiley & Sons, London.

Baarschers, W.H. (1996) *Eco-facts and eco-fiction: understanding the environmental debate.* Routledge, London.

Baird, A.J. and R.L. Wilby (1999) *Eco-hydrology.* Routledge, London.

Baker, S., M. Kousis, D. Richardson and S. Young (eds) (1996) *The politics of sustainable development.* Routledge, London.

Ballantyne, C.K. and C. Harris (1994) *The periglaciation of Great Britain.* Cambridge University Press, London.

Barbier, E.B., J.C. Burgess and C. Folke (1994) *Paradise lost? The ecological economics of biodiversity.* Earthscan, London.

Barraclough, S.L. and K.B. Ghimire (2000) *Agricultural expansion and tropical deforestation: poverty, international trade and land use.* Earthscan, London.

Barrow, C.J. (1995) *Developing the environment: problems and management.* Longman, Harlow.

Barry, J. (1999) *Environment and social theory.* Routledge, London.

Barry, R.G. (1992) *Mountain weather and climate.* Routledge, London.

Barry, R.G. and R. Chorley (1998) *Atmosphere, weather and climate.* Routledge, London.

Beatley, T. (2000) *Green urbanism: learning from European cities.* Island Press, Washington.

Beaumont, P. (1993) *Drylands: environmental management and development.* Routledge, London.

Bell, F. (1999) *Geological hazards: their assessment, avoidance and mitigation.* E & FN Spon, London.

Bell, M. and M.J.C. Walker (1992) *Late Quaternary environmental change: physical and human perspectives.* Longman, Harlow.

Benito, G., V.R. Baker and K.J. Gregory (eds) (1998) *Palaeohydrology and environmental change.* John Wiley & Sons, London.

Benn, D.I. and D.J.A. Evans (1997) *Glaciers and glaciation.* Edward Arnold, London.

Bennett, M.R. and P. Doyle (1997) *Environmental geology.* John Wiley & Sons, London.

Bennett, M.R. and N.F. Glasser (1996) *Glacial geology: ice sheets and landforms.* John Wiley & Sons, London.

Bennison, G.M. and K.A. Moseley (1997) *Introduction to geological structures and maps.* Edward Arnold, London.

Berkhaut, F. (1991) *Radioactive waste: politics and technology.* Routledge, London.

Best, G., E. Niemirycz and T. Bogacka (1997) *International river water quality.* E & FN Spon, London.

Bird, E.C.F. (1996) *Beach management.* John Wiley & Sons, London.

Black, R. (1998) *Refugees, environment and development.* Longman, Harlow.

Bland, W. and D. Rolls (1998) *Weathering.* Edward Arnold, London.

Boehmer-Christiansen, S. and J. Skea (1991) *Acid politics.* John Wiley & Sons, London.

Boggs, S. (1995) *Principles of sedimentology and stratigraphy.* Prentice Hall, Hemel Hempstead.

Bolin, R. and L. Stanford (1998) *The Northridge earthquake: vulnerability and disaster.* Routledge, London.

Bolt, B.A. (1993) *Earthquakes: a primer.* W.H. Freeman, Oxford.

Brack, D. (1999) *International trade and climate change policies.* Earthscan, London.

Brack, D., M. Grubb and C. Vrolijk (1999) *The Kyoto Protocol: a guide and assessment.* Earthscan, London.

Bradbury, I.K. (1998) *The biosphere.* John Wiley & Sons, London.

Bradbury, M., J. Boyle and A. Morse (2000) *Science for physical geography and environment.* Longman, Harlow.

Bradley, R. and P. Jones (eds) (1994) *Climate since AD 1500.* Routledge, London.

Bridges, E.M. (1990) *World geomorphology.* Cambridge University Press, London.

Bridgman, H. (1990) *Global air pollution: problems for the 1990s.* John Wiley & Sons, London.

Briggs, D.J., P. Smithson, K. Addison and K. Atkinson (1997) *Fundamentals of the physical environment.* Routledge, London.

Bromley, D. (ed.) (1995) *The handbook of environmental economics.* Blackwell, Oxford.

Brown, A.G. (ed.) (1995) *Geomorphology and groundwater.* John Wiley & Sons, London.

Brown, L., G. Gardner and B. Halweil (1999) *Beyond Malthus: nineteen dimensions of the population challenge.* Worldwatch Institute, Washington.

Brown, L.R. (ed.) (2000) *State of the world 2000.* Earthscan, London.

Brown, L.R., M. Renner and B. Halweil (eds) (2000) *Vital signs 2000–2001: the environmental trends that are shaping our future.* Worldwatch Institute, Washington.

Bruce, D. and A. Bruce (eds) (1999) *Engineering Genesis: the ethics of genetic engineering in non-human species.* Earthscan, London.

Brum, G., G. Karp and L. McKane (1993) *Biology.* John Wiley & Sons, London.

Brumbaugh, D.S. (1998) *Earthquakes: science and society.* Prentice Hall, Hemel Hempstead.

Buckingham-Hatfield, S. (2000) *Environment and gender.* Routledge, London.

Burton, I., R.W. Kates and G.F. White (1994) *The environment as hazard.* Longman, Harlow.

Butcher, S.S., G.H. Orians, R.J. Charlson and G.V. Wolfe (1992) *Global biogeochemical cycles.* Academic Press, London.

Carley, M. and I. Christie (2000) *Managing sustainable development.* Earthscan, London.

Carlson, T. (1991) *Mid-latitude weather systems.* Routledge, London.

Carter, R.W.G. (1990) *Coastal environments: an introduction to the physical, ecological and cultural systems of coastlines.* Academic Press, London.

Chatterjee, P. and M. Finger (1994) *The Earth brokers.* Routledge, London.

Chester, D.K. (1993) *Volcanoes and society.* Edward Arnold, London.

Christopherson, R.W. (2000) *Geosystems.* Prentice Hall, Hemel Hempstead.

Cline, W.R. (1992) *The economics of global warming.* Longman, Harlow.

Coleman, D.C. and D.A. Crossley (eds) (1996) *Fundamentals of soil ecology.* Academic Press, London.

Colinvaux, P. (1993) *Ecology.* John Wiley & Sons, London.

Collins, D. (2000) *Contaminated land; managing legal liabilities.* Earthscan, London.

Collins, R.O. (1990) *The waters of the Nile: hydropolitics and the Jonglei Canal, 1900–1988.* Clarendon Press, Oxford.

Colls, J. (1996) *Air pollution.* E & FN Spon, London.

Condie, K.C. and R.E. Sloan (1998) *Origin and evolution of the Earth: principles of historical geology.* Prentice Hall, Hemel Hempstead.

Connelly, J. and G. Smith (1999) *Politics and the environment.* Routledge, London.

Cook, H.F. (1998) *The protection and conservation of water resources: a British perspective.* John Wiley & Sons, London.

Cooper, D.E. and J.A. Palmer (eds) (1998) *Spirit of the environment: religion, value and environmental concern.* Routledge, London.

Cosgrove, W.J. and F.R. Rijsberman (2000) *World water vision: making water everybody's business.* Earthscan, London.

Coull, J.R. (1993) *World fisheries resources.* Routledge, London.

Cunningham, A.B. (2000) *Applied ethnobotany: people, wild plant use and conservation.* WWF/UNESCO/Earthscan, London.

Cutter, S.L. and W.H. Renwick (1998) *Exploitation, conservation, preservation.* John Wiley & Sons, London.

Davidson, D.A. (1992) *The evaluation of land resources.* Longman, Harlow.

Davis, R.A. (1993) *The evolving coast.* W.H. Freeman, Oxford.

Dawson, A. (1991) *Ice age Earth.* Routledge, London.

De Wall, L.C., A. Large and M. Wade (eds) (1998) *Rehabilitation of rivers: principles and implementation.* John Wiley & Sons, London.

Deacon, M., T. Rice and C. Summerhays (eds) (1999) *Understanding the oceans.* UCL Press, London.

Decker, R. and B. Decker (1995) *Volcanoes.* W.H. Freeman, Oxford.

Deer, W.A., R.A. Howie and J. Zussman (1992) *An introduction to rock forming minerals.* Longman, Harlow.

Degens, E.T., S. Kempe and J.E. Richey (1990) *Biogeochemistry of major world rivers.* John Wiley & Sons, London, for Scientific Committee on Problems of the Environment of ICSU/UNEP; SCOPE 42.

Demeny, P. and G. McNicholl (eds) (1996) *The Earthscan reader in population and development.* Earthscan, London.

Dickinson, G. and K. Murphy (1998) *Ecosystems.* Routledge, London.

Dikau, R., L. Schrott, D. Brunsden and M.L. Ibsen (eds) (1996) *Landslide recognition.* John Wiley & Sons, London.

Djuric, D. (1994) *Weather analysis.* Prentice Hall, Hemel Hempstead.

Dodds, F. (ed.) (2000) *Earth Summit 2002: a new deal.* Earthscan, London.

Donovan, S.K. and C.R.C. Paul (eds) (1998) *The adequacy of the fossil record.* John Wiley & Sons, London.

Doornbos, M., A. Saith and B. White (2000) *Forests: nature, people, power.* Blackwell, Oxford.

Downing, T. and R.S.J. Tol (eds) (1998) *Climate, change and risk.* Routledge, London.

Doyle, H. (1995) *Seismology.* John Wiley & Sons, London.

Doyle, P. and M. Bennett (eds) (1998) *Unlocking the stratigraphic record: advances in modern stratigraphy.* John Wiley & Sons, London.

Doyle, P., M.R. Bennett and A.N. Baxter (1994) *The key to Earth history: an introduction to stratigraphy.* John Wiley & Sons, London.

Drever, J.I. (1997) *The geochemistry of natural waters.* Prentice Hall, Hemel Hempstead.

Dyer, K. (1997) *Estuaries: a physical introduction.* John Wiley & Sons, London.

Dyson, T. (1999) *Population and food.* Routledge, London.

Edwards, M. (2000) *Future positive: international co-operation in the twenty-first century.* Earthscan, London.

Egorov, N.N., V.M. Novikov, F.L. Parker and V.K. Popov (eds) (2000) *The radiation legacy of the Soviet nuclear complex.* IIASA/Earthscan, London.

Ehlers, J. (1996) *Quaternary and glacial geology.* John Wiley & Sons, London.

Elliott, J. (1993) *An introduction to sustainable development.* Routledge, London.

Ellis, S. and T. Mellor (1996) *Soils and environment.* Routledge, London.

Elsom, D. (1992) *Atmospheric pollution: a global problem.* Blackwell, Oxford.

Elsom, D. (1996) *Smog alert: managing urban air quality.* Earthscan, London.

Emiliani, C. (1992) *Planet Earth: cosmology, geology, and the evolution of life and environment.* Cambridge University Press, Cambridge.

Engel, J.R. and J.G. Engel (eds) (1992) *Ethics of environment and development: global challenge and international response.* John Wiley & Sons, London.

Evans, D.J.A. (ed.) (1994) *Cold climate landforms.* John Wiley & Sons, London.

Evans, I.S. (1996) *The geography of glaciers.* Edward Arnold, London.

Fairhead, J. and M. Leach (1998) *Reframing deforestation: global analysis and local realities – studies in West Africa.* Routledge, London.

Farmer, A. (1997) *Managing environmental pollution.* Routledge, London.

Faure, G. (1998) *Principles and applications of geochemistry.* Prentice Hall, Hemel Hempstead.

Foth, H. (1991) *Fundamentals of soil science.* John Wiley & Sons, London.

Frederick, K. (1997) *Water resources and climate change.* Resources for the Future, Washington.

French, H. (1996) *The periglacial environment.* Longman, Harlow.

French, P. (1997) *Coastal and estuarine management.* Routledge, London.

Fry, N. (1991) *The field description of metamorphic rocks.* John Wiley & Sons, London.

Gardner, E.J., M. Simmons and D. Snustad (1991) *Principles of genetics.* John Wiley & Sons, London.

Gerrard, J. (2000) *Fundamentals of soil.* Routledge, London.

Gerrard, J. (1992) *Soil geomorphology.* Chapman & Hall, London.

Gleick, P.H. (2000) *The world's water 2000–2001.* Island Press, Washington.

Goldring, R. (1999) *Field palaeontology.* Longman, London.

Goldsmith, F.B. and A. Warren (eds) (1993) *Conservation in progress.* John Wiley & Sons, London.

Gordon, N., T. McMahon and B. Findlayson (1994) *Stream hydrology: an introduction for ecologists.* John Wiley & Sons, London.

Gore, A. (2000) *Earth in the balance: forging a new common purpose.* Earthscan, London.

Goudie, A. (1993) *The nature of the environment.* Blackwell, Oxford.

Goudie, A. (ed.) (1997) *The human impact reader: readings and case studies.* Blackwell, Oxford.

Goudie, A. (1999) *The human impact on the natural environment.* Blackwell, Oxford.

Goudie, A. and H. Viles (1997) *The Earth transformed: an introduction to human impacts on the environment.* Blackwell, Oxford.

Goudie, A.S. and H.A. Viles (1997) *Salt weathering hazard.* John Wiley & Sons, London.

Graedel, T.E. and P.J. Crutzen (1993) *Atmospheric change: an Earth system perspective.* W.H. Freeman, Oxford.

Graedel, T.E. and P.J. Crutzen (1995) *Atmosphere, climate and change.* W.H. Freeman, Oxford.

Graf, W.H. (1998) *Fluvial hydraulics.* John Wiley & Sons, London.

Graham, D.T. and N.K. Poku (1999) *Migration, globalisation and human security.* Routledge, London.

Gray, N.F. (1994) *Drinking water quality.* John Wiley & Sons, London.

Gregory, K.J., L. Starkel and V.R. Baker (eds) (1995) *Global continental palaeohydrology.* John Wiley & Sons, London.

Gren, I.M., R.K. Turner and F. Wulff (2000) *Managing a sea: the ecological economics of the Baltic.* Earthscan, London.

Gurnell, A. and G. Petts (eds) (1995) *Changing river channels.* John Wiley & Sons, London.

Hall, A. (1995) *Igneous petrology.* Longman, Harlow.

Hanson, J. and J. Gordon (1998) *Antarctic environments and resources.* Longman, Harlow.

Hardman, D., S. McEldowney and S. Waite (1995) *Pollution: ecology and biotreatment.* Longman, Harlow.

Harper, D. and M. Benton (1996) *Basic palaeontology.* Longman, Harlow.

Harvey, D. (2000) *Climate and global environmental change.* Longman, Harlow.

Harvey, D. (2000) *Global warming.* Longman, Harlow.

Hasan, S.E. (1996) *Geology and hazardous waste management.* Prentice Hall, Hemel Hempstead.

Haslam, S.M. (1992) *River pollution: an ecological perspective.* John Wiley & Sons, London.

Hatcher, R.D. (1995) *Structural geology: principles, concepts and problems.* Prentice Hall, London.

Hewitt, K. (1997) *Regions of risk: a geographical introduction to disasters.* Longman, Harlow.

Heywood, I., S. Cornelius and S. Carver (1998) *Introduction to Geographical Information Systems.* Longman, Harlow.

Hickin, E.J. (ed.) (1995) *River geomorphology.* John Wiley & Sons, London.

Hidore, J.J. (1996) *Global environmental change: its nature and impact.* Prentice Hall, Hemel Hempstead.

Hinrichs, R.A. (1996) *Energy.* Saunders College Publishing, Philadelphia.

Holdgate, M. (1995) *From care to action: making a sustainable world.* Earthscan, London.

Holland, C. (1999) *The idea of time.* John Wiley & Sons, London.

Holton, J.R. (1992) *An introduction to dynamic meteorology.* Academic Press, London.

Honari, M. and T. Boleyn (1999) *Health ecology: nature, culture and human–environment interaction.* Routledge, London.

Hooke, R.LeB. (1998) *Principles of glacier mechanics.* Prentice Hall, Hemel Hempstead.

Howitt, R. (2000) *Rethinking resource management.* Routledge, London.

Huggett, R. (1997) *Environmental change: the evolving ecosphere.* Routledge, London.

Huggett, R. (1998) *Fundamentals of biogeography.* Routledge, London.

Hulme, M. and E. Barrow (eds) (1997) *Climates of the British Isles: present, past and future.* Routledge, London.

Hussein, A.M. (1999) *Principles of environmental economics.* Routledge, London.

Hutton, J. and B. Dickson (1999) *Endangered species, threatened convention: the past, present and future of CITES.* Earthscan, London.

Huxham, M. and D. Sumner (2000) *Science and environmental decision making.* Prentice Hall, London.

Hviding, E. and T. Bayliss-Smith (2000) *Islands of rainforest: agroforestry, logging and eco-tourism in Solomon Islands.* Ashgate, London.

Iudicello, S., M. Weber and R. Wieland (1999) *Fish, markets and fishermen: the economics of over-fishing.* Earthscan, London.

Jackson, S. (1998) *Britain's population: demographic issues in a contemporary society.* Routledge, London.

Jackson, T., K. Begg and S. Parkinson (eds) (2000) *Flexibility in climate policy: making the Kyoto mechanisms work.* Earthscan, London.

Jeffries, M. (1997) *Biodiversity and conservation.* Routledge, London.

Jensen, J.R. (2000) *Remote sensing of the environment.* Prentice Hall, London.

Jepma, C.J. (1995) *Tropical deforestation: a socio-economic approach.* Earthscan, London.

Jobin, W. (1999) *Dams and diseases: ecological design and health impact of large dams and irrigation schemes.* E & FN Spon, London.

Jones, A. (1997) *Environmental biology.* Routledge, London.

Jones, E.J.W. (1999) *Marine geophysics.* John Wiley & Sons, London.

Jones, J.A.A. (1996) *Global hydrology: process and management implications.* Longman, Harlow.

Kate, K.T. and S.A. Laird (1999) *The commercial use of biodiversity.* Earthscan, London.

Kaufmann, W.J. (1993) *Discovering the universe.* W.H. Freeman, Oxford.

Kay, B. (1998) *Water resources: health, environment and development.* E & FN Spon, London.

Kay, R. and J. Alder (1998) *Coastal planning and management.* E & FN Spon, London.

Keller, E.A. (2000) *Environmental geology.* Prentice Hall, London.

Keller, E.A. and N. Pinter (1996) *Active tectonics: earthquakes, uplift and landscape.* Prentice Hall, Hemel Hempstead.

Kellert, S.R. (1996) *The value of life: biological diversity and human society.* Island Press, New York.

Kellman, M. and R. Tackaberry (1997) *Tropical environments: the functioning and management of tropical ecosystems.* Routledge, London.

Kemp, D. (1998) *The environment dictionary.* Routledge, London.

Kemp, D. (1994) *Global environmental issues: a climatological approach.* Routledge, London.

Keulartz, J. (1998) *The struggle for nature: a critique of environmental philosophy.* Routledge, London.

Keys, D. (1999) *Catastrophe: a quest for the origins of the modern world.* Ballantine, New York.

Kliot, N. (1993) *Water resources and conflict in the Middle East.* Routledge, London.

Knighton, D. (1997) *Fluvial forms and processes: a new perspective.* Edward Arnold, London.

Komar, P.D. (1998) *Beach processes and sedimentation.* Prentice Hall, Hemel Hempstead.

Kondratyev, K.Y., A. Buznikov and O. Pokrovsky (1996) *Global change and remote sensing.* John Wiley & Sons, London.

Laird, S.A. (ed.) (2000) *Biodiversity and traditional knowledge: equitable partnerships in practice.* WWF/UNESCO/ Earthscan, London.

Lamb, H.H. (1995) *Climate, history and the modern world.* Routledge, London.

Lamb, R. (1996) *Promising the Earth.* Routledge, London.

Lancaster, N. (1995) *Geomorphology of desert dunes.* Routledge, London.

Lappé, M. and B. Bailey (1999) *Against the grain: the genetic transformation of global agriculture.* Earthscan, London.

Leggett, J. (2000) *Global warming and the end of the oil era.* Penguin Books, London.

Levin, H.L. (1995) *The Earth through time.* Saunders College Publishing, Philadelphia.

Lewis, J.S., M.S. Matthews and M.L. Guerriri (1993) *Resources of near-Earth space.* University of Arizona Press, Tucson.

Lewis, L.A. and D.L. Johnson (1994) *Land degradation.* Blackwell, Oxford.

Lillesand, T.M. and R.W. Kiefer (1999) *Remote sensing and image interpretation.* John Wiley & Sons, London.

Linacre, E. and B. Geerts (1997) *Climates and weather explained.* Routledge, London.

Livi-Bacci, M. (1997) *A concise history of world population.* Blackwell, Oxford.

Livi-Bacci, M. (2000) *The population of Europe.* Blackwell, Oxford.

Livingstone, I. and A. Warren (1996) *Aeolian geomorphology: an introduction.* Longman, Harlow.

Lohnert, B. and H. Geist (eds) (2000) *Coping with changing environments: social dimensions of endangered ecosystems in the developing world.* Ashgate, London.

Lomnitz, C. (1994) *Fundamentals of earthquake prediction.* John Wiley & Sons, London.

Lonsdale, P. (1997) *Deep sea geomorphology.* Blackwell, Oxford.

Lovelock, J.E. (1979) *Gaia: a new look at life on Earth.* Oxford University Press, Oxford.

Lovelock, J.E. (1988) *The ages of Gaia: a biography of our living Earth.* W.W. Norton, New York.

Lowe, J.L. and M.J.C. Walker (1996) *Reconstructing Quaternary environments.* Longman, Harlow.

Mabey, N., S. Hall, C. Smith and S. Gupta (1997) *Argument in the greenhouse: the international economics of controlling global warming.* Routledge, London.

McClay, K. (1991) *The mapping of geological structures.* John Wiley & Sons, London.

McCormick, J. (1995) *The global environmental movement.* John Wiley & Sons, London.

McGinnis, M.V. (ed.) (1998) *Bioregionalism.* Routledge, New York.

Macklin, M. (1997) *Holocene river environments.* Blackwell, Oxford.

McNamara, K. and J. Long (1998) *The evolution revolution.* John Wiley & Sons, London.

Maddison, D. (2000) *The amenity value of global climate.* Earthscan, London.

Mairota, P., J.B. Thornes and N. Geeson (eds) (1997) *Atlas of Mediterranean environments in Europe: the desertification context.* John Wiley & Sons, London.

Manning, J.C. (1997) *Applied principles of hydrology.* Prentice Hall, Hemel Hempstead.

Mannion, A.M. and S.R. Bowlby (eds) (1992) *Environmental issues in the 1990s.* John Wiley & Sons, London.

Mannion, A.M. (1995) *Agriculture and environmental change.* John Wiley & Sons, London.

Mannion, A.M. (1999) *Natural environmental change.* Routledge, London.

Markham, A. (1994) *A brief history of pollution.* Earthscan, London.

Marsh, W.M. and J.M. Grossa (1996) *Environmental geography.* John Wiley & Sons, London.

Mason, C.F. (1996) *Biology of freshwater pollution.* Longman, Harlow.

Mather, A. (1992) *Global forest resources.* John Wiley & Sons, London.

Mather, A.S. and K. Chapman (1995) *Environmental resources.* Longman, Harlow.

Merrett, S. (1997) *Introduction to the economics of water resources.* UCL Press, London.

Middleton, N. (1999) *The global casino: an introduction to environmental issues.* Edward Arnold, London.

Middleton, N. and D. Thomas (1992) *World atlas of desertification.* Edward Arnold, London.

Milliken, W. and J. Ratter (1998) *Marácа: the biodiversity and environment of an Amazonian rainforest.* John Wiley & Sons, London.

Millington, A. and K. Pye (eds) (1994) *Environmental change in drylands.* John Wiley & Sons, London.

Mistry, J. (2000) *World savannas.* Prentice Hall, London.

Mitchell, B. (1997) *Resource and environmental management.* Longman, Harlow.

Mooney, H.A. and R.J. Hobbs (eds) (2000) *Invasive species in a changing world.* Island Press, Washington.

Moores, E.M. (ed.) (1990) *Shaping the Earth: tectonics of continents and oceans.* W.H. Freeman, Oxford.

Morgan, M.D. and J.P. Moran (1997) *Weather and people.* Prentice Hall, Hemel Hempstead.

Morgan, R.P.C. (1995) *Soil erosion and conservation.* Longman, Harlow.

Morley, R.J. (1999) *Origin and evolution of tropical rain forests.* John Wiley & Sons, London.

Mortimore, M.J. and W.M. Adams (1999) *Working the Sahel.* Routledge, London.

Moss, N. (2000) *Managing the planet: the politics of the new millennium.* Earthscan, London.

Murck, B.W., B.J. Skinner and S.C. Porter (1996) *Dangerous Earth.* John Wiley & Sons, London.

Mursky, G. (1996) *Introduction to planetary volcanism.* Prentice Hall, Hemel Hempstead.

Newson, M. (1997) *Land, water and development: sustainable management of river basin systems.* Routledge, London.

Nordstrom, K.F. and C.T. Roman (eds) (1996) *Estuarine shores: evolution, environments and human alterations.* John Wiley & Sons, London.

Nunn, P. (1994) *Oceanic islands.* Blackwell, Oxford.

Oliver, C. and B.C. Larson (1996) *Forest stand dynamics.* John Wiley & Sons, London.

Ollier, C. and C. Pain (1995) *Regolith, soils and landforms.* John Wiley & Sons, London.

Ollier, C. and C. Pain (2000) *The origin of mountains.* Routledge, London.

O'Riordan, T. (ed.) (1999) *Environmental science for environmental management.* Longman, Harlow.

O'Riordan, T. (ed.) (2000) *Globalism, localism and identity: new perspectives on the transition to sustainability.* Earthscan, London.

O'Riordan, T. and J. Jager (eds) (1995) *Politics of climate change: a European perspective.* Routledge, London.

Owen, O.S., D.D. Chiras and J.P. Reganold (1998) *Natural resource conservation: management for a sustainable future.* Prentice Hall, Hemel Hempstead.

Park, C.C. (1987) *Acid rain: rhetoric and reality.* Methuen, London.

Park, C.C. (1989) *Chernobyl: the long shadow.* Routledge, London.

Park, C.C. (1991) *Environmental hazards.* Macmillan, London.

Park, C.C. (1992) *Tropical rainforests.* Routledge, London.

Parker, D. and J. Handmer (eds) (2000) *Floods.* Routledge, London.

Parry, M. and R. Duncan (eds) (1995) *The economic implications of climate change in Britain.* Earthscan, London.

Pasachoff, J.M. (1995) *Astronomy.* Saunders College Publishing, Philadelphia.

Paterson, M. (1996) *Global warming and global politics.* Routledge, London.

Pavlinek, P. and J. Pickles (2000) *Environmental transitions: transformation and ecological defense in Central and Eastern Europe.* Routledge, London.

Peacock, W.G., B.H. Morrow and H. Gladwin (1997) *Hurricane Andrew: ethnicity, gender and the sociology of disasters.* Routledge, London.

Pearce, D. and E.B. Barbier (2000) *Blueprint for a sustainable economy.* Earthscan, London.

Pentecost, A. (1999) *Analysing environmental data.* Longman, Harlow.

Pepper, D. (1996) *Modern environmentalism.* Routledge, London.

Pereira, L.S. and J. Gowing (1998) *Water and the environment: innovation issues in irrigation and drainage.* E & FN Spon, London.

Perkins, D. (1998) *Mineralogy.* Prentice Hall, Hemel Hempstead.

Perry, A. and R. Thompson (1997) *Applied climatology: principles and practice.* Routledge, London.

Pickering, K.T. and L.A. Owen (1997) *An introduction to global environmental issues* (second edition). Routledge, London.

Pielke, R.A. and R.A. Pielke (1997) *Hurricanes: their nature and impacts on society.* John Wiley & Sons, London.

Pielke, R.A. and R.A. Pielke (eds) (1999) *Storms.* Routledge, London.

Pilson, M.E.Q. (1998) *Introduction to the chemistry of the sea.* Prentice Hall, Hemel Hempstead.

Pirazzoli, P.A. (1996) *Sea-level changes: the last 20,000 years.* John Wiley & Sons, London.

Porteous, A. (1992) *Dictionary of environmental science and technology.* John Wiley & Sons, Chichester.

Power, T.M. (1996) *Extraction and the environment: the economic battle to control our natural landscapes.* Island Press, New York.

Press, F. and R. Seiver (1993) *Understanding Earth.* W.H. Freeman, Oxford.

Preston, S., P. Heiveline and M. Guillot (2000) *Demography: measuring and modelling population processes.* Blackwell, Oxford.

Pritchard, P. (2000) *Environmental risk management.* Earthscan, London.

Pugh, C. (2000) *Sustainable cities in developing countries.* Earthscan, London.

Rao, P.K. (2000) *Sustainable development: economics and policy.* Blackwell, Oxford.

Reading, A.J., R.D. Thompson and A.C. Millington (1995) *Humid tropical environments.* Blackwell, Oxford.

Retallack, G.J. (1997) *A colour guide to paleosols.* John Wiley & Sons, London.

Rice, T. and P. Owen (1999) *Decommissioning the Brent Spar.* E & FN Spon, London.

Rietbergen-McCracken, J. and H. Abaza (eds) (2000) *Economic instruments for environmental management.* UNEP/Earthscan, London.

Rijsberman, F.R. (ed.) (2000) *World water scenarios: analysing global water resources and use.* Earthscan, London.

Ripley, E.A., R.E. Redman and A.A. Crowder (1995) *Environmental effects of mining.* St Lucie Press, New York.

Roberts, N. (ed.) (1993) *The changing global environment.* Blackwell, Oxford.

Roberts, N. (1998) *The Holocene: an environmental history.* Blackwell, Oxford.

Robinson, D.A. and R.B.G. Williams (eds) (1994) *Rock weathering and landform evolution.* John Wiley & Sons, London.

Robinson, P. and A. Henderson-Sellers (2000) *Contemporary climatology.* Longman, Harlow.

Ross, D.A. (1995) *Introduction to oceanography.* Longman, Harlow.

Roth, H. (1991) *Fundamentals of soil science.* John Wiley & Sons, London.

Rowell, D.L. (1994) *Soil science: methods and applications.* Longman, Harlow.

Saiko, T. (2000) *Environmental crises.* Prentice Hall, London.

Sarewitz, D., R.A. Pielke and R. Byerly (eds) (2000) *Prediction: science, decision-making and the future of nature.* Island Press, Washington.

Satterthwaite, D. (ed.) (1999) *The Earthscan reader in sustainable cities.* Earthscan, London.

Savage, D. (ed.) (1995) *The scientific and regulatory basis for the geological disposal of nuclear waste.* John Wiley & Sons, London.

Schlesinger, W.H. (1991) *Biogeochemistry: an analysis of global change.* Academic Press, London.

Schwab, G., D. Fangmeier and W. Elliot (1996) *Soil and water management systems.* John Wiley & Sons, London.

Scragg, A. (1999) *Environmental biotechnology.* Longman, Harlow.

Secretariat of the Convention on Biological Diversity (2000) *The handbook of the Convention on Biological Diversity.* Earthscan, London.

Seitz, J.L. (1995) *Global issues: an introduction*. Blackwell, Oxford.

Simmonds, M. and J. Hutchinson (eds) (1996) *The conservation of whales and dolphins*. John Wiley & Sons, London.

Simmons, I.G. (1993) *Interpreting nature: cultural constructions of the environment*. Routledge, London.

Simmons, I.G. (1996) *Changing the face of the Earth: culture, environment, history*. Blackwell, Oxford.

Simon, J. (1995) *The state of humanity*. Blackwell, Oxford.

Simpson-Housley, P. (1992) *Antarctica: exploration, perception and metaphor*. Routledge, London.

Skeldon, R. (1997) *Migration and development*. Longman, Harlow.

Skinner, B.J. and S.C. Porter (1995) *Dynamic Earth*. John Wiley & Sons, London.

Skinner, B.J. and S.C. Porter (1995) *The blue planet: an introduction to Earth systems science*. John Wiley & Sons, London.

Slack, P. (ed.) (1999) *Environments and historical change*. Oxford University Press, Oxford.

Slaymaker, O. (1995) *Steepland geomorphology*. John Wiley & Sons, London.

Slaymaker, O. (ed.) (1996) *Geomorphic hazards*. John Wiley & Sons, London.

Slaymaker, O. and T. Spencer (1998) *Physical geography and global environmental change*. Longman, Harlow.

Smith, J. (ed.) (2000) *The daily globe: environmental change, the public and the media*. Earthscan, London.

Smith, K. (1996) *Environmental hazards: assessing risk and reducing disaster*. Routledge, London.

Smith, K. and R. Ward (1998) *Floods: physical processes and human impacts*. John Wiley & Sons, London.

Smith, M. (ed.) (1999) *Thinking through the environment: classic and contemporary readings*. Routledge, London.

Snape, W.J. (ed.) (1996) *Biodiversity and the law*. Island Press, New York.

Snustad, D.P., M.J. Simmons and J.B. Jenkins (1999) *Principles of genetics*. John Wiley & Sons, London.

Spellerberg, I.F. (ed.) (1996) *Conservation biology*. Longman, Harlow.

Spellerberg, I.F. *et al.* (eds) (1991) *The scientific management of temperate communities for conservation*. Blackwell Scientific, Oxford.

Spencer, E.W. (2000) *Geological maps: a practical guide to the interpretation and preparation of geological maps*. Prentice Hall, London.

Stanley, S.M. (1993) *Exploring Earth and life through time*. W.H. Freeman, Oxford.

Strahler, A.H. and A.N. Strahler (1994) *Introducing physical geography*. John Wiley & Sons, London.

Summerfield, M.A. (1991) *Global geomorphology: an introduction to the study of landforms*. Longman, Harlow.

Summerfield, M.A. (ed.) (1999) *Geomorphology and global tectonics*. John Wiley & Sons, London.

Tacconi, L. (2000) *Biodiversity and ecological economics: participatory approaches to resource management*. Earthscan, London.

Teisseyre, R., J. Leliwa Kopystynski and B. Lang (1992) *Evolution of the Earth and other planetary bodies*. Elsevier, New York.

Thomas, D.S.G. (ed.) (1997) *Arid zone geomorphology*. John Wiley & Sons, London.

Thomas, D.S.G. and N.J. Middleton (1994) *Desertification: exploding the myth*. John Wiley & Sons, London.

Thomas, M. (1994) *Geomorphology in the tropics: a study of weathering and denudation in low latitudes*. John Wiley & Sons, London.

Thompson, R.D. (1998) *Atmospheric processes and systems*. Routledge, London.

Thorne, C.R. and N. Newson (eds) (1997) *Applied fluvial geomorphology*. John Wiley & Sons, London.

Thorpe, R. and G. Brown (1991) *The field description of igneous rocks*. John Wiley & Sons, London.

Thurman, H.V. and A.P. Trujillo (1999) *Essentials of oceanography*. Prentice Hall, London.

Tivy, J. (1993) *Biogeography: a study of plants in the ecosphere*. Longman, Harlow.

Tóth, F.L. (ed.) (1999) *Fair weather? Equity concerns in climate change*. Earthscan, London.

Trudgill, S.T. (1990) *Barriers to a better environment: what stops us solving environmental problems?* John Wiley & Sons, London.

Tucker, M. (1996) *Sedimentary rocks in the field*. John Wiley & Sons, London.

Turk, J. and G.R. Thompson (1995) *Environmental geoscience*. Saunders College Publishing, Philadelphia.

Turner, R.K. (ed.) (1993) *Sustainable environmental economics and management*. John Wiley & Sons, London.

Twort, A.C., D.D. Ratnayaka and M.J. Brandt (1999) *Water supply*. Edward Arnold, London.

United Nations Environment Program (1993) *Environmental data report 1993–94*. Blackwell, Oxford.

United Nations Environment Program (1999) *Global environment outlook 2000*. UNEP/Earthscan, London.

Upton, D. (1996) *Waves of fortune: past, present and future of the UK continental shelf*. John Wiley & Sons, London.

Vig, N.J. and R.S. Axelrod (eds) (1999) *The global environment: institutions, law and policy.* Earthscan, London.

Viles, H. and T. Spencer (1995) *Coastal problems: geomorphology, ecology and society at the coast.* Edward Arnold, London.

Voet, D., J.G. Voet and C. Pratt (1999) *Fundamentals of biochemistry.* John Wiley & Sons, London.

Vogler, J. (2000) *The global commons: environmental and technological governance.* John Wiley & Sons, London.

Wadsworth, R. and J. Treweek (1999) *GIS for ecology.* Longman, Harlow.

Wall, D. (ed.) (1993) *Green history: a reader in environmental literature, philosophy and politics.* Routledge, London.

Wang, G.T. (1999) *China's population: problems, thoughts and policies.* Ashgate, London.

Ward, P.D. (1991) *On Methuselah's trail: living fossils and the great extinctions.* W.H. Freeman, Oxford.

Watts, S. (ed.) (1996) *Essential environmental science: methods and techniques.* Routledge, London.

Webb, P. and J. Von Braun (1994) *Famine and food security in Ethiopia: lessons for Africa.* John Wiley & Sons, London.

Wellburn, A. (1994) *Air pollution and climate change: the biological impact.* Longman, Harlow.

Wells, N. (1997) *The atmosphere and ocean.* John Wiley & Sons, London.

Wheeler, D.A. and J.C. Mayes (eds) (1997) *Regional climates of the British Isles.* Routledge, London.

Whitmore, T.C. and J.A. Sayer (eds) (1991) *Tropical deforestation and species extinction.* Chapman & Hall, London.

Whyte, I. (1995) *Climatic change and human society.* Edward Arnold, London.

Wilhite, D.A. (ed.) (1999) *Drought: a global assessment.* Routledge, London.

Williams, M. (ed.) (1993) *Wetlands: a threatened landscape.* Blackwell, Oxford.

Williams, M.A. and R.C. Balling (1996) *Interactions of desertification and climate.* Edward Arnold, London.

Williams, M.A.J., D.L. Dunkerley, P. De Deckker, A.P. Kershaw and T. Stokes (1993) *Quaternary environments.* Edward Arnold, London.

Wilson, R.C.L., S.A. Drury and J.L. Chapman (1999) *The Great Ice Age.* Routledge, London.

Windley, B.F. (1995) *The evolving continents.* John Wiley & Sons, London.

Wood, A., P. Stedman-Edwards and J. Mang (eds) (2000) *The root causes of biodiversity loss.* WWF/Earthscan, London.

Wood, J.A. (1997) *The solar system.* Prentice Hall, London.

Yardley, B.W.D., W.S. Mackenzie and C. Guilford (1990) *Atlas of metamorphic rocks and their textures.* Longman, Harlow.

Yarnal, B. (2000) *Local water supply and climate extremes.* Ashgate, London.

Glossary

Ablation The process by which snow or ice is lost from glacier e.g. melting.

absorption The process by which a substance retains radiant energy.

acid rain The precipitation of dilute solutions of strong mineral acids from the atmosphere. Produced by the mixing of various industrial pollutants with oxygen and water vapour.

advection The movement of air, water and other fluids in a horizontal plane.

aerosols Minute liquid droplets or solid particles suspended in the atmosphere.

Agenda 21 A programme of global-local action on environmental and development issues arising from the Rio conference 1992. Aimed at governments, organizations and institutions that make policy or seek to influence policy-making.

AIDS A condition caused by a virus in which the body loses the ability to protect itself against disease.

albedo The ratio of the amount of light reflected by a surface to the amount of incident light.

ana-front A front along which warmer air mass rises over a layer of cold air.

anticline An arched upfold in the Earth's crust.

anticyclone An extensive region of relatively high pressure. Low level winds spiral out clockwise in the northern hemisphere and anti-clockwise in the southern hemisphere.

aquifer An underground layer of water bearing rock.

artesian well A well sunk into a confined aquifer. The release of hydrostatic pressure forces the water to the surface.

atmosphere The envelope of air that surrounds the earth, consisting principally of a mixture of the gases: oxygen (21%); nitrogen (78%) and carbon dioxide (0.03%).

atoll A ring of coral reef enclosing a lagoon. Often develops in conjunction with a subsiding volcanic island.

avalanche The rapid downslope movement of snow and ice or rock debris in steep mountain areas.

bajada The gentle, sloping surface from the mountain front to an inland basin in arid areas.

Bhopal The location, in India, of the world's worst industrial disaster in terms of loss of life. The escape of a pesticide vapour resulted in over 2500 direct fatalities.

biodiversity The variability among living organisms from all ecosystems and ecological complexes of which they are part. This includes diversity within species, between species and of ecosystems.

biogeochemical cycle The constant transfer of essential nutrients from living organisms to the physical environment and back to the organisms via a cyclical pathway.

biome A major climax community of plants and animals. It generally corresponds to a climatic region.

biosphere This is the collective ecosystem of Earth comprising the oceans, atmosphere and terrestrial ecosystems.

calcification A soil forming process in which calcium carbonate accumulates in the lower horizons. Particularly characteristic of arid climates.

carbonation A type of chemical weathering of rock by weak carbonic acid from dissolved carbon dioxide in rainwater. Very significant in limestone areas.

carbon cycle The natural circulation of carbon in the biosphere involving interconnected cycles on land, sea and the atmosphere.

carbon dioxide An atmospheric gas capable of absorbing radiation in wavelengths similar to those emitted by the Earth. Thus, it prevents excessive loss of terrestrial radiation and accompanying heat loss.

carrying capacity The maximum number of individuals that can be supported by a given part of the environment. This sustained use of an environmental resource should not result in its destruction or unacceptable deterioration.

catalytic converter A device fitted to the exhaust systems of motor vehicles which chemically changes noxious pollutants into less harmful emissions.

Chernobyl The site, in the Ukraine, of the world's worst nuclear power accident.

Chinook A warm, dry adiabatic wind which flows on the eastern side of the Rockies in the USA.

chlorofluorocarbons (CFCs) A group of inert gases used in aerosols, refrigerators, solvent cleaners and foam plastics. They have damaging effects on stratospheric ozone as they degrade to yield chlorine atoms.

cirque A hollow with a downstream opening formed by glacial erosion in upland areas.

Climatic Optimum A term used to describe the warm and moist climatic phase between 7,500 and 5,200 B.P. Commonly known as the Atlantic Period.

closed system Denotes a system that can exchange energy but not matter with its surroundings.

clouds A mass of water droplets formed throughout the troposphere by condensation of water vapour around nuclei such as dust, salt and soil particles.

cold front The front boundary of a mass of cold air where it undercuts and replaces a slower moving warm air mass.

commons Resources that are perceived to be inexhaustible and freely available to all without any individual/communal responsibility for them. The atmosphere, clean water and the oceans are perceived as commons.

competence The ability of a stream to move particles of a particular size. The maximum size of particle carried at a given stream velocity.

condensation The process by which energy is released from a gas when it undergoes a change of state to a liquid.

conduction The transfer of heat either within a substance or between two substances in direct physical contact.

constructive boundary A divergent plate boundary where new lithospheric rocks are being formed. Usually associated with sea floor spreading and mid-oceanic ridges.

continental drift A theory popularized and refined by Alfred Wegener in 1915, which suggested that the continents move around the Earth's surface because of the weakness of the suboceanic crust. Replaced by the more sophisticated plate tectonic theory.

continental shelf A gently sloping submerged plain bordering a continental landmass.

convection The vertical movement of air or water which follows the transference of heat from a warmer body via conduction.

core The central part of the Earth's interior. It is subdivided into a solid inner core and a liquid outer core.

Coriolis force A deflecting motion or force caused by the rotation of the Earth. The deflection is to the right in the northern hemisphere and to the left in the southern hemisphere.

craton A stable continental area that has undergone little internal deformation since the Precambrian (c. 570 million years ago). Normally refers to the ancient core of a shield area.

creep The extremely slow continuous movement of soil or rock debris down a slope in response to gravity.

crust The outer layers of the Earth's structure, varying between 6 and 48 km in thickness.

cyclogenesis The development of atmospheric low pressure systems.

Darwin, Charles Put forward the theory of evolution by means of natural selection in 1859.

delta A fan shaped alluvial deposit at a river mouth. Built up into a landform by successive layers of sediment.

demographic transition A general model describing the evolution of levels of fertility and mortality over time. Devised with reference to developing countries that have experienced the processes of industrialization and urbanization.

demography The study of numbers of organisms in a population and their variance over time.

dendritic A type of drainage pattern that develops as a random network and has a branched pattern.

denudation The removal of overlying material to expose rocks or strata. Commonly includes all processes that wear down the surface of the Earth.

deposition The laying down of material after it has been eroded and transported.

depression A low pressure atmospheric system. Also referred to as cyclonic.

desertification The spread of desert-like conditions due to human influence and/or climatic change.

destructive boundary A convergent plate boundary where one plate overrides the other in a subduction zone. Usually associated with volcanic and earthquake activity.

determinism A concept relating to the inevitability of a set of events under the influence of physical factors, e.g. drought conditions promoting migration.

diagenesis The compaction and cementation of unconsolidated sediment to form sedimentary rock. Also known as lithification.

diastrophism The deformation and movement of rock in the Earth's crust.

discharge The rate of flow of a river at a particular moment in time. Related to the river's velocity and volume.

doldrums An equatorial zone of low air pressure, developed over the ocean. Characterized by high temperatures and humidity, and calm conditions.

doubling time The time it takes for a population growing exponentially to increase twofold.

drainage basin The part of the land surface drained by a unitary river system.

drought An extended and continuous period of very dry weather. Water supply becomes insufficient to meet usual domestic, industrial and agricultural demands.

drumlin A glacial depositional landform which has been streamlined into a low hill by the passage of overlying ice.

earthquake A series of shocks and tremors resulting from sudden pressure release along active faults and in areas of volcanic activity.

ecology The study of the interrelationships between organisms and their environment.

ecosystem A system in which there is interdependence upon, and interaction between living organisms and their immediate environment.

emigration The movement of people out of a country or area. Opposite of immigration.

energy budget The balance of energy input and use in a biological system. May be analysed for each of the system components.

entropy A thermodynamic term describing the disorder or randomness of a system.

epicentre The point on the Earth's surface located immediately above the focus of an earthquake. At this point the most severe shock waves are usually experienced.

equilibrium A state of balance in a system produced by a variety of forces remaining unchanged through time.

ergs Deserts characterized by sand sheets and dunes.

erosion The group of processes whereby rock material is loosened or dissolved and removed from its original location on the surface. This includes weathering, solution and transportation.

esker A sinuous ridge of course gravel deposited by a subglacial meltwater stream.

ethnicity An affiliation to a common racial, cultural, religious or linguistic group whose characteristics distinguish it from a larger population.

eutrophication The enrichment of bodies of water by inorganic plant nutrients, e.g. nitrate and phosphate.

evaporation Process by which a liquid changes into a gas.

evaporites Sedimentary rocks formed by the precipitation of minerals dissolved in water a result of evaporation.

externalities Effects, positive or negative, that arise from an economic choice but are not reflected in market prices, e.g. a trained workforce (positive) or pollution (negative).

faulting A fracturing of the Earth's crustal rocks producing a line along which significant movement takes place. This involves the displacement of the rock on one side of the fault plane with respect to the rock on the other side.

fecundity The potential level of births over time for a population.

feedback A system mechanism in which once a component reaches a certain level it inhibits or promotes further action.

fertility The number of live births produced by a female and one of the fundamental influences on population size in any area.

firn Snow which has survived the summer melt season and has yet to become glacial ice.

Flandrian A term for the post-glacial stage of the Quaternary in N.-W. Europe commencing 10,300 B.P. Marked by an amelioration of climate.

flooding The inundation by water of land not normally covered with water.

fog Produced by the cooling of moist near the ground which condenses to produce a lowered visibility of 1 km or less.

folding The process by which rocks bend and buckle in response to compressional forces. Often associated with layered sedimentary rocks.

fracturing The process which produces a clean break in a rock due to stress and strain from faulting or folding.

free trade International trade that is free of such government interference as quotas, subsidies and tariffs.

fronts Sharp transition zones separating air of different temperatures and origins.

frost A form of precipitation consisting of the deposition of ice particles onto a surface. The severity of a frost depends on the moisture content and temperature of the air.

Gaia hypothesis Formulated in 1979 by James Lovelock. He proposed that the Earth's systems act as a totality producing feedback that seeks to optimise the environment for life. The presence of life impacts back on the condition of the biosphere and its systems.

geodesy The science associated with the size and shape of the Earth. It also studies the Earth's rotation, gravitational field and tidal variation.

geostrophic wind A wind that flows parallel to isobars. Produced when the pressure gradient is balanced by the Coriolis force; usually at higher altitudes where topographic friction is reduced.

geothermal gradient The natural increase in temperature of the Earth's crust with increasing depth.

gestalt A pattern or structure whose whole qualities exceed the sum of its constituent parts.

geyser A violent ejection of superheated water and steam from a hole in the ground. The underground reservoir consists of water filled chambers connected by a central pipe.

glacials Describes the occasions when ice sheets expanded during Ice Ages. Average global climates were colder and drier.

global change May be a change at the global spatial scale or a cumulative change based on localised change occurring world-wide.

Gondwanaland A large former continent, lying mainly in the southern hemisphere which rifted apart in the late Palaeozoic, to form parts of Africa, Australia, Antarctica, South America and India.

graben A valley or trough produced by faulting and uplift or subsidence of adjacent blocks of the Earth's crust.

green A descriptor that has become applied to anything that is seen to be connected with concern for the environment such as green politics.

greenhouse effect The mechanism where re-radiation from the Earth's surface is trapped by the presence of greenhouse gases, e.g. carbon dioxide in the atmosphere. This heats the underlying surface and maintains temperatures conducive to biotic habitation. Has been enhanced by human activity leading to accelerated global warming.

Green Revolution The development of high yielding cereal crops and their introduction to the less economically developed world from the 1960s. Also incorporated chemical and water supply initiatives to reduce the gap between population growth and food output in these areas.

Gross Domestic Product Total monetary value of all the goods and services produced in a national economy, usually over a period of one year.

Gross National Product Used as measure of economic success and relates to all goods and services produced by a country plus trade figures. Normally calculated over a period of one year.

groundwater Water than occupies spaces in the crustal rocks. Normally used in relation to subsurface water that participates in the hydrological cycle.

gyre A huge open-ocean circulation cell.

habitat The place where an organism or species generally lives.

Hadley cells An extensive circulatory system where air rises near the equator before flowing polewards at high altitudes before descending at about 30 degrees north and south to return towards the equator at low altitudes as trade winds.

hail Solid precipitation that falls as ice particles from cumulonimbus clouds.

HIV Human immunodeficiency virus that may lead to Aids.

holistic Explanations that attempt to explain complex phenomena in terms of the properties of the system as a whole.

Holocene Recent geological epoch beginning 10,000 years ago.

horst An upstanding block of the Earth's crust that is bounded by faults and has been uplifted by tectonic forces.

humanitarian Striving to promote the welfare of humankind.

humidity The amount of water vapour in the atmosphere. The proportion of water vapour present relative to the maximum quantity possible (controlled by temperature) is called relative humidity. The actual quantity of moisture held in the air is the absolute humidity level.

hurricane An intense tropical non-frontal depression generated by atmospheric instability. They are a major atmospheric hazard.

hydroelectricity Electricity generated from flowing water via turbines.

hydrolysis The main type of chemical weathering in which water combines with rock minerals to form an insoluble precipitate. Carbonation is the exception.

hydrosphere The surface water component of the Earth's system. Interlinked to the atmosphere by the hydrological cycle.

igneous Type of rock formed when magma solidifies either on the surface or within the Earth's crust.

immigration The movement of people into a country or area.

Intergovernmental Panel on Climate Change (IPCC) An organization of over 400 scientists established in 1988 that produced a consensus that deforestation and increased greenhouse gas emissions were resulting in global warming.

International Geosphere-Biosphere Programme (IGBP) A major programme, instigated in the early 1980s, focusing on the Earth's regulatory systems and the influence on them of human activities.

intertropical convergence zone (ITCZ) The zone of converging trade winds along the equator which cause rising air currents and low atmospheric pressure.

irrigation The application of water by sprinklers, ditches or canals over a land area in order to offset aridity. Usually related to agricultural applications.

isotope One of two or more forms of the same element who vary by having different numbers of neutrons.

jet stream A narrow band of extremely fast moving air located high in the troposphere.

kame An undulating mound of sorted glacial drift. Deposited in an irregular pattern by meltwater adjacent to an ice sheet.

kata-front A depressed frontal zone in which warm air is given little opportunity to rise over cold air because of prevailing downward movement of air from a higher level.

kettle hole An enclosed depression in glacial drift deposits accumulated in a recently glaciated area.

lag The time period elapsing between the occurrence of a causative event and its resulting impact.

laminar flow A type of flow in which the movement of each fluid element is along a specific path with uniform velocity. There is no mixing between adjacent layers.

Laurasia The northern protocontinent believed to be part of the fragmentation of the Pangea supercontinent. It is thought to have comprised North America, Greenland and all of Eurasia.

lava The term given to magma when it reaches the Earth's surface.

laws of thermodynamics Two physical laws related to the conservation of energy and the production of waste heat. They apply to all systems.

levée An elevated bank flanking the channel of a river and standing above the flood plain.

lightning A visible flash produced by electrical discharge within the clouds of a thunderstorm or between the clouds and the ground.

Limits to Growth Report produced in 1972, using computer models, which predicted future global collapse as the limits imposed by the finite global resource base on population growth and industrial development would be exceeded if current trends continued.

lithosphere The Earth's crust comprising layers of rock and soil.

longshore drift The processes by which material is transported along the coast.

magma Molten rock found beneath the Earth's crust from which igneous rocks are formed. May contain some solid particles and gases.

Malthus, Thomas In 1798 proposed a theory on the interaction of population growth and food supply. Malthus's basic proposition was that as population increased exponentially and food supply only arithmetically then unchecked population growth would outrun the growth in food supply leading to poverty.

mantle The zone within the Earth's interior between the thin surface crust and the partially molten core.

mega-city An urban area with a population exceeding ten million.

Mercalli scale A scale between one and twelve for measuring the intensity of earthquakes based on the amount of structural damage they cause.

mesosphere One of the concentric layers of the atmosphere lying above the stratosphere and below the ionosphere. It lies between 50 and 80 km above the Earth's surface.

metamorphic A rock type produced by the application of great temperature and pressure to igneous or sedimentary rocks over long periods of time. Alters the texture, structure and composition of the original rock.

methane The main constituent of natural gas and a contributor to the greenhouse effect.

mid-ocean ridge Underwater volcanic mountain ranges, which link together to form a global network. They are associated with divergent plate boundaries.

mineral A naturally occurring material with a local constant chemical composition.

migration Permanent or semi-permanent change of residence of an individual or group of people. Migration may be temporary or permanent, often involving movement across international frontiers.

mistral A strong, dry, cold northwesterly wind which blows offshore along the Mediterranean coasts of France and Spain.

monsoon The seasonal reversal of winds and air pressure systems over continental land masses and adjacent oceans.

Montreal Protocol An international protocol drawn up in 1987 and came into force in 1989 to protect the ozone layer from depletion. Concentrated primarily on stabilizing and then reducing the consumption and production of CFCs.

moraine An accumulation of mixed material that has been transported and deposited by a glacier or ice sheet.

mortality Death of individuals within a population. An essential determinant of population structure and growth.

natural hazard Any extreme event or condition in the natural environment that causes harm to people or property.

net primary productivity The rate at which chemical energy is manufactured by green plants per unit area in a given time minus losses due to respiration.

nitrogen cycle The sum total of processes by which nitrogen circulates between the atmosphere and the biosphere or any subsidiary cycles within this overall process.

nitrogen oxides Compounds of nitrogen and oxygen. Three of which, collectively called NOx, are significant atmospheric pollutants.

non-governmental organization (NGO) An organization that is separate from government and that operates without distributing a profit. Examples include the World Wide Fund for Nature and the Red Cross.

nuclear waste Radioactive waste materials produced as a by-product of nuclear power generation and research.

occlusion An atmospheric process where the cold front overtakes the warm front of a depression. This results in the warm sector being lifted away from the ground surface.

ontogeny The history of the development and growth of an individual.

open system Denotes a system that can exchange both matter and energy with its surroundings.

orogenesis The process by which intensely folded and faulted mountain ranges are formed.

osmosis The movement of a solvent (normally water) through a membrane from a dilute solution to a more concentrated one.

oxidation A type of chemical weathering in which oxygen dissolved in water reacts with certain rock minerals to form oxides and hydroxides.

pedogenesis The natural process of soil formation.

pedology The scientific study of soils.

periglacial Conditions, processes and landforms found at the periphery of glaciers and ice sheets. May also be used to describe cold climate areas where frost weathering is operative.

permafrost A layer, usually subsurface, of permanently frozen ground.

persistence The non-degradable ability of pesticides or other chemicals; thus allowing them to remain in the environment.

photochemical smog An atmospheric haze often found above large urban/industrial centres. It is the product of reactions between pollutants, often related to motor vehicles, in the presence of sunlight energy to produce a visible discoloration of the lower atmosphere.

photosynthesis A complex process occurring within the cells of green plants where sunlight is utilized in combination with carbon dioxide and water to produce oxygen and simple sugars or foods.

phylogeny The evolutionary history and line of descent of a species.

phytoplankton Free floating microscopic aquatic plants.

plate tectonics A widely accepted theory that the Earth's surface is divided into rigid crustal plates. The plastic nature of the underlying upper mantle allows these plates to move relative to each other.

playa A closed depression in an arid or semi-arid region that is periodically inundated by surface runoff.

Pleistocene The glacial and postglacial epoch lasting from about 2 million to 10,000 years ago.

pluton A mass of rock which has solidified underground from intrusions of magma.

pluvial A lengthy period of time when rainfall is considerably heavier than in preceding and succeeding stages.

polluter pays principle The polluter should pay the cost of anti-pollution measures.

population A group of individuals, usually of the same species, that inhabit a given area at a point in time.

population explosion The rapid increase in the population size of a biological species. May result from its introduction to a new, more favourable, environmental location.

poverty The condition of being without adequate food or economic resource.

precautionary principle The principle that scientific uncertainty on a particular issue is no excuse for postponing the implementation of measures to prevent environmental degradation.

proximity principle Waste must be disposed of in the nearest suitable facility using the most appropriate technology to guarantee a high level of protection to the public and environment.

Quaternary Geological period that comprises the Pleistocene and Holocene epochs (2 million years B.P. to present).

race A term which classifies members of the same species who differ in certain secondary characteristics, e.g. skin colour or blood type.

radiation The transfer of heat and other energy by means of electromagnetic waves.

radioactivity The ability of some elements to undergo spontaneous disintegration of their nuclei associated with the emission of ionizing particles or electromagnetic radiation.

radon A colourless, odourless gas about eight times denser than air. It is derived largely from uranium that is present in rocks such as granite.

rainsplash Raindrop impact during high intensity rainstorms. Results in soil erosion.

remote sensing The observation and measurement of an object without touching it. Usually associated with satellite imagery of the Earth's surface.

Richter scale A logarithmic scale of earthquake magnitude indicating the amount of energy generated by an earthquake. It ranges along a progression from zero to nine.

Rio Earth Summit The United Nations Conference on Environment and Development held in 1992.

risk assessment The identification of hazards and the evaluation of their economic and social costs.

Rossby waves Horizontal wave-like motions in the mid-latitude westerly circumpolar air stream. Occur in the upper troposphere and form and decay over one or two months.

runoff Water that moves across the surface of the land rather than being absorbed by the soil.

sedimentary Type of rock produced by the compression, compaction and cementation of sediments.

seismic waves Earthquake shock waves generated within the Earth's crust. They comprise a variety of types with different travel properties and hazard potential.

sere A successional series of plant communities. May also define a stage in succession.

shanty-town A peripheral section of a town or city inhabited by the very poor. Usually consisting of crude dwellings and little or no infrastructure/services.

shear strength The maximum resistance of a material to applied stress. Practically the degree to which soil and rock can resist mass movement.

shear stress A disturbing force that may result in the mass movement of soil or rock.

sial The part of the Earth's crust which is composed of granitic rocks rich in silica and alumina. Has a lower density than sima and forms the continental crust.

sill An horizontal intrusion of magna between existing rock strata.

sima The part of the Earth's crust composed of material rich in silica and magnesium. It is denser than sial that covers it in places. Where it is not covered by sial it comprises the ocean bed.

sovereignty The claim by a state to have authority over the people within its jurisdiction and over the course of events within its domain. Operationally this is the legal freedom of the state to act under international law.

storm surge An abnormally high wall of water driven ashore by high winds normally associated with the approach of a hurricane.

stratosphere The layer of the atmosphere that lies between the troposphere (below) and the mesosphere (above). Its base altitude varies with latitude being 9 km at the poles and 16 km near the equator. It extends to an altitude of around 50 km.

subduction The process that takes place when two lithospheric plates converge and one plate is forced below the other.

succession The sequence of different communities developing over time in the same area.

sulphur cycle The cycle of biological processes by which sulphur circulates within the biosphere.

sulphur dioxide A gaseous pollutant of the atmosphere. It has both natural, e.g. volcanic, and human e.g. fossil fuels, sources. Important component in the formation of acid precipitation.

sustainable yield The rate at which a renewable resource may be used without reducing its supply.

symbiotic A relationship between members of two different species which results in a mutual benefit.

syncline A downfold, or basin, of crustal rocks in which the strata dip inwards towards a central axis.

synergy The combination of two substances to produce an impact greater than the sum of the separate effects. May also be applied to greater potential effects by combining biotic communities.

techno-fix The reliance on technological processes to solve human problems.

teleconnections Simultaneous atmospheric events in areas that are remote from each other.

thermosphere The layer of the atmosphere above the mesosphere, some 80 km above the Earth's surface, in which temperature increases with altitude.

threshold This is the level or value that must be reached before an event occurs.

thunderstorm A storm generated by extreme atmospheric instability. Associated conditions include thunder, lightning and very heavy precipitation, e.g. hail.

till Unsorted glacial drift that is deposited directly from the ice.

trophic level A level in a food chain defined by the method of obtaining food. Also defines the number of energy transfers away from the original source of energy.

tropopause Boundary between the troposphere and the stratosphere. Located at an altitude of about 20 km in the tropics falling stepwise with latitude to 10 km at the poles. The steep rises are associated with jetstreams.

troposphere The lower portion of the atmosphere from the Earth's surface up to the lower stratospheric boundary (tropopause). In the troposphere temperature usually decreases with height.

tsunami A sea surface wave generated by submarine, earthquake and volcanic activity.

turbulent flow A fluid flow pattern where the flow is broken up.

UNCED The United Nations Conference on Environment and Development (also known as Rio Earth Summit) held in 1992. It was the largest intergovernmental conference to be held on the environment.

viscous The internal flow resistance of a fluid.

volcano A vent or fissure in the Earth's surface through which lava and volatiles are extruded.

vulcanism The movement of magma or molten rock and associated volatiles onto or towards the Earth's surface. Often used to refer to the extrusion of material onto the surface only.

wadi An ephemeral river channel in desert areas.

warm front A frontal zone in the atmosphere where, from its direction of movement, cool air is being replaced by rising warm air.

warping The gentle deformation of the Earth's crust over a large area. The process does not involve folding or faulting.

water table The upper level of the zone of groundwater in permeable rocks.

weathering The decomposition and disintegration of rocks in situ by the action of external factors, e.g. rain.

wind A horizontal movement of air in relation to the Earth's surface. Caused by variations of pressure related to differential global surface heating.

World Conservation Strategy A policy document published in 1980 which presented a single integrated approach to global problems. This was based on the concepts of helping species and populations; conserving planetary life support systems and maintaining genetic diversity.

Index

Note: Page numbers in **bold** type refer to figures
Page numbers in *italic* type refer to tables
Page numbers followed by 'P' refer to plates